# The Collected Works of J. Richard Büchi

J. Richard Büchi, 1983

Saunders Mac Lane Dirk Siefkes
Editors

# The Collected Works of J. Richard Büchi

With 60 Illustrations

Springer-Verlag
New York Berlin Heidelberg
London Paris Tokyo Hong Kong

Saunders Mac Lane
Department of Mathematics
University of Chicago
Chicago, Illinois 60637
USA

Dirk Siefkes
Technische Universität Berlin
Fachberich Informatik
D-1000 Berlin 10
Federal Republic of Germany

Library of Congress Cataloging-in-Publication Data
Büchi, J. Richard.
[Works. 1990]
The collected works of J. Richard Büchi / Saunders Mac Lane, Dirk Siefkes, editors.
p. cm.
ISBN 0-387-97064-9 (alk. paper)
1. Machine theory. 2. Logic, symbolic and mathematical.
3. Büchi, J. Richard. I. Mac Lane, Saunders.
II. Siefkes, Dirk. III. Title.
QA267.B79 1990
511.3—dc20 89-21769

Printed on acid-free paper.



Printed and bound by R.R. Donnelley & Sons, Harrisonburg, Virginia.
Printed in the United States of America.

9 8 7 6 5 4 3 2 1

ISBN 0-387-97064-9 Springer-Verlag New York Berlin Heidelberg
ISBN 3-540-97064-9 Springer-Verlag Berlin Heidelberg New York

# Preface

J. Richard Büchi is well known for his work in mathematical logic and theoretical computer science. (He himself would have sharply objected to the qualifier "theoretical," because he more or less identified science and theory, using "theory" in a broader sense and "science" in a narrower sense than usual.) We are happy to present here this collection of his papers.

I (DS)[1] worked with Büchi for many years, on and off, ever since I did my Ph.D. thesis on his Sequential Calculus. His way was to travel locally, not globally: When we met we would try some specific problem, but rarely discussed research we had done or might do. After he died in April 1984 I sifted through the manuscripts and notes left behind and was dumbfounded to see what areas he had been in. Essentially I knew about his work in finite automata, monadic second-order theories, and computability. But here were at least four layers on his writing desk, and evidently he had been working on them all in parallel.

I am sure that many people who knew Büchi would tell an analogous story. Therefore when Saunders Mac Lane asked me to help him edit a volume of collected papers of Büchi, I gladly accepted, although I was afraid too. Walter Kaufmann-Bühler from Springer-Verlag, who had struggled with Büchi for many years to get him to finish his book on a theory of automata, grammars, and terms, encouraged us. When Walter sadly died, Lynn Montz took over, so both books are now Springer volumes. We are grateful to the Springer people for that, and thank them for their help.

I (SM)[1] first met Richard in 1947. I was then spending six months in study at the Eidgenössische Technische Hochschule in Zürich. Richard and I found and discussed common interests in Boolean algebras and in the foundations of mathematics. One week early in the spring of 1948 we went on a skiing trip together to St. Moritz, where Richard knew all the ropes and found a good pension for our stay. We skied up and down the slopes and down the glacier, talking the while about logic. Since then it has been my privilege to

[1] The preface is written by both editors. Where necessary we add initials to indicate who is writing.

follow his career and now to share in this presentation of his varied and decisive work.

At our request, several people wrote comments on different parts of Büchi's work, to introduce and evaluate what he did in an area. All of them knew Büchi personally, be it through a single visit or as a year-long colleague. We made it a policy, however, not to have coworkers of Büchi write comments—except for some help from doctoral students. Thus nine comments appear herein, each with the corresponding section of Büchi's work. These comments turned out quite differently, some involving a personal attitude, some expositing a whole area. We feel this is good, since it reflects the different reactions Büchi could draw from people. We thank all of them for the work they did. Many people helped by contributing to the volume in other ways—you will see their names while reading—or from the background.

We are personally most indebted to Sylvia Büchi and Leonard Lipshitz, who gave time and space, advice, memories, material support, and a cheerful spirit when necessary. Walter Schnyder helped as both a colleague and a friend of Büchi. Beat Glaus from the ETH library in Zurich indexed and stored the many boxes of Büchi's notes and papers; through his immense effort all material is now publicly available.[2] Also I (DS) thank the people in Berlin who would sometimes rather have seen me doing my regular work at home or in my office than "Büchi again." I thank especially Helga Barnewitz for her excellent typing and my wife for loving patience. At the same time I (SM) thank Ann Kauth and Lynette Whalum for quickly typing the numerous letters needed to keep this project on track.

SAUNDERS MAC LANE
DIRK SIEFKES

[2] ETH-Bibliothek, Rämistr. 101, CH-8092 Zürich, Schweiz.

# Contents



---

* Numbers in brackets refer to the Publications section on pp. xi–xiii.

# Publications


[1] Die Boole'sche Partialordnung und die Paarung von Gefügen. *Port. Math.* 7 (1950), pp. 119–178. Doctoral dissertation.

[2] Representation of complete lattices by sets. *Port. Math.* 11 (1952), pp. 151–167.

[3] Investigation of the equivalence of the axiom of choice and Zorn's lemma from the viewpoint of the hierarchy of types. *J. Symbolic Logic* 18 (1953), pp. 125–135.

[4] On the existence of totally heterogeneous spaces. *Fund. Math.* 41 (1954), pp. 97–102.

[5] (with Jesse B. Wright) The theory of proportionality as an abstraction of group theory. *Math. Annalen* 130 (1955), pp. 102–108.

[6] (with Jesse B. Wright) Invariants of the anti-automorphisms of a group. *Proc. AMS* 8 (1957), pp. 1134–1140.

[7] Weak second-order arithmetic and finite automata. *Zeitschrift Math. Logik und Grundlagen der Mathematik* 6 (1960), pp. 66–92.

[8] On a decision method in restricted second order arithmetic. *Proc. Int. Congress Logic, Methodology, and Philosophy of Science, Berkeley 1960*. Stanford University Press, Stanford, Calif. (1962), pp. 1–11. Invited address.

[9] Turing machines and the Entscheidungsproblem. *Math. Annalen* 148 (1962), pp. 201–213.

[10] Mathematische Theorie des Verhaltens endlicher Automaten. *Zeitschrift Angew. Mathematik und Mechanik* 42 (1962), T9–T16. Invited address, yearly meeting of the German Soc. Appl. Math. and Mech.

[11] Regular canonical systems. *Archiv Math. Logik und Grundlagenforschung* 6 (1964), pp. 91–111.

[12] Algebraic theory of feedback in discrete systems. In E. Caianello (Ed.), *Automata Theory*. First Course on Automata Theory, Ravello, Italy, 1964. Academic Press, New York (1966), pp. 70–101.

[13] Transfinite automata recursions and weak second order theory of ordinals. *Proc. Int. Congress Logic, Methodology, and Philosophy of Science*, Jerusalem 1964. North-Holland Publ. Co., Amsterdam (1965), pp. 2–23. Invited address.

[14] Decision methods in the theory of ordinals. *Bull. AMS* 71 (1965), pp. 767–770.

[15] (with Lawrence H. Landweber) Solving sequential conditions by finite-state strategies. *Trans. AMS* 138 (1969), pp. 295–311.

[16] (with Lawrence H. Landweber) Definability in the monadic second-order theory of successor. *J. Symbolic Logic* 34 (1969), pp. 166–170.

[17] (with William H. Hosken) Canonical systems which produce periodic sets. *Math. System Theory* 4 (1970), pp. 81–90.

[18] Algorithmisches Konstruieren von Automaten und die Herstellung von Gewinnstrategien nach Cantor-Bendixson. In J. Dörr and G. Hotz (Eds.), *Automatentheorie und Formale Sprachen*, Tagung Math. Forschungsinst. Oberwolfach, 1969. Mannheim, F.R. Germany (1970), pp. 385–398. Invited address.

[19] (with Gary Haggard) Jordan circuits of a graph. *J. Combinatorial Theory* 10 (1971), pp. 185–197.

[20] (with Stephen Klein) On the presentation of winning strategies via the Cantor-Bendixson Method. *Report Purdue University* CSD TR-81 (1972), 14 pp.

[21] (with Kenneth J. Danhof) Model theoretic approaches to definability. *Zeitschrift Math. Logik und Grundlagen der Mathematik* 18 (1972), pp. 61–70.

[22] (with Kenneth J. Danhof) Definability in normal theories. *Israel Journal of Mathematics* 14 (1973), pp. 248–256.

[23] (with Kenneth J. Danhof) Variations on a theme of Cantor in the theory of relational structures. *Zeitschrift Math. Logik und Grundlagen der Mathematik* 19 (1973), pp. 411–426.

*[24] The monadic second order theory of $\omega_1$. In J.R. Büchi and D. Siefkes, *The Monadic Second Order Theory of all Countable Ordinals*. Lect. Notes Math. vol. 328 (1973), Springer-Verlag, pp. 1–127.

*[25] (with Dirk Siefkes) Axiomatization of the monadic second order theory of $\omega_1$. *The Monadic Second Order Theory of All Countable Ordinals*. Lect. Notes Math. vol. 328 (1973), Springer-Verlag, pp. 129–217.

[26] (with Dirk Siefkes) The complete extensions of the monadic second order theory of countable ordinals. *Report Forschungsinstitut Mathematik*, ETH Zürich (1974). *Zeitschrift Math. Logik und Grundlagen der Mathematik* 29 (1983), pp. 289–312.

[27] Using determinancy of games to eliminate quantifiers. In M. Karpinski (Ed.), *Fundamentals of Computation Theory*, 1977. Lecture Notes Comp. Science vol. 56, Springer-Verlag, pp. 367–378. Invited address.

[28] (with Charles Zaiontz) Deterministic automata and the monadic theory of ordinals $<\omega_2$. *Zeitschrift Math. Logik und Grundlagen der Mathematik* 29 (1983), pp. 313–336.

[29] State-strategies for games in $F_{\sigma\delta} \cap G_{\delta\sigma}$. *J. Symbolic Logic* 48 (1983), pp. 1171–1198.

*[30] (with Bernd Mahr, Dirk Siefkes) Manual on REC—A language for use and cost analysis of recursion over arbitrary data structures. *Techn. Univ. Berlin, FB Informatik* (20), Bericht Nr. 84-06 (1984), 79 pp.

[31] (with Bernd Mahr and Dirk Siefkes) Recursive definition and complexity of functions over arbitrary data structures. In G. Wechsung (Ed.), *Proc. 2. Frege Conference*. Akademie-Verlag Berlin (1984), pp. 303–308.

[32] (with Steven Senger) Coding in the existential theory of concatenation. *Archiv Math. Logik und Grundlagenforschung* 26 (1986/87), pp. 101–106.

[33] (with Steven Senger) Definability in the existential theory of concatenation and undecidable extensions of this theory. *Zeitschrift Math. Logik und Grundlagen der Mathematik* 34 (1988), pp. 337–342.

[34] (with William E. Fenton) Large convex sets in oriented matroids. *Journal Combinatorial Theory*, series B 45 (1988), pp. 293–304.

---

* Does not appear in this book.

*[35] (with William E. Fenton) *Directed Circuits of a Graph with an Application to Series-Parallel Graphs*. Manuscript.

*[36] *Implicative Boolean algebras, Copeland's theory of conditional probability*. Manuscript.

[37] (with T. Michael Owens) Skolem rings and their varieties. *Report Purdue University* CSD TR-140 (1975).

*[38] (with T. Michael Owens) *Complemented Monoids and Hoops*. To be submitted.

*[39] *Finite Automata, Their Algebras and Grammars: Towards a Theory of Formal Expressions*. D. Siefkes (Ed.). Springer-Verlag, New York (1989).

[40] *Relatively Categorical and Normal Theories*. Manuscript (1963–1965).

---

* Does not appear in this book.

# Permissions

[1]* Reprinted with permission.
[2] Reprinted with permission.
[3] Reprinted with permission.
[4] Reprinted with permission.
[5] Reprinted with permission from *Math. Annal.* Volume 130, 102–108. © 1955 Springer-Verlag.
[6] Reprinted from *Proceedings of the American Mathematical Society* (1957), "Invariants of the Anti-Automorphism of a Group," J. Richard Büchi and Jesse B. Wright, Volume 8, pages 1134–1140, with permission of the American Mathematical Society and Jesse B. Wright.
[7] Reprinted with permission.
[8] Reprinted from *Logic, Methodology, and Philosophy of Science*, Proceedings of the 1960 International Congress, edited by Ernest Nagel, Patrick Suppes, and Alfred Tarski, with the permission of the publishers, Stanford University Press. © 1962 by the Board of Trustees of the Leland Stanford Junior University.
[9] Reprinted with permission from *Math. Annal. Bd.* Volume 148, 201–213. © 1962 Springer-Verlag.
[10] Reprinted with permission.
[11] Reprinted with permission.
[12] Reprinted from *Automata Theory.* © 1966 Academic Press, New York.
[13] Reprinted with permission from *Proc. Int. Congress Logic, Method., Philos. of Sci. 1964, Jerusalem.* © 1965 North-Holland Publishing Company.
[14] Reprinted from *Bulletin of the American Mathematical Society* Volume 71 (1965) 767–770. Reprinted with permission of the American Mathematical Society.
[15] Reprinted from *Trans. of the American Mathematical Society* Volume 138 (1969) 295–311. Reprinted with permission of the American Mathematical Society.
[16] Reprinted with permission from the Association for Symbolic Logic and Lawrence H. Landweber.
[17] Reprinted with permission from *Mathematical Systems Theory* Volume 4, 81–90. © 1970 Springer-Verlag.
[18] Reprinted with permission.
[19] Reprinted from *Journal of Combinatorial Theory* Volume 10, 185–197. © 1971 Academic Press, New York.

---

*Numbers in brackets refer to the Publications section on pp. xi–xiii.

[20] Reprinted with permission of Stephen J. Klein.
[21] Reprinted with permission.
[22] Reprinted with permission.
[23] Reprinted with permission.
[26] Reprinted with permission.
[27] Reprinted with permission from *Fundamentals of Computation Theory* Lecture Notes in Computer Science Volume 56, 367–378. Editor: Marek Karpinski. © 1977 Springer-Verlag.
[28] Reprinted with permission from V.E.B. Deutscher Verlag der Wissenshaften and Charles Zaiontz.
[29] Reprinted with permission.
[31] Reprinted with permission.
[32] Reprinted with permission.
[33] Reprinted with permission.
[34] Reprinted with permission from *Journal of Combinatorial Theory* Series B, Volume 45, Number 3, December 1988, pp. 293–304. © 1988 Academic Press.
[37] Reprinted with permission.
Abstract 11 Reprinted with permission from *Proceedings of the Symposium on the Theory of Models.* © 1965 North-Holland Publishing Company, Amsterdam.

# Part 1 The Person and His Work

# Association for Symbolic Logic

ASL NEWSLETTER NOVEMBER 1984

**J. Richard Büchi (1924–1984)*** was born January 31, 1924, in Porto Alegre, Brazil, a citizen of Zell, Switzerland. He received his Doctorat in Mathematics at the E.T.H., Zürich, 1948, under Paul Bernays, and then moved to the United States. For several years he worked at the University of Michigan, as an instructor and then as the cofounder of the Logic of Computers Group with J.B. Wright, C.C. Elgot, R. McNaughton, H. Wang, and others. In 1958 he married Sylvia Duncan Hall. He held positions at the University of Illinois, Notre Dame University, the University of Mainz in Germany as a Fulbright Lecturer, and at Ohio State University. In 1963 he was appointed Professor of Mathematics and Computer Science at Purdue University. He was the first logician there, and built up a significant group in mathematical logic and theoretical computer sciences; among his colleagues were Paul Young, Jean Rubin, John Doner, Michael Machtey, and Leonard Lipshitz. He also drew a number of visitors from different countries and had 10 Ph.D. students in both fields, mathematics and computer science.

The period 1958–63 at Ann Arbor, Michigan, was particularly fruitful in his research. He became well known by 1958 when he used finite automata as a combinatorial tool to establish a decision procedure for a monadic fragment of arithmetic. He then introduced these ideas into infinite combinatorics and descriptive set theory, extending the decidability results to much stronger theories of numbers and ordinals. Following a contribution by McNaughton he transferred the approach to infinite game theory to prove constructive determinacy results. In this way he established a new field. People like M. Rabin, M. Magidor, S. Shelah, and Y. Gurevich have since successfully adapted and extended his methods: here his last and technically most difficult paper appeared in 1983 in the *Journal of Symbolic Logic*.

In teaching and research, finite automata had long been very close to him. He identified automata with algebras, and thus opened up a mathematical theory of automata. This he blended with his earlier work in lattice theory and formal language theory (Post's canonical systems) to form a theory of discrete

* Reprinted with permission of the Association for Symbolic Logic and the authors.

systems that uniformizes all these different areas. Much of this work, which he continued throughout his life, is published in the book FINITE AUTOMATA, THEIR ALGEBRAS AND GRAMMARS: TOWARDS A THEORY OF FORMAL EXPRESSIONS (Springer-Verlag, 1989).

In 1961, Richard Büchi introduced a new method for describing computations in logical theories to prove that the prefix class $\exists^\wedge\forall\exists\forall$ is a reduction type for the predicate calculus. Many consider this paper a milestone in the development of this field. Originally his interest came from Hilbert's 10th problem on the solvability of diophantine equations. This interest later led him to investigate Gauss' theory of quadratic forms, of which he constructed his own beautiful algorithmic version. This work in number theory will, we fear, remain unpublished. It is connected with results on definability and decidability in the existential theory of concatenation of words, which will be published jointly with S. Senger.

Starting with his doctoral thesis, Richard Büchi focused on describing mathematical structures axiomatically. Together with J.B. Wright he introduced a concept of "abstraction" to characterize structures by their automorphism groups. Although he used the concept for strong results in model theory, it remained remarkably unrecognized. Combined with his work in lattice theory it led him to an axiomaic theory of convexity, which will be published jointly with W. Fenton. How much of the lattice theoretic results will be published is not yet decided.

Richard Büchi mostly worked alone: those who worked with him were deeply influenced. He never could talk or let one talk *about* a subject: rather he had to create the subject fresh each time. (This applied to any field, not just mathematics.) Most of the rest of us are not like this; many have, therefore, failed to see his insight and influence. He demanded the utmost of himself and others. This made him a great mathematician and a great, though difficult, friend. It also made him forget his limits—he overworked himself in the effort to complete his work.

Contributed by: Dirk Siefkes, Technische Universität, Berlin; Paul Young, University of Washington; and Leonard Lipshitz, Purdue University.

# The Life of J. Richard Büchi

*Sylvia Büchi*

Richard was born in Porto Alegre, Brazil, on January 31, 1924, the second son of Jacob and Hedwig Büchi-Eigensatz. The family, including his one-year-older brother, James, returned to St. Gallen, Switzerland, when Richard, christened Julius Richard, was two years old. There his younger brother, Edgar, was born. The Büchis came from the Canton of Zürich, where the family can be traced back to the Reformation in the church records from Elgg and Zell. Richard had an ideal childhood. The household included, at various times, three grandparents and an aunt. They formed a close-knit, happy family. All three brothers became good tennis players and skiers, encouraged by their father, who had played soccer as a young man in St. Gallen and Porto Alegre.

Jacob Büchi would have liked to study at the university, but was financially unable to do so. He had been apprenticed in the textile industry in St. Gallen, but left for the United States, and a while later for Argentina, where with two American partners, he went to Porto Alegre and set up the Ford Agency. He was the business manager of the Agency. He decided to return to Switzerland so that his children could have a good education. Hedwig Büchi was educated at the seminary at Wil and then at Menzigen, in Switzerland. She wanted to see the world and became a private tutor, first for a year in Sicily, and later in Rio de Janeiro. She helped her sons with their education and encouraged them to go out into the world. Richard came to the United States, and his two brothers went to Brazil.

Richard decided at the age of 14 to become a mathematician. In school he did not like memorizing things, but enjoyed mathematics and the sciences. All three brothers went to the E.T.H., the Swiss Federal Institute of Technology in Zürich. James studied geology, Edgar chemistry. After Richard received his diploma in mathematics and theoretical physics, he returned to St. Gallen for eight months to work on a problem of his own. He returned to the E.T.H. to show the result to Professor Hopf, with whom he had done his diploma work. Hopf suggested that he show it to Professor Bernays. After a few conversations with Bernays, the work was accepted for his doctorate in mathematics.

During the war Richard had to spend some time working on a farm in the Rheintal. He later did his military service in the artillery, where it was suggested that he go for officer training, after he had worked out some equations to improve the calculating of trajectories. He did not do so, disliking regimentation, as it would have required him to remain longer in active service.

After receiving his doctorate, he came to the United States on a Swiss Rotary Club scholarship and spent a few months at the University of Chicago. He had wanted to come to the United States a year earlier, but was not allowed to do so as he did not know English (he studied French and Italian in school instead). After teaching at Ripon college in Wisconsin for a year, he went to the University of Michigan as an instructor in mathematics. I was doing graduate work in geology when we met there. After we were introduced, my first impressions of Richard were of a quiet, intelligent, kind and attractive man. We were married in Niles, Michigan, in 1958.

For three years we lived in an old farmhouse outside Ann Arbor, while Richard had a research appointment with the Logic of Computers Group and I worked on my dissertation. Because of the lack of interruptions, we both started to work at night, a habit Richard continued for the rest of his life. Later he talked to students and lectured in the afternoon, then ate supper, listened to music or read for an hour or so, then went to his desk and worked through until about 5 in the morning. On weekends, he also worked at night, but spent a few hours outside, walking in the woods, working with his tractor, or in the summer, relaxing at our small pond. Because of being up at night, we got to know the foxes and racoons that came regularly to eat with our cats.

We returned to Europe to visit his parents in St. Gallen and, later, after his father died in 1964, to visit his mother. Richard missed Switzerland, and we often thought of living in Europe. Although I grew up in the United States, I was born in Geneva (of Australian parents) and also spent my first nine years in Switzerland. But when the offer came from Kiel for Schütte's chair in logic in 1967, we had just finished building our house and were settling in, and it was just not the right time. Also we both enjoyed living in the country, and that would have been more difficult to do in Europe.

Richard was intensely involved in everything he did, no matter what. He expected the best from people, as he did from himself, and thus was sometimes hurt. In some respects he was very Swiss. He lived a moderate life, was frugal, critical yet kind, and completely without malice. When confronted with a problem, he kept at it until he solved it. He was very good with his hands, doing all the household repairs, partly to save the time that the repairmen would have wasted and to get the job done right. He was very conscious of time and disliked meetings and committee work because it cut into time that should be devoted to mathematics.

As he worked so much at home, we spent most of our time together. In later years we read and discussed books together, such as Jane Austen, Darwin, and early scientific writers. He liked to read such writers as Morgenstern, Kleist,

Schopenhauer, Hölderlin, the Bronte sisters, and Lewis Carroll. As a child he very much liked stories about animals, and his whole life he liked having animals around and observing them. In reading he analyzed what he was reading, and read very slowly thinking about it as he read. In fact, the most characteristic thing about him was that he really looked at and thought about the world around him. He was happiest when doing mathematics, and remarked once that no sooner had he sat down to work than it seemed he had to stop. At times he fretted that there was so little time to work on all the ideas he had in his head. He very much enjoyed doing mathematics with others, such as his friend, Jesse Wright, Dirk Siefkes, and his students, and thought of mathematics as a very social occupation, with mathematicians from one- or two-hundred years ago being just as alive for him as his contemporaries.

His health was adversely affected by the stress he was under in his last three years. His drive to do mathematics was sometimes at the expense of his own welfare and in the last few weeks before his sudden death, he was working later and later every night, on the book he was writing on closure spaces. He had a ticket to visit his 93-year-old mother and was trying to get as much as possible done before he left. His family loved him and we all miss him very much.

# The Work of J. Richard Büchi*†

*Dirk Siefkes*‡

## Abstract

J. Richard Büchi has done influential work in mathematics, logic, and computer sciences. He is probably best known for using finite automata as combinatorial devices to obtain strong results on decidability and definability in monadic second-order theories and extending the method to infinite combinatorial tools. Many consider his way of describing computations in logical theories as seminal in the area of reduction types. Treating automata as algebras, he opened up a mathematical theory of automata, which he generalized to a theory of discrete systems; this later work will hopefully be published. Less recognized is his concept of "abstraction" for characterizing structures by their automorphism groups, which he considered basic for a theory of definability. An axiomatic theory of convexity, which originated thereof, will be published jointly with W. Fenton. Joint work on formalizing computing and complexity on abstract data types is being published by the current author. Unpublished is, and likely will be, his continued work on an algorithmic version of Gauss' theory of quadratic forms, which stemmed from his interest in Hilbert's 10th problem. Results in the existential theory of concatenation, which link the two areas, will be published jointly with S. Senger.

Working with J. Richard Büchi has been strenuous, frustrating, and beautifully rewarding. Our meetings always lasted through the night. We would start with a heated political discussion or inquire into a biological question; he had read most of Darwin. Then we would head into mathematics. If he brought up the subject, he would be the teacher, I the student, and I would learn a lot. He never, however, told me what he knew. He would work on some problem, in this way trying to make me see the things he saw. When I brought up the

---

* Invited talk, AMS spring meeting, Chicago, 1985.

† Reprinted with permission from *Perspectives on the History of Mathematical Logic*. Ed. L. Drucker. Birkhäuser Boston. In press.

‡ Technische Universität Berlin, August 1985.

I did this work with the help of travel grants from Purdue University and from the Deutsche Forschungsgemeinschaft. They enabled me to visit Purdue University twice after Richard Büchi's death to look through his notes. I also thank the Deutsche Forschungsgemeinschaft, the Stiftung Volkswagenwerk, and the Fulbright Commission for earlier travel grants to work with Richard Büchi.

subject, it was worse. He would ask questions that might sound silly; I would answer; he would ask more questions; I would realize that I had better answer carefully; but soon my answers would not fit and my ideas would crumble away. Then we would start together afresh, and something new and beautiful would evolve, although it might take long.

This is the way he dealt with mathematics or any other field. He could not talk, or hear, *about* a subject. He had to work *through* it, each time again, to make it his own. (For this reason he quarreled with so many people: after a conversation both parties would think that the other had learned from himself.) This nonrelenting way of working made him one of the great mathematicians of our time; he never stopped short of creating a clear and complete picture.

He did not, however, get stuck in the details either. He gave new directions to many fields. (I learned of many of them only after his death when I looked through his notes.) He always worked into the history of the area and was deeply involved with the philosophical, social, and personal problems of people in mathematics and science in general.

Now you see the double squeeze I am in: I am supposed to write about Richard Büchi's work, although he himself would have hated that; and much of it I even do not know well. So instead of writing about subjects and results and giving a complete overview, let me show you some of the things he liked to work with and I'll just mention some others. This still will not do but it is the best thing that came to my mind.

## Trees, Algebras, Automata, and Languages

Richard Büchi liked trees. So do I. Thus let me start with the trees that played a role in much of his mathematics. The full binary tree has a root $e$ and successors $w1$ and $w2$ for each node $w$:

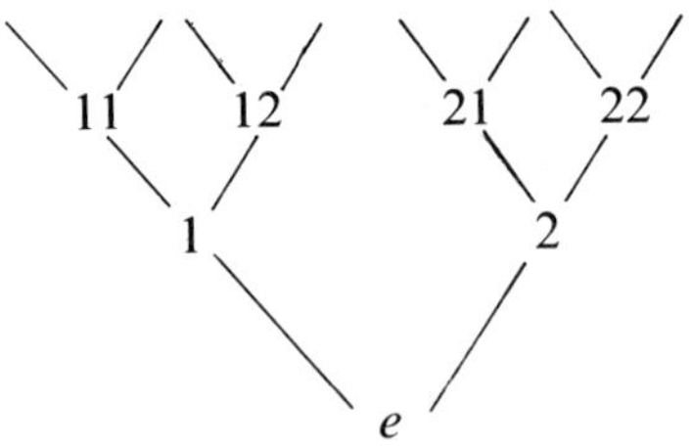


It represents the set $\{1, 2\}^*$ of words over the alphabet $\{1, 2\}$ with empty word $e$ and writing to the right of words. Or in another formulation: it represents the free algebra generated by the element $e$ and the unary operations $^\frown 1$ and $^\frown 2$. Every algebra with the same generators is a quotient (homomorphic image) of this one. Every such finite algebra is also a finite automaton over the alphabet $\{1,2\}$: nodes are states, connected by paths in tree (input words).

For an example take a finite automaton that accepts the language (set) $L$ of words with an even number of 2s:

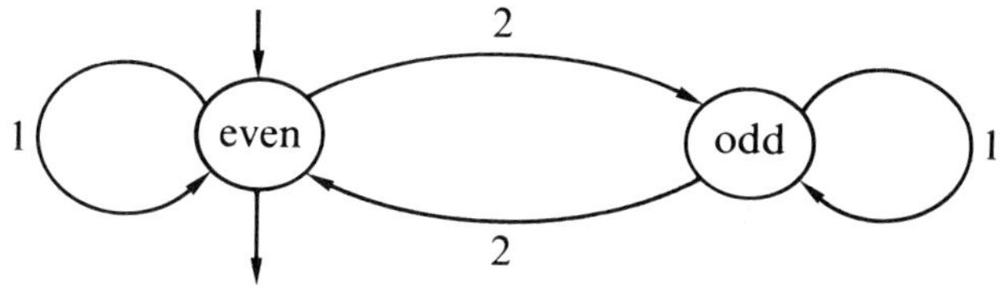


The two states tell whether the input has had an even or an odd number of 2s so far. Map the automaton into the tree by putting the initial state at the root and closing a path when you hit a state the second time:

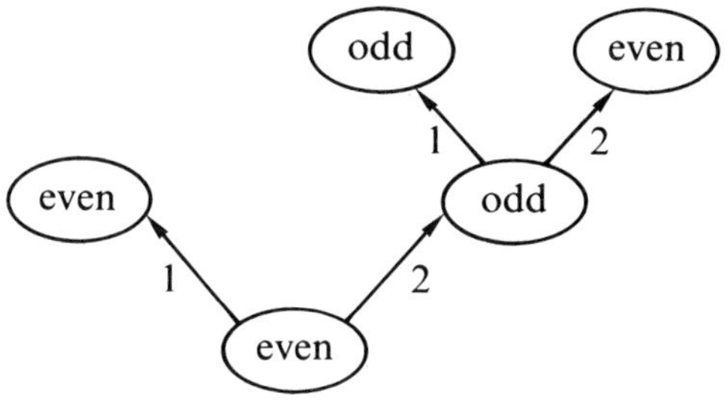


Identifying states gives back the automaton. So automata and unary algebras are really the same. Given a language $L$ of words define

$$u \equiv v \quad \text{iff} \quad \forall w \in \{1,2\}^* \,(uw \in L \Leftrightarrow vw \in L);$$

then $\{1,2\}^*/\equiv$ yields a minimal automaton accepting $L$. Given an automaton with a set $S$ of states and final states $F$ define

$$u \equiv v \quad \text{iff} \quad \forall s \in S \,(s \xrightarrow{u} F \Leftrightarrow s \xrightarrow{v} F);$$

then $\{1,2\}^*/\equiv$ yields a minimal equivalent automaton.

In the years 1952–1962 Richard Büchi and his coworkers in the Logic of Computers Group at the University of Michigan, especially Jesse Wright, Calvin Elgot, Jim Thatcher, Robert McNaughton, Michael Harrison, and Hao Wang, established and worked out these connections:

$$\text{languages} \leftrightarrow \text{automata} \leftrightarrow \text{unary algebras} \leftrightarrow \text{trees}.$$

The relevant publications are 10–12 in the bibliography, as well as 17. For years he worked on a book that contains all this material and generalizes it to the more-than-unary case:

term languages $\leftrightarrow$ tree automata $\leftrightarrow$ general algebras $\leftrightarrow$ trees.

This book, 39, hopefully will be published soon.

## Discrete Spaces

A finite automaton is a discrete deterministic system. Starting with his doctoral dissertation and continuing through all his life Richard Büchi also used the more general setting of discrete spaces to investigate discrete systems. Some of this work on set spaces, lattices, discrete linear and convex closure, directed graphs, and matroids is contained in publications 1–4, 19, and 34–38. William Fenton gave a talk on his work with Richard Büchi on convexity theory at this conference. Other material will likely remain unpublished.

## Finite Automata on Infinite Input, Monadic Theories, and Determinacy

The (weak) monadic second-order theory of a class of structures is the elementary theory enriched by quantifiable variables for (finite) predicates, i.e., sets. It was Richard Büchi's idea to express statements on finite automata as formulas of (weak) monadic second-order theories, and in this way to get decidability and definability results for such theories by using combinatorial statements expressed in terms of finite automata. For an example write T(true) for the state "odd" and the input 1 and F(false) for the state "even" and the input 2 of the automaton in section 1:

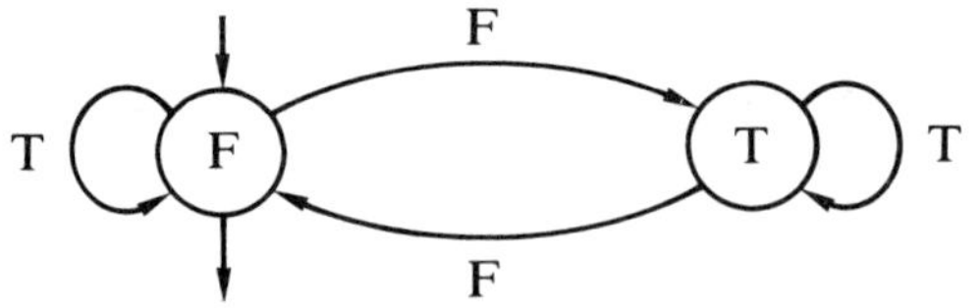


Then the automaton accepts the word $w \in \{F, T\}^*$ iff there is an accepting "run," i.e., a sequence of states $v \in \{F, T\}^*$ beginning and ending with F and following the transition condition. This statement can be read as a formula of the weak monadic second-order theory of the successor on the natural numbers:

$$\exists Z. \quad \neg Z_o \wedge \forall t(Zt' \leftrightarrow (Zt \leftrightarrow Xt)),$$

where $X = wFF\ldots$; $Z = vFF\ldots$, and $'$ is the successor function. Richard Büchi showed that "automata formulas" of this type constitute a normal form and in this way proved that the theory is decidable; see publication 7. The same method works for arbitrary ordinals; see publication 13. Using this method Calvin Elgot obtained related results; see also the joint abstract 5.

The input $X$ and the run $Z$ in the preceding automata formula are actually infinite, though eventually constant, sequences. Two years later Richard Büchi had generalized his method to automata on $\omega$-input ($\omega$-automata) and to the strong monadic second-order successor theory (publications 8 and 16). For example, the formula

$$\exists Z. \quad \neg Z_o \wedge \forall t(Zt' \leftrightarrow (Zt \leftrightarrow Xt)) \wedge \exists^\omega t\, Zt$$

says that this automaton accepts those sequences $X$ that contain an even number of $F$s or infinitely many of them. Automata formulas for nondeterministic $\omega$-automata yield a normal form:

$$\exists Z. \quad A[Z_o] \wedge \forall t\, B[Xt, Zt, Zt'] \wedge \exists^\omega t\, C[Zt].$$

Robert McNaughton proved that every such automaton can be changed into a deterministic one if one replaces the acceptance condition

$$\exists^\omega t\, C[Zt] \quad \text{by} \quad \sup Z \in D \quad \text{where} \quad \sup Z := \{s; \exists^\omega t\, Zt = s\}$$

is the set of states that occur infinitely often.

Similar ideas with more and more sophisticated automata and thus more and more help from infinite combinatorics and set theory work for all ordinals $<\omega_2$ (see publications 14, 24, and 28). They also work for nonstandard structures like $\omega + \omega^*$, and then allow to axiomatize the theories in question and to characterize their complete extensions by prime models (publications 25 and 26; see also the paper by Charles Zaiontz in the same volume as 28). The central part of the normal form construction is always a complementation lemma, which says that the automaton acceptable sets are closed under complement, and thus allows to rewrite the negation of an automata formula as an automata formula. It is crucial to find the right transition and final conditions for the automata, which make them jump not only from one state to the next, but also to the limit, for example from $\omega$ to $\omega + 1$, or to $\omega + \omega^*$. Yuri Gurevich, Menachem Magidor, and Saharon Shelah have shown that at $\omega_2$ decidability depends on statements about big ordinals; so what is the natural theory there?

Robert McNaughton was the first to observe the close connection of the area to problems on infinite two-person games with complete information. Such a game is given by a condition $C(X, Y)$ on $\omega$-sequences over two finite sets $H$ and $K$. Two players $I$ and $J$ alternately choose from $H$ and $K$ respectively. Their moves depend on the previous choices of both players and constitute two $\omega$-sequences $X$ and $Z$. Player $J$ wins if the result satisfies condition $C$; otherwise player $I$ wins. A winning strategy for player $J$ ($I$) is an operation $W$ ($V$), which to any play of the opponent yields a play that beats him:

$$\forall X\, C(X, WX) \quad \text{or} \quad \forall Y \neg C(VY, Y) \quad \text{respectively.}$$

A game is determinate if one of the players has a winning strategy. A finite state winning strategy is one that can be realized by an $\omega$-automaton.

Richard Büchi and Lawrence Landweber proved that all games definable in the monadic second-order theory of successor are determinate; one can decide who wins and can construct a finite state winning strategy (see publications 15 and 18). This had been stated as a problem by Richard Büchi in 8, and conjectured by Robert McNaughton. By another result of Büchi and Landweber (see 16) the games in question are just the regular sets in the Boolean algebra over $F_\sigma$, i.e., definable by Boolean combinations of formulas

$$\exists y \forall t \geq y \quad (\bar{X}t, \bar{Y}t) \in L$$

where $(X, Y)$ is thought as a path in an $H \times K$-tree, $(\bar{X}t, \bar{Y}t)$ is the corresponding node at height $t$, and the congruence induced by $L$ has finite rank (section 1). In the early 1970s Richard Büchi removed the restriction to regular sets (see publication 27). The last part of the result then reads: one can decide who wins and can construct a winning strategy that is finite state over the given game. He published the proof in 29 for the slightly bigger class of $F_{\sigma\delta} \cap G_{\delta\sigma}$-games (one more quantifier change).

In a very difficult and complicated construction Michael Rabin extended Richard Büch's method of handlung monadic second-order theories with the help of finite automata, to the binary tree, i.e., to the case of two successors. He observed that his proof, together with McNaughton's result on deterministic $\omega$-automata, yields the second half of the Büchi-Landweber result, namely, decidability and construction; it does not yield determinacy. In 27 Richard Büchi introduced determinacy as a general tool for eliminating quantifiers, and proved especially that his determinacy for Boolean $F_\sigma$-games together with McNaughton's result yields Rabin's decision procedure. For this he showed that in the monadic second-order theory of two successors the complementation lemma follows from a determinacy statement. Namely, the automata normal form here is

$$\exists W. \quad A[We] \wedge \forall u\, B[Vu, Wu, Wu1, Wu2] \wedge \forall X \sup WX \in D \tag{1}$$

where the input $V$ and the run $W$ are finite state trees, $X$ is a path through the tree, and sup $WX$ is the sup (as defined earlier) of the run $W$ along the path $X$. Using McNaughton's result he brought formula (1) into the form

$$\exists W \forall X \exists Z. \quad Z_o = c \wedge \forall t(Zt' \leftrightarrow B[Zt, Xt, W\bar{X}t, V\bar{X}t]) \wedge \sup Z \in \tilde{D} \tag{2}$$

where $V$, $W$, and $X$ are as defined earlier, and $Z$ is the run of a deterministic $\omega$-automaton along $X$. Now the formula

$$\exists Z. \quad Z_o = c \wedge \forall t(Zt' \leftrightarrow B[Zt, Xt, Yt, V\bar{X}t]) \wedge \sup Z \in \tilde{D}$$

defines a special Boolean $F_\sigma$-game where player $I$ chooses from the set $\{1, 2\}$, and $V$ is a parameter tree. Therefore formula (2) says that player $J$ has a winning strategy for that game. By determinacy the negation of (2) can be brought into the same form; thus this holds for (1) as well.

## Quadratic Forms, the Five-Squares Problem, and Diophantine Equations

For many years Richard Büchi worked on quadratic forms. He presented Gauss' class field theory in an algorithmic form, rather in the spirit of Gauss himself, but more thoroughly organized in a way that was suitable for his purpose. It seems that his interest in quadratic forms stems from Hilbert's 10th problem, which asks for a procedure to decide whether a given set of Diophantine (i.e., integer coefficient) equations has a solution. He tried to access it by reducing Diophantine equations to equations on square numbers in the following way.

Let $x_1 < \cdots < x_n$ be positive integers, $n \geq 3$. If they are consecutive, i.e., $x_{i+1} = x_i + 1$, then the second difference of their squares is 2, namely

$$(x_{i+1}^2 - x_i^2) - (x_i^2 - x_{i-1}^2) = (x_i + 1)^2 - 2x_i^2 + (x_i - 1)^2 = 2.$$

Richard Büchi raised the question: Are there nonconsecutive integers with this property? He knew a procedure to generate infinitely many such sequences in the case $n = 4$ (and thus $n = 3$); an example is 6, 23, 32, 39. For $n = 5$ the question is open; his students called it the five-squares problem. He proved: If for some $n$ the answer to the $n$-squares problem is "no," then to any Diophantine equation $P$ one can construct a set $Q$ of Diophantine equations with only quadratic terms *s.t.* $P$ has an integral solution iff $Q$ has:

$$\exists z_1 \ldots z_m\, P(z_1, \ldots, z_m) = O \quad \text{iff} \quad \exists y_1 \ldots y_k\, A \cdot \begin{pmatrix} y_1^2 \\ \vdots \\ y_k^2 \end{pmatrix} = B$$

where $P$ is a polynomial with integer coefficients, $A$ is an integer matrix, and $B$ is an integer vector. The idea of the proof is to define the relations

$$y = x + 1,\; y = x^2 \text{ and finally } z = x \cdot y$$

with the help of squared variables and existential quantifiers. Thus if the answer is "no," Hilbert's 10th problem can be stated in this much simpler form.

Since by now Hilbert's 10th problem is solved in the negative, the result amounts to the following: If the answer to some $n$-squares problem is "no," then there is no procedure to decide whether a given set of Diophantine equations with only squared variables has a solution. Slightly more general, then the existential theory of addition over the integers with the predicate "is a square" is undecidable.

Carl Friedrich Gauss in his "Disquisitiones Arithmeticae" (1800) investigates which numbers $M$ can be represented by quadratic forms, i.e., as

$$M = ax^2 + bxy + cy^2$$

where $a$, $b$, $c$ and $x$, $y$ are integers. He denotes a form $F$ by its coefficients $(a, b, c)$, and classifies forms by the determinant of their matrix

$$D := ac - b^2/4 = \begin{vmatrix} a & b/2 \\ b/2 & c \end{vmatrix},$$

or, as one does it today, by their discriminant $b^2 - 4ac = -4D$. He calls two forms $F$ and $F'$ equivalent if they can be transformed into each other by a unimodular transformation:

$$F' = A^T FA = \begin{pmatrix} h & k \\ i & j \end{pmatrix} \begin{pmatrix} a & b/2 \\ b/2 & c \end{pmatrix} \begin{pmatrix} h & i \\ k & j \end{pmatrix} \text{where } |A| = hj - ki = \pm 1.$$

Equivalent forms have the same discriminant and represent the same numbers, and conversely: two forms that represent the same numbers are equivalent. For each discriminant there are only finitely many classes of forms (class field theorem). Gauss devotes much of the book to studying which types of classes occur for which types of discriminants.

For these investigations Richard Büchi introduces three structures. First he notes that the positive unimodular matrices (of dimension 2 by 2) are freely generated from the unit matrix $e$ through multiplication from the right by the matrices 1 and 2, where

$$e := \begin{pmatrix} 1 & 0 \\ 0 & 1 \end{pmatrix}, \qquad 1 := \begin{pmatrix} 1 & 0 \\ 1 & 1 \end{pmatrix}, \qquad 2 := \begin{pmatrix} 1 & 1 \\ 0 & 1 \end{pmatrix}.$$

Thus they can be represented by the full binary tree of section 1. To generate all unimodular matrices he adds the inverses

$$\dot{1} = \begin{pmatrix} 1 & 0 \\ -1 & 1 \end{pmatrix} \quad \text{and} \quad \dot{2} = \begin{pmatrix} 1 & -1 \\ 0 & 1 \end{pmatrix}$$

of 1 and 2. In this way he gets the structure $\mathfrak{U}$

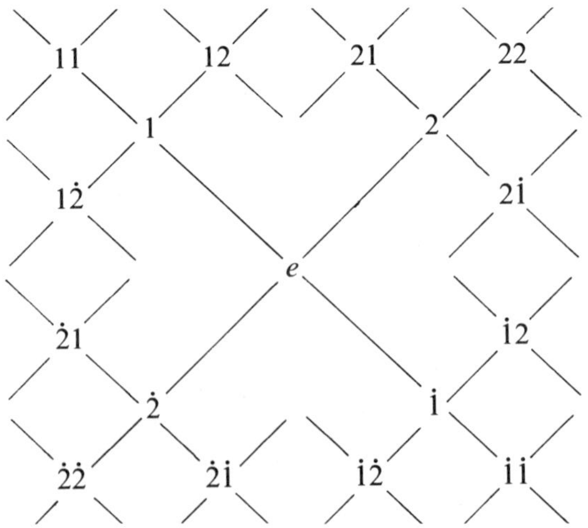

of unimodular matrices. It is not free; for example, $L^2 = -e$, and thus $L^4 = e$, where

$$L := 2\dot{1}2 = \begin{pmatrix} 0 & 1 \\ -1 & 0 \end{pmatrix}.$$

Gauss uses $L$ instead of 1 to generate the unimodular matrices. Therefore he misses the connection to the word algebra, and thus to the other two structures.

Richard Büchi gets the second structure from an algorithm which Nicomachus introduces around 100 A.D. in his "Arithmetics." Nicomachus states that it generalizes the Euclidean algorithm for computing the greatest common divisor of a pair of integers. It applies to triples of integers and yields at each step four new triples. Thus it results in the structure NIC:

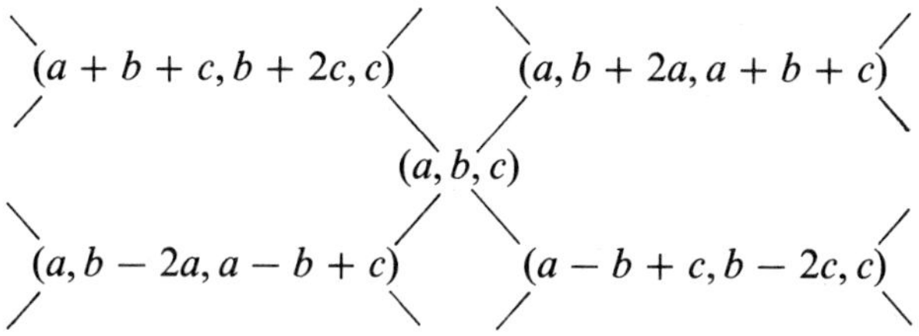


If $F = (a, b, c)$ is a form, then the NIC-structure consists of the forms equivalent to $F$. In fact NIC is a homomorphic image of the structure $\mathfrak{U}$ of unimodular matrices if we map the matrix $A$ into the form $A^T F A$ (see earlier). For example, $e^T F e = F$ and

$$\begin{aligned} 1^T(a,b,c)1 &= \begin{pmatrix} 1 & 1 \\ 0 & 1 \end{pmatrix}\begin{pmatrix} a & b/2 \\ b/2 & c \end{pmatrix}\begin{pmatrix} 1 & 0 \\ 1 & 1 \end{pmatrix} = \begin{pmatrix} a + b/2 & b/2 + c \\ b/2 & c \end{pmatrix}\begin{pmatrix} 1 & 0 \\ 1 & 1 \end{pmatrix} \\ &= \begin{pmatrix} a+b+c & b/2 + c \\ b/2 + c & c \end{pmatrix} = (a+b+c, b+2c, c), \end{aligned}$$

which is the upper right successor of $(a, b, c)$ in NIC. The connection to $\mathfrak{U}$ also explains in which sense the Nicomachus algorithm generalizes the Euclidean algorithm. Namely, in $\mathfrak{U}$

$$\begin{pmatrix} h & i \\ k & j \end{pmatrix} \cdot \dot{1} = \begin{pmatrix} h-i & i \\ k-j & j \end{pmatrix}, \begin{pmatrix} h & i \\ k & j \end{pmatrix} \cdot \dot{2} = \begin{pmatrix} h & i-k \\ k & j-k \end{pmatrix},$$

and therefore in the positive unimodular matrices the predeccessor operations apply the Euclidean algorithm to both rows simultaneously.

The third structure consists of the conjugates $\dot{A}XA$ of some given matrix $X$ for unimodular matrices $A$. With the generators 1, 2, $\dot{1}$, $\dot{2}$ from before, the conjugacy structure is

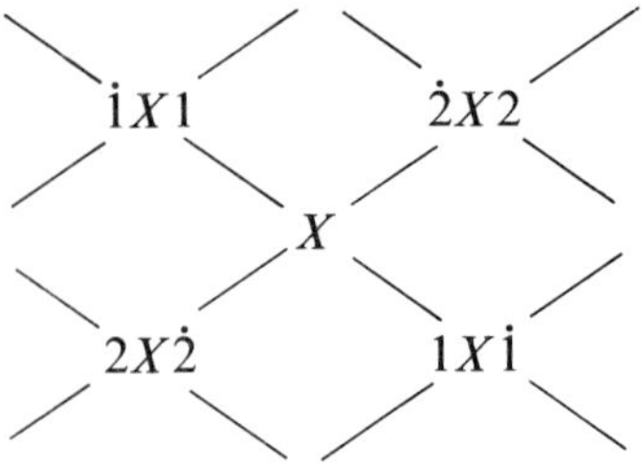


It is obviously again a homomorphic image of the structure $\mathfrak{U}$. In fact it is isomorphic to the NIC-structure if we map the form $F = (a, b, c)$ into the matrix

$$X := \begin{pmatrix} \frac{1}{2}(\iota - b) & -c \\ a & \frac{1}{2}(\iota + b) \end{pmatrix}$$

where $\iota = 0, 1$ is the residuo modulo 4 of the discriminant of $F$. Thus also the conjugacy structure can be viewed as representing the class of forms equivalent to $F$.

Going back and forth between the three structures and using whichever he found convenient, Richard Büchi reconstructed Gauss' results on quadratic forms, and worked on the five-squares problem. Nothing of this work is published; it is preserved only in very many notes and in two sets of lecture notes taken by the author and Steven Senger.

## Words, Turing Machines, and Recursive Functions

Richard Büchi refers often to Hilbert's 10th problem in his early notes. It may well be that more of his work was stimulated by this problem and it is surely the case with the existential theory of words over a finite alphabet. Since we represent numbers by words, Hilbert's 10th problem would be solved in the negative if the existential theory of words were undecidable. He therefore was long interested in that problem, but was overtaken by Makanin who proved in 1976 that this theory is decidable. An offspin of this interest is the work he did with Steven Senger on definability and decidability in extensions of that theory (see publications 32 and 33). Steven Senger gave a talk on these results at this conference.

Besides the first accounts of his work on finite automata and algebra (10) and on monadic second-order theories (8) Richard Büchi published a third landmark paper in 1962 (9) in which he proves that the prefix $\exists \wedge \forall\exists\forall$ is a reduction type. In the proof he expresses the halting problem for Turing machines in formulas of this type, and thereby—although he did not invent "domino" or "tiling" problems—he is the first to use them in such a reduction, which is now the standard method.

Also in the area of computability, in 1980 Richard Büchi (together with Bernd Mahr) joined the current author in studying recursive definitions on arbitrary partial algebras. In publications 30 and 31 we show that recursion is a natural tool to formulate algorithms over any data structure and to analyze their cost.

## Abstraction

The relation of proportionality in a multiplicative group,

$$\rho(x, y, u, v) :\leftrightarrow x \cdot y^{-1} = u \cdot v^{-1}$$

can be characterized by simple axioms. Let $T$ be the elementary theory of $\rho$ with these axioms, let $T_1$ be the elementary theory of groups with multiplication ($\cdot$), inverse ($^{-1}$), and unit $e$ as primitives, and let $T_2$ be $T$ plus the constant $e$ without additional axioms. Then $T_1$ and $T_2$ are strongly equivalent in that their basic notions are interdefinable, and thus they have the same models up to this change of primitives. Richard Büchi and Jesse B. Wright in publications 5 and 6 call $T$ an abstraction of $T_2$ (and thus of group theory) in either of the following senses:

(i) $T$ is weaker than $T_2$ in expressive power, namely, the constant $e$ is not definable in $T$; but $T$ and $T_2$ have the same models up to adding or deleting $e$.

(ii) The automorphism group of $T$ extends that of $T_2$, namely, by the translations (which are not group automorphisms).

Another example of abstraction appears already in Richard Büchi's doctoral thesis (1). These observations were the starting point for his long-lasting work towards a theory of definability, incorporating Craig's lemma and Beth's definability theorem. Part of the work was done together with Kenneth Danhof, and is published in (21–23). The manuscript announced in abstract 11 was not published, because he refused to shorten it.

*Acknowledgment.* I am grateful to Sylvia Büchi, Leonard Lipshitz, and Walter Schnyder who helped me through the ups and downs of discovering what Richard Büchi had left behind. I also thank William Fenton and Steven Senger, his two last doctoral students, for helpful conversations, and Ernst Specker for arranging with the E.T.H. Zürich to store the Büchi material.

# The Role of Büchi's Automata in Computing Science

*E. Allen Emerson**

Automata on infinite objects (strings) were introduced in the early 1960s by Büchi (1), motivated by issues in mathematical logic, that is, the decidability of his monadic second-order theory of one successor (S1S). This logical theory was actually referred to by Büchi as the Sequential Calculus because of its potential utility for describing the behavior of sequential circuits. Somewhat later, Muller (7) proposed that automata on infinite strings might themselves be applicable to describing the behavior of circuits, in particular, nonstabilizing asynchronous digital circuits. By the end of the 1960s, Rabin (9) had shown how automata on infinite trees, a natural but very powerful generalization of Büchi's automata on strings, could be used to show decidability of a surprisingly rich class of logical theories including the basic monadic second-order theory of multiple successors (SnS).

We emphasize that these automata, while they accept input strings of infinite length $\omega$ (or input trees with paths of length $\omega$), are still finite-state devices. For example, a Büchi string automaton with acceptance condition specified by a set of state GREEN, accepts an input string if there is a path through the automaton starting from its initial state leading through a GREEN state infinitely often. For this reason, testing nonemptiness of such an automaton (i.e., whether there exists some infinite input string the automaton accepts) is a decidable problem that amounts to checking if there is a path through the automaton leading to a cycle that includes a GREEN state.

In the 1970s there was relatively little interest in these automata. There was some theoretical work on automata with infinite state spaces such as pushdown tree automata. However, the decision problems usually became undecidable. Thus, while of some theoretical interest, it did not appear to have major impact on computing science.

However, over the past 12 years (1977–89) there has been a significant resurgence of interest in finite automata on infinite strings and trees (henceforth referred to simply as automata), owing to their close relationship with nonterminating concurrent computer programs and formalisms for specifying and reasoning about such programs.

* University of Texas at Austin.

An ordinary sequential program has a transformational character. It starts executing in an initial state corresponding to a given set of input data values. It then executes for some finite number of steps, whereupon it terminates in a particular final state corresponding to a certain set of output data values. Its semantics can be described as a relation from initial to final states. Most such programs are deterministic, so the relation is actually a function (or transformation). It is thus possible to specify correct behavior using precondition/postcondition pairs that assert that if the program starts executing in a state satisfying the precondition it will terminate in a final state satisfying the postcondition. Examples of such programs abound and include most data processing programs; for a trivial example consider the program that takes as input a postive integer and outputs the integer squared.

In contrast, a continuously operating concurrent program exhibits ongoing behavior that is ideally nonterminating. Examples of such programs include computer operating systems, network communication protocols, air traffic control systems, and nonstabilizing asynchronous digital circuits. For these programs, termination represents an abnormal condition while normal behavior entails executing forever. Thus, such programs are not naturally described using the initial state/final state semantics used for transformational programs. The semantics of these programs is instead described in terms of infinite computation sequences recording the states the program goes through as it executes. For a given initial state, there are in general infinitely many computation sequences beginning with that state, because the programs are usually highly nondeterministic. This reflects our modeling of the execution of the parallel program $P_1//\ldots//P_k$ composed of the sequential programs $P_1, \ldots, P_k$ in the usual way, which is to treat it as the nondeterministic interleaving of the atomic steps (individual indivisible instructions) of each sequential program $P_i$. A concurrent program currently in a particular initial state nondeterministically selects which sequential program to execute a step of next. After executing that step the program is in a new state from which it makes another choice of next step. And so on. Thus, the set of all possible computation sequences the program may follow starting from a given state can be seen to in general contain infinitely many sequences. In many cases, additional nondeterminism arises from the program's interaction with the external world (e.g., is a character input from the keyboard or not?). Finally, one additional requirement that we might wish to impose is that each computation sequence be "fair" in the sense that each sequential program $P_i$ is executed infinitely often along the sequence.

In 1977 Pnueli (8) proposed the use of temporal logic for reasoning about continuously operating concurrent programs. Temporal logic is a type of modal logic that provides a formalism for describing how the truth values of assertions vary over time. While there are a variety of different systems of temporal logic, typical temporal operators or modalities include $Fp$ ("sometime $p$"), which is true now provided there is a future moment where $p$ holds, and $Gp$ ("always $p$"), which is true now provided that $p$ holds at all future moments. As Pnueli argued, temporal logic seems particularly well-suited to describing correct behavior of continuously operating concurrent programs. For example, to assert that two concurrently executing programs are not in their respective critical sections at the same time, one can write $G(\neg inCS_1 \vee \neg inCS_2)$, where $inCS_i$, $i = 1, 2$, is an atomic proposition true exactly when process $P_i$ is in its critical section. This mutual exclusion request for service is eventually granted, one can write $G(\text{req} \Rightarrow F\text{grant})$. This type of liveness requirement is typical of what might be expected of "resource controller" that monitors access to a shared resource.

Following the publication of Pnueli's 1977 paper, temporal logic has become an active area of research interest. There is now a consensus among a large body of practicioners as well as theoreticians that the various forms of temporal logic comprise a very promising approach for specifying and reasoning about continuously operating concurrent programs. Concommittant with the rising interest in temporal logic, there has also been a resurgence of interest in automata on infinite objects. This interest arises from the intimate connection between the two formalisms. Today we find finite-state automata on infinite objects used or proposed for use in virtually every capacity for reasoning about ongoing concurrent programs. Automata provide not only important technical machinery, but also a convenient and uniform conceptual framework.

One ambitious application is the mechanical synthesis of a concurrent program from a specification of its correct behavior formulated in a system of propositional temporal logic (cf. 4). These propositional logics are surprisingly expressive, and are sufficient for a large variety of applications. Synthesis is possible because these logics are decidable, since they possess the small model property: if a formula is satisfiable, then it has a finite model of bounded size. This finite model can be viewed as the global state graph of a concurrent program meeting the specification, from which its constituent sequential subprograms can be obtained. An unsatisfiable formula, on the other hand, indicates that the specification is inconsistent and must be reformulated.

For simple logics, the decision procedure can be obtained using various tableau constructions, similar to those used in classical modal logic. A tableau is a graph constructed in a systematic way from the syntax of the formula that essentially encodes all potential models of a formula. A uniform method for developing decision procedures can be obtained by observing that a tableau defines the transition diagram of an automaton with the Büchi acceptance condition. The formula is satisfiable if the Büchi automaton accepts some input. So the decision procedure is obtained by reduction to the nonemptiness problem for Büchi automata.

For richer logics, the tableau construction is no longer applicable. However, it is still possible to reduce satisfiability to testing nonemptiness of a finite-state automaton on infinite strings or trees by appealing to such delicate automata-theoretic constructions as McNaughton's or Safra's algorithm for determinizing a Büchi automaton on infinite strings (cf. 5, 10) and by allowing a more complicated acceptance condition (essentially boolean combinations of Büchi's acceptance condition). An alternative way to get a decision procedure is by encoding the logics in SnS. Unfortunately, this yields a decision procedure of nonelementary complexity, i.e., not bounded by the composition of a fixed number of exponential functions, $\exp(n) = c^n$, for some constant $c > 1$ (cf. 6). The only known way to get to elementary time decision procedures is by directly reducing from the logic to the automaton. In fact, it was this application—to improve the complexity of decision procedures for temporal logics—that was the primary motivation in computing science for resuming the study of automata on infinite objects (cf. 11).

If we restrict our attention to finite-state concurrent programs, the problem of determining whether a given program meets a given specification expressed in temporal logic (or a related formalism) is always decidable, and in many cases quite practical and useful. It amounts to a brute force graph reachability analysis. Very efficient algorithms can be developed, which run in time linear in the size of the program and in many cases linear or polynomial in the size of the specification to check whether the program defines a model of the specification formula. This "model checking" approach to mechanical verification of finite-state concurrent programs is still of wide applicability, since many solutions to concurrent programming problems in the literature are finite state, as are many network communication protocols (cf. 2, 3).

Automata have also been proposed as a useful and general specification language themselves. Some would argue that it is more convenient and comprehensible for humans to write specifications as automata rather than in temporal logic. In fact, automata provide strictly more expressive power than (ordinary) temporal logic. The property $G_2 p$, meaning that at all even moments $p$ holds, is easily described by an automaton, but not by any temporal logic formula. Moreover, automata provide a sort of yardstick for measuring the expressive power of various systems of temporal logic and related formalisms for describing sequences. Ordinary temporal logic is equivalent to counter-free automata, star-free $\omega$-regular expressions, and the first-order language of linear order. Büchi automata, on the other hand, provide us with the full power of $\omega$-regular expressions, the second-order language of linear order, and S1S.

Additionally, automata are useful as a representation of concurrent programs, with the global state graph corresponding to the automaton's transition diagram in an obvious way, and any needed fairness requirements captured by the automaton's acceptance condition. Since both programs and specifications can be naturally represented using automata, it becomes possible to formulate and compare different approaches to mechanical program synthesis, automatic program verification, and model checking within the uniform framework of automata (cf. 12).

To sum up, we recall the twin origins of automata on infinite objects. First, Büchi introduced them for an essentially theoretical purpose: to give a decision procedure for testing the validity of sentences of S1S. There was also a more practical motivation: describing the behavior of digital circuits. Today, Büchi's and related automata are studied from both a theoretical and a practical viewpoint. But now, the theoretical and practical have blended together to a large extent. A noteworthy point is that the central technical use of automata by Büchi—to provide a decision procedure for a logical theory by reduction to the emptiness problem for the automata—remains today the main use of such automata in connection with logical theories, such as temporal logic, for reasoning about program correctness.

## *References*


1. Büchi, J.R. (1962). On a decision method in restricted second order arithmentic. *Proc. 1960 International Congress on Logic, Methodology, and Philosophy of Science*, 1–11.
2. Clarke, E.M., Emerson, E.A., and Sistla, A.P. (1986). Automatic verification of finite-state concurrent systems using temporal logic specifications. *ACM Transactions on Prog. Lang. and Sys.*, 8 (2), 244–263.
3. Clarke, E.M., and Grumberg, O. (1987). Research on automatic verification of finite state concurrent systems. *Annual Reviews in Computer Science*, 2, 269–290.
4. Emerson, E.A., and Clarke, E.M. (1982). Using branching time logic to synthesize synchronization skeletons. *Science of Computer Programming*, 2, 241–266.
5. McNaughton, R. (1966). Testing and generating infinite sequences by a finite automaton. *Information and Control*, 9, 521–530.
6. Meyer, A.R. (1974). Weak monadic second order theory of successor is not elementary recursive. *Boston Logic Colloquium*. Lecture Notes in Mathematics 453, Springer-Verlag.
7. Müller, D.E. (1963). Infinite sequences and finite machines. *Proc. 4th Annual IEEE Symposium on Switching Circuit Theory and Logical Design*, 3–16.
8. Pnueli, A. (1977). The temporal logic of programs. *19th IEEE Symp. on Foundations of Computer Science*, 1–14.
9. Rabin, M.O. (1969). Decidability of second order theories and automata on infinite trees. *Trans. AMS*, 141, 1–35.
10. Safra, S. (1988). On complexity of $\omega$-automata. *29th IEEE Symposium on Foundations of Computer Science*, 319–327.
11. Streett, R.S. (1982). Propositional dynamic logic of looping and converse is elementary decidable. *Information and Control*, 54, 121–141.
12. Vardi, M.Y., and Wolper, P.L. (1986). An automata-theoretic approach to automatic program verification. *Proc. IEEE Symp. on Logic in Computer Science*, 332–346.

# J. Richard Büchi's Doctoral Students

William Henry Hosken (1966)—Canonical Systems Which Produce Regular Sets.

Lawrence Hugh Landweber (1967)—A Design Algorithm for Sequential Machines and Definability in Monadic Second Order Arithmetic.

Gary Martin Haggard (1968)—Embedding of Graphs in Surfaces.

Kenneth Joe Danhof (1969)—On Definability and the Cantor Method in Model Theory.

Peng-Siu Mei (1971)—Linear Closure Spaces and Matroids, Convex Closure Spaces and Paramatroids.

Jean-Louis Lassez (1973)—On the Relationship between Prefix Codes, Trees and Automata.

Charles Zaiontz (1974)—Automata and the Monadic Theory of Ordinals $<\omega_2$.

Terrence Michael Owens (1975)—Varieties of Skolem Rings.

William Ellis Fenton (1982)—Axiomatic Convexity Theory.

Steven Orville Senger (1982)—The Existential Theory of Concatenation over a Finite Alphabet.

# Abstracts Published by J. Richard Büchi


1. (with Jesse B. Wright) 1954, Abstraction versus generalization. *Proc. Int. Congress of Mathematics*, North-Holland: Amsterdam p. 398.
2. (with Jesse B. Wright) 1956, A fundamental concept in the theory of models. *J. Symb. Logic*, 21, p. 110.
3. (with William Craig) 1956, Notes on the family $PC_\Delta$ of sets of models. *J. Symb. Logic*, 21, pp. 222–223.
4. (with Calvin C. Elgot and Jesse B. Wright) 1958, The nonexistence of certain algorithms of finite automata theory. *Notices AMS*, 5, p. 98.
5. (with Calvin C. Elgot) 1958, Decision problems of weak second order arithmetic and finite automata, I. *Notices AMS*, 5, p. 834.
6. 1959, Regular canonical systems and finite automata. *Notices AMS*, 6, p. 618.
7. 1960, On the hierarchy of monadic predicate quantifiers. *Notices AMS*, 7, p. 381.
8. 1960, On a problem of Tarski. *Notices AMS*, 7, p. 382.
9. 1961, Turing machines and the Entscheidungsproblem. *Notices AMS*, 8, p. 354.
10. 1961, Validity in finite domains. *Notices AMS*, 8, p. 354.
11. 1963, Relatively categorical and normal theories. *Proc. Int. Symp. on Theory of Models*, Berkeley, Calif., North-Holland: Amsterdam, pp. 424–426.
12. 1965, Transfinite automata recursions. *Notices AMS*, 12, p. 371.
13. 1965, Restricted second order theory of ordinals. *Notices AMS*, 12, p. 457.
14. (with Lawrence H. Landweber) 1967, Definability in the monadic second order theory of successor. *Notices AMS*, 14, p. 852.
15. 1968, Affine definability of affine invariants of Euclidean geometry. *Notices AMS*, 15, p. 932.

16. (with Kenneth J. Danhof) 1968, A strong form of Beth's definability theorem. *Notices AMS*, 15, p. 932.

17. (with Dirk Siefkes) 1971, Axiomatization of the monadic second-order theory of countable ordinals. *Tarski Symposium*, Berkeley, Calif.

18. 1971, The monadic second order theory of $\omega_1$. *Notices AMS*, 19, p. 662.

# Part 2 The Publications, with Comments

# Section 1 Boolean Algebras

*With comments by Saunders Mac Lane, The University of Chicago*



*

# Büchi's Thesis: Die Boole'sche Partialordnung und die Paarung von Gefügen

(The Boolean Partial Order and the Pairing of Frames)

*Saunders Mac Lane*

This thesis is designed to be an introduction to atom-free mathematics—that is, mathematics carried out not with the elements of a set $K$, but with the subsets of $K$, using their inclusion *and* disjointness relations. A Boolean partial order $B_0$ is first defined to be a poset $B_0$ with a least element 0 and one additional axiom: If not $X \subset Y$ then there is an element $U \subset X$ with $U \not\subset 0$ and $U$ disjoint from $Y$. Here $U$ disjoint from $Y$, in symbols $U \circ Y$, is defined by the requirement that every $V$ with $V \subset U$ and $V \subset Y$ has $V \subset 0$. Let $B$ be the set $B_0$ with the least element 0 removed. There are two relations defined on elements $X$, $Y$ of $B$: inclusion $X \subset Y$ and disjointness $X \circ Y$, and these relations satisfy four axioms ($T_1$ to $T_4$, on page 4).

A set $B$ with two such binary relations that satisfy these axioms is called a "Gefüge;" we translate this as "frame." Each of these relations can be defined in terms of the other, and every such frame $B$ arises, by the omission of 0, from a Boolean partial order. Conversely, for any frame $B$ the symbol $B_0$ denotes the same poset with a (new) bottom element 0 adjoined. There are numerous examples of frames; in particular, the set $B$ of all nonempty subsets of a set $K$ is a frame—a so-called complete frame. The first half of the thesis is then a systematic axiomatic study of frames, with definitions of (infinite) intersection and of a sharpened infinite union, and the consideration of complete frames. The use of suitable orthocomplements provides a completion theorem: Every Boolean partial order can be embedded in a (unique minimal) complete Boolean algebra (Theorems $S_1$ and $S_2$, pp. 25, 27). This is analogous to the well-known MacNeille completion of a poset by cuts.

The second part of the thesis comes to grips with the idea of "atom-free" mathematics. The central question is that of replacing a function $\varphi : K \to L$ between sets by something acting between the corresponding frames $B$ and $C$. The most general such thing is a "pairing" of frames: two functions $f: B_0 \to C_0$ and $g: C_0 \to B_0$ such that $v \circ fu$ if and only if $u \circ gv$, for all $u \in B_0$ and $v \in C_0$; note that the definition uses only the disjointness relation. Any binary relation $F$ between sets $K$ and $L$ yields such a pairing between the corresponding frames, and the usual properties of binary relations can be formulated in terms of pairings. In particular, a pairing $(f, g)$ of frames $B$ and $C$ is said to be functional in $B$ when $g$ preserves disjointness (for nonzero elements) and $u \neq 0$ implies $fu \neq 0$. When $B$ and $C$ are the frames of all nonzero subsets of sets $K$ and $L$, a functional pairing $(f, g)$ corresponds exactly to a function $\varphi$ from $K$ to $L$, and in the correspondence $f$ sends each $X \in B$ (each subset $X \subset K$) into its image $\varphi X$, while $g$ is the operation "inverse image."

This then provides a framework for an atom-free mathematics, formulated in terms of inclusion and disjointness. There are some applications to point-set topology, in the context of Kuratowski's axiomatics on the closure operation on subsets. There is also reference to an earlier paper by Schoenflies, who provided atom-free axioms for the cardinal numbers aspects of Cantorian set theory.

This atom-free approach has, as best I known, not been carried further. There are currently active subjects that might be regarded as atom-free—for instance, topos-theoretic axioms for sets, not using elements, or the typed $\lambda$-calculus, or the study of topological spaces as locales (of open sets). These, however, do not fall under Büchi's use of inclusion and disjointness. Sikorski's well-known book on Boolean algebra (*Ergebnisse der Mathematik*, Vol. 25, 1964) does quote one theorem of Büchi's thesis: If $S$ is a poset such that whenever $A$ is not contained in $B$ there is $C \subset A$ disjoint from $B$, then $S$ is a dense subjset of a Boolean algebra in which the partial order is an extension of that of $S$. (That is, every Boolean partial order is a dense subset of the partial order of a Boolean algebra.) This is a version of Büchi's completion theorem.

*

# Representation of Complete Lattices by Sets

*Saunders Mac Lane*

This paper of Büchi's is a systematic study of the following problem: When is a complete lattice $L$ isomorphically represented by a complete ring of sets? This problem and its extension turn out to have some fascinating and non-obvious aspects.

By definition, a complete lattice $L$ is a poset with a largest element $e$ in which arbitrary subsets $a_i$ of $L$ have g.l.b.'s $\cap a_i$. This is known to imply that arbitrary subsets also have l.u.b.'s $\cup a_i$. An example is a "complete set lattice" $L$: For any set $e$, a collection $L$ of subsets of $e$ closed under arbitrary intersection (set-intersection). However, in such a complete set lattice, l.u.b. in the lattice sense need not be the set-theoretic union. (There are easy examples; for instance, omit all two-element subsets from the Boolean algebra of all subsets of $\{1, 2, 3\}$.) In a complete set lattice $L$ when all the *finite* l.u.b.'s are exactly the set-theoretic unions, call $L$ a complete $F$-set lattice, with $F$ for "finite," and similarly for other classes ($C$-set lattice for all countable unions; $A$-set lattice when *all* l.u.b.s are set-theoretic unions). Thus a complete $A$-set lattice is just a complete ring of sets (a collection of subsets of a set $E$ closed under arbitrary set-theoretic union and intersection).

Call an element $u$ of $L$ *subirreducible* (or $F$- or $C$-subirreducible) if whenever $u \subset$ l.u.b.$a_i$, then $u$ is contained in some one $a_i$ (for $F$, for a finite set of $a_i$; for $C$, for a denumerable set). Then Büchi's Theorem 15 asserts that a complete lattice $L$ is isomorphic to a complete $F$-set lattice (or a complete $C$-set lattice) if and only if every element of $L$ is an l.u.b. of $F$- or $C$-subirreducible elements; also that $L$ is isomorphic to a complete ring of sets if and only if every element of $L$ is the l.u.b. of subirreducible elements. The latter case (complete ring of sets) of this theorem was also proved independently by Raney in a 1952 paper cited below and by V.K. Balachandran (1954, *Fundamenta Math*, 41, 38–41). This paper was submitted on December 9, 1952, three months after Büchi's paper was submitted.

Büchi then formulated a related theorem assuming distributivity. Call an element $u$ in $L$ *irreducible* when $u = \text{l.u.b.}a_i$ implies $u$ equal to some $a_i$ (or $F$-irreducible, or $C$-irreducible if this holds for finite or denumberable l.u.b.). Consider the distributive law $b \cap (\text{l.u.b.}a_i) = \text{l.u.b.}(b \cap a_i)$. Then (Theorem 19) a complete lattice $L$ is isomorphic to a complete ring of sets (or to a $F$- or $C$-set lattice) if and only if it satisfies the corresponding distributive law and every element is a union of irreducible (or $F$-irreducible or $C$-irreducible) elements. Actually, his Theorem 19 is formulated more generally for $F$ and $C$ replaced by a much more general $N$. In this general form (which is perhaps of less interest) the theorem is wrong ("only if" fails), because of the counter-example stated in the review by M. Novotny (1953, *Mathematical Reviews*, Vol. 14, p. 940). The corresponding correction is written in, in Büchi's handwriting, on his copy of this paper.

For subsets, the union and intersection operation satisfy the most general distributive law—in particular, arbitrary g.l.b.'s distribute over arbitrary l.u.b.'s as in Büchi's equation $\bar{D}$, page 158, foot. However, this distributive law for a complete lattice is not enough to give an isomorphism theorem. Büchi proves (Theorem 35) that a complete lattice that satisfies this distributive law $\bar{D}$ is the homomorphic image of a complete ring of sets (called by Büchi an "A"-set lattice). This theorem (communicated by Büchi in September 1952) had also been proved by G.N. Raney (1952, *Proc. Amer. Math. Soc.*, 3, 677–680) in a paper communicated on January 7, 1952. These two discoveries of this result seem to be independent. This theorem is well known; it appears for example on page 120 in the third edition of Garrett Bürkhoff's *Lattice Theory* (1967, Amer. Math. Soc. Colloquium Publication, Vol. XXV, 3d ed.) as Raney's theorem. It might well be called the Raney-Büchi theorem.

To summarize, this paper presents difficult and interesting theorems about complete lattices. It represents a real advance on Büchi's earlier work in his thesis, in that the present paper is closer to mainstream concerns (complete lattices) and reaches a number of definite results.

Shortly after this work was completed, Büchi's research interests turned to problems concerning automata, formal language, and logic.

The Raney-Büchi theorem has played an important role in the recent active development of the subject of continuous lattices. It is used in the compendium of Gierz et. al. [2] and has appeared in a number of related papers (there are citations of Raney but not Büchi in J.D. Lawson [3] and D. Novak [4]). Both authors are cited in a current paper by Fawcett and Wood [1], who make extensive use of Büchi's distributivity of infs over various classes of sups. Galois connections and other adjunctions play a prominent role in several of these studies.

For example (Fawcett-Wood), if $DL$ is a class of subsets of the lattice $L$, and if $L$ has a supremum for each subset in $DL$, then the axiom of choice and complete distributivity for $DL$, in the sense of Büchi, implies that the join map $v$: $DL \to L$ has both a left and a right adjoint, and a resulting map of $DL$ into the power set of $L$ exhibits $DL$ as a complete ring of sets.

## *References*

1. Fawcett, B., and Wood, R.J. Constructive complete distributivity I. *Proc. Camb. Phil. Soc.*, to appear.
2. Gierz, G., Hofmann, K.H., Keimel, K., Lawson, J.D., Mislove, M., and Scott, D.S. (1980). *A compendium of continuous lattices*. Berlin, Heidelberg, New York: Springer, 371 pp.
3. Lawson, J.D. (1979). The duality of continuous posets. *Houston J. Math.*, 5, 351–386.
4. Novak, D. (1988). Generalizations of continuous posets. *Trans. AMS*, 272, 645–667.

# DIE BOOLE'SCHE PARTIALORDNUNG UND DIE PAARUNG VON GEFÜGEN

VON DER

EIDGENÖSSISCHEN TECHNISCHEN HOCHSCHULE IN ZÜRICH

ZUR ERLANGUNG

DER WÜRDE EINES DOKTORS DER MATHEMATIK

GENEHMIGTE

PROMOTIONSARBEIT

VORGELEGT VON

J. RICHARD BÜCHI

AUS ZELL, ZÜRICH UND PORTO ALEGRE, BRASILIEN

Referent:
Prof. Dr. P. Bernays

Korreferent:
Prof. Dr. F. Gonseth

1950

PORTUGALIAE MATHEMATICA — LISSABON

# VORWORT

Ich will diese Gelegenheit nicht versäumen Herrn Professor Dr. Paul Bernays meinen herzlichen Dank auszusprechen für das Interesse, das er meiner Arbeit entgegengebracht hat. In eingehenden Besprechungen durfte ich sein reichhaltiges Wissen in Anspruch nehmen, und seine aufbauende Kritik hat die Arbeit in vielen Teilen gefördert.

Auch allen Mitgliedern der Mathematischen Abteilung der E. T. H., deren Vorlesungen ich während meiner Studienzeit besuchen durfte, fühle ich mich zu grösstem Dank verpflichtet. Insbesondere war es Herr Professor Dr. Heinz Hopf, der mir immer wieder Verständnis entgegenbrachte und mir mit mannigfachen Ratschlägen zur Seite stand. Herrn Professor Dr. Ferdinand Gonseth möchte ich die Arbeit verdanken, die ein Korreferat mit sich bringt und die er in zuvorkommender Weise auf sich nahm.

Widmen will ich meine Doktorarbeit meinen lieben Eltern. Ihrer stets treuen und aufopfernden Liebe verdanke ich es vor allem, wenn ich heute auf eine erfolgreiche Schul und Studienzeit zurückblicken darf.

St. Gallen. Dezember 1948.

J. Richard Büchi

# DIE BOOLE'SCHE PARTIALORDNUNG UND DIE PAARUNG VON GEFUEGEN [1]

von J. Richard Büchi (St. Gallen)

## EINLEITUNG

Der Begriff des Teiles ist in der heutigen Mathematik eng an denjenigen der Menge von Elementen gebunden. Jedes mathematische Ganze ist eine Menge, jeder Teil des Ganzen eine Teilmenge.

Diese Beschränkung des Teilbegriffes auf denjenigen der Teilmenge liegt nun aber gar nicht im Wesen der Sache. Es besteht daher das Bedürfnis nach einer mathematischen Theorie, die der Wechselbeziehung zwischen dem Ganzen und seinen Teilen die volle Freiheit lässt.

Versucht man eine solche mathematische Teillehre axiomatisch zu begründen, so stellt man zunächst fest, dass der Teilbegriff aufs Engste mit demjenigen der partiellen Ordnung verbunden ist. Ebenso elementar, und das wird leicht verkannt, ist aber der Begriff der Fremdheit von Teilen.

Im 1. Teil der Arbeit wird nun eine Axiomatik gegeben, die die allgemeinste sinnvolle Koppelung von Inklusion und Fremdheitsbeziehung gewährleistet.

Diese Axiomatik hat an und für sich als allgemeine Teillehre ihre Bedeutung. Darüber hinaus wird die abstrakte Theorie des 1. Teiles zeigen, dass sie die Grundlage zur Behandlung der Boole'schen Algebren in ähnlicher Weise ist, wie die Theorie der partiellen Ordnung diejenige der Verbandstheorie.

[1] Von der Eidgenössisch Technischen Hochschule in Zürich genehmigte Promotionsarbeit. Eingereicht im Dezember 1948.

Dass schliesslich der Begriff der Boole'schen Partialordnung auch in anderen mathematischen Disziplinen fruchtbar ist, zeigt die in Teil I. § 8 gegebene Anwendung auf die Theorie der topologischen Räume.

Im 2. Teil soll das Problem der Abbildung von allgemeinen Teilen zur Diskussion stehen. Da ein Teil nicht aus Atomen zu bestehen braucht versagt der klassische Abbildungsbegriff, der ja nur für Mengen definiert ist und sich wesentlich auf die Elemente stützt. Im Begriff der Paarung von Teilen ist eine Verallgemeinerung der «Relation zwischen Elementen» auf den Fall nicht atomarer Objekte gefunden. Mit Hilfe der Paarung wird dann ausser anderem die Definition der Abbildung atomfreier Objekte gelingen.

So zeigt die Arbeit die Bedeutung des «Gefüges» und seiner «Paarungen» in verschiedenen mathematischen Disziplinen. Es scheint aber, dass den genannten Begriffen ein viel umfassenderes Anwendungsgebiet offen steht.

Wegen genauerer Beschreibung der Resultate sei auf die Vorworte zu den einzelnen Teilen sowie auf das Schlusswort verwiesen.

# 1. TEIL

# DAS GEFÜGE UND DIE BOOLE'SCHE PARTIALORDNUNG

Die Boole'sche Partialordnung gehört, wie auch etwa der Verband oder die Boole'sche Algebra in die Theorie der partiellen Ordnung. An Stelle der für den Verband typischen Existenzforderungen (kleinstes gemeinsames Grösseres usw) tritt dann ein Axiom, welches die Inklusionsrelation näher umschreibt. Es wird die «Fremdheit» zweier Elemente definiert.

In § 1 werden wir von dem zur «Boole'schen Partialordnung» äquivalenten Begriff des «Gefüges» ausgehen. Dabei werden zwei Relationen «sub» und «dis» postuliert, die etwa als Inklusion und Fremdheitsbeziehung veranschaulicht werden können. Dieses Vorgehen ist dadurch gerechtfertigt, dass die Relation «dis» in der ganzen Theorie eine der Inklusion mindestens ebenbürtige Rolle spielen wird.

Die Theorie der Gefüge hat als typisches Beispiel einer in sich geschlossenen Axiomatik an und für sich ihre Bedeutung. Durch die Vervollständigungstheorie, wie sie in § 5 gegeben wird, tritt sie in enge Beziehung zu den Boole'schen Algebren. Ja, es darf gesagt werden, dass die Boole'sche Partialordnung die Grundlage der Theorie der Boole'sche Algebren, in analoger Weise ist, wie die teilweise Ordnung diejenige der allgemeinen Verbandstheorie.

Die Resultate des 1. Teiles lassen sich etwa so beschreiben: Ueber die Boole'sche Partialordnung gelangt man in natürlicher Weise zu einem Axiomenystem für vollständige, d. h. total additive Boole'sche Algebren. Sodann gelingt es, einen Satz über die Einbettung einer Boole'sch partiellgeordneten Menge in eine vollständige Boole'sche Algebra zu beweisen, der ein Analogon zum entsprechenden MacNeille'schen Satz über die

Einbettung von partiell geordneten Mengen in vollständige Verbände ist. Auch auf den Begriff des Atoms in Boole'schen Algebren wird die Theorie einiges Licht werfen. So entpuppt sich ein Kriterium für atomistische Boole'sche Algebren von Tarski als Spezialfall eines solchen für Gefüge.

Durch diese Resultate ist der Begriff des Gefüges zur Genüge legitimiert. Es bleibt vielleicht noch zu erwähnen, dass er, allerdings in impliziter Weise, schon vielfach verwendet wurde. So hat z. B. Huntington schon 1924 ein Axiomensytem für Boole'sche Algebren angegeben, dessen erste Axiome gerade eine Boole'sche Partialordnung gewährleisten. Schliesslich zeigt § 8, dass das Gefüge auch in der Theorie der topologischen Räume seine Bedeutung hat.

## § 1 DAS GEFÜGE UND DIE BOOLE'SCHE PARTIALORDNUNG

Den nachfolgenden Betrachtungen liegt immer ein vorgegebener Bereich von Dingen zu grunde. Zwar wird gelegentlich die Teilklassenbildung zur Konstruktion neuer Elemente herbeigezogen; dies geschieht aber höchstens in zweiter Stufe. Man hat daher keine Schwierigkeiten in der Form mengentheoretischer Antinomien zu befürchten.

1. Axiome des Gefüges: Ein Bereich B von Dingen $(a, b, \ldots)$ heisst ein Gefüge, wenn zwei Relationen «$x$ sub $y$» (in Zeichen $x \subset y$) und «$x$ dis $y$» (in Zeichen $x \circ y$) zwischen den Elementen von B gegeben sind, die folgende Axiome erfüllen: Für alle $x, y, z$ aus B gilt [1]

T$_1$) $$\overline{(x \circ y \wedge y \subset x)}$$

T$_2$) $$(x \circ y \wedge z \subset y) \rightarrow (x \circ z)$$

T$_3$) $$\bigwedge_u (u \circ y \rightarrow u \circ x) \rightarrow (x \subset y)$$

T$_4$) $$\bigwedge_u (u \mathrel{\overline{\subset}} x \vee u \mathrel{\overline{\subset}} y) \rightarrow (x \circ y).$$

Zur Veranschaulichung der Axiome, sowie der nachfolgenden Formeln stelle man sich die Relationen sub und dis als Inklusion und Fremdheitsbeziehung zwischen Mengen vor.

[1] Es werden die üblichen logischen Zeichen gebraucht. d. h.: $\wedge$ Konjunktion; $<$ Disjunktion; $\rightarrow$ Implikation; — Negation; weiter bedeutet $\bigwedge_u$ «für alle $u$ aus B» und $\bigvee_u$ «es existiert $u$ in B.» Sodann werden die Zeichen $\overline{\subset}$ und $\bar{o}$ für die Negation von $\subset$ und $o$ benützt.

Es sollen nun die folgenden Formeln für beliebige Elemente $x, y, z$ des Gefüges B nachgewiesen werden:

$c_1)$ $(x \subset x)$

$c_2)$ $(x \subset y \wedge y \subset z) \rightarrow (x \subset z)$

$o_1)$ $(x \,\bar{\circ}\, x)$

$o_2)$ $(x \circ y) \rightarrow (y \circ x)$

$S_1)$ $(x \subset y) \rightarrow \bigwedge_u (u \circ y \rightarrow u \circ x)$

$S_2)$ $(x \circ y) \rightarrow \bigwedge_u (u \,\bar{\subset}\, x \vee u \,\bar{\subset}\, y)$

$S_3)$ $(x \,\bar{\subset}\, y) \rightarrow \bigvee_u (u \subset x \wedge u \circ y)$.

*Beweise:*

$S_1$: Es sei $x \subset y$. Dann folgt wegen $T_2$ aus $u \circ y$ immer auch $u \circ x$. Dies ist aber gerade der Inhalt der Formel $S_1$.

$S_2$: Es sei $x \circ y$. Dann gilt wegen $T_2$ mit $u \subset y$ auch $x \circ u$ und daher wegen $T_1$ $u \,\bar{\subset}\, x$. Somit gilt die Formel $S_2$.

$c_1$: Trivialer Weise gilt für jedes $u$ aus B: $(u \circ x \rightarrow u \circ x)$. Nach $T_3$ gilt also $x \subset x$.

$c_2$: Es sei $(x \subset y \wedge y \subset z)$. Dann folgt durch zweimalige Anwendung von $T_2$ aus $u \circ z$ auch $u \circ x$. Nach $T_3$ gilt dann $x \subset z$.

$o_1$: Wegen $c_1$ gilt $x \subset x$ also nach $T_1$ auch $x \not\circ x$.

$o_2$: $T_4$ und $S_2$ geben zusammen ein Kriterium für $x \circ y$. Der Ausdruck $\bigwedge_u (u \,\bar{\subset}\, x \vee u \,\bar{\subset}\, y)$ ist aber symmetrisch in $x$ und $y$, also ist es auch der äquivalente $x \circ y$.

$S_3$: Es sei $x \,\bar{\subset}\, y$. Wegen $T_3$ existiert dann ein $v$ in B, sodass $(v \,\bar{\circ}\, x \wedge v \circ y)$. Wegen $T_4$ gibt es auch ein $u$ in B, sodass $(u \subset v \wedge u \subset x)$. Nach $T_2$ folgt schliesslich aus $(u \subset v \wedge v \circ y)$ auch $u \circ y$. Damit ist ein $u$ in B gefunden, welches die Formel $(u \subset x \wedge u \circ y)$ befriedigt.

$S_1$ und $S_2$ sind die Umkehrungen der Implikationen $T_3$ und $T_4$. Sie zeigen zusammen mit diesen, dass sich die Relationen sub und dis gegenseitig eindeutig bestimmen. Dadurch wird nahe gelegt, nur eine der Relationen z. B. sub zu postulieren und dann dis mit Hilfe der Implikationen $T_4$ und $S_2$ zu definieren. Man wird dann nahe liegender Weise die Formeln $c_1$ und $c_2$ als Axiome für sub wählen; d. h. voraussetzen, dass sub den Bereich B partiell ordnet. Diese Axiome allein sind aber nicht hinreichend. Die Formel $S_3$ braucht beispielsweise nicht erfüllt zu sein. Man kann sie aber als drittes Axiom postulieren und kommt so zur Boole'schen Partialordnung.

**2.** Die Boole'sche Partialordnung: $B_0$ sei ein Bereich von Elementen $a, b, \ldots, x, \ldots$ der bezüglich einer Relation sub partiell geordnet ist, sodass ein kleinstes Element $o$ existiert. D. h. es gelten für alle $x, y, z$ aus $B_0$ die Formeln:

$c_0$) $$(o \subset x)$$

$c_1$) $$(x \subset x)$$

$c_2$) $$(x \subset y \wedge y \subset z) \rightarrow (x \subset z).$$

Durch die folgende Formel ($f$) ist dann eine Relation dis in $B_0$ definiert:

$f$) $$(x \subset y) \longleftrightarrow \bigwedge_u (u \subset x \wedge u \subset y \rightarrow u \subset 0).$$

Die Relation sub heisst dann eine Boole'sche Partialordnung des Bereiches $B_0$, wenn das folgende Axiom erfüllt ist:

$c_3$) $$(x \mathrel{\bar{\subset}} y) \rightarrow \bigvee_u (u \mathrel{\bar{\subset}} o \wedge u \subset x \wedge u \circ y).$$

Zunächst ist nicht jede Partialordnung auch Boole'sch, wie die Beispiele der linear, das heisst im gewöhnlichen Sinne geordneten Mengen zeigen. Dann gilt:

*a)* Der Bereich B aller Elemente $x$ eines Boole'sch geordneten Bereiches $B_0$, die nicht sub o sind, bilden ein Gefüge bezüglich der Relation sub und der nach der Formel ($f$) definierten Relation dis.

*Beweis*: Es gilt: ($i$) $(x \in B) \longleftrightarrow (x \in B_0 \wedge x \mathrel{\bar{\subset}} 0)$. Schon aus den Formeln $c_0, c_1, c_2$ und $f$ folgert man dann die Axiome $T_1, T_2, T_4$ wie folgt:

$T_1$: Aus $x \circ y \wedge y \subset x$ folgt wegen $f$: $(y \mathrel{\bar{\subset}} y \wedge y \subset 0)$. Wegen $c_1$ gilt also $y \subset 0$ und damit wegen $i$ auch $y \notin B$. Die Formel $x \circ y \wedge y \subset x$ kann also niemals für Elemente aus B zutreffen.

$T_2$: Es sei $x \circ y \wedge z \subset y$. Ist dann $u \subset x \wedge u \subset z$, so gilt wegen $z \subset y$ nach $c_2$ auch $u \subset x \wedge u \subset y$ und daher wegen $x \circ y$ nach $f$ schliesslich $u \subset 0$. Für jedes $u$ aus $B_0$ ist damit die Formel $(u \subset x \wedge u \subset z) \rightarrow (u \subset 0)$ nachgewiesen. Wegen $f$ gilt daher $x \circ z$. Zusammenfassend ist dann die Formel $T_2$ im Bereich $B_0$ nachgewiesen. Sie gilt dann um so mehr auch in B.

$T_4$; Gilt für alle $u$ in B die Formel $(u \mathrel{\bar{\subset}} x \vee u \mathrel{\bar{\subset}} y)$, so gilt wegen $i$ und $c_0$ für alle $u$ in $B_0$ die Formel $(u \subset x \wedge u \subset y \rightarrow u \subset 0)$. Wegen $f$ ist dann $x \circ y$. Damit ist das Axiom $T_4$ nachgewiesen.

Es bleibt nur noch das Axiom $T_3$ nachzuweisen. Dazu braucht man aber auch die Formel $c_3$.

$T_3$: Es sei $x \,\bar{\subset}\, y$. Nach $c_3$ existiert dann ein $u$ in $B_0$, sodass $u \,\bar{\subset}\, 0 \wedge u \subset x \wedge u \circ y$. Wegen $i$ folgt aus $u \in B_0 \wedge u \,\bar{\subset}\, 0$ auch $u \in B$ und aus $u \subset x \wedge u \,\bar{\subset}\, 0$ wegen $f$ und $c_1$ auch $u \,\bar{\circ}\, x$. Zusammenfassend gilt also für $u$: $u \in B \wedge u \not\circ x \wedge u \circ y$. Damit ist dann auch $T_3$ in B nachgewiesen.

Weiter gilt die folgende Umkehrung von $(a)$:

*b)* Ist B ein Gefüge und $B_0$ der Bereich aller Elemente von B vermehrt um ein Element $o$ mit der Eigenschaft: [1]

(O) $$(o \subset x \wedge o \circ x)$$

so ist:

1) Die Relation sub eine Boole'sche Partialordnung des Bereiches $B_0$ und;

2) genügt die Relation dis der Formel $f$.

*Beweis:* Zunächst ist B, wie die Formeln $c_1$ und $c_2$ aus 1. zeigen, partiell geordnet. Dasselbe gilt dann auch von $B_0$ und $o$ erfüllt die Formel $c_0$. Es bleibt also nur die Formel $c_3$ zu zeigen, um zu beweisen, dass $B_0$ Boole'sch geordnet ist. Es sei daher $x, y \in B_0$. Ist dann zunächst $y$ von $o$ verschieden, so gilt wegen $x \,\bar{\subset}\, y$ nach 0 auch $x$ ungleich $o$; Es ist also $x, y \in B$. Da aber B ein Gefüge ist, existiert nach 1. $S_3$ ein $u$ in B, sodass $u \subset x \wedge u \circ y$. Es gilt also $u \,\bar{\subset}\, 0 \wedge u \subset x \wedge u \circ y$. Ist anderseits $y$ das Element $o$, so gilt $x \subset x \wedge x \circ y$ und nach Voraussetzung $x \,\bar{\subset}\, 0$. In beiden Fällen existiert also ein $u$ in $B_0$ mit der Eigenschaft $u \,\bar{\subset}\, 0 \wedge u \subset x \wedge u \circ y$, womit auch $c_3$ nachgewiesen ist.

Damit ist $b$ 1) nachgewiesen, es bleibt $b$ 2):

Nach *a)* ist B ein Gefüge bezüglich sub und der nach $f$ definierten Relation DIS. B ist aber nach Voraussetzung auch Gefüge bezüglich der Relationen sub und dis. In einem Gefüge bestimmen sich aber die Relationen sub und dis gegenseitig eindeutig. Die Relation DIS muss daher mit der ursprünglichen Relation dis übereinstimmen. Es genügt dann also auch dis der Formel $f$.

Die Sätze *a)* und *b)* zeigen die völlige Aequivalenz der Begriffe «Boole'sche Partialordnung» und «Gefüge». Aus terminologischen Grün-

[1] Das Element o kann niemals dem Gefüge B angehören. Die Formel O widerspricht ja dem Axiom $T_1$.

den werden wir auch in Zukunft beide Begriffe verwenden. Es soll jedoch ein für alle Mal festgesetzt werden:

Ist B ein Gefüge, so bedeute $B_0$ immer den Bereich B vermehrt um das Element $o$ mit der Eigenschaft 0. Ist umgekehrt die Relation sub eine Bool'sche Partialordnung des Bereiches $B_0$, so bedeutet B immer das nach *a)* definierte Gefüge.

**3.** Die Gleichheit: Wie dies auch in partiell geordneten Bereichen üblich ist, definiert man in Gefügen eine Gleichheit zwischen den Elementen:

G) $$(x = y) \leftarrow\rightarrow (x \subset y \wedge y \subset x).$$

Die Relation «$=$» erfüllt dann alle Bedingungen, die man von einer Gleichheit verlangen muss:

$G_1$) $$(x = x)$$

$G_2$) $$(x = y) \rightarrow (y = x)$$

$G_3$) $$(x = y \wedge y = z) \rightarrow (x = z)$$

$G_4$) $$(x = y \wedge y \subset z) \rightarrow (x \subset z)$$

$G_5$) $$(x = y \wedge z \subset y) \rightarrow (z \subset x)$$

$G_6$) $$(x = y \wedge y \circ z) \rightarrow (x \circ z).$$

Die Beweise dieser Formeln sind leicht und sollen übergangen werden.

Anstatt die Gleichheit im Gefüge durch die Formel G zu definieren, hätten wir sie als logische Identität annehmen können. Damit wäre aber ausser der Postulierung einer weiteren Relation «$=$» auch eine Erweiterung des Axiomensystems um die Formel G verbunden.

## §2 BEISPIELE VON GEFÜGEN

**1.** Das volle Klassengefüge: Es sei K eine Menge von Dingen $\alpha, \beta, \ldots$ und $a, b, \ldots$ die nicht leeren Teilklassen von K. Ferner sei nach Definition:

*a)* $$(x \subset y) \leftarrow\rightarrow \bigwedge_{\alpha} (\alpha \in x \rightarrow \alpha \in y)$$
$$(x \circ y) \leftarrow\rightarrow \bigwedge_{\alpha} (\alpha \notin x \rightarrow \alpha \notin y).$$

Für die Operationen sub und dis aus a weist man dann leicht die Axiome $T_1 \ldots T_4$ nach. Es gilt also der

*Satz:* Der Bereich B aller nicht leeren Teilklassen einer Klasse K bildet ein Gefüge bezüglich der natürlichen Teil—und Fremdheitsbeziehung.

B heisst das volle Gefüge über der Klasse K.

Jedes volle Gefüge besitzt «kleinste Teile», sogenannte Atome; nämlich die Klassen, welche nur aus einem einzigen Element bestehen. Die Atome umfassen keine anderen Elemente des Gefüges. Ferner sind die Relationen sub und dis im vollen Gefüge allein aus der Zugehörigkeit der Atome zu den Klassen aus B entscheidbar (wie die Formeln a zeigen). Solche Gefüge heissen atomar (§ 7). Wir können also festhalten:

Jedes volle Gefüge ist atomar. Die Atome sind die Einermengen.

Man könnte nun denken, dass überhaupt jedes Gefüge atomar sei. Dem ist aber nicht so, wie die folgenden Beispiele zeigen:

**2.** Die verallgemeinerte Boole'sche Algebra: Es stellt sich die Frage, welche bekannten Gebilde aus der Theorie der partiellen Ordnung sogar Boole'sch angeordnet sind.

Zunächst braucht bei einer Boole'schen Partialordnung von $B_0$ das kleinste Umfassende und das grösste Gemeinsame zweier Elemente nicht zu existieren; $B_0$ braucht kein Verband zu sein. Anderseits ist ein Verband $V_0$ mit $o$-Element zwar partiell geordnet bezüglich der folgenden Relation sub:

$$(x \subset y) \leftarrow\rightarrow (x \cup y = y) \leftarrow\rightarrow (x \cap y = x).$$

Diese Relation sub braucht aber im allgemeinen keine Boole'sche Partialordnung zu sein. So zeigen die folgenden Beispiele von Verbänden, dass die Formel $c_3$ nicht erfüllt zu sein braucht:

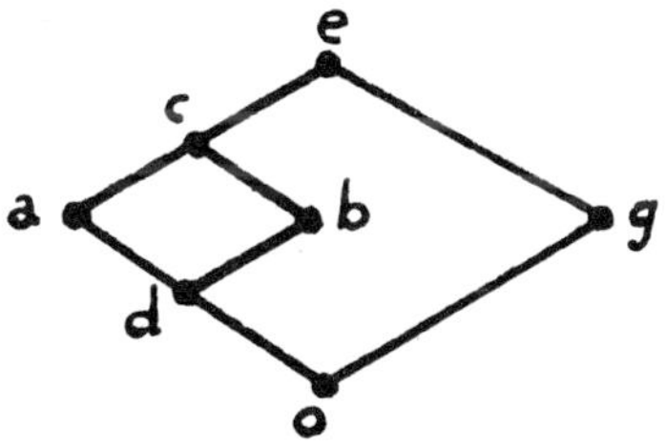


$(b \overline{\subset} a) \wedge \bigwedge_u (u \subset b \wedge u \circ a \rightarrow u = 0)$ $(a \overline{\subset} b) \wedge \bigwedge_u (u \subset a \wedge u \circ b \rightarrow u = 0)$

Es gibt nun aber spezielle Verbände, die Boole'sch angeordnet sind. Zu diesen gehören vor allem die Boole'schen Algebren:

Ein Verband $V_0$ mit $o$-Element $o$ heisst eine verallgemeinerte Boole'sche Algebra, wenn er distributiv ist und wenn zu jedem Elementepaar $b$ sub $a$

aus $V_0$ ein Element $(a - b)$ in $V_0$ existiert, welches folgende Eigenschaften besitzt:

$k_1)$ $$(a - b) \cup b = a$$

$k_2)$ $$(a - b) \cap b = 0$$

$(a - b)$ heisst das Komplement von $b$ relativ $a$.

Die Boole'schen Algebren sind dann diejenigen verallgemeinerten Algebren, die ein 1-Element $e$ besitzen.

Für diese speziellen Verbände gilt nun:

*Satz:* Die Inklusion vermittelt in der verallgemeinerten Boole'schen Algebra eine Boole'sche Partialordnung, oder:
Die von $o$ verschiedenen Elemente einer verallgemeinerten Boole'schen Algebra $V_0$ bilden ein Gefüge V bezüglich der Relationen:

$$(x \subset y) \leftarrow\rightarrow (x \cup y = y) \leftarrow\rightarrow (x \cap y = x)$$
$$(x \circ y) \leftarrow\rightarrow (x \cap y = 0).$$

*Beweis:* Es ist nur das Axiom $c_3$ in $V_0$ nachzuweisen. Zu diesen Zweck setze man:

$$u = (x \cup y) - y.$$

Wegen $k_2$ gilt dann zunächst $u \cap y = 0$.

Wegen $k_1$ gilt ferner: $(i)$ $(u \cup y) = (x \cup y)$. Daraus folgt $u \subset x \cup y$ und daher $u \cap (x \cup y) = (u \cap x) \cup (u \cap y) = u$ und daher wegen $u \cap y = 0$ auch $u \cap x = u$ also $u \subset x$.

Ist nun $x \,\bar{\subset}\, y$, so gilt $x \cup y \neq y$ also wegen $i$ auch $u \cup y \neq y$, d. h. $u \,\bar{\subset}\, y$. Es gilt also sicher $u \neq 0$.

Zusammenfassend ist damit unter der Voraussetzung $x \,\bar{\subset}\, y$ ein $u$ gefunden, welches die Formel $(u \neq 0 \wedge u \subset x \wedge u \cap y = 0)$ befriedigt. Da aber wegen der Formel $(f)$ aus § 1 im Verband gilt: $(x \circ y) \leftarrow\rightarrow (x \cap y = 0)$ ist damit die Formel $c_3$ nachgewiesen; $V_0$ ist Boole'sch geordnet. Zugleich ist dann auch die zweite Formulierung des Satzes bewiesen.

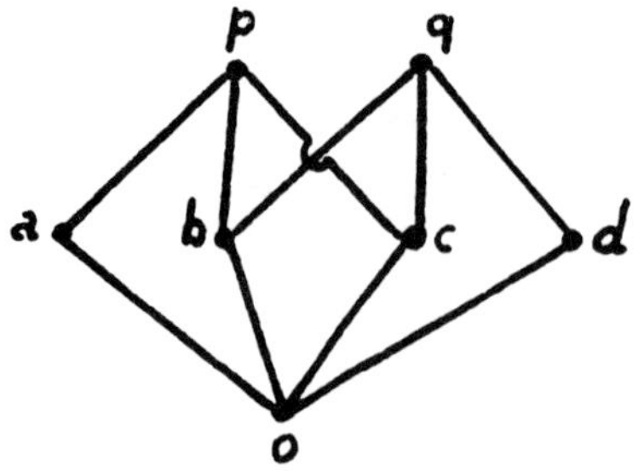


Vielleicht ist es noch interessant zu bemerken, dass es endliche Boole'sche Partialordnungen gibt, die keine Verbände sind. Ein solches Beispiel wird durch die nebenstehende Figur symbolisiert.

**3.** In 1 und 2 haben wir Gefüge kennengelernt, die sich in einfacher Weise von Boole'schen Algebren herleiten lassen. Dass nun aber auch Gefüge vorkommen, in denen Vereinigung, Durchschnitt und Komplement nicht oder nicht immer existieren, sei durch die folgenden Beispiele gezeigt:

*a)* B sei die Klasse aller offenen Kreisscheiben der euklidischen Ebene, sub und dis seien die mengentheoretischen Teil — und Fremdheitsrelationen.
*b)* B sei die Klasse aller offenen Quadrate mit achsenparallelen Seiten und sub und dis wieder die mengentheoretischen Relationen.

Man verifiziert in beiden Fällen leicht die Axiome $T_1 \ldots T_4$. Ferner stellt man fest, dass bei a weder Durchschnitt, Vereinigung noch Komplement von irgendwelchen Teilen existieren. Im Falle *b* kann der Durchschnitt unter günstigen Umständen existieren, nämlich dann, wenn die Quadrate eine gemeinsame Diagonale besitzen.

Schliesslich gibt der § 8 ein Beispiel eines Gefüges aus der Theorie der topologischen Räume.

## § 3 DURCHSCHNITT, VEREINIGUNG, KOMPLEMENT IM GEFÜGE

Die in § 1 gegebene Axiomatik kann als mathematische Teillehre bezeichnet werden. Eng verbunden mit dem Teil sind die Begriffe: Durchschnitt, Vereinigung, Komplement von Teilen. Es soll nun gezeigt werden, wie diese im Gefüge definiert werden können.

Bekanntlich werden diese Begriffe schon in partiell geordneten Mengen eingeführt. Es sei aber schon hier bemerkt, dass unsere Definition der Vereinigung schärfer ist als die übliche.

**1.** Durchschnitt und Vereinigung: N sei ein Bereich von Indizes $\nu, \mu, \ldots$ und B ein Gefüge. Ferner sei jedem $\nu \in$ N ein $x_\nu \in B_0$ zugeordnet. Ist dann $v$ ein Element aus $B_0$ mit den Eigenschaften:

$d_1$) $$\bigwedge_\nu (v \subset x_\nu)$$

$d_2$) $$\bigwedge_\nu (a \subset x_\nu) \rightarrow (a \subset v)$$

respektive

$v$) $$\bigwedge_\nu (x_\nu \circ u) \leftarrow\rightarrow (v \circ u)$$

so heisst $v$ der Durschschnitt, respektive die Vereinigung der Elemente $x_\nu$, und wird mit $\bigcap_{\nu \in N} x_\nu$, respektive $\bigcup_{\nu \in N} x_\nu$, bezeichnet.

Zur Symbolik sei noch bemerkt, dass häufig an Stelle des Bereiches N ein Prädikat $\varphi(x)$ vom Umfange N Verwendung findet. Wir schreiben dann anstatt $\bigcap\limits_{\nu \in N} x_\nu$ auch $\bigcap\limits_{\varphi(x)} x$ und desgleichen $\bigcup\limits_{\varphi(x)} x$ an Stelle von $\bigcap\limits_{\nu \in N} x_\nu$. Ferner kann der Bereich N aus endlich vielen Elementen $1, 2, \ldots, n$ bestehen. Wir schreiben dann für $\bigcap\limits_{\nu \in N} x_\nu$, respektive $\bigcup\limits_{\nu \in N} x_\nu$, auch $x_1 \cap x_2 \cap \ldots \cap x_n$, respektive $x_1 \cup x_2 \cup \ldots \cup x_n$. Schliesslich bedeutet $x \cap y$, respektive $x \cup y$, den Durchschnitt, respektive die Vereinigung der Elemente $x_1 = x$ und $x_2 = y$.

Während nun der Durchschnitt genau so definiert ist, wie es in der Theorie der partiellen Ordnung üblich ist, stimmt das bei der Vereinigung nicht. Dort werden ja von der Vereinigung $v$ nur die Formeln:

$v_1$) $$\bigwedge_\nu (x_\nu \subset v)$$

$v_2$) $$\bigwedge_\nu (x_\nu \subset a) \rightarrow (v \subset a)$$

gefordert, die, wie folgende Ueberlegungen zeigen, schwächer sind als die Formel ($v$).

Dass die Formel ($v$) nicht eine Folge von $v_1$ und $v_2$ ist, zeigt das folgende Beispiel: B sei das in der Figur angedeutete Gefüge der Elemente $a, b, u, e$. $e$ ist dann zwar die «Vereinigung» der Elemente $a$ und $b$ im schwächeren Sinne der Formeln $v_1$ und $v_2$. Trotzdem gilt $(u \circ a \wedge u \circ b \wedge u \bar{\circ} e)$; $e$ ist also nicht Vereinigung von $a$ und $b$ im Sinne der Gefügetheorie. Umgekehrt sind nun aber die Formeln $v_1$ und $v_2$ leichte Folgen aus ($v$):

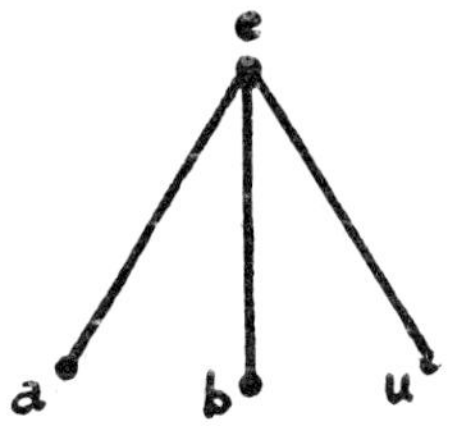


Ist nämlich $u \circ a$, so gilt wegen der Prämisse von $v_2$ auch $\bigwedge\limits_\nu (u \circ x_\nu)$ und daher wegen ($v$) auch $u \circ \bigcup\limits_\nu x_\nu$. Für jedes $u$ aus B gilt daher: $(u \circ a) \rightarrow (u \circ \bigcup\limits_\nu x_\nu)$ d. h. $\bigcup\limits_\nu x_\nu \subset a$. Aus ($v$) folgt also die Formel $v_2$. Ferner folgt aus ($v$) für alle $u$ und $\mu$: $u \circ \bigcup\limits_\nu x_\nu \rightarrow u \circ x_\mu$ also $x_\mu \subset \bigcup\limits_\nu x_\nu$, was die Formel $v_1$ ist.

Die Vereinigung im Sinne des Gefüges ist also ein engerer Begriff, als diejenige im Sinne der partiellen Ordnung im Gefüge. Es können daher auch für die Vereinigung alle Regeln, wie man sie von der Verbandstheorie her kennt, übernommen werden. Insbesondere sind Durchschnitt und Vereinigung eindeutig bestimmt, sofern sie existieren. Der scharfe Vereinigungsbegriff kommt der intuitiven Vorstellung näher als der schwache. Darüber hinaus leistet er Gewähr für die folgende Distributivität:

*Satz:* Ist B ein Gefüge und existieren beide Seiten der nachfolgenden Gleichung, so gilt:

$$\text{D)} \qquad y \cap \bigcup_\nu x_\nu = \bigcup_\nu (y \cap x_\nu).$$

*Beweis:* Es gilt für alle $\nu$: $(y \cap x_\nu) \subset x_\nu$ also $\bigcup_\nu (y \cap x_\nu) \subset \bigcup_\nu x_\nu$. Ferner ist auch $y \cap x_\nu \subset y$ also gilt auch $\bigcup_\nu (y \cap x_\nu) \subset y$. Wir haben damit die Formel:

$$i) \qquad \bigcup_\nu (y \cap x_\nu) \subset y \cap \bigcup_\nu x_\nu .$$

Anderseits folgt aus $u \circ \bigcup_\nu (y \cap x_\nu)$ für alle $\nu$: $u \circ (y \cap x_\nu)$. Es gilt dann auch für alle $\nu$: $(u \cap y) \circ x_\nu$ und daher wegen $(v)$: $(u \cap y) \circ \bigcup_\nu x_\nu$, also auch $u \circ (y \cap \bigcup_\nu x_\nu)$. Zusammenfassend gilt für alle $u$ aus B: $u \circ \bigcup_\nu (y \cap x_\nu) \to u \circ (y \cap \bigcup_\nu x_\nu)$, also:

$$ii) \qquad y \cap \bigcup_\nu x_\nu \subset \bigcup_\nu (y \cap x_\nu).$$

Mit $i$ und $ii$ ist aber die Formel D nachgewiesen. Es gilt dann auch:

*Satz:* Ist B ein Gefüge und existiert zu jedem Paar $x$, $y$ aus $B_0$ sowohl $x \cap y$ als auch $x \cup y$ (und zwar im scharfen Sinn der Formel $v$) so ist $B_0$ ein distributiver Verband.
Existiert auch die unbeschränkte Vereinigung in $B_0$, so gilt sogar das Distributivgesetz D.

Verbandstheoretisch lässt sich dieser Satz auch so formulieren:

*Satz:* Ist $V_0$ ein Verband mit $o$-Element, der die folgenden Eigenschaften besitzt:

$$(1) \qquad (x \mathrel{\overline{\subset}} y) \to \bigvee_u (u \neq 0 \wedge u \subset x \wedge u \cap y = 0)$$

so ist $V_0$ dann und nur dann distributiv, wenn auch die folgende Formel erfüllt ist:

$$(2) \qquad (u \cap x = 0 \wedge u \cap y = 0) \to (u \cap (x \cup y) = 0).$$

Es ist nämlich V wegen (1) ein Gefüge. Wegen des vorhergehenden Satzes und da die Vereinigung im Sinne des Verbandes nach (2) auch Vereinigung im Sinne des Gefüges ist, muss dann $V_0$ distributiv sein. Umgekehrt ist (2) eine leichte Folge der Distributivität.

Solche Verbände, die zugleich distributiv und Boole'sch angeordnet sind, haben eine gewisse Bedeutung in der Topologie. Wie H. Wallman [1] zeigt, lassen sie sich als Basis abgeschlossener Mengen eines bikompakten topologischen Raumes darstellen.

**2.** Das Komplement: B sei wieder ein Gefüge und $x$ eines seiner Elemente oder das $o$-Element 0. Existiert dann ein Element $y$ in $B_0$, sodass:

$k_1$) $$u \subset x \rightarrow u \circ y$$

$k_2$) $$u \circ x \rightarrow u \subset y$$

so heisst $y$ das Komplement von $x$ und wird mit $x'$ bezeichnet.

Es ist dann auch $x$ das Komplement von $x'$ d. h. es gelten die Formeln:

$k_3$) $$u \subset x \leftrightarrow u \circ x'$$

$k_4$) $$u \circ x \leftrightarrow u \subset x'$$

oder also $x'' = x$. Ferner bestimmen sich die Komplemente eindeutig.

*Beweise*: Um die Formeln $k_3$ und $k_4$ nachzuweisen, sind nur die Umkehrungen von $k_1$ und $k_2$ für $y = x'$ zu zeigen. Es sei also $u \mathrel{\bar{\subset}} x$. Dann existiert ein $v$ in B, sodass $(v \subset u \wedge v \circ x)$. Nach $k_2$ gilt dann auch $(v \subset u \wedge v \subset x')$, also ist $u \mathrel{\bar{\circ}} x'$. Aus $u \mathrel{\bar{\subset}} x$ folgt also immer $u \mathrel{\bar{\circ}} x'$, womit die Umkehrung von $k_1$ nachgewiesen ist.

Ist anderseits $u \mathrel{\bar{\circ}} x$, so existiert ein $v$ in B, sodass $(v \subset u \wedge v \subset x)$. Wegen $k_1$ gilt dann auch $(v \subset u \wedge v \circ x')$ also ist $u \mathrel{\bar{\subset}} x'$. Damit ist dann auch die Umkehrung von $k_0$ bewiesen.

Es bleibt noch zu zeigen, dass $x'$ eindeutig bestimmt ist. Es sei also $x^*$ ein zweites Komplement von $x$. $u \circ x'$ ist dann nach $k_3$ äquivalent mit $u \subset x$ und dies wiederum nach $k_3$ äquivalent mit $u \circ x^*$. Es gilt also für alle $u$ aus B die Formel: $u \circ x' \leftrightarrow u \circ x^*$, d. h.: $x' = x^*$.

Ganz ähnlich zeigt man, dass auch jede der Formeln $k_3$ und $k_4$ allein zur Charakterisierung des Komplements hinreichend ist. Zum Vergleich mit dem verbandstheoretischen Komplementbegriff sind folgende Bemerkungen nützlich:

1) Existiert $x \cup x'$, so existiert auch ein Supremum $e$ in B und es gilt: $x \cup x' = e$.
2) Existiert ein Supremum $e$ in B, so existiert mit $x'$ immer auch $x \cup x'$ und es ist $x \cup x' = e$.

[1] Wallman (1). Die Bedingung 1 d. h $c_3$ heisst dort Disjunktion Property.

3) Es gilt immer $x \cap x' = 0$, wenn nur $x'$ existiert.

4) Existiert das Supremum $e$ in B und ist $y$ ein Element von $B_0$ mit den Eigenschaften:

$$x \cup y = e$$

$$x \cap y = 0$$

so ist $y = x'$.

5) Es gilt immer:

$$x' = \bigcup_{u \circ x} u.$$

Die Beweise sind leicht und sollen übergangen werden. Schliesslich wird folgender Satz im nächsten § 4 von Nutzen sein:

*Satz:* Existieren alle vorkommenden Ausdrücke, so gilt die Formel:

$$\bigcap_{\nu} x_{\nu} = (\bigcup_{\nu} x'_{\nu})'.$$

*Beweis:* Es sei $v = (\bigcup_{\nu} x'_{\nu})'$. Wegen $(v_1)$ gilt für jedes $\nu$: $x'_{\nu} \subset \bigcup_{\nu} x'_{\nu}$ und daher wegen $k_3$: $x'_{\nu} \circ v$, also wiederum wegen $k_3$ die Formel:

$$(i) \qquad \bigwedge_{\nu} (v \subset x_{\nu}).$$

Es sei für alle $\nu$: $z \subset x_{\nu}$. Wegen $k_3$ gilt dann ebenfalls für alle $\nu$: $z \circ x'_{\nu}$ und daher wegen $(v)$ $z \circ \bigcup_{\nu} x'_{\nu}$. Nach $k_3$ folgt daraus aber $z \subset v$; es gilt also zusammenfassend:

$$(ii) \qquad \bigwedge_{\nu} (z \subset x_{\nu}) \rightarrow (z \subset v).$$

Die Formeln $i$ und $ii$ zeigen aber, dass $v$ der Durchschnitt aller $x_{\nu}$ ist, womit der Satz bewiesen ist.

## § 4 DAS VOLLSTÄNDIGE GEFÜGE

Für manche Zwecke ist das Axiomensystem $T_1 \ldots T_4$ zu schwach. So brauchen ja beispielsweise nicht einmal Durchschnitt und Vereinigung zweier Elemente zu existieren. Es soll daher ein weiteres Axiom postuliert werden:

$T_5$) A und A' seien zwei Klassen von Elementen des Gefüges B, sodass für alle $a$ und $b$ aus B gilt:

1) $$(a \in A \wedge b \in A') \rightarrow (a \circ b).$$

Dann existiert ein $v$ in $B_0$, sodas für alle $a$ aus B gilt:

2) $$\begin{matrix} (a \in A\ ) \rightarrow (a \subset v) \\ (a \in A') \rightarrow (a \circ v) \end{matrix}.$$

Ein Gefüge B, welches das Axiom $T_5$ befriedigt, heisst vollständig.

Obwohl das Axiom $T_5$ logisch nicht sehr stark zu sein scheint, hat es doch weittragende Konsequenzen. Zunächst ist $T_5$ äquivalent mit:

$T'_5$) Zu jedem Indexbereich N existiert die Vereinigung $\bigcup_{\nu \in N} x_\nu$. (Es sei ausdrücklich bemerkt, dass die Vereinigung im Sinne, wie sie in § 3 definiert wurde, gemeint ist.)

Sind nämlich A und $A'$ zwei Klassen, welche die Bedingung 1) aus $T_5$ erfüllen, so genügt das Element $v = \bigcup_{x \in A} x$ den Bedingungen 2) aus $T_5$. Aus $T'_5$ folgt also $T_5$.

Ist umgekehrt N irgend ein Indexbereich, so setze man

A $=$ Klasse aller $x$, die einem $x_\nu$ gleich sind.

$A' =$ Klasse aller $x$, die zu allen $x_\nu$ fremd sind.

Wegen $T_5$ existiert dann ein $v$ mit den Eigenschaften:

$$\bigwedge_\nu (x_\nu \subset v)$$
$$\bigwedge_\nu (u \circ x_\nu) \rightarrow (u \circ v)$$

$v$ ist daher nach § 3 die Vereinigung aller $x_\nu$. Aus $T_5$ folgt also auch $T'_5$.

Im vollständigen Gefüge existiert also die Vereinigung beliebiger Elemente. Dasselbe gilt dann aber auch von Komplement und Durchschnitt. Man hat nur zu setzen:

($a$) $$x' = \bigcup_{u \circ x} u$$

($b$) $$\bigcap_\nu x_\nu = (\bigcup_\nu x'_\nu)'.$$

Um zu zeigen dass die Formel ($a$) das Komplement zum Element $x$ liefert, hat man die Formeln $k_1$ und $k_2$ aus § 3. 2. nachzuweisen:

$k_1$: Ist $u \subset x$, so gilt für alle $z$ aus B ($z \circ x \rightarrow z \circ u$). Wegen der Formel ($v$) aus § 3 und wegen ($a$) gilt dann auch $u \circ x'$. Aus $u \subset x$ folgt also $u \circ x'$.

$k_2$: Ist $u \circ x$, so gilt wegen der Formel ($v_1$) aus § 3 und wegen ($a$) auch $u \subset x'$.

Es bleibt zu zeigen, dass ($b$) den Durchschnitt der Elemente $x_\nu$ ergibt. Dazu sind die Formeln $d_1$ und $d_2$ aus § 3 nachzuweisen. Dies wurde am Ende von § 3 durchgeführt.

Damit sind mit den Elementen eines vollständigen Gefüges alle Operationen definiert, wie sie zwischen den Teilklassen einer Klasse bestehen. Darüber hinaus ist aber durch die Axiome $T_1 \ldots T_5$ auch ein sinnvoller Kalkül mit diesen Operationen gewährleistet. Es gelten nämlich fast alle Regeln, wie sie vom Klassenkalkül her bekannt sind [1]. Dies kommt schärfer im folgenden Satz zum Ausdruck:

*Satz:* Ist B ein vollständiges Gefüge, so ist $B_0$ nicht nur Boole'sch partiell geordnet, sondern sogar eine vollständige (d. h. total additive und total multiplikative) Boole'sche Algebra bezüglich der partiellen Ordnungsrelation sub.
Als Summe und Produkt der Elemente $x_\nu$ und als Komplement zu $x$ hat man nur die Elemente $\bigcup_\nu x_\nu$, $\bigcap_\nu x_\nu$ und $x'$ zu wählen. Oder anders formuliert:
Die Klasse $B_0$, bestehend aus den Elementen des vollständigen Gefüges B und dem $o$-Element $o$, erfüllt folgende Gesetze:

I: $B_0$ ist bezüglich der Relation sub partiell geordnet.

Ist N irgend ein Indexbereich und $x_\nu \in B_0$, so existiert

II: ein Element $v$ in $B_0$, nämlich $v = \bigcup_\nu x_\nu$, mit den Eigenschaften

1) $\bigwedge_\nu (x_\nu \subset v)$

2) $\bigwedge_\nu (x_\nu \subset z) \rightarrow (v \subset z)$ und

III: ein Element $w$ in $B_0$, nämlich $w = \bigcap_\nu x_\nu$, mit den Eigenschaften

1) $\bigwedge_\nu (w \subset x_\nu)$

2) $\bigwedge_\nu (z \subset x_\nu) \rightarrow (z \subset w)$

IV: Es existieren zwei Elemente $o$ und $e$ in $B_0$, nämlich $o = o$-Element von B und $e =$ Vereinigung aller Elemente von B, sodass

1) $\bigwedge_u (o \subset u)$

2) $\bigwedge_u (u \subset e)$

1 Es fahlt nur eine verschärfte Distributivität. Tarski (1).

V: Zu jedem $x$ aus $B_0$ existiert ein $y$ aus $B_0$, nämlich $y = x'$, sodass

1) $$(x \cup y) = e$$

2) $$(x \cap y) = 0$$

VI: Es gelten die Distributivgesetze

1) $$a \cap \bigcup_\nu x_\nu = \bigcup_\nu (a \cap x_\nu)$$

2) $$a \cup \bigcap_\nu x_\nu = \bigcap_\nu (a \cup x_\nu).$$

*Beweis*: In § 1 wurde gezeigt, dass $B_0$ partiell geordnet ist, womit I nachgewiesen ist. Nadh Axiom $T_5'$ existiert $v = \bigcup_\nu x_\nu$ immer und nach § 3 hat es die Eigenschaften II, 1 und 2. Dass auch $w = \bigcap_\nu x_\nu$ existiert, wurde unmittelbar vor dem Satz gezeigt. Nach seiner Definition in § 3 erfüllt es die Formeln III, 1 und 2. IV ist fast trivial, ebenso die Formel V. Es bleiben die Formeln VI nachzuweisen. VI, 1 ist aber die in § 3 nachgewiesene Formel (D). Ferner ist, wie man in der Theorie der Boole'schen Algebren zeigt VI, 2 eine Folge der Axiome I ... V und VI, 1. Damit ist der Satz bewiesen.

Umgekehrt wurde in § 2 gezeigt, dass die von $o$ verschiedenen Elemente einer Boole'schen Algebra $B_0$ immer ein Gefüge B bilden. Ist nun $B_0$ sogar vollständige Boole'sche Algebra, so erfüllt die Summe $v = \sum_\nu x_\nu$ der Elemente $x_\nu$ die Formel ($v$) aus § 3. Die folgenden Formeln sind ja äquivalent:

$$\bigwedge_\nu (u \cap x_\nu = 0) \quad ; \quad \sum_\nu (u \cap x_\nu) = 0 \quad ; \quad u \cap \sum_\nu x_\nu = 0$$

also gilt:

$$\bigwedge_\nu (u \circ x_\nu) \longleftrightarrow u \circ \sum_\nu x_\nu .$$

Das Element $v = \sum_\nu x_\nu$ ist also die Vereinigung der Elemente $x_\nu$ im scharfen Sinne von § 3. Damit ist gezeigt, dass im Gefüge B das Axiom $T_5'$ zutrifft, d. h. B ist vollständiges Gefüge. Wir können also festhalten:

*Satz:* Die von $o$ verschiedenen Elemente der vollständigen Boole'schen Algebra $B_0$ bilden ein vollständiges Gefüge B.

In § 1 wurde gezeigt, dass die Begriffe «Boole'sche Partialordnung» und «Gefüge» äquivalent sind. Die beiden letzten Sätze zeigen, dass auch die Begriffe «vollständige Boole'sche Algebra» und «vollständiges Gefüge» äquivalent sind. Somit kommt man zu neuartigen Axiomensystemen für vollständige Boole'sche Algebren:

*Satz:* Die bezüglich einer Relation sub partiell geordnete Menge $B_0$ ist dann und nur dann eine vollständige Boole'sche Algebra, wenn
1. sub eine Boole'sche Ordnung ist und
2. das Axiom $T_5$ zutrifft.

Ganz ähnliche Axiomensysteme sind zwar schon 1924 von Huntington und dann in neuerer Zeit auch von Tarski behandelt worden [1]. Die Beweise ergeben sich aber bei Verwendung der Boole'schen Partialordnung viel natürlicher. Dies hat wohl seinen Grund in der konsequenten Benutzung der Fremdheitsrelation.

Ueber die Huntington'-und Tarski'schen Resultate hinaus führt nun aber die Theorie der Gefüge zur folgenden Erkenntnis:

Die Forderung 2 (d. h. das Axiom $T_5$) ist unwesentlich, in dem Sinne, als sie durch die Adjunktion von Elementen erzwungen werden kann.

Der Schlüssel zum Beweis dieser Behauptung ist der Begriff der Spaltung von Gefügen.

## § 5 ORTHOKOMPLEMENTE UND SPALTUNG

Da im folgenden häufig die Teilklassen von B Verwendung finden, sei die Symbolik folgendermassen festgelegt:

Die Teilklassen werden mit den Lettern A, . . . , X, Y, . . . bezeichnet. Es bedeutet auch etwa $(a, b, c, \ldots)$ die Klasse bestehend aus den Elementen $a, b, c, \ldots$ Speziell ist $(a)$ die Klasse mit dem einen Element $a$. Für sub — und dis — Relationen zwischen Klassen, so wie für ihre Verknüpfungen sollen die den bisherigen entsprechenden eckigen Zeichen verwendet werden. Es bedeutet also z. B. : $X < Y$ soviel wie «X ist (nicht notwendig echte) Teilklasse von Y»; $X \vee Y$ die Vereinigung der Klassen X und Y; $\bigwedge_\nu X_\nu$ den Durchschnitt aller Klassen $X_\nu$. [2] Schliesslich sei CX das Komplement von X in B.

1. Orthokomplement und S-Klassen: Zu jeder Klasse X von Elementen des Gefüges B gehört eindeutig eine Klasse X′ definiert durch:

$$(k) \qquad (a \in X') \longleftrightarrow \bigwedge_u (u \in X \rightarrow u \circ a)$$

[1] Huntington (1); Tarski (1).

[2] Dass dieselben Zeichen auch für die logischen Verknüpfungen Verwendung finden, kann nicht zu Verwechslungen Anlass geben.

$X'$ heisst die zu X orthokomplementäre Klasse oder kurz das Orthokomplement von X. Es gelten dann folgende Formeln:

1) $$(a \in X' \wedge b \subset a) \rightarrow (b \in X')$$

2) $$(X < Y) \rightarrow (Y' < X')$$

3) $$(X < X'').$$

Ist nämlich $a \in X'$ so gilt wegen $k$ für alle $u$ aus B:

$$(u \in X \rightarrow u \circ a)$$

also wegen $T_2$ und $b \subset a$ auch $(u \in X \rightarrow u \circ b)$.

Nach $k$ ist dann $b \in X'$, womit die Formel 1) nachgewiesen ist.

Ist $a \in Y'$, so ist es zu allen $u$ aus Y fremd, also wegen $X < Y$ auch zu allen $u$ aus X. Es gilt dann auch $a \in X'$. Damit ist 2) nachgewiesen.

Die Elemente von X sind zu allen Elementen von $X'$ fremd, es gilt also $X < X''$, womit schliesslich auch 3) nachgewiesen ist.

Aus $X < X''$ folgt nun wegen 2): $X''' < X'$. Wegen 3) gilt aber auch $X' < X'''$ also:

4) $$X''' = X'.$$

Es sei nun definiert:

*Definition:* Eine Klasse X von Elementen des Gefüges B heisst spaltend oder kurz S-Klasse, wenn $X = X''$ ist.

Wegen 4) gilt dann:

*Satz:* Die S-Klassen des Gefüges B sind die Klassen von der Form $X'$ oder kurz:
Die S-Klassen sind die Orthokomplemente.

Ist nun A irgend eine S-Klasse, die die Klasse X umfängt, so ist wegen 2): $A' < X'$ und $X'' < A''$. Da A S-Klasse ist, gilt also $X'' < A$. Anderseits ist aber auch $X''$ eine S-Klasse, die wegen 3) ebenfalls X umfängt. Es gilt also:

*Satz:* $X''$ ist die kleinste S-Klasse, die X umfängt. Sie heisst die von X erzeugte S-Klasse.

Ist speziell $X = (a)$, so gilt wegen $k$: $u \in (a)' \leftarrow\rightarrow u \circ a$ und nochmals wegen $k$: $x \in (a)'' \leftarrow\rightarrow \bigwedge_u (u \circ a \rightarrow u \circ x)$. Wegen $T_3$ und $S_1$ aus § 1 gilt also: $x \in (a)'' \leftarrow\rightarrow x \subset a$ also:

*Satz:* Die von $(a)$ erzeugte S-Klasse besteht aus den Elementen $u$ sub $a$. Sie heisst die vom Element $a$ erzeugte Hauptklasse oder kurz H-Klasse.

Es sollen noch die folgenden Hilfsformeln bewiesen werden, die im nächsten § nützlich sind:

Für alle S-Klassen $X, Y$ und für beliebige Klassen $X_\nu$ gilt:

$(h_1)$ $$X < Y \longleftrightarrow X \diamond Y'$$

$(h_2)$ $$X \diamond Y \longleftrightarrow X < Y'$$

$(h_3)$ $$(\bigvee_\nu X_\nu)' = \bigwedge_\nu X'_\nu .$$

*Beweise:*

$h_2$: Ist $X \overline{\diamond} Y$, so existiert ein $u$ in B, sodass $u \in X \wedge u \in Y$. Es gilt dann auch $u \in X \wedge u \notin Y'$, also $X \nless Y'$.

Ist umgekehrt $X \nless Y'$, so existiert ein $u$ in B, sodass $u \in X \wedge u \notin Y'$. Wegen der Formel $k$ existiert dann ein $v$ in B, sodass $v \in Y \wedge u \not\phi\, v$ und daher ein $z$ in B, sodass $z \subset u \wedge z \subset v$. Nach 1) und da $X, Y$ S-Klassen sind, folgt aus $u \in X \wedge z \subset u$ bzw. $v \in Y \wedge z \subset v$ auch $z \in X \wedge z \in Y$. Es ist also $X \overline{\diamond} Y$. Zusammenfassend ist die Formel $X \overline{\diamond} Y \longleftrightarrow X \nless Y'$ nachgewiesen, die zu $h_2$ äquivalent ist.

$h_1$: Aus $X < Y$ folgt nach 2): $Y' < X'$, also wegen $h_2$ auch $X \diamond Y'$. Ist umgekehrt $X \diamond Y'$, so gilt wegen $h_2$ auch $Y' < X'$ und daher wegen 2): $X'' < Y''$, also da X und Y S-Klassen sind auch $X < Y$.

$h_3$: Zunächst gilt für jedes $\nu$: $X_\nu < \bigvee_\nu X_\nu$ also wegen 2): $(\bigvee_\nu X_\nu)' < X'_\nu$ und daher:

$(i)$ $$(\bigvee_\nu X_\nu)' < \bigwedge_\nu X'_\nu .$$

Ist anderseits $u \in \bigwedge_\nu X'_\nu$, so gilt für alle $\nu$: $u \in X'_\nu$, also wegen 2) und 3) auch $X_\nu < (u)'$. Daraus folgt aber $\bigvee_\nu X_\nu < (u)'$. Wegen 2) gilt dann $(u)'' < (\bigvee_\nu X_\nu)'$, also sicher $u \in (\bigvee_\nu X_\nu)'$. Zusammenfassend folgt also aus $u \in \bigwedge_\nu X'_\nu$ immer auch $u \in (\bigvee_\nu X_\nu)'$ also:

$(ii)$ $$\bigwedge_\nu X'_\nu < (\bigvee_\nu X_\nu)'$$

$i$ und $ii$ ergeben aber die zu beweisende Formel $h_3$.

**2.** Spaltende Klassen vollständiger Gefüge: In den vollständigen Gefügen sind die Verhältnisse besonders einfach. Es gilt zunächst:

$S_1$: Jede S-Klasse A des vollständigen Gefüges B wird vom Element $v = \bigcup_{x \in A} x$ erzeugt und ist also sogar H-Klasse.

*Beweis:* A sei S-Klasse. Weil B vollständig ist, existiert dann $v = \bigcup_{x \in A} x$, und es gilt für jedes $u$ aus B:

$$(u \circ v) \longleftrightarrow \bigwedge_x (x \in A \rightarrow x \circ u).$$

Wegen der Formel $k$ aus 1. gilt dann auch $(u \circ v) \longleftrightarrow (u \in A')$ und daher für jedes $x$ aus B.

$$\bigwedge_u (u \circ v \rightarrow u \circ x) \longleftrightarrow \bigwedge_u (u \in A' \rightarrow u \circ x).$$

Wegen $T_3$ und $S_1$ aus § 1 einerseits und wegen ($k$) aus 1. andererseits gilt dann auch $(x \subset v) \longleftrightarrow (x \in A'')$ und daher wegen des letzten Satzes aus 1. schliesslich $(v)'' = A$. Damit ist $S_1$ nachgewiesen. Umgekehrt gilt aber auch

$S_2$: Ein Gefüge B in dem jede S-Klasse sogar H-Klasse ist, ist vollständig.

*Beweis:* Wir zeigen, dass in B das Axiom $T_5'$ zutrifft. Es sei also N irgend ein Indexbereich und $x_\nu \in B$. Ist dann A die von allen $x_\nu$ erzeugte S-Klasse, so existiert nach Voraussetzung ein $v$ in B, sodass $A = (v)''$ ist. Wegen des letzten Satzes aus 1. gilt also $(x \subset v \longleftrightarrow x \in A)$. Es gilt also:

$$(u \circ v) \longleftrightarrow \bigwedge_x (x \in A \rightarrow x \circ u).$$

Also wegen $k$ aus 1.: $u \circ v \longleftrightarrow u \in A'$. Daraus folgt schliesslich, weil $A'$ das Orthokomplement der Klasse aller $x_\nu$ ist:

$$(u \circ v) \longleftrightarrow \bigwedge_\nu (u \circ x_\nu).$$

Das Element $v$ ist also nach § 3 die Vereinigung aller $x_\nu$. Zusammenfassend existiert also zu jedem Bereich N das Element $\bigcup_{\nu \in N} x$. Wegen $T_5'$ ist also B vollständig.

Die Sätze $S_1$ und $S_2$ zusammenfassend ist folgendes Kriterium für die Vollständigkeit eines Gefüges gefunden:

*Satz:* Es sind diejenigen Gefüge vollständig, in denen jede S-Klasse sogar H-Klasse ist. Oder:
Ein Boole'sch partial geordneter Bereich ist dann und nur dann eine vollständige Boole'sche Algebra, wenn jede spaltende Klasse von einem einzigen Element erzeugt wird.

**3.** Die Spaltung: Es gibt einen zweiten Weg die S-Klassen einzuführen. Er beruht auf folgender Definition:

Das Paar (A A′) zweier Klassen von Elementen des Gefüges B heisst eine Spaltung, wenn die folgenden Bedingungen zutreffen:

$$(s_1) \qquad (x \in A \wedge y \in A') \rightarrow (x \circ y)$$

$$(s_2) \qquad \bigwedge_u (u \in A \rightarrow u \circ x) \rightarrow (x \in A')$$

$$(s_3) \qquad \bigwedge_u (u \in A' \rightarrow u \circ x) \rightarrow (x \in A).$$

Aus den Formeln $s_1, s_2, s_3$ und der Definitionsformel ($k$) in 1. schliesst man leicht:

*a)* Ist (A A′) Spaltung des Gefüges B, so ist A′ das Orthokomplement von A und umgekehrt A dasjenige von A′. Wegen des ersten Satzes aus 1. sind dann weiter A und A′ S-Klassen.

Weiter gilt aber:

*b)* Ist A eine S-Klasse des Gefüges B und A′ das Orthokomplement von A, so ist (A A′) eine Spaltung von B. Zunächst gelten wegen ($k$) aus 1. die Formeln $s_1$ und $s_2$. Sodann ist, weil A S-Klasse ist: $A'' = A$ es ist also auch A das Orthokomplement von A′. Damit gilt wegen der Formel ($k$) auch $s_3$.

Die Bemerkungen *a)* und *b)* zusammenfassend kann also der folgende Satz formuliert werden:

*Satz:* (A A′) ist dann und nur dann Spaltung des Gefüges B, wenn A und A′ orthokomplementäre S-Klassen sind.

Die Spaltung erinnert stark an die bekannte Dedekind'sche Idee des Schnittes in geordneten Mengen. Der Begriff des Schnittes wurde von MacNeille [1] in die Theorie der partiellen Ordnung übernommen:

Sind A und $\bar{A}$ zwei Klassen von Elementen des partiell geordneten Bereiches $B_0$, so heisst $(A\bar{A})$ ein Schnitt, wenn folgende Bedingungen zutreffen:

$$(r_1) \qquad (x \in A \wedge y \in \bar{A}) \rightarrow (x \subset y)$$

$$(r_2) \qquad \bigwedge_u (u \in A \rightarrow u \subset x) \rightarrow (x \in \bar{A})$$

$$(r_3) \qquad \bigwedge_u (u \in \bar{A} \rightarrow x \subset u) \rightarrow (x \in A).$$

[1] MacNeille (1).

Im Falle der Boole'schen Algebra sind nun die Begriffe Spaltung und Schnitt äquivalent. Man zeigt nämlich leicht: Ist $B_0$ eine Boole'sche Algebra, so werden durch die Formel [1].

$$x \in A' \longleftrightarrow x' \in \bar{A}$$

die Spaltungen $(AA')$ von B ein-eindeutig den Schnitten $(A\bar{A})$ von $B_0$ zugeordnet.

Im allgemeinen Falle, wo $B_0$ zwar Boole'sch partialgeordnet, aber nicht Boole'sche Algebra ist, liegen die Dinge anders. Es existieren dann im allgemeinen mehr Spaltungen als Schnitte, wie man etwa so einsieht:

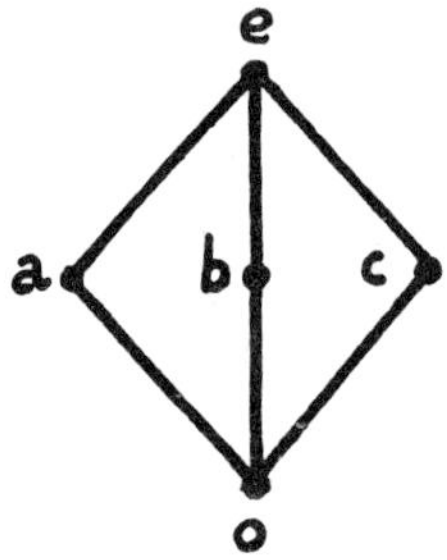


Zunächst zeigt man leicht, dass die Klasse A aus dem Schnitt $(A\bar{A})$ eine S-Klasse sein muss; dass also eine Spaltung $(AA')$ existiert. Umgekehrt zeigt aber das folgende Beispiel, dass nicht zu jeder Spaltung $(AA')$ ein Schnitt $(A\bar{A})$ zu existieren braucht:

$B_0$ sei die in der Figur angedeutete partiell geordnete Menge. Man verifiziert dann leicht die Formel $c_3$ aus § 1, $B_0$ ist also sogar Boole'sch geordnet. Ist nun $A = (a, b)$ und $A' = (c)$, so ist zwar $(AA')$ eine Spaltung, es existiert aber trotzdem keine Klasse $\bar{A}$, sodass $(A\bar{A})$ ein Schnitt wäre.

4. S-Klassen und Ideale in Boole'schen Algebren: Ist $B_0$ nicht nur Boole'sch geordnet, sondern sogar eine verallgemeinerte Boole'sche Algebra, so fällt das Orthokomplement einer Klasse X, wie es in 1. definiert wurde, mit dem entsprechenden Stone'schen Begriff zusammen [2]. Die S-Klassen sind dann die Stone'schen normalen Ideale von $B_0$.

In diesem Zusammenhange stellt sich das Problem, wie weit die bekannten Stone'schen Resultate von der verallgemeinerten Boole'schen Algebra auf die Boole'sche Partialordnung übernommen werden können. Dass in dieser Richtung Resultate zu erwarten sind, zeigt der nächste §. Es wird dort der folgende Satz verallgemeinert:

Die normalen Ideale einer verallgemeinerten Boole'schen Algebra $B_0$ bilden eine vollständige Boole'sche Algebra V mit folgenden Eigenschaften:

1) V umfängt eine zu $B_0$ isomorphe Algebra $\bar{B}$. $\Phi$ sei der Isomorphismus.
2) V ist bis auf Isomorphie die kleinste vollständige Algebra mit der Eigenschaft 1.
3) Existiert das Element $\bigcup_\nu x_\nu$ in $B_0$, so gilt $\Phi(\bigcup_\nu x_\nu) = \bigcup_\nu \Phi(x_\nu)$.

[1] Mit $x'$ wird das Komplement zu $x$ in der Boole'schen Algebra $B_0$ bezeichnet.
[2] Stone (1).

Es scheint sogar, dass ein schöner Teil der Stone'schen Idealtheorie auf die Gefüge übernommen werden kann, wenn man definiert:

Die Klasse I heisst ein Ideal des Gefüges B, wenn sie folgende Bedingung erfüllt:

$$\text{(I)} \qquad (x_1 \in \mathrm{I} \wedge \ldots \wedge x_n \in \mathrm{I}) \rightarrow (x_1, \ldots, x_n)'' < \mathrm{I}.$$

Diese Verallgemeinerung dürfte auch im Hinblick auf die Zusammenhänge der Idealtheorie mit der Theorie der topologischen Räume [1] interessant sein.

## § 6 DIE VERVOLLSTÄNDIGUNG EINES GEFÜGES

Der Schritt vom Gefüges zum vollständigen mag vorerst etwas gross erscheinen. Man könnte daran denken zuerst Durchschnitt und Vereinigung nur für endliche Klassen von Elementen zu postulieren; d. h., nicht eine vollständige, sondern nur eine gewöhnliche Boole'sche Algebra vorauszusetzen.

Im Folgenden soll nun aber gezeigt werden, dass jedes Gefüge zu einem kleinsten vollständigen ergänzt werden kann. Im Lichte dieser Tatsache erscheint die gewöhnliche Boole'sche Algebra nur als Zwischenstufe auf dem Wege von der Boole'schen Partialordnung zur vollständigen Boole'schen Algebra.

Der Schlüssel zum Problem der Vervollständigung ist der Begriff der Spaltung. Es gilt nämlich;

$S_1$: Die Klasse S(B) aller S-Klassen von B bilden bezüglich der klassentheoretischen Relationen "$<$" und "$<>$" ein vollständiges Gefüge.
Die leere Klasse 0 spielt dabei die Rolle des Nullelementes.

*Beweis*: Es sind für S(B) die Axiome $T_1 \ldots T_5$ nachzuweisen. $T_1$ und $T_2$ sind trivialer Weise erfüllt, da wir es mit den klassentheoretischen sub — und dis — Relationen zu tun haben.

$T_3$: Es seien X und Y zwei S-Klassen und $X \not< Y$. Dann gilt wegen der Formel $h_1$ aus § 5, 1. auch $X \overline{<>} Y'$. Es existiert also ein $u$ in B, sodass $u \in X \wedge u \in Y'$. Durch zweimalige Anwendung der Formel 2) aus § 5, 1. folgert man daraus $(u)'' < X'' \wedge (u)'' < Y'''$, also da X und Y'

[1] Stone (2).

S-Klassen sind auch $(u)'' < X \wedge (u)'' < Y'$. Wegen $h_2$ aus § 5 gilt dann schliesslich $(u)'' < X \wedge (u)'' <> Y$.

Zusammenfassend ist damit unter Voraussetzung $X \not< Y$ ein $U = (u)'' \in S(B)$ so gefunden, dass $U \overline{<>} X \wedge U <> Y$. Damit ist aber die Formel $T_3$ in $S(B)$ nachgewiesen.

$T_4$: Es seien wieder X und Y zwei S-Klassen und jetzt $X \overline{<>} Y$. Dann existiert ein $u$ in B, sodass $u \in X \wedge u \in Y$. Durch zweimalige Anwendung der Formel 2) aus § 5, 1. schliesst man dann $(u)'' < X'' \wedge (u)'' < Y''$, also da X und Y S-Klassen sind auch $(u)'' < X \wedge (u)'' < Y$. Damit ist ein $U = (u)''$ in $S(B)$ gefunden, welches die Formel $U < X \wedge U < Y$ erfüllt. Es trifft also auch das Axiom $T_4$ in $S(B)$ zu. Es bleibt noch die Vollständigkeit des Gefüges $S(B)$ zu zeigen. Es genügt zu diesem Zweck das Axiom $T'_5$ nachzuweisen.

$T'_5$: Es sei $\mu, \nu, \ldots$ irgend ein Indexbereich und jedes $X_\nu$ eine S-Klasse. Man setze:

$$V = \bigvee_\nu X_\nu$$

V braucht dann keine S-Klasse zu sein. Es soll aber gezeigt werden, dass die von V erzeugte S-Klasse $V''$ die Rolle der Vereinigung (und zwar im scharfen Sinn von § 3) im Gefüge $S(B)$ spielt. Dazu ist nur die Formel:

$$(i) \qquad U <> V'' \longleftrightarrow \bigwedge_\nu (U <> X_\nu)$$

für jede S-Klasse U nachzuweisen.

Es sei zunächst $U <> V''$. Wegen der Formel 3) aus § 5, 1. gilt dann auch $U <> V$ und daher für alle $\nu$: $U <> X_\nu$. Damit ist der eine Teil der Formel ($i$) nachgewiesen. Ist umgekehrt $\bigwedge_\nu (U <> X_\nu)$, so gilt wegen der Formel $h_2$ aus § 5, 1. auch $\bigwedge_\nu (U < X'_\nu)$ also: $U < \bigwedge_\nu X'_\nu$. Wegen $h_3$ aus § 5, 1. gilt also auch $U < (\bigvee_\nu X_\nu)'$, d. h. $U < V'$. Daraus folgt dann wieder nach $h_2$ auch $U <> V''$. Damit ist der 2. Teil der Formel ($i$) nachgewiesen.

Zu jedem Indexbereich existiert also eine Vereinigung $\bigcup_\nu X_\nu = (\bigvee_\nu X_\nu)''$. Es gilt also im Gefüge $S(B)$ das Axiom $T'_5$, d. h. $S(B)$ ist vollständiges Gefüge.

Wegen der Resultate aus § 4 und durch genauere Auswertung des vorigen Beweises lässt sich $S_1$ auch so formulieren:

$S'_1$: Der Bereich $S_0(B)$ aller S-Klassen des Gefüges B und der leeren Klassen 0 ist eine vollständige Boole'sche Algebra bezüglich der klassentheoretischen Inklusionsrelation "$<$". Die Operationen in $S_0(B)$ sind dabei folgendermassen zu wählen:

$$\text{Summe aller S-Klassen } X_\nu = (\bigvee_\nu X_\nu)''$$
$$\text{Produkt aller S-Klassen } X_\nu = \bigwedge_\nu X_\nu$$
$$\text{Komplement der S-Klasse } X = \text{Orthokomplement } X'.$$

Weiter existiert nun im vollständigen Gefüge S(B) ein zu B isomorphes Teilgefüge. Ordnet man nämlich jedem $x$ aus B die von $x$ erzeugte H-Klasse $(x)''$ aus S(B) zu, so ist dies wegen der Formeln

$$(i) \qquad (x \subset y) \longleftrightarrow (x)'' < (y)'' \quad ; \quad (x \circ y) \longleftrightarrow (x)'' \diamond (y)''$$

deren nachweis dem Leser überlassen sei, ein Isomorphismus von B auf die Klasse H(B) aller H-Klassen von B.

Es gilt also:

$S_2$: Die Klasse H(B) aller H-Klassen von B bilden ein zu B isomorphes Teilgefüge des vollständigen Gefüges S(B) aller S-Klassen von B. Oder, wenn man $x$ mit $(x)''$ identifiziert:
S(B) ist eine vollständige Erweiterung des Gefüges B.

Das Spaltungsgefüge S(B) besitzt nun aber noch eine weitere charakteristische Eigenschaft. Es ist unter allen vollständige Erweiterungen von B die kleinste, d. h.:

$S_3$: Ist V irgendeine vollständige Erweiterung des Gefüges B, so existiert ein zu S(B) isomorphes Teilgefüge $\overline{B}$ von V, welches wiederum B als Teilgefüge umfängt.

*Beweis:* Es sei A eine S-Klasse von B. A ist dann auch Teilklasse von V, es existiert also in V die Vereinigung $v$ aller Elemente von A. Es sei nun folgende Zuordnung $f$ betrachtet:

$$f(A) = \bigcup_{x \in A} x.$$

Dies ist eine eindeutige Abbildung von S(B) in V. $\overline{B}$ sei der Bildbereich. Es gelten nun die Formeln:

$$(i) \qquad A \diamond D \rightarrow f(A) \circ f(D)$$
$$(ii) \qquad A \,\overline{\diamond}\, D \rightarrow f(A) \,\bar{\circ}\, f(D)$$
$$(j) \qquad A < D \rightarrow f(A) \subset f(D)$$
$$(jj) \qquad A \not< D \rightarrow f(A) \,\overline{\subset}\, f(D).$$

Sie werden etwa so nachgewiesen:

$i$: Ist $A \diamond D$, so gilt nach $h_2$ aus § 5, 1.: $A < D'$, also $\bigwedge_x (x \in A \rightarrow x \in D')$ und daher wegen der Formel $k$ aus § 5, 1.: $\bigwedge_x ((x \in A) \rightarrow \bigwedge_y (y \in D \rightarrow x \circ y))$. Durch zweimalige Anwendung der Formel ($v$) aus § 3, 1. schliesst man dann $\bigcup_{x \in A} x \circ \bigcup_{y \in D} y$, also $f(A) \circ f(D)$.

$ii$: Ist $A \overline{\diamond} D$, so existiert ein $u$ in B, sodass $u \in A \wedge u \in D$. Es gilt dann $u \subset \bigcup_{x \in A} x \wedge u \subset \bigcup_{x \in D} x$, also $f(A) \bar{\circ} f(D)$.

$j$: Ist $A < D$, so gilt auch $\bigcup_{x \in A} u \subset \bigcup_{x \in D} x$, also $f(A) \subset f(D)$.

$jj$: Ist $A \not< D$, so haben wir nach $h_1$ aus § 5, 1. auch $A \overline{\diamond} D'$. Es existiert also ein $u$ in B, sodass $u \in A \wedge u \in D'$. Aus $u \in A$ folgt aber $u \subset \bigcup_{x \in A} x$ also $u \subset f(A)$. Anderseits folgt wegen $k$ aus § 5, 1. aus $u \in D'$: $\bigwedge_x (x \in D \rightarrow x \circ u)$, und daher wegen ($v$) aus § 3, 1. auch $u \circ \bigcup_{x \in D} x$ d. h. $f(D) \circ u$. Aus $u \subset f(A) \wedge u \circ f(D)$ folgt dann schliesslich $f(A) \overline{\subset} f(D)$.

Damit sind die Formeln $i \ldots jj$ nachgewiesen. Da aus $j$ und $jj$ die Ein-Eindeutigkeit von $f$ folgt, ist also $f$ ein Isomorphismus von S(B) auf $\overline{B}$. Von $S_3$ bleibt dann schliesslich nur noch zu beweisen, dass $\overline{B}$ das Gefüge B umfängt. Es ist aber leicht einzusehen, dass die Abbildung $f$ jede H-Klasse $(a)''$ von B auf das Element $a$ in B abbildet. Es ist also sicher $B < \overline{B}$.

Die Sätze $S_1 S_2$ und $S_3$ zusammenfassend gilt dann der.

*Einbettungssatz*: Ist B ein Gefüge, so bildet die Klasse S(B) aller S-Klassen von B ein vollständiges Gefüge mit folgenden Eigenschaften:

1) S(B) umfängt ein zu B isomorphes Gefüge, nämlich dasjenige H(B) aller H-Klassen von B.
2) Ist V irgend eine vollständige Erweiterung von B, so existiert ein zu S(B) isomorphes Gefüge $\overline{B}$ in V, das wiederum B umfängt.

Identifiziert man isomorphe Gefüge, so gilt wegen § 4:

Jeder Boole'sch partiell geordnete Bereich $B_0$ lässt sich so in eine kleinste vollständige Boole'sche Algebra $V_0$ einbetten, dass

1) Die Ordnung in $B_0$ erhalten bleit und
2) Immer dann, wenn die Vereinigung $\bigcup_\nu x_\nu$ in $B_0$ existiert, dieses Element auch die Summe aller $x_\nu$ in der Algedra $V_0$ ist [1].

[1] Dieser 2. Teil ist eine leichte Folge der verschärften Definition der Vereinigung im § 3.

Wie man leicht einsieht, bleibt obiger Satz bestehen, wenn man unter S(B) die Klasse aller Spaltungen von B und unter H(B) die Klasse aller Hauptspaltungen von B versteht. Man hat nur zu definieren:

$$(AA') \subset (DD') \longleftrightarrow A < D$$

$$(AA') \circ (DD') \longleftrightarrow A \diamond D.$$

Es ist daher naheliegend S(B) die Spaltungsstruktur über B zu nennen.

Die Einbettung allgemeiner Gebilde in speziellere ist ein Problem, das in fast jeder mathematischen Disziplin eine Rolle spielt. In der Theorie der partiellen Ordnung hat vor allem MacNeille solche Probleme gelöst [1]) Unter anderem ist es ihm gelungen, mit Hilfe des Schnittbegriffes (§ 5, 3.. jede partiell geordnete Menge M in einen kleinsten vollständigen Verband V einzubetten. MacNeille zeigt dann weiter, dass die Klasse aller Schnitte einer Boole'schen Algebra $B_0$ eine kleinste vollständige Boole'sche Algebra $S(B_0)$ über $B_0$ bilden. Dieser Satz ist ein Spezialfall unseres Einbettungssatzes für Gefüge. Im Falle der Boole'schen Algebra ist ja, wie in § 5, 3. bemerkt wurde der Spaltungsbegriff äquivalent zu demjenigen des Schnittes.

Schnitt und Spaltung sind also in gleicher Weise zur Vervollständigung einer Boole'schen Algebra $B_0$ geeignet. Ist aber $B_0$ nicht Algebra, sonder nur Boole'sch geordnet, so liegen die Dinge anders. Wie in § 5, 3. gezeigt wurde, existieren nämlich dann im Allgemeinen mehr Spaltungen als Schnitte. Weil aber das Spaltungsgefüge die kleinste vollständige Erweiterung ist, braucht dann die Klasse aller Schnitte kein vollständiges Gefüge zu sein.

Zusammenfassend erweist sich also die Spaltung als der geeignete Begriff in der Theorie der Boole'schen Algebren und der Boole'schen Ordnung, während der Schnitt in die Verbandstheorie und die Lehre der allgemeinen partiellen Ordnung gehört. Ueberhaupt kann in Form einer Analogie festgehalten werden:

| | | | | |
|---|---|---|---|---|
| Partialordnung | : | Schnitt | : | Verband |
| Boole'che Part. Ordg. | : | Spaltung | : | Boole'sche Algebra. |

Die Boole'sche Partialordnung kann also als Grundlage der Boole'schen Algebra bezeichnet werden, in analoger Weise, wie die allgemeine Partialordnung die Grundlage der allgemeinen Verbandstheorie ist.

1 MacNeille (1).

## § 7 ATOME UND ATOMARE GEFÜGE

**1. Das Atom:** Das Element $\dot{a}$ eines Gefüges B heisst ein Atom wenn es eine der folgenden äquivalenten Bedingungen für alle $u$ aus B erfüllt:

$a_1$) $(\dot{a} \subset u \vee \dot{a} \circ u)$

$a_2$) $(u \subset \dot{a} \rightarrow u = \dot{a})$.

Atome seien immer durch Punktierung gekennzeichnet. $\dot{a} \in B$ heisst also: «$\dot{a}$ ist Atom von B». Die Aequivalenz der Formeln $a_1$ und $a_2$ zeigt man etwa so:

Es sei $u \subset \dot{a}$. Wegen $a_1$ gilt dann $\dot{a} \subset u$. Unter Voraussetzung von $a_1$ folgt also aus $u \subset \dot{a}$ immer auch $u = \dot{a}$.

Ist anderseits $\dot{a} \overline{\subset} u$, so existiert $v$ in B, sodass $v \subset \dot{a} \wedge v \circ u$. Wegen $a_2$ gilt dann $v = \dot{a}$ und also $\dot{a} \circ u$. Aus $a_2$ folgt also für jedes $u$ die Formel $\dot{a} \overline{\subset} u \rightarrow \dot{a} \circ u$; was mit $a_1$ äquivalent ist.

In § 2 haben wir Gefüge kennen gelernt, die keine Atome enthalten. Hier sollen nun im Gegensteil solche Gefüge behandelt werden, die so viele Atome enthalten, dass sie überhaupt isomorph zu Klassengefügen sind.

**2. Das atomare Gefüge:** Das Gefüge B heisst atomar, wenn jedes Element $x$ aus B ein Atom $\dot{a}$ besitzt [1].

Diese Bedingung hat, trotzdem sie schwach scheint, weitgehende Folgen. Es gelingt nämlich zu zeigen, dass für das Bestehen der Relationen $x$ sub $y$ und $x$ dis $y$ nur die Lage der Atome entscheidend ist, d. h. genau:

In atomaren Gefügen gelten die Formeln:

$a$) $$(x \subset y) \longleftrightarrow \bigwedge_{\dot{a} \in B} (\dot{a} \subset x \rightarrow \dot{a} \subset y)$$

$b$) $$(x \circ y) \longleftrightarrow \bigwedge_{\dot{a} \in B} (\dot{a} \overline{\subset} x \wedge \dot{a} \overline{\subset} y).$$

Die Schlüsse von links nach rechts sind trivial. Es bleiben die umgekehrten zu beweisen:

Es sei $x \overline{\subset} y$, dann existiert wegen $T_3$ ein $u$ in B, sodass $u \bar{\circ} x \wedge u \circ y$. Nach $T_4$ gibt es dann ein $v$ in B mit $v \subset u \wedge v \subset x$. Es gilt also wegen $T_2$: $v \in B \wedge v \subset x \wedge v \circ y$. Da nun B atomar ist, existiert ein Atom $\dot{a}$ sub $v$. Es gilt also $\dot{a} \subset x \wedge \dot{a} \circ y$ und damit $\dot{a} \subset x \wedge \dot{a} \overline{\subset} y$. Zusammenfassend

1 Dabei heisst das Atom $\dot{a}$ ein Atom von $x$, wenn es sub $x$ ist.

ist die zu *a)* äquivalente Formel $(x \mathrel{\overline{\subset}} y) \longleftrightarrow \bigvee_{\dot a \in B} (\dot a \subset x \wedge \dot a \mathrel{\overline{\subset}} y)$ bewiesen. Ist nun $x \,\bar{\circ}\, y$, so existiert wegen $T_4$ ein $u$ in B mit $u \subset x \wedge u \subset y$. Weil B atomar ist, existiert ein Atom $\dot a$ sub $u$, für welches dann gilt $\dot a \subset x \wedge \dot a \subset y$. Damit ist auch *b)* nachgewiesen.

Die Formeln *a)* und *b)* lassen sich auch im folgenden Satz aussprechen:

*Satz:* Das atomare Gefüge B ist isomorph zum Gefüge B* aller Klassen X, die aus der Gesamtheit der Atome eines Elementes $x$ von B bestehen.
Der Isomorphismus wird vermittelt durch die Zuordnung: $x \Longleftrightarrow$ Klasse aller Atome von $x$.

Jedes atomare Gefüge B ist also isomorph zu einem Gefüge von Klassen. Umgekert braucht aber ein Klassengefüge nicht notwendig atomar zu sein. Dies drückt sich deutlich in der folgenden ganz speziellen Eigenschaft von B* aus:

B* besitzt die Einermengen des zu Grunde liegenden Bereiches K aller Atome als Elemente.

Da jedes endliche Gefüge atomar ist, gilt das folgende Korollar zum vorigen Satz:

Jedes endliche Gefüge ist isomorph zum Gefüge aller Klassen, die aus den Atomen eines Elementes bestehen.

**3. Vollständige atomare Gefüge:** In § 4 wurden die vollständigen Gefüge behandelt. Trotzdem in diesen fast alle Rechenregeln gelten, wie in den vollen Klassengefügen, brauchen sie doch keine Atome zu besitzen. Verlangt man nun aber weiter, dass das Gefüge atomar sei, so gelangt man zum vollen Klassengefüge. Dies zu zeigen ist die Aufgabe des vorliegenden Abschnittes.

Es sei zunächst das folgende Kriterium bewiesen:

*Satz:* Ein atomares Gefüge ist dann und nur dann isomorph einem vollen Gefüge (nämlich demjenigen über der Klasse aller Atome) wenn folgendes Axiom erfüllt ist:

*g)* Zu jeder nicht leeren Klasse A von Atomen des Gefüges B existiert ein Element $v_A$ in B, das die Atome der Klasse A und nur diese besitzt. d. h. $v_A$ erfüllt die Formeln:

$$\dot a \in A \longleftrightarrow \dot a \subset v_A$$

oder

$$\dot a \notin A \longleftrightarrow \dot a \circ v_A .$$

*Beweis:* Dass die vollen Gefüge das Axiom $g$ erfüllen ist klar. Ist umgekehrt B atomares Gefüge und $A_x$ die Klasse aller Atome von $x$, so bilden nach 2. die $A_x$ ein Klassengefüge B*. Es bleibt also nur zu zeigen, dass B* voll ist, sofern B das Axiom $g$ erfüllt. Mit andern Worten: Es ist zu zeigen, dass jede Klasse A von Atomen aus B ein Element von B* ist, wenn in B nur das Axiom $g$ gilt.

Es sei also A eine Klasse von Atomen aus B. Wegen $g$ existiert dann ein $v_A$, sodass $\dot{a} \subset \dot{v}_A \longleftrightarrow a \in A$. A ist also die Klasse aller Atome von $v_A$, es gilt daher $A \in B^*$.

Da $g$ eine leichte Folge von $T_5$ ist gilt:

*Satz:* Ein Gefüge B ist dann und nur dann isomorph einem vollen Klassengefüge, wenn es atomar und vollständig ist.

Da nach § 4 die Forderung «B ist vollständiges Gefüge» äquivalent ist mit «$B_0$ ist vollständige Boole'sche Algebra», folgt aus vorigem Satz, die schon von Tarski [1] bewiesene Tatsache:

Eine vollständige Boole'sche Algebra, in welcher jedes Element ein Atom besitzt, ist immer isomorph einer vollen Algebra bestehend aus allen Teilklassen einer festen Klasse.

## § 8 DIE VOLLEN MENGEN EINES TOPOLOGISCHEN RAUMES

Es soll hier eine Anwendung der Boole'schen Partialordnung in der Theorie der topologischen Räume gemacht werden. Wohl eine der schönsten Methoden zur Einführung einer abstrakten Topologie in der Menge R ist die Auszeichnung der offenen Mengen:

**1.** Definition des topologischen Raumes: Eine Menge R heisst ein topologischer Raum, wenn eine Klasse U von Teilmengen $a, b, \ldots$ aus R so ausgezeichnet ist, dass folgende Axiome gelten:

$0_1$) $$o \in U \wedge R \in U$$

$0_2$) $$x \in U \wedge y \in U \rightarrow x \cap y \in U$$

$0_3$) $$X < U \rightarrow \bigcup_{x \in X} x \in U.$$

Dabei stehen die Zeichen $\cup$, $\cap$ für die Mengenvereinigung und -durch schnitt.

[1] Tarski (1).

Die Elemente von U heissen dann die offenen Mengen von R, ihre Komplemente die abgeschlossenen Mengen von R.

Eine Teilklasse B offener Mengen $b \neq 0$ heisst eine Basis offener Mengen, wenn sich jedes $x$ aus U in der Form

$$x = \bigcup_{u \in B \wedge u \subset x} u$$

darstellen lässt [1]. Es gilt dann der Satz:

$S_1$: Eine Klasse B von nicht leeren Teilmengen der Menge R kann dann und nur dann als Basis offener Mengen einer Topologie in R ge ·ählt werden, wenn folgende Bedingungen zutreffen:

$b_1$) $$\bigcup_{u \in B} u = R$$

$b_2$) $$(x \in B \wedge y \in B \wedge a \in x \wedge a \in y) \rightarrow \bigvee_u (u \in B \wedge u \subset x \wedge u \subset y \wedge a \in u).$$

Die offenen Mengen sind dann diejenigen, die sich als Vereinigung von (beliebig vielen) Elementen aus B darstellen lassen.

**2.** Die Operationen H und I: Es sei B eine Basis offener Mengen des topologischen Raumes R. Man definiert dann für jede Teilmenge $a$ von R:

(I) $$\mathrm{I}\,a = \bigcup_{u \subset a \wedge u \in B} u$$

und nennt I $a$ das Innere der Menge $a$. Es gelten dann, wie man leicht nachweist, die Formeln:

$I_1$) $$\mathrm{I}\,e = e$$

$I_2$) $$\mathrm{I}\,a \subset a$$

$I_3$) $$\mathrm{II}\,a = \mathrm{I}\,a$$

$I_4$) $$\mathrm{I}(a \cap b) = \mathrm{I}\,a \cap \mathrm{I}\,b$$

$I_5$) $$a \subset b \rightarrow \mathrm{I}\,a \subset \mathrm{I}\,b$$

wobei $e = R$ gesetzt ist. Weiter definiert man die Operation H nach der Formel:

(H) $$\mathrm{H}\,a = \mathrm{CIC}\,a,$$

1 Gewöhnlich wird auch die leere Menge $o$ als Element einer Basis B zugelassen. In unserem Fall ist dies jedoch nicht zweckmässig.

wobei C die Komplementbildung in der Menge R bedeutet. Ha heisst die abgeschlossene Hülle der Menge $a$ und es gelten wegen $\mathrm{CC}x = x$ die Formeln:

$$\text{(U)} \qquad \mathrm{HC}\,a = \mathrm{CI}\,a$$
$$\mathrm{IC}\,a = \mathrm{CH}\,a$$

und daher die dualen zu $\mathrm{I}_1 \ldots \mathrm{I}_5$:

$$\mathrm{H}_1) \qquad \mathrm{H}\,o = o$$
$$\mathrm{H}_2) \qquad a \subset \mathrm{H}\,a$$
$$\mathrm{H}_3) \qquad \mathrm{HH}\,a = \mathrm{H}\,a$$
$$\mathrm{H}_4) \qquad \mathrm{H}(a \cup b) = \mathrm{H}\,a \cup \mathrm{H}\,b$$
$$\mathrm{H}_5) \qquad a \subset b \rightarrow \mathrm{H}\,a \subset \mathrm{H}\,b$$

$\mathrm{H}_1 \ldots \mathrm{H}_4$ ergeben das bekannte Kuratowski'sche Axiomen-System des topologischen Raumes. $a$ ist schliesslich dann und nur dann offen bezw. abgeschlossen, wenn $\mathrm{I}\,a = a$ bezw. $\mathrm{H}\,a = a$ ist.

**3.** Die vollen Mengen eines topologischen Raumes. Es wurde schon verschiedentlich auf die Bedeutung der Mengen von der Form $\mathrm{IH}\,a$ eines topologischen Raumes hingewiesen [1]. Es soll gezeigt werden, dass sie in gewissen Räumen eine ausgezeichnete Rolle spielen. Wir definieren:

Eine nicht leere Teilmenge $x$ des topologischen Raumes R heisst voll, wenn sie von der Form $x = \mathrm{IH}\,u$ ist.

Zunächst braucht eine offene Menge nicht voll zu sein. Das charakteristische Gegenbeispiel ist die offene Kreisscheibe der Euklidischen Ebene, von der ein eindimensionales Stück fehlt. Sodann gelten die folgenden Hilfsformeln:

$$(i) \qquad \mathrm{I}\,a \subset \mathrm{IHI}\,a$$
$$(ii) \qquad \mathrm{IH}\,a = \mathrm{IHIH}\,a.$$

*Beweis:* Wegen $\mathrm{H}_2$ gilt $\mathrm{I}a \subset \mathrm{HI}\,a$ und daher wegen $\mathrm{I}_5$ auch $\mathrm{II}\,a \subset \mathrm{IHI}\,a$. Daraus folgt dann wegen $\mathrm{I}_3$ die Formel $i$.

Um $ii$ nachzuweisen setze man für $a$ in der Formel $i$ die Menge $\mathrm{H}\,a$. Es gilt dann $\mathrm{IH}\,a \subset \mathrm{IHIH}\,a$. Anderseits gilt wegen $\mathrm{I}_2$: $\mathrm{IH}\,a \subset \mathrm{H}\,a$ und daher wegen $\mathrm{H}_5$ und $\mathrm{I}_5$ auch $\mathrm{IHIH}\,a \subset \mathrm{IHH}\,a$. Wegen $\mathrm{H}_3$ folgt daraus

[1] Kuratowski (1) (2). In (2) verwendet der Autor die Operation HI in gewissen kurventheoretischen Betrachtungen.

Lebesgue (1). Der Autor bezeichnet dort die Ränder der Mengen $\mathrm{IH}\,a$ einnes $\mathrm{R}^n$ als «$(n-1)$ dimensionale Grenzen» und zeigt, dass dieser Begriff eine brauchbare Verallgemeinerung desjenigen der $(n-1)$ dimensionalen Mannigfaltigkeiten ist.

auch $\mathrm{IHIH}\,a \subset \mathrm{IH}\,a$. Zusammen mit der vorigen Formel ist dann auch *ii* nachgewiesen.

Aus der Formel *ii* folgt nun sofort folgendes Kriterium für volle Mengen:

*Satz:* Die Menge $a \neq 0$ ist dann und nur dann voll, wenn $\mathrm{IH}\,a = a$ gilt.

Ferner lässt sich beweisen:

$S_2$: Der Durchschnitt $a \cap b$ zweier nicht fremder voller Mengen $a$ und $b$ ist ebenfalls voll.

Nach vorigem Satz gilt nämlich $a = \mathrm{IH}\,a$ und $b = \mathrm{IH}\,b$, also wegen $I_4$ auch $a \cap b = \mathrm{I}(\mathrm{H}\,a \cap \mathrm{H}\,b)$. $\mathrm{H}\,a \cap \mathrm{H}\,b$ ist aber als Durchschnitt abgeschlossener Mengen wieder abgeschlossen, also von der Form $\mathrm{H}\,u$. Es ist daher $a \cap b = \mathrm{IH}\,u$, also wegen der Voraussetzung $a \cap b \neq 0$, die Menge $a \cap b$ voll.

Alle diese Eigenschaften voller Mengen hat schon Kuratowski bewiesen. Zu einer weitern interessanten Eigenschaft der vollen Mengen führt nun aber der Begriff der Boole'schen Partialordnung. Ist nämlich R ein beliebiger topologischer Raum, so braucht zunächst die Klasse aller nicht leeren offenen Mengen kein Gefüge [1] zu sein. Ist $a$ eine offene Kreisscheibe des $R^2$ und $b$ dieselbe Kreisscheibe mit Ausnahme eines Punktes, so gilt zwar $a \mathrel{\bar{\subset}} b$ und trotzdem existiert kein offenes $u \neq 0$ mit der Eigenschaft $u \mathrel{\bar{\circ}} a \wedge u \circ b$. Das Axiom $T_3$ ist also nicht erfüllt.

Ganz anders verhält sich die Klasse V aller vollen Mengen. Es gilt nämlich:

$S_3$: Die Gesamtheit V aller vollen Mengen eines topologischen Raumes R bildet ein Gefüge.

*Beweis:* Die Axiome $T_1$ und $T_2$ sind sicher erfüllt, da V eine Klasse von Mengen ist. $T_3$ und $T_4$ werden in V etwa so nachgewiesen:

$T_3$: Es seien $x$ und $y$ Elemente von V und es sei $x \mathrel{\bar{\subset}} y$. Setzt man dann $u = \mathrm{CH}\,y$, so gilt:

*a)* $u \in \mathrm{V} \vee u = 0$. Da $y$ voll ist, gilt ja $y = \mathrm{IH}\,y$, also $u = \mathrm{CHIH}\,y$. Wegen der Formeln (U) aus 2. ist dann $u = \mathrm{IHIC}\,y$ und daher $u = \mathrm{IH}u$. Es ist also $u$ in V oder dann leer.

*b)* $u \circ y$. Es ist ja wegen $H_2$ aus 2. $y \subset \mathrm{H}y$, also $\mathrm{CH}\,y \subset \mathrm{C}\,y$, also $u \subset \mathrm{C}\,y$ und damit $u \circ y$.

[1] Es ist in der Folge immer das Gefüge bezüglich der klassentheoretischen sub- und dis-Relation gemeint, sofern es sich um einen Bereich B von Klassen handelt.

*c)* $u \bar{\circ} x$. Denn wäre $u \circ x$, so hätten wir $x \circ \mathrm{CH}\,y$, also $x \subset \mathrm{H}\,y$ und damit wegen der Formeln $\mathrm{H}_5$ und $\mathrm{I}_5$ aus 2. $\mathrm{IH}\,x \subset \mathrm{IHH}\,y$, also wegen $\mathrm{H}_3$ auch $\mathrm{IH}\,x \subset \mathrm{IH}\,y$. Weil $x$ und $y$ voll sind, ist $\mathrm{IH}\,x = x$ und $\mathrm{IH}\,y = y$, also wäre $x \subset y$. Dies ist aber der gesuchte Widerspruch mit der Voraussetzung $x \bar{\subset} y$.

Schliesslich folgt dann aus $u \bar{\circ} x$ speziell $u \neq 0$, also gilt zusammenfassend die Formel $\mathrm{T}_3$ im Bereich V.

$\mathrm{T}_4$: Sind wieder $x$ und $y$ in V und jetzt $x \bar{\circ} y$, so setze man $u = x \cap y$. Nach $\mathrm{S}_2$ ist dann $u$ in V. Ferner gilt $u \subset x \wedge u \subset y \wedge u \neq 0$. Damit ist auch die Formel $\mathrm{T}_4$ in V nachgewiesen.

4. Der G-Raum: B sei eine Basis offener Mengen des topologischen Raumes R. Es sind dann in B die Axiome $\mathrm{T}_1$ und $\mathrm{T}_2$ mit Bezug auf die gewöhnlichen klassentheoretischen sub — und dis — Relationen erfüllt. Darüber hinaus zeigt aber die Formel $b_2$ aus 1., dass B auch das Axiom $\mathrm{T}_4$ befriedigt. Trotzdem braucht nun B kein Gefüge zu sein; das Axiom $\mathrm{T}_3$ ist nämlich nicht notwendig erfüllt:

Beispiel eines topologischen Raumes der kein Gefüge als Basis offener Mengen besitzt:

R sei die Menge bestehend aus den Elementen $\alpha$ und $\beta$. Als offene Mengen seien $0, (\alpha), (\alpha, \beta)$ ausgezeichnet. Es sind dann die Axiome $\mathrm{O}_1, \mathrm{O}_2, \mathrm{O}_3$ erfüllt; R ist topologischer Raum. Auf Grund der Definition in 1. sieht man leicht ein, dass die einzige Basis von R die Klasse B bestehend aus den Mengen $(\alpha)$ und $(\alpha, \beta)$ ist. Anderseits ist B kein Gefüge, denn es gilt zwar $(\alpha, \beta) \bar{\subset} (\alpha)$, trotzdem existiert kein Basiselement $u$, welches zu $(\alpha)$ aber nicht zu $(\alpha, \beta)$ fremd ist. Das Axiom $\mathrm{T}_3$ ist also in B nicht erfüllt.

Ein weiteres solches Beispiel ist die unendliche Menge R, in der man die endlichen Mengen als abgeschlossene Mengen auszeichnet.

*Definition:* Eine Basis B offener Mengen eines topologischen Raumes R heisst G-Basis, wenn sie ein Gefüge ist.

Der Raum R heisst G-Raum, wenn er eine G-Basis besitzt.

Mit dem vorstehenden Beispiel ist dann gezeigt:

A) Nicht jede Basis offener Mengen braucht G-Basis zu sein und noch mehr.

B) Nicht jeder topologische Raum ist ein G-Raum.

Anderseits gehören aber wichtige Räume in die Klasse der G-Räume. So bilden z. B. die offenen Kugeln des $\mathrm{R}^n$ und allgemeiner diejenigen eines metrischen Raumes eine G-Basis; die metrischen Räume sind also G-Räume. Man zeigt sogar leicht, dass alle regulären Räume G-Räume sind.

5. Die vollen Mengen eines G-Raumes: Die in 3. definierten vollen Mengen spielen besonders im G-Raum eine wichtige Rolle. Zunächst gilt:

$S_4$: Ist B eine Basis offener Mengen des topologischen Raumes R und besteht B aus lauter vollen Mengen, so ist es eine G-Basis und also R ein G-Raum.

*Beweis:* Als Basis erfüllt der Bereich B die Axiome $T_1, T_2, T_4$ bezüglich der natürlichen sub — und dis — Relationen für Mengen. Es bleibt also nur $T_3$ in B zu verifizieren:

Es seien $x$ und $y$ in B und es sei $x \mathrel{\overline{\subset}} y$. Nach Voraussetzung sind dann $x$ und $y$ volle Mengen, nach $S_3$ existiert also wegen $x \mathrel{\overline{\subset}} y$ eine weitere volle Menge $v$, sodass $v \mathrel{\bar{\circ}} x \wedge v \circ y$. Weil B Basis und $v$ offen ist gilt $v = \bigcup_{u \in B \wedge u \subset v} u$. Wegen $v \mathrel{\bar{\circ}} x$ existiert also ein $u$, sodass $u \in B \wedge u \subset v \wedge u \mathrel{\bar{\circ}} x$. Wegen $v \circ y \wedge u \subset v$ gilt schliesslich auch $u \circ y$. Zusammenfassend ist also die Formel

$$(x \in B \wedge y \in B \wedge x \mathrel{\overline{\subset}} y) \rightarrow \bigvee_u (u \in B \wedge u \mathrel{\bar{\circ}} x \wedge u \circ y)$$

nachgewiesen. Sie ist zu $T_3$ in B äquivalent.

Umgekehrt sind nun aber mit dem Satz $S_4$ schon alle G-Räume charakterisiert. Es gilt nämlich:

$S_5$: Ist B eine G-Basis des topologischen Raumes R, so sind alle ihre Elemente volle Mengen. Oder:
Damit eine Basis offener Mengen eines topologischen Raumes ein Gefüge sei, muss sie notwendig aus vollen Mengen bestehen.

*Beweis:* Es sei B eine G-Basis des Raumes R. Es gilt dann für alle $x$ und $y$ aus B die Hilfsformel

$$(i) \qquad x \subset y \longleftrightarrow x \subset \mathrm{H}\, y.$$

Zunächst folgt ja wegen $H_2$ aus 2. aus $x \subset y$ auch $x \subset \mathrm{H}\, y$.

Ist anderseits $x \mathrel{\overline{\subset}} y$, so existiert, da B Gefüge ist, ein $u$ in B, sodass $u \mathrel{\bar{\circ}} x \wedge u \circ y$. Aus $u \circ y$ folgt aber $y \subset \mathrm{C}\, u$ und nach $H_5$ aus 2. gilt dann auch $\mathrm{H}\, y \subset \mathrm{H}\mathrm{C}\, u$. Nach (U) aus 2. und da $u$ als Element von B offen ist, gilt aber $\mathrm{H}\mathrm{C}\, u = \mathrm{C}\mathrm{I}\, u = \mathrm{C}\, u$, also ist $\mathrm{H}\, y \subset \mathrm{C}\, u$ und damit $\mathrm{H}\, y \circ u$. Wir hatten aber $u \mathrel{\bar{\circ}} x$, also gilt sicher auch $x \mathrel{\overline{\subset}} \mathrm{H}\, y$. Zusammenfassend ist damit auch der zweite Teil der Formel $i$, nämlich $x \mathrel{\overline{\subset}} y \rightarrow x \mathrel{\overline{\subset}} \mathrm{H}\, y$, nachgewiesen.

Ist nun $y$ irgend ein Element aus B, so gilt, da $y$ und $\mathrm{IH}\,y$ offene Mengen und B eine Basis sind:

$$y = \bigcup_{x \in \mathrm{B} \wedge x \subset y} x \qquad\qquad \mathrm{IH}\,y = \bigcup_{x \in \mathrm{B} \wedge x \subset \mathrm{H}\,y} x .$$

Wegen der Formel *i* gilt daher $y = \mathrm{IH}\,y$, $y$ ist also volle Menge.

Fassen wir $S_4$ und $S_5$ zusammen, so erhalten wir folgendes Resultat:

*Satz:* Die Basis B des topologischen Raumes R ist dann und nur dann ein Gefüge (d. h. G-Basis), wenn sie aus lauter vollen Mengen besteht.

Die G-Räume sind diejenigen topologischen Räume, die eine Basis von vollen Mengen besitzen.

Es ist dann speziell auch die Klasse V aller vollen Mengen eine Basis von R.

# 2. TEIL

# DIE PAARUNG VON GEFÜGEN

Eine Paarung ist ein Paar von Zuordnungen der Elemente zweier Gefüge, die einer charakteristischen Bedingung genügen.

Im Falle des vollständigen Gefüges, wo also eine vollständige Boole'sche Algebra zwischen den Elementen besteht, stehen die Paarungen in enger Beziehung zu den Homomorphismen bezüglich der totalen Vereinigung. Die Paarung ist eine Verallgemeinerung dieser Vereinigungshomomorphismen auf das beliebige Gefüge.

Noch besser kann die Paarung als Verallgemeinerung des Begriffes «Relation zwischen den Elementen zweier Gebilde (Mengen)» auf den Fall elementeloser Gebilde (Gefüge) charakterisiert werden. Im Falle zweier vollständiger und atomarer Gefüge stehen nämlich die Paarungen in ein-eindeutiger Beziehung zu den Relationen zwischen den Atomen der beiden Gefüge.

In § 3 soll gezeigt werden, wie sich Begriffe wie «Reflexität, Symmetrie, Transitivität einer Relation» auf den Paarungsbegriff übertragen lassen.

In § 5 wird auf Grund der Paarung die Abbildung nicht atomarer Gebilde definiert.

## § 1 DEFINITION DER PAARUNG

B und $\bar{B}$ seien zwei Gefüge. $f$ ordne jedem $x$ aus $B_0$ genau ein $f(x)$ aus $\bar{B}_0$ zu Genau so ordne $\bar{f}$ jedem $y$ aus $\bar{B}_0$ genau ein $\bar{f}(y)$ aus $B_0$ zu [1].

[1] Mit $B_0$ wird, wie im 1. Teil, der Bereich B vermehrt um das Null-Element $o$ bezeichnet.

$(f\bar{f})$ heisst dann eine Paarung der Elemente von B und $\bar{B}$ (oder kurz von B und $\bar{B}$), wenn folgende Bedingung für alle $u$ aus $B_0$ und $v$ aus $\bar{B}_0$ zutrifft:

(P) $$v \circ f(u) \longleftrightarrow u \circ \bar{f}(v).$$

$f$ und $\bar{f}$ heissen dann die Glieder der Paarung; $f(u)$ und $\bar{f}(v)$ der Nach — bezw. Vor — Bereich von $u$ und $v$.

Wegen P gilt für jedes $v$ aus $\bar{B}$ die Formel $v \circ f(o)$, es ist also $f(o) = 0$. Genau so gilt $\bar{f}(o) = 0$. Ferner sei festgehalten: Wenn $x$ in B und $f(x) = 0$, so kann $x$ nicht Wert von $\bar{f}$ sein.

Interessanterweise tritt in der Definition der Paarung nur die Relation dis auf. Dies liegt im Wesen der Sache. Die Fremdheit spielt also hier die Hauptrolle und nicht die Teilrelation. Es seien nun einige Eigenschaften von Paarungen festgehalten:

1. Die Glieder einer Paarung bestimmen sich gegenseitig eindeutig, sie heissen konjugierte Zuordnungen.
2. $(f\bar{f})$ ist dann und nur dann Paarung von B und $\bar{B}$, wenn $(\bar{f}f)$ Paarung von $\bar{B}$ und B ist. Die beiden Paarungen heissen dann konjugiert.
3. Ist $(f\bar{f})$ Paarung von B mit $\bar{B}$, so existiert genau eine Paarung $(\bar{f}\bar{\bar{f}})$ von $\bar{B}$ und B mit dem Vorderglied $\bar{f}$. Das zweite Glied $\bar{\bar{f}}$ ist dann identisch mit $f$.
4. Für jedes Glied $f$ einer Paarung und für alle Elemente $x$ und $y$ gilt:

   $P_0$) $$x \subset y \rightarrow f(x) \subset f(y).$$

   Die Glieder einer Paarung sind also ordnungstreue Zuordnungen.

*Beweis:* $(f\bar{f})$ und $(ff^*)$ seien Paarungen von B und B mit demselben Vorderglied $f$. Dann gilt wegen P

$$u \circ \bar{f}(v) \longleftrightarrow v \circ f(u) \longleftrightarrow u \circ f^*(v).$$

Da diese Formeln für alle $u$ gelten, haben wir für jedes $v$: $\bar{f}(v) = f^*(v)$. Damit ist die eindeutige Bestimmtheit von $\bar{f}$ durch $f$ gezeigt. Genau so zeigt man, dass $f$ durch $\bar{f}$ eindeutig bestimmt ist, womit 1. nachgewiesen ist.

Die Behauptung 2. folgt leicht aus der Symmetrie der Formel P in $f$ und $\bar{f}$. Schliesslich ist 3. eine Folge aus 1. und 2., es bleibt also nur noch 4. nachzuweisen:

Es sei $(f\bar{f})$ eine Paarung von B und $\bar{\mathrm{B}}$ und es seien ferner $x \subset y$ zwei Elemente aus B. Ist dann $u \in \bar{\mathrm{B}}$, so gilt wegen P mit $u \circ f(y)$ auch $y \circ \bar{f}(u)$, also wegen $x \subset y$ auch $x \circ \bar{f}(u)$. Daraus folgt wegen P auch $u \circ f(x)$. Zusammenfassend gilt also für jedes $u$ aus $\bar{\mathrm{B}}$ die Formel: $u \circ f(y) \rightarrow u \circ f(x)$. Aus $x \subset y$ folgt also immer auch $f(x) \subset f(y)$. Damit ist $\mathrm{P}_0$ nachgewiesen.

Ausser den Sätzen 1 . . . . 4. werden später auch folgende Bemerkungen von Nutzen sein:

*a)* Ist $b \in \bar{\mathrm{B}}$ und $x \in \mathrm{B}$, so gilt: $x \subset \bar{f}(b) \rightarrow f(x) \neq 0$.
Ist $a \in \mathrm{B}$ und $y \in \bar{\mathrm{B}}$, so gilt: $y \subset f(a) \rightarrow \bar{f}(y) \neq 0$.

*b)* Für alle $x$ und $y$ gilt: $\bar{f}(y) \subset \bar{f}f\bar{f}(y)$.
$f(x) \subset f\bar{f}f(x)$.

Es genügt etwa die ersten Formeln nachzuweisen:

*a:* Ist $x \subset \bar{f}(b)$, so gilt $x \,\bar{\circ}\, \bar{f}(b)$, also auch $b \,\bar{\circ}\, f(x)$ und damit $f(x) \neq 0$.

*b:* Wäre $\bar{f}(y) \not\subset \bar{f}f\bar{f}(y)$, so existierte ein $u$ in B. sodass $u \subset \bar{f}(y) \wedge u \circ \bar{f}f\bar{f}(y)$. Wegen P und $\mathrm{P}_0$ gilt dann auch $f(u) \subset f\bar{f}(y) \wedge f(u) \circ f\bar{f}(y)$, es müsste also $f(u) = 0$ sein, was wegen *a)* ein Widerspruch mit $u \subset \bar{f}(y)$ ist.

Die Formeln *a)* und *b)* lassen sich folgendermassen verallgemeinern:

*a')*
$$x \in \bar{f}(\bar{\mathrm{B}})' \longleftrightarrow f(x) = 0$$
$$y \in f(\mathrm{B})' \longleftrightarrow \bar{f}(y) = 0$$

*b')*
$$x \in \bar{f}(\bar{\mathrm{B}})'' \rightarrow x \subset \bar{f}f(x)$$
$$y \in f(\mathrm{B})'' \rightarrow y \subset f\bar{f}(y).$$

Dabei bedeutet $f(\mathrm{X})$ die Klasse aller $f(x)$, wenn $x$ in X. Weiter bedeutet X' das Orthokomplement der Klasse X (siehe 1. Teil § 5). Die Beweise der Formeln gehen etwa so:

*a':* Ist $x$ in $\bar{f}(\bar{\mathrm{B}})'$, so gilt für alle $u$ aus $\bar{\mathrm{B}}$ die Beziehung $x \circ \bar{f}(u)$, also wegen P auch $u \circ f(x)$. Es kann also nur $f(x) = 0$ sein.

*b':* Die Formel $b'$ folgt genau so aus $a'$, wie $b$ aus $a$.

## § 2 BEISPIELE VON PAARUNGEN

Zuerst soll ein Anschauungsbeispiel gegeben werden. Es kann dazu dienen, die Definitionen und Sätze zu prüfen.

1. Veranschaulichung des Paarungsbegriffes am Koordinatenkreuz: Es sei B das Gefüge aller Teilmengen der $x$-Achse und $\bar{\mathrm{B}}$ dasjenige aller Teilmengen der $y$-Achse. Die Zuordnungen $f$ und $\bar{f}$ seien durch eine

Menge in der $xy$-Ebene (z.B. die Kurve $c$) so bestimmt, wie es die Figur andeutet. Man verifiziert dann leicht das Kriterium P; d.h. $(f\bar{f})$ ist eine Paarung der Gefüge B und $\bar{\mathrm{B}}$.

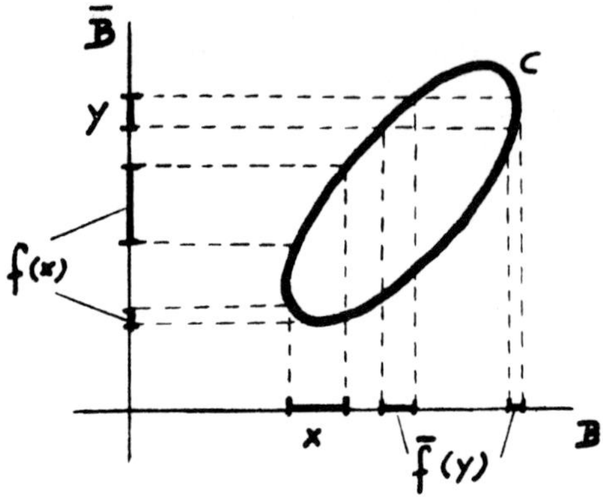


Im Falle vollständiger und atomarer Gefüge kann also jede Paarung durch eine Teilmenge des Produktes der Mengen aller Atome der beiden Gefüge repräsentiert werden.

**2.** Die normalen Paarungen: Sind B und $\bar{\mathrm{B}}$ zwei Gefüge und $a \in \mathrm{B}$, $b \in \bar{\mathrm{B}}$ irgend zwei ihrer Elemente, so wird durch die folgenden Formeln eine Paarung von B und $\bar{\mathrm{B}}$ definiert:

$$f(x) = \begin{cases} b & \text{für alle } x \,\bar{\circ}\, a \\ o & \text{für alle } x \circ a \end{cases}$$

$$\bar{f}(y) = \begin{cases} a & \text{für alle } y \,\bar{\circ}\, b \\ o & \text{für alle } y \circ b\,. \end{cases}$$

$(f\bar{f})$ heisst die von den Elementen $a$ und $b$ erzeugte normale Paarung von B und $\bar{\mathrm{B}}$. Veranschaulicht man sie nach der in 1. beschriebenen Weise, so entspricht ihr die Rechtecksfläche mit den Seiten $a$ und $b$ in der $xy$-Ebene.

Das Beispiel 1. weist ganz deutlich auf einen Zusammenhang zwischen Paarung und Relation hin. Ist nämlich $\mathrm{F}(\xi, \eta)$ die folgende Relation zwischen den Punkten der $x$-und $y$-Achse:

$\mathrm{F}(\xi, \eta) \longleftrightarrow$ die Koordinatenlinien durch die Punkte $\xi$ und $\eta$ schneiden sich auf der Menge $c$,

so ist $f(x)$ nichts anderes als der Nachbereich der Teilmenge $x$ bezüglich der Relation F. $\bar{f}(y)$ ist der Vorbereich von $y$ bezüglich F. Umgekehrt ist die Relation $\mathrm{F}(\xi, \eta)$ gleichbedeutend mitder Formel $\eta \in f((\xi))$ [1].

Diese Zusammenhänge im Allgemeinen zu klären ist Aufgabe des nächsten Beispieles.

**3.** Die Paarungen voller Klassengefüge: Es seien K und $\bar{\mathrm{K}}$ zwei Bereiche von Dingen $\xi, \eta, \zeta, \ldots .\, \alpha, \ldots$ Ihre nicht leeren Teilklassen seien mit $a, b, \ldots, x \ldots$ bezeichnet.

1 $(\xi)$ bedeutet die Klasse bestehend aus dem einen Element $\xi$. In der Folge wird aber $f((\xi))$ kurz durch $f(\xi)$ symbolisiert.

Im ersten Teil wurde gezeigt, dass die Bereiche B und $\overline{B}$ aller nicht leeren Teilklassen von K respektive $\overline{K}$ ein vollständiges und atomares Gefüge bezüglich der natürlichen sub — und dis — Relationen zwischen Klassen bilden. B und $\overline{B}$ hiessen dort die vollen Gefüge über den Klassen K und $\overline{K}$. Was sind nun die Paarungen dieser speziellen Gefüge? Um dieses Problem zu lösen soll eine Erzeugung von Paarungen aus Relationen behandelt werden:

Es sei $F(\xi, \eta)$ eine Relation zwischen den Elementen von K und $\overline{K}$; d.h. für jedes Paar $\xi \in K$ und $\eta \in \overline{K}$, ist gesagt, ob $F(\xi, \eta)$ zutrifft oder nicht zutrifft. Man definiere dann für jede Teilmenge $x$ von K und $y$ von $\overline{K}$:

$$\text{I)} \quad \begin{aligned} \eta \in f(x) &\longleftrightarrow \bigvee_{\xi} (\xi \in x \wedge F(\xi, \eta)) \\ \eta \in \overline{f}(y) &\longleftrightarrow \bigvee_{\xi} (\xi \in y \wedge F(\eta, \xi)) \end{aligned} .$$

$f(x)$ ist also der Nachbereich der Teilklasse $x$ von K bezüglich der Relation F; $\overline{f}(y)$ der Vorbereich der Klasse $y$ aus $\overline{K}$ bezüglich F. Es gilt dann.

$S_1$: Die nach I definierten Zuordnungen $f$ und $\overline{f}$ bilden eine Paarung $(f\overline{f})$ der vollen Gefüge B und $\overline{B}$. Sie heisst «die von der Relation F erzeugte Paarung».

*Beweis:* Es ist für $f$ und $\overline{f}$ das Paarungskriterium P nachzuweisen. Es sei daher $\eta \in f(x)$. Dann existiert wegen I ein $\xi \in x$, sodass $F(\xi, \eta)$ gilt. Ist dann $x \circ \overline{f}(y)$, so gilt $\xi \notin \overline{f}(y)$. Wegen I und $F(\xi, \eta)$ folgt daraus $\eta \notin y$. Unter der Voraussetzung $x \circ \overline{f}(y)$ gilt also für alle $\eta$ aus $\overline{K}$ die Formel $\eta \in f(x) \rightarrow \eta \notin y$, es ist also $y \circ f(x)$. Damit ist die Formel P in einer Richtung nachgewiesen. Genau so zeigt man auch die andere Richtung.

Jede Relation F zwischen den Elementen der Klassen K und $\overline{K}$ erzeugt also eine Paarung $(f\overline{f})$ der vollen Gefüge B und $\overline{B}$. Damit sind nun bereits alle Paarungen von B und $\overline{B}$ erschöpft, es gilt nämlich:

*Satz:* Jede Paarung $(f\overline{f})$ der vollen Gefüge B und $\overline{B}$ über den Bereichen K und $\overline{K}$ wird durch eine Relation F zwischen den Elementen von K und $\overline{K}$ vermittels der Formeln I erzeugt.
Die Relation F ist durch die folgende Formel eindeutig bestimmt:

$$\text{II)} \qquad F(\xi, \eta) \longleftrightarrow \eta \in f(\xi) \longleftrightarrow \xi \in \overline{f}(\eta) .$$

*Beweis*: Die Gültigkeit der Formel II und die Eindeutigkeit von F sind Trivialitäten. Es bleibt nur noch zu beweisen, dass $(f\overline{f})$ durch die

Relation F erzeugt wird. Wegen der völligen Symmetrie in $f$ und $\bar{f}$ genügt es zu zeigen, dass $f(x)$ der Nachbereich $N(x)$ von $x$ bezüglich F ist:

Es sei $\eta \in N(x)$. Dann existiert ein $\xi$, sodass $\xi \in x \wedge F(\xi, \eta)$. Es gilt also wegen II und weil $f$ Glied einer Paarung ist: $f(\xi) \subset f(x) \wedge \eta \in f(\xi)$. Daraus folgt $\eta \in f(x)$. Zusammenfassend ist also gezeigt:

*(i)* $$N(x) \subset f(x).$$

Ist anderseits $\eta \in f(x)$, so gilt: $(\eta)\, \bar{o}\, f(x)$, also da $(f\bar{f})$ Paarung ist auch $x\, \bar{o}\, \bar{f}(\eta)$. Es existiert also ein $\xi$, sodass $\xi \in x \wedge \xi \in \bar{f}(\eta)$. Wegen II gilt dann auch $\xi \in x \wedge F(\xi, \eta)$. $\eta$ ist also auch Element des Nachbereiches $N(x)$. Zusammenfassend gilt daher:

*(ii)* $$f(x) \subset N(x).$$

Die Formeln *i* und *ii* zeigen, dass $(f\bar{f})$ von der Relation F erzeugt wird.

Mit diesem Satz sind die Paarungen voller Klassengefüge charakterisiert. Er kann aber auch etwa so, ohne Verwendung des Paarungsbegriffes, formuliert werden:

*Satz:* K und $\bar{K}$ seien zwei Bereiche von Dingen. $f$ ordne jeder Teilklasse $x$ von K genau eine Teilklasse $f(x)$ von $\bar{K}$ und $\bar{f}$ jeder Teilklasse $y$ von $\bar{K}$ genau enie solche $\bar{f}(y)$ aus K zu.
$f$ und $\bar{f}$ bedeuten dann und nur dann Nach — bezw. Vor — Bereich bezüglich einer Relation $F(\xi, \eta)$ zwischen den Elementen von K und $\bar{K}$, wenn sie das Kriterium P erfüllen.

Im Teil 1. wurde gezeigt, dass jedes vollständige und atomare Gefüge B isomorph ist zum ganzen Gefüge B* über der Klasse K aller Atome von B. In leicht verständlicher Weise gilt daher:

Jede Paarung zweier vollständiger und atomarer Gefüge wird erzeugt durch eine Relation zwischen ihren Atomen.

Schliesslich, und das ist wohl der interessanteste Aspekt, lässt sich der Satz auch noch so formulieren:

Die Paarung ist eine Verallgemeinerung des Begriffes «Relation zwischen den Elementen zweier Gebilde (Mengen)» auf den Fall elementefreier, d.h. nicht atomarer Gebilde (Gefüge).

## § 3 PAARUNGEN EINES GEFÜGES MIT SICH

Im vorigen § wurde gezeigt, dass die Paarung eine Verallgemeinerung des Relationsbegriffes ist. Hier soll nun gezeigt werden, wie spezielle

Eigenschaften von Relationen, wie etwa Transitivität oder Symmetrie, auch für Paarungen definiert werden können.

Es sei K ein Bereich von Dingen und F eine Relation in K. F heisst, dann

| | | | |
|---|---|---|---|
| (1) | reflexiv | wenn: | $F(\xi,\xi)$ |
| (2) | symmetrisch | wenn: | $F(\xi,\eta)\rightarrow F(\eta,\xi)$ |
| (3) | disjunktiv | wenn: | $F(\xi,\eta)\vee F(\eta,\xi)$ |
| (4) | transitiv | wenn: | $F(\xi,\eta)\wedge F(\eta,\zeta)\rightarrow F(\xi,\zeta)$. |

Ist nun B ein Gefüge und $(f\bar{f})$ eine Paarung von B mit sich selber, so definiert man die entsprechenden Begriffe wie folgt: $(f\bar{f})$ heisst

| | | | |
|---|---|---|---|
| I | reflexiv | wenn: | $x\subset f(x)$ |
| II | symmetrisch | wenn: | $f(x)=\bar{f}(x)$ |
| III | disjunktiv | wenn: | $x\subset f(y)\vee y\subset f(x)$ |
| IV | transitiv | wenn: | $ff(x)\subset f(x)$. |

Dass diese Definitionen sinnvoll sind, zeigt folgender

*Satz:* Die Paarung $(f\bar{f})$ des vollen Gefüges B über der Klasse K ist dann und nur dann reflexiv, symmetrisch, disjunktiv oder transitiv, wenn es die erzeugende Relation F zwischen den Elementen von K ist.

*Beweis:* Die Paarung $(f\bar{f})$ sei von der Relation F erzeugt. Es gelten also die Formeln:

*(i)* $$F(\xi,\eta)\longleftrightarrow\eta\in f(\xi)\longleftrightarrow\xi\in\bar{f}(\eta)$$

*(ii)* $$\eta\in f(x)\longleftrightarrow\bigvee_{\xi}(\xi\in x\wedge F(\xi,\eta))$$
$$\eta\in\bar{f}(x)\longleftrightarrow\bigvee_{\xi}(\xi\in x\wedge F(\eta,\xi)).$$

Ist dann

1 *a:* F reflexiv, so folgt aus $\xi\in x$ wegen der Gültigkeit von $F(\xi,\xi)$ und wegen *ii* auch $\xi\in f(x)$. Es gilt also $x\subset f(x)$, wegen I ist daher auch $(f\bar{f})$ reflexiv.

1 *b:* $(f\bar{f})$ reflexiv, so gilt wegen I die Formel $(\xi)\subset f(\xi)$ und daher wegen *i* auch $F(\xi,\xi)$. Nach (1) ist also auch F reflexiv.

2 *a:* F symmetrisch, so fallen Vor- und Nachbereiche zusammen, also gilt $f(x)=\bar{f}(x)$. $(f\bar{f})$ ist also wegen II ebenfalls symmetrisch.

2 *b:* $(f\bar{f})$ symmetrisch, so folgt aus $F(\xi,\eta)$ wegen *i* auch $\eta\in f(\xi)$. Wegen II gilt also auch $\eta\in\bar{f}(\xi)$ und wegen *i* $F(\eta,\xi)$. Nach (2) ist daher auch F symmetrisch.

3 *a*: F disjunktiv, und $x \overline{\subset} f(y)$, so existiert ein $\xi$ in K, sodass $\xi \in x \wedge \xi \notin f(y)$. Wegen *ii* gilt dann $\bigwedge_{\eta} (\eta \in y \rightarrow \overline{F(\eta, \xi)})$. Wegen (3) folgt aber aus $\overline{F(\eta, \xi)}$ immer $\overline{F(\xi, \eta)}$. Wegen $\xi \in x$ und *ii* gilt daher $\bigwedge_{\eta} (\eta \in y \rightarrow \eta \in f(x))$, d. h. $y \subset f(x)$. Zusammenfassend folgt also aus $x \overline{\subset} f(y)$ immer $y \subset f(x)$ Nach III ist daher auch $(f\bar{f})$ disjunktiv.

3 *b*: $(f\bar{f})$ disjunktiv, so folgt aus $\overline{F(\xi, \eta)}$ wegen *i* auch $(\eta) \overline{\subset} f(\xi)$, also wegen III auch $(\xi) \subset f(\eta)$. Wegen *i* gilt daher $\overline{F(\eta, \xi)}$. Nach (3) ist also auch F disjunktiv.

4 *a*: F transitiv, so folgt aus $\xi \in ff(x)$ wegen *ii* die Existenz eines $\eta$ in K, sodass $\eta \in f(x) \wedge F(\eta, \xi)$. Ebenfalls wegen *ii* existiert also ein $\xi$ in K, sodass $\xi \in x \wedge F(\zeta, \eta)$. Aus $F(\zeta, \eta)$ und $F(\eta, \xi)$ folgt wegen (4) auch $F(\zeta, \xi)$. Es ist aber $\xi \in x$, also wegen *ii* $\xi \in f(x)$. Aus $\xi \in ff(x)$ folgt also zusammenfassend $\xi \in f(x)$. Es gilt daher $ff(x) \subset f(x)$. Nach IV ist also $(f\bar{f})$ transitiv.

4 *b*: $(f\bar{f})$ transitiv, so folgt aus $F(\xi, \eta) \wedge F(\eta, \zeta)$ wegen *i* $\eta \in f(\xi) \wedge \zeta \in f(\eta)$. Weil $f$ Glied einer Paarung ist, gilt dann $f(\eta) \subset ff(\xi)$, also wegen IV auch $f(\eta) \subset f(\xi)$. Wegen $\xi \in f(\eta)$ gilt daher $\zeta \in f(\xi)$. Nach *i* haben wir also $F(\xi, \zeta)$. Zusammenfassend ist damit die Formel aus (4) nachgewiesen; F ist also transitiv.

## § 4 PAARUNG VOLLSTÄNDIGER GEFÜGE

**1.** Die Isomorphie-Eigenschaften von Paarungen vollständiger Gefüge: Die Paarungen vollständiger Gefüge besitzen gewisse Isomorphie-Eigenschaften bezüglich der Boole'schen Algebra zwischen den Elementen.

Es seien B und $\bar{B}$ zwei vollständige Gefüge und $(f\bar{f})$ eine Paarung derselben. Dann ist das Paarungskriterium P gleichbedeutend mit den Formeln P′ oder P″:

(P′) $$f(x) \subset y' \leftarrow\rightarrow \bar{f}(y) \subset x'$$

(P″) $$f(x) \cap y = 0 \leftarrow\rightarrow \bar{f}(y) \cap x = 0.$$

Damit lassen sich nun leicht folgende Eigenschaften für die Glieder einer Paarung vollständiger Gefüge nachweisen:

($P_1$) $$f(\bigcup_{\nu} x_{\nu}) = \bigcup_{\nu} f(x_{\nu})$$

($P_2$) $$f(\bigcap_{\nu} x_{\nu}) \subset \bigcap_{\nu} f(x_{\nu}).$$

$f$ ist also Homomorphie bezüglich $\bigcup$, nicht aber notwendig bezüglich $\bigcap$.

*Beweis:* Es sei $a' = \bigcup_\nu f(x_\nu)$. Dann gilt für alle $\nu : f(x_\nu) \subset a'$, also wegen P′ auch $\bar{f}(a) \subset x'_\nu$. Daraus folgt aber $\bigcap_\nu x'_\nu \supset \bar{f}(a)$ und damit $(\bigcup_\nu x_\nu)' \supset \bar{f}(a)$, also wegen $P' : f(\bigcup x_\nu) \subset a'$. Damit ist zusammenfassend die folgende Formel bewiesen:

*(i)*
$$f(\bigcup_\nu x_\nu) \subset \bigcup_\nu f(x_\nu).$$

Anderseits gilt für alle $\mu : x_\mu \subset \bigcup_\nu x_\nu$. Wegen der Formel $P_0$ aus § 1 (Teil 2.) gilt damit auch $f(x_\mu) \subset f(\bigcup_\nu x_\nu)$. Daraus folgt die Formel

*(ii)*
$$\bigcup_\nu f(x_\nu) \subset f(\bigcup_\nu x_\nu).$$

Mit *i* und *ii* ist aber die Formel $P_1$ nachgewiesen.

Es gilt für alle $\mu$: $\bigcap_\nu x_\nu \subset x_\mu$. Wegen $P_0$ aus § 1 gilt dann auch $f(\bigcap_\nu x_\nu) \subset f(x_\mu)$ und damit $f(\bigcap_\nu x_\nu) \subset \bigcap_\nu f(x_\nu)$. Damit ist die Formel $P'_2$ nachgewiesen.

**2.** Das konjugierte Glied: In § 1 wurde gezeigt, dass sich die Glieder einer Paarung gegenseitig eindeutig bestimmen. Im vollständigen Gefüge lässt sich nun eine Formel angeben, die das konjugierte Glied explizite bestimmt:

(K)
$$\bar{f}(b) = \bigcap_{f(x') \subset b'} x.$$

Ist nämlich $u \circ \bar{f}(b)$, so gilt wegen P auch $f(u) \circ b$, d. h. $f(u) \subset b'$. Es gilt dann auch $f(u'') \subset b'$, also auch $\bigcap_{f(x') \subset b'} x \subset u'$ und schliesslich $u \circ \bigcap_{f(x') \subset b'} x$. Für jedes $u$ folgt also aus $u \circ \bar{f}(b)$ immer auch $u \circ \bigcap_{f(x') \subset b'} x$. Es gilt daher:

*(i)*
$$\bigcap_{f(x') \subset b'} x \subset \bar{f}(b).$$

Ist anderseits $u \bar{\circ} \bar{f}(b)$, so existiert ein $z$ in B, sodass $z \subset u \wedge z \subset \bar{f}(b)$. Es gilt nun für alle $x : f(x') \subset b' \rightarrow \bar{f}(b) \subset x$, also wegen $z \subset \bar{f}(b)$ auch $f(x') \subset b' \rightarrow z \subset x$ und daher $z \subset \bigcap_{f(x') \subset b'} x$. Wegen $z \subset u$ gilt daher auch $u \bar{\circ} \bigcap_{f(x') \subset b'} x$. Zusammenfassend gilt also für jedes $u$ die Formel $u \bar{\circ} \bar{f}(b) \rightarrow u \bar{\circ} \bigcap_{f(x') \subset b'} x$, womit die folgende Beziehung nachgewiesen ist:

*(ii)*
$$\bar{f}(b) \subset \bigcap_{f(x') \subset b'} x.$$

Aus *i* und *ii* folgt aber die Formel K.

3. Paarungskriterien in vollständigen Gefügen: Die Paarungskriterien P, P′, P″ haben alle den Mangel, dass man die konjugierte Zuordnung $\bar{f}$ kennen muss, um entscheiden zu können, ob $f$ Glied einer Paarung sei. Im vollständigen Gefüge erlaubt nun aber die Boole'sche Algebra zwischen den Elementen die Aufstellung von Kriterien, in denen nur $f$ selber auftritt.

*Satz:* Sind B und $\bar{B}$ vollständige Gefüge und ordnet $f$ jedem Element $x$ aus B genau ein Element $f(x)$ aus $\bar{B}_0$ zu, so ist $f$ dann und nur dann das Glied einer Paarung $(f\bar{f})$ von B und $\bar{B}$, wenn es das folgende Kriterium erfüllt:

$P_1$) $$f(\bigcup_\nu x_\nu) = \bigcup_\nu f(x_\nu)$$

oder was dasselbe ist:

$P_0$) $$x \subset y \rightarrow f(x) \subset f(y)$$

$P'_1$) $$f(\bigcup_\nu x_\nu) \subset \bigcup_\nu f(x_\nu).$$

Die konjugierte Zuordnung $\bar{f}$ berechnet sich dann nach der Formel K:

K) $$\bar{f}(b) = \bigcap_{f(x') \subset b'} x.$$

*Beweis:* Zunächst ist es klar, dass die Kriterien $P_1$, $P_0$ und $P'_1$ notwendig sind, sie wurden ja in 1. nachgewiesen. Ebenso wurde die Formel K in 2. bewiesen. Es bleibt also nur zu zeigen, *a)* dass aus $P_0$, $P'_1$, K das Paarungskriterium P für $f$ und $\bar{f}$ folgt und, *b)* dass dann aus $P_0$ auch die Formel $j$ folgt:

*(j)* $$\bigcup_\nu f(x_\nu) \subset f(\bigcup_\nu x_\nu).$$

*a)* Es sei $\bar{f}(b) \circ a$. Dann gilt $a \subset (\bar{f}(b))'$, also wegen K: $a \subset (\bigcap_{f(x') \subset b'} x)' = \bigcup_{f(x') \subset b'} x' = \bigcup_{f(x) \subset b'} x$. Nach $P_0$ gilt dann auch $f(a) \subset f(\bigcup_{f(x) \subset b'} x)$ und wegen $P'_1$ auch $f(a) \subset \bigcup_{f(x) \subset b'} f(x)$. Aus der letzten Formel folgt dann $f(a) \subset b'$, also auch $f(a) \circ b$. Zusammenfassend ist also gezeigt:

*(i)* $$\bar{f}(b) \circ a \rightarrow f(a) \circ b.$$

Ist nun anderseits $f(a) \circ b$, so gilt $f(a) \subset b'$. Daraus folgt aber $a' \supset \bigcap_{f(x') \subset b'} x$, denn es genügt ja $a' = x$ der Formel $f(x') \subset b'$. Wegen K gilt also $a' \supset \bar{f}(b)$, also auch $\bar{f}(b) \circ a$. Zusammenfassend ist dann gezeigt:

*(ii)* $$f(a) \circ b \rightarrow \bar{f}(b) \circ a.$$

*i* und *ii* ergeben aber das zu beweisende Paarungskriterium P. Es bleibt nur noch *b)* nachzuweisen:

*b)* Es gilt für alle $\mu$: $x_\mu \subset \bigcup_\nu x_\nu$ und daher wegen $P_0$ auch $f(x_\mu) \subset f(\bigcup_\nu x_\nu)$, also schliesslich: $\bigcup_\nu f(x_\nu) \subset f(\bigcup_\nu x_\nu)$. Damit ist die Formel $j$ nachgewiesen.

Die Paarungen zweier vollständiger Gefüge sind also nichts anderes als die Homomorphismen bezüglich der unbeschränkten Vereinigungsoperation:

*Satz:* $B_0$ und $\overline{B}_0$ seien zwei vollständige Boole'sche Algebren und $f$ sei Abbildung von $B_0$ in $\overline{B}_0$.
$f$ ist dann und nur dann Homomorphismus bezüglich der unbeschränkten Vereinigung, wenn es Glied einer Paarung ist.
Die Formel K bestimmt also zu jedem Vereinigungshomomorphismus $f$ von $B_0$ in $\overline{B}_0$ einen konjugierten Vereinigungshomomorphismus $\overline{f}$ von $\overline{B}_0$ in $B_0$.

4. Die Paarungen der Spaltungsgefüge: Es seien B und $\overline{B}$ zwei Gefüge; S(B) und S($\overline{B}$) die zugehörigen vollständigen Gefüge aller S-Klassen von B resp. $\overline{B}$ (Teil 1. § 6). Welche Zusammenhänge bestehen nun zwischen den Paarungen von B mit $\overline{B}$ und denjenigen von S(B) mit S($\overline{B}$)?

Es sei $(f\overline{f})$ Paarung von B und $\overline{B}$. Man weist dann folgende Formeln für beliebige Elemente $x$ in B und beliebige S-Klassen Y aus S($\overline{B}$) nach:

$$(j) \qquad f((x)'')'' = (f(x))''$$

$$(jj) \qquad f(x) \in Y' \to x \in \overline{f}(Y)'.$$

*Beweise:* Ist $u \in f((x)'')$, so existiert ein $z$ in B, sodass $u = f(z) \wedge z \subset x$. Wegen $P_0$ aus § 1 folgt daraus $u \subset f(x)$, also $u \in (f(x))''$. Es ist also $f((x))'' < (f(x))''$ und daher wegen 2) und 4) aus § 5 Teil 1 auch $f((x)'')'' < (f(x))''$. Ist umgekehrt $u \in (f(x))''$, so gilt $u \subset f(x)$, also wegen $f(x) \in f((x)')$ auch $u \in f((x)'')''$. Damit ist die Formel $j$ nachgewiesen.
Der Beweis von $jj$ geht etwa so: aus $f(x) \in Y'$ folgt nach der Formel $k$ aus § 5, 1. des 1. Teiles für jedes $y$ aus $\overline{B}$: $y \in Y \to y \circ f(x)$, also da $(f\overline{f})$ Paarung ist, auch $y \in Y \to x \circ \overline{f}(y)$. Wiederum wegen der Formel $k$ gilt also $x \in \overline{f}(Y)'$.

Nun seien die folgenden Zuordnungen F und $\overline{F}$ der S-Klassen $X \in S(B)$ und $Y \in S(\overline{B})$ definiert:

$$(d) \qquad \begin{aligned} F(X) &= f(X)'' \\ \overline{F}(Y) &= \overline{f}(Y)''. \end{aligned}$$

F(X) ist also von der Klasse $f$(X) erzeugte S-Klasse von $\overline{B}$ und F ordnet also jedem Element X aus $S(B)_0$ genau ein Element aus $S(\overline{B})_0$ zu. Entsprechendes gilt von $\overline{F}$. Es soll nun gezeigt werden, dass F und $\overline{F}$ eine

Paarung bilden. Dazu ist die folgende Formel *i* für alle S-Klassen X aus S(B) und Y aus S($\overline{\mathrm{B}}$) nachzuweisen:

$$(i) \qquad \mathrm{Y} <> \mathrm{F}(\mathrm{X}) \leftarrow\rightarrow \mathrm{X} <> \overline{\mathrm{F}}(\mathrm{Y}).$$

Es genügt, die Formel in der einen Richtung zu beweisen. Es sei also etwa Y <> F(X). Wegen der Formel $h_2$ aus Teil 1 § 5, 1. und wegen *(d)* gilt dann $f(\mathrm{X})'' < \mathrm{Y}'$, also wegen 3) aus Teil 1 § 5, 1. auch $f(\mathrm{X}) < \mathrm{Y}'$ und wegen der Hilfsformel *(jj)* $\mathrm{X} < \overline{f}(\mathrm{Y})'$. Daraus folgt aber schliesslich nach $h_1$ aus Teil 1 § 5, 1. auch $\mathrm{X} <> \overline{f}(\mathrm{Y})''$, also $\mathrm{X} <> \overline{\mathrm{F}}(\mathrm{Y})$.

(F $\overline{\mathrm{F}}$) ist also Paarung der Gefüge S(B) und S($\overline{\mathrm{B}}$). Weiterhin folgt aber aus den beiden Formeln *j* und *d* für jedes $x$ aus B: $\mathrm{F}((x)'') = (f(x))''$. Es wird also jede H-Klasse $(x)''$ aus S(B) durch F auf die H-Klasse $(f(x))''$ aus S($\overline{\mathrm{B}}$) abgebildet und Analoges gilt von $\overline{\mathrm{F}}$.

Damit ist aber gezeigt, dass (F $\overline{\mathrm{F}}$) auch Paarung der Gefüge H(B) und H($\overline{\mathrm{B}}$) ist und, dass diese Paarung bei der Identifikation von $x$ mit $(x)''$ mit $(f\overline{f})$ zusammenfällt. Zusammenfassend sind damit die Punkte 1) und 2) des folgenden Satzes bewiesen.

*Satz:* Es sei $(f\overline{f})$ Paarung der Gefüge B und $\overline{\mathrm{B}}$, und es seien S(B) und S($\overline{\mathrm{B}}$) die Gefüge aller S-Klassen von B resp. $\overline{\mathrm{B}}$. Setzt man dann für die S-Klassen X aus S(B) und Y aus S($\overline{\mathrm{B}}$):

$$\mathrm{F}(\mathrm{X}) = f(\mathrm{X})''$$
$$\overline{\mathrm{F}}(\mathrm{Y}) = \overline{f}(\mathrm{Y})''.$$

So gilt:

1) (F $\overline{\mathrm{F}}$) ist Paarung der Gefüge S(B) und S($\overline{\mathrm{B}}$).
2) Identifiziert man B und $\overline{\mathrm{B}}$ mit den Hauptspaltungsgefügen H(B) und H($\overline{\mathrm{B}}$), d.h. indentifiziert man das Element $x$ mit der H-Klasse $(x)''$, so fällt (F $\overline{\mathrm{F}}$) auf B und $\overline{\mathrm{B}}$ mit $(f\overline{f})$ zusammen.
3) (F $\overline{\mathrm{F}}$) ist die einzige Paarung von S(B) und S($\overline{\mathrm{B}}$) mit der Eigenschaft 2).

Ausserdem existieren natürlich noch Paarungen von S(B) und S($\overline{\mathrm{B}}$), die nicht die Hauptklassen einander zuordnen. Zu beweisen bleibt nur der Punkt 3) des Satzes.

Es sei (G $\overline{\mathrm{G}}$) eine Paarung von S(B) und S($\overline{\mathrm{B}}$), die ebenfalls die Bedingung 2) erfüllt. Dann fällt also (G $\overline{\mathrm{G}}$) auf H(B) und H($\overline{\mathrm{B}}$) mit (F $\overline{\mathrm{F}}$) zusammen, d.h. es gilt für alle $x$ aus B und alle $y$ aus $\overline{\mathrm{B}}$:

$$(i) \qquad \begin{aligned} \mathrm{G}((x)'') &= \mathrm{F}((x)'') \\ \overline{\mathrm{G}}((y)'') &= \overline{\mathrm{F}}((y)''). \end{aligned}$$

Es ist zu zeigen, dass auch für alle S-Klassen die folgende Formel gilt:

$$(ii) \qquad \begin{aligned} G(X) &= F(X) \\ \overline{G}(Y) &= \overline{F}(Y). \end{aligned}$$

Wäre etwa $G(X) \overline{<} F(X)$ so wäre nach $h_2$ aus Teil 1, § 5, 1, auch $G(X) \overline{<>} F(X)'$, es existierte also ein $u$ in $\overline{B}$, sodass $u \in G(X) \wedge u \in F(X)'$. Daraus folgt $(u)'' < G(X) \wedge (u)'' < F(X)'$. Wiederum nach $h_2$ folgt aus $(u)'' < F(X)'$ auch $(u)'' <> F(X)$ und da $(F\overline{F})$ Paarung ist $X <> F((u)'')$. Wegen $i$ und da auch $(G\overline{G})$ Paarung ist, gilt also auch $(u)'' <> G(X)$. Vorhin hatten wir aber $(u)'' < G(X)$, womit ein Widerspruch aufgedeckt ist. Es gilt also: $G(X) < F(X)$. Genau so zeigt man $F(X) < G(X)$, womit dann die erste Formel aus *ii* nachgewiesen ist. Der Beweis der zweiten geht genau so.

## ANWENDUNG: DER DISKRETE TOPOLOGISCHE RAUM

Als Anwendung von § 3 und § 4 soll hier der Zusammenhang zwischen den diskreten Topologien in einer Menge R einerseits und den reflexiven und transitiven Relationen zwischen den Elementen von R anderseits aufgedeckt werden.

R sei ein topologischer Raum bezüglich der Abschliessungs — Operation $h$. D.h. $h$ ordne jeder Teilmenge $x$ von R eine Teilmenge $h(x)$ von R zu, sodass die vier Kuratowski'schen Axiome erfüllt sind:

$$\text{I} \qquad h(o) = 0$$

$$\text{II} \qquad x \subset h(x)$$

$$\text{III} \qquad h(x \cup y) = h(x) \cup h(y)$$

$$\text{IV} \qquad hh(x) \subset h(x).$$

Der Raum R heisst dann diskret, wenn die folgende Verschärfung von III gilt:

$$\text{V} \qquad h(\bigcup_{\nu} x_{\nu}) = \bigcup_{\nu} h(x_{\nu}).$$

Wegen des Paarungskriteriums $P_1$ in vollständigen Gefügen ist also $h$ das Glied einer Paarung $(h\overline{h})$ des Gefüges B aller nicht leeren Teilmengen von R mit sich selber. Weil aber B ein volles Mengengefüge ist, wird $(h\overline{h})$ von einer Relation $H(\xi, \eta)$ zwischen den Punkten von R erzeugt. Schliesslich folgt nach § 3 aus II und IV die Reflexivität und Transitivität der Relation H.

Umgekehrt erzeugt jede Relation H eine Paarung $(h\bar{h})$. Nach § 4 gilt dann für $h$ die Formel V. Ist ferner H reflexiv und transitiv, so gilt nach § 3 für $h$ auch II und IV. Zusammenfassend kann folgender Satz formuliert werden:

*Satz:* Ist $H(\xi, \eta)$ eine reflexive und transitive Relation zwischen den Elementen der Menge R und ist $h(x)$ der Nachbereich der Teilmenge $x$ von R bezüglich der Relation H, so ist R ein diskreter topologischer Raum mit der Hüllen-Operation $h$.
Umgekehrt kann jede diskrete Topologie in R in der beschriebenen Weise aus einer reflexiven und transitiven Relation erzeugt werden.

## § 5 DIE FUNKTIONELLE PAARUNG

Die Paarung ist, wie in § 2 gezeigt wurde eine Verallgemeinerung der «Relation zwischen den Elementen zweier Klassen» auf den Fall nicht atomarer Objekte. Auf der Relation beruht aber der mengentheoretische Abbildungsbegrift. Es ist daher zu hoffen, dass auf Grund der Paarung auch die Abbildung nicht atomarer Gebilde definiert werden kanu.

**1. Definition der funktionellen Paarung:** Das Glied $f$ einer Paarung $(f\bar{f})$ der Gefüge B und $\bar{B}$ heisse eindeutig, wenn eine der folgenden äquivalenten Bedingungen erfüllt ist:

$e_1$) $\quad x \circ y \rightarrow \bar{f}(x) \circ \bar{f}(y) \quad$ für alle $x, y$ aus $\bar{B}$

$e_2$) $\quad f\bar{f}(y) \subset y \quad$ für alle $y$ aus $\bar{B}$.

$f$ heisst «überall in B definiert», wenn eine der folgenden äquivalenten Bedingungen zutrifft:

$u_1$) $\quad f(x) \circ f(y) \rightarrow x \circ y \quad$ für alle $x, y$ aus B

$u_2$) $\quad x \subset \bar{f}f(x) \quad$ für alle $x$ aus B

$u_3$) $\quad f(x) \neq 0 \quad$ für alle $x$ aus B.

Die Aequivalenz der verschiedenen Formeln zeigt man so:

$u_3 \rightarrow u_1$: Es sei $x \bar{\circ} y$. Dann existiert ein $u$ in B, sodass $u \subset x \wedge u \subset y$. Es ist dann auch $f(u) \subset f(x) \wedge f(u) \subset f(y)$. Wegen $u_3$ gilt aber $f(u) \neq 0$, also ist $f(x) \bar{\circ} f(y)$. Zusammenfassend folgt also aus $u_3$ die Formel $x \bar{\circ} y \rightarrow f(x) \bar{\circ} f(y)$, welche mit $u_1$ äquivalent ist.

$u_1 \to u_2$: Es sei $u \circ \bar{f} f(z)$. Dann gilt wegen P auch $f(u) \circ f(z)$ und daher wegen $u_1$ auch $u \circ z$. Wegen $T_3$ ist also $z \subset \bar{f} f(z)$. Aus der Formel $u_1$ folgt also auch $u_2$.

$u_2 \to u_3$: Es sei $f(x) = 0$. Dann gilt auch $\bar{f} f(x) = 0$ und wegen $u_2$ auch $x = 0$. Aus der Formel $u_2$ folgt also $f(x) \neq 0$ für alle $x$ aus B.

Damit ist die Aequivalenz der Formeln $u_1, u_2, u_3$ nachgewiesen. Es bleibt diejenige von $e_1$ und $e_2$ zu zeigen:

$e_1 \to e_2$: Es sei $z$ in $\bar{B}$ und $u \circ z$. Wegen $e_1$ ist dann auch $\bar{f}(u) \circ \bar{f}(z)$, also wegen P auch $u \circ f \bar{f}(z)$. Aus $u \circ z$ folgt also immer $u \circ f \bar{f}(z)$, es gilt also $f \bar{f}(z) \subset z$. Aus $e_1$ folgt daher $e_2$.

$e_2 \to e_1$: Es sei $\bar{f}(x) \bar{\circ} \bar{f}(y)$. Dann existiert ein $u$ in B, sodass $u \subset \bar{f}(x) \wedge u \subset \bar{f}(y)$. Es gilt also $f(u) \subset f \bar{f}(x) \wedge f(u) \subset f \bar{f}(y)$ und wegen $e_2$ auch $f(u) \subset x \wedge f(u) \subset y$. Wegen $u \subset \bar{f}(x)$ gilt nach § 1, $a$ aber $f(u) \neq 0$ also ist $f(u)$ in $\bar{B}$ und daher $x \bar{\circ} y$. Zusammenfassend gilt also unter Voraussetzung von $e_2$ die zu $e_1$ äquivalente Formel $\bar{f}(x) \bar{\circ} \bar{f}(y) \to x \bar{\circ} y$.

Wir wollen nun eine Paarung der Gefüge B und $\bar{B}$ in B funktionell nennen, wenn das erste Glied $f$ überall in B definiert und eindeutig ist. Das Glied $f$ heisse dann auch etwa eine Funktion von B in $\bar{B}$.

Diese Benennung wird durch die Resultate in 3. gerechtfertigt. Es wird dort gezeigt, dass die so definierte Funktion eine Verallgemeinerung des klassischen Abbildungsbegriffes ist.

2. Funktionelle Paarungen vollständiger Gefüge: In vollständigen Gefügen gilt das folgende einfache Kriterium für die Funktionalität einer Paarung:

*Satz:* Die Paarung $(f\bar{f})$ des vollständigen Gefüges B mit dem ebenfalls vollständigen Gefüge $\bar{B}$ ist dann und nur dann funktionell in B, wenn für alle $y$ aus $\bar{B}$ gilt:

F) $$\bar{f}(y') = \bar{f}(y)'$$

genauer: Die Paarung $(f\bar{f})$ ist dann und nur dann eindeutig in B, wenn gilt:

*(i)* $$\bar{f}(y') \subset \bar{f}(y)'.$$

Sie ist dann und nur dann überall in B definiert, wenn gilt:

*(ii)* $$\bar{f}(y)' \subset \bar{f}(y').$$

*Beweis:* $(f\bar{f})$ sei im ersten Glied $f$ eindeutig. Wegen $e_1$ aus 1. folgt dann aus $y \circ y'$ immer auch $\bar{f}(y) \circ \bar{f}(y')$ und damit die Formel $i$.

Ist umgekehrt $\bar{f}(y') \subset \bar{f}(y)'$, so gilt $\bar{f}(y') \circ \bar{f}(y)$, also wegen P auch $y' \circ f\bar{f}(y)$ und damit $f\bar{f}(y) \subset y$. Dies ist aber die Formel $e_2$ aus 1., also ist $f$ eindeutig. Damit ist der eine Teil des Satzes nachgewiesen. Der Beweis des anderen geht etwa so:

Es sei $f$ überall in B definiert. Ist dann $u \circ \bar{f}(y')$, so gilt wegen P auch $f(u) \circ y'$, also $f(u) \subset y$. Daraus folgt nach $P_0$ $\bar{f}f(u) \subset \bar{f}(y)$. Nach $u_2$ aus 1. gilt dann auch $u \subset \bar{f}(y)$, also $u \circ \bar{f}(y)'$. Zusammenfassend folgt also aus $u \circ \bar{f}(y')$ immer auch $u \circ \bar{f}(y)'$, d.h. es gilt die Formel $ii$.

Ist umgekehrt die Formel $ii$ erfült und ist $x$ ein beliebiges Element von B, so gilt $f(x)' \circ f(x)$, also wegen P auch $\bar{f}(f(x)') \circ x$. Wegen $ii$ haben wir aber $\bar{f}f(x') \subset \bar{f}(f(x)')$, also ist auch $\bar{f}f(x) \circ x$ und daher $x \subset \bar{f}f(x)$. Damit ist die Formel $u_2$ aus 1. nachgewiesen, $f$ ist also über all in B definiert.

Zusammen mit den Resultaten aus § 4 lässt sich der Satz auch so formulieren:

*Satz:* Eine Zuordnung $f$ ist dann und nur dann eine Funktion des Vollständigen Gefüges B in das ebenfalls vollständige Gefüge $\bar{B}$, wenn ein Homomorphismus $\bar{f}$ der vollständigen Boole'schen Algebren $\bar{B}_0$ in $B_0$ existiert und $f$ sich nach folgender Formel berechnet:

K′) $$f(a) = \bigcap_{a \subset \bar{f}(y)} y.$$

Nac § 3 ist ja $\bar{f}$ dann und nur dann Glied einer Paarung, wenn es Vereinigungs-Homomorphismus ist und die konjugierte Zuordnung $f$ berechnet sich dann nach der Formel

K) $$f(a) = \bigcap_{\bar{f}(y') \subset a'} y.$$

Soeben wurde gezeit, dass $(f\bar{f})$ dann und nur dann funktionell ist in B, wenn $\bar{f}$ auch Homomorphimus bezüglich der Komplementbildung ist. Es bleibt also vom Satz nur noch die Formel K′ zu beweisen. Sie folgt aber aus der Formel K, wenn der Nachweis von $(v)$ gelingt:

*(v)* $$\bar{f}(y') \subset a' \leftarrow\rightarrow a \subset \bar{f}(y).$$

Diese Formel ist aber sichergestellt, da $(f\bar{f})$ funktionell ist und also, gilt: $\bar{f}(y') = \bar{f}(y)'$.

Der Funktionsbegriff tällt also in der vollständigen Boole'schen Algebra zusammen mit demjenigen des (vollständigen) Homomorphismus. Es sei aber nochmals ausdrücklich bemerkt, dass nicht die Funktion $f$ selber, sondern das konjugierte Glied $\bar{f}$ der Homomorphismus ist. $f$ besitzt im Allgemeinen nur die schwächere Homomorphie-Eigenschaft $P_1$.

Im Zusammenhang mit den Resultaten aus § 4.4. über die Paarungen der Spaltungsgefüge sei noch der folgende Satz bemerkt:

*Satz:* Die Erweiterung $(F\bar{F})$ der Paarung $(f\bar{f})$ von B und $\bar{B}$ auf die Spaltungsgefüge $S(B)$ und $S(\bar{B})$ ist dann und nur dann funktionell, wenn es schon $(f\bar{f})$ ist.

**3.** Die funktionellen Paarungen voller Klassengefüge: Es seien K und $\bar{K}$ zwei Bereiche von Elementen $\alpha, \beta, \ldots$, dann bilden die Bereiche B und $\bar{B}$ aller nicht leeren Teilklassen $a, b, \ldots$ von K resp. $\bar{K}$ zwei Gefüge, die sogenannten vollen Gefüge über K resp. $\bar{K}$.

Es sei nun weiter $\eta = \varphi(\xi)$ eine Abbildung der Klasse K in $\bar{K}$, d.h. durch die Formel $\eta = \varphi(\xi)$ ist eine Relation $F(\xi, \eta)$ zwischen den Elementen $\xi$ aus K und jenen $\eta$ aus $\bar{K}$ mit den Eigenschaften *i* und *ii* symbolisiert.

(*i*) $$(\xi \in K) \rightarrow \bigvee_{\eta} (\eta \in \bar{K} \wedge F(\xi, \eta))$$

(*ii*) $$F(\xi, \mu) \wedge F(\xi, \zeta) \rightarrow \eta = \zeta.$$

Mit $f(x)$ und $\bar{f}(y)$ seien Nach — und Vorbereich bezüglich der Relation F bezeichnet, d.h. es gilt

(*j*) $f(x) =$ Klasse aller $\eta$ sodass $\xi$ in $x$ mit $\eta = \varphi(\xi)$ existiert.

(*jj*) $f(y) =$ Klasse aller $\xi$ sodass $\eta$ in $y$ mit $\eta = \varphi(\xi)$ existiert.

In der gebräuchlichen Ausdrucksweise ist also $f(x)$ das Bild der Klasse $x$ und $\bar{f}(y)$ das Urbild der Klasse $y$ bezüglich der Abbildung $\varphi$.

Schon in § 2 wurde gezeigt, dass dann $(f\bar{f})$ eine Paarung der Gefüge B und $\bar{B}$ ist. Darüber hinaus folgt nun aber aus den Bedingungen *i* und *ii*, dass $(f\bar{f})$ in B funktionell ist:

Ist nämlich $x$ in B, so existiert ein $\xi$ in $x$. Wegen *i* gibt es dann ein $\eta$, sodass $\eta = \varphi(\xi)$. Wegen *j* gilt daher $\eta \in f(x)$. Aus *i* folgt also die Formel $u_3$ aus 1., $f$ ist daher überall in B definiert.

Sind ferner $x$ und $y$ Elemente von $\bar{B}$ und gilt $\bar{f}(x) \,\bar{o}\, \bar{f}(y)$, so existiert ein $\xi$ in K, sodass $\xi \in \bar{f}(y) \wedge \xi \in \bar{f}(x)$. Wegen *jj* gibt es also auch zwei Elemente $\eta \in y$ und $\zeta \in x$, sodass $F(\xi, \eta) \wedge F(\xi, \zeta)$. Wegen *ii* ist dann aber $\eta = \zeta$, also $y \,\bar{o}\, x$. Aus *ii* folgt also die Formel $e_1$ aus 1. Damit ist folgender Satz bewiesen:

$S_1$ Ist $\varphi$ eine Abbildung von K in $\overline{K}$, so ist durch die Formel $j$ eine Funktion $f$ des vollen Gefüges B über K in das volle Gefüge $\overline{B}$ über $\overline{K}$ bestimmt.
$f$ heisst die von der Abbildung $\varphi$ erzeugte Funktion. Die konjugierte Zuordnung $\overline{f}$ ist dann durch die Formel $jj$ bestimmt.

Weiter gilt nun aber:

$S_2$ Jede Funktion $f$ von B in $\overline{B}$ wird von einer Abbildung $\varphi$ von K in $\overline{K}$ erzeugt.
$\varphi$ ist dabei eindeutig durch die folgende Formel $k$ bestimmt:

$$(k) \qquad \eta = \varphi(\xi) \longleftrightarrow \eta \in f(\xi) \longleftrightarrow \xi \in \overline{f}(\eta).$$

*Beweis:* Schon in § 2 wurde gezeigt, dass die durch $k$ definierte Relation $F(\xi, \eta) \longleftrightarrow \eta = \varphi(\xi)$ die Paarung $(f\overline{f})$ erzeugt. Es bleibt also nur zu zeigen, dass dann $F(\xi, \eta)$ die Bedingungen $i$ und $ii$ erfüllt, sobald $(f\overline{f})$ funktionell ist.

Es sei daher $\xi \in K$. Wegen $u_3$ aus 1. gilt dann $f(\xi) \neq 0$. Es existiert also ein $\eta \in f(\xi)$. Wegen $k$ gilt dann $F(\xi, \eta)$. Zusammenfassend existiert also zu jedem $\xi$ in K ein $\eta$ in $\overline{K}$, sodass $F(\xi, \eta)$ zutrifft. F erfüllt also die Formel $i$.

Ist ferner $F(\xi, \eta)$ und $F(\xi, \zeta)$, so haben wir wegen $k$ auch $\xi \in \overline{f}(\eta) \wedge \xi \in \overline{f}(\zeta)$. Es ist also $\overline{f}(\eta)\, \ddot{o}\, \overline{f}(\zeta)$. Wegen $e_1$ aus 1. gilt daher auch $(\eta)\, \ddot{o}\, (\eta)$ also $\eta = \zeta$. Zusammenfassend folgt also aus $F(\xi, \eta) \wedge F(\xi, \zeta)$ immer $\eta = \zeta$. F erfüllt auch die Formel $ii$.

Damit ist $S_2$ nachgewiesen. $S_1$ und $S_2$ ergeben aber folgenden Satz:

*Satz:* Sind B und $\overline{B}$ die vollen Gefüge aller nicht leeren Teilklassen der Bereiche K und $\overline{K}$, so sind die Funktionen $f$ von B in $\overline{B}$ genau die von den Abbildungen $\varphi$ von K in $\overline{K}$ erzeugten Zuordnungen. Oder genauer:

1) Jede Abbildung $\varphi$ von K in $\overline{K}$ erzeugt eine Funktion $f$ von B in $\overline{B}$. Man hat nur zu setzen
$f(x) =$ Bildbereich von $x$ bezüglich $\varphi$
$\overline{f}(y) =$ Urbild von $y$ bezüglich $\varphi$.
2) Jede Funktion $f$ von B in $\overline{B}$ wird gemäss 1 durch eine Abbildung $\varphi$ von K in $\overline{B}$ erzeugt. Dabei ist $\varphi$ eindeutig durch die folgende Formel bestimmt:

$$\eta = \varphi(\xi) \longleftrightarrow (\eta) = f(\xi).$$

Da jedes vollständige und atomare Gefüge isomorph zum vollen Gefüge über seinen Atomen ist, gilt also auch in leicht verständlicher Ausdrucksweise:

> Jede Funktion des vollständigen und atomaren Gefüges B in das vollständige und atomare Gefüge $\overline{B}$ wird erzeugt von einer Abbildung der Klasse K aller Atome von B in die Klasse $\overline{K}$ aller atome von $\overline{B}$.

Die funktionelle Paarung ist also eine Verallgemeinerung des klassischen Begriffes der Abbildung atomarer Gebilde (Mengen) auf den Fall nicht atomarer Gebilde (Gefüge).

## SCHLUSSBETRACHTUNG: DIE ATOMFREIE MATHEMATIK

Die Theorie der Gefüge ist die der Intuition entsprechende mathematische Teillehre.

Eine solche vom Teilmengenbegriff losgelöste allgemeine Teillehre hat nicht nur an und für sich ihre Bedeutung. Vielmehr eröffnet sie eine weitgehendere, die gesamte Mathematik betreffende Perspektive:

Der heutigen Mathematik liegt der Begriff der Menge von Elementen zu Grunde. Alle mathematischen Objekte ($R^n$, abstrakte Räume, Gruppen) sind Mengen, also aus kleinsten Teilen (Atomen) aufgebaut. Es ist sogar gelungen typisch kontinuierliche Begriffe wie die Stetigkeit, den Zusammenhang, das Kontinuum im Rahmen der Mengenlehre zu definieren.

Durchgeht man aber die Bergriffsbildungen und Sätze verschiedener mathematischer Disziplinen (Mächtigkeitslehre, Theorie der topologischen Räume, Mass-und Integrations-Theorie), so ist es geradezu auffallend, wie selten das Element explizite Verwendung findet. Vielmehr dient es nur zur Definition der Teilmengen sowie von Relationen (enthalten in, fremd zu) und Verknüphfungen (Vereinigung, Durchschnitt, Komplement) zwischen denselben. In der Folge findet das Element dann nur noch implizite Verwendung.

Diese Verhältnisse legen es nahe, gewisse Gebiete der Mathematik direkt auf der allgemeinen Teillehre (Gefügetheorie) zu begründen. Damit gelangt man zu einer atomfreien Mathematik.

Dass es klassische Mathematische Disziplinen gibt, die sich in dieser Richtung verallgemeinern lassen, zeigen die folgenden Beispiele:

1. C. Carathéodory [1] zeigt, wie sich die Mass-und Integrations-Theorie in natürlicher Weise auf Grund der allgemeinen Teillehre behandeln lässt. Er sagt von seiner atomfreien Theorie:

« Einmal wird das Feld der möglichen Anwendungen beträchtlich erweitert. Dann aber zeigt sich, dass die Theorie eher einfacher wird, weil nur ihr scharfumrissenes Skelett übrig bleibt. Durch den Zwang, den man sich auferlegt, Objekte zu handhaben, die nur wenig differentiiert sind, fallen bei vielen Beweisen diejenigen komplizierten Teile von selbst weg, die bei der Verarbeitung eines reichhaltigeren Materials aus Versehen haften geblieben waren. »

In neuerer Zeit sind dann viele ähnliche und verbesserte Arbeiten in dieser Richtung entstanden [2].

2. Die Theorie der topologischen Räume ist zur Verallgemeinerung auf die Teillehre geradezu prädestiniert. Mann denke nur etwa an die Kuratowski'schen Axiome, die unter alleiniger Verwendung des Teilmengenbegriffes formuliert sind.

In neuester Zeit wurde denn auch der Begriff des topologischen Raumes in die atomfreie Mathematik übernommen [3].

3. A. Schöenflies [4] hat schon 1921 ein Postulatensystem angegeben, auf Grund dessen er eine Mächtigkeitstheorie entwickelte, ohne den Begriff des Elementes zu verwenden. Er schreibt von seiner Theorie:

« Der Vergleich der Mengen bezüglich ihres Grössencharakters ist nichts, was dem Mengenbegriff allein eigentümlich ist; er betrifft allgemeiner alle Objekte, für die man das ganze und den Bestandteil unterscheiden kann... Der Begriff des Elementes wird gar nicht benutzt; immer nur bilden die an sich möglichen Beziehungen zwischen dem Ganzen und seinen Teilen den Gegenstand der Untersuchung. »

Diese Beispiele zeigen, wie sich auch mit atomfreien Gebilden sehr wohl Mathematik treiben lässt. Neben den Mengen selber spielt in der Mathematik wohl ihre Abbildung die fundamentalste Rolle [5]. Es muss daher auffallen, dass in den Theorien von Carathéodory und Schöenflies wohl neue atomfreie Gebilde an Stelle der Mengen vorgeschlagen werden,

[1] Carathéodory (1). Carathéodory setzt zwar eine abzählbar additive Boole'sche Algebra zwischen den Teilen (genannt Somen) voraus. Dies ist aber, wie die Vervollständigungstheorie aus Teil 1. zeigt, nur eine unwesentliche Beschränkung der Allgemeinheit.

[2] MacNeille (2), Olmsted (1), Sikorski (1), Horn-Tarski (1).

[3] Sikorski (2), Gomes (1).

[4] Schöenflies (1).

[5] Ueberall dort, wo zwei mathematische Strukturen miteinander verglichen werden, geschieht ja dieser Vergleich durch Abbildung.

nirgends aber die Abbildung dieser Gebilde definiert wird. Sogar Schöenflies, der ja einen Vergleich seiner Gebilde («bezüglich ihres Grössencharakters») anstrebt, umgeht den Abbildungsbegriff [1].

Dieser Mangel bei der atomfreien Mathematik wird nun durch den im 2. Teil der Arbeit definierten Paarungsbegriff behoben. Die funktionelle Paarung ist ja eine Verallgemeinerung des klassischen Abbildungbegriffes auf den Fall nicht atomarer Gebilde.

Es stellt sich nun die Frage, wie weitgehend die funktionelle Paarung die Rolle der Abbildung in der atomfreien Mathematik zu übernehmen vermag. Dass in dieser Richtung einiges zu hoffen ist, zeigt die folgende von R. Sikorski [2] bewiesene Verallgemeinerung des Bernstein'schen Satzes:

Es seien $B_0$ und $\bar{B}_0$ zwei abzählbar additive Boole'sche Algebren und es existiere sowohl eine funktionelle Paarung $(ff)$ von B in B, als auch eine solche $(g\bar{g})$ von $\bar{B}$ in B. Dann sind B und $\bar{B}$ isomorph.

Aber auch in anderen Beziehungen vertritt die Paarung die Rolle der Relation zwischen Elementen. So gelingt es, auf Grund der Paarung, das Produkt zweier Boole'scher Algebren zu definieren, während ja das Produkt zweier Mengen die Menge aller Relationen zwischen ihren Elementen ist.

[1] Schöenflies postuliert die «Gleichmächtigkeit» seiner Gebilde als undefinierte Aquivalenzrelation, anstatt sie wie Cantor mit Hilfe eines Funktionsbegriffes zu definieren.

[2] Sikorski (3).

# LITERATURANGABEN


WALLMAN (1) Lattices and topological Spaces. Annals of Mathematics **39** (1938) p. 112.

TARSKI (1) Zur Grundlegung der Boole'schen Algebra I. Fundamenta Mathematicae **24** (1935) p. 177.

HORN-TARSKI (1) Measure in Boolean algebras. TAMS[1] **64** (1948) p. 467.

HUNTIGTON (1) New sets of independent postulates for the algebra of logic TAMS **35** (1933) p. 557.

MACNEILLE (1) Partially ordered sets. TAMS **42** (1937) p. 416.

STONE (1) The theory of representation of Boolean algebras. TAMS **40** (1936) p. 37.

(2) Application of the theory of Boolean algebras to general topology. TAMS **41** (1937) p. 375.

KURATOWSKI (1) Topologie. Mongrafje Matematyczne Tom III.

(2) Contribution à l'étude de continu de Jordan. Fundamenta Mathematicae **5** (1924) p. 113.

LEBESGUE (1) Sur les correspondences entre les points de deux éspaces. Fundamenta Mathematicae **2** (1921) p. 256.

CARATHÉODORY (1) Entwurf für eine Algebraisierung des Integralbegriffes. Sitzungsberichte der bayrischen Academie der Wissenschaften 1938 pp. 27, 175.

OLMSTED (1) Lebesgue theory in a Boolean algebra TAMS **51** (1944) p. 164

SIKORSKI (1) The integral on a Boolean algebra. Colloquium Mathematicum Wroclaw **1** (1948).

(2) Sur les corps de Boole topologiques. Fundamenta Mathematicae **36** (1949).

(3) On a generalisation of theorems of Banach and Cantor-Bernstein. Colloquium Mathematicum Wroclaw **1** (1948).

GOMES (1) Introdução ao estudo duma noção de funcional em espaços sem pontos. Portugaliae Mathematica **5** (1946) p. 1.

SCHÖNFLIES (1) Zur Axiomatik der Mengenlehre. Mathematische Annalen **83** (1921) p. 173.


[1] Transactions of the American Mathematical Society.

# INHALTSVERZEICHINIS

# ERRATA

ZU

## DIE BOOLE'SCHE PARTIALORDNUNG UND DIE PAARUNG VON GEFUEGEN,

von J. RICHARD BÜCHI (St. Gallen)

PORTUGALIAE MATHEMATICA VOL. 7, FASCS. 3-4

Seite 3, Zeile 14 von oben : statt *ehenbürtige* lies *ebenbürtige.*
» 3, » 4 von unten : statt *Axiomenystem* lies *Axiomensystem.*
» 4, » 3 von unten : statt $<$ *Disjunktion* lies $\vee$ *Disjunktion.*
» 6, » 10 von oben : statt $(x \subset y)$ lies $(x \circ y)$.
» 17, » 1 von unten : statt *fahlt* lies *felht.*
» 18, » 8 von oben : statt *Nadh* lies *Nach.*
» 25, » 10 von oben : statt *Gefüges* lies *Gefüge.*
» 27, » 10 von oben : statt *nachweis* lies *Nachweis.*
» 27, » 15 von unten : statt *vollständige* lies *vollständigen.*
» 28, » 5 von unten : statt *bleit* lies *bleibt.*
» 28, » 3 von unten : statt *Algedra* lies *Algebra.*
» 29, » 14 von oben : statt *bilden* lies *bildet.*
» 30, » 15 von oben : statt *Gegensteil* lies *Gegenteil.*
» 31, » 13 von oben : statt *Umgekert* lies *Umgekehrt.*
» 34, » 3 von unten : statt *einnes* lies *eines.*
» 54, » 17 von unten : statt *Vollständigen* lies *vollständigen.*
» 54, » 13 von unten : statt *Nac* lies *Nach.*
» 54, » 9 von unten : statt *gezeit* lies *gezeigt.*
» 58, » 14 von oben : statt *Mann* lies *Man.*

# LEBENSLAUF

Ich wurde am 31. Januar 1924 in Porto Alegre, der Hauptstadt des brasilianischen Staates Rio Grande do Sul, geboren. Meine Eltern, die sich dort eine Existenz geschaffen hatten, verliessen Brasilien nach meinem zweiten Lebensjahr, um sich in der Schweiz niederzulassen.

Die obligatorische Schulzeit von 9 Jahren absolvierte ich in St. Gallen und trat dann in die dortige Kantonsschule ein. Nach der regulären Studienzeit von 4 1/2 Jahren erreichte ich die Maturität Typus C. Damit war der Weg zur Hochschule frei.

Die Wahl der Studienrichtung war mir kein Problem. Ich immatrikulierte mich im Herbst 1943 an der mathematisch-physikalischen Abteilung der Eidgenössischen Technischen Hochschule in Zürich. Im Herbst 1947 schloss ich den normalen Studiengang mit dem Diplom in Mathematik ab. Das Jahr 1948 widmete ich dann ganz der Ausarbeitung der Dissertation.

PORTUGALIAE MATHEMATICA
Vol. 11 — Fas. 4 — 1952

# REPRESENTATION OF COMPLETE LATTICES BY SETS*

BY J. RICHARD BUCHI
*University of Michigan, U. S. A.*

Throughout this paper L denotes a complete lattice. That is a set of elements $a, b, c, \cdots$ partly ordered by a relation «$\subset$» such that to every class $(a_\nu)$ of elements $a_\nu$ in L there exists a greatest lower bound $\bigcap_\nu a_\nu$ in L. Then also the least upper bound $\bigcup_\nu a_\nu$ exists in L and furthermore there is a greatest element $e$ in L, namely the g. l. b. of the empty subclass of L, and there exists the smallest element 0 in L, the g. l. b. of all elements of L.

$\mathcal{M}$ denotes a class of subclasses $(a_\nu)$ of L, which includes the one-element-subclasses. In particular $\mathcal{M}$ may represent the class $\mathfrak{O}$ of all one-element-subclasses of L. In view of applications the following choises for $\mathcal{M}$ are interesting: the class $\mathfrak{F}$ of all finite subclasses of L; the class $\mathfrak{D}$ of all directed subclasses of L (the class $(a_\nu)$ is called directed if to any two of its elements $a_\nu$ and $a_\mu$ there exists a third $a_\rho$ such that $a_\nu \subset a_\rho$ and $a_\mu \subset a_\rho$); the class $\mathfrak{K}$ of all chains in L; the class $\mathfrak{C}$ of all countable subclasses of L; the class $\mathfrak{A}$ of all subclasses of L.

In section 1 we shall find properties of $\mathcal{M}$-set-lattices, that is lattices of sets in which g. l. b. means set-intersection and l. u. b. of classes belonging to $\mathcal{M}$ means set-union. The purpose of section 2 is to determine those complete lattices L which are isomorphic to $\mathcal{M}$-set-lattices. First we shall find a characterization of these lattices by a pure decomposition property. Later we shall point out the connection between distributive properties of a lattice L and the possibility of representing it as $\mathcal{M}$-set-lattice. In contrast to ordinary lattices and complete complemented lattices, it turns out that in general the complete $\mathcal{M}$-set-lattices are not fully characterized by their distributive properties. This is shown in section 3, where we exhibit completely distributive complete lattices which are not isomorphic to $\mathfrak{A}$-set-lattices and not

* Received September, 1952.

even to $\mathfrak{K}$-set-lattices. However the homomorphic images of $\mathfrak{N}$-set-lattices can be characterized by distributive properties at least in the two cases where $\mathfrak{N}=\mathfrak{A}$ and $\mathfrak{N}=\mathfrak{D}$. This is shown in section 4. In section 5 we give a complete description of all possible meet-representations of a complete lattice with respect to their economy (number of elements required to form the sets for a representation). Special attention is given to the question of the existence of most economic meet-representations.

Applications of these results will be collected in another paper. Here we only point out that the case $\mathfrak{N}=\mathfrak{F}$ will lead to conclusions about the lattices of all closed subsets of a topological space. The case $\mathfrak{N}=\mathfrak{D}$ has its importance in algebra. It leads to conclusions about the lattices of all subalgebras and the lattices of all congruence relations of universal algebras. The case $\mathfrak{N}=\mathfrak{A}$ has a close connection to totally disconnected spaces.

**1. Complete lattices of sets.** In this part L represents a class of sets $a, b, c, \cdots$ with element $\xi, \eta, \zeta, \cdots$. We assume that there is a largest set $e$ in L. $<, \wedge, \vee$ denote respectively set-inclusion, set-intersection and set-union. We furthermore make the convention that $\bigwedge_\nu a_\nu = e$ if $(a_\nu)$ is the empty subclass of L. If L now is closed under the operation $\wedge$ it is a complete lattice with the largest element $e$. However l. u. b. in L need not be set-union.

DEFINITION 1: *The class* L *of sets is called a* $\mathfrak{N}$*-set-lattice if it contains a largest set* e, *if for any subclass* $(a_\nu)$ *of* L, $\bigwedge_\nu a_\nu$ *belongs to* L, *and if for any subclass* $(a_\nu)$ *of* L *belonging to* $\mathfrak{N}$, $\bigvee_\nu a_\nu$ *belongs to* L.

It is a well known fact that the $o$-set-lattices L with largest set $e$ are in one-to-one correspondence with the closure operators H in the set $e$. The closure of a subset $x<e$ is defined as $Hx = \bigwedge_{x<u\in L} u$.

We now shall investigate those elements $H(\xi) = \bigwedge_{\xi\in a\in L} a$ of the $\mathfrak{N}$-set-lattice L which are generated by a single element $\xi$ of $e$.

DEFINITION 2. *A subset* B *of the complete lattice* L *is called a* U*-basis of* L *if every element* a *of* L *can be expressed as a l. u. b.* $a=\bigcup_\nu u_\nu$ *of elements* $u_\nu$ *of* B.

LEMMA 3. *The subset* B *of the complete lattice* L *is a* U*-basis of* L *if and only if to any two elements* $a \not\subset b$ *of* L, *there exists an element* u *in* B *such that* $u \subset a$ *and* $u \not\subset b$.

Proof: Let B be a U-basis of L. Then $a = \bigcup_\nu u_\nu$ where all $u_\nu$ are in B. $a \not\subset b$ then implies the existence of a $u_\nu$ such that $u_\nu \not\subset$ b. But $u_\nu$ is in B and $u_\nu \subset a$, which shows that B has the property stated in the lemma. Conversely let B be a class of elements of L which has this property and let $a$ be any element of L. Now define $b = \bigcup_{u \subset a,\, u \in B} u$. Then $b \subset a$. $a \not\subset b$ would imply the existence of an element $u$ in B such that $u \subset a$ and $u \not\subset b$, but this is in contradiction with the definition of $b$. Thus we have $a = b$, which proves that any element of L is a l. u. b. of elements in B.

Proposition 4. *The elements* $u_\xi = H(\xi)$ *of a* $\mathfrak{N}$*-set-lattice* L *which are generated by a single elements* $\xi$ *of* e *form a* U*-basis of* L.

Proof: Let $a$ and $b$ bo two different elements of L. Then $a \not< b$ or $b \not< a$. Suppose for example $a \not< b$. Then there exists $\xi$ such that $\xi \in a$ and $\xi \notin b$, which implies $H(\xi) < a$ and $H(\xi) \not< b$. Thus by Lemma 3 the class B of all $H(\xi)$ is a U-basis of L.

Definition 5. *An element* u *of a complete lattice* L *is* $\mathfrak{N}$*-subirreducible if for any class* $(a_\nu)$ *belonging to* $\mathfrak{N}$, $u \subset \bigcup_\nu a_\nu$ *implies the existence of an* $a_\nu$ *such that* $u \subset a_\nu$.

Proposition 6. *The set* $u_\xi = H(\xi)$ *generated by a single element* $\xi$ *of* e *is a* $\mathfrak{N}$*-subirreducible element of the* $\mathfrak{N}$*-set-lattice* L.

Proof: Let $u_\xi = H(\xi) < \bigcup_\nu a_\nu$. Because $(a_\nu)$ is in $\mathfrak{N}$ and because L is a $\mathfrak{N}$-set-lattice $\bigcup_\nu a_\nu = \bigvee_\nu a_\nu$. We thus have $\xi \in \bigvee_\nu a_\nu$, which implies the existence of a set $a_\nu$ such that $\xi \in a_\nu$. This again implies $u_\xi = H(\xi) < a_\nu$. We conclude that $u_\xi$ is $\mathfrak{N}$-subirreducible.

Propositions 4 and 6 show that a $\mathfrak{N}$-set-lattice has a U-basis of $\mathfrak{N}$-subirreducible elements. In part 2 we shall see that this fact characterizes the $\mathfrak{N}$-set-lattices among all complete lattices.

Definition 7. *A* $\mathfrak{O}$*-set-lattices* L *is called reduced if the smallest set* 0 *in* L *is the empty set and if one of the following equivalent conditions is satisfied:*

(1) *If* $\xi$ *and* $\eta$ *are different elements of the largest set* e *in* L *then there exists a set* u *in* L *which contains either* $\xi$ *or* $\eta$ *but not both.*

(2) *Different elements* $\xi$ *and* $\eta$ *of the largest set* e *in* L *generate different elements* $H(\xi)$ *and* $H(\eta)$ *of* L.

*Two* $\mathfrak{O}$*-set-lattices* $L_1$ *and* $L_2$ *with the largest set* $e_1$ *and* $e_2$ *are called essentially equal if there exists a one-to-one mapping of* $e_1$ *onto* $e_2$ *which generates an isomorphism between* $L_1$ *and* $L_2$.

PROPOSITION 8. *Every $\mathfrak{O}$-set-lattice* L *is isomorphic with a reduced $\mathfrak{O}$-set-lattice* L*. L* *is essentially unique and is a $\mathfrak{N}^*$-set-lattice if and only if* L *is a $\mathfrak{N}$-set lattice. (Here $\mathfrak{N}^*$ is generated by $\mathfrak{N}$ and the isomorphism between* L *and* L* *in an obvious manner).*

L* *can be constructed from* L *by identifying elements $\xi$ and $\eta$ of* e *for which* $H(\xi)=H(\eta)$ *and by dropping elements $\xi$ in* e *for which* $H(\xi)$ *is the smallest element* 0 *of* L.

The proof of proposition 8 and that of the equivalence of conditions (1) and (2) is straight forward and will be omited.

**2.** Representation by sets. We now shall investigate isomorphisms R of a complete lattice L with $\mathfrak{O}$-set-lattices R(L), which carry g. l. b. in L into set-intersections in R(L). Such a representation R clearly preserves inclusions and l. u. b. However the l. u. b. $\bigcup_\nu R(a_\nu)$ in R(L) need not to be the set-union of the $R(a_\nu)$.

DEFINITION 9. *By a $\mathfrak{N}$-representation of a complete lattice* L *we understand a one-to-one mapping* R, *which assigns to every element* a *in* L *a set* R(a) *in such a way that g. l. b.* $\bigcap_\nu a_\nu$ *in* L *are carried into set--intersections* $\bigwedge_\nu R(a_\nu)$ *and all l. u. b.* $\bigcup_\nu a_\nu$ *of classes* $(a_\nu)$ *belonging to $\mathfrak{N}$ are carried into set-unions* $\bigvee_\nu R(a_\nu)$.

DEFINITION 10. *An $\mathfrak{O}$-representation* R *of* L *is called reduced if the $\mathfrak{O}$-set-lattice* R(L) *of all sets* R(a) *is reduced.*

*Two representations* $R_1$ *and* $R_2$ *are essentially equal if the corresponding set-lattices* $R_1(L)$ *and* $R_2(L)$ *are essentially equal.*

We now shall exhibit a close connection between the U-basis of a complete lattice L and its representations.

PROPOSITION 11. *Let* B *be a* U*-basis of the complete lattice* L *and let* B(a) *be the class of all elements* $u \neq 0$ *of* B *which are contained in* a. *Then* B(a) *is a reduced* o*-representation of* L.

*If furthermore* $\sigma$ *is a function which assigns to every* u *in* B *a set* $\sigma u$ *in such a way that* $\sigma u$ *is not empty if* $u \neq 0$ *and such that* $\sigma u$ *and* $\sigma v$ *are disjoint whenever* u *and* v *are different then* $B^\sigma(a)$ *is also a* o*-representation of* L, *where* $B^\sigma(a)$ *is defined as the set-union of all sets* $\sigma u$ *for which* $u \subset a$. *If* $\sigma_1$ *and* $\sigma_2$ *are two functions attached to the* U*-basis* B *and if the powers of the sets* $\sigma_1 u$ *and* $\sigma_2 u$ *are equal for all* u *in* B *then the representation* $B^{\sigma_1}$ *is essentiallly equal to the representation* $B^{\sigma_2}$.

PROOF: Let B be a U-basis L, let $\sigma$ be a function as described in the proposition and let $B^\sigma(a) = \bigvee_{u \subset a} \sigma u$. To show that $B^\sigma$ is a $\mathfrak{D}$-representation of L we first prove:

(I) $a \subset b$ implies $B^\sigma(a) < B^\sigma(b)$, for all $a$ and $b$ in L.
(II) $a \not\subset b$ implies $B^\sigma(a) \not< B^\sigma(b)$, for all $a$ and $b$ in L.
(III) $B^\sigma(\bigcap_\nu a_\nu) < \bigwedge_\nu B^\sigma(a_\nu)$, for all subclasses $(a_\nu)$ of L.
(IV) $\bigwedge_\nu B^\sigma(a_\nu) < B^\sigma(\bigcap_\nu a_\nu)$, for all subclasses $(a_\nu)$ of L.

PROOF OF (I): If $a \subset b$ then $u \subset a$ implies $u \subset b$ and therefore $B^\sigma(b)$ is a union of at least as many sets $\sigma u$ as $B^\sigma(a)$. We conclude $B^\sigma(a) < B^\sigma(b)$.

PROOF OF (II): Because B is a U-basis, $a \not\subset b$ implies the existence of an element $u$ in B such that $u \subset a$ and $u \not\subset b$. $u \not\subset b$ implies $u \not\subset o$ and by hypothesis $\sigma u$ is not empty. Therefore $\sigma u < B^\sigma(b)$ would imply the existence of an element $v$ in B such that $v \subset b$ and $\sigma v$ not disjoint to $\sigma u$. By the hypothesis for $\sigma$ we then would have $u = v$, which is impossible because $u \not\subset b$ and $v \subset b$. Thus $\sigma u \not< B^\sigma(b)$. Furthermore, $u$ in B and $u \subset a$ implies $\sigma u < B^\sigma(a)$. We conclude $B^\sigma(a) \not< B^\sigma(b)$

PROOF OF (III): For all $\nu$ we have $\bigcap_\nu a_\nu \subset a_\nu$ and by (I), $B^\sigma(\bigcap_\nu a_\nu) < B^\sigma(a_\nu)$. We conclude $B^\sigma(\bigcap_\nu a_\nu) < \bigwedge_\nu B^\sigma(a_\nu)$.

PROOF OF (IV): Let $\xi$ be any element of $\bigwedge_\nu B^\sigma(a_\nu)$. Then for every $\nu$ there exists a $u_\nu$ in B such that $\xi \in \sigma u_\nu$ and $u_\nu \subset a_\nu$. Because all the $\sigma u_\nu$ have the element $\xi$ in common all ellements $u_\nu$ must be equal to a single element $u$ of B. Because all $u_\nu \subset a_\nu$ we have $u \subset \bigcap_\nu a_\nu$. Furthermore $\xi \in \sigma u$. We conclude that $\xi$ also belongs to $B^\sigma(\bigcap_\nu a_\nu)$, which proves (IV).

Now (I) and (II) show that B is a one-to-one mapping. (III) and (IV) show furthermore that B is a $\mathfrak{D}$-representation of the lattice L, which proves the second part of proposition 11. To prove the first part we define $\sigma$ as follows: If $o$ is in B then $\sigma o$ is the empty set. If $u \neq 0$ is in B then $\sigma u$ consists of one element, $u$. $B^\sigma(a)$ then clearly is the mapping $B(a)$, which therefore is a representation. It is easy to see that $B(a)$ is reduced. The third part of proposition 11 is easily established, also.

Next we show that the representations described in proposition 11 are all possible $\mathfrak{D}$-representation of L.

PROPOSITION 12. *Let* R *be a* $\mathfrak{D}$*-representation of the complete lattice* L *and let* $\xi$ *be any element of the set* U=R(e). *If* $u_\xi$ *is the g. l. b. of all elements* a *in* L *for which* $\xi$ *is in* R(a), *then the class of all elements* $u_\xi$ *constitutes a* U*-basis* B *of* L.

*If furthermore for any* u *in* B, $\sigma$u *is the class of all element* $\xi$ *for which* $u_\xi$=u *then the representation* R(a) *is just the representation* $B^\sigma(a)$ *generated by* B *and* $\sigma$ *in the sense of proposition 11.*

PROOF: Let $a \not\subset b$ be two elements of L. Because R is a representation we conclude $R(a) \not< R(b)$, or there exists an element $\xi$ in $R(a)$ which does not belong to $R(b)$. Now consider the element $u_\xi$ of L, as defined in the proposition. Then we have clearly $u_\xi \subset a$ and $u_\xi \not\subset b$. By lemma 3 we conclude that the collection B of all $u_\xi$ is a U-basis of L.

To prove the second part let $\sigma$ be defined as in proposition 12. $\sigma$ then clearly satisfies the conditions in proposition 11 and $B^\sigma$ is well-defined. It is sufficient to show that $R(a) = B^\sigma(a)$ for all $a$ in L. Let therefore $\xi$ be an element of $R(a)$. Then $u_\xi \subset a$ and it follows that there exists an element $u$ in B such that $\xi$ in $\sigma u$ and $u \subset a$. This again implies that $\xi$ belongs to $B^\sigma(a)$. All these steps are reversible, which shows that $B^\sigma$ and R are identical.

Proposition 11 together with the fact that L is a U-Basis of L shows that every complete lattice L is isomorphic to a o-set-lattice. Because proposition 4 and 6 clearly impose a condition on $\mathfrak{N}$-set-lattices we can not expect that $\mathfrak{N}$-representations always exist. The following proposition gives a complete description of all possible $\mathfrak{N}$-representations.

PROPOSITION 13. *The* $\mathfrak{D}$*-representation* $B^\sigma(a)$ *described in proposition 11 is a* $\mathfrak{N}$*-representation of the complete lattice* L *if and only if all elements of the* U*-basis* B *are* $\mathfrak{N}$*-subirreducible.*

PROOF: Let B be a U-basis which consists of $\mathfrak{N}$-subirreducible elements and let $\sigma$ have the properties described in proposition 11. Let furthermore $(a_\nu)$ be a subclass of L which belongs to $\mathfrak{N}$. Then we have $a_\nu \subset \bigcup_\nu a_\nu$ and $B^\sigma(a_\nu) < B^\sigma(\bigcup_\nu a_\nu)$, for all $\nu$. Therefore $\bigvee_\nu B^\sigma(a_\nu) < B^\sigma(\bigcup_\nu a_\nu)$. Furthermore if $\xi$ belongs to $B^\sigma(\bigcup_\nu a_\nu)$ then there exists an element $u$ in B such that $\xi$ is in $\sigma u$ and $u \subset \bigcup_\nu a_\nu$. Because $u$ is $\mathfrak{N}$-subirreducible and $(a_\nu)$ belongs to $\mathfrak{N}$, there must be a $\nu$ such that $u \subset a_\nu$. We conclude $\xi$ is in $B^\sigma(a_\nu)$ and therefore in $\bigvee_\nu B^\sigma(a_\nu)$. Therefore we have shown $\bigvee_\nu B^\sigma(a_\nu) = B^\sigma(\bigcup_\nu a_\nu)$, which proves the first part of proposition 13.

To prove the second part we now suppose that $B^\sigma$ is a $\mathfrak{N}$-representation, $(a_\nu)$ belongs to $\mathfrak{N}$ and $u \neq 0$ is an element of B such that $u \subset \bigcup_\nu a_\nu$. We have to show the existence of a $\nu$ such that $u \subset a^\nu$. Because $B^\sigma$ is a $\mathfrak{N}$-representation, $u \subset \bigcup_\nu a_\nu$ implies $B^\sigma(u) < \bigvee_\nu B^\sigma(a_\nu)$ or $\sigma u < \bigvee \sigma v$, where $v$ ranges over all elements in B which are contained in some $a_\nu$. This implies either $\sigma u$ is empty or the existence of an element $a_\nu$ and an element $v$ in B such that $v \subset a_\nu$ and $\sigma u$ and $\sigma v$ are not disjoint. By the hypothesis for $\sigma$ this implies either $u = o$ or $u = v$. Because $u \neq 0$ and $v \subset a_\nu$ we have shown the existence of a $\nu$ such that $u \subset a_\nu$, which completes the proof.

DEFINITION 14. *A complete lattice* L *satisfies the* $\mathfrak{N}$*-subdecomposition property if anyone of the following equivalent conditions is satisfied.*

(1) *The* $\mathfrak{N}$*-subirreducible elements of* L *form a* U*-basis of* L.

(2) *Every element of* L *is a l. u. b. of* $\mathfrak{N}$*-subirreducible elements.*

(3) *To any two elements* $a \not\subset b$ *of* L *there exists a* $\mathfrak{N}$*-subirreducible element* u *such that* $u \subset a$ *and* $u \not\subset b$.

Condition (2) is just a reformulation of (1). The equivalence of (1) and (3) is a consequence of lemma 3. Propositions 11, 12, 13 now can be combined as follows:

THEOREM 15. *The subcomplete lattice* L *is* $\mathfrak{N}$*-representable if and only if it satisfies the* $\mathfrak{N}$*-decomposition property.*

*If* B *is a* U*-basis of* L *which contains* $\mathfrak{N}$*-subirreducible elements only, then the representation* $B^\sigma(a)$ *defined in proposition 11 is a* $\mathfrak{N}$*-representation.*

*All possible* $\mathfrak{N}$*-representations of* L *are generated in the described manner by a* U*-basis of* $\mathfrak{N}$*-irreducible elements.*

$\mathfrak{N}$-subirreducibility can be weakened to the following more familiar notion:

DEFINITION 16. *An element* $u \neq o$ *of the complete lattice* L is $\mathfrak{N}$*-irreducible if for any class* $(a_\nu)$ *beloging to* $\mathfrak{N}$ *the equation* $u = \bigcup_\nu a_\nu$ *implies the existence of a* $\nu$ *such that* $u = a^\nu$.

DEFINITION 17. *The complete lattice* L *satisfies the* $\mathfrak{N}$*-decomposition property if one of the following equivalent conditions holds:*

(1) *There exists a* U*-basis of* L *which contains* $\mathfrak{N}$*-irreducible elements only.*

(2) *Every element of* L *is a l. u. b. of* $\mathfrak{N}$*-irreducible elements of* L.

(3) *To any elements* $a \not\subset b$ *of* L *there exists an* $\mathfrak{N}$*-irreducible element* u *such that* $u \not\subset a$ *and* $u \not\subset b$.

In general the $\mathfrak{N}$-decomposition property is weaker that the $\mathfrak{N}$-subdecomposition property and therefore is not sufficient for $\mathfrak{N}$--representability. We shall show now that for lattices which satisfy a distributive law this situation is improved.

($d_{\mathfrak{N}}$) If the class $(a_\nu)$ belongs to $\mathfrak{N}$ then the class $(b \cap a_\nu)$ belongs to $\mathfrak{N}$ also, and furthermore the formula $b \cap \bigcup_\nu a_\nu = \bigcup_\nu b \cap a_\nu$ holds.

We then have:

Lemma 18. *Every* $\mathfrak{N}$*-subirreducible element* u *of a complete lattice* L *is* $\mathfrak{N}$*-irreducible. If* L *satifies* ($d_{\mathfrak{N}}$) *then the two notions are equivalent.*

Proof: The first part of the lemma is trivial. Let us therefore assume that L satisfies ($d_{\mathfrak{N}}$), $(a_\nu)$ is in $\mathfrak{N}$ and $u$ is $\mathfrak{N}$-irreducible. $u \subset \bigcup_\nu a_\nu$ then implies $u = \bigcup_\nu (u \cap a_\nu)$ and there exists $\nu$ such that $u = u \cap a_\nu$. Thus there exists $\nu$ such that $u \subset a_\nu$, which shows that $u$ is $\mathfrak{N}$-subirreducible.

Now we note that a $\mathfrak{N}$-representable L automatically satisfies $d_{\mathfrak{N}}$.* Considering Theorem 15, definition 17 and lemma 18 we therefore conclude:

Theorem 19. *A complete lattice* L *is isomorphic to a* $\mathfrak{N}$*-set-lattice if and only if* it satisfies the distributive law* $d_{\mathfrak{N}}$ *and the* $\mathfrak{N}$*-decomposition property.*

*All* $\mathfrak{N}$*-representatious of* L *then are the representations* $B^\sigma(a)$ *induced by a* U*-basis* B *of* $\mathfrak{N}$*-irreducible elements.*

**3.** Distributive complete lattices. Set-intersection and set-union are distributive with each other in the strongest sense. $\mathfrak{A}$-representable complete lattices L therefore satisfy the distributive laws:

(D) $$\bigcup_\mu \bigcap_\nu a_{\mu,\nu} = \bigcap_\vartheta \bigcup_\mu a_{\mu,\vartheta(\mu)}$$

($\bar{D}$) $$\bigcap_\mu \bigcup_\nu a_{\mu,\nu} = \bigcup_\vartheta \bigcap_\mu a_{\mu,\vartheta(\mu)}.$$

Here $(a_{\mu,\nu})$ is any indexed subset of L and $\vartheta$ varies over all mappings of the index sets $(\mu)$ into $(\nu)$.

We now shall describe a large class of lattices satisfying these strong distributive laws. Especially we shall exhibit lattices satisfying D and $\bar{D}$ which nevertheless are not $\mathfrak{A}$-representable. More generally let us consider the conditions:

*The "only if" holds, when $\mathfrak{N}$ satisfies the condition: if $(a_\nu) \in \mathfrak{N}$ then $(b n a_\nu) \in \mathfrak{N}$. This is in particular the case if $\mathfrak{N}$ is either one of $\mathfrak{F}$, $\mathfrak{D}$, $\mathfrak{R}$, $\mathfrak{C}$, or $\mathfrak{U}$.
(In Büchi's hand copy this line is deleted, and the footnote to the theorem is added in his handwriting; cf. Mac Lane's comment. The Editors.)

($D_{\mathfrak{M}}$) The formula D holds for any class $(a_{\mu,\nu})$ such that the class $(a_{\mu,\vartheta(\mu)})$ belongs to $\mathfrak{M}$ for any fixed $\vartheta$.

($\bar{D}_{\mathfrak{M}}$) The formula $\bar{D}$ holds for any class $(a_{\mu,\nu})$ such that for every fixed $\mu$ the class $(a_{\mu,\nu})$ belongs to $\mathfrak{M}$.

In a $\mathfrak{M}$-set-lattice both $D_{\mathfrak{M}}$ and $\bar{D}_{\mathfrak{M}}$ are satisfied. We conclude: A complete lattice L which satisfies the $\mathfrak{M}$-subdecomposition property also satisfies $D_{\mathfrak{M}}$ and $\bar{D}_{\mathfrak{M}}$.

We now shall improve this statement considerably.

*Notation:* If the elements $v \subset u$ of a lattice L are different we shall write $v \sqsubset u$.

DEFINITION 20. *An element* u *of a complete lattice* L *is called weakly* $\mathfrak{M}$*-subirreducible if for any class* $(a_\nu)$ *belonging to* $\mathfrak{M}$ *the formula* $u \sqsubset \bigcup_\nu a_\nu$, *implies the existence of an element* $a_\nu$ *such that* $u \subset a_\nu$.

Obviously $\mathfrak{M}$-irreducibility does not imply weak $\mathfrak{M}$-subirreducibility. But $\mathfrak{M}$-irreducibility together with weak $\mathfrak{M}$-subirreducibility is equivalent to $\mathfrak{M}$-subirreducibility.

DEFINITION 21. *A complete lattice* L *satisfies the weak* $\mathfrak{M}$*-subde composition property if one of the following equivalent conditions holds:*

(1) *There exists a* U*-basis of weakly* $\mathfrak{M}$*-subirreducible elements.*

(2) *Every element of* L *is a l. u. b. of weakly* $\mathfrak{M}$*-subirreducible elements.*

(3) *To any elements* $a \not\subset b$ *of* L *there exists a weakly* $\mathfrak{M}$*-subirreducible element* u *such that* $u \subset a$ *and* $u \not\subset b$.

The equivalence of (1) and (2) in trivial. That of (1) and (3) follows by lemma 3.

LEMMA 22. *In a complete lattice* L *satisfying the weak* $\mathfrak{M}$*-subdecomposition property every* $\mathfrak{A}$*-irreducible element* u *is* $\mathfrak{M}$*-subirreducible.*

PROOF: Because L satisfies the weak $\mathfrak{M}$-subdecomposition property $u$ can be expressed as $u = \bigcup_\nu u_\nu$ where the $u_\nu$ are weakly $\mathfrak{M}$-subirreducible. Because $u$ is $\mathfrak{A}$-irreducible we infer that $u = u_\nu$ for some $\nu$, $u$ therefore is weak $\mathfrak{M}$-subirreducible, which together with its $\mathfrak{A}$-irreducibility implies its $\mathfrak{M}$-subirreducibility.

DEFINITION 23. b *is a direct successor of a if* $a \sqsubset b$ *and if* $u \sqsubset b$ *implies* $u \subset a$.

LEMMA 24. *In a complete lattice* L *the direct successors are the* $\mathfrak{A}$*-irreducible elements.*

PROOF: Let $b$ be a direct successor of $a$ and suppose $b$ is not $\mathfrak{A}$-irreducible. Then $b = \bigcup_\nu b_\nu$ where $b \neq b_\nu$ for all $\nu$. We then have $b_\nu \sqsubset b$ and therefore $b_\nu \subset a$ for all $\nu$. But this leads to $b \subset a$, which is a contradiction with $a \sqsubset b$. Thus, every direct successor is $\mathfrak{A}$-irreducible. Conversely let $b$ be $\mathfrak{A}$-irreducible and $a = \bigcup_{u \sqsubset b} u$. Then $a \neq b$ and $b$ is clearly a direct successor of $a$.

THEOREM 25. *A complete lattice* L *satisfying the weak* $\mathfrak{N}$*-subdecomposition property also satisfies the distributive laws* $D_{\mathfrak{N}}$ *and* $\overline{D}_{\mathfrak{N}}$.

PROOF: Let $a = \bigcup_\mu \bigcap_\nu a_{\mu,\nu}$ and $b = \bigcap_\vartheta \bigcup_\mu a_{\mu,\vartheta(\mu)}$ and let us assume that the class $(a_{\mu,\vartheta(\mu)})$ belongs to $\mathfrak{N}$ for any $\vartheta$. To prove $D_{\mathfrak{N}}$ we have to show $a = b$. The inclusion $a \subset b$ is trivial, therefore it is sufficient to show that $a \sqsubset b$ leads to a contradiction. We shall do this by considering two cases:

1. Let $b$ be a direct successor of $a$. Then, by lemma 22 and 24 $b$ is $\mathfrak{N}$-subirreducible. Because $b \subset \bigcup_\mu a_{\mu,\vartheta(\mu)}$ and because the class $(a_{\mu,\vartheta(\mu)})$ belongs to $\mathfrak{N}$, we conclude that for all $\vartheta$ there exists a $\mu$ such that $b \subset a_{\mu,\vartheta(\mu)}$. This implies the existence of a $\mu$ such that for all $\nu$, $b \subset a_{\mu,\nu}$. We conclude $b \subset \bigcup_\mu \bigcap_\nu a_{\mu,\nu} = a$ which is a contradiction with the assumption $a \sqsubset b$.

2. Suppose now $b$ is not a direct successor of $a$. Then there exists an element $v$ such that $v \sqsubset b$ but $v \not\subset a$. Because L satisfies the weak $\mathfrak{N}$-subdecomposition property $v$ can be expressed as $v = \bigcup_\rho v_\rho$ where all $v_\rho$ are weakly $\mathfrak{N}$-subirreducible. For all $\rho$ and $\vartheta$ we then have $v_\rho \subset v \sqsubset b \subset \bigcup_\mu a_{\mu,\vartheta(\mu)}$ and because $v_\rho$ is weakly $\mathfrak{N}$-subirreducible and because the class $(a_{\mu,\vartheta(\mu)})$ belongs to $\mathfrak{N}$, we can conclude that for every $\vartheta$ there exists a $\mu$ such that $v_\rho \subset a_{\mu,\vartheta(\mu)}$. This implies the existence of a $\mu$ such that for all $\nu$, $v_\rho \subset a_{\mu,\nu}$ and therefore $v_\rho \subset \bigcup_\mu \bigcap_\nu a_{\mu,\nu} = a$. We conclude $v = \bigcup_\rho v_\rho \subset a$ which is in contradiction with $v \not\subset a$.

The distributive law $\overline{D}_{\mathfrak{N}}$ can be proved in an analogous way.

Comparing Theorems 15 and 25 one could expect that complete lattices satisfying the distributive laws $D_{\mathfrak{N}}$ and $\overline{D}_{\mathfrak{N}}$ are always representable. In the case $\mathfrak{N} = \mathfrak{A}$ this expectation is supported by two well known results:

1. A finite distributive lattice is $\mathfrak{A}$-representable.

2. A complete and complemented lattice satisfying the distributive law D is isomorphic to the Boolean algebra of all subsets of a set (and therefore $\mathfrak{A}$-representable).

Nevertheless the expectation is false. We shall now exhibit complete lattices satisfying D and $\overline{\mathrm{D}}$, which in spite of that are not even $\mathfrak{N}$-representable.

DEFINITION 26. *A partly ordered set* L *is a chain if* $a \not\subset b$ *implies* $b \subset a$, *for all elements* a *and* b *in* L.

*A partly ordered set* L *is dense if for elements* a *and* b *of* L $a \sqsubset b$ *implies the existence of an element* u *in* L *such that* $a \sqsubset u \sqsubset b$.

LEMMA 27. *Every element* u *of a complete chain* L *is weakly* $\mathfrak{A}$*-subirreducible. No element* $u \neq o$ *of a dense complete lattice* L *is* $\mathfrak{A}$*-irreducible.*

PROOF: Suppose $u \sqsubset \bigcup_{\nu} a_{\nu}$. Then $\bigcup_{\nu} a_{\nu} \not\subset u$ and there exists $a_{\nu}$ such that $a_{\nu} \not\subset u$. In a chain this implies $u \subset a_{\nu}$. Therefore every element $u$ of a complete chain is weakly $\mathfrak{A}$-subirreducible.

Suppose now that L is a dense complete lattice. To show that no element a of L is $\mathfrak{A}$-irreducible we prove that $a$ is equal to the element $b = \bigcup_{u \sqsubset a} u$ and we do that by showing that $a \neq b$ leads to a contradiction. First we have trivially $b \subset a$. $a \neq b$ then implies $b \sqsubset a$ and because L is dense there exists an element $u$ in L such that $b \sqsubset u \sqsubset a$. But $u \sqsubset a$ implies $u \subset b$ which is a contradiction with $b \sqsubset u$.

As a corollary to Theorem 25 and Lemma 27 we have:

THEOREM 27. *Every complete chain satisfies the strongest distributive laws* D *and* $\overline{\mathrm{D}}$.

As a corollary to Theorem 15, Theorem 25, and Lemma 27 we have:

THEOREM 29. *There are complete lattices* L *which satisfy the strongest distributive laws* D *and* $\overline{\mathrm{D}}$ *and in spite of that are not isomophic to* $\mathfrak{A}$*-set-lattices. This is in particular the case if* L *is a dense complete chain.*

Because every subclass of a chain is directed we also conclude: There are complete lattices L which satisfy the distributive laws $\mathrm{D}_{\mathfrak{D}}$ and $\overline{\mathrm{D}}_{\mathfrak{D}}$ and still are not isomorphic with a $\mathfrak{D}$-set-lattice.

In the following section we shall see however that distributive lattices are homomorphic images of set-lattices.

Another attempt of characterizing completely distributive lattices is suggested by Theorem 24. It fails because there exist complete lattices which satisfy D and $\overline{\mathrm{D}}$ and which nevertheless do not contain any weakly $\mathfrak{A}$-subirreducible elements. Examples of such lattices are the direct composition of dense chains (i. e., the pairs of real numbers between 0 and 1, partly ordered by the relation $(x, y) \subset (u, v)$ if and only if $x \leqq u$ and $y \leqq v$).

**4.** Homomorphic images of set-latices. In the preceeding section we have seen that in general distributivity is not characteristic for set-lattices. Here our aim is to show that in the special cases $\mathfrak{N} = \mathfrak{A}$ and $\mathfrak{N} = \mathfrak{D}$ however, the homomorphic images of $\mathfrak{N}$-set-lattices are essentially all $\overline{\mathrm{D}}_{\mathfrak{N}}$-distributive complete lattices.

DEFINITION 30. *The complete lattice* L *is a homomorphic image of the complete lattice* L* *if there exists a mapping* h *of* L* *onto* L *which maps g. l. b. into g. l. b. and l. u. b. into l. u. b.*

DEFINITION 31. *A $\mathfrak{D}$-ideal of a complete lattice* L *is a subclass* I *of* L *which is directed and furthermore has the property that whenever* x *is in* I *and* y⊂x *then also* y *is in* I.

LEMMA 32. *The class $\mathfrak{D}(\mathrm{L})$ of all $\mathfrak{D}$-ideals of a complete lattice* L *is a $\mathfrak{D}$-set-lattice.*

PROOF: Let $(\mathrm{I}_\mu)$ be a class of $\mathfrak{D}$-ideals of L. Then $\bigwedge_\mu \mathrm{I}_\mu$ is clearly again a $\mathfrak{D}$-ideal of L. If $(\mathrm{I}_\mu)$ is furthermore direct then $\bigvee_\mu \mathrm{I}_\mu$ is a $\mathfrak{D}$-ideal, also.

*Notation:* If the class $(\mathrm{I}_\mu)$ of $\mathfrak{D}$-ideals of L is not directed then its l. u. b. still exists. However it is not necessarily the set union of the $\mathrm{I}_\mu$. This l. u. b. is denoted by $\bigcup_\mu \mathrm{I}_\mu$. We then have

$$(1) \qquad \bigcup_\mu \mathrm{I}_\mu = \bigwedge \mathrm{I}, \text{ where I ranges over all } \mathfrak{D}\text{-ideals such that for all } \mu,\ \mathrm{I}_\mu < \mathrm{I}.$$

Every element $u$ of L determines the class of all elements $z \subset u$. This class is clearly a $\mathfrak{D}$-ideal of L. We shall denote it by $[u]$ and call it the principal ideal generated by $u$.

LEMMA 33. *Let* L *be a complete lattice and define* $\mathrm{h}(\mathrm{I}) = \bigcup_{\mathrm{u} \in \mathrm{I}} \mathrm{u}$ *for any $\mathfrak{D}$-ideal* I *of* L. *If* L *satisfies the distributive law* $\overline{\mathrm{D}}_{\mathfrak{D}}$ *then for any class* $(\mathrm{I}_\mu)$ *of $\mathfrak{D}$-ideals:*

(2) $$h(\bigwedge_{\mu} I_{\mu}) = \bigcap_{\mu} h(I_{\mu})$$

(3) $$h(\bigcup_{\mu} I_{\mu}) = \bigcup_{\mu} h(I_{\mu}).$$

PROOF: Let $a=h(\bigwedge_{\mu} I_{\mu})$ and $b=\bigcap_{\mu} h(I_{\mu})$. Then by the definition of $h$, $b=\bigcap_{\mu}\bigcup_{u\in I_{\mu}} u$. Because $I_{\mu}$ is directed we can apply $\overline{D}_{\mathfrak{D}}$ to get $b=\bigcup_{\vartheta}\bigcap_{\mu}\vartheta(\mu)$ where $\vartheta$ varies over all functions with the property $\vartheta(\mu)\in I_{\mu}$. Because $I_{\mu}$ is $\mathfrak{D}$-ideal and $\bigcap_{\mu}\vartheta(\mu)\subset\vartheta(\mu)$ we have $\bigcap_{\mu}\vartheta(\mu)\in I_{\mu}$, for every $\vartheta$ and every $\mu$. We conclude $\bigcap_{\mu}\vartheta(\mu)\in\bigwedge_{\mu} I_{\mu}$. If conversely $u\in\bigwedge_{\mu} I_{\mu}$ we can define $\vartheta(u)=u$. Then $\bigcap_{\mu}\vartheta(\mu)=u$. This shows that the class of all $\bigcap_{\mu}\vartheta(\mu)$ is equal to $\bigwedge_{\mu} I_{\mu}$. Therefore $b=\bigcup_{u\in\bigwedge_{\mu} I_{\mu}} u$, which is by the definition of $h$ equal to $a$. Thus $a=b$ which proves (2).

As an easy consequence of (2) we have:

(4) For any $\mathfrak{D}$-ideals $I<J$ implies $h(I)\subset h(J)$.

Let now $x=h(\bigcup_{\mu} I_{\mu})$ and $y=\bigcup_{\mu} h(I_{\mu})$. To prove (3) we have to show $x\subset y$ and $y\subset x$: For any $\mu$, $I_{\mu}<\bigcup_{\mu} I_{\mu}$ and by (4), $h(I_{\mu})<h(\bigcup_{\mu} I_{\mu})$. We conclude $y\subset x$. By formula (1) and (2), $x=\bigcap_{I_{\mu}<I} h(I)$. If in this g. l. b. we drop all $h(I)$ where I is not a principal ideal we get $x\subset\bigcap_{I_{\mu}<[u]} h([u])$. But $h([u])=u$ and furthermore $I_{\mu}<[v]$ is equivalent with $h(I_{\mu})\subset u$. Therefore $x\subset\bigcap_{h(I_{\mu})\subset u} u=\bigcup_{\mu} h(I_{\mu})=y$. This completes the prove of Lemma 32.

Every $u$ in L is the image $h([u])$ of a $\mathfrak{D}$-ideal $[u]$ of L. As a corollary to Lemmas 32 and 33 we therefore have:

THEOREM 34. *Every complete lattice* L *which satisfies the distributive law* $\overline{D}_{\mathfrak{D}}$ *is the homomorphic image of a* $\mathfrak{D}$ *set-lattice* L*.

Instead of operating with $\mathfrak{D}$-ideals of L we can consider classes I which have the property that $x$ in I and $x\subset y$ implies $y$ in I. A procedure analogous to the one just given then leads to:

THEOREM 35 (1). *Every complete lattice* L *which satisfies the distributive law* $\overline{D}$ *is the homomorphic image of a* $\mathfrak{A}$ *set-lattice* L*.

*Essentially all possible* L* *are generated by a* U*-basis* B *of* L *in*

---

(1) This result was found independently by G. N. RANEY, «Proc. Am. Math. Soc.» **3**, 1952, pp. 677-680.

*the sense that* $L^*$ *is the class of all subsets* $I < B$ *where* $I$ *is a subclass of* $L$ *having the property that* $y$ *in* $I$ *and* $x \subset y$ *implies* $x$ *in* $I$.

Because of duality reasons we also have:

COROLLARY 36. *A complete lattice* $L$ *which satisfies the distributive law* $\overline{D}$ *also satisfies the distributive law* $D$ *and conversely.*

**6.** Economy of representations. If a representation $R$ and a complete lattice $L$ requires less elements to form the sets $R(a)$ than another representation $R'$ of $L$ then we can say that $R$ is more economic than $R'$. More exactly we define:

DEFINITION 37. *Let* $R_1$ *and* $R_2$ *be* $\mathfrak{D}$*-representations of the complete lattice* $L$ *and let* $U_1 = R_1(e)$ *and* $U_2 = R_2(e)$ *respectively be the sets of elements used for the representation. We shall say that* $R_1$ *is at least as economic as* $R_2$ *and we shall write in symbols* $R_1 < R_2$ *if there exists a one-to-one mapping* $f$ *of* $U_1$ *into* $U_2$ *such that for all elements* $a$ *of* $L$ *the equation* $R_1(a) = f^{-1}(R_2(a))$ *holds.*

Our aim is to give a complete description of the relation $R_1 < R_2$ between the representations of a complete lattice $L$.

DEFINITION 38. *Let* $B$ *be a* $U$*-basis of the complete lattice* $L$, *let* $\sigma$ *be a function which assigns to every element* $u \neq 0$ *in* $B$ *a not empty set* $\sigma u$, *in such a way that* $\sigma u$ *and* $\sigma v$ *are disjoint if* $u$ *and* $v$ *are different and let* $\tau u$ *be the function which assigns to every element* $u$ *in* $B$ *the power* $\tau u$ *of the set* $\sigma u$. *Then the pair* $(B, \tau)$ *is called the type of the representation* $B^\sigma$, *described in proposition 11.*

Proposition 12 then says that the type of any representation $R$ is well defined. By proposition 11 we conclude that there are representations of any type and that two representations are essentially equal if and only if they have the same type.

PROPOSITION 39. *Let* $R_1$ *and* $R_2$ *be representations of the complete lattice* $L$ *and let* $(B_1, \tau_1)$ *and* $(B_2, \tau_2)$ *respectively be the types of* $R_1$ *and* $R_2$. *Then* $R_1$ *is at least as economic as* $R_2$ *if and only if the* $U$*-basis* $B_1$ *is a subclass of* $B_2$ *and if for any* $u$ *in* $B_1$ *the cardinal number* $\tau_1 u$ *is smaller than or equal the cardinal number* $\tau_2 u$.

*In symbols we can state* $R_1 < R_2$ *if and only if* $(B_1, \tau_1) < (B_2, \tau_2)$, *where the relation «$<$» between types is defined in an obvious manner.*

PROOF: By proposition 12 we can assume that $R_1 = B_1^{\sigma_1}$ and $R_2 = B_1^{\sigma_2}$ where $\sigma_1 u$ and $\sigma_2 u$ are sets of power $\tau_1 u$ and $\tau_2 u$ respectively. We then have:

($i_1$) $\xi \in R_1(a)$ is equivalent with the existence of an element $u$ in $B_1$ such that $\xi \in \sigma_1 u$ and $u \subset a$.

($i_2$) $\xi \in R_2(a)$ is equivalent with the existence of an element $u$ in $B_2$ such that $\xi \in \sigma_2 u$ and $u \subset a$.

Let now $(B_1, \tau_1) < (B_2, \tau_2)$. Then $B_1 < B_2$ and for every $u$ in $B_1$ there exists a one-to-one mapping $f_u$ of $\sigma_1 u$ into $\sigma_2 u$. If $\xi$ is any element of any $\sigma_1 u$ we define $f(\xi) = f_u(\xi)$. $f$ then is clearly a mapping of $R_1(e) = \bigvee_{u \in B_1} \sigma_1 u$ into $R_2(e) = \bigvee_{u \in B_2} \sigma_2 u$. Furthermore $f$ is one-to-one because the $f_u$ are one-to-one and the $\sigma_2 u$ are mutually disjoint. Considering ($i_1$) and ($i_2$) one finally shows that $\xi \in R_1(a)$ is equivalent with $f(\xi) \in R_2(a)$. Thus $(B_1, \tau_1) < (B_2, \tau_2)$ implies the existence of a one-to-one mapping of $R_1(e)$ into $R_2(e)$ for which $R_1(a) = f^{-1}(R_2(a))$, and by definition 37 the relation $R_1 < R_2$.

Let now coversely $R_1 < R_2$. Then by definition 37 there exists a one-to-one mapping $f$ of $R_1(e) = \bigvee_{u \in B_1} \sigma_1 u$ into $R_2(e) = \bigvee_{u \in B_2} \sigma_2 u$ for which the condition ($ii$) holds:

($ii$) $\xi \in R_1(a)$ is equivalent with $f(\xi) \in R_2(a)$.

Suppose now $\xi \in \sigma_1 u$ and $f(\xi) \in \sigma_2 x$, where $u$ is in $B_1$ and $x$ in $B_2$. By ($i_1$) we infer $\xi \in R_1(u)$ and by ($ii$), $f(\xi) \in R_2(u)$. By ($i_2$) we infer the existence of an element $y$ in $B_2$ such that $y \subset u$ and $f(\xi) \in \sigma_2 y$. Because $f(\xi)$ belongs to both sets, $\sigma_2 x$ and $\sigma_2 y$ are not disjoint, therefore $x = y$. But $y \subset u$ and therefore $x \subset u$. A similar argument will show $u \subset x$. Summing up, we have proved the following statement.

($iii$) If $\xi \in \sigma_1 u$ and $f(\xi) \in \sigma_2 x$ then $u = x$.

Now let $u \neq 0$ be any element of $B_1$. Then $\sigma_1 u$ is not empty and we can choose an element $\xi \in \sigma_1 u$. $f(\xi)$ then lies in some set $\sigma_2 x$, where $x$ is in $B_2$. But ($iii$) shows $x = u$ and therefore $u$ also belongs to $B_2$. We conclude $B_1 < B_2$. By ($iii$) it follows furthermore that $f$ maps every element $\xi$ of $\sigma_1 u$ into an element $f(\xi)$ of $\sigma_2 u$. Because $f$ is one-to-one we conclude that the power $\tau_1 u$ of the set $\sigma_1 u$ is smaller than or equal to the power $\tau_2 u$ the set $\sigma_2 u$. The conclusion is that $R_1 < R_2$ implies $(B_1, \tau_1) < (B_2, \tau_2)$, which ends the proof of proposition 39.

The relation $R_1 < R_2$ is clearly reflexive and transitive. Because cardinal numbers have the property that $m \leqq n$ and $n \leqq m$ implies $m = n$, theorem 32 shows furthermore that $R_1 < R_2$ and $R_2 < R_1$ implies the essential equality of $R_1$ and $R_2$. Thus we have:

COROLLARY 40. *The classification of the $\mathfrak{O}$-representations of a complete lattice* L *whith respect to their economy is essentially a partial ordering.*

It should be noted also, that if $R_1$ is at least as economic as $R_2$ then the U-basis $B_1$ generating $R_1$ must be a subset of the U-basis $B_2$ generating $R_2$.

Proposition 39 reduces the description of the economy relationship between representations R to a description of all possible types $(B, \tau)$. But the ordering of all types is just a direct composition in which the first component is the ordering of all U-basis with respect to set-inclusion and all remaining components are well-orderings. Because we have full command over the rest, we now shall concentrate on the first component.

LEMMA 41. *Let* B *be a* U*-basis of the complete lattice* L *and let* a *be an element of* B *which is not* $\mathfrak{A}$*-irreducible. Then* B − a *is also a* U*-basis of* L.

PROOF: Because $a$ is not $\mathfrak{A}$ irreducible there exists a representation $a = \bigcup_\nu a_\nu$ where all $a_\nu \neq a$. Because B is a U-basis every element $a_\nu$ is a of the form $a_\nu$ is of the form $a = \bigcup_\mu u_{\mu,\nu}$ where all $u_{\mu,\nu}$ belong to B. Because $a_\nu \neq a$ we also have $u_{\mu,\nu} \neq a$. Thus $a$ can be expressed in the form $a = \bigcup_\rho u_\rho$ where $u_\rho$ is in $(B - a)$. Now let $b$ be any element of L. Because B is a U-basis we have $b = \bigcup_\sigma u_\sigma$ where $u^\sigma$ is in B. If one or the other of the $u_\sigma$ is equal to $a$ we replace it by $\bigcup_\rho u_\rho$, which shows that $b$ is of the form $b = \bigcup_\tau u_\tau$ when all $u_\tau$ belong to $B - a$. We conclude that $B - a$ is still a U-basis of L.

PROPOSITION 42. *Let* L *be a complete lattice, let* $\mathfrak{B}$ *be the class of all* U*-basis of* L *and let* $B_0$ *be the class of all* $\mathfrak{A}$*-irreducible elements of* L. *Then the structure of* $\mathfrak{B}$ *relative to set-inclusion can be described as follows:*

1. case. L *satisfies the* $\mathfrak{A}$*-decomposition property or in other words,* $B_0$ *is a* U*-basis of* L. *Then* $\mathfrak{B}$ *consists of all subsets* X *of* L *for which* $B_0 < X < L$. $\mathfrak{B}$ *is a complete atomistic Boolean algebra, in particular there exists a smallest element* $B_0$ *in* $\mathfrak{B}$.

2. case. L *does not satisfy the* $\mathfrak{A}$*-decomposition property or,* $B_0$ *is not a* U*-basis of* L. *Then to every* U*-basis* B *there exist next smaller ones. In particular there can not be a smallest and not even a minimal* U*-basis.* $\mathfrak{B}$ *still has the following properties: If* B *is in* $\mathfrak{B}$ *then* $B_0 < B$. *If* B *is in* $\mathfrak{B}$ *and* $B < X < L$ *then* X *is in* $\mathfrak{B}$.

PROOF: If B is a U-basis and $B < X < L$ then clearly X is also a U-basis. Furthermore it is obvious that every $\mathfrak{A}$-irreducible element is in B. Therefore $B_0 < B$.

These remarks imply everything stated under case 1. In case 2 it remains to be shown that to every U-basis there exists a next smaller one. Let therefore B be a U-basis of L. By supposition $B_0$ is not a U-basis and therefore B is not contained in $B_0$. Then there exists an element $a$ in B which is not $\mathfrak{A}$-irreducible. By Lemma 41, $B - a$ is still a U-basis, which concludes the proof of proposition 42.

The Journal of Symbolic Logic
Volume 18, Number 2, June 1953

# INVESTIGATION OF THE EQUIVALENCE OF THE AXIOM OF CHOICE AND ZORN'S LEMMA FROM THE VIEWPOINT OF THE HIERARCHY OF TYPES

J. RICHARD BÜCHI

It is a well known fact that Zermelo's Axiom of Choice and Zorn's Lemma are equivalent logical assumptions. However, an investigation from the viewpoint of the hierarchy of types reveals a complication in the nature of this equivalence. In a type-theoretical formalism both postulates enter as spectra ($ZA^{\alpha}$) and ($ZL^{\alpha}$) of formulas, where $ZA^{\alpha}$ is Zermelo's Axiom and $ZL^{\alpha}$ is Zorn's Lemma stated for variables of a fixed type $\alpha$. The complication has its origin in the fact that, although the formula $ZA^{\alpha}$ implies $ZL^{\alpha}$, in order to deduce $ZA^{\alpha}$ it seems to be necessary to assert Zorn's Lemma $ZL^{\beta}$ for a type $\beta$ which is higher than $\alpha$. In other words, we don't know a proof assuring the equivalence of the two formulas $ZA^{\alpha}$ and $ZL^{\alpha}$. Therefore the equivalence of Zermelo's Axiom and Zorn's Lemma has to be understood in the sense that the assertion of the formulas $ZA^{\alpha}$ for all types $\alpha$ implies every one of the formulas $ZL^{\beta}$ and conversely the assertion of the spectrum of formulas ($ZL^{\alpha}$) implies every one of the formulas $ZA^{\beta}$.

In § 3 we shall indicate a type $\beta$ which makes the deduction of $ZA^{\alpha}$ from $ZL^{\beta}$ possible. For this purpose a proof by G. Birkhoff [3] is translated into type-theoretical language.

In § 2 it is shown that $ZA^{\alpha}$ implies $ZL^{\alpha}$. In this proof no use is made of the notion of well-ordering. Specifically, we do not first deduce the Well-Ordering Theorem as it is usually done in proving Zorn's Lemma from Zermelo's Axiom. However, the method is suggested by Zermelo's second proof of the Well-Ordering Theorem [2].

All proofs are formalized to an extent which allows an exact investigation of the logical assumptions employed. The results are collected in § 4.

## 1. Introduction.

As a logical base we use the formalism developed by A. Church in his paper [1]. The reader is presumed to be acquainted with pp. 56–63 of this paper. We use the same notation throughout.

Remarks which intend to supply the intuitive motivation are separated from the formal proofs by the signs ⟨⟨...⟩⟩.

In addition to the familiar symbols for the logical connectives we make use of the following abbreviations throughout this paper.

$T_{o(oaa)} \rightarrow \lambda r_{oaa}(x_a)(y_a)(z_a)[rxy \wedge ryz \supset rxz]$.
$P_{o(oaa)} \rightarrow \lambda r_{oaa}[Tr \wedge (x_a)rxx]$.
$C_{o(oaa)(oa)} \rightarrow \lambda c_{oa}\lambda r_{oaa}(x_a)(y_a)[cx \wedge cy \supset rxy \vee ryx]$.

Received October 13, 1952.

<<$Tr$ expresses that $r_{oaa}$ is a transitive relation between the elements of type $a$. $Pr$ expresses that $r_{oaa}$ quasi-orders the type $a$. $Ccr$ expresses that $c_{oa}$ (or rather the class determined by the propositional function $c_{oa}$, to which we shall also refer as $c$) is a chain with respect to the relation $r_{oaa}$.>>

$$V_{o(oaa)(oa)a} \rightarrow \lambda y_a \lambda c_{oa} \lambda r_{oaa} (x_a)[cx \supset rxy].$$
$$U_{o(oaa)(oa)a} \rightarrow \lambda y_a \lambda c_{oa} \lambda r_{oaa} \,.\, Vycr \wedge (z_a)[Vzcr \supset ryz].$$
$$W_{o(oaa)} \quad \rightarrow \lambda r_{oaa}(c_{oa})[Ccr \supset (\exists y_a) Vycr].$$

<<$Vycr$ expresses that $y_a$ is an upper bound of the class $c_{oa}$ with respect to the relation $r_{oaa}$. $Uycr$ stands for: $y_a$ is a least upper bound of $c_{oa}$ with respect to $r_{oaa}$. A relation $r_{oaa}$ has the property $W$ if to every $r$-chain there exists a $r$-upper bound.>>

$$\mathrm{ZA}^a \rightarrow (\exists h_{a(oa)})(a_{oa})[ax \supset a(ha)].$$
$$\mathrm{ZL}_a \rightarrow (r_{oaa}) \,.\, [Pr \wedge Wr] \supset (\exists x_a)(u_a)[rxu \supset rux].$$

<<$\mathrm{ZA}^a$ is Zermelo's proposition for elements of type $a$. It states the existence of a function $h_{a(oa)}$, which to every propositional function which can be satisfied and has arguments of type $a$, selects one particular element $ha$ from all the elements for which the proposition $ax$ holds. $\mathrm{ZL}^a$ is Zorn's proposition formulated for elements of type $a$. It states: To every relation $r$ which quasi-orders the type $a$ and which has the property $W$, there exists an element $x_a$ which is maximal in the sense that $rxu$ always implies that $x$ and $u$ are equivalent relative to $r$.>>

We shall make use of the following formal axiom schemata.

1. $p_o \vee p_o \supset p_o$.
2. $p_o \supset p_o \vee q_o$.
3. $p_o \vee q_o \supset q_o \vee p_o$.
4. $[p_o \supset q_o] \supset [r_o \vee p_o \supset r_o \vee q_o]$.

$5^a$. $\Pi_{o(oa)} a_{oa} \supset a_{oa} x_a$.

$6^a$. $(x_a)[p_o \vee a_{oa} x_a] \supset p_o \vee \Pi_{o(oa)} a_{oa}$.

$7^a$. $(x_a)[a_{oa} x_a \equiv b_{oa} x_a] \supset [f_{a(oa)} a_o = f_{a(oa)} b_{oa}]$.

$8^a$. $a_{oa} x_a \wedge (y_a)[a_{oa} y_a \supset x_a = y_a] \supset a_{oa}(\iota_{a(oa)} a_{oa})$.

The axioms 1 to $6^a$ for a particular type $a$ lead to a logical functional calculus (hereafter called l.f.c.) in which the variables of type $a$ are the individual variables. These parts of the proofs which are based on such a l.f.c. are done informally.

The axioms $7^a$ are axioms of extensionality. <<The formula $7^a$ says that propositional functions $a_{oa}$ and $b_{oa}$ with the same extension (associated class) are mapped into equal elements by any function $f_{a(oa)}$. This makes it possible to interpret the elements of type $a(oa)$ as functions of classes.>>

The axiom $8^a$ is a restricted choice principle. <<The primitive constant $\iota_{a(oa)}$ becomes a selection-operator which chooses a particular element $\iota a$

with the property $a$, provided that there exists an element $x$ with the property $a$ and provided that any two elements having the property $a$ are identical.⟫

The proofs of the following theorems, which are consequences of the axioms 1 to $6^a$ and the definition of the identity, are left to the reader.

$9^a$. $x_a = x_a$.

$10^a$. $x_a = y_a \supset [a_{oa}x_a \supset a_{oa}y_a]$.

$11^a$. $x_a = y_a \supset y_a = x_a$.

$12^a$. $x_a = y_a \wedge y_a = z_a \supset x_a = z_a$.

**2. Deduction of Zorn's Lemma from Zermelo's Axiom.** The following abbreviations are used in this paragraph only.

$E_{o(oaa)} \rightarrow \lambda r_{oaa}(c_{oa})[Ccr \supset (\exists z_a)Uzcr]$ ⟪to every $r$-chain there exists a $r$–l.u.b.⟫.

$G_{o(oaa)(aa)} \rightarrow \lambda f_{aa}\lambda r_{oaa}(x_a)(y_a)[rxy \wedge ryx \supset r(fx)(fy)]$ ⟪$r$-equivalence is preserved by $f$⟫.

$S_{o(oaa)(aa)} \rightarrow \lambda f_{aa}\lambda r_{oaa}(x_a)rx(fx)$.

$D_{o(oaa)(aa)} \rightarrow \lambda f_{aa}\lambda r_{oaa}(x_a)(y_a)[rxy \wedge ry(fx) \supset ryx \vee r(fx)y]$. ⟪$Sfr \wedge Dfr$ expresses that $fx$ is either $r$-equivalent to $x$ or an immediate $r$-successor of $x$.⟫

$H_{o(aa)(oa)} \rightarrow \lambda q_{oa}\lambda f_{aa}(x_a)[qx \supset q(fx)]$ ⟪$q$ is closed with respect to $f$⟫.

$I_{o(oaa)(oa)} \rightarrow \lambda q_{oa}\lambda r_{oaa}(c_{oa})(y_a) \,.\, Ccr \wedge Uycr \wedge (x_a)[cx \supset qx] \supset qy$ ⟪$q$ is closed with respect to the formation of $r$–l.u.b. of $r$-chains⟫.

$L_{o(oa)} \rightarrow \lambda a_{oa} \,.\, (y_a)(c_{oa}) \,.\, Uycr \wedge Ccr \wedge (x_a)[cx \supset ax] \supset ay$ ⟪all l.u.b. of chains of $a$ are in $a$.⟫.

$M_{o(oa)} \rightarrow \lambda a_{oa} \,.\, (x_a)[qx \wedge ax \supset a(fx)]$ ⟪$f$ maps the part of $a$ belonging to $q$ into $a$⟫.

$N_{o(oa)} \rightarrow \lambda a_{oa}(x_a)(y_a)[rxy \wedge ryx \wedge ax \supset ay]$ ⟪if $x$ and $y$ are $r$-equivalent then $x$ in $a$ implies $y$ in $a$.⟫.

$\bar{A}_{oa} \rightarrow \lambda x_a(a_{oa})[La \wedge Ma \wedge Na \supset ax]$ ⟪intersection of all classes $a$ having the properties $L$, $M$ and $N$⟫.

$A_{oa} \rightarrow \lambda u_a \,.\, \bar{A}u \wedge [ru(fv) \vee r(fv)u]$ ⟪class of all $u$ in $\bar{A}$ which are comparable with $fv$⟫.

$A'_{oa} \rightarrow \lambda v_a \,.\, qv \wedge (u)(\bar{A}u \supset ruv \vee rvu]$ ⟪class of all $v$ in $q$ which are comparable with all $u$ in $\bar{A}$⟫.

⟪Our first task will be to show that $Tr \wedge Gfr \wedge Sfr \wedge Dfr$ together with the hypothesis that $v$ is comparable with all elements in $A$ implies the equality of $\bar{A}$ and $A$ (this is formula $m$ below). The formal proof can be outlined as follows. $\bar{A}$ is the smallest class having the properties $L$, $M$ and $N$ (see its definition and formulas (i), (ii), (iii)). $A$ is by its definition a subclass of $\bar{A}$; furthermore $A$ has the properties $L$, $M$ and $N$ (see formulas (j), (jj), (jjj)). We therefore conclude that the classes $A$ and $\bar{A}$ must be equal.⟫

Making use of l.f.c., it is easy to derive the formulas

(i) $L\bar{A}$, (ii) $M\bar{A}$, (iii) $N\bar{A}$,

and the formulas

$$(x)[cx \supset rx(fv) \vee r(fv)x] \supset (x)[cx \supset rx(fv)] \vee (\exists x)[cx \wedge r(fv)x],$$
$$Uycr \wedge (x)[cx \supset rx(fv)] \supset ry(fv),$$
$$Tr \wedge Uycr \wedge (\exists x)[cx \wedge r(fv)x] \supset r(fv)y,$$

which in turn yield

$$Tr \wedge Uycr \wedge (x)[cx \supset rx(fv) \vee r(fv)x] \supset ry(fv) \vee r(fv)y.$$

Because of (i) and l.f.c., this yields

$$Tr \wedge Uycr \wedge Ccr \wedge (x)[cx \supset Ax] \supset Ay$$

and

(j) $Tr \supset LA$.

By l.f.c. $Ax \supset [\bar{A}x \wedge rx(fv)] \vee r(fv)x$, which yields

$$(u)[\bar{A}u \supset ruv \vee rvu] \wedge Ax \supset [rvx \wedge rx(fv)] \vee rxv \vee r(fv)x.$$

Furthermore by l.f.c.

$$(u)[\bar{A}u \supset ruv \vee rvu] \wedge \bar{A}(fx) \supset r(fx)v \vee rv(fx),$$

which, together with the last formula, yields

(k) $(u)[\bar{A}u \supset ruv \vee rvu] \wedge Ax \wedge \bar{A}(fx) \supset [rvx \wedge rx(fv)] \vee r(fx)v \vee [rxv \wedge rv(fx)] \vee r(fv)x$.

By l.f.c. one derives the two formulas

$$Dfr \supset. [rvx \wedge rx(fv)] \supset [rvx \wedge rxv] \vee r(fv)x,$$
$$Dfr \supset. [rxv \wedge rv(fx)] \supset [rvx \wedge rxv] \vee r(fx)v.$$

Together with $k$ this yields

(h) $Dfr \wedge (u)[\bar{A}u \supset ruv \wedge rvu] \wedge Ax \wedge \bar{A}(fx) \supset [rvx \wedge rxv] \vee r(fv)x \vee r(fx)v$.

By l.f.c. one gets the two formulas

$$Gfr \supset [rvx \wedge rxv \supset r(fx)(fv)],$$
$$Tr \wedge Sfr \supset [r(fv)x \supset r(fv)(fx)].$$

Together with (h) this yields

$$Tr \wedge Gfr \wedge Sfr \wedge Dfr \wedge (u)[\bar{A}u \supset ruv \vee rvu] \wedge Ax \wedge \bar{A}(fx) \supset r(fv)(fx) \vee r(fx)(fv).$$

But $Ax \supset \bar{A}x$, which together with (ii) yields $Ax \supset \bar{A}(fx)$. By l.f.c., this together with the last formula leads to

(jj) $Tr \wedge Cfr \wedge Sfr \wedge Dfr \wedge (u)[\bar{A}u \supset ruv \vee rvu] \supset MA$.

By l.f.c., $Tr \wedge rxy \wedge ryx \wedge [rx(fv) \vee r(fv)x] \supset ry(fv) \vee r(fv)y$.

Because of (iii), this yields by l.f.c.

(jjj) $$Tr \supset N\bar{A}.$$

By l.f.c., the formulas (j), (jj), (jjj), and the definition of $\bar{A}$, one finally derives

(m) $$Tr \wedge Gfr \wedge Sfr \wedge Dfr \wedge (u)[\bar{A}u \supset ruv \vee rvu] \supset (u)[\bar{A}u \equiv Au].$$

<<Next we use the same idea to derive the equality of the classes $A'$ and $\bar{A}$ (formula (*s*)). Because by its definition $A'$ is a chain contained in $q$, we conclude that $\bar{A}$ is a chain contained in $q$ (formula (*n*)).>>

Considering the definition of $A$, $A'$ and $H$, the formula (m) yields by an application of l.f.c.

(ll) $$Tr \wedge Gfr \wedge Sfr \wedge Dfr \wedge Hfq \supset MA'.$$

By l.f.c. one derives the three formulas

$$(x)[cx \wedge (u) \,.\, \bar{A}u \supset rux \vee rxu] \wedge \bar{A}w \supset (x)[cx \supset rxw] \vee (\exists x)[cx \wedge rxw],$$
$$Uycr \wedge (x)[cx \supset rxw] \supset ryw,$$
$$Tr \wedge Uycr \wedge (\exists x)[cx \wedge rxw] \supset rwy,$$

which yield

$$Tr \wedge Uycr \wedge (x)[cx \supset (u) \,.\, \bar{A}u \supset rux \vee rxu] \supset (w)[\bar{A}w \supset ryw \vee rwy];$$

and together with the formula $Iqr \wedge Uycr \wedge Ccr \supset qy$ this yields

(l) $$Tr \wedge Iqr \supset LA'.$$

Finally, as an easy consequence of l.f.c.,

(lll) $$Tr \supset NA'.$$

Now (l), (ll) and (lll) together with the definition of $\bar{A}$ lead to the formula

(s) $$Tr \wedge Gfr \wedge Sfr \wedge Dfr \wedge Hfq \wedge Iqr \supset (x)[A'x \equiv \bar{A}x].$$

Considering the definition of $A'$ and $C$, the formula (s) yields

(n) $$Tr \wedge Gfr \wedge Sfr \wedge Dfr \wedge Hfq \wedge Iqr \supset C\bar{A}r \wedge (x)[\bar{A}x \supset qx].$$

<<Here it might seem surprising that (n) holds even though $\bar{A}$ can be empty. This is simply due to the fact that according to our definition the empty class is a chain. Using (n) we now easily derive the important lemma ($1^{oa}$).>>

From (i) and $\bar{A}x \supset \bar{A}x$ one derives by l.f.c. $Uz\bar{A}r \supset \bar{A}z$, and by (ii), $Uz\bar{A}r \supset \bar{A}(fz)$, which together with $Uz\bar{A}r \wedge \bar{A}(fz) \supset r(fz)z$ yields

(p) $$Uz\bar{A}r \supset r(fz)z.$$

Furthermore $Er \wedge C\bar{A}r \supset (\exists z)Uz\bar{A}r$ and $Iqr \wedge (x)[\bar{A}x \supset qx] \wedge Uz\bar{A}r \supset qz$. From these two formulas and (p) we derive by l.f.c.

$$Er \wedge Iqr \wedge C\bar{A}r \wedge (x)[\bar{A}x \supset qx] \supset (\exists z)[r(fz)z \wedge qz].$$

This, together with (n), leads to the formula

$(1^{oa})$ $\quad Tr \wedge Er \wedge Gfr \wedge Sfr \wedge Dfr \wedge Hqf \wedge Iqr \supset (\exists u)[r(fu)u \wedge qu].$

<<This is a fixed point theorem about a function $f_{aa}$. Informally it says: Let $r$ be a transitive relation such that to every chain there exists a l.u.b. Let $f$ be a mapping such that $r$-equivalent elements $x$ and $y$ (elements with the property $rxy \wedge ryx$) are mapped into $r$-equivalent elements $fx$ and $fy$, and such that $fx$ is either an immediate successor of $x$ or $fx$ is $r$-equivalent to $x$. Let $q$ be a class which is closed relative to the formation of l.u.b. of chains in $q$ and which is mapped into itself by $f$. Then there exists a fixed point of $f$ in $q$ (a point $u$ in $q$ which is mapped into an $r$-equivalent point $fu$). Note that by our definition the empty class is a chain, and $Er \wedge Iqr$ therefore assure the existence of an element in $q$.>>

Now consider the abbreviations,

$B_{o(oaa)(a(oa))} \to \lambda f_{a(oa)} \lambda r_{oaa} (c_{oa})(x_a)[Ccr \wedge cr \supset rx(fc)]$ <<for every chain $c$, $fc$ is an upper bound>>,

$Q_{o(oa)} \to Ccr$ <<class of all $r$-chains>>,
$R_{o(oa)(oa)} \to \lambda a_{oa} \lambda b_{oa} (x_a)[ax \supset bx]$ <<class inclusion>>,
$F_{oa(oa)} \to \lambda a_{oa} \lambda x_a [ax \vee x = fa]$ <<$a$ enlarged by $fa$>>,
$Z_{oa} \to \lambda x_a (\exists a_{oa})[p_{o(oa)} a_{oa} \wedge ax]$ <<union of all classes $a$ in $p$>>.

<<Now we are going to show that $R$, $F$ and $Q$ satisfy the hypotheses of lemma $(1^{oa})$. The application of $(1^{oa})$ then leads directly to the formula $(h^a)$ below.>>

By l.f.c., and observing $UZaR$, one sees easily

(a) $\quad TR,$ (b) $\quad ER.$

By l.f.c. and axiom $7^a$,

$$Rab \wedge Rba \supset (z)[az \equiv bz] \wedge fa = fb;$$

by l.f.c. and theorem $(12^a)$ this yields

$$Rab \wedge Rba \supset [ax \vee x = fa \supset bx \vee x = fb].$$

Considering the definition of $F$ and $G$, this yields

(c) $\quad GFR.$

$Rc(Fc)$ is easy to prove. It yields

(d) $\quad SFR.$

By l.f.c., $Ra(Fb) \wedge \sim Rab \supset (\exists x)[ax \wedge x = fb]$. By $12^a$ this yields the formula $Ra(Fb) \wedge \sim Rab \wedge z = fb \supset az$. Therefore

$$Ra(Fb) \wedge \sim Rab \supset (z)[z = fb \supset az].$$

Furthermore, by l.f.c.,

$$Rba \wedge \sim R(Fb)a \supset (\exists z)[z = fb \wedge \sim az].$$

These two formulas yield by l.f.c. $Rba \wedge Ra(Fb) \supset Rab \vee R(Fb)a$, which in turn yields

(e) $DFR$.

By l.f.c.,

$$Ccr \wedge Bfr \wedge cy \supset ry(fc),$$
$$Ccr \wedge Bfr \wedge cx \supset rx(fc),$$
$$(x)rxx \supset rxx.$$

By $10^a$ one infers

$$Ccr \wedge Bfr \wedge cy \wedge x = fc \supset ryx,$$
$$Ccr \wedge Bfr \wedge cy \wedge y = fc \supset rxy,$$
$$(x)rxx \wedge x = y \supset rxy.$$

Furthermore, by l.f.c. and $10^a$ one obtains the two formulas

$$Ccr \wedge cx \wedge cy \supset rxy \vee ryx,$$
$$(Fc)x \wedge (Fc)y \supset [cx \wedge cy] \vee [x = fc \wedge cy] \vee [y = fc \wedge cx] \vee x = y.$$

The last five formulas yield $(x)rxx \wedge Bfr \supset . \ Ccr \supset [(Fc)x \wedge (Fc)y \supset rxy \vee ryx]$ or $(x)rxx \wedge Bfr \supset . \ Ccr \supset C(Fc)r$ or

(f) $(x)rxx \wedge Bfr \supset HQF$.

By l.f.c., $(x)[ax \equiv bx] \wedge Car \supset Cbr$ and $UapR \wedge UbpR \supset (x)[ax \equiv bx]$. This yields

$$UapR \wedge UbpR \wedge CbR \supset Car.$$

Furthermore, by l.f.c., $UZpR$ and $CpR \wedge (c)[pc \supset Ccr] \supset CZr$. The last three formulas yield by l.f.c.

$$UapR \wedge CpR \wedge (c)[pc \supset Ccr] \supset Car.$$

Because $CCr$ and $Car$ may be written as $Qc$ and $Qa$, this yields

(g) $IQR$.

The formulas (a) to (g) together with lemma ($1^{oa}$) and l.f.c. yield

$$(x)rxx \wedge Bfr \supset (\exists c)[R(Fc)c \wedge Qc],$$

or, considering the definitions of $R$, $F$ and $Q$,

$$(x)rxx \wedge Bfr \supset (Ec) \ . \ Ccr \wedge (x)[x = fc \supset cx].$$

But by theorem ($9^a$) we have $fc = fc$. This yields the formula

($h^a$) $(x_a)rxx \wedge Bfr \supset (\exists c_{oa})[Ccr \wedge c(fc)]$.

<<So far no application of any choice principle has been necessary. The existence of the needed selection functions was always presupposed. Thus in the formula ($h^a$) the function $f_{a(oa)}$ assigns to every chain a particular upper bound. Making use now of the Choice Axiom it is clearly possible to

replace the assumption of the existence of such a selection function $f$ by the assumption that to every chain there exists an upper bound. Zorn's Lemma then will be a corollary to ($h^a$).>>

First consider the abbreviations

$$\overline{V}_{oa} \to \lambda v_a(u_a)[cu \supset. ruv \wedge \sim rvu] \text{ <<"strict" u.b. of } c\text{>>},$$
$$X_{oa} \to \lambda v_a Vvcr \wedge [(\exists z)\overline{V}z \supset \overline{V}v].$$

<<If there exists a strict u.b. of $c$, then $X_{oa}$ is the class of all strict u.b.; in the other case, $X_{oa}$ is the class of all u.b. of $c$. Next I shall show that $X_{oa}$ is not empty. A Zermelo selection function $h_{a(oa)}$ then allows the selection of a particular element $hX$ in $X$.>>

By l.f.c., $Vvcr \wedge \sim(\exists z)\overline{V}z \supset Xv$ and $\overline{V}z \supset Xz$. Therefore $Vvcr \supset (\exists u)Xu$, and this yields by l.f.c.

$$(a_{oa})[au \supset a(ha)] \wedge Vvcr \supset X(hX).$$

Consider now the abbreviation

$$F_{a(oa)} \to \lambda c_{oa} h_{a(oa)} X_{oa}.$$

By the second rule of $\lambda$-conversion $hX$ can be replaced by $Fc$ in the last formula. Considering the definition $W$, this yields by l.f.c.

(r) $\quad (a)[au \supset a(ha)] \wedge Wr \wedge Ccr \supset X(Fc).$

Because $X(Fc) \supset V(Fc)cr$, we have

$$(a)[au \supset a(ha)] \wedge Wr \supset. [Ccr \wedge cx] \supset rx(Fc);$$

and because of the definition of $B$ in lemma ($h^a$),

$$(a)[au \supset a(ha)] \wedge Wr \supset Bfr.$$

An application of ($h^a$) then yields

(t) $\quad (a)[au \supset a(ha)] \wedge (x)rxx \wedge Wr \supset (\exists c)[Ccr \wedge c(Fc)].$

<<$c(Fc)$ implies that $Fc$ is not a strict u.b. of $c$. On the other hand (r) shows $X(Fc)$, which implies that $c$ can not have any strict u.b. at all. This again implies that the u.b. $Fc$ of $c$ is a maximal element relative to the relation $r$. Formally these conclusions are as follows.>>.

By l.f.c., $r(Fc)w \supset rw(Fc) \vee [r(Fc)w \wedge \sim rw(Fc)]$ and $X(Fc) \supset (x)[cx \supset rx(Fc)]$.

This yields

$$Tr \wedge X(Fc) \wedge r(Fc)w \supset rw(Fc) \vee (x)[cx \supset rxw \wedge \sim rwx].$$

Because $(x)[cx \supset rxw \wedge \sim rwx] \supset (\exists z)\overline{V}z$, this yields by l.f.c.

$$Tr \wedge X(Fc) \wedge r(Fc)w \supset rw(Fc) \vee (u)(v)[Xu \wedge cv \supset \sim ruv].$$

Because $X(Fc) \wedge c(Fc) \wedge r(Fc)(Fc) \supset (\exists u)(\exists v)[Xu \wedge cv \wedge ruv]$, one concludes by l.f.c.

$$(x)rxx \wedge Tr \wedge X(Fc) \wedge c(Fc) \wedge r(Fc)w \supset rw(Fc).$$

Making use of (r) a second time, this yields

$$(a)[au \supset a(ha)] \land Pr \land Wr \land Ccr \land c(Fc) \land r(Fc)w \supset rw(Fc).$$

Together with (t) and introducing the abbreviation $U \to Fc$, this yields

$$(a)[au \supset a(ha)] \land Pr \land Wr \supset (\exists c)(w)[rUw \supset rwU],$$

which finally yields by l.f.c.

$$ZA^{a} \supset. Pr \land Wr \supset (\exists u)(w)[ruw \supset rwu].$$

This is the formula $ZA^{a} \supset ZL^{a}$ which was to be proved.

**3. Deduction of Zermelo's Axiom from Zorn's Lemma.** In the following proof of the Formula $ZL^{a(oa)} \supset ZA^{a}$ we make use of the logical functional calculus for several types $\beta$, of the selection operator $\iota_{a(oa)}$, and of axiom $8^{a}$.

First consider the abbreviation

$$R_{o(a(oa))(a(oa))} \to \lambda f_{a(oa)} \lambda g_{a(oa)} (c_{oa})[c(fc) \supset fc=gc].$$

<<This is a relation between elements of type $a(oa)$. The idea of the proof is to show:

(1) $R$ quasi-orders the type $a(oa)$.

(2) An $R$-maximal element is a Zermelo selection function.

(3) There exists an $R$-maximal function $h_{a(oa)}$.

The existence of a Zermelo selection function follows.>>

(1) By l.f.c., $[c(fc) \supset c(gc)] \land [c(gc) \supset lc=gc] \supset [c(fc) \supset lc=gc]$; and by theorem $10^{a}$ and l.f.c., $[c(fc) \supset fc=gc] \supset [c(fc) \supset c(gc)]$. This yields

$$[c(fc) \supset fc=gc] \land [c(gc) \supset lc=gc \supset [c(fc) \supset lc=gc \land fc=gc],$$

and because of $12^{a}$, $Rfg \land Rgl \supset Rfl$, hence $TR$. Furthermore $Rff$ and therefore

(a) $PR$.

(2) Consider now the abbreviation

$$L_{oa} \to \lambda y_{a} . [c(hc) \supset y=hc] \land [\sim c(hc) \supset y=u].$$

Because of $12^{a}$ and l.f.c., $c(hc) \land Lx \land Ly \supset x=y$ and $\sim chc \land Lx \land Ly \supset x=y$.

Therefore

(i) $Lx \land Ly \supset x=y$.

By l.f.c., $c(hc) \supset L(hc)$ and $\sim c(hc) \supset Lu$. Therefore

(ii) $(\exists x)\, Lx$.

The formulas (i) and (ii) make the application of the selection-axiom $8^{a}$ possible. It yields the formula $L(\iota L)$. Consider now the abbreviation

$$G_{a(oa)} \to \lambda c_{oa}(\iota L).$$

By the rules of $\lambda$-conversion, $\iota L$ then may be replaced by $Gc$ in the formula $L(\iota L)$. This yields $L(Gc)$, or

(b) $[c(hc) \supset Gc{=}hc] \wedge [\sim c(hc) \supset Gc{=}u]$.

<<Suppose now that $h$ is not a Zermelo selection function. The Formula (b) then shows as follows that $G$ is greater than $h$ and therefore $h$ is not maximal.>>

By (b) and l.f.c., $\sim a(ha) \supset Ga{=}u$; and by $10^{\alpha}$ and l.f.c., $au \wedge \sim a(ha) \supset \sim[ha{=}u]$.
Therefore by l.f.c. and $12^{\alpha}$, $au \wedge \sim a(ha) \supset \sim[Ga{=}ha]$. Furthermore by (b) and l.f.c., $au \wedge \sim a(ha) \supset a(Ga)$. This together with the last formula yields $au \wedge \sim a(ha) \supset \sim RGh$. $RhG$ follows directly from (b). Hence

$$(\exists a)(\exists u)[au \wedge \sim a(ha)] \supset (\exists g)[Rhg \wedge \sim Rgh],$$

or by l.f.c.

(c) $(g)[Rhg \supset Rgh] \supset (a)[au \supset a(ha)]$.

(3) Consider now the abbreviation

$$H_{o\alpha} \rightarrow \lambda x_\alpha \centerdot c_{o\alpha}x_\alpha \wedge (\exists g_{\alpha(o\alpha)})[k_{o(\alpha(o\alpha))}g_{\alpha(o\alpha)} \wedge gc{=}x \wedge c(gc)].$$

By l.f.c.,

(i) $CkR \supset [Hx \wedge Hy \supset x{=}y]$,

and $kh \wedge c(hc) \supset c(hc) \wedge [hc{=}hc \wedge kh \wedge c(hc)]$, which leads to

(ii) $kh \wedge c(hc) \supset (\exists x)Hx$.

Now (i) and (ii) make an application of the axiom $8^{\alpha}$ possible. This leads to

$$CkR \wedge kh \wedge c(hc) \supset H(\iota H).$$

Now consider the abbreviation

$$F_{\alpha(o\alpha)} \rightarrow \lambda c_{o\alpha}(\iota_{\alpha(o\alpha)}H_{o\alpha}).$$

By the rules of $\lambda$-conversion, $\iota H$ may be replaced by $Fc$ in the last formula. This yields

$$CkR \wedge kh \wedge c(hc) \supset (\exists g)[kg \wedge c(gc) \wedge gc{=}Fc].$$

By definition of $C$ and $R$, we have

$$CkR \wedge kh \wedge c(hc) \supset [kg \wedge c(gc) \supset hc{=}gc].$$

The last two formulas yield

$$CkR \wedge kh \wedge c(hc) \supset hc{=}Fc$$

and

$$CkR \supset [kh \supset RhF]$$

and

(d) $CkR \supset (\exists f)(h)[kh \supset Rhf]$.

Then (a) and (d) together with the definition of $ZL^{\alpha(o\alpha)}$ lead to the formula

$$ZL^{\alpha(o\alpha)} \supset (\exists h)(g)[Rhg \supset Rgh],$$

and together with (c) we obtain the formula $ZL^{\alpha(o\alpha)} \supset ZA^{\alpha}$, which was to be proved.

**4. Concluding remarks.** In § 2, the following was shown. Let $\alpha$ be any type symbol. Then Zorn's Lemma $ZL^{\alpha}$ stated for variables of the type $\alpha$ is a consequence of axioms 1 to $6^{\beta}$ for several types $\beta$, axiom $7^{\alpha}$, and Zermelo's Axiom $ZA^{\alpha}$ stated for variables of type $\alpha$.

Because $ZA^{\alpha}$ asserts the existence of a stronger selection function, it is understandable that no use has to be made of $\iota_{\alpha(o\alpha)}$ and axiom $8^{\alpha}$. The question arises whether the extensionality axiom $7^{\alpha}$ can be avoided.

It seems that this is impossible without a considerable complication of the formulas $ZA^{\alpha}$ and $ZL^{\alpha}$. The result of § 3 is as follows. Let $\alpha$ be any type symbol. Then Zermelo's Axiom $ZA^{\alpha}$ stated for variables of type $\alpha$ is a consequence of axioms 1 to $6^{\beta}$ for several types $\beta$, axiom $8^{\alpha}$ and Zorn's Lemma $ZL^{\alpha(o\alpha)}$ for variables of the higher type $\alpha(o\alpha)$.

Here no use has been made of any axiom of extensionality. On the other hand, the given proof depends heavily upon the selection operator $\iota_{\alpha(o\alpha)}$ and axiom $8^{\alpha}$, by which it is regulated. There seems to be little chance of avoiding selections from one-element classes, at least if contextual definitions are not allowed.

Making use of the results in § 2 and § 3, we reach the following conclusions.

First, on the base of the infinite list of axioms 1 to $8^{\alpha}$, the infinite collection of formulas ($ZA^{\alpha}$) is equivalent to the infinite collection of formulas ($ZL^{\alpha}$).

Second, Zorn's Lemma $ZL^{\alpha(\alpha o)}$ for elements of type $\alpha(o\alpha)$ implies Zorn's Lemma $ZL^{\alpha}$ for elements of type $\alpha$. A similar statement holds for Zermelo's Axiom.

It has not been shown that the formula $ZA^{\alpha}$, where $\alpha$ is a fixed type, is equivalent to the formula $ZL^{\alpha}$ nor to any formula $ZL^{\beta}$, where $\beta$ is another particular type. There seems to be little hope that such an equivalence holds.

## REFERENCES

[1] Church, Alonzo, *A formulation of the simple theory of types*, this Journal, vol. 5 (1940), pp. 56–68.

[2] Zermelo, Ernst, *Neuer Beweis für die Möglichkeit einer Wohlordnung*, ***Mathematische Annalen***, vol. 65 (1908), pp. 107–128.

[3] Birkhoff, Garrett, ***Lattice Theory***, American Mathematical Society Colloquium Publications, vol. 25 (1940), v + 155 pp.

THE UNIVERSITY OF MICHIGAN

# Section 2 Discrete Spaces

*With comments by Don Pigozzi, Iowa State University, Ames*



*

# Richard Büchi's Work on Discrete Spaces


*Don Pigozzi*


Richard Büchi was interested throughout his career in the representation problem for abstract lattices and the dual characterization property. He had highly original ideas on the subject that are not always fully appreciated.

To Büchi a *closure space* $\mathscr{S}$ is a family of subsets of a set $U$ that contains $U$ and is closed under arbitrary intersection. A closed set $X$ of $\mathscr{S}$ is *finitely generated* if there is a finite set $F \subseteq U$ such that $X = \bigcap\{Y \in \mathscr{S} : F \subseteq Y\}$. Every closure space $\mathscr{S}$ can also be viewed as a complete lattice where, for every $\mathscr{M} \subseteq \mathscr{S}$, $\bigwedge\mathscr{M} = \cap\mathscr{M}$ (set theoretical intersection). Conversely, every complete lattice of subsets of $U$ that contains $U$ and for which $\bigwedge\mathscr{M} = \cap\mathscr{M}$ can be identified with its closure space.

In the representation problem the special character of the closure spaces that can be found to represent various kinds of complete lattices is investigated. The paradigm for this problem is Stone's result that the lattice of ideals of a Booleam algebra is isomorphic to the lattice of closed sets of a totally disconnected topological space. In the dual problem one tries to characterize abstractly the lattices representable by closure spaces with special properties.

In the complete lattice of closed sets of a closure space the join $\bigvee\mathscr{M}$ does not in general coincide with the set-theoretical union, and closure spaces can be classified in terms of the $\mathscr{M}$ for which we have $\bigvee\mathscr{M} = \cup\mathscr{M}$. In [2] Büchi develops a general representation theory based on this classification. Our concern here is with one particular kind of closure space in this classification, the *discrete spaces*. A space $\mathscr{S}$ is *discrete* if $\bigvee\mathscr{M} = \cup\mathscr{M}$ for any $\mathscr{M}$ directed by inclusion. ($\mathscr{M}$ is *directed* if, for every $X, Y \in \mathscr{M}$, there exists a $Z \in \mathscr{M}$ such that $X, Y \subseteq Z$.) A complete lattice $\mathbf{L}$ is representable (i.e., isomorphic) to the lattice of closed subsets of a discrete closure space iff every element of $\mathbf{L}$ is the join of compact elements of $\mathbf{L}$ ($u$ is *compact* if $u \leq \bigvee M$ implies $u \leq \bigvee M'$ for some finite $M' \subseteq M$); such lattices are called *algebraic*. In a discrete closure space the finitely generated closed sets coincide with the compact elements of the lattice of closed sets. (This property in fact serves to characterize discrete closure spaces.) Algebraic lattices and discrete closure spaces arise most naturally in finitary mathematics: geometry, algebra, logic, etc. For example, the linear (and the affine) subspaces of Euclidean space and the congruence relations on an algebra form discrete spaces. The same is true of the theories of most logical systems; in fact in his development of general metamathematics Tarski defines the fundamental notion of consequence in terms of a countable discrete closure space (15). The representation of algebraic lattices as discrete spaces of this kind, and the dual problem of giving an intrinsic characterization of lattices that arise from such spaces, played an important role in the development of lattice theory, and was as a major factor in the development of universal algebra. An arbitrary algebraic lattice can be represented both as the lattice of subalgebras of some algebra (Birkhoff and Frink, 1) and as a lattice of congruence relations (Grätzer and Schmidt, 6).

The set of compact elements of any complete lattice $\mathbf{L}$ forms a join-semilattice with smallest element: $\mathbf{Cp}\,\mathbf{L} = \langle Cp\,\mathbf{L}, \vee, 0\rangle$ (the smallest element 0 is the smallest element of $\mathbf{L}$). If $\mathbf{L}$ is algebraic, then $\mathbf{L}$ is isomorphic to $\mathbf{Id}\,\mathbf{Cp}\,\mathbf{L}$, the lattice of ideals of $\mathbf{Cp}\,\mathbf{L}$ (Nachbin, 10). So in representing $\mathbf{L}$ as the lattice of closed sets of a discrete space it suffices to show that $\mathbf{Cp}\,\mathbf{L}$ can be represented as the semilattice of finitely generated closed sets of $\mathscr{S}$, i.e., that $\mathbf{Cp}\,\mathbf{L} \cong \mathbf{Fg}\,\mathscr{S}$.

The two papers [37] and [38] that Büchi wrote with his student T.M. Owens, and which were never published, contain his most ambitious work in this area.

In both papers the underlying structure is a semilattice. In [37] they deal with *Skolem semilattices*. These are upper semilattices $\langle S, \vee, 0\rangle$ that are relatively pseudo-complemented in the sense that, for every pair of elements $x$, $y$ of $S$, the set $\{u : x \leq u \vee y\}$ contains a smallest element; let $x\backslash y$ denote this element. $x\backslash y$ is called the *pseudo-complement of y relative to x*. The structure $\langle S, \vee, \backslash, 0\rangle$ (strictly speaking, its dual, a lower semilattice) is commonly called a *Brouwerian semilattice*. Büchi and Owens consider the *pseudo-symmetric difference*. In a relatively pseudo-complemented semilattice the set $\{u : u \vee x = u \vee y\}$ also has a smallest element, definable in terms of $\backslash$ and $\vee$ by the term $(x\backslash y) \vee (y\backslash x)$. Büchi and Owens denote the pseudo-symmetric difference by $x + y$; the enriched structure $\langle A, +, \vee, 0\rangle$ is called a *Skolem semilattice* also. Since $x\backslash y = y + (x \vee y)$, Skolem semilattices and Brouwerian semilattices are term-definitionally equivalent, and consequently their representation and structure theories are interchangeable.

The Boolean structure $\langle B, +, \vee, 0\rangle$, where $x + y$ is the symmetric difference of $x$ and $y$, is the paradigm for Skolem semilattices. Replacing join by meet we get Boolean rings $\langle B, +, \wedge, 0\rangle$. In an arbitrary Skolem semilattice one can define a pseudo-meet operation by

$$x \cdot y = (y\backslash(y\backslash x)) \vee (x\backslash(x\backslash y)),$$

which coincides with the actual meet in the case of Boolean structures. It is not a real meet operation in general since an arbitrary Skolem semilattice is in general not a lattice. Büchi and Owens call the structure $\langle A, +, \cdot, 0\rangle$ a *Skolem ring*; it generalizes the Boolean ring but in general it is not a ring in the traditional sense. Skolem rings are term-definitionally equivalent to Skolem semilattices (and hence also to Brouwerian semilattices) by means of the definition

$$x \vee y = (x + y) + (x \cdot y).$$

At the time Büchi and Owens were beginning their work on Skolem rings the theory of Brouwerian semilattices was being extensively developed by a number of different investigators (8, 11, 12, 13; see 8 for additional references). Büchi and Owens became aware of this work only when theirs was at an advanced stage. (This may be why they never attempted to publish theirs.) But the paper contains some new and highly original ideas that deserve exposure.

The most novel is the choice of the pseudo-symmetric difference as primitive operation in place of relative pseudo-complementation. This had not been done before, which is suprising since symmetric difference is the Abelian group operation of Boolean rings. It is also natural when one considers the Grätzer–Schmidt representation theorem. Büchi and Owens call a universal **A** algebra *nice* if there is an element 0 such that every finitely generated congruence relation on **A** is of the form $\Theta(x, 0)$ for some $x \in A$. ($\Theta(x, y)$ is the congruence relation generated by the pair $\langle x, y\rangle$.) Let **S** be a semilattice. A representation $\mathbf{S} \cong \mathbf{Fg\,Co\,A}$ (equivalently, $\mathbf{Id\,S} \cong \mathbf{Co\,A}$) is called *nice* if **A** is nice. (**Co A** is the lattice of all congruence relations on **A**.) Büchi and Owens observe that the Grätzer–Schmidt representation is nice, so that every algebraic lattice has a nice representation.

A nice algebra and a nice representation are *fission-free* if $\Theta(x,0) = \Theta(y,0)$ implies $x = y$. Every Skolem semilattice $\mathbf{S} = \langle S, \vee, 0\rangle$ is a fission-free nice algebra that is representable in its own congruence lattice in the sense that $x \mapsto \Theta(x,0)$ induces an isomorphism from $\mathbf{S} \cong \mathbf{Fg\,Co\,S}$. Conversely, and unexpectedly, Büchi and Owens were able to show that under quite general conditions the converse holds. One can identify each finitely generated congruence $\Phi$ on a fission-free nice algebra $\mathbf{A}$ with the unique element $x$ such that $\Phi = \Theta(x,0)$. Then binary operations $\vee$ and $+$ are uniquely defined by the conditions $\Theta(x,0) \vee \Theta(y,0) = \Theta(x \vee y, 0)$ (the first join is in the congruence lattice) and $\Theta(x,y) = \Theta(x + y, 0)$. If $\vee$ is compatible with the congruences of $\mathbf{A}$ (in particular if $\vee$ is a term or polynomial operation), then $\langle A, +, \vee, 0\rangle$ is a Skolem semilattice. The key to this result is in showing $+$ is a pseudo-symmetric difference, i.e.,

$$x + y \leq z \Leftrightarrow x \vee z = y \vee z.$$

But this is a simple consequence of the properties of congruence generation. For, suppose $x \vee z = y \vee z$. Then

$$x = x \vee 0 \equiv_{\Theta(z,0)} x \vee z = y \vee z \equiv_{\Theta(z,0)} y \vee 0 = y,$$

i.e., $\Theta(x + y, 0) = \Theta(x,y) \subseteq \Theta(z,0)$, i.e., $x + y \leq z$. Now assume conversely that $x + y \leq z$, that is, $x \equiv y \pmod{\Theta(z,0)}$. Then $y \equiv 0 \pmod{\Theta(x,0) \vee \Theta(z,0)}$, i.e., $y \leq x \vee z$. Similarly, $x \leq y \vee z$, and together these inclusions give $x \vee z = y \vee z$.

Büchi and Owens exploit this relationship to obtain an equational axiomatization of Skolem semilattices consisting of the standard semilattice axioms together with

$$x \leq x + 0,$$

$$x + y \leq (x + z) \vee (y + z),$$

$$(x \vee y) + y \leq x.$$

Varieties of algebras all of whose members are nice (but not necessarily fission-free) are investigated in (2) in connection with the property of having definable principal congruences.

The other novelty of the paper is the notion of a regular element in a Skolem ring. Regular elements are intended to capture the abstract notion of regular closed sets in a topological space. Other important features of the paper: A characterization of Skolem rings that are subdirect products of chains; an investigation of part of the lattice of subvarieties of Skolem rings; axiomatization results for varieties of Skolem rings including the construction of a variety that fails to be finitely axiomatizable. Almost all these results have appeared elsewhere in the literature of Brouwerian semilattices previously cited.

In [38] Büchi and Owens extend their program from Skolem rings to a much wider class of algebraic structures, the so-called *complementary monoids*. Only a preliminary draft of this paper exists, but as opposed to the Skolem ring paper almost all of the results are new and cannot be found elsewhere in the literature.

Let $\langle A, \cdot, 1\rangle$ be a monoid ($x \cdot y$ is associative and 1 is a two-sided unit). The relation of *right divisibility* is defined by $x \leq y$ iff $\exists u(x \cdot u = y)$. The enriched structure $\mathbf{A} = \langle A, \cdot, /, 1\rangle$ is a *right-complemented monoid* if $\leq$ is a partial ordering and $x/y$ is the smallest element of the set $\{u : x \leq y \cdot u\}$, i.e., if

$$\mathbf{A} \models \forall u(x \leq y \cdot u \Leftrightarrow x/y \leq u).$$

*Left-complemented monoids* are defined similarly. A *complemented monoid* (CM) $\langle A, \cdot, /, \backslash, 1\rangle$ is a right- and left-complemented monoid satisfying the identity

$$x \cdot (y/x) = (y \backslash x) \cdot x.$$

In a right (or left) CM the partial ordering $x \leq y$ is always an upper semilattice ordering with $x \vee y = x \cdot (y/x) = y \cdot (x/y)$; similarly for left CMs. Thus, in a complemented monoid the left and right divisibility orderings coincide. If $x \cdot y$ is commutative, then $x \backslash y = x/y$ and the notions of right-complemented, left-complemented, and complemented monoid collapse into one. Büchi and Owens call a commutative CM a *hoop*. Hoops $\langle A, \cdot, \backslash, 1\rangle$ in which $x \cdot y = x \vee y$ are exactly the (dual) Brouwerian semilattices.

The notion of a residuated lattice was introduced by Krull (9) and Ward and Dilworth (16) as an abstract form of the algebra of ideals of a commutative ring. Roughly speaking, CMs bear the same relation to residuated lattices as Brouwerian semilattices bear to Heyting algebras. Right CMs form a variety and are axiomatized by the following identities due to Bosbach (3).

$$x/x = 1,$$

$$x \cdot 1 = 1 \cdot x = x,$$

$$x \cdot (y/x) = y \cdot (x/y),$$

$$x/(y \cdot z) = (x/y)/z.$$

Left CMs have a similar axiomatization, and this gives an equational axiomatization of hoops. The most important examples of hoops are Skolem (particularly Boolean) rings; finite chains $\mathbf{C}_n$ with universe $\{1, a^1, a^2, \ldots, a^n\}$ and $a^k \cdot a^l = a^{\mathrm{MIN}\{k+l,n\}}$ and $a^k/a^l = a^{\mathrm{MAX}\{k-l,1\}}$; the infinite chain $\mathbf{C}_\omega$; and the ideal algebras of Dedekind domains. The last three examples satisfy the *cancellation laws* $(x \cdot y)/y = x$ and $(x \cdot y)\backslash y = x$. (In the case of $\mathbf{C}_n$ the cancellation laws hold for all $x$, $y$ such that $x \cdot y$ is not maximal; this is called *semicancellation*.) In contrast Skolem rings are idempotent ($x \cdot x = x$) and in fact can be characterized as idempotent CMs.

The main feature of [38] is an analysis of subdirectly irreducible CMs. This is applied to study the structure of CMs satisfying various natural conditions, such as the cancellation laws, and the structure of the lattice of subvarieties of

CMs. The starting point for this investigation is Bosbach's (4) characterization of the 1-equivalence classes of congruences on CMs. A subset $N$ of a CM $\mathbf{A}$ is *normal* if (i) $x \cdot y \in N$ for all $x, y \in A$, and (ii) $x \cdot N = N \cdot x$ for all $x \in A$.

The normal sets coincide with the 1-equivalence classes of congruences, and thus the subdirectly irreducible CMs are characterized by having a smallest nontrivial normal set. A CM is *k-potent* if it satisfies the identity $x^{k+1} = x^k$. Büchi and Owens show that the minimal normal subset of a subdirectly irreducible $k$-potent CM $\mathbf{A}$ is a chain $\mathbf{C}_n$ with $n \leq k$. Moreover, $\mathbf{A}$ can be decomposed into $\mathbf{C}_n$ and a naturally determined subalgebra $\mathbf{A}'$ that lies over $\mathbf{C}_n$ with respect to the divisibility ordering. By iterating the decomposition alternatively with subdirect product decomposition the structure of every finitely generated $k$-potent CM can be determined. As a consequence they show that every $k$-potent CM is a hoop, and that every finitely generated $k$-potent CM is finite. More generally, they prove that the minimal normal set $N$ of any hoop is *Archimedean* in the sense that, if $x, y \in N$ and $x^n \leq y$ for all $n < \omega$, then $x = 1$; hence the minimal normal subset of any hoop is linearly ordered. However the structure of an arbitrary subdirectly irreducible hoop cannot be easily described in terms of its minimal normal subset as it can in the $k$-potent case. Although it can be decomposed into a chain and a naturally determined subalgebra, the chain is not finite or even Archimedian; it is, however, embeddable in an ultraproduct of finite chains. A consequence of this structure theory is the identification of the varieties generated by $\mathbf{C}_1$ and $\mathbf{C}_\omega$ as the only two equationally complete varieties of hoops. They also prove that the variety generated by $\mathbf{C}_\omega$ is axiomatized by the hoop identities together with the cancellation law $(x \cdot y)/y = x$.

The main result of [38] is a characterization in terms of the term operation

$$x * y = (y/(y/x)) \vee (x/(x/y))$$

of those hoops that are subdirect products of chains. This operation is the analog of the pseudo-meet operation in Skolem rings, and Büchi and Owens characterize the property of being a subdirect product of chains in terms of several equivalent conditions on the pseudo-meet. The most interesting condition is that the divisibility ordering is a lattice ordering in which the lattice meet coincides with the pseudo-meet. Another equivalent condition is that $x * y$ is an associative operation.

In another preliminary manuscript [36] Büchi considers a related representation problem for the *implicative Boolean algebras* introduced by Copeland (5). These algebraic structures have, in addition to the usual Boolean operations, two binary operations $\times$ and $\supset$. The operation $x \supset y$ is intended to represent the implication "if $x$ then $y$" interpreted as a statement whose probability is the conditional probability of $y$ with respect to $x$. The class of implicative Boolean algebras is a universal class but not a variety.

In [36] Büchi obtains a Stone-like representation for these structures, where each element $X$ of the underlying Boolean set algebra is associated with a one-one mapping $\varphi x$ of the universe $U$ onto $X$ satisfying certain conditions; the implication $X \supset Y$ is represented by the element $\varphi_X^{-1}(Y)$. The result is analogous to the representation theorem for Boolean algebras with operators (7).

Finally, we want to discuss briefly two other published papers of Büchi ([4], [19]) that deal with representations of a somewhat different character.

A set $M$ of real numbers of cardinality $2^{\aleph_0}$ is *totally heterogeneous* if, for every Borel-measurable function $f$ of a subset $X \subseteq M$ into $M$, the set $\{f(x) : f(x) \neq x\}$ is of cardinality $< 2^{\aleph_0}$. In the main result of [4] Büchi proves that under the assumption of the continuum hypothesis (CH), there exists a totally heterogeneous set of cardinality $2^{\aleph_0}$. It follows that under CH, there exist $2^{\aleph_0}$ totally heterogeneous sets of reals. Moreover, every Borel set contains a totally heterogeneous set. There are totally heterogeneous sets of measure zero and of the first category, but no Borel set has the property.

Without assuming the CH Büchi shows that there are $2^{2^{\aleph_0}}$ sets $M_\xi$ of real numbers such that $M_\xi$ is not generalized-homeomorphic to any subset of $M_\eta$ with $\eta \neq \xi$. (This generalizes a result of Sierpinski (14).) Another consequence of this kind is that under CH, there exists a $\sigma$-complete Boolean algebra **A** with $\aleph_1$ atoms that is heterogeneous in the following sense: If $x$ and $y$ are any pair of disjoint elements of **A** such that $x$ contains $\aleph_1$ atoms, then there is no $\sigma$-homomorphism from the principal ideal generated by $y$ onto the principal ideal generated by $x$.

In the joint paper [19] with Gary Haggard, Büchi considers the following graph-theoretic problem: Let $G$ be a finite, connected, planar graph, and let $C$ be a circuit of $G$. Does there exist an embedding $\xi$ of $G$ on $T_2$ (the 2-sphere) such that $\xi(C)$ bounds a component of $T_2 \backslash \xi(C)$? A characterization of the circuits of $G$ that have this property is given.

The following is taken from the abstract of [19]. Let $G$ be a finite, connected graph and $C$ a circuit of $G$. $\beta(C)$, the *strong bridge graph* of $C$ in $G$, is defined as follows: (i) the vertices of $\beta(C)$ are the bridges of $C$ in $G$, and (ii) there is an edge in $\beta(C)$ joining a pair of vertices $B_1$ and $B_2$ iff $B_1$ and $B_2$ separate each other relative $C$.

**Theorem.** *Let G be a finite, connected graph. G is planar iff $\beta(C)$ is bipartite for each circuit C in G.*

**Lemma.** *Let G be a finite, connected graph. G is not planar iff there is a circuit C of G for which $\beta(C)$ contains a loop or a triangle.*

This lemma isolates the crucial step in a new proof of the Kuratowski theorem.

## *References*


1. Birkhoff, G., & Frink, O. (1948). Representations of lattices by sets. *Trans. Amer. Math. Soc.*, 64, 299–316.
2. Blok, W.J., & Pigozzi, D. (1984). On the structure of varieties with equationally definable principal congruences II. *Algebra Universalis*, 18, 334–379.

3. Bosbach, B. (1969). Komplementäre Halbgruppen. Axiomatik und Arithmetik. *Fund. Math.*, 64, 257–287.
4. Bosbach, B. (1970). Komplementäre Halbgruppen. Kongruenzen and Quotienten. *Fund. Math.*, 69, 1–14.
5. Copeland, A.H. Sr. (1950). Implicative Boolean algebra. *Math. Zeit.*, 53, 285–290.
6. Grätzer, G., and Schmidt, E.T. (1962). Characterizations of congruence lattices of abstract algebras. *Acta. Sci. Math. (Szeged)*, 24, 34–59.
7. Jónsson, B., and Tarski, A. (1951). Boolean algebras, Part I. *Amer. J. Math.*, 73, 891–939.
8. Khöler, P. (1981). Brouwerian semilattices. *Trans. Amer. Math. Soc.*, 268, 103–126.
9. Krull, W. (1924). Axiomatische Begründung der allgemeinen Idealtheorie. *Sitzungsberichte der physikalisch-medicinischen Societät Erlangen*, 56, 47–63.
10. Nachbin, L. (1949). On a characterization of the lattice of all ideals of a Boolean ring. *Fund Math.*, 36, 137–142.
11. Nemitz, W.C. (1965). Implicative semi-lattices. *Trans. Amer. Math. Soc.*, 117, 128–142.
12. Nemitz, W.C., and Whaley, T. (1971). Varieties of implicative semi-lattices. *Pacific. J. Math.*, 37, 759–769.
13. Nemitz, W.C., and Whaley, T. (1973). Varieties of implicative semi-lattices II. *Pacific J. Math.*, 45, 303–311.
14. Sierpiński, W. (1950). Sur les types d'ordre des ensembles linéaires. *Fund. Math.*, 37, 253–264.
15. Tarski, A. (1983). Fundamental concepts of the methodology of the deductive sciences. In *Logic, semantics, metamathematics*, 2d ed., A. Tarski, Hackett Pub. Co., Indianapolis, Ind.
16. Ward, M., and Dilworth, R.P. (1939). Residuated lattices. *Trans. Amer. Math. Soc.*, 45, 335–354.

FUNDAMENTA
MATHEMATICAE
XLI (1954)

# On the existence of totally heterogeneous spaces

by

**J. R. Büchi** (Ann Arbor)

The main purpose of this note is to prove the existence of a set $M$ of real numbers, which is heterogeneous in the sense that every Borel-function defined on a subset $X$ of $M$ into $M$ is trivial. Some consequences and related facts are pointed out in notes at the end of the paper.

We first state the following fact:

(1) *Let $f$ be a real valued measurable function defined on a measurable set $X$ of real numbers. Then the set $D$ of all $y$, for which $f^{-1}(y)$ is of positive measure, is at most of cardinality $\aleph_0$.*

Now we prove,

LEMMA 1. *Let $F$ be a class of real valued measurable functions, defined on measurable sets of real numbers, and suppose the cardinality of $F$ is $\aleph_1$. Then there exists a set $M$ of real numbers, which is of cardinality $\aleph_1$, such that the sets $[f(x)|x \in M, f(x) \in M, f(x) \neq x]$ are at most of cardinality $\aleph_0$, for all members $f$ of $F$.*

Proof. Let $\omega_1$ be the first ordinal of cardinality $\aleph_1$. By hypothesis the class $F$ can be arranged into a $\omega_1$-series $[f_\xi|\xi<\omega_1]$. Let $D_\xi=[y|f_\xi^{-1}(y)$ of positive measure$]$ and define a $\omega_1$-series of real numbers $x_\xi$ by the following induction.

Choose any real number as $x_1$. If the $x_\eta$ are already defined for all $\eta<\xi$, then choose $x_\xi$ such that the following conditions are satisfied:

$$(\alpha) \qquad x_\xi \neq x_\eta \quad \text{for all} \quad \eta<\xi,$$

$$(\beta) \qquad x_\xi \neq f_\nu(x_\eta) \quad \text{for all} \quad \eta<\xi \quad \text{and} \quad \nu<\xi,$$

$$(\gamma) \qquad f_\nu(x_\xi) \neq x_\eta \quad \text{or} \quad f_\nu(x_\xi) \in D_\nu \quad \text{for all} \quad \eta<\xi \quad \text{and} \quad \nu<\xi.$$

That such an element $x_\xi$ exists one shows as follows. To realize $(\alpha)$ and $(\beta)$ one has to avoid a set of cardinality less than $\aleph_1$ only. As for the realization of $(\gamma)$ note first that in case $x_\eta \in D_\nu$ the condition $(\gamma)$ is void. In the alternative case the pair $(\eta,\nu)$ is such that $x_\eta \notin D_\nu$. Then, by definition of $D_\nu$, $f_\nu^{-1}(x_\eta)$ is of measure 0. Therefore, for any pair $(\eta,\nu)$, one can satisfy ($\gamma$) by avoiding a set of measure 0 only. But $\xi<\omega_1$, and therefore the conditions ($\alpha$), ($\beta$), ($\gamma$) can be realized simultaneously by avoiding a set of measure 0. Thus, the $\omega_1$-series $[x_\xi|\xi<\omega_1]$ is well-defined.

Now by ($\beta$), $x_\eta=f_\nu(x_\xi)$ implies $\eta\leqslant\nu$ or $\eta=\xi$ or $\eta<\xi$. By ($\gamma$), $x_\eta=f_\nu(x_\xi)$ and $\eta<\xi$ implies $f_\nu(x_\xi)\in D_\nu$ or $\eta\leqslant\nu$. We conclude that $x_\eta=f_\nu(x_\xi)$ implies $\eta\leqslant\nu$ or $x_\eta=x_\xi$ or $f_\nu(x_\xi)\in D_\nu$. Or, if we now define $M=[x_\xi|\xi<\omega_1]$, $x\in M$ and $f_\nu(x)\in M$ and $f_\nu(x)\neq x$ implies $f_\nu(x)\in[x_\eta|\eta<\nu]$ or $f_\nu(x)\in D_\nu$. But both sets $D_\nu$ and $[x_\eta|\eta<\nu]$ are at most of cardinality $\aleph_0$, as it follows from (1) and $\nu<\omega_1$. Thus the set $M$ clearly satisfies the conditions in lemma 1.

THEOREM 1. *If $2^{\aleph_0}=\aleph_1$, there exists a set $M$ of real numbers, such that $M$ is of cardinality $2^{\aleph_0}$, and such that the sets $[f(x)|f(x)\neq x]$ are at most of cardinality $\aleph_0$, for all Borel-measurable functions $f$: $X\to M$, defined on arbitrary subsets $X$ of $M$.*

Proof. Let $F$ be the class of all real valued Borel-measurable functions, defined on Borel-sets of real numbers. The cardinality of $F$ is $2^{\aleph_0}$. Thus, by $2^{\aleph_0}=\aleph_1$ and lemma 1, there exists a set $M$ of cardinality $2^{\aleph_0}$, such that $[g(x)|x\in M, g(x)\in M, g(x)\neq x]$ is at most of cardinality $\aleph_0$, for all members $g$ of $F$. Now suppose $X$ is a subset of $M$ and $f$ is any Borel-measurable function of $X$ into $M$. It is known, (see [4] and [8]), that such an $f$ can be extended to a function $g$, which is a member of $F$. Since $[f(x)|f(x)\neq x]$ is a subset of $[g(x)|x\in M, g(x)\in M, g(x)\neq x]$, it follows that the cardinality of $[f(x)|f(x)\neq x]$ is at most $\aleph_0$. This proves theorem 1.

## Some notes and further results

**1.** Theorem (1) strengthens a result of B. Dushnik and E. W. Miller [1], who proved it with *Borel-measurable functions* replaced by *strictly monotonic functions*. A similar result can be proved without assuming the continuum-hypothesis. (See theorem 2).

**2.** It does not seem to be easy to eliminate the assumption $2^{\aleph_0}=\aleph_1$ from theorem 1. We do not know how to do this, even if we restrict our attention to continuous functions. However, by an obvious variation of the proof of lemma 1 we can get, without assuming $2^{\aleph_0}=\aleph_1$.

LEMMA 2. *Let $F$ be a class of real valued measurable functions, defined on measurable sets of real numbers. Suppose the cardinality of $F$ is $2^{\aleph_0}$ and, for every $f$ in $F$ and every real number $y$, $f^{-1}(y)$ is either of positive measure or at most of cardinality $\aleph_0$. Then there exists a set $M$ of cardinality $2^{\aleph_0}$, such that the sets $[f(x)|x\in M, f(x)\in M, f(x)\neq x]$ are of cardinality less than $2^{\aleph_0}$, for every member $f$ of $F$.*

An example of a class $F$ which satisfies the conditions in lemma 2 consists of all real valued weakly monotonic functions, defined for all real numbers. Furthermore, every weakly monotonic function defined on any set of reals can be extended to a member of this class $F$. We obtain at once

THEOREM 2. *Without assuming* $2^{\aleph_0}=\aleph_1$, *it is possible to prove the existence of a set* $M$ *of real numbers, such that the cardinality of* $M$ *is* $2^{\aleph_0}$, *but the cardinality of* $[f(x)|f(x)\neq x]$ *is less than* $2^{\aleph_0}$, *for every weakly monotonic function* $f$, *which maps a subset of* $M$ *into* $M$.

Another class $F$ which satisfies all conditions of lemma 2 is the set of all generalized homeomorphisms (one-to-one mappings, which are Borel-measurable in both ways) between Borel-sets of real numbers. According to a result of C. Kuratowski [5], every generalized homeomorphism between any two sets of real numbers can be extended to a member of this class $F$. It follows that

(2) *Without assuming* $\aleph_1=2^{\aleph_0}$ *one can show the existence of a set* $M$ *of real numbers, such that the cardinality of* $M$ *is* $2^{\aleph_0}$, *and such that the set* $[x|f(x)\neq x]$ *is of cardinality less than* $2^{\aleph_0}$, *for every generalized homeomorphism* $f$ *between two subsets of* $M$.

This theorem has been proved by W. Sierpiński [9], with the word *generalized* removed. By taking two disjoined subsets of $M$, both of cardinality $2^{\aleph_0}$, it follows that

(3) *Without assuming* $\aleph_1=2^{\aleph_0}$ *one can show the existence of two sets* $N_1$ *and* $N_2$ *of real numbers, both of cardinality* $2^{\aleph_0}$, *such that there is no set* $Z$ *of cardinality* $2^{\aleph_0}$ *which can be mapped into* $N_1$ *and* $N_2$ *by generalized homeomorphisms.*

**3.** The set $M$ of theorem 1 has in particular the property that no two exclusive subsets of it are homeomorphic, except, when they are of power less than $2^{\aleph_0}$. This suggests

Definition. *A set* $M$ *of real numbers which has cardinality* $2^{\aleph_0}$ *is totally heterogeneous, if for every Borel-function* $f$ *of a subset* $X\subseteq M$ *into* $M$ *the set* $[f(x)|f(x)\neq x]$ *is of cardinality less than* $2^{\aleph_0}$.

Now, a subset of cardinality $2^{\aleph_0}$ of a heterogeneous set is clearly heterogeneous. Thus by theorem 1 we have

THEOREM 3. *If* $2^{\aleph_0}=\aleph_1$ *there exist* $2^{\aleph_1}$ *totally heterogeneous sets of real numbers.*

Next we note,

(4) *Every perfect set contains an order-isomorphic image of the set of all reals. Every Borel-set of cardinality greater than* $\aleph_0$ *contains a perfect set* (see Hausdorff [2]).

Now, Cantor's set is a perfect set of measure zero and of first category. Together with (4) this yields another improvement of theorem 1.

THEOREM 4. *If $2^{\aleph_0}=\aleph_1$, there exists a totally heterogeneous set in every Borel-set of power greater than $\aleph_0$ (in every perfects set). Furthermore, there exist totally heterogeneous sets which are of measure zero and of first category.*

But (4) implies the following negative results, also.

THEOREM 5. *A totally heterogeneous set cannot be a Borel-set and is always of inner measure zero.*

Nevertheless in the sense of outer measure, a heterogeneous set may be thick.

THEOREM 6. *If $2^{\aleph_0}=\aleph_1$, there exists a totally heterogeneous set $M$ of real numbers which has the outer measure $\infty$, and even stronger, the outer measure of $M\cap E$ is equal to the measure of $E$, for every measurable set $E$. Such a set $M$ is automatically of second category.*

Proof. Since $2^{\aleph_0}=\aleph_1$, we can arrange the Borel-sets of real numbers which have a positive measure into a $\omega_1$-series $[B_\xi|\xi<\omega_1]$. Now we refine the proof of lemma 1 by choosing $x_\xi$ in $B_\xi$, which can be done, because $B_\xi$ is of positive measure and the conditions ($\alpha$), ($\beta$), ($\gamma$) eliminate a set of measure 0, only. The heterogeneous set $M=[x_\xi|\xi<\omega_1]$ then clearly intersects every set $A$ of positive measure. Now let $E$ be any measurable set. If $M\cap E\subseteq X\subseteq E$, then $(E-X)\cap M=0$. Therefore, if $X$ is measurable, we conclude by the property of $M$ that the measure of $E-X$ must be 0, and therefore the measure of $X$ is equal to the measure of $E$. It follows that the outer measure of $M\cap E$ is equal to the measure of $E$.

To prove or disprove the existence of totally heterogeneous sets of second category appears to be difficult.

**4.** We say that the set $V$ is a Borel-image of the set $U$, if there is a Borel-measurable function $f$ defined on $U$, such that $f(U)=V$. Now, every Borel-measurable function $f:X\to Y$ can be extended to a Borel-measurable function $f':X'\to Y'$ where $X'$ and $Y'$ are Borel-sets of reals. Furthermore, if $Y$ is of cardinality greater than $\aleph_0$, then $Y'$ is automatically of cardinality $2^{\aleph_0}$ (see Hausdorff [2]). By a theorem of C. Kuratowski [7], and theorem 1, we conclude:

THEOREM 7. *Without assuming $2^{\aleph_0}=\aleph_1$, we can show the existence of $2^{2^{\aleph_0}}$ sets $M_\nu$ of real numbers, such that no Borel-image $f(M_\nu)$ of cardinality greater than $\aleph_0$ of any $M_\nu$ can be contained in a different $M_\nu$.*

We define a Borel-invariant to be a class $I$ of sets, which together with any set contains all its Borel-images. As a corollary to theorem 7 we then get the following improvement of a theorem by C. Kuratowski [6].

COROLLARY. *Without assuming* $2^{\aleph_0}=\aleph_1$ *one can show the existence of* $2^{2^{2^{\aleph_0}}}$ *Borel-invariants of the space of real numbers.*

As another obvious corollary to theorem 7 we get the following improvement of a theorem by W. Sierpiński [11].

COROLLARY. *Without assuming* $2^{\aleph_0}=\aleph_1$, *one can show the existence of* $2^{2^{\aleph_0}}$ *sets* $M_\nu$ *of real numbers, such that none of the* $M_\nu$ *is generalized homeomorphic to any subset of any different one of the* $M_\nu$*'s.*

To get W. Sierpiński's result, replace "$2^{2^{\aleph_0}}$" by "more than $2^{\aleph_0}$" and replace "generalized homeomorphic" by "order-isomorphic".

If we now assume the continuum-hypothesis, we can improve theorem 7 to

THEOREM 8. *If* $2^{\aleph_0}=\aleph_1$ *and* $2^{\aleph_1}=\aleph_2$, *then there exist* $\aleph_2$ *sets of real numbers* $M_\nu$, *such that no subset of cardinality* $\aleph_1$ *of* $M_{\nu_1}$ *is a Borel-image of any subset of* $M_{\nu_2}$, *where* $M_{\nu_1}\neq M_{\nu_2}$.

Proof. By a theorem of W. Sierpiński [10] (see A. Tarski [13]), if $M$ is a set of cardinality $m$, then there exists a class $K$ of subsets $M_\nu$ of $M$, such that $K$ is of cardinality greater than $m$ and such that $M_{\nu_1}\cap M_{\nu_2}$ is of cardinality less than $m$, whenever $M_{\nu_1}$ and $M_{\nu_2}$ are different members of $K$. Thus, under the hypotheses $2^{\aleph_0}=\aleph_1$ and $2^{\aleph_1}=\aleph_2$, theorem 8 clearly follows from theorem 1.

Note that similarly we can prove from theorem 2 and (2), without assuming $2^{\aleph_0}=\aleph_1$ or $2^{\aleph_1}=\aleph_2$, that there exist more than $2^{\aleph_0}$ sets of real numbers $M_\nu$, such that no subset of $M_{\nu_1}$ of cardinality $2^{\aleph_0}$ is a monotonic image of (generalized homeomorphic to) any subset of $M_{\nu_2}$.

**5.** In the proof of lemma 1 we can choose all numbers $x_\xi$ from a given set of positive outer measure. Thus, in every set of real numbers $X$ which has positive outer measure, there exists a totally heterogeneous subset $M$.

It is clear that all our results can be proved if we replace the set of real numbers by a complete separable metric space, for which there is a $\sigma$-measure on the Borel-sets, which is not identically 0.

**6.** The following fact has been proved by R. Sikorski [12]. Let $B_1$ and $B_2$ be the $\sigma$-complete fields of all Borel-sets of separable metric spaces $X_1$ and $X_2$ and let $h$ be a $\sigma$-homomorphism of $B_2$ into $B_1$; then there exists a mapping $f$ of $X_1$ into $X_2$, such that $h(U)=f^{-1}(U)$ for every member $U$ of $B_2$. In other words, every $\sigma$-homomorphism $h$ of $B_2$ into $B_1$ is generated by a Borel-measurable function $f$ of $X_1$ into $X_2$. From theorem 1 we obtain at once

THEOREM 9. *If* $2^{\aleph_0}=\aleph_1$, *there exists a* $\sigma$*-complete Boolean algebra with* $\aleph_1$ *atoms, which is heterogeneous in the following sense. If* $x$ *and* $y$ *are* *any elements of B, such that* $x \cap y = 0$ *and y contains* $\aleph_1$ *atoms, then there is no* $\sigma$*-homomorphism defined on the Boolean algebra* $[u|u \in B, u \subseteq y]$ *onto the Boolean algebra* $[u|u \in B, u \subseteq x]$.

It is possible that the quotient-algebra $Q$ of $B$ in theorem 9, modulo the $\sigma$-ideal of all elements $x \in B$, which are the union of at most $\aleph_0$ atoms, does not admit any $\sigma$-homomorphisms. (This would follow from a result of R. Sikorski [12] if the heterogeneous set $M$ which generates $B$ were a Borel-set of real numbers. But, by theorem 4 there is no such $M$). In this connection note the ingenious construction of B. Jónsson [3] of a Boolean algebra which admits no automorphism except the identity. His algebra is of very high cardinality.

**7.** The rather ingenious use of well-orderings, employed to prove the fundamental lemma 1, has often been used to derive pseudo-antinomious results about the continuum. It seems to originate with G. Hamel, who devised it to show the existence of a base for the reals.

## References

[1] B. Dushnik and E. W. Miller, *Partially ordered sets*, Am. Jour. of Math. 63 (1941), p. 600-610.

[2] F. Hausdorff, *Die Mächtigkeit der Borelschen Mengen*, Math. Ann. 77 (1916), p. 430-437.

[3] B. Jónsson, *A Boolean algebra without proper automorphisms*, Proc. Am. Math. Soc. 2 (1951), p. 766-770.

[4] C. Kuratowski, *Sur les théorèmes topologiques de la théorie des fonctions de variables réelles*, Comptes Rendus 197 (1933), p. 19-20.

[5] — *Sur le prolongement de l'homéomorphie*, Comptes Rendus 197 (1933) p. 1090-1091.

[6] — *Sur la puissance de l'ensemble des nombres de dimension au sens de M. Fréchet*, Fund. Math. 8 (1926), p. 201-208.

[7] — *Sur l'extension de deux théorèmes topologiques à la théorie de ensembles*, Fund. Math. 34 (1947), p. 34-38.

[8] W. Sierpiński, *Sur l'extension des fonctions de Baire définies sur les ensembles linéaires quelconques*, Fund. Math. 16 (1930), p. 81-89.

[9] — *Sur un problème concernant les types de dimension*, Fund. Math. 19 (1932), p. 65-71.

[10] — *Sur une décomposition d'ensembles*, Monatsheft für Math. und Phys. 35 (1928), p. 239-242.

[11] — *Sur les types d'ordre des ensembles linéaires*, Fund. Math. 37 (1950), p. 253-264.

[12] R. Sikorski, *On the inducing of homomorphisms by mappings*, Fund. Math. 36 (1949), p. 7-22.

[13] A. Tarski, *Sur la décomposition des ensembles en sous-ensembles presque disjoints*, Fund. Math. 12 (1928), p. 188-205.

*Reçu par la Rédaction le 8. 7. 1953*

Reprinted from JOURNAL OF COMBINATORIAL THEORY
Vol. 10, No. 3, May 1971
Printed in Belgium


# Jordan Circuits of a Graph*

J. RICHARD BÜCHI

*Purdue University, Lafayette, Indiana 47907*

AND

GARY HAGGARD

*University of California, Santa Cruz, California 95060*

*Communicated by W. T. Tutte*

Received December 20, 1968

Let $G$ be a connected, finite graph. Let $C$ be a circuit of $G$. $\beta(C)$, the strong bridge graph of $C$ in $G$, is defined as follows: (1) the vertices of $\beta(C)$ are the bridges of $C$ in $G$, and (2) there is an edge in $\beta(C)$ joining a pair of vertices $B_1$ and $B_2$ if and only if $B_1$ and $B_2$ separate each other relative $C$.

THEOREM. *Let $G$ be a finite, connected graph. $G$ is planar if and only if $\beta(C)$ is bipartite for each circuit $C$ in $G$.*

LEMMA. *Let $G$ be a finite, connected graph. $G$ is not planar if and only if there is a circuit $C$ of $G$ for which $\beta(C)$ contains a loop or a triangle.*

This LEMMA isolates the crucial step in a new proof of the Kuratowski Theorem.

## 1. INTRODUCTION

Let $G$ be a finite, connected, planar graph. Let $C$ be a circuit of $G$. Does there exist an embedding $\xi$ of $G$ on $T_2$ (2-sphere) such that $\xi(C)$ bounds a component of $T_2 - \xi(G)$? A characterization of the circuits of $G$ which have this property is given. This result corrects a false answer to this question found in [1] and a false answer to a similar question found in [2].

The remainder of this paper is devoted to proving that the following seven statements are equivalent for a finite, connected graph $G = \langle V, E, I\rangle$.

* This work was supported in part by N.S.F. grant GJ-120. The authors would also like to thank Mr. Peng-Siu Mei for comments on the manuscript.

(I) $\gamma(G) = 0$, i.e., $G$ is planar.

(II) For any circuit $C$ in $G$ (a) $\gamma(G \cdot E(B \cup C)) = 0$ for each bridge $B$ of $C$ in $G$ and (b) $C$ is weakly Jordan, i.e., $\bar{\beta}(C)$ is bipartite.

(III) $G$ is weakly Jordan, i.e., $\bar{\beta}(C)$ is bipartite for each circuit $C$ in $G$.

(IV) $G$ is Jordan, i.e., $\beta(C)$ is bipartite for each circuit $C$ in $G$.

(V) For no circuit $C$ of $G$ does $\beta(C)$ contain an odd circuit.

(VI) For no circuit $C$ of $G$ does $\beta(C)$ contain a loop or a triangle.

(VII) $G$ does not contain any subgraph which is isomorphic to $K_{3,3}$ or $K_5$ up to replacing suspended chains by single edges.

(II) corrects a statement found in [1]. At (III) the planarity of a graph becomes equivalent to a purely combinatorial condition. At (VI) it becomes apparent that the class of 1-irreducible graphs is finite.

## 2. Definitions and Notations

All graphs considered will be assumed to be finite, connected, and undirected. A graph $G$ will be denoted by $\langle V, E, I\rangle$ where $V$ is the set of vertices of $G$, $E$ is the set of edges of $G$, and $I \subseteq V \times E$ tells which edges and vertices are adjacent. For a graph $G$, $E(G)$ will denote the set of edges in $G$. If $S \subseteq E(G)$ for some graph $G$, then we denoted by $G \cdot S$ that subgraph of $G$ whose edges are the members of $S$ and whose vertices are the ends in $G$ of the members of $S$. A graph $G = \langle V, E, I\rangle$ is bipartite if and only if the vertex set decomposes into two disjoint sets $V'$ and $V''$ such that each edge connects a vertex of $V'$ with a vertex of $V''$. For a graph $G$ $\gamma(G)$, the genus of $G$, is the smallest of the numbers $\gamma(N)$ for orientable 2-manifolds $N$ in which $G$ can be embedded. A graph $G$ is $n$-irreducible if and only if $\gamma(G) = n$ and for any proper subgraph $G'$ of $G$ $\gamma(G') < n$. Notions for which definitions are assumed are defined in [2] or [3].

Definition 1. Let $G = \langle V, E, I\rangle$ be a graph. Let $H = \langle V', E', I'\rangle$ be a connected subgraph of $G$. A way $W = (v_0, e_0, v_1, \ldots, v_n, e_n, v_{n+1})$ in $G$ is *H-avoiding* if and only if

(1) $\{e_0, e_1, \ldots, e_n\} \cap E' = \varnothing$, and

(2) $\{v_1, \ldots, v_n\} \cap V' = \varnothing$.

Definition 2. Let $G = \langle V, E, I\rangle$ be a graph. Let $E' \subseteq E$ such that $G \cdot E'$ is connected. $\sim$ is defined for members of $E - E'$ as follows:

$(\forall e_1, e_2)_{E-E'}\, e_1 \sim e_2 . \equiv.$ There is a $G \cdot E'$ avoiding way in $G$ which contains both $e_1$ and $e_2$.

$\sim$ is an equivalence relation. The equivalence classes of $E - E'/\sim$ are called the *pseudo bridges* of $G \cdot E'$ in $G$. If $A \in E - E'/\sim$, then $G \cdot A$ is defined to be a *bridge of* $G \cdot E'$ *in* $G$. One sees that the bridges of $G \cdot E'$ in $G$ partition $G \cdot (E - E')$ into a set of edge disjoint, connected subgraphs.

To clarify the notion of a bridge Figure 1 shows a graph $G$, a connected subgraph $H$, and the bridges of $H$ in $G$.

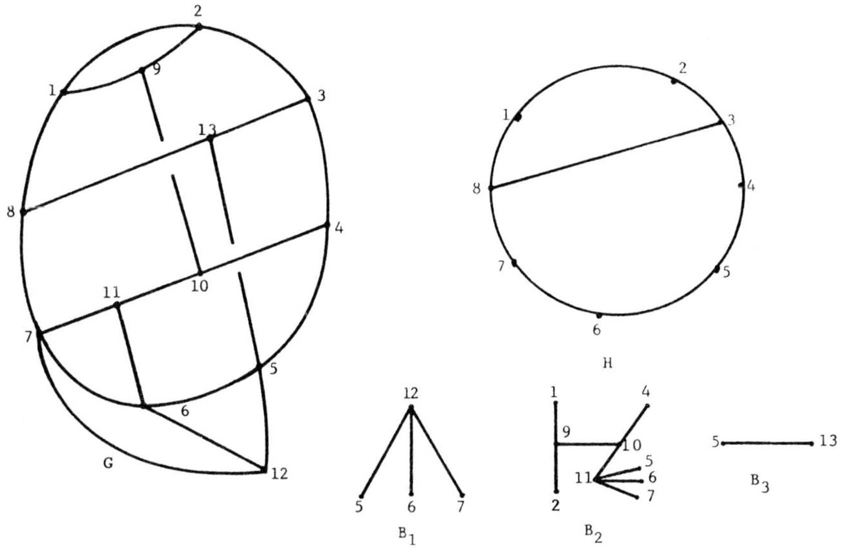


FIGURE 1

Let $G = \langle V, E, I\rangle$ be a graph. Let $C = \langle V', E', I'\rangle$ be a circuit of $G$. Let $B_1 = \langle V_1, E_1, I_1\rangle$ and $B_2 = \langle V_2, E_2, I_2\rangle$ be bridges of $C$ in $G$.

DEFINITION 3. The vertices of $V' \cap V_1$ are called the *vertices of attachment* of $B_1$ relative $C$. $\omega(B_1) = | V_1 \cap V' |$ is called the *attachment number* of $B_1$ relative $C$. $B_1$ and $B_2$ are called *equivalent n-bridges* if and only if (1) $\omega(B_1) = \omega(B_2) = n$ and (2) $V_1 \cap V' = V_2 \cap V'$. $B_1$ and $B_2$ are called *strongly equivalent* 3-*bridges* if and only if $B_1$ and $B_2$ are equivalent 3-bridges and there are connected subgraphs $P_1 = \langle V_{11}, E_{11}, I_{11}\rangle$ and $P_2 = \langle V_{22}, E_{22}, I_{22}\rangle$ such that

(1) $P_i \subseteq B_i$ for $i = 1, 2$ (read $P_i$ is a subgraph of $B_i$),

(2) $E_{11} \cap E_{22} = \varnothing$, and

(3) $V_{11} \cap V_{22} = V_1 \cap V'$.

The point of DEFINITION 3 is to allow a 3-bridge to be strongly equivalent to itself. Figure 2 gives an example of such an occurrence.

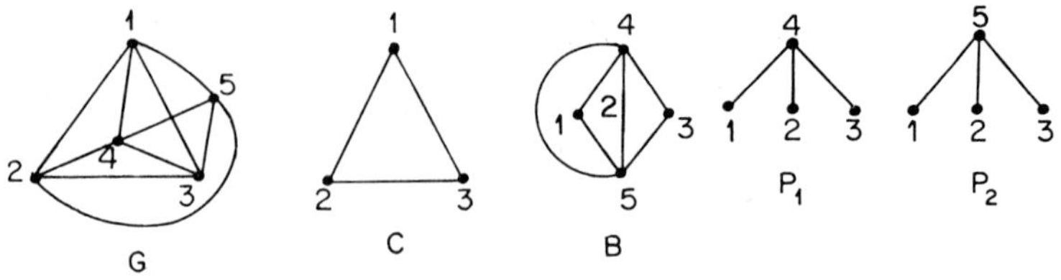


FIGURE 2

DEFINITION 4. Let $G$ be a graph. Let $C$ be a circuit of $G$. A pair of points $\langle a, b\rangle$ on $C$ *separate* a pair of points $\langle a_1, b_1\rangle$ on $C$ if and only if the four points $a, b, a_1, b_1$ are distinct and these points occur on $C$ in the cyclic order $a, a_1, b, b_1$. Let $W$ and $W_1$ be distinct ways of $G$ (i.e., no common vertices and no common edges) such that $W$ meets $C$ in exactly two vertices $a$ and $b$ and $W_1$ meets $C$ in exactly two vertices $a_1$ and $b_1$. Then $\langle W, W_1\rangle$ are said to separate each other on $C$ if and only if $\langle a, b\rangle$ separates $\langle a_1, b_1\rangle$.

DEFINITION 5. Let $G$ be a graph. Let $C$ be a circuit of $G$. Let $B_1$ and $B_2$ be bridges of $C$ in $G$. $B_1$ and $B_2$ *separate each other on* $C$ if and only if there is a way $P_i \subseteq B_i$ such that $P_i$ meets $C$ at its end-points $a_i$ and $b_i$ for $i = 1, 2$ and such that $\langle P_1, P_2\rangle$ separate each other on $C$.

It is possible for a bridge to separate itself as seen in Figure 3.

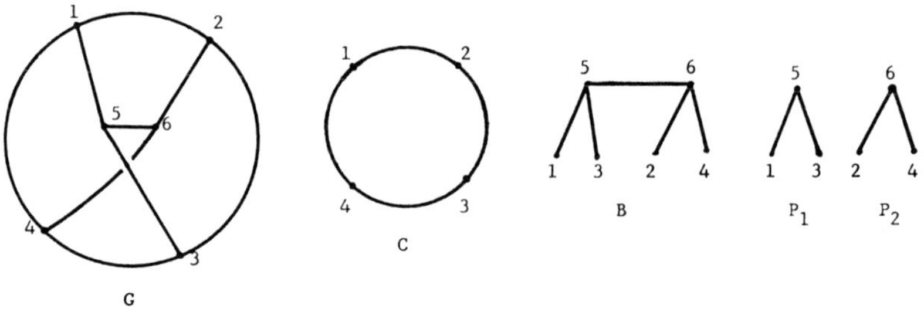


FIGURE 3

DEFINITION 6. Let $G$ be a graph. Let $C$ be circuit of $G$. Let $B_1$ and $B_2$ be bridges of $C$ in $G$. $B_1$ and $B_2$ *overlap on* $C$ if and only if (1) $B_1$ and $B_2$ separate each other on $C$, or (2) $B_1$ and $B_2$ are strongly equivalent 3-bridges.

The notion of overlap was found in [4] in the context of matroid theory.

DEFINITION 7. Let $S = (v_0, v_1, ..., v_n)$ be a sequence of vertices of a graph $G$. $S$ is called a *minimal vertex cycle* of $G$ if and only if $v_0, ..., v_n$ are the vertices, in cyclic order, of a circuit of $G$ and no proper subset of $\{v_0, ..., v_n\}$ with two or more elements is the set of vertices of a circuit of $G$. $S$ is said to have length $n + 1$.

DEFINITION 8. Let $G$ be a graph. Let $C$ be a circuit of $G$. The *weak bridge graph of $C$ in $G$*, denoted by $\bar{\beta}(C)$, is defined as follows: (1) the vertices of $\bar{\beta}(C)$ are the bridges of $C$ in $G$ and (2) there is an edge in $\bar{\beta}(C)$ joining vertices $B_1$ and $B_2$ if and only if $B_1$ and $B_2$ overlap on $C$.

DEFINITION 9. Let $G$ be a graph. Let $C$ be a circuit of $G$. $C$ is called *weakly Jordan in $G$* if and only if there is a division of the bridges of $C$ in $G$ into two sets $I$ and $J$ such that no pair of bridges in $I$ overlap on $C$ and no pair of bridges in $J$ overlap on $C$. $G$ is called *weakly Jordan* if and only if each circuit of $G$ is weakly Jordan in $G$.

*Remark* 1. Let $G$ be a graph. Let $C$ be any circuit of $G$. $C$ is *weakly Jordan* if and only if $\bar{\beta}(C)$ is bipartite. $G$ is *weakly Jordan* if and only if $\bar{\beta}(C)$ is bipartite for each circuit $C$ of $G$.

It is now possible to answer the question posed at the beginning of this paper.

LEMMA A. *Let $G$ be a planar graph. Let $C$ be any circuit of $G$. There is an embedding $\xi$ of $G$ on $T_2$ (2-sphere) such that $\xi(C)$ bounds a region of $\xi$, i.e. component of $T_2 - \xi(G)$, if and only if $\bar{\beta}(C)$ contains no edges, i.e., no two bridges of $C$ in $G$ overlap on $C$.*

*Proof.* ($\rightarrow$) Obvious by Jordan's Theorem.

($\leftarrow$) Let $C$ be any circuit in $G$ for which $\bar{\beta}(C)$ contains no edges. The conclusion follows from an induction on $k$, the number of vertices in $\bar{\beta}(C)$, i.e., the number of bridges of $C$ in $G$.

For $k = 0$ the result is obvious.

Suppose the theorem is true for all circuits of $G$ which satisfy the hypothesis and which have fewer than $k$ bridges in $G$.

Let $C$ be a circuit of $G$ for which $\bar{\beta}(C)$ has $k$ vertices and no edges. Let the brigdes of $C$ in $G$ be $B_1, ..., B_{k-1}, B_k$. By the inductive assumption there is a $\xi : (C \cup B_1, \cup \cdots \cup B_{k-1}) \rightarrow T_2$ such that $\xi(C)$ bounds a region of $\xi$. Since $\gamma(C \cup B_k) \leqslant \gamma(G) = 0$, it is possible to embed $B_k$ in the region $C$ bounds under an embedding $\xi'$.

*Case* 1. $\omega(B_k) = 1$. In this case it is obvious how to embed $B_k$ exterior to $\xi(C)$. Let $\xi''$ be the resulting embedding. $\xi''$ has the required property.

*Case* 2. $\omega(B_k) > 1$. $\xi$ is a quasi-disc embedding (the regions of $\xi$ are homeomorphic to the interior of a disc) since $G$ is connected and planar. Since $B_k$ does not overlap $B_1$ or $\cdots$ or $B_{k-1}$ some region of $T_2 - \xi(C \cup B_1 \cup \cdots \cup B_{k-1})$ besides $\text{int}(\xi(C))$ contains all the vertices of attachment of $B_k$ in its closure. Call this region $R$. Let $\xi'' : G \to T_2$ be defined such that $\xi'' |_{C \cup B_1 \cup \cdots \cup B_{k-1}} = \xi$ and $\xi''$ embeds $B_k$ in $R$. Then $\xi''$ has the required property.

DEFINITION 10. Let $G$ be a graph. Let $C$ be a circuit of $G$. The *strong bridge graph of* $C$ *in* $G$, denoted by $\beta(C)$, is defined as follows: (1) the vertices of $\beta(C)$ are the bridges of $C$ in $G$, and (2) there is an edge in $\beta(C)$ joining vertices $B_1$ and $B_2$ if and only if $B_1$ and $B_2$ separate each other on $C$.

DEFINITION 11. Let $G$ be a graph. Let $C$ be a circuit of $G$. $C$ is called *Jordan in* $G$ if and only if there is a division of the bridges of $C$ in $G$ into two sets $I$ and $J$ such that no pair of bridges in $I$ separate each other on $C$ and no pair of bridges in $J$ separate each other on $C$. $G$ is called *Jordan* if and only if each circuit of $G$ is Jordan in $G$.

*Remark* 2. Let $G$ be a graph. Let $C$ be any circuit of $G$. $C$ is Jordan in $G$ if and only if $\beta(C)$ is bipartite. $G$ is Jordan if and only if $\beta(C)$ is bipartite for each circuit $C$ of $G$.

## 3. PROOF OF THE EQUIVALENCE OF I, II, III, IV, V, VI, VII

$(II) \to (I)$. This will be proved by induction.

Let $G$ be a graph satisfying II(a) and containing a circuit with exactly one bridge. $G$ is planar by II(a).

Now suppose that any graph $G$ satisfying II(a) and II(b) and containing a circuit with fewer than $k (k > 1)$ bridges is planar.

Let $G$ be a graph satisfying II(a) and II(b) and containing no circuit with fewer than $k$ bridges. Let $C$ be a circuit of $G$ with $k$ bridges.

Since II(b) holds, there is a division of the $k$ bridges of $C$ in $G$ into two sets $I$ and $J$ such that no pair of bridges in $I$ overlap on $C$ and no pair of bridges in $J$ overlap on $C$. Without loss of generality it can be assumed $0 < |I|, |J| < k$. For suppose $|I| = 0$. Let $B$ be any element of $J$. Let $I' = \{B\}$ and $J' = J - \{B\}$. $I'$ and $J'$ form a division of the bridges of $C$ in $G$ with the required property.

Since $0 < |I| < k$ by the inductive hypothesis $\gamma(C \cup (\bigcup_{B \in I} B)) = 0$. Further, since no pair of bridges of $I$ overlap relative $C$, by LEMMA A there is an embedding $\xi_1$ of $(C \cup (\bigcup_{B \in I} B))$ on $T_2$ such that $\xi_1(C)$ bounds a region of $\xi_1$, i.e., a component of $T_2 - \xi_1(G)$. Similarly, there is an embedding $\xi_2$ of $(C \cup (\bigcup_{B \in I} B))$ on $T_2$ such that $\xi_2(C)$ bounds a region of $\xi_2$. It is now obvious how these two embeddings may be combined to form an embedding of $G$ on $T_2$.

*(III)* → *(II)*. Let $G$ be weakly Jordan. The conclusion follows from an induction on $|E|$, the number of edges in $G$.

If $|E| = 1$ the result is obvious.

Suppose the conclusion holds for all weakly Jordan graphs with $|E| < k$.

Let $G$ be weakly Jordan with $|E| = k$. Let $C$ be a circuit of $G$ and let $\mathscr{B}(C) = \{B \mid B \text{ is a bridge of } C \text{ in } G\}$. If $|\mathscr{B}(C)| = 0$, the result is obvious.

*Case* 1. $|\mathscr{B}(C)| = 1$ for all circuits $C$ of $G$. By Theorem 3 of [1] either the bridge is a tree for some circuit or the bridge meets each circuit of $G$ exactly one point. Moreover, by Theorem 3 $G$ is planar.

*Case* 2. $|\mathscr{B}(C)| \geqslant 2$ for some circuit $C$ of $G$. For each $B \in \mathscr{B}(C)$, $(B \cup C)$ has fewer than $k$ edges and is a weakly Jordan graph. Therefore, by the inductive assumption $\gamma(B \cup C) = 0$. Since $G$ is weakly Jordan, $C$ is weakly Jordan. Therefore, the conclusion follows.

*(IV)* → *(III)*. Let $G$ be Jordan but not weakly Jordan. Let $C$ be a circuit of $G$ which is Jordan but not weakly Jordan. Let $I$ and $J$ form a division of the bridges of $C$ in $G$ such that no pair of bridges of $I$ separate each other on $C$ and no pair of bridges of $J$ separate each other on $C$.

Since the obvious modification of $I$ and $J$ to separate pairs of distinct strongly equivalent 3-bridges does not result in $C$ becoming weakly Jordan, then:

i) $C$ has a 3-bridge which is strongly equivalent to itself,

ii) $C$ has 3 distinct strongly equivalent 3-bridges, or

iii) $C$ has a third bridge which separates on $C$ each of a pair strongly of equivalent 3-bridges.

In any case, Figure 4 shows how to find a circuit $C'$ in $G$ which is not Jordan. It then follows that $G$ is not Jordan, which is a contradiction.

*(V)* → *(IV)*. This follows from the following well-known result: A graph $G$ is bipartite if and only if every circuit of $G$ consists of an even number of edges.

*(VI)* → *(V)*. This implication follows from a series of lemmas that will now be proved.

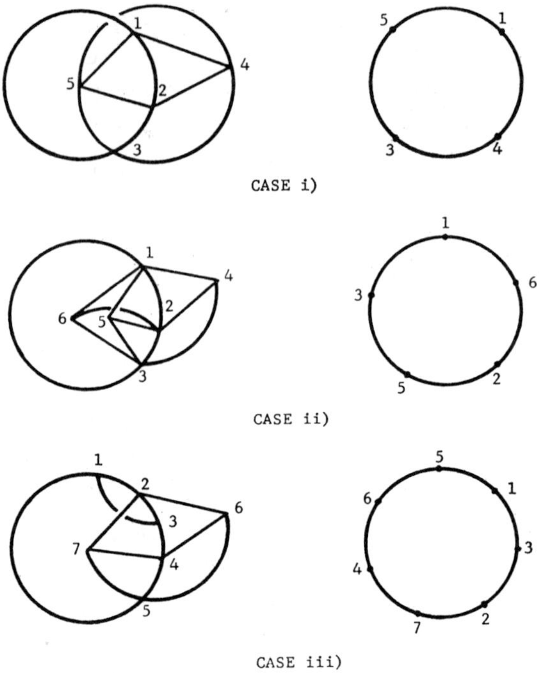


FIGURE 4

LEMMA 1. *Let G be a graph. Let C be a circuit in G. Let X, Y, and Z be bridges of C in G. Suppose X and Z do not separate on C. Let* $W_1$, $W_2$, $A$, $W_3$, $W_4$ *be ways in G such that* $W_1 \subseteq X$; $W_2, A, W_3 \subseteq Y$; $W_4 \subseteq Z$. *Suppose A connects* $W_2$ *and* $W_3$ *and that* $\langle W_1, W_2 \rangle$ *and* $\langle W_3, W_4 \rangle$ *separate each other on C. Then there is a way B in Y such that* $B \subseteq W_2 \cup A \cup W_3$ *and such that* $\langle W_1, B \rangle$ *and* $\langle B, W_3 \rangle$ *separate each other on C.*

*Proof.* Since $X$ and $Z$ do not separate each other on $C$, $W_1$ and $W_4$ and $C$ can occur only in one of the four configurations appearing in FIGURE 5. In each case, it is clear how to construct a suitable way $B$.

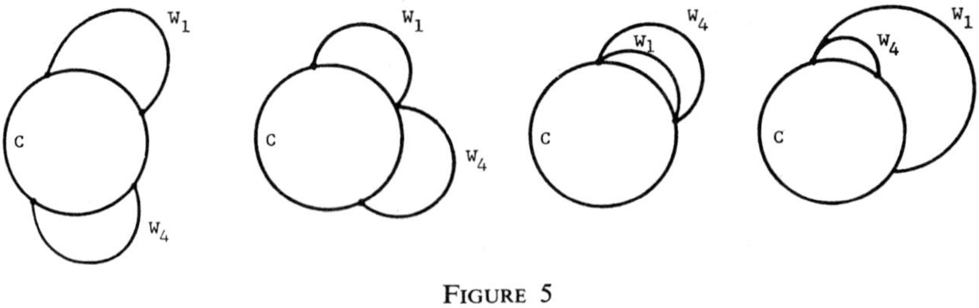


FIGURE 5

LEMMA 2. *Let $G$ be a graph. Let $C$ be a circuit of $G$. Let $(B_0, \dots, B_n)$ be a minimal vertex cycle of $\beta(C)$ with $n \geqslant 3$. For each $i$, $0 \leqslant i \leqslant n$, there is a way $W_i \subseteq B_i$ such that, for $0 \leqslant i < n$, $\langle W_i, W_{i+1}\rangle$ separate each other on $C$ and $\langle W_n, W_0\rangle$ separate each other on $C$.*

*Proof.* By $n$ applications of Lemma 1, we clearly can obtain ways $W_0' \subseteq B_0, \dots, W_n \subseteq B_n$, $W_0'' \subseteq B_0$ such that $\langle W_0', W_1\rangle, \dots, \langle W_{n-1}, W_n\rangle$, and $\langle W_n, W_0''\rangle$ all separate each other on $C$. A further application of Lemma 1 will replace $W_0'$ and $W_0''$ by a single way $W_0 \subseteq B_0$ such that $\langle W_n, W_0\rangle$ and $\langle W_0, W_1\rangle$ separate each other on $C$.

LEMMA 3. *Let $G$ be a graph. Let $C$ be a circuit of $G$. Let $(B_0, \dots, B_n)$ be a minimal vertex cycle of $\beta(C)$ with $n \geqslant 4$. There is a subgraph $G'$ of $G$ containing a circuit $C'$ whose bridge graph relative to $G'$ has a minimal vertex cycle of length $n - 2$. (In fact $G' = C \cup B_0 \cup \cdots \cup B_n$ will do.)*

*Proof.* Apply LEMMA 2 to $(B_0, \dots, B_n)$ and $C$ to obtain $W_0, W_1, \dots, W_n$ such that $\langle W_i, W_{i+1}\rangle$ separate each other on $C$ for $0 \leqslant i \leqslant n$ (subscripts are to be regarded as modulo $n$, of course). Let $G' = C \cup W_0 \cup \cdots \cup W_n$. From here on we work on $G'$. Take any $j$, $0 \leqslant j \leqslant n$. Let $x_j$ and $y_j$ be the vertices of attachment of $W_j$ relative $C$. The pair of vertices $x_j$ and $y_j$ naturally divide $C$ into two parts $C_1$ and $C_2$ (cf. Figure 6).

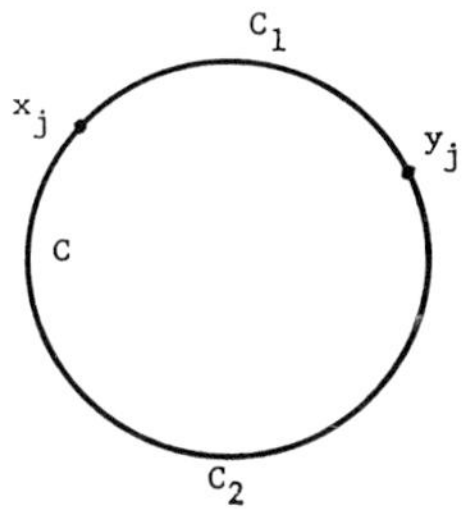


FIGURE 6

Since $W_j$ separates both $W_{j+1}$ and $W_{j-1}$ on $C$ and no other bridge of $C$, one of $C_1$ and $C_2$, say $C_1$, contains only vertices of attachment of $W_{j-1}$ and $W_{j+1}$. Let $C' = W_j \cup C_2$ and $A = W_{j-1} \cup W_{j+1} \cup C_1$. $(W_0, W_1, \dots, W_{j-2}, A, W_{j+2}, \dots, W_n)$ is then a minimal vertex cycle of $\beta(C')$ relative $G'$.

LEMMA 4. *Let $G$ be a graph. Let $C$ be a circuit of $G$. If $C$ consists of an odd number of edges, then $G$ has a minimal circuit $C'$ consisting of an odd number of edges whose vertices are among those of $C$.*

*Proof.* Obvious.

*Proof of* (*VI*) → (*V*). Let $C$ be a circuit of a graph $G$ such that $\beta(C)$ has a circuit $S$ of odd length. If $S$ is of length 1 or 3, we are finished. Otherwise, let $(B_0, ..., B_n)$ be the vertices of $S$ in their order of occurrence in $S$. By LEMMA 4 since the length of $S$ is odd, there is an odd length minimal vertex cycle of $\beta(C)$, say $(A_0, ..., A_m)$, where $\{A_0, ..., A_m\} \subseteq \{B_0, ..., B_n\}$. If the length $m + 1$ is three, we are finished. Otherwise, starting with the subgraph $C \cup A_0 \cup \cdots \cup A_m$ and applying LEMMA 3, we obtain $G'$ and $C'$ of the conclusion of LEMMA 3. If necessary iterate the application of LEMMA 4 to $G'$ and $C'$ to arrive at a subgraph $H$ of $G$ with a circuit $D$ where $D$ has three bridges $E_1$, $E_2$, and $E_3$ in $H$ and such that $\langle E_1, E_2\rangle$, $\langle E_2, E_3\rangle$, and $\langle E_3, E_1\rangle$ all separate each other on $D$.

$D$ is clearly a circuit of $G$ and $E_1$, $E_2$, $E_3$ are, respectively, subgraphs of bridges $F_1$, $F_2$, $F_3$ of $D$ in $G$. If all of the $F$'s are disjoint, then $\beta(D)$ contains a triangle. Otherwise $\beta(D)$ contains a loop.

(*VII*) → (*VI*). Let $G$ be a graph. Let $C$ be a circuit of $G$ for which $\beta(C)$ contains a loop or a triangle. The proof will be completed if we can show $G$ must contain a subgraph which is isomorphic to either $K_{3,3}$ or $K_5$, up to replacing suspended chains by single edges.

*Case* 1. $\beta(C)$ contains a loop. In this case $G$ must have a subgraph isomorphic (to within replacing suspended chains by single arcs) to the graph pictured in Figure 7. This graph is seen to be $K_{3,3}$.

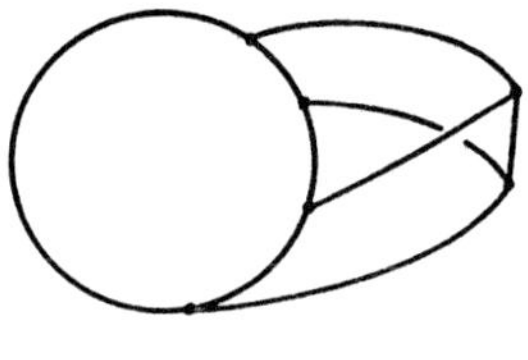

FIGURE 7

*Case* 2. $\beta(C)$ contains a triangle. Let $C = [r_1, ..., s_1, ..., r_2, ..., s_2, ...]$. Let $R, S, T$ be the bridges of $C$ in $G$ which form a triangle in $\beta(C)$. Suppose $R$ meets $C$ at $r_1$ and $r_2$ and $S$ meets $C$ at $s_1$ and $s_2$ such that $r_1$ and $r_2$ separate $s_1$ and $s_2$ on $C$. The proof will be completed by analyzing where $T$ can meet $C$ and still have the property that $T$ separates both $R$ and $S$. One notes that $T$ must meet $C$ in at least two points.

*Case* 2a. $\omega(T) = 2$ and $T$ meets $C$ at $t_1$ and $t_2$. $T$ cannot separate both $R$ and $S$ if either $t_1$ or $t_2$ is contained in $\{r_1, r_2, s_1, s_2\}$. Therefore, some $t_i$, say $t_1$, is either (a) between $r_1$ and $s_1$ or (b) between $s_1$ and $r_2$. In case (a), $t_2$ must be between $r_2$ and $s_2$. In case (b), $t_2$ must be between $s_2$ and $r_1$. In either case, $G$ contains $K_{3,3}$ as a subgraph as seen in Figure 8.

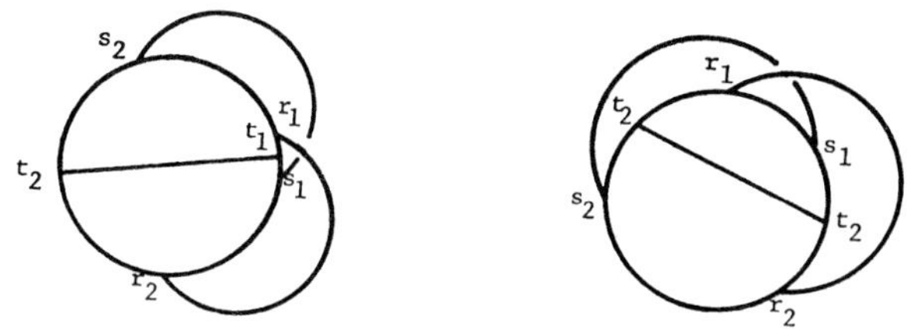


FIGURE 8

*Case* 2b. $\omega(T) = 3$ and $T$ meets $C$ at $t_1$, $t_2$, and $t_3$. Onc notes that

$$|L| = |\{r_1, s_1, r_2, s_2\} \cap \{t_1, t_2, t_3\}| \leqslant 2.$$

By analysis of how the vertices $t_1$, $t_2$, and $t_3$ can be distributed around $C$ such that $|L| = 2$ and such that $T$ separates both $R$ and $S$, it is seen that only the graph pictured in Figure 9 has the required properties (up to obvious permutations of $r_1$, $r_2$, $s_1$, $s_2$, $t_1$, $t_2$, $t_3$). Again one sees that $G$ contains $K_{3,3}$ as a subgraph.

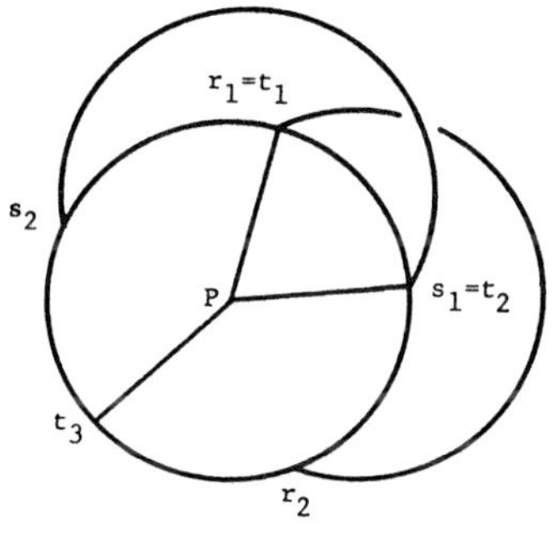


FIGURE 9

If $|L| = 1$ then $R$, $S$, $T$ must occur in $G$ as pictured in Figure 10. Again $G$ is seen to contain $K_{3,3}$ as a subgraph.

If $|L| = 0$ then Case 2a applies.

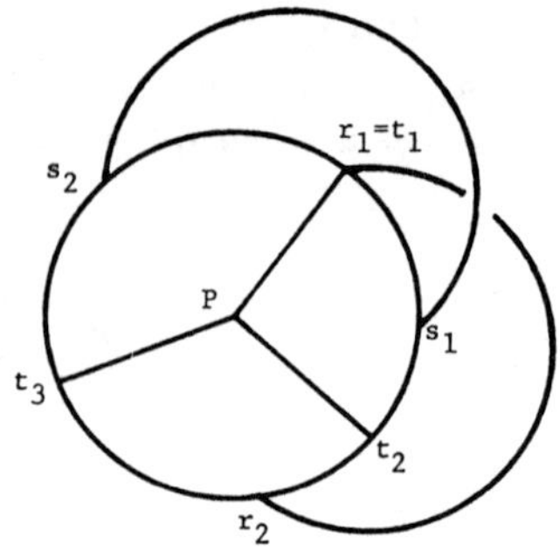


FIGURE 10

*Case* 2c. $\omega(T) = 4$ and $T$ meets $C$ at $t_1$, $t_2$, $t_3$, and $t_4$. One sees that unless $\{r_1, r_2, s_1, s_2\} = \{t_1, t_2, t_3, t_4\}$ one of the previous cases applies. When $\{r_1, r_2, s_1, s_2\} = \{t_1, t_2, t_3, t_4\}$ it is seen that $G$ contains a subgraph of the form (a) or (b), pictured in Figure 11. In case (a) it is seen that $G$ contains $K_{3,3}$ as a subgraph. In case (b) it is seen that $G$ contains $K_5$ as a subgraph.

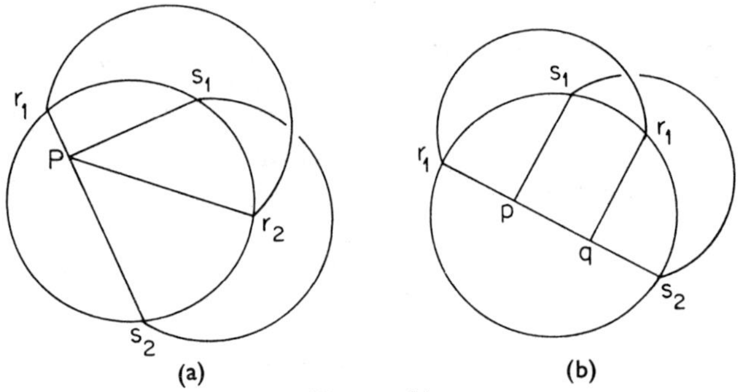


FIGURE 11

*Case* 2d. If $\omega(T) > 4$, then one of the previous cases applies.

$(I) \to (VII)$. This follows from the Euler formula.

COROLLARY TO THE PROOF. *Let G be a cubic graph.* $\gamma(G) > 0$ *if and only if G contains* $K_{3,3}$ *as a subgraph up to replacing suspended chains by single edges.*

## 4. CONCLUSION

It is hoped that this new proof of the Kuratowski Theorem may contain ideas which generalize to surfaces of higher genus. As there are many

2-irreducible graphs, nobody will want to list all of them. Now, in case of a surface of genus $n \geqslant 1$, and adaptation of our proof could stop at stage (VI), yielding the finiteness of the class of $n$-irreducible graphs without actually listing them all, which (VII) does. Such a proof would certainly eliminate much combinatorial horror—the question is: how much is left?

## References

1. L. Auslander and S. V. Parter, On Imbedding Graphs in the Sphere, *J. Math. Mech.* **10** (1961), 517–524.
2. O. Ore, *The Four Color Problem*, Academic Press, New York, 1967.
3. G. Ringel, *Färbungsprobleme auf Flächen und Graphen*, VEB Deutscher Verlag der Wissenschaften, Berlin, 1959.
4. W. T Tutte, Matroids and Graphs, *Trans. Amer. Math. Soc.* **90** (1959), 527–552.

Printed in Belgium by the St. Catherine Press Ltd., Tempelhof 37, Bruges

# Skolem Rings and Their Varieties

J.R. Büchi and T.M. Owens
Purdue University, Lafayette, Indiana 47906 CSD TR 140

## INTRODUCTION

Skolem semi-lattices $\langle A,\vee,o\rangle$ may be defined as those which admit a binary operation $+$ (the symmetric pseudo-complement), whereby $x+y$ is the smallest element $u$, $x \vee u = y \vee u$. Define $x{*}y = (x \vee y) + y$, and $x \cdot y = (x{*}y) \vee (y{*}x)$. One can recover $\vee$ from $+$, $\cdot$ by: $x \vee y = (x+y)+(x \cdot y)$. So the structures $\langle A,+,\vee,o\rangle$ and $\langle A,+,\cdot,o\rangle$ are equivalent up to equational translation. We call these structures Skolem rings.

The lattice of closed sets of a topological space and the algebra of formulas of intuitionistic propositional calculus provide the best known examples of Skolem lattices $\mathcal{A} = \langle A,+,\vee,\wedge,o\rangle$ (see Curry [4], Rasiowa [14]). More natural still is the notion of a Skolem semi-lattice. We add a further reason for studying these structures; Theorem 1.1. The semi-lattice $\mathcal{A} = \langle A,\vee,o\rangle$ admits a nice representation $\mathrm{Id}\ell\mathcal{A} \cong \mathrm{Cgr}\,\mathcal{B}$ ($\mathrm{Id}\ell\,\mathcal{A}$ is the ideal lattice of $\mathcal{A}$, $\mathrm{Cgr}\,\mathcal{B}$ is the congruence lattice of an algebra $\mathcal{B}$), just in case $\mathcal{A}$ is Skolem. In this case the representation $U \to \equiv(U)$ is given by $x \equiv y(U)$ $.\equiv.$ $x+y \in U$, and the representing algebra $\mathcal{B}$ may be taken as the Skolem ring $\langle A,+,\vee,o\rangle$. Here "nice" means that 1) the class of 0 modulo $\equiv(U)$ is just the ideal $U$, and 2) $\equiv(U)$ is a congruence of $\vee$.

The representation problem will be discussed in section 1. As a bonus to this work an equational axiom system for Skolem rings is developed and Theorem 1.2: (Monteiro [ 9 ]) the class of all Skolem rings forms a variety. The laws for semi-lattices plus and the laws $x \leq x + 0$, $x + y \leq (x+z) \vee (y+z)$, and $(x \vee y) + y \leq x$ are an equational axiom system.

In section 3 we will clarify somewhat the relationship of Skolem rings to Topology. The semi-lattice of closed sets of a topological space is easily seen to be Skolem, here the $+$ operation is obtained by defining $A + B$ as the closure of the symmetric set difference $(A-B) \cup (B-A)$. We will prove in theorem 3.1 that every Skolem ring may be embedded as a subalgebra of the Skolem ring of closed sets of a topological space. The notion of a regular closed set has a natural analogue in Skolem rings, that of a regular element. We have the following counterpart to the Topological result; theorem 3.2: The regular elements of a Skolem ring $\mathcal{A}$ form a subalgebra of $\mathcal{A}$ and a Boolean ring.

The remainder of this paper is concerned with the subvariety lattice, or dually the lattice of equational extensions of Skolem rings. Theorem 4.1 asserts that those Skolem rings in which the $\cdot$ operation is a meet may be characterized as subdirect products of chains. This class of Skolem rings then forms a subvariety which Whalen and Nemitz [13 ] have shown is characterized by the associative law for $\cdot$, $x\cdot(y\cdot z) = (x\cdot y)\cdot z$. A characterization of the subdirectly irreducible Skolem rings developed in section 2 is used in theorem 5.1 to show that every finitely generated Skolem ring is finite. This result is in contrast to the situation for Skolem lattices. We conclude by showing that the Variety lattice of Skolem rings is completely distributive (theorem 5.2) and has the power of the continuum (theorem 5.3).

1.1. Discrete Systems

The closely related notions of a closure space, and a complete lattice are found throughout mathematics. The two spaces which most frequently arise are characterized by their $\bigvee$-bases. Let S be a closure space (complete lattice), call $a \in S$ strongly-$\vee$-irreducible if whenever $a \in cl\{b_1,\ldots,b_n\}$ $(a \leq b_1 \vee \ldots \vee b_n)$, there is $i \leq n$ and $a \in cl\{b_i\}(a \leq b_i)$; a is called strongly-directed-$\bigvee$-irreducible (or compact) if whenever $a \in clB$, $B \subseteq S$ $(a \leq \bigvee B)$, there are $b_1,\ldots,b_n \in B$ and $a \in cl\{b_1,\ldots,b_n\}$ $(a \leq b_1 \vee \ldots \vee b_n)$. Closure spaces and complete lattices which have $\bigvee$-bases of strongly-$\vee$-irreducible elements usually occur in continuous mathematics and are called topological. Those which have $\bigvee$-bases of strongly-directed-$\vee$-irreducible (compact) elements are most often met with in finitary mathematics, Logic, Geometry, Algebra, etc., and are called discrete.

As examples of the discrete complete lattices we mention Cgr$\mathcal{B}$ (the congruence lattice of the algebra $\mathcal{B}$), Sba$\mathcal{B}$ (the subalgebra lattice of $\mathcal{B}$), and Idl$\mathcal{A}$ (the ideal lattice of a semi-lattice $\mathcal{A} = \langle A,\vee,0\rangle$). Of course these systems may equally well be described by closure spaces. In each case the set of compact elements, $Cgr_0\mathcal{B}$, $Sba_0\mathcal{B}$, $Idl_0\mathcal{A}$, may be described as the finitely generated congruences, subalgebras, and ideals respectively.

These examples suggest a number of ways of representing a discrete complete lattice L, namely:

a) As all closed sets of a discrete space $\mathcal{S} = \langle S, \mathrm{Cld} \rangle$: Let $S = L_0$ = all compact elements of L, for $X \subseteq S$ define,

$$\mathrm{cl}\, X = \{u \in S;\ u \leq \bigvee X\} = \bigcup \{\mathrm{cl}\{x_1, \ldots, x_n\};\ x_1, \ldots, x_n \in X\}$$

b) As all ideals of a semi-lattice $\mathcal{A} = \langle A, \vee, 0 \rangle$: (Nachbin [12]). Let $A = L_0$, let $\vee$ be the join in L, 0 be the least element in L. Then $L \cong \mathrm{Idl}\, \mathcal{A}$ where $\mathcal{A} = \langle A, \vee, 0 \rangle$.

c) As all subalgebras of an algebra $\mathcal{B}$: (Büchi [ 3 ]). By Nachbin [12] we may assume $L = \mathrm{Idl}\, \mathcal{A}$, $\mathcal{A} = \langle A, \vee, 0 \rangle$ a semi-lattice. For $a \in A$ define

$$f_a(x) = \begin{cases} x & \text{if } a \nleq x \\ a & \text{if } a \leq x \end{cases}$$

Let $\mathcal{B} = \langle A, \vee, 0, \ldots, f_a, \ldots \rangle_{a \in A}$. Then $L \cong \mathrm{Sba}\, \mathcal{B}$.

d) As all congruences of an algebra $\mathcal{B}$: (Grätzer-Schmidt [ 6 ]). This remarkable result involves a rather intricate construction. We will only mention several interesting features of this construction here. Starting from a semi-lattice $\mathcal{A} = \langle A, \vee, 0\rangle$ an isomorphism $u \to \equiv(u)$ is constructed which takes $\mathcal{A}$ onto $Cgr_0\mathcal{B}$. The algebra $\mathcal{B}$ is constructed in such a way that $Cgr_0\mathcal{B} = Pgr\mathcal{B}$, i.e. finitely generated congruences are principal, and also every principal congruence $Cg_{\mathcal{B}}\{xy\}$ (the least congruence identifying $x$ with $y$) is of the form $Cg_{\mathcal{B}}\{z0\}$ for a fixed element $0 \in B$.

Following from this construction we define a representation $Idl\,\mathcal{A} \cong Cgr\,\mathcal{B}$ to be <u>nice</u> if it has the above features of the Grätzer-Schmidt construction: that is if $u \to \equiv(u)$ is a map of $\mathcal{A}$ onto $Cgr\,\mathcal{B}$ satisfying,

1) $u \leq v \;.\!\equiv\!.\; \equiv(u) \subseteq \equiv(v)$ for $u, v \in A$

2) $(\forall u)_A(\exists x)_B \quad \equiv(u) = Cg\{x0\}$ for fixed $0 \in B$

3) $(\forall x_1, y_1, x_2, y_2)_B(\exists u)_A \quad Cg\{x_1y_1, x_2y_2\} = \equiv(u)$ .

It is immediate from 3 that nice representations satisfy,

3a) $(\forall x)_B \; (\exists u)_A \quad \equiv(u) = Cg\{x0\}$

3b) $(\forall xy)_B(\exists z)_B \quad Cg\{xy\} = Cg\{z0\}$

3c) $(\forall xy)_B(\exists z)_B \quad Cg\{x0, y0\} = Cg\{z0\}$ .

Notice that 3a) together with 2) give an onto map $x \to \bar{x}$ of $B$ onto $A$ by the rule,

$$Cg\{x0\} = \equiv(\bar{x}).$$

Furthermore 3b) and 3c) show that operations + and v may be defined on B (not necessarily in a unique way) so that,

$$Cg\{xy\} = Cg\{(x+y)0\}$$

and

$$Cg\{x0,y0\} = Cg\{(x \vee y)0\}.$$

Suppose now that $u \to \equiv(u)$ is a nice representation of $\mathcal{A}$ onto Cgr $\mathcal{B}$. Let $u = \bar{y}$ for $y \in B$, then

$$\begin{aligned} x \equiv 0(u) \ &.\equiv.\ x \equiv 0\ (Cg\{y0\}) \\ &.\equiv.\ Cg\{x0\} \subseteq Cg\{y0\} \\ &.\equiv.\ \equiv(\bar{x}) \subseteq \equiv(\bar{y}) \\ &.\equiv.\ \bar{x} \le \bar{y} = u \end{aligned}$$

and thus for the congruence class of 0 modulo u, $0^{\equiv(u)}$, we have

4) $$0^{\equiv(u)} = \{x;\ \bar{x} \in id\{u\}\}$$

where id{u} denotes the least ideal in $\mathcal{A}$ containing u. As the map $x \to \bar{x}$ is onto,

5) $$0^{\equiv(u)} = 0^{\equiv(v)} \;.\supset.\; (\bar{x} \le u \;.\equiv.\; \bar{x} \le v)$$

$$.\supset.\; u = v \quad .$$

Also,

$$\begin{aligned} \equiv(x \vee y)^{-} &= Cg\{(x \vee y)0\} \\ &= Cg\{x0\} \vee Cg\{y0\} \\ &= \equiv(\bar{x}) \vee \equiv(\bar{y}) \\ &= \equiv(\bar{x} \vee \bar{y}) \quad \text{by } 1 \end{aligned}$$

and thus,

6) $$(x \vee y)^{-} = \bar{x} \vee \bar{y} \quad .$$

Finally,

$$\begin{aligned} x \equiv y(u) \;&.\equiv.\; Cg\{xy\} \subseteq \equiv(u) \\ &.\equiv.\; x + y \equiv 0 \;(u) \end{aligned}$$

and thus by 4).

7) $$x \equiv y(u) \;.\equiv.\; (x+y)^{-} \le u.$$

Formulas 4) and 5) shows that nice representations are of the kind usually associated with ring theory, i.e., $id\{u\} \to 0^{\equiv(u)}$. 6) indicates the parallels between the v operator on B and the $\vee$ on $\mathcal{A}$, while 7) states that $\equiv(u)$ is uniquely determined by the + and -.

1.2. Fission-free representation, Skolem semi-lattices.

Let $u \to \equiv(u)$ be a nice representation of $\mathcal{A} = \langle \Lambda, \vee, 0 \rangle$ as $Cgr_0 \mathcal{B}$, $\mathcal{B} = \langle B, F \rangle$. Define:

$$x \sim y \;.\!\equiv\!.\; Cg\{x0\} = Cg\{y0\}$$

$$.\!\equiv\!.\; \equiv(\bar{x}) = \equiv(\bar{y})$$

$$.\!\equiv\!.\; \bar{x} = \bar{y} \quad .$$

If $\sim$ is not the equality on B, we say that the representation uses <u>fission</u>.

If no fission is used in the representations, we have the fission-free condition,

$$Cg\{x0\} = Cg\{y0\} \;.\!\supset\!.\; x = y$$

that is,

$$\bar{x} = \bar{y} \quad .\!\supset\!.\; x = y.$$

The representation given by the Grätzer-Schmidt construction, being very general, will not usually be the best representation of $\mathcal{A}$. In particular, their construction always involves infinite fission. That this can be improved can be seen in,

<u>The Boolean Case</u>: Let $\mathcal{B} = \langle A,+,\cdot,0\rangle$ be a Boolean ring, let $\vee$ be the join in $\mathcal{B}$. We call $\mathcal{A} = \langle A,\vee,0\rangle$ a Boolean semi-lattice, and it is well-known that $\mathrm{Idl}\,\mathcal{A} = \mathrm{Cgr}\,\mathcal{B} = \mathrm{Cgr}\langle A,\vee,+\rangle$. Furthermore this representation is given by

$$x \equiv y(u) \;.\!\equiv\!.\; x + y \leq u$$

is nice, and without fission.

There are many features of the Boolean case that are common to all nice, fission-free representations. Namely, the $\bar{\ }$ operation on B is uniquely defined, and hence so are the + and v, moreover the map $x \to \bar{x}$ is one to one from B onto A and we may therefore assume A = B, $x = \bar{x}$. So, for the nice fission-free representation we have,

1. $u \leq v \;.\!\equiv\!.\; \equiv(u) \leq \equiv(v)$, hence congruences on $\mathcal{B}$ preserve the $\vee$ on $\mathcal{A}$, ie., $\mathrm{Cgr}\langle A,F\rangle = \mathrm{Cgr}\langle A,\vee,F\rangle$.

2. $\mathrm{Cg}\{x0\} = \equiv(x)$.

3. $\mathrm{Cg}\{xy\} = \mathrm{Cg}\{(x+y)0\}$, hence + is uniquely defined.

4. $Cg\{x0,y0\} = Cg\{(x\vee y)0\}$, hence $v$ is uniquely defined, moreover,

$$Cg\{x0,y0\} = Cg\{x0\} \cup Cg\{y0\}$$
$$= \equiv(x) \cup \equiv(y) \qquad \text{by 2}$$
$$= \equiv(x\vee y) \qquad \text{by 1}$$
$$= Cg\{(x\vee y)^{-}\} \text{ and } v = \vee \text{ by 2}$$

5. $x \equiv 0(u) .\equiv. x \leq u$, ie., $0^{\equiv(u)} = id\{u\}$

6. $0^{\equiv(u)} = 0^{\equiv(v)} .\supset. u = v$

7. $x \equiv y(u) .\equiv. x + y \leq u$, hence $\equiv(u)$ is completely determined by $+$.

Using these properties, we may prove the following laws for the $+$ and $\vee$,

i) $x \leq x + 0$ this follows by 7,5

ii) $x + y \leq (x + z) \vee (y + z)$ as $Cg\{xy\} \subseteq Cg\{xz,yz\}$

iii) $(x\vee y) + y \leq x$ as $Cg\{(x\vee y)y\} \subseteq Cg\{x0\}$.

Definition 1.1: $\mathcal{A} = \langle A,\vee,0\rangle$ is a <u>Skolem semi-lattice</u> if $\mathcal{A}$ admits a <u>symmetric pseudo-complement</u> $+$. That is if there is a binary operation $+$ on $\mathcal{A}$ such that $x + y =$ smallest $u$, $x \vee u = y \vee u$.

Lemma 1.1: Let $\mathcal{A} = \langle A, \vee, 0\rangle$ be a semi-lattice, and let $+$ be an operation on $\mathcal{A}$ satisfying i), ii), iii) above, then for all $x, y, u \in A$,

1. $x + x = 0$
2. $x + y = y + x$
3. $x + 0 = x$
4. $x \vee u = y \vee u \;.\equiv.\; x + y \leq u$

Proof:

1. $x + x = (x \vee 0) + x \leq 0$ by ii)

2. $x + y \leq (x + x) \vee (y + x) = (y + x)$ by 1.

The result follows by symmetry.

3. $x + 0 = (x \vee 0) + 0 \leq x$ by iii), and $x + 0 = x$ by i).

4. $x + 0 \leq (x + y) \vee (y + 0)$ by ii) and $x \leq (x + y) \vee y$ by 3). Hence $x \vee (x + y) \leq y \vee (x + y)$ and $x \vee (x + y) = y \vee (x + y)$ by symmetry. Conversely suppose $x \vee u = y \vee u$, then let $z = (x \vee u) = (y \vee u)$ in ii) to get,
$x + y \leq (x + (x \vee u)) \vee (y + (y \vee u)) = ((x \vee u) + x) \vee ((y \vee u) + y)$ by 2, and using iii) we get $x + y \leq u$. Q.E.D.

By Lemma 1.1, a semi-lattice $\mathcal{A}$ has a nice, fission-free representation as $\mathrm{Cgr}_0\mathcal{B}$ only if it is Skolem. In fact every Skolem semi-lattice has such a representation as we see in the following,

Lemma 1.2: Every Skolem semi-lattice $\mathcal{A}$ admits a nice representation without fission. This representation is given by $x \equiv y\ (u)\ .\equiv.\ x + y \leq u$ and $\mathcal{B} = \langle A, \vee, + \rangle$ will do.

Proof: Consider a Skolem semi-lattice $\mathcal{A}$ with symmetric pseudo-complement $+$, let $\mathcal{B} = \langle A, \vee, + \rangle$. Then

$$x \equiv y\ .\supset.\ x + x \equiv x + y$$

$$.\supset.\ 0 \equiv x + y$$

and

$$x + y \equiv 0\ .\supset.\ x \vee (x + y) \equiv x\ \&\ y \vee (x + y) \equiv y$$

$$.\supset.\ x \vee y \equiv x\ \&\ x \vee y \equiv y$$

$$.\supset.\ x \equiv y\ .$$

Thus congruences on $\mathcal{B}$ are completely determined by the congruence classes of $0$. And as $x \equiv 0\ \&\ y \equiv 0\ .\supset.\ x \vee y \equiv 0$, and $x \equiv 0\ \&\ y \leq x\ .\supset.\ y = (y \vee 0) \equiv (y \vee x) = x \equiv 0$, the congruence classes of $0$ form ideals. The required map $u \to \equiv(u)$ from $\mathcal{A}$ to $Cgr_0\mathcal{B}$ is then given by

$$x \equiv y\ (u)\ .\equiv.\ x + y \leq u\ .\equiv.\ x + y \in id\{u\}.$$

Q.E.D.

Lemmas 1.2 and 1.2 are combined in the following theorems.

Theorem 1.1: Skolem semi-lattices $\mathcal{A}$ and only such admit nice representations without fission as $\mathrm{Cgr}_0\mathcal{B}$. The representation is given by

$$x \equiv y \ (u) \ .\equiv.\ x + y \leq u$$

$$.\equiv.\ x \vee u = y \vee u$$

and $\mathcal{B} = \langle A, \vee, + \rangle$ will do.

Theorem 1.2: All Skolem semi-lattices $\langle A, +, \vee, 0 \rangle$ form an equational class (variety). An equational axiom system is given by the equations which make $\langle A, \vee, 0 \rangle$ a semi-lattice with zero, together with the axioms.

i) $x \leq x + 0$

ii) $x + y \leq (x + z) \vee (y + z)$

iii) $(x \vee y) + y \leq x$

2.1. Relative Pseudo-complementation.

Skolem Rings may be defined alternatively as semi-lattices with a binary operation $/$, the relative pseudo-complement, which satisfies the rule

$$y \vee u \geq x \quad .\equiv. \quad u \geq x/y.$$

Then as

$$\begin{aligned} y \vee u = x \vee u \quad &.\equiv. \quad y \vee u \geq x \ \& \ x \vee u \geq y \\ &.\equiv. \quad u \geq x/y \ \& \ u \geq y/x \\ &.\equiv. \quad u \geq (x/y) \vee (y/x). \end{aligned}$$

The symmetric pseudo-complement $+$ may be defined by

a) $$x + y = (x/y) \vee (y/x).$$

Similarly for a Skolem Ring $\mathcal{A} = \langle A, +, \vee, 0 \rangle$,

$$\begin{aligned} y \vee u \geq x \quad &.\equiv. \quad y \vee u = x \vee y \vee u \\ &.\equiv. \quad u \geq (x \vee y) + y \end{aligned}$$

and thus the relative pseudo-complement $/$ is defined as

b) $$x/y = (x \vee y) + y$$

The following lemma gives a list of useful equations for Skolem Rings involving the $/$ .

Lemma 2.1: Let $\mathcal{A}$ be a Skolem Ring, let $/$ be defined as in b). Then,

1. $x \leq y \vee u \quad .\equiv. \quad u \leq x/y$

2. $x \vee (y/x) = x \vee y$

3. $x \leq y \ .\equiv. \ x/y = 0$

4. $x/y \leq x$

5. $x \geq y \quad .\supset. \quad x/z \geq y/z \quad \& \quad z/x \leq z/y$

6. $x/(x/y) \leq y$

7. $(x \vee y)/z = (x/z) \vee (y/z)$

8. $(x/y)/z = x/(y \vee z) = (x/z)/y.$

Proof: 1) has already been shown, and 2) through 6) are obvious from 1). 7) is obtained by 1) as follows,

$$\begin{aligned} u \geq (x \vee y)/z \ &.\equiv. \ z \vee u \geq x \vee y \\ &.\equiv. \ z \vee u \geq x \ \& \ z \vee u \geq y \\ &.\equiv. \ u \geq x/z \ \& \ u \geq y/z \\ &.\equiv. \ u \geq (x/z) \vee (y/z) \end{aligned}$$

8) follows similarly using the fact that $z \vee u \geq x/y$ .≡. $y \vee z \vee u \geq x$. Q.E.D.

2.2. Distributive Semi-lattices.

Finite distributive lattices form the simplest examples of Skolem lattices and semi-lattices. More generally complete lattices in which finite joins distribute over complete meets, i.e., in which $a \vee \bigwedge_{i \in I}\{b_i\} = \bigwedge_{i \in I}\{a \vee b_i\}$, are Skolem and in this case $a/b = \bigwedge\{c;\ b \vee c \geq a\}$. Conversely it is not hard to show that every Skolem lattice is distributive, and that a complete lattice is Skolem only if it is $\vee$-$\bigwedge$ distributive in the above sense.

Definition 2.1 below extends to the notion of distributivity to semi-lattices. As we will show in Lemma 2.4, every Skolem semi-lattice is distrubitive in this sense.

Definition 2.1: A semi-lattice $\mathcal{A}$ is distributive if for any $x, y, z \in A$, whenever $z \leq x \vee y$, there are $u, v \in A$, $u \leq x$, $v \leq y$, and $z = u \vee v$.

Lemma 2.3 below gives evidence that definition 1 correctly extends the notion of a distributive lattice.

Lemma 2.3: The semi-lattice $\mathcal{A}$ is distributive if and only if $\mathrm{Idl}\mathcal{A}$ is a distributive lattice. $\langle A,\vee,\wedge,0\rangle$ is a distributive lattice if and only if the semi-lattice $\langle A,\vee,0\rangle$ is distributive.

Proof: Suppose $\mathcal{A}$ is a distributive semi-lattice, let $I,J,K \in \mathrm{Idl}\,\mathcal{A}$, and let $a \in (I \vee K) \wedge (I \vee J)$. Then there are $i_1, i_2 \in I$, $k \in K$, $j \in J$ and $a \leq i_1 \vee k,\ i_2 \vee j$. As $\mathcal{A}$ is distributive, there are $i' \leq i$, $k' \leq k$, and $a = i' \vee k'$. Thus $k' \in K$ and as $k' \leq a \leq j$, $k' \in J$, and $k' \in (J \wedge K)$. Then $a \in I \vee (J \wedge K)$ and $(I \vee J) \wedge (I \vee K) \leq I \vee (J \wedge K)$. The reverse inequality is obvious.

Conversely if $\mathrm{Idl}\,\mathcal{A}$ is distributive, and $a \leq b \vee c$ then $\mathrm{id}\{a\} = \mathrm{id}\{a\} \wedge (\mathrm{id}\{b\} \vee \mathrm{id}\{c\}) = (\mathrm{id}\{a\} \wedge \mathrm{id}\{b\}) \vee (\mathrm{id}\{a\} \wedge \mathrm{id}\{c\})$, and there are $b' \leq a,b$, $c' \leq a,c$ and $a \leq b' \vee c'$. As $b',c' \leq a$ we have $a = b' \vee c'$.

The second statement follows from the first, and the well-known fact that $\langle A,\vee,\wedge,0\rangle$ is a distributive lattice if and only if $\mathrm{Idl}\langle A,\vee,\wedge,0\rangle \cong \mathrm{Idl}\langle A,\vee,0\rangle$ is distributive.

Q.E.D.

Lemma 2.4: Let $\mathcal{A} = \langle A,+,\vee,0\rangle$ be a Skolem semi-lattice then the semi-lattice $\langle A,\vee,0\rangle$ is distributive.

Proof: By Lemma 2.1,

$$z \leq x \vee y \;.\supset.\; z/x \leq y \;\;\&\;\; z \leq x \vee (z/x)$$

$$.\supset.\; z/(z/x) \leq x \;\;\&\;\; z \leq (z/(z/x)) \vee (z/x)$$

as $z/x,\ z/(z/x) \leq z$, we have $z = u \vee v$ for $u = z/(z/x) \leq x$ and $v = (z/x) \leq y$.

Q.E.D.

Corollary: If $\mathcal{A}$ is a Skolem Ring, then $\mathrm{Idl}\,\mathcal{A}$ is distributive.

2.3. Homomorphisms, Subalgebras, $^+$ - Process.

Theorem 1.1 gives a strong grip on the congruence relations and hence the homomorphisms of Skolem semi-lattices. If $\mathcal{A}$ and $\mathcal{B}$ are Skolem semi-lattices and $\phi:\mathcal{A}\to\mathcal{B}$ is a homomorphism of $\mathcal{A}$ onto $\mathcal{B}$, then $U = \{a \in A;\ \phi(a) = 0\}$ is an ideal on $\mathcal{A}$ and

a) $\phi(x) = \phi(y)\ .\equiv.\ \phi(x) + \phi(y) = 0\ .\equiv.\ \phi(x + y) = 0\ .\equiv.$

$$x + y \in U.$$

Conversely every ideal $U$ on $\mathcal{A}$ determines a homomorphism $\phi$ of $\mathcal{A}$ by the rule $\phi(x) = \phi(y)\ .\equiv.\ x + y \in U$. Let $\mathcal{A}/U$ denote the homomorphic image of $\mathcal{A}$ determined by $U \in \mathrm{Idl}\,\mathcal{A}$, then every homomorphic image of $\mathcal{A}$ is of the form $\mathcal{A}/U$ for some ideal $U$.

Lemma 2.5: Let $\mathcal{A}$ be a Skolem semi-lattice, then $\mathrm{Cgr}\,\mathcal{A} \cong \mathrm{Idl}\,\mathcal{A}$, with the isomorphism given by a).

For Skolem semi-lattice $\mathcal{A}$ and fixed $u \in A$, the set $\{a/u\colon a \in A\}$ forms a subalgebra of $\mathcal{A}$, as by Lemma 2.1 $(a/u) \vee (b/u) = (a \vee b)/u$ and $(a/u)/(b/u) = a/(u \vee (b/u)) = a/(b \vee u) = (a/b)/u$. We denote this subalgebra of $\mathcal{A}$ by $\mathcal{A}_u$. Then $\mathcal{A}_u \cong A/\mathrm{id}\{u\}$. To see this, note that $x/(x/u) \le u$ and

$(x/u)/x = 0 \leq u$ and thus $x + (x/u) \leq u$. The map $x \to x/u$ is then a homomorphism of $\mathcal{A}_u$ induced by id{u} and $\mathcal{A}_u \cong \mathcal{A}/\text{id}\{u\}$. We have shown

Lemma 2.6: Every homomorphic image of a Skolem ring $\mathcal{A}$ induced by a principal ideal id{u}, is isomorphic to the subalgebra $\mathcal{A}_u$ of $\mathcal{A}$. If every ideal in $\mathcal{A}$ is principal, in particular if $\mathcal{A}$ is finite, every homomorphic image of $\mathcal{A}$ is isomorphic to a subalgebra of $\mathcal{A}$.

In addition to the formation of subalgebras, homomorphic images, and direct products, there is another important method of constructing Skolem semi-lattices which is often useful. Let $\mathcal{A}$ and $\mathcal{B}$ be Skolem semi-lattices, $\mathcal{A} = \langle A, +_A, \vee_A, 0_A \rangle$, $\mathcal{B} = \langle B, +_B, \vee_B, 0_B \rangle$, and assume $A \cap B = \phi$. Let $A' = A - \{0_A\}$ and define a new Skolem semi-lattice $\mathcal{A} {}^+ \mathcal{B}$ on $A' \cup B$ as follows:

$$a_1 \vee a_2 = a_2 \vee a_1 = \begin{cases} a_1 \vee_A a_2 & \text{if } a_1, a_2 \in A' \\ a_1 \vee_B a_2 & \text{if } a_1, a_2 \in B \\ a_1 & \text{if } a_1 \in A', \quad a_2 \in B \end{cases}$$

$$a_1 + a_2 = a_2 + a_1 = \begin{cases} 0_B & \text{if } a_1 = a_2 \\ a_1 +_A a_2 & \text{if } a_1, a_2 \in A',\ a_1 \neq a_2 \\ a_1 +_B a_2 & \text{if } a_1, a_2 \in B \\ a_1 & \text{if } a_1 \in A',\ a_2 \in B \end{cases} .$$

It is easy to verify that $\mathcal{A}^+\mathcal{B} = \langle A' \cup B, +, \vee, 0_B \rangle$ is a Skolem semi-lattice. In the follwoing example $\mathcal{A}$ is a four element Skolem semi-lattice, and $\mathcal{B}$ is the (unique) two element Skolem semi-lattice.

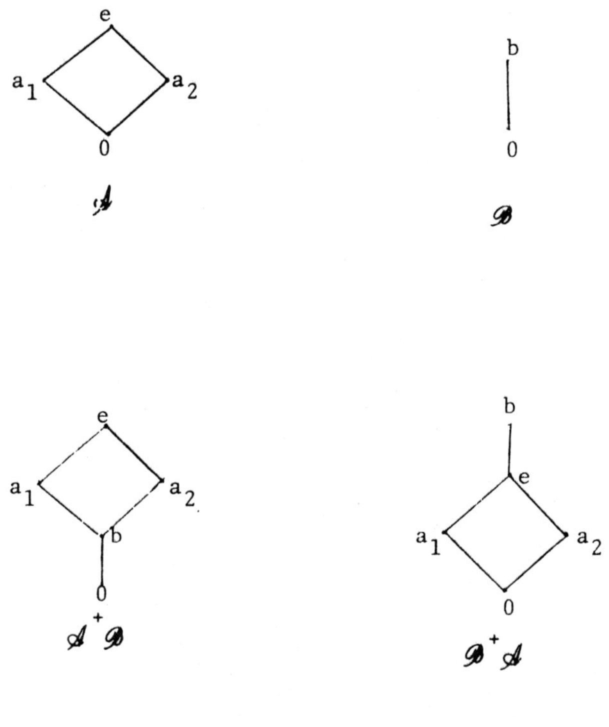


Figure 2.1

An example $\mathcal{A}^+\mathcal{B}$, $\mathcal{B}^+\mathcal{A}$

When $\mathcal{B}$ is the two element Skolem semi-lattice we will usually write $\mathcal{A}^+$ in place of $\mathcal{A}^+\mathcal{B}$.

We will write $\mathcal{A} \subseteq_s \mathcal{B}$ if $\mathcal{A}$ is isomorphic to a subalgebra of $\mathcal{B}$. The following Lemma is easily shown.

Lemma 2.7: Let $\mathcal{A}, \mathcal{B}$ be Skolem semi-lattices, then $\mathcal{A}^+\mathcal{B}$ is a Skolem semi-lattice. The following statements are true,

i) $\mathcal{A} \subseteq_s \mathcal{A}^+\mathcal{B}$

ii) $\mathcal{A} \subseteq_s \mathcal{C}$ , $\mathcal{B} \subseteq_s \mathcal{D}$ .⊃. $\mathcal{A}^+\mathcal{B} \subseteq_s \mathcal{C}^+\mathcal{D}$

iii) $\mathcal{B} \varepsilon$ Idl $\mathcal{A}^+\mathcal{B}$

iv) $\mathcal{A} \cong \mathcal{A}^+ \mathcal{B}/\mathcal{B}$

Definition 2.2: An ideal U of a Skolem semi-lattice $\mathcal{A}$ is called <u>relatively maximal</u> if there is an a ε A, a ∉ U, and for all V ε Idl $\mathcal{A}$ , U ⊆ V and U ≠ V imply a ε V.

The Relatively maximal ideals of $\mathcal{A}$ are thus the completely ∩-irreducible memebers of Idl $\mathcal{A}$ , and by Lemma 2.5, correspond to the ∩-irreducible members of Cgr $\mathcal{A}$. It follows that

Lemma 2.8: $\mathcal{A}$/U is a subdirectly irreducible factor of $\mathcal{A}$ precisely when U is relatively maximal. $\mathcal{A} \cong \mathcal{A}$/id{0} is subdirectly irreducible just in case there is a smallest non-zero element in $\mathcal{A}$, i.e., when $\mathcal{A} \cong \mathcal{B}^+$ for $\mathcal{B} \subseteq_s \mathcal{A}$.

3.1. Topological Skolem Semi-lattices

Topological lattices, those which have a $\bigwedge$-base of strongly-$\vee$-irreducible elements, are easily seen to be $\vee$-$\bigwedge$ distributive. They provide the most important examples of Skolem lattices and semi-lattices. In particular for Cld S, the Skolem semi-lattice of closed sets of a topological space S, the + operation may be defined as the closure of the symmetric set difference, i.e., $A + B = \bigcap\{C : (A-B) \subseteq C \;\&\; (B-A) \subseteq C\}$. It is natural to look to these topological structures, Cld S, for counter examples to possible Skolem equations. As we will see in Theorem 3.1, it is sufficient in such questions to confine our attention to those structures. In fact every Skolem semi-lattice may be embedded as a sub-algebra of Cld S, for a suitable space S. To see this, let $\mathcal{A} = \langle A, +, \vee, 0 \rangle$ be a Skolem semi-lattice and let $\mathcal{D}$ be an arbitrary dense set of proper ideals on $\mathcal{A}$ (i.e., $\mathcal{D} \subseteq \mathrm{Idl}\,\mathcal{A}$ and for $a, b \in A$, $a \nleq b$, there is $I \in \mathcal{D}$, $b \in I$ and $a \notin I$). For $a \in A$, define $P_a = \{I \in \mathcal{D};\; a \notin I\}$. Then

$$P_a \cup P_b = \{I \in \mathcal{D};\; a \notin I \text{ or } b \notin I\}$$

$$= \{I \in \mathcal{D};\; a \vee b \notin I\}$$

$$= P_{a \vee b}$$

The set $\{P_a;\ a \in A\}$ then forms a $\cap$-base for a topology $S$ on $\mathcal{D}$. As $\mathcal{D}$ is dense, $a \neq b \ .\supset.\ P_a \neq P_b$. The map $a \to P_a$ is then an embedding of $A$ into $\mathrm{Cld}(S)$ which preserves the $\vee$. We have only to check now that $P_a + P_b = P_{a+b}$. By definition,

$$P_a + P_b = \cap\{P_c : P_a - P_b \subseteq P_c \ \&\ P_b - P_a \subseteq P_c\}\ .$$

Now

$$\begin{aligned} P_a - P_b \subseteq P_c\ &.\equiv.\ (\forall I)_{\mathcal{D}}(a \notin I \ \&\ b \in I\ .\supset.\ c \notin I) \\ &.\equiv.\ (\forall I)_{\mathcal{D}}(b \in I \ \&\ c \in I\ .\supset.\ a \in I) \\ &.\equiv.\ (\forall I)_{\mathcal{D}}(b \vee c \in I\ .\supset.\ a \in I) \\ &.\equiv.\ a \le b \vee c, \quad \text{as } \mathcal{D} \text{ is dense.} \end{aligned}$$

Similarly $P_b - P_a \subseteq P_c\ .\equiv.\ b \le a \vee c$. And

$$\begin{aligned} P_a + P_b \subseteq P_c\ &.\equiv.\ a \vee c = b \vee c \\ &.\equiv.\ a + b \le c \\ &.\equiv.\ P_{a+b} \subseteq P_c \end{aligned}$$

we have $P_a + P_b = P_{a+b}$. This proves

Theorem 3.1: Every Skolem semi-lattice may be embedded as a subalgebra of $\mathrm{Cld}(S)$, for some topological space $S$.

Corollary: Every Skolem semi-lattice has an embedding into a Skolem lattice which preserves the $+$ and $\vee$ .

3.2. Regular elements

A regular closed set in topology is defined as a set which is the closure of its interior. In order to understand the appropriate analogue for Skolem semi-lattices, we will first study the regular closed sets of the topology $S$ of Theorem 3.1.

Define $\bar{P}_a = \{I \in \mathcal{D}; a \in I\}$, the set $\{\bar{P}_a : a \in A\}$ is then a $\cup$-base for the open sets of $S$. For $T \in S$, let $T^0$ denote the interior of $T$, i.e., $T^0 = \cup\{\bar{P}_b; \bar{P}_b \subseteq T\}$. Then for $a \in A$, $P_a^0 = \cup\{\bar{P}_b; \bar{P}_b \subseteq P_a\}$ and

$$\begin{aligned} \bar{P}_b \subseteq P_a \ &.\equiv.\ (\forall I)_{\mathcal{D}}(b \in I\ .\supset.\ a \notin I) \\ &.\equiv.\ (\forall I)_{\mathcal{D}}(b \vee a \notin I) \\ &.\equiv.\ (\forall c)_A\ c \le b \vee a \end{aligned}$$

where the last equivalence follows since $\mathcal{D}$ is dense. Thus $\bar{P}_b \subseteq P_a$, and there are non-trivial regular closed sets, just in case there is a largest element $1 \in \mathcal{A}$, and $b \vee a = 1$. And,

$$\begin{aligned} \bar{P}_b \subseteq P_a \ &.\equiv.\ b \vee a = 1 \\ &.\equiv.\ 1/a \le b \\ &.\equiv.\ \bar{P}_b \subseteq \bar{P}_{1/a} \\ &.\equiv.\ \bar{P}_{1/a} = P_a^0\ . \end{aligned}$$

In this case $P_a$ is regular closed if and only if $P_a = \bigcap\{P_b; P_a^0 \subseteq P_b\} = \bigcap\{P_b; \bar{P}_{1/a} \subseteq P_b\}$, i.e. if and only if

$$\bar{P}_{1/a} \subseteq P_b \ .\equiv.\ P_a \subseteq P_b$$

and this last statement is equivalent to,

$$b \vee (1/a) = 1 \ .\equiv.\ a \le b$$

$$.\equiv.\ a \le 1/(1/a) \quad .$$

As $1/(1/a) \le a$ by Lemma 2.1, we have $P_a$ is regular closed just in case $a = 1/(1/a)$. If for some element $b \in \mathcal{A}$, $a = 1/b$, then it is easy to show that $1/(1/a) = 1/(1/(1/b)) = 1/b = a$. This leads to the following definition.

Definition 3.1: In a Skolem semi-lattice $\mathcal{A}$ with identity 1 an element $a \in \mathcal{A}$ is <u>regular</u> if there is $b \in \mathcal{A}$, and $a = 1/b$.

Lemma 3.1. In a Skolem semi-lattice with identity, the regular elements form a subalgebra and a Boolean ring.

Proof: This follows from theorem 3.1; and the well known fact that the regular closed sets form a subalgebra of Cld(S) and a Boolean ring.

We will see now that definition 3.1 may be modified to extend the concept of a regular element to Skolem semi-lattices without a 1. The new definition should satisfy two conditions: 1) it should agree with definition 3.1 on Skolem semi-lattices with identity, and 2) the regular elements of an arbitrary Skolem semi-lattice should form a subalgebra and a Boolean ring. With this in mind we make the following definition.

Definition 3.2: An element $a$ of a Skolem semi-lattice $\mathcal{A}$ is <u>regular</u> if there is an ideal $I_a \in \mathrm{Idl}\,\mathcal{A}$ and $\mathrm{id}\{a\}$ is the least ideal such that $A = I_a \vee \mathrm{id}\{a\}$.

It is easy to verify that Definition 3.1 and 3.2 agree for Skolem semi-lattices with identity. Before we show that definition 3.2 meets condition 2), we will give an equivalent formulation of regularity.

Lemma 3.2: $a$ is a regular element of a Skolem semi-lattice $\mathcal{A}$ if and only if $a = a/(b/a)$ for all $b \in A$.

Proof: Let $a$ be a regular element of a Skolem semi-lattice $\mathcal{A}$. Then

$$b \in A = I_a \vee \mathrm{id}\{a\} \;.\supset.\; (\exists i)_{I_a} \; b \leq i \vee a$$

$$.\supset.\; (b/a) \leq i \in I_a$$

and thus $b/a \in I_a$ for all $b \in A$. Then $b/a \in I_a$ and $a \leq (b/a) \vee (a/(b/a)) \;.\supset.\; A = I_a \vee \mathrm{id}\{a\} = I_a \vee \mathrm{id}\{a/(b/a)\}$. As $a$ is regular, $a \leq a/(b/a)$, and $a \geq a/(b/a)$ by Lemma 2.1, thus for all $b \in A$, $a = a/(b/a)$.

Conversely suppose $a = a/(b/a)$ for all $b \varepsilon A$, and let $I = id\{(b/a); b \varepsilon A\}$. It is clear that $A = I \vee id\{a\}$. Now suppose $A = I \vee J$ for some $J \varepsilon Idl\, \mathcal{A}$, then $a \varepsilon I \vee J .\supset. a \leq i \vee c$ for $i \varepsilon I$, $c \varepsilon J$, and thus $a \leq (c/a) \vee c$ for $c/a \varepsilon I$. Then $a = a/(c/a) \leq c$, and $id\{a\} \subseteq J$. Q.E.D.

This new formulation is useful in proving the following Lemma.

Lemma 3.3: The regular elements of a Skolem semi-lattice $\mathcal{A}$ form a subalgebra of $\mathcal{A}$.

Proof: Let a, c be regular in $\mathcal{A}$. By Lemmas 3.2 and 2.1 it is sufficient to show for $b \varepsilon A$, $(a \vee c)/(b/(a \vee c)) \geq a \vee c$ and $(a/c)/(b/(a/c)) \geq a/c$.

Using Lemma 2.1,

$$\begin{aligned}(a \vee c)/(b/(a \vee c)) &= a/(b/(a \vee c)) \vee c/(b/(a \vee c)) \\ &\geq a/(b/a) \vee c/(b/c) \\ &= a \vee c \quad \text{as } a,c \text{ are regular.}\end{aligned}$$

Again by Lemma 2.1,

$$\begin{aligned}(a/c)/(b/(a/c)) &= (a/(b/(a/c))/c \\ &\geq (a/(b/a))/c \\ &= a/c \quad \text{by regularity of } a.\end{aligned}$$

Q.E.D.

Let $\operatorname{Reg}\mathcal{A}$ denote the subalgebra of regular elements of $\mathcal{A}$. Let $a, b, c \in \operatorname{Reg}\mathcal{A}$, and suppose $c \leq a, b$. Then $c \leq b \;.\supset.\; a/c \geq a/b$ and $(a/c) \vee (a/b) \geq (a/b) \vee (a/(a/b)) = a \geq c$, thus $(a/(a/b)) \geq c/(a/c) = c$ as $c$ is regular. As $(a/(a/b)) \leq a,b$ by Lemma 2.1, it follows that $a/(a/b)$ is the meet $a \cdot b$ of $a, b$ in $\operatorname{Reg}\mathcal{A}$, and $\operatorname{Reg}\mathcal{A}$ is a lattice. Moreover for $a, b \in \operatorname{Reg}\mathcal{A}$, $a \cdot (b/a) = a/(a/(b/a)) = a/a = 0$, and $a \vee (b/a) = a \vee b$. Thus $(b/a)$ is a complement for $a$ in the interval $(a \vee b, 0)$. It follows that $\operatorname{Reg}\mathcal{A}$ is generalized Boolean algebra, and thus a Boolean ring. Combined with Lemma 3.3, this proves

Theorem 3.2: The regular elements of a Skolem semi-lattice $\mathcal{A}$ form a subalgebra of $\mathcal{A}$ and a Boolean ring.

Corollary: A Boolean ring may be defined as a Skolem semi-lattice which satisfies the additional axiom $x/(y/x) = x$.

It is easy to check that $\operatorname{Reg}(\mathcal{A}^{+}\mathcal{B}) \cong \operatorname{Reg}(\mathcal{A})$, $\operatorname{Reg}(\mathcal{B}) \subseteq_{s} \mathcal{B} \subseteq_{s} \mathcal{A}^{+}\mathcal{B}$. In general $\operatorname{Reg}(\mathcal{A})$ need not be the largest boolean subalgebra of $\mathcal{A}$.

4.1 Skolem Rings

We have seen in Section 1 how the classical representation of Boolean semi-lattices $\langle A,+,\vee,0\rangle$ naturally extends to Skolem semi-lattices. The Boolean structure is usually presented and is more familiar in the ring form $\langle A,+,\cdot,0\rangle$ where

(a) $$x \cdot y = (x \vee y) + (x + y),$$

the two structures being equivalent up to equational transformation. We will therefore try to introduce a corresponding $\cdot$ operation also for the more general Skolem semi-lattice; of course the two operations should coincide in the Boolean case. It turns out that (a) is not the appropriate definition, we define

(b) $$x * y = ((x \vee y)+x)+y = y/(y/x)$$
$$x \cdot y = (x*y) \vee (y*x).$$

Definition 4.1: Let $\mathcal{A} = \langle A,+,\vee,0\rangle$ be a Skolem semi-lattice and let $\cdot$ be defined by (b). Then the structure $\langle A,+,\cdot,0\rangle$ is called a <u>Skolem ring</u>.

A Skolem ring is not generally a ring in the usual algebraic sense. The term is used here to indicate the parallels with Boolean rings.

Lemma 4.1: The following hold in any Skolem ring;

1. $x \cdot y = y \cdot x \leq x$

2. $x \leq y \;.\supset.\; x \cdot y = x$

3. $x/y \leq (x/y)/(y/x)$

4. $(x/y) \cdot (y/x) = 0$

5. $(x+y) \vee (x \cdot y) = x \vee y$

6. $x \cdot (y \vee z) \leq x \cdot y \vee x \cdot z$

Proof:

1. Using Lemma 2.1 $x/(x/y) \leq x$ and $y/(y/x) \leq x$, thus $x \cdot y = (x/(x/y)) \vee (y/(y/x)) \leq x$, $x \cdot y = y \cdot x$ follows by symmetry.

2.
$$\begin{aligned} x \leq y \;&.\supset.\; x/y = 0 \qquad \text{Lemma 2.1} \\ &.\supset.\; x/(x/y) = x \\ &.\supset.\; x \leq (x/(x/y)) \vee (y/(y/x)) = x \cdot y \\ &.\supset.\; x = x \cdot y \quad \text{by } 1). \end{aligned}$$

3. It suffices to show $x \leq y \vee ((x/y)/(y/x))$. By Lemma 2.1, $y/x \leq y$ and thus $(x/y)/y \leq (x/y)/(y/x)$. Then $x \leq y \vee (x/y) = y \vee ((x/y)/y) \leq y \vee ((x/y)/(y/x))$.

4. By 3) and Lemma 2.1, $(x/y)/((x/y)/(y/x)) = 0$ and $(y/x)/((y/x)/(x/y)) = 0$, thus

$$(x/y)\cdot(y/x) = (x/y)/((x/y)/(y/x)) \vee (y/x)/((y/x)/(x/y))$$
$$= 0.$$

5. As $x+y = (y/x) \vee (x/y)$ we have

$$(x+y) \vee (x\cdot y) = ((y/x) \vee (x/y)) \vee ((x/(x/y)) \vee (y/(y/x))$$
$$= ((y/x) \vee (y/(y/x))) \vee ((x/y) \vee (x/(x/y))$$
$$= y \vee x.$$

6. By Lemma 2.1, $x/(x/(y\vee z)) \le x/(x/y)$ and

$$(y \vee z)/(y \vee z)/x) = y/((y \vee z)/x) \vee z/((y \vee z)/x)$$
$$\le y/(y/x) \vee z/(z/x)$$

thus,

$$x\cdot(y \vee z) = x/(x/(y \vee z)) \vee (y \vee z)/((y \vee z)/x)$$
$$\le y/(y/x) \vee z/(z/x) \vee x/(x/y)$$
$$\le (y/(y/x) \vee x/(x/y)) \vee (z/(z/x) \vee x/(x/z))$$
$$= (x\cdot y) \vee (x\cdot z)$$

Q.E.D.

Lemma 4.2: Let $\mathscr{A} = \langle A,+,\vee,0\rangle$ be a Skolem semi-lattice and let $\cdot$ be defined by (b). Then $\vee$ can be defined in the Skolem ring $\langle A,+,\cdot,0\rangle$ by

(c) $$x \vee y = (x+y) + (x \cdot y).$$

The notions of a Skolem semi-lattice and a Skolem ring are equivalent.

Proof: The second statement follows directly from (c). To prove (c) let $u = (x+y) + (x \cdot y)$; as $x+y, x \cdot y \leq x \vee y$ it is clear that $u \leq x \vee y$. It suffices to show $x, y \leq u$. By Lemma 2.1 and 4.1

$$(x+y) \vee u = (x \cdot y) \vee u = (x+y) \vee (x \cdot y) = x \vee y$$

and thus $x \leq (x+y) \vee u, (x \cdot y) \vee u$. Using the distributivity of $\mathscr{A}$ we may write $x = z_1 \vee u_1 = z_2 \vee u_2$ for $z_1 \leq x+y$, $z_2 \leq x \cdot y$, $u_1, u_2 \leq u$. Again by distributivity $z_1 \leq z_2 \vee u_2$ implies $z_1 = z \vee u'$ for $z \leq z_2$, $u' \leq u_2$. Let $v_1 = u_1 \vee u'$, then $x = z \vee v_1$ for $z \leq x+y, x \cdot y$ and $v_1 \leq u$. Similarly $y = w \vee v_2$ for $w \leq x+y, x \cdot y$ and $v_2 \leq u$. As $w \leq x \cdot y \leq x$, $x = (z \vee w) \vee v_1$ and similarly $y = (z \vee w) \vee v_2$. Then, $(x+y) \leq v_1 \vee v_2$ and thus $(z \vee w) \leq (x+y) \leq v_1 \vee v_2 \leq u$. As $z, w, v_1 v_2 \leq u$, we have $x, y \leq u$.

Q.E.D.

As Skolem semi-lattices and Skolem rings are equivalent up to equational translation, we will use the two terms interchangeably.

## 4.2. Correct Skolem Rings

The $\cdot$ operation in a Boolean ring $\mathscr{A}$ is the meet operation of the lattice, that is $x \cdot y = x \wedge y =$ the greatest lower bound of $x,y$ in $\mathscr{A}$. This contrasts with the general Skolem case where the meet $x \wedge y$ may not exist, or if it exists may not be equationally definable. As we have seen the $\cdot$ operation for Skolem rings was derived by modifying the Boolean definition, and as formulas (1) and (2) of Lemma 4.1 show, it retains some of the characteristics of a meet. The question arises: for which Skolem rings is the $\cdot$ operation a meet? More generally: for which Skolem rings is a meet equationally definable?

Whalen and Nemitz [13] have shown that the $\cdot$ operation is the meet precisely when it satisfies the associative law $x\cdot(y\cdot z) = (x\cdot y)\cdot z$. This result is included in Theorem 4.1 below. It is easily seen that these "associative" structures are the only ones in which the meet is equationally definable.

Linear ordered sets with smallest element $0$ are the simplest examples of Skolem rings in which $x\cdot y = x \wedge y$. We call these structures chains and use $C_n$, $n \in \omega$ to denote the n-element chain.

Definition 4.2: Let $U \in \mathrm{Idl}\,\mathscr{A}$, for Skolem ring $\mathscr{A}$. $U$ is said to be prime if $x\cdot y \in U$ implies $x \in U$ or $y \in U$. $x$ and $y$ are said to be 0-divisors if $x,y \neq 0$ and $x\cdot y = 0$.

Lemma 4.3: Let $P$ be an ideal in the Skolem ring $\mathscr{A}$. Then

$$P \text{ is prime } .\equiv.\ \mathscr{A}/P \text{ is a chain.}$$

Proof: Suppose $P$ is prime, let $x,y \in A$, and $\phi$ be the homomorphism $\phi: A \to \mathscr{A}/P$. As $(x/y)\cdot(y/x) = 0 \in P$ by 4. of

Lemma 4.4, either $x/y \in P$ or $y/x \in P$. Suppose $x/y \in P$ then $0 = \phi(x/y) = \phi(x)/\phi(y)$ and thus $\phi(x) \leq \phi(y)$, and $\mathcal{A}/P$ is a chain.

Conversely suppose $\mathcal{A}/P$ is a chain, $x,y \in A$, $x \cdot y \in P$. As $\mathcal{A}/P$ is a chain we may assume $\phi(x) \geq \phi(y)$. Then by Lemma 4.1, $\phi(y) = \phi(x) \cdot \phi(y) = \phi(x \cdot y) = 0$, and thus $y \in P$.

Q.E.D.

Corollary: Let $\mathcal{A}$ be a Skolem ring. Then,

$$\mathcal{A} \text{ has no 0-divisors } .\equiv. \ \mathcal{A} \text{ is a chain.}$$

If the $\cdot$ operation is a meet in some Skolem ring $\mathcal{A}$, then by Lemmas 2.3 and 2.4 $\langle A, \vee, \cdot, 0 \rangle$ is a distributive lattice and thus every relatively maximal ideal is prime. The proof goes as follows: Let $P$ be relatively maximal in $\mathcal{A}$, say $a \notin P$ and for all $b \in P$, $a \in P \vee id\{b\}$. Let $b,c \notin P$, then $a \in P \vee id\{b\}$, $P \vee id\{c\}$ and thus $a = p_1 \vee b' = p_2 \vee c'$ for $p_1, p_2 \in P$, $b' \leq b$, $c' \leq c$ by distributivity. Then as $\cdot$ is the $\wedge$, by Lemmas 2.3, and 2.4,

$$(*) \qquad a = a \cdot a = (p_1 \vee b') \cdot (p_2 \vee c')$$

$$= (p_1 \cdot p_2 \vee p_1 \cdot c' \vee p_2 \cdot b') \vee b' \cdot c'$$

thus $a \in P \vee id\{b' \cdot c'\}$ and,

(**) $\quad b' \leq b,\ c' \leq c\ .\supset.\ b'\cdot c' \leq b\cdot c$

then $a \in P \vee id\{b\cdot c\}$, $b\cdot c \notin P$ and $P$ is prime.

Then $x\cdot y = x \wedge y$ in $\mathcal{A}$ implies that every relatively maximal ideal is Prime, which implies by Lemma 4.2, that every subdirectly irreducible factor of $\mathcal{A}$ is a chain, and $\mathcal{A}$ is a subdirect product of chains. Conversely as $x\cdot y$ is the meet of $x,y$ in every chain, and as the statement "$\cdot$ is meet" is equational (e.g. replace $\wedge$ with $\cdot$ in the usual lattice equations) we have the following:

Lemma 4.4: For a Skolem ring $\mathcal{A}$, the following statements are equivalent:

i) the $\cdot$ operation is the meet in $\mathcal{A}$

ii) $\mathcal{A}$ is a subdirect product of chains.

Remark 1: A closer look at the proof just given for Lemma 4.4 shows that only two properties need be assumed for the $\cdot$ operation on $\mathcal{A}$ in order for the relatively maximal ideals of $\mathcal{A}$ to be prime. Namely at (*) it is enough to assume the distributive law

(1) $x\cdot(y \vee z) = x\cdot y \vee x\cdot z$, and at (**) it is sufficient that

(2) $x \leq y\ .\supset.\ x\cdot z \leq y\cdot z$.

In fact it is easily seen that (1) is equivalent with (2), and both are equivalent to the associative law (3) $x\cdot(y\cdot z) = (x\cdot y)\cdot z$. This is summarized in the following theorem.

Theorem 4.1: The following are equivalent statements for a Skolem ring $\mathcal{A} = \langle A,+,\cdot,0\rangle$.

a) $x\cdot(y\cdot z) = (x\cdot y)\cdot z$

b) $x \leq y \;.\supset.\; x\cdot z \leq y\cdot z$

c) $x\cdot(y \vee z) = x\cdot y \vee x\cdot z$

d) every relatively maximal ideal of $\mathcal{A}$ is prime.

e) $\mathcal{A}$ is a subdirect product of chains.

f) $\cdot$ is a meet operation on $\mathcal{A}$.

Proof: (b) $\leftrightarrow$ (c) is trivial, (d) $\leftrightarrow$ (e) $\leftrightarrow$ (f) by Lemmas 4.3 and 4.4. (f) $\rightarrow$ (a) is trivial. We will prove (a) $\rightarrow$ (b) and (c) $\rightarrow$ (d).

(a) $\rightarrow$ (b): Using 2 of Lemma 4.1, we have,

$$\begin{aligned} x \leq y \;&.\equiv.\; x\cdot y = x \\ &.\supset.\; (x\cdot y)\cdot z = x\cdot z \\ &.\supset.\; (x\cdot y)\cdot(z\cdot z) = x\cdot z \end{aligned}$$

by (a) and 1 of Lemma 4.1 we may rewrite this last expression as $(x \cdot z) \cdot (y \cdot z) = x \cdot z$ and conclude again by 2 of Lemma 4.1 that $x \cdot z \leq y \cdot z$.

(c) → (d): Using the fact that c) → b) we may assume

$$x \leq y \ .\supset.\ x \cdot z \leq y \cdot z \quad \text{and} \quad x \cdot (y \vee z) = (x \cdot y) \vee (x \cdot z)$$

and conclude by applying Remark 1 to the proof of Lemma 4.3.

Q.E.D.

Definition 4.3: We will call a Skolem ring satisfying any of the equivalent statements of Theorem 4.1 a <u>Correct Skolem ring</u>.

Theorem 4.1 may be read as asserting that the Variety of correct Skolem rings is identical with the Variety generated by chains. As every n-element subset of a chain $\mathscr{C}$ which includes the 0 of $\mathscr{C}$ forms a subalgebra isomorphic to $C_n$, an equation fails in $\mathscr{C}$ only if it fails in some finite subchain of $\mathscr{C}$. We may conclude,

Corollary 1: (Whaley, Nemitz [13]). The Variety of Correct Skolem rings is the Variety generated by the finite chains.

Moreover as Whalen and Nemitz point out, every subdirectly irreducible Skolem ring $\mathscr{A}$ which is not a chain must have a subalgebra of the form in Figure 4.1.

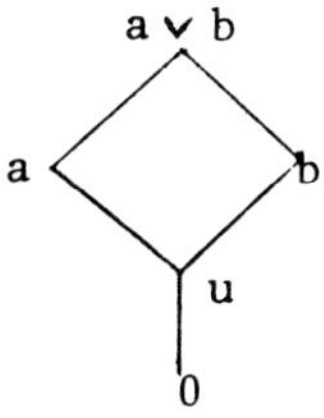


Figure 4.1
An example of a non-correct Skolem Ring

To see this, let $x,y \in \mathcal{A}$ such that $x \nleq y$ and $y \nleq x$, then for $a = (x/y)$, $b = (y/x)$ and $u$ the smallest non-zero element in $\mathcal{A}$. The set $\{(a \vee b),a,b,u,0\}$ is easily seen to be a subalgebra of $\mathcal{A}$. Note that in this case $a \cdot b = 0$ and hence $\cdot$ is not the meet. Further the four elements $a,b,(a \vee b),0$ form a subalgebra of $\mathcal{A}$ which does not include $a \wedge b = u$, and thus the meet is not equationally definable in $\mathcal{A}$. We have the following corollary to Theorem 4.1,

Corollary 2: The Variety generated by the finite chains is the only Variety of Skolem rings for which a meet may be equationally defined, and in this case $x \wedge y = x \cdot y$.

5.1. The Subvariety Lattice of Skolem Rings.

Let A denote the class of all algebras of a given similarity type, and $\mathcal{E}$ the class of all identities or equations of that type. For $K \subseteq A$, $M \subseteq \mathcal{E}$, define $K^{\tau} \subseteq \mathcal{E}$, $M^{\mu} \subseteq A$ by: $K^{\tau}$ is the set of all identities which hold for every algebra in K, $M^{\mu}$ is the set of all algebras which satisfy the identities in M. This describes a Galois connection (see Birkhoff [2 ]), between two important closure spaces, the one over universal algebras, the other over equational theories.

In the first space, the map $K \to K^{\tau\mu}$ defines the closure operator. Closed sets, those of the form $K^{\tau\mu}$ for $K \subseteq A$, are called Varieties. $K^{\tau\mu}$ is called the variety generated by K, denoted V(K). If $K = \{\mathcal{A}\}$ for an algebra $\mathcal{A} \in A$, we will write $V(\mathcal{A})$ in place of $V(\{\mathcal{A}\})$. By a well-known result of Birkhoff[1], V(K) = HSP(K) where H,S,P denote the closure respectively under the taking of homomorphisms, subalgebras and direct products. If $K_1, K_2$ are varieties, $K_1 \subset K_2$ then $K_1$ is said to be a subvariety of $K_2$. The subvarieties of a given variety K, ordered by inclusion, thus form a complete lattice VL(K) called the subvariety lattice of K. Th(K), the complete lattice of equational theories extending $K^{\tau}$ is defined dually over $\mathscr{L}$. Th(K) and VL(K) are dually isomorphic by the Galois connection.

The class of all Skolem Rings, (S.R.), forms a Variety by theorem 1.2, in this section we will give a description of VL(S.R) and dually Th(S.R.). We will use the fact that varieties of the form $V(\mathscr{A})$ for $\mathscr{A}$ a subdirectly irreducible Skolem ring form a $\bigvee$-base for VL(S.R).

Let $\mathscr{C}_n$ denote the n-element chain and $\mathscr{D}$ the Skolem Ring of Figure 4.1. The one element algebra $\mathscr{C}_1$ is a homomorphic image of every Skolem Ring and thus $V(\mathscr{C}_1)$ is the zero of the lattice VL(S.R): it corresponds to the largest theory in Th(S.R.), the inconsistent theory $x = y$. As every non-trivial subdirectly irreducible Skolem ring has $\mathscr{C}_2$ as a subalgebra, $V(\mathscr{C}_2)$ is the smallest non-trivial subvariety, and $\mathscr{C}_2^{\tau}$ is the maximally consistent or complete theory in Th(S.R.). Let $\mathscr{A}$ be a subdirectly irreducible Skolem ring, and let $u$ be the smallest non-zero element in $\mathscr{A}$. If $\mathscr{A}$ is not $\mathscr{C}_2$, then there is $a \in \mathscr{A}$, $a > u$ and the set $\{a,u,0\}$ forms a subalgebra of $\mathscr{A}$ isomorphic to $\mathscr{C}_3$. As the Boolean identity of theorem 3.2 $x = x/(y/x)$ holds in $\mathscr{C}_2$ but fails in $\mathscr{C}_3$, $V(\mathscr{C}_3)$ properly contains $V(\mathscr{C}_2)$. Similarly it may be shown that every variety which properly contains $V(\mathscr{C}_3)$ contains either $V(\mathscr{C}_4)$ or $V(\mathscr{D})$. The base of the lattice VL(S.R.) thus looks like,

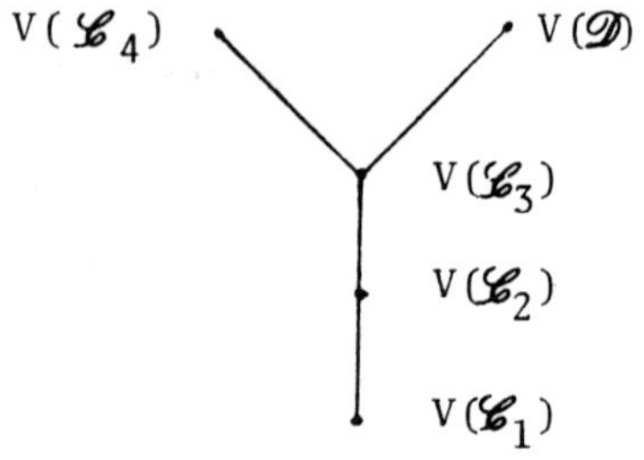


Figure 5.1
Base of VL(S.R.)

It follows further that V($\mathscr{C}_2$) is the variety of all Boolean rings, and hence Stones' result: every Boolean ring is a subdirect product of $\mathscr{C}_2$. Any identity such as $x = x/(y/x)$ which holds in $\mathscr{C}_2$ but fails in $\mathscr{C}_3$ together with the Skolem ring identities of theorem 1.2, will serve as an equational set of axioms for $\mathscr{C}_2^{\tau}$.

The Whalen-Nemitz result in section 4 shows that the variety of correct Skolem rings is the join of varieties generated by finite Skolem rings. It is not difficult to show that this is also the case for the variety of all Skolem rings. This follows from a result due to McKinsey and Tarski [9,10], using theorem 3.1. To see this suppose $\phi(x_1,\ldots,x_n)$ is an equation which fails in some Skolem ring $\mathscr{A} = \langle A,+,\vee,0\rangle$. We may assume by theorem 3.1 that A is a distributive lattice. Say $\phi$ fails for $x_1 = a_1,\ldots,x_n = a_n$, for $a_i \in A$. Let B' be the finite subset of A containing $a_1,\ldots,a_n$ and all subterms of the expression $\phi(a_1,\ldots,a_n)$

in A. Let B be the finite distributive sublattice of A generated by B' and let $\mathcal{B} = \langle B, +', \vee, 0\rangle$ be the Skolem ring on B. For $a, b \in B$, if $a + b \leq B$ then $a +' b = \bigwedge\{c \in B, a \vee c = b \vee c\} = a + b$. Hence $\phi(x_1, \ldots, x_n)$ fails in $\mathcal{B}$ for $x_i = a_i$.

Theorem 5.1 below shows that in fact the varieties of the form $V(\mathcal{A})$ for $\mathcal{A}$ finite and subdirectly irreducible form a $\bigvee$-base for VL(S.R.). This follows from the more general fact that the finitely generated Skolem rings are finite. This result is rather surprising in view of the example given by McKinsey and Tarski [9,10] of an infinite Skolem lattice generated by only two elements.

We will let $\mathcal{F}_n$ denote the free Skolem ring generated by n elements. A simple argument shows that $\mathcal{F}_n$ is finite. To see this, let x,y be the two generators of $\mathcal{F}_2$, and let $\mathcal{A}^+$ be a subdirectly irreducible factor of $\mathcal{F}_2$. Let $\phi$ be the onto projection $\phi: \mathcal{F}_2 \to \mathcal{A}^+$, and let u be the smallest non-zero element of $\mathcal{A}^+$. As $\phi$ is onto $\mathcal{A}^+$, $\phi(x)$, $\phi(y)$ generate $\mathcal{A}^+$ and as $u \notin \mathcal{A} \subseteq_s \mathcal{A}^+$, either $\phi(x) = u$ or $\phi(y) = u$. Say $\phi(x) = u$, then $\mathcal{A} \cong \mathcal{C}_3$ or $\mathcal{C}_2$ according as $\phi(y) > \phi(x)$ or $\phi(y) \leq \phi(x)$. In either case $\mathcal{A} \in V(\mathcal{C}_3)$.

Lemma 5.1: $\mathcal{F}_2 \in V(\mathcal{C}_3)$, $\mathcal{F}_2$ is finite.

Proof: The first statement has already been proven, the second statement follows from the first and following general fact,

Lemma 5.2: Let $\mathcal{A}, \mathcal{B}$ be algebras of the same type and $A \in V(\mathcal{B})$. If $\mathcal{B}$ is finite and $\mathcal{A}$ is finitely generated, then $\mathcal{A}$ is finite.

Proof: Suppose $\mathcal{A}$ is generated by n-elements, then as $\mathcal{A} \in HSP\{\mathcal{B}\}$ there is an $\mathcal{A}' \in SP\{B\}$, and $\mathcal{A} \in H\{A'\}$. We may assume $\mathcal{A}'$ is generated by n-elements, it suffices to show $\mathcal{A}'$ is finite. Suppose $\mathcal{A}'$ is generated by $\{a_1, \ldots, a_n\}$ and $\mathcal{A}'$ is a subalgebra of $\prod_{i \in I} \mathcal{B}_i$ where $\mathcal{B}_i = \mathcal{B}$ for all i. Let $\phi_i : \mathcal{A}' \to \mathcal{B}_i$ be the ith projection function. We may define an equivalence relation $\theta$ on I as follows: $i \sim j(\theta)$ if and only if $\phi_i(a_k) = \phi_j(a_k)$ for all $k \le n$. As B is finite, say $|B| = m$, there are at most $m^n$ different equivalence classes $i^\theta$, $i \in I$. Pick a representative for each such equivalence class and let $I' \subseteq I$ be this set of representatives. Then clearly $\mathcal{A}'$ is isomorphic to a subalgebra of $\prod_{i \in I'} \mathcal{B}_i$, and I' and B are finite, $\mathcal{A}'$ and $\mathcal{A}$ are finite and $|A| \le m \cdot m^n$.

Q.E.D.

It follows that $|\mathcal{F}_2| \le 3 \cdot 3^2 = 27$. Direct computation shows that $\mathcal{F}_2 \cong \mathcal{L}_3 \times \mathcal{L}_3 \times \mathcal{L}_2$ and has 18 elements. This argument is extended for all $\mathcal{F}_n$, $n \in \omega$ in the following:

Theorem 5.1: $\mathcal{F}_{n+1} \in V(\mathcal{F}_n^+)$, $\mathcal{F}_{n+1}$ is finite for all $n \in \omega$.

Proof: The proof is by induction, the case $n = 1$ having been covered in Lemma 2.1, where $\mathcal{L}_3 \cong \mathcal{F}_1^+$. Assume for all $k < n$, $\mathcal{F}_{k+1} \in V(\mathcal{F}_k^+)$ and $\mathcal{F}_{k+1}$ is finite. Let $\mathcal{A}^+$ be a subdirectly irreducible factor of $\mathcal{F}_{n+1}$ and $\phi:\mathcal{F}_{n+1} \to \mathcal{A}^+$ the onto projection. As $\phi$ is onto, $\mathcal{A}^+$ is generated by $\leq n+1$ elements, thus $\mathcal{A} \subseteq_s \mathcal{A}^+$ is generated by $\leq n$ elements and $\mathcal{A} \varepsilon H(\mathcal{F}_n)$. As $\mathcal{F}_n$ is finite by induction hypothesis, $\mathcal{A} \subseteq_s \mathcal{F}_n$ by lemma 2.6. Then $\mathcal{A}^+ \subseteq_s \mathcal{F}_n^+$ by lemma 2.7, and $\mathcal{F}_{n+1} \varepsilon V(\mathcal{F}_n^+)$. $\mathcal{F}_{n+1}$ is then finite by Lemma 5.2, and the induction hypothesis. Q.E.D.

Corollary 1: Every finitely generated Skolem ring is finite.

Corollary 2: Varieties of the form $V(\mathcal{A})$ for $\mathcal{A}$ finite and subdirectly irreducible form a $\vee$-base for VL(SR).

Proof of Corollary 2: If an equation $\phi$ fails in a variety K of Skolem Rings, it fails in some finitely generated and hence finite member of K. Q.E.D.

As Lemma 5.3, below shows, varieties of the form $V(\mathcal{A})$, for $\mathcal{A}$ finite, are strongly-directed-$\vee$-irreducible in VL(S.R.).

Lemma 5.3: Let K be a set of Skolem rings. Let $\mathcal{A} \varepsilon V(K)$. If $\mathcal{A}$ is finite, then there is a finite subset $K' \subseteq K$ and $\mathcal{A} \varepsilon V(K')$.

Proof: As $\mathcal{A} \varepsilon HSP(K)$, there are $K_i \varepsilon K$, $\mathcal{B} \subseteq_s \prod_{i \varepsilon I} K_i$, and $\phi: \mathcal{B} \to \mathcal{A}$ a homomorphism of $\mathcal{B}$ onto $\mathcal{A}$. Since $\mathcal{A}$ is finite we may assume $\mathcal{B}$ is finitely generated and thus by theorem 5.1, $\mathcal{B}$ is finite. Let $B = \{b_1, \ldots, b_n\}$ and for each $i < j \le n$ pick $k_{ij} \in I$ such that $\pi_{k_{ij}}(b_i) \neq \pi_{k_{ij}}(b_j)$ where $\pi_k$ denotes the k-th projection $\mathcal{B} \to K_k$. Let $I' = \{k_{ij} : i < j \le n\} \subseteq I$ and let $\mathcal{B}_{I'}$ be the projection $\pi_{I'}$ of $\mathcal{B}$ onto $\prod_{i \in I'} K_i$. Then $\pi_{I'}$ is an isomorphism of $\mathcal{B}$ onto $\mathcal{B}_{I'}$ and $\mathcal{A} \in HSP\{K_i : i \in I'\}$.

Q.E.D.

Closure Spaces $\mathcal{S}$ which are both discrete and topological are called singulary. They have the property that for $B \subseteq \mathcal{S}$, $\text{cl } B = \bigvee_{b \in B} \text{cl}\{b\} = \bigcup_{b \in B} \text{cl}\{b\}$; that is, the closure of a set is the union of the closure of its point subsets. It follows that singulary complete lattices (the lattices of closed sets of a singulary space) are completely distributive. We will show by Lemma 5.3, and a result due to B. Jonsson [ 8 ], that VL(S.R.) is also topological and hence singulary.

Lemma: (Jonsson [8]). Let K be a class of algebras of the same type such that $Cgr(\mathcal{B})$ is distributive for each $\mathcal{B} \in V(K)$. If $\mathcal{A}$ is subdirectly irreducible and $\mathcal{A} \in V(K)$, then $\mathcal{A} \in HSP_u(K)$ where $P_u$ denotes closure under the taking of ultraproducts.

Corollary: Let $\mathcal{A}$, K be as above, if $K_1, K_2 \subseteq K$, and $\mathcal{A} \varepsilon V(K_1 \cup K_2)$, then $\mathcal{A} \varepsilon V(K_1)$ or $\mathcal{A} \varepsilon V(K_2)$.

As we have noted, for any variety K, varieties of the form $V(\mathcal{A})$ for $\mathcal{A}$ subdirectly irreducible form a $\bigvee$-base for VL(K). If in addition K has the property that for every algebra $\mathcal{A} \varepsilon K$, $Cgr(\mathcal{A})$ is distributive, then $\mathcal{A}$ subdirectly irreducible implies $V(\mathcal{A})$ is strongly-$\vee$-irreducible, and VL(K) is topological. For in this case by the corollary to Jonsson's Lemma, if $\mathcal{A} \varepsilon (\mathcal{B}_1) \vee \ldots \vee (\mathcal{B}_n)$ for $\mathcal{B}_1, \ldots, \mathcal{B}_n \varepsilon K$, then there is $i \leq n$ and $\mathcal{A} \varepsilon \vee(\mathcal{B}_i)$.

Lemma 5.4. VL(SR) is topological.

Proof: By Jonsson's Lemma, we have only to show that $Cgr\,\mathcal{A}$ is distributive for every Skolem ring $\mathcal{A}$, and this follows as $Cgr\,\mathcal{A} \cong Idl\,\mathcal{A}$ (theorem 1.1, lemma 2.5) and $Idl(\mathcal{A})$ is distributive (lemmas 2.3 and 2.4).

Combining Lemmas 5.3 and 5.4 we get the following

Theorem 5.2: VL(S.R.) is singulary. Varieties of the form $V(\mathcal{A})$ for $\mathcal{A}$ finite and subdirectly irreducible are both strongly-$\vee$-irreducible and strongly-directed-$\bigvee$-irreducible and form a $\bigvee$-base for VL(SR).

Corollary: VL(S.R.) is completely distributive.

5.2. Non-finitely Based Theories of Skolem Rings.

Definition 5.1: An equational theory is <u>finitely based</u> if it is axiomatized by a finite number of identities.

Theorem 1.2 shows that the theory of Skolem rings is finitely based, and theorem 3.2 shows that $\mathscr{C}_2^\tau$ is finitely based. Similarly by picking identities $\phi_1, \phi_2 \in \mathscr{C}_3^\tau$ such that $\phi_1$ fails in $\mathscr{C}_4$ and $\phi_2$ fails in $\mathscr{D}$ we see by figure 4.1 that $\phi_1, \phi_2$ together with the Skolem ring identities of theorem 1.2 form a finite base for $\mathscr{C}_3^\tau$. This argument is easily generalized by means of theorems 4.2 & 4.4, to show that the theory of every finite Skolem ring is finitely based.

Definition 5.2: Let $K \in VL(S.R.)$, we will call $\{\mathscr{A}_i : i \in I\}$ where each $\mathscr{A}_i$ is a finite subdirectly irreducible Skolem ring, a <u>cover</u> for $K$, if for each $i$, $K \vee V(\mathscr{A}_i)$ covers $K$ in $VL(S.R.)$, and if whenever $K' \supseteq K$, for $K' \in VL(S.R.)$ either $K' = K$ or there is an $i \in I$ and $\mathscr{A}_i \in K'$. The set $\{\mathscr{A}_i : i \in I\}$ is a <u>minimal cover</u> for $K$ if it is a cover for $K$, and for each proper subset $I' \subset I$, $\{\mathscr{A}_i : i \in I'\}$ is not a cover for $K$.

Lemma 5.5: Every variety $K$ of Skolem rings has a minimal cover.

Proof: This follows directly from Theorem 5.1, Corollary 2. For each $K' \in VL(S.R.)$ which properly contains $K$, there is a finite subdirectly irreducible $\mathscr{A} \in K'$, $\mathscr{A} \notin K$. The minimal cover for $K$ consists of all such $\mathcal{A}$ of minimal cardinality.

Q.E.D.

It is easily seen that for every variety $K$ of Skolem rings which has a finite cover, $K^{\tau}$ is finitely based. For if $\{\mathcal{A}_i : i \leq n\}$ is a cover for $K$, we may pick an identity $\phi_i \in K^{\tau}$ for each $i \leq n$ such that $\phi_i \notin (\mathcal{A}_i)^{\tau}$, the set $\{\phi_i : i \leq n\}$ together with the Skolem ring identities of theorem 1.2 will form a finite generating set for $K^{\tau}$.

Lemma 5.6: Let $\mathcal{A}$ be any finite Skolem ring, then $\mathcal{A}^{\tau}$ is finitely based.

Proof: We will show that $VL(\mathcal{A})$ has a finite cover. As $\mathcal{A}$ is finite, we may write $V(\mathcal{A}) = V(\{\mathcal{A}_i : i \leq n\})$ for some finite set of subdirectly irreducible Skolem rings $\mathcal{A}_i$. Let $m$ be the maximum cardinality of the $\mathcal{A}_i$, the result will follow from theorem 5.1, by showing that each member $\mathcal{B}$ of a minimal cover for $V(\mathcal{A})$ is generated by $\leq m + 1$ elements. Let $\mathcal{B}$ be a member of a minimal cover for $V(\mathcal{A})$, and let $\mathcal{B}_1$ be a maximal proper subalgebra of $\mathcal{B}$. As $\mathcal{B}$ is finite, we have $|\mathcal{B}_1| < |\mathcal{B}|$ and thus $\mathcal{B} \notin HS(\mathcal{B}_1) = HSP_u(\mathcal{B}_1)$. By the Jonsson Lemma this gives $\mathcal{B} \notin V(\mathcal{A}) \vee V(\mathcal{B}_1)$ and thus $V(\mathcal{A}) = V(\mathcal{A}) \vee V(\mathcal{B}_1)$. Again by the Jonsson Lemma we have $\mathcal{B}_1 \in HSP_u(\mathcal{A}_i) = HS(\mathcal{A}_i)$ for some $i$ and hence $|\mathcal{B}_1| \leq m$. It follows that $\mathcal{B}$ has $\leq m + 1$ generators.

Q.E.D.

Remark: It is easily seen that every theory of Skolem Rings generated by two identities $\phi_1:\sigma_1 = \tau_1$ and $\phi_2:\sigma_2 = \tau_2$ is equivalent to the theory generated by the single identity $\phi:(\sigma_1 + \tau_1) \vee (\sigma_2 + \tau_2) = 0$. Thus we have a result analogous to that of first-order sentences, namely every finitely based theory is one-based, i.e., is the theory generated by a single equation.

Definition 5.3: A set of identities $\{\phi_i : i \in I\}$ of the same type is <u>independent</u> if each distinct subset generates a distinct theory. That is if for $i \notin I' \subset I$, we have $\phi_i \notin \{\phi_j : j \in I'\}^{\mu\tau}$.

Lemma 5.7: The following are equivalent statements:

1. There exists a non-finitely based theory of Skolem rings.
2. There exists an infinite set of finite Skolem rings $\{\mathcal{A}_i : i \in \omega\}$ with the property that $\mathcal{A}_i \not\subseteq_s \mathcal{A}_j$ for $i \neq j$.
3. There is an infinite independent set $\{\phi_i : i \in \omega\}$ of Skolem ring identities.
4. There are $2^{\aleph_0}$ distinct theories of Skolem rings.

Proof:

$1 \rightarrow 2$: Let $K \in VL(S.R.)$ and $K^\tau$ be non-finitely based. Let $\{\mathcal{A}_i : i \in I\}$ be a minimal cover for $K$. As $K^\tau$ is non-finitely based, $|I| = \aleph_0$, and any countable subset of the $\mathcal{A}_i$ will do for 2.

$2 \to 3$: Let $\{\mathscr{A}_i : i \in \omega\}$ be a countable set of finite Skolem rings with the property that $i \neq j \;.\supset.\; \mathscr{A}_i \not\subseteq_s \mathscr{A}_j$. Then for $i \neq j$, $\mathscr{A}_i \not\subseteq_s \mathscr{A}_j$ which implies by Lemma 2.6, $\mathscr{A}_i \notin HS(\mathscr{A}_j)$ and by Jonssons Lemma $\mathscr{A}_i \notin V(\mathscr{A}_j)$. Let $K_i = V(\{\mathscr{A}_j : j \neq i, j \in \omega\})$. It follows from Lemma 5.4, that $\mathscr{A}_i \notin K_i$. For each $i \in \omega$ we may pick $\phi_i \in K_i^\tau$, $\phi_i \notin \mathscr{A}_i^\tau$. The set $\{\phi_i : i \in \omega\}$ picked in this manner is clearly independent.

$3 \to 4$, and $4 \to 1$ are clear. Q.E.D.

In order to answer the question of the existence of a non-finitely based theory of Skolem rings, it seems easiest to look for an infinite set $\{\mathscr{A}_i : i \in \omega\}$ of Skolem rings as in 2 of Lemma 5.7. We will now exhibit such a set together with an infinite Skolem ring $\mathscr{A}_\omega^+$ whose theory is easily shown to be non-finitely based.

We will let $\mathscr{A}_i$, $i \in \omega$ denote the Skolem rings associated with the distributive lattices of figure 5.2. Each of the structures $\mathscr{A}_n$ may be described as being generated by two chains $a > b > b_2 > \ldots > b_{2n} > d$ and $b_1 > b_3 > \ldots > b_{2n+1} > c$ such that (1) a,b are incomparable with $b_1$, (2) c is incomparable $b_{2n}$,d, (3) $b_i$ is incomparable with $b_{i+1}$ all $i \leq 2n$, (4) $b > b_3$, $b_{2n+1} > d$, and $b_i > b_{i+3}$ all $i \leq 2n - 2$, and (5) $b_1$ and $b_{2n}$ are the only two elements of $\mathscr{A}_n$ incomparable with four distinct elements. Where by "x incomparable with y" we mean $x \nleq y$ and $y \nleq x$.

We will say that $\mathcal{A}_n$ has rank n. This construction may be extended to form similar Skolem rings with ranks of arbitrarily large cardinality. $\mathcal{A}_\omega$ will be used to denote such a structure with an $\omega + \omega^*$ sequence of $b_i$.

The important feature of this construction is the interlocking nature of the $b_i$-sequence. Namely $b_i > b_{i+3}$ but $b_i$ incomparable with $b_{i+1}$. Let $\phi$ be an embedding of $\mathcal{A}_n$ into $\mathcal{A}_m$, $n,m \in \omega$. It follows that if $b_i \in \mathcal{A}_n$, $b'_j \in \mathcal{A}_m$ and $\phi(b_i) = b'_j$, then $\phi(b_{i+1}) = b'_{j+1}$. By (5) $\phi(b_1) = b'_1$ and $\phi(b_{2n}) = b'_{2m}$. Thus $\phi$ is an embedding of $\mathcal{A}_n$ into $\mathcal{A}_m$ if and only if $n = m$ and $\phi$ is an isomorphism of $\mathcal{A}_n$. We have

Lemma 5.8: For each of the Skolem rings $\mathcal{A}_i$, $i \in \omega$ described in Figure 4.2, we have $i \neq j \supset \mathcal{A}_i \not\subseteq_s \mathcal{A}_j$.

Lemma 5.9: Let $\mathcal{A}$ be any non-trivial ultraproduct of the set $\{\mathcal{A}_i^+ : i \in \omega\}$ of Lemma 5.8. Then $\mathcal{A}^\tau = (\mathcal{A}_\omega^+)^\tau$ and the equational theory of $\mathcal{A}_\omega^+$ is non-finitely based.

Proof: The features (1)-(5) of the $\mathcal{A}_i$ can be stated in sentences of first order, it follows that any ultraproduct of the $\mathcal{A}_i^+$ must be similarly constructed, with non-trivial ultraproducts having infinite (in fact uncountable) rank. It is clear that any two such Skolem rings of infinite rank have the same finite subalgebras and thus the same equational theory.

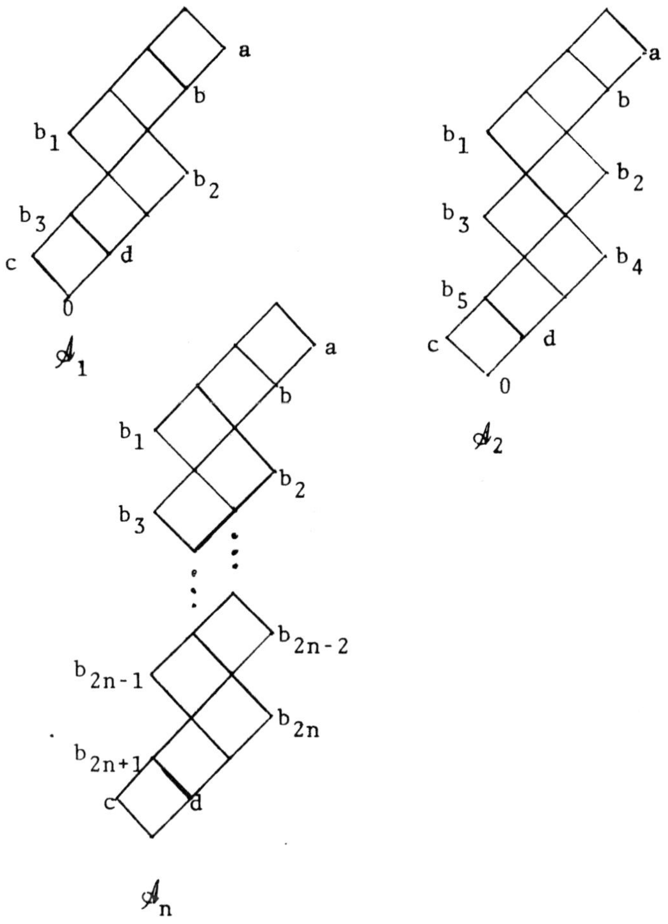


Figure 5.2

Skolem Rings $\mathcal{A}_n$ of Lemma 5.8

As no proper homomorphism of $\mathcal{A}_\omega$ will satisfy feature (5) of the construction, by the Jonsson lemma $\mathcal{A}_i^+ \notin V(\mathcal{A}_\omega^+)$. Now suppose $\mathcal{A}_\omega^\tau$ is finitely based, then by the remark following Lemma 5.6 $\mathcal{A}_\omega^\tau = \{\phi\}^{\mu\tau}$ for a single identity $\phi$. As $\mathcal{A}_i^+ \notin V(\mathcal{A}_\omega^+)$ we have $\phi \notin (\mathcal{A}_i^+)^\tau$ for all i. $\phi$ then fails in every ultraproduct of the $\mathcal{A}_i$, and thus $\phi \notin (\mathcal{A}_\omega^+)^\tau$. It follows that $(\mathcal{A}_\omega^+)^\tau$ is non-finitely based. Q.E.D.

Remark: If we consider $\mathcal{A}_\omega$ as a Skolem lattice, that is as an algebra $\langle A_\omega, +, \vee, \wedge, 0\rangle$, any two elements $(b_i \wedge b_{i+1})$, $b_{i+2}$ generate an infinite subalgebra. This is in contrast to the situation for Skolem rings, Lemma 5.1.

5.3. Positive Implication Algebras, Skolem Lattices

Let $\mathcal{P}$ be a propositional calculus of the intuitionistic logic of Heyting and Brouwer (see Heyting [7]). Let P be the set of all formulas of $\mathcal{P}$. We may define a quasi-ordering $\prec$ on P by $A \prec B$ .≡. $\vdash_{\mathcal{P}} A \to B$. Let & and $\vee$ be the conjunction and disjunction on $\mathcal{P}$, let $(\equiv)$ denote the equivalence relation on P induced by $\prec$, and $1 = [A \to A]^{(\equiv)}$, $0 = [\neg A \mathbin{\&} A]^{(\equiv)}$. The structure $\langle P/(\equiv), \to, \vee, \&, 1\rangle$ forms a Skolem lattice in the dual form, that is, $\vee$ and & are join and meet operations for the partial ordering $\le$ on $P/(\equiv)$, and $(A \mathbin{\&} B) \le C$ .≡. $\vdash_{\mathcal{P}} (A \mathbin{\&} B) \to C$ .≡. $\vdash_{\mathcal{P}} B \to (A \to C)$ .≡. $B \le (A \to C)$. Algebras of this form are also called relatively pseudo-complemented lattices (Birkhoff [2]). Relatively pseudo-complemented lattices with the zero element

0 are called Heyting algebras, pseudo-complemented algebras, and in their dual form Brouwerian algebras, (Rasiowa [14]).

The equational theory of implication, that is, the equational theory of the structures $\langle P/(\equiv), \rightarrow, 1\rangle$ has also been studied; Diego [ 5 ], in particular, has proven a result for these structures corresponding to our theorem 5.1. Following Rasiowa we will call such algebras Positive implication algebras.

In this section we will compare the variety (and dually the theory) lattices of positive implication algebras (P.I.), Skolem rings (S.R.) and Skolem lattices (S.L.).

Recall from Lemma 2.1 that

1) $x/(y \vee z) = (x/y)/z$

2) $(x \vee y)/z = (x/z) \vee (y/z)$

we use these two facts to prove a result for Skolem rings analogous to the disjunctive normal form in propositional calculus.

Lemma 5.10: Let $\phi$ be any term in the Skolem ring operations $/$, $\vee$, 0. Then there are terms $\psi_1,\dots,\psi_n$ in $/$, 0 such that $\phi = \bigvee_{i=1}^{n} \psi_i$ is a Skolem ring identity.

Proof: By induction on the length of the expression $\phi$. If $\phi$ is of the form $\phi_1 \vee \phi_2$ the result follows immediately from the induction assumption. Otherwise $\phi$ is of the form $\phi_1/\phi_2$ and we may assume there are terms $\psi_i$, $\psi_i'$ such that

$\phi_1 = \psi_1 \vee \ldots \vee \psi_n$ and $\phi_2 = \psi'_1 \vee \ldots \vee \psi'_m$. Then $\phi = \phi_1/\phi_2 = (\psi_1 \vee \ldots \vee \psi_n)/\phi_2 = (\psi_1/\phi_2) \vee \ldots \vee (\psi_n/\phi_2)$ using 1). Each $\psi_i/\phi_2 = (\ldots(\psi_i/\psi'_1)/\psi'_2)/\ldots/\psi'_m)$ and

$$\phi = \bigvee_{i=1}^{n} (\ldots(\psi_i/\psi'_1)/\psi'_2)/\ldots/\psi'_m)$$

Q.E.D.

Corollary 1: Let $\phi$ be any Skolem ring identity. Then there are identities $\psi_1,\ldots,\psi_n$ in /, 0 such that the Skolem ring theory generated by $\phi$ is the Skolem ring theory generated by $\{\psi_1,\ldots,\psi_n\}$.

Proof: $\phi$ is equivalent to an identity of the form $\tau = 0$. By the Lemma, there are $\psi'_1,\ldots,\psi'_n$ in /, 0 such that $\tau = \bigvee_{i=1}^{n} \psi'_n$, then $\tau = 0 \;.\equiv.\; (\bigvee_{i=1}^{n} \psi'_i) = 0 \;.\equiv.\; \psi'_i = 0$ all $i \leq n$, then let $\psi_i$ be the identity $\psi'_i = 0$.

Corollary 2: Every Skolem ring theory can be axiomatized by identities in /, 0, together with the two identities 1) and 2) above.

Proof: The partial ordering on a Skolem ring is completely determined by / , 0, that is, $a \leq b \;.\equiv.\; a/b = 0$. It is easily verified that 1) and 2) suffice to make $\vee$ a join for this partial ordering, while $x \leq y \vee z \;.\equiv.\; x/(y \vee z) = 0 \;.\equiv.\; (x/y)/z = 0 \;.\equiv.\; (x/y) \leq z$.

Q.E.D.

Ideals on a positive implication algebra $\mathcal{A}$ are defined as subsets $I \subseteq A$ such that $a \in I,\ b/a \in I \;.\supset.\; b \in I$.

Ideals on Skolem lattices are defined in the usual way. It follows as in Lemma 2.5 that for P.I. structures or S.L. structures $\mathcal{A}$, $Cgr(\mathcal{A}) \cong Idl(\mathcal{A})$ and $Idl(\mathcal{A})$ is distributive. It follows from theorem 5.1 and Jonssons Lemma that V.L.(P.I) (and th(P.I.)) is completely distributive.

Each finite subdirectly irreducible Skolem ring $\mathcal{A} = \langle A,/,\vee,0\rangle$ has associated with it a unique finite, subdirectly irreducible P.I. algebra $\mathcal{A}' = \langle A,/,0\rangle$, and as $\mathcal{A}$ is finite, a unique finite, subdirectly irreducible Skolem lattice $\mathcal{A}^\wedge = \langle A,/,\vee,\wedge,0\rangle$, where $\wedge$ is the meet operation on $\mathcal{A}$. By corollary 2 above, $\mathcal{A}^\tau$ is generated by $(\mathcal{A}')^\tau$ together with the identities 1) and 2). To each variety $K \in$ V.L.(P.I) there corresponds a variety $K^\vee \in$ V.L.(S.R.) given by the rule $\mathcal{A} \in K^\vee \;.\!\equiv.\; \mathcal{A}' \in K$ where $\mathcal{A}$ is finite and subdirectly irreducible.

Lemma 5.11: The map $\phi: K \to K^\vee$ defines a homomorphism of V.L.(P.I.) onto V.L.(S.R.).

Proof: Let $K_1, K_2 \in$ V.L.(P.I.). That $\phi(K_1 \cap K_2) = \phi(K_1) \cap \phi(K_2)$ is clear, we must show $\phi(K_1 \vee K_2) = \phi(K_1) \vee \phi(K_2)$. Let $\mathcal{A}$ be a finite subdirectly irreducible Skolem ring and suppose $\mathcal{A} \in \phi(K_1 \vee K_2)$, then the corresponding P.I. algebra $\mathcal{A}' \in K_1 \vee K_2$. As $\mathcal{A}$ is subdirectly irreducible, $\mathcal{A}'$ is also, hence by the distributivity of V.L.(P.I.) and the Jonsson Lemma $\mathcal{A}' \in K_1$ or $\mathcal{A}' \in K_2$ and $\mathcal{A} \in \phi(K_1)$ or $\mathcal{A} \in \phi(K_2)$. Then $\phi(K_1 \vee K_2) \subseteq \phi(K_1) \vee \phi(K_2)$. The reverse inclusion is obvious. That $\phi$ is onto follows by Corollary 2 above. Q.E.D.

For $K \varepsilon$ V.L.(S.R.) let $K' \varepsilon$ V.L.(P.I.) and $K^\wedge \varepsilon$ V.L.(S.L.) be defined as follows: $K'$ is generated by the algebras $\mathcal{A}' = \langle A,/,0\rangle$ where $\mathcal{A}$ is finite, subdirectly irreducible, $\mathcal{A} \varepsilon K$, and $K^\wedge$ is similarly generated by the algebras $\mathcal{A}^\wedge$.

Lemma 5.12: The map $\phi: K \to K'$ defines an embedding of V.L.(S.R.) as a complete sub-lattice of V.L.(P.I.). The map $\psi: K \to K^\wedge$ defines an embedding of V.L.(S.R.) as a complete sub-lattice of V.L.(S.L.).

Proof: As V.L.(P.I.) is distributive, $\phi$ is an embedding of V.L.(S.R.) as a sub-lattice by theorem 5.1 and Jonssons Lemma. It is clear that $\phi$ preserves arbitrary meets, and Lemma 5.3 is used to show $\phi$ preserves arbitrary joins. The second statement is proved similarly. Q.E.D.

By theorem 5.2 for any independent set $\{\phi_i : i \varepsilon I\}$ of Skolem ring identities there are finite Skolem rings $\mathcal{A}_i$ such that $\phi_j \varepsilon \mathcal{A}_i^\tau$ .≡. $j \neq i$. Since each $\mathcal{A}_i$ is finite there is a $\wedge$-operation on $\mathcal{A}_i$ and $\phi_j \varepsilon \langle A_i,/,\vee,\wedge,0\rangle^\tau$ .≡. $j \neq i$. The identities are independent when considered as identities in S.L.

Theorem 5.3: There is a countable set of independent Skolem ring and Skolem lattice identities. There is a countable set of independent identities for positive implication algebras.

Proof: The first statement follows from lemmas 5.7 and 5.8 and the above remarks. The second statement follows by generalizing lemmas 5.7 and 5.8 to the P.I-algebras.

Q.E.D.

BIBLIOGRAPHY

1. Birkhoff, G., "On the structure of abstract algebras", Proc. Cambridge Phil. Soc., 31(1935) pp. 433-453.

2. ____________, Lattice Theory, 2nd Ed., AMS Coll. Publ. 25, Providence (1948).

3. Buchi, J. R., "Representation of complete lattices by sets," Portugalia Math., 11(1952) pp. 151-167.

4. Curry, H. B., Foundations of mathematical Logic, New York, McGraw-Hill, 1963.

5. Diego, A., Les algebres de Hilbert, Paris, 1966.

6. Gratzer, G. and Schmidt, E. T., "Characterizations of congruence lattices of abstract algebras", Acta Sci. Math., 24(1963) pp 34-59.

7. Heyting, A., Intuitionism, an Introduction, Studies in Logic and Foundations of Mathematics, Amsterdam, 1956.

8. Jonsson, B., "Algebras whose congruence lattices are distributive," Math. Scand., 21(1967) pp 110-121.

9. McKinsey, J. C. C. and Tarski, A., "The algebra of topology," Annals of Math., 45(1944) pp 141-191.

10. ____________________, "On closed elements in closure algebras", Annals of Math., 47(1946) pp 122-162.

11. Monteiro, A., "Axiomes independents pour les algebres de Brouwer," Rev. Un. Mat. Argentina, 17(1955) pp 149-160.

12. Nachbin, L, "On a characterization of the lattice of all ideals of a Boolean ring," Fund. Math. 36(1949) pp 137-142.

13. Nemitz, W. and Whaley, T., "Varieties of implicative semi-lattices," Pacific Journal of Mathematics, Vol. 37, No. 3, (1971) pp 759-769.

14. Rasiowa, H., An algebraic approach to non-classical logics, Studies in Logic and Foundations of Mathematics, Amsterdam, 1974.

This work was supported by NSF Grant No. GJ 40891.

# Section 3 Towards a Theory of Definability

*With comments by Ernst Specker, ETH Zürich, with assistance from Kenneth Danhof, Southern Illinois University at Carbondale*



*

* This paper was written for the *Proceedings of the Symposium on the Theory of Models*, Berkeley, Calif., 1963, but not included due to its length; only the abstract (11) appeared instead. There exists a later copy of the paper with many changes and additions in Büchi's handwriting, with the title changed to "Galois Theory of Elementary Axiom Systems." [*The Editors*]

# Büchi's Work in Definability

*Comments by Ernst Specker and Kenneth Danhof*

In papers [5] and [6] (and later in [40] and [22]), an attempt is made to generalize (and to make more precise) the fundamental ideas of Klein's Erlanger program (see reference 10 in [22]) and those of Galois. Klein's starting point is as follows: Given a structure $A$ (say, Euclidian space) with automorphism group $G_o(A)$ (in Euclid's case, group of similitudes) and a group extension $G(A)$ of $G_o(A)$ (say, group of affine transformations) find invariants characterizing the transformations of $G(A)$ (solution: ratio of segments on a line).

In both [5] and [6], the structure $A$ is a group. As extensions of the group of automorphisms the following groups are considered: In [5], the group generated by automorphisms and translations $T_a$ ($a \in A, T_a(x) = x \cdot a$); in [6], the group of automorphisms and anti-automorphisms. The following solution is presented: [5] The quaternary relation $\pi$ (called proportionality relation) defined by $\pi(x, y, u, v) := x \cdot y^{-1} = u \cdot v^{-1}$ is a characterizing invariant. [6] The relation $\alpha$ defined by $(x \cdot y = z \vee y \cdot x = z)$ is a characterizing invariant.

The main technical result of [6] is the proof that the disjunction in the definition of $\alpha$ is essential; more precisely there is no set of relations expressible as equations between words characterizing the group of automorphisms and anti-automorphisms.

In [5], the proportionality relation is studied in its own right; necessary and sufficient conditions for a quaternary relation to be the proportionality relation of a group are stated and these properties of a relation are used to define the notion of a "regular bi-equivalence." The main technical result is that a quaternary relation is the proportionality relation of some group iff it is a regular bi-equivalence.

It is, however, quite obvious that the real interest is not supposed to lie in these technical results—13 notes added to paper [5] clearly testify to that. (There are seven notes at the end of Klein's paper exposing the Erlanger program.) What the authors seem to hope for is that their work is a first step toward "adequate metamathematics" (cf. note 5), where, e.g., a group is not just a structure on a set but something more general (which in "real mathematics" it certainly is).

In [40] (a fundamental paper that has never received the exposure it deserves), Büchi moves from the specific examples of [5] and [6] toward a more general study of the invariants of elementary concepts in elementary theories. The ideas of a theory being categorical relative to a primitive concept and normal relative to a concept are introduced. Two results are established. The first result asserts that in a theory $T(R)$, which is *categorical relative to* the concept (primitive) $R$ (i.e., any two models of $T(R)$ in which $R$ takes the same value are isomorphic), every $R$-invariant concept is definable from $R$. This theorem is shown to follow from Beth's definability theorem and has as a simple consequence a fundamental result of Klein's Erlanger program—roughly speaking, every Euclidian concept that is affine invariant can be represented by an affine concept.

The second result involves a theory $T(R, S)$ that is *normal relative* to $R$, i.e., a theory in which any two $R$-conjugate models (models $\langle R, S_1 \rangle$ and $\langle R, S_2 \rangle$ are *R-conjugate* if they are isomorphic) have the same automorphism groups. It is shown that if $T(R, S)$ is normal relative to $R$ then there is an integer $n$ such that every model of $T(R, S)$ has at most $n$ $R$-conjugates. This result is shown to follow from a theorem of Svenonius and has as a consequence the fact that Galois groups of elementary theories are finite.

These results suggest a new branch of model theory that is concerned with the classification of theories by means of automorphism groups on models.

Paper [22] attempts to extend the ideas introduced in [5], [6] and [40] and in particular focuses on the case of complete theories. In this situation Beth's definability theorem (relative to a complete theory $T$) results in an upper semilattice having as elements (concepts) the interdefinable formulas of $T$ (i.e., formulas having equal automorphism groups in all models). It is shown that in the case of a normal concept in a complete theory, the upper semi-lattice of all concepts that define the given normal concept is a finite lattice—anti-isomorphic to the lattice of subgroups of the corresponding anti-isomorphism group. Connections with the Galois theory of fields are discussed.

Papers [21] and [23] represent something of a departure from the theme of the papers just discussed. Paper [21] investigates methods or approaches available when considering the problems of mathematical logic from a model-theoretic point of view. The particular approaches considered are based on elementary embeddings, strong elementary embeddings, and ultrapowers. These various approaches are organized into a series of levels. The underlying uniformity of these different levels is illustrated by using the methods on each level to obtain a corresponding series of definability results. Extensions of Craig's lemma and Beth's definability theorem occur as consequences of the analysis.

Paper [23] considers the universality of the methods used by Cantor in characterizing the order types of the linear continuum and the rationals. An axiomatic theory designed to capture this universality is proposed. The paper notes several applications of the resulting abstraction of Cantor's methods to the model-theoretic study of relational systems. It is suggested that the concepts introduced might serve as the beginnings of an axiomatic treatment of model theory.

In his careful review of [6], R.C. Lyndon (*Math. Review*, 20, p. 395) points out the following errata: p. 1136, last line, read "invariance." p. 1138, sixth line from bottom, read "four." He also sketches an alternative (and rather short) proof of what has been called the main technical result.

BÜCHI, J. R., and J. B. WRIGHT
Math. Annalen, Bd. 130, S. 102—108 (1955).

# The Theory of Proportionality as an Abstraction of Group Theory.

By

J. RICHARD BÜCHI and JESSE B. WRIGHT in Ann Arbor, Michigan.

In this paper we present an axiomatic theory denoted by $T$ which is an abstraction of group theory in the sense that projective geometry is an abstraction of affine geometry. Extending to group theory, F. KLEIN's idea, that a geometry is determined by its group of automorphisms, we describe $T$ as the theory whose group of automorphisms is generated by the automorphisms and translations of group theory. In axiomatic terms the relationship between the two theories is characterized by the fact that from $T$ we can arrive at group theory by distinguishing an element, i. e., by introducing a name for the group identity. This situation is analogous to that in geometry, where affine geometry is derived from projective geometry by the introduction of a name for the lines at infinity.

The theory $T$ is of interest because it turns out to be an axiomatization of the notion of proportionality. Furthermore, this paper represents a step in our current investigation of a more or less familiar metamathematical concept which we will call abstraction. The relation between $T$ and group theory is studied here because it affords a useful example of the process of abstraction. In section 4 we put forward a series of notes suggested by this and other instances of abstraction.

## 1. A Characterizing Invariant of the Holomorph of a Group.

Because group multiplication is not invariant under translations, it clearly does not belong to the theory $T$. It is our first task to find, in group theory, a notion which is invariant under group automorphisms and group translations and which is invariant only under transformations which are composed of group automorphisms and group translations.

Let $A$ be a set and let $x, y, z, \ldots$ be variables ranging over $A$. Let $x \cdot y$, $\dot{x}$ and $e$ denote the product, inverse and identity of a group $\bar{A} = [A, \cdot, \dot{}, e]$ on $A$.

**Definition 1:** The group $H$ of transformations of $A$ generated by the automorphisms and translations of a group $\bar{A}$ is called the holomorph of $\bar{A}$. The elements of $H$ are called holomorphisms.

**Definition 2:** The quaternary relation $\pi$ on $A$, defined by:

$$\pi(x, y, u, v) \quad \equiv \quad x \cdot \dot{y} = u \cdot \dot{v},$$

is called the proportionality relation of the group $\bar{A}$.

**Theorem 1:** *The proportionality relation $\pi$ of a group $\bar{A}$ is an invariant of the holomorph $H$ of $\bar{A}$, and every transformation of $A$ which has the invariant $\pi$ is a holomorphism of $\bar{A}$.*

**Proof:** It is easily seen that $\pi$ is invariant under automorphisms as well as translations. This proves the first part of Theorem 1.

Next suppose $F$ is a transformation of $A$, which has the invariant $\pi$. Now, if $x \cdot y = z$, then $\pi(x, e, z, y)$ and because $\pi$ is an invariant of $F$, $\pi(Fx, Fe, Fz, Fy)$. If $Fe = e$ it follows, $Fx \cdot Fy = Fz$. Thus $F$ is an automorphism of $\bar{A}$. If $Fe = a$ we define $S$ by, $Sx = \dot{a} \cdot Fx$. Then $S$ obviously has the invariant $\pi$ and $Se = e$. It therefore is an automorphism of $\bar{A}$. But $Fx = a \cdot Sx$.

We conclude, every transformation $F$ of $A$, which has $\pi$ as an invariant is a holomorphism. Q.E.D.

By further analysis of this proof we get:

**Theorem 2:** *Every holomorphism of a group $\bar{A}$ is of the form $LS$, where $L$ is a left translation of $\bar{A}$ and $S$ is an automorphism of $\bar{A}$. Every holomorphism of $\bar{A}$ which leaves e fixed is an automorphism of $\bar{A}$. The proportionality relation $\pi$ of a group $\bar{A}$ together with its identity e are characterizing invariants for the group of automorphisms of $\bar{A}$.*

Geometric intuition about the relation $\pi$ can be gained by considering a parallelogram in the affine plain with an origin 0. The vectors drawn from the origin to the four vertices of the parallelogram are in relation $\pi$. This example will also provide an intuitive background to the next section.

## 2. Characterizing Properties of the Proportionality Relation.

In this section we will find necessary and sufficient conditions for a quaternary relation $\varrho$ on a set $A$ to be the proportionality relation of some group $\bar{A}$ on $A$. Because of our intuition about the notion of proportionality it becomes evident that the proportionality relation of a group on $A$ has the following properties:

(1) $$\varrho(x, y, x, y)$$

(2) $$\varrho(x, y, u, v) \rightarrow \varrho(u, v, x, y)$$

(3) $$\varrho(x, y, u, v) \wedge \varrho(u, v, p, q) \rightarrow \varrho(x, y, p, q)$$

(4) $$\varrho(x, x, y, y)$$

(5) $$\varrho(x, y, u, v) \rightarrow \varrho(y, x, v, u)$$

(6) $$\varrho(x, y, u, v) \wedge \varrho(y, z, v, w) \rightarrow \varrho(x, z, u, w)$$

(7) $$\varrho(x, x, y, x) \rightarrow x = y$$

(8) $$(\mathrm{Eu})\ \varrho(x, y, u, v)$$

The relation between $(x, y)$ and $(u, v)$ which holds in case $\varrho(x, y, u, v)$, is by (1), (2) and (3) an equivalence relation on $A \times A$. Similarly a second equivalence relation arises in connection with (4), (5) and (6). Also, note that the duality between $x \cdot y$ and $y \cdot x$ is reflected in a duality of $\varrho$.

**Definition 3:** A relation $\varrho$ on the set $A$ which satisfies the conditions (1) to (6) is called a bi-equivalence on $A$. If $\varrho$ also satisfies (7) and (8), it is called a regular bi-equivalence.

**Lemma:** If $\varrho$ is a regular bi-equivalence on the set $A$ and if $x, e$ and $y$ are any elements of $A$, then there exists exactly one element $z$ of $A$, such that $\varrho(x, e, z, y)$.

**Proof:** That such $a$ $z$ exists follows from (8). Suppose now, $\varrho(x, e, z, y)$ and $\varrho(x, e, z' y)$. Then by (2), $\varrho(z, y, x, e)$ and by (3), $\varrho(z, y, z', y)$. By (1), $\varrho(y, z, y, z)$. Thus by (6) it follows, $\varrho(z, z, z', z)$, and by (7), $z = z'$. Q.E.D.

**Theorem 3:** *A quaternary-relation $\varrho$ on a set $A$ is the proportionality relation of some group $\bar{A}$ on $A$, if and only if $\varrho$ is a regular bi-equivalence.*

**Proof:** The "only if" can be shown by obvious computations in a group $\bar{A}$. We now assume that $\varrho$ is a regular bi-equivalence on $A$, and we choose an element $e$ in $A$. By the lemma, to any pair $x, y$ of elements of $A$, there exists exactly one $z$ in $A$, such that $\varrho(x, e, z, y)$. We define this $z$ to be $x \cdot y$. Also by the lemma, to any element $x$ in $A$ there exists exactly one element $y$ in $A$, such that $\varrho(x, e, e, y)$. We define this $y$ to be $\dot{x}$. Now let $\bar{A}$ be the algebraic structure $[A, \cdot, \dot{}, e]$. To end the proof of Theorem 3 we have to show first, $\bar{A}$ is a group, and second, $\varrho$ is the proportionality relation of $\bar{A}$.

By (5) and (6), $\varrho(v, e, r, w)$ and $\varrho(p, e, q, w)$ implies $\varrho(p, v, q, r)$, and by (3), $\varrho(u, e, p, v)$ and $\varrho(p, v, q, r)$ implies $\varrho(u, e, q, r)$. Thus, $\varrho(u, e, p, v)$ and $\varrho(p, e, q, w)$ and $\varrho(v, e, r, w)$ implies $\varrho(u, e, q, r)$. Translated according to our definition this means, $p = u \cdot v$ and $q = p \cdot w$ and $r = v \cdot w$ implies $q = u \cdot r$, or, $(u \cdot r) \cdot w = u \cdot (v \cdot w)$. Thus, the operation $x \cdot y$ is associative. The remaining group-axioms are easily established for $\bar{A}$.

Now suppose $\varrho(x, y, u, v)$. Then by (3), $\varrho(z, e, x, y)$ implies $\varrho(z, e, u, v)$. By our definition this means, $z \cdot y = x$ implies $z \cdot v = u$, or, because $\bar{A}$ is a group, $z = x \cdot \dot{y} = u \cdot \dot{v}$. Next, suppose $x \cdot \dot{y} = u \cdot \dot{v}$. Because $\bar{A}$ is a group it follows that there is a $z$, such that $x = z \cdot y$ and $u = z \cdot v$. By our definition this means, $\varrho(z, e, x, y)$ and $\varrho(z, e, u, v)$. By (2) and (3) we conclude, $\varrho(x, y, u, v)$. Thus, $\varrho(x, y, u, v)$ is equivalent with $x \cdot \dot{y} = u \cdot \dot{v}$. Q.E.D.

The significance of theorem 3 becomes more evident if we summarize it as follows: *The conditions* (1) *to* (8) *provide an axiomatic characterization of the notion of proportionality.*

### 3. The Relationship Between Group Theory and the Theory of Regular Bi-equivalences.

Here it is our aim to re-interpret the results of sections one and two. We will discuss the relationship between group theory and the theory of regular bi-equivalences. This relationship can be perceived more clearly only when each of the subject matters is presented as an applied first order functional calculus. However, we hope that the informal presentation which follows suffices to convey the fundamental ideas.

For later convenience, let us introduce the notation "$(\mu z) \ldots$" to stand for "the $z$ for which ...". Furthermore, let (a) and (b) be the following formulas:

(a) $$\varrho(x, y, u, v) \quad \equiv \quad x \cdot \dot{y} = u \cdot \dot{v}$$

(b) $$x \cdot y = (\mu z)\, \varrho(x, e, z, y)$$
$$\dot{x} = (\mu z)\, \varrho(x, e, e, z).$$

Now it is easy to verify:

**I.** If $T_1$ is an applied first order functional calculus with the primitives $A, \cdot, {}^{\cdot}, e$ and a conventional set of axioms for group theory, and if $\varrho$ is defined by (a), then $\varrho$ satisfies the axioms for a regular bi-equivalence, and moreover the statements (b) become theorems of $T_1$.

Conversely, by an analysis of the proof to Theorem 3 one verifies:

**II.** Let $T_2$ be an applied first order functional calculus with the primitives $A$, $\varrho$, $e$, which has as axioms the statements which make $\varrho$ a regular bi-equivalence on $A$ together with the statement that $e$ belongs to $A$. If $\cdot$ and ${}^{\cdot}$ are defined by (b), then $A, \cdot, {}^{\cdot}, e$ satisfy the group axioms, and moreover, (a) becomes a theorem of $T_2$.

Although it is difficult to define equivalence of logical systems in general, it will be recognized that I and II are sufficient for $T_1$ and $T_2$ to be called equivalent. Thus, in the light of I and II, $T_2$ can be regarded as an unconventional formulation of group theory.

From the point of view of models the equivalence of $T_1$ and $T_2$ expresses itself in the fact that every model of one theory is a model of the other theory. Finally, the equivalence of $T_1$ and $T_2$ is reflected in the fact that a transformation of $A$ is an automorphism of $T_1$ if and only if it is an automorphism of $T_2$. This is part of theorem 2.

Now let $T$ be the applied first order functional calculus with the primitives $A$ and $\varrho$, which has as axioms the statements making $\varrho$ into a regular bi-equivalence on $A$. I and II show that $T_1$ and $T_2$ have the same expressive power. However, the expressive power of $T$ is weaker than that of $T_2$, because there is no primitive constant of the type of individuals in $T$ nor can such a constant be defined in $T$. (If such a constant $e$ could be defined in $T$ it would be an invariant of every automorphism of $T$, but this contradicts theorem 1.) Therefore, from the equivalence of $T_1$ and $T_2$ it follows that $T$ is weaker than $T_1$, and in particular no constants $\cdot$ and ${}^{\cdot}$ can be defined in $T$. From Theorem 1 it follows that $T$ is weaker than $T_1$, also from the standpoint of groups of transformations.

The series of steps which leads from group theory $T_1$ to the theory $T$ of regular bi-equivalences is a typical example of what we call the process of abstraction. To better understand this process we considered a new formulation $T_2$ of group theory. $T$ might be called a direct abstraction of $T_2$.

## 4. Some Notes on the Process of Abstraction.

In the preceding sections we introduced the theory of regular bi-equivalences as an abstraction of group theory. There is a need for a systematic metamathematical theory which will explain the characteristics of this example as well as the characteristics of other examples of abstraction. The purpose of this section is to present in the form of informal notes some of the problems, conjectures and examples concerned with the notion of abstraction. A precise definition of abstraction can be given only as part of a metamathematical theory of the relationships between mathematical subject matters.

**Note 1:** The hierarchy of geometries affords the classical examples of abstraction. In the introduction we have already pointed out the close analogy between the relationship of affine to projective geometry and that of group theory to regular bi-equivalences. But the field of geometry still has room for new theories. One of us has formulated the abstraction of projective geometry whose automorphisms are the colineations and correlations [4].

**Note 2:** In the light of section 3 we may say that the theory of regular bi-equivalences differs from group theory only in its vocabulary, i. e., in its expressive power. From the formal point of view the weakening of the expressive power is characteristic for the relationship between subject matters which interests us. Our use of the term abstraction should be considered in this light.

**Note 3:** Abstraction can also be studied with reference to automorphisms. One of the characteristics of the process of abstraction is that the automorphism group is enlarged. Our investigations started with a consideration of the holomorph $H$, a super group of the automorphism group $G$ of group theory. The fact that an invariant of $H$ automatically is an invariant of $G$ together with the fact that $H$ is definable in group theory suggests that there is a characterizing invariant of $H$ definable in group theory. This consideration leads to the discovery of the proportionality relation $\varrho$ in section one. The fruitfulness of the automorphism approach to the theory of regular bi-equivalences suggests its use in defining and investigating other problems of abstraction.

**Note 4:** Although it is not true that the models of the theory of regular bi-equivalences and the models of group theory are exactly the same, nevertheless, every model of the abstracted theory can be extended to a model of the stronger theory by a slight variation, and every model of the stronger theory is an extention of a model of the abstracted theory. This suggests a third way of defining the concept of abstraction.

**Note 5:** The equivalence of theories $T_1$ and $T_2$ was discussed in section 3. An adequate metamathematics must provide a rigorous definition of the concept of equivalence. It should be noted that the equivalence and non-equivalence of two theories is not determined until their logical frameworks are specified. For example, let us consider the situation which arises in section 3 if in the logical frame the description operator $\mu$ is replaced by a full power selection operator. Now we can define a constant of the type of individuals in $T$. As a matter of fact, many such constants can be defined in $T$ and $T_1$ and the two theories become equivalent and of much stronger expressive power. There are many examples which demonstrate that this situation is not brought about by any peculiar properties of the operators in question.

**Note 6:** we used heuristically the principle that whatever is an invariant of the automorphism group of a formal theory must be definable in that theory. In section 3 we used the converse of this principle. These metamathematical propositions should be rigorously formulated and their valididy investigated.

**Note 7:** Besides the theory of regular bi-equivalences other abstractions of group theory arise if other extentions $G'$ of the group $G$ of group automorphisms are considered. We hope to present in another paper the theory whose group $G'$ consists of the automorphisms and anti-automorphisms of group theory.

**Note 8:** Let $\varrho$ be a regular bi-equivalence on the set $A$ and let $e$ be an element of $A$. Then by Theorem 3 we have a group $\bar{A}$ on $A$ which has $\varrho$ as its proportionality relation and $e$ as its identity. An element $e'$ different from $e$ will give rise to a different group $A'$ with the same proportionality relation. However, $\bar{A}$ and $A'$ clearly are isomorphic. Because this situation also appears in the case of affine and projective geometry and many other instances of abstraction the question arises whether or not it should be taken as characteristic for the notion of abstraction.

**Note 9:** It is possible to subject a regular bi-equivalence to further symmetry conditions. If $\varrho$ is a regular bi-equivalence on $A$ we define, $\varrho$ *to be Abelian* if $\varrho(x, y, u, v) \to \varrho(x, u, y, v)$, and, $\varrho$ *is Boolean* if $\varrho(x, y, u, v) \to \to \varrho(x, v, u, y)$. The theory of Abelian regular bi-equivalences is an abstraction of the theory of commutative groups and the theory of Boolean regular bi-equivalences is an abstraction of the theory Boolean groups (groups in which $x = \dot{x}$ identically). It might be noted that in the Boolean case $\varrho$ is completely symmetric and therefore Abelian.

**Note 10:** Like the proportionality relation, also the ternary operation $f(x, y, z) = x \cdot \dot{y} \cdot z$ is a characterizing invariant for the holomorph. The operator $f$ has been studied by J. Certaine [2], who however is not careful in making the separation between group theory and the abstracted theory. R. Baer [1] also made use of the operator $f$, the paper in question contains many interesting ideas concerning a theory of general invariants.

**Note 11:** We are interested in regular bi-equivalences primarily because of their relationship to groups and in particular to proportionality. Nevertheless, the intrinsic theory of bi-equivalences shows many interesting features of which we mention but one.

All significant ideas connected with subgroups and in particular normal subgroups can without difficulty be reformulated in the theory of regular bi-equivalences. Moreover these ideas take an especially attractive form in the abstracted theory. For example, let $\underline{P}$ be the partition of $A$ generated by a normal subgroup $N$. From the group point of view $\underline{P}$ has a preferred element $N$. However, the proportionality relation $\pi$ of the group is a regular bi-equivalence on every co-set of $N$, furthermore, all these co-sets are $\pi$-isomorphic to $N$. Thus, in the theory of regular bi-equivalences every such partition $\underline{P}$ is homogeneous.

**Note 12:** There are models of regular bi-equivalences which are not groups. For example, the set $A$ of correlations of projective geometry is not a group under composition, however, the proportionality relations of composition is a regular bi-equivalence on $A$. $A$ similar remark holds for any co-set $A$ of a subgroup of a group.

Affine geometry affords another example. The relation $\pi(x, y, u, v)$, defined by $xy \parallel uv \wedge xu \parallel yv$, is an Abelian regular bi-equivalence on the set of all points. Since it is an important feature of affine geometry that no point is preferred, it follows from the discussion in section 3 that in affine geometry it is impossible to define a group of which $\pi$ is the proportionality relation. However, if an origin is added an affine geometry becomes a vector space and $\pi$ becomes the proportionality relation of the vector addition.

**Note 13:** Let $\varrho$ be a regular bi-equivalence on $A$. As in proportionality theory we then define ratios as follows: $(xy) \equiv (uv)$ defined by $\varrho(x, y, u, v)$ is an equivalence relation on $A \times A$. Let $\underline{L}$ be the corresponding partitioning into equivalence classes $\overline{xy}$. Now $\underline{L}$ is a group under the operation $\overline{xy} \cdot \overline{yz} = \overline{xz}$, the inverse of $\overline{xy}$ is $\overline{yx}$ and the identity is $\overline{xx}$. The class $\overline{xy}$ might be called the left ratio of $x$ and $y$.

Also, the concept of a group translation can be defined already in the abstracted theory. A transformation $T$ of $A$ is called a left translation if $\varrho(x, Tx, y, Ty)$ holds for all $x$ and $y$ in $A$. Note that if a translation is regarded as a set of pairs, the group $\underline{L}$ defined before turns out to be the group of left translations. Clearly, the group $\underline{L}$ is isomorphic to the group $[A, \varrho, e]$ which arises from $\varrho$ if an element $e$ of $A$ is distinguished.

Although a group structure isomorphic to $[A, \varrho, e]$ has been defined within the theory of regular bi-equivalences, it must be realized that this does not imply the equivalence of the theory of regular bi-equivalences with group theory. On the basis of our construction of $\underline{L}$ one might attempt to set up a translation of the theory $T_1$ in section 3 into the theory $T$. However, in trying to formulate the analogs to I and II in section 3, one will run into difficulties with logical types.

## *Bibliography.*

[1] R. BAER: Zur Einführung des Scharbegriffes. Crelles J. **160**, 199—207 (1929). — [2] J. CERTAINE: The ternary operation $(abc) = ab^{-1}c$ of a group. Bull. Amer. Math. Soc. **49**, 869—877 (1943). — [3] A. A. GRAU: Ternary Boolean algebra. Bull. Amer. Math. Soc. **53**, 567—572 (1947). — [4] J. B. WRIGHT: Quasi projective geometry. Mich. Math. J. **2**, 115—122 (1953/4).

*(Eingegangen am 3. Oktober 1954.)*

Druck: Brühische Universitätsdruckerei Gießen

# INVARIANTS OF THE ANTI-AUTOMORPHISMS OF A GROUP

J. RICHARD BÜCHI AND JESSE B. WRIGHT

1. **Introduction.** The background for this paper is provided by Klein's work presented in the Erlangerprogram [**1**], and more recent developments of these ideas as well as their application outside the field of geometry [**2**; **3**; **4**; **5**; **6**].

Klein deals with the Euclidean space $\bar{A}$ as a *fundamental structure.* Its group of automorphisms $G_0(\bar{A})$ consists of the similitudes. Let $G(\bar{A})$ denote any group arrived at by adjoining to $G_0(\bar{A})$ new transformations of the set $A$. The basic problems then are of the following type: *Given an extension* $G(\bar{A})$ *of* $G_0(\bar{A})$, *find invariants of* $G(\bar{A})$ *which characterize it.*

Analogous problems can be formulated in the theory of abstract groups [**2**]. Now the fundamental structure itself is an abstract group $\bar{A}$. Its automorphism-group $G_0(\bar{A})$ may be extended to a group $G(\bar{A})$, by adjoining new transformations of the set $A$. The group-multiplication of $\bar{A}$, the characterizing invariant of $G_0(\bar{A})$, is not invariant under $G(\bar{A})$. The basic problems, stated above for geometry, take the same form in group theory, namely, what are characterizing invariants for $G(\bar{A})$.

A study of this type has already been made in the case $G(\bar{A})$ is taken to be the holomorph $H(\bar{A})$, i.e., the group of transformations obtained by adjoining to $G_0(\bar{A})$ the translations of $\bar{A}$ [**4**]. The present paper deals with the case $G(\bar{A})=G_1(\bar{A})$, the group consisting of the automorphisms and anti-automorphisms of $\bar{A}$. Characterizing invariants of $G_1(\bar{A})$ are investigated.

In a group $\bar{A}$ on a set $A$ one can define two multiplications $p(x, y) = x\cdot y$ and $q(x, y) = y\cdot x$. The automorphism group $G_0(\bar{A})$ consists of the automorphisms of the operation $p$, while the anti-automorphisms are those transformations $T$ of the set $A$ which interchange $p$ and $q$. It follows that the set $\{p, q\}$ is invariant under all transformations belonging to the group $G_1(\bar{A})$ consisting of the automorphisms and anti-automorphisms of $\bar{A}$. Furthermore, this invariant characterizes $G_1(\bar{A})$, i.e., every transformation $T$ of the set $A$ which keeps the set $\{p, q\}$ invariant belongs to $G_1(\bar{A})$. However, there are simpler characterizing invariants for $G_1(\bar{A})$, namely relations whose arguments

Received by the editors September 13, 1955, and, in revised form, December 11, 1955.

range over the set $A$. For example $G_1(\bar{A})$ clearly is the group of automorphisms of the 6-term relation $\beta$ defined as follows:

$$\beta(x, y, z, u, v, w) : (xy = z \wedge uv = w) \vee (yx = z \wedge vu = w).$$

Theorem 1 shows that even a 3-term relation will serve to characterize $G_1(\bar{A})$.

THEOREM 1. *The automorphisms and anti-automorphisms of a group $\bar{A}$ constitute the group of automorphisms of the relation*

$$\alpha(x, y, z) : xy = z \vee yx = z$$

*i.e., the relation $\alpha$ is a characterizing invariant for $G_1(\bar{A})$.*

The significance of this theorem is that it shows how to replace the rather complex relation $\beta$ by a simpler relation $\alpha$ which is a characterizing invariant for the same group of transformations. It is natural to ask whether this result can be further improved. The relation $\alpha$ is a disjunction of two equations. The question is whether there is a relation expressible in the form of a single equation, which characterizes the group $G_1(\bar{A})$ of automorphisms and anti-automorphisms. The answer is negative, it is possible to prove.

THEOREM 2. *There is no finite or infinite set of relations expressible as equations between words, which would constitute a system of invariants characterizing the group $G_1(\bar{A})$ of automorphisms and anti-automorphisms. I.e., there is a group $\bar{A}_0$ and a transformation $T$ on $A_0$ such that every equation between words which in every group $\bar{A}$ is invariant under $G_1(\bar{A})$ is also invariant in $\bar{A}_0$ under $T$, and such that $T$ is not a member of $G_1(\bar{A}_0)$.*

In algebra one prefers to deal with operations rather than relations. An operation on the set $A$, which characterizes $G_1(\bar{A})$, is given in §3. However, such an operation is neither definable explicitly by a word, nor is it definable implicitly as a solution of an equation of grouptheory. This follows as a corollary to Theorem 2.

2. PROOFS. Theorem 1 states: If $\bar{A}$ is a group and $T$ is a transformation of the set $A$ having the property that for all $x, y \in A$, $T(x \cdot y)$ is either equal to $Tx \cdot Ty$ or equal to $Ty \cdot Tx$, and $T^{-1}(x \cdot y)$ is either equal to $T^{-1}x \cdot T^{-1}y$ or equal to $T^{-1}y \cdot T^{-1}x$, then $T$ must be an automorphism or an anti-automorphism of $\bar{A}$. In this form the theorem was independently obtained by W. R. Scott [8]. As his proof appears in this journal, our proof of Theorem 1 will be omitted.

Let us define a semi-automorphism of a group $\bar{A}$ to be a mapping which preserves $e$ and the functions $Sx = x^{-1}$ and $s(x, y) = xyx$. Let

$G_1(\overline{A})$ and $G_2(\overline{A})$ denote respectively the group of automorphisms plus anti-automorphisms, and the group of semi-automorphisms. The proof of Theorem 2 now proceeds as follows: first a complete description of all equations invariant under $G_1$ in all groups $\overline{A}$ is given. It then can be seen easily that, in the abelian case, all these equations are invariant also under $G_2$. The proof is completed by displaying abelian groups $\overline{A}_0$ for which $G_2(\overline{A}_0)$ is not contained in $G_1(\overline{A}_0)$.

An equation of grouptheory will be called *reduced* if it is either the equation $e=e$ or then is of the type $a_1a_2 \cdots a_n=e$, whereby every $a_i$ is of the form $x$ or $x^{-1}$, $x$ being a variable, and none of the pairs $a_ia_{i+1}$ and $a_1a_n$ is of the form $xx^{-1}$ or $x^{-1}x$. Clearly, to every equation $f=g$ one can find a reduced equation $h=e$, such that $(f=g)\leftrightarrow(h=e)$ holds in all groups. It follows that every relation expressible by an equation $f=g$ can also be expressed by a reduced equation $h=e$. In describing the equational invariants of $G_1$ and $G_2$ it therefore is sufficient to deal with reduced equations only. This procedure will be followed in the sequel. Furthermore, the following notations will be used: Let $x$ be a variable, then $[x]$ stands for $x^{-1}$ and $[x^{-1}]$ stands for $x$. Let $w$ be a word of grouptheory, i.e., an expression $a_1a_2 \cdots a_n$, whereby every $a_i$ is of the form $x$ or $x^{-1}$. Then $w^*$ stands for the word $a_n \cdots a_2a_1$, and $[w]$ stands for the word $[a_1][a_2] \cdots [a_n]$. The symbol "$\approx$" is used to denote syntactic identity of words.

L1: If $g=e$ and $h=e$ are reduced equations such that $(g=e)\leftrightarrow(h=e)$ is true in all groups, then $h$ results by a cyclic permutation of the constituents of either $g$ or $[g^*]$.

To prove this one best uses Gödel's completeness theorem for first order predicate calculus. It says that in L1 one can replace "true in all groups" by "provable in first-order group theory." Although the validity of the resulting meta-group-theoretic statement is fairly obvious on intuitive grounds, its proof is rather lengthy and therefore it is omitted.

Next we define a reduced equation $g=e$ to be *regular*$_1$ in case $g^*$ results from $g$ by a cyclic permutation, and to be *regular*$_2$ in case $[g]$ results from $g$ by a cyclic permutation.

L2: If the reduced equation $g=e$ in all groups $\overline{A}$ is invariant under $G_1(\overline{A})$, then it is either regular$_1$ or regular$_2$, or $g$ is $e$.

Proof. Suppose $g=e$ is reduced and invariant under $G_1$ and $g$ is not $e$. Then $g=e$ is invariant under $Sx=x^{-1}$, i.e., $(g=e)\leftrightarrow([g]=e)$ holds in all groups. Therefore by L1, $[g]$ results from $g$ or $[g^*]$ by cyclic permutation. Consequently $g^*$ or $[g]$ results from $g$ by cyclic permutation, i.e., $g$ is regular. Q.E.D.

The next step is to investigate the invariants under $G_2$ of regular equations. For this purpose the structure of regular equations has to be described. This is done in L4.

L3: Let $g$ be a word of length $n$, and let $P$ be the cyclic permutation of $n$ objects through $m$ places. If $Pg \approx g$, then there is a word $w$, such that $g \approx ww \cdots w$ and $m$ is a multiple of the length $l$ of $w$.

PROOF. Let $g$ be the word $a_1 \cdots a_n$. The equation

$$a_i \approx a_{i+n}, \qquad \text{for all integers } i, \tag{1}$$

clearly defines a function $i \to a_i$ of the integers into the set $\{a_1, \cdots, a_n\}$, which is periodic with period $n$. Because $Pg \approx g$, it follows that the function $i \to a_i$ is also periodic with period $m$, i.e.,

$$a_i \approx a_{i+m}, \qquad \text{for all integers } i. \tag{2}$$

Let $l$ be the largest common divisor of $n$ and $m$. Then, $l = pm + qn$ for some integers $p$ and $q$. Therefore, by (1) and (2), the function $i \to a_i$ is also periodic with period $l$, i.e.,

$$a_i \approx a_{i+l}, \qquad \text{for all integers } i. \tag{3}$$

Let $w$ be the word $a_1 \cdots a_l$. Because $l$ divides $n$, $g$ is of the form $ww \cdots w$. Because $l$ divides $m$, $m$ is a multlple of the length $l$ of $w$. Q.E.D.

L4: If the equation $g = e$ is regular$_1$, then the word $g$ must be of the form $g_1 g_2$, whereby both $g_1$ and $g_2$ are symmetric words, i.e., $g_1 = g_1^*$ and $g_2 = g_2^*$.

If the equation $g = e$ is regular$_2$, then the word $g$ must be of the form $v[v]v[v] \cdots v[v]$, whereby $v$ is some word.

PROOF. Let $g = e$ be regular$_1$. Then there is a number $i$ such that a cyclic permutation of $g$ through $i$ places yields $g^*$. It may be assumed that $i$ is less or equal to half of the length of $g$, so that $g$ is of the form $a_1 \cdots a_j b_1 \cdots b_i$ whereby $i \leqq j$. The cyclic permutation of $g$ through $i$ places then yields $b_1 \cdots b_i a_1 \cdots a_j$, while $g^*$ is the word $b_i \cdots b_1 a_j \cdots a_1$. Because these two words are identical it follows that $b_1 \cdots b_i$ is identical with $b_i \cdots b_1$, and $a_1 \cdots a_j$ is identical with $a_j \cdots a_1$. Therefore $g$ is of the form $g_1 g_2$, whereby both $g_1$ and $g_2$ are symmetric.

Next let $g = e$ be regular$_2$. Then there is a number $i$ such that the cyclic permutation $P$ through $i$ places takes $g$ into $[g]$, i.e., $Pg \approx [g]$. It follows that $PPg \approx P[g] \approx [Pg] \approx [[g]] \approx g$, i.e., $PPg \approx g$. By L3, there is a word $w$ of length $l$, such that $g$ is of the form $w \cdots w$ and $2i$ is a multiple of $l$, say $2i = s \cdot l$. Suppose first that $s$ is even. Then $i$ would be a multiple of $l$, and therefore, $Pg$ would be identical to $g$. Because $Pg$ is identical with $[g]$, it would follow that $g$ and $[g]$ are identical, which is impossible. Consequently $s$ must be odd, and therefore it follows from $2i = s \cdot l$, that $l$ is even, and $w$ is of the form $w_1w_2$, whereby both $w_1$ and $w_2$ are of length $l/2$. Thus, the situation is as follows:

$$g \approx aa, \text{ whereby } a \approx w_1w_2w_1w_2 \cdots w_1w_2,$$

$$Pg \approx bb, \text{ whereby } b \approx w_2w_1w_2w_1 \cdots w_2w_1,$$

$$\lfloor g] \approx cc, \text{ whereby } c \approx [w_1][w_2][w_1][w_2] \cdots [w_1][w_2].$$

Because $Pg \approx [g]$ it follows that $w_2 \approx [w_1]$, and therefore

$$g \approx w_1[w_1]w_1[w_1] \cdots w_1[w_1]. \qquad \text{Q.E.D.}$$

L5: If the equation $g = e$ is regular$_1$, then in all groups $\overline{A}$ it is invariant under $G_2(\overline{A})$.

If the equation $g = e$ is regular$_2$, then in all abelian groups $\overline{A}$ it is invariant under $G_2(\overline{A})$.

PROOF. Suppose $g = e$ is regular$_1$. Then by L4, $g = e$ must be of the form $g_1g_2 = e$, whereby $g_1$ and $g_2$ are both symmetric. It is easily seen that every symmetric word is provably equal to an expression composed from $s(x, y) = xyx$ and $Sx = x^{-1}$, furthermore, $g_1g_2 = e$ is provably equivalent to $g_1 = S(g_2)$. It follows that there are expressions $E_1$ and $E_2$ in $e$, $S$ and $s$, such that $(g = e) \leftrightarrow (E_1 = E_2)$ holds in all groups. Because $E_1$ and $E_2$ are defined from $e$, $S$ and $s$, the equation $E_1 = E_2$ must be invariant under the automorphism group $G_2(\overline{A})$ of $e$, $S$ and $s$. It follows that $g = e$ is invariant under $G_2(\overline{A})$.

Suppose the equation $g = e$ is regular$_2$. Then by L3 it must be of the form $v[v]v[v] \cdots v[v] = e$. In every abelian group $\overline{A}$ this equation is identically satisfied, and therefore invariant under $G_2(\overline{A})$. Q.E.D.

L6: There are abelian groups $\overline{A}_0$ for which $G_2(\overline{A}_0)$ is not contained in $G_1(\overline{A}_0)$.

PROOF. Let $\overline{A}_0$ be a Boolean group, i.e., a group which satisfies the equation $x^2 = e$ identically. In this group $Sx = x$ and $s(x, y) = y$. It follows that $G_2(\overline{A}_0)$ consists of all transformations of the set $A_0$ which keep $e$ fixed. On the other hand, because $\overline{A}_0$ is abelian, $G_1(\overline{A}_0)$ consist of all automorphisms of $\overline{A}_0$. Clearly $G_2(\overline{A}_0)$ is not contained in $G_1(\overline{A}_0)$, when $A_0$ has more than two elements. (For other examples see Dinkines [7].) Q.E.D.

By L2. and L5. it follows that, for abelian groups $\overline{A}$, if an equation is invariant under $G_1(\overline{A})$, then it is also invariant under $G_2(\overline{A})$. Because of L6, this yields that the equations invariant under $G_1(\overline{A})$ cannot characterize $G_1(\overline{A})$. This concludes the proof of Theorem 2.

3. *Remarks.* Since by Theorem 1, $\alpha$ and $\beta$ have the same group of automorphisms in any $\overline{A}$, they may be said to be equivalent in Klein's sense [1]. This suggests that a stronger sort of equivalence may be established by finding a definition of $\beta$ in terms of $\alpha$. That this is possible will be shown elsewhere by use of the following stronger form of Theorem 1: If two groups $\langle A, \cdot \rangle$ and $\langle A, * \rangle$ have the same $\alpha$, then they must either be identical or anti-groups of each other. From this it also follows that the $\alpha$-theory is an abstraction ([4]; [5]) of group theory, and that every concept of group theory which is invariant under anti-automorphisms is definable in terms of $\alpha$.

The notion of an anti-automorphism applies to any algebraic system $\overline{A} = \langle A, \cdot \rangle$ consisting of a set $A$ and a binary operation $x \cdot y$. While the relation $\beta$ will still be a characterizing invariant for the group $G_1(\overline{A})$ consisting of all automorphisms and anti-automorphisms of $\overline{A}$, this will in general not be the case for $\alpha$. However, our proof for Theorem 1 as well as W. R. Scott's makes use of the associative-law and both cancellation-laws only. Therefore, if $\overline{A}$ is a cancellation-semi-group, then $\alpha$ is a characterizing invariant for $G_1(\overline{A})$. The following example shows that cancellation-semi-groups still do not exhaust all systems $\langle A, \cdot \rangle$ for which Theorem 1 holds: Let $A$ be any set and let $x \cdot y = x$. Then $\overline{A} = \langle A, \cdot \rangle$ violates one of the cancellation-laws, however, $G_1(\overline{A})$ and the group of automorphisms of $\alpha$ are identical, they both consist of all transformations of the set $A$.

In connection with Theorem 2 it should be noted that it is a statement about invariants which are "uniformally" defined for all groups (general invariants in the sense of Baer [2]). In particular groups it may well happen that the anti-automorphisms may be characterized by an equational relation. Thus, as it is shown by F. Dinkines [7], there are many groups in which the semi-automorphisms are exactly the automorphisms and anti-automorphisms. For these groups the equations $x = e$, $z = xyx$ clearly constitute a system of characterizing invariants for the group of automorphisms and anti-automorphisms.

As a corollary to Theorem 2 it follows that there is no word $w$ in grouptheory, such that in every group $\overline{A}$ the operation $w_{\overline{A}}$ defined by $w$ is a characterizing invariant for $G_1(\overline{A})$. However, there are other ways of uniformly defining operations by the use of expressions in grouptheory. For example consider the function $f_{\overline{A}}(a, b, c)$ which takes the value $c$ or $e$ according to whether $\alpha(a, b, c)$ holds or does not hold in $\overline{A}$. One can recover the relation $\alpha(a, b, c)$ from $f$, $Sx = x^{-1}$ and $e$, by defining: $\alpha(a, b, c)$, if and only if, $(c \neq e \wedge f(a, b, c) = c) \vee (c = e \wedge Sa = b)$. It follows that $(e, S, f)$ is a system of characterizing invariants for $G_1(\overline{A})$.

Theorem 2 belongs into meta-group theory, i.e., it is a statement about a first order functional calculus $F[e, \cdot, ^{-1}]$ with extralogical primitives $e$, $\cdot$, and $^{-1}$, and extralogical axioms corresponding to conventional group-axioms. The statement may become false if a different formalization of grouptheory is used, for example the rather nonconventional formalization $F[e, \cdot, ^{-1}, f]$ with an additional primitive $f$ and an additional axiom, $f(x, y, z) = n \leftrightarrow ((xy = z \vee yx = z) \wedge n = z) \vee (xy \neq z \wedge yx \neq z \wedge n = e)$.

## Bibliography


**1.** F. Klein, *Vergleichende Betrachtungen über neuere geometrische Forschungen*, Erlangen (1872), Verlag von Andreas Deichert.

**2.** R. Baer, *Zur Einführung des Scharbegriffs*, Crelle's Journal vol. 160 (1929) pp. 199–207.

**3.** F. I. Mautner, *An extension of Klein's Erlangerprogram: Logic as invariant theory*, Amer. J. Math. vol. 68 (1946) pp. 345–384.

**4.** J. R. Büchi and J. B. Wright, *The theory of proportionality as an abstraction of group theory*, Math. Ann., vol. 130 (1955) pp. 102–108.

**5.** ———, *Abstraction versus generalization*, Proceedings of the International Congress of Mathematics, vol. 2, 1954, p. 398.

**6.** J. B. Wright, *Quasi-projective geometry of two dimensions*, Michigan Mathematical Journal, vol. 2 (1953–1954) pp. 115–122.

**7.** F. Dinkines, *Semi-automorphisms of symmetric and alternating groups*, Proc. Amer. Math. Soc. vol. 2 (1951) pp. 478–486.

**8.** W. R. Scott, *Half-homomorphisms of groups*, Proc. Amer. Math. Soc. vol. 8 (1957) pp. 1141–1144.


University of Illinois and
University of Michigan

Reprinted from the
Proceedings of the American Mathematical Society
Vol. 8, No. 6, pp. 1134-1140
December, 1957

*Zeitschr. f. math. Logik und Grundlagen d. Math.*
*Bd. 18, S. 61–70 (1972)*

# MODEL THEORETIC APPROACHES TO DEFINABILITY

by J. RICHARD BÜCHI in Lafayette, Indiana,
and KENNETH J. DANHOF in Carbondale, Illinois (U.S.A.)[1])

## 1. Introduction

When approaching the problems of mathematical logic from a model-theoretic point of view, one has at his disposal a variety of 'methods of attack.' We have in mind, for example, methods based on the compactness theorem and utilized by ROBINSON, SVENONIUS and others (see [18] and [20]) or methods based on the ultraproduct construction (perhaps more precisely, the Łos theorem [16]) and its generalizations and employed by KOCHEN, KEISLER and others (see [8], [11], [12], [13], [14]). Some of the more important of these methods are considered here with a view toward revealing the underlying uniformity. The analysis is made in the context of the problems of definability. Particularly in Section 3, an attempt is made to indicate basis similarities between the various methods considered while addressing each approach to a proof of the BETH theorem [1]. These different proofs of the BETH theorem invoke CRAIG's lemma (see [6], or [7]). Although much of Section 3 is expository in nature, we feel that the material should be available in the unified presentation.

Regarding definability results, it is natural to seek an extension of CRAIG's lemma which would give a direct proof of SVENONIUS' theorem [20], i.e., is there a lemma which performs the same role in the proof of SVENONIUS' theorem as does CRAIG's lemma in the proof of BETH's theorem? One might also desire such a lemma for proving the strong form of the BETH theorem [5]. Lemmas for these purposes are given in 4, resulting in new versions of the proofs of the theorems mentioned. Generalizations of ROBINSON's consistency lemma [18] and of KOCHEN's basic ultralimit result [14] follow.

## 2. Preliminaries

Ordinal numbers are generally denoted by the Greek letters $\alpha$, $\beta$, $\gamma$, $\eta$, $\lambda$, $\xi$, $\sigma$, while $\mu$, $\pi$, $\sigma$, $\tau$, $\varphi$, $\chi$, $\psi$ are usually functions. Familiarity with the notion of a relational system $\boldsymbol{A} = \langle A, R, S\rangle$, where $R$ and $S$ are relation (or sequences of relations) on $A$ is assumed. To avoid cumbersome notation, we will assume $R$ and $S$ are binary relations; the general case requires no additional ideas. The first-order predicate logic $L(\boldsymbol{R}, \boldsymbol{S})$ contains, in addition to the usual logical connectives and the identity '=' predicate letters $\boldsymbol{R}$ and $\boldsymbol{S}$ corresponding to $R$ and $S$. $\tau(\boldsymbol{A})$ (the *elementary theory* of $\boldsymbol{A}$) will represent the sentences of $L(\boldsymbol{R}, \boldsymbol{S})$ which are true in $\boldsymbol{A}$ and $\boldsymbol{A} \equiv \boldsymbol{B}$ (*elementary equivalence*) indicates that $\tau(\boldsymbol{A}) = \tau(\boldsymbol{B})$. For a sentence $\Sigma$, $\mu(\Sigma)$ (the *models* of $\Sigma$) is the set of systems in which $\Sigma$ is true and $\Sigma_1 \models \Sigma_2$

[1]) This work was partially supported by N.S.F. Grant No. GJ-120.

(*logical consequence*) indicates that $\mu(\Sigma_1) \subseteqq \mu(\Sigma_2)$. The maps $\mu$ and $\tau$ are extended to sets in the usual fashion. The compactness theorem is frequently needed:

Theorem. *If every finite subset of a set $\Gamma$ of sentences has a model, then $\Gamma$ has a model.*

For a system $\boldsymbol{A} = \langle A, R, S\rangle$, $\boldsymbol{A_R}$ denotes the system $\langle A, R\rangle$ and if $B \subseteqq A$, $\boldsymbol{A}/B$ the system obtained by restricting $R$ and $S$ to $B$. For a class of systems $K$ we often write $K = K(\boldsymbol{R}, \boldsymbol{S})$ to indicate the type of systems of $K$ and then $K_{\boldsymbol{R}} = \{\langle A, R\rangle; \boldsymbol{A} \in K\}$. As usual we write $K \in EC$ ($K$ is an *elementary class*) if $K = \mu(\Sigma)$ for some sentence $\Sigma$ and $K \in EC_\Delta$ if $K = \bigcap_\eta M_\eta$ where each $M_\eta \in EC$. Similarly $K \in PC$ if for some $M(\boldsymbol{R}, \boldsymbol{S}) \in EC$, $K = M_{\boldsymbol{R}}$ and $K \in PC_\Delta$ if there is an $M(\boldsymbol{R}, \boldsymbol{S}) \in EC_\Delta$ such that $K = M_{\boldsymbol{R}}$. Classes of systems are always assumed to be closed under isomorphism. For $K(\boldsymbol{R}, \boldsymbol{S})$, $w_{\boldsymbol{R}}$ denotes the function defined on $K$ by $w_{\boldsymbol{R}}(\boldsymbol{A}) = \boldsymbol{A}_R$ and we write $w_{\boldsymbol{R}} \in EF(K)$ ($w_{\boldsymbol{R}}$ is an *elementary concept* on $K$) if there is a formula $\Phi(\boldsymbol{R}, x_0, x_1)$ such that $(\forall x_0, x_1)\,[\Phi(\boldsymbol{R}, x_0, x_1) \leftrightarrow \boldsymbol{S}(x_0, x_1)] \in \tau(K)$.

For systems $\boldsymbol{A}$ and $\boldsymbol{A}'$, $\varphi$ is called an *elementary embedding* of $\boldsymbol{A}$ into $\boldsymbol{A}'$ if $\varphi$ maps $A$ into $A'$ ($\varphi\colon A \to A'$) and for any formula $\Phi(x_0, \ldots, x_{n-1})$ any $a_0, \ldots, a_{n-1}$ in $A$, $a_0, \ldots, a_{n-1}$ satisfies $\Phi$ in $\boldsymbol{A}$ iff $\varphi(a_0), \ldots, \varphi(a_{n-1})$ satisfies $\Phi$ in $\boldsymbol{A}'$ and we write $\varphi : \boldsymbol{A} \prec \boldsymbol{A}'$. $\boldsymbol{A} \prec \boldsymbol{A}'$ indicates that for some $\varphi$, $\varphi : \boldsymbol{A} \prec \boldsymbol{A}'$. We see that if $\boldsymbol{A} \prec \boldsymbol{A}',$ then $\boldsymbol{A} \equiv \boldsymbol{A}'$; if $\boldsymbol{A} \prec \boldsymbol{A}'$ and $\boldsymbol{A}' \prec \boldsymbol{A}''$, then $\boldsymbol{A} \prec \boldsymbol{A}''$; and we always have $i: \boldsymbol{A} \prec \boldsymbol{A}$ ($i$ the identity map.) The following result is due to TARSKI [21].

Theorem T 1. *If $\{\boldsymbol{A}_n; n < \omega_0\}$ is a class of systems such that $i: \boldsymbol{A}_j \prec \boldsymbol{A}_{j+1}$ $(j < \omega_0)$, then $i\colon \boldsymbol{A}_0 \prec \bigcup_{k<\omega_0} \boldsymbol{A}_k$.*

That $\varphi$ is an isomorphic injection from $\boldsymbol{A}$ into $\boldsymbol{B}$ is indicated by writing $\varphi$: $\boldsymbol{A} \mathrel{\overset{\cong}{\to}} \boldsymbol{B}$; if $\varphi$ is also onto we write $\varphi$: $\boldsymbol{A} \cong \boldsymbol{B}$. A map $\sigma$ defined on $A_0 \subseteqq A$ is called a *partial $R$-automorphism* on $\boldsymbol{A} = \langle A, R, S\rangle$ if any two $n$-tuples of elements of $A$ which correspond to each other by $\sigma$ satisfy the same formulas $\Phi(\boldsymbol{R}, x_0, \ldots, x_{n-1})$. If $A_0 = A$, $\sigma$ is called an *$R$-automorphism on $A$*. $D\sigma$ and $\text{Ɑ}\sigma$ indicate the domain and range of $\sigma$ respectively. $\varkappa(\boldsymbol{A})$ denotes the group of automorphisms of $\boldsymbol{A}$.

We also consider briefly the ultraproduct construction (see [14] or [16]). Let $\boldsymbol{A}_i = \langle A_i, R_i\rangle$ for $i \in I$ where, e.g., $R_i \subseteqq A_i^2$, and let $\prod_I A_i = \{f : I \to \bigcup_I A_i; f(i) \in A_i$ for $i \in I\}$. If $D$ is a filter on $I$, then the relation $f \equiv_D g$ iff $\{i \in I; f(i) = g(i)\} \in D$ is an equivalence relation on $\prod_I A_i$ and the resulting set of equivalence classes is denoted by $\prod_I A_i/D$. The *reduced product* of the family $\{\boldsymbol{A}_i; i \in I\}$ relative $D$ is the system $\prod_I \boldsymbol{A}_i/D = \langle \prod_I A_i/D, R_D\rangle$ where $R_D$ is defined by $\langle f/D, g/D\rangle \in R_D$ iff $\{i \in I; \langle f(i), g(i)\rangle \in R_i\} \in D$. If the reduced product is taken relative an ultrafilter, the result is called an *ultraproduct*, and if furthermore, $\boldsymbol{A}_i = \boldsymbol{A}$ for $i \in I$, the result is called the *ultrapower of $\boldsymbol{A}$ relative $\boldsymbol{D}$* and written $\boldsymbol{A}_D^I$. $\prod(K)$ denotes the class of ultraproducts over $K$, $P(\boldsymbol{A})$ the class of ultrapowers of $\boldsymbol{A}$. The diagonal function $\delta$: $A \to \prod_I A/D$ is defined as follows: for $a \in A$, $\delta(a) = \{\langle i, a\rangle; i \in I\}/D$.

The following fact is basic in the study of ultraproducts:

Theorem (Łos [16]). *For $D$ an ultrafilter on $I$, $\Sigma$ a sentence; $\Sigma \in \tau(\prod_I \boldsymbol{A}_i/D)$ iff $\{i \in I; \Sigma \in \tau(\boldsymbol{A}_i)\} \in D$. In particular, for the ultrapower, $\delta : \boldsymbol{A} \prec \boldsymbol{A}_D^I$.*

### 3. Various Approaches of Beth's Theorem

An important observation on definability, which is often used by mathematicians, was first stated explicitly in PADOA [17]. In our terminology, PADOA's method might be stated as follows: in order to show that $\boldsymbol{S}$ is not syntactically definable from $\boldsymbol{R}$ in a theory $\Sigma(\boldsymbol{R}, \boldsymbol{S})$ it suffices to find two models $\boldsymbol{A}$ and $\boldsymbol{A}'$ of $\Sigma(\boldsymbol{R}, \boldsymbol{S})$ such that $\boldsymbol{A}_{\boldsymbol{R}} = \boldsymbol{A}'_{\boldsymbol{R}}$ but $\boldsymbol{A} \neq \boldsymbol{A}'$. BETH's theorem [1] is the converse to PADOA's remark. More precisely, if

i) for $\boldsymbol{A}$, $\boldsymbol{A}'$ models of $\Sigma(\boldsymbol{R}, \boldsymbol{S})$, if $\boldsymbol{A}_{\boldsymbol{R}} = \boldsymbol{A}'_{\boldsymbol{R}}$, then $\boldsymbol{A} = \boldsymbol{A}'$,

then the following syntactic statement holds for $\Sigma(\boldsymbol{R}, \boldsymbol{S})$:

ii) there is a formula $\Phi(\boldsymbol{R}, x_0, x_1)$ such that

$$\Sigma(\boldsymbol{R}, \boldsymbol{S}) \models (\forall x_0, x_1)\, [\Phi(\boldsymbol{R}, x_0, x_1) \leftrightarrow \boldsymbol{S}(x_0, x_1)].$$

Thus BETH's theorem insures that the two systems required in PADOA's method will always exist. We note that i) above is equivalent to the following:

i′) For $\boldsymbol{A}$, $\boldsymbol{A}'$ models of $\Sigma(\boldsymbol{R}, \boldsymbol{S})$, if $\varphi\colon \boldsymbol{A}_{\boldsymbol{R}} \cong \boldsymbol{A}'_{\boldsymbol{R}}$, then $\varphi\colon \boldsymbol{A} \cong \boldsymbol{A}'$.

We also remark that BETH's theorem and all following definability results can be stated in more generality; see CRAIG's extension of BETH's theorem to elementary concepts and to $\mathrm{PC}_\Delta$ [7]. However, in order not to obscure the main ideas with additional notation, we will restrict our attention to primitive concepts.

We will consider four 'levels' and on each level list a sequence of results culminating in a proof of BETH's theorem. The results are labeled in a manner which indicates the analogous statements on each of the other levels.

Level 1. We have in mind here the results and methods of ROBINSON as presented in [18]. The statements are listed below and followed by a discussion

$\mathrm{F}_1$ a) Given $\boldsymbol{A} = \langle A, R, S\rangle$ and $\boldsymbol{A}' = \langle A', R'\rangle$ such that $\boldsymbol{A}_{\boldsymbol{R}} \equiv \boldsymbol{A}'$ there are maps $\varphi$ and $\pi$ and a system $\boldsymbol{A}'' = \langle A'', R'', S''\rangle$ such that $\varphi\colon \boldsymbol{A} \prec \boldsymbol{A}''$ and $\pi\colon \boldsymbol{A}' \prec \boldsymbol{A}''$.
b) If $\boldsymbol{A} = \langle A, R, S\rangle$, $\boldsymbol{A}' = \langle A, R'\rangle$ and $\varphi\colon \boldsymbol{A}_{\boldsymbol{R}} \prec \boldsymbol{A}'$, then there are maps $\psi$ and $\pi$ and a system $\boldsymbol{A}'' = \langle A'', R'', S''\rangle$ such that $\psi\colon \boldsymbol{A} \prec \boldsymbol{A}''$, $\pi\colon \boldsymbol{A}' \prec \boldsymbol{A}''_{\boldsymbol{R}}$ and $\pi \cdot \varphi = \psi$.

$\mathrm{R}_1$ If $\boldsymbol{A} \equiv \boldsymbol{A}'$ where $\boldsymbol{A} \in K$, $\boldsymbol{A}' \in M$ and $K$, $M \in \mathrm{PC}_\Delta$, then $K \cap M \neq 0$.

$\mathrm{L}_1$ If $K \in \mathrm{PC}_\Delta$, $\tau(K) \subseteq \tau(\boldsymbol{A})$, then there is $\boldsymbol{A}' \in K$ such that $\boldsymbol{A}' \equiv \boldsymbol{A}$.

$\mathrm{C}_1^1$ If $\tau(K) \cup \tau(K')$ is consistent and $K$, $K' \in \mathrm{PC}_\Delta$, then $K \cap K' \neq 0$.

$\mathrm{C}_1^2$ If $K$, $K' \in \mathrm{PC}_\Delta$ and $K \cap K' = 0$, then there is $N \in \mathrm{EC}$ such that $K \subseteq N$ and $N \cap K' = 0$.

$\mathrm{B}_1$ For a class of systems $K = K(\boldsymbol{R}, \boldsymbol{S}) \in \mathrm{PC}_\Delta$, $K_{\boldsymbol{R}} = w_{\boldsymbol{R}}(K)$, the following are equivalent: i) $w_{\boldsymbol{R}} \in \mathrm{EF}(K)$; ii) $w_{\boldsymbol{R}}$ is one-one on $K$.

$\mathrm{R}_1$ is a restatement of ROBINSON's consistency lemma [18] and $\mathrm{F}_1$ can be abstracted from the proof of that result. (The label $\mathrm{F}_1$ is used because of the formal similarity of this statement to FRAYNE's result $\mathrm{F}_3$ on level 3.) $\mathrm{F}_1$ is proved using the compactness theorem. (The proof of the fact that if $\boldsymbol{A}' \equiv \boldsymbol{A}''$, then there is an $\boldsymbol{A}$ such that $\boldsymbol{A}' \prec \boldsymbol{A}$ and $\boldsymbol{A}'' \prec \boldsymbol{A}$ (KEISLER) might be viewed as a prototype of this sort of proof.) $R_1$ is proved using repeated applications of $F_1$ to form an 'alternating-chain' of systems and then applying T 1. We call this the 'outside CANTOR process'. $L_1$ is again proved using the compactness theorem and $C_1^1$ (a version of CRAIG's lemma) results from $R_1$ and $L_1$. $C_1^2$ is a restatement of $C_1^1$ as a separation property of members of $PC_\Delta$ (see [4] or [8]). $B_1$ (specifically the implication from ii) to i)) is a statement of BETH's theorem and follows as in [7] or [18]. We give the details of these various proofs in a more general form in section 4.

The proofs of BETH's theorem and CRAIG's lemma discussed here are not the original proofs. In fact, BETH published his result in 1953 and CRAIG had a proof of his lemma before this time (Doctoral Dissertation). Each of these proofs had its origins in HERBRAND's theorem (or GENTZEN's Hauptsatz, see [10]). The basic nature of his lemma became clear in a 1956 correspondence in which CRAIG realized that BETH's result is a simple corollary to his lemma, and BÜCHI found the model-theoretic forms $C_1$ of the lemma.

Level 2. On level 1, the basic relation between systems was that of an elementary embedding. We now consider a restriction of that notion, the strong elementary embedding. We first define this relation on sets.

Definition. *A strong elementary embedding of $\boldsymbol{A}$ into $\boldsymbol{B}$* is a mapping $\varphi$ defined on $A \cup \bigcup_{0<k<\omega} (2^{A^k})$ such that;

i) for all $a \in A$, $\varphi(a) \in B$ and for all $R \subseteqq A^k$, $\varphi(R) \subseteqq B^k$,
ii) $\varphi(E_A) = E_B$ where $E_A$ and $E_B$ are the respective equality relations on $A$ and $B$,
iii) for any sentence $\Phi(\boldsymbol{a}_0, \boldsymbol{a}_1, \boldsymbol{R}_0, \boldsymbol{R}_1)$ and any $a_0, a_1 \in A$, $R_0 \subseteqq A^{k_0}$, $R_1 \subseteqq A^{k_1}$ if $\Phi \in \tau(\langle A, a_0, a_1, R_0, R_1\rangle)$, then $\Phi \in \tau(\langle B, \varphi(a_0), \varphi(a_1), \varphi(R_0), \varphi(R_1)\rangle)$.

If $\varphi$ is a strong elementary embedding of $A$ into $B$, we write $\varphi : A \lll B$. For systems $\boldsymbol{A} = \langle A, a, R\rangle$ and $\boldsymbol{A}' = \langle A', a', R'\rangle$, we say *$\varphi$ is a strong elementary embedding of $\boldsymbol{A}$ into $\boldsymbol{A}'$* (written $\varphi : \boldsymbol{A} \lll \boldsymbol{A}'$) if $\varphi : A \lll A'$, $\varphi(a) = a'$ and $\varphi(R) = R'$. (See also [12] in this connection, especially the discussion of complete systems.) One quickly sees that the notion of strong elementary embedding satisfies those properties noted for elementary embedding, and particularly the obvious analogue to theorem T 1.

We now extend this relation to classes of systems. Again, this is first done for sets.

Definition. $\varphi : \{A_j; j \in J\} \lll A'$ means that $\varphi$ is defined for all maps $r$ on $J$ with the property that for $j \in J$ either $r j \in A_j$ or $r j \in A_j^k$ $(k < \omega)$ and for all such maps $r_0$, $r_1$ and sentences $\Phi(\chi_0, \chi_1)$ (fitting the type of $r_0$, $r_1$), if for all $j \in J$, $\Phi \in \tau(\langle A_j, r_0 j, r_1 j\rangle)$, then $\Phi \in \tau(\langle A', \varphi(r_0), \varphi(r_1)\rangle)$.

As before, if $K$ is a class of systems, $\varphi : K \lll \boldsymbol{A}' = \langle A', r_0', r_1'\rangle$ means that $\varphi : \{A; \boldsymbol{A} \in K\} \lll A'$ and $\varphi(r_i) = r_i'$ for $i = 0, 1$ where for all $\boldsymbol{A} \in K$ and $i = 0, 1$, $r_i(\boldsymbol{A}) = i^{\text{th}}$ distinguished element of $\boldsymbol{A}$. $\Lambda(K)$ will denote the class of strong elementary extensions over $K$. i.e., those $\boldsymbol{A}$ such that for some $K_0 \subseteqq K$, $K_0 \lll \boldsymbol{A}$.

We can now give the results on level 2 as follows:

$F_2$ a) If $A \equiv A'$, then there are maps $\varphi$ and $\pi$ and a system $A''$ such that $\varphi : A \lll A''$ and $\pi : A' \prec A''$.

b) If $\varphi : A \prec A'$, there are maps $\psi$ and $\pi$ and a system $A''$ such that $\psi : A \lll A''$, $\pi : A' \prec A''$ and $\pi \cdot \varphi = \psi$.

$R_2$ If $A \equiv A'$, then $\Lambda(A) \cap \Lambda(A') \neq 0$.

$L_2$ If $\tau(K) \subseteqq \tau(A)$, then there is a system $A' \in \Lambda(K)$ such that $A \equiv A'$.

$C_2^1$ If $\tau(K) \cup \tau(K')$ is consistent, $K$ and $K'$ are closed under $\Lambda$, then $K \cap K' \neq 0$.

$C_2^2$ If $K$ and $K'$ are closed under $\Lambda$ and $K \cap K' = 0$, then there is an $N \in \mathrm{EC}$ such that $K \subseteqq N$ and $N \cap K' = 0$.

$B_2$ For a class of systems $K = K(\boldsymbol{R}, \boldsymbol{S})$ closed under $\Lambda$, $K_{\boldsymbol{R}} = w_{\boldsymbol{R}}(K)$, the following are equivalent: i) $w_{\boldsymbol{R}} \in \mathrm{EF}(K)$; ii) $w_{\boldsymbol{R}}$ is one-one on $K$.

$R_2$ is a restatement of the consistency lemma and is proved using $F_2$ and T 1 just as in the case of level 1. We now give the proof of $L_2$.

Let $K$, $A$ be as in $L_2$. For each appropriate map $r$ on $K$ we select a letter $x_r$. Let $\Gamma_1 = \{\Phi(\ldots x_r \ldots); \Phi \in \tau(\langle B, \ldots r(B) \ldots \rangle)$ for all $B \in K\}$ and let $\Gamma_2 = \tau(A)$, with $A = \langle A, R\rangle$. We assert that $\Gamma_1 \cup \Gamma_2$ is satisfiable, for if not, by the compactness theorem, for some $\Phi_1 \in \Gamma_1$, $\Phi_2 \in \Gamma_2$, $\Phi_1 \wedge \Phi_2$ is not satisfiable. We would then have $\sim \Phi_2 \in \Gamma_1$ and as a result, $\sim \Phi_2 \in \tau(K)$. But this is a contradiction, since $\tau(K) \subseteqq \tau(A)$ by assumption. Let $A' = \langle A', R', \ldots, R'_r, \ldots\rangle$ be a model of $\Gamma_1 \cup \Gamma_2$. Then clearly $A'_R \equiv A$. Also, if $\Phi \in \tau(\langle B, \ldots, r B, \ldots\rangle)$ for all $B \in K$. then since $\Gamma_1 \subseteqq \tau(A')$, we have $\Phi \in \tau(A')$. It follows that if $\varphi$ is defined by $\varphi(r) = R'_r$, then $\varphi : K \lll A'$.

Using $L_2$, $C_2^1$ follows from $R_2$ and again, one easily shows that $C_2^1$ and $C_2^2$ are equivalent. $B_2$, our version of BETH's theorem on this level, can now be derived using $C_2^2$. We also remark that each result on level 2 implies its counterpart on level 1 since, e.g., if $A \in K \in \mathrm{PC}_\Delta$ and $A \lll A'$, then $A' \in K$.

In [12], KEISLER introduces the notion of a limit ultrapower – a generalization of the notion of ultrapower. The following relationship between limit ultrapowers and strong elementary extensions is established: $B$ is isomorphic to a limit ultrapower of $A$ iff $A \lll B$. On level 3 below, we consider another generalization of the ultrapower, the ultralimit. It is also shown in [12] that every ultralimit of a system $A$ is isomorphic to a limit ultrapower of $A$. Thus, in spite of the fact that the results and methods of level 2 are closely related to those of level 1, the role of level 2 in passing from level 1 to level 3 is apparent.

For the purpose of extending the notion of strong elementary extensions to classes, KEISLER [13] makes available an alternative to that given above. The resulting construction is called the *limit ultraproduct* and we refer the reader to [13] for more information in this regard.

Level 3. As mentioned previously, the basic construction to be considered here is that of the ultralimit. Roughly speaking, the ultralimit of a system $A$ is the directed limit of a sequence $\langle \boldsymbol{A}_i \rangle_{i<\omega}$ where $\boldsymbol{A} = \boldsymbol{A}_0$ and $\boldsymbol{A}_{i+1}$ is an ultrapower of $\boldsymbol{A}_i$ for $i < \omega$ (for a precise definition see [14]). $\Omega(\boldsymbol{A})$ will denote the class of ultralimits of a system $\boldsymbol{A}$. For level 3, our sequence of results is as follows:

$F_3$ a) If $\boldsymbol{A} \equiv \boldsymbol{A}'$, then there is a system $\boldsymbol{A}'' \in \mathrm{P}(A)$ and a map $\varphi$ such that $\varphi : \boldsymbol{A}' \prec \boldsymbol{A}''$.
b) If $\varphi : \boldsymbol{A} \prec \boldsymbol{A}'$. then there is $\boldsymbol{A}'' \in \mathrm{P}(A)$ and a map $\psi$ such that $\psi : \boldsymbol{A}' \prec \boldsymbol{A}''$ and $\psi \cdot \varphi = \delta$.

$R_3$ If $\boldsymbol{A} \equiv \boldsymbol{A}'$, then $\Omega(\boldsymbol{A}) \cap \Omega(\boldsymbol{A}') \neq 0$.

$L_3$ If $\tau(K) \subseteqq \tau(\boldsymbol{A})$, there is $\boldsymbol{A}' \in \Pi(K)$ such that $\boldsymbol{A} \equiv \boldsymbol{A}'$.

$C_3^1$ If $\tau(K) \cup \tau(K')$ is consistent and $K$ and $K'$ are closed under $\Pi$ and $\Omega$, then $K \cap K' \neq 0$.

$C_3^2$ If $K$ and $K'$ are closed under $\Pi$ and $\Omega$ and if $K \cap K' = 0$, then there is $N \in \mathrm{EC}$ such that $K \subseteqq N$ and $N \cap K' = 0$.

$B_3$ For a class of systems $K = K(\boldsymbol{R}, \boldsymbol{S})$, closed under $\Pi$ and $\Omega$, $K_{\boldsymbol{R}} = w_{\boldsymbol{R}}(K)$, the following are equivalent: (i) $w_{\boldsymbol{R}} \in \mathrm{EF}(K)$; (ii) $w_{\boldsymbol{R}}$ is one-one on $K$.

$F_3$ above is due to FRAYNE [8] and theorem T 3 again follows from $F_3$ by the 'alternating-chain' method. The proof of $L_3$ can be found, e.g., in [14] and the remaining results follow as in the previous cases. In view of the fact that an ultralimit of a system is isomorphic to a limit ultrapower of that system, each result here again implies its counterpart on level 2.

Level 4. Using the generalized continuum hypothesis (GCH), KEISLER [11] has shown that $R_3$ can be proved directly, without passing through the results $F_3$ and the directed limit process. In place of using the 'alternating' chain (outside CANTOR method) suggested by $F_3$ and directed limit, he applies the alternating chain from within (inside CANTOR method) as suggested by CANTOR's proof of the $\omega_0$-categoricity of the order type $\eta$. The results might be listed as follows:

$R_4$ If GCH and $\boldsymbol{A} \equiv \boldsymbol{A}'$, then $\mathrm{P}(\boldsymbol{A}) \cap \mathrm{P}(\boldsymbol{A}') \neq 0$.

$L_4$ If $\tau(K) \subseteqq \tau(\boldsymbol{A})$, there is $\boldsymbol{A}' \in \Pi(K)$ with $\boldsymbol{A} \equiv \boldsymbol{A}'$.

$C_4^1$ If GCH, $K$ and $K'$ are closed under $\Pi$ and $\tau(K) \cup \tau(K')$ is consistent, then $K \cap K' \neq 0$.

$C_4^2$ If GCH, $K$ and $K'$ are closed under $\Pi$ and $K \cup K' = 0$, then there is $N \in \mathrm{EC}$ with $K \subseteqq N$ and $N \cap K' = 0$.

$B_4$ If GCH, $K(\boldsymbol{R}, \boldsymbol{S})$ is closed under $\Pi$ and $w_{\boldsymbol{R}}(K) = K_{\boldsymbol{R}}$, the following are equivalent: (i) $w_{\boldsymbol{R}} \in \mathrm{EF}(K)$; (ii) $w_{\boldsymbol{R}}$ is one-one on $K$.

It has been pointed out that the statements on a given level imply those on prior levels. That these are proper implications has been shown by KEISLER [12]. In particular, he shows that there are systems $\boldsymbol{A}$ such that, i) $\mathrm{P}(\boldsymbol{A})$ is properly contained in $\Omega(\boldsymbol{A})$; ii) if GCH, then $\Omega(\boldsymbol{A})$ is properly contained in $\Lambda(\boldsymbol{A})$; and iii) $\Lambda(\boldsymbol{A})$ is properly contained in the class of elementary extensions of $\boldsymbol{A}$.

A survey of the methods used in this section shows that, on levels 3 and 4, the Łos theorem plays much the same role as does the compactness theorem on levels 1 and 2. In view of the absence of any syntactic development for ultraproducts in logic, this suggests such an approach beginning perhaps with a 'completeness theorem' having the Łos theorem as a semantic corollary. Considering the syntactic proofs of CRAIG's lemma (such as CRAIG's [6]), one might require that such a 'completeness theorem' be extendable to a syntactic proof of say, $C_4$. A further problem we wish to mention is the following: Prove $R_2$ – i.e. show that two elementarily equivalent systems have a common limit ultrapower (strong elementary extension) – by the *inside* CANTOR method and *without* using GCH.

## 4. Further Results on Definability

The hypothesis of BETH's theorem (condition i) of section 3) can be weakened in various ways. One such possibility is as follows:

iii) If $\boldsymbol{A}$ and $\boldsymbol{A}'$ are models of $\Sigma(\boldsymbol{R}, \boldsymbol{S})$, $\boldsymbol{A}_{\boldsymbol{R}} = \boldsymbol{A}'_{\boldsymbol{R}}$ and $\boldsymbol{A} \cong \boldsymbol{A}'$, then $\boldsymbol{A} = \boldsymbol{A}'$ or equivalently:

iii′) For any model $\boldsymbol{A}$ of $\Sigma(\boldsymbol{R}, \boldsymbol{S})$, $\varkappa(\boldsymbol{A}_{\boldsymbol{R}}) \subseteqq \varkappa(\boldsymbol{A})$.

SVENONIUS [20] proved (using BETH's theorem) that if iii) holds for $\Sigma(\boldsymbol{R}, \boldsymbol{S})$, then there are finitely many formulas $\Delta_1, \ldots, \Delta_n$ where each $\Delta_i$ is of the form $(\forall x_0, x_1)\,[\Phi_i(\boldsymbol{R}, x_0, x_1) \leftrightarrow \boldsymbol{S}(x_0, x_1)]$ and $\Sigma(\boldsymbol{R}, \boldsymbol{S}) \models \bigvee_{i=1}^{n} \Delta_i$.

A theory $\Sigma(\boldsymbol{R}, \boldsymbol{S})$ will be called *categorical relative* $\boldsymbol{R}$ if

iv) For $\boldsymbol{A}$, $\boldsymbol{A}'$ models of $\Sigma(\boldsymbol{R}, \boldsymbol{S})$, $\boldsymbol{A}_{\boldsymbol{R}} = \boldsymbol{A}'_{\boldsymbol{R}}$ implies $\boldsymbol{A} \cong \boldsymbol{A}'$.

It was observed by BÜCHI [3] that conditions iii) and iv) are equivalent to i). Thus iv) is just what is required in addition to iii) in order to derive the conclusion of BETH's theorem.

We will say that a theory $\Sigma(\boldsymbol{R}, \boldsymbol{S})$ is *complete relative* $\boldsymbol{R}$ if

v) For any sentence $\Phi(\boldsymbol{R}, \boldsymbol{S})$ there is a sentence $\Phi'(\boldsymbol{R})$ such that

$$\Sigma(\boldsymbol{R}, \boldsymbol{S}) \models \Phi(\boldsymbol{R}, \boldsymbol{S}) \leftrightarrow \Phi'(\boldsymbol{R}).$$

Using CRAIG's lemma (e.g., $C_1^2$) it can be shown that v) is equivalent to,

v′) For $\boldsymbol{A}$, $\boldsymbol{A}'$ models of $\Sigma(\boldsymbol{R}, \boldsymbol{S})$, $\boldsymbol{A}_{\boldsymbol{R}} = \boldsymbol{A}'_{\boldsymbol{R}}$ implies $\boldsymbol{A} \equiv \boldsymbol{A}'$.

It was noted in [5] that the hypothesis of BETH's theorem can be weakened by replacing iii) and iv) by iii) and v). The conclusion of the BETH theorem then remains valid. BÜCHI's proof of this strong form of BETH's theorem requires both BETH's theorem and SVENONIUS' theorem. (For an interesting application of SVENONIUS' theorem (or the strong form of the BETH theorem) see BÜCHI [2].) We give here a sequence of results (following the pattern of section 3) that yield direct proofs of the definability theorems just mentioned.

$\mathrm{R}_1$ Let $\boldsymbol{A} = \langle A, R, S, T\rangle$, $\boldsymbol{A}' = \langle A', R', S', T'\rangle$ where $\boldsymbol{A} \in K$, $\boldsymbol{A}' \in K'$, $K, K' \in \mathrm{PC}_\Delta$. Then if $\boldsymbol{A}_{\boldsymbol{R},\boldsymbol{T}} \equiv \boldsymbol{A}'_{\boldsymbol{R},\boldsymbol{T}}$ and $\tau(\boldsymbol{A}_{\boldsymbol{R},\boldsymbol{S}}) \cap \tau(\boldsymbol{A}'_{\boldsymbol{R},\boldsymbol{S}})$ is complete relative $\boldsymbol{R}$, there is a system $\boldsymbol{A}'' = \langle A'', R'', S'', T''\rangle$ and $\sigma \in \varkappa(\boldsymbol{A}''_{\boldsymbol{R}})$ such that $\boldsymbol{A}'' \in K$ and $\langle A'', R'', \sigma S'', T''\rangle \in K'$.

5*

Before proving $\bar{\mathrm{R}}_1$, we note its consequences.

$\bar{\mathrm{C}}_1^1$ For $K(\boldsymbol{R}, \boldsymbol{S}, \boldsymbol{T})$, $K'(\boldsymbol{R}, \boldsymbol{S}, \boldsymbol{T}) \in \mathrm{PC}_\Delta$ with $\tau(K_{\boldsymbol{R},\boldsymbol{S}}) \cap \tau(K'_{\boldsymbol{R},\boldsymbol{S}})$ complete relative $\boldsymbol{R}$, if $\tau(K_{\boldsymbol{R},\boldsymbol{T}}) \cap \tau(K'_{\boldsymbol{R},\boldsymbol{T}})$ is consistent, there are $\boldsymbol{A} \in K$ and $\sigma \in \varkappa(\boldsymbol{A}_R)$ such that $\langle A, R, \sigma S, T\rangle \in K'$.

Proof. Let $\langle B, R, T\rangle$ be a model of $\tau(K_{\boldsymbol{R},\boldsymbol{T}}) \cup \tau(K'_{\boldsymbol{R},\boldsymbol{T}})$. By $\mathrm{L}_1$ there are $\boldsymbol{A}_1 \in K_{\boldsymbol{R},\boldsymbol{T}}$, $\boldsymbol{A}'_1 \in K'_{\boldsymbol{R},\boldsymbol{T}}$ such that $\boldsymbol{A}_1 \equiv \boldsymbol{A}'_1$. $\boldsymbol{A}_1$ $(\boldsymbol{A}'_1)$ can be extended to $\boldsymbol{A} \in K$ $(\boldsymbol{A}' \in K')$. Now $\tau(K_{\boldsymbol{R},\boldsymbol{S}}) \cap \tau(K'_{\boldsymbol{R},\boldsymbol{S}}) \subseteqq \tau(\boldsymbol{A}_{\boldsymbol{R},\boldsymbol{S}}) \cap \tau(\boldsymbol{A}'_{\boldsymbol{R},\boldsymbol{S}})$ and hence by our assumption, the latter is complete relative $\boldsymbol{R}$, The conclusion follows from $\bar{\mathrm{R}}_1$.

$\bar{\mathrm{C}}_1^2$ For $K(\boldsymbol{R}, \boldsymbol{S}, \boldsymbol{T})$, $K'(\boldsymbol{R}, \boldsymbol{S}, \boldsymbol{T}) \in \mathrm{PC}_\Delta$ with $\tau(K_{\boldsymbol{R},\boldsymbol{S}}) \cap \tau(K'_{\boldsymbol{R},\boldsymbol{S}})$ complete relative $\boldsymbol{R}$, if for all $\boldsymbol{A} \in K$ and $\sigma \in \varkappa(\boldsymbol{A}_R)$, $\langle A, R, \sigma S, T\rangle \notin K'$, then there is $N(R, T) \in \mathrm{EC}$ with $K \subseteqq N$ and $K' \cap N = 0$.

$\bar{\mathrm{C}}_1^1$ is, as before, equivalent to $\bar{\mathrm{C}}_1^2$.

$\bar{\mathrm{B}}_1$ For $K(\boldsymbol{R}, \boldsymbol{S}) \in \mathrm{PC}_\Delta$, $K_{\boldsymbol{R}} = w_{\boldsymbol{R}}(K)$, the following are equivalent: (i) $w_{\boldsymbol{R}} \in \mathrm{EF}(K)$; (ii) $\tau(K)$ is complete relative $\boldsymbol{R}$ and for $\boldsymbol{A} \in K$, $\varkappa(\boldsymbol{A}_{\boldsymbol{R}}) \subseteqq \varkappa(\boldsymbol{A})$.

Proof. For (ii)$\Rightarrow$(i), let $a$ and $b$ be individual constants and let $K' = K \cap \mu(\boldsymbol{S}(a,b))$, $K'' = K \cap \mu(\sim \boldsymbol{S}(a,b))$. Then $K'$, $K'' \in \mathrm{PC}_\Delta$ and $\tau(K'_{\boldsymbol{R},\boldsymbol{S}}) \cap \tau(K''_{\boldsymbol{R},\boldsymbol{S}})$ is complete relative $\boldsymbol{R}$. Using our assumption, for $\langle A, R, S, a, b\rangle \in K'$, $\sigma \in \varkappa(\boldsymbol{A}_{\boldsymbol{R}})$, $\langle A, R, \sigma S, a, b\rangle \notin K''$. Hence by $\bar{\mathrm{C}}_1^2$, there is $N(\boldsymbol{R}, \boldsymbol{a}, \boldsymbol{b}) \in \mathrm{EC}$ with $K' \subseteqq N$ and $N \cap K'' = 0$. $w_{\boldsymbol{R}}$ is then defined by the formula $\Phi(\boldsymbol{R}, x_1, x_2)$ where $N$ is defined by $\Phi(\boldsymbol{R}, a, b)$.

Using $\bar{\mathrm{B}}_1$ and the compactness theorem, one now has:

$\bar{\mathrm{S}}_1$ For $K(\boldsymbol{R}, \boldsymbol{S}) \in \mathrm{PC}_\Delta$, $K_{\boldsymbol{R}} = w_{\boldsymbol{R}}(K)$, the following are equivalent: (i) There is a finite partition $\{G_1, \ldots, G_r\}$ of $K$ such that $w_{\boldsymbol{R}}$ restricted to $G_i$ is in $\mathrm{EF}(G_i)$, $i = 1, \ldots, r$. (ii) For $\boldsymbol{A} \in K$, $\varkappa(\boldsymbol{A}_{\boldsymbol{R}}) \subseteqq \varkappa(\boldsymbol{A})$.

Note that $\bar{\mathrm{B}}_1$ implies the strong form of BETH's theorem and $\bar{\mathrm{S}}_1$ implies SVENONIUS' theorem – moreover on this level they differ only by an application of the compactness theorem.

We now prove $\bar{\mathrm{R}}_1$. The reader should note the presence of $\bar{\mathrm{F}}_1$ and T 1 in the proof.

Proof of $\bar{\mathrm{R}}_1$. We are to find, for each $n$, systems $\langle A_n, R_n, S_n, T_n\rangle$ and $\langle A'_n, R'_n, S'_n, T'_n\rangle$ and maps $\sigma_n$, $\sigma'_n$, $\varphi_n$, $\psi_n$, $\pi_n$, and $\chi_n$ such that:

(1) $\sigma_n(\sigma'_n)$ is a partial $R$-automorphism on $\boldsymbol{A}_n$ $(\boldsymbol{A}'_n)$ with $\mathrm{C\!\!I}\sigma_n = \pi_n A'_{n-1}$ $(\mathrm{D}\sigma'_{n-1} = \chi_n A_n)$ and extends $\sigma'_{n-1}(\sigma_{n-1})$

(2) $\sigma_n^{-1}\pi_n S'_{n-1} = S_n \cap \sigma_n^{-1}\pi_n(A'_{n-1})^2$ and $\sigma'_n\chi_n S_n = S'_n \cap \sigma'_n\chi_n(A_n)^2$

(3) $\varphi_{n+1}: \boldsymbol{A}_n \prec \boldsymbol{A}_{n+1}$, $\psi_{n+1}: \boldsymbol{A}'_n \prec \boldsymbol{A}'_{n-1}$, $\pi_{n+1}: \boldsymbol{A}'_{n_{\boldsymbol{R},\boldsymbol{T}}} \prec \boldsymbol{A}_{n+1_{\boldsymbol{R},\boldsymbol{T}}}$, $\chi_n: \boldsymbol{A}_{n_{\boldsymbol{R},\boldsymbol{T}}} \prec \boldsymbol{A}'_{n_{\boldsymbol{R},\boldsymbol{T}}}$

(4) $\varphi_{n+1} = \pi_n \cdot \chi_n$, $\psi_{n+1} = \chi_{n+1} \cdot \pi_{n+1}$.

Given these systems and maps, condition (4) permits formation of the limits of the sequences by theorem T 1. (1) and (2) insure that the limit $\sigma$ of the $\sigma_n$ will serve as the desired $\boldsymbol{R}$-automorphism on the resulting system.

In i) below it is shown how to arrive at $\boldsymbol{A}_1$, $\sigma_1$ satisfying (1), (2), (3) and (4). The remaining steps are justified by ii) below.

i) Let $a_0, a_1, \ldots, a_\nu, \ldots, \nu < \eta$ be a well-ordering of $A$, $b_0, b_1, \ldots, b_\mu, \ldots$, $\mu < \lambda$ be a well-ordering of $B$. Let $\Sigma_1 = \tau(\langle \boldsymbol{A}, a_0, a_1, \ldots, a_\nu, \ldots\rangle_{\nu<\eta})$ representing $a_\nu$ by $\boldsymbol{a}_\nu$, $\Sigma_2 = \tau(\langle A', R', T', b_0, b_1, \ldots, b_\mu, \ldots\rangle_{\mu<\lambda})$ representing $b_\mu$ by $\boldsymbol{b}_\mu$. Let $\langle \boldsymbol{c}_\mu\rangle_{\mu<\lambda}$ be a new list of individual constants symbols and form $\Sigma_3 = \tau(\langle A', R', b_0, b_1, \ldots, b_\mu, \ldots\rangle_{\mu<\lambda})$ representing $b_\mu$ by $\boldsymbol{c}_\mu$. Let $\Sigma_4 = \{\boldsymbol{S}(\boldsymbol{c}_\xi, \boldsymbol{c}_{\xi'}), \sim\boldsymbol{S}(\boldsymbol{c}_\delta, \boldsymbol{c}_{\delta'});\ \langle b_\xi, b_{\xi'}\rangle \in S',\ \langle b_\delta, b_{\delta'}\rangle \notin S'$ respectively$\}$ and $\Delta = \bigcup_{i=1}^{4} \Sigma_i$. If $\Delta$ has a model, it will serve as $\boldsymbol{A}_1$ and thus, by the compactness theorem, it suffices to show $\Delta$ is finitely satisfiable. We consider a typical finite subset $\Delta_1 = \{\Phi_1(\boldsymbol{R}, \boldsymbol{S}, \boldsymbol{T}, \boldsymbol{a}_\nu), \Phi_2(\boldsymbol{R}, \boldsymbol{T}, \boldsymbol{b}_\mu), \Phi_3(\boldsymbol{R}, \boldsymbol{c}_\xi, \boldsymbol{c}_{\xi'}), \boldsymbol{S}(\boldsymbol{c}_\xi, \boldsymbol{c}_{\xi'})\}$. In $\Delta_1, \Phi_2(\boldsymbol{R}, \boldsymbol{T}, \boldsymbol{b}_\mu)$ may be replaced by $(\exists x)\Phi_2(\boldsymbol{R}, \boldsymbol{T}, x) = \Phi_2'$. Now $\Phi_2' \in \tau(\boldsymbol{A}'_{\boldsymbol{R},\boldsymbol{T}}) = \tau(\boldsymbol{A}'_{\boldsymbol{R},\boldsymbol{T}})$ and thus $\Phi_2' \in \Sigma_1$. $\Phi_3$ and $\boldsymbol{S}(\boldsymbol{c}_\xi, \boldsymbol{c}_{\xi'})$ may be replaced by $(\exists x_1, x_2)[\Phi_3(\boldsymbol{R}, x_1, x_2) \wedge \boldsymbol{S}(x_1, x_2)] = \Phi_5$ where $\Phi_5 \in \tau(\boldsymbol{A}'_{\boldsymbol{R},\boldsymbol{S}})$. Since $\tau(\boldsymbol{A}'_{\boldsymbol{R},\boldsymbol{S}}) \cap \tau(\boldsymbol{A}_{\boldsymbol{R},\boldsymbol{S}})$ is complete relative $\boldsymbol{R}$ and $\tau(\boldsymbol{A}_{\boldsymbol{R}}) = \tau(\boldsymbol{A}'_{\boldsymbol{R}})$ it follows that $\Phi_5 \in \tau(\boldsymbol{A}_{\boldsymbol{R},\boldsymbol{S}}) \subseteqq \Sigma_1$ and hence $\Delta_1$ (and $\Delta$) is satisfiable. Thus the existence of $\boldsymbol{A}_1$ and $\sigma_1$ follows; $\sigma_1$ given by associating the interpretation of $\boldsymbol{c}_\mu$ with that of $\boldsymbol{b}_\mu$ for $\mu < \lambda$.

ii) Assume $\boldsymbol{A}'$, $\boldsymbol{A}_1$, $\sigma_1$, $\pi_1$ as required and let $a_0, a_1, \ldots, a_\nu, \ldots, \nu < \eta$, and $b_0, b_1, \ldots, b_\mu, \ldots, \mu < \lambda$ be well-orderings of $A'$ and $A_1 \setminus \pi_1 A'$ respectively. We form $\Sigma_1 = \tau(\langle \boldsymbol{A}', a_0, \ldots, a_\nu, \ldots\rangle_{\nu<\eta})$ representing $a_\nu$ by $\boldsymbol{a}_\nu$ and $\Sigma_2 = \tau(\langle A_1, R_1, T, \pi(a_0), \ldots, \pi(a_\nu), \ldots, b_0, \ldots, b_\mu, \ldots\rangle_{\nu<\eta,\, \mu<\lambda})$ representing $\pi(a_\nu)$ by $\boldsymbol{a}_\nu$ and $b_\mu$ by $\boldsymbol{b}_\mu$. Let $\langle \boldsymbol{c}_\nu\rangle_{\nu<\eta}$ and $\langle \boldsymbol{d}_\mu\rangle_{\mu<\lambda}$ be new lists of individual constant symbols and form $\Sigma_3 = \tau(\langle A_1, R_1, \pi(a_0), \ldots, \pi(a_\nu), \ldots, b_0, \ldots, b_\mu, \ldots\rangle_{\nu<\eta,\mu<\lambda})$ representing $\pi(a_\nu)$ by $\boldsymbol{c}_\nu$ and $b_\mu$ by $\boldsymbol{d}_\mu$. As before let $\Sigma_4 = \{\boldsymbol{S}(\boldsymbol{c}_\xi, \boldsymbol{c}_{\xi'}), \sim\boldsymbol{S}(\boldsymbol{c}_\varrho, \boldsymbol{c}_{\varrho'}), \boldsymbol{S}(\boldsymbol{d}_\xi, \boldsymbol{d}_{\xi'}), \sim\boldsymbol{S}(\boldsymbol{d}_\varrho, \boldsymbol{d}_{\varrho'});\ \langle \pi(a_\xi), \pi(a_{\xi'})\rangle \in S_1,\ \langle \pi\, a_\varrho, \pi\, a_{\varrho'}\rangle \notin S_1,\ \langle b_\xi, b_{\xi'}\rangle \in S_1,\ \langle b_\varrho, b_{\varrho'}\rangle \notin S_1$ respectively$\}$. For each $\boldsymbol{a}_\nu$, $\nu < \eta$ let $\boldsymbol{a}_\nu'$ denote its preimage under $\sigma_1$, let $\boldsymbol{c}_\nu'$ denote the corresponding elements among the $\{\boldsymbol{c}_\nu, \boldsymbol{d}_\mu\}$ and let $\Sigma_5 = \{\boldsymbol{c}_\nu' = \boldsymbol{a}_\nu;\ \nu < \eta\}$. Again, it suffices to show that $\Delta = \bigcup_{i=1}^{5} \Sigma_i$ is finitely satisfiable. We consider a typical finite subset $\Delta_1 = \{\Phi_1(\boldsymbol{R}, \boldsymbol{S}, \boldsymbol{T}, \boldsymbol{a}_\xi, \boldsymbol{a}_{\xi'}), \Phi_2(\boldsymbol{R}, \boldsymbol{T}, \boldsymbol{a}_\xi, \boldsymbol{a}_{\xi'}, \boldsymbol{b}_\mu), \Phi_3(\boldsymbol{R}, \boldsymbol{c}_\xi \boldsymbol{c}_{\xi'}, \boldsymbol{d}_\mu), \boldsymbol{S}(\boldsymbol{c}_\xi, \boldsymbol{c}_{\xi'}), \boldsymbol{c}_\xi' = \boldsymbol{a}_\xi, \boldsymbol{c}_{\xi'}' = \boldsymbol{a}_{\xi'}\}$. Clearly we may suppose $\boldsymbol{c}_\xi = \boldsymbol{c}_\xi'$ and $\boldsymbol{c}_{\xi'} = \boldsymbol{c}_{\xi'}'$, i.e., that $\Delta_1 = \{\Phi_1(\boldsymbol{R}, \boldsymbol{S}, \boldsymbol{T}, \boldsymbol{a}_\xi, \boldsymbol{a}_{\xi'}), \Phi_2(\boldsymbol{R}, \boldsymbol{T}, \boldsymbol{a}_\xi, \boldsymbol{a}_{\xi'}, \boldsymbol{b}_\mu), \Phi_3(\boldsymbol{R}, \boldsymbol{a}_\xi, \boldsymbol{a}_{\xi'}, \boldsymbol{d}_\mu), \boldsymbol{S}(\boldsymbol{a}_\xi, \boldsymbol{a}_{\xi'})\}$. Since $\boldsymbol{c}_\xi = \boldsymbol{c}_{\xi'}$ and $\boldsymbol{c}_{\xi'} = \boldsymbol{c}_{\xi'}'$, we have that $\boldsymbol{a}_\xi = \boldsymbol{a}_\xi'$ and $\boldsymbol{a}_{\xi'} = \boldsymbol{a}_{\xi'}'$, i.e. $\sigma_1 \pi_1 a_\xi = \pi_1 a_\xi$ and $\sigma_1 \pi_1 a_{\xi'} = \pi_1 a_{\xi'}$. Also, $\boldsymbol{S}(\boldsymbol{c}_\xi, \boldsymbol{c}_{\xi'}) \in \Delta$ implies $\langle \pi_1 a_\xi, \pi_1 a_{\xi'}\rangle \in S_1$. Now if $\sim\boldsymbol{S}(a_\xi, \boldsymbol{a}_{\xi'}) \in \Sigma_1$, then $\langle a_\xi, a_{\xi'}\rangle \notin S'$. But since $\sigma_1^{-1}\pi_1 S' = S_1 \cap \sigma_1^{-1}\pi_1(A')^2$, we would then have that $\langle \sigma_1^{-1}\pi_1 a_\xi, \sigma_1^{-1}\pi_1 a_{\xi'}\rangle \notin S_1$ which implies $\langle \pi_1 a_\xi, \pi_1 a_{\xi'}\rangle \notin S_1$. This is a contradiction and thus $\boldsymbol{S}(a_\xi, a_{\xi'}) \in \Sigma_1$. That $\Delta_1$ (and hence $\Delta$) is satisfiable now follows as in the case of i).

Repeated applications of ii) (reversing the partial automorphisms as necessary) now permit completion of the proof.

$\mathrm{R}_1$ also admits extensions to levels 2 and 3. For example, on level 3, the result can be stated as follows:

$\mathrm{R}_3$ For $\boldsymbol{A} = \langle A, R, S, T\rangle$, $\boldsymbol{A}' = \langle A', R', S', T'\rangle$ with $\boldsymbol{A}_{\boldsymbol{R},\boldsymbol{T}} \equiv \boldsymbol{A}'_{\boldsymbol{R},\boldsymbol{T}}$ and $\tau(\boldsymbol{A}_{\boldsymbol{R},\boldsymbol{S}}) \cap \tau(\boldsymbol{A}'_{\boldsymbol{R},\boldsymbol{S}})$ complete relative $\boldsymbol{R}$, there are $A'' \in \Omega(\boldsymbol{A})$ and $\sigma \in \varkappa(\boldsymbol{A}''_{\boldsymbol{R}})$ such that $\langle A'', R'', \sigma S'', T''\rangle \in \Omega(\boldsymbol{A}')$.

The proof of $\bar{R}_3$ is similar to that of $R_1$, incorporating features of the proof of $R_3$. We again have $\bar{C}_3^1$ and $\bar{C}_3^2$ and from this:

$\bar{B}_3$ For a class $K(\boldsymbol{R}, \boldsymbol{S})$ closed under $\Pi$ and $\Omega$, the following are equivalent: (i) $\omega_{\boldsymbol{R}} \in \mathrm{EF}(K)$; (ii) $\tau(K)$ is complete relative $\boldsymbol{R}$ and for $\boldsymbol{A} \in K$, $\varkappa(\boldsymbol{A}_{\boldsymbol{R}}) \subsetneqq \varkappa(\boldsymbol{A})$.

To get $\bar{S}_3$ from $\bar{B}_3$, a generalization of the compactness theorem is needed – the following result of FRAYNE, MOREL, and SCOTT [8] suffices:

Lemma. *For U a family of classes of systems, each of which is closed under $\Pi$, if the intersection over U is empty, then the intersection over some finite subfamily of U is empty.*

From this lemma and $\bar{B}_3$, one now has:

$\bar{S}_3$ For a class $K(\boldsymbol{R}, \boldsymbol{S})$ closed under $\Pi$ and $\Omega$, the following are equivalent: (i) There is a partition $\{G_1, \ldots, G_r\}$ of $K$ such that $\omega_{\boldsymbol{R}}$ restricted to $G_i$ belongs to $\mathrm{EF}(G_i)$ for $i = 1, \ldots, r$. (ii) For $\boldsymbol{A} \in K$, $\varkappa(\boldsymbol{A}_{\boldsymbol{R}}) \subsetneqq \varkappa(\boldsymbol{A})$.

$\bar{S}_3$ was noted by KOCHEN [15] and proved there using a fact about universal ultrafilter sequences.

## Bibliography


[1] BETH, E. W., On Padoa's method in the theory of definition. Indagationes Mathematicae **15** (1953).

[2] BÜCHI, J. R., Affine definability of affine invariants of Euclidian geometry. Notices Amer. Math. Soc. **15**, No. 6, 68T-E23.

[3] BÜCHI, J. R., Relatively categorical and normal theories. In: The Theory of Models. Amsterdam 1965.

[4] BÜCHI, J. R., and W. CRAIG, Notes on the family $PC_\Delta$ of sets of models. J. Symb. Logic **21** (1965).

[5] BÜCHI, J. R., and K. DANHOF, A strong form of Beth's definability theorem. Notices Amer. Math. Soc. **15**, No. 6, 68T-E22.

[6] CRAIG, W., Linear reasoning. A new form of the Herbrand-Gentzen theorem. J. Symb. Logic **22** (1957).

[7] CRAIG, W., Three uses of the Herbrand-Gentzen theorem in relating proof theory and model theory. J. Symb. Logic **22** (1957).

[8] FRAYNE, T., A. MOREL and D. SCOTT, Reduced direct products. Fund. Math. **51** (1962).

[9] HENKIN, L., The completeness of the first-order functional calculus. J. Symb. Logic **14** (1949).

[10] HILBERT, D., und P. BERNAYS, Grundlagen der Mathematik, Band 1. Berlin 1934.

[11] KEISLER, H. J., Ultraproducts and elementary classes. Indag. Math. **23** (1961).

[12] KEISLER, H. J., Limit Ultrapowers. Trans. Amer. Math. Soc. **108** (1963).

[13] KEISLER, H. J., Limit Ultraproducts. J. Symb. Logic **30** (1965).

[14] KOCHEN, S., Ultraproducts in the theory of models. Ann. of Math. **74** (1961).

[15] KOCHEN, S., Topics in the theory of definition. In: The Theory of Models. Amsterdam 1965.

[16] ŁOS, J., Quelques remarques, theoremes, et problemes sur les classes definissables d'algebras. In: Mathematical Interpretations of Formal Systems. Amsterdam 1965.

[17] PADOA, A., Un nouveau systems irreductible de postulats pour l'algebre. Compte rendu du deuxième congres international des mathematiciens. Paris 1902.

[18] ROBINSON, A., A result on consistency and its application to the theory of definition. Indag. Math. **18** (1956).

[19] ROBINSON, A., Complete theories. Amsterdam 1965.

[20] SVENONIUS, L., A theorem on permutations in models. Theoria (Lund) **25** (1959).

[21] TARSKI, A., and R. L. VAUGHT, Arithmetical extensions of relational systems. Comp. Math. **13** (1957).


(Eingegangen am 20. Februar 1971)

# DEFINIBILITY IN NORMAL THEORIES

BY

J. RICHARD BUCHI AND KENNETH J. DANHOF

ABSTRACT

This paper initiates an investigation which seeks to explain elementary definability as the classical results of mathematical logic (the completeness, compactness and Löwenheim-Skolem theorems) explain elementary logical consequence. The theorems of Beth and Svenonius are basic in this approach and introduce automorphism groups as a means of studying these problems. It is shown that for a complete theory $T$, the definability relation of Beth (or Svenonius) yields an upper semi-lattice whose elements (concepts) are interdefinable formulas of $T$ (formulas having equal automorphism groups in all models of $T$). It is shown that there are countable models $A$ of $T$ such that two formulae are distinct (not interdefinable) in $T$ if and only if they are distinct (have different automorphism groups) in $A$. The notion of a concept $h$ being normal in a theory $T$ is introduced. Here the upper semi-lattice of all concepts which define $h$ is proved to be a finite lattice — anti-isomorphic to the lattice of subgroups of the corresponding automorphism group. Connections with the Galois theory of fields are discussed.

## 1. Introduction

In 1932 all the basic results concerning the notion of elementary logical deductions were available; namely, the theorems of Skolem-Löwenheim, Herbrand and Gödel. Four additional years of investigation of the notion of algorithm brought about the solution of Hilbert's decision problem of mathematics by Church and (independently and almost simultaneously) Turing. A few years later the compactness theorem for elementary logic must have been clear to several people.

At this time it might have seemed that most of the basic problems of elementary axiom systems were solved. A more careful observer however, upon reading the papers of Tarski [13, 14], might have wondered about the existence of general theorems which would explain elementary definability as the above theorems

Received April 2, 1972 and in revised form August 21, 1972

explain the basic properties of elementary logical consequence. One such theorem, the completeness, in the sense of definability, of elementary logic was proved by Beth in 1953 [1]. Also, around this time, Craig, in his doctoral dissertation, proved what is now known as the interpolation lemma (or separation lemma).

Craig's lemma [6] and Beth's theorem went unnoticed for a time. In 1956 the basic importance of Craig's lemma became apparent: first, it admits Beth's theorem as an easy corollary and second, it sounds more impressive in its model-theoretic version. About this time also, Robinson's consistency lemma [11], which is intimately related to Craig's lemma, appeared. The relationship between these definability results may be traced on various levels, i.e., there are stronger forms which may be proved by using model-operators like ultrapowers, etc. For a discussion of these various levels, see Buchi and Danhof [3].

In 1959 Svenonius [12] published a further result on elementary definability. Just as with the earlier results of Beth and Craig, logicians seem slow in recognizing Svenonius' theorem as a basic tool in the theory of definability, perhaps because it is not generally known to be available.

With the appearance of Klein's Erlangerprogramm in 1872 [10], it became apparent that automorphism groups are a most useful means of studying mathematical theories. In a more rigorous model-theoretic manner, these ideas have been discussed in the case of various elementary mathematical theories by Buchi and Wright (see [4, 5, 15]). It is not surprising that both the theorems of Beth and Svenonius are about the automorphism groups of the models of a theory. In Buchi [2], Beth's theorem is restated as a general result on relative categoricity and Svenonius' theorem is used to establish a basic result on the Galois group of normal concepts. Here these matters are carried out in detail and further results leading toward a theory of elementary definability are added.

## 2. Preliminaries

We assume familiarity with the notion of an *elementary theory* (*class*) $T = T(\boldsymbol{R})$ with primitives $\boldsymbol{R}$ and equality. We frequently identify a theory $T(\boldsymbol{R})$ with the class of models $\langle A, R\rangle$ of $T$ (and assume this class to be closed under isomorphism). For a system $\boldsymbol{A}$, $\tau(A)$ denotes the set of all sentences true in $\boldsymbol{A}$ (*elementary theory of* $\boldsymbol{A}$). A theory $T(\boldsymbol{R})$ is *pseudo-elementary* if there is an elementary theory $T(\boldsymbol{R}, \boldsymbol{S})$ such that $T(\boldsymbol{R})$ is the class of systems $\langle A, R\rangle$ for which there is an $S$ with $\langle A, R, S\rangle \in T(\boldsymbol{R}, \boldsymbol{S})$. For a system $\boldsymbol{A}$, $\kappa \boldsymbol{A}$ denotes the group of automorphisms of

$A$; $A \cong A'$ indicates that $A$ and $A'$ are isomorphic. A formula is called an *$R'$-condition* if the predicate $R'$ (primitive or defined) is its only extralogical constant. For a system $A$, a one-one mapping $\phi$ between two subsets of $A$ is called a *partial $R'$-automorphism* if any two sequences of elements which correspond to each other by $\phi$ satisfy the same $R'$-conditions. "$(\forall A)_T \cdots$" should be read as "for all models $A$ of $T$, $\cdots$". An elementary formula $\Phi = \Phi(R, x_1, \cdots, x_n)$ of $T(R)$ defines an *elementary concept* of $T$ as follows: For any system $A = \langle A, R\rangle$,

$$\Phi(\langle A, R\rangle) = \langle A, (\hat{x}_1, \cdots, \hat{x}_n)\Phi\rangle.$$

Thus, an elementary concept $c$ of $T$ ($c \in \overline{ec}(T)$) is a mapping defined by an elementary formula from a species into a (possibly different) species. Note that $A$ and $cA$ have the same domain. Moreover, for any such concept and any transformation $\phi$ of $A$ we have:

i) $$\phi c A = c \phi A.$$

For the purpose of comparing elements of $\overline{ee}(T)$, we have the following two quasi-orders:

1) $c \leqq_1 d(T) .\equiv. (\forall A, A')_T\, dA = dA' \to cA = cA'$

2) $c \leqq_2 d(T) .\equiv. (\forall A)_T\, \kappa dA \subseteq \kappa cA$.

It is easy to see that $c \leqq_1 d(T)$ implies $c \leqq_2 d(T)$. For $T$ pseudo-elementary, Craig's extension [7] of Beth's theorem may be stated as follows:

ii) $c \leqq_1 d(T) .\equiv.$ there is an elementary concept $d'$ such that

$$(\forall A)_T\, cA = d'\, dA.$$

Similarly, with $T$ again pseudo-elementary, Svenonius' theorem [12] may be generalized to:

iii) $c \leqq_2 d(T) .\equiv.$ there are elementary concepts $d_1, \cdots, d_n$ such that

$$(\forall A)_T\, (cA = d_1 dA \vee \cdots \vee cA = d_n dA).$$

Note in particular that if $T$ is a complete elementary theory, then for all $c$, $d \in \overline{ec}(T)$, $c \leqq_1 d(T) .\equiv. c \leqq_2 d(T)$. In the sequel, we restrict our attention to complete theories and let $\leqq$ denote $\leqq_1$ (or equivalently $\leqq_2$). For $h \in \overline{ec}(T)$, let $\overline{ec}(T, h) = \{c \in \overline{ec}(T);\, h \leqq c(T)\}$. $\leqq$ is a quasi-order on $\overline{ec}(T, h)$ and if $\sim$ is defined by $c \sim d .\equiv. c \leqq d(T) \wedge d \leqq c(T)$, then $\sim$ is an equivalence relation on $\overline{ec}(T, h)$. $ec(T, h)$ will denote the set of equivalence classes (relative to the relation $\sim$). $\leqq$ induces a partial order on these classes. The partially ordered set has (0−) the concept $h$ and $(1-)$ the primitive concept $\boldsymbol{R}$. Hereafter, we often identify a concept with its equivalence class. For $c$, $c' \in ec(T,h)$ defined by $\Phi(x_1,\cdots,x_n)$ and $\Phi'(y_1,\cdots,y_m)$, we let $c \otimes c' = (\hat{x}_1 \cdots \hat{x}_n \hat{y}_1 \cdots \hat{y}_m)\ (\Phi \wedge \Phi')$. $\otimes$ is a l.u.b. in $ec(T,h)$ and consequently $ec(T,h)$ is an upper semi-lattice with 0 and 1. Note that if the concept '$=$' is defined by the formula $x_1 = x_2$, then for any concept $c$, '$=$' $\leqq c(T)$. We denote $ec(T,$ '$=$'$)$ by $ec(T)$.

For a structure $\boldsymbol{A}$ of $T$, we write $c \leqq d(A)$ if $\kappa d A \subseteq \kappa c \boldsymbol{A}$. Note $c \leqq d(T)$ implies $c \leqq d(A)$. By imitating the above construction, we get, for each model $\boldsymbol{A}$ of $T$ and each $h \in \overline{ec}(T)$, an upper semi-lattice $ec(\boldsymbol{A},h)$—a subsystem of $ec(T,h)$.

It can be shown that $ec(T)$ need not be a lattice. For example, let

$$T = \tau\langle A, f_1, f_2, a\rangle$$

where $a \in A$, $f_i$ is a partial one-one unary function from $A_i = \{f_i^n(a);\ n \geqq 0\}$ onto $A_i - \{a\}$ for $i = 1,2$, and $f_1^n(a) = f_2^m(a)$ iff $n = m$ and $n$ is even (i.e., $A_1 \cup A_2 = A$ and $A_1 \cap A_2 = \{f_i^n(a);\ n \text{ even}\}$). One can then show g.l.b. $\{f_1(x) = y, f_2(x) = y\}$ does not exist (such a g.l.b. would be definable from each of $f_1$ and $f_2$).

Several questions suggest themselves at this point: What is the relation between $ec(T)$ and the theory $T$? When is $ec(T,h)$ finite?, etc. In the following section we describe a sufficient condition for the finiteness of $ec(T,h)$.

**3.** The following lemma expands only slightly the basic result used in extending Beth's theorem to Svenonius' theorem and the proof employs the ideas of that proof (see Svenonius [12]).

LEMMA 1. *Given elementarily equivalent systems $\boldsymbol{A}$ and $\boldsymbol{A}'$, concepts $c$ and $c'$ and partial automorphisms $\phi$ ($\phi'$) of $c\boldsymbol{A}$ ($c'\boldsymbol{A}'$), there is an elementary extension $\boldsymbol{B}$ of $\boldsymbol{A}$ and $\boldsymbol{A}'$ and partial automorphisms $\sigma$ ($\sigma'$) of $c\boldsymbol{B}$ ($c'\boldsymbol{B}$) which extend $\phi$ ($\phi'$) such that the domain (or range) of $\sigma$ ($\sigma'$) is the image of $A(A')$ in $B$.*

Our first theorem shows that for $T$ complete, the study of $ec(T,h)$ reduces to that of $ec(A,h)$ for certain countable models of $A$ of $T$.

THEOREM 1. *If $T$ is complete and $h \in ec(T)$, there is a countable model $\boldsymbol{A}$ of $T$ such that $ec(T,h) = ec(\boldsymbol{A},h)$.*

OUTLINE OF PROOF. $ec(T,h)$ is countable (its elements are represented by formulas of $T$). Let $\{\alpha_i\}_{i<\omega}$ ennumerate all pairs $\langle c,c'\rangle \in (ec(T,h))^2$ such that $c < c'(T)$. Then for each $i < \omega$ and $\alpha_i = \langle c_i, c_i'\rangle$ there is a model $\boldsymbol{A}_i$ of $T$ such that $\kappa c_i'\, \boldsymbol{A}_i \subset \kappa c_i \boldsymbol{A}_i$; moreover we may assume $\boldsymbol{A}_i$ countable. Now using Lemma 1 one constructs models $\boldsymbol{B}_n$ of $T$ and partial mappings $\phi_i^n$ such that $\boldsymbol{B}_0 = \boldsymbol{A}_0$, $\boldsymbol{B}_{n+1}$

is an elementary extension of $\boldsymbol{B}_n$ and $\boldsymbol{A}_{n+1}$, and $\phi_i{}^{n+1}(i = 0, \cdots n+1)$ is a partial automorphism of $c_i\boldsymbol{B}_{n+1}$ but not of $c_i'\boldsymbol{B}_{n+1}$. As $n$ is even (odd) the domain (range) of $\phi_i^{n+1}$ $(i \leqq n)$ is the image of $\boldsymbol{B}_n$ in $\boldsymbol{B}_{n+1}$ and the domain (range) of $\phi_{n+1}^{n+1}$ is the image of $\boldsymbol{A}_{n+1}$ in $\boldsymbol{B}_{n+1}$. $\boldsymbol{A} = \cup \boldsymbol{B}_n$ is now the desired system since for $\phi_i = \cup \phi_i^n$ we have $\phi_i \in \kappa c_i \boldsymbol{A} \,/\, \kappa c_i' \boldsymbol{A}$ for $i < \omega$. Thus $c < d(T) \;.\!\equiv\!.\; c < d(\boldsymbol{A})$ and consequently $ec(T,h) = ec(\boldsymbol{A}, h)$.

Definition. The theory $T$ is said to be *normal relative the elementary concept* $h$ $(h \lhd T)$ if:

$$(\forall \boldsymbol{A}, \boldsymbol{A}')_T \boldsymbol{A} \cong \boldsymbol{A}' \wedge h\boldsymbol{A} = h\boldsymbol{A}' \;.\!\rightarrow\!.\; \kappa \boldsymbol{A} = \kappa \boldsymbol{A}'.$$

Models $\boldsymbol{A}$ and $\boldsymbol{A}'$ such that $\boldsymbol{A} \cong \boldsymbol{A}'$ and $h\boldsymbol{A} = h\boldsymbol{A}'$ (i.e., there is an automorpism of $h\boldsymbol{A} = h\boldsymbol{A}'$ which takes $\boldsymbol{A}$ onto $\boldsymbol{A}'$) are called *h-conjugate* and we write $\boldsymbol{A} \simeq \boldsymbol{A}'(h)$.

Lemma 2. $h \lhd T \;.\!\equiv\!.\; (\forall \boldsymbol{A})_T \kappa \boldsymbol{A} \lhd \kappa h \boldsymbol{A}$ (i.e., $\kappa \boldsymbol{A}$ *is a normal subgroup of* $\kappa h \boldsymbol{A}$).

Proof. Note that by definition,

$$h \lhd T \;.\!\equiv\!.\; (\forall \boldsymbol{A}, \boldsymbol{A}')_T \boldsymbol{A} \simeq \boldsymbol{A}'(h) \;.\!\rightarrow\!.\; \kappa \boldsymbol{A} = \kappa \boldsymbol{A}',$$

or equivalently, $(\forall \boldsymbol{A})_T \; \phi h \boldsymbol{A} = h\boldsymbol{A} \wedge \psi \boldsymbol{A} = \boldsymbol{A} \;.\!\rightarrow\!.\; \psi\phi \boldsymbol{A} = \phi \boldsymbol{A}$. However, this last condition may be restated as $(\forall \boldsymbol{A})_T \; \phi \in \kappa h \boldsymbol{A} \wedge \psi \in \kappa \boldsymbol{A} \;.\!\rightarrow\!.\; \phi^{-1}\psi\phi \in \kappa \boldsymbol{A}$.

Theorem 2. *If $h \lhd T$, there is a number $n$ and elementary concepts $d_1, \cdots, d_n$ such that for any model $\boldsymbol{A}$ of $T$,*

$$\boldsymbol{A} \simeq \boldsymbol{A}'(h) \;.\!\rightarrow\!.\; \boldsymbol{A}' = d_1 \boldsymbol{A} \vee \cdots \vee \boldsymbol{A}' = d_n \boldsymbol{A}$$

*and in particular, $G_h(\boldsymbol{A}) = \kappa h \boldsymbol{A} / \kappa \boldsymbol{A}$ has order $\leqq n$.*

Proof. Since $h \lhd T$, we have

a) $$(\forall \boldsymbol{A})_T \boldsymbol{A} \simeq \boldsymbol{A}'(h) \;.\!\rightarrow\!.\; \phi \boldsymbol{A} = \boldsymbol{A} \rightarrow \phi \boldsymbol{A}' = \boldsymbol{A}'.$$

Recall that $\boldsymbol{A} \simeq \boldsymbol{A}'(h)$ means $h\boldsymbol{A} = h\boldsymbol{A}' \wedge (\exists \phi)\; \phi \boldsymbol{A} = \boldsymbol{A}'$. Since $h$ is elementary, this is a pseudo-elementary proposition. Since $T = T(\boldsymbol{R})$ is elementary, it follows that $\boldsymbol{A} \in T \wedge \boldsymbol{A} \simeq \boldsymbol{A}'(h)$ is a pseudo-elementary theory $T'(\boldsymbol{R}, \boldsymbol{R}')$. Now (a) is just the other assumption of (iii) (Svenonius' theorem) and hence there are elementary concepts $d_1, \cdots, d_n$ such that

$$A \in T \wedge A \simeq A'(h) . \rightarrow . A' = d_1 A \vee \cdots \vee A' = d_n A.$$

This establishes the first part of Theorem 2.

To see that $G_h(A)$ is finite, let $A \in T$. Since, by the above, there are at most $n$ different $h$-conjugates $A_1, \cdots, A_n$ of $A$, we have

b) $$\phi \in \kappa h A \rightarrow \phi A = A_1 \vee \cdots \vee \phi A = A_n.$$

Let $\sigma$, $\rho \in \kappa h A$. Then by (b), we have

c) $$\sigma^{-1}\rho \in \kappa A . \equiv . \sigma A = \rho A = A_1 \vee \cdots \vee \sigma A = \rho A = A_n.$$

From (c), one sees the relation $\sigma^{-1}\rho \in \kappa A$ is an equivalence relation on $\kappa h A$, of index $\leqq n$. Moreover, it is the congruence on $\kappa h A$ relative to the normal subgroup $\kappa A$. Therefore, $G_n(A)$ has $\leqq n$ members.

We give two examples where $h \lhd T$. $T_1$ has $n$ unary predicate letters $R_1, \cdots, R_n$. Axioms for $T_1$ insure that in any model of $T_1$, the interpretations of $R_1, \cdots, R_n$ partition the domain of the model into $n$ disjoint equivalence classes. Let $h$ be given by the formula which defines the resulting equivalence relation. It is easy to show $h \lhd T_1$. Each model of $T_1$ has $\leqq n!$ $h$-conjugates and equality holds if all of the equivalence classes have the same cardinality. In the latter case $G_h(A)$ is just the symmetric group on $n$ objects.

As a second example, let $F$ be a finite (algebraic) extension of its prime subfield and $G = F(\theta_1, \cdots, \theta_n)$ a finite extension of $F$. Let $T_2$ be all sentences true in $\mathbf{G} = \langle G, F(G), \theta_1, \cdots, \theta_n \rangle$ where $F(G)$ consists of $0, 1, +, \cdot$ and the elements needed to generate $F$. For $\mathbf{A} = \langle A, F(A), a_1, \cdots, a_n \rangle$ a model of $T_2$, let $h\mathbf{A} = \langle A, F(A) \rangle$. Using basic facts about fields, we again have $h \lhd T_2$.

THEOREM 3. *If $h \lhd \tau(A)$, then there is an order anti-isomorphism from $ec(\mathbf{A}, h)$ onto the lattice of subgroups of $G_h(A)$. Consequently, applying Theorem 2, $ec(\mathbf{A}, h)$ is a finite lattice.*

PROOF. For $c \in ec(\mathbf{A}, h)$, $c \rightarrow \kappa c A / \kappa A \subseteq G_h(A)$. This correspondence is clearly one-one, for $\kappa c A / \kappa A = \kappa c' A / \kappa A . \equiv . \kappa c A = \kappa c' A . \equiv . c = c'$. Also, $c \leqq c'(A)$ $. \equiv . \kappa c' A \subseteq \kappa c A . \equiv . \kappa c' A / \kappa A \subseteq \kappa c A / \kappa A$. We are to show that the mapping is onto. By Theorem 2, there are elementary concepts $d_1, \cdots, d_n$ such that

1) $$A \cong A' \wedge hA = hA' . \rightarrow . A' = d_1 A \vee \cdots \vee A' = d_n A.$$

Moreover, we may take $d_iA = \phi_iA$ where $G_h(A) = \{\phi_i; i = 1,\cdots,n\}$. One also has (see (i)) that if $\phi_i\phi_j = \phi_k$, then $d_jd_iA = d_kA$ and if $\phi^{-1} = \phi_j$, then $d_jd_iA = d_id_jA = A$ (we take $d_1A = A$, i.e., $\phi_1$ is the identity). Thus, $D = \{d_i; i = 1,\cdots,n\}$ restricted to $\{d_iA; i = 1,\cdots,n\}$ forms a group anti-isomorphic to $G_h(A)$.

For $d_i$, $d_j \in D$, $\sim d_i$, $d_i \wedge d_j$ and $d_i \vee d_j$ are defined in the obvious manner. For each $f\colon \{d_2,\cdots,d_n\} \to \{d_2, \sim d_2,\cdots,d_n, \sim d_n\}$ such that $f(d_j) \in \{d_j, \sim d_j\}$ for $j = 2,\cdots,n$, let

2) $$d_i^f = d_i \wedge (\wedge_{j=2,\dots,n} f(d_j)d_i); \qquad i \leqq n.$$

We have,

3) $$a \in d_jd_iA \;.\equiv.\; \phi_i^{-1}\ (a) \in d_jA.$$

From (3), $d_i^f(A) = 0 \;.\equiv.\; d_j^f(A) = 0$. We say $f \neq 0$ if $d_1^f(A) \neq 0$. For $i = 1,\cdots,n$; $d_i^* = \otimes_{f \neq 0}\ d_i^f$ and for $H \subseteq G_h(A)$,

4) $$d_H = \left(\bigvee_{\phi_i \in H} d_i^*\right) \otimes h.$$

Note $\kappa d_HA \subseteq \kappa hA \to d_H \in ec(A,h)$. We assert $H = \kappa d_HA/\kappa A$. If $\sigma \in H$, then $\sigma = \phi_k$ for some $k \leqq n$. Also, $\sigma d_HA = d_HA$ since if $\phi_i \in H$, $\sigma d_i^*A = \phi_k d_i^*A = d_i^* d_kA = (d_id_k)^*A = d_j^*A$ for $\phi_j = \phi_k\phi_i \in H$. Thus, $H \subseteq \kappa_{dH}A\, d\kappa A$. For the converse, note that $d_j^*A \cap d_i^*A = 0$ for $i \neq j$. If $\sigma \in \kappa d_HA/\kappa A$, then $\sigma = \phi_j$ for some $j \leqq n$. Now for $a \in d_1^*A$,

$$\sigma a \in \left(\bigvee_{\phi_i \in H} d_i^*\right)$$

i.e., $\sigma a \in d_i{}^*A$ for some $\phi_i \in H$. By (3), $a \in d_k^*A$ where $\phi_k = \phi_j^{-1}\phi_i$ and hence $k = 1$ by the disjointness property noted above. Thus, $\phi_j = \phi_i$ and $\sigma = \phi_j \in H$. This proves the theorem.

Combining Theorems 1 and 3, we have;

THEOREM 4. *If $h \lhd T$ ($T$ complete), then $ec(T,h)$ is anti-isomorphic to the lattice of subgroups of $G_h(A)$ for some countable model $A$ of $T$ (and hence is a finite lattice).*

Returning to the earlier example $T_2$ from fields, since $h \lhd T_2$, Theorem 3 applies here. Moreover, if $G$ is assumed to be the splitting field of a separable polynomial over $F$, then $G = F(\theta)$, a simple extension of $F$, and $G_h(G)$ is just the group of automorphisms of $G$ fixing $F$. In this case also, there is a one-one correspondence between $ec(\boldsymbol{G},h)$ and subfields $J$, $F\subseteq J\subseteq G$. For if $F\subseteq J\subseteq G$, then $J=F(\delta)$ for some $\delta\in G$. We let $c_J\boldsymbol{G}=\langle G,F(G),\delta\rangle$; then $\kappa c_J\boldsymbol{G}$ is the group of automorphisms of $G$ over $J$. Also, if $c_J=c_{J'}$, then $J=J'$. Conversely, given $c\in ec(\boldsymbol{G},h)$, let $J=\{a\in G;\ \phi a=a$ for all $\phi\in\kappa cG=H\}$. Then $c_J\leqq c(\boldsymbol{G})$ and from the proof of Theorem 3, we may write $c=\bigvee_H(x=\phi_i(\theta))\otimes h$. Now in $T_2$, $\bigvee_H(x=\phi_i(\theta))$ $.\equiv.f(x)=0$; it follows that the coefficients of $f(x)$ are in $J$. Any element of $\kappa c_J\boldsymbol{G}$ fixes $f(x)$ and hence fixes $c$. Thus, $c\leqq c_J(\boldsymbol{G})$. It follows that there is a one-one correspondence between the subgroups of $\kappa h(G)$ and subfields $J,F\subseteq J\subseteq G$. (cf. Fundamental theorem of Galois theory).

Remark. In an attempt to keep the presentation as smooth as possible, our definition of elementary concept in §2 was less general than it might have been. As an example of how this notion might be extended, for $\Phi=\Phi(\boldsymbol{R},\boldsymbol{S},x_1,\cdots,x_n)$ a formula of $T(\boldsymbol{RS})$, we can let $\Phi(\langle A,R,S\rangle)=\langle A,R,(\hat{x}_1,\cdots,\hat{x}_n)\Phi\rangle$. This version subsumes the earlier notion and the theorems stated above remain intact. In this setting the restriction in example 2 that $F$ be a finite extension of its prime field can be removed; $F$ can now be any field.

We conclude with a final observation reminiscent of the Galois groups of fields. Let $h\lhd T,\boldsymbol{A}$ be a model of $T$ and $c\in ec(T,h)$. Then for $\sigma\in\kappa h\boldsymbol{A}$, $\kappa c\sigma\boldsymbol{A}=\sigma(\kappa c\boldsymbol{A})\sigma^{-1}$. Now $\kappa c\boldsymbol{A}=\sigma(\kappa c\boldsymbol{A})\sigma^{-1}$ for all $\sigma\in\kappa h\boldsymbol{A}\ .\equiv.\ \kappa c\boldsymbol{A}\lhd\kappa h\boldsymbol{A}\ .\equiv.\ \boldsymbol{A}\simeq\boldsymbol{A}'(h)\ .\rightarrow.\ \kappa c\boldsymbol{A}=\kappa c\boldsymbol{A}'$. Accordingly, we write

$$h\lhd c(T)\ .\equiv.\ (\forall\boldsymbol{A},\boldsymbol{A}')_T\boldsymbol{A}\simeq\boldsymbol{A}'(h)\ .\rightarrow.\ \kappa c\boldsymbol{A}=\kappa c\boldsymbol{A}'.$$

From the above, $h\lhd c(T)\ .\equiv.\ (\forall\boldsymbol{A})_T\ \kappa c\boldsymbol{A}\lhd\kappa h\boldsymbol{A}\ .\equiv.\ (\forall\boldsymbol{A})_T(\forall\sigma)_{\kappa h\bar{A}}\ \kappa c\sigma\boldsymbol{A}=\kappa c\boldsymbol{A}$. Now if $\kappa c\boldsymbol{A}\lhd\kappa h\boldsymbol{A}$, then $\kappa c\boldsymbol{A}/\kappa\boldsymbol{A}\lhd\kappa h\boldsymbol{A}/\kappa\boldsymbol{A}$. In this case, by the law of homomorphism for groups,

$$\kappa h\boldsymbol{A}/\kappa c\boldsymbol{A}\cong(\kappa h\boldsymbol{A}/\kappa\boldsymbol{A})/(\kappa c\boldsymbol{A}/\kappa\boldsymbol{A}).$$

## References


1. E. W. Beth, *On Padoa's method in the theory of definitions*, Indag. Math. **15** (1953).
2. J. R. Buchi, *Relatively Categorical and Normal Theories*, in The Theory of Models, Amsterdam, 1965.
3. J. R. Buchi and K. J. Danhof, *Model Theoretic Approaches to Definability*, Z. Math. Logik Grundlagen Math. **18** (1972), 61–70.
4. J. R. Buchi and J. B. Wright, *The theory of proportionality as an abstraction of group theory*, Math. Ann. **130** (1955).

5. J.R. Buchi and J.B. Wright, *Invariants of the anti-automorphisms of a group*, Proc. Amer. Math. Soc. **8** (1957).

6. W. Craig, *Linear reasoning. A new form of the Herbrand-Gentzen theorem*, J. Symbolic Logic **22** (1957).

7. W. Craig, *Tree uses of the Herband-Gentzen theorem in relating model theorey and proof theory*, J. Symbolic Logic **22** (1957).

8. K.J. Danhof, *Concepts in normal theories*, Notices Amer. Math. Soc. **71 T-E26** (1971).

9. D. Hilbert and P. Bernays, *Grundlagen der Mathematik*, Vol. 1, Berlin, 1934.

10. F. Klein, *Vergleichende Betrachutngen über neuere geometrische Forschungen*, Verlag A. Deicher, Erlangen, 1872.

11. A. Robinson, *A result on consistency and its application to the theory of definition*, Indag. Math. **18** (1956), 47-58.

12. L. Svenonius, *A theorem on permutations in models*, Theoria (Lund) **25** (1959).

13. A. Tarski, *Der Wahrheitsbegriff in den formalisierten Sprachen*, Studia Philosophica **1**, (1936).

14. A. Tarski, *Einige Methodologische Untersuchungen über die Definierbarkeit der Begriffe* Erkenntnis **5** (1935).

15. J.B. Wright, *Quasi-projective geometry of two dimensions*, Michigan Math. J. **2** (1953-4).

PURDUE UNIVERSITY
AND
SOUTHERN ILLINOIS UNIVERSITY

*Reprinted from*
Israel Journal of Mathematics
Vol. 14, No. 3, 1973

Zeitschr. f. math. Logik und Grundlagen d. Math
Bd. 19 S. 411—426 (1973)

# VARIATIONS ON A THEME OF CANTOR IN THE THEORY OF RELATIONAL STRUCTURES

by J. RICHARD BÜCHI in Lafayette, Indiana
and KENNETH J. DANHOF in Carbondale, Illinois (U.S.A.)[1])

## 1. Introduction

In the process of characterizing the order type $\lambda$ of the linear continuum, CANTOR [2] proved the following facts about the order type $\eta$ of the rationals:

Theorem 1. *For linearly ordered sets $A$ and $B$, if $A$ is finite or countable and $B$ is dense and unbounded, then $A$ is similar to a subset of $B$.*

Theorem 2. *Any two countable dense unbounded linearly ordered sets are isomorphic.*

It is peculiar to CANTOR's methods of proof that they surpass the context and are useful in a variety of situations. In particular, variations of the above theorems have occurred in the literature on ordered fields [5] and model theory (note that theorem 2 is a result on $\omega_0$-categoricity). We will present, in sections 2 and 3, an axiomatic theory (based on "$C$-relations") which does justice to the universality of CANTOR's proofs of theorems 1 and 2. In section 4 we consider the derivatives of our $C$-relations and discuss various applications of the general versions of the CANTOR theorems. Further applications are noted in section 5. It is suggested that the basic notions introduced in this abstract approach play a very pervasive role in the model-theoretic study of relational systems. We hope that these notions might serve as a basis for a similar treatment of CANTOR's results on the relationship between $\eta$ and $\lambda$ (denseness of $\eta$ in $\lambda$). Also the CANTOR-BENDIXSON argument on perfect sets has recently appeared in model theory and can be extended to the present setting. This paper might be viewed as the beginning of an axiomatic treatment of model theory. We would like to thank John Doner for several helpful suggestions and conversations regarding the manuscript.

## 2. Preliminaries

The objects of investigation are relational systems of a fixed type. We will present the general definitions and proofs for the case of one binary relation. (It will be clear how to extend the development to general relational systems of finite type; the case of infinite types requires some modification.) Thus a relational system $\boldsymbol{A} = \langle A, R\rangle$ consists of a domain $A$ and a binary relation $R \subseteq A^2$.

[1]) This work was supported in part by N.S.F. grant No. GJ-980.

The ordinal number $\xi$ is identified with the set $\{\eta; \eta < \xi\}$ of smaller ordinals. On is the class of all ordinals. $\omega_\xi$ denotes both the $\xi$-th initial ordinal and the $\xi$-th infinite cardinal number; we write $\omega_0$ as $\omega$. If $\alpha = \omega_\xi$, $\alpha^+ = \omega_{\xi+1}$. $|X|$ represents the cardinal number of the set $X$. A $\xi$-sequence $u$ of elements of $A$ is a member of $A^\xi$, i.e., a map from $\nu < \xi$ to $u_\nu \in A$. For $u \in A^\xi$, we say the *length of* $u$ ($lg\, u$) is $\xi$. If $u \in A^\xi$ and $v \in A^\eta$, then $uv$ denotes that $w \in A^{\xi+\eta}$ such that $w_\nu = u_\nu$ for $\nu < \xi$ and $w_{\xi+\nu} = v_\nu$ for $\nu < \eta$. If $\boldsymbol{A}$ is a system and $\mu \in A^\xi$, then $\boldsymbol{A}u$ is the *extended system* $\langle A, R, u_0, \ldots, u_\eta, \ldots\rangle_{\eta<\xi}$. For $u \in A^\eta$ and $x \subseteq \eta$, $\bar{u}_x$ denotes the restriction of $u$ to $x$. In particular for $\xi < \eta$, $\bar{u}_\xi$ is the initial segment of $u$ of length $\xi$. For $u \in A^\xi$ and $\varphi \in \xi^\xi$, $u\varphi \in A^\xi$ is defined by $(u\varphi)_\nu = u_{\varphi(\nu)}$ for $\nu < \xi$. The expression $(\forall x)_A \Sigma$ is to be read $(\forall x)\,[x \in A \Rightarrow \Sigma]$, $(\exists x)_A \Sigma$ means $(\exists x)\,[x \in A \wedge \Sigma]$ and $(\mu x)_A \Sigma$ is used to mean "the first $x \in A$ such that $\Sigma$" if such an $x$ exists. Thus, e.g., $(\forall \nu)_\xi \Sigma$ means "for all $\nu < \xi, \Sigma$". When a well-ordering is indicated by an application of the $\mu$-operator, we assume that one has been selected. $\mathrm{D}f$ and $\mathfrak{A}f$ denote the domain and range respectively, of the function $f$.

For an ordinal number $\varkappa \geqq 2$, we let $A_\varkappa = \bigcup_{0<\nu<\varkappa} A^\nu$. Thus $x \in A_\varkappa^\xi$ means that $x$ is a $\xi$-string of sequences $x_\eta$ each of length less than $\varkappa$. If $lg\, x_\eta = \varrho_\eta$ for $\eta < \xi$, we set $\hat{x} = x_0 x_1 \ldots x_\eta \ldots$, $\eta < \xi$. Thus $\hat{x}$ is a sequence in $A^\nu$ where $\nu = \sum_{\eta<\xi} \varrho_\eta$. We write $x \in A_{\varkappa\alpha}$ if $x \in A_\varkappa^\xi$ for some $\xi < \alpha$. The cardinal number $|A|_\varkappa$ is defined by $|A|_\varkappa = (\mu\xi)\,[(\exists x)_{A_\varkappa^\xi}\, \mathfrak{A}\hat{x} = A]$.

Definition. A *C-relation* $\theta$ on a class $K$ of relation systems is a relation between extensions $\boldsymbol{A}x$ of systems $\boldsymbol{A} \in K$ such that

($C_0$) $\boldsymbol{A}x\,\theta\,\boldsymbol{B}y \mathrel{.{\Rightarrow}.} lg\, x = lg\, y$,

($C_1$) $\theta$ is an equivalence relation on extensions of systems in $K$,

($C_2$) $\boldsymbol{A}x\,\theta\,\boldsymbol{B}y \wedge lg\, x = \xi \wedge \varphi \in \xi^\xi \mathrel{.{\Rightarrow}.} \boldsymbol{A}x\varphi\,\theta\,\boldsymbol{B}y\varphi$,

($C_3$) $\boldsymbol{A}x\,\theta\,\boldsymbol{B}y \wedge \eta \leqq lg\, x \mathrel{.{\Rightarrow}.} \boldsymbol{A}\bar{x}_\eta\,\theta\,\boldsymbol{B}\bar{y}_\eta$,

($C_4$) $\boldsymbol{A}x\,\theta\,\boldsymbol{B}y \wedge \sigma, \gamma < lg\, x \mathrel{.{\Rightarrow}.} [x_\sigma = x_\gamma \mathrel{.{\equiv}.} y_\sigma = y_\gamma]$.

For systems $\boldsymbol{A} = \langle A, R\rangle$ and $\boldsymbol{B} = \langle B, S\rangle$, $x \in A^\xi$ and $y \in B^\xi$ we define $\boldsymbol{A}x \sim \boldsymbol{B}y$ ($\boldsymbol{A}x$ is *partially isomorphic* to $\boldsymbol{B}y$) by

$$\boldsymbol{A}x \sim \boldsymbol{B}y \mathrel{.{\equiv}.} (\forall \sigma, \gamma)_\xi\,[x_\sigma = x_\gamma \equiv y_\sigma = y_\gamma \mathrel{.{\wedge}.} Rx_\sigma x_\gamma \equiv Sy_\sigma y_\gamma].$$

Partial isomorphism is a simple example of a $C$-relation. Elementary equivalence is another example. These and other $C$-relations satisfy an additional axiom.

($C_F$) $x \in A^\xi \wedge y \in B^\xi \wedge (\forall F)\,[F \leqq \xi \wedge |F| < \omega \mathrel{.{\Rightarrow}.} \boldsymbol{A}\bar{x}_F\,\theta\,\boldsymbol{B}\bar{y}_F] \mathrel{.{\Rightarrow}.} \boldsymbol{A}x\,\theta\,\boldsymbol{B}y$.

$C$-relations satisfying ($C_F$) are called *finitary*. Note that ($C_F$) may be restated as

($C_F'$) $\lambda$ a limit ordinal $\wedge\, x \in A^\lambda \wedge y \in B^\lambda \wedge (\forall \eta)_\lambda\,[\boldsymbol{A}\bar{x}_\eta\,\theta\,\boldsymbol{B}\bar{y}_\eta] \mathrel{.{\Rightarrow}.} \boldsymbol{A}x\,\theta\,\boldsymbol{B}y$.

With any $C$-relation $\theta$, we associate its *finitary extension* $\tilde{\theta}$ defined, for $x \in A^\xi$, $y \in B^\xi$ by $\boldsymbol{A}x\,\tilde{\theta}\,\boldsymbol{B}y \mathrel{.{\equiv}.} (\forall F)\,[F \subseteqq \xi \wedge |F| < \omega \mathrel{.{\Rightarrow}.} \boldsymbol{A}\bar{x}_F\,\theta\,\boldsymbol{B}\bar{y}_F]$.

Note that if $\theta$ is a $C$-relation, $\tilde{\theta}$ is a finitary $C$-relation.

Let $f$ be a partial function from $A$ into $B$ and choose $x \in A^{\xi}$ such that $\mathrm{D}f = \mathrm{Q}x$. Define $fx \in B^{\xi}$ by $(fx)_{\nu} = fx_{\nu}$ for $\nu < \xi$. If $\boldsymbol{A}x\,\theta\,\boldsymbol{B}fx$ we call $f$ a *partial $\theta$-injection from* $\boldsymbol{A}$ into $\boldsymbol{B}$; in symbols $f: \boldsymbol{A} \rightharpoonup \boldsymbol{B}(\theta)$. If in addition, $\mathrm{D}f = A$, we call $f$ a *$\theta$-injection from* $\boldsymbol{A}$ *into* $\boldsymbol{B}$ and write $f: \boldsymbol{A} \to \boldsymbol{B}(\theta)$. If also $\mathrm{Q}f = B$, $f$ is called a *$\theta$-isomorphism* and we write $f: \boldsymbol{A} \cong \boldsymbol{B}(\theta)$. $\boldsymbol{A} \to \boldsymbol{B}(\theta)$ denotes the existence of a $\theta$-injection from $A$ into $\boldsymbol{B}$, and $\boldsymbol{A} \cong \boldsymbol{B}(\theta)$ indicates that $\boldsymbol{A}$ and $\boldsymbol{B}$ are $\theta$-isomorphic. Note that $\boldsymbol{A} \cong \boldsymbol{B}(\sim)$ means that $\boldsymbol{A}$ and $\boldsymbol{B}$ are isomorphic in which case we write $\boldsymbol{A} \cong \boldsymbol{B}$. For mappings $f$ and $g$, $f < g$ indicates that $g$ extends $f$.

From the notion "partial $\theta$-injection" one can recover our primitive relation $\theta$. Namely $\boldsymbol{A}x\,\theta\,\boldsymbol{B}y$ denotes that the function $fx_{\nu} = y_{\nu}$ is a partial $\theta$-injection. Hence in place of the relation $\theta$, we might have used the notion "partial $\theta$-injection" as primitive in our axiomatic theory. An axiom system similar to that of a category would result (the difference being the absence of many-one maps and the presence of partial maps). The reader may restate our main results in the category version.

Let $\theta$ be a $C$-relation and $\varkappa$ a cardinal number, $2 \leqq \varkappa$. We define the relation $\theta_{\varkappa}$ as follows: for $x \in A_{\varkappa}^{\xi}$, $y \in B_{\varkappa}^{\xi}$

$$\boldsymbol{A}x\,\theta_{\varkappa}\,\boldsymbol{B}y \;.\!\equiv.\; (\forall\nu)_{\xi}\,[lg\, x_{\nu} = lg\, y_{\nu}] \wedge \boldsymbol{A}\hat{x}\,\theta\,\boldsymbol{B}\hat{y}\,.$$

With appropriate modifications ($x$ and $y$ are no longer sequences of elements, but sequences of sequences, if $2 < \varkappa$), $\theta_{\varkappa}$ again satisfies $(\mathrm{C}_0), \ldots, (\mathrm{C}_4)$. We say $\theta$ is *$\varkappa$-finitary* if $\theta_{\varkappa}$ is finitary. Note that if $\theta$ is finitary, then it is $\varkappa$-finitary. We conclude the section with a final definition. For $x \in A_{\varkappa}^{\xi}$, $y \in B_{\varkappa}^{\xi}$

$$\boldsymbol{A}x\,\theta'_{\varkappa}\,\boldsymbol{B}y \;.\!\equiv.\; (\forall a)_{A_{\varkappa}}\,(\exists b)_{B_{\varkappa}}\,[\boldsymbol{A}xa\,\theta_{\varkappa}\,\boldsymbol{B}yb] \wedge (\forall b)_{B_{\varkappa}}\,(\exists a)_{A_{\varkappa}}\,[\boldsymbol{A}xa\,\theta_{\varkappa}\,\boldsymbol{B}yb]\,.$$

$\theta'_{\varkappa}$ is called the *derivative* of $\theta_{\varkappa}$. We consider derivatives of $C$-relations more carefully in section 4.

## 3. The generalized Cantor theorems

In this section we present our general forms of CANTOR's theorems 1 and 2. In the sequel $\theta$ is assumed to be a $C$-relation. Also, from here on, $\alpha$ and $\varkappa$ will be cardinals with $\omega \leqq \alpha$ and $2 \leqq \varkappa$ (we will not always repeat this assumption). The relation $\lhd_{\varkappa}^{\alpha}(\theta)$ is defined as follows:

$$\boldsymbol{A} \lhd_{\varkappa}^{\alpha} \boldsymbol{B}(\theta) \;.\!\equiv.\; \boldsymbol{A}\,\theta\,\boldsymbol{B} \wedge (\forall x)_{A_{\varkappa\alpha}}(\forall y)_{B_{\varkappa\alpha}}[\boldsymbol{A}x\,\theta_{\varkappa}\,\boldsymbol{B}y \Rightarrow (\forall a)_{A_{\varkappa}}(\exists b)_{B_{\varkappa}}\boldsymbol{A}xa\,\theta_{\varkappa}\,\boldsymbol{B}yb]\,.$$

It is easy to show the relation $\lhd_{\varkappa}^{\alpha}(\theta)$ is transitive, but not (necessarily) reflexive. Moreover, $\boldsymbol{A} \lhd_{\varkappa}^{\alpha} \boldsymbol{B}(\theta)$ implies $\boldsymbol{A} \lhd_{\tau}^{\beta} \boldsymbol{B}(\theta)$ for $|\beta| \leqq |\alpha|$ and $|\tau| \leqq |\varkappa|$ so that there is no loss of generality in assuming $\alpha$ and $\varkappa$ cardinals. Closely related to the relation $\lhd_{\varkappa}^{\alpha}(\theta)$ is the relation $\diamond_{\varkappa}^{\alpha}(\theta)$ defined as follows:

$$\boldsymbol{A} \diamond_{\varkappa}^{\alpha} \boldsymbol{B}(\theta) \;.\!\equiv.\; \boldsymbol{A}\,\theta\,\boldsymbol{B} \wedge (\forall x)_{A_{\varkappa\alpha}}(\forall y)_{B_{\varkappa\alpha}}[\boldsymbol{A}x\,\theta_{\varkappa}\,\boldsymbol{B}y \Rightarrow \boldsymbol{A}x\,\theta'_{\varkappa}\,\boldsymbol{B}y]\,.$$

Remark 1. Note that $\boldsymbol{A} \diamond_{\varkappa}^{\alpha} \boldsymbol{B}(\theta) \;.\!\equiv.\; \boldsymbol{A} \lhd_{\varkappa}^{\alpha} \boldsymbol{B}(\theta) \wedge \boldsymbol{B} \lhd_{\varkappa}^{\alpha} \boldsymbol{A}(\theta)$ and it follows, since $\lhd_{\varkappa}^{\alpha}(\theta)$ is transitive, that the relation $\diamond_{\varkappa}^{\alpha}(\theta)$ is symmetric and transitive.

Lemma 1. *Let $\theta$ be a $\varkappa$-finitary C-relation, then*

(i) $\boldsymbol{A} \lhd_\varkappa^\alpha \boldsymbol{B}(\theta) .\Rightarrow. (\forall x)_{A_{\varkappa\alpha}} (\forall y)_{B_{\varkappa\alpha}} \boldsymbol{A} x \,\theta_\varkappa\, \boldsymbol{B} y \Rightarrow (\forall a)_{A_\varkappa^\alpha} (\exists b)_{B_\varkappa^\alpha} \boldsymbol{A} xa \,\theta_\varkappa\, \boldsymbol{B} yb,$

(ii) $\boldsymbol{A} \lhd_\varkappa^\alpha \boldsymbol{B}(\theta) .\Rightarrow. (\forall a)_{A_\varkappa^\alpha} (\exists b)_{B_\varkappa^\alpha} \boldsymbol{A} a \,\theta_\varkappa\, \boldsymbol{B} b,$

(iii) $\boldsymbol{A} \lhd_\varkappa^\alpha \boldsymbol{B}(\theta) \wedge |A|_\varkappa \leqq \alpha .\Rightarrow. \boldsymbol{A} \to \boldsymbol{B}(\theta).$

Proof. For (i), suppose $x \in A_{\varkappa\alpha}$, $y \in B_{\varkappa\alpha}$, $\boldsymbol{A} x \,\theta_\varkappa\, \boldsymbol{B} y$ and $a \in A_\varkappa^\alpha$. Define $b \in B_\varkappa^\alpha$ by $b_\nu = (\mu c)_{B_\varkappa} \boldsymbol{A} x \bar{a}_\nu a_\nu \,\theta_\varkappa\, \boldsymbol{B} y \bar{b}_\nu c$, for $\nu < \alpha$. By induction on $\nu < \alpha$ we show (a) $\boldsymbol{A} x \bar{a}_\nu \,\theta_\varkappa\, \boldsymbol{B} y \bar{b}_\nu$ and $b_\nu$ exists. We have $\boldsymbol{A} x \,\theta_\varkappa\, \boldsymbol{B} y$ and since $\boldsymbol{A} \lhd_\varkappa^\alpha \boldsymbol{B}$, for $a_0$, there is $c \in B_\varkappa$ such that $\boldsymbol{A} x a_0 \,\theta_\varkappa\, \boldsymbol{B} y c$. Thus (a) holds for $\nu = 0$. Assuming (a) holds for $\nu$, we have $\boldsymbol{A} x \bar{a}_\nu \,\theta_\varkappa\, \boldsymbol{B} y \bar{b}_\nu$ and $b_\nu$ exists, i.e., $\boldsymbol{A} x \bar{a}_\nu a_\nu \,\theta_\varkappa\, \boldsymbol{B} y \bar{b}_\nu b_\nu$ or $\boldsymbol{A} x \bar{a}_{\nu+1} \,\theta_\varkappa\, \boldsymbol{B} y \bar{b}_{\nu+1}$. Since $\alpha$ is an infinite cardinal and $\nu < \alpha$, $lg\, x \bar{a}_{\nu+1} = lg\, x + \nu + 1 < \alpha$. Then because $\boldsymbol{A} \lhd_\varkappa^\alpha \boldsymbol{B}(\theta)$, there is a $c \in B_\varkappa$ such that $\boldsymbol{A} x \bar{a}_{\nu+1} a_{\nu+1} \,\theta_\varkappa\, \boldsymbol{B} y \bar{b}_{\nu+1} c$. This shows that (a) holds at $\nu + 1$. If $\nu$ is a limit, (a) follows by the inductive assumption and the fact that $\theta$ is $\varkappa$-finitary. Now by (a) we have $\boldsymbol{A} x \bar{a}_\nu \,\theta_\varkappa\, \boldsymbol{B} y \bar{b}_\nu$ for all $\nu < \alpha$ and since $\theta$ is $\varkappa$-finitary, it follows that $\boldsymbol{A} x a \,\theta_\varkappa\, \boldsymbol{B} y b$. (ii) follows from (i) since $\boldsymbol{A} \lhd_\varkappa^\alpha \boldsymbol{B}(\theta)$ implies $\boldsymbol{A} \,\theta_\varkappa\, \boldsymbol{B}$ and (iii) follows from (ii) using the fact that if $|A|_\varkappa \leqq \alpha$, for some $\xi \leqq \alpha$, we can pick $a \in A_\varkappa^\xi$ with $\mathrm{Cl}\, \hat{a} = A$.

Corollary 1. *If $\theta$ is a C-relation, then*

$$\boldsymbol{A} \lhd_\varkappa^\omega \boldsymbol{B}(\theta) \wedge |A|_\varkappa \leqq \omega .\Rightarrow. \boldsymbol{A} \to \boldsymbol{B}(\tilde{\theta}).$$

Proof. By assumption, there are $\xi \leqq \omega$ and $x \in A_\varkappa^\xi$ with $\mathrm{Cl}\, \hat{x} = A$. As in the proof of lemma 1, we can define $b \in B_\varkappa^\xi$ such that $\boldsymbol{A} a_\nu \,\theta_\varkappa\, \boldsymbol{B} \bar{b}_\nu$ for all $\nu < \xi$. We may then conclude $\boldsymbol{A} a \,\tilde{\theta}_\varkappa\, \boldsymbol{B} b$ and hence $\boldsymbol{A} \to \boldsymbol{B}(\tilde{\theta})$.

From the definition of $\diamond$, we have

Remark 2. $\boldsymbol{A} \diamond_\varkappa^\alpha \boldsymbol{B}(\theta) \wedge x \in A_{\varkappa\alpha} \wedge y \in B_{\varkappa\alpha} \wedge \boldsymbol{A} x \,\theta_\varkappa\, \boldsymbol{B} y .\Rightarrow.$
$(\forall a)_{A_\varkappa} (\forall b)_{B_\varkappa} (\exists c)_{A_\varkappa} (\exists d)_{B_\varkappa} \boldsymbol{A} x a c \,\theta_\varkappa\, \boldsymbol{B} y d b.$

Lemma 2. *Let $\theta$ be a $\varkappa$-finitary C-relation, then*

(i) $\boldsymbol{A} \diamond \boldsymbol{B}(\theta) .\Rightarrow. (\forall x)_{A_{\varkappa\alpha}} (\forall y)_{B_{\varkappa\alpha}} \boldsymbol{A} x \theta_\varkappa \boldsymbol{B} y \Rightarrow (\forall a)_{A_\varkappa^\alpha} (\forall b)_{B_\varkappa^\alpha} (\exists c)_{A_\varkappa^\alpha} (\exists d)_{B_\varkappa^\alpha} \boldsymbol{A} x a c \theta_\varkappa \boldsymbol{B} y d b,$

(ii) $\boldsymbol{A} \diamond_\varkappa^\alpha \boldsymbol{B}(\theta) .\Rightarrow. (\forall a)_{A_\varkappa^\alpha} (\forall b)_{B_\varkappa^\alpha} (\exists c)_{A_\varkappa^\alpha} (\exists d)_{B_\varkappa^\alpha} \boldsymbol{A} a c \,\theta_\varkappa\, \boldsymbol{B} d b,$

(iii) $\boldsymbol{A} \diamond_\varkappa^\alpha \boldsymbol{B}(\theta) \wedge |A|_\varkappa, |B|_\varkappa \leqq \alpha .\Rightarrow. \boldsymbol{A} \cong \boldsymbol{B}(\theta).$

Proof. For (i), suppose $x \in A_{\varkappa\alpha}$, $y \in B_{\varkappa\alpha}$, $\boldsymbol{A} x \,\theta_\varkappa\, \boldsymbol{B} y$, $a \in A_\varkappa^\alpha$ and $b \in B_\varkappa^\alpha$. We define $c \in A_\varkappa^\alpha$ and $d \in B_\varkappa^\alpha$ by $(c_\nu, d_\nu) = (\mu u, v)_{A_\varkappa, B_\varkappa} \boldsymbol{A} x \bar{a}_\nu a_\nu \bar{c}_\nu u \,\theta_\varkappa\, \boldsymbol{B} y \bar{d}_\nu v \bar{b}_\nu b_\nu$. By induction on $\nu < \alpha$, we show (a) $\boldsymbol{A} x \bar{a}_\nu \bar{c}_\nu \,\theta_\varkappa\, \boldsymbol{B} y \bar{d}_\nu \bar{b}_\nu$ and $c_\nu$ and $d_\nu$ exist. Since $\boldsymbol{A} \diamond_\varkappa^\alpha \boldsymbol{B}(\theta)$, we have by remark 2 that for $a_0$, $b_0$ there are $u \in A_\varkappa$, $v \in B_\varkappa$ such that $\boldsymbol{A} x a_0 u \,\theta_\varkappa\, \boldsymbol{B} y v b_0$. Thus (a) holds for $\nu = 0$. Assuming (a) holds for $\nu$, we have $\boldsymbol{A} x \bar{a}_\nu \bar{c}_\nu \,\theta_\varkappa\, \boldsymbol{B} y d_\nu \bar{b}_\nu$ and $c_\nu$ and $d_\nu$ exist, i.e., $\boldsymbol{A} x \bar{a}_\nu a_\nu \bar{c}_\nu c_\nu \,\theta_\varkappa\, \boldsymbol{B} y d_\nu d_\nu \bar{b}_\nu b_\nu$ or $\boldsymbol{A} x \bar{a}_{\nu+1} \bar{c}_{\nu+1} \,\theta_\varkappa\, \boldsymbol{B} y \bar{d}_{\nu+1} \bar{b}_{\nu+1}$. Since $\alpha$ is an infinite cardinal and $\nu < \alpha$, $lg\, x \bar{a}_{\nu+1} \bar{c}_{\nu+1} = lg\, x + lg\, \bar{a}_{\nu+1} \bar{c}_{\nu+1} < \alpha$. Then since $\boldsymbol{A} \diamond_\varkappa^\alpha \boldsymbol{B}(\theta)$, we have by remark 2 that there are $u \in A_\varkappa$ and $v \in B_\varkappa$ such that $\boldsymbol{A} x \bar{a}_{\nu+1} \bar{c}_{\nu+1} a_{\nu+1} u \,\theta_\varkappa\, \boldsymbol{B} y d_{\nu+1} \bar{b}_{\nu+1} v\, b_{\nu+1}$. Using axiom C2, it follows that $\boldsymbol{A} x \bar{a}_{\nu+1} a_{\nu+1} \bar{c}_{\nu+1} u \,\theta_\varkappa\, \boldsymbol{B} y \bar{d}_{\nu+1} v\, \bar{b}_{\nu+1} b_{\nu+1}$. Thus (a) holds at $\nu + 1$. If $\nu$ is a limit, (a) follows from the inductive assumption and the fact that $\theta$ is $\varkappa$-finitary. Hence (a) is established. By (a), we have $\boldsymbol{A} x \bar{a}_\nu \bar{c}_\nu \, \theta_\varkappa \, \boldsymbol{B} y \bar{d}_\nu \bar{b}_\nu$ for all $\nu < \alpha$ and since $\theta$ is $\varkappa$-finitary, it follows that $\boldsymbol{A} x a c \, \theta_\varkappa \, \boldsymbol{B} y d b$. (ii) follows from (i) since $\boldsymbol{A} \diamond_\varkappa^\alpha \boldsymbol{B}(\theta)$ implies $\boldsymbol{A} \, \theta \, \boldsymbol{B}$ and (iii) follows from (ii) for because $|A|_\varkappa, |B|_\varkappa \leqq \alpha$, for some $\eta, \xi \leqq \alpha$, we can pick $a \in A_\varkappa^\eta$, $b \in B_\varkappa^\xi$ with $\mathrm{D}\hat{a} = A$ and $\mathrm{D}\hat{b} = B$. By (ii) then there are $c$ and $d$ such that $\boldsymbol{A} a c \, \theta_\varkappa \, \boldsymbol{B} d b$. Since $\mathrm{D}\widehat{ac} = A$ and $\mathrm{D}\widehat{db} = B$, we then have $\boldsymbol{A} \cong \boldsymbol{B}(\theta)$.

Corollary 2. *If $\theta$ is a C-relation, then*

$$\boldsymbol{A} \diamond_\varkappa^\omega \boldsymbol{B}(\theta) \wedge |A|_\varkappa, |B|_\varkappa \leqq \omega \;.\Rightarrow.\; \boldsymbol{A} \cong \boldsymbol{B}(\tilde{\theta}).$$

Corollary 2 follows from lemma 2 in the same manner as which corollary 1 followed from lemma 1. The ideas used in proving lemmas 1 and 2 are just those used by CANTOR in proving his theorems 1 and 2. However these lemmas are not the true generalizations of CANTOR's theorems. To get these, we introduce the following concepts:

A $\binom{\xi}{\varkappa}$*-type* $(\theta)$ is an equivalence class, modulo $\theta_\varkappa$, of a system $\boldsymbol{A} x$ $x \in A_\varkappa^\xi$. The STONE-*space* $St_\varkappa^\alpha(\boldsymbol{A}, \theta)$ is the set of all $\binom{\xi}{\varkappa}$-types $(\theta)$, $\xi < \alpha$, which are *realized* in $\boldsymbol{A}$. We call $\boldsymbol{A}$ $\binom{\alpha}{\varkappa}$*-younger* $(\theta)$ *than* $\boldsymbol{B}$ if $\boldsymbol{B}$ realizes all $\binom{\xi}{\varkappa}$-types $(\theta)$, $\xi < \alpha$, which are realized in $\boldsymbol{A}$. In symbols,

$$\boldsymbol{A} \leqq_\varkappa^\alpha \boldsymbol{B}(\theta) : (\forall x)_{A_{\varkappa\alpha}} (\exists y)_{B_{\varkappa\alpha}} \boldsymbol{A} x \, \theta_\varkappa \, \boldsymbol{B} y, \quad \text{i.e.,} \quad St_\varkappa^\alpha(\boldsymbol{A}, \theta) \subseteq St_\varkappa^\alpha(\boldsymbol{B}, \theta).$$

The following fundamental notions are suggested by CANTOR's theorems 1 and 2:

$\boldsymbol{B}$ is $\binom{\alpha}{\varkappa}$*-saturated* $(\theta)$ $.\equiv.\; \boldsymbol{B} \, \theta \, \boldsymbol{B} \wedge (\forall \boldsymbol{A}) [\boldsymbol{A} \, \theta \, \boldsymbol{B} \Rightarrow \boldsymbol{A} \lhd_\varkappa^\alpha \boldsymbol{B}(\theta)]$.

$\boldsymbol{B}$ is $\binom{\alpha}{\varkappa}$*-homogeneous* $(\theta)$ $.\equiv.\; \boldsymbol{B} \lhd_\varkappa^\alpha \boldsymbol{B}(\theta)$.

$\boldsymbol{B}$ is $\binom{\alpha}{\varkappa}$*-universal* $(\theta)$ $.\equiv.\; (\forall A) [\boldsymbol{A} \, \theta \, \boldsymbol{B} \Rightarrow A \leqq_\varkappa^\alpha \boldsymbol{B}(\theta)]$.

$\boldsymbol{B}$ is called $\varkappa$*-saturated* $(\theta)$ if it is $\binom{\alpha}{\varkappa}$-saturated for $\alpha = |B|_\varkappa$.

The terms $\varkappa$*-homogeneous* and $\varkappa$*-universal* are used similarly.

These notions have been discussed in the literature for the case $\varkappa = 2$ and $\theta$ = elementary equivalence. Many terms have been used where in fact there is just the one notion—$\alpha$-universal. Both "$\alpha$-weakly-saturated" and "$\alpha$-weakly-universal" have been used for our $\binom{\alpha}{2}$-universal and "$\alpha$-universal" has been used for our $\binom{\alpha^+}{2}$-universal. The definitions in the literature differ from ours in yet another way. We now show that this difference disappears in view of the fact that elementary equivalence satisfies the following SKOLEM-LÖWENHEIM condition:

$$(\mathrm{SL})^{\alpha}_{\varkappa} \qquad (\forall \boldsymbol{A})\,(\forall x)_{A_{\varkappa\alpha}}(\exists \boldsymbol{C})\,(\exists z)_{C_{\varkappa\alpha}}[|C|_{\varkappa} \leqq \alpha \wedge \boldsymbol{A}x\,\theta_{\varkappa}\,\boldsymbol{C}z].$$

Lemma 3.1. *If the C-relation $\theta$ satisfies* $(\mathrm{SL})^{\alpha}_{\varkappa}$, *then*

(i) $\boldsymbol{B}$ *is* $\binom{\alpha}{\varkappa}$*-saturated* $(\theta)$ $.\equiv.$ $(\forall \boldsymbol{A})\,[\boldsymbol{A}\,\theta\,\boldsymbol{B} \wedge |A|_{\varkappa} \leqq \alpha \,.\!\Rightarrow\!.\, \boldsymbol{A} \lhd^{\alpha}_{\varkappa} \boldsymbol{B}(\theta)]$,

(ii) $\boldsymbol{B}$ *is* $\binom{\alpha}{\varkappa}$*-universal* $(\theta)$ $.\equiv.$ $(\forall \boldsymbol{A})\,[\boldsymbol{A}\,\theta\,\boldsymbol{B} \wedge |A|_{\varkappa} \leqq \alpha \,.\!\Rightarrow\!.\, \boldsymbol{A} \leqq^{\alpha}_{\varkappa} \boldsymbol{B}(\theta)]$.

*The implications from left to right hold without* $(\mathrm{SL})^{\alpha}_{\varkappa}$.

Proof. The implications from left to right are trivial. To prove (i) from right to left, suppose $\boldsymbol{A}\,\theta\,\boldsymbol{B}$, $x \in A_{\varkappa\alpha}$, $y \in B_{\varkappa\alpha}$, $a \in A_{\varkappa}$ and $\boldsymbol{A}x\,\theta_{\varkappa}\,\boldsymbol{B}y$. Since $\alpha$ is infinite, $xa \in A_{\varkappa\alpha}$ and so by $(\mathrm{SL})^{\alpha}_{\varkappa}$, there are $\boldsymbol{C}$ and $zc \in C_{\varkappa\alpha}$ with $|C|_{\varkappa} \leqq \alpha$ and $\boldsymbol{A}xa\,\theta_{\varkappa}\,\boldsymbol{C}zc$. By $(\mathrm{C}_3)$, $\boldsymbol{A}x\,\theta_{\varkappa}\,\boldsymbol{C}z$ and by $(\mathrm{C}_1)$, $\boldsymbol{C}z\,\theta_{\varkappa}\,\boldsymbol{B}y$. Since $|C|_{\varkappa} \leqq \alpha$, we have $\boldsymbol{C} \lhd^{\alpha}_{\varkappa} \boldsymbol{B}(\theta)$ and thus there is $b \in B_{\varkappa}$ for which $\boldsymbol{C}zc\,\theta_{\varkappa}\,\boldsymbol{B}yb$. By $(\mathrm{C}_1)$, $\boldsymbol{A}xa\,\theta_{\varkappa}\,\boldsymbol{B}yb$. Thus $\boldsymbol{A} \lhd^{\alpha}_{\varkappa} \boldsymbol{B}(\theta)$. The proof of (ii) from right to left is similar.

Note that for $|A|_{\varkappa} < \alpha$, $\boldsymbol{A} \leqq^{\alpha}_{\varkappa} \boldsymbol{B}(\theta)$ is equivalent with $\boldsymbol{A} \to \boldsymbol{B}(\theta)$. It remains to be shown that our notion of homogeneous is equivalent to that occurring in the literature. This will be our first use of lemma 2.

Lemma 3.2. *If $\theta$ is a $\varkappa$-finitary C-relation, then*

(i) $\boldsymbol{B}$ $\binom{\alpha}{\varkappa}$*-homogeneous* $(\theta)$ $.\equiv.$

$$\boldsymbol{B}\,\theta\,\boldsymbol{B} \wedge (\forall f, X, Y)\,[f\colon \boldsymbol{B} \twoheadrightarrow B(\theta) \wedge |\mathrm{D}f|_{\varkappa} < \alpha \wedge X, Y \subseteqq B \wedge |X|_{\varkappa}, |Y|_{\varkappa} \leqq \alpha$$
$$.\!\Rightarrow\!.\, (\exists g)\,(g\colon \boldsymbol{B} \twoheadrightarrow \boldsymbol{B}(\theta) \wedge f < g \wedge X \subseteqq \mathrm{D}g \wedge Y \subseteqq \mathrm{G}g)],$$

(ii) $\boldsymbol{B}$ *$\varkappa$-homogeneous* $(\theta)$ $.\equiv.$

$$\boldsymbol{B}\,\theta\,\boldsymbol{B} \wedge (\forall f)\,[f\colon \boldsymbol{B} \twoheadrightarrow \boldsymbol{B}(\theta) \wedge |\mathrm{D}f|_{\varkappa} < |B|_{\varkappa} \,.\!\Rightarrow\!.\, (\exists g)\,(g\colon B \cong B(\theta) \wedge f < g)].$$

Proof. From, left to right, (i) follows directly from the partial $\theta$-injection version of lemma 2(i). Since, by $(\mathrm{C}_0)$ and $(\mathrm{C}_4)$, $\boldsymbol{B}x\,\theta_{\varkappa}\,\boldsymbol{B}y$ implies the existence of a partial $\theta$-injection from $\hat{x}$ to $\hat{y}$, the right hand side of (i) implies $\boldsymbol{B} \lhd^{\alpha}_{\varkappa} \boldsymbol{B}(\theta)$. (ii) follows from (i) by choosing $X = Y = B$.

Lemma 3.2 connects our notion of homogeneity with the more usual notion—namely, that "sufficiently partial" automorphisms can be extended to automorphisms. We now state and prove our generalizations of CANTOR's theorems 1 and 2.

Theorem 1. *If $\theta$ is a $\varkappa$-finitary C-relation, then*

(i) $\boldsymbol{B}$ $\binom{\alpha}{\varkappa}$*-saturated* $(\theta)$ $.\!\Rightarrow\!.$ $\boldsymbol{B}$ $\binom{\alpha^{+}}{\varkappa}$*-universal* $(\theta)$,

(ii) $\boldsymbol{B}$ $\binom{\alpha}{\varkappa}$*-saturated* $(\theta) \wedge \boldsymbol{A}\,\theta\,\boldsymbol{B} \wedge |A|_{\varkappa} \leqq \alpha \,.\!\Rightarrow\!.\, \boldsymbol{A} \to \boldsymbol{B}(\theta)$,

(iii) $\boldsymbol{B}$ *$\varkappa$-saturated* $(\theta) \wedge \boldsymbol{A}\,\theta\,\boldsymbol{B} \wedge |A|_{\varkappa} \leqq |B|_{\varkappa} \,.\!\Rightarrow\!.\, \boldsymbol{A} \to \boldsymbol{B}(\theta)$.

Proof. To prove (i), suppose $\boldsymbol{A}\,\theta\,\boldsymbol{B}$ and $x \in A_{\varkappa\alpha^+}$, i.e. $x \in A_\varkappa^\xi$ for some $\xi < \alpha^+$. We may, if necessary, rearrange $x$ to $x' \in A_\varkappa^{\xi'}$ where $\xi' \leqq \alpha$. Since $\boldsymbol{B}$ is $\binom{\alpha}{\varkappa}$-saturated $(\theta)$, we have $\boldsymbol{A} \lhd_\varkappa^\alpha \boldsymbol{B}(\theta)$ and hence by lemma 1(ii), $\boldsymbol{A}x'\,\theta_\varkappa\,\boldsymbol{B}y'$ for some $y' \in B_\varkappa^{\xi'}$. Using $(C_2)$, it follows that $\boldsymbol{A}x\,\theta_\varkappa\,\boldsymbol{B}y$ for the appropriate rearrangement $y$ of $y'$. Hence $\boldsymbol{A} \leqq_\varkappa^{\alpha^+} \boldsymbol{B}(\theta)$. (ii) and (iii) are direct consequences of lemma 1(iii).

Theorem 2. *If $\theta$ is a $\varkappa$-finitary C-relation, then*

(i) $\boldsymbol{A}, \boldsymbol{B}$ $\binom{\alpha}{\varkappa}$*-saturated* $(\theta) \wedge \boldsymbol{A}\,\theta\,\boldsymbol{B} \wedge |A|_\varkappa, |B|_\varkappa \leqq \alpha .\Rightarrow. \boldsymbol{A} \cong \boldsymbol{B}(\theta)$,

(ii) $\boldsymbol{A}, \boldsymbol{B}$ $\varkappa$*-saturated* $(\theta) \wedge |A|_\varkappa = |B|_\varkappa .\Rightarrow. \boldsymbol{A} \cong \boldsymbol{B}(\theta)$.

Proof. For (i), since $\boldsymbol{A}$ and $\boldsymbol{B}$ are $\binom{\alpha}{\varkappa}$-saturated $(\theta)$ and $\boldsymbol{A}\,\theta\,\boldsymbol{B}$, we have $\boldsymbol{A} \diamondsuit_\varkappa^\alpha \boldsymbol{B}(\theta)$. The conclusion now follows by lemma 2(iii). (ii) is a special case of (i).

By way of example we note that these theorems subsume CANTOR's theorems 1 and 2. For let $K$ be the class of all non-empty linear orders and let $\theta$ be $\sim$ (partial-isomorphism) restricted to $K$. Let $\alpha = \omega$ and $\varkappa = 2$. Then for $\boldsymbol{B} \in K$, $\boldsymbol{B}$ is $\binom{\omega}{2}$-saturated $(\sim)$ if and only if $\boldsymbol{B}$ is dense and unbounded (and hence $\omega \leqq |B|$). Now if $\boldsymbol{A} \in K$ and $|A|_2 = |A| \leqq \omega$, then by theorem 1(ii) $\boldsymbol{A} \to \boldsymbol{B}(\sim)$, i.e., $\boldsymbol{A}$ is similar to a subset of $\boldsymbol{B}$. Also by theorem 2, if $\boldsymbol{A}$ and $\boldsymbol{B}$ are countable dense and unbounded (i.e., $\binom{\omega}{2}$-saturated $(\sim)$) members of $K$, then $\boldsymbol{A} \cong \boldsymbol{B}(\sim)$, i.e. $\boldsymbol{A} \cong \boldsymbol{B}$. In a similar fashion, if $\boldsymbol{A}$ is an $\eta_\alpha$-order (see HAUSDORFF [8]) we have that $\boldsymbol{A}$ is $\binom{\omega_\alpha}{2}$-saturated $(\sim)$. Consequently for $\boldsymbol{A}$ and $\boldsymbol{B}$ $\eta_\alpha$-orders, if $|A| = |B| = \omega_\alpha$, we have, by Theorem 2, HAUSDORFF's result that $\boldsymbol{A} \cong \boldsymbol{B}$.

We have seen that $\boldsymbol{B}$ $\binom{\alpha}{\varkappa}$-saturated $(\theta)$ implies $\boldsymbol{B}$ $\binom{\alpha^+}{\varkappa}$-universal $(\theta)$ and it trivially implies $\boldsymbol{B}$ $\binom{\alpha}{\varkappa}$-homogeneous $(\theta)$. To prove the converse in the general setting, we make use of the following lemma:

Lemma 3.3. *For a C-relation* $(\theta)$,

$$\boldsymbol{B} \binom{\alpha}{\varkappa}\text{-}homogeneous\ (\theta) \wedge \boldsymbol{A} \leqq_\varkappa^\alpha \boldsymbol{B}(\theta) .\Rightarrow. \boldsymbol{A} \lhd_\varkappa^\alpha \boldsymbol{B}(\theta).$$

Proof. Suppose $x \in A_{\varkappa\alpha}$, $y \in B_{\varkappa\alpha}$, $a \in A_\varkappa$ and $\boldsymbol{A}x\,\theta_\varkappa\,\boldsymbol{B}y$. Since $\boldsymbol{A} \leqq_\varkappa^\alpha \boldsymbol{B}(\theta)$, for some $z \in B_{\varkappa\alpha}$ and $c \in B_\varkappa$, $\boldsymbol{A}xa\,\theta_\varkappa\,\boldsymbol{B}zc$. By $(C_3)$, $\boldsymbol{A}x\,\theta_\varkappa\,\boldsymbol{B}z$ and by $(C_1)$, $\boldsymbol{B}y\,\theta_\varkappa\,\boldsymbol{B}z$. Since $\boldsymbol{B} \lhd_\varkappa^\alpha \boldsymbol{B}(\theta)$, $\boldsymbol{B}zc\,\theta_\varkappa\,\boldsymbol{B}yb$ for some $b \in B_\varkappa$ and thus by $(C_1)$, $\boldsymbol{A}xa\,\theta_\varkappa\,\boldsymbol{B}yb$. This proves the lemma.

Theorem 3.4. *For a $\varkappa$-finitary C-relation* $(\theta)$,

$$\boldsymbol{B} \binom{\alpha}{\varkappa}\text{-}saturated\ (\theta) .\equiv. \boldsymbol{B} \binom{\alpha}{\varkappa}\text{-}homogeneous\ (\theta) \wedge \boldsymbol{B} \binom{\alpha}{\varkappa}\text{-}universal\ (\theta).$$

Proof. The validity of the conclusion from left to right was noted above. The converse follows directly from lemma 3.3.

Theorem 3.4 is known in the special case $\theta$ = elementary equivalence, and $\varkappa = 2$ (see KEISLER [9]) and plays an important role in recent model theory. A second important instance of this theorem is the case $\varkappa = \omega$ and $\theta$ = EHRENFEUCHT's relation [4] (see the relation $\approx_\omega^\omega$ in section 4).

## 4. Derivatives of C-relations

Let $\theta$ be a $C$-relation. Recall that in section 2, the derivative $\theta'_\varkappa$ of $\theta$ was defined as follows, for $x \in A_{\varkappa\alpha}$, $y \in B_{\varkappa\alpha}$:

$$\boldsymbol{A}x\,\theta'_\varkappa\,\boldsymbol{B}y \;.\!\equiv\!.\; (\forall a)_{A_\varkappa}(\exists b)_{B_\varkappa}[\boldsymbol{A}xa\,\theta_\varkappa\,\boldsymbol{B}yb] \wedge (\forall b)_{B_\varkappa}(\exists a)_{A_\varkappa}[\boldsymbol{A}xa\,\theta_\varkappa\,\boldsymbol{B}yb].$$

It is easy to check that $\theta'_\varkappa$ is again a $C$-relation and $\theta'_\varkappa \subseteqq \theta_\varkappa$. Moreover if $\theta_0 \supseteqq \theta_1 \supseteqq \cdots \supseteqq \theta_\xi \supseteqq \cdots$ are $C_\varkappa$-relations for $\xi < \lambda$, then so is $\theta_\lambda = \bigcap_\lambda \theta_\xi$. Starting from a $C$-relation $\theta$, we obtain therefore for $2 \leqq \varkappa$, a transfinite sequence of *derived* $C$-relations $\theta^\xi_\varkappa$ as follows:

$$\theta^0_\varkappa = \theta_\varkappa, \qquad \theta^{\xi+1}_\varkappa = (\theta^\xi_\varkappa)'_\varkappa, \qquad \theta^\lambda_\varkappa = \bigcap_\lambda \theta^\xi_\varkappa \quad \text{for } \lambda \text{ a limit ordinal.}$$

With each of the $C$-relations $\theta^\xi_\varkappa$ we can associate its finitary extension $\tilde\theta^\xi_\varkappa$. $\tilde\theta^\xi_\varkappa$ is again a $C$-relation however it is not in general true that $(\widetilde{\theta^{\xi+1}_\varkappa})$ is equal to $(\tilde\theta^\xi_\varkappa)'_\varkappa$. Moreover, if $\theta$ is finitary $\theta'_\varkappa$ need not be finitary. (These remarks are easily verified by letting $\theta$ be partial isomorphism.)

For the case $\theta = \sim$ (partial isomorphism), the hierarchy $\sim^i_2$, $i \leqq \omega$, was introduced by FRAÏSSE [7] and $\sim^i_\omega$ was introduced by EHRENFEUCHT [4, 6]. FRAÏSSE proved that $\approx^\omega_2$ is just elementary equivalence and EHRENFEUCHT showed that $\approx^\omega_\omega$ implies monadic second-order equivalence.

We now make some observations on the relationship between the relations $\lhd^\alpha_\varkappa(\theta)$ and $\diamond^\alpha_\varkappa(\theta)$ and the derived sequence $\{\theta^\xi_\varkappa\}$. As before $\alpha$ and $\varkappa$ will be cardinals with $\omega \leqq \alpha$ and $2 \leqq \varkappa$.

Lemma 4.1. *If $\theta$ is a $C$-relation, $\boldsymbol{A} \diamond^\alpha_\varkappa \boldsymbol{B}(\theta)$ and $\eta \in On$, then*

(i) $x \in A_{\varkappa\alpha} \wedge y \in B_{\varkappa\alpha} \wedge \boldsymbol{A}x\,\theta_\varkappa\,\boldsymbol{B}y \;.\!\Rightarrow\!.\; \boldsymbol{A}x\,\theta^\eta_\varkappa\,\boldsymbol{B}y$,

(ii) $\boldsymbol{A}\,\theta^\eta_\varkappa\,\boldsymbol{B}$,

(iii) $\boldsymbol{A} \diamond^\alpha_\varkappa \boldsymbol{B}(\theta^\eta_\varkappa)$.

Proof. (i) is proved by induction on $\eta$. If $\eta = 0$, the conclusion holds trivially. Assuming (i) holds for $\eta$, suppose $x \in A_{\varkappa\alpha}$, $y \in B_{\varkappa\alpha}$, $a \in A_\varkappa$ and $\boldsymbol{A}x\,\theta_\varkappa\,\boldsymbol{B}y$. Since $\boldsymbol{A} \diamond^\alpha_\varkappa \boldsymbol{B}(\theta)$, we have $\boldsymbol{A}xa\,\theta_\varkappa\,\boldsymbol{B}yb$ for some $b \in B_\varkappa$. By the induction assumption (recall $\omega \leqq \alpha$) $\boldsymbol{A}xa\,\theta^\eta_\varkappa\,\boldsymbol{B}yb$. The argument is symmetric if we start with $b \in B_\varkappa$. Hence $\boldsymbol{A}x\,\theta^{\eta+1}_\varkappa\,\boldsymbol{B}y$. Assuming (i) holds for $\eta < \lambda$, $\lambda$ a limit, it follows trivially that it holds at $\lambda$. Thus (i) is established for all $\eta \in On$. (ii) now follows from (i) since the assumption implies $\boldsymbol{A}\,\theta_\varkappa\,\boldsymbol{B}$. To prove (iii) note that we have $\boldsymbol{A}\,\theta_\varkappa^\eta\,\boldsymbol{B}$ by (ii). If $x \in A_{\varkappa\alpha}$, $y \in B_{\varkappa\alpha}$ and $\boldsymbol{A}x\,\theta_\varkappa^\eta\,\boldsymbol{B}y$, then $\boldsymbol{A}x\,\theta_\varkappa\,\boldsymbol{B}y$ and hence by (i) $\boldsymbol{A}x\,\theta_\varkappa^{\eta+1}\,\boldsymbol{B}y$. Thus $\boldsymbol{A} \diamond_\varkappa^\alpha \boldsymbol{B}(\theta_\varkappa^\eta)$.

Lemma 4.2. *If $\theta$ is a C-relation, then*

(i) *For $\boldsymbol{A}$ such that* $\omega \leqq |A|$, $(\exists\xi)_{|A_\varkappa|^+}[A \lhd_\varkappa^\omega \boldsymbol{A}(\theta_\varkappa^\xi)]$,

(ii) *For $\boldsymbol{A}$ such that* $|A| \leqq \omega$, $(\exists\xi)_{\omega_1}[\boldsymbol{A} \lhd_\varkappa^\omega \boldsymbol{A}(\theta_\varkappa^\xi)]$.

Proof. We give the proof of (i). (ii) is proved in a similar manner. For $x, y \in A_{\varkappa\omega}$, define

$$f(x, y) = \begin{cases} (\mu\xi)\,[\neg\,\boldsymbol{A}x\,\theta_\varkappa^\xi\,\boldsymbol{A}y], & \text{if } \neg\,\boldsymbol{A}x\,\theta_\varkappa^{|A_\varkappa|^+}\,\boldsymbol{A}y \\ 0, & \text{otherwise} \end{cases}$$

and let $\xi_0 = \sup\{f(x, y); x, y \in A_{\varkappa\omega}\}$. Note that the number of such pairs $(x, y)$ is $\sum_\omega |A_\varkappa|^n \leqq |A_\varkappa| \cdot \omega = |A_\varkappa|$ (recall $\omega \leqq |A|$) and thus since $f(x, y) < |A_\varkappa|^+$ for each such pair $(x, y)$, we have $\xi_0 < |A_\varkappa|^+$. We now assert $\boldsymbol{A} \lhd_\varkappa^\omega \boldsymbol{A}(\theta_\varkappa^{\xi_0+1})$. For if $x, y \in A_{\varkappa\omega}$, $a \in A_\varkappa$ and $\boldsymbol{A}x\,\theta_\varkappa^{\xi_0+1}\,\boldsymbol{A}y$, then there is $b \in A_\varkappa$ such that $\boldsymbol{A}xa\,\theta_\varkappa^{\xi_0}\,\boldsymbol{A}yb$. Hence by definition of $f$, $f(xa, yb) > \xi_0$ or $f(xa, yb) = 0$. The first possibility contradicts the definition of $\xi_0$ and thus $f(xa, yb) = 0$. By definition of $f$ then, $\boldsymbol{A}xa\,\theta_\varkappa^{\xi_0+1}\,\boldsymbol{A}yb$. Therefore $\boldsymbol{A} \lhd_\varkappa^\omega \boldsymbol{A}(\theta_\varkappa^{\xi_0+1})$.

Lemma 4.3. *Let $\theta$ be a C-relation, then*

$$\boldsymbol{A} \lhd_\varkappa^\omega \boldsymbol{A}(\theta) \wedge \boldsymbol{B}\,\theta_\varkappa^\omega\,\boldsymbol{A} .\Rightarrow. \boldsymbol{B} \diamond_\varkappa^\omega \boldsymbol{A}(\theta).$$

Proof. Suppose $x \in A_{\varkappa\omega}$, $y \in B_{\varkappa\omega}$ and $\boldsymbol{A}x\,\theta_\varkappa\,\boldsymbol{B}y$. Since $\boldsymbol{B}\,\theta_\varkappa^\omega\,\boldsymbol{A}$, $\boldsymbol{B}y\,\theta_\varkappa'\,\boldsymbol{A}z$ for some $z \in A_{\varkappa\omega}$. Hence $\boldsymbol{B}y\,\theta_\varkappa\,\boldsymbol{A}z$ and by $(C_1)$, $\boldsymbol{A}z\,\theta_\varkappa\,\boldsymbol{A}x$. Since $\boldsymbol{A} \lhd_\varkappa^\omega \boldsymbol{A}(\theta)$, we have $\boldsymbol{A}z\,\theta_\varkappa'\,\boldsymbol{A}x$ and again by $(C_1)$, $\boldsymbol{A}x\,\theta_\varkappa'\,\boldsymbol{B}y$. Hence $\boldsymbol{B} \diamond_\varkappa^\omega \boldsymbol{A}(\theta)$.

Corollary 4.4. *Let $\theta$ be a C-relation, then*

(i) $(\forall\boldsymbol{A})\,[\omega \leqq |A| \Rightarrow (\exists\xi)_{|A_\varkappa|^+}(\forall\boldsymbol{B})\,(\boldsymbol{B}\,\theta_\varkappa^\xi\,\boldsymbol{A} \Rightarrow \boldsymbol{B} \diamond_\varkappa^\omega \boldsymbol{A}(\theta_\varkappa^\xi))]$,

(ii) $(\forall\boldsymbol{A})\,[|A| \leqq \omega .\Rightarrow. (\exists\xi)_{\omega_1}(\boldsymbol{B}\,\theta_\varkappa^\xi\,\boldsymbol{A} \Rightarrow \boldsymbol{B} \diamond_\varkappa^\omega \boldsymbol{A}(\theta_\varkappa^\xi))]$.

Proof. For (i), note that by lemma 4.2, there is $\xi \in |A_\varkappa|^+$ such that $\boldsymbol{A} \lhd_\varkappa^\omega \boldsymbol{A}(\theta_\varkappa^\xi)$. The conclusion now follows from lemma 4.3. (ii) follows similarly.

We define $\psi(A, \varkappa) = \sum_{\nu<\varkappa} |A^\nu|^+$. Consequently, $\psi(A, \varkappa) \leqq |A_\varkappa|^+$. If $A \diamond_\varkappa^\alpha \boldsymbol{B}(\theta)$ for all cardinals $\alpha$, we write $\boldsymbol{A} \diamond_\varkappa^{On} \boldsymbol{B}(\theta)$; similarly $\boldsymbol{A}\,\theta_\varkappa^{On}\,\boldsymbol{B}$ means that for all $\xi \in On$, $\boldsymbol{A}\,\theta_\varkappa^\xi\,\boldsymbol{B}$. In the following, $\psi$ denotes an infinite cardinal number.

Lemma 4.5. *Let $\theta$ be a C-relation;*

$$\textit{if} \quad \psi(A, \varkappa), \psi(\boldsymbol{B}, \varkappa) \leqq \psi, \quad \textit{then} \quad \boldsymbol{A}\,\theta_\varkappa^\psi\,\boldsymbol{B} .\equiv. \boldsymbol{A} \diamond_\varkappa^{On} \boldsymbol{B}(\theta_\varkappa^\psi).$$

Proof. The implication from right to left is trivial. For the converse, we are to show if $\alpha \in On$, $x \in A_{\varkappa\alpha}$, $y \in B_{\varkappa\alpha}$ and $\boldsymbol{A}x\,\theta_\varkappa^\psi\,\boldsymbol{B}y$, then $Ax\,\theta_\varkappa^{\psi+1}\,\boldsymbol{B}y$. Now $a \in A_\varkappa$ means that $a \in A^\nu$ for some $\nu < \varkappa$. For such $a$, since $\boldsymbol{A}x\,\theta_\varkappa^\psi\,\boldsymbol{B}y$, we have for each $\gamma < \psi$, $b_\gamma \in B^\nu$ such that $\boldsymbol{A}xa\,\theta_\varkappa^\gamma\,Byb_\gamma$. The number of such $b$ is $|B|^\nu < \psi(B, \varkappa) \leqq \psi$. Hence for some $b \in B^\nu$, $\boldsymbol{A}xa\,\theta_\varkappa^\gamma\,\boldsymbol{B}yb$ for all $\gamma < \psi$ or $\boldsymbol{A}xa\,\theta_\varkappa^\psi\,\boldsymbol{B}yb$. Since the roles of $\boldsymbol{A}$ and $\boldsymbol{B}$ can clearly be reversed, we have $\boldsymbol{A}x\,\theta_\varkappa^{\psi+1}\,\boldsymbol{B}y$.

Corollary 4.6. *For a C-relation* $\theta$,

(i) $(\forall \boldsymbol{A})\,[\omega \leqq |A| \Rightarrow (\exists \xi)_{|A_\varkappa|^+}\, \forall \boldsymbol{B}(\boldsymbol{A}\,\theta^{\xi}_{\varkappa}\,\boldsymbol{B} \equiv \boldsymbol{A}\,\theta^{On}_{\varkappa}\,\boldsymbol{B})]$,

(ii) $(\forall \boldsymbol{A}, \boldsymbol{B})\,[\psi(A, \varkappa), \psi(\boldsymbol{B}, \varkappa) \leqq \psi \mathrel{.\Rightarrow.} \boldsymbol{A}\,\theta^{\psi}_{\varkappa}\,\boldsymbol{B} \mathrel{.\equiv.} \boldsymbol{A}\,\theta^{On}_{\varkappa}\,\boldsymbol{B}]$.

Proof. These statements follow directly from lemmas 4.4, 4.5 and 4.1.

The preceding lemmas are of particular interest in the case $\varkappa = 2 - |A|_2 = |A|$ and $|A|_2^+ = |A|^+$. In this case for $|A| = \omega$, lemma 4.2 states that for some $\xi < \omega_1$, $\boldsymbol{A}$ is $\binom{\omega}{2}$-homogeneous $(\theta^{\xi}_{2})$. We summarize the results of combining these lemmas with corollary 2 in the following two theorems:

Theorem 4.7. *Let* $\theta$ *be a C-relation,* $\boldsymbol{A}$ *and* $\boldsymbol{B}$ *systems with* $\psi(A, \varkappa), \psi(B, \varkappa) \leqq \psi$. *Then*

$$\boldsymbol{A}\,\theta^{\psi}_{\varkappa}\,\boldsymbol{B} \mathrel{.\equiv.} \boldsymbol{A}\,\theta^{On}_{\varkappa}\,\boldsymbol{B} \mathrel{.\equiv.} \boldsymbol{A} \diamond^{On}_{\varkappa} \boldsymbol{B}(\theta^{\psi}_{\varkappa}).$$

*Moreover, if* $|A|_\varkappa, |B|_\varkappa \leqq \omega$, *then any of the above imply* $A \cong \boldsymbol{B}(\tilde{\theta}^{\xi}_{\varkappa})$.

Theorem 4.8. *Let* $\theta$ *be a C-relation and* $\boldsymbol{A}$ *an infinite system. Then there is a* $\xi < |A_\varkappa|^+$ *such that for any* $\boldsymbol{B}$,

$$\boldsymbol{A}\,\theta^{\xi}_{\varkappa}\,\boldsymbol{B} \mathrel{.\equiv.} \boldsymbol{A}\,\theta^{On}_{\varkappa}\,\boldsymbol{B} \mathrel{.\equiv.} \boldsymbol{A} \diamond^{\omega}_{\varkappa} \boldsymbol{B}(\theta^{\xi}_{\varkappa}).$$

*Moreover if* $|A|_\varkappa, |B|_\varkappa \leqq \omega$, *then any of the above imply* $\boldsymbol{A} \cong \boldsymbol{B}(\tilde{\theta}^{\xi}_{\varkappa})$.

As noted earlier, FRAÏSSE [7] introduced the hierarchy $\sim^{i}_{2}$ $(i \leqq \omega)$ to provide an alternative to the study of systems by means of elementary sentences. We now introduce a class of infinitary languages which correspond to the transfinite hierarchy $\sim^{\xi}_{\varkappa}$. Note however that these languages are not really needed. The basic results in trans-elementary model theory are more naturally stated and proved in terms of the hierarchies $\sim^{\xi}_{\varkappa}$ and $\approx^{\xi}_{\varkappa}$ (see theorems 4.7 and 4.8).

The language $\mathrm{L}_\varkappa$ consists of sentences only. The primitives occurring in these sentences are the equality symbol, (finitely many) predicate symbols and individual (constant) symbols corresponding to the primitives in the systems $\boldsymbol{A}, \boldsymbol{B}, \ldots$ Furthermore, the sentences may contain individual symbols from a well-ordered set $\Delta = \{\delta_0, \delta_1, \ldots, \delta_\nu, \ldots;\ \nu \in On\}$. Regarding interpretation, in a system $\boldsymbol{A}a$ $(a \in A^{\xi})$, we will always use $\delta_\nu$ to represent $a_\nu$ for $\nu < \xi$. Members of the well-ordered set $V = \{w_0, w_1, \ldots, w_\nu, \ldots;\ \nu \in On\}$ will serve as bound individual variables. If $\Phi$ is a sentence, $\beta \in \Delta^{\xi}$ and $v \in V^{\xi}$, $\Phi^{\beta}_{v}$ denotes the expression obtained from $\Phi$ by replacing $\beta_\nu$ by $v_\nu$ for $\nu < \xi$. The sentences of $\mathrm{L}_\varkappa$ are built up by starting with atomic sentences and applying the logical operations $\neg\Phi$, $\wedge S$ for $S$ a set of sentences and $(\exists v)\,\Phi^{\beta}_{v}$ where $v \in V^{\gamma}$ for $\gamma < \varkappa$. We now state the production rules for $\mathrm{L}_\varkappa$ in a manner which simultaneously yields a *ranking* $\mathrm{L}^{\xi}_{\varkappa}$ of $\mathrm{L}_\varkappa = \bigcup_{\xi \in On} \mathrm{L}^{\xi}_{\varkappa}$ according to the depth of quantification.

$$\Phi \text{ an atomic sentence} \Rightarrow \Phi \in \mathrm{L}^{0}_{\varkappa}, \qquad \xi < \eta, \Phi \in \mathrm{L}^{\xi}_{\varkappa} \mathrel{.\Rightarrow.} \Phi \in \mathrm{L}^{\eta}_{\varkappa},$$

$$\Phi \in \mathrm{L}^{\xi}_{\varkappa} \mathrel{.\Rightarrow.} \neg\Phi \in \mathrm{L}^{\xi}_{\varkappa}, \qquad S \subseteqq \mathrm{L}^{\xi}_{\varkappa} \mathrel{.\Rightarrow.} \wedge S \in \mathrm{L}^{\xi}_{\varkappa},$$

$$\xi < \varkappa, \beta \in \Delta^{\xi}, v \in V^{\xi}, \Phi \in \mathrm{L}^{\xi}_{\varkappa} \mathrel{.\Rightarrow.} (\exists v)\,\Phi^{\beta}_{v} \in \mathrm{L}^{\xi+1}_{\varkappa}.$$

The notion $\bigwedge_{i \in I} \Phi_i$ represents $\bigwedge\{\Phi_i;\ i \in I\}$. Note that if only finitely many of the $\Phi_i$ are different, this conjunction is finite. The disjunction $\bigvee S$ and the universal quantifier $(\forall v)$ are introduced as abbreviations in the usual manner. It is now easy to introduce, by recursion on $\xi \in On$, the sets $\tau^\xi_\varkappa(\boldsymbol{A}a) \subseteq \mathrm{L}^\xi_\varkappa$ which have exactly the models $\boldsymbol{B}b$ such that $\boldsymbol{B}b \sim^\xi_\varkappa \boldsymbol{A}a$. Recall that the individual symbol $\delta_\nu$ represents $a_\nu$ and that $A_2 = A$.

For $a \in A^\xi$

$\tau^0_\varkappa(\boldsymbol{A}a) =$ the set of all atomic sentences and negations of atomic sentences true in $\boldsymbol{A}a$,

$\bar\tau^\xi_\varkappa\begin{pmatrix} & b \\ \boldsymbol{A} a & v\end{pmatrix} = [\bigwedge \tau^\xi_\varkappa(\boldsymbol{A}ab)]^\beta_v$ where $b \in A^\sigma$ and $\beta, v$ denote the sequences $\delta_\nu, w_\nu$, $\xi \leqq \nu < \xi + \sigma$,

$$\tau^{\xi+1}_\varkappa(\boldsymbol{A}a) = \left\{(\exists v)\, \bar\tau^\xi_\varkappa\begin{pmatrix} & b \\ \boldsymbol{A} a & v\end{pmatrix};\ b \in A_\varkappa\right\} \cup \left\{(\forall v) \bigvee_{b \in A^{lg\, v}} \bar\tau^\xi_\varkappa\begin{pmatrix} & b \\ \boldsymbol{A} a & v\end{pmatrix};\ lg\ v < \varkappa\right\},$$

$\tau^\lambda_\varkappa(\boldsymbol{A}a) = \bigcup_{\xi<\lambda} \tau^\xi_\varkappa(\boldsymbol{A}a)$ for $\lambda$ a limit ordinal.

For $a \in A^\varrho_\varkappa$

$\tau^\xi_\varkappa(\boldsymbol{A}a) = \tau^\xi_\varkappa(\boldsymbol{A}\hat a)$, $\quad \tilde\tau^\xi_\varkappa(\boldsymbol{A}a) = \bigcup\{\tau^\xi_\varkappa(\boldsymbol{A}\bar a_F);\ F \subseteq \varrho \wedge |F| < \omega\}$.

For a sentence $\Phi \in \mathrm{L}^\xi_\varkappa$ and $a \in A^\varrho_\varkappa$, $\boldsymbol{A}a$ is a *model of* (satisfies) $\Phi$ if and only if $\boldsymbol{A}\hat a$ is a model of $\Phi$. Note that for finite $a \in A_{\varkappa\omega}$, $\tilde\tau^\xi_\varkappa(\boldsymbol{A}a) = \tau^\xi_\varkappa(\boldsymbol{A}a)$. If furthermore $\varkappa$ and $i$ are finite, $\tau^i_\varkappa(\boldsymbol{A}a)$ is a finite set of sentences and $\tau^\omega_\varkappa(\boldsymbol{A}a)$ describes $\boldsymbol{A}a$ up to elementary equivalence. The following theorem suggests that $\tau^\xi_\varkappa(\boldsymbol{A}a)$ is properly called the $\begin{pmatrix}\xi\\ \varkappa\end{pmatrix}$*-theory* of $\boldsymbol{A}a$.

Theorem 4.9. *For systems $\boldsymbol{A}$ and $\boldsymbol{B}$, $a \in A_{\varkappa\alpha}$ and $b \in B_{\varkappa\alpha}$ $\tau^\xi_\varkappa(\boldsymbol{A}a) \subseteq \mathrm{L}^\xi_\varkappa$. Moreover if $lg\ a = lg\ b$ and $lg\ a_\nu = lg\ b_\nu$ for $\nu < lg\ a$, the following are equivalent*

(i) $\boldsymbol{A}a \sim^\xi_\varkappa \boldsymbol{B}b$,

(ii) *$\boldsymbol{A}a$ and $\boldsymbol{B}b$ are $\mathrm{L}^\xi_\varkappa$-equivalent,*

(iii) $\tau^\xi_\varkappa(\boldsymbol{A}a) = \tau^\xi_\varkappa(\boldsymbol{B}b)$,

(iv) *$\boldsymbol{B}b$ is a model of $\tau^\xi_\varkappa(\boldsymbol{A}a)$.*

Proof. The theorem is proved by induction on $\xi \in On$. We show only (i) implies (ii), i.e., if $\boldsymbol{A}a \sim^\xi_\varkappa \boldsymbol{B}b$, then for $\Phi \in \mathrm{L}^\xi_\varkappa$, $\boldsymbol{A}a$ satisfies $\Phi$ if and only if $\boldsymbol{B}b$ satisfies $\Phi$. This follows directly from the definitions if $\xi = 0$ and if true for $\xi < \lambda$, $\lambda$ a limit, it is clearly true for $\xi = \lambda$. Assuming the statement true for $\xi$, we show it true for $\xi + 1$. Suppose $\boldsymbol{A}a \sim^{\xi+1}_\varkappa \boldsymbol{B}b$ and $\Phi \in \mathrm{L}^{\xi+1}_\varkappa$. If $\Phi \in \mathrm{L}^\xi_\varkappa$, the result follows by the inductive assumption. If $\Phi \notin \mathrm{L}^\xi_\varkappa$, we argue by induction on the number of occurrences of $\neg$, $\bigwedge$ and $\exists$ in $\Phi$. The argument is straight-forward in the cases where $\Phi$ is of the form $\neg\Psi$ or $\bigwedge S$. We consider the case where $\Phi$ has the form $\exists v\, \Psi^\beta_v$, for $v \in V^\xi$, $b \in \Delta^\xi$ and $\xi < \varkappa$. Note that $\Psi \in \mathrm{L}^\xi_\varkappa$. Then if $\boldsymbol{A}a$ satisfies $\Phi$, for some $a' \in A^\xi$, $\boldsymbol{A}aa'$ satisfies $\Psi$. Since $\boldsymbol{A}a \sim_\varkappa^{\xi+1} \boldsymbol{B}b$, there is $b' \in B^\xi$ for which $\boldsymbol{A}aa' \sim_\varkappa^\xi \boldsymbol{B}bb'$. By the inductive assumption, $\boldsymbol{B}bb'$ satisfies $\Psi$ and hence $\boldsymbol{B}b$ satisfies $\Phi$. Similarly, if $\boldsymbol{B}b$ satisfies $\Phi$, so does $\boldsymbol{A}a$. This completes the argument.

Observe that theorem 4.9 implies FRAÏSSE's result [7] that $\approx_2^\omega$ is elementary equivalence. Furthermore, we now have a well-defined language, $\bigcup_{i<\omega} \mathrm{L}_\omega^i$ which corresponds to EHRENFEUCHT's relation $\approx_\omega^\omega$ [4] exactly as elementary logic $\bigcup_{i<\omega} \mathrm{L}_2^i$ corresponds to elementary equivalence.

Using theorem 4.9 we can now restate theorems 4.7 and 4.8 in semantic terminology. Note that $\boldsymbol{A} \cong \boldsymbol{B}\ (\approx_\varkappa^\xi)$ implies $\boldsymbol{A} \cong \boldsymbol{B}$.

Corollary 4.10. *For systems $\boldsymbol{A}$ and $\boldsymbol{B}$, if $\psi(A, \varkappa)$, $\psi(B, \varkappa) \leqq \psi$, then*

$$\boldsymbol{A} \text{ and } \boldsymbol{B} \text{ are } \mathrm{L}_\varkappa^\psi\text{-equivalent} .\equiv. \boldsymbol{A} \text{ and } \boldsymbol{B} \text{ are } \mathrm{L}_\varkappa\text{-equivalent}.$$

*Moreover, either of these statements implies that if $|A|_\varkappa, |B|_\varkappa \leqq \omega$, then $\boldsymbol{A} \cong \boldsymbol{B}$.*

Corollary 4.11. *For any system $\boldsymbol{A}$, there is a $\xi < |A_\varkappa|^+$ such that for any system $\boldsymbol{B}$, the following are equivalent:*

(i) *$\boldsymbol{B}$ is a model of $\tau_\varkappa^\xi(\boldsymbol{A})$,*
(ii) *$\boldsymbol{A}$ and $\boldsymbol{B}$ are $\mathrm{L}_\varkappa^\xi$-equivalent,*
(iii) *$\boldsymbol{A}$ and $\boldsymbol{B}$ are $\mathrm{L}_\varkappa$-equivalent.*

*Moreover, any of these statements implies that if $|A|_\varkappa, |B|_\varkappa \leqq \omega$, then $\boldsymbol{A} \cong \boldsymbol{B}$.*

Corollary 4.10 implies CHANG's result [3] that for $\psi(A, \varkappa) + \psi(B, \varkappa) \leqq \xi$, $\boldsymbol{A}$ and $\boldsymbol{B}$ are $\mathrm{L}_\varkappa^\xi$-equivalent if and only if they are $\mathrm{L}_\varkappa$-equivalent and if $|A| = |B| = \varkappa, \omega$ cofinal with $\varkappa$ and $\psi(A, \varkappa) \leqq \xi$, then $\boldsymbol{A}$ and $\boldsymbol{B}$ are $\mathrm{L}_\varkappa^\xi$-equivalent if and only they are isomorphic.

Corollary 4.11 yields CHANG's result [3] regarding the characterization of $\boldsymbol{A}$ up to $\mathrm{L}_\varkappa$-equivalence (and isomorphism if $\omega$ cofinal with $|A|$) by a single sentence. For in this case the cardinality of the set of sentences $\tau_\varkappa^\xi(\boldsymbol{A})$ is less than $|A_\varkappa|^+$. In particular, if $\boldsymbol{A}$ is countable and $\varkappa = 2$, we have $\xi < \omega_1$ and $\tau_2^\xi(\boldsymbol{A})$ is a countable set containing only countable $\bigwedge$. The "SCOTT sentence", $\bigwedge \tau_2^\xi(\boldsymbol{A})$, describes $\boldsymbol{A}$ up to isomorphism (see SCOTT [13]).

As noted earlier, an important (finitary) $C$-relation is elementary equivalence, i.e., $\approx_2^\omega$. Consequently many of the categoricity theorems of elementary model theory are encompassed by our generalized CANTOR theorems. For example, let $\Sigma$ be a complete elementary theory without finite models and for $n < \omega$, let $St_n(\Sigma)$ denote the set of all $\binom{n}{2}$-types $(\approx_2^\omega)$ realized in models $\boldsymbol{A}$ of $\Sigma$. If for each $n$, $St_n(\Sigma)$ is finite, it is easy to show that $\boldsymbol{A} \diamond_2^\omega \boldsymbol{B}\ (\approx_2^\omega)$ for any two models $\boldsymbol{A}$ and $\boldsymbol{B}$ of $\Sigma$. Hence by Lemma 2 (or theorem 2) $\boldsymbol{A} \cong \boldsymbol{B}$ if $|A| = |B| = \omega$, i.e., $\Sigma$ is $\omega$-categorical (see RYLL-NARDZEWSKI [12]).

Note in particular that if $\Sigma$ is an elementary theory such that $\boldsymbol{A} \diamond_2^\omega \boldsymbol{B}\ (\approx_2^\omega)$ for all models $\boldsymbol{A}$ and $\boldsymbol{B}$ of $\Sigma$, then $\Sigma$ is $\omega$-categorical. Conversely if $\Sigma$ is $\omega$-categorical and $\boldsymbol{A}$ and $\boldsymbol{B}$ are models of $\Sigma$, then $\boldsymbol{A} \diamond_2^\omega \boldsymbol{B}\ (\approx_2^\omega)$. For if not, by the LÖWENHEIM-

SKOLEM-TARSKI theorem [14], one may assume that $\boldsymbol{A}$ and $\boldsymbol{B}$ are countable. This leads to a contradiction for since $\Sigma$ is $\omega$-categorical, $\boldsymbol{A} \cong \boldsymbol{B}$ and then clearly, $\boldsymbol{A} \diamond_2^\omega \boldsymbol{B}\ (\approx_2^\omega)$.

Less natural $C$-relations are often useful. In working with objects such as real closed fields, for example, one might take $\boldsymbol{A}x\,\theta\,\boldsymbol{B}y$ to mean that the sequences of elements elementarily definable in $\boldsymbol{A}x$ and $\boldsymbol{B}y$ are partially isomorphic. (It is understood that these elements must be in one-one correspondence and ordered by their defining formulas.) It can then be shown, with the aid of the ARTIN-SCHREIER theorem (see [5]), that if $\boldsymbol{A}$ and $\boldsymbol{B}$ are real closed fields which are $\eta_1$-orders relative their order, then $\boldsymbol{A} \diamond_2^{\omega_1} \boldsymbol{B}(\theta)$. Hence by lemma 2, $\boldsymbol{A} \cong \boldsymbol{B}$ if $|A| = |B| = \omega_1$ (see also ERDÖS, GILLMAN and HENRIKSEN [5]). Similar remarks apply to divisible ordered abelian groups.

## 5. Further applications

In this section we conclude with some additional notions which generalize naturally to our present context.

Recall that by lemma 3.2, if $\theta$ is a $\varkappa$-finitary $C$-relation, $|A|_\varkappa = \alpha$ and $\boldsymbol{A}\,\theta\,\boldsymbol{A}$, then $\boldsymbol{A}$ is $\varkappa$-homogeneous $(\theta)$ if and only if for $x, y \in A_{\varkappa\alpha}$, $\boldsymbol{A}x\,\theta_\varkappa\,\boldsymbol{A}y$ implies there is an $f$ such that $f: \boldsymbol{A} \cong \boldsymbol{A}(\theta)$ and $fx = y$ (i.e., every "sufficiently partial" $\theta$-automorphism of $\boldsymbol{A}$ can be extended to a $\theta$-automorphism of $\boldsymbol{A}$). We compare this with the following weaker notion: For $\alpha = |A|_\varkappa$, $\boldsymbol{A}$ is *$\varkappa$-pseudo-homogeneous* $(\theta)$ if $\boldsymbol{A}\,\theta\,\boldsymbol{A}$ and $(\forall x)_{A_{\varkappa\alpha}}(\exists y)_{A_\varkappa}(\forall z)_{A_{\varkappa\alpha}}[\boldsymbol{A}xy\,\theta_\varkappa\,\boldsymbol{A}z\ .\!\Rightarrow\!.\ (\exists f)\,(f: \boldsymbol{A} \cong \boldsymbol{A}(\theta) \wedge fxy = z)]$.

A system may be $\varkappa$-pseudo-homogeneous $(\theta)$ and not $\varkappa$-homogeneous $(\theta)$. For example, if $\varkappa = \omega$ and $\theta = \sim$, then $\langle I, S\rangle$, where $I$ is the set of integers and $S$ is the graph of the successor function, has these properties. CALAIS [1] introduced the notion of pseudo-homogeneity in the case $\varkappa = \omega$ and $\theta = \sim$.

Remark 3. If $\theta$ is a $C$-relation, $f: \boldsymbol{A} \cong \boldsymbol{A}(\theta)$ and for $x, y \in A_{\varkappa\alpha}$, $fx = y$, then $\boldsymbol{A}x\,\theta_\varkappa^\xi\,\boldsymbol{A}y$ for all $\xi \in On$.

This remark is easily proved by induction on $\xi$. Our next lemma relates homogeneity and pseudo-homogeneity.

Lemma 5.1. *If $\theta$ is a $C$-relation, then*

(i) *$\boldsymbol{A}$ $\varkappa$-pseudo-homogeneous $(\theta)$ $.\!\Rightarrow\!.$ $\boldsymbol{A}$ $\varkappa$-homogeneous $(\theta_\varkappa^\xi)$ for $2 \leqq \xi$.*

(ii) *If $A$ and $\boldsymbol{B}$ are $\varkappa$-pseudo-homogeneous $(\theta)$ and $\alpha = |A|_\varkappa = |B|_\varkappa$, then $St_\varkappa^\alpha(\boldsymbol{A}, \theta) = St_\varkappa^\alpha(\boldsymbol{B}, \theta) \Rightarrow St_\varkappa^\alpha(\boldsymbol{A}, \theta_\varkappa^\xi) = St_\varkappa^\alpha(\boldsymbol{B}, \theta_\varkappa^\xi)$ for $\xi \in On$.*

Proof. In (i) we will assume $\omega \leqq |A|_\varkappa$. If $|A|_\varkappa < \omega$, the following argument requires only minor modifications. Likewise in (ii), we assume $\omega \leqq \alpha$. For $x \in A_{\varkappa\alpha}$, we call $x$ *closed in* $\boldsymbol{A}$ if $\boldsymbol{A}x\,\theta_\varkappa\,\boldsymbol{A}y$ implies $f: \boldsymbol{A} \cong \boldsymbol{A}(\theta)$ with $fx = y$ for some $f$. To prove (i), let $|A|_\varkappa = \alpha$. It suffices, by lemma 4.1, to prove (i) for the case $\xi = 2$. Suppose $x, y \in A_{\varkappa\alpha}$, $a \in A_\varkappa$ and $\boldsymbol{A}x\,\theta_\varkappa^2\,\boldsymbol{A}y$. Since $xa \in A_{\varkappa\alpha}$ and $\boldsymbol{A}$ is $\varkappa$-pseudo-homogeneous $(\theta)$, there is $a' \in A_\varkappa$ such that $xaa'$ is closed in $\boldsymbol{A}$. Since $\boldsymbol{A}x\,\theta_\varkappa^2\,\boldsymbol{A}y$, for some $b, b' \in A_\varkappa$, $\boldsymbol{A}xaa'\,\theta_\varkappa\,\boldsymbol{A}ybb'$. Because $xaa'$ is closed, there is an $f$, $f: \boldsymbol{A} \cong \boldsymbol{A}(\theta)$ and $fxa = yb$. By remark 3, $\boldsymbol{A}xa\,\theta_\varkappa^2\,\boldsymbol{A}yb$ so that $\boldsymbol{A}$ is $\varkappa$-homogeneous $(\theta)$.

28*

To prove (ii), it clearly suffices to show $\boldsymbol{A} \leqq_\varkappa^\alpha \boldsymbol{B}(\theta_\varkappa^\xi)$ for $1 \leqq \xi$. We do this for $\xi = 1$ – a simple induction argument replaces 1 by $\xi \in On$. Since $\boldsymbol{A} \leqq_\varkappa^\alpha \boldsymbol{B}(\theta)$, we may thus establish (ii) by showing for $x \in A_{\varkappa\alpha}$, $y \in B_{\varkappa\alpha}$, $\boldsymbol{A}x\,\theta_\varkappa\,\boldsymbol{B}y$ implies $\boldsymbol{A}x\,\theta_\varkappa^1\,\boldsymbol{B}y$. Moreover since $\boldsymbol{A}$ is $\varkappa$-pseudo-homogeneous $(\theta)$ and $\alpha = |A|_\varkappa$, it follows by $(C_3)$ that we need only do this for the case in which $x$ is closed in $\boldsymbol{A}$. Suppose then for such $x$ and $y$, $\boldsymbol{A}x\,\theta_\varkappa\,By$. Let $a \in A_\varkappa$. Since $\boldsymbol{B}$ is $\varkappa$-pseudo-homogeneous $(\theta)$ and $\alpha = |B|_\varkappa$, there is $y' \in B_\varkappa$ such that $yy'$ is closed in $\boldsymbol{B}$ and $\boldsymbol{B} \leqq_\varkappa^\alpha \boldsymbol{A}(\theta)$ implies $\boldsymbol{A}vv'\,\theta_\varkappa\,\boldsymbol{B}yy'$ for some $v \in A_{\varkappa\alpha}$ and $v' \in A_\varkappa$. By $(C_3)$ and $(C_1)$, $\boldsymbol{A}x\,\theta_\varkappa\,\boldsymbol{A}v$ and because $x$ is closed in $\boldsymbol{A}$, we have by remark 3 that $\boldsymbol{A}x\,\theta_\varkappa^1\,\boldsymbol{A}v$. Hence $\boldsymbol{A}xx'\,\theta_\varkappa\,\boldsymbol{A}vv'$ for some $x' \in A_\varkappa$. Since $\boldsymbol{A}$ is $\varkappa$-pseudo-homogeneous $(\theta)$, for some $x'' \in A_\varkappa$, $xx'ax''$ is closed in $\boldsymbol{A}$ and as above, $\boldsymbol{A}xx'ax''\,\theta_\varkappa\,\boldsymbol{B}uu'b'u''$ for some $u \in B_{\varkappa\alpha}$ and $u', b', u'' \in B_\varkappa$. Then by $(C_3)$ and $(C_1)$, $\boldsymbol{B}uu'\,\theta_\varkappa\,\boldsymbol{B}yy'$ and because $yy'$ is closed in $\boldsymbol{B}$, we have (using remark 3) that $\boldsymbol{B}uu'b'u''\,\theta_\varkappa\,\boldsymbol{B}yy'by''$ for some $b, y'' \in B_\varkappa$. Then by $(C_1)$ and $(C_3)$, $\boldsymbol{A}xa\,\theta_\varkappa\,\boldsymbol{B}yb$. Now let $b \in B_\varkappa$. By assumption, for some $y' \in B_\varkappa$, $yby'$ is closed in $\boldsymbol{B}$ and because $\boldsymbol{B} \leqq_\varkappa^\alpha \boldsymbol{A}(\theta)$, $\boldsymbol{B}yby'\,\theta_\varkappa\,\boldsymbol{A}ua'u'$ for appropriate $u \in A_{\varkappa\alpha}$ and $a', u' \in A_\varkappa$. As before $\boldsymbol{A}x\,\theta_\varkappa\,\boldsymbol{A}u$ and $x$ closed in $\boldsymbol{A}$ implies $\boldsymbol{A}ua'u'\,\theta_\varkappa\,\boldsymbol{A}xax'$ for some $a, x' \in A_\varkappa$. It follows that $\boldsymbol{A}xax'\,\theta_\varkappa\,\boldsymbol{B}yby'$ and hence $\boldsymbol{A}xa\,\theta_\varkappa\,\boldsymbol{B}yb$. Thus we have $\boldsymbol{A}x\,\theta_\varkappa^1\,\boldsymbol{B}y$.

As a direct consequence of lemmas 5.1 and 3.3 and corollary 2, we have

Corollary 5.2. *If $\theta$ is a C-relation, $\boldsymbol{A}$ and $\boldsymbol{B}$ are $\varkappa$-pseudo-homogeneous $(\theta)$, $\alpha = |A|_\varkappa = |B|_\varkappa$, and $St_\varkappa^\alpha(\boldsymbol{A}, \theta) = St_\varkappa^\alpha(\boldsymbol{B}, \theta)$, then*

(i) $\boldsymbol{A} \diamond_\varkappa^\alpha \boldsymbol{B}(\theta_\varkappa^\xi)$ *for* $2 \leqq \xi$, (ii) $|A|_\varkappa = |B|_\varkappa \leqq \omega .\Rightarrow. \boldsymbol{A} \cong \boldsymbol{B}(\bar{\theta}_\varkappa^2)$.

In [1] Calais proved that if $\boldsymbol{A}$ and $\boldsymbol{B}$ are $\omega$-pseudo-homogeneous $(\sim)$, $St_\omega^\omega(\boldsymbol{A}, \sim) = St_\omega^\omega(\boldsymbol{B}, \sim)$ and $|A|, |B| \leqq \omega$, then $\boldsymbol{A} \cong \boldsymbol{B}$. Moreover, his proof is based on a Cantor-type argument which is apparently somewhat more sophisticated than that used in the proof of lemma 2. Note however that Calais' result is a consequence of corollary 5.2. The main point here is contained in lemma 5.1 – pseudo-homogeneity relative $\theta$ implies homogeneity relative the second derivative of $\theta$.

A final notion we wish to consider in this abstract setting is that of an atomic system. We will call $\boldsymbol{A}$ $\binom{\alpha}{\varkappa}$*-atomic* $(\theta_\varkappa^\xi)$ if $\boldsymbol{A}\,\theta\,\boldsymbol{A}$ and

$$(\forall x)_{A_{\varkappa\alpha}}(\exists \nu)_\xi((\forall \boldsymbol{B})\,\forall y)_{B_{\varkappa\alpha}}[\boldsymbol{B}\,\theta_\varkappa^\xi\,\boldsymbol{A} \wedge \boldsymbol{A}x\,\theta_\varkappa^\nu\,\boldsymbol{B}y .\Rightarrow. \boldsymbol{A}x\,\theta_\varkappa^\xi\,\boldsymbol{B}y].$$

Lemma 5.3. *If $\theta$ is a C-relation, then*

(i) $\boldsymbol{A}$ $\binom{\alpha}{\varkappa}$*-atomic* $(\theta_\varkappa^\xi) \wedge \boldsymbol{B}\,\theta_\varkappa^\xi\,\boldsymbol{A} .\Rightarrow. \boldsymbol{A} \lhd_\varkappa^\alpha \boldsymbol{B}(\theta_\varkappa^\xi)$,

(ii) $\boldsymbol{A}$ $\binom{\alpha}{\varkappa}$*-atomic* $(\theta_\varkappa^\xi) \Rightarrow \boldsymbol{A}$ $\binom{\alpha}{\varkappa}$*-homogeneous* $(\theta_\varkappa^\xi)$.

Proof. For (i), let $x \in A_{\varkappa\alpha}$, $y \in B_{\varkappa\alpha}$, $a \in A_\varkappa$ and $\boldsymbol{A}x\,\theta_\varkappa^\xi\,\boldsymbol{B}y$. Since $\boldsymbol{A}$ is $\binom{\alpha}{\varkappa}$-atomic $(\theta_\varkappa^\xi)$, there is $\nu < \xi$ such that for any $z \in B_{\varkappa\alpha}$, if $\boldsymbol{A}xa\,\theta_\varkappa^\nu\,\boldsymbol{B}z$, then $\boldsymbol{A}xa\,\theta_\xi^\varkappa\,\boldsymbol{B}z$. Because $\boldsymbol{A}x\,\theta_\varkappa^\xi\,\boldsymbol{B}y$ and $\nu < \xi$, we have $\boldsymbol{A}x\,\theta_\varkappa^{\nu+1}\,\boldsymbol{B}y$ and thus $\boldsymbol{A}xa\,\theta_\varkappa^\nu\,Byb$ for some $b \in B_\varkappa$. Therefore $\boldsymbol{A}xa\,\theta_\varkappa^\xi\,\boldsymbol{B}yb$ as was to be shown. (ii) is a special case of (i).

It is easy to check that a system $\boldsymbol{A}$ is atomic in the usual sense (i.e., as in VAUGHT [15]) just in case $\boldsymbol{A}$ is $\binom{\omega}{2}$-atomic ($\approx_2^\omega$). Combining lemma 5.3 and corollaries 1 and 2, we can specialize to the following results of VAUGHT [15]:

If $\boldsymbol{A}$ is $\binom{\omega}{2}$-atomic ($\approx_2^\omega$), $\boldsymbol{A} \approx_2^\omega \boldsymbol{B}$ and $|A| \leqq |B|$, then $\boldsymbol{A}$ is elementarily embeddable in $\boldsymbol{B}$ (i.e., an atomic, denumerable system is *prime*);

If $\boldsymbol{A}$ and $\boldsymbol{B}$ are $\binom{\omega}{2}$-atomic ($\approx_2^\omega$), denumerable and $\boldsymbol{A} \approx_2^\omega \boldsymbol{B}$, then $\boldsymbol{A} \cong \boldsymbol{B}$.

Note that the above notions are again applicable in the case of the EHRENFEUCHT relation.

One feature of the development of model theory via the hierarchies $\sim_\varkappa^\xi$ and $\approx_\varkappa^\xi$ is the fact that all results are naturally stated and proved independent of formulas. That is, model theory can be developed without using semantic methods! For example, the FRAÏSSE relation $\approx_2^\omega$ means elementary equivalence, but is defined without reference to formulas. Investigation of its first derivative becomes a study of the STONE-space of elementary types. It therefore seems natural to step up this derivation process in search of more refined methods. How this CANTOR-BENDIXSON process is related to the one introduced by MORLEY [10] remains to be investigated. We also would like to see a neat proof of the compactness theorem for $\approx_2^\omega$:

Theorem. *Given the sequence* $\boldsymbol{A}_0 \sim_2^1 \boldsymbol{A}_1 \sim_2^2 \boldsymbol{A}_2 \sim_2^3 \boldsymbol{A}_3 \sim_2^4 \ldots$, *there exists a countable system* $\boldsymbol{A}$ *for which* $\boldsymbol{A} \sim_2^n \boldsymbol{A}_n$, *for all* $n < \omega$.

At higher ranks $\lambda > \omega$, there will be chains, $\nu < \mu < \lambda$ and $\boldsymbol{A}_\nu \sim_2^\nu \boldsymbol{A}_\mu$ which end at $\lambda$. Using the methods of MORLEY [10], one can show:

Theorem. *Let* $\lambda$ *be a countable ordinal and let* $S_\lambda$ *be the space of all* $\lambda$*-chains of countable systems* $\boldsymbol{A}_\nu$, $\nu < \lambda$. *Let* $T_\lambda$ *be the subset of* $S_\lambda$ *consisting of those chains which admit an extension* $\boldsymbol{A}$ *at rank* $\lambda$. *Then* $T_\lambda$ *is a Suslin set in* $S_\lambda$.

We would like to see such results proved without the use of languages.

## Bibliography


[1] CALAIS, J. P., Relation et multirelation pseudo-homogènes. C.R. Acad. Sci. Paris **265** (1967), A2–A4.

[2] CANTOR, G., Beiträge zur Begründung der transfiniten Mengenlehre. Math. Ann. **46** (1895), 481–512.

[3] CHANG, C., Some remarks on the model theory of infinitary languages. In: Syntax and Semantics on Infinitary Languages, Berlin–Heidelberg–New York, 1968, p. 36–63.

[4] EHRENFEUCHT, A., Application of games to the completeness problem. Fund. Math. **49** (1961), 129–141.

[5] ERDÖS, P., GILLMAN, L., and HENRIKSEN, M., An isomorphism theorem for real closed fields. Ann. of Math. **61** (1955), 542–554.

[6] FEFERMAN, S., Some recent work of Ehrenfeucht and Fraïsse. Cornell Summer Logic Institute, 1957.

[7] FRAÏSSE, R., Sur les classifications des systems de relations. Pub. Sc. de l'Université d'Alger **1** (1954).
[8] HAUSDORFF, F., Grundzüge der Mengenlehre. Leipzig, 1914.
[9] KEISLER, J. H., Ultraproducts and elementary classes. Indag. Math. **23** (1961), 477–495.
[10] MORLEY, M., The number of countable models. J. Symb. Log. **35** (1970), 14–18.
[11] MORLEY, M., and VAUGHT, R., Homogeneous universal models. Math. Scand. **11** (1962), 37–57.
[12] RYLL-NARDZEWSKI, C., On theories categorical in power $\leqq \aleph_0$. Bull. Acad. Pol. Sci. Sér. Math. Astr. Phys. **7** (1959), 545–548.
[13] SCOTT, D., Logic with denumerably long formulas and finite strings of quantifiers. In: Theory of Models, Amsterdam, 1965, p. 329–341.
[14] TARSKI, A., and VAUGHT, R., Arithmetical extensions of relational systems. Comp. Math. **13** (1957), 81–102.
[15] VAUGHT, R., Denumerable models of complete theories. In: Infinitistic Methods, Warsaw 1959, p. 303–321.

(Eingegangen am 28. Mai 1972)

# Relatively Categorical and Normal Theories

J. Richard Büchi, Purdue University

We present here two remarks, preliminary to a more serious study of invariants of elementary concepts in elementary theories. The ideas of Klein and Galois suggest the basic questions in this field. To render them more precise we introduce two very natural model-theoretic notions: "the theory $\mathbf{A}(\mathcal{R}, \mathcal{S})$ *is categorical relative* to the primitive concept $\mathcal{R}$" and "$\mathbf{A}(\mathcal{R}, \mathcal{S})$ is *normal relative* $\mathcal{R}$". Once this is accomplished, it turns out that a fundamental result in Klein's program applied to elementary theories (theorem 1'), is a simple corollary to Beth's [1] definability theorem (or its extension by Craig [4]). Similarly, the fact that Galois groups of elementary theories are finite (theorem 2) follows at once from a definability theorem of Svenonius [6].

That the notion of relative categoricity might be the key to a more precise formulation of the Erlangerprogramm [5] was suggested in Büchi and Wright [2, 3]. These papers and also Wright [7] provide further examples which serve to illustrate the present general discussion. Of course, Klein's ideas are by no means limited to the hierarchy of geometries. In fact they suggest a new branch of general model theory, which is concerned with the classification of theories via automorphism groups on models. It has the following two features:
1) Relations between theories of different grammar (models of different species) come into play.
2) Formulas (concepts) are in the foreground rather than sentences (relational systems); consequently definability theorems are needed, rather than results about the model-relation between sentences and relational systems.
It would be nice, if both of these underdeveloped areas of model theory were to be stimulated by the ideas of Galois and Klein.

## 1. A basic result in Klein's Erlangerprogramm

Theorem 1 is a general model-theoretic result, of which the following is an application.

(a) Every concept $\mathbf{c}(\|, \perp)$ of Euclidean geometry $\mathbf{E}(\|, \perp)$, which (in all models of $\mathbf{E}$) is invariant under all affine transformations (i.e. automorphism of $\|$) is in fact definable from $\|$ (i.e. there is a concept $\mathbf{d}(\|)$ such that $\mathbf{E} \models \mathbf{c} = \mathbf{d}$).

In particular this implies:

(b) To every proposition $\mathbf{P}(\|, \perp)$ of Euclidean geometry $\mathbf{E}$ one can find a proposition $\mathbf{D}(\|)$ of affine geometry, such that $\mathbf{E} \models \mathbf{P} = \mathbf{D}$.

Roughly speaking: (b) Affine geometry $\mathbf{A}(\|)$ (equivalent $(\exists \mathcal{X})\ \mathbf{E}(\|\mathcal{X})$) is deductively just as strong as Euclidean geometry $\mathbf{E}(\|, \perp)$, and (a) $\mathbf{A}(\|)$ is as strong in expressive power as one may reasonably expect it to be: every Euclidean concept $\mathbf{c}(\|, \perp)$ that is an affine invariant is represented by an affine concept $\mathbf{d}(\|)$. Of course, corresponding results hold for other pairs of geometries in Klein's hierarchy, and it seems strange that this very fundamental observation concerning the Erlanger-programm has not received more attention. In particular the following question is in order.

(Q) Which property that the theory $\mathbf{E}(\|, \perp)$ possesses relative to the concept $\|$, is used to establish (a)?

The answer is given by theorem 1. We now present the basic definitions, beginning with some indications concerning notational matters.

$\mathcal{R}$; a *relational structure*, i.e., a sequence of relations on the *universe* $D(\mathcal{R})$ = first member of $\mathcal{R}$. Requirements on the species (similarity type) of $\mathcal{R}$ and on $D(\mathcal{R})$, if any, should be inferred from the context.

**a**: a function with relational structures as arguments and values. The domain is always a species, the values belong to a (possibly different) species. Furthermore, $\mathcal{R}$ and $\mathbf{a}\mathcal{R}$ have the same universe.

A: a set of relational structures, all of which are of equal species. **A** will be construed as being a special sort of **a**, namely a two-valued function. $\mathbf{A}\mathcal{R}$, $\mathcal{R} \in \mathbf{A}$, $\mathbf{A}\mathcal{R} = D(\mathcal{R})$ have the same meaning and so do $\sim\mathbf{A}\mathcal{R}$, $\mathcal{R} \notin \mathbf{A}$, $\mathbf{A}\mathcal{R} =$ empty set.

$\varphi$: a transformation, i.e., a one-to-one mapping of some universe $D_1$ onto $D_2$. Requirements on the universes of $\varphi$ are not explicitly mentioned and should be clear from the context.

$\varphi\mathcal{R}$: the isomorphic image of the relational structure $\mathcal{R}$ under the transformation $\varphi$.

$\mathcal{R} \simeq \mathcal{S}$: $(\exists\varphi)\ (\varphi\mathcal{R} = \mathcal{S})$, i.e. $\mathcal{R}$ and $\mathcal{S}$ are *isomorphic*, and therefore have identical species.

$\mathcal{K}\mathcal{R}$: $\{\varphi: \varphi\ \mathcal{R} = \mathcal{R}\}$, i.e., the *automorphism group* of $\mathcal{R}$.

A *(relational) concept* is a function **a** which possesses the following invariance property

(1) $\varphi\, \mathbf{a}\, \mathcal{R} = \mathbf{a}\, \varphi\, \mathcal{R},$

for all relational structures $\mathcal{R}$ of the argument species of **a** and for all transformations $\varphi$ of the universe D($\mathcal{R}$). We remark that (1) is equivalent with

(1') $\varphi\, \mathbf{a}\, \mathcal{R} \subseteq \mathbf{a}\, \varphi\, \mathcal{R},$

and also inclusion in the other direction has the same effect. Furthermore, (1) implies as a special case,

(2) $\mathcal{K}\mathcal{R} \subseteq \mathcal{K}\mathbf{a}\, \mathcal{R}.$

I.e., if **a** is a concept, then every automorphism of $\mathcal{R}$ is also an automorphism of **a**$\mathcal{R}$. Finally we note that a concept **a** maps isomorphic structures into isomorphic structures, i.e.,

(3) $\mathcal{R}_1 \simeq \mathcal{R}_2 \supset \mathbf{a}\mathcal{R}_1 \simeq \mathbf{a}\mathcal{R}_2.$

Note that (1) is stronger than (2), it requires that the particular isomorphism between $\mathcal{R}_1$ and $\mathcal{R}_2$ also works between **a**$\mathcal{R}_1$ and **a**$\mathcal{R}_2$. Thus, a concept is more than a mapping of isomorphism types of relational structures.

A two valued concept **A** is also called a *proposition*, and sometimes a *theory*. The defining condition (1), in this case, is equivalent to (3). **A** is a proposition if and only if it has the invariance property,

(3') $\mathcal{R}_1 \simeq \mathcal{R}_2 \wedge \mathbf{A}\mathcal{R}_1 .\supset. \mathbf{A}\mathcal{R}_2.$

Thus, a theory is simply a class **A** of relational structures, closed under isomorphisms. If **A**$\mathcal{R}$ we will also say that $\mathcal{R}$ *is a model of the theory* **A**, or that the *proposition* **A** *holds in* $\mathcal{R}$.

If we speak of a *concept* **c** *of the theory* **A**, we mean to indicate that we are actually interested only in the restriction of **c** to arguments $\mathcal{R} \in \mathbf{A}$. In particular $\mathbf{c}_1 = \mathbf{c}_2(\mathbf{A})$ ($\mathbf{c}_1$ and $\mathbf{c}_2$ are *equal concepts of* **A**) means $\mathbf{A}\mathcal{R} \supset.\ \mathbf{c}_1\mathcal{R} = \mathbf{c}_2\mathcal{R}$. Of course this is just a formality; one might prefer to define a concept of **A** to be a function **c** whose domain is **A**, satisfying (1) for all $\mathcal{R} \in \mathbf{A}$.

Note that such a **c** can always be extended to the whole species, preserving property (1). The *primitive concept of a theory* **A** is the identity function $\mathbf{e}\mathcal{Q} = \mathcal{Q}$ on the argument species of **A**.

For propositions we have the important relation of "logical consequence":

$\mathbf{P} \models \mathbf{Q}$ or $\mathbf{P}\mathcal{R} \models \mathbf{Q}\mathcal{R}$ simply means $\mathbf{P} \subseteq \mathbf{Q}$. We are more interested here in some basic relations between concepts. One of these is,

$$\mathbf{c} \leq \mathbf{b}\ (\mathbf{A}):\ (\forall \mathcal{R}_1\ \mathcal{R}_2).\ \mathbf{A}\mathcal{R}_1 \wedge \mathbf{A}\mathcal{R}_2 \wedge \mathbf{b}\mathcal{R}_1 = \mathbf{b}\mathcal{R}_2 .\!\supset \mathbf{c}\mathcal{R}_1 = \mathbf{c}\mathcal{R}_2 .$$

In words, *in the theory* **A** *the concept* **c** *is definable from* **b**. It is quite easy to prove that definability is equivalent with "uniform definability" in the following sense,

$$\mathbf{c} \leq \mathbf{b}\ (\mathbf{A})\ .\!\equiv\!.\ (\exists \text{ concept } \mathbf{h})(\forall \mathcal{R}).\mathbf{A}\mathcal{R} \supset \mathbf{c}\mathcal{R} = \mathbf{h}\mathbf{b}\mathcal{R} .$$

Compare this with the discussion of Padoa's method in Beth [1]. Another principle of classifying the concepts of a theory is via automorphism groups, and was proposed by Klein.

$$\mathbf{c} \precsim \mathbf{b}\ (\mathbf{A}):\ (\forall \mathcal{R}).\mathbf{A}\mathcal{R} \supset \mathcal{K}\mathbf{b}\mathcal{R} \subseteq \mathcal{K}\mathbf{c}\mathcal{R},$$

i.e., in every model $\mathcal{R}$ of the theory **A**, every automorphism $\varphi$ of $\mathbf{b}\mathcal{R}$ is also an automorphism of $\mathbf{c}\mathcal{R}$. We will express this also thus, *in the theory* **A** *the concept* **c** *is* ***b****-invariant.* Note that trivially

$$\mathbf{c} \leq \mathbf{b}\ (\mathbf{A}) \supset \mathbf{c} \precsim \mathbf{b}\ (\mathbf{A})$$

and the question arises whether reversely every **b**-invariant concept is definable from **b**, i.e., whether the classification of concepts according to definability is identical to Klein's classification by automorphism groups. It is not clear that Klein considered this problem in some generality. However, the particular hierarchy of geometries and geometric concepts discussed in the Erlangerprogramm is in fact both a definability and an invariance hierarchy. Theorem 1 answers the general problem.

**Definition 1:**
A theory **A** is *categorical relative* to the concept **r** if any two models $Q_1$ and $Q_2$ of **A** in which **r** takes the same value are isomorphic, i.e.,

$$(\forall Q_1 Q_2)\ .\ \mathbf{A}Q_1 \wedge \mathbf{A}Q_2 \wedge \mathbf{r}Q_1 = \mathbf{r}Q_2 .\!\supset Q_1 \simeq Q_2$$

In this case the theory **B** defined by $\mathbf{B}\mathcal{R} \equiv (\exists Q).\ \mathbf{A}Q \wedge \mathcal{R} = \mathbf{r}Q$ is called the *abstraction* of **A** relative **r**.

Note that $\mathbf{r}Q_1 = \mathbf{r}Q_2$ implies in particular that $Q_1$ and $Q_2$ are on the same universe. An isomorphism between $Q_1$ and $Q_2$ is therefore an automorphism of $\mathcal{R} = \mathbf{r}Q_1 = \mathbf{r}Q_2$; one might call $Q_1$ and $Q_2$ *conjugates modulo* $\mathcal{R}$. In this terminology, **A** is categorical relative **r** means: for a given $\mathcal{R}$ there is at most one conjugate class of models $\mathcal{Q}$ of **A**, $\mathbf{r}\mathcal{Q} = \mathcal{R}$.

**Theorem 1:**
Let **A** be a theory which is categorical relative to the concept **r** in **A**. Every concept **c** that in **A** is **r**-invariant is definable from **r** in **A**. I.e., if in all models of **A** the concept **c** is invariant under all automorphisms of **r**, then there is a concept **c**' such that in all models of **A**, **c** = **c'r**. I.e., if **B** is the abstraction of **A** relative **r**, then every **r**-invariant concept **c** of **A** can be recovered in **B**, in the sense that there is a unique concept **c**' in **B** such that for any proposition **P**,

$\mathbf{A} \models \mathbf{Pc}$ if and only if $\mathbf{B} \models \mathbf{Pc}'$.

**Proof**:
Suppose $\mathbf{A}\mathcal{Q}_1$, $\mathbf{A}\mathcal{Q}_2$, and $\mathbf{r}\mathcal{Q}_1 = \mathbf{r}\mathcal{Q}_2$. Because **A** is categorical relative **r** it follows that there is a transformation $\varphi$ such that $\varphi\mathcal{Q}_1 = \mathbf{r}\mathcal{Q}_2$, and therefore $\mathbf{r}\varphi\mathcal{Q}_1 = \mathbf{r}\mathcal{Q}_2$ and $\mathbf{c}\varphi\mathcal{Q}_1 = \mathbf{c}\mathcal{Q}_2$. Because **r** and **c** are concepts we use (1) to infer $\varphi\mathbf{r}\mathcal{Q}_1 = \mathbf{r}\mathcal{Q}_2$ and $\varphi\mathbf{c}\mathcal{Q}_1 = \mathbf{c}\mathcal{Q}_2$. Because $\mathbf{r}\mathcal{Q}_1 = \mathbf{r}\mathcal{Q}_2$ we obtain $\varphi\mathbf{r}\mathcal{Q}_1 = \mathbf{r}\mathcal{Q}_1$. Because **c** is **r**-invariant and $\mathbf{A}\mathcal{Q}_1$, $\varphi\mathbf{r}\mathcal{Q}_1 = \mathbf{r}\mathcal{Q}_1$, it follows that $\varphi\mathbf{c}\mathcal{Q}_1 = \mathbf{c}\mathcal{Q}_1$. Thus, $\mathbf{c}\mathcal{Q}_1 = \varphi\mathbf{c}\mathcal{Q}_1 = \mathbf{c}\mathcal{Q}_2$, so that by using the assumptions of theorem 1 we have shown,

$$\mathbf{A}\mathcal{Q}_1 \wedge \mathbf{A}\mathcal{Q}_2 \wedge \mathbf{r}\mathcal{Q}_1 = \mathbf{r}\mathcal{Q}_2 \,.\supset \mathbf{c}\mathcal{Q}_1 = \mathbf{c}\mathcal{Q}_2,$$

i.e., in **A** the concept **c** is definable from **r**. That the existence of a concept **c**' that uniformly defines **c** from **r**, follows, we have claimed before. That in the abstraction **B** this **c**' represents **c**, in the sense described in theorem 1, also is easy to see.

Q.E.D.

Let us now return to the statement (a) at the beginning of this section. To fix the ideas, let us assume that $\mathbf{E}(\|, \perp)$ is given by an axiom system on the primitive concepts $\|$ (interpreted as: the points x, y and u, v determine parallel lines) and $\perp$ (the angle x, y, z is right). Of course it is not essential for our argument that $\|$ be chosen as primitive concept. That $\mathbf{E}(\|, \perp)$ is categorical relative $\|$ means: If $[D, \|, \perp_1]$ and $[D, \|, \perp_2]$ are two models of **E**, then there is an affine transformation $\varphi$ (i.e. an automorphism of $[D, \|]$ which takes $\perp_1$ onto $\perp_2$. This clearly holds for any axiom system $\mathbf{E}(\|, \perp)$ that would reasonably be called "Euclidean geometry". The statement (a) then is a consequence of theorem 1. To establish (b), a much weaker assumption is sufficient, namely, that **E** be complete relative $\|$ in the following sense.

**Definition 2:**
The theory **A** is *complete relative* to the concept **r**, if for any proposition **P**,

$$\mathbf{A}\mathcal{Q}_1 \wedge \mathbf{A}\mathcal{Q}_2 \wedge \mathbf{r}\mathcal{Q}_1 = \mathbf{r}\mathcal{Q}_2 \,.\supset .\, \mathbf{P}\mathcal{Q}_1 \equiv \mathbf{P}\mathcal{Q}_2.$$

It is easy to see that this is equivalent to: to every proposition **P** there is a proposition **P'**, such that $\mathbf{A}\mathcal{Q} \supset . \mathbf{P}\mathcal{Q} \equiv \mathbf{P}'\mathbf{r}\mathcal{Q}$. This is the general form of (b).

## 2. Elementary concepts

If $\Gamma(\mathcal{R}, x_1,...,x_n)$ is a formula of elementary predicate calculus, define $\mathbf{c}\mathcal{R} = \hat{x}_1...\hat{x}_n\Gamma$. By induction on the structure of $\Gamma$ one shows that **c** is a concept. We call it the *elementary concept* defined by $\hat{x}_1...\hat{x}_n\Gamma$. **A** is an *elementary theory* in case there are elementary sentences $\Sigma_i$ such that $\mathbf{A}\mathcal{Q} . \equiv . (\forall i)\Sigma_i\mathcal{Q}$. **A** is a *pseudo-elementary theory* in case there is an elementary **B** such that $\mathbf{A}\mathcal{Q} \equiv (\exists \mathcal{X})\mathbf{B}(\mathcal{Q},\mathcal{X})$. Furthermore, a concept **a** is *elementarily definable from* **b** in the theory **A**, if there is an elementary concept **c** such that $\mathbf{A}\mathcal{Q} \supset \mathbf{a}\mathcal{Q} = \mathbf{c}\mathbf{b}\mathcal{Q}$. In this terminology, the extension by Craig [4] of Beth's [1] theorem takes the following form.

**Lemma 1:**
Let **A** be any pseudo-elementary theory, let **a** and **b** be elementary concepts of **A**. If **a** is definable in **A** from **b**, then **a** is elementarily definable in **A** from **b.**

**Theorem 1':**
Let **A** be a pseudo-elementary theory which is categorical relative to the elementary concept **r**. Let **c** be an elementary concept on **A** which in **A** is **r**-invariant. Then **c** is elementarily definable from **r** in **A**.

Clearly this follows by theorem 1 and lemma 1. Of course, theorem 1' is much more interesting than theorem 1. It should be viewed as an application of Beth's definability theorem to the theory of invariance of elementary concepts. The significance of theorem 1 is not in its mathematical depth; rather, it requires a clear definition of "concept" and suggests a more precise formulation of Klein's ideas.

*Erlangerprogramm for a pseudo-elementary theory of* **A**: Study the hierarchy of pseudo-elementary abstractions of **A**, i.e., investigate the partial order $\mathbf{r}_1 \precsim \mathbf{r}_2$ (**A**) on the set of all elementary concepts relative to which **A** is categorical.

The significance of theorem 1' is to show that, in this context, the strongly model-theoretic relation $\precsim$ has the same meaning as the more syntactic relation "$\mathbf{r}_1$ is elementarily definable from $\mathbf{r}_2$". Of course, this is only a first step towards a general theory of elementary abstraction.

In Büchi and Wright [2, 3] some elementary abstractions of the group-axioms are discussed; we mention that also Post's multi-groups fit into this particular hierarchy. It seems that a rather complete survey over all abstractions of group theory relative to equational concepts might be given. Finally we point out that, as a consequence of theorem 1', assertion (a) holds even if $\mathbf{E}(\parallel, \perp)$ stands for an elementary approximation of Euclidean geometry. In fact "definable" may now be replaced by "elementarily definable", if only elementary concepts are considered.

## 3. Galois groups of normal extensions of a theory

The relational structures $\mathcal{Q}_1$ and $\mathcal{Q}_2$ are called *conjugate relative to the concept* **r** if $\mathbf{r}\mathcal{Q}_1 = \mathbf{r}\mathcal{Q}_2$ and $\mathcal{Q}_1 \simeq \mathcal{Q}_2$, i.e., if there is an automorphism $\varphi$ of $\mathbf{r}\mathcal{Q}_1 = \mathbf{r}\mathcal{Q}_2$ which takes $\mathcal{Q}_1$ onto $\mathcal{Q}_2$. We will use the notation

$$\mathcal{Q}_1 \simeq \mathcal{Q}_2\,(\mathbf{r}) \;:\; \mathcal{Q}_1 \text{ is conjugate of } \mathcal{Q}_2 \text{ relative } \mathbf{r}\,.$$

There is a second equivalence relation on relational structures over the same universe:

$$\mathcal{Q}_1 \approx \mathcal{Q}_2 : \; \mathcal{K}\mathcal{Q}_1 = \mathcal{K}\mathcal{Q}_2, \text{ i.e., } \mathcal{Q}_1 \text{ and } \mathcal{Q}_2 \text{ have the same automorphisms.}$$

**Definition:**
The theory **A** is *normal relative to the concept* **r** in case any **r**-conjugate models of **A** have the same automorphism group, i.e.,

$$\mathbf{A}\mathcal{Q}_1 \wedge \mathcal{Q}_1 \simeq \mathcal{Q}_2\,(\mathbf{r}) \;.\!\supset\; \mathcal{Q}_1 \approx \mathcal{Q}_2.$$

Note that this condition may also be stated as follows:

$$\mathbf{A}\mathcal{Q} \wedge \varphi\mathbf{r}\mathcal{Q} = \mathbf{r}\mathcal{Q} \wedge \psi\mathcal{Q} = \mathcal{Q} \;.\!\supset\; \psi\varphi\mathcal{Q} = \varphi\mathcal{Q}$$

or equivalently,

$$\mathbf{A}\mathcal{Q} \wedge \varphi \in \mathcal{K}\mathbf{r}\mathcal{Q} \wedge \psi \in \mathcal{K}\mathcal{Q} \;.\!\supset\; \varphi^{-1}\psi\varphi\mathcal{K}\mathcal{Q};$$

therefore,

**Lemma 2:**
The theory **A** is normal relative to the concept **r** just in case, in every model $\mathcal{Q}$ of **A** the automorphism group $\mathcal{K}\mathcal{Q}$ is a normal subgroup of $\mathcal{K}\mathbf{r}\mathcal{Q}$.

If **A** is normal relative **r**, and $\mathcal{Q}$ is a model of **A**, we define

$$\mathcal{G}^{\mathbf{r}}{}_{\mathbf{A}}\mathcal{Q} : \; \mathcal{K}\mathbf{r}\mathcal{Q}/\,\mathcal{K}\mathcal{Q}, \textit{ Galois group on } \mathcal{Q}, \text{ of } \mathbf{r} \text{ relative } \mathbf{A}\,.$$

We will now show that for pseudo-elementary theories **A** and elementary **r** the Galois group is finite. This follows quite simply from a result of Svenonius [6] that in our terminology can be stated as follows.

**Lemma 3:**
Let $\mathbf{T}(\mathcal{R}, \mathcal{S})$ be a pseudo-elementary theory, such that in every model $(\mathcal{R}, \mathcal{S})$ of $\mathbf{T}$ every $\mathcal{R}$-automorphism is also an $\mathcal{S}$-automorphism. Then there is a number n, and there are elementary concepts $\mathbf{d}_1,\ldots,\mathbf{d}_n$ such that

$$\mathbf{T}(\mathcal{R}, \mathcal{S}) \supset .\ \mathcal{S} = \mathbf{d}_1\mathcal{R} \vee \ldots \vee \mathcal{S} = \mathbf{d}_n\mathcal{R}.$$

**Theorem 2:**
Let $\mathbf{A}$ be a pseudo-elementary theory which is normal relative to the elementary concept $\mathbf{r}$. Then there is a number n, and there are elementary concepts $\mathbf{d}_1,\ldots,\mathbf{d}_n$ such that

$$\mathbf{A}\mathcal{Q} \wedge \mathcal{Q}' \simeq \mathcal{Q}(\mathbf{r}) . \supset .\ \mathcal{Q}' = \mathbf{d}_1\mathcal{Q} \vee \ldots \vee \mathcal{Q}' = \mathbf{d}_n\mathcal{Q}$$

I.e., to every model $\mathcal{Q}$ of $\mathbf{A}$ there exist at most n models $\mathcal{Q}'$ that are $\mathbf{r}$-conjugates of $\mathcal{Q}$. Furthermore, each one of these $\mathcal{Q}'$ is elementarily definable from $\mathcal{Q}$. In particular it follows that the Galois group $\mathcal{G}^{\mathbf{r}}{}_{\mathbf{A}}\mathcal{Q}$ is finite, and in fact has $\leq$n members, for every model $\mathcal{Q}$ of $\mathbf{A}$.

**Proof:**
Because $\mathbf{A}$ is normal relative $\mathbf{r}$ we have,

$$\text{(j)} \quad \mathbf{A}\mathcal{Q} \wedge \mathcal{Q}' \simeq \mathcal{Q}(\mathbf{r}) . \supset .\ \varphi\mathcal{Q} = \mathcal{Q} \supset \varphi\mathcal{Q}' = \mathcal{Q}'.$$

Note that $\mathcal{Q}' \simeq \mathcal{Q}(\mathbf{r})$ means $\mathbf{r}\mathcal{Q}' = \mathbf{r}\mathcal{Q} \wedge (\exists\varphi)\mathcal{Q}' = \varphi\mathcal{Q}$. Because $\mathbf{r}$ is elementary this is clearly a pseudo-elementary proposition. Because $\mathbf{A}$ is pseudo-elementary it follows that $\mathbf{A}\mathcal{Q} \wedge \mathcal{Q}' \simeq \mathcal{Q}(\mathbf{r})$ is a pseudo-elementary theory $\mathbf{T}(\mathcal{Q}, \mathcal{Q}')$. Because (j) is just the other assumption of lemma 3 on $\mathbf{T}$, we conclude the existence of elementary concepts $\mathbf{d}_1,\ldots,\mathbf{d}_n$ such that $\mathbf{T}(\mathcal{Q}, \mathcal{Q}') \supset .\ \mathcal{Q}' = \mathbf{d}_1\mathcal{Q} \vee \ldots \vee \mathcal{Q}' = \mathbf{d}_n\mathcal{Q}$. This establishes the first part of theorem 1. That the Galois group is finite is seen as follows.
Let $\mathcal{Q}$ be a model of $\mathbf{A}$. As we have just seen, there are at most n different $\mathbf{r}$-conjugates $\mathcal{Q}_1,\ldots,\mathcal{Q}_m$ of $\mathcal{Q}$. Therefore,

$$\text{(jj)} \quad \varphi \in \mathcal{K}\mathbf{r}\mathcal{Q} \supset .\ \varphi\mathcal{Q} = \mathcal{Q}_1 \vee \ldots \vee \varphi\mathcal{Q} = \mathcal{Q}_m.$$

Suppose now that $\alpha\beta \in \mathcal{K}\mathbf{r}\mathcal{Q}$. Note that $\alpha^{-1}\beta \in \mathcal{K}\mathcal{Q}$ means $\alpha\mathcal{Q} = \beta\mathcal{Q}$. Therefore, by (jj),

$$\alpha^{-1}\beta \in \mathcal{K}\mathcal{Q} . \equiv .\ \alpha\mathcal{Q} = \beta\mathcal{Q} = \mathcal{Q}_1 \vee \ldots \vee \alpha\mathcal{Q} = \beta\mathcal{Q} = \mathcal{Q}_m.$$

Now, the formula on the right side clearly determines an equivalence relation $\alpha \sim \beta$ of index $m \leq n$. Therefore, $\alpha^{-1}\beta \in \mathcal{K}\mathcal{Q}$ is an equivalence relation on $\mathcal{K}\mathbf{r}\mathcal{Q}$, of finite index $m \leq n$. But this is just the congruence on $\mathcal{K}\mathbf{r}\mathcal{Q}$, relative the normal subgroup $\mathcal{K}\mathcal{Q}$. It follows that the Galois group $\mathcal{K}\mathbf{r}\mathcal{Q}/\mathcal{K}\mathcal{Q}$ has $m \leq n$ members. Q.E.D.

Let us return to Svenonius' lemma 3. The following is but a trivial generalization of it.

**Theorem 3:**
Let **A** be a pseudo-elementary theory, let **r** and **c** be elementary concepts in **A**. If **c** is **r**-invariant in **A**, then there are elementary concepts $\mathbf{d}_1,\ldots,\mathbf{d}_n$ such that

$$\mathbf{A}\mathcal{Q} \supset . \; \mathbf{c}\mathcal{Q} = \mathbf{d}_1\mathbf{r}\mathcal{Q} \vee \ldots \vee \mathbf{c}\mathcal{Q} = \mathbf{d}_n\mathbf{r}\mathcal{Q}.$$

This shows what remains of theorem 1', in case the assumption "**A** is categorical relative **r**" is dropped. The alternative definitions $\mathbf{d}_i$ of **c** from **r** clearly are related to the many **r**-conjugate classes which **A** now possesses.

Unfortunately, that part of elementary model theory that deals with concepts rather than sentences and for which the results of Beth, Craig, and Svenonius are typical, is rather underdeveloped. The study of automorphism groups of elementary concepts, along the ideas of Galois and Klein, would fill a gap in this part of model theory. We feel that much more could be done, even with the already available tools. As a source of ideas and problems we tentatively suggest the following dictionary.

| | |
|---|---|
| +,•, elements of ground field K | concepts elementarily definable in $\mathbf{B}\mathcal{R}$ |
| polynomial equation p(x) = 0 with coefficients in K | elementary sentence $\mathbf{E}(\mathcal{R},\mathcal{S})$ |
| p(x) = 0 has solution in some extension of K | $\mathbf{B}\mathcal{R} \supset (\exists \mathcal{S})\, \mathbf{E}(\mathcal{R},\mathcal{S})$ |
| p(x) is irreducible in K | $\mathbf{B}\mathcal{R} \wedge \mathbf{E}(\mathcal{R},\mathcal{S})$ is categorical relative $\mathcal{R}$ |
| elements of the algebraic extension K(x) by p | concepts elementarily definable in the theory $\mathbf{B}\mathcal{R} \wedge \mathbf{E}(\mathcal{R},\mathcal{S})$ |
| K(x) is normal extension of K | $\mathbf{B}\mathcal{R} \wedge \mathbf{E}(\mathcal{R},\mathcal{S})$ is normal relative $\mathcal{R}$ |
| Galois group of p over K | Galois group of $\mathcal{R}$ relative $\mathbf{B} \wedge \mathbf{E}$ |

## Bibliography


[1] Beth, E.W.: On Padoa's method in the theory of definition. Indag. math. 15 (1953), pp. 330-339

[2] Büchi and Wright: The theory of proportionality as an abstraction of group theory. Math. Ann. 130 (1955), pp. 102-108

[3] Büchi and Wright: Invariants of the anti-automorphisms of a group. Proc. Am. Math. Soc. 8 (1957), pp. 1134-1140

[4] Craig, William: Three uses of the Herbrand-Gentzen theorem in relating model theory and proof theory. Journ. Symb. Log. 22 (1957), pp. 269-285

[5] Klein, Felix: Vergleichende Betrachtungen über neuere geometrische Forschungen. Verlag A. Deicher, Erlangen 1872

[6] Svenonius, Lars: A theorem on permutations in models. Theoria, Swed. Journ. Phil. and Psych., 25 (1959), pp. 173-178

[7] Wright, J.B.: Quasi-projective geometry of two dimensions. Mich. Math. Journ. 2 (1953-4), pp. 115-122

*Symposium on the Theory of Models; North-Holland Publ. Co., Amsterdam (1965)*

# RELATIVELY CATEGORICAL AND NORMAL THEORIES

J. RICHARD BUCHI

*Purdue University, Lafayette, Indiana, U.S.A.*

Let $\Sigma(Q, x_1, \ldots, x_n)$ be an elementary formula on the predicate letters $Q$ and free variables $x_1, \ldots, x_n$. The function $\mathfrak{c}Q$, which to the relational system $Q$ assigns the relation $\hat{x}_1 \ldots \hat{x}_n\Sigma$ as value, is called an *elementary concept*. As a consequence of our general model-theoretic Theorem 1 we obtain,

Theorem. *Let* $\mathfrak{c}(\|, \perp)$ *be an elementary concept of elementary Euclidean geometry* $\mathfrak{E}(\|, \perp)$. *If, in every model of* $\mathfrak{E}$, $\mathfrak{c}$ *is an affine invariant (i.e., is invariant under automorphisms of* $\|$*), then* $\mathfrak{c}$ *is elementarily definable in affine geometry (i.e., there is an elementary concept* $\mathfrak{c}'(\|)$ *such that* $\mathfrak{c}(\|, \perp) = \mathfrak{c}'(\|)$ *holds in* $\mathfrak{E}$*). Similar statements hold for any pair of geometries in Klein's hierarchy* [1872].

It should be noted that the accent in this theorem lies on "elementary". If this word is dropped and we deal with a categorical situation, as Klein does in the Erlangerprogramm, the theorem becomes rather trivial. In fact, it is our main concern here to indicate how Klein's ideas can be made compatible with the axiomatic method, and may eventually lead to a general theory of invariants of elementary concepts in elementary model theory. Theorems 1 and 2 should be viewed as basic contributions to this new branch of model theory.

Definition 1. *A theory* $\mathfrak{A}(R, S)$ *is* categorical relative *to the primitive concepts* $R$, *if any two models* $\langle R, S_1\rangle$ and $\langle R, S_2\rangle$ of $\mathfrak{A}$ *(on the same domain and with identical* $R$*) are isomorphic.*

That this notion may be the key to a more precise formulation of Klein's program was suggested by Büchi and Wright [55], [57]. These papers and Wright [53] provide further illustrations of Theorem 1.

Theorem 1. *Let* $\mathfrak{A}(R, S)$ *be an elementary theory which is categorical relative* $R$. *Let* $\mathfrak{c}(R, S)$ *be an elementary concept. If, in every model of* $\mathfrak{A}$, $\mathfrak{c}$ *is invariant under all automorphisms of* $R$, *then* $\mathfrak{c}$ *is elementarily definable in* $\mathfrak{A}$ *from* $R$ *alone (i.e., there is an elementary concept* $\mathfrak{c}'(R)$ *such that* $\mathfrak{c}(R, S) = \mathfrak{c}'(R)$ *holds in* $\mathfrak{A}$).

If $\mathfrak{A}(R, S)$ is categorical relative $R$, we shall call the theory $\mathfrak{B}(R) = (\exists S)\mathfrak{A}(R, S)$ an *abstraction* of $\mathfrak{A}$. In this terminology Theorem 1 says: If $\mathfrak{B}(R)$ is an abstraction of the elementary theory $\mathfrak{A}(R, S)$, then every elementary concept $\mathfrak{c}(R, S)$ which is $R$-invariant in $\mathfrak{A}$ is represented by an elementary concept $\mathfrak{c}'(R)$ in $\mathfrak{B}$, in the sense that for every elementary proposition $\mathfrak{P}(C)$, $\mathfrak{A}(R, S) \models \mathfrak{P}(\mathfrak{c}(R, S))$ if and only if $\mathfrak{B}(R) \models \mathfrak{P}(\mathfrak{c}'(R))$.

Theorem 1 is a simple consequence of the definability theorem of Beth [53], or Craig's lemma [57]. An important task is to obtain general results on the hierarchy of abstractions of an elementary theory $\mathfrak{A}(Q)$, i.e., to give a survey of elementary concepts $\mathfrak{r}(Q)$ relative to which $\mathfrak{A}$ is categorical.

$S_1$ and $S_2$ may be called *conjugate relative* $R$, in symbols $S_1 \cong S_2(R)$, if $<R, S_1>$ is isomorphic to $<R, S_2>$, i.e., if there is an automorphism of $R$ which takes $S_1$ onto $S_2$. That $\mathfrak{A}(R, S)$ is categorical relative $R$ is to say that, in every model of $\mathfrak{A}$, there is relative to $R$ but one conjugate class of $S$'s; in Galois' terminology one would say that $\mathfrak{A}(R, S)$ is irreducible over $(\exists S)\,\mathfrak{A}(R, S)$. There is a second equivalence relation $S_1 \backsim S_2(R)$, which is even more fundamental for Galois theory, and which is defined as "$<R, S_1>$ and $<R, S_2>$ have the same automorphisms".

Definition 2. *The theory* $\mathfrak{A}(R, S)$ *is normal relative to the primitive concepts* $R$ *in case any* $R$*-conjugate models of* $\mathfrak{A}$ *have the same automorphism group, i.e.,* $\mathfrak{A}(R, S_1) \wedge S_1 \cong S_2(R) \supset S_1 \backsim S_2(R)$.

If $\mathscr{K}(Q)$ denotes the automorphism group (*Klein-group*) of the relational system $Q$ one easily proves;

Lemma. *The theory* $\mathfrak{A}(R, S)$ *is normal relative* $R$ *just in case in every model of* $\mathfrak{A}$, $\mathscr{K}(R, S)$ *is a normal subgroup of* $\mathscr{K}(R)$.

The quotient $\mathscr{K}(R)/\mathscr{K}(R, S) = \mathscr{G}(R)$ (which exists in the normal case, and is independent of $S$) is properly called the *Galois group* of $\mathfrak{A}$ relative $R$, in the model $<R, S>$ of $\mathfrak{A}$.

Theorem 2. *Let* $\mathfrak{A}(R, S)$ *be an elementary theory which is normal relative* $R$. *There is a number* $n$, *and there are elementary concepts* $\mathfrak{d}_1(R, S), \ldots, \mathfrak{d}_n(R, S)$ *such that in every model* $<R, S>$ *of* $\mathfrak{A}$ *every conjugate* $S'$ *of* $S$ *relative* $R$ *is one of the relations* $\mathfrak{d}_1(R, S), \ldots, \mathfrak{d}_n(R, S)$, *i.e.,* $\mathfrak{A}(R, S) \wedge S' \cong S(R) . \supset . S' = \mathfrak{d}_1(R, S) \vee \ldots \vee S' = \mathfrak{d}_n(R, S)$.

Corollary. *Let* $\mathfrak{A}(R, S)$ *be an elementary theory which is normal relative* $R$. *Then in every model of* $\mathfrak{A}$ *the Galois group* $\mathcal{G}(R)$ *is finite, and in fact has* $\leqslant n$ *members, and hence* $n$ *depends on* $\mathfrak{A}$ *only.*

Theorem 2 is a consequence of the definability theorem of Svenonius [59a]. The definability results of Beth and Svenonius thus are basic for a theory of invariance of elementary concepts. Do these tools suffice to translate a substantial part of classical Galois theory? If not, a careful investigation might lead to additional basic results on elementary definability.

# Section 4 Automata and Grammars

*With comments by Peter Deussen, Universität Karlsruhe*



*

# J. Richard Büchi: Automata and Grammars

*Peter Deussen**

JRB's mental position with regard to computer science is best characterized by a citation from [10] in the bibliography of Büchi:

> This theory is, however, a branch (and to be sure a young and modest one) of a highly developed mathematical discipline which as a whole could be that for the computer specialist what for the physicist is the theory of differential equations and what for the statistician is measure theory.

And in fact, he employed all his logical and algebraic knowledge and skill to form a streamlined and mathematically satisfactory presentation of what others then called automata theory.

The commentator, a then freshly baked research assistant, had the honor, the pleasure, and the burden to listen to JRB's two semester course on automata theory, which he gave during his sabbatical stay at the Institute for Applied Mathematics of the University of Mainz in 1961–62. The burden, because a draft lecture manuscript was intended to be written (it has never been finished), and JRB never allowed any deviation from his sometimes strange and unusual notions and notations.

And it was this attitude that rendered it more difficult for others to read his papers and appreciate his excellent ideas sufficiently. Accordingly, one finds bitter remarks thereon in many of his papers commented in this article. This attitude also rendered it more difficult for him to keep track of developments related to his own work.

Inspired by both Post's canonical systems and Thue's work, and by finite automata through Jesse B. Wright (see the acknowledgment in the introduction to [39]), JRB made three steps that were revolutionary in the beginning 1960s and that have unfortunately been recognized by only a few people.

First, he considered finite automata deliberately as semi-Thue systems, or (word-) rewrite systems, or nowadays fashionably called rule-based systems, rather than as some sort of "machine," which they definitely are not. So he considered a natural "indeterminism" of such "machines" mathematically as an existential quantifier rather than as a more or less dubious probabilistic gnome.

His rewrite rules are of the form $a\xi \to b\xi$ where $a$, $b$ are letter strings from some given alphabet and where $\xi$ is a variable for letter strings. The rule means that if in some word the part $a$ occurs at its left end then this $a$ may (!) be replaced by $b$. (Another way of writing such rules without variables for letter

* Technische Universität Karlsruhe, Federal Republic of Germany.

strings is $\blacksquare a \to \blacksquare b$ and considering only letter strings of the form $\blacksquare u$ where $\blacksquare$ does not occur in $a$, $b$, and $u$.)

JRB called these rules *regular* (as opposed to the *normal* rules $a\xi \to \xi b$, which Post had inspected). The reason for this terminology was that in the late 1950s so-called *regular events* had been discussed as a means for characterizing (abstract models of) neuronal networks. In his paper [11] JRB proved that regular events are in a sense governed by regular rules. In fact, he gave three different characterizations for sets of letter strings that can be produced from finitely many letter strings by applying regular rules only: these sets are regular events; they are the behaviors of finite automata; they are "the union of some of the equivalence classes of a congruence of concatenation of finite partition" (cited from [11]).

Rules with more than one premise $a\xi$ have also been considered in [11] and [39]. In [17] JRB attempted to carry over these neat regularity results to certain types of canonical productions studied by Post.

In the context of using semi-Thue systems, it is worthwhile to cite JRB once more:

> Inspection will show that such automata [namely, two-way nonerasing automata], as well as other special types of Turing machines, may be interpreted as reduced general-right-regular systems. ([11] p. 108)

If this note were recognized by more computer scientists the research direction on multiway/multiband/multihead, etc., automata could have been considerably different.

Second, JRB interpreted the letters of a letter string as symbols for unary functions (from whence the strange terminology 'unary letters' in [39]). Hence a letter string represents a repeated application of its constituent functions to a given zeroary function or constant. Consequently, JRB considered an automaton with $k$ input letters and state set $S$ as an algebra with $k$ unary functions and carrier $S$ ($k$-algebra), with the above rewrite rules being the (finitely many) generators of the congruence that defines this algebra from the corresponding free algebra.

This viewpoint opened the way for employing all the heavy machinery of algebra: "the universal algebraic concepts of subalgebra, homomorphisms, congruence relations, direct products, and free algebra, in the special case of $k$-algebras are discussed" (from [12]), although JRB has "refrained from bringing categories into the picture. [... because] category theory doesn't solve problems in automata theory ..." (from [39] Introduction).

Similarly the lattice of all homomorphic images of a $k$-algebra (structure lattice) is inspected and brought into relation with the lattice of its congruences ([12]).

The intuitive notion of state merging turned out to be the factorization modulo a congruence ([10], [12], [39]), and what had been called behavior of an automaton was in fact the interpretation of terms (of a free $k$-algebra) in a given $k$-algebra ([10], [11]).

Third: it was only a small but important step from JRB's unary letters to function symbols with other arities than just one (see [12] and [39] Chapter VI) and additionally to admit types ([39] Chapter VII).

Since "terms are trees and trees are terms" ([39] Chapter VI, title of Section 2), this is a very unusual but likewise elegant way to cope with tree languages, tree automata, and tree acceptors. It would therefore be very interesting, and surely profitable, to compare JRB's generalized regularity results for his tree acceptors with the huge mass of papers that have been published on the subject.

This third step also provided an elegant way to obtain a rigorous and clear theory not only for context-free languages and their corresponding (pushdown-) acceptors but (based on a paper by H. Langmaack) also for $LR(k)$-languages, which play a paramount role in the parsing business for languages, be they natural or artificial like programming languages. And again it is interesting and typical that JRB has reinvented the notion of $LL(k)$-languages without considering the big heap of literature on the parsing subject: look at exercise 4b in Section 4 of Chapter VII in [39].

To come back to the introductory citation: JRB believed in the unity of science, he advocated that one first should look around for already existing (mathematical) tools before inventing new ones, and he railed at dillettants (see [10]) whose knowledge is not broad enough to provide for the necessary mathematical background.

He trusted in a theoretical building that he began to erect and whose stories should range at least from logic over algebra up to the design of digital systems ([12] p. 71), and he hoped that this building would be practically applicable in a higher, surely not naive sense.

Really applicable, however, in many areas of computer science is that seemingly theoretical construction to which he gave its most general and final form, the subset construction (see [11]): for example, it occurs basically in every parser generating system, and is therefore a basic step in the construction of each compiler; also the well-known linear algorithm for pattern matching provided by Knuth, Morris, and Pratt is in fact a cleverly refined version of a subset construction.

His idea of a theoretical building for computer science was materialized as a book project [39], which he was not given the time to complete. Did he already feel his end coming when he wrote at the end of the introduction to this book "... in this life you don't have too much time to waste"?

Chapters I–V of it are essentially the book version of his papers [10], [11], and [12], and consist of JRB's lecture notes for his courses on finite automata, which he taught at Purdue and elsewhere. What was earlier herein named the third step is treated extensively in chapters VI and VII. The book has now been worked through and edited by D. Siefkes, who, as its editor, in his preface shows up all the various currents and influences that contributed to the book. It is published by Springer with the title *Finite Automata, Their Algebras and Grammars: Towards a Theory of Formal Expressions.*

# Mathematische Theorie des Verhaltens endlicher Automaten*)

Von J. Richard Büchi**)

Der Begriff des endlichen Automaten entstand als mathematische Abstraktion der deterministischen-diskreten-sequentiellen Systeme, wie es z. B. die modernen Rechenanlagen sind. Das Charakteristische am Funktionieren solcher Systeme liegt in der Rückkoppelung oder Selbst-Steuerung. Die Theorie der reinen Schaltungen (Booleesche Algebra) genügt nicht zur Erfassung dieses Phänomens und muß erweitert werden zur Theorie der endlichen Automaten (Kleenesche Algebra), die geradezu als Theorie der Schaltung durch Rückkoppelung bezeichnet werden kann. Es ist zu hoffen, daß eine streng mathematische Entwicklung von Begriffen und Sätzen über endliche Automaten auch dem Praktiker einen Einblick in das Verhalten und die Möglichkeiten digitaler Systeme gewährt, der sich dem nur empirischen Probieren auf dem Gebiete der Konstruktion und des Gebrauches von Schaltwerken komplementär zur Seite stellt. Allerdings ist es fraglich, ob die schon vorliegende Theorie im mehr naiven direkten Sinn praktisch anwendbar sei. Andererseits aber werden leistungsfähigere und vor allem neuartige Systeme schon in nächster Zukunft nur entstehen können auf Grund tiefgreifender Einsichten von allgemeiner Natur, wie sie nur bei der konsequenten Entwicklung mathematischer Theorien entstehen. Es wäre daher wünschenswert, daß sich ein weiter Kreis mathematisch denkender Forscher der verschiedensten Richtungen mit endlichen Automaten, Turing-Maschinen und ähnlichen Strukturen beschäftigte, und daß in Zukunft diese Dinge nicht nur einigen Logikern und vielen Dilettanten überlassen blieben. Seit einiger Zeit ist ein großes Gewimmel von Schlagwörtern (wie etwa Selbstreproduzierende, Selbstorganisierende, Lernende Systeme, Operations Research, Communication Science usw.) charakteristisch für einen ganzen Komplex von Aktivitäten, der sich um die modernen Rechenanlagen entwickelt hat. Ob es dabei immer bei großen Worten und anregenden Spekulationen bleiben soll? Daß das nicht so zu sein braucht, zeigt das Beispiel des Wörtchens „Verhalten". Im Folgenden werden wir sehen, daß sich diesem, wenigstens mit Bezug auf eine ganz einfache Sorte von Systemen, eine exakte Bedeutung zulegen läßt, die dann auch Anlaß gibt zu einer fruchtbaren Theorie.

| | A | B | C | D | E |
|---|---|---|---|---|---|
| 1 | B | D | C | B | B |
| 2 | C | C | D | E | E |
| W | F | T | F | F | T |

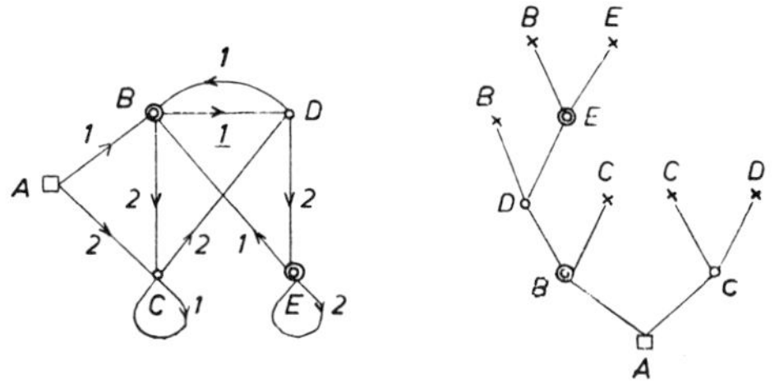


(a) Transit-Tafel von $\mathfrak{A}_0$ (b) Transit-Graph von $\mathfrak{A}_0$ (c) Transit-Baum von $\mathfrak{A}_0$

Bild 1

Ein Automat mit $k$ Eingabezuständen ist ein algebraisches System $\mathfrak{A} = \langle \underline{S}, f_i, \ldots, f_k, \underline{W} \rangle$ bestehend aus einer Menge $\underline{S}$, deren Elemente die Zustände von $\mathfrak{A}$ heißen, einem Element $A$ von $\underline{S}$ genannt Anfangszustand, Abbildungen $f_i$ von $\underline{S}$ in $\underline{S}$ genannt Transit-Operatoren und einer Teilmenge $\underline{W}$ von $\underline{S}$, genannt die Ausgabe von $\mathfrak{A}$. Die Eingabezustände werden durch die Symbole $1, 2, \ldots, k$ bezeichnet, jeder von ihnen entspricht einem Transit-Operator von $\mathfrak{A}$. Der Automat heißt endlich, wenn die Menge $\underline{S}$ seiner Zustände endlich ist.

Ein Beispiel eines endlichen Automaten $\mathfrak{A}_0$ mit zwei Eingabezuständen ist durch Bild 1(a) gegeben. Seine Zustände sind $A$, $B$, $C$, $D$ und $E$. Die Tafel ist so zu verstehen, daß z. B. $f_1 B = D$, $f_2 B = C$, $B \in \underline{W}$ da $W(B) = T$, $C \notin W$ da $W(C) = F$. In selbstsprechender Weise ist derselbe Automat $\mathfrak{A}_0$ auch durch die Bilder 1(b) oder 1(c) bestimmt. Diese graphischen Darstellungen eines endlichen Automaten sind natürlich für die Theorie überflüssig, erweisen sich aber beide als nützliche Stützen der Anschauung. Auf Eingabe etwa des Signals 1121 reagiert der Automat $\mathfrak{A}_0$ indem er vom Anfangszustand $A$ ausgehend sukzessiv die Zustände $f_1 A = B$, $f_1 B = D$, $f_2 D = E$, $f_1 E = B$ durchläuft. Der schlußendliche Zustand $B$ heißt die interne Reaktion von $\mathfrak{A}_0$ auf das Eingabesignal 1121 und sei mit $\mathfrak{A}_0(1121)$ bezeichnet. Das Signal 1121 erregt die Ausgabe

*) Auf Einladung der Tagungsleitung gehaltener Hauptvortrag.

**) University of Michigan, Ann Arbor und Gutenberg-Universität, Mainz.

von $\mathfrak{A}_0$, weil $\mathfrak{A}_0(1121) \in \underline{W}$. Hingegen gilt z. B. $\mathfrak{A}_0(212) \notin W$, also erregt das Signal 212 die Ausgabe von $\mathfrak{A}_0$ nicht. Genauer wird die Arbeitsweise eines Automaten $\mathfrak{A}$ mit $k$ Eingabezuständen so definiert:

Die Wörter über dem Alphabet $1, \ldots, k$ heißen die Eingabesignale von $\mathfrak{A}$. Sie bilden die Menge $N_k$, zu der auch das leere Wort $o$ gerechnet wird. Die interne Reaktion $\mathfrak{A}(x)$ ist die für jedes $x \in N_k$ durch folgende einfache Rekursion definierte Funktion;

$$\mathfrak{A}(o) = A$$
$$\mathfrak{A}(x\,i) = f_i\,\mathfrak{A}(x) \quad \text{für} \quad i = 1, \ldots, k\,.$$

Für jedes Eingabesignal $x$ bestehen nun folgende zwei Alternativen

$$\mathfrak{A}(x) \in \underline{W} : x \text{ erregt die Ausgabe von } \mathfrak{A}$$
$$\mathfrak{A}(x) \notin \underline{W} : x \text{ erregt die Ausgabe von } \mathfrak{A} \text{ nicht.}$$

Ein Automat $\mathfrak{A}$ ist also die mathematische Abstraktion eines konkreten Systems, dessen Zweck es ist, die Eingabesignale in zwei komplementäre Mengen aufzulösen. Allerdings spielen in dieser Hinsicht nur endliche Automaten eine Rolle, und auf solche werden wir uns daher im folgenden konzentrieren. Dabei handelt es sich aber nicht darum, über diese und jene endliche Automaten Aussagen adhoc zu machen. Vielmehr soll eine allgemeine mathematische Theorie, bestehend aus Definitionen, Beweisen und Sätzen, entwickelt werden, mit dem Zweck, das Verhalten endlicher Automaten verstehen zu lernen. Dazu tut es vor allem not, geeignete Konzepte durch präzise Definitionen hervorzuheben. Wir beginnen mit der von Kleene [10] stammenden Definition des fundamentalen Begriffs des Verhaltens.

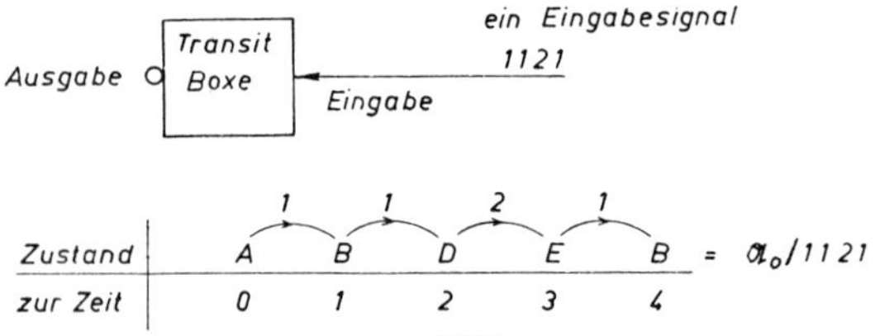

Bild 2

Seien $\mathfrak{A} = \langle S, A, f_1, \ldots, f_k, \underline{W}\rangle$ und $\mathfrak{B} = \langle \underline{T}, B, g_1, \ldots, g_k, \underline{V}\rangle$ zwei Automaten. Man wird sagen „$\mathfrak{A}$ verhält sich gleich wie $\mathfrak{B}$", wenn die beiden Automaten auf gleiche Eingabesignale mit gleicher Ausgabe reagieren, d. h., wenn für jedes $x$ aus $N_k$ gilt, $\mathfrak{A}(x) \in \underline{W}$ genau dann, wenn $\mathfrak{B}(x) \in \underline{V}$. Diesen Sachverhalt kann man aber auch so ausdrücken:

$$\{x \mid \mathfrak{A}(x) \in \underline{W}\} = \{x \mid \mathfrak{B}(x) \in \underline{V}\}\,.$$

Es ist daher zweckmäßig, die Menge $\{x \mid \mathfrak{A}(x) \in \underline{W}\}$ das Verhalten von $\mathfrak{A}$ zu nennen und mit $v\,r\,h(\mathfrak{A})$ zu bezeichnen. Damit kann nun der Zweck unserer Theorie schon ziemlich scharf formuliert werden.

Hauptaufgabe: Sei $\Pi_k$ die Menge aller Verhalten endlicher Automaten mit $k$ Eingabezuständen. Es ist die Art derjenigen Teilmengen $\beta \subseteq N_k$ zu erforschen, die zu $\Pi_k$ gehören. Auch sind Eigenschaften von $\Pi_k$ zu untersuchen. Insbesondere gilt es Kriterien zu finden, die für irgendwelches $\beta \subseteq N_k$ aussagen, ob $\beta$ zu $\Pi_k$ gehört oder nicht gehört.

Zunächst ist zu bemerken, daß nur abzählbar viele der Teilmengen von $N_k$ durch endliche Automaten realisierbar sind. Fernerhin sind diese Mengen rekursiv; ein endlicher Automat $\mathfrak{A}$ präsentiert ja geradezu ein effektives Verfahren, das darüber entscheidet, ob ein vorgelegtes $x \in N_k$ zu $v\,r\,h(\mathfrak{A})$ gehört. Es gibt aber andererseits ganz einfache rekursive Teilmengen von $N_k$, die nicht das Verhalten eines endlichen Automaten sind. Z. B. zeigt man leicht, daß die Menge $\{yy \mid y \in N_2\}$, bestehend aus allen Wörtern aus $N_2$, die aus zwei gleichen Stücken zusammengesetzt sind, nicht zu $\Pi_2$ gehört.

Gegenüber den in der Literatur vorkommenden Varianten hat unsere Definition der Automaten den Vorteil, daß so fundamentale Begriffe wie Homomorphie, direktes Produkt und Kongruenz gleich aus der abstrakten Algebra übernommen werden können. Damit ist dann erstens eine systematische Strukturtheorie der Automaten eröffnet und zweitens ein Beitrag zum Hauptproblem gewonnen. Auch die Resultate von Moore [13] über minimale Automaten ergeben sich sozusagen als Beiprodukt. Diese Dinge können in Kürze wie folgt angedeutet werden. Dabei wollen wir uns auf solche Automaten beschränken, in welchen die Menge $\underline{S}$ aller Zustände vom Anfangszustand $A$ durch die Transit-Operatoren erzeugt wird. Von $A$ aus nicht „erreichbare" Zustände haben ja keinen Einfluß auf das Verhalten von $\mathfrak{A}$.

Eine Abbildung $h$ von $S$ auf $T$ heißt Homomorphie des Automaten $\mathfrak{A} = \langle S, A, f_1, \ldots, f_k, W\rangle$ auf $\mathfrak{B} = \langle T, B, g_1, \ldots, g_k, V\rangle$ wenn $h\,A = B$, $h(f_i\,X) = g_i(h\,X)$ und $(X \in W) \leftrightarrow (h\,X \in V)$. Bedeutet $\mathfrak{B} \leqq \mathfrak{A}$, daß $\mathfrak{B}$ homomorphes Bild von $\mathfrak{A}$ ist, dann gilt

$$\mathfrak{B} \leqq \mathfrak{A} \rightarrow v\,r\,h(\mathfrak{A}) = v\,r\,h(\mathfrak{B})$$

$$(\exists\,\mathfrak{C})\,(\mathfrak{A} \leqq \mathfrak{C} \wedge \mathfrak{B} \leqq \mathfrak{C}) \leftarrow v\,r\,h\,(\mathfrak{A}) = v\,r\,h\,(\mathfrak{B})$$

Für jedes $\beta \subseteq N_k$ ist die Menge $V_\beta$ aller Automaten vom Verhalten $\beta$ ein vollständiger Verband bezüglich der Ordnungsrelation $\leqq$. Dieser Strukturverband $(V_\beta, \leqq)$ hat als gröbstes Element den freien Automaten $\langle N_k, o, \varphi_1, \ldots, \varphi_k, \beta\rangle = \mathfrak{F}_\beta$, wobei $\varphi_1\,x = x\,1, \ldots, \varphi_k\,x = x\,k$. Sein kleinstes Element ist der minimale Automat $\mathfrak{M}_\beta$ des Verhaltens $\beta$. Gibt es endliche Automaten vom Verhalten $\beta$, so liegen sie unten im Strukturverband; $\mathfrak{M}_\beta$ ist dann auch der minimale in der Zahl von Zuständen unter allen Automaten vom Verhalten $\beta$.

Bild 3 zeigt den Strukturverband bestehend aus allen homomorphen Bildern von $\mathfrak{A}_1$. Wie man dem Beispiel entnimmt, brauchen solche Verbände nicht distributiv zu sein. Es brauchen auch keine Kettensätze zu gelten, und die irreduziblen Elemente können sehr unregelmäßig angeordnet sein. Die Strukturtheorie endlicher Automaten scheint daher eine fast hoffnungslos schwierige Angelegenheit zu sein.

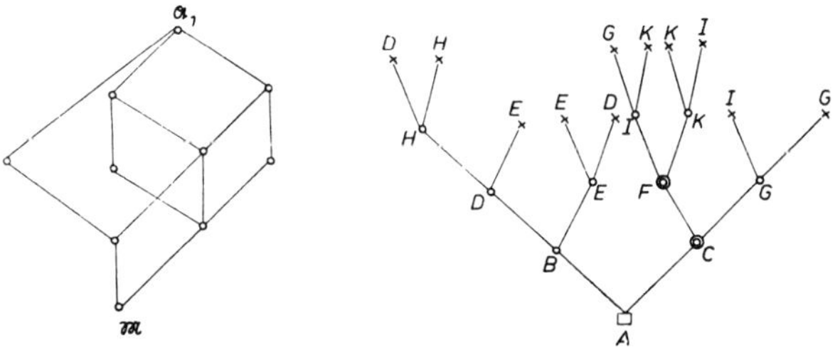


Strukturverband von $\mathfrak{A}_1$ Automat $\mathfrak{A}_1$

Bild 3

Eine Äquivalenzrelation $\backsim$ über der Menge $S$ aller Zustände wird man dann eine Kongruenz des Automaten $\mathfrak{A}$ nennen, wenn gilt $(X \backsim Y) \rightarrow (f_i\,X \backsim f_i\,Y)$ und $(X \backsim Y \wedge X \in W) \rightarrow (Y \in W)$. In bekannter Weise bildet man den Quotienten $\mathfrak{A}/\backsim$ von $\mathfrak{A}$ bezüglich einer Kongruenz. Wie man das von der Algebra her weiß, besteht nun der Strukturverband von $\mathfrak{A}$ aus allen Quotienten von $\mathfrak{A}$ und ist isomorph zum Verband der Kongruenzen von $\mathfrak{A}$. Insbesondere liefert die gröbste (das Supremum) $\simeq$ aller Kongruenzen von $\mathfrak{A}$ den minimalen Automaten des Verhaltens $v\,r\,h(\mathfrak{A})$. Daraus ergeben sich leicht effektive Methoden zur Minimierung; es handelt sich ja einfach um das Auffinden der gröbsten Kongruenz eines vorgelegten endlichen Automaten.

Bis auf Isomorphie sind also die Automaten vom Verhalten $\beta$ genau die Quotienten $\mathfrak{F}_\beta/\backsim$ des freien Automaten $\mathfrak{F}_\beta$. Insbesondere erhalten wir daher die folgenden Kriterien für die Realisierbarkeit von $\beta \subseteq N_k$ als Verhalten eines endlichen Automaten:

1. Die Menge $\beta \subseteq N_k$ gehört zu $\Pi_k$ genau dann, wenn der freie Automat $\mathfrak{F}_\beta$ eine Kongruenz von endlichem Index (Zahl der Kongruenzklassen) zuläßt.

2. Die Menge $\beta \subseteq N_k$ gehört zu $\Pi_k$ genau dann, wenn die gröbste Kongruenz von $\mathfrak{F}_\beta$ endlichen Index hat.

Zwecks prägnanter Formulierung lohnt es sich, folgende Definitionen anzubringen. Eine Kongruenz der rechts-Nachfolgerfunktionen $\varphi_1, \ldots, \varphi_k$ nennen wir rechts-Kongruenz von $N_k$. Die gröbste Kongruenz von $\mathfrak{F}_\beta$ bezeichnen wir mit $\simeq(\beta)$ und nennen sie die von $\beta$ induzierte rechts-Kongruenz. Der Index von $\simeq(\beta)$ heißt auch rechts-Rang von $\beta$. Obige Kriterien lassen sich nun auch so formulieren:

1′. Die Menge $\beta \subseteq N_k$ gehört zu $\Pi_k$ genau dann, wenn sie Vereinigung von Kongruenzklassen einer rechts-Kongruenz von $N_k$ von endlichem Index ist.

(II) Die Menge $\beta \subseteq N_k$ gehört zu $\Pi_k$ genau dann, wenn der rechts-Rang von $\beta$ endlich ist, d. h., wenn die von $\beta$ induzierte rechts-Kongruenz $\simeq(\beta)$ nur endlich viele Klassen besitzt.

Mit Hilfe der Formel

$$\text{(a)} \qquad x \simeq y(\beta) \leftrightarrow (\forall\,u)\,(x\,u \in \beta \leftrightarrow y\,u \in \beta)$$

gelingt es oft, die induzierte rechts-Kongruenz genügend konkret zu ermitteln (oder doch ihren Index abzuschätzen), sodaß dann Kriterium (II) wirklich angewandt werden kann. Z. B. zeigt man so leicht, daß die induzierte rechts-Kongruenz der Menge $\{x\,x \mid x \in N_2\}$ die Gleichheitsrelation ist, und sie daher nicht durch einen endlichen Automaten realisiert werden kann. Zur Bedeutung von Kriterium (1′) ist folgendes zu sagen:

$N_1 = \{0, 1, 11, 111, \ldots\}$ ist doch einfach die Menge der natürlichen Zahlen. Die Nachfolgerfunktion ist $\varphi_1 x = x\,1 = (x + 1)$, ihre Kongruenzen sind, nebst der Gleichheit, die wohlbekannten elementaren Kongruenzen

$$x \equiv y(l, q) \leftrightarrow (x < l \wedge x = y) \vee (l \leqq x \wedge l \leqq y \wedge x \equiv y \ (\mathrm{mod}\ q))\,,$$

wobei $0 \leqq l$ und $1 \leqq q$. Sie sind alle von endlichen Index $(l + q)$, und Vereinigungen ihrer Kongruenzklassen pflegt man (schlußendlich-) periodische Mengen von Zahlen zu nennen. Es scheint daher angebracht, auch im Falle $k > 1$, die Vereinigungen von Kongruenzklassen modulo einer rechts-Kongruenz von endlichem Index, als rechts-periodische Teilmengen von $N_k$ zu bezeichnen. Das Kriterium (1′) kann dann auch so formuliert werden:

(I) Die Menge $\beta$ gehört zu $\Pi_k$ genau dann, wenn sie rechts-periodisch ist.

Übrigens kann $N_k$ auch im Falle $k > 1$ als die Menge der natürlichen Zahlen interpretiert werden. Nur sind jetzt die Zahlen nicht mehr durch eine einzige, aber durch $k$ freie Funktionen $\varphi_1 x = x\,1 = (k \cdot x + 1), \ldots, \varphi_k x = x\,k = (k \cdot x + k)$ aus 0 erzeugt (siehe Bild 4 für den Fall $k = 2$). Mit einer Variante dieses $k$-ären Nummernsystems $\langle N_k, 0, \varphi_1, \ldots, \varphi_k \rangle$ lernt man zwar schon auf der Schule zu manipulieren (Algorithmen für $k$-äre Addition und Multiplikation), und es wird doch gerade dieses Umgehen mit Wörtern von altersher Rechnen genannt. Warum aber sind eigentlich diese $k$-Kongruenzen (Kongruenzen des Funktionssystems $k \cdot x + 1, \ldots, k \cdot x + k$) von endlichem Index und die zugehörigen $k$-periodischen Mengen von Zahlen nicht schon früher systematisch behandelt worden? Die Theorie der endlichen Automaten kann geradezu als Beitrag zu diesem Teil der elementaren Zahlentheorie bezeichnet werden. — Einige Bemerkungen über $k$-periodische Mengen von Zahlen sind bei Büchi [2] zu finden. — Ein sehr anspruchsvolles Problem ist es, irgendwelche Übersicht über alle $k$-Kongruenzen (von endlichem Index) der Zahlen zu gewinnen. Daß die Verbandtheorie hier nicht weiter hilft, wurde schon angedeutet. — Im Kontrast zum Falle $k = 1$ gibt es unter den $k$-Kongruenzen von unendlichem Index sehr komplizierte (nämlich nicht rekursive) Relationen (Post und Markow, rekursive Unlösbarkeit des Wortproblems für Halbgruppen).

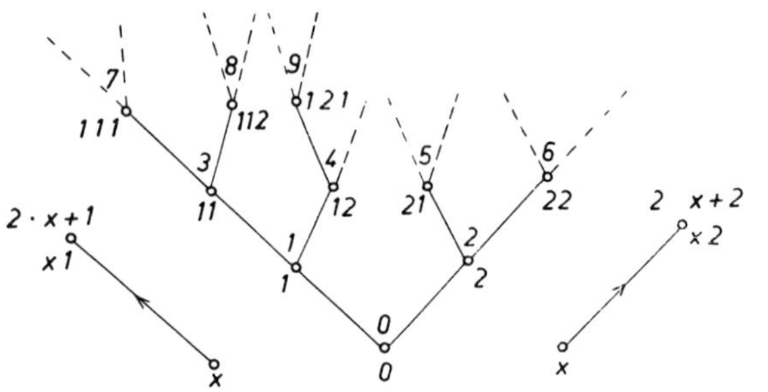


Bild 4

Über $N_k$ gibt es auch die links-Nachfolgerfunktionen $\lambda_1 x = 1\,x, \ldots, \lambda_k x = k\,x$, und ihre Kongruenzen, die links-Kongruenzen von $N_k$. Eine Menge $\beta \subseteq N_k$ heißt links-periodisch, wenn die von $\beta$ induzierte links-Kongruenz $\approx(\beta)$ endlichen Index hat. Dieser heißt auch der links-Rang von $\beta$, und es gilt,

(b) $$x \approx y(\beta) \leftrightarrow (\forall u)(u\,x \in \beta \leftrightarrow u\,y \in \beta)\,.$$

Schließlich nennen wir eine Relation, die links- und rechts-Kongruenz ist, einfach eine Kongruenz von $N_k$. Es sind dies die Kongruenzen der freien Halbgruppe $\langle N_k, \hat{} \rangle$. Eine Menge $\beta \subseteq N_k$ heißt periodisch, wenn die induzierte Kongruenz $\eqsim(\beta)$ endlichen Index hat. Dieser Index heißt auch Rang von $\beta$, und es gilt

(c) $$x \eqsim y(\beta) \leftrightarrow (\forall u, v)(u\,x\,v \in \beta \leftrightarrow u\,y\,v \in \beta)\,.$$

Unter Verwendung von (a), (b), (c) ist es nun möglich zu zeigen, daß die Ränge $r_1$, $r_2$, $r$ einer Menge $\beta \subseteq N_k$ den Abschätzungen $r_1 \leqq 2^{r_2}$, $r_2 \leqq 2^{r_1}$ und $r \leqq r_1^{r_2}, r_2^{r_1}$ genügen. Da trivialerweise $r_1, r_2, \leqq r$, folgt also aus der Endlichkeit einer der drei Ränge die Endlichkeit der beiden anderen, d. h., „rechts-periodisch", „links-periodisch" und „periodisch" haben alle denselben Sinn. Wegen (I) gilt also,

Satz 1: Die Verhaltensmengen endlicher Automaten sind genau die periodischen Mengen von Wörtern. Diese wiederum sind identisch mit den rechts-periodischen und auch den links-periodischen Mengen.

Trotzdem brauchen natürlich die drei Ränge einer Menge $\beta \in \Pi_k$ nicht identisch zu sein, sie genügen aber den oben zitierten Ungleichungen. Übrigens ist der rechts-Rang $r_1$ einer periodischen Menge $\beta$ gerade die Minimalzahl von Zuständen die zur Realisierung von $\beta$ durch einen

Automaten nötig sind; es ist ja $\mathfrak{F}_\beta / \simeq(\beta)$ der minimale Automat des Verhaltens $\beta$. Das links-rechts Vertauschte eines Wortes $x$ heiße das Konverse zu $x$ und sei mit $\overleftarrow{x}$ bezeichnet. Trivialerweise ist $\beta$ periodisch genau dann, wenn es $\overleftarrow{\beta}$ ist, und der links-Rang von $\beta$ ist gleich dem rechts-Rang von $\overleftarrow{\beta}$. Aus Satz 1 folgt also: Ist $\beta \subseteq N_k$ realisierbar durch einen endlichen Automaten, dann ist es auch $\overleftarrow{\beta}$. Mit anderen Worten, bei der Frage der Realisierbarkeit von $\beta$ spielt es keine Rolle, ob man sich dazu entscheidet, die Signale immer von links nach rechts, oder immer von rechts nach links einzugeben. Apriori (ohne Satz 1) ist dies ein erstaunliches Resultat über endliche Automaten, und die Zahl der zur Realisierung von $\beta$ nötigen Zustände kann auch für die eine Art des Eingebens viel größer sein als für die andere (rechts-Rang und links-Rang von $\beta$). Das Resultat kann übrigens auch so formuliert werden: $\Pi_k$ ist abgeschlossen unter Konversion. Unter Verwendung der Formeln (a), (b) zur Abschätzung von Rängen ist es nicht schwer zu zeigen, daß Periodizität auch erhalten bleibt unter den BOOLEschen Operationen $\cup$, $\cap$, $\sim$ und den weiteren, sogenannten regulären Operationen $\alpha^\frown\beta = \{xy \mid x \in \alpha \wedge y \in \beta\}$ und $\alpha^* = \{x_1 x_2 \ldots x_n \mid x_1, x_2, \ldots, x_n \in \alpha\} = \alpha \cup (\alpha^\frown\alpha) \cup (\alpha^\frown\alpha^\frown\alpha) \cup \cdots$. Aus Satz 1 folgt dann

Satz 2: Die Menge $\Pi_k$ der Verhalten endlicher Automaten ist abgeschlossen unter den Operationen $\cup$, $\cap$, $\sim$, $\leftarrow$, $^\frown$, $*$.

Wir werden sehen, daß die abstrakte Algebra $\mathfrak{K}_k = \langle \Pi_k, \cup, \cap, \sim, \leftarrow, ^\frown, * \rangle$ das Einmaleins der Theorie der Schaltung durch Rückkoppelung ist. Fälschlich wird dies oft von der BOOLEschen Algebra behauptet, die aber nur das reine Schalten beherrscht[1]). Dies wurde zuerst von KLEENE [10] erkannt, und wir schlagen deshalb vor, $\mathfrak{K}_k$ die KLEENE'schen Algebren zu nennen. Über sie ist leider nicht viel bekannt; es bleibt die wichtige Aufgabe, ihre Theorie auf den Stand derjenigen der BOOLEschen Algebra zu bringen. Z. B. sind die Probleme der Axiomatisierung der KLEENEschen Algebra ungelöst.

Der oben angedeutete Beweis von Satz 2 ist nicht derjenige von KLEENE [10]. Diese Arbeit enthält vieles von vielleicht sekundärer Bedeutung und ist daher schwer verdaulich. Eine bessere Darstellung der Hauptresultate ist bei COPI, ELGOT und WRIGHT [8] zu finden. Im folgenden werde ein eleganter und sehr aufschlußreicher Beweis von MYHILL [12] des Satzes 2 wiedergegeben. Er beruht auf der folgenden Verallgemeinerung der Automaten, die auch aus ganz anderen Gründen beachtenswert sein dürfte. Ein Transit-System (auch Flußdiagramm und nicht deterministischer Automat genannt) $\mathfrak{C} = \langle \underline{S}, \underline{A}, F_1, \ldots, F_k, F, \underline{W} \rangle$ besteht aus einer Menge $S$ von Zuständen, der Anfangsmenge $\underline{A} \subseteq \underline{S}$, den Transit-Relationen $F_1, \ldots, F_k$, $F \subseteq (\underline{S} \times \underline{S})$ und der Endmenge $W \subseteq S$. Die Relationen $F_i$ entsprechen den Eingabezuständen $i = 1, \ldots, k$. $F$ heißt auch Relation der spontanen Übergänge und entspricht dem leeren Signal o. In Bild 5 ist ein Transit-System $\mathfrak{C}_0$ mit den Zuständen $A, B, C, D, E$ durch seinen Transit-Graphen gegeben. Man sieht gleich, daß $\mathfrak{C}_0$ kein Automat ist, denn erstens gibt es zwei Anfangszustände, zweitens sind die Relationen $F_1$ und $F_2$ nicht funktionell, und drittens treten spontane Übergänge auf.

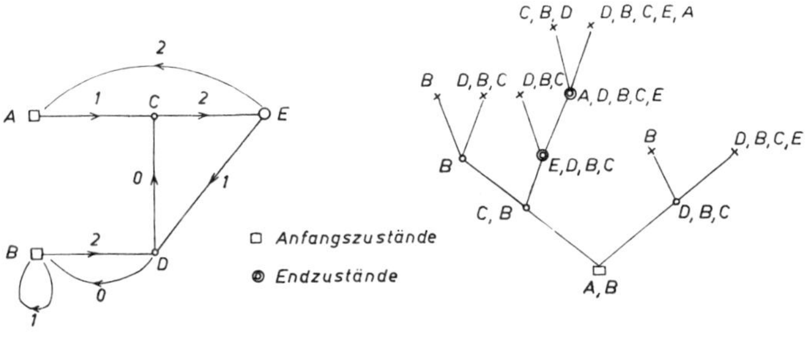


(a) Das Transit-System $\mathfrak{C}_0$ (b) Teilmengenkonstruktion Tm ($\mathfrak{C}_0$)

Bild 5

Bemerkung zu Bild 5(a): $E$ ist Endzustand

Unter der Kontaktmenge eines Transit-Systems $\mathfrak{C}$ verstehen wir die Menge $k\,n\,t(\mathfrak{C})$, bestehend aus allen Signalen $x \in N_k$, die einen gerichteten Weg vom Anfang nach dem Ende markieren. Z. B. gehört also das Signal $121122 = 121o12o2$ zu $k\,n\,t(\mathfrak{C}_0)$, da in Bild 5 der gerichtete Weg $A \to C \to E \to D \to B \to B \to D \to C \to E$ in der Anfangsmenge $\underline{A} = \{A, B\}$

[1]) Eigentlich ist es nur die zweiwertige Schaltung, die der BOOLEschen Algebra entspricht. Die $n$-wertigen Schaltungen werden von den POSTschen Algebren [14] beherrscht, die leider wenig Beachtung finden. Für eine kurze Darstellung siehe ROSENBLOOM [17].

beginnt, in der Endmenge $W = \{E\}$ endet und seine Kanten durch 1, 2, 1, o, 1, 2, o, 2 markiert sind. Es ist leicht einzusehen, daß das Verhalten eines Automaten gerade die Kontaktmenge des zugehörigen Transit-Graphen ist. Umgekehrt gilt aber auch das

Lemma: Zu jedem endlichen Transit-System $\mathfrak{C}$ (mit $n$ Zuständen) kann man einen endlichen Automaten $\mathfrak{A} = Tm(\mathfrak{C})$ (mit höchstens $2^n$ Zuständen) konstruieren, so daß $v\,r\,h(\mathfrak{A}) = k\,n\,t(\mathfrak{C})$.

Die Teilmengen-Konstruktion $Tm(\mathfrak{C})$ wurde von Myhill [12] erfunden und ist in Bild 5(b) im Beispiel $\mathfrak{C}_0$ durchgeführt. Die Zustände des Automaten $Tm(\mathfrak{C}_0)$ sind gewisse Teilmengen aller Zustände von $\mathfrak{C}_0$. Als Anfangszustand wählt man die Anfangsmenge $A = \{A, B\}$. Sei die Menge $X$ schon als Zustand von $Tm(\mathfrak{C}_0)$ gefunden. Dann schlägt man zum Nachfolger $f_1(X)$ zunächst alle 1-Nachfolger von $\mathfrak{C}_0$ von Elementen $X \in X$ und schließt die so entstandene Menge ab, durch sukzessives Zufügen aller ihrer spontanen Nachfolger in $\mathfrak{C}_0$. Ähnlich konstruiert man $f_2(X)$. Schließlich wählt man als Elemente der Ausgabe von $Tm(\mathfrak{C}_0)$ alle diejenigen Mengen, die Zustände von $Tm(\mathfrak{C}_0)$ sind und sich mit der Endmenge $W = \{E\}$ von $\mathfrak{C}_0$ überschneiden. Es ist nun ganz leicht einzusehen, daß das Verhalten des Automaten $Tm(\mathfrak{C}_0)$ gerade mit der Kontaktmenge $k\,n\,t(\mathfrak{C}_0)$ übereinstimmt. Zunächst ergibt das Lemma von Myhill ein weiteres Kriterium für das Verhalten endlicher Automaten:

(III) Eine Menge $\beta$ gehört zu $\Pi_k$ genau dann, wenn sie Kontaktmenge eines endlichen Transit-Systems ist.

Daraus folgt nun wieder ganz leicht die Abgeschlossenheit von $\Pi_k$ unter den Operationen $\sim$, $\cup$, $\frown$, $*$; es braucht ja nur gezeigt zu werden, daß mit $\alpha$ und $\beta$ auch $\tilde{\alpha}$, $\alpha \cup \beta$, $\alpha^\frown\beta$ und $\beta^*$ Kontaktmengen sind. Aus einem Transit-System für $\alpha$ erhält man aber eines für $\tilde{\alpha}$, indem man alle Transitionen umdreht und Anfang mit Ende vertauscht. Ferner ist in Bild 6 angedeutet, wie aus Transit-Systemen für $\alpha$ und $\beta$ neue Transit-Systeme mit den Kontaktmengen $\alpha \cup \beta$, $\alpha^\frown\beta$, $\beta^*$ konstruiert werden können. Die Bilder erklären wohl diese Konstruktionen zur Genüge (die Doppelpfeile bedeuten jeweils ganze Kabel von spontanen Transitionen). Damit ist dann Satz 2 nach der Methode von Myhill bewiesen (die Abgeschlossenheit von $\Pi_k$ unter $\sim$ zeigt man ja leicht direkt; das Komplement der Ausgabe eines Automaten ergibt doch das komplementäre Verhalten). Übrigens zeigt die Konstruktion für $\beta^*$ in Bild 6 ganz deutlich die fundamentale Bedeutung des Operators $*$. Er verdient geradezu als Rückkoppelungsoperator bezeichnet zu werden.

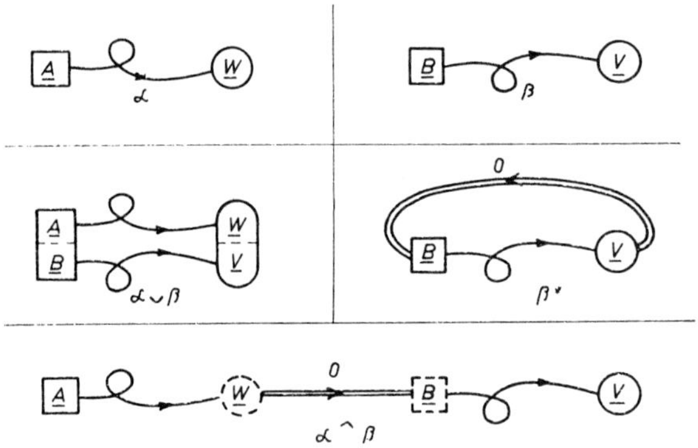


Bild 6

Leicht konstruiert man zu jeder endlichen Menge $\beta \subseteq N_k$ ein endliches Transit-System (Automaten), dessen Kontaktmenge (Verhalten) gleich $\beta$ ist. Aus Satz 2 folgt daher:

(d) Jede aus endlichen Teilmengen von $N_k$ durch die Operationen $\cup$, $\frown$, $*$ aufgebaute Menge $\beta$ ist das Verhalten eines endlichen Automaten.

Eine Formel, die aus den Symbolen $1, \ldots, k$ durch Anwendung der Operationssymbole $\cup$, $\frown$, $*$ aufgebaut ist, wenn man einen regulären Ausdruck (dabei ist oft $(\mathfrak{E}^\frown\mathfrak{F})$ durch $(\mathfrak{E}\,\mathfrak{F})$ abgekürzt). Jeder solchen Formel $\mathfrak{E}$ wird, in augenscheinlicher Weise, ein Wert $|\mathfrak{E}| \subseteq N_k$ zugeordnet. Z. B. bezeichnet der Ausdruck $(1 \cup 2)^*\,2$ die Menge aller Wörter in $N_2$, die auf eine 2 enden. Es ist nun leicht einzusehen, daß (d) in folgender stärkerer Form gilt:

Synthesen-Satz: Es gibt ein effektives Verfahren, das zu jedem regulären Ausdruck $\mathfrak{E}$ einen endlichen Automaten $\mathfrak{A}$ vom Verhalten $|\mathfrak{E}|$ liefert.

Ein solches Verfahren kann doch etwa so beschrieben werden: 1. Durch Anwendung der Konstruktionen aus Fig. 6 stelle man ein Transit-System $\mathfrak{C}$ her, dessen Kontaktmenge gleich dem Wert des vorgelegten Ausdrucks $\mathfrak{E}$ ist. 2. Auf $\mathfrak{C}$ wende man die Teilmengenkonstruktion an; der so gewonnene Automat $\mathfrak{A}$ hat das gewünschte Verhalten $|\mathfrak{E}|$. In Bild 7 wurde diese Konstruktion im Beispiel des Ausdruckes $(12^*1 \cup 22)^*22$ durchgeführt.

Auch die Umkehrung zum Synthesen-Satz wurde von Kleene [10] gefunden; für den Beweis verweisen wir auf diese Arbeit, sowie auf Myhill [12], Copi, Elgot und Wright [8],

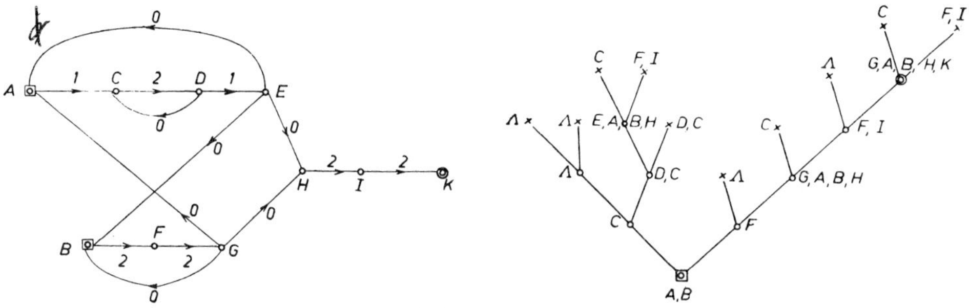

Bild 7

Rabin und Scott [16]. Zu den regulären Ausdrücken ist jetzt aber noch das Symbol $\Lambda$ für die leere Menge, zu rechnen.

Analysen-Satz: Es gibt ein effektives Verfahren, das zu jedem endlichen Automaten $\mathfrak{A}$ einen regulären Ausdruck $\mathfrak{E}$ liefert, der das Verhalten von $\mathfrak{A}$ bezeichnet.

Die beiden Sätze zusammen zeigen, daß die regulären Ausdrücke genau das Verhalten endlicher Automaten beschreiben. Ihre Bedeutung für die Theorie der Schaltung durch Rückkoppelung entspricht also gerade der Bedeutung Boolescher Ausdrücke in der Theorie der reinen Schaltung. Endlich ergibt der Analysen-Satz die Umkehrung zu (d), so daß wir als weiteres Kriterium für das Verhalten endlicher Automaten erhalten:

(IV) Eine Menge $\beta \subseteq N_k$ gehört zu $\Pi_k$, genau dann, wenn sie der Wert eines regulären Ausdruckes ist, d. h., wenn sie durch die Operationen $\cup$, $\frown$, * aus endlichen Mengen aufgebaut werden kann.

Wegen Satz 2 können in (IV) auch die Operationen $\cap$, $\sim$, $\leftarrow$ zugefügt werden. Über reguläre Ausdrücke bleiben viele wichtige Fragen ungelöst, die mit den Problemen der Axiomatisierung der Kleeneschen Algebren verwandt sind. So folgt zwar aus dem Synthesensatz, daß das Zutreffen von Gleichungen $\mathfrak{E} = \mathfrak{F}$ zwischen regulären Ausdrücken algorithmisch entschieden werden kann. Man kennt aber keine endliche Menge von Identitäten, aus denen alle zutreffenden Gleichungen $\mathfrak{E} = \mathfrak{F}$ herleitbar sind. Sodann fehlt für reguläre Ausdrücke ein brauchbares Konzept der Normalform.

Wir haben damit die fundamentalen Resultate über das Verhalten endlicher Automaten besprochen. Tiefgreifendere Ergebnisse in dieser Richtung sind bei Büchi [3] zu finden. Es bleibt uns noch, vor allzu stur-einseitiger Beurteilung der Bedeutung dieser Dinge zu warnen. Es ist doch einfach undenkbar, daß Einsichten über so anschauliche konkrete Strukturen, wie es z. B. die endlichen Transit-Systeme sind, nicht auch in ganz anderen Zusammenhängen von praktischem Wert sein können. Zudem verdient die Arithmetik der Wörter (oder $k$-ären Zahlsysteme) im allgemeinen und der periodischen Mengen von Wörtern ($k$-äre Kongruenzen der Zahlen) im speziellen auch die Beachtung des reinen Mathematikers. Daß, auch von diesem Standpunkt aus gesehen, die Theorie des Verhaltens endlicher Automaten nicht inhaltsleer ist, zeigt vielleicht am schönsten ihre Anwendung, Büchi [1], auf ein Problem von Tarski, das sich immerhin von anderer Seite her als unzugänglich erwiesen hat.

Eine ganz andere Fragestellung über endliche Automaten ist wohl am ausführlichsten bei Church [7] formuliert; siehe auch Büchi, Elgot und Wright [4]. Es handelt sich hier um Probleme der Existenz von Algorithmen zur Konstruktion von endlichen Automaten, die vorgelegten Anforderungen genügen sollen. Solche Probleme sind natürlich erst dann fixiert, wenn man sich für eine genau umschriebene, also eine formalisierte Sprache $S$ entschieden hat, in der die Anforderungen zu formulieren sind. Abhängig vom Reichtum der Sprache $S$ existieren dann Konstruktions-Algorithmen oder sie existieren nicht. Ein sehr starker Algorithmus dieser Art findet sich bei Büchi [1]. Weiter sind zu diesem Thema folgende Arbeiten zu nennen: Friedman [9], Büchi [2], Wang [19].

Für gewisse Zwecke ist es wohl nützlich, digitale Rechenanlagen als Turingsche Maschinen zu interpretieren. Mit seiner universellen Maschine ist Turing ja der geniale Erfinder der heute so wichtigen Programmsteuerung geworden. Allerdings darf man nicht stur auf der praktischen Unbrauchbarkeit seiner linearen Bänder herumreiten; es kann ja leicht Turings Konzeption der Maschine in vielen Richtungen flexibler gestaltet werden. (Schließlich hatte Turings Arbeit den mehr theoretischen Zweck der exakten Definition der Berechenbarkeit und Lösung des HILBERTschen Entscheidungsproblems.) Als sehr weitreichende Idealisierung verbleibt aber die Unbegrenztheit des Rechenbandes. Im Unterschied zur allgemeineren TURING-Maschine mit endlichem, aber unbegrenztem Gedächtnis ist das Gedächtnis des endlichen Automaten beschränkt durch die Zahl seiner Zustände. Es gibt nun eine Mannigfaltigkeit von Begriffen, die zwischen denjenigen des endlichen Automaten und der TURING-Maschine fallen. Siehe z. B. RABIN und SCOTT [16]. Eine wichtige Aufgabe bleibt es zu erforschen, ob einer dieser Begriffe sich besser eigne, als mathematische Abstraktion der programmgesteuerten Rechenmaschine zu dienen.

Zum Schluß sei noch ganz deutlich gesagt, daß hier nicht der Eindruck erweckt werden soll, als ob die Automatentheorie alle Aspekte der Konstruktion und des Programmierens von digitalen Rechenanlagen und Datenverarbeitungssystemen erfasse. Diese Theorie ist aber ein Zweig (und zwar ein ganz junger und bescheidener) einer hoch entwickelten mathematischen Disziplin, die als Ganzes für den Computerspezialisten sehr wohl sein könnte, was für den Physiker die Theorie der Differentialgleichungen und was für den Statistiker die Maßtheorie ist. Gemeint natürlich ist die formale Logik; allerdings nicht mit Akzent auf Grundlagenforschung, sondern als eine Art von Mathematik gewisser endlicher Strukturen. Der angehende Computermann sollte (vielleicht sogar auf Kosten der Analysis) lernen, was der Logiker über exakte Sprachen weiß und wie man mit solchen umgeht. Vor allem sollte er sich ein tiefes Verständnis aneignen des Begriffes der Berechenbarkeit, wie er in den 30er Jahren seine Klärung fand mit der Entwicklung der Theorie der rekursiven Funktionen, TURING-Maschinen und Algorithmen.

## Literatur


[1] J. R. BÜCHI, A Decision Method for Sequential Calculus, Proceedings of the International Congress for Logic, Methodology, and the Philosophy of Science, Stanford University, 1960, S. 1—11.

[2] J. R. BÜCHI, Weak Second Order Arithmetic and Finite Automata, Zeitschrift für Mathematische Logik und Grundlagen der Mathematik **6** (1960), S. 66—92.

[3] J. R. BÜCHI, Regular Canonical Systems, University of Michigan Research Institute Technical Report 2794–7–T, 1959. To appear in Archiv für Mathematische Logik, 1962*

[4] BÜCHI, ELGOT and WRIGHT, The Non-existence of Certain Algorithms of Finite Automata Theory, Abstract, Notices of the American Mathematical Society **5** (1958), S. 98.

[5] A. W. BURKS and J. B. WRIGHT, Sequence Generators and Digital Computers, Proceedings, Symposium on Recursive Functions, Am. Math. Soc., New York 1961.

[6] A. W. BURKS and J. B, WRIGHT Theory of Logical Nets, Proc. IRE **41** (1953), S. 1357—1365.

[7] A. CHURCH, Application of Recursive Arithmetic to the Problem of Circuit Synthesis, Proceedings of the Cornell Logic Conference, Cornell University, 1957. Also see Application of Recursive Arithmetic in the Theory of Computing and Automata, Notes for a summer course, The University of Michigan, 1959.

[8] I. M. COPI, C. C. ELGOT and J. B. WRIGHT, Realization of Events by Logical Nets, Journal Ass. Comp. Mach, **5** (1958), S. 181—196.

[9] J. FRIEDMAN, Some Results in Church's Restricted Recursive Arithmetic, The Journal of Symbolic Logic **22** (1957), S. 337—342.

[10] S. C. KLEENE, Representation of Events in Nerve Nets and Finite Automata, Automata Studies, Princeton University Press, 1956, S. 3—41.

[11] I. T. MEDVEDEV, On a Class of Events Representable in a Finite Automaton, MIT Lincoln Laborators Group Report, S. 34—73, translated from the Russian by J. Schorr–Kon, 1958.

[12] J. MYHILL, Finite Automata and Representation of Events, WADC Report TR 57—624, Fundamental Concepts in the Theory of Systems, 1957, S. 112—137.

[13] E. F. MOORE, Gedanken-Experiments on Sequential Machines, Automata Studies, Princeton 1956, S. 129 bis 153.

[14] E. L. POST, Introduction to a General Theory of Elementary Propositions, Amer. Jour. Math. **43** (1921), S. 163—185.

[15] E. L. POST, Recursive Unsolvability of a Problem of Thue, Journal of Symbolic Logic **12** (1947), S. 1—11.

[16] M. RABIN and D. SCOTT, Finite Automata and Their Decision Problems, IBM Journal, April 1959, S. 114 bis 125.

[17] P. C. ROSENBLOOM, The Elements of Mathematical Logic, Dover 1950.

[18] A. THUE, Probleme über Veränderungen von Zeichenreihen nach gegebenen Regeln, Skrifter utgit av Videnskapsselskapet i Kristiania, I. Matematisk-naturvidens-kabelig klasse, No. 10 (1914), 34 S.

[19] HAO WANG, Circuit Synthesis by Solving Sequential Boolean Equations, Zeitsch. für Math. Logik und Grundl. der Math. **5** (1959), S. 291—322.


*Anschrift:* Prof. Dr. J. BÜCHI, Mainz, Mathematisches Institut der Universität

* Actually in vol. 6 (1964), pp. 91–111. [Editors.]

Reprinted from Zeit für Angew. Math. Mech., vol. 42, 1962, T9-16.

**Mathematische Theorie des Verhaltens endlicher Automaten**

von J.Richard Büchi

Zeit. für Angew. Math. Mech., vol. 42, 1962, T9-16.

Translated by Sylvia Büchi, Peter Deussen, and Dirk Siefkes

# Mathematical Theory of the Behavior of Finite Automata*

by J.R. Büchi**

The concept of finite automata originated as a mathematical abstraction of deterministic discrete sequential systems, such as modern computers. Characteristic for the functioning of such systems is feedback or self-control. The theory of pure switching (Boolean algebra) does not suffice for the comprehension of these phenomena, and must be extended to the theory of finite automata (Kleenean algebra), which can indeed be called the theory of switching through feedback. It is to be hoped that a strongly mathematical development of concepts and theorems on finite automata will provide also the practitioner with an insight into the behavior and the possibilities of digital systems, which complements the purely empirical trying out in the areas of construction and use of switching networks. It seems questionable, however, whether the theory developed so far is practical in the more naive direct sense. On the other hand,more powerful and, above all, innovative systems will only be developed in the near future on the basis of deep-reaching insights of a general nature, as they originate solely from a persistent development of mathematical theories. It would therefore be desirable that a wide circle of mathematically thinking researchers from different fields should occupy themselves with finite automata, Turing machines, and similar structures, and that in the future these things were not just left to a few logicians and the many dilettantes. For some time now there has been a big confusion of catchwords (such as, for example, self-reproducing, self-organizing, or learning systems, operations research, communications science, etc.), which are characteristic for a whole complex of activities which has developed around the modern computers. Will big words and stimulating speculations be all there is? That this need not be the case is shown by the example of the little word "behaviour". As we will see in the following, at least with regard to a very simple type of system, this word can be given a precise meaning, which also then gives rise to a fruitful theory.

---

* Invited main address, yearly meeting German Soc. Appl. Math. and Mech.

** University of Michigan, Ann Arbor; Gutenberg-Universität Mainz

An *automaton* with k input states is an algebraic system **A** = <**S**, A, $f_1$,...,$f_k$,W> consisting of a set S, whose elements are the *states* of A, an element A of S, called *initial state*, mappings $f_i$ from S into S, called *transition operators*, and a subset W of S, called the *output* of **A**. The *input states* are denoted by the symbols 1, 2,..., k; each of them corresponds to a transition operator of **A**. The automaton is called *finite* if the set S of its states is finite.

| | A | B | C | D | E |
|---|---|---|---|---|---|
| 1 | B | D | C | B | B |
| 2 | C | C | D | E | E |
| W | F | T | F | F | T |

(a) Transition table of $\mathbf{A}_o$

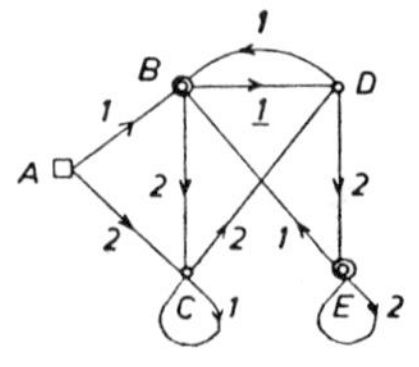

(b) Transition graph of $\mathbf{A}_o$

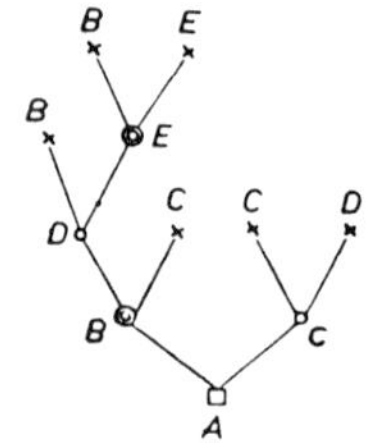

(c) Transition tree of $\mathbf{A}_o$

Figure 1

An example of a finite automaton $\mathbf{A}_o$ with two input states is given by figure 1(a). Its states are A, B, C, D, and E. The table is to be understood to say that, for example, $f_1B = D$, $f_2B = C$, B∈W since W(B) = T, C∉W since W(C) = F. (W(X) means the predicate X∈W.) In an obvious way the automaton $\mathbf{A}_o$ is also determined by the figures 1(b) or 1(c). These graphic representations of a finite automaton are clearly not necessary for the theory; they provide, however, useful supports for the understanding. On input of, say, the signal 1121, the automaton $\mathbf{A}_o$ reacts by running, starting with the initial state A, through the successive states $f_1A = B$, $f_1B = D$, $f_2D = E$, $f_1E = B$. The resulting state B is called the internal reaction of $\mathbf{A}_o$ on the input signal 1121, and is denoted by $\mathbf{A}_o(1121)$. The signal 1121 excites the outputs of $\mathbf{A}_o$, because $\mathbf{A}_o$ (1121)∈W. On the other hand $\mathbf{A}_o(212)$∉W, thus the signal 212 does not excite the output of $\mathbf{A}_o$.

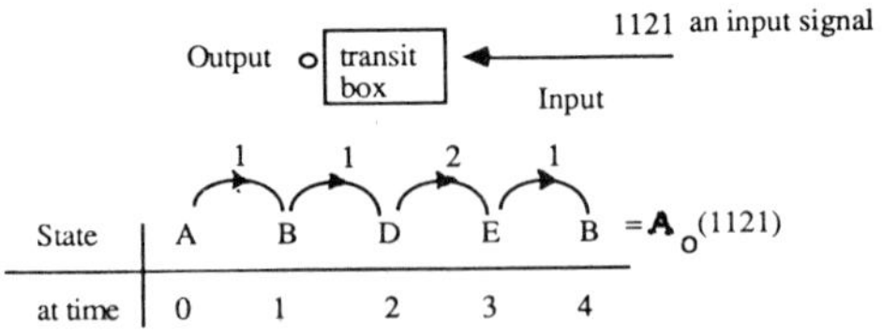


Figure 2

More precisely, the working of an automaton **A** with k input states is defined as follows:

The words over the alphabet 1,...,k are called the *input signals* of **A**. They form the set $N_k$ to which also the empty word 0 belongs. The *internal reaction* **A**(x) is the function which for every $x \in N_k$ is defined by the following simple recursion:

$$\mathbf{A}(0) = A$$
$$\mathbf{A}(xi) = f_i\, \mathbf{A}(x) \quad \text{for} \quad i = 1,\ldots,k.$$

For every input signal x there are now the following two alternatives:

$\mathbf{A}(x) \in \underline{W}$: x *excites* the output of **A**,
$\mathbf{A}(x) \notin \underline{W}$: x does not *excite* the output of **A**.

An automaton **A** is thus the mathematical abstraction of a concrete system which has the purpose to separate the set of input signals into two complementary sets. Only finite automata, however, play a role in this respect, and we will concentrate on such in the following. The idea here is not to make ad-hoc statements on this or that finite automaton. Rather, a generalized mathematical theory will be developed, consisting of definitions, proofs, and theorems, with the purpose of learning to understand the behaviour of finite automata. For that it is above all necessary to carve out appropriate concepts through precise definitions. We begin with the fundamental concept of behaviour as defined by Kleene [10].

Let $\mathbf{A} = \langle \underline{S}, A, f_1,\ldots,f_k,\underline{W} \rangle$ and $\mathbf{B} = \langle \underline{T}, B, g_1,\ldots,g_k,\underline{V} \rangle$ be two automata. One says "**A** *behaves like* **B**" when the two automata react to the same input signals with the same outputs, that is, if for every x of $N_k$ $\mathbf{A}(x) \in \underline{W}$ holds if and only if $\mathbf{B}(x) \in \underline{V}$. This fact can also be expressed as:

$$\{x \mid \mathbf{A}(x) \in \underline{W}\} = \{x \mid \mathbf{B}(x) \in \underline{V}\}.$$

It is therefore appropriate to call the set $\{x \mid \mathbf{A}(x) \in \underline{W}\}$ *the behaviour* of **A**, and to denote it by bh(**A**). With this notion the purpose of our theory can already be expressed quite sharply:

Main task: *Let $\Pi_k$ be the set of all behaviours of finite automata with* k *input states. The type of those subsets $\beta \subseteq N_k$ that belong to $\Pi_k$ is to be examined. Also the properties of $\Pi_k$ are to be investigated. Especially, criteria are to be found that for any $\beta \subseteq N_k$ tell us whether $\beta$ does or does not belong to $\Pi_k$.*

First it should be noted that only countably many subsets of $N_k$ are realizable by finite automata. Furthermore these sets are recursive; indeed, a finite automaton **A** directly presents an effective procedure which decides whether a given $x \in N_k$ belongs to bh(**A**). On the other hand there are very simple recursive subsets of $N_k$ which are not the behaviour of a finite automaton. For example, one easily shows that the set $\{yy \mid y \in N_2\}$, consisting of all words from $N_2$ which are composed of two equal pieces, does not belong to $\Pi_2$.

Compared to the variants found in the literature our definition of automata has the advantage that such fundamental concepts as homomorphism, direct product, and congruence can be taken over immediately from abstract algebra. Thereby first a systematic structural theory of automata is started and second a contribution to the main problem (the main task above) is won. Also the results of Moore [13] on minimal automata follow as by-products as it were. These matters can briefly be indicated as follows. In doing so we will restrict ourselves to automata where the set $\underline{S}$ of states is generated from the initial state A through the transition operators. States not "reachable" from A obviously have no influence on the behaviour of **A**.

A mapping h from $\underline{S}$ onto $\underline{T}$ is called *homomorphism of the automaton* $\mathbf{A} = \langle \underline{S}, A, f_1,\ldots,f_k,\underline{W} \rangle$ *onto* $\mathbf{B} = \langle \underline{T}, B, g_1,\ldots,g_k,\underline{V} \rangle$ if $hA = B$, $h(f_iX) = g_i(hX)$ and $X \in \underline{W}$ iff $hX \in \underline{V}$. Let $\mathbf{B} \leq \mathbf{A}$ mean that **B** is a homomorphic image of **A**. Then

$$\mathbf{B} \leq \mathbf{A} \rightarrow \mathrm{bh}(\mathbf{A}) = \mathrm{bh}(\mathbf{B})$$
$$(\exists \mathbf{C})(\mathbf{A} \leq \mathbf{C} \wedge \mathbf{B} \leq \mathbf{C}) \leftarrow \mathrm{bh}(\mathbf{A}) = \mathrm{bh}(\mathbf{B})$$

For every $\beta \subseteq N_k$ the set $V_\beta$ of all automata of behavior $\beta$ is a complete lattice with respect to the order relation $\leq$. This *structure lattice* $(V_\beta, \leq)$ has as its maximal element the *free automaton* $\langle N_k, 0, \varphi_1,\ldots,\varphi_k,\beta \rangle = \mathbf{F}_\beta$ where $\varphi_1 x = x1,\ldots,\varphi_k x = xk$. Its smallest element is the *minimal automaton* $\mathbf{M}_\beta$ of behaviour $\beta$. If there exist finite automata of behavior $\beta$, they are located in the lower part of the structure lattice; among all automata of behaviour $\beta$, $\mathbf{M}_\beta$ is also the minimal one with respect to the number of states.

Figure 3 shows <that part of> the structure lattice consisting of all homomorphic images of $\mathbf{A}_1$. As one sees from the example such lattices need not be distributive. Also chain theorems do not necessarily hold, and the irreducible elements might be placed in a very irregular manner. The structure theory of finite automata therefore appears to be a nearly hopelessly difficult business.

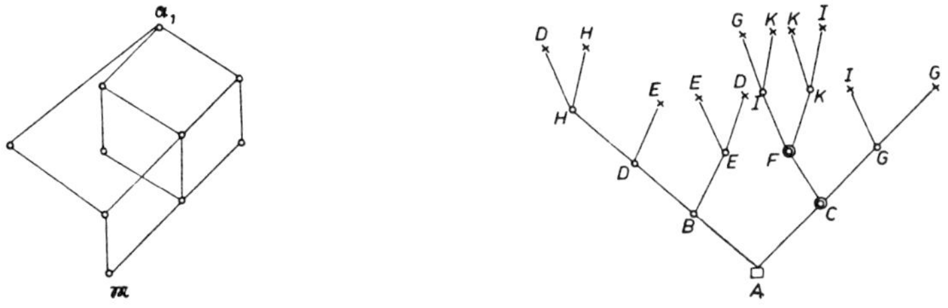

Structure lattice of $A_1$ automaton $A_1$

Figure 3

One will call an equivalence relation ~ over the set $\underline{S}$ of all states a *congruence of the automaton* **A** if both $(X \sim Y) \to (f_i X \sim f_i Y)$ and $(X \sim Y \wedge X \in \underline{W}) \to (Y \in \underline{W})$ hold. In the familiar way one forms the quotient **A**/~ of **A** with respect to a congruence. As one knows from algebra, the structure lattice of **A** now consists of all quotients of **A**, and is isomorphic to the lattice of congruences of **A**. In particular the coarsest (the supremum) $\simeq$ of all congruences of **A** yields the minimal automaton of behaviour bh(**A**). From there result easily *effective methods for minimization*; indeed, this amounts simply to finding the coarsest congruence of a given finite automaton.

Thus, up to isomorphism the automata of behaviour $\beta$ are exactly the quotients $\mathbf{F}_\beta/\sim$ of the free automaton $\mathbf{F}_\beta$. In particular, we obtain thereby the following criteria for the realizability of $\beta \subseteq N_k$ as behaviour of a finite automaton:

1. The set $\beta \subseteq N_k$ belongs to $\Pi_k$ if and only if the free automaton $\mathbf{F}_\beta$ admits a congruence of finite index (number of congruence classes).
2. The set $\beta \subseteq N_k$ belongs to $\Pi_k$ if and only if the coarsest congruence of $\mathbf{F}_\beta$ has finite index.

For a terse formulation it is worthwhile to introduce the following definitions. We call a congruence of the right-successor functions $\varphi_1,\ldots,\varphi_k$ a *right-congruence* on $N_k$. We denote the coarsest congruence of $F_\beta$ by $\simeq(\beta)$, and call it the *right-congruence induced by* $\beta$. The index of $\simeq(\beta)$ is also called the *right-rank* of $\beta$. The above criteria can now also be formulated as follows:

1 . The set $\beta \subseteq N_k$ belongs to $\Pi_k$ if and only if it is the union of some congruence classes of a right-congruence on $N_k$ of finite index.

(II) *The set $\beta \subseteq N_k$ belongs to $\Pi_k$ if and only if the right-rank of $\beta$ is finite; that is, if the right-congruence $\simeq(\beta)$ induced by $\beta$ has only finitely many classes.*

With the help of the formula

(a) $x \simeq y(\beta) \leftrightarrow (\forall u)(xu \in \beta \leftrightarrow yu \in \beta)$

it is often possible to obtain the induced right-congruence sufficiently concretely (or at least to calculate a bound for its index), so that criterion II can really be used. For example one easily shows that the induced right-congruence of the set $\{xx \mid x \in N_2\}$ is the equality relation, and therefore cannot be realized through a finite automaton. The following can be said about the importance of criterion (1'):

Obviously $N_1 = \{0, 11, 11, 111,...\}$ is nothing but the set of natural numbers. The successor function is $\varphi_1 x = x1 = (x+1)$. Its congruences are, besides the equality, the well-known elementary congruences

$$x \equiv y(p,q) \leftrightarrow (x < p \wedge x = y) \vee (p \leq x \wedge p \leq y \wedge x \equiv y(\mathrm{mod}\, q)),$$

where $0 \leq p$ and $1 \leq q$. They are all of finite index $(p + q)$, and one usually calls the unions of their congruence classes the (ultimately) periodic sets of numbers. It seems therefore appropriate, also in case $k > 1$, to call unions of congruence classes modulo a right-congruence of finite index *right-periodic* subsets of $N_k$. The criterion (1') can then also be formulated as:

(I) *The set $\beta$ belongs to $\Pi_k$ if and only if it is right-periodic.*

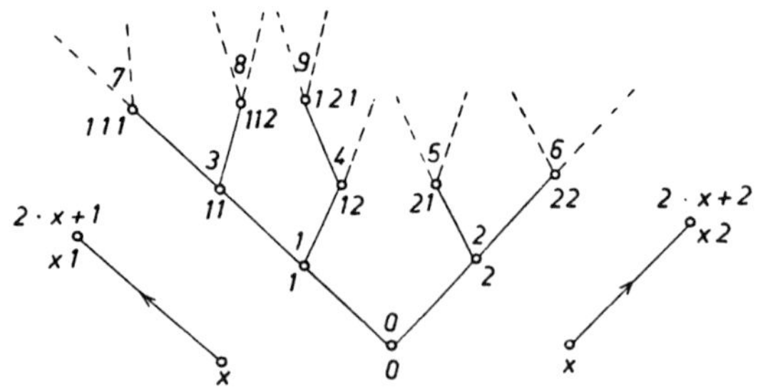


Figure 4

Incidentally, also in the case $k > 1$, $N_k$ can be interpreted as the set of natural numbers. Numbers are now generated from 0 no longer through a single, but through k free functions $\varphi_1 x = x1 = (k \cdot x+1),\dots,\varphi_k x = xk = (k \cdot x+k)$ (see figure 4 for the case $k = 2$). Already in school one learns to manipulate a variant of this k-ary number system $<N_k,0,\varphi_1,\dots,\varphi_k>$ (algorithms for k-ary addition and multiplication), and it is exactly this handling of words which from ancient times is called computing. Why have these k-congruences (congruences of the system of functions $k \cdot x+1,\dots,k \cdot x+k$) of finite index and their corresponding k-periodic sets of numbers not been systematically treated before? The theory of finite automata can indeed be termed a contribution to this part of elementary number theory. - A few remarks about k-periodic sets of numbers are to be found in Büchi [2]. - A very demanding problem is to classify somehow all k-congruences of finite index. That lattice theory does not help here has already been hinted at. - In contrast to the case $k = 1$, among the k-congruences of infinite index there are very complicated (in fact, nonrecursive) relations (Post and Markov, recursive unsolvability of the wordproblem for semigroups).

Over $N_k$ there are also the *left-successor functions* $\lambda_1 x = 1x,\dots,\lambda_k x = kx$, and their congruences, the *left-congruences on* $N_k$. A set $\beta \subseteq N_k$ is called *left-periodic* if the *left-congruence induced by* $\beta$, $\sim(\beta)$, has finite index. This index is also called the *left-rank* of $\beta$, and we have:

(b) $\quad x \sim y(\beta) \leftrightarrow (\forall u)(ux \in \beta \leftrightarrow uy \in \beta)$.

Finally we call a relation which is a left- and a right-congruence, simply a *congruence on* $N_k$. These are the congruences of the free semigroup $<N_k,^\wedge>$. A set $\beta \subseteq N_k$ is called *periodic* if the *induced congruence* $\sim(\beta)$ has finite index. This index is also called the *rank* of $\beta$, and we have:

(c) $\quad x \sim y(\beta) \leftrightarrow (\forall u,v)(uxv \in \beta \leftrightarrow uyv \in \beta)$.

Using (a), (b), and (c) it is now possible to show that the ranks $r_1$, $r_2$, $r$ of a set $\beta \subseteq N_k$ are bound by $r_1 \le 2^{r_2}$, $r_2 \le 2^{r_1}$ and $r \le r_1{}^{r_2}, r_2{}^{r_1}$. Therefore, since trivially $r_1, r_2 \le r$, from the finiteness of one of the three ranks follows the finiteness of the two others. Thus "right-periodic", "left-periodic" and "periodic" all have the same meaning. On account of (I) one therefore has:

Theorem 1: *The sets of behaviour of finite automata are exactly the periodic sets of words. These in turn are identical with the right-periodic sets and also with the left-periodic sets.*

Nonetheless of course the three ranks of a set $\beta \in \Pi_k$ need not be identical; they satisfy, however, the above cited inequalities. By the way the right rank $r_1$ of a periodic set $\beta$ is exactly the minimal number of states that are necessary for the realization of $\beta$ by an automaton. Indeed, $F_{\beta}/\simeq(\beta)$ is the minimal automaton of behaviour $\beta$. The left-right exchange of a word x is called the *converse* to x, and is denoted by x. Trivially $\beta$ is periodic if and only if $\beta$ is, and the left-rank of $\beta$ is equal to the right-rank of $\beta$. From theorem 1 thus follows: *If* $\beta \subseteq N_k$ *is realizable by a finite automaton, then* $\beta$ *is also.* In other words, as far as the question of realizability of $\beta$ is concerned, it makes no difference whether one decides to feed in the signals always from left to right, or always from right to left. A priori (without theorem 1) this is an astonishing result on finite automata, and indeed the number of states necessary for the realization of $\beta$ may be much greater for one kind of input than for the other (right-rank and left-rank of $\beta$). Incidentally the result can also be formulated thus: $\Pi_k$ is closed under conversion. Using the formulas (a), (b) for estimating the ranks it is not hard to show that periodicity is also preserved under the Boolean operations $\cap$, $\cup$, $\neg$, and the further, so called *regular operations*

$$\alpha \wedge \beta = \{xy \mid x \in \alpha \wedge y \in \beta\} \text{ and}$$

$$\alpha^* = \{x_1 x_2 \ldots x_n \mid x_1, x_2, \ldots, x_n \in \alpha\} = \alpha \cup (\alpha \wedge \alpha) \cup (\alpha \wedge \alpha \wedge \alpha) \cup \ldots .$$

From theorem 1 then follows:

Theorem 2: *The set* $\Pi_k$ *of the behaviours of finite automata is closed under the operations* $\cup$, $\cap$, $\neg$, $\leftarrow$, $\wedge$, $*$.

We will see that the abstract algebra $\mathbf{K}_k = <\Pi_k, \cup, \cap, \neg, \leftarrow, \wedge, *>$ is the multiplication table of the theory of switching circuits with feedback. This often is falsely asserted of Boolean algebra, which, however, rules pure switching only [1]. This was first discovered by Kleene [10] and we therefore propose to name $\mathbf{K}_k$ the *Kleenean algebras*. Unfortunately little is known about them; the important task remains to bring their theory to equal footing with that of Boolean algebra. For example the problems of axiomatization of the Kleenean algebra are unsolved.

The proof of theorem 2 sketched above is not the one from Kleene [10]. That paper contains much of perhaps secondary significance, and is therefore hard to digest. A better presentation of the main results is found in Copi, Elgot, and Wright [8]. In the following an elegant and very informative proof by Myhill [12] of theorem 2 is reproduced. It is based on the following generalization of automata, which may be also worth considering for completely different reasons. A *transition system* (also called flow diagram, or non-deterministic automaton) $\mathbf{C} = <\underline{S}, \underline{A}, F_1, \ldots, F_k, F, \underline{W}>$ consists of a set $\underline{S}$ of *states*, the *initial set* $\underline{A} \subseteq \underline{S}$, the *transition*

[1] Actually it is only two-valued switching which corresponds to Boolean algebra. n-valued switching is ruled by Post algebras [14], which unfortunately are rarely considered. For a short presentation see Rosenbloom [17].

*relations* $F_1,\dots,F_k$, $F \subseteq (\underline{S} \times \underline{S})$ and the *final set* $\underline{W} \subseteq \underline{S}$. The relations $F_i$ correspond to the input states $i = 1,\dots,k$. F is also called the relation of *spontaneous* transitions, and corresponds to the empty signal 0. In figure 5 a transition system $\mathbf{C_o}$ with the states A, B, C, D, E is given by its transition graph. One sees immediately that $\mathbf{C_o}$ is not an automaton, for first it has two initial states, second the relations $F_1$ and $F_2$ are not functions, and third spontaneous transitions occur.

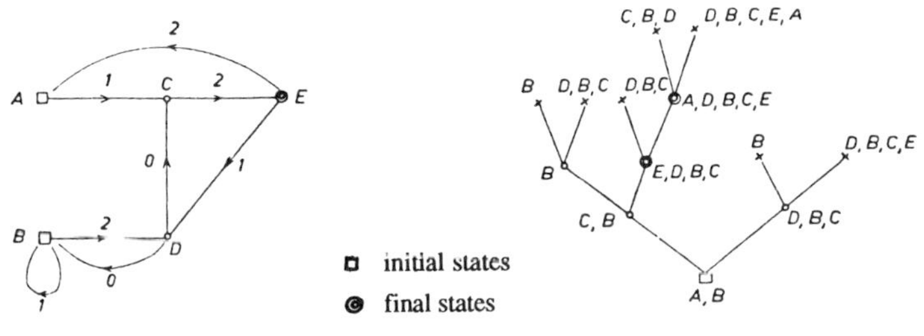


(a) The transition system $\mathbf{C_o}$ (b) Subset construction $\mathrm{Sb}(\mathbf{C_o})$

Figure 5

By the *contact set* of a transition system **C** we understand the set ct(**C**) consisting of all signals $x \in N_k$ which mark a directed path from some initial state to some final state. For example the signal 121122 = 12101202 belongs to ct($\mathbf{C_o}$), because in figure 5 the directed path $A \to C \to E \to D \to B \to B \to D \to C \to E$ starts from the initial set $\underline{A} = \{A, B\}$ and ends in the final set $\underline{W} = \{E\}$, and its edges are marked by 1, 2, 1, 0, 1, 2, 0, 2. It is easily seen that the behaviour of a finite automaton is exactly the contact set of the corresponding transition graph. Conversely also the following holds:

Lemma: *To every finite transition system* **C** *(with n states) one can construct a finite automaton* **A** = Sb(**C**) *(with at most* $2^n$ *states) such that* bh(**A**) = ct(**C**).

The subset construction Sb(**C**) was discovered by Myhill [12], and is carried out in figure 5(b) for the example $\mathbf{C_o}$. The states of the automaton Sb($\mathbf{C_o}$) are certain subsets of the states of $\mathbf{C_o}$. As initial state one chooses the initial set $\underline{A} = \{A, B\}$. Let the set $\underline{X}$ already be found as a state of Sb($\mathbf{C_o}$). Then as successor $f_1(\underline{X})$ one chooses first all 1-successors (in $\mathbf{C_o}$) of elements $X \in \underline{X}$, and closes the resulting set by successively adding all their spontaneous successors in $\mathbf{C_o}$. Similarly one constructs $f_2(\underline{X})$. Finally one chooses as elements of the output of Sb($\mathbf{C_o}$) all those sets which are states of Sb($\mathbf{C_o}$) and have common elements with the final set $\underline{W} = \{E\}$ of $\mathbf{C_o}$. It is now very easy to see that the behaviour of the automaton Sb($\mathbf{C_o}$) coincides just with the contact set ct($\mathbf{C_o}$). First the lemma of Myhill gives an additional criterion for the behaviour of finite automata:

(III) *A set* $\beta$ *belongs to* $\Pi_k$ *if and only if it is the contact set of a finite transition system.*

From this follows again very easily the closure of $\Pi_k$ under the operations $\leftarrow$, $\cup$, $\wedge$, $*$. It only has to be shown that with $\alpha$ and $\beta$ also $\alpha$, $\alpha \cup \beta$, $\alpha \wedge \beta$ and $\beta^*$ are contact sets. Now from a transition system for $\alpha$ one obtains one for $\alpha$ by reversing all transitions and by exchanging initial set and final set. Further it is indicated in figure 6 how from transition systems for $\alpha$ and $\beta$ new transition systems can be constructed with contact sets $\alpha \cup \beta$, $\alpha \wedge \beta$ and $\beta^*$, respectively. The figures hopefully explain these constructions sufficiently well (the double arrows indicate whole bundles of spontaneous transitions).

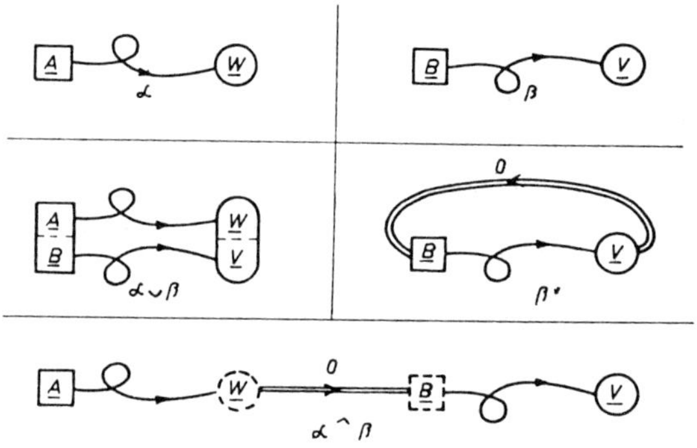


Figure 6

Thereby theorem 2 is now proved by the method of Myhill. (The closure of $\Pi_k$ under $\neg$ is in fact easily shown directly; the complement of the output of an automaton gives, of course, the complementary behaviour.) By the way, the construction for $\beta^*$ in figure 6 shows quite clearly the fundamental meaning of the operator $*$. It justly deserves to be called *feedback operator*.

For every finite set $\beta \subseteq N_k$ one easily constructs a finite transition system (automaton) whose contact set (behaviour) is equal to $\beta$. From theorem 2 therefore follows:

(d) Every set built from finite subsets of $N_k$ by the operations $\cup$, $\wedge$, $*$, is the behaviour of a finite automaton.

A formula that is built up from the symbols 1,...,k using the operation symbols $\cup$, $\wedge$, $*$, is called *regular expression* (there often **(E** $\wedge$ **F)** is abbreviated by **(E F)**). To every such formula

**E** a value |**E**| $\subseteq N_k$ is assigned in an obvious way. For example, the expression $(1 \cup 2)^*2$ denotes the set of all words in $N_2$ that end with 2. It is now easy to see that (d) holds in the following stronger form:

Synthesis Theorem: *There is an effective procedure that to every regular expression* **E** *gives a finite automaton* **A** *of behavior* |**E**|.

Such a method can be described in this way:

1. Using the construction from figure 6 one sets up a transition system **C** whose contact set is the value of the given expression **E**.
2. To **C** one applies the subset construction. The automaton **A** so obtained has the desired behavior |**E**|.

In figure 7 this construction is carried out for the expression $(12^*1 \cup 22)^*22$.

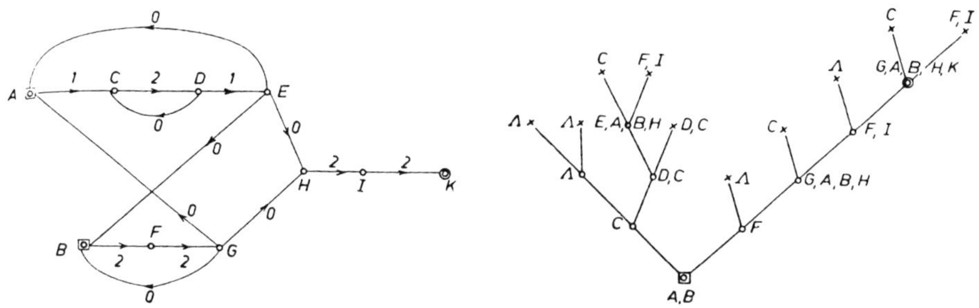


Figure 7

The reversal of the synthesis theorem was also found by Kleene [10]. For the proof we refer to that paper, as well as to Myhill [12], Copi, Elgot, and Wright [8], Rabin and Scott [16]. Now also the symbol $\Lambda$ for the empty set is to be counted to the regular expressions.

*Analysis theorem: There is an effective procedure that to every finite automaton* **A** *yields a regular expression* **E** *whose value* |**E**| *is the behaviour of* **A**.

The two theorems together show that the regular expressions describe exactly the behaviour of finite automata. Their significance for the theory of switching circuits with feedback corresponds thus exactly to that of Boolean expressions in the theory of pure switching. Finally the analysis theorem gives the converse of (d), so that we obtain a further criterium for the behaviour of finite automata:

(IV) *A set $\beta \subseteq N_k$ belongs to $\Pi_k$ if and only if it is the value of a regular expression; that is, if it can be built up by the operations $\cup$, $\wedge$, * from finite sets.*

On account of theorem 2, in (IV) also the operations $\cap$, $\neg$, $\leftarrow$ can be added. Many important questions for regular expressions, which are related to the problems of the axiomatization of Kleenean algebras, remain unsolved. On one side, from the synthesis theorem it follows that the validity of equations $\mathbf{E} = \mathbf{F}$ between regular expressions can be decided algorithmically. On the other side, however, one does not know any finite set of identities from which all the valid equations $\mathbf{E} = \mathbf{F}$ can be derived. Furthermore, for regular expressions there is no useful concept of normal forms.

We have now discussed the fundamental results on the behaviour of finite automata. Deeper reaching results in this direction are to be found in Büchi [3]. It still remains for us to warn against an all too rigid and one-sided judgement on the significance of these matters. Surely it is simply unthinkable that insights into such intuitive and concrete structures, as are e.g. finite transition systems, should not be of practical value also in quite different contexts. In addition the arithmetic of words (or k-ary number systems) in general and of periodic sets of words (or k-ary congruences of numbers) in particular, deserves the attention of the pure mathematician. Also from this standpoint the theory of behaviour of finite automata is not lacking in content, as shown perhaps most beautifully by its application to a problem of Tarski, which after all was not approachable from other sides; Büchi [1].

A completely different inquiry into finite automata is formulated in a perhaps most detailed way by Church [7]; see also Büchi, Elgot, and Wright [4]: One is concerned here with problems of the existence of algorithms for the construction of finite automata which have to meet given requirements. Naturally such problems are fixed only when one has settled for an exactly described, thus formalized, language S in which these requirements can be formulated. Depending on the expressive power of the language S construction algorithms do or do not exist. A very strong algorithm of this kind is to be found in Büchi [1]. Further to this subject the following papers should be named: Friedman [9], Büchi [2], Wang [19].

For certain purposes it might be useful to interpret digital computers as Turing machines. With his universal machine Turing indeed has become the ingenious inventor of program control, which is so important today. Only one should not always be harping stubbornly on the practical uselessness of his linear tapes. Turing's conception of the machine can easily be formed more flexibly in many directions. (After all Turing's work had the more theoretical purpose of an exact definition of computability and of solving Hilbert's decision problem.) As a very far reaching idealization, however, remains the unboundedness of the tape. In contrast to the general Turing machine with finite, but unlimited, memory the memory of a finite automaton is limited by the number of its states. There is now a great variety of concepts between finite automata and Turing machines. See for example Rabin and Scott [16]. The important task remains to investigate whether one of these concepts serves better as a mathematical abstraction of the program controlled computer.

In conclusion it should be very clearly stated that it is not intended to raise here the impression that automata theory could comprise all aspects of constructing and programming digital computers and data processing systems. This theory is, however, a branch (and to be sure a quite young and modest one) of a highly developed mathematical discipline that as a whole could very well be, for the computer specialist, what the theory of differential equations is for the physicist, and measure theory for the statistician. Meant here of course is formal logic; not with an emphasis on foundations, but as a kind of mathematics of certain finite structures. The beginning computer man should (perhaps even at the expense of calculus) learn what the logician knows on precise languages and how one deals with such. Above all he should acquire a deep understanding of the concept of computability as it found its clarification in the 30s with the development of the theory of recursive functions, Turing machines, and algorithms.

## Literature

[1] J.R. Büchi, A Decision Method for Sequential Calculus, Proceedings of the International Congress for Logic, Methodology, and the Philosophy of Science, Stanford University, 1960, S. 1-11.

[2] J.R. Büchi, Weak Second Order Arithmetic and Finite Automata, Zeitschrift für Mathematische Logik und Grundlagen der Mathematik 6 (1960), S. 66-92.

[3] J.R. Büchi, Regular Canonical Systems, University of Michigan Research Institute Technical Report 2794-7-T, 1959. To appear in Archiv für Mathematische Logik, 1962. (Note added: Actually in vol. 6 (1964) pp. 91–111.)

[4] Büchi, Elgot and Wright, The Non-existence of Certain Algorithms of Finite Automata Theory, Abstract, Notices of the American Mathematical Society 5 (1958), S. 98.

[5] A. W. Burks and J.B. Wright, Sequence Generators and Digital Computers, Proceedings, Symposium on Recursive Functions, Am. Math. Soc., New York 1961.

[6] A.W. Burks and J.B Wright, Theory of Logical Nets, Proc. IRE 41 (1953), S. 1357-1365.

[7] A. Church, Application of Recursive Arithmetic to the Problem of Circuit Synthesis, Proceedings of the Cornell Logic Conference, Cornell University, 1957. Also see Application of Recursive Arithmetic in the Theory of Computing and Automata, Notes for a summer course, The University of Michigan, 1959.

[8] I.M. Copi, C.C. Elgot and J.B. Wright, Realization of Events by Logical Nets, Journal Ass. Comp. Mach. 5 (1958), S. 181-196.

[9] J. Friedman, Some Results in Church's Restricted Recursive Arithmetic, The Journal of Symbolic Logic 22 (1957), S. 337-342.

[10] S.C. Kleene, Representation of Events in Nerve Nets and Finite Automata, Automata Studies, Princeton University Press, 1956, S. 3-41.

[11] I.T. Medvedev, On a Class of Events Representable in a Finite Automaton, MIT Lincoln Laboratory Group Report, S. 34-73, translated from the Russian by J. Schorr-Kon, 1958.

[12] J. Myhill, Finite Automata and Representation of Events, WADC Report TR 57-624, Fundamental Concepts in the Theory of Systems, 1957, S. 112-137.

[13] E.F. Moore, Gedanken-Experiments on Sequential Machines, Automata Studies, Princeton 1956, S. 129-153.

[14] E.L. Post, Introduction to a General Theory of Elementary Propositions, Am. Jour. Math. 43 (1921), S. 163-185.

[15] E.L. Post, Recursive Unsolvability of a Problem of Thue, Journal of Symbolic Logic 12 (1947), S. 1-11.

[16] M. Rabin and D. Scott, Finite Automata and Their Decision Problems, IBM Journal, April 1959, S. 114-125.

[17] P.C. Rosenbloom, The Elements of Mathematical Logic, Dover 1950.

[18] A. Thue, Probleme über Veränderungen von Zeichenreihen nach gegebenen Regeln, Skrifter utgit av Videnskapsselskapet i Kristiania, I. Matematisk-naturvidenskabelig klasse, No. 10 (1914), 34 S.

[19] Hao Wang, Circuit Synthesis by Solving Sequential Boolean Equations, Zeitschrift für Math. Logik und Grundl. der Math. 5 (1959), S. 291-322.

# REGULAR CANONICAL SYSTEMS* **

J. RICHARD BÜCHI, Ann Arbor und Mainz

A special type of canonical systems, called *normal systems* and containing rules of production of form $a\underline{x} \rightarrow \underline{x}b$ only, plays a role in Post's version of the theory of effectiveness. One's attention is therefore naturally drawn to another type of canonical system, whose rules are of form $a\underline{x} \rightarrow b\underline{x}$. We shall call these *regular systems*.

As Post [2] has shown, normal systems, in spite of their deductive simplicity, produce all canonical (i.e. recursively generable) sets of words. At the same place Post claims that in contrast, regular systems produce only recursive sets of words. While this much is not difficult to prove, it requires a more careful investigation of deductions by regular rules to characterize, among all recursive sets of words, those which can be produced by regular systems. It turns out that they are of surprisingly simple nature. We shall prove that for a set $\beta$ of words the following are equivalent conditions:

(1) $\beta$ may be produced from a finite set of axioms by regular rules $a\underline{x} \rightarrow b\underline{x}$.

(2) $\beta$ is periodic in the sense that it may be produced from a finite set of axioms by rules of form $a\underline{x} \rightarrow ap\underline{x}$.

(3) $\beta$ is periodic in the sense that it is the union of some of the equivalence classes of a congruence of concatenation of finite partition.

To fully appreciate (2), the „expansive character" of rules of form $a\underline{x} \rightarrow ap\underline{x}$ should be noted ($apx$ is longer than $ax$). Furthermore, both (2) and (3) are very natural generalizations of the concept of an (ultimately) periodic set of natural numbers, from ordinary (1-ary) arithmetic to the $k$-ary arithmetic over the words on $k$ letters.

Another equivalent characterization of sets $\beta$ satisfying (1) is the following:

(4) $\beta$ is the behaviour of (i.e. the set accepted by) a finite automaton.

This shows that regular systems are related to finite automata in much the same way as normal systems are to Turing machines. The characterization (4) in fact is very similar to (2); finite automata are but another very special type of expansive regular systems.

---

* Eingegangen am 12. 3. 1962

** The work was done under contracts with the US. Office of Naval Research and Army Signal Corps and appeared as Technical Report 03105, 2794-7-T, University of Michigan, 1959.

*1. Terminology and notations.*

*Words* are finite strings formed from the *letters* $1_1, 1_2, 1_3, \ldots$, and *auxiliary letters* usually taken from the list $S_1, S_2, S_3, \ldots$. We will use as syntactic variables with or without subscripts:

| | |
|---|---|
| $A, B, C, \ldots$ | ranging over letters |
| $a, b, c, \ldots$ | ranging over words |
| $\underline{A}, \underline{B}, \underline{C}, \ldots$ | ranging over finite sets and families of words |
| $\alpha, \beta, \gamma, \ldots$ | ranging over sets of words |

A notation $\{a_i\}_r$ will be used ambiguously for the set, and also the family of words $a_1, a_2, \ldots, a_r$. *Concatenation* of words is denoted by $x^\frown y$, and often abbreviated as $xy$. $lg(x)$ denotes the *length* of the word $x$, $o$ denotes the word of length zero. An *alphabet* is a finite set of letters. $\underline{I}_k$ denotes the alphabet consisting of $1_1, \ldots, 1_k$, and $\underline{S}_n$ denotes the alphabet consisting of $S_1, \ldots, S_n$. $N_k$ denotes the set of all words on the alphabet $\underline{I}_k$, so that $N_1$ may be interpreted to be the set of natural numbers (zero included).
We add some remarks on the structure of $N_k$, which generalizes that of the natural number system $N_1$. The algebra $[N_k, o, ^\frown]$ is the *free semigroup with identity* [i.e., free relative to the equations $x^\frown 0 = x$, $0^\frown x = x$, $x^\frown(y^\frown z) = (x^\frown y)^\frown z$)] on $k$ generators $1_1, \ldots, 1_k$. More elementary is the following characterization of $N_k$: If $l_i(x)$ denotes the $i$-the *left successor* $1_i x$ of the word $x \in N_k$, then $(N_k, l_1, \ldots, l_k)$ is the *absolutely free algebra* (with $k$ unary operations) and one generator $o$. The same remark, of course, holds for the *right-successor functions* $r_i(x) = x1_i$. Further important relations on $N_k$ are the *left-segment relation* $x \leqq y$, and the *right-segment relation* $y \geqq x$, defined by

$$x \leqq y\colon (\exists u)\,[x^\frown u = y]$$
$$y \geqq x\colon (\exists u)\,[y = u^\frown x]$$

The *proper left- and right-segment* relations $x < y$, and $y > x$ are defined by $x \leqq y \wedge x \neq y$, and $y \geqq x \wedge x \neq y$. We will call $[N_k, <]$ the *left-tree* on $N_k$; valuable intuition is added by interpreting $[N_k, <]$ as the graph which is obtained by starting with a root $o$ and joining to every vertex $v$, on the level $s = lg(v)$, new vertices $v1_1, \ldots, v1_k$ on the next higher level s + 1. For later reference we introduce the following notions. If $\underline{X} \subseteq N_k$ then the *interior* int $(\underline{X})$, and the exterior ext$(\underline{X})$ are defined by:

$$u \in int(\underline{X})\colon (\exists x)\,[x \in \underline{X} \wedge u < x]$$
$$u \in ext(\underline{X})\colon (\exists x)\,[x \in \underline{X} \wedge x < u]$$

Futhermore, $\underline{X}$ will be called a *frontier* of the left-tree on $N_k$ if every $u \in N_k$ belongs to exactly one of the sets $\underline{X}$, $int(\underline{X})$, $ext(\underline{X})$.
If $a$ and $b$ are words then the notation $a\underline{x} \to b\underline{x}$ is called *a regular production.* This production *directly produces* from the word $u$ a word $v$, if there is a word $x$ such that $u = ax$ and $v = bx$. Thus regular productions are a very

simple type of canonical productions, studied by Post (see also Rosenbloom [7]).

*Definition 1.* A *regular system* on the alphabet $\underline{I}_k$ is a finite collection $\Sigma$ of regular productions, whereby the words $a$ and $b$ occurring in the productions are words on an alphabet $\underline{I}_k \cup \underline{S}$. If $\underline{S} \cap \underline{I}_k = \Lambda$ then the letters $S \in \underline{S}$ are called the *auxiliary letters* of $\Sigma$. If $\underline{S}$ is empty $\Sigma$ is called a *pure regular system* on $\underline{I}_k$.
If $\underline{A} = \{a_i\}_r$ and $\underline{B} = \{b_i\}_r$ are families of words then $\Sigma(\underline{A}, \underline{B})$ denotes the regular system whose productions are $a_1\underline{x} \to b_1\underline{x}, \ldots, a_r\underline{x} \to b_r\underline{x}$.
Let $\Sigma$ be a regular system on $\underline{I}_k$ with auxiliary alphabet $\underline{S}$. A sequence $u_1, \ldots, u_r$ of words on the alphabet $\underline{I}_k \cup \underline{S}$ is called a *$\Sigma$-deduction*, if, for any $i = 1, \ldots, r-1$ some production in $\Sigma$ directly produces from $u_i$ the word $u_{i+1}$. If $u$ and $v$ are words on $\underline{I}_k \cup \underline{S}$ for which there is a $\Sigma$-deduction $u, \ldots, v$ then we will write $u \mid \Sigma \vdash v$, and say that $\Sigma$ *produces from* $u$ the word $v$.

Without reference we will often use the following property which is characteristic for regular system $\Sigma$:

$$(*) \qquad u \mid \Sigma \vdash v \,.\supset.\, u^\frown y \mid \Sigma \vdash v^\frown y.$$

*Definition 2.* Let $\Sigma$ be a regular system on $\underline{I}_k$ with auxiliary alphabet $\underline{S}$, and let $\underline{U}$ and $\underline{V}$ be finite sets of words on the alphabet $\underline{I}_k \cup \underline{S}$. The set $\tau(\underline{U}, \Sigma, \underline{V})$ of words *produced by* $[\underline{U}, \Sigma, \underline{V}]$, and the set $\beta(\underline{U}, \Sigma, \underline{V})$ of words *accepted by* $[\underline{U}, \Sigma, \underline{V}]$ are defined by

$$x \,\varepsilon\, \tau(\underline{U}, \Sigma, \underline{V}) \quad .\equiv. \quad x \in N_k \wedge (\exists\, uv)\,[u \in \underline{U} \wedge u \mid \Sigma \vdash vx \wedge v \in \underline{V}]$$
$$x \,\varepsilon\, \beta(\underline{U}, \Sigma, \underline{V}) \quad .\equiv. \quad x \in N_k \wedge (\exists\, uv)\,[u \in \underline{U} \wedge ux \mid \Sigma \vdash v \wedge v \in \underline{V}]$$

The set $\tau(\underline{U}, \Sigma, \{o\})$ is also denoted by $\tau(\underline{U}, \Sigma)$, and will be called the *set of theorems* of $[\underline{U}, \Sigma]$. The elements $u$ of $\underline{U}$ are called the *axioms*.

We remark that even if $\Sigma$ has auxiliary letters which may come to use in $\Sigma$-deductions, $\tau$ and $\beta$ still consist of words $x$ on $\underline{I}_k$ only. Usually in the literature sets of theorems $\tau(\underline{U}, \Sigma)$ of canonical systems are studied. However, also the notion of accepted sets occurs in particular in the theory of Turing machines. If, as in regular systems, all productions of $\Sigma$ have but one datum, it is a matter of conveniency whether one prefers to work with $\tau$ or $\beta$. This is a consequence of the following remark. By the *converse production* to $a\underline{x} \to b\underline{x}$ we mean $b\underline{x} \to a\underline{x}$, and by the *converse system* to $\Sigma$ we mean the system $\overleftarrow{\Sigma}$ consisting of all converses to productions in $\Sigma$. Then it is clear that $(\beta(\underline{U}, \Sigma, \underline{V}) = \tau(\underline{V}, \overleftarrow{\Sigma}, \underline{U})$, and if $\Sigma_1 = \overleftarrow{\Sigma}$ then $\Sigma = \overleftarrow{\Sigma}_1$.

*Definition 3.* A *finite automaton* with *input states* $\underline{I}_k$ and *transit states* $\underline{S}_n$, is a regular system $\Sigma$ on $\underline{I}_k$ with auxiliary alphabet $\underline{S}_n$ which consist of $nk$ productions of form $S_i 1_p \underline{x} \to S_j \underline{x}$, and such that to every pair $S_i \in \underline{S}_n$, $1_p \in \underline{I}_k$, there is exactly one production $S_i 1_p \underline{x} \to S_j \underline{x}$ in $\Sigma$.
The functions $tr_p\colon \underline{S}_n \to \underline{S}_n$, defined by $tr_p(S_i) = Sj$ if and only if $S_i l_p \underline{x} \to S_j \underline{x}$

belongs to $\Sigma$, are called the *transition functions* of the finite automaton $\Sigma$; $[\underline{S}_n, tr_1, \ldots, tr_k]$ is called its *transition system*. The *rank* of $\Sigma$ is the number $n$ of transit states.

For any transit state $X \in \underline{S}_n$ and any sequence of input states $z \in N_k$ of a finite automaton $\Sigma$ on $\underline{I}_k$ there is exactly one transit state $Y \in \underline{S}_n$ such that $Xz \mid \Sigma \vdash Y$. This $Y$ is called the *response* of $\Sigma$ to $z$ with respect to the initial state $X$, and will be denoted by $X/z$ (or more precisely $X/_{\Sigma}z$).

We remark that a finite automaton $\Sigma$ is uniquely determined by its transition system, and that every system $[\underline{S}, tr_1, \ldots, tr_k]$ of function $tr_1$ from $\underline{S}$ to $\underline{S}$ is the transition system of a finite automaton $\Sigma$ on $\underline{I}_k$, namely the one whose productions are $S1_i x \rightarrow tr_i(S)\, x$. This shows that our concept of a finite automaton is essentially equivalent to others occurring in the literature. Note in this connection that the binary transition function $tr(1_i, S) = tr_i(S)$ could be used in place of $tr_1, \ldots, tr_k$. The response operation $X/z$ can be obtained by the following recursion

$$X/0 = X, \qquad X/1_i z = tr_i(X)/z$$

which perhaps more closely reflects the intuitive idea of how an automaton, originally in transit state $X$, responds to injection of signals $z = 1_i 1_j \ldots 1_p$ by going successively through the transition states $X$, $tr_i(X)$, $tr_j tr_i(X), \ldots X/z$.

*Definition 4.* A (*binary*) *output* of a finite automaton $\Sigma$ is a subset $\underline{V}$ of its transition states $\underline{S}$. The behavior of a finite automaton $\Sigma$ on $\underline{I}_k$ with initial state $U$ and output $\underline{V}$ is the set $\beta(U, \Sigma, \underline{V})$ of input words accepted by $[\{U\}, \Sigma, \underline{V}]$, i.e.,

$$x \in \beta(U, \Sigma, \underline{V}) \equiv U/x \in \underline{V}.$$

The intended interpretation is that if the automaton is in transit state $S$ then the output is either on or off according to whether $S$ does or does not belong to $\underline{V}$. The behavior $\beta(U, \Sigma, \underline{V})$ therefore consists of all input words $x$ which injected into the automaton initially in state $U$ ultimately produce a state $S$ which activates the output $\underline{V}$. In a somewhat modified form this concept of behavior is due to *Kleene* [4].

Let $\Phi$ be a finite automaton on $\underline{I}_k$ with transit states $\underline{S}_n$, and let $A \in \underline{S}_n$. We will say that a state $X \in \underline{S}_n$ is *A-accessible* if there is an input word $z \in N_k$ such that $A/z = X$, i.e., $Az \mid \Phi \vdash X$; let $\underline{S}_n(A)$ denote the set of $A$-accessible states. Suppose that $X_1 I_1 .. I_s, \ldots, X_s I_s, X_{s+1}$ is a $\Phi$-deduction. If $s \geqq n =$ rank of $\Phi$, then a repetition $Xp = Xq$. $1 \leqq p < q \leqq s + 1$ must occur in the deduction. It follows that $X_1 I_1 .. I_p I_{q+1} \ldots I_s, \ldots, X_p I_q I_{q+1} .. I_s, \ldots, X_s I_s, X_{s+1}$ is still a $\Phi$-deduction. Thus, if $Az \mid \Phi \vdash X$ then there must be a $y$ of length $lg(y) \leqq n$ such that $Ay \mid \Phi \vdash Y$. Consequently, $X$ is $A$-accessible if and only if there is a $y \in N_k$ such that $lg(y) \leqq n =$ rank of $\Phi$ and $A/y = X$. Therefore for any finite automaton $\Phi$ and transit state $A$ one can effectively select the set $\underline{S}(A)$ of $A$-accessible states. Furthermore, $\underline{S}(A)$ is closed under the transition functions $tr_1, \ldots, tr_k$ of $\Phi$, and the automaton $\Phi(A)$ obtained by retaining only $\underline{S}(A)$ as transit states has the properties:

(a) Every transit state of $\Phi(A)$ is $A$-accessible.

(b) $\Phi$ and $\Phi(A)$ yield the same response $A/z \in \underline{S}(A)$ to every input word $z \in N_k$.

*2. Reduced regular systems.*

It turns out that regular systems may be reduced to a simple form. In this section we will therefore present our basic result for this simple type of regular system.

*Definition 5.* A regular system $\Sigma$ on $\underline{I}_k$ with auxiliary alphabet $\underline{S}_n$ is called reduced if all its productions are either of the forms:

$$S_i 1_p \underline{x} \to S_j \underline{x} \qquad \textit{contraction}$$
$$S_i \underline{x} \to S_j \underline{x} \qquad \textit{neutration}$$
$$S_i \underline{x} \to S_j 1_p \underline{x} \qquad \textit{expansion}$$

The number $n$ of auxiliary letters will be called the *rank* of $\Sigma$.

Of special interest are those reduced regular systems $\Sigma$ which, like finite automata, do not contain expansions. They may be interpreted as *transition-graphs*: Take $\underline{S}_n$ as set of vertices. Join $S_i$ to $S_j$ by an arrow labelled by $1_h$ (by $o$), in case $S_i 1_h \underline{x} \to S_j \underline{x}$ ($S_i \underline{x} \to S_j \underline{x}$) belongs to $\Sigma$. The set $\beta(\underline{U}, \Sigma, \underline{V})$ consist of those words $u$, whose letters occur in order as labels on some path through the transition-graph, starting in $\underline{U}$ and ending in $\underline{V}$. Clearly $\beta(\underline{U}, \Sigma, \underline{V})$ is a recursive subset of $N_k$.
We will now show that to every reduced system $\Sigma$ there exists a reduced system $\Sigma_0$ without expansions, such that $\beta(\underline{U}, \Sigma, \underline{V}) = \beta(\underline{U}, \Sigma_0, \underline{V})$. It follows that $\beta(\underline{U}, \Sigma, \underline{V})$ is recursive for arbitrary reduced systems $\Sigma$.

*Definition 6.* Let $\Sigma$ be a reduced regular system on $\underline{I}_k$ with auxiliary alphabet $\underline{S}_n$. The system $ctr(\Sigma)$ in the same alphabets $\underline{I}_k$ and $\underline{S}_n$ consist of all contractions and neutrations which are derived rules of $\Sigma$, i.e., if $X, Y \in \underline{S}_n$ and $I \in \underline{I}_k$ then

$$XI\underline{x} \to Y\underline{x} \quad \text{belongs to } ctr(\Sigma) \textit{ if } XI \mid \Sigma \vdash Y$$
$$X\underline{x} \to Y\underline{x} \quad \text{belongs to } ctr(\Sigma) \text{ if } X \mid \Sigma \vdash Y$$

*Lemma 1.* If $\Sigma$ is a reduced regular system on $\underline{I}_k$ with auxiliary alphabet $\underline{S}_n$, and $\Sigma_0 = ctr(\Sigma)$ then for any $X, Y \in \underline{S}_n$ and $z \in N_k$,

$$Xz \mid \Sigma \vdash Y \quad .\equiv. \quad Xz \mid \Sigma_0 \vdash Y$$

*Corollary.* For any reduced regular $\Sigma$, $\underline{U} \subseteq \underline{S}_n$ and $\underline{V} \subseteq \underline{S}_n$, $\beta(\underline{U}, \Sigma, \underline{V}) = \beta(\underline{U}, ctr(\Sigma), \underline{V})$, and $\beta(\underline{U}, \Sigma, \underline{V})$ is a recursive subset of $N_k$.

Proof: That $Xz \mid \Sigma_0 \vdash Y$ implies $Xz \mid \Sigma \vdash Y$ clearly follows from the definition of $\Sigma_0 = ctr(\Sigma)$. Suppose now that $Xz \mid \Sigma \vdash Y$. In case $z = 0$ we have

$X \mid \Sigma \vdash Y$ so that, by definition of $\Sigma_0 = \text{ctr}(\Sigma)$, $X \mid \Sigma_0 \vdash Y$, i.e., $Xz \mid \Sigma_0 \vdash Yz$. In case $z = I_1 \ldots I_h \neq o$ there is a $\Sigma$-deduction $XI_1 \ldots I_h, \ldots, Y$. It is clear that this deduction must be of form:
(i) $X_1 I_1 ..I_h, \ldots, U_1 I_1 ..I_h, X_2 I_2 ..I_h, ----, U_r I_r ..I_h, X_{r+1} I_{r+1} ..I_h, \ldots, U_{r+1} I_{r+1} ..I_h, X_{r+2} I_{r+2} ..I_h, ----, U_h I_h, X_{h+1}, \ldots, Y$ whereby $X_1 = X$, and for $r = 1, \ldots, h\text{-}1$ $(r = h)$ the indicated contraction $U_r I_r ..I_h, X_{r+1} I_{r+1} ..I_h$ $(U_h I_h, X_{h+1})$ is the first to the left of any contraction $AI_r ..I_h, BI_{r+1} ..I_h$ $(AI_h, B)$ occurring in (i). It follows that for $r = 1, \ldots, h$, $X_r \mid \Sigma \vdash U_r$, and of course $U_r I_r \mid \Sigma \vdash X_{r+1}$, $X_{h+1} \mid \Sigma \vdash Y$. Thus, by definition of $\Sigma_0 = ctr(\Sigma)$, $X_r \underline{x} \to U_r \underline{x}$, $U_r I_r \underline{x} \to X_{r+1} \underline{x}$, and $X_{h+1} \underline{x} \to Y \underline{x}$ are productions belonging to $\Sigma_0$. Consequently the modified deduction,
(ii) $X_1 I_1 ..I_h, U_1 I_1 ..I_h, X_2 I_2 ..I_h, ----, U_r I_r ..I_h, X_{r+1} I_{r+1} ..I_h, U_{r+1} I_{r+1} ..I_h, X_{r+2} I_{r+2} ..I_h, ----, U_h I_h, X_{h+1}, Y$ is a $\Sigma_0$-deduction. Thus we have shown that $Xz \mid \Sigma \vdash Y$ implies $Xz \mid \Sigma_0 \vdash Y$, which concludes the proof of lemma 1.
The first part of the corollary follows by lemma 1 and definition 2. Furthermore by definition of $\Sigma_0 = ctr(\Sigma)$ it is clear that if $U 1_p \underline{x} \to V \underline{x}$, $V \underline{x} \to W \underline{x}$, $W \underline{x} \to Z \underline{x}$ are productions in $\Sigma_0$, then also $U 1_p \underline{x} \to W \underline{x}$, $V \underline{x} \to Z \underline{x}$ belong to $\Sigma_0$. It follows that every $\Sigma_0$-deduction $Xz, \ldots, Y$ can be modified to a $\Sigma_0$-deduction consisting of $h = lg(z)$ contraction followed by one neutration. Consequently, to check whether or not $x \in \beta(\underline{U}, \Sigma_0, \underline{V})$ one only has to investigate the finite number of $\Sigma_0$-deductions of length $\leqq lg(x) + 2$. Therefore $\beta(\underline{U}, \Sigma, \underline{V}) = \beta(\underline{U}, \Sigma_0, \underline{V})$ is recursive.
This corollary may be considerably improved in two ways. First, the system $ctr(\Sigma)$ can be effectively constructed (see corollary to lemma 4). Second, one can obtain a finite automaton which accepts the set $\beta(\underline{U}, \Sigma, \underline{V})$. We begin with the second point.

*Definition 7.* Let $\Sigma$ be a reduced regular system on $I_k$ with auxiliary alphabet $\underline{S}$, and *Let* $\underline{\underline{S}}$ be the set of all subsets $\underline{X} \subseteq \underline{S}$. Define the functions $sp$, $tr_1$, $\ldots$, $tr_k$ on $\underline{\underline{S}}$, and the subset $\underline{\underline{C}} \subseteq \underline{\underline{S}}$ as follows:

$$Y \in sp(\underline{x}) \quad .\equiv. \quad (\exists X)[X \in \underline{X} \wedge X \mid \Sigma \vdash Y)$$

$$Y \in tr_i(\underline{x}) \quad .\equiv. \quad (\exists X)[X \in X \wedge X1_i \mid \Sigma \vdash Y], \quad i = l, \ldots, k$$

$$Y \in \underline{\underline{C}} \quad .\equiv. \quad sp(\underline{X}) = \underline{X}$$

sub $(\Sigma)$ is the finite automaton on $\underline{I_k}$ whose transit states are the subsets $\underline{X} \in \underline{\underline{C}}$ of $\underline{S}$, and whose productions are:

$$\underline{Y}\, 1_i \underline{x} \to tr_i(\underline{Y})\, \underline{x}, \qquad \underline{Y} \in \underline{\underline{C}},\ i = 1, \ldots, k$$

To see that sub$(\Sigma)$ is well-defined it is necessary to establish that $tr_i(\underline{Y}) \in \underline{\underline{C}}$ for $\underline{Y} \in \underline{\underline{C}}$. This follows by (6) below. One migth object to using subsets of $\underline{S}$ as letters of an alphabet, in which case one would have to modify the definition

of sub$(\Sigma)$ by using some other alphabet whose letters are brought into one-to-one correspondence with the elements of $\underline{\underline{C}}$.
The following properties of $sp$, $tr_i$, $\underline{\underline{C}}$ are easily established:

(1) $\underline{X} \subseteq sp(\underline{X})$

(2) $\underline{X} \subseteq \underline{Y} \quad .\supset. \quad sp(\underline{X}) \subseteq sp(\underline{Y})$

(3) $sp(sp(\underline{X})) = sp(\underline{X})$

(4) $\underline{X} \subseteq \underline{Y} \quad .\supset. \quad tr_i(\underline{X}) \subseteq tr_i(\underline{Y})$

(5) $tr_i(sp(\underline{X})) = tr_i(\underline{X})$

(7) $\underline{X} \in \underline{\underline{C}} \quad .\equiv. \quad (\exists\, \underline{Y})\, [\underline{X} = sp(\underline{Y})]$

(6) $sp(tr_i(\underline{X})) = tr_i(\underline{X})$

Thus, $sp$ is a closure operation on $\underline{S}$, and $\underline{\underline{C}}$ is the collection of closed sets. Furthermore it was pointed out in section 1, following definition 3, that the response operation $\underline{X}/z$ of the automaton sub $(\Sigma)$ satisfies the recursion:

(8) $$\underline{X}/o = \underline{X}, \qquad \underline{X}/1_i z = tr_i(\underline{X})/z, \quad \textit{for } \underline{X} \in \underline{\underline{C}} \textit{ and } z \in N_k.$$

We prove next by an induction on $z$ that,

(9) $$\underline{X} \in \underline{\underline{C}} \wedge Y \in \underline{X}/z \quad .\supset. \quad (\exists X)\, [X \in \underline{X} \wedge Xz \mid \Sigma \vdash Y], \quad \text{for all } z \in N_k.$$

That (9) holds for $z = o$ is clear if one notes that $\underline{X}/o = \underline{X}$, $Yo \mid \Sigma \vdash Y$. Now we make the inductive assumption that (9) holds for $z = y$, and assume that $\underline{X} \in \underline{\underline{C}}$ and $Y \in \underline{X}/1_i y$. Then by (8), $Y \in tr_i(\underline{X})/y$, and by (6), $tr_i(\underline{X}) \in \underline{\underline{C}}$. Therefore, by the inductive assumption, there is a $U$ such that $U \in tr_i(\underline{X})$ and $Uy \mid \Sigma \vdash Y$. It follows that there is a $X \in \underline{X}$ such that $X1_i \mid \Sigma \vdash U$, and therefore $X1_i y \mid \Sigma \vdash Uy$. Together with $Uy \mid \Sigma \vdash Y$ this yields $X1_i y \mid \Sigma \vdash Y$. Thus we haves hown, $\underline{X} \in \underline{\underline{C}} \wedge Y \in \underline{X} \mid 1_i y \;.\supset.\; (\exists\, X)\, [X \in \underline{X} \wedge X1_i y \mid \Sigma \vdash Y]$, which completes the proof by induction of (9).
Again by induction on $z$ we show,

(10) $$X \in \underline{X} \in \underline{\underline{C}} \wedge Xz \mid \Sigma \vdash Y \quad .\supset. \quad Y \in \underline{X}/z, \text{ for } z \in N_k.$$

For $z = o$ this is established thus. $X \in \underline{X} \in \underline{\underline{C}}$ and $Xo \mid \Sigma \vdash Y$ implies $Y \in sp(\underline{X}) = \underline{X}$, which by (8) implies $Y \in \underline{X}/o$. Now we make the inductive assumption that (10) holds for $z = y$, and suppose that $X \in \underline{X} \in \underline{\underline{C}}$ and $X1_i y \mid \Sigma \vdash Y$. Then there is a $\Sigma$-deduction (i) $X1_i y, \ldots, U1_i y, Vy, \ldots, Y$. Here it may be assumed that $U1_i y$, $Vy$ is the first to the left of any occurence of a contraction of form $A1_i y$, $By$ in (i). This has the effect that the part $X1_i y, \ldots, U1_i y, Vy$ of (i) remains a $\Sigma$-deduction if one takes of the terminal segment $y$ from each word. Consequently, $X1_i \mid \Sigma \vdash V$, which together with $X \in \underline{X}$ yields $V \in tr_i(\underline{X})$. Furthermore the second part $Vy, \ldots, Y$ of (i) shows that $Vy \mid \Sigma \vdash Y$, so that by inductive assumption $Y \in tr_i(\underline{X})/y$, and by (8), $Y \in \underline{X}/1_i y$. Thus we have shown, $X \in \underline{X} \in \underline{\underline{C}} \wedge X1_i y \mid \Sigma \vdash Y \;.\supset.\; Y \in \underline{X}/1_i y$. This completes the proof by induction of (10).

*Lemma 2.* Let $\Sigma$ be a reduced regular system on $\underline{I}_k$ with auxiliary alphabet $\underline{S}$, and let $\underline{U}$ and $\underline{V}$ be any subsets of $\underline{S}$. Define the finite automaton $\Sigma_1 = \text{sub}(\Sigma)$, as in definition 7, take $\underline{U}' = sp(\underline{U})$ as initial state, and define the output $\underline{\underline{V}}$ by $\underline{Y} \in \underline{\underline{V}}$ if and only if $\underline{Y} \in \underline{\underline{C}}$ and $\underline{Y} \cap \underline{V} \neq \Lambda$ ($\Lambda$ denotes the empty set). Then $\beta(\underline{U}, \Sigma, \underline{V}) = \beta(\underline{U}', \Sigma_1, \underline{\underline{V}})$.

*Corollary.* If $\Sigma$ is reduced regular, then every set of words $\beta(\underline{U}, \Sigma, \underline{V})$, accepted by $\Sigma$ and sets $\underline{U}$ and $\underline{V}$ of auxiliary letters, is equal to the behavior of an output to the finite automaton $\Phi = \text{sub}(\Sigma)$, for proper choice of the initial state. If the rank of $\Sigma$ is $n$, then $\Phi$ may be taken to have rank $\leqq 2^n$.

Proof: Suppose $z \in \beta(\underline{U}, \Sigma, \underline{V})$. Then by definition 2 there are $U \in \underline{U}$ and $V \in \underline{V}$ such that $Uz \mid \Sigma \vdash V$. By (1) it follows that $U \in sp(\underline{U}) = \underline{U}'$ and by (10), $V \in \underline{U}'/z$. Because $V \in \underline{V}$ and $V \in \underline{X}/z$ it follows by the definition of $\underline{\underline{V}}$ that $\underline{U}'/z \in \underline{\underline{V}}$. Therefore, by definition 4, $z \in \beta(\underline{U}, \Sigma_,, \underline{\underline{V}})$. Thus we have shown $\beta(\underline{U}, \Sigma, \underline{V}) \subseteq \beta(\underline{U}', \Sigma_1, \underline{\underline{V}})$.
Suppose next that $z \in \beta(\underline{U}', \Sigma_1, \underline{\underline{V}})$. By definition 4 it follows that $\underline{U}'/z \in \underline{\underline{V}}$, i.e. there is a $V \in \underline{S}$ such that $V \in \underline{V}$ and $V \varepsilon \underline{U}'/z$. Therefore by (9) there is a $X \in \underline{U}'$ such that $Xz \mid \Sigma \vdash V$. Because $X \in \underline{U}' = sp(\underline{U})$ there is a $U \in \underline{U}$ such that $U \mid \Sigma \vdash X$, which together with $Xz \mid \Sigma \vdash V$ implies $Uz \mid \Sigma \vdash V$. Because $U \in \underline{U}$, $V \in V$ this yields, by definition 2, $z \in \beta(\underline{U}, \Sigma, \underline{V})$. Thus we have shown that also $\beta(\underline{U}', \Sigma, \underline{\underline{V}}) \subseteq \beta(\underline{U}, \Sigma, \underline{V})$, which concludes the proof of lemma 2. The corollary is an obvious consequence.

Our definition 7 of the automaton sub$(\Sigma)$ is a refinement of a subset-construction first used by Myhill [6], and later by Medvedev [8], and Rabin and Scott [1]. In these papers lemma 2 and its corollary are proved in the special case where $\Sigma$ consists of contractions only. While in their case it is obvious that sub$(\Sigma)$ can be effectively obtained, the situation is quite different if $\Sigma$ also contains expansions. We will still be able to show that sub$(\Sigma)$ is effectively constructable from $\Sigma$, however this requires a rather more careful investigation of $\Sigma$-deductions. Incidentally we will also obtain the result that *ctr*$(\Sigma)$, of definition 6 and lemma 1, may be effectively constructed. Because sub$(\Sigma)$ = sub(*ctr*$(\Sigma)$) the constructibility of sub$(\Sigma)$ actually comes to the same as that of *ctr*$(\Sigma)$.
We remark that for a reduced system any $\Sigma$-deduction contains words of form $Sx$, $S \in \underline{S}$ and $x \in N_k$ only. We define the *height of a $\Sigma$-deducation* to be the maximum of all length $lg(x)$ such that $Sx$ occurs in the deduction. By the *length of a $\Sigma$-deduction* we mean the number of words occurring as terms in the deduction. The important result in the process of proving constructibility of *ctr*$(\Sigma)$ and *sub*$(\Sigma)$ is:
*Lemma 3.* Let $\Sigma$ be a reduced regular system on $\underline{I}_k$ and of rank n. If $X$ and $Y$ are auxiliary letters of $\Sigma$ and $X \mid \Sigma \vdash Y$, then there is a $\Sigma$-deduction $X, \ldots, Sz, \ldots, Y$ of height $h \leqq kn^2$, and of length $s \leqq n(k^0 + k^1 + \ldots + k^h)$.

Proof: Suppose $X, \ldots, Wy, \ldots, Y$ is a $\Sigma$-deducation and suppose that $Wy = WI_h \ldots I_1$ is chosen such that $lg(y) = h =$ heights of the deduction. It is easy to verify that the deduction must be of form:
(i) $X, \ldots, U_1, X_1I_1, ----, U_rI_{r-1}..I_1, X_rI_r..I_1, ----, U_hI_{h-1}..I_1, X_h I_h..I_1, \ldots, WI_h..I_1, \ldots, Y_hI_h..I_1, V_hI_{h-1}, ..I_1, ----, Y_rI_r..I_1, V_rI_{r-1}..I_1, ----, Y_1I_1, V_1, \ldots, Y.$
Whereby for every $r = 1, \ldots, h$ the indicated expansion $U_rI_r..I_1, X_rI_{r-1}.. I_1$ (contraction $Y_rI_r..I_1, V_rI_{r-1}..I_1$) is the first one occurring to the left of (right of) the indicated occurence of $WI_h..I_1$, of any expansion of form $AI_{r-1}..I_1, BI_r..I_1$ (contraction of form $AI_r..I_1, BI_{r-1}..I_1$) occurring in (i). We note that this has the effect that for any $r = 1, \ldots, h$, and any $z \in N_k$ the deduction
(ii) $X_rI_rz, ----, U_hI_{h-1}..I_rz, X_hI_h..I_rz, \ldots, WI_h..I_rz, \ldots, Y_hI_h..I_rz, V_hI_{h-1}..I_rz, ----, X_rI_rz$, obtained from (i) by the indicated modifications is still a $\Sigma$-deduction.
Now suppose that $h > kn^2$. Then two of the triples $(X_1, Y_1, I_1), \ldots, (X_h, Y_h, I_h)$ must be identical, say $X_p = X_q$, $Y_p = Y_q$, $I_p = I_q$ for $1 \leqq p < q \leqq h$. Using (ii) with $r = q$, $z = I_{p-1}..I_1$ one obtains that the following modification of (i) still is a $\Sigma$-deduction:

(iii) $X, \ldots, U_1, X_1I_1, ----, U_pI_{p-1}..I_1, X_qI_qI_{p-1}..I_1, ----, U_hI_{h-1}.. I_qI_{p-1}..I_1, X_hI_h..I_qI_{p-1}..I_1, \ldots, WI_h..I_qI_{p-1}..I_1, \ldots, Y_hI_h..I_qI_{p-1}..I_1, V_hI_h..I_qI_{p-1}..I_1, ----, Y_qI_qI_{p-1}..I_1, V_pI_{p-1}..I_1, ----, Y_1I_1, V_1, \ldots, Y$

Clearly the height of the deduction (iii) is not larger than that of (i), and because $p < q$, (iii) is shorter than (i). Thus we have shown that to every $\Sigma$-deduction (i) $X, \ldots, Y$ of height $h > kn^2$ and length $s$ one can construct a $\Sigma$-deduction ($i_1$) $X, \ldots, Y$ of height $h_1 \leqq h$ and length $s_1 < s$. Now if $h_1$ still is larger than $kn^2$ one can iterate the procedure to obtain a $\Sigma$-deduction ($i_2$) $X, \ldots, Y$ of height $h_2 \leqq h_1$ and length $s_2 < s_1$, etc. Because $s > s_1 > s_2 > \ldots > 0$ it is clear that the construction must come to an end, say in $m$ steps. Clearly this is possible only if ($i_m$) is of heights $h_m \leqq kn^2$. Thus we have shown:

(a) If $X \mid \Sigma \vdash Y$ then there is a $\Sigma$-deduction $X, \ldots, Y$ of height $h \leqq kn^2$.

Suppose next that (j) $X = X_1u_1, X_2u_2, \ldots, X_su_s = Y$ is a $\Sigma$-deduction of height $h$. Then $lg(u_1), lg(u_2), \ldots, lg(u_s) \leqq h$. But there are just $r = k^0 + k^1 + \ldots + k^h$ words $z$ in $N_k$ such that $lg(z) \leqq h$, and therefore there are just $nr$ words $Sz$ which may occur in (j). Consequently if the length $s$ of (j) is larger than $nr$ there will be a repetition $X_pu_p = X_qu_q$, $p < q$ in (j). Then clearly ($j_1$) $X, X_2u_2, \ldots, X_pu_p, X_{q+1}\, u_{q+1}, \ldots, X_su_s, = Y$ is still a $\Sigma$-deduction of height $h_1 \leqq h$ and length $s_1 < s$. By iteration of this argument, if necessary, one finally arrives at a $\Sigma$-deduction $X, \ldots, Y$ of height $\leqq h$ and length $\leqq n \,.\, r = n(k^0 + k^1 \ldots + k^h)$. Together with (a) this establishes lemma 3.

7*

*Lemma 4.* For any reduced regular system $\Sigma$ on $\underline{I}_k$, and any auxiliary letters $X$, $Y$ one can effectively decide ($a$) whether or not $X \mid \Sigma \vdash Y$, ($b$) whether or not $X1_i \mid \Sigma \vdash Y$.

*Corollary.* To every reduced regular system $\Sigma$ one can effectively construct the system *ctr* ($\Sigma$), and the finite automaton sub($\Sigma$), of definitions 6 and 7.

Proof: Let $n$ be the rank of $\Sigma$, let $r = kn^2$. Then clearly there are but a finite number of $\Sigma$-deductions $X, \ldots, Y$ of height $\leqq r$ and length $\leqq n(k^0 + k^1 + \ldots + k^r)$. Using lemma 3 this means that to decide whether or not $X \mid \Sigma \vdash Y$ one has to investigate but a finite number of $\Sigma$-deductions. This establishes (a).

Suppose next $X1_i \mid \Sigma \vdash Y$. Then there is a $\Sigma$-deduction $X1_i, \ldots, U1_i, V, \ldots, Y$ whereby the indicated pair $U1_i$, $V$ is the first from the left of any occurrence of a contraction of from $A1_i$, $B$. Then clearly $X \mid \Sigma \vdash U$, $U1_i\underline{x} \to V\underline{x}$ is a contraction belonging to $\Sigma$, and $V \mid \Sigma \vdash Y$. Thus we have shown: ($c$) If $X1_i \mid \Sigma \vdash Y$ then there are auxiliary letters $U$ and $V$ such that $X \mid \Sigma \vdash U$, $U1_i\underline{x} \to V\underline{x}$ belongs to $\Sigma$, and $V \mid \Sigma \vdash Y$. The converse to ($c$) is obvious. Therefore, to decide whether or not $X1_i \mid \Sigma \vdash Y$ it is sufficient to check whether or not among the contractions $U1_i\underline{x} \to V\underline{x}$ of $\Sigma$ there is one such that $X \mid \Sigma \vdash U$ and $V \mid \Sigma \vdash Y$. Because of ($a$) this can be done effectively, which establishes the remaining part ($b$) of the lemma.

By definition 6 and lemma 4 it is clear that $\Sigma_0 = ctr(\Sigma)$ can be effectively obtained. Furthermore, by definition 7 and lemma 4 it is clear that to every set $\underline{X}$ of auxiliaries one can effectively find the sets $sp(\underline{X})$ and $tr_i(\underline{X})$. Consequently the system sub($\Sigma$) also can be effectively constructed from $\Sigma$. This establishes the corollary.

*Theorem 1.* To every reduced regular system $\Sigma$ on $\underline{I}_k$, and any subsets $\underline{U}$ and $\underline{V}$ of its auxiliary alphabet $S_n$ one can effectively construct a finite automaton $\Phi$ on $\underline{I}_k$ with initial state $C$ and output $\underline{D}$ such that the behavior $\beta(C, \Phi, \underline{D})$ is exactly the set $\beta(\underline{U}, \Sigma, \underline{V})$ of accepted words. Moreover $\Phi$ can be so constructed that its rank is at most $2^n$, whereby $n$ is the rank of $\Sigma$.

This statement remains true if $\beta(\underline{U}, \Sigma, \underline{V})$ is replaced by $\tau(\underline{U}, \Sigma, \underline{V})$.

Proof: The first part of this assertion follows by the corollaries to lemmas 2 and 4. To obtain the second part we note (see remark in section 1) that $\tau(\underline{U}, \Sigma, \underline{V}) = \beta(\underline{V}, \overleftarrow{\Sigma}, \underline{U})$, whereby the converse system $\overleftarrow{\Sigma}$ still is reduced regular. Using the first part of theorem 1 we construct the automaton $\Phi_1$ with initial state $C$ and output $\underline{D}_1$ such that $\beta(\underline{V}, \Sigma, \underline{U}) = \beta(C_1, \Phi_1, \underline{D}_1)$. Then clearly, $\tau(\underline{U}, \Sigma, \underline{V}) = \beta(C_1, \Phi_1, \underline{D}_1)$, which proves the second part of theorem 1.

## 3. *Pure regular Systems*

According to definition 1 a regular system $\Sigma$ on $\underline{I}_k$ is pure if it does not make use of any auxiliary letters. We will first show that the theorems $\tau(\underline{C}, \Sigma)$ of a pure system $\Sigma$ with respect to a finite set of axioms $\underline{C}$ must always be the behavior of some finite automaton. The method consists in constructing a reduced system red$(\Sigma)$ such that $\tau(\underline{C}, \Sigma) = \tau(\underline{U}, \text{red } (\Sigma), \underline{V})$, the assertion then follows by theorem 1.

*Lemma 5.* To every pure regular system $\Sigma$ on $\underline{I}_k$ and every finite set of axioms $\underline{C} \subseteq N_k$ one can effectively construct a reduced regular system red$(\Sigma)$ on $\underline{I}_k$ with auxiliary alphabet $\underline{S}$, and subsets $\underline{U} \subseteq \underline{S}$ and $\underline{V} \subseteq \underline{S}$, such that $\tau(\underline{C}, \Sigma) = \tau(\underline{U}, \text{red } (\Sigma), \underline{V})$.

Proof: From $[\underline{C}, \Sigma]$ we first construct $[\underline{C}_1, \Sigma_1]$ by introducing the auxiliary letter $S_1$ and putting $S_1c$ into $\underline{C}_1$, and $S_1a\underline{x} \rightarrow S_1\, b\underline{x}$ into $\Sigma_1$ just in case $c \in C$, and $a\underline{x} \rightarrow b\underline{x}$ is in $\Sigma$. If $[\underline{C}_m, \Sigma_m]$ with auxiliary letters $S_1, S_2, ----, S_{f(m)}$ has already been obtained, we construct $[\underline{C}_{m+1}, \Sigma_{m+1}]$ according to the following specifications:

(a) If $S_i1_jc \in \underline{C}_m$, then introduce the next auxiliary letter $S$ which has not yet come to use. Put $Sc$ into $\underline{C}_{m+1}$, and put the production $S\underline{x} \rightarrow S_i1_j\underline{x}$ into $\Sigma_{m+1}$.

(b) If $S_i1_ja\underline{x} \rightarrow u\underline{x}$ is in $\Sigma^m$, then introduce the next auxiliary letter $S$ which has not yet come to use. Put the productions $Sa\underline{x} \rightarrow u\underline{x}$ and $S_i1_j\underline{x} \rightarrow S\underline{x}$ into $\Sigma_{m+1}$.

(c) If $u\underline{x} \rightarrow S_i1_ja\underline{x}$ is in $\Sigma_m$, then introduce the next auxiliary letter $S$ which has not yet come to use. Put the productions $u\underline{x} \rightarrow Sa\underline{x}$ and $S\underline{x} \rightarrow S_i1_j\underline{x}$ into $\Sigma_{m+1}$.

It is clear that these constructions must come to an end with some $[\underline{C}_s, \Sigma_s]$, with auxiliary letters $S_1, S_2, ----, S_{f(s)}$, and having the properties that $\underline{C}_s$ consists of auxiliary letters only (else instruction (a) could still be applied), and that $\Sigma_s$ consists of reducéd regular rules only (else either instruction (b) or (c) could still be applied). Furthermore it is easily seen that $\tau(\underline{C}, \Sigma) - \tau(\underline{C}_1, \Sigma_1, S_1)$, and $\tau(\underline{C}_i, \Sigma_i, S_1) = \tau(\underline{C}_{i+1}, \Sigma_{i+1}, S_1)$ for $i = 1, \ldots, s\text{-}1$. Therefore $\underline{U} = \underline{C}_s$, $red(\Sigma) = \Sigma_s$, $\underline{V} = \{S_1\}$ is a triple as required in lemma 5.

*Theorem 2.* To every pure regular system $\Sigma$ on $\underline{I}_k$ and any finite set of axioms $\underline{C} \subseteq N_k$ one can effectively construct a finite automaton $\Phi$ on $I_k$, an initial state $A$, and on output $\underline{U}$, such that the set of theorem $\tau(\underline{C}, \Sigma)$ is exactly the behavior $\beta(A, \Phi, \underline{U})$.

Proof: Construct $[\underline{X}, \text{red}(\Sigma), \underline{Y}]$ according to lemma 5, so that $\tau(\underline{C}, \Sigma) = \tau(\underline{X}, \text{red}(\Sigma), \underline{V})$. Then construct $[A, \Phi, \underline{U}]$ from $\underline{X}$, red$(\Sigma)$, $\underline{Y}$ according to the second part of theorem 1, so that $\tau(\underline{X}, \text{red}(\Sigma), \underline{Y}) = \beta(A, \Phi, \underline{U})$. Then clearly $[A, \Phi, U]$ is as required in theorem 2.

We shall next establish the converse to theorem 2: every behavior $\beta(A, \Phi, \underline{U})$ of a finite automaton on $\underline{I}_k$ can be generated from a finite set $\underline{C} \subseteq N_k$ of axioms by a pure regular system $\Sigma$ on $\underline{I}_k$. This comes to showing how the auxiliary letters (transit states) can be eliminated from $\Phi$. We refer to the discussion of the lefttree $[N_k, <]$ at the beginning of section 1, and begin with an investigation of deductions in a particular type of pure systems.

*Definition 8.* Let $\underline{A} = \{a_i\}_r$ and $\underline{B} = \{b_i\}_r$ be families of elements of $N_k$, and let $\Phi$ be a finite automaton on $\underline{I}_k$ with transitstate $A$. If $\underline{B}$ is a frontier of $[N_k, <]$ and $\underline{A} \subseteq \operatorname{int}(\underline{B})$ then the pure regular system $\Sigma(\underline{A}, \underline{B})$ is called a *frontier system.* If in addition $A/a_i = A/b_i$ for $i = 1, \ldots, r$, then $\Sigma(\underline{A}, \underline{B})$ is called a $[A, \Phi]$ – *frontier system.*

*Lemma 6.* If $\Sigma = \Sigma(\underline{A}, \underline{B})$ is a frontier system on $\underline{I}_k$ then to every $z \in N_k$ there is a $z_0 \in \operatorname{int}(\underline{B})$ such that $z_0 \mid \Sigma \vdash z$.

Proof: To every $y \in N_k$ define the excess $ec(y)$ over the frontier $\underline{B}$ as follows:

$$(1) \qquad \begin{aligned} ec(y) &= -1 \quad , \quad \text{if } y \in \operatorname{int}(\underline{B}) \\ ec(y) &= lg(u) \quad , \quad \text{if } y = bu \text{ and } b \in \underline{B} \end{aligned}$$

Note that (1) unambiguously defines $ec(y)$, because by assumption $\underline{B}$ is a frontier. The proof of the lemma now goes by induction on $ec(z)$.
If $ec(z) = -1$ then by (1), $z \in \operatorname{int}(\underline{B})$. Therefore if $z_0 = z$ we have $z_0 \in \operatorname{int}(\underline{B})$ and $z_0 \mid \Sigma \vdash z$. This establishes the base of the induction. Next assume that $ec(z) = s > -1$, and make the inductive assumption:

$$(2) \qquad \begin{aligned} &\text{If } ec(y) < s \text{ then there is a } y_0 \in \operatorname{int}(\underline{B}) \\ &\text{such that } y_0 \mid \Sigma \vdash y. \end{aligned}$$

Because $ec(z) = s > -1$ it follows by (1) that there are $y$ and $b_i$ such that $z = b_i y$ and $s = lg(y)$. Because $a_i\underline{x} \to b_i\underline{x}$ is a production of $\Sigma$, $a_i y \mid \Sigma \vdash b_i y$, i.e., $a_i y \mid \Sigma \vdash z$. There are two cases to be considered, namely $(a)$ $ec(a_i y) = -1$, and $(b)$ $ec(a_i y) > -1$. In case $(a)$ it follows by (1) that $a_i y \in \operatorname{int}(\underline{B})$. Therefore, if $z_0 = a_i y$, we have $z_0 \in \operatorname{int}(\underline{B})$ and $z_0 \mid \Sigma \vdash z$. In case $(b)$ it follows by (1) that there are $v \in N_k$ and $b_p \in \underline{B}$ such that $ec(a_i y) = lg(v)$ and $a_i y = b_p v$. By assumption $a_i \in \operatorname{int}(\underline{B})$,an d therefore $b_p \nleq a_i$. Because $a_i y = b_p v$ this implies that $y > v$, and therefore $s = lg(y) > lg(v)$. Because $lg(v) = ec(a_i y)$ it follows that $ec(a_i y) < s$. Therefore, by inductive assumption (2), there is a $z_0 \in$ *int*$(\underline{B})$ such that $z_0 \mid \Sigma \vdash a_i y$. Because $a_i y \mid \Sigma \vdash z$, it follows that $z_0 \mid \Sigma \vdash z$. Thus also in case $(b)$ we have found a $z_0 \in$ *int*$(B)$, $z_0 \mid \Sigma \vdash z$. This concludes the proof of lemma 6.

*Lemma 7.* Let $\Phi$ be a finite automaton on $\underline{I}_k$ with transistate $A$ and output $\underline{U}$. And let $\Sigma = \Sigma(\underline{A}, \underline{B})$ be a $[A, \Phi]$ – frontier system. Then the behavior $\beta(A, \Phi, \underline{U})$ is equal to a set of theorems $\tau(\underline{C}, \Sigma)$. The axioms set is given by $c \in \underline{C} \;.\!\equiv\!.\; c \in int(\underline{B}) \wedge A/c \in \underline{U}$.
Proof: Suppose $u \mid \Sigma \vdash v$. Then there is a deduction $u = p_0 z_0$, $q_0 z_0 = p_1 z_1$, $q_1 z_1 = p_2 z_2, \ldots, q_{s-1} z_{s-1} = p_s z_s$, $q_s z_s = v$, whereby $p_i\underline{x} \to q_i\underline{x}$ belongs to $\Sigma$. Because $\Sigma$ is a $[A, \Phi]$–frontier system it follows that $A/p_i = A/q_i$, and by (*) of section (1), $A/p_i z_i = A/q_i z_i$. Therefore, $A/u = A/p_0 z_0 = A/q_0 z_0 = \ldots = A/p_s z_s = A/q_s z_s = A/v$. Thus we have shown

$$(a) \qquad u \mid \Sigma \vdash v \quad .\!\supset\!. \quad A/u = A/v$$

Suppose now that $y \in \tau(\underline{C}, \Sigma)$. Then there is a $y_0 \in \underline{C}$ such that $y_0 \mid \Sigma \vdash y$. By (*a*) it follows $A/y_0 = A/y$, and by definition of $\underline{C}$, $A/y_0 \in \underline{U}$. Therefore $A/y \in \underline{U}$, i.e., $y \in \beta(A, \Phi, \underline{U})$. This proves the part $\tau(\underline{C}, \Sigma) \subseteq \beta(A, \Phi, \underline{U})$ of the lemma.

Suppose next that $y \in \beta(A, \Phi, \underline{U})$, i.e., $A/y \in \underline{U}$. By lemma 6 there is a $y_0 \in int(B)$ such that $y_0 \mid \Sigma \vdash y$. It follows by (*a*) that $A/y_0 = A/y$. Thus $A/y_0 \in \underline{U}$ and $Y_0 \in int(\underline{B})$, i.e., by definition of $\underline{C}$, $y_0 \in \underline{C}$. Thus $y_0 \in \underline{C}$ and $y_0 \mid \Sigma \vdash y$ which means that $y \in \tau(\underline{C}, \Sigma)$. This proves the remaining part, $\beta(A, \Phi, \underline{U}) \subseteq \tau(\underline{C}, \Sigma)$, of lemma 7.

*Theorem 3.* To every finite automaton $\Phi$ on $\underline{I}_k$ with transit state $A$ and output $\underline{U}$ one can effectively construct a pure regular system $\Sigma$ on $\underline{I}_k$ and a finite set of axiom $\underline{C} \subseteq K_k$ such that the behavior $\beta(A, \Phi, \underline{U})$ is exactly the set of theorem $\tau(\underline{C}, \Sigma)$.

Proof: Let $n$ be the rank of $\Phi$, let $m = k^n$, and let $z_1, \ldots, z_m$ be an enumeration of all words $z \in N_k$, such that $lg(z) = n$. Now choose $(a_i, p_i)$ such that $a_i p_i$ is the first left-segment of $z_i$ with a repetition $A/a_i = A/a_i p_i$. That this is possible follows by the accessibility argument at the end of section 1. Furthermore, it is clear that for $i \neq j$ it is not possible that $a_i p_i < a_j p_j$, because else $a_j p_j$ would not be the first segment of $z_j$ of the kind considered. It is also clear that to every $u \in N_k$ there is $a_i p_i$ such that either $u < a_i p_i$ or $u = a_i p_i$ or $a_i p_i < u$. Consequently, $\underline{B} = \{a_i p_i\}_m$ is a frontier of $[N_k, <]$. Furthermore, $a_i \in int(\underline{B})$ and $A/a_i = A/a_i p_i$. Therefore $\Sigma = \Sigma(\underline{A}, \underline{B})$ is a $[A, \Phi]$ – frontier system. Thus we have indicated how one can effectively construct a $[A, \Phi]$ – frontier system to any finite automaton $\Phi$ and transitstate $A$. Theorem 3 now follows by lemma 7.

*Remark.* The system $\Sigma(\underline{A}, \underline{B})$ constructed in the proof of theorem 3 consist of as many as $m = k^n$ productions. However many among the pairs $(a_1, p_1), \ldots, (a_m, p_m)$ will usually turn out to be equal to each other, so that $\Sigma$ can be greatly reduced to $\Sigma_0$ by dropping repetitions (see end of next section for improved construction). In fact, only for more trivial automata $\Phi$ this does not happen. That the construction leads to a rather economical $\Sigma_0$ is indicated by the fact that $\Sigma_0$ is independent in the sense that none of its productions may be omitted without changing $\tau(\underline{C}, \Sigma_0)$. If one is not interested in this additional feature of economy one can obtain theorem 3 more directly, without using lemma 6 and frontier systems.

Note also that $\Sigma$ constructed in the proof of theorem 3 is a frontier system, which has the effect that it is *strictly expanding* in the sense that $u \mid \Sigma \vdash v$, $u \neq v$ implies that the excess (see proof of lemma 6 )of $v$ is larger than that of $u$. In section 5 we will analyze sets of theorems of such systems $\Sigma$. Via theorem 3 this leads to a better understanding of the structure of behaviors of finite automata.

It is possible to modify the construction as to obtain a $[A, \Phi]$ – frontier system $\Sigma(\underline{A}, \underline{B})$ with the property that $A/u \neq A/v$ for $u, v \in int(\underline{B})$. This has the effect that $\Sigma(\underline{A}, \underline{B})$ becomes minimal in the sense that the lengths of words a, $b$ of productions $a\underline{x} \rightarrow b\underline{x}$ in the system are small. Also the number of axioms needed will be minimal.

One may consider a triple $[\underline{C}, \underline{A}, \underline{B}]$, consisting of finite families $\underline{C}, \underline{A}, \underline{B}$ of words in $N_k$, to be a syntactic objet denoting the set $\tau(\underline{C}, \Sigma(\underline{A}, \underline{B})) \subseteq N_k$. Theorem 2 then is a *synthesis result*: to every notation $[\underline{C}, \underline{A}, \underline{B}]$ one can construct a finite automaton $\Phi$ with initial state $A$ and output $\underline{U}$ such that the behavior $\beta(A, \Phi, \underline{U})$ is just the set denoted by $[\underline{C}, \underline{A}\ \underline{B}]$. Theorem 3 is the corresponding *analysis result*.

*4. Periodic sets of words.*

In its own right the following seems to be a reasonable extension of the concept of a (ultimately) periodic set of natural numbers to the arithmetic of words in $N_k$:

*Definition 9.* A set $\beta \subseteq N_k$ is *closed* with respect to the production $a\underline{x} \rightarrow b\underline{x}$ if for any $u \in N_k$, $au \in \beta$ .>. $bu \in \beta$. If $\beta$ is closed with respect to $a\underline{x} \rightarrow ap\underline{x}$, we will say that $(a, p)$ is a *phase-period* of $\beta$.

A set $\beta \subseteq N_k$ is called *k-periodic* if there are a finite set $\underline{C} \subseteq N_k$ and a finite set of pairs $\underline{P} \subseteq N_k \times N_k$, such that $\beta$ is the smallest subset of $N_k$ containing $\underline{C}$ and having the phase-periods $(a, p) \in P$. The pair $(\underline{C}, \underline{P})$ then is called a *periodic description* of $\beta$.

A *periodic regular system* on $\underline{I}_k$ is a pure regular system $\Sigma$ on $\underline{I}_k$ all of whose productions are of the form $a\underline{x} \rightarrow ap\underline{x}$.

Later we will indicate more evidence in the direction of showing that $k$-periodicity is a natural generalization of periodicity on $N_1$.

*Lemma* 8. A set $\beta \subseteq N_k$ is $k$-periodic if and only if there is a periodic regular system $\Sigma$ and a finite set of axioms $\underline{C} \subseteq N_k$ such that $\beta$ is the set of theorems $\tau(\underline{C}, \Sigma)$. Furthermore, if $(\underline{C}, P)$ is a periodic description of $\beta$, one can take $\Sigma$ to consist of all productions $a\underline{x} \rightarrow ap\underline{x}$, $(a, p) \in \underline{P}$.

Proof: If $\Sigma$ is a system of productions on $\underline{I}_k$ and $\underline{C} \subseteq N_k$ is a system of axioms, then $\tau(\underline{C}, \Sigma)$ is the intersection of (smallest of) all set $\gamma \subseteq N_k$ such that $\underline{C} \subseteq \gamma$ and $\gamma$ is closed under the productions in $\Sigma$. (This is the well known fact used already by Dedekind in obtaining explicit set-theoretic definitions for recursively defined sets and relations on natural numbers). With this remark in mind one easily obtains the lemma from definition 9.

*Theorem 4.* A set $\beta \subseteq N_k$ is $k$-periodic if and only if it is the behavior $\beta(A, \Phi, \underline{U})$ of some finite automaton $\Phi$ on $\underline{I}_k$, initial state $A$, and output $\underline{U}$. Furthermore from a periodic description $(\underline{C}, \underline{P},)$ of $\beta$ one can effectively construct $[A, \Phi, \underline{U}]$ and conversely.

Proof: Suppose that $(\underline{C}, \underline{P})$ is a periodic description of $\beta$. Then by lemma 8 and theorem 2 one can obtain an automaton $\Phi$, initial state $A$, and output $U$ such that $\beta = \beta(A, \Phi, \underline{U})$. This establishes one direction of theorem 4.

Suppose now that $[A, \Phi, \underline{U}]$ is given. In the proof to theorem 3 we constructed $\Sigma = \Sigma(\underline{A}, \underline{B})$ and $\underline{C}$ such that $\beta(A, \varnothing, \underline{U}) = \tau(\underline{C}, \Sigma)$, and furthermore the family $\underline{B}$ is of form $\{a_i p_i\}_r$, whereby $\underline{A} = \{a_i\}_r$. Therefore, if $\underline{P}$ is the set consisting of the pairs $(a_1, p_1), \ldots, (a_r, p_r)$, then by lemma 8 the pair $(\underline{C}, \underline{P})$ is a periodic description of $\beta(A, \Phi, \underline{U})$. This establishes theorem 4 in the other direction.

*Remark.* A periodic description $(\underline{C}, \underline{P})$ of a set $\beta$ of words seems to give a ratther clear picture of how the set $\beta$ is built up. Therefore the value of theorem 4 is that it provides a good idea of how behaviors of finite automata look like. This is not the case for other characterizations of behaviors of finite automata, which have appeared in the literature. Let us therefore recapitulate the construction of $(C, P)$, in the proof of theorem 3, in more intuitive terminology:

*Construction of a periodic description for the behavior* $\beta(A, \Phi, \underline{U})$. Let $n$ be the rank of $\Phi$. Construct the finite part of the left tree $[N_k, <)$ starting with o at level o, and ending with level $n$, consisting of the words $z_1 = 1_1 1_1 .. 1_1, \ldots, z_m = 1_k 1_k .. 1_k$ of length $n$, in lexicographic order. Attach the label $A/u$ to each vertex $u$ in the tree. Find $y_1, y_2, \ldots,$ and $(a_1, p_1), (a_2, p_2), \ldots$ according to the following instructions

(1) Take $y_1$ to be $z_1$.

(2) If $y_i$ has been found then search along the path $o$ to $y_i$ for the first repetition $A/a = A/ap$ of a label, and let $(a_i, p_i) = (a, p)$.

(3) If $(a_i, p_i)$ has been found then search for the first element $y$ in the list $y_i = z_j, z_{j+1}, \ldots, z_m$ such that $a_i p_i < y$, and let $y_{i+1} = y$.

This procedure clearly comes to an end in $s \leqq m$ steps, and $\underline{B} = \{a_1 p_1, \ldots, a_s p_s\}$ is a frontier. Among the vertices $c$ occurring between $o$ and the frontier $\underline{B}$ (i. e., $c \in int(\underline{B})$) select those which carry a label $A/c \in \underline{U}$, and put them into the set $\underline{C}$. Let $\underline{P} = \{(a_1, p_1), \ldots, (a_s, p_s)\}$. Then $(\underline{C}, \underline{P})$ is a periodic description of $\beta(A, \Phi, \underline{U})$.

For a discussion of economy of this $(\underline{C}, \underline{P})$ see the remark following theorem 3. Going through, in an example, the construction of $(\underline{C}, \underline{P})$ in case $\Phi$ is given by a transition diagram $[\underline{S}, tr_1, \ldots, tr_k]$ will show the workability of this procedure, and will provide an idea of how much better one knows the behavior $\beta(A, \Phi, \underline{U})$ once the periodic description $(\underline{C}, \underline{P})$ is found.

## 5. *The structure of the set of theorems of a frontier system.*

In this section we will analyze how the sets $\tau(\underline{C}, \Sigma)$ for frontier systems are built up. Our analysis applies equally well to expansive systems (definition 11) which are slightly more general than frontier systems. The fundamental concept in our analysis is that of an elementary set:

*Definition 10.* Let $R$ be a binary relation on the numbers $1, 2, \ldots, r$. By an $R$-*family* $\underline{E} = \{e_{i,j}\}_R$ in $N_k$ we mean a doubly indexed family of words $e_{i,j} \in N_k$, whereby $e_{i,j}$ is defined just in case $iRj$ holds.

Let $\underline{E} = \{e_{i,j}\}_R$ be an *R-family* in $N_k$. Then the *$\underline{E}$-elementary sets* $\varrho_{p,q}(\underline{E}) \subseteq N_k$ are defined, for $p, q = 1, \ldots, r$, as follows:
$\varrho_{p,q}(\underline{E})$ consist of all words of form $e_{i_0, i_1} e_{i_1, i_2} \ldots e_{i_{s-2}, i_{s-1}} e_{i_{s-1}, i_s}$ whereby $i_0 = p$, $i_s = q$, and s $= 1, 2, 3, \ldots$ The word $o$ is included in $\varrho_{p,q}(\underline{E})$ just in case $p = q$
Note that a sequence $i_0, i_1, \ldots, i_s$ gives rise to a word $e_{i_0, i_1} e_{i_1, i_2} \ldots e_{i_{s-1}, i_s}$ just in case it is an R-sequence, i. e., in case $i_0 R i_1, \ldots, i_{s-1} R i_s$ holds.

*Definition 11.* Let $\underline{A} = \{a_i\}_r$ and $\underline{B} = \{b_i\}r$ be families in $N_k$. The pure regular system $\Sigma = \Sigma(\underline{A}, \underline{B})$ on $\underline{I}_k$ is called *expansive* in case, $b_i \leqq a_j$ for $i, j = 1, \ldots, r$. The R-family $\underline{E} = \{e_{i,j}\}$ of such a system $\Sigma$ is defined as follows:

$$iRj \quad .\equiv. \quad a_i < b_j \qquad , \text{ for } i, j = l, \ldots, r$$
$$iRj \quad .\supset. \quad a_i e_{i,j} = b_j$$

The elementary sets of $\Sigma$ are the $\underline{E}$-elementary sets, and will also be denoted by $\varrho_{i,j}(\Sigma)$.

*Lemma 9.* Let $\Sigma = \Sigma(\underline{A}, \underline{B})$ be an expansive system on $\underline{I}_k$, and let $\underline{E} = \{e_{i,j}\}_R$ be its R-family. Suppose that $c, \ldots, y$ is a $\Sigma$-deduction of length greater than 2. Then there is a $d \epsilon N_k$ and *a* $(i, j)$ such that $c = a_i d$ and $y \in b_j{}^\frown \varrho_{j,i}{}^\frown d$.

Proof: The $\Sigma$-deduction clearly must be of form $c = a_{i_0} d_0$, $b_{i_0} d_0 = a_{i_1} d_1$, $b_{i_1} d_1 = a_{i_2} d_2, \ldots, b_{i_{s-1}} d_{s-1} = a_{i_s} d_s$, $b_{i_s} d_s = y$. Because $\Sigma$ is expansive it follows from $b_{i_0} d_0 = a_{s_1} d_1$ that $a_{i_1} < b_{i_0}$, and therefore, $b_{i_0} = a_{i_1} e_{i_1, i_0}$ and $e_{i_1, i_0} d_0 = d_1$. Similarly one obtains $e_{i_2, i_1} d_1 = d_2, \ldots, e_{i_s, i_{s-1}} d_{s-1} = d_s$. Therefore $d_s = e_{i_s, i_{s-1}} \ldots e_{i_2, i_1} e_{i_1, i_0} d_0$. By definition, of $\varrho_{j,i}$ this implies $d_s \in \varrho_{j,i}{}^\frown d_0$, if $j = i_s$ and $i = i_0$. Because $y = b_{i_s} d_s$ we therefore have $y \in b_j{}^\frown \varrho_{j,i}{}^\frown d_0$. Because $c = a_i d_0$ $(i = i_0)$, this completes the proof of the lemma.

*Theorem 5.* If $\Sigma$ is an expansive regular system on $\underline{I}_k$ and $\underline{C}$ is any finite set of axioms then the set of theorems $\tau(\underline{C}, \Sigma)$ is a finite union of sets of form $b^\frown \varrho^\frown d$, whereby $\varrho$ is elementary. More exactly, if $\Sigma = \Sigma(\underline{A}, \underline{B})$, $\underline{A} = \{a_i\}_r$, $\underline{B} = \{b_i\}_r$, and $\varrho_{i,j} = p_{i,j}(\Sigma)$ are the elementary sets of $\Sigma$, then

$$\tau(\underline{C}, \Sigma) = \underline{C} \cup \bigvee_{\substack{i = 1, \ldots, r \\ j = 1, \ldots, r}} b_i{}^\frown \varrho_{i,j}{}^\frown \underline{D}_j$$

whereby $\underline{D}_j$ consists of all $d$ such that $a_j d \epsilon \underline{C}$.

Proof: Let $\{e_{i,j}\}_R$ be the R-family of $\Sigma$ (see definition 11). Then clearly $\Sigma$ contains (among others) the following productions:

(1) $\qquad a_i \underline{x} \to a_j e_{j,i} \underline{x}$, $\qquad$ whenever $jRi$

Suppose first that $y$ belongs to the indicated union. If $y \epsilon \underline{C}$ then clearly $y \epsilon \tau(\underline{C}, \Sigma)$. If $y \notin \underline{C}$ then $y \in b_{i_0}{}^\frown \varrho_{i_0, i_s}{}^\frown d$ for some $i_0, i_s$ and some $d$ such that $a_{i_s} d \epsilon \underline{C}$. Therefore, by definition of $\varrho_{i,j}$, either $i_0 = i_s$ and $y = b_{i_s} d$. or $y = b_{i_0} e_{i_0, i_1} e_{i_1, i_2} \ldots e_{i_{s-1}, i_s} d$. In thefi rst case it is clear that $a_i\, d \mid \Sigma \vdash b_{i_s} d$, and because $a_{i_s} d \epsilon \underline{C}$, $y = b_{i_s} d \epsilon \tau(\underline{C}, \Sigma)$. In the second case it follows by (1) that

$$a_{i_s}d,\ a_{i_{s-1}}e_{i_{s-1}},\ {}_{i_s}d,\ \ldots\ldots,\ a_{i_0}e_{i_0},\ {}_{i_1}\ldots e_{i_{s-1}},\ {}_{i_s}d$$

is a $\Sigma$-deduction. Because $a_{i_s}\, d \in \underline{C}$, and $a_{i_0}\underline{x} \to b_{i_0}\underline{x}$ is a production of $\Sigma$, this implies that $y = b_{i_0}e_{i_0},\ {}_{i_1}\ldots e_{i_{s-1}},\ {}_{i_s}d \in \tau(\underline{C}, \Sigma)$. Thus we have shown that the union indicated in theorem 5 is contained in $\tau(\underline{C}, \Sigma)$.

Assume now that $y \in \tau(\underline{C}, \Sigma)$, i.e., $\underline{C} \mid \Sigma \vdash y$. Then either $y \in \underline{C}$ or there is an $c \in \underline{C}$ and a $\Sigma$-deduction $c, \ldots, y$ of length greater than one. In case the length of the deduction is 2, it must be of form $c = a_i d$, $b_i d = y$. By definition of $\underline{D_i}$ it follows that $d \in \underline{D_i}$, and because $o \in \varrho_{i,i}$ we obtain $y \in b_i{}^\frown \varrho_{i,i}{}^\frown \underline{Di}$. In case the length of the deduction, is greater than 2 it follows by lemma 9 that $c = a_i d$ and $y \in b_j{}^\frown \varrho_{j,i}{}^\frown d$, for some $i, j$ and $d \in N_k$. By definition of $\underline{D_i}$, $c = a_i d$ implies $d \in \underline{D_i}$, so that also in this case $y$ belong to some $b_j{}^\frown \varrho_{j,i}{}^\frown d$. Consequently $\tau(\underline{C}, \Sigma)$ is contained in the union indicated in theorem 5, which completes the proof.

*Remark.* By their very definition one has a clear idea of how elementary sets are constructed. Theorem 5 therefore explains much about the structure of sets of theorems of expansive systems. Because the construction of theorem 3 (see also end of section 4) yields an expansive system, this explanation carries over to behaviors of finite automata, and, via theorems 1 and 2, to sets of theorems of other regular systems.

For an elementary set $\varrho_{i,j}(\underline{E})$ it is easy to set up a regular system $\Sigma$ and axioms such that $\varrho_{i,j}(\underline{E})$ becomes the set of theorems. More generally, every finite union of terms of form $a^\frown \varrho^\frown b$ with elementary $\varrho$, is easily seen to be the set of theorems of some regular system and finitely many axioms.

If $\underline{E} = \{e_{i,j}\}_R$ is an R-family on $N_k$, one may introduce a different letter $1_{i,j}$ corresponding to every pair $(i, j)$ such that $iRj$. The R-family $\underline{E}_0 = \{1_{i,j}\}_R$ then clearly is such that $\varrho_{i,j}(\underline{E})$ may be obtained from $\varrho_{i,j}(\underline{E}^0)$ by replacing in each word $u \in \varrho_{i,j}(\underline{E}^0)$ every occurrence of $1_{p,\,q}$ by the word $e_{i,j}$.

## 6. *Concluding remarks.*

a) It remains to generalize theorem 2 to arbitrary regular systems. This however does not pose new difficulties and can be handled thus:

Let $\Sigma$ be a regular system on $\underline{I}_k$ with auxiliary alphabet $\underline{S}_n$, and let $\underline{C}$ be a finite set of words on $\underline{I}_k \cup \underline{S}_n$. We take $\Sigma_1$ to be the pure regular system on the alphabet $\underline{I}_k \cup \underline{S}_n$ and consisting of the same productions as $\Sigma$. Then clearly

$$y \in \tau(\underline{C}, \Sigma) \quad .\equiv. \quad a \in N_k \wedge y \in \tau(\underline{C}, \Sigma_1)$$

By theorem 2 one can construct a finite automaton $\Phi_1$ with input alphabet $\underline{I}_k \cup \underline{S}_n$, transition alphabet $\underline{A} = \{A_1, \ldots, A_m\}$, such that

$$y \in \tau(\underline{C}, \Sigma_1) \quad .\equiv. \quad y \in \beta(A_1, \Phi_1, \underline{U})$$

whereby $\underline{U} \subseteq \underline{A}$. Now let $\Phi$ be the automaton on $\underline{I}_k$ with transit alphabet $\underline{A}$ obtained from $\Phi_1$ by simply dropping the productions $A_i S_p \underline{x} \to A_j \underline{x}$. Then clearly:

$$y \epsilon \beta(A_1, \Phi, \underline{U}) \quad .\equiv. \quad y \epsilon N_k \wedge y \epsilon \beta(A_1, \Phi_1, \underline{U})$$

It is now clear that $\beta(A_1, \Phi, \underline{U}) = \tau(\underline{C}, \Sigma)$.

b) Up to this point we have called regular, what should more appropriately be termed *right- regular*. Of course our results also hold for *left-regular systems*, whose rules are of form $\underline{x}a \to \underline{x}b$. However, additional investigation is necessary to prove the following stronger form of theorem 2:
If the system $\Sigma$ consists of right- and left-regular rules and $\underline{C}$ is a finite set of axioms, then the set of theorems $\tau(\underline{C}, \Sigma)$ is the behavior $\beta(A, \Phi, \underline{U})$ of a finite automaton $\Phi$ with initial state $A$ and output $U$. Again, $[A, \Phi, \underline{U}]$ can be effectively constructed.

c) Theorem 2 can also be generalized to systems $\Sigma$ which contain productions of form $a_1\underline{x}, \ldots, a_n\underline{x} \to b\underline{x}$. Let us call these *general-right-regular systems*. However, it does not seem possible to derive this result from theorem 2, one rather has to modify the proof, which is mainly contained in the lemmas of section 1. The difficulty is that deductions now are not any more linear, but of tree-form. This does not really require new ideas, however, the presentation of the proofs gets clumsy.
The fact that, also for general-right-regular systems, $\tau(C, \Sigma)$ is the behavior of a finite automaton, generalizes a result of Rabin and Scott [1] concerning "two way non erasing automata". Inspection will show that such automata, as well as other special types of Turing machines, may be interpreted as reduced general-right-regular systems.

d) In view of remarks *b*) and *c*) one might except that also systems $\Sigma$ containing both general-right-regular and general-left-regular rules produce recursive sets of words. However, as noted by Post [2], this is not the case. In fact every recursively generable set of words can be produced from a single axiom by rules of form $a_1\underline{x}, a_2\underline{x} \to b\underline{x}$ and $\underline{x}a_1, \underline{x}a_2 \to \underline{x}b$.
We intend to present elsewhere our proofs of the assertions in *b*, *c*, and *d*. It seems likely that *b* and *c* may be improved to,
*Conjecture*: If the system $\Sigma$ consist of right-regular and general-left regular rules and $\underline{C}$ is finite, then the set of theorems $\tau(\underline{C}, \Sigma)$ is the behavior $\beta(A, \Phi, \underline{U})$ of a finite automaton $\Phi$. Furthermore $[A, \Phi, \underline{U}]$ can be effectively constructed from $[\underline{C}, \Sigma]$.

e) A right-regular system $\Sigma$ on $\underline{I}_k$ may be called *symmetric* if with every production $a\underline{x} \to b\underline{x}$ it also contains the converse $b\underline{x} \to a\underline{x}$. If $\Sigma$ is pure, note that the relation $x \mid \Sigma \vdash y$ is a *right-congruence* on $[\overline{N}_k, ^\frown]$, i.e., it is an equivalence relation and

$$x \mid \Sigma \vdash y \quad .\supset. \quad x^\frown u \mid \Sigma \vdash y^\frown u$$

In contrast to the analogous word problem for semi-groups, formulated by Thue [10], and solved in the negative sense by Post [3], our results show that $x \mid \Sigma \vdash y$ is recursive, for any symmetric regular system. We have shown even more: there is a general method applying to all $x$, $\Sigma$, $y$ and yielding a decision as to whether or not $x \mid \Sigma \vdash y$.

f) Let us say that a right-congruence $x \sim y$ on $N_k$ is of finite rank $n$, if its partition consists of $n$ classes. The (right-) congruences of finite rank play the same role in the artihmetic on $N_k$ as do the relations $x \equiv y$ (*mod n*) in ordinary arithmetic. In particular:
A set $\beta \subseteq N_k$ is $k$-periodic if and only if it is the union of congruence classes of some (right-) congruence of finite rank.
An equivalent condition is that the *induced* right-congruence

$$x \sim y \ (\text{mod}\ \beta) \quad .\equiv. \quad (\forall u)\ (x^\frown u \in \beta \equiv y^\frown u \in \beta)$$

has finite rank. These assertions follow by theorem 4 and the remark that to every right congruence $\sim$ of finite rank $n$ there is a finite automaton $\Phi$ (of rank $n$) with transit state $A$, such that $A/x = A/y$ just in case $x \sim y$. (This has been noted also by Rabin and Scott [7]).
Using our results one can show that to every (right-) congruence $\sim$ of finite rank on $N_k$ there is a symmetric pure regular system $\Sigma$ on $\underline{I}_k$ such that $x \sim y$ if and only if $x \mid \Sigma \vdash y$.
Perhaps of special interest are the congruences $x \sim y$ (mod $\varrho$) for elementary sets $\varrho$. Then there are the meet-irreducible congruences of finite rank, which essentially correspond to congruences modulo powers of primes. To investigate in which manner the fundamental facts about primes generalize to the $k$-array arithmetic on words, would seem to be of importance for a better understanding of $k$-periodicity, i.e., the behavior of finite automata. Some remarks on $k$-array arithmetic are contained in Buchi [9].

g) Simplifying the work of Kleene [4], Myhill [6], and Copi, Elgot, and Wright [5] have shown that:

*Synthesis*: To every regular expression $E$ one can construct a finite automata $\Phi$, $A$, and $\underline{U}$ such that $E$ denotes the behavior $\beta(A, \Phi, \underline{U})$.

*Analysis*: To every finite automaton $\Phi$, $A$, and $\underline{U}$ one can construct a regular expression $E$ which denotes $\beta(A, \Phi, \underline{U})$.

Let $\Sigma_1$ and $\Sigma_2$ be reduced regular system without expansions on $\underline{I}_k$ and auxiliary alphabets $\underline{S}_1$ respectively $\underline{S}_2$. If $\underline{S}_1 \cap \underline{S}_2 = \Lambda$, $\underline{U}_1, \underline{V}_1 \subseteq \underline{S}_1$ and $\underline{U}_2, \underline{V}_2 \subseteq \underline{S}_2$, one can construct $\Sigma$, $\underline{U}$, $\underline{V}$ such that

(a) $\beta(\underline{U}, \Sigma, V) = \beta(\underline{U}_1, \Sigma_1, \underline{V}_1) \cup \beta(\underline{U}_2, \Sigma_2, \underline{V}_2)$

(b) $\beta(\underline{U}, \Sigma, V) = \beta(\underline{U}_1, \Sigma_1, \underline{V}_1)^\frown \beta(\underline{U}_2, \Sigma_2, \underline{V}_2)$

(c) $\beta(\underline{U}, \Sigma, V) = \beta(\underline{U}_1, \Sigma_1, \underline{V}_1)^*$

Namely as follows:

(a) Let $\Sigma = \Sigma_1 \cup \Sigma_2$, $\underline{U} = \underline{U}_1 \cup \underline{U}_2$, $\underline{V} = \underline{V}_1 \cup \underline{V}_2$

(b) Let $\Sigma$ consist of $\Sigma_1$, $\Sigma_2$, and $V\underline{x} \to U\underline{x}$ for $V \in \underline{V}_1$, $U \in \underline{U}_2$.
Let $\underline{U} = \underline{U}_1$, and $\underline{V} = \underline{V}_1$.

(c) Let $\Sigma$ consist of $\Sigma_1$ and $V\underline{x} \to U\underline{x}$, for $V \in \underline{V}$ and $U \in \underline{U}$
Let $\underline{U} = \underline{U}_1$, and $\underline{V} = \underline{V}_1$.

These remarks together with theorem 1 can be used to prove the synthesis-result.
By theorem 5 it is sufficient to prove the analysis result for elementary sets $\varrho_{i,j}(\underline{E})$. That to every elementary set $\varrho_{i,j}(\underline{E})$ one can obtain a regular expression is best shown by an induction on the cardinality of $\underline{E}$. This result also follows by the remark concerning $\underline{E}^0$ at the end of section 5, and the fact that the sets $\varrho_{i,j}(\underline{E}^0)$ are regular, which is simply arest atement of lemma 1 of Copi, Elgot, and Wright. (This lemma was also used by Myhill [6] and is implicit already in Kleene [4]).

h) Among the right- and left-regular systems there are right- and left-automata. Let $u^0$ denote the word obtained from $u$ by a left-right-inversion, and let $\beta^0$ consist of all $u^0$, $u \in \beta$. Then clearly $\beta$ is the behavior of a left-automaton just in case $\beta^0$ is the behavior of the corresponding right-automaton. However, a priori it seems likely that a behavior $\beta$ of some right-automaton might not be the behavior of a left-automaton, i.e., right-periodicity might not coincide with left-periodicity. That this is not actually the case follows from the theory of regularity; the primitive operations $\cup$, $\frown$, $^*$ on regular sets do not distinguis right from left.
While for automata behavior this remark is non-trivial, it is on the other hand quite easy to construct a reduced left-regular system $\Sigma_l$ to any right-automaton $\Phi$ such that $\beta(A, \Phi, \underline{U}) = \beta(\underline{U}, \Sigma_l, A)$, namely

$$\underline{x}\, 1_\varrho S_\mu \quad \to \quad \underline{x} S_\nu \text{ in } \Sigma_l$$

if and only if

$$S_\nu 1_\varrho \underline{x} \quad \to \quad S_\mu \underline{x} \text{ in } \Phi.$$

The desired result then follows by theorem 1.

*Theorem.* The following condition on a set $\beta$ of words in $N_k$ are equivalent:

(1) $\beta$ is the behavior $\beta(A, \Phi, \underline{U})$ of some finite right-automaton $\Phi$ $\underline{I}_k$ on with initial state $A$ and output $\underline{U}$.

(2) $\beta$ is the set $\beta(\underline{U}, \Sigma, \underline{V})$ accepted by a reduced right-regular system $\Sigma$ on $\underline{I}_k$, and sets $\underline{U}$ and $\underline{V}$ of auxiliary letters of $\Sigma$.

(3) $\beta$ is the set of theorems $\tau(\underline{C}, \Sigma)$ of a pure right-regular system $\Sigma$ on $\underline{I}_k$ and a finite number of axioms $\underline{C} \subseteq N_k$.

(4) $\beta$ is right-$k$-periodic, i.e., $\beta$ has a right-periodic description $(\underline{C}, \underline{P})$.

(5) The right-congruence generated by $\beta$ is of finite rank.

(6) $\beta$ is the union of equivalence classes modulo a right-congruence of finite rank.

(7) $\beta$ is the set of theorems $\tau(\underline{C}, \Sigma)$ of a general right-regular system $\Sigma$ on $\underline{I}_k$ and finite set $\underline{C}$ of axioms.

(8) $\beta$ is a regular set, i.e., $\beta$ is the set $\beta(E)$ denoted by a regular expression $E$.

$(i')$ standing for $(i)$ with "right" replaced by "left"; $i = 1, \ldots, 7$.

## BIBLIOGRAPHY


[1] Rabin, M., and Scott, D., "Finite Automata and Their Decision Problems", *IBM Journal*, April, 1959, pp. 114–125.

[2] Post, Emil L., "Formal Reductions of the General Combinatorial Decision Problem". *American Journal of Mathematics* Vol. 65, pp. 197–215. (1943).

[3] Post, Emil L., "Recursive Unsolvability of a Problem of Thue", *Journal of Symbolic Logic*, Vol 12, pp. 1–11 (1947).

[4] Kleene, S. C., "Representation of Events in Nerve Nets and Finite Automata", *Automata Studies*, Princeton University Press, 1956, pp. 3–41.

[5] Copi, I. M., Elgot, C. C., and Wright, J. B. "Realization of Events by Logical Nets", *Journal of the Association for Computing Machinery*, 5, pp. 181–196 (1958).

[6] Myhill, John, "Finite Automata and Representation of Events", WADC Report TR 57–624, *Fundamental Concepts in the Theory of Systems*, October, 1957, pp. 112–137.

[7] Rosenbloom, P. C., "*The Elements of Mathematical Logic*". Dover (1950).

[8] Medvedev, I. T., "On a Class of Events Representable in a Finite Automaton", MIT Lincoln Laboratory Group Report, 34–73, translated from the Russian by J. Schorr-Kon, June 30, 1958.

[9] Büchi, J. R., "Weak Second Order Arithmetic and Finite Automata", *Zeitschrift für Mathematische Logik und Grundlagen der Mathematik*, Vol. 6 (1960), pp. 66–92.

[10] Thue, Axel, "Probleme über Veränderungen von Zeichenreihen nach gegebenen Regeln", *Skrifter utgit av Videnskaps selskapet i Kristiania*, I. Matematisknaturvidenskabelig klasse 1914, no. 10, 34 pp.


Reprinted from Archiv für Math. Logik und Grundlagenforschung 6 (1964).

Reprinted from
AUTOMATA THEORY


# Algebraic Theory of Feedback in Discrete Systems, Part I[1]

J. RICHARD BÜCHI

*Department of Mathematics, Ohio State University*
*Columbus, Ohio*

## I. Introduction

Our world, both natural and technological, abounds in systems which may be thought of, at least in a first approximation, as operating in accordance with the following specifications:

1. *Finite number of internal states.* At each instance the system is in one, out of a finite number $n$, of well distinguishable internal configurations.

2. *Finite number of input and output states.* The system is connected to the environment by an input channel, through which at each instance one out of a finite number $k$ of well distinguishable stimuli can be imposed on the system. In turn, the system can influence the environment through an output channel, capable of taking on but a finite number $n$ of states.

Because of 1 and 2 the input, internal, and output states can change only at discrete time instances $t = 0, 1, 2, \ldots$.

3. *Determinism.* At time $t = 0$ the system is in a specific state $A$, called the *initial state.* The internal state at time $t + 1$ is uniquely determined by the pair consisting of the internal state at time $t$ and the input state at time $t$. The output state at time $t$ is uniquely determined by the internal state at time $t$.

Mechanical devices, or parts of machines working on mechanical principles, provide obvious examples of such discrete deterministic systems. (Clocks may be mentioned as input-free examples; a combination lock is clearly meant to operate according to specifications 1, 2, and 3.) It is rather tempting to consider certain biological systems (nerve nets, interaction among organs) from this point of view. Last but not least, we mention electronic devices, such as digital computers and their components.

[1] These notes were prepared under a grant from the National Science Foundation.

We are presenting here an outline of a mathematical theory of such discrete deterministic systems and their behavior. *Finite automaton* is the mathematical concept which renders precise the intuitive idea of *discrete deterministic system.* In his pioneering work of 1952, Kleene [1] gave a rigorous definition of "behavior," and he proved two theorems about the behavior of finite automata, which provide a clear understanding of what discrete deterministic systems can, and what they cannot, do. Much work has since been done in this field, so that it is now possible to present concisely the rudiments of a mathematical theory, which is appealing both to the practitioner and the mathematician:

(a) The rigorous development of basic concepts and theorems on automata provides a kind of understanding of discrete deterministic systems which cannot be obtained by empirical methods and experience alone.

(b) There are many intriguing solved and unsolved problems which well deserve attention, especially the attention of the mathematician with an interest in discrete questions. There is a strong intuitive background, and there are obvious connections to algebra (finite semigroups), graph theory, and logic.

Most of the material presented here is not new. However, we have chosen a strongly algebraic presentation. If the definition of "finite automata" is appropriately chosen, it turns out that all the basic concepts and results concerning structure and behavior of finite automata are in fact just special cases of the fundamental concepts (homomorphism, congruence relation, free algebra) and facts of abstract algebra. Automata theory is simply the theory of universal algebras (in the sense of Birkhoff [2]) with unary operations, and with emphasis on finite algebras. In turn, all the material presented here can be generalized to universal algebras with $n$-ary operations, and in part leads to novel conceptions in this field. From another point of view, the theory of finite automata may be viewed as a chapter in the arithmetic of words. It is a study of congruences of finite index on words, and these are a very natural generalization of the elementary congruences on natural numbers.

Finite automata and their theory are clearly of interest to the designer of digital systems; in fact, such a theory might well be called the "theory of switching through feedback." As a contribution to the study of words (i.e., sequences of symbols) the theory is of interest to formal linguistics (finite-state grammars are but another version of finite automata). But also for more theoretical purposes, finite automata have proved to be useful. The solution of a decision problem of logic, posed by Tarski and solved by

Büchi [3], and some preliminary results of Büchi [4], are based on the understanding of regular sets of words. We do not discuss these matters here, but refer to Church [5], where the history of results on design algorithms for finite automata is presented.

## II. The Transition Algebra of a Logical Net

The purpose of this section is to motivate the claim that the study of finite algebras with unary operations is the study of feedback in discrete

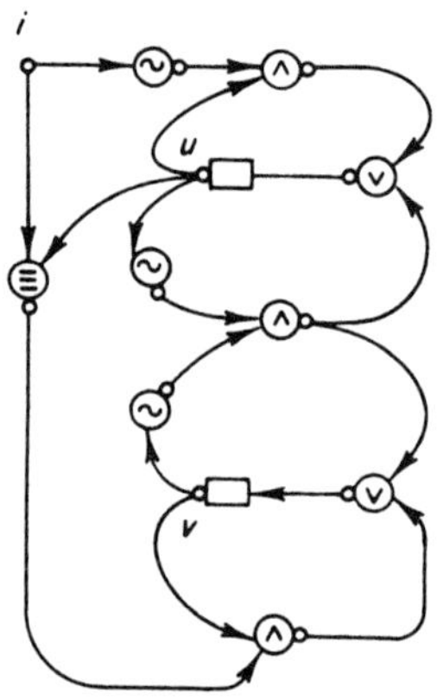

FIG. 1.

systems. We shall therefore discuss the concepts of logical nets and restricted recursion, by way of examples. For extensive studies of these matters we refer to Burks and Wright [6], Kleene [1], and Church [7].

**Definition 1.** A $k$-algebra is a system $\mathfrak{A} = \langle \mathbf{A}, A, f_1, \ldots, f_k \rangle$ consisting of a set $\mathbf{A}$, an element $A$ of $\mathbf{A}$, and unary operations $f_1, \ldots, f_k$ which map $\mathbf{A}$ into $\mathbf{A}$.

The value of $f_i$ at $U \in \mathbf{A}$ will be denoted by $U^{f_i}$.

Figure 1 represents an example of a (*well formed*) *logical net* with one input junction, labeled $i$; various *switch junctions*, marked by small circles; and two *delay junctions*, labeled $u$ and $v$. The *logical elements* $\sim$, $\wedge$, $\vee$, and $\equiv$ behave in the usual manner. The *delay elements* behave thus: At time 0 the junction of a delay is $F$ (inactive); at time $t' = t + 1$ the junction is $F$, if the input wire at time $t$ is $F$ [otherwise the junction is $T$ (active)]. It is now clear how the net transforms an input predicate $i = i0, i1, i2, \ldots$ into delay-output predicates $u$ and $v$. In fact, this transformation $i \to u, v$ is also defined by the *restricted recursion*

$$\begin{aligned} u0 &\equiv F \\ v0 &\equiv F \\ ut' &\equiv [\widetilde{it} \wedge ut] \vee [\widetilde{ut} \wedge \widetilde{vt}] \\ vt' &\equiv [[it \equiv ut] \wedge vt] \vee [\widetilde{ut} \wedge \widetilde{vt}] \end{aligned} \tag{1}$$

This recursion clearly provides the same information as the net. According to taste, one may prefer to deal either with logical nets or restricted recursions. In the sequel we shall discuss the recursions, simply because printing of nets is more expensive. Let us now consider the general form of a restricted recursion:

$$\begin{aligned} u0 &= A \\ ut' &= B[it, ut] \end{aligned} \tag{2}$$

Here $i$ and $u$ stand for vectors $i_1, \ldots, i_n$ and $u_1, \ldots, u_m$ of predicates, i.e., for sequences $i0, i1, i2, \ldots$ and $u0, u1, u2, \ldots$ of $n$-vectors ($m$-vectors) of truth values $T$ and $F$. $A$ stands for a given $m$-vector of truth values $A_1, \ldots, A_m$. $B[it, ut]$ is a given $m$-vector of Boolean expressions $B_1[it, ut], \ldots, B_m[it, ut]$, and "$=$" in (2) stands for componentwise equivalence. The recursion (2) is uniquely described by the pair $\langle A, B[X, Y]\rangle$.

Let $k = 2^n$, and let $C_1, \ldots, C_k$ be an enumeration of all $n$-vectors of truth values; call these the *input states* of the recursion (2). Let $\mathbf{A}$ be the set of all $m$-vectors of truth values, called the *internal states* of (2). For $i = 1, \ldots, k$, let $f_i$ be the unary operation on $\mathbf{A}$, defined by the vector of expressions $B[C_i, Y]$ (i.e., $Y^{f_i} = B[C_i, Y]$ for all $Y \in \mathbf{A}$). The $k$-algebra $\mathfrak{A} = \langle \mathbf{A}, A, f_1, \ldots, f_k\rangle$ may be called the *transition algebra* of the recursion (2) and of the corresponding logical net.

From the transition algebra $\mathfrak{A}$ of a recursion (2), one can clearly reconstruct the recursion, at least up to equivalent expressions $B'[X, Y]$. More exactly, if the transition operators $f_1, \ldots, f_k$ are given in some fashion, one can find a vector $B'[X, Y]$ of Boolean expressions, such that $Y^{f_i} = B'[C_i, Y]$ for all $i = 1, \ldots, k$ and all $Y \in \mathbf{A}$. While these expressions $B'$ may be different from $B$, they will define the same Boolean functions, i.e., $B'[X, Y] \equiv B[X, Y]$ must be valid, and the recursion $\langle A, B'[X, Y]\rangle$ defines the same input-to-output transformation $i \to u$ as (2). Suppose we do not possess the algebra $\mathfrak{A}$ itself but only an isomorphic $k$-algebra $\mathfrak{A}'$. In this case there is, in addition to the matter just discussed, an arbitrary choice of a one-to-one correspondence between the set of elements of $\mathfrak{A}'$ and $\mathbf{A}$, i.e., of a coding of the elements of $\mathfrak{A}'$ as vectors of truth values. We may express this situation as follows: *The theory of (isomorphism types of) finite k-algebras is the theory of logical nets, modulo matters of coding and switching.*

Because the study of logical nets (or restricted recursions) is the study of switching through feedback (in finite discrete systems) we may also say:

*The theory of finite k-algebras is the theory of feedback in discrete systems.*

We hope that this discussion will be clearer after the reader has worked out the following exercises.

*Exercise 1.*

(a) Find the transition algebra of the net in Fig. 1. Let $C_1 = T$, $C_2 = F$, and make up a table for the transition functions $f_1$ and $f_2$. Find expressions $B_1'[X, U, V]$, different from those occurring in (1), which yield a recursion, defining the same transformation $i \to u, v$ as (1).

(b) Find a logical net whose transition algebra $\mathfrak{A}$ is isomorphic to the algebra on the set $\mathbf{A} = \{1, \ldots, 8\}$, whose distinguished element is $A = 1$, and whose operators $f_1, f_2, f_3, f_4$ are given by Table 1.

TABLE 1

| | 1 | 2 | 3 | 4 | 5 | 6 | 7 | 8 |
|---|---|---|---|---|---|---|---|---|
| $f_1$ | 2 | 2 | 2 | 5 | 5 | 5 | 5 | 5 |
| $f_2$ | 2 | 2 | 4 | 7 | 7 | 8 | 7 | 8 |
| $f_3$ | 1 | 2 | 3 | 4 | 6 | 6 | 8 | 8 |
| $f_4$ | 1 | 3 | 3 | 8 | 8 | 8 | 8 | 8 |

(c) Find a logical net with two input functions $i, j$, which realizes the algebra $\mathfrak{A}$ given by $\mathbf{A} = \{1, 2, 3, 4, 5\}$, $A = 1$, and $f_1, f_2, f_3$ as shown in Table 2. Note that three delay junctions are required, so that the net will have to possess eight internal states and four input states. Choose the "don't-care" states and transitions wisely, to obtain a simple net.

TABLE 2

| | 1 | 2 | 3 | 4 | 5 |
|---|---|---|---|---|---|
| $f_1$ | 2 | 2 | 3 | 3 | 3 |
| $f_2$ | 1 | 2 | 3 | 4 | 4 |
| $f_3$ | 5 | 4 | 3 | 2 | 1 |

### III. The Response Function of a $k$-Algebra

Let $N_k$ denote the set consisting of all *words* (i.e., finite sequences) over the alphabet $1,\ldots,k$. Thus, examples of members of $N_3$ are 122, 221, 33213, and also the *empty word* 0. The *length* of a word $x$ will be denoted by $\ln(x)$. Thus, $\ln(122)=3$ and $\ln(0)=0$. The result of juxtaposing the word $x$ to the left of the word $y$ will be denoted by $x^\frown y$. $^\frown$ thus becomes a binary operation on $N_k$, called *concatenation*. If no confusion arises we shall sometimes abbreviate "$x ^\frown y$" by "$xy$." The members of $N_k$ may be called $k$-ary *input signals*.

Let $\mathfrak{A} = \langle \mathbf{A}, A, f_1, \ldots, f_k \rangle$ be a $k$-algebra as defined in Sec. II. Compare this mathematical concept with the idea of a discrete deterministic system, as described in Sec. I. To preserve the intuitive background we shall call $A$ the *initial state* of $\mathfrak{A}$; the numerals $1,\ldots,k$ are called the *input states* of $\mathfrak{A}$; the elements $U \in \mathbf{A}$ are called the (*internal*) *states* of $\mathfrak{A}$; and for $i=1,\ldots,k$ the map $f_i$ is called the *transition operator* of the input state $i$. Note that we shall later have to add additional structure to $\mathfrak{A}$, to provide for an output channel.

Words $x$ in $N_k$ are called *input signals* (or input histories) of $\mathfrak{A}$. The intended interpretation is as follows: The transition algebra $\mathfrak{A}$ at first is in its initial state $A$, and whenever it is in state $U$ and the input state $j$ is applied it will go into state $V = U^{f_j}$. Thus, $\mathfrak{A}$ reacts to the input signal 2113 by successively passing through the states $A$, $A^{f_2}$, $A^{f_2 f_1}$, $A^{f_2 f_1 f_1}$, $A^{f_2 f_1 f_1 f_3}$. The final state $rp(2113) = A^{f_2 f_1 f_1 f_3}$ may be called the *response* of $\mathfrak{A}$ to the input signal 2113. More precisely this may be put as follows.

**Definition 2.** Let $\mathfrak{A} = \langle \mathbf{A}, A, f_1, \ldots, f_k \rangle$ be a $k$-algebra. Then its binary *response function* $V = U/x$ is the function with arguments $U \in \mathbf{A}$, $x \in N_k$ defined by the recursion

$$\begin{aligned} U/0 &= U & & U \in \mathbf{S} \\ U/xj &= (U/x)^{f_j} & & j = 1,\ldots,k,\ x \in N_k \end{aligned}$$

The *response function* $rp(x)$ of $\mathfrak{A}$ is defined by $rp(x) = A/x$. For a fixed $a \in N_k$ the *a-transition operator* $f_a$ on $\mathbf{A}$ is defined by $U^{f_a} = U/a$.

Somewhat less precisely, $U/i_1 i_2 \cdots i_l = U^{f_{i_1} f_{i_2} \cdots f_{i_l}}$, and $U^{f_{221}} = U^{f_2 f_2 f_1}$. Thus, $f_{221}$ is but the result of *composing* the operators $f_2, f_2, f_1$. Note also that $f_0$ is the identity operator of $\mathbf{A}$. Thus the transition operators $f_a$, $a \in N_k$ form a *monoid* (semigroup with identity) $G(\mathfrak{A})$ of mappings from $\mathbf{A}$ into $\mathbf{A}$. $G(\mathfrak{A})$ is generated by the direct transition operators $f_1,\ldots,f_k$. Because a monoid $G(\mathfrak{A})$ may be attached to the transition part of an automaton, it is sometimes claimed that monoids should be used to provide an algebraic approach to finite automata. This, however, appears somewhat strained; unary algebras serve the purpose much more appropriately, and, in addition, they are in a sense simpler algebraic objects than monoids.

*Note:* The operators $f_a$ and $f_b$ may be equal, even for $a \neq b$. In fact if $\mathfrak{A}$ is finite then $G(\mathfrak{A})$ must be finite also.

Often without reference we shall use the following elementary properties of $U/x$ and $rp(x)$. The proofs are left to the reader. The binary response function of a $k$-algebra $\mathfrak{A}$ may also be calculated by the following recursions "from the left":

$$\text{(a)} \qquad \begin{array}{ll} U/0 = U & U \in \mathbf{S} \\ U/jx = U^{f_j}/x, & j = 1,\ldots,k, \;\; x \in N_k \end{array}$$

From this it follows more generally that

$$\text{(b)} \qquad \begin{array}{ll} (\forall U)_{\mathbf{A}}(\forall x, y)_{N_k} & U/(x \frown y) = (U/x)/y \\ (\forall x, y)_{N_k} & rp(x \frown y) = rp(x)/y \end{array}$$

This in turn yields

$$\text{(c)} \qquad \begin{array}{ll} (\forall U)_{\mathbf{A}}(\forall x, y, v)_{N_k} & U/x = U/y \supset U/(x \frown v) = U/(y \frown v) \\ (\forall x, y, v)_{N_k} & rp(x) = rp(y) \supset rp(x \frown v) = rp(y \frown v). \end{array}$$

Clearly the direct transition operators $f_1,\ldots,f_k$ of a $k$-algebra $\mathfrak{A} = \langle \mathbf{A}, A, f_1,\ldots,f_k \rangle$ may be recovered from the *binary transition operator* $\mathrm{tr}(U, X) = U/X$, defined for $U \in \mathbf{A}, X \in \mathbf{I} = \{1,\ldots,k\}$, and taking values in $\mathbf{A}$. Namely, $U^{f_j} = \mathrm{tr}(U, j)$. Therefore, in place of our unary transition operators $f_1,\ldots,f_k$ one could take the binary transition operator tr as primitive. In place of transition algebras one would then have to investigate systems of the form $\langle \mathbf{I}, \mathbf{A}, A, \mathrm{tr} \rangle$ whereby $A \in \mathbf{A}$ and tr maps $\mathbf{A} \times \mathbf{I}$ into $\mathbf{A}$. This actually is more commonly done in the literature on automata theory, and various other variants of the concept of a $k$-algebra are also in use. However, our $k$-algebras are just as natural intuitively and have the advantage of being algebraic systems of a more conventional sort. Consequently, it will be possible to apply standard ideas concerning such fundamental algebraic concepts as homomorphisms, congruence relations on free algebras, and direct products (see the following sections).

Note that *time* enters implicitly into our discussion through the definition of the response function $rp$. This time is discrete: At time 0 the algebra is in state $A$, at time 1 it is in a new state $A^{f_j}$ corresponding to an input state $j$ which has been injected at some instance between 0 and 1, etc. Whether or not a physical realization of $\mathfrak{A}$ is such that the time intervals $0-1, 1-2, \ldots$ must be of equal length does not concern us, and our theory also abstracts from the question whether the input state is to be injected at time $t$, or sometimes between $t$ and $t'$. It might be good here to consider the following two universal principles:

1. The more abstract the theory, the wider the range of applications.
2. The more abstract the theory, the less it will say in any one application.

The positive side of (2) can be brought out thus: If properly carried out an abstraction emphasizes the essential aspects of a more concrete situation.

Concluding this section we shall now introduce a graphical representation of a finite $k$-algebra $\mathfrak{A}$. In essence it is equivalent to the operation table. However, the graph is much more suggestive, and often provides a very

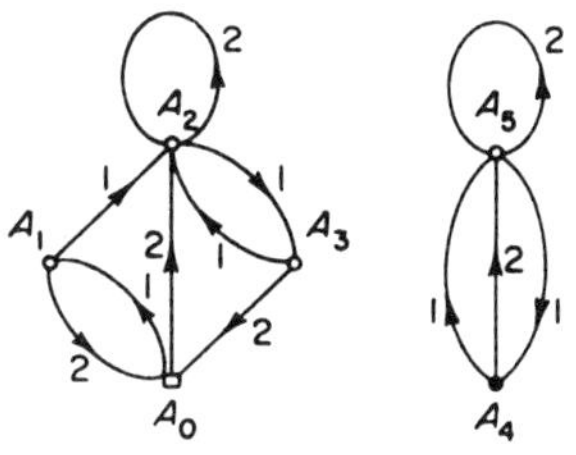


FIG. 2.

handy method of dealing with $k$-algebras. Instead of stating the general definition of the *transition graph* of $\mathfrak{A}$, we shall show it for an example of a 2-algebra.

*Example 1.* Let $\mathfrak{A}$ be the finite 2-algebra whose states are $\mathbf{A} = \{A_0, \ldots, A_5\}$, whose initial state is $A = A_0$, and whose transition operators $f_1, f_2$ are given by Table 3.

TABLE 3

| | $A_0$ | $A_1$ | $A_2$ | $A_3$ | $A_4$ | $A_5$ |
|---|---|---|---|---|---|---|
| $f_1$ | $A_1$ | $A_2$ | $A_3$ | $A_2$ | $A_5$ | $A_4$ |
| $f_2$ | $A_2$ | $A_0$ | $A_2$ | $A_0$ | $A_5$ | $A_5$ |

The transition graph of $\mathfrak{A}$ is the directed graph with labeled edges and marked root, shown in Fig. 2. Clearly from it one can recover the algebra $\mathfrak{A}$. Note that we could drop the labeling of vertices by states; the graph would still describe the algebra $\mathfrak{A}$ up to an isomorphism. In this example the transition graph splits into two disconnected components.

*Exercise 2.*

(a) Let $\mathfrak{A}$ be as in Example 1. Calculate $rp(122)$ and $rp(221)$. Calculate $A_2/122$, first by using the right recursions of Definition 2, and second by using the left recursion of (a). Make up a table and graph for the transition operator $f_{12}$. Find two input signals $a$ and $b$ such that $rp(a) = rp(b)$ but $f_a \neq f_b$. Find two different input signals $a$ and $b$ such that $f_a = f_b$.

(b) Let $\mathbf{A} = \{A_0, A_1, A_2, A_3\}$, $\mathbf{I} = \{1,2,3\}$, and let the binary transition operator tr: $\mathbf{A} \times \mathbf{I} \to \mathbf{A}$ be defined by $\mathrm{tr}(A_i, j) = A_{i+j}$, whereby $i + j$ stands for adding $i$ and $j$ modulo 4. If $\mathfrak{A}$ is the 3-algebra corresponding to the system $\langle \mathbf{I}, \mathbf{A}, A_1, \mathrm{tr}\rangle$, draw the transition graph of $\mathfrak{A}$.

(c) Suppose $G$ is a directed graph, on the set of vertices $\mathbf{A}$, each of whose edges is labeled by either 1 or 2. State necessary and sufficient conditions for $G$ to be the transition graph of a transition algebra.

(d) Make up a table for the product operation of the monoid of transition operators of the algebra $\mathfrak{A}$ of Example 1.

(e) Construct the transition graphs of the algebras of Exercise 1, (b) and (c).

### IV. Accessible States of a Transition Algebra

Note that not every state $U$ of a $k$-algebra need occur as a response to an input signal. Clearly such states are inessential, in the sense that they do not enter into the calculation of $rp(x)$

**Definition 3.** A state $U$ of a $k$-algebra $\mathfrak{A}$ is called accessible if it occurs as a response to some input signal. The set of accessible states of $\mathfrak{A}$ will be denoted by $as(\mathfrak{A})$. Thus, $as(\mathfrak{A}) = rp(N_k)$. A $k$-algebra is called *reduced* if all its states are accessible, i.e., if $as(\mathfrak{A}) = \mathbf{A}$. The reduced *transition algebra* $rd(\mathfrak{A})$ of $\mathfrak{A}$ is the algebra $\langle as(\mathfrak{A}), A, g_1, \ldots, g_k\rangle$, whereby the operators $g_1, \ldots, g_k$ are obtained by restricting $f_1, \ldots, f_k$ to the subset $as(\mathfrak{A})$ of $\mathbf{A}$.

We shall also say that $U$ is *accessible from* $V$, if $U/x = V$ for some $x$ in $N_k$. In more algebraic terminology, $\mathfrak{A} = \langle \mathbf{A}, A, f_1, \ldots, f_k\rangle$ is reduced just in case $A$ is a *generator* of $\mathfrak{A}$; i.e., a reduced $k$-algebra is but an "algebra with only monadic operators and one generator."

Note that $rd(\mathfrak{A})$ is reduced. Furthermore, $\mathfrak{A}$ and $rd(\mathfrak{A})$ have identical response functions. Because we are using algebras to generate response functions, this means that in essence we need only be concerned with reduced

$k$-algebras. We shall now discuss a simple method for reducing a finite $k$-algebra.

It is hoped that Fig. 3 sufficiently explains what is meant by the *right tree* over $N_k$ for $k = 3$. Its vertices correspond one-to-one to the elements of $N_k$. The vertices in the $m$th *level* correspond to the $x \in N_k$ of length $\ln(x) = m$. The *response tree* of a $k$-algebra $\mathfrak{A}$ is obtained by attaching the state $rp(x)$ as a label to the vertex $x$ of the right tree over $N_k$. Thus the response tree is but a graphical description of the response function $rp$. Note that only accessible states occur as labels.

The *transition tree* of $\mathfrak{A}$ is obtained by labeling the vertices of the right tree just as in the response tree, except that in case $U$ occurs a second time as a label, say at $x$, we delete all vertices $y \geqslant x$ which stand over $x$ in the right

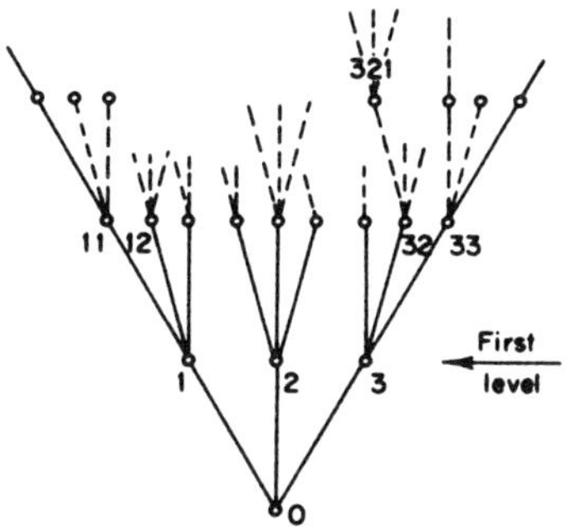


FIG. 3.

tree. Furthermore, the labeling is done by starting at 0 and running through successive levels from left to right.

*Example 2.* Let $\mathfrak{A}$ be the 3-algebra on the states $\mathbf{A} = \{A, B, C, D, E, F\}$, with initial state $A$ and transit operators given by Table 4. Fig. 4 shows the transition tree.

TABLE 4

| | $A$ | $B$ | $C$ | $D$ | $E$ | $F$ |
|---|---|---|---|---|---|---|
| $f_1$ | $B$ | $C$ | $A$ | $D$ | $F$ | $A$ |
| $f_2$ | $C$ | $B$ | $C$ | $B$ | $B$ | $E$ |
| $f_3$ | $A$ | $D$ | $D$ | $C$ | $C$ | $F$ |

It is clear that just the accessible states of $\mathfrak{A}$ occur in the transition tree. Furthermore, the transition tree fully describes the $rd(\mathfrak{A})$; it is but a modified version of the transition graph of $rd(\mathfrak{A})$. In particular, the initial state of $rd(\mathfrak{A})$ can be read from the transition tree; it is the root. To further clarify the idea we add the following comments: (a) The labeling of edges is omitted in the transition tree; it is intended that the branches, in left-to-right order, correspond to the input states $1, 2, \ldots, k$. Thus our trees include an orientation of branches, coming off a vertex. (b) The tree of a $k$-algebra is is $k$-branching (c) Each accessible state of $\mathfrak{A}$ occurs exactly once as interior vertex; it may or may not recur at the frontier.

Transition trees turn out to be a useful device for describing and handling finite reduced $k$-algebras. In particular, constructing the transition tree of $\mathfrak{A}$ is the most efficient method to obtain the accessible states, and at the same time one obtains $rd(\mathfrak{A})$.

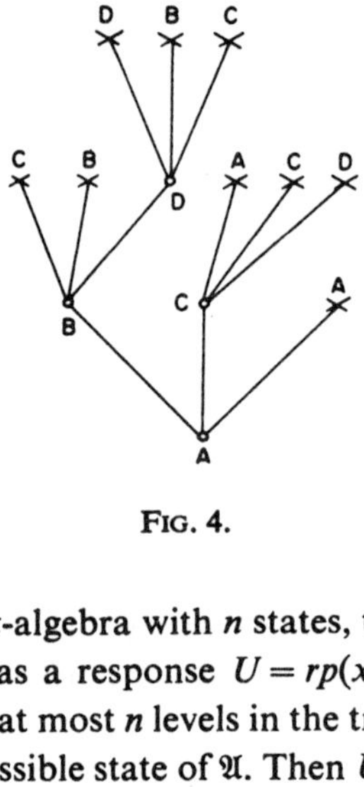


FIG. 4.

**Lemma 1.** If $\mathfrak{A}$ is a finite $k$-algebra with $n$ states, then every accessible state $U$ of $\mathfrak{A}$ can be obtained as a response $U = rp(x)$ to an input signal $x$ of length $< n$; i.e., there are at most $n$ levels in the transition tree of $\mathfrak{A}$.

*Proof.* Let $U$ be an accessible state of $\mathfrak{A}$. Then $U = rp(y)$ for some $y \in N_k$. To prove the lemma it is clearly sufficient to show that if $U = rp(y)$ and $\ln(y) \geqslant n$, then there is a $z$ such that $U = rp(z)$ and $\ln(y) > \ln(z)$.

Suppose therefore that $U = rp(y)$, $y = X_1 X_2 \cdots X_l$, $l \geqslant n$, and $X_1, \ldots, X_l$ are input states. Let $U_0, U_1, \ldots, U_l$ be the states $\mathfrak{A}$ passes through if the input states $X_1, X_2, \ldots, X_l$ are applied in succession, i.e., $U_0 = A$, $U_1 = U_0/X_1$, $U_2 = U_1/X_2, \ldots, U_l = U_{l-1}/X_l = rp(y) = U$. Because there are but $n$ states in $\mathfrak{A}$ and $l \geqslant n$, there must be a repetition $U_p = U_q$ for some $0 \leqslant p < q \leqslant l$. Thus we have $U_0 = A$, $U_1 = U_0/X_1, \ldots, U_p = U_{p-1}/X_p$, $U_{q+1} = U_p/X_{q+1}, \ldots$, $U_l = U_{l-1}/X_l = U$. Thus $A = U_0, U_1, \ldots, U_p, U_{q+1}, \ldots, U_l = U$ is the succession of states $\mathfrak{A}$ passes through if the input signal $z = X_1 X_2 \cdots X_p X_{q+1} X_{q+2} \cdots X_l$ is applied. Therefore, $U = rp(z)$. But $\ln(z) < \ln(y)$, because $p < q$, and we have found a $z$ such that $rp(z) = U$ and $\ln(z) < \ln(y)$. Q.E.D.

This simple proof is typical for many similar but more sophisticated arguments used in the theory of finite automata.

*Exercise 3*

(a) Let $\mathfrak{A}$ be as in Example 1. Find $as(\mathfrak{A})$ by constructing the transition tree. Note that in this particular case there is a better bound on the number of levels than the one predicted by Lemma 1.

(b) Let $\mathfrak{A}$ be as in Example 2. Construct the transition graph of $rd(\mathfrak{A})$ from the transition tree.

(c) Find a function $h: N_2 \to \{A_0, A_1, A_2\} = \mathbf{A}$ which is not the response function of an algebra on the states $\mathbf{A}$. If $g$ is a function from $N_k$ into a finite set $\mathbf{S}$, find a necessary and sufficient condition for existence of an algebra on the set $\mathbf{S}$ whose response function is $g$.

(d) Construct the transition tree of the algebra of Exercise 1, b and c, and of Exercise 2b.

(e) Give a survey over all reduced 1-algebras, up to isomorphism. How many of these are infinite? Find two nonisomorphic infinite 2-algebras.

(f) Show that if a $k$-algebra has $n$ states, then to every word $y$ there is a word $z$ such that $\ln(z) \leqslant n^2$ and the transition operators $f_y$ and $f_z$ are equal.

## V. The Basic Concepts of Algebra

In this section we shall discuss the universal algebraic concepts of *subalgebra*, *homomorphism*, *congruence relation*, *direct product*, and *free algebra*, in the special case of $k$-algebras. We shall see that each one of these general notions is of significance for the understanding of discrete deterministic systems. Had not the algebraists invented them, these ideas would sooner or later have independently grown out of the study of states transitions. In fact, something of the sort has actually happened during the past decade.

We have already used the notion of subalgebra of a $k$-algebra $\mathfrak{A}$. The reduced $rd(\mathfrak{A})$ of $\mathfrak{A}$ is simply the smallest among all subalgebras of $\mathfrak{A}$. $\mathfrak{A}$ is reduced in case it admits no proper subalgebra. In the sequel we shall often deal with reduced algebras only, the reason for this having been stated in Sec. IV.

A $k$-algebra $\mathfrak{A} = \langle \mathbf{A}, A, f_1, \ldots, f_k \rangle$ is called (*totally*) *free* if it satisfies the "generalized Peano axioms":

$$U^{f_i} \neq A \qquad \text{for } U \in \mathbf{A}, i = 1, \ldots, k \tag{3}$$

$$U^{f_i} \neq V^{f_j} \qquad \text{for } U, V \in \mathbf{A}, i < j = 1, \ldots, k \tag{4}$$

$$U^{f_i} = V^{f_i} \supset U = V \qquad \text{for } U, V \in \mathbf{A}, i = 1, \ldots, k \tag{5}$$

For any $\mathbf{X} \subseteq \mathbf{A}$, if $A \in \mathbf{X}$ and $(\forall U).\ U \in \mathbf{X} \supset U^{f_1}, \ldots, U^{f_k} \in \mathbf{X}$, then $\mathbf{X} = \mathbf{A}$ (6)

Note that (4) and (5) may be expressed thus: $A^{f_{i_1} \cdots f_{i_n}} = A^{f_{j_1} \cdots f_{j_m}}$ holds only in case $n = m$ and $i_1 = j_1, \ldots, i_n = j_n$, and condition (6) simply states that $\mathfrak{A}$ is reduced. On the set $N_k$ of words one defines the *right-successor functions* $\sigma_1, \ldots, \sigma_k$; thus

$$x^{\sigma_i} = xi \qquad \text{for } x \in N_k, i = 1, \ldots, k \tag{7}$$

It is easy to see that the $k$-algebra $\mathfrak{F}_k = \langle N_k, 0, \sigma_1, \ldots, \sigma_k \rangle$ is free, and that every free $k$-algebra $\mathfrak{A}$ must be isomorphic to $\mathfrak{F}_k$. The transition tree of $\mathfrak{F}_k$ is the right tree of words, discussed in Sec. IV. In fact, to say that a reduced $k$-algebra $\mathfrak{A}$ is free simply means that no feedback occurs in its transition tree (or graph).

A *homomorphism* of the $k$-algebra $\mathfrak{A} = \langle \mathbf{A}, A, f_1, \ldots, f_k \rangle$ onto the $k$-algebra $\mathfrak{B} = \langle \mathbf{B}, B, g_1, \ldots, g_k \rangle$ is a mapping $h$ from $\mathbf{A}$ onto $\mathbf{B}$ such that $hA = B$ and

$$h(U^{f_i}) = (hU)^{g_i} \qquad \text{for } U \in A, i = 1, \ldots, k \tag{8}$$

An *isomorphism* is a one-to-one homomorphism; the inverse $h^{-1}$ of an isomorphism of $\mathfrak{A}$ onto $\mathfrak{B}$ is an isomorphism of $\mathfrak{B}$ onto $\mathfrak{A}$. We use the symbol $\mathfrak{A} \Leftarrow \mathfrak{B}$ to denote that $\mathfrak{A}$ is the *homomorphic image* of $\mathfrak{B}$, and $\mathfrak{A} \simeq \mathfrak{B}$ means that $\mathfrak{A}$ and $\mathfrak{B}$ are *isomorphic.*

We have already dealt with homomorphisms. The response function $rp$ of a reduced $k$-algebra $\mathfrak{A}$ is, by its very definition, a homomorphism of the free algebra $\mathfrak{F}_k$ onto $\mathfrak{A}$. Thus $\mathfrak{F}_k \Rightarrow \mathfrak{A}$ holds for any reduced algebra. In fact $rp$ is the only homomorphism of $\mathfrak{F}_k$ onto $\mathfrak{A}$; more generally one easily shows:

> If $\mathfrak{A}$ is reduced and $\mathfrak{A} \Rightarrow \mathfrak{B}$, then $\mathfrak{B}$ is reduced and the homomorphism from $\mathfrak{A}$ onto $\mathfrak{B}$ is uniquely determined. (9)

Isomorphic algebras $\mathfrak{A} \simeq \mathfrak{B}$ have identical response functions. Conversely, if $\mathfrak{A}$ and $\mathfrak{B}$ are reduced and $rp_{\mathfrak{A}} = rp_{\mathfrak{B}}$, then $\mathfrak{A} \simeq \mathfrak{B}$. The transition graphs and trees of isomorphic algebras are identical up to renaming of the vertices. The significance of the notion of isomorphism is that it provides a rigorous definition of "$\mathfrak{A}$ *is structurally identical to* $\mathfrak{B}$." The *isomorphism type* $\mathfrak{A}$, i.e., the class of all $k$-algebras $\mathfrak{B}$ which are isomorphic to $\mathfrak{A}$, may well be called the structure of $\mathfrak{A}$. We are usually interested in a $k$-algebra $\mathfrak{A}$ only up to isomorphism. However, we shall often follow the common practice of talking about $\mathfrak{A}$, when in fact we mean $\mathfrak{A}^{\simeq}$. The intuitive meaning of $\mathfrak{A} \Rightarrow \mathfrak{B}$ is "$\mathfrak{A}$ is *structurally stronger* than $\mathfrak{B}$."

Let $\mathfrak{A} = \langle \mathbf{A}, A, f_1, \ldots, f_k \rangle$ be a $k$-algebra, and let $h: \mathbf{A} \to \mathbf{B}$ be a mapping of $\mathbf{A}$ onto some set $\mathbf{B}$. We will say that $h$ is *compatible with* $f_1, \ldots, f_k$ (or with $\mathfrak{A}$) if

$$hU = hV \supset h(U^{f_i}) = h(V^{f_i}) \qquad \text{for } U, V \in \mathbf{A}, i = 1, \ldots, k \tag{10}$$

Clearly a homomorphism $h$ of $\mathfrak{A}$ onto $\mathfrak{B}$ is compatible with $\mathfrak{A}$, i.e., (10) is a simple consequence of (8). Conversely, if $h$ is a mapping from $\mathbf{A}$ onto $\mathbf{B}$ which is compatible with $\mathfrak{A}$, then on $\mathbf{B}$ one can introduce a $k$-algebra $\mathfrak{B} = \langle \mathbf{B}, B, g_1, \ldots, g_k \rangle$ such that $h$ is a homomorphism of $\mathfrak{A}$ onto $\mathfrak{B}$. Namely, let $B = hA$, and note that, because (10) is assumed to hold, formula (8) properly defines mappings $g_i$ from $\mathbf{B}$ onto $\mathbf{B}$. Thus,

> A mapping $h: \mathbf{A} \to \mathbf{B}$ is compatible with $\mathfrak{A} = \langle \mathbf{A}, A, f_1, \ldots, f_k \rangle$, if and only if on $\mathbf{B}$ there is an algebra $\mathfrak{B} = \langle \mathbf{B}, B, g_1, \ldots, g_k \rangle$ such that $h$ is homomorphism of $\mathfrak{A}$ onto $\mathfrak{B}$. $\mathfrak{B}$, of course, is uniquely determined by $h$. (11)

In other words, finding homomorphic images of $\mathfrak{A}$ is the same as finding mappings which are compatible with $\mathfrak{A}$. This analysis can be carried one step further, and leads to the concept of congruence relation. We first note that many homomorphic images of $\mathfrak{A}$ will be equal up to an isomorphism; in particular,

> If $h_1: \mathfrak{A} \Rightarrow \mathfrak{B}_1$, $h_2: \mathfrak{A} \Rightarrow \mathfrak{B}_2$ are homomorphisms such that the equivalence relations $h_1 U = h_1 V$ and $h_2 U = h_2 V$ on $\mathbf{A}$ are identical then $\mathfrak{B}_1 \simeq \mathfrak{B}_2$. (12)

We therefore call an equivalence relation $\backsim$ on $\mathbf{A}$ a *congruence* (*relation*) on $\mathfrak{A}$ if

$$U \backsim V \supset U^{f_i} \backsim V^{f_i} \qquad \text{for } U, V \in \mathbf{A}, i = 1, \ldots, k \tag{13}$$

Inspection of (10) and (13) shows the following: To say that $\backsim$ is a congruence of $\mathfrak{A}$ is the same as saying that the *canonical map* $U \to \overset{\frown}{U}$ of $\mathbf{A}$ onto $\overset{\frown}{\mathbf{A}} = \{\overset{\frown}{U}, U \in A\}$ is compatible with $\mathfrak{A}$. Thus, by (11), to a congruence $\backsim$ on $\mathfrak{A}$ there corresponds an algebra $\mathfrak{A}/\backsim$ on $\overset{\frown}{\mathbf{A}}$, namely, the homomorphic image of $\mathfrak{A}$ under the canonical map $U \to \breve{U}$. This algebra $\mathfrak{A}/\backsim = \langle \breve{\mathbf{A}}, \breve{A}, \breve{f}_1, \ldots, \breve{f}_k \rangle$ is called the quotient of $\mathfrak{A}$ by $\backsim$; its initial state is the congruence class $\breve{A}$ of the initial state $A$ of $\mathfrak{A}$, and its operators $\breve{f}_i$ are given by

$$\breve{U}^{\breve{f}_i} = (U^{f_i})^{\backsim} \qquad \text{for } U \in \mathbf{A},\ i = 1, \ldots, k \tag{14}$$

Note that, owing to (13), formula (14) unambiguously defines operators $\breve{f}_i$ on $\breve{\mathbf{A}}$. Reconsidering (12) we see the significance of congruences and their quotients:

> The homomorphisms $h$ of a $k$-algebra $\mathfrak{A}$ come in classes $K(\backsim)$ which correspond one-to-one to the congruences $\backsim$ on $\mathfrak{A}$. Namely, $h$ belongs to $K(\backsim)$ just in case $hU = hV$ means $U \backsim V$. All images $h(\mathfrak{A})$ of congruences $h \in K(\backsim)$ are isomorphic among each other and isomorphic to the quotient $\mathfrak{A}/\backsim$. (15)

In other words, the canonical map $U \to \breve{U}$ of a congruence $\backsim$ on $\mathfrak{A}$ may be viewed as a normalized version of any one of the homomorphisms $h$ in $K(\backsim)$. Every homomorphic image of $\mathfrak{A}$ is isomorphic to some quotient of $\mathfrak{A}$. In the case of reduced algebras, which are of special interest to us, the relationship between homomorphisms and congruences is particularly simple. Using (9), one shows that for reduced algebras the converse to (12) also holds, i.e.,

> Let $\mathfrak{A}$ be a reduced $k$-algebra. Two homomorphic images $h_1(\mathfrak{A})$ and $h_2(\mathfrak{A})$ are isomorphic if and only if $h_1$ and $h_2$ belong to the same class $K(\backsim)$; i.e., if $\mathfrak{A}/\backsim_1 \simeq \mathfrak{A}/\backsim_2$, then $\backsim_1 = \backsim_2$. (16)

For later reference we restate this as

**Lemma 2.** The isomorphism types of homomorphic images of a reduced $k$-algebra $\mathfrak{A}$ correspond one-to-one to the congruence relations on $\mathfrak{A}$. If $\backsim$ is a congruence on $\mathfrak{A}$, then $\mathfrak{A}/\backsim$ represents the corresponding isomorphism type. If $h: \mathfrak{A} \Rightarrow \mathfrak{B}$ is a homomorphism, then $hU = hV$ is the congruence corresponding to $\mathfrak{B}$.

In papers concerned with digital systems one often encounters a process called *merging of states*. In fact, this always means passing from a $k$-algebra (or similar algebraic system) $\mathfrak{A}$ to some quotient $\mathfrak{A}/\backsim$. The point is: The basic facts about congruences (which we are here discussing, and which are well known to the algebraist) yield a clear and satisfactory understanding of the idea of merging of states. See in particular the discussion of Moore's minimality theory in Sec. VI.

A congruence relation on the free $k$-algebra $\mathfrak{F}_k = \langle N_k, 0, \sigma_1, \ldots, \sigma_k \rangle$ we shall also call a *right congruence* on the set $N_k$ of words. Thus a right congruence on $N_k$ is an equivalence relation on $N_k$ satisfying

$$(x \backsim y) \supset (xu \backsim yu) \qquad \text{for } x, y, u \in N_k \tag{17}$$

As we have already remarked, every reduced $k$-algebra $\mathfrak{A}$ is a homomorphic image of $\mathfrak{F}_k$, and the (unique) homomorphism is the response function $rp$ of $\mathfrak{A}$. The corresponding congruence relation on $\mathfrak{F}_k$ we denote by $\perp(\mathfrak{A})$, and we call it the *equi-response* relation of $\mathfrak{A}$. Thus,

$$x \perp y(\mathfrak{A}) \equiv rp_{\mathfrak{A}}\, x = rp_{\mathfrak{A}}\, y \qquad \text{for } x, y \in N_k \tag{18}$$

Note that $\perp(\mathfrak{A})$ has a very natural intuitive meaning; $x \perp y(\mathfrak{A})$ simply means that the input signals $x$ and $y$ produce the same internal state, when fed through the input channel of $\mathfrak{A}$. Note also,

> If $\mathfrak{A}$ is a reduced $k$-algebra, then the index[1] $id(\mathfrak{A})$ of the equi-response relation $\perp$ of $\mathfrak{A}$ is equal to the number of states of $\mathfrak{A}$. (19)

Using this terminology we can state the following corollary to Lemma 2:

**Lemma 2′.** The isomorphism types of reduced $k$-algebras correspond one-to-one to the right-congruence relations on $N_k$. This correspondence and its inverse are given by

$$\backsim \text{ right congruence} \rightarrow \mathfrak{F}_k/\backsim \text{ reduced } k\text{-algebra}$$
$$\mathfrak{A} \text{ reduced } k\text{-algebra} \rightarrow \perp(\mathfrak{A}) \text{ right congruence}$$

Furthermore, if $\mathfrak{A} \leftrightarrow \backsim$, then $id(\backsim)$ = number of states of $\mathfrak{A}$. Thus, the study of (finite) reduced $k$-algebras is equivalent to a study of right congruences (of finite index) on the set $N_k$ of words.

The *direct product* $\mathfrak{A} \times \mathfrak{B}$ of two $k$-algebras, $\mathfrak{A} = \langle \mathbf{A}, A, f_1, \ldots, f_k \rangle$ and $\mathfrak{B} = \langle \mathbf{B}, B, g_1, \ldots, g_k \rangle$, is the $k$-algebra $\langle \mathbf{A} \times \mathbf{B}, (A, B), h_1, \ldots, h_k \rangle$, whose states are the pairs $(U, V), U \in \mathbf{A}, V \in \mathbf{B}$, whose initial state is $(A, B)$ and whose transition functions are defined by

$$(U, V)^{h_i} = (U^{f_i}, V^{g_i}) \qquad \text{for } U \in \mathbf{A}, V \in \mathbf{B}, i = 1, \ldots, k$$

[1] By the *index* $id(\backsim)$ of an equivalence relation $\backsim$ we mean the number of its equivalence classes. Clearly, $id(\backsim)$ is the number of states of the quotient $\mathfrak{A}/\backsim$ of $\mathfrak{A}$ by a congruence $\backsim$ on $\mathfrak{A}$.

The *projections* $p$ and $q$, defined below, clearly are homomorphisms of $\mathfrak{A} \times \mathfrak{B}$ onto $\mathfrak{A}$ and $\mathfrak{B}$, respectively.

$$\left.\begin{array}{l} p(U,V) = U \\ q(U,V) = V \end{array}\right\} \qquad \text{for } U \in \mathbf{A},\ V \in \mathbf{B}$$

Note that $\mathfrak{A} \times \mathfrak{B}$ does not need to be reduced, even in case $\mathfrak{A}$ and $\mathfrak{B}$ are reduced. We define $\mathfrak{A} \otimes \mathfrak{B}$ to be the reduction of $\mathfrak{A} \times \mathfrak{B}$ and we call this the *reduced product*. A very natural construction will yield the transition tree of $\mathfrak{A} \otimes \mathfrak{B}$ (without going through $\mathfrak{A} \times \mathfrak{B}$) from the trees of $\mathfrak{A}$ and $\mathfrak{B}$. We show this construction in an example; see Fig. 5.

Clearly $\mathfrak{A} \otimes \mathfrak{B}$ has at most $n \cdot m$ states if $\mathfrak{A}$ has $n$ and $\mathfrak{B}$ has $m$ states. One easily verifies that the projections $p$ and $q$, restricted to $\mathfrak{A} \otimes \mathfrak{B}$, are homomorphisms of $\mathfrak{A} \otimes \mathfrak{B}$ onto $\mathfrak{A}$ and $\mathfrak{B}$, respectively, if $\mathfrak{A}$ and $\mathfrak{B}$ are

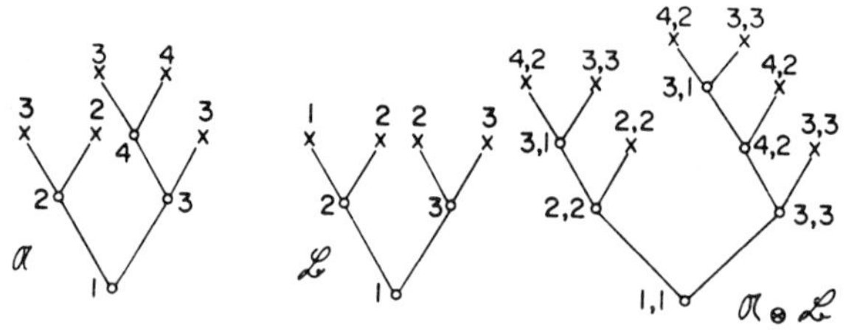

FIG. 5.

reduced. We now show that, for reduced algebras, $\mathfrak{A} \otimes \mathfrak{B}$ is the smallest $k$-algebra having $\mathfrak{A}$ and $\mathfrak{B}$ as homomorphic images.

**Lemma 3.** Let $\mathfrak{A}$ and $\mathfrak{B}$ be reduced $k$-algebras. Then $\mathfrak{A} \otimes \mathfrak{B} \Rightarrow \mathfrak{A}$, $\mathfrak{A} \otimes \mathfrak{B} \Rightarrow \mathfrak{B}$ Furthermore, if $\mathfrak{C}$ is a reduced $k$-algebra such that $\mathfrak{C} \Rightarrow \mathfrak{A}$, $\mathfrak{C} \Rightarrow \mathfrak{B}$, then $\mathfrak{C} \Rightarrow \mathfrak{A} \otimes \mathfrak{B}$; i.e., among all reduced $k$-algebras $\mathfrak{A} \otimes \mathfrak{B}$ is the smallest having $\mathfrak{A}$ and $\mathfrak{B}$ as homomorphic images.

*Proof.* Let $\mathfrak{C}$ be reduced $k$-algebras. Let $h_1 : \mathfrak{C} \Rightarrow \mathfrak{A}$ and $h_2 : \mathfrak{C} \Rightarrow \mathfrak{B}$. Define the mapping $h : \mathbf{C} \to (\mathbf{A} \times \mathbf{B})$ by

$$hQ = (h_1 Q, h_2 Q) \qquad \text{for } Q \in \mathbf{C}$$

Because $\mathfrak{C}$ is reduced, $h$ clearly is a homomorphism of $\mathfrak{C}$ into $\mathfrak{A} \otimes \mathfrak{B}$. That all states of $\mathfrak{A} \otimes \mathfrak{B}$ are actually values of $h$ follows because $h_1$ is onto $\mathfrak{A}$ and $h_2$ is onto $\mathfrak{B}$. Thus, $h$ is a homomorphism of $\mathfrak{C}$ onto $\mathfrak{A} \otimes \mathfrak{B}$, and therefore $\mathfrak{C} \Rightarrow \mathfrak{A} \otimes \mathfrak{B}$. Q.E.D.

If $\mathbf{A}_s, s \in S$ is a family of sets, let $\times_{s \in S} \mathbf{A}_s$ denote the set of all functions $U$, defined on $S$, and such that $U_s \in \mathbf{A}_s$ for all $s \in S$. For every $s \in S$ let $\mathfrak{A}_s = \langle \mathbf{A}_s, A_s, f_{1s}, \ldots, f_{ks} \rangle$ be a $k$-algebra. The *direct product* $\underset{s \in S}{\times} \mathfrak{A}_s$ is the algebra whose states are $\underset{s \in S}{\times} \mathbf{A}_s$, whose initial state is the function $A$ which takes $s$ into $A_s$, and whose transition operators $f_i$ are defined by $(U^{f_i})_s = U_s^{f_{is}}$. Reducing the direct product yields the reduced *product*, denoted by $\underset{s \in S}{\otimes} \mathfrak{A}_s$. Lemma 3 easily generalizes to arbitrary reduced products.

*The case* $k = 1$. It is quite instructive to reconsider the discussion of this section in the special case $k = 1$, which is properly termed the *input-free case*. The free algebra $\mathfrak{F}_1$ is $\langle N, 0, ' \rangle$, whereby $N = N_1$ is the set of natural numbers (including 0) and $x' = x + 1$ is the ordinary successor function on $N$. This turns out to be, up to isomorphism, the only infinite reduced 1-algebra. Every finite reduced 1-algebra is, up to isomorphism, uniquely characterized by two numbers $s \geqslant 0$ (called the *phase*) and $p \geqslant 1$ (called the *period*). Figure 6 shows the transition graph of the 1-algebra $\mathfrak{A}_{s,p}$ of phase

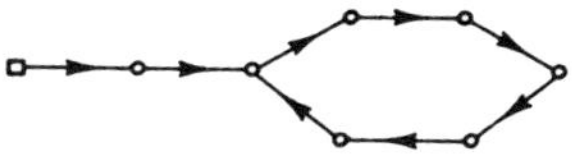

FIG. 6.

$s = 2$ and period $p = 6$. The congruence relation $\approx (s,p)$ on $\mathfrak{F}_1$, corresponding to the algebra $\mathfrak{A}_{s,p}$, is given by

$$x \approx y(s,p) \mathrel{.{\equiv}.} [x < s \wedge x = y] \wedge [s \leqslant x \wedge s \leqslant y \wedge x \equiv y (\operatorname{mod} p)] \quad (20)$$

We may extend these notations to $p = 0$; for any $s$ $\mathfrak{F}_1 = \mathfrak{A}_{s,0}$ and $\approx (s,0)$ is the equality relation on $N$. Thus, the study of 1-algebras simply means the study of the usual congruences of elementary number theory.

*k-ary notation for natural numbers.* Also, the words on $k$ symbols $1, \ldots, k$ may be interpreted as natural numbers. Note that the free algebra $\mathfrak{F}_k = \langle N_k, 0, \sigma_1, \ldots, \sigma_k \rangle$ is isomorphic to the algebra $\mathfrak{F}'_k = \langle N, 0, k \cdot x + 1, \ldots, k \cdot x + k \rangle$; the isomorphism $h: N \to N_k$ yields a $k$-ary notation for natural numbers. (This is not the standard $k$-ary notation! It has the advantage of using every word in a one-to-one fashion as the notation for a natural number). Results on finite $k$-algebras and finite automata may therefore be viewed as contributions to elementary number theory with emphasis on $k$-ary, rather than 1-ary notation. The first task in this field is to provide a survey over all congruences of finite index of the functions $k \cdot x + 1, \ldots, k \cdot x + k$ on $N$ (i.e., of all congruences of finite index on $\mathfrak{F}_k$). As a distant goal one may hope to eventually reach an understanding of these congruences which compares to the understanding we possess of the ordinary congruences (20) on the natural numbers. The results on finite automata which are available today should be considered as a humble contribution to such an understanding. That the step from $k = 1$ to $k = 2$ is not a trivial one is most clearly shown by the lattice-theoretic difficulties we shall discuss in Sec. VI. That things become much more interesting for $k \geqslant 2$ is also clear from the following.

*Congruences on $\mathfrak{F}_k$ of infinite index.* While there is but one infinite reduced 1-algebra, it is clear that there are many nonisomorphic infinite reduced 2-algebras. In fact, there are congruences of $2x + 1$, $2x + 2$ which are of infinite index and are intricate in a much deeper sense. The result of Post and Markov, which states that the word problem for semigroups is unsolvable, may be restated thus: For $k \geqslant 2$, there is a congruence relation $\backsim$ on $\mathfrak{F}_k$ which is a non-recursive relation.

*Commutative k-algebras.* A $k$-algebra $\mathfrak{A}$ is called *commutative* if its transition operators $f_1,\ldots,f_k$ commute with each other, i.e., $U^{f_i f_j} = U^{f_j f_i}$. From the viewpoint of sequential systems, finite commutative $k$-algebras may be called *modular counters.* Note that 1-algebras are trivially commutative. The *free commutative* $k$-algebra $\mathfrak{C}_k$, of which all other reduced commutative $k$-algebras are homomorphic images, is a sort of product $\mathfrak{C}_k = \mathfrak{F}_1 \times \cdots \times \mathfrak{F}_1$ of $k$ copies of $\mathfrak{F}_1$. Thus the states of $\mathfrak{C}_k$ are the $k$-tuples $x_1,\ldots,x_k$ of natural numbers; the initial state is $0,\ldots,0$; and the transition operators are $\langle x_1,\ldots,x_i,\ldots,x_k\rangle^{\lambda_i} = \langle x_1,\ldots,x_i + 1,\ldots,x_k\rangle$. We leave it to the reader to give a satisfactory survey over all finite reduced commutative $k$-algebras, i.e., to classify the congruences of finite index of $\mathfrak{C}_k$. This can actually be done, using well-known facts of elementary number theory or the basic facts about finite commutative monoids.

*Exercise 4.*

(a) Let $\mathfrak{A}$ be as in Example 1. Does $122 \perp 2211(\mathfrak{A})$ hold? Is there an input signal $x \neq 0$ such that $x \perp 0(\mathfrak{A})$? Find a set of representatives for the congruence classes of $\perp$.

TABLE 5

| | 1 | 2 | 3 | 4 | 5 | 6 |
|---|---|---|---|---|---|---|
| $g_1$ | 2 | 3 | 4 | 2 | 4 | 1 |
| $g_2$ | 3 | 5 | 4 | 3 | 3 | 3 |
| $g_3$ | 4 | 6 | 6 | 1 | 6 | 3 |

(b) Construct the transition tree of a 3-algebra $\mathfrak{C}$ which has $\mathfrak{A}$ and $\mathfrak{B}$ as homomorphic images. $\mathfrak{A}$ is the algebra of Fig. 4; $\mathfrak{B}$ is given by Table 5.

(c) Let $T$ be the transition tree of the finite $k$-algebra $\mathfrak{A}$. What does it mean for $T$ that $\backsim$ is a congruence of $\mathfrak{A}$? From $T$, how does one construct the transition tree of the quotient $\mathfrak{A}/\backsim$? Find all congruences of the algebra given by Table 6. Find the transition trees of the corresponding quotients.

TABLE 6

| | 1 | 2 | 3 | 4 | 5 |
|---|---|---|---|---|---|
| $f_1$ | 2 | 4 | 5 | 4 | 5 |
| $f_2$ | 3 | 4 | 5 | 4 | 5 |

(d) Find all homomorphic images of the algebra $\mathfrak{A}$ given by Table 7. Represent $\mathfrak{A}$ as a reduced product of a two-state algebra and a four-state algebra.

TABLE 7

| | 1 | 2 | 3 | 4 | 5 | 6 |
|---|---|---|---|---|---|---|
| $f_1$ | 2 | 4 | 6 | 6 | 4 | 6 |
| $f_2$ | 3 | 5 | 2 | 5 | 5 | 5 |

(e) Describe the relations $\approx(0,1)$, $\approx(s,1)$, $\approx(s,0)$, $\approx(0,p)$. Find the index of $\approx(s,p)$. Show that $\approx(s,p)$ is a congruence relation of the operations $x+y$ and $x{\cdot}y$.

(f) Show that $\approx(3,12)$ is the equi-response relation of $\mathfrak{A}_{3,6} \otimes \mathfrak{A}_{1,4}$. Which is the equi-response relation of $\mathfrak{A}_{s,p} \otimes \mathfrak{A}_{r,q}$? Find $u,v$ such that $\approx(u,v)$ is the intersection of $\approx(s,p)$ and $\approx(r,p)$. State a theorem concerning these matters.

(g) Prove that every congruence relation $\backsim$ of $\langle N,0,'\rangle$ is one of the relations $\approx(s,p)$. Do not use Lemma 2 but only the defining property $(x \backsim y) \supset (x' \backsim y')$.

(*h*) Find an infinite reduced commutative 2-algebra which is not isomorphic to the free commutative algebra $\mathfrak{C}_2$; i.e., find a congruence of infinite index on $\mathfrak{C}_2$ which is not the equality relation.

(i) Find all reduced 2-counters with six states, and the corresponding congruences on $\mathfrak{C}_2$.

(j) Show that the 3-algebra $\mathfrak{F}_3' = \langle N, 0, 3x+1, 3x+2, 3x+3 \rangle$ is totally free. Draw its transition tree, and compare the corresponding ternary notation for natural numbers with the standard ternary notation. Set up an algorithm for adding two natural numbers in this modified ternary notation.

(k) Which of the following equivalence relations on $N_2$ are right congruences and left congruences; find the indices. For notation see below.

1. $\ln_1 x \approx \ln_1 y(1,2) \wedge \ln_2 x \approx \ln_2 y(2,4)$
2. $(12 \leqslant x) \equiv (12 \leqslant y)$
3. $\ln_1 x \approx \ln_1 y(1,2) \wedge (\ln_2 x \geqslant 1) \equiv (\ln_2 y \geqslant 1)$
4. $(\ln_1 x + 2\ln_2 x) \approx (\ln_1 y + 2\ln_2 y)(1,3)$

(l) Show that the binary relations on $N_2$ defined by the following expressions are right congruences on $N_2$, of finite index. In each case construct a 2-algebra which has the relation as its equi-response relation. For notation see below.

1. $(12 \leqslant x) \equiv (12 \leqslant y) \wedge (x = 0) \equiv (y = 0) \wedge (x = 1) \equiv (y = 1)$
2. $(x \geqslant 12) \equiv (y \geqslant 12) \wedge (x \geqslant 1) \equiv (y \geqslant 1)$
3. $\ln_1 x \approx \ln_1 y(1,4)$
4. $\ln_{12} x \approx \ln_{12} y(1,2)$

(m) Find the right congruences $\backsim$ on $N_2$ of smallest index such that $\backsim$ is contained in the relation.

$$1\colon (112 \leqslant x) \equiv (112 \leqslant y) \qquad 2\colon (x \geqslant 112) \equiv (y \geqslant 112)$$

Construct the transition trees of the quotients.

*Notation.* If $x, y$ are words in $N_k$, define

$x \leqslant y\colon (\exists u)\, xu = y$ "$x$ is left segment of $y$"
$y \geqslant x\colon (\exists u)\, ux = y$ "$x$ is right segment of $y$"
$x \preceq y\colon (\exists uv)\, uxv = y$ "$x$ occurs in $y$"
$x < y\colon (x \leqslant y) \wedge (x \neq y)$ "$x$ is proper left segment of $y$"
$y > x\colon (y \geqslant x) \wedge (x \neq y)$ "$x$ is proper right segment of $y$"
$\ln_x y$: number of occurrences of $x$ in $y$

## VI. The Structure Lattice of $k$-Algebras

It seems clear that mathematical results concerning the following general problems will be of interest for the understanding of deterministic digital systems.

*Problem A.* To give a survey over all isomorphism types of finite reduced $k$-algebras.

*Problem B.* To decompose finite reduced $k$-algebras, i.e., to find operations of composition of $k$-algebras, such that every finite reduced $k$-algebra $\mathfrak{A}$ can be represented as a composite of $k$-algebras $\mathfrak{P}$, which are of simple structure.

As the algebraist would say, we would like to see significant contributions to the *structure theory* of finite reduced $k$-algebras. Of course these problems are not precisely stated. In fact, an essential part of the task at hand is to invent significant principles of ordering (A) and operations of composition (B) of (isomorphism types of) $k$-algebras. In this section we shall present the basic results which are available in case we choose $\mathfrak{A} \Rightarrow \mathfrak{B}$ (the homomorphism or quotient relation) as ordering, and $\mathfrak{A} \otimes \mathfrak{B}$ (the reduced product) as composition. It turns out that $\mathfrak{A} \Rightarrow \mathfrak{B}$ is a lattice order on reduced $k$-algebras, and $\mathfrak{A} \otimes \mathfrak{B}$ is the "join" operation in this lattice. Furthermore, every reduced $k$-algebra $\mathfrak{A}$ is decomposable into $\otimes$-indecomposable components.

*1. Lattices*

A partially ordered set $\langle L, \leqslant \rangle$ is called a *lattice order* if the *meet* (greatest lower bound) $x \wedge y$, and the *join* (smallest upper bound) $x \vee y$ exist for any

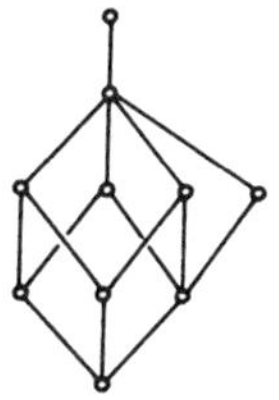

FIG. 7.

members $x, y$ of $L$. It is called a *complete lattice order* if the meet $\wedge_\gamma x_\gamma$ and the join $\wedge_\gamma x_\gamma$ exist for arbitrary families $\{x_\gamma\}$ of members of $L$. Note that

$$\bigvee_\gamma x_\gamma = \bigwedge_{(\forall \gamma) x_\gamma \leqslant y} y \tag{21}$$

Thus, if all $\wedge$'s exist, so do all $\vee$'s, and $\langle L, \leqslant \rangle$ is a complete lattice. The join (meet) of the empty family of members of $L$, if it exists, is the smallest (largest) member of $L$. Thus a complete lattice order must possess a smallest element 0 and a largest element $e$. A finite lattice order is always complete, and can be represented by a diagram, as exemplified in Figs. 7 and 10. It is left to the reader to learn to understand such diagrams, i.e., to find out in which manner they represent a partial order $x \leqslant y$ on the vertices, and how one goes about finding $x \wedge y$ and $x \vee y$. A *lattice* $\langle L, \wedge, \vee \rangle$ is an algebra with two binary operations $\wedge$, $\vee$, which are meet and join of some partial order $\leqslant$ on $L$. Note that $x \leqslant y$ means $x = x \wedge y$ $(y = x \vee y)$.

A *complete lattice* $\langle L, \bigwedge, \bigvee \rangle$ is an algebra with infinitary operations, which are the operations of meet and join of a partial order on $L$.

A one-to-one mapping $h$ of $L_1$ onto $L_2$ is called an *anti-isomorphism* of the partial order $\langle L_1, \leqslant_1 \rangle$ onto $\langle L_2, \leqslant_2 \rangle$ if $(x \leqslant_1 y) \equiv (hy \leqslant_2 hx)$ holds for all $x, y \in L_1$. Clearly an anti-isomorphism $h$ takes meets into joins and vice versa, i.e.,

$$h(x \wedge y) = (hx \vee hy)$$
$$h(x \vee y) = (hx \wedge hy)$$

and similarly for $\bigwedge$ and $\bigvee$. We shall now discuss lattices of equivalence relations.

## 2. *Algebraic Closure Lattices*

Let $S$ be a set; let $\mathbf{L}$ be a set of subsets of $S$ such that $S \in \mathbf{L}$ and $\mathbf{L}$ is closed under intersection, i.e.,

$$\bigcap_{\gamma} X_{\gamma} \in \mathbf{L} \qquad \text{for any family } \{X_{\gamma}\} \text{ of members of } \mathbf{L} \tag{22}$$

Such a pair $\langle S, \mathbf{L} \rangle$ is called a *closure space*. The operation $U^c$, defined by

$$U^c = \bigcap_{U \subseteq X \in \mathbf{L}} X \qquad \text{for } U \subseteq S \tag{23}$$

is a closure operator on $S$, i.e., $U \subseteq U^c$, $(U \subseteq V) \supset (U^c \subseteq V^c)$, $U^{cc} = U^c$. $\mathbf{L}$ just consists of the closed sets $X = X^c$.

By remark (21) it is clear that $\langle \mathbf{L}, \subseteq \rangle$ is a complete lattice order. Its meet operation is the operation $\bigcap$ of set intersection; its join operation $\bigvee$ is defined by

$$\bigvee_{\gamma} X_{\gamma} = \Big(\bigcup_{\gamma} X_{\gamma}\Big)^c \qquad \text{for any family } \{X_{\gamma}\} \text{ of members of } \mathbf{L} \tag{24}$$

By a *closure lattice* we understand a complete lattice $\langle \mathbf{L}, \bigcap, \bigvee \rangle$ whose elements are subsets of a set $S$, whose largest element is $S$, and whose meet operation is set intersection; i.e., a closure lattice is the lattice of all closed sets of a closure space $\langle S, \mathbf{L} \rangle$. A closure lattice (and the corresponding space) is called *algebraic* if it satisfies either one of the following equivalent conditions:

$$\bigvee_{\gamma} X_{\gamma} = \bigcup_{\gamma_1, \ldots, \gamma_l} (X_{\gamma_1} \vee \cdots \vee X_{\gamma_l}) \qquad \text{for any family } \{X_{\gamma}\} \text{ in } \mathbf{L} \tag{25a}$$

$$\bigvee_{\gamma} X_{\gamma} = \bigcup_{\gamma} X_{\gamma} \qquad \text{for any directed family in } \mathbf{L} \tag{25b}$$

A family of sets $X_\gamma$ is called *directed* if to any two members $X_{\gamma_1}$ and $X_{\gamma_2}$ there is a third member $X_\gamma$, such that $(X_{\gamma_1} \cup X_{\gamma_2}) \subseteq X_\gamma$. It is left to the reader to show that (25a) and (25b) are equivalent, and in fact are equivalent to the assertion that $U^c$ is the union of closures $\{u_1, \ldots, u_i\}^c$ of all finite subsets of $U$.

Algebraic closure lattices naturally arise in algebra and logic. For example, one easily sees that the subalgebras of any algebra form an algebraic closure lattice. More important yet is the fact that all congruences on an algebra $\mathfrak{A}$ form an algebraic closure lattice $Cg(\mathfrak{A})$; below we shall show this for $k$-algebras. We shall now discuss a basic result of Birkhoff's and provide some hints to the proof.

**Definition 4.** Let $X$ be an element of the closure lattice $\langle \mathbf{L}, \bigcap, \vee \rangle$ over the set $S$. $X$ is called $\bigcap$-*irreducible* if $X \neq S$, and $X = \bigcap_\gamma X_\gamma$ implies that $X = X_\gamma$, for some member of the family $\{X_\gamma\}$ in $\mathbf{L}$. $X$ is called maximal relative $a \in S$ if $a \notin X$ and there is no $U \in \mathbf{L}$ such that $a \notin U$, $X \subset U$. $X$ is called maximal if $S$ is the only element of $\mathbf{L}$ which properly contains $X$.

Note that every maximal $X$ is relatively maximal. Furthermore, it is not hard to establish that

In every closure lattice $\langle \mathbf{L}, \bigcap, \vee \rangle$, the relatively maximal elements of $\mathbf{L}$ are exactly the $\bigcap$-irreducibles of $\mathbf{L}$. (26)

Birkhoff's theorem says that in an algebraic closure lattice every $X \in \mathbf{L}$ is an intersection of $\bigcap$-irreducibles; i.e.,

**Lemma 4.** In an algebraic closure lattice $\langle \mathbf{L}, \bigcap, \vee \rangle$ the $\bigcap$-irreducible (i.e., the relatively maximal elements) form a $\bigcap$-basis.

*Proof.* Let $Y \in \mathbf{L}$; let $M = \{X;\ Y \subseteq X;\ X \text{ relatively maximal}\}$; let $Z$ be the intersection of all $X \in M$. By (26) it is sufficient to show that $Y = Z$. But $Y \subseteq Z$ is obvious; thus it remains to show that $(a \notin Y) \supset (a \notin Z)$ for any $a \in S$. This is proved as follows (using the axiom of choice):

Suppose $a \notin Y$. Let $\mathbf{P} = \{X;\ Y \subseteq X \wedge a \notin X \in \mathbf{L}\}$. Clearly $Y \in \mathbf{P}$, so that $\mathbf{P}$ is not empty. Furthermore, from the assumption (25b) that the space is algebraic, it follows that $\mathbf{P}$ is closed under directed union. Therefore, by Zorn's lemma, $\mathbf{P}$ contains a maximal member $X_0$; i.e., (a) $Y \subseteq X_0$, (b) $a \notin X_0 \in \mathbf{L}$, and (c) $Y \subseteq X$, $a \notin X \in \mathbf{L}$ implies $X_0 \not\subset X$. Because of (b) and (c) $X_0$ is maximal relative $a$. Therefore by ($a$), $X_0 \in M$. Thus we have $a \notin X_0 \in M$, so that $a \notin Z$. Q.E.D.

### 3. *Lattices of Equivalence Relations*

It is easy to see that the intersection $\bigcap_\gamma E_\gamma$ of a family of equivalence relations $\{E_\gamma\}$ on a set **A** is again an equivalence relation on **A**. The set of all equivalence relations on **A** therefore forms a closure lattice $Eq(\mathbf{A})$. The smallest member of $Eq(\mathbf{A})$ is the equality relation $=(\mathbf{A})$ on the set **A**; the largest member is the universal relation $\mathbf{A} \times \mathbf{A}$. The join operation on equivalences will be denoted by $\bigsqcup$, while $X^e$ is used to denote the equivalence closure of the relation $X \subseteq \mathbf{A} \times \mathbf{A}$. Thus, by (23) and (24),

$$\begin{aligned} X^e &= \text{intersection of all equivalence relations } E \text{ on } \mathbf{A},\ X \subseteq E \\ \bigsqcup_\gamma E_\gamma &= (\bigcup_\gamma E_\gamma)^e \qquad \text{for any family of equivalences on } \mathbf{A} \end{aligned} \tag{27}$$

$X^e$ is usually called the *equivalence relation generated* by the relation $X \subseteq \mathbf{A} \times \mathbf{A}$.

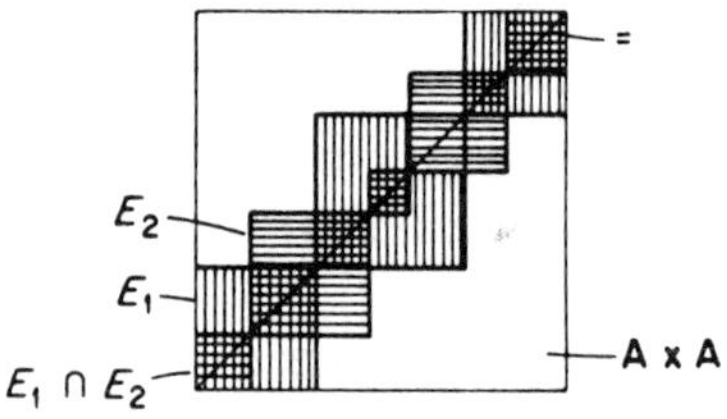


FIG. 8.

As shown in Fig. 8, there is a nice graphic way of thinking of equivalence relations and their intersection. (*Caution:* The picture depends on the order which is arbitrarily introduced among the members of **A**!) Also, it is well known that instead of talking about equivalences $E$ on **A**, one may talk about the corresponding partition of **A**. The reader should establish the meaning of $\subseteq, \cap$, and $\bigsqcup$ in terms of partition. The following are obvious remarks on the index of equivalence relations.

$$\begin{aligned} &E_1 \subseteq E_2 . \supset . \operatorname{ind}(E_2) \leqslant \operatorname{ind}(E_1) \\ &\operatorname{ind}(E_1 \cap E_2) \leqslant \operatorname{ind}(E_1)\cdot\operatorname{ind}(E_2) \end{aligned} \tag{28}$$

We leave it to the reader to show that the closure operator $X^e$ on the set $\mathbf{A} \times \mathbf{A}$ is algebraic, i.e., $X^e = \bigcup U^e$, whereby $U$ ranges over all finite subrelations of $X$. Actually, we need this only in case $X$ is a symmetric relation; for these the assertion easily follows from

$$xR^ey \,.\equiv.\, (\exists x_0 \cdots x_m)[x_0 = x \wedge x_0 R x_1 \wedge \cdots \wedge x_{m-1} R x_m \wedge x_m = y] \quad \text{if } R \text{ is symmetric} \tag{29}$$

To prove (29) one has to prove three facts about the relation $xSy$ defined by the expression on the right. Namely, (a) $S$ is an equivalence relation, (b) $R \subseteq S$, and (c) if $E$ is an equivalence and $R \subseteq E$, then $S \subseteq E$. These are easy to establish (and it seems clear how the expression for $S$ has to be modified in case $R$ is not symmetric). Using (29) and (27), and the fact that $\bigcup_\gamma E_\gamma$ is symmetric if the $E_\gamma$'s are equivalences, we obtain a more constructive method for finding the join of a family of equivalences:

$$x(\bigsqcup_\gamma E_\gamma)y \;.\equiv.\; (\exists_{\gamma_1}^{x_0}\cdots\exists_{\gamma_m}^{x_m})[x = x_0 \wedge x_0 E_{\gamma_1} x_1 \wedge \cdots \wedge x_{m-1} E_{\gamma_m} x_m \wedge x_m = y] \tag{30}$$

From (30) one easily proves that $Eq(\mathbf{A})$ is an algebraic closure space, i.e.,

$$\bigsqcup_\gamma E_\gamma = \bigcup_{\gamma_1,\ldots,\gamma_m} (E_{\gamma_1} \sqcup \cdots \sqcup E_{\gamma_m}) \qquad \text{for any family of equivalences} \tag{31}$$

We will now establish a second important consequence of (30). Let $\mathfrak{A} = \langle \mathbf{A}, A, f_1, \ldots, f_k \rangle$ be a $k$-algebra, and let $\{E_\gamma\}$ be a family of equivalence relations on $\mathbf{A}$ which are congruences of $\mathfrak{A}$, i.e., $(XE_\gamma Y) \supset (X^{f_i} E_\gamma Y^{f_i})$. From (30) it clearly follows that $\bigsqcup_\gamma E_\gamma$ is again a congruence relation of $\mathfrak{A}$.

**Theorem 1.** The congruence relations on a $k$-algebra $\mathfrak{A}$ form a closure lattice $Cg(\mathfrak{A})$. The join operation of this lattice is $\sqcup$, i.e., $Cg(\mathfrak{A})$ is a sublattice of $Eq(\mathbf{A})$. Furthermore, formula (31) holds in $Cg(\mathfrak{A})$, so that it is an algebraic closure lattice. In particular, the right congruences on the set of words $N_k$ form a closure lattice $Cg_k$. It is algebraic, and its join operation is $\sqcup$.

*Proof.* That the intersection $\bigcap_\gamma E_\gamma$ of congruences is again a congruence is easily seen. Therefore $Cg(\mathfrak{A})$ is a closure lattice. That $\bigsqcup_\gamma E_\gamma$ is a congruence, if the $E_\gamma$'s are, was shown above. Thus $\sqcup$ is the join operation of $Cg(\mathfrak{A})$, and by (31) the lattice is algebraic. Q.E.D.

As a consequence of Lemma 4 and Theorem 1 we obtain

**Corollary 1.** Every congruence $E$ of a $k$-algebra $\mathfrak{A}$ is an intersection $E = \bigcap_\gamma P_\gamma$ of $\bigcap$-irreducible congruences $P_\gamma$ of $\mathfrak{A}$. In particular, every right congruence $E$ on $N_k$ is the intersection of $\bigcap$-irreducible right congruences.

It remains to discuss the congruence-closure-operator $R^c$ = intersection of all congruence relations $X$ on $\mathfrak{A}$, $R \subseteq X$. $R^c$ is also called the *congruence generated* by the relation $R \subseteq \mathbf{A} \times \mathbf{A}$. We leave it to the reader to show that $R^c$ can be calculated as follows. If $R_i, E_i$ are defined inductively by

$$R_0 = R \qquad E_i = R_i^e \tag{32}$$

$$XR_{i+1}Y \;.\equiv.\; XE_iY \vee X^{f_1}E_iY^{f_1} \vee \cdots \vee X^{f_k}E_iY^{f_k}$$

then

$$R^c = \bigcup_{i=0}^{\infty} E_i$$

Note that if $\mathfrak{A}$ is finite and has $n$ states, then for some $j \leqslant 2^{(n^2)}$, $E_{j+1} = E_j$. If this holds for $j$, then $R = E_j$.

*Example 3.* Let $\mathfrak{A}$ be the $k$-algebra, on the set of states $\mathbf{A} = \{1,2,3,4,5\}$, Exercise 4(c). Using the method described above one easily finds the congruences induced by the relations $R \subseteq \mathbf{A} \times \mathbf{A}$, starting with one-member relations. This process yields 10 congruences of $\mathfrak{A}$. Inspection will show that the lattice $Cg(\mathfrak{A})$ is that of Fig. 7. We leave it to the reader to mark the five $\bigcap$-irreducible elements (just one of them is maximal), and to verify Corollary 1. Note that the decomposition into $\bigcap$-irreducibles is not unique!

*4. The Structure Lattice of k-Algebras*

In the sequel we shall follow the common practice of talking about the $k$-algebra $\mathfrak{A}$, when in fact we mean its isomorphism-type $\mathfrak{A}^{\simeq}$. In other words, we "identify" isomorphic $k$-algebras. Note that this ambiguity is even less dangerous than usual, because we are dealing with reduced algebras (the isomorphism between reduced algebras is unique if it exists, and a reduced algebra cannot be isomorphic to a proper subalgebra).

As announced at the beginning of this section, we wish to study the relation $\mathfrak{B} \Rightarrow \mathfrak{A}$ (i.e., $\mathfrak{A}$ is a homomorphic image of $\mathfrak{B}$) between all homomorphic images of a reduced $k$-algebra $\mathfrak{C}$. It is easy to see that this is a partial order. Furthermore, Lemma 3 (which easily extends to general reduced products) says that $\mathfrak{A} \otimes \mathfrak{B}$ is the join operation in this partial order. Thus all homomorphic images of a reduced $k$-algebra $\mathfrak{A}$ form a complete lattice $SL(\mathfrak{A})$ relative $\Rightarrow$, and the reduced product $\otimes$ is the join operation in this lattice. We call $SL(\mathfrak{A})$ the *structure lattice* of $\mathfrak{A}$. We already know from Lemma 2 that $E \to \mathfrak{A}/E$ establishes a one-to-one map of the congruences of $\mathfrak{A}$ onto the homomorphic images of $\mathfrak{A}$. This map is in fact an anti-isomorphism of $Cg(\mathfrak{A})$ onto $SL(\mathfrak{A})$.

**Theorem 2.** The structure lattice $SL(\mathfrak{A})$ of a reduced $k$-algebra $\mathfrak{A}$ is anti-isomorphic to the lattice $Cg(\mathfrak{A})$ of all congruences on $\mathfrak{A}$. The anti-isomorphism from $Cg(\mathfrak{A})$ onto $SL(\mathfrak{A})$ is given by $E \to \mathfrak{A}/E$, i.e.

$$\begin{aligned} &E_1 \subseteq E_2 \,.\!\equiv\!.\, (\mathfrak{A}/E_2) \Leftarrow (\mathfrak{A}/E_1) && \text{for congruences } E_1, E_2 \text{ on } \mathfrak{A} \\ &\mathfrak{A}/(\bigcap_\gamma E_\gamma) = \bigotimes_\gamma (\mathfrak{A}/E_\gamma) && \text{for a family } \{E_\gamma\} \text{ of congruences on } \mathfrak{A} \end{aligned} \tag{33}$$

In particular, the structure lattice $SL_k$ of the free $k$-algebra $\mathfrak{F}_k$ is anti-isomorphic to the lattice $Cg_k$ of all right congruences on $N_k$. The anti-isomorphism is given by $\mathfrak{C} \to \perp(\mathfrak{C})$; its inverse is $E \to \mathfrak{F}_k/E$; i.e.,

$$\mathfrak{C}_1 \Leftarrow \mathfrak{C}_2 \,.\!\equiv\!.\, \perp(\mathfrak{C}_2) \subseteq \perp(\mathfrak{C}_1) \qquad \text{for reduced } k\text{-algebras}$$
$$E_1 \subseteq E_2 \,.\!\equiv\!.\, (\mathfrak{F}_k/E_2) \Leftarrow (\mathfrak{F}_k/E_1) \qquad \text{for right congruences on } N_k \quad (34)$$

*Proof.* Suppose $E_1 \subseteq E_2$ are congruences on the reduced $k$-algebra $\mathfrak{A}$. Let $X^{E_i}$ denote the congruence class, modulo $E_i$, of the state $X$ of $\mathfrak{A}$. Then clearly $X^{E_1} \to X^{E_2}$ is a homomorphism of $\mathfrak{A}/E_1$ onto $\mathfrak{A}/E_2$. Thus $E_1 \subseteq E_2$ implies $(\mathfrak{A}/E_2) \Leftarrow (\mathfrak{A}/E_1)$. Suppose next that $(\mathfrak{A}/E_2) \Leftarrow (\mathfrak{A}/E_1)$. Let $h$ be the homomorphism. Then $gX = h(X^{E_1})$ is a homomorphism from $\mathfrak{A}$ onto $\mathfrak{A}/E_2$. Because $\mathfrak{A}$ is reduced, there is but one homomorphism of $\mathfrak{A}$ onto $\mathfrak{A}/E_2$, namely, $X \to X^{E_2}$. Consequently, $gX = X^{E_2}$, i.e., $h(X^{E_1}) = X^{E_2}$. Therefore we have $(X^{E_1} = Y^{E_1}) \supset (X^{E_2} = Y^{E_2})$, i.e., $(XE_1\,Y) \supset (XE_2\,Y)$, i.e., $E_1 \subseteq E_2$. Thus, $(\mathfrak{A}/E_2) \Leftarrow (\mathfrak{A}/E_1)$ implies $E_1 \subseteq E_2$. This establishes the first formula (33); i.e., $Cg(\mathfrak{A})$ is anti-isomorphic to $SL(\mathfrak{A})$.

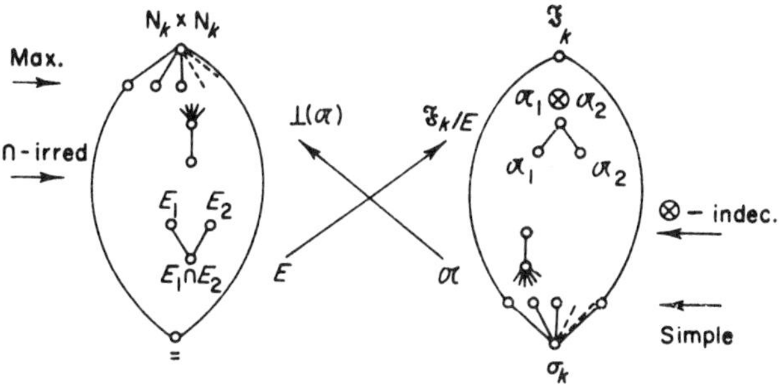


FIG. 9.

The second part of (33) follows, because an anti-isomorphism takes meets into joins, and we have already seen (Lemma 2) that $\otimes$ is the join in the lattice $SL(\mathfrak{A})$. The rest of Theorem 2 is just a restatement of the first part, in the special case $\mathfrak{A} = \mathfrak{F}_k$. Q.E.D.

A reduced $k$-algebra $\mathfrak{A}$ is called $\otimes$-*indecomposable* (reduced directly indecomposable) if $\mathfrak{A} = \otimes_\gamma \mathfrak{C}_\gamma$ implies $\mathfrak{A} \simeq \mathfrak{C}_\gamma$ for some member of the family $\{\mathfrak{C}_\gamma\}$.

**Corollary 2.** Every reduced $k$-algebra $\mathfrak{A}$ is isomorphic to a reduced product $\otimes_\gamma \mathfrak{P}_\gamma$ of $\otimes$-indecomposable reduced $k$-algebras $\mathfrak{P}_\gamma$.

This clearly follows from Theorem 1, Corollary 1, and the remark that a right congruence $E$ is $\bigcap$-irreducible if and only if its quotient $\mathfrak{F}/_kE$ is $\otimes$-indecomposable. We remark further that the maximal right congruences

$E$ on $N_k$ correspond to the *simple* reduced $k$-algebras. A reduced $k$-algebra $\mathfrak{A}$ is called simple if the one-state algebra $\mathfrak{O}_k$ and $\mathfrak{A}$ are the only homomorphic images of $\mathfrak{A}$. In other words, $\mathfrak{A}$ is simple if it admits no congruence, except the equality and universal relations. Figure 9 represents the anti-isomorphism between the lattices $Cg_k$ and $SL_k$.

*Example 4*. Figure 10 shows the structure lattice $SL(\mathfrak{B})$ of the 2-algebra $\mathfrak{B}$ given by Table 8; the $\otimes$-indecomposable elements are marked. We leave it to the reader to develop his own streamlined method of finding all homomorphic images of $\mathfrak{B}$. It seems quite clear that Theorem 2 is of much help; it is easier to manipulate congruences than homomorphisms. We have

TABLE 8

| | 0 | 1 | 2 | 3 | 4 | 5 | 6 | 7 | 8 | 9 |
|---|---|---|---|---|---|---|---|---|---|---|
| $f_1$ | 1 | 3 | 5 | 7 | 4 | 8 | 8 | 3 | 6 | 9 |
| $f_2$ | 2 | 4 | 6 | 4 | 3 | 9 | 6 | 7 | 9 | 8 |

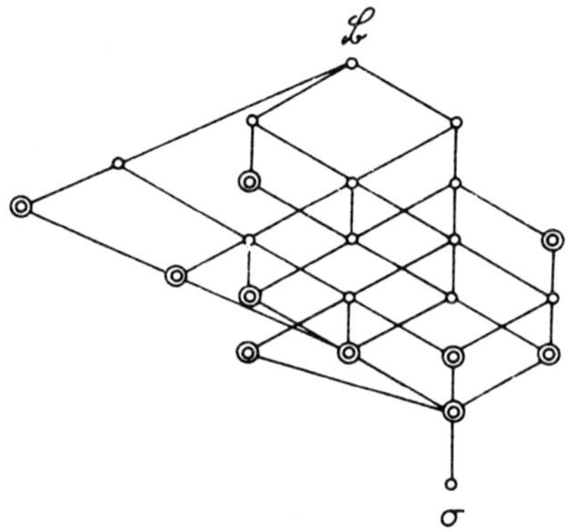


FIG. 10.

chosen this example because it shows how bad structure lattices of reduced $k$-algebras can get. Note the following features:

a. The representation of $\mathfrak{B}$ as a reduced product of indecomposable algebras is far from unique. There are very different irredundant decompositions of $\mathfrak{B}$.

b. Some of the indecomposable members of $SL(\mathfrak{B})$ are not small, in any obvious sense. Compare their strange distribution in $SL(\mathfrak{B})$ with the orderly one in $SL_1$.

c. There are maximal chains from $\mathfrak{D}$ to $\mathfrak{B}$, which are of different length. The lattice is not modular nor even semimodular.

Of course, the idea of studying the structure of an algebra via congruences is not new. What we have done here is to show that, by choosing the proper mathematical formulation, these classical algebraic methods become available in the study of finite automata. To the algebraist our presentation of Corollary 2 may be of interest. This is just another version of Birkhoff's result: Every universal algebra is a subdirect product of subdirectly indecomposable algebras.

We do not want to give the impression that the basic lattice-theoretic results, discussed in this section, settle the structure theory of finite reduced $k$-algebras. On the contrary, there are good reasons to suspect that this structure theory is just as difficult and interesting a field as the structure theory of finite monoids and groups. Much solid mathematical work is needed to make the lattice-theoretic approach more significant (see Problems 1, 2, 3 below). Furthermore, there are very natural non-lattice-theoretic operations on $k$-algebras, which might be the key to a satisfactory structure theory (see Problem 3 below). We mention only the idea of cascading, where an output of the $k$-algebra $\mathfrak{A}$ is used to drive (part of) the input of a $k$-algebra $\mathfrak{B}$. Finally, we would like to stress that Corollary 2 is trivial for finite algebras. Thus, as far as the lattice $SL_k^0$ of all finite reduced $k$-algebras is concerned, we are left with only the most elementary results.

In spite of the present lack of deeper results concerning the structure of finite reduced $k$-algebras, investigating the lattice $SL(\mathfrak{A})$ of a particular finite $\mathfrak{A}$ might still yield practical information on $\mathfrak{A}$. Furthermore, in the case of finite commutative $k$-algebras, the lattice-theoretic approach can actually be worked out to yield a satisfactory structure theory, and it is quite reasonable to expect that this systematic survey should be of interest to people who design counters. Finally, we remark that congruences and the forming of quotients are the basis for a clear understanding of minimality algorithms for finite automata. Minimality and Kleene's theory of behavior will be presented as Part II of these lectures.

## PROBLEMS

1. Classify the simple and $\otimes$-indecomposable finite reduced $k$-algebras; i.e., give a survey over all maximal and $\bigcap$-irreducible right congruences on $N_k$.

2. Investigate the nonuniqueness of $\otimes$-decompositions for finite reduced $k$-algebras.

3. Characterize those lattices which are isomorphic to $SL(\mathfrak{A})$, for some finite reduced $k$-algebra $\mathfrak{A}$.

4. Even in the case $k = 1$, the relationship between simple (primes) and $\otimes$-indecomposables (prime powers) can be better understood if nonlattice operations (product, powers) are considered. Is there a reasonable operation on $k$-algebras, which would clear up these matters for $k \geqslant 2$?

*Exercise 5*

(a) Draw the lattice $Eq$ of all equivalence relations on the set $\{1,2,3,4\}$.

(b) Find the relation $(s,p) \Leftarrow (r,q)$ and the operation $(s,p) \otimes (r,q)$ on pairs of natural numbers, which correspond to the relation $\mathfrak{A}_{s,p} \Leftarrow \mathfrak{A}_{r,q}$ and the operation $\mathfrak{A}_{s,p} \otimes \mathfrak{A}_{r,q}$ on finite reduced 1-algebras. Draw the structure lattice $SL(\mathfrak{A}_{2,12})$. Which $\mathfrak{A}_{s,p}$ are simple, which are $\otimes$-indecomposable? Show that the structure lattice $SL_1$ (all reduced 1-algebras) is isomorphic to the lattice of divisibility on natural numbers, i.e., isomorphic to the infinite direct power of the lattice $\langle N, \leqslant \rangle$. (Consequently, $SL_1$ is distributive!) Which is the irredundant $\otimes$-representation of $\mathfrak{A}_{s,p}$?

(c) Let the 2-algebra $\mathfrak{A}$ be given by Table 9. Find the congruence $E$ generated by (the relation whose only element is) the pair $BC$. Find the transition tree of $\mathfrak{A}/E$. This process is called *merging of the states $B, C$* of $\mathfrak{A}$.

TABLE 9

| | $A$ | $B$ | $C$ | $D$ | $E$ | $F$ | $G$ | $H$ |
|---|---|---|---|---|---|---|---|---|
| $f_1$ | $B$ | $D$ | $F$ | $H$ | $C$ | $H$ | $H$ | $F$ |
| $f_2$ | $C$ | $E$ | $G$ | $D$ | $B$ | $D$ | $C$ | $G$ |

(d) Construct a 2-algebra $\mathfrak{A}$ whose equi-response relation is

$$E = E_1 \cap E_2 \cap E_4.$$

$E_1, E_2, E_4$ are given by the corresponding expressions in Exercise 4b. *Hint:* Use the reduced-product operation.

(e) Work out Examples 2 and 4. Note that $\mathfrak{A}$ occurs in $SL(\mathfrak{B})$.

## References

1. S. C. Kleene, Representation of events in nerve nets and finite automata, *Automata Studies*, Princeton University Press, Princeton, N.J., 1956, pp. 3–41.
2. G. Birkhoff, *Lattice Theory*, Am. Math. Soc., New York, 1948.

3. J. R. Büchi, On a decision method in restricted second order arithmetic, in *Proc. Intern. Congr. Logic Phil. Sci.*, Stanford University Press, Stanford, Calif., 1962, pp. 1–11.
4. J. R. Büchi, Weak second order arithmetic and finite automata, *Z. Math. Logik und Grundlagenforschung der Math.* **6**, 66–92 (1960).
5. A. Church, Logic, Arithmetic, and Automata. *Proc. Intern. Congr. Math.*, Stockholm, 1962, pp. 23–35.
6. A. W. Burks and J. B. Wright, Theory of logical nets, *Proc. IRE* **41**, 1357–65 (1953).
7. A. Church, Application of recursive arithmetic to the problem of circuit synthesis, in *Proc. Logic Conf.*, Cornell University, Ithaca, New York, 1957.

# Canonical Systems which Produce Periodic Sets

by

J. R. BÜCHI

Mathematics Department
Purdue University

and

W. H. HOSKEN

Computer Science Department
Pennsylvania State University

ABSTRACT

Canonical systems with productions of the form $a\underline{x}$, $b\underline{x} \to c\underline{x}$ and $\underline{x}a \to \underline{x}c$ are shown to produce only the periodic sets. Thus, these systems are equivalent to finite-state grammars.

## 1. Introduction

Certain canonical systems discussed by Post in his paper on normal systems [1] are investigated.†

These systems produce periodic sets and, as pointed out in Büchi [2], are closely related to finite automata, i.e., finite-state grammars (see [2] for details). Kratko [3] contains results equivalent to Theorems 1 and 2. The methods used here are quite different and should be of interest. Also, a proof of Theorem 3 has not been published before.

*Words* are finite (possibly null) strings formed from the *letters* 1, 2, 3,$\cdots$. $o$ denotes the null string. $I_k$ denotes the set (alphabet) $\{1, 2, \cdots, k\}$. $N_k$ denotes the set of all words with letters from $I_k$. If $y$ is a word, $\lg(y)$ denotes the length of $y$. The concatenation of words $x$ and $y$ is written $x^\wedge y$ and is often abbreviated by $xy$. Let $a, a_1, a_2, \cdots, a_n, b$ be words. The notation $\underline{x}a_1, \underline{x}a_2, \cdots, \underline{x}a_n, \cdots, \underline{x}a_n \to \underline{x}b$ is called an *n-left* regular production. Such a production *directly produces* a word $v$ from words $u_1, u_2, \cdots, u_n$ if there is a word $x$ such that $u_i = xa_i$ $(i = 1, 2, \cdots, n)$ and $v = \underline{x}b$. Similarly for *n-right regular productions* $a_1\underline{x}, a_2\underline{x}, \cdots, a_n\underline{x} \to b\underline{x}$ and normal productions $a\underline{x} \to \underline{x}b$. Thus, these are types of canonical productions studied by Post [1]. (See also Rosenbloom [4].)

**Definition 1.** An *n-m*-system on the alphabet $I_k$ is a finite collection $\Sigma$ of $n$-left and $m$-right regular productions, where the $a_1, a_2, \cdots, a_n, b$ occurring in the productions of $\Sigma$ are words on $I_k$. In case all productions are on the same side in $\Sigma$, we call $\Sigma$ an $n$-system. A normal system $\Sigma$ on $I_k$ is a finite collection of normal productions where the $a$, $b$ occurring in the production are words on $I_k$.

†This work was supported by NSF Grant No. GP 06120. Theorem 1 was reported in Büchi [2] and Theorem 2 in Hosken [5].



MATHEMATICAL SYSTEMS THEORY, Vol. 4, No. 1.
Published by Springer-Verlag New York Inc.

**Definition 2.** Let $\Sigma$ be an $n$-$m$-system or a normal system on $I_k$ and $U \subseteq N_k$. The set $\tau[\Sigma, U]$ of theorems of $[\Sigma, U]$ is defined as follows:

$\tau[\Sigma, U]$ is the intersection of all sets $\alpha \subseteq N_k$ such that $U \subseteq \alpha$ and if $x_1, x_2, \cdots, x_r \in \alpha$ and $y$ is directly produced from $x_1, x_2, \cdots, x_r$ by some production of $\Sigma$, then $y \in \alpha$.

This might be called the *outside* definition of $\tau[\Sigma, U]$. $\tau[\Sigma, U]$ may also be defined from the inside using the notion of a deduction.

**Definition 3.** A $k$-transition system $\underline{S} = \langle S, A, R_1, R_2, \cdots, R_k, B\rangle$ is a $(k+3)$-tuple where $S$ is a finite set of states, $A \subseteq S$ is called the initial set, $B \subseteq S$ is called the terminal set and $R_1, R_2, \cdots, R_k$ are binary relations on $S$.

$k$-transition systems are often called "non-deterministic finite automata". The behavior, $\beta_{\underline{S}}$, of a $k$-transition system is defined as follows:

$$o \in \beta_{\underline{S}} \cdot\equiv\cdot A \cap B \neq \varnothing;$$

if $x_1, x_2, \cdots, x_m \in I_k$ then

$$x_1x_2\cdots x_m \in \beta_{\underline{S}} \cdot\equiv\cdot (\exists y_0y_1\cdots y_m)\,[y_0 \in A \wedge R_{x_1}[y_0, y_1] \wedge \cdots \wedge R_{x_m}[y_{m-1}y_m] \wedge Y_m \in B].$$

**Definition 4.** A $k$-finite automaton $\underline{A}$ is a $k$-transition system in which $A = S_0$ has only one element and $R_i$, $i = 1, \cdots, k$, are functions from $S$ into $S$. If $\beta_{\underline{A}}$ is the behavior of a $k$-finite automaton $\underline{A}$, then $\beta_{\underline{A}}$ is called a periodic set in $I_k$. The following is a well known fact from automata theory.

**LEMMA 1.** (Myhill [6], Medvedey [7], Rabin and Scott [8]). *There is an effective method which, given a k-transition system $\underline{S}$, produces a k-finite automaton $\underline{A}$ such that $\beta_{\underline{S}} = \beta_{\underline{A}}$.*

## 2. One-sided Productions

In the proof of Theorem 1, the outside definition of the set of theorems is used.

**THEOREM 1.** *If $\Sigma$ is an n-system on $I_k$ and $U \subseteq N_k$ is a finite set, then $\tau[\Sigma, U]$ is a periodic set in $I_k$. Moreover, there is an effective method, which, given $\Sigma$ and U, produces a k-finite automaton $\underline{A}$ such that $\beta_{\underline{A}} = \tau[\Sigma, U]$.*

Before giving the proof we first prove the following lemmas. We work only with right-regular systems since the modifications for left-regular systems are obvious.

The notation $\underline{\alpha}$ stands for an $n$-tuple $(\alpha_1, \alpha_2, \cdots, \alpha_n)$ of variables ranging over sets of words on $I_k$. $\underline{A}, \underline{B}, \cdots$ range over propositional formulas.

For words $x$ and $y$, we write $x<y$ if there is a word $z$ such that $xz = y$.

Variables $a, b, \cdots$ range over words from $N_k$.

**LEMMA 2.** *If $\Sigma$ is an n-system on $I_k$ and $C = \{c_1, c_2, \cdots, c_\ell\} \subseteq N_k$, then there is a formula $U(y)$ of the form*

$$(\forall\alpha)\cdot(\forall x)\ \underline{C}\begin{vmatrix}\underline{\alpha}(c_1) & \underline{\alpha}(xa_1)\\ \vdots & \vdots\\ \underline{\alpha}(c_p) & \underline{\alpha}(xa_q)\end{vmatrix} \supset D\begin{vmatrix}\underline{\alpha}(d_1) & \underline{\alpha}(ye_1)\\ \vdots & \vdots\\ \underline{\alpha}(d_r) & \underline{\alpha}(ye_s)\end{vmatrix}$$

*such that $y \in \tau[\Sigma, C) \equiv U(y)$.*

*Proof.* If $\Sigma$ is the $n$-system

$$\begin{matrix} \underset{\sim}{x}a_{11}, \cdots, \underset{\sim}{x}a_{1n} \to \underset{\sim}{x}b_1 \\ \vdots \quad\quad \vdots \quad\quad \vdots \\ \underset{\sim}{x}a_{m1}, \cdots, \underset{\sim}{x}a_{mn} \to \underset{\sim}{x}b_m \end{matrix}$$

then $\tau[\Sigma, U]$ is the intersection of all sets $\alpha$ such that

$$\bigwedge_{v=1,\cdots,\ell} c_v \in \underset{\sim}{\alpha} \quad \bigwedge_{v=1,\cdots,m} (\forall x)\,[xa_{v1} \in \underset{\sim}{\alpha} \wedge \cdots \wedge xa_{vm} \in \underset{\sim}{\alpha} \cdot\supset\cdot xv \in \underset{\sim}{\alpha}].$$

Therefore the following formula $U(y)$ defines $\tau[\Sigma, C]$

$$(\forall \underset{\sim}{\alpha})_{v=1,\cdots,\ell}\, \underset{\sim}{\alpha}(c_v)\, (x)_{v=1,\cdots,m}\, [\underset{\sim}{\alpha}(xa_{v1})\cdots\underset{\sim}{\alpha}(xa_{vn}) \cdot\supset\cdot \underset{\sim}{\alpha}(xb_v)]$$

This formula is of the type required. Note that the proof of Theorem 1 requires only a very special form of the formula given in Lemma 2.

**LEMMA 3.** *Every set definable by a formula $U(y)$ of the form given in Lemma 2 is also definable by a formula $V(y)$ of the following form:*

$$(\forall \underset{\sim}{\alpha}) \cdot \underline{K}(\underset{\sim}{\alpha}(o))\, (\forall x)\underline{B}[\underset{\sim}{\alpha}(x), \underset{\sim}{\alpha}(x1), \underset{\sim}{\alpha}(x2), \cdots, \underset{\sim}{\alpha}(x_k)] \cdot\supset\cdot \underline{A}[\underset{\sim}{\alpha}(y)].$$

*Proof.* New set variables $\alpha_2, \alpha_3, \cdots, \alpha_n$ are to be introduced as illustrated in the following example. If $\alpha_1(x12)$ occurs in $U(y)$, replace $\alpha_1(x12)$ by $[\alpha_2(x) \equiv \alpha_1(x2)]\alpha_2(x1)$.

**LEMMA 4.** *Every set definable by a formula $U(y)$ of the form given in Lemma 3 is also definable by a formula $V(y)$ of the following form:*

$$(V\underset{\sim}{\alpha})_{\text{finite}} \cdot \underline{K}[\underset{\sim}{\alpha}(o)]\, (Vx)_{x<y}\underline{B}^*[\underset{\sim}{\alpha}(x), \underset{\sim}{\alpha}(x1), \cdots, \underset{\sim}{\alpha}(xk)] \supset \underline{A}[\underset{\sim}{\alpha}(y)],$$

where "$(\forall\underset{\sim}{\alpha})_{\text{finite}}\cdots$" stands for "for all partial valuations which extend at least from $o$ to $y$ . . . ".

*Proof.* It seems somewhat more intuitive to prove the dual assertion: If $U(y)$ is

$$(\forall\alpha)_{\text{finite}} \cdot \underline{K}[\underset{\sim}{\alpha}(o)]\, (\forall x)_{x<y}\underline{B}[\underset{\sim}{\alpha}(x), \underset{\sim}{\alpha}(x1), \cdots, \underset{\sim}{\alpha}(xk)] \wedge \underline{A}(\underset{\sim}{\alpha}(y)),$$

then one can find a formula $V(y)$ of the form

$$(\exists\underset{\sim}{\alpha})_{\text{finite}} \cdot K[\underset{\sim}{\alpha}(o)] \wedge (\forall x)_{x<y}\underline{B}^*[\underset{\sim}{\alpha}(x), \underset{\sim}{\alpha}(x1), \cdots, \underset{\sim}{\alpha}(xk)] \wedge \underline{A}(\underset{\sim}{\alpha}(y))$$

such that $(\forall y)U(y) \equiv V(y)$.

Let $\underline{E}(I)$ be a propositional formula ($I$ ranges over $n$-tuples of truth values when $\underset{\sim}{\alpha}$ is an $n$-tuple) such that

$$\underline{E}[I] \equiv (\exists\underset{\sim}{\alpha})\,[\underset{\sim}{\alpha}(o) \equiv I \wedge (\forall x)\underline{B}[\underset{\sim}{\alpha}(x), \underset{\sim}{\alpha}(x1), \cdots, \underset{\sim}{\alpha}(xk)]].$$

Note that: (1) If $\underline{E}[I]$ and if $\underset{\sim}{\alpha}$ is a partial valuation which extends from $o$ through $u$ and satisfies $\underline{B}$ through $u$ and if $\underset{\sim}{\alpha}(u) \equiv I$, then the partial valuation $\underset{\sim}{\alpha}$ can be modified and extended to all successors $ui$ in such a way that $\underline{B}$ will remain satisfied through $ui$.

(2) Every total infinite valuation which satisfies $\underline{B}$ makes use only of $n$-tuples $I$ for which $\underline{E}[I]$. From these remarks it is clear that

$$\begin{aligned} U(y) \equiv (\exists\underset{\sim}{\alpha})_{\text{finite}} \cdot \underline{K}(\underset{\sim}{\alpha}(o)) \wedge (\forall x)_{x<y}[\underline{B}[\underset{\sim}{\alpha}(x), \underset{\sim}{\alpha}(x1), \cdots, \underset{\sim}{\alpha}(xk)] \\ \wedge \underline{E}[\alpha(x1)] \wedge \underline{E}[\alpha(x2) \wedge \ldots \wedge \underline{E}[\underset{\sim}{\alpha}(xk)] \wedge \underline{A}(\underset{\sim}{\alpha}(y))]; \end{aligned}$$

thus we have found the required formula. To show that $V(y)$ may be effectively found, we show that $\underline{E}[I]$ may be taken to be

$$\underline{E}[I]: \bigvee_{\substack{\text{Valuation of } \underset{\sim}{\alpha} \text{ on} \\ \{u \mid \lg(u) \leq 2^n\}}} [\underset{\sim}{\alpha}(o) \equiv I \bigwedge_{\lg(x) < 2^n} \underline{B}[\underset{\sim}{\alpha}(x), \underset{\sim}{\alpha}(x1), \cdots, \underset{\sim}{\alpha}(xk)]].$$

Clearly

$$(\exists \underset{\sim}{\alpha})\, [\underset{\sim}{\alpha}(o) \equiv I \wedge (\forall x)\underline{B}[\underset{\sim}{\alpha}(x), \underset{\sim}{\alpha}(x1), \cdots, \underset{\sim}{\alpha}(xk)]] \supset \underline{E}[I].$$

To show the converse, suppose $\underline{E}[I]$. Thus there is a partial valuation of $\underset{\sim}{\alpha}$, through all $u$ of length $2^n$, which satisfies $\underline{B}$. Now in each path from $o$ to a word $u$ of length $2^n$ there are $2^n+1$ values of $\underset{\sim}{\alpha}$ occurring; therefore, there must be a repetition. Thus the valuation may be modified (if necessary) and extended to the successors of $u$ in such a way that $\underline{B}$ still holds.

**LEMMA 5.** *Every set definable by a formula $U(y)$ of the form in Lemma 4 is also definable by a formula $V(y)$ of the following form*

$$\begin{aligned}(\forall \underset{\sim}{\alpha})_y \,.\, \underline{K}[\underset{\sim}{\alpha}(o)] \wedge [(\forall x)_{x1<y}&\underline{B}_1[\underset{\sim}{\alpha}(x), \underset{\sim}{\alpha}(x1)] \\ (\forall x)_{x2<y}&\underline{B}_2[\underset{\sim}{\alpha}(x), \underset{\sim}{\alpha}(x2)] \\ &\vdots \\ (\forall x)_{xk<y}&\underline{B}_k[\underset{\sim}{\alpha}(x), \underset{\sim}{\alpha}(xk)]] \supset A[\underset{\sim}{\alpha}(y)],\end{aligned}$$

*where "$(\forall \underset{\sim}{\alpha})_y \cdots$" stands for all partial valuations $\underset{\sim}{\alpha}$ defined exactly on $\{u \mid u<y\}$ . . .".*

*Proof.* Given $\underline{B}[I_1, I_2, \cdots, I_{k+1}]$ from $U(y)$, $\underline{B}_i[I_1, I_i]$, $i = 1, 2, \cdots, k+1$ are defined as follows

$$\underline{B}_i[I_1, I_i] \equiv \bigvee_{J_2, \cdots, J_{i-1}, J_{i+1}, \cdots, J_{k+1}} \underline{B}[I_1, J_2, \cdots, J_{i-1}, I_i, J_{i+1}, J_{k+1}].$$

Then clearly $(\forall y)U(y) \equiv V(y)$.

Note that at this point we can show that $\tau[\Sigma, C]$ is recursive. It follows from Lemmas 2, 3, 4, 5 that $\tau[\Sigma, C]$ is definable by a formula $U(y)$ of the form given in Lemma 6. Thus, to determine if $b \notin \tau[\Sigma, C]$, it is simply necessary to check whether among the finite number of valuations of $\underset{\sim}{\alpha}$ on $\{u \mid u<b\}$ there is one for which

$$\begin{aligned}\underline{K}[\underset{\sim}{\alpha}(o)] &\bigwedge_{x1 \leq b} \underline{B}_1[\underset{\sim}{\alpha}(x), \underset{\sim}{\alpha}(x1)] \\ &\bigwedge_{x2 \leq b} \underline{B}_2[\underset{\sim}{\alpha}(x), \underset{\sim}{\alpha}(x2)] \\ &\vdots \\ &\bigwedge_{xk \leq b} \underline{B}_k[\underset{\sim}{\alpha}(x), \underset{\sim}{\alpha}(xk)] \\ &\wedge A[\underset{\sim}{\alpha}(b)].\end{aligned}$$

We will now proceed to show that $\tau[\Sigma, C]$ must be periodic.

**LEMMA 6.** *Every set $\beta$ definable by a formula $U(y)$ of the form given in Lemma 5 is the behavior of a k-transition system.*

*Proof.* Given $U(y)$ as in Lemma 6, we define a $k$-transition system $\underline{S}$ such that $\beta_{\underline{S}} = \tilde{\beta}$. The states of $\underline{S}$ are $n$-tuples of truth values. Define relations $R_i[Y_1, Y_2]$ $(i = 1, 2, \cdots, k)$ on the set $S$ of $n$-tuples of truth values as follows:

$$R_i[Y_1, Y_2] \equiv \underline{B}_i[Y_1, Y_2], \qquad i = 1, 2, \cdots, k.$$

Let $A = \{Y|\underline{K}(Y)\}$, $B = \{Y|\underline{A}(Y)\}$; then $\tilde{\beta}$ is the behavior of the $k$-transition system $\langle S, A, R_1, \cdots, R_K, B\rangle$, i.e.,

$$o \notin \beta \equiv \bigvee_Y \underline{K}(Y) \wedge \tilde{\underline{A}}[Y]$$

$$x_1 x_2 \cdots x_m \notin \beta \equiv \exists y_0 y_1 \cdots y_m \cdot [\underline{K}[Y_0] \wedge R_{x_1}[Y_0, Y_1] \wedge \cdots \wedge R_{x_m}[Y_{m-1}, Y_m] \wedge \tilde{\underline{A}}[Y_m]].$$

The proof of Theorem 1 is now a matter of combining the procedures of Lemmas 1–6 in the order 2–6, 1, and the remark that the class of periodic sets is closed under complementation.

## 3. Two-sided Productions

In extending Theorem 1 to $n-1$ systems we make use of the inside definition of $\tau[\Sigma, U]$ and make a careful investigation of the deduction trees.

**Definition 5.** Let $\Sigma$ and $U$ be as in Definition 2, and let $x \in N_k$. A deduction tree $[\Sigma, U, x]$ is a finite tree whose vertices are labeled with words from $N_k$ satisfying:

(1) $x$ is the label of the root.
(2) If $y$ is the label of any vertex, then $y \in U$ and there are no vertices above $y$, or there are $n$ vertices directly above $y$ labeled with $x_1, x_2, \cdots, x_n$ and $y$ is directly produced from $x_1, x_2, \cdots, x_n$ by some $n$-premise production of $\Sigma$.

A branch of a deduction tree is a maximal length sequence $V_1, V_2, \cdots, V_\ell$ of vertices such that $V_i$ is the root, $V_\ell$ is labeled with $y \in U$, and $V_{i+1}$ is directly above $V_i$, $i = 1, 2, \cdots, \ell-1$. The notation $U|\Sigma \vdash x$ means that there is a deduction tree $[\Sigma, U, x]$.

It is easy to see that $\tau[\Sigma, U]$ may be defined as

$$\tau[\Sigma, U] = \{x|U|\Sigma \vdash x\}.$$

**THEOREM 2.** *If $\Sigma$ is an $n-1$ $(1-n)$ system on $I_k$ and $U \subseteq N_k$ is a finite set, then $\tau[\Sigma, U]$ is a regular set in $I_k$. Moreover, there is an effective method which, given $[\Sigma, U]$, produces a k-finite automaton $A$ such that $\beta_{\underline{A}} = \tau[\Sigma, U]$.*

*Proof.* The proof is contained in the following two lemmas. We handle $n-1$ systems only. The proof for $1-n$ systems is entirely similar.

**LEMMA 7.** *If $\Sigma$ is an $n-0$ system on $I_k$ and $U \subseteq N_k$ is a periodic set, then $\tau[\Sigma, U]$ is a periodic set. Moreover, there is an effective method which, given $\Sigma$ and a k-finite automaton $\underline{A}$ of behavior $U$, produces a k-finite automaton $A$ such that $\beta_{\underline{A}} = \tau[\Sigma, U]$.*

*Proof.* We prove Lemma 7 for the case in which $\Sigma$ is an $n$-left regular system. Büchi [2] gives a method by which, from a $k$-finite automaton of behavior $U$, a 1-system $\Sigma_1$ on $I_k$ and a finite set $U_1 \subseteq N_k$ may be constructed such that $U = \tau[\Sigma_1, U_1]$. Define $[\Sigma_2, U_2]$ on $I_{k+1}$ as follows:

if $u \in U_1$ then $u\hat{}k+1 \in U_2$;

if $\underset{\sim}{x}a \to \underset{\sim}{x}b \in \Sigma_1$, then $\underset{\sim}{x}a\hat{}k+1 \to \underset{\sim}{x}b\hat{}k+1 \in \Sigma_2$, $x\hat{}k+1 \to x \in \Sigma_2$;

if $\underset{\sim}{x}a_1, \underset{\sim}{x}a_2, \cdots, \underset{\sim}{x}a_n \to \underset{\sim}{x}b \in \Sigma$, then $\underset{\sim}{x}a_1, \underset{\sim}{x}a_2, \cdots, \underset{\sim}{x}a_n \to \underset{\sim}{x}b \in \Sigma_2$.

Clearly $\tau[\Sigma_2, U_2] \cap N_k = \tau[\Sigma, U]$. $\Sigma_2$ is an $n$-system in $I_{k+1}$ and thus by Theorem 1, $\tau[\Sigma_2, U_2]$ is a periodic set in $I_{k+1}$. If $\underset{\sim}{S} = \langle S, s, f_1, f_2, \cdots, f_{k+1}, B\rangle$ is a $k+1$ finite automaton such that $\beta_S = \tau[\Sigma_2, U_2]$, then $A = \langle S, s, f_1, \cdots, f_k, B\rangle$ is a $k$-finite automaton such that $\beta_A = \beta_S \cap N_k = \tau[\Sigma, U]$.

**LEMMA 8.** Let $\Sigma_L$ *be a n-left regular system on* $I_k$, $\Sigma_R$ *a* 1-*right regular system on* $I_k$ *and* $U \subseteq N_k$ *a finite set, and let* $\Sigma_{2i} = \Sigma_R$ *and* $\Sigma_{2i+1} = \Sigma_L$, $i = 0, 1, \cdots$. *Then there is an N which may be effectively found, given* $\Sigma_L$ *and* $\Sigma_R$, *such that*

$$\tau[\Sigma_L \cup \Sigma_R, U] = \tau[\Sigma_N, \tau[\Sigma_{N-1}, \cdots, \tau[\Sigma_0, U]\cdots]].$$

*Proof.* Let $M = 2\cdot\max\{\lg(a); a \text{ occurs in a production of } \Sigma_L \text{ or } \Sigma_R\}$. Let $N_1$ be the number of words of length less than $M$. Let $N = 2N_1+1$.

It is clear that

$$\tau[\Sigma_N, \tau[\Sigma_{N-1}, \cdots, \tau[\Sigma_0, U], \cdots] \subseteq \tau[\Sigma_L \cup \Sigma_R, U].$$

To show the opposite inclusion we will show that every $x \in [\Sigma_L \cup \Sigma_R, U]$ has a deduction tree in a special form. Let $T$ be a deduction tree for $x$ in $\tau[\Sigma_L \cup \Sigma_R, U]$.

We may assume that the following conditions are satisfied.

(1) On any branch, a word occurs at most once as a label. If not, the tree above the lower occurrence could be replaced by the tree above the upper occurrence.

(2) If $y$ and $z$ are successive labels on any branch, then either $\lg(y) < M$ or, if $z$ is directly produced by a left-regular production, then so is $y$.

If not, $\lg(y) \geq M$ and $z$ is directly produced from some $w_1, w_2, \cdots, w_n$ by a left-regular production $\underset{\sim}{x}a_1, \underset{\sim}{x}a_2, \cdots, \underset{\sim}{x}a_n \to \underset{\sim}{x}b$ and $y$ is directly produced by $z$ by a right-regular production $c\underset{\sim}{x} \to d\underset{\sim}{x}$. Thus $w_i = cea_i$, $i = 1, 2, \cdots, n$, $z = ceb$ and $y = deb$. Thus $T$ may be modified as follows: add a vertex labeled $dea_i$ below the vertices labeled $w_i$, $i = 1, 2, \cdots n$, and directly above the vertex labeled $y$. Delete the vertex labeled $z$. In this way a deduction tree $T'$ for $x$ in $[\Sigma_L \cup \Sigma_R, U]$ for which (2) holds may be produced.

Thus if $y$ and $z$ are successive labels on a branch of $T'$, then if $z$ is directly produced by a left-regular production and $y$ by a right-regular production, $\lg(y) < M$. We call such an occurrence, for short, an $L$-$R$-point. By (1) there can be at most $N_1$ $L$-$R$-points on any branch of $T'$. Let each label of $T'$ be assigned a number $Ht(y)$ called the $L$-$R$ height.

$Ht(y)$ = the maximum number of $L$-$R$ points occur on any branch in the tree above $y$ with $y$ as root.

It is easy to see by induction on the $L$-$R$ height that for every word $w$ occurring in $T'$,

$$w \in \tau[\Sigma_{2Ht(w)+1}, \tau[\Sigma_{2Ht(w)}\cdots\tau[\Sigma_0, U]\cdots]].$$

Thus $w \in \tau[\Sigma_{2N_1+1}, \tau[\Sigma_{2N_1}, \cdots, \tau[\Sigma_0, U]]]$.

Theorem 2 now follows from Lemma 8 if we choose the $n$-left regular productions of $\Sigma$ as $\Sigma_L$ and the 1-right regular productions of $\Sigma$ as $\Sigma_R$ and apply Lemma 7 $N$ times.

## 4. 2-2 Systems

For completeness we include a proof of the following theorem, which was stated without proof in Post [1]. It shows that every recursively enumerable set in $N_k$ can be obtained as $\tau[\Sigma, U] \cap N_k$, where $\Sigma$ is a 2-2 system on $I_e$, $e>k$, and $U \subseteq N_e$, $U$ finite.

**THEOREM 3.** *Given a normal system* $\Sigma_1$ *on* $I_k$ *and* $g \in N_k$, *a 2-2 system* $\Sigma_2$ on $I_{k+m}$ *and a finite set* $U \subseteq N_{k+m}$ *can effectively be found such that*

$$\tau[\Sigma_1, \{g\}] = \tau[\Sigma_2, U] \cap N_k.$$

*Proof.* Given a normal system $\Sigma_1$, and an axiom $g$:

$$\Sigma_1 = \{a_i\underset{\sim}{x} \to \underset{\sim}{x}b_i\}_{i=1,\cdots,s}; \; g, a_i, b_i \in N_k,$$

introduce additional letters $A$, $B_i$, $C_i$, $D$, $E$, $F$, and let $\Sigma_2$ and the new axioms $\underline{G}$ be as follows:

| Rules of $\Sigma_2$, $i = 1, \cdots, s$, | | $r = 1, \cdots, k$ | | $\underline{G}$ |
|---|---|---|---|---|
| ($\mathrm{I}_i$) | $\underset{\sim}{x}, \underset{\sim}{x}D \to \underset{\sim}{x}B_i$ | ($1_r$) | $\underset{\sim}{x}D \to \underset{\sim}{x}rD$ | $D$ |
| ($\mathrm{II}_i$) | $a_i\underset{\sim}{x}, C_i\underset{\sim}{x} \to A\underset{\sim}{x}$ | ($2_{i,r}$) | $C_i\underset{\sim}{x} \to C_ir\underset{\sim}{x}$ | $D_iB_i$ |
| ($\mathrm{III}_i$) | $\underset{\sim}{x}E, \underset{\sim}{x}B_i \to \underset{\sim}{x}b_i$ | ($3_r$) | $\underset{\sim}{x}E \to \underset{\sim}{x}rE$ | $AE$ |
| (IV) | $F\underset{\sim}{x}, A\underset{\sim}{x} \to \underset{\sim}{x}$ | ($4_r$) | $F\underset{\sim}{x} \to Fr\underset{\sim}{x}$ | $F$ |

**LEMMAS:** (a) $g, \underline{G}|\Sigma_2 \vdash uE \cdot\equiv\cdot u \in A^\wedge N_k$.

(b) $g, \underline{G}|\Sigma_2 \vdash Fu \cdot\equiv\cdot u \in N_k$.

(c) $g, \underline{G}|\Sigma_2 \vdash C_iu \cdot\equiv\cdot u \in N_k{}^\wedge B_i$.

(d) $g, \underline{G}|\Sigma_2 \vdash uD \cdot\equiv\cdot u \in N_k$.

*Proof.* These follow easily from the definitions.

**LEMMA A.** *If* $u \in N_k$ *then* $\underline{G}$, $a_iu|\Sigma_2 \vdash ub_i$.
*Proof.*

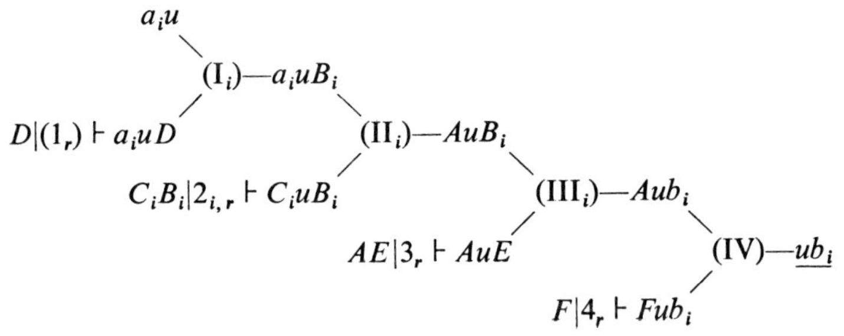


because $u \in N_k$.

**LEMMA B.** *Let $z \in N_k$, $z \neq g$ and $g, \underline{G}|\Sigma_2 \vdash z$. In any $\Sigma_2$-deduction of $z$ from $g, \underline{G}$ there must occur a word $a_i u$ such that $z = ub_i$.*

*Proof.* The deduction cannot consist of $z$ alone, else $z = g$ or $z \in \underline{G}$ and therefore $z \notin N_k$. Furthermore, the last step in the deduction cannot be by any of the rules (1), (2), (3), (4), (I), (II), else $z \notin N_k$. It can also not be by the rule (III$_i$), else $z = vb_i$ and $g, \underline{G}|\Sigma_2 \vdash vE$; by Lemma (a) it would follow that $v \in A^\wedge N_k$, which contradicts $z \in N_k$ and $z = vb_i$. Thus, the given $\Sigma_2$-deduction is of the form

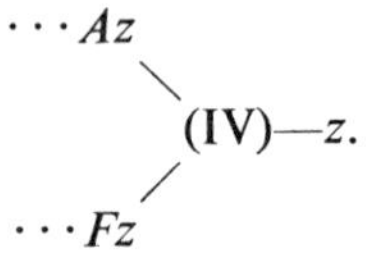


Again $Az$ can neither be in $g, \underline{G}$ nor can it be the result of a step using (1), (2), (3), (4), (I). Also (II$_i$) is out, else $g, \underline{G}|\Sigma_2 \vdash C_i z$, which contradicts Lemma (c) and $z \in N_k$. The same for (IV), $g, \underline{G}|\Sigma_2 \vdash FAz$, which contradicts Lemma (b). It remains that $Az$ follows by use of (III$_i$), and thus $z = ub_i$ and the given $\Sigma_2$-deduction is of the form

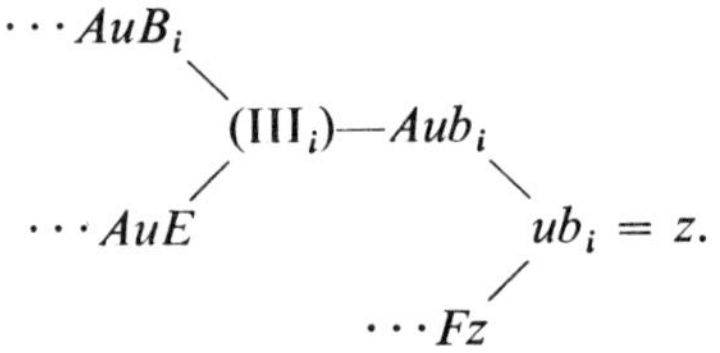


Again $AuB_i$ is not a member of $g, \underline{G}$, nor can it be obtained by (1), (2), (3), (4), (III). Also:

(I$_i$) is out, else $g, \underline{G}|\Sigma_2 \vdash AuD$ and by Lemma (d), $Au \in N_k$, which is contradictory.

(IV) is out, else $g, \underline{G}|\Sigma_2 \vdash FAuB_i$ and by Lemma (b), $AuB_i \in N_k$, which is contradictory.

It remains that $AuB_i$ results if we apply (II$_j$); that is, the $\Sigma_2$-deduction has the form

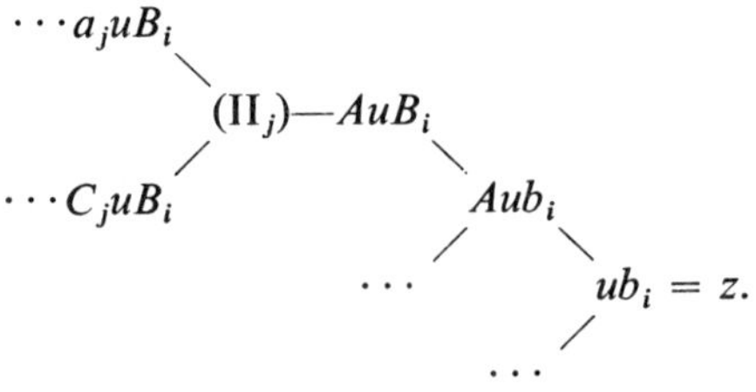


It follows that $g, \underline{G}|\Sigma|C_j uB_i$. Therefore, by Lemma (c), $uB_i \in N_k{}^\wedge B_j$. Therefore $i = j$.

Again $a_i uB_i$ is not a member of $g, \underline{G}$, nor can it be obtained by (1), (2), (3),

(4), $(\mathrm{II}_i)$, $(\mathrm{III}_i)$. Also, (IV) is out; else $g$, $\underline{G}|\Sigma_2|F_i uB_i$, which contradicts Lemma (b). There remains the rule $(\mathrm{I}_i)$ to obtain $a_i uB_i$, so that the deduction is as follows:

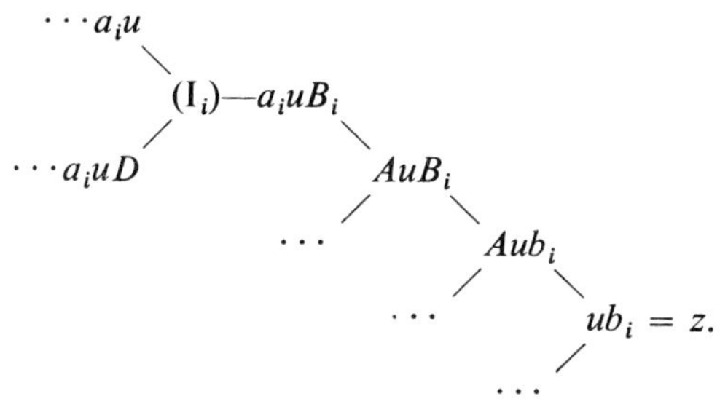


This ends the proof of Lemma B.

Now $\tau(\Sigma_1, \{g\}) = \tau(\Sigma_2, \underline{G}\{g\}) \cap N_k$.

1. Trivially $g \in \tau(\Sigma_2\{g, \underline{G}\}) \cap N_k$. Furthermore, by Lemma A, if $a_1 u \in \tau(\Sigma_2, \{g, \underline{G}\})$ then also $ub_i \in \tau(\Sigma_2, \{g, \underline{G}\})$, i.e., $\tau(\Sigma_2, \{g, \underline{G}\})$ contains $g$ and is closed under the rules of $\Sigma_1$. Therefore

$$\tau(\Sigma_1, g) \subseteq \tau(\Sigma_2, \{g, \underline{G}\}) \cap N_k.$$

2. Let $\tau_L = \{z \in N_k|$ there is a $\Sigma_2$-deduction of height $\leq L$ from $\{g, \underline{G}\}$ to $z\}$. We will show by induction that for any $L = 1, 2, 3, \cdots, \tau_L \subseteq \tau(\Sigma_1, g)$.

$L = 1$: $\qquad \tau_L = \tau_1 = \{g\} \subseteq \tau(\Sigma_1, g)$.

$L = n+1$: Suppose $z \in \tau_1$ and $z \in \tau_{n+1}$. Then by Lemma B there are $u$ and $i$ such that $a_i u \in \tau_n$, and $z = ub_i$. By the inductive hypothesis we have $\tau_n \subseteq \tau(\Sigma_1, g)$. Thus $g|\Sigma_1 \vdash z$, i.e., $z \in \tau(\Sigma_1, g)$.

This shows that $\tau_{n+1} \subseteq \tau(\Sigma_1, g)$. Now

$$\tau(\Sigma_2, \{g, \underline{G}\}) \cap N_k = \bigcup_{L = 1, 2, \cdots} \tau_L \subseteq \tau(\Sigma_1, g).$$

## 5. Concluding Remarks

We feel that Theorems 2 and 3 present an interesting contrast, and hope that further results on special multi-premise Post systems will be forthcoming. It seems strange that so little about multi-premise systems has been published. For example, the straightforward relationship between systems (multi-premise) with rules of the form $A\underline{x}_1, B\underline{x}_2, \cdots, C\underline{x} \rightarrow D\omega$ (where $\omega$ is a word in $\{1, 2, \cdots, \ell, \underline{x}_1, \underline{x}_2, \cdots, \underline{x}_n\}$ where for each $i$, $\lg_{x_i}(\omega) = 1$, and $A, B, \cdots, C, D$ are auxiliary letters) and phase structure grammar [9] does not seem to be generally recognized.

REFERENCES


[1] E. Post, Formal reductions of the general combinatorial decision problem, *Amer. J. Math.* **65** (1943), 197–215.

[2] J. R. Büchi, Regular canonical systems, *Arch. Math. Logik Grundlagenforsch.* **6** (1964), 91–111.

[3] M. Kratko, Formal'nye isčislenija Posta i konečnye avtomaty (Russian), *Problemy Kibernet.* **17** (1966), 41–65.

[4] P. C. ROSENBLOOM, *The Elements of Mathematical Logic*, Dover, New York, 1950.
[5] W. HOSKEN, Combinatorial systems which produce regular sets, *Notices Amer. Math. Soc.* **13** (1966), 732.
[6] JOHN MYHILL, Finite automata and representations of events, WADO Report TR 57-624, *Fundamental Concepts in the Theory of Systems* (1957), 112–137.
[7] I. T. MEDVEDEV, On a class of events representable in a finite automaton, MIT Lincoln Laboratory Group Report, 34–73 (1958).
[8] M. RABIN and D. SCOTT, Finite automata and their decision problems, *IBM J. Res. Develop.* **3** (1959), 114–125.
[9] N. CHOMSKY, Three models for the description of language, *IRE Trans. Information Theory* **IT2** (1956), 113–124

(*Received 12 August 1969*)

# Section 5 Automata and Monadic Theories

*With comments by Robert McNaughton, Rensselaer Polytechnic Institute*



*

# Büchi's Sequential Calculus

*Robert McNaughton**

From one point of view Richard Büchi's sequential calculus (SC) is a restricted version of the arithmetic of the nonnegative integers studied as a logical system and proved to be decidable; the restriction is that neither addition nor multiplication is definable in it. From another point of view, it is a formalism for describing the finite and infinite behavior of finite automata. Actually, it is the second point of view that provides the groundwork for the decision procedure. This paper will offer an elementary exposition of the sequential calculus, explain its use in describing finite automata, and present a detailed exposition of the decision procedure. The last section will examine related papers by Büchi and others. It is my hope that my exposition will be close to Büchi's original train of thought and at the same time more accessible than the original paper [8] (as numbered in Publications, p. xi).

Büchi's paper was, I think, the first example of a problem in pure logic being solved by reference to the theory of automata. For it answered two questions that had been posed by Tarski some years before: whether addition is definable in SC, and whether SC is decidable. (We shall discuss Büchi's negative answer to the first question in Section 5.) Especially interesting was Büchi's reference to (11) (references to the list on pp. 396–397 in parentheses), which describes these problems.

## The Formalism

SC has two kinds of variables: $t, u, v, w, x, y, z$, these lower-case Roman letters with numerical subscripts standing for nonnegative integers; and the lower-case letters $p$ through $s$ standing for monadic predicates of nonnegative integers. It has the constant term 0, and the constant function symbol $'$. $x'$ is the successor of $x$, and $0, 0', 0'', \ldots$ are numerals for all the nonnegative integers. A *term* is either (1) a number variable, (2) 0, or (3) a term followed by the function symbol $'$. Thus, terms are the syntactic entities that denote nonnegative integers.

Truth-functional connectives are $\vee$ (or), $\wedge$ (and), $\neg$ (not), $\rightarrow$ (if then), and $\leftrightarrow$ (if and only if). Quantifiers $\forall$ and $\exists$ are allowed over both types of variables.

An atomic (well-formed) formula is a predicate variable followed by a term enclosed in parentheses, e.g., $p(x)$, $p(x')$, $p(0)$, $p(0')$. If $\Phi$ and $\Psi$ are formulas then $(\Phi) \vee (\Psi)$, $(\Phi) \wedge (\Psi)$, $\neg(\Phi)$, $(\Phi) \rightarrow (\Psi)$, and $(\Phi) \leftrightarrow (\Psi)$ are formulas (in practice we sometimes omit the parentheses). Finally, the result of quantifying a formula either by a number variable or a predicate variable, either existen-

*Department of Computer Science, Rensselaer Polytechnic Institute, Troy, NY 12180-3590.

tially or universally, is a formula, e.g., $(\forall x)p(x)$, $(\exists x)p(x)$, $(\forall p)p(x)$, and $(\exists p)p(x)$ are all formulas.

We introduce definitions:

| | | |
|---|---|---|
| $x = y$ | for | $(\forall p)(p(x) \leftrightarrow p(y))$ |
| $x \neq y$ | for | $\neg(x = y)$ |
| $x \leq y$ | for | $(\forall p)([p(x) \wedge (\forall z)(p(z) \rightarrow p(z'))] \rightarrow p(y))$ |
| $x \geq y$ | for | $y \leq x$ |
| $x > y$ | for | $x \geq y \wedge x \neq y$ |
| $x < y$ | for | $y < x$ |
| $(\exists^{\omega} x)\Psi$ | for | $(\forall y)(\exists x)(x > y \wedge \Psi)$ |

(In the last definition, we issue the customary caution that the variable $y$ is selected to be new; it is assumed not to occur in $\Psi$.)

SC is an interpreted formalism: 0 means zero and $'$ means successor. The truth functions and quantifiers have their customary meanings. Thus, if $\Psi$ is a formula, the formulas (1) $(\forall x)\Psi$, (2) $(\exists x)\Psi$, (3) $(\forall p)\Psi$, and (4) $(\exists p)\Psi$ are true, respectively, if and only if (1) $\Psi$ is true for every interpretation of $x$ (as a nonnegative integer), (2) $\Psi$ is true for at least one interpretation of $x$, (3) $\Psi$ is true for all interpretations of $p$ (as a predicate of nonnegative integers), and (4) $\Psi$ is true for at least one interpretation of $p$.

It is to be noted that the formulas $(\forall x)p(x)$ and

$$(\forall q)(q(0) \wedge (\forall y)[q(y) \rightarrow q(y')] \rightarrow (\forall x)[q(x) \rightarrow p(x)])$$

are equivalent in the sequential calculus. Any interpretation of $p$ as a predicate over the nonnegative integers making one of them true also makes the other true. But the two formulas are not logically equivalent, since there are interpretations in some universes that separate them. For example if the universe contains all the integers, negative as well as nonnegative, then $p$ can be interpreted to mean "is nonnegative," making the first formula false and the second true.

Because SC is an interpreted formalism, we do not write axioms for it. We simply declare that the valid formulas are those that are true under all interpretations (specified earlier) of its free predicate and number variables. This was Büchi's point of view in his paper on SC [8], although in later similar papers he and his collaborators did become interested in axiomatizations (e.g., [25] and [26]).

## Finite Automata

For the purposes of this paper we define a *deterministic finite automaton* algebraically as an ordered triple $\langle Q, A, \delta \rangle$ such that $Q$ and $A$ are finite sets and $\delta\colon Q \times A \rightarrow Q$ is a function. $Q$ is the set of states, $A$ is the input alphabet, and $\delta$ is the transition function. The meaning of $\delta$ is extended so that its domain is $Q \times A^*$ ($A^*$ is the set of all finite words over $A$) by the following recursive definition:

$$\delta(k, \lambda) = k, \qquad k \in Q$$

$$\delta(k, wa) = \delta(\delta(k, w), a), \qquad k \in Q,\ w \in A^*,\ a \in A.$$

To say $\delta(k, w) = r$ is to say the word $w$ takes the automaton from state $k$ to state $r$. (In the literature, a deterministic finite automaton also includes certain other things: a subset $F \subseteq Q$ [the set of *final* or *acceptance* states] and an initial state $k_0 \in Q$. The language of the automaton is defined to be the set of all words $w$ such that $\delta(k_0, w) \in F$.)

A *nondeterministic finite automaton* is an ordered triple $\langle Q, A, R\rangle$ where $Q$ and $A$ are as they were but $R$ is now a ternary relation: $R \subseteq Q \times A \times Q$. Again, $R$ is extended so that its domain is $Q \times A^* \times Q$: (1) $(k_1, \lambda, k_2) \in R$ if and only if $k_1 = k_2$; and (2) for $w \in A^*$ and $a \in A$, $(k_1, wa, k_3) \in R$ if and only if, for some $k_2 \in Q$, $(k_1, w, k_2) \in R$ and $(k_2, a, k_3) \in R$.

Now for given $k_1 \in Q$, $w \in A^*$, there may be several $k_2$ such that $(k_1, w, k_2) \in R$, there may be exactly one, or there may be none at all. But, for any $k_2$, if $(k_1, w, k_2) \in R$ we say the input word $w$ may take the automaton from state $k_1$ to $k_2$. The relation $R$ specifies which transitions are possible. Note that a deterministic finite automaton is a special case of a nondeterministic finite automaton.

Finite automata can be described in SC by taking enough state predicates $q_1, \ldots, q_n$. It suffices to take $n$ to be the smallest integer not less than $\log_2|Q|$. Similarly, we can take input predicates $p_1, \ldots, p_m$, where $m$ is the smallest integer not less than $\log_2|A|$. We can then arbitrarily identify the states with some (possibly all) of the truth conditions of the state predicates and the input letters with truth conditions of the input predicates.

For example, the automaton in Figure 2.1 has $Q = \{k_0, k_1, \ldots, k_5\}$, $A =$

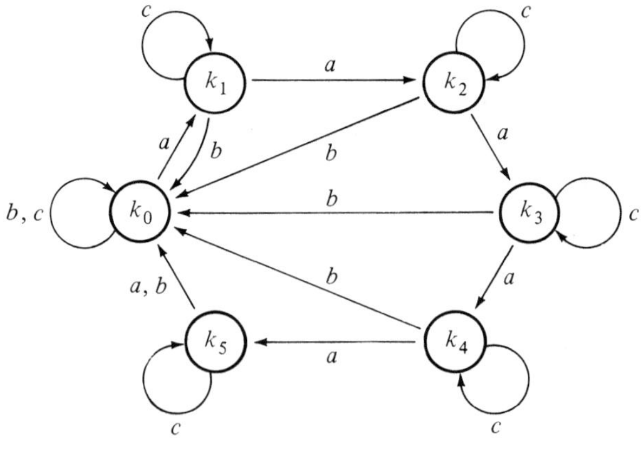


FIGURE 2.1

| | $q_1(x)$ | $q_2(x)$ | $q_3(x)$ |
|---|---|---|---|
| $k_0$ | $T$ | $T$ | $T$ |
| $k_1$ | $T$ | $T$ | $F$ |
| $k_2$ | $T$ | $F$ | $T$ |
| $k_3$ | $T$ | $F$ | $F$ |
| $k_4$ | $F$ | $T$ | $T$ |
| $k_5$ | $F$ | $T$ | $F$ |

| | $p_1(x)$ | $p_2(x)$ |
|---|---|---|
| $a$ | $T$ | $T$ |
| $b$ | $T$ | $F$ |
| $c$ | $F$ | $T$ |

FIGURE 2.2

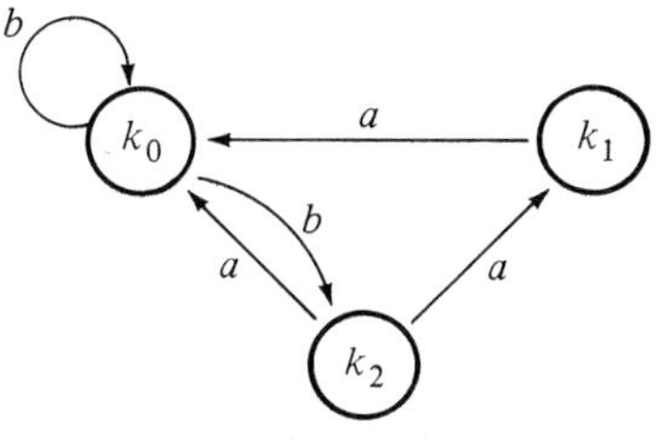


FIGURE 2.3

$\{a, b, c\}$, and $\delta$ is as shown. For this deterministic automaton, we can take state predicates $q_1, q_2, q_3$ and input predicates $p_1, p_2$. At time $x$, if the automaton is in state $k_h$, $0 \le h \le 5$, then we assign truth and falsity to $q_1(x)$, $q_2(x)$, $q_3(x)$ according to the first table in Fig. 2.2. If the input at time $x$ is $a$, $b$, or $c$ then we assign truth or falsity to $p_1(x)$, $p_2(x)$ according to the second table in that figure.

With this identification Fig. 2.1 can be depicted as a large formula in SC

$$(\forall t)(\Psi_1 \vee \Psi_2 \vee \cdots \vee \Psi_{18})$$

in which each $\Psi_h$ represents a transition (arrow) in Fig. 2.1. For example, $\Psi_1$, $\Psi_2$, and $\Psi_3$, representing the three transitions out of state $k_0$, are as follows:

$$\Psi_1\colon [q_1(t) \wedge q_2(t) \wedge q_3(t) \wedge p_1(t) \wedge p_2(t) \wedge q_1(t') \wedge q_2(t') \wedge \neg q_3(t')]$$

$$\Psi_2\colon [q_1(t) \wedge q_2(t) \wedge q_3(t) \wedge p_1(t) \wedge \neg p_2(t) \wedge q_1(t') \wedge q_2(t') \wedge q_3(t')]$$

$$\Psi_3\colon [q_1(t) \wedge q_2(t) \wedge q_3(t) \wedge \neg p_1(t) \wedge p_2(t) \wedge q_1(t') \wedge q_2(t') \wedge q_3(t')]$$

We are deliberately chosing the variable $t$ here to suggest the discrete moments of time that comprise the history of the automaton: time 0, time 1, time 2, etc.

Figure 2.3 is a properly nondeterministic automaton, which can be represented in exactly the same manner, namely, by the formula:

$$\begin{aligned}(\forall t)([q_1(t) \wedge q_2(t) \wedge \neg p(t) \wedge q_1(t') \wedge q_2(t')]\\ \vee [q_1(t) \wedge q_2(t) \wedge \neg p(t) \wedge \neg q_1(t') \wedge q_2(t')]\\ \vee [q_1(t) \wedge \neg q_2(t) \wedge p(t) \wedge q_1(t') \wedge q_2(t')]\\ \vee [\neg q_1(t) \wedge q_2(t) \wedge p(t) \wedge q_1(t') \wedge \neg q_2(t')]\\ \vee [\neg q_1(t) \wedge q_2(t) \wedge p(t) \wedge q_1(t') \wedge q_2(t')]).\end{aligned}$$

Here $p(t)$ means input $a$ while $\neg p(t)$ means input $b$. The formula $q_1(t) \wedge q_2(t)$ means state $k_0$, while $q_1(t) \wedge \neg q_2(t)$ means state $k_1$ and $\neg q_1(t) \wedge q_2(t)$ means state $k_2$.

The two formulas we have just considered are state-to-state formulas. They describe, or prescribe, the way an input at any time affects the transition to a new state. In the case of a deterministic automaton the next state is determined precisely, while in the case of a nondeterministic automaton the next state is merely limited (in general) to a number of possibilities.

Having given two examples we shall henceforth be forced to talk about formulas in the abstract, since the formulas become rather large even for small automata. We shall use capital Greek letters for formulas. Following Büchi's paper, we shall use capital Roman letters often followed by one or more formulas enclosed in brackets to represent truth functions of those formulas as arguments, e.g.,

$$(\forall t)B[q_1(t), q_2(t), p(x), q_1(t'), q_2(t')]$$

could represent the formula given earlier for Fig. 2.3. The general state-to-state formula is, for some $m \geq 0$ and $n \geq 1$,

$$(\forall t)B[q_1(t), \ldots, q_n(t), p_1(t), \ldots, p_m(t), q_1(t'), \ldots, q_n(t')].$$

It should be noted that a state-to-state formula tells not only which transitions between states are possible for given input conditions, but also restricts the set of input conditions and the set of state conditions that occur. The formula for Fig. 2.1, for example, tells that $\neg p_1(t) \wedge \neg p_2(t)$ does not hold for any time $t$, and thus that in the history of the device, at most three input possibilities occur. It also tells that $\neg q_1(t) \wedge \neg q_2(t)$ does not hold at any time and, thus, that at most six states ever occur.

By a possible history of an automaton we mean an infinite sequence

$$k_{i_0}, a_{i_0}, k_{i_1}, a_{i_1}, \ldots$$

where each $k_{i_t}$ is a state, each $a_{i_t}$ is an input, and there is a transition from $k_{i_t}$ to $k_{i_{t+1}}$ labeled $a_{i_t}$. We observe that, for every state-to-state formula there is an automaton such that the formula is true for an infinite sequence of this kind if and only if the sequence is a possible history of the automaton.

It is important to note what this means in a degenerate case, e.g., to the formula $(\forall t)(q(t) \wedge \neg q(t))$. For this we could take the nondeterministic automaton with one state and no transitions; the formula describes the empty set of behaviors, i.e., the empty set of possible histories.

## Input-Acceptance Formulas

Our discussion of automata has included inputs and states but has not included outputs. We assume the rather common point of view that the function of an automaton is to classify its possible input histories into those it accepts and those it rejects. We shall use SC to write *acceptance formulas* in which the input predicates $p_1, p_2, \ldots, p_m$ are the only free predicate variables; the formula is to be true when these are interpreted as predicates describing an accepted input history, and false when the predicates describe an unaccepted input history. State variables used in the formula must be bound by quantifiers: existential quantifiers initially placed are generally appropriate.

Consider first a formula of the form:

$$(\exists q_1)\ldots(\exists q_n)(H[q_1(0),\ldots,q_n(0)] \wedge (\forall t)(0 \le t < x \to K[q_1(t),\ldots,q_n(t),$$
$$p_1(t),\ldots,p_m(t),q_1(t'),\ldots,q_n(t')]) \wedge L[q_1(x),\ldots,q_n(x)]).$$

Note that the $K$-clause in this formula describes, or prescribes, the state-to-state behavior of a finite automaton. The formula refers to this automaton, stating that the input predicates have values between time 0 and time $x - 1$ inclusive so as to provide a possible history of the automaton in which the state satisfies the $H$-clause at time 0, and the $L$-clause at time $x$. Thus the formula as a whole makes a statement about the input history between time 0 and time $x - 1$ inclusive. Such formulas would give us a basis for the well-studied finite automata theory.

However, such formulas are not helpful in formulating our decision procedure since they do not capture the full power of SC. It turns out that formulas of the following form are sufficiently powerful:

$$(\exists q_1)\ldots(\exists q_n)(H[q_1(0),\ldots,q_n(0)] \wedge (\forall t)K[q_1(t),\ldots,q_n(t),$$
$$p_1(t),\ldots,p_m(t),q_1(t'),\ldots,q_n(t')] \wedge (\exists^{\omega} t)L[q_1(t),\ldots,q_n(t)]].$$

The middle clause corresponds to some nondeterministic automaton, and we can therefore interpret the whole part of the formula apart from the initial quantifiers as saying something about the infinite behavior of that automaton. Let the $H$-states and $L$-states be the states for which the $q$-predicates have truth values making the truth functions $H$ and $L$, respectively, true. Then the formula says that the input predicates are such that the automaton begins its history in an $H$-state and infinitely often assumes an $L$-state. Let us call formulas of this kind *Büchi formulas.*

The key to Büchi's decision procedure for SC is the fact that every formula of SC is effectively reducible to an equivalent Büchi formula. Thus, every formula of SC says something about the infinite behavior of a finite automaton and nothing more.

Let us call formulas of the following form *general behavioral formulas*:

$$(\exists q_1)\ldots(\exists q_n)(K[q_1(0),\ldots,q_n(0)]$$
$$\wedge\ (\forall t)H[q_1(t),\ldots,q_n(t),p_1(t),\ldots,p_m(t),q_1(t'),\ldots,q_n(t')]$$
$$\wedge\ M[(\exists^{\omega}t)K_1[q_1(t),\ldots,q_n(t)],\ldots,(\exists^{\omega}t)K_k[q_1(t),\ldots,q_n(t)]]).$$

These formulas are more expressive than Büchi formulas for several reasons: First, they can stipulate that certain conditions occur only finitely often. Second, they can make several assertions at once, each being either that some condition occurs infinitely often or that some condition occurs only finitely often. For example, it can assert that states $k_1$ and $k_2$ both occur infinitely often, but states $k_3$ and $k_4$ each occur only finitely often.

In any infinite history of an automaton there is a set $\gamma$ of precisely those states that occur infinitely often. A general behavioral formula is capable of making any assertion about $\gamma$ that can be expressed in any way.

General behavioral formulas lead us to an important concept introduced by David E. Muller. We define a *Muller automaton* [(7); see also (6)] to be a deterministic finite automaton, with a designated initial state, and a list of designated *acceptance* subsets of the set of states. The *omega-language* of the automaton is the set of all infinite words over the automaton's alphabet that determine an infinite history in which the set of states assumed infinitely often is one of the acceptance subsets. We leave it to the reader to observe that, given a Muller automaton, one can write a general behavioral formula for it. In fact, the Muller automata correspond exactly to those general behavioral formulas in which the state-to-state clauses are deterministic.

For example, consider the Muller automaton of Fig. 3.1 in which $k_0$ is the initial state and the designated acceptance subsets are $\{k_0, k_1\}$ and $\{k_0, k_1, k_2, k_3\}$. The omega-language of this automaton is the set of all infinite words over the alphabet $\{a, b\}$ having no subword *aaaa*, and having either (1) infinitely many *a*s and only finitely many subwords *aa* or else (2) infinitely many subwords *aaa*. We let $\Sigma^{\omega}$ denote the set of all infinite words over an alphabet $\Sigma$.

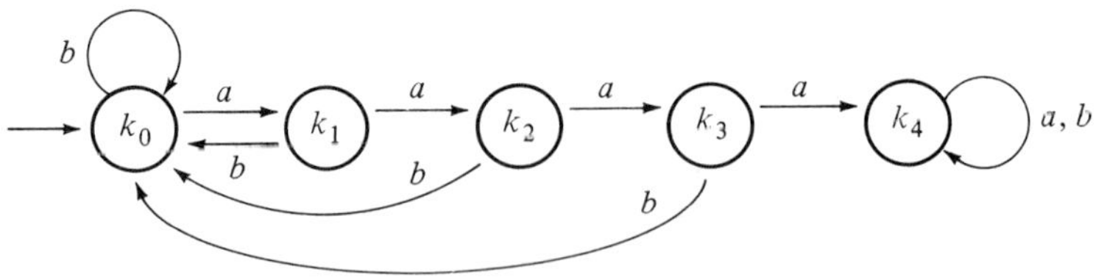


FIGURE 3.1

**Theorem 3.1.** *If $\Gamma$ is the omega-language of a Muller automaton M with alphabet $\Sigma$, then $\Sigma^{\omega} - \Gamma$ is the omega-language of a Muller automaton effectively obtained from M.*

PROOF. Let $Q$ be the set of states of $M$ and let $Q_1, Q_2, \ldots, Q_h$ be the acceptance subsets of $Q$. Let $Q_{h+1}, \ldots, Q_k$ be all the subsets of $Q$ other than $Q_1, \ldots, Q_h$ (so that $k = 2^{|Q|}$). Then $M'$ can be taken as the Muller automaton like $M$ in every way except that the acceptance subsets of $Q$ are $Q_{h+1}, \ldots, Q_k$. Since $M$ and $M'$ are deterministic it follows that the omega-language of $M' = \Sigma^\omega - \Gamma$. □

**Theorem 3.2.** *An omega-language is the set of all input histories satisfying a Büchi formula of SC if and only if it is the omega-language of some Muller automaton. The Büchi formula and the Muller automaton are each effectively obtained from the other.*

The first proof of this difficult theorem (6) was a machine-like proof. Subsequent proofs [see, e.g., (1), Chapter XIV of (2), (10), Chapter I of (15), (16)] were in varying degrees more algebraic in style. We shall not prove Theorem 3.2 but we shall use it in the next section at a vital point in our development of a decision procedure for SC. The Muller automaton is useful chiefly because it is deterministic.

## The Decision Procedure for SC

In this section we shall develop an algorithm for constructing, from an arbitrary formula of SC, an equivalent Büchi formula. This construction is the major part of the decision procedure. Note that a Büchi formula has no individual-variable quantifiers except for the two $t$ quantifiers. Since formulas of SC generally have individual-variable quantifiers, we shall devise a way of replacing these by predicate-variable quantifiers. To this end, we convert a formula containing a free variable $x$ to a formula without $x$ but, in its place, the predicate $p_x$ with a special meaning: $p_x(x)$ is true, and $p_x(y)$ is false for $y \neq x$. $p_x$ is not a free predicate variable; its meaning depends on the assigned meaning of the variable $x$.

This extension of the SC formalism will not cause any trouble. We shall stipulate that a Büchi formula not contain any terms containing free individual variables; however, we shall allow it to have these $p_x$ predicates.

Thus a *Büchi formula* is a formula

$$Q(H \wedge (\forall t)K \wedge (\exists^\omega t)L)$$

where $Q$ is a string of existential predicate-variable quantifiers; $H$ is a quantifier-free formula in which 0 is the only term and in which only predicates quantified in $Q$ may occur; $K$ is a quantifier-free formula in which all terms are either $t$ or $t'$, and in which any free predicate variable or predicate of the type $p_x$ may have only term $t$ as argument; and $L$ is a quantifier-free formula in which only quantified predicates may occur and $t$ is the only permissible term. (Any one or two of $H$, $(\forall t)K$ and $(\exists^\omega t)L$ may, with neighboring conjunction signs, be omitted.) The quantified predicates we call *state predicates*; the free predicate variables we call the *input predicates*.

**Lemma 1.** *A quantifier-free formula of SC has an equivalent Büchi formula effectively obtained from it.*

PROOF. Without loss of generality, assume that $t$ does not occur in the given formula $\Phi$. All the atomic constituents of $\Phi$ are of the form $q(0^{(i)})$ or $q(y^{(i)})$—where $q$ is a predicate variable or a $p_x$ predicate; $y$ is an individual variable, and superscript $(i)$ is a string of $i$ apostrophes, $i \geq 0$. Assume no term of $\Phi$ has more than $k$ apostrophes and assume that $q_1, \ldots, q_n$ are all the predicate variables and $p_x$ predicates occurring in $\Phi$.

We construct the Büchi formula

$$Q(H \wedge (\forall t)K \wedge (\exists^\omega t)L)$$

from $\Phi$ by constructing $Q$, $H$, $K$, and $L$. For each term $\tau$ that is a constituent of a term in $\Phi$ we shall introduce a predicate that we shall write as $p_\tau$. Unless $\tau$ is an individual variable, $p_\tau$ will be a predicate variable and will have an existential quantifier in $Q$ (in which case the subscript $\tau$ will have no semantical significance and, for conformity with the syntax of SC, will ultimately be replaced by a numerical subscript). For each $\tau$ that is not an individual variable, we shall put clauses into $H$ and $K$ to insure that, for all $t$, $p_\tau(t)$ is true for $t = \tau$ and false for $t \neq \tau$.

For example, suppose the terms $0''$ (or 2) and $x'$ occur. Then $p_0, p_1, p_2, p_x$ and $p_{x'}$ must occur; the clause $p_0(0)$ must occur in $H$ and $\neg p_0(t')$ must occur in $K$. Similarly, the clauses $\neg p_1(0)$, $p_1(t') \leftrightarrow p_0(t)$, $\neg p_2(0)$, $p_2(t') \leftrightarrow p_1(t)$, $\neg p_{x'}(0)$, $p_{x'}(t') \leftrightarrow p_x(t)$ must all occur (some in $H$, some in $K$). Note that no such clauses are appropriate for $p_x$ since $x$ is a free variable, and the meaning of $p_x$ is semantically related to the meaning of $x$.

If $q_i(\tau)$ occurs in $\Phi$ we shall use the predicate variable $\dot{q}_{i,\tau}$ with existential quantifier in $Q$ with the idea that

$$\dot{q}_{i,\tau}(t) \leftrightarrow q_i(\tau) \wedge t > \tau.$$

Thus we shall have the clause $\neg \dot{q}_{i,\tau}(0)$ in $H$ and the following clause in $K$:

$$\dot{q}_{i,\tau}(t') \leftrightarrow \dot{q}_{i,\tau}(t) \wedge (q_i(t) \wedge p_\tau(t)).$$

We leave it to the reader to verify that a Büchi formula can be constructed to be equivalent to $\Phi$, aided by the following example:

Suppose $\Phi$ is $F[q_3(2), q_4(x')]$. Then $Q$ and $H$ are, respectively, as follows:

$$(\exists p_0)(\exists p_1)(\exists p_2)(\exists p_{x'})(\exists \dot{q}_{3,2})(\exists \dot{q}_{4,x'})$$

$$p_0(0) \wedge \neg p_1(0) \wedge \neg p_2(0) \wedge \neg p_{x'}(0) \wedge \neg \dot{q}_{3,2}(0) \wedge \neg \dot{q}_{4,x'}(0).$$

$K$ is the conjunction of the following

$$\neg p_0(t') \qquad p_1(t') \leftrightarrow p_0(t)$$

$$p_2(t') \leftrightarrow p_1(t) \qquad p_{x'}(t') \leftrightarrow p_x(t)$$

$$\dot{q}_{3,2}(t') \leftrightarrow \dot{q}_{3,2}(t) \vee (q_3(t) \wedge p_2(t))$$

$$\dot{q}_{4,x'}(t') \leftrightarrow \dot{q}_{4,x'}(t) \vee (q_4(t) \wedge p_{x'}(t)).$$

Finally, $L$ is $F[\dot{q}_{3,2}(t), \dot{q}_{4,x'}(t)]$.

Explanation of the example: If values are assigned to the predicate variables $q_3$ and $q_4$ and individual variable $x$, then to say that $L$ is true for infinitely many $t$ is to say, given the nature of the predicates $\dot{q}_{3,2}$ and $\dot{q}_{4,x'}$ as stipulated in $H$ and $K$, that $L$ is true for $t = \max(2, x') + 1$. But the latter is true if and only if, given the stipulations of $H$ and $K$, that $F[q_3(2), q_4(x')]$ is true. □

**Lemma 2.** *If $\Phi$ is a Büchi formula then $(\exists p)\Phi$ is a Büchi formula.*

The Proof is immediate from the definition of Büchi formula.

**Lemma 3.** *If $\Phi$ is a Büchi formula then there is a Büchi formula equivalent to $(\exists x)\Phi$ effectively obtained from $\Phi$.*

PROOF. Let $\Phi = Q(H \wedge (\forall t)K \wedge (\exists^\omega t)L)$ and assume that the predicate variables $p$ and $s$ do not occur in $\Phi$. Let $H'$, $K'$, and $L'$ be $H$, $K$, and $L$, respectively, with $p_x$ replaced by $p$. We claim that the following is a Büchi formula equivalent to $(\exists x)\Phi$:

$$\begin{aligned}(\exists p)(\exists s)Q(&\neg s(0) \wedge H' \\ &\wedge (\forall t)[(s(t') \leftrightarrow p(t) \vee s(t)] \wedge \neg[p(t) \wedge s(t)] \wedge K') \\ &\wedge (\exists^\omega t)(s(t) \wedge L'))\end{aligned}$$

Note that the parts of this formula after $Q$ but outside of $H'$, $K'$, and $L'$ are enough to insure that there be exactly one $x$ such that $p(x)$ is true: for if $p(x)$ is false for all $x$ then $s(x)$ is false for all $x$; this possibility is excluded because of the last clause. If $x_0$ is the smallest $x$ making $p(x)$ true then we have

$$s(t) \leftrightarrow t > x_0.$$

Thus $p(y)$ cannot be true for $y > x_0$ by virtue of the clause $\neg[p(t) \wedge s(t)]$.

Thus our formula is equivalent to $(\exists x)\Phi$. □

**Lemma 4.** *If $\Phi$ is a Büchi formula then there is a Büchi formula equivalent to $\neg\Phi$, effectively obtained from $\Phi$.*

PROOF. By Theorem 3.2, the set of all input histories making $\Phi$ true is the omega-language $\Gamma$ of a Muller automaton $M$. If $\Sigma$ is the input alphabet of $M$, then by Theorem 3.1 there is a Muller automaton $M'$ for $\Sigma^\omega - \Gamma$. By Theorem 3.2 again there is a Büchi formula $\Psi$ satisfying just thoes input histories represented by $\Sigma^\omega - \Gamma$; which means that $\Psi$ is equivalent to $\neg\Phi$. □

The proof of the step in Büchi's proof corresponding to our Lemma 4 (i.e., Lemma 9 of [8]) was quite a difficult one. Theorem 3.2 had not yet been discovered when Büchi wrote his paper.

Recently there has been an interest in the question of how large a Büchi formula has to be, as a function of the size of the given Büchi formula $\Phi$, to be equivalent to $\neg\Phi$. Two recent references in this problem area are (14) and (12).

**Theorem 4.1.** *Every formula $\Phi$ of SC is equivalent to some Büchi formula, effectively obtained from $\Phi$.*

Proof. Put $\Phi$ into prenex normal form. Then convert every universal quantifier $(\forall \ldots)$ to $\neg(\exists \ldots)\neg$. Lemmas 1, 2, 3, and 4 are then sufficient to convert this result into an equivalent Büchi formula. □

If the given formula $\Phi$ has free individual variables then the Büchi formula that results from the construction described in the proof of Theorem 4.1 will have $p_x$ predicates. These are necessary, since any formula equivalent to a formula with a free variable $x$ (in an occurrence that is not vacuous logically) must also have some manifestation of that free variable $x$. To say that SC is decidable is to say that there is a decision procedure for the truth or falsity of any formula of SC without free variables. (Formulas with free variables are neither true nor false until an interpretation is placed on all the free variables.)

**Theorem 4.2.** *SC is decidable.*

Proof. Every formula without free variables is equivalent to a Büchi formula without free variables. So we must prove that we can decide the truth or falsity of a Büchi formula without free variables.

But such a Büchi formula refers to an automaton without inputs. The state-to-state part of the formula, $K$, simply tells, given the state at time $t$, what states are possible for time $t'$. The $H$ part of the formula tells which states it may assume at time 0. And the $L$ part gives a set of states such that the automaton is required to be in at least one of those states infinitely often.

To decide whether this formula is true or false, one could simply draw a directed graph with the states as nodes, drawing arcs from the information in $K$, designating the possible initial nodes by reference to $H$ ($H$-nodes), and designating the states satisfying $L$ ($L$-nodes). The Büchi formula is true if and only if there is a path from an $H$-node to some $L$-node, and a nonnull loop from that $L$-node back to itself. Algorithms to decide this for such a directed graph are well known. □

Readers who are interested in computational complexity will note that the parts of this decision procedure corresponding to Lemmas 1, 2, and 3, as well as the part explicitly described in the proof of Theorem 4.2, require only polynomial time. There seems to be no hope, however, that the part corresponding to Lemma 4 can be done in polynomial time. See (14) and (12).

## Comments on Related Papers

Büchi's paper on the sequential calculus appeared about the same time as his paper [7] on the weak second-order arithmetic (W2A). W2A and SC are syntactically identical; the semantic difference is that the predicate variables of W2A are interpreted as predicates that are true only of finitely many nonnegative integers (which is the signficance of the term "weak" as Büchi used it). The result is that a formula of W2A can say nothing interesting about an infinite history, since the only predicates to which we can refer are predicates that are false for all integers beyond a certain point. W2A is a formalism in which one can express the usual content of finite automata theory.

Büchi proves that every formula of W2A can effectively be put into a certain normal form, the proof being quite similar to the proof that every formula of SC is equivalent to some Büchi formula. From the normal form, the automaton is easily constructed. The decision procedure is based on the constructive proof of this theorem.

The normal form is as follows:

$$Q(H \wedge (\forall t)K \wedge L_1 \wedge \ldots \wedge L_k)$$

where $Q$ is a string of existential predicate-variable quantifiers; $H$ and $K$ are like the $H$ and $K$ of the Büchi formula; and $L_1, \ldots, L_k$ are quantifier-free formulas, each $L_i$ having only the term $x_i$, where $x_1, \ldots, x_k$ are all the free individual variables of the original formula (cf. the first example of Section 3, which is essentially in this normal form for $k = 1$).

(The appearance of the clauses $L_i$, in which $x_i$ is the only term, may make this normal form seem more unlike the Büchi formula of SC than it really is. If we were to introduce the predicates $p_{x_i}$ into W2A as we did for SC then the clauses $L_i$ would not be necessary; the content of the $L_i$'s would then be part of the $K$ clause. On the other hand, if we had chosen to do without such predicates in SC we would have ended up needing the $L_i$ clauses in Büchi formulas in addition to the clauses already there, as Büchi notes in remark 3, p. 7, of [8]. The only esential difference between the Büchi formula and the normal form for W2A is the presence in the former of the clause $(\exists^{\omega} t)L$.)

SC is richer than W2A in expressive power, in that whatever we can say in W2A we can also say in SC, but not vice versa. However, in some respects W2A and SC are equal in expressive power: for example, any property of nonnegative integers is definable in SC if and only if it is definable in W2A. For Büchi shows in [7] and [8] that any property is definable in either system if and only if it is ultimately periodic. (A property $p$ is *ultimately periodic* if there are nonnegative integers $e$ [the *period*] and $m$ [the *bound*] such that, for all $x$, if $x \geq m$ then $P(x)$ holds if and only if $P(x + e)$ holds.) Chapter II of (13) has a more detailed discussion of this issue.

From this result the undefinability of addition in SC easily follows. For if "$x + y = z$" were definable then "$x$ is a multiple of $y$" would be definable, and then "$x$ is a prime" would be definable, which is a property that is not ultimately periodic. (We leave it to the reader to fill in the details of this argument.)

A paper that Büchi wrote with Landweber [16] contains a proof that predicates over nonnegative integers and sets thereof are quite low in the Kleene hierarchy. We regret that we cannot explain this result here.

From my present point of view, Büchi's work on the sequential calculus was his greatest achievement, for which the paper on W2A [7] provided important background. In my review (5) of these two papers, I gave them equal treatment.

Much of Büchi's work and that of his students and other collaborators was an attempt to extend his result. But if an enumeration of the extensions to this great result is to be made, first place on the list should go to Michael Rabin's proof that the formalism that is like SC except for having two successor functors is decidable [(9) and (10)]. The intended interpretation of this formalism is the infinite binary tree: the objects denoted by individual variables are the nodes of the tree, while the two successor functors applied to any node determine its left successor and right successor. Rabin's Büchi-like analysis of this formalism proved to be quite fruitful. His proof of decidability was quite an accomplishment, as certain notions in Büchi's analysis turned out to be difficult to generalize to the two-sucessor problem.

Most of Büchi's own efforts at extending his results of SC and W2A were directed at formalisms representing various systems of ordinals. This interest occupied him for many years.

To get the kind of decision procedure for theories of ordinals that he got for theories of nonnegative integers, he extended the notion of time used in describing the history of an automaton to allow any ordinal to be a temporal measure. (Actually, Büchi did not use the word time in his papers on ordinals, although he did write about automata, states, and transitions.)

For example, if all the ordinals less than $\omega^2$ are to be moments of time, then we have time 0, time 1, ... time $\omega$, time $\omega + 1$, ... time $\omega \cdot 2$, time $\omega \cdot 2 + 1$, .... The general moment of time is time $\omega \cdot i + j$ where $i$ and $j$ are any nonnegative integers. Time $\omega \cdot i + (j + 1)$ is the successor of time $\omega \cdot i + j$. For $i \geq 1$, $\omega \cdot i$ is not the successor of any ordinal but is the limit of all ordinals preceding it.

If $p_1, \ldots, p_m$ are the input predicates and $q_1, \ldots, q_n$ are the state predicates, then the general statement about the automaton says something about the state at time 0, and something about how the state at time $t + 1$ relates to the state and input at time $t$, as before. But the general statement must now also say something about the state at those moments of time that are limit ordinals. It turns out to be appropriate to relate the state at time $\omega \cdot i$, $i \geq 1$, to the set of states occurring infinitely often during the time interval after time $\omega \cdot (i - 1)$ and before time $\omega \cdot i$. More generally, for any limit ordinal $t$, let $\Sigma_t$ be the set of states $\sigma$ such that, for all $y < t$, there is an $x$, $y < x < t$, such that state $\sigma$ occurs at time $x$. Then the general behavioral formula says something about the relation between the state at time $t$ and $\Sigma_t$.

Generalizing SC to the set of ordinals less than a certain ordinal $\alpha$, we get the interpreted formal system $SC_\alpha$. Syntactically, the only difference is the addition of $<$ as a primitive relation, since $\geq$, $\leq$, $>$, and $<$ can no longer be defined as they were when the intended interpretation was the set of nonnegative integers. We define $x = y$ (as before) as $(\forall p)(p(x) \leftrightarrow p(y))$. Then with $<$ as primitive, $>$, $\geq$, and $\leq$ are easily defined. Semantically, the individual variables range over the set of all ordinals less than $\alpha$; the predicate variables range over the set of all predicates of these ordinals. For the purposes of this exposition, we shall assume that $\alpha$ is a limit ordinal.

The generalization of the Büchi formula is

$$Q(H \wedge (\forall t)K \wedge \Psi \wedge \Phi)$$

where $Q$, $H$, and $K$ are as before. We informally describe $\Psi$ and $\Phi$: Recall that the states of an automaton of a formula are given by the predicate variables that are existentially quantified in $Q$. The clause $\Psi$ tells, for any limit ordinal $t < \alpha$, how the state at time $t$ depends on $\Sigma_t$, as discussed earlier. The clause $\Phi$ is a condition on $\Sigma_\alpha$; it is a generalization of the third clause $(\exists^\omega t)L$ of the original Büchi formula.

In any result of decidability of $SC_\alpha$ for an ordinal $\alpha$, the proof proceeds by establishing the generalized Büchi formula as a normal form for $SC_\alpha$.

Büchi's first paper on ordinals [13] considered the generalization of W2A to ordinals rather than the generalization of SC: thus predicate variables ranged over the predicates true only of finitely many ordinals. Let $W2A_\alpha$ be this extension to the system of ordinals less than $\alpha$, for any ordinal $\alpha$. Büchi proved that $W2A_\alpha$ is decidable for any ordinal $\alpha$. Furthermore, he was able to prove that for every $\alpha \geq \omega^\omega \cdot 2$, the set of all true sentences of $W2A_\alpha$ is exactly the same as the set of all true sentences of $W2A_\beta$ for some $\beta$, $\omega^\omega \leq \beta < \omega^\omega \cdot 2$. This last result essentially showed that the systems $W2A_\alpha$ are quite unsatisfactory as a means of studying the theory of ordinals.

On the other hand, the systems $SC_\alpha$ proved to be quite expressive, sufficient to permit making many important distinctions that need to be made in the general theory of ordinals. Along with that, the problem of establishing decision procedures for $SC_\alpha$ for various ordinals $\alpha$ showed itself to be substantially more difficult than his sweeping result for $W2A_\alpha$ for all ordinals $\alpha$.

In a rather lengthy and well-written paper [24], Büchi gives a good account of the results by himself and others in the decidable second-order theories of ordinals. I recommend this paper as first reading for anyone wishing to get a well-rounded understanding of these theories and their decision procedures. In it he carries through the decision procedure for $SC_\alpha$ for all denumerable ordinals $\alpha$, and for $SC_{\omega_1}$, where $\omega_1$ is the first nondenumerable ordinal. However, in the paper [28] the decidability result is extended to $SC_\alpha$ for $\alpha$ any ordinal that has the same cardinality as $\omega_0$ or $\omega_1$. (These last two papers seem to have been written nine years before they appeared in print. Because of the delay in publication, other authors came forth claiming the same result in the intervening years.) In (4) it is proved that for $\alpha \geq \omega_2$ one needs strong assumptions on the underlying set theory, which indicates that the decidability result cannot be extended to $SC_\alpha$ for $\alpha \geq \omega_2$. Regrettably, the interesting technical issues in the papers mentioned in this paragraph cannot be explained here.

By now there are many applications of Büchi's work with SC and related systems. One such application is to temporal logic, a topic of investigation in theoretical computer science. We are unable to give any explanation here; it must suffice to give the title of one reference: "The complementation problem for Büchi automata with applications to temporal logic" (14).

In May 1984 a five-day conference was held in Le Mont Dore in France on automata and infinite words. Several times every day Büchi's name was mentioned, yet apparently no one at the conference knew that Richard Büchi had died a few weeks before. The published proceedings of the conference (8) reveals how the problem of the infinite history of finite automata, as conceived originally by Büchi and Muller, has been responsible not only for many research results, but also for spawning other problem areas as well.

## *References*


1. Choueka, Y. (1974). Theories of automata on omega-tapes: A simple approach. *J. Computer and Systems Science*, 8, 117–141.
2. Eilenberg, S. (1974). *Automata, languages and machines*, vol. A. New York: Academic Press.
3. Gurevich, Y., & Harrington, L. (1982). Trees, automata and games. *Proc. 14th ACM Symposium on Theory of Computing*, 60–65.
4. Gurevich, Y., Magidor, M., & Shelah, S. (1983). The monadic theory of $\omega_2$. *Journal of Symbolic Logic*, 48, 387–398.
5. McNaughton, R. (1963). Review of Büchi [7] and [8]. *J. Symbolic Logic*, 28, 100–102.
6. McNaughton, R. (1966). Testing and generating infinite sequences by a finite automaton. *Information and Control*, 9, 521–530.
7. Muller, D. (1963). Infinite sequences and finite machines. *Switching Circuit Theory and Logical Design: Proceedings of the Fourth Annual Symposium*, IEEE, 3–16.
8. Nivat, M., & Perrin, D. (1985). *Automata on infinite words: Ecole de Printemps d'Informatique Théoretique*, Mai 1984. Lecture Notes in Computer Science, 192, New York: Springer-Verlag.
9. Rabin, M.O. (1969). Decidability of second-order theories and automata on infinite trees. *Trans. Am. Math. Soc.*, 141, 1–35.
10. Rabin, M.O. (1972). *Automata on infinite objects and Church's problem*. Regional conference series in mathematics, 13, Amer. Math. Soc., Providence, R.I.
11. Robinson, R.M. (1958). Restricted set-theoretical definitions in arithmetic. *Proc. Am. Math. Soc.*, 9, 238–242.
12. Safra, S. (1988). On the complexity of omega-automata. *Proc. 29th Annual Symposium on Foundations of Computer Science*, IEEE Computer Society, 319–327.
13. Siefkes, D. (1970). *Büchi's monadic second-order successor arithmetic*. In Decidable Theories, G.H. Müller, ed. Lecture Notes in Mathematics, 120, Heidelberg: Springer-Verlag.
14. Sistla, A.P., Vardi, M.Y., & Wolper, P. (1987). The complementation problem for Büchi automata with applications to temporal logic. *Theoretical Computer Science*, 49, 217–237.

15. Trakhtenbrot, B.A., & Barzdin, Y.M. (1973). *Finite automata behavior and synthesis*. Amsterdam: North-Holland; New York: American Elsevier.
16. Thomas, W. (1981). A combinatorial approach to the theory of omega-automata. *Information and Control*, 48, 261–283.

I would like to thank Sylvia Büchi, Yuri Gurevich, Saunders Mac Lane, and Dirk Siefkes for many valuable comments and suggestions used in the final draft of this article.

*Zeitschr. f. math. Logik und Grundlagen d. Math.*
*Bd. 6, S. 66—92 (1960)*

# WEAK SECOND-ORDER ARITHMETIC AND FINITE AUTOMATA[1])

By J. RICHARD BÜCHI in Ann Arbor, Michigan

## Introduction

The formalism of regular expressions was introduced by S. C. KLEENE [6] to obtain the following basic theorems.

Synthesis. *To every regular expression* $\mathfrak{E}$ *one can effectively obtain a finite automata* $\mathfrak{A}$ *with binary output* $\mathfrak{U}$ *such that* $\mathfrak{E}$ *denotes the behavior of* $\langle \mathfrak{A}, \mathfrak{U} \rangle$.

Analysis. *To every finite automaton* $\mathfrak{A}$ *with binary output* $\mathfrak{U}$ *one can effectively construct a regular expression* $\mathfrak{E}$ *such that the behavior of* $\langle \mathfrak{A}, \mathfrak{U} \rangle$ *is denoted by* $\mathfrak{E}$.

For simplified expositions of KLEENE's theory see COPI-ELGOT-WRIGHT [4], RABIN-SCOTT [13], and MYHILL [8].

It will be shown here that a more conventional formalism, a weak second-order arithmetic, can be used in place of the formalism of regular expressions. Our theorems 1, 2 section 4 are equivalent to KLEENE's synthesis and analysis theorems. This result is of interest for automata theory because formulas of weak second-order arithmetic seem to be more convenient than regular expressions for formalizing conditions on the behavior of automata. In addition, our synthesis and analysis theorems yield rather complete information on the strength of weak second-order arithmetic (see Section 5), thus providing an example of applying automata theory to logic.

## § 1. Notations and Terminology

The following letters (possibly with subscripts) will be used as syntactic variables with range as indicated,

| | |
|---|---|
| $x, y, z, t$ | denote *individual variables* |
| $X, Y, Z$ | denote *propositional variables* |
| $i, j, r, s, u$ | denote *monadic predicate variables* |
| $A, B, C, D, \ldots$ | denote *formulas of propositional calculus* |
| $\mathfrak{A}, \mathfrak{B}, \mathfrak{C}, \mathfrak{D}$ | denote *formulas.* |

1) The author wishes to thank Dr. J. B. WRIGHT for many stimulating discussions. Some the results were announced in the *Notices*, American Mathematical Society. (Decision Problem Weak Second-Order Arithmetics and Finite Automata. Prelim. Report, Part I, Vol. 5, No December 1958.) This work was done on contract with the Office of Naval Research, Offi of Ordnance Research, Army Signal Corps, and with the assistance of a grant from the Natio Science Foundation.

Besides the already mentioned (countable) lists of variables our formalism contains the primitive symbols $\Lambda$, $\mathrm{V}$, $\wedge$, $\vee$, $\sim$, $\supset$, $\equiv$, [, ], $\forall$, $\exists$, (, ), 0, $'$. The *formulas of propositional calculus* (p. f.'s) are obtained in the conventional manner from $\Lambda$, $\mathrm{V}$ and propositional variables by means of the connectives and [, ]. Examples of *atomic formulas* are $i(0'')$, $i(x)$, $r_2(t''')$. *Matrices* are obtained by replacing all propositional variables in a p. f. $A$ by atomic formulas. *Formulas* (of restricted, i. e. monadic, second-order-predicate calculus) are obtained in the conventional manner from matrices by applying individual — and predicate — quantifiers, and propositional connectives. A *sentence* is a formula without free variables.

A method for indicating occurrences of free variables is explained by the following examples. "$D[X, Y, Z]$" denotes a p. f. in which the indicated propositional variables but no others may occur. "$D[i(0''), i(x), s(t')]$" denotes the matrix obtained by making the obvious substitutions in $D[X, Y, Z]$, and may be abbreviated as "$D(0, x, t)$". "$\mathfrak{C}(i_1, i_2, t)$" denotes a formula in which the variables $i_1$, $i_2$, $t$ but no others may have free occurrences. We will also abbreviate "$[\sim A]$" by "$\tilde{A}$", and "$[A \wedge B]$" by "$AB$".

We will often deal with *n-tuples* of objects. The symbol "$\frown$" denotes concatenation of $n$-tuples. Thus for example "$H_1{}^\frown H_2{}^\frown H_1$" and "$i_1{}^\frown i_2{}^\frown i_3{}^\frown i_4$" denote respectively a 3-tuple of matrices and a 4-tuple of predicate variables. The second *component* of the 3-tuple $a^\frown b^\frown c$ is $b$. The $n$-tuple $2^\frown 2^\frown 2 \ldots {}^\frown 2$ will also be denoted by "$\boldsymbol{n}$". We next explain the use of a vector-notation which will considerably condense the presentation and make it more comprehensible.

| | |
|---|---|
| $\boldsymbol{X}^n$, $\boldsymbol{Y}^n$, $\boldsymbol{Z}^n$ | denote *n-tuples of propositional variables* |
| $\boldsymbol{i}^n$, $\boldsymbol{j}^n$, $\boldsymbol{r}^n$, $\boldsymbol{s}^n$ | denote *n-tuples of monadic predicate variables* |
| $\boldsymbol{A}^n$, $\boldsymbol{B}^n$, ... | denote *n-tuples of propositional formulas.* |

For example, "$\boldsymbol{r}^m(t') \equiv \boldsymbol{H}^m[\boldsymbol{r}^m(t), \boldsymbol{i}^n(t)]$" stands for the $m$-tuples of matrices whose components are

$$r_1(t') \equiv H_1[r_1(t), \ldots, r_m(t), i_1(t), \ldots, i_n(t)],$$
$$\vdots$$
$$r_m(t') \equiv H_m[r_1(t), \ldots, r_m(t), i_1(t), \ldots, i_n(t)].$$

"$\boldsymbol{X}^n{}^\frown Y$" stand for the $(n+1)$-tuple $X_1{}^\frown X_2{}^\frown \ldots {}^\frown X_n{}^\frown Y$. At some places a notation for a $n$-tuple of propositional formulas is used ambiguously to denote the conjunction of the p. f.'s occurring in the $n$-tuple. Furthermore the superscripts on notations for $n$-tuples are often omitted; in such cases it is clear from the context how they are to be restored.

*Interpretation:* We will not make use of any deductive structure on the syntactic frame. However, propositional formulas and formulas will be interpreted. Because the interpretations are quite conventional we will not state rigorous definitions. Furthermore, we will make ambiguous use of the syntactic notations, using them at times also for reference to the interpretation. The following list will explain additional notations and terminology.

| | |
|---|---|
| $\Lambda$, V | truth values false, true |
| $\mathbf{\Lambda}^n$ | the $n$-tuple $\Lambda^\frown\ldots{}^\frown\Lambda$ |
| predicate | function from natural numbers to $\{\Lambda, \mathrm{V}\}$ (set of natural numbers) |
| special predicate | predicate which is ultimately false (finite set of natural numbers) |
| $\boldsymbol{n}$-predicate | $n$-tuple of predicates |
| special $\boldsymbol{n}$-predicates | $\boldsymbol{n}$-predicate with special components ($\boldsymbol{n}$-predicate which is ultimately $\mathbf{\Lambda}^n$). |

Further notations are introduced in the various sections.

## § 2. Weak second-order arithmetic

We now consider the following interpretation of the primitive symbols of the restricted second-order system described in § 1.

| | |
|---|---|
| $\Lambda$, V | false, true |
| $\wedge, \vee, \sim, \supset, \equiv$ | the usual truth functions |
| $0$, $'$ | the natural number zero, the successor-function on natural numbers |
| individual variables | range over natural numbers |
| predicate variables | range over *special* predicates (finite sets of natural numbers) |
| $(\exists x)$, $(\forall x)$ | there is a n. n. $x$, for all n. n.'s $x$ |
| $(\exists i)$, $(\forall i)$ | there is a *special* predicate $i$, for all *special* predicates $i$. |

This leads in a conventional manner to a definition of *satisfaction* for formulas, and *truth* for sentences. The resulting interpreted system will be called *weak second-order arithmetic* (W. 2. A). The notation "$\hat{i}\mathfrak{C}(i)$" will be used for the set of all special predicates $i$ on natural numbers which satisfy the formula $\mathfrak{C}(i)$. The notations "$\hat{x}\,\mathfrak{C}(x)$", "$\hat{\boldsymbol{i}}\mathfrak{C}(\boldsymbol{i}^n)$" are used similarly. The formula $\mathfrak{C}(i)$ *defines the set* $\hat{i}\,\mathfrak{C}(i)$ in W. 2. A. $\mathfrak{C}(i, x)$ is *equivalent* to $\mathfrak{B}(i, x)$ if $\hat{i}\,\hat{x}\,\mathfrak{C}(i, x) = \hat{i}\,\hat{x}\,\mathfrak{B}(i, x)$, i. e., if $(\forall i)\,(\forall x).\ \mathfrak{C}(i, x) \equiv \mathfrak{B}(i, x)$ holds in W. 2. A.

Lemma 1. *In* W. 2. A., *to every formula* $\mathfrak{C}(\boldsymbol{i}^n)$ *one can effectively construct a formula* $\mathfrak{C}^*(\boldsymbol{i}^n)$ *such that* $\hat{\boldsymbol{i}}\mathfrak{C}^*(\boldsymbol{i}^n) = \hat{\boldsymbol{i}}\mathfrak{C}(\boldsymbol{i}^n)$ *and* $\mathfrak{C}^*(\boldsymbol{i}^n)$ *is of the form*

$$(\boldsymbol{j}) \cdot K[\boldsymbol{j}(0)] \wedge (\forall t)\, B[\boldsymbol{i}(t),\ \boldsymbol{j}(t),\ \boldsymbol{j}(t')],$$

*whereby* $\boldsymbol{j}$ *is a r-tuple* $j_1{}^\frown\ldots{}^\frown j_r$ *of pedicate variables and* $(\boldsymbol{j})$ *is a prefix of quantifiers* $(\exists j_\nu)$ *and* $(\forall j_\nu)$.

Proof. It is clear that one can construct a formula $\mathfrak{Z}(x)$ such that 0 is the only number satisfying $\mathfrak{Z}(x)$. Now if 0 occurs in $\mathfrak{C}(\boldsymbol{i})$ and $\mathfrak{D}(\boldsymbol{i}, x)$ is obtained by substituting $x$ for 0 in $\mathfrak{C}(\boldsymbol{i})$, then

$$\mathfrak{C}(\boldsymbol{i}) \cdot\!\equiv\!\cdot (\exists x)\,[\mathfrak{Z}(x) \wedge \mathfrak{D}(\boldsymbol{i}, x)]$$

holds for $\boldsymbol{i}$. Thus we have obtained a formula $\mathfrak{C}_1(\boldsymbol{i})$ equivalent to $\mathfrak{C}(\boldsymbol{i})$ in which 0 does not occur. The next step is to construct from $\mathfrak{C}_1(\boldsymbol{i})$ an equivalent formula $\mathfrak{C}_2(\boldsymbol{i})$

in which no iterations of "$'$" occur and no "$'$" occurs in argument places of $\boldsymbol{i}$. This is easily accomplished by introduction of new predicate variables which are appropriately quantified. Then $\mathfrak{C}_3(\boldsymbol{i})$ is obtained by passing to prenex form.

Next we repeatedly apply to $\mathfrak{C}_3(\boldsymbol{i})$ identities of the form

$$\begin{aligned}&(\text{prefix } 1)\ (\exists x)\ (\text{prefix } 2)\ A(x, \ldots)\\ &\cdot\equiv\cdot\ (\text{prefix } 1)\ (\exists j)\ (\forall x)\ (\text{prefix } 2)\ (\exists y)\ [j(y) \wedge j(x) \supset A(x, \ldots)]\end{aligned}$$

and their duals, to obtain a formula $\mathfrak{C}_4(\boldsymbol{i})$ equivalent to $\mathfrak{C}(\boldsymbol{i})$ which is of the form

$\mathfrak{C}_4(\boldsymbol{i})$: (predicate prefix) (individual prefix) [Matrix].

Because the matrices are constructed from monadic predicate variables only we can make use of BEHMANN's [1] device of moving individual quantifiers into the matrix of $\mathfrak{C}_4(\boldsymbol{i})$. The result is a formula $\mathfrak{C}_5(\boldsymbol{i})$ equivalent to $\mathfrak{C}(\boldsymbol{i})$, and of form

$\mathfrak{C}_5(\boldsymbol{i})$: (predicate prefix) $\bigwedge_\nu[(\exists t)\ D_\nu(t) \vee (\forall t)\ E_{\nu,1}(t) \vee \cdots \vee (\forall t)\ E_{\nu,n}(t)]$.

Now we note that predicate quantification is over special predicates, and therefore identities of the form

$$(\forall t)\ E(t) \cdot\cdot\equiv\cdot\cdot\ (\forall j) \cdot \tilde{j}(0) \vee (\exists t)\ [j(t)\ \tilde{j}(t')\ E(t)]$$

hold in W. 2. A. If we accordingly replace each constituent $(\exists t)\ E_{\nu,\mu}(t)$ in $\mathfrak{C}_5(\boldsymbol{i})$, making use of different $j$'s, and then move the newly introduced predicate-quantifiers into the prefix, the result is a formula $\mathfrak{C}_6(\boldsymbol{i})$ equivalent to $\mathfrak{C}(\boldsymbol{i})$ and of the form,

$\mathfrak{C}_6(\boldsymbol{i})$: (predicate prefix) $\bigwedge_\nu [A_\nu(0) \vee (\exists t) B_\nu(t)]$.

Now we remark again that predicate quantification is over special predicates, and therefore identities of the form

$$A(0) \vee (\exists t)\ B(t) \cdot\cdot\equiv\cdot\cdot\ (\exists j) \cdot [\tilde{A}(0) \supset j(0)] \wedge (\forall t)\ [j(t)\ \tilde{j}(t') \supset B(t)]$$

hold in W. 2. A. Accordingly we replace each $A_\nu(0) \vee (\exists t)\ B_\nu(t)$ in $\mathfrak{C}_6(\boldsymbol{i})$, making use of different $j$'s, and then move the new predicate quantifiers into the prefix. The result is a formula $\mathfrak{C}^*(\boldsymbol{i})$ equivalent to $\mathfrak{C}(\boldsymbol{i})$ and of the form required in lemma 1.

## § 3. Finite automata

We will now define automata as syntactic entities. We refer to § 6 for motivation. In § 6, our concept also is compared with other definitions of "finite automaton".

Definition 1. A (*finite*) *$n/m$-automaton* with *input* $\boldsymbol{X}^n$ and *transit* $\boldsymbol{Y}^m$ is a $2m$-tuple $\boldsymbol{E}^m{}^\frown\boldsymbol{H}^m[\boldsymbol{Y}^m, \boldsymbol{X}^n]$ of propositional formulas. (Note that no propositional variables occur in $\boldsymbol{E}^m$.) A (binary) *output* of a $n/m$-automaton with transit $\boldsymbol{Y}^m$ is a propositional formula $U[\boldsymbol{Y}^m]$.

The *transition-recursion* of an automaton $\boldsymbol{E}^m{}^\frown\boldsymbol{H}^m$ is the $2m$-tuple of matrices

$$\begin{aligned}\boldsymbol{r}^m(0) &\equiv \boldsymbol{E}^m,\\ \boldsymbol{r}^m(t') &\equiv \boldsymbol{H}^m[\boldsymbol{r}^m(t),\ \boldsymbol{i}^n(t)],\end{aligned}$$

it defines recursively a functional which will be denoted by $\boldsymbol{r} = \zeta(\boldsymbol{E}, \boldsymbol{H}, \boldsymbol{i})$.

The *output-recursion* of an automaton with output $\boldsymbol{E}^m \frown \boldsymbol{H}^m \frown U$, is obtained by adding the matrix $u(t) \equiv U[\boldsymbol{r}(t)]$ to the transition-recursion. It defines recursively a functional which will be denoted by $u = \psi(\boldsymbol{E}, \boldsymbol{H}, U, \boldsymbol{i})$.

The "input to output" functional $u = \psi(\boldsymbol{E}, \boldsymbol{H}, U, \boldsymbol{i})$ might be defined to be the behavior of the automaton with output. However, this is inconvenient for establishing relations to W. 2. A., because in general $u = \psi(\boldsymbol{i})$ will not be special even for special $\boldsymbol{i}$. We therefore will deal with a special sort of output only. It will be seen in § 6 that this is not essentially a restriction.

Definition 2. $U[\boldsymbol{Y}^m]$ is a *special output* of the automaton $\boldsymbol{E}^m \frown \boldsymbol{H}^m$ if $U[\boldsymbol{Y}] \equiv U[\boldsymbol{H}[\boldsymbol{Y}, \boldsymbol{\Lambda}]]$.

It is easy to see that in case $U$ is special the output recursion defines an operator $\psi$ such that $\psi(\boldsymbol{i})$ or $\sim\psi(\boldsymbol{i})$ is special whenever $\boldsymbol{i}$ is special. This makes the operator definition of behavior manageable in W. 2. A. However, it is more convenient to work with the set $\beta$ consisting of all special $\boldsymbol{i}$ for which $\sim\psi(\boldsymbol{i})$ is special, rather than with $\psi$ directly. This leads to the following definition of behavior, which is closely related to KLEENE's [6] (see § 7).

Definition 3. The *behavior* $\beta(\boldsymbol{E}^m, \boldsymbol{H}^m, U)$ of a $\boldsymbol{n}/\boldsymbol{m}$-automaton with special output is the set of all special $\boldsymbol{n}$-predicates $\boldsymbol{i}^n$ for which the predicate $u = \psi(\boldsymbol{E}, \boldsymbol{H}, U, \boldsymbol{i})$, determined by the output-recursion, is ultimately true, i. e., $(\exists x)(\forall t)_x^\infty u(t)$.

By the *length* of a special $\boldsymbol{n}$-predicate $\boldsymbol{i}^n$ we mean the smallest number $x$ such that $\boldsymbol{i}^n(t) \equiv \boldsymbol{\Lambda}^n$ for $t > x$. By definitions 2, 3 one easily proves,

(*) *If $U$ is a special output of the automaton $\boldsymbol{E}^m \frown \boldsymbol{H}^m$, and if $\boldsymbol{i}^n$ is a $\boldsymbol{n}$-predicate of length $l$, then $\boldsymbol{i} \in \beta(\boldsymbol{E}, \boldsymbol{H}, U) \cdot\equiv\cdot U[\boldsymbol{r}(l+1)]$ whereby $\boldsymbol{r} = \zeta(\boldsymbol{E}, \boldsymbol{H})$ is given by the transition-recursion of $\boldsymbol{E} \frown \boldsymbol{H}$.*

Lemma 2. *If $U$ is a special output of the $\boldsymbol{n}/\boldsymbol{m}$-automaton $\boldsymbol{E}^m \frown \boldsymbol{H}^m$ then so is $\sim U$. Furthermore the behaviors $\beta(\boldsymbol{E}, \boldsymbol{H}, U)$ and $\beta(\boldsymbol{E}, \boldsymbol{H}, \sim U)$ are complementary subsets of the set of all special $\boldsymbol{n}$-predicates.*

Proof. That $\sim U$ again is special follows directly by definition 2. That the behaviors of $U$ and $\sim U$ are complementary is best seen by referring to (*).

Definition 4. An $\boldsymbol{n}/\boldsymbol{m}$-automaton with output, $\boldsymbol{E}^m \frown \boldsymbol{H}^m \frown U$, is in *expanded form* if $\boldsymbol{H}^m$ and $U$ are of the form

$$\boldsymbol{H}^m[Y_1, \ldots, Y_m, \boldsymbol{X}^n] : Y_1 \boldsymbol{K}_1^m[\boldsymbol{X}^n] \vee \cdots \vee Y_m \boldsymbol{K}_m^m[\boldsymbol{X}^n],$$

$$U[Y_1, \ldots, Y_m] \qquad : Y_1 \vee Y_2 \vee \cdots \vee Y_k.$$

Lemma 3. *To every $\boldsymbol{n}/\boldsymbol{m}$-automaton $\boldsymbol{E}^m \frown \boldsymbol{H}^m \frown U$ one can construct a $\boldsymbol{n}/\boldsymbol{k}$-automaton $\boldsymbol{G}^k \frown \boldsymbol{L}^k \frown W$ which is in expanded form, and such that if $U$ is special output of $\boldsymbol{E}^m \frown \boldsymbol{H}^m$ then $W$ is special output of $\boldsymbol{G}^k \frown \boldsymbol{L}^k$ and the behavior $\beta(\boldsymbol{G}, \boldsymbol{L}, W)$ is equal to $\beta(\boldsymbol{E}, \boldsymbol{H}, U)$.*

Proof. We indicate the construction of $\boldsymbol{G}, \boldsymbol{L}, W$ in case $\boldsymbol{m} = 2 \frown 2$, and $U[Y_1, Y_2] \equiv Y_1$. The given automaton then consists of p. f.'s $E_1$, $E_2$, $H_1[Y_1, Y_2, \boldsymbol{X}^n]$, $H_2[Y_1, Y_2, \boldsymbol{X}^n]$.

The first step in the construction is to obtain disjunctive forms for $H_1$, $H_2$. and $U$. Let this be,

$$(1)\quad \begin{cases} H_1[Y_1,\ Y_2,\ \mathbf{X}] \cdot\equiv\cdot\ Y_1 Y_2 A_{11}[\mathbf{X}] \vee Y_1 \tilde{Y}_2 A_{12}[\mathbf{X}] \vee \\ \qquad\qquad \vee\ \tilde{Y}_1 Y_2 A_{13}[\mathbf{X}] \vee \tilde{Y}_1 \tilde{Y}_2 A_{14}[\mathbf{X}], \\ H_2[Y_1,\ Y_2,\ \mathbf{X}] \cdot\equiv\cdot\ Y_1 Y_2 A_{21}[\mathbf{X}] \vee Y_1 \tilde{Y}_2 A_{22}[\mathbf{X}] \vee \\ \qquad\qquad \vee\ \tilde{Y}_1 Y_2 A_{23}[\mathbf{X}] \vee \tilde{Y}_1 \tilde{Y}_2 A_{24}[\mathbf{X}], \end{cases}$$

$$(2)\qquad U[Y_1,\ Y_2] \cdot\equiv\cdot\ Y_1 Y_2 \vee Y_1 \tilde{Y}_2$$

Now we let $\boldsymbol{k} = 2^\frown 2^\frown 2^\frown 2$ and construct $\boldsymbol{G}^k$, $\boldsymbol{H}^k$, $W$ as follows,

$$(3)\quad \begin{cases} G_1 : E_1 E_2, \qquad G_3 : \tilde{E}_1 E_2, \\ G_2 : E_1 \tilde{E}_2, \qquad G_4 : \tilde{E}_1 \tilde{E}_2, \end{cases}$$

$$(4)\quad \begin{cases} L_1[Z_1, Z_2, Z_3, Z_4, \mathbf{X}] : Z_1 A_{11}[\mathbf{X}]\, A_{21}[\mathbf{X}] \vee Z_2 A_{12}[\mathbf{X}]\, A_{22}[\mathbf{X}] \vee \\ \qquad\qquad \vee\ Z_3 A_{13}[\mathbf{X}]\, A_{23}[\mathbf{X}] \vee Z_4 A_{14}[\mathbf{X}]\, A_{24}[\mathbf{X}], \\ L_2[Z_1, Z_2, Z_3, Z_4, \mathbf{X}] : Z_1 A_{11}[\mathbf{X}]\, \tilde{A}_{21}[\mathbf{X}] \vee Z_2 A_{12}[\mathbf{X}]\, \tilde{A}_{22}[\mathbf{X}] \vee \\ \qquad\qquad \vee\ Z_3 A_{13}[\mathbf{X}]\, \tilde{A}_{23}[\mathbf{X}] \vee Z_4 A_{14}[\mathbf{X}]\, \tilde{A}_{24}[\mathbf{X}], \\ L_3[Z_1, Z_2, Z_3, Z_4, \mathbf{X}] : Z_1 \tilde{A}_{11}[\mathbf{X}]\, A_{21}[\mathbf{X}] \vee Z_2 \tilde{A}_{12}[\mathbf{X}]\, A_{22}[\mathbf{X}] \vee \\ \qquad\qquad \vee\ Z_3 \tilde{A}_{13}[\mathbf{X}]\, A_{23}[\mathbf{X}] \vee Z_4 \tilde{A}_{14}[\mathbf{X}]\, A_{24}[\mathbf{X}], \\ L_4[Z_1, Z_2, Z_3, Z_4, \mathbf{X}] : Z_1 \tilde{A}_{11}[\mathbf{X}]\, \tilde{A}_{21}[\mathbf{X}] \vee Z_2 A_{12}[\mathbf{X}]\, \tilde{A}_{22}[\mathbf{X}] \vee \\ \qquad\qquad \vee\ Z_3 \tilde{A}_{13}[\mathbf{X}]\, \tilde{A}_{23}[\mathbf{X}] \vee Z_4 \tilde{A}_{14}[\mathbf{X}]\, \tilde{A}_{24}[\mathbf{X}], \end{cases}$$

$$(5)\qquad W[Z_1, Z_2, Z_3, Z_4] : Z_1 \vee Z_2.$$

By definition 4, and (3), (4), (5) it is clear that $\boldsymbol{G}^\frown \boldsymbol{L}^\frown W$ is in expanded form. Next we obtain from (1) the identities

$$\tilde{H}_\nu[Y_1,\ Y_2,\ \mathbf{X}] \cdot\equiv\cdot\ Y_1 Y_2 \tilde{A}_{\nu 1}[\mathbf{X}] \vee Y_1 \tilde{Y}_2 \tilde{A}_{\nu 2}[\mathbf{X}] \vee Y_1 \tilde{Y}_2 \tilde{A}_{\nu 3}[\mathbf{X}] \vee \tilde{Y}_1 \tilde{Y}_2 \tilde{A}_{\nu 4}[\mathbf{X}].$$

Together with (1) and (4) this yields,

$$(6)\quad \begin{cases} L_1[Y_1 Y_2,\ Y_1 \tilde{Y}_2,\ \tilde{Y}_1 Y_2,\ \tilde{Y}_1 \tilde{Y}_2,\ \mathbf{X}] \cdot\equiv\cdot H_1[Y_1,\ Y_2,\ \mathbf{X}]\ H_2[Y_1,\ Y_2,\ \mathbf{X}], \\ L_2[Y_1 Y_2,\ Y_1 \tilde{Y}_2,\ \tilde{Y}_1 Y_2,\ \tilde{Y}_1 \tilde{Y}_2,\ \mathbf{X}] \cdot\equiv\cdot H_1[Y_1,\ Y_2,\ \mathbf{X}]\ \tilde{H}_2[Y_1,\ Y_2,\ \mathbf{X}], \\ L_3[Y_1 Y_2,\ Y_1 \tilde{Y}_2,\ \tilde{Y}_1 Y_2,\ \tilde{Y}_1 \tilde{Y}_2,\ \mathbf{X}] \cdot\equiv\cdot \tilde{H}_1[Y_1,\ Y_2,\ \mathbf{X}]\ H_2[Y_1,\ Y_2,\ \mathbf{X}], \\ L_4[Y_1 Y_2,\ Y_1 \tilde{Y}_2,\ \tilde{Y}_1 Y_2,\ \tilde{Y}_1 \tilde{Y}_2,\ \mathbf{X}] \cdot\equiv\cdot \tilde{H}_1[Y_1,\ Y_2,\ \mathbf{X}]\ \tilde{H}_2[Y_1,\ Y_2,\ \mathbf{X}] \end{cases}$$

and from (2) and (5) we get,

$$(7)\qquad W[Y_1 Y_2,\ Y_1 \tilde{Y}_2,\ \tilde{Y}_1 Y_2,\ \tilde{Y}_1 \tilde{Y}_2] \cdot\equiv\cdot\ U[Y_1 Y_2].$$

Now assume that $U$ is a special output of $\boldsymbol{E}^\frown \boldsymbol{H}$. Let $\boldsymbol{C}[Y_1,\ Y_2]$ stand for $Y_1 Y_2{}^\frown Y_1 \tilde{Y}_2{}^\frown \tilde{Y}_1 Y_2{}^\frown \tilde{Y}_1 \tilde{Y}_2$. Then using in order (7), definition 2, (7), (6) one obtains

$W[C[Y_1, Y_2]] \equiv W[L[C[Y_1\, Y_2], \wedge]]$. This means that $W[Z] \equiv W[L[Z, \wedge]]$ holds for the values $\vee^\frown\wedge^\frown\wedge^\frown\wedge$, $\wedge^\frown\vee^\frown\wedge^\frown\wedge$, $\wedge^\frown\wedge^\frown\vee^\frown\wedge$, and $\wedge^\frown\wedge^\frown\wedge^\frown\vee$ of $Z$. Because of the particular form (5) of $W$ the restriction of the range of $Z$ can be omitted, which by definition 2 means that $W$ is special output of $G^\frown L$.

Next let $i$ be any $n$-predicate and let $r_1{}^\frown r_2 = \zeta(E, H, i)$, $s_1{}^\frown s_2{}^\frown s_3{}^\frown s_4 = \zeta(G, L, i)$, so that

$$(8)\quad \begin{cases} r(0) \equiv E, \\ r(t') \equiv H[r(t),\, i(t)], \end{cases} \qquad (9)\quad \begin{cases} s(0) \equiv G, \\ s(t') \equiv L[s(t),\, i(t)]. \end{cases}$$

Using (3), (6), (8), (9) one shows by an induction on $t$ taht $s_1(t) \equiv r_1(t)\, r_2(t)$, $s_2(t) \equiv r_1(t)\, \tilde{r}_2(t)$, $s_3(t) \equiv \tilde{r}_1(t)\, r_2(t)$, $s_4(t) \equiv \tilde{r}_1(t)\, \tilde{r}_2(t)$ hold for all $t$. Because of (7) this yields, $(\forall t) \cdot W[s(t)] \equiv U[r(t)]$. By definition 3 it follows that $\beta(G, L, W) = \beta(K, H, U)$.

Lemma 4. *Let $E^m{}^\frown H^m$ be an $n^\frown 2/m$-atomaton in expanded form and let $L^m$ be defined by,* $L^m[Y^m, X^n] : H^m[Y^m, X^{n\frown}\vee] \vee H^m[Y^m, X^{n\frown}\wedge]$.

(a) *If $i^{n\frown}j$ is any $n^\frown 2$-predicate and if $r^m = \zeta(E, H, i^{m\frown}j)$, and $s^m = \zeta(E, L, i^n)$, then* $(t)\,[r(t) \supset s(t)]$.

b) *If $i^n$ is any $n$-predicate, and $s^m = \zeta(E, L, i^n)$, and $s_\nu(x') \equiv \vee$, then there exists a special predicate $j$ of length $\leq x$ such that also $r_\nu(x') \equiv \vee$, in case* $r^m = \zeta(E, H, i^{n\frown}j)$.

Proof. By assumption $H^n$ is of form

$$(1)\qquad H^m[Y^m, X^{n\frown}Z] : Y_1 K_1^m[X^{n\frown}Z] \vee \cdots \vee Y_m K_m^m[X^{n\frown}Z].$$

Now let $i^\frown j$ be any $n^\frown 2$-predicate and let $r = \zeta(E, H, i^\frown j)$, $s = \zeta(E, L, i)$. Then by definition 3 and the construction of $L$,

$$(2)\qquad \begin{cases} r(0) \equiv E, \\ r(t') \equiv H[r(t),\, i(t)^\frown j(t)], \end{cases}$$

$$(3)\qquad \begin{cases} s(0) \equiv E, \\ s(t') \equiv H[s(t),\, i(t)^\frown\vee] \vee H[s(t),\, i(t)^\frown\wedge] \end{cases}$$

hold for all $t$. Making use of (1), (2), (3) it follows by induction on $t$ that $(\forall t)\,[r(t) \supset s(t)]$. Thus part (a) of the lemma is established.

Next let $i$ be any $n$-predicate, let $s = \zeta(E, L, i)$, and suppose that $s_\nu(x') \equiv \vee$. Then by definition 3 the predicates $i$ and $s$ satisfy (3) for all $t$, so that by (1) we obtain

$$(4)\qquad \begin{cases} s_\varrho(0) \equiv E_\varrho, & \varrho = 1, \ldots, m \\ s_\varrho(t') \equiv \bigvee\limits_{\substack{\nu = 1, \ldots, m \\ Y = \wedge, \vee}} s_\nu(t)\, K_{\varrho,\nu}[i(t')^\frown Y], & \varrho = 1, \ldots, m \end{cases}$$

Now we stepwise choose a sequence of truth values $j(x), \ldots, j(0)$ and a sequence of indices $\nu_{x+1}, \ldots, \nu_0$ according to the following specifications. I. Let $\nu_{x+1}$ be $\nu$. II. If $\nu_{y+1}$ has already been obtained choose $\nu_y$, $j(y)$ such that $s_{\nu_y}(y)$ and $K_{\nu_{y+1}, \nu_y}[i^n(y)^\frown j(y)]$ hold. That these choices are possible clearly follows by the assumption $s_\nu(x) \equiv \vee$ and (4). Let now the values $r(0), \ldots, r(x')$ be defined from $\boldsymbol{i}(0), \ldots, \boldsymbol{i}(x), j(0), \ldots, j(x)$ by the recursions (2), so that by (1), $r_\varrho(0) \equiv E_\varrho$; $r_\varrho(t') \equiv \bigvee_{\mu=1,\ldots,m} r_\mu(t) K_{\varrho,\mu}[\boldsymbol{i}(t)^\frown j(t)]$ for $t \leq x$. Then by using II one stepwise obtains $r_{\nu_0}(0)$, $K_{\nu_1,\nu_0}[\boldsymbol{i}(0)^\frown j(0)], \ldots, r_{\nu_x}(x)$, $K_{\nu_{x+1},\nu_x}[\boldsymbol{i}(x)^\frown j(x)]$, $r_{\nu_{x+1}}(x+1)$. Therefore by I, $r_\nu(x') \equiv \mathsf{V}$. Thus by extending the sequence $j(0), \ldots, j(x)$ letting $j(t) \equiv \mathsf{\Lambda}$ for $t > x$, we obtain a special predicate $j$ such that $j$ is of length $\leq x$, and $r_\nu(x') \equiv \mathsf{V}$ in case $\boldsymbol{r} = \zeta(\boldsymbol{E}, \boldsymbol{H}, \boldsymbol{i}^n)$. Therefore also part (b) of the lemma is established.

Lemma 5. *For every $n/m$-automaton with output $\boldsymbol{E}^m{}^\frown \boldsymbol{H}^m{}^\frown U$ one can construct a number $h$ and an output $W$ such that*

(a) $U[\boldsymbol{Y}] \supset W[\boldsymbol{Y}]$,

(b) $\boldsymbol{Z}_1 \equiv \boldsymbol{Y} \wedge \boldsymbol{Z}_2 \equiv \boldsymbol{H}[\boldsymbol{Z}_1, \mathsf{\Lambda}] \wedge \cdots \wedge \boldsymbol{Z}_h \equiv \boldsymbol{H}[\boldsymbol{Z}_{h-1}, \mathsf{\Lambda}] \cdot\supset\cdot U[\boldsymbol{Z}_1] \vee \cdots \vee U[\boldsymbol{Z}_h]$,

(c) *If $U[\boldsymbol{Y}] \supset U[\boldsymbol{H}[\boldsymbol{Y}, \mathsf{\Lambda}]]$ then $W$ is special output of $\boldsymbol{E}^\frown \boldsymbol{H}$.*

Proof. The construction of $h$ and $W[\boldsymbol{Y}^m]$ from $\boldsymbol{H}^m[\boldsymbol{Y}^m, \boldsymbol{X}^m]$ and $U[\boldsymbol{Y}]$ proceeds by the following rules,

I. Let $U_0[\boldsymbol{Y}]$ be $\mathsf{V}$, let $U_1[\boldsymbol{Y}]$ be $U[\boldsymbol{Y}]$.

II. If $U_k[\boldsymbol{Y}] \equiv U_{k+1}[\boldsymbol{Y}]$ is not tautologous,
let $U_{k+2}[\boldsymbol{Y}]$ be $U_{k+1}[\boldsymbol{Y}] \vee U_{k+1}[\boldsymbol{H}[\boldsymbol{Y}, \mathsf{\Lambda}]]$.

III. If $U_k[\boldsymbol{Y}] \equiv U_{k+1}[\boldsymbol{Y}]$ is tautologous, let $h$ be $k$, let $W$ be $U_k$, and stop.

From this construction it is clear that $U_0[\boldsymbol{Y}] \supset U_1[\boldsymbol{Y}]$, $U_1[\boldsymbol{Y}] \supset U_2[\boldsymbol{Y}], \ldots$. Because there are only $a = 2^{(2^m)}$ truth-functions on $m$ arguments the construction must reach a stage $k \leq a$ such that $U_k$ and $U_{k+1}$ denote the same truth function, i. e., such that $U_k[\boldsymbol{Y}] \equiv U_{k+1}[\boldsymbol{Y}]$. The next step in the construction then clearly must use the stoprule III. Thus the construction always ends in $k \leq a$ steps, yielding an $h$ and a $W$.

Suppose now for example that $h$ turns out to be 3, so that $W$ is $U_3$ and $U_3 \equiv U_4$. Letting $\boldsymbol{K}[\boldsymbol{Y}]$ stand for $\boldsymbol{H}[\boldsymbol{Y}, \mathsf{\Lambda}]$ it follows from the construction,

(1) $$U_2[\boldsymbol{Y}] \equiv\cdot\ U[\boldsymbol{Y}] \vee U[\boldsymbol{K}[\boldsymbol{Y}]],$$

and

(2) $$U_3[\boldsymbol{Y}] \equiv\cdot\ U_2[\boldsymbol{Y}] \vee U_2[\boldsymbol{K}[\boldsymbol{Y}]],$$

and

(3) $$U_4[\boldsymbol{Y}] \equiv U_3[\boldsymbol{Y}] \vee U_3[\boldsymbol{K}[\boldsymbol{Y}]].$$

By (1), $U[\boldsymbol{Y}] \supset U_2[\boldsymbol{Y}]$ and by (2), $U_2[\boldsymbol{Y}] \supset U_3[\boldsymbol{Y}]$. Therefore $U[\boldsymbol{Y}] \supset U_3[\boldsymbol{Y}]$. Because $W$ is $U_3$ this established part (a) of the lemma.

Next suppose $\boldsymbol{Z}_1 \equiv \boldsymbol{Y}_1$, $\boldsymbol{Z}_2 \equiv \boldsymbol{K}[\boldsymbol{Z}_1]$, $\boldsymbol{Z}_3 \equiv \boldsymbol{K}[\boldsymbol{Z}_2]$, and $U_3[\boldsymbol{Y}_1]$. Then by (2), $U_2[\boldsymbol{Z}_1] \vee U_2[\boldsymbol{Z}_2]$. This yields by (1), $U[\boldsymbol{Z}_1] \vee U[\boldsymbol{Z}_2] \vee U[\boldsymbol{Z}_3]$. Because $U_3$ is $W$ and $\boldsymbol{K}[\cdot]$ is $\boldsymbol{H}[\cdot, \mathsf{\Lambda}]$ this established part (b) of the lemma.

Now suppose $U[\boldsymbol{Y}] \supset U[\boldsymbol{K}[\boldsymbol{Y}]]$. Then by (1), $U_2[\boldsymbol{Y}] \supset U[\boldsymbol{K}[\boldsymbol{Y}]]$, and also by (1), $U[\boldsymbol{K}[\boldsymbol{Y}]] \supset U_2[\boldsymbol{K}[\boldsymbol{Y}]]$. Therefore $U_2[\boldsymbol{Y}] \supset U_2[\boldsymbol{K}[\boldsymbol{Y}]]$. From this, by using (2), $U_3[\boldsymbol{Y}] \supset U_3[\boldsymbol{K}[\boldsymbol{Y}]]$ is similarly obtained. But also $U_3[\boldsymbol{K}[\boldsymbol{Y}]] \supset U_3[\boldsymbol{Y}]$, because of (3) and $U_3 \equiv U_4$. Thus, $U_3[\boldsymbol{Y}] \equiv U_3[\boldsymbol{K}[\boldsymbol{Y}]]$. Because $U_3$ is $W$ and $\boldsymbol{K}[\boldsymbol{Y}]$ is

$\boldsymbol{H}[\boldsymbol{Y}, \boldsymbol{\Lambda}]$ this means that $W$ is a special output of $\boldsymbol{E}^\frown\boldsymbol{H}$. This establishes part (c) of the lemma.

Lemma 6. *To every* $\boldsymbol{n}^\frown 2/\boldsymbol{m}$*-automaton with special output* $\boldsymbol{E}^{m\frown}\boldsymbol{H}^{m\frown}U$ *in expanded form one can construct a* $\boldsymbol{n}/\boldsymbol{m}$*-automaton with special output* $\boldsymbol{E}^{m\frown}\boldsymbol{L}^{m\frown}W$ *such that for every special* $\boldsymbol{n}$*-predicate* $\boldsymbol{i}^n$, $\boldsymbol{i} \in \beta(\boldsymbol{G}, \boldsymbol{L}, W) \equiv (\exists j)[\boldsymbol{i}^\frown j \in \beta(\boldsymbol{E}, \boldsymbol{H}, U)]$ *and* $\boldsymbol{E}^\frown\boldsymbol{L}^\frown W$ *is again in expanded form.*

Proof. The construction of $\boldsymbol{L}^m$, $W$ is as follows,

$$\text{(1)} \qquad \boldsymbol{L}[\boldsymbol{Y}, \boldsymbol{X}] : \boldsymbol{H}[\boldsymbol{Y}, \boldsymbol{X}^\frown\mathsf{V}] \vee \boldsymbol{H}[\boldsymbol{Y}, \boldsymbol{X}^\frown\Lambda].$$

$$\text{(2)} \qquad h, W[Y] : \text{apply construction of lemma 4 to } \boldsymbol{E}^\frown\boldsymbol{L}^\frown U.$$

Assume now that $\boldsymbol{E}^\frown\boldsymbol{H}^\frown U$ is in expanded form and $U$ is special. Then by definitions 2, 4,

$$\text{(3)} \qquad U[\boldsymbol{Y}] : Y_1 \vee Y_2 \vee \cdots \vee Y_k,$$

$$\text{(4)} \qquad U[\boldsymbol{Y}] \equiv {}^{\boldsymbol{\cdot}}U[\boldsymbol{H}[\boldsymbol{Y}, \boldsymbol{\Lambda}^\frown\Lambda]].$$

By using (4), (3), (1), (3) in this order one shows $U[\boldsymbol{Y}] \supset U[\boldsymbol{L}, [\boldsymbol{Y}, \boldsymbol{\Lambda}]]$. Therefore, by (2) and lemma 5 (c), $W$ is a special output of $\boldsymbol{E}^\frown\boldsymbol{L}$.

Next assume $\boldsymbol{i}^\frown j \in \beta(\boldsymbol{E}, \boldsymbol{H}, U)$, and let $\boldsymbol{r} = \zeta(\boldsymbol{E}, \boldsymbol{H}, \boldsymbol{i}^\frown j)$ and $\boldsymbol{s} = \zeta(\boldsymbol{E}, \boldsymbol{L}, \boldsymbol{i})$. Then by definition 3, $(\exists x)(\forall t)_x^\infty U[\boldsymbol{r}(t)]$, and by (1), lemma 4 (a), $(\forall t)[\boldsymbol{r}(t) \supset \boldsymbol{s}(t)]$. Therefore it follows by (3) that $(\exists x)(\forall t)_x^\infty U[\boldsymbol{s}(t)]$. By (2) and lemma 5 (a) this yields $(\exists x)(\forall t)_x^\infty W[\boldsymbol{s}(t)]$. Therefore, by definition 3, $\boldsymbol{i} \in \beta(\boldsymbol{E}, \boldsymbol{L}, W)$. Thus we have shown that $(\exists j)[\boldsymbol{i}^\frown j \in \beta(\boldsymbol{E}, \boldsymbol{H}, U)] \supset \boldsymbol{i} \in \beta(\boldsymbol{E}, \boldsymbol{L}, W)$.

Assume now that $\boldsymbol{i} \in \beta(\boldsymbol{E}, \boldsymbol{L}, W)$, and let $\boldsymbol{s} = \zeta(\boldsymbol{E}, \boldsymbol{L}, \boldsymbol{i})$. Then if $l$ is the length of $\boldsymbol{i}$ it follows by (*) that $W[\boldsymbol{s}(l+1)]$. Because of (3) and lemma 5 (b) this implies that $U[\boldsymbol{s}(l+1)] \vee \cdots \vee U[\boldsymbol{s}(l+h+1)]$, and by (3), there are $\nu, p \leqq h$ such that $s_\nu(l+p+1) \equiv \mathsf{V}$. Because $\boldsymbol{E}^\frown\boldsymbol{H}$ is in expanded form and because of (1) we therefore can apply lemma 4 (b) to conclude that there is a $j$ of length $\leqq l+p$ such that $r_\nu(l+p+1) \equiv \mathsf{V}$, for $\boldsymbol{r} = \zeta(\boldsymbol{E}, \boldsymbol{H}, \boldsymbol{i}^\frown j)$. Using (3) we obtain $U[\boldsymbol{r}(l+p+1)]$. Now observe that $\boldsymbol{i}(t) \equiv \boldsymbol{\Lambda}$, and $j(t) \equiv \Lambda$, for $t > l+p$. Because $U$ is special output of $\boldsymbol{E}^\frown\boldsymbol{H}$ it therefore follows that $(\forall t)_{l+p+1}^\infty U[\boldsymbol{r}(t)]$. By definition 3 this means that $\boldsymbol{i}^{n\frown} j \in \beta(\boldsymbol{E}, \boldsymbol{H}, U)$. Thus we have concluded the proof of lemma 6 by showing that $\boldsymbol{i} \in \beta(\boldsymbol{E}, \boldsymbol{L}, W) \supset (\exists j)[\boldsymbol{i}^\frown j \in \beta(\boldsymbol{E}, \boldsymbol{H}, U)]$.

## § 4. Analysis and synthesis

We begin by establishing a synthesis result for formulas of W. 2. A. which do not contain predicate-quantifiers. Using the lemmas of § 2 and § 3 one then easily extends the result.

Lemma 7. *To every formula* $\mathfrak{C}(\boldsymbol{i}^n)$ *of the form* $K[\boldsymbol{i}(0)] \wedge (\forall t)\, B[\boldsymbol{i}(t), \boldsymbol{i}(t')]$ *one can construct a* $\boldsymbol{n}/\boldsymbol{m}$*-automaton* $\boldsymbol{E}^{m\frown}\boldsymbol{H}^m$ *and a special output* $U[\boldsymbol{Y}^m]$ *such that* $\beta(\boldsymbol{E}, \boldsymbol{H}, U) = \hat{\boldsymbol{i}}\, \mathfrak{C}(\boldsymbol{i})$.

Proof. We first determine whether or not $B[\boldsymbol{\Lambda}, \boldsymbol{\Lambda}]$. In case $\widetilde{B}[\boldsymbol{\Lambda}, \boldsymbol{\Lambda}]$, we take for $\boldsymbol{E}^\frown\boldsymbol{H}$ any automaton and for $U$ the output $\Lambda$. Then clearly $U$ is special output and $\beta(\boldsymbol{E}, \boldsymbol{H}, U) = \hat{\boldsymbol{i}}\, \mathfrak{C}(\boldsymbol{i})$ are both empty. Thus in this case Lemma 7 is established.

Let us next consider the case $B[\mathbf{\Lambda}, \mathbf{\Lambda}]$. Then we take $m$ to be $n+2$, and $\boldsymbol{E}^m$ to be $\mathbf{V}^m$. $\boldsymbol{H}^m$ we define as follows.

$$H_\nu[\boldsymbol{Y}^n{}^\frown Z_1{}^\frown Z_2, \boldsymbol{X}^n] : X_\nu, \qquad \text{for } \nu = 1, \ldots, n$$

$$H_{n+1}[\boldsymbol{Y}^n{}^\frown Z_1{}^\frown Z_2, \boldsymbol{X}^n] : \Lambda,$$

$$H_{n+2}[\boldsymbol{Y}^n{}^\frown Z_1{}^\frown Z_2, \boldsymbol{X}^n] : Z_1\, K[\boldsymbol{X}^n] \vee \tilde{Z}_1\, Z_2\, B[\boldsymbol{Y}^n, \boldsymbol{X}^n].$$

As output we choose the formula

$$U[\boldsymbol{Y}^n{}^\frown Z_1{}^\frown Z_2] : Z_1 K[\mathbf{\Lambda}^n] \vee \tilde{Z}_1\, Z_2\, B[\boldsymbol{Y}^n, \mathbf{\Lambda}^n].$$

Noting that $B[\mathbf{\Lambda}, \mathbf{\Lambda}]$, and using definition 2, one easily checks that $U$ is a special output of $\boldsymbol{E}^\frown \boldsymbol{H}$. Furthermore, the transition-recursion of $\boldsymbol{E}^\frown\boldsymbol{H}$ is clearly equivalent to the recursion

$$r_\nu(0) \equiv \mathsf{V}, \qquad \nu = 1, \ldots, n+2$$

$$r_{n+2}(1) \equiv K[\boldsymbol{i}[0)],$$

$$r_\nu(t') \equiv i_\nu(t), \qquad \nu = 1, \ldots, n$$

$$r_{n+1}(t') \equiv \Lambda,$$

$$r_{n+2}(t'') \equiv r_{n+2}(t')\, B[\boldsymbol{i}(t), \boldsymbol{i}(t')]$$

and therefore the operator $\boldsymbol{r}^m = \zeta(\boldsymbol{E}, \boldsymbol{H}, \boldsymbol{i}^n)$ can also be defined by

$$r_\nu(0) \equiv \mathsf{V}, \qquad \nu = 1, \ldots, n+2$$

$$r(t') \equiv i_\nu(t), \qquad \nu = 1, \ldots, n$$

$$r_{n+1}(t') \equiv \Lambda,$$

$$r_{n+2}(t') \equiv K[\boldsymbol{i}(0)] \wedge (\forall x)_0^{t-1}\, B[\boldsymbol{i}(x), \boldsymbol{i}(x')],$$

Consequently, the output-operator $u = \psi(\boldsymbol{E}, \boldsymbol{H}, U)$ is defined by

$$u(0) \equiv K[\mathbf{\Lambda}],$$

$$u(t') \equiv K[\boldsymbol{i}(0)] \wedge (\forall x)_0^{t-1}\, B[\boldsymbol{i}(x), \boldsymbol{i}(x')] \wedge B[\boldsymbol{i}(t), \mathbf{\Lambda}].$$

Because $B[\mathbf{\Lambda}, \mathbf{\Lambda}]$ it follows that for any special $\boldsymbol{i}$, $u = \psi[\boldsymbol{E}, \boldsymbol{H}, U)$ is ultimately $\mathsf{V}$ just in case $K[\boldsymbol{i}(0)] \wedge (\forall t)\, B[\boldsymbol{i}(t), \boldsymbol{i}(t')]$. I. e., by definition 3, $\beta(\boldsymbol{E}, \boldsymbol{H}, U) = \hat{\boldsymbol{i}}\, \mathfrak{C}(\boldsymbol{i})$.

Theorem 1 (Synthesis). *For every formula* $\mathfrak{C}(\boldsymbol{i}^n)$ *of* W. 2. A. *one can construct a* $\boldsymbol{n}/\boldsymbol{m}$*-automaton* $\boldsymbol{E}^{m\frown}\boldsymbol{H}^m$ *with special output* $U[\boldsymbol{Y}^m]$ *such that* $\beta(\boldsymbol{E}, \boldsymbol{H}, U) = \hat{\boldsymbol{i}}\, \mathfrak{C}(\boldsymbol{i})$, *i. e., such that the behavior of* $\boldsymbol{E}^\frown\boldsymbol{H}^\frown U$ *is just the set of special* $\boldsymbol{n}$*-predicates which satisfy* $\mathfrak{C}(\boldsymbol{i})$.

Proof. Using Lemma 1 we first construct the formula $\mathfrak{C}^*(\boldsymbol{i})$ equivalent to $\mathfrak{C}(\boldsymbol{i})$. Let us for example assume that $\mathfrak{C}^*(\boldsymbol{i})$ is as follows, $\mathfrak{C}^*(\boldsymbol{i})$:

$$(\forall \boldsymbol{r})\, (\exists \boldsymbol{s})\, K[\boldsymbol{r}(0), \boldsymbol{s}(0)] \wedge (\forall t)\, B[\boldsymbol{i}(t), \boldsymbol{r}(t), \boldsymbol{s}(t), \boldsymbol{r}(t'), \boldsymbol{s}(t')].$$

Next we use the construction of Lemma 7 to obtain an automaton $\boldsymbol{E}_1{}^\frown\boldsymbol{H}_1$ with special output $U_1$ such that for special $\boldsymbol{i}$, $\boldsymbol{r}$, $\boldsymbol{s}$,

(1) $$\boldsymbol{i}^\frown\boldsymbol{r}^\frown\boldsymbol{s} \in \beta(\boldsymbol{E}_1, \boldsymbol{H}_1, U) \cdot\!\equiv\!\cdot\, K(0) \wedge (\forall t)\, B(t).$$

Using Lemma 3 we next construct $E_2{}^\frown H_2{}^\frown U_2$ in expanded form and such that

(2) $$\beta(E_2, H_2, U_2) = \beta(E_1, H_1, U_1).$$

By repeated application of the construction of Lemma 6 starting with $E_2{}^\frown H_2{}^\frown U$ one obtains $E_3{}^\frown H_3{}^\frown U_3$ such that for special $i$, $r$,

$$i^\frown r \in \beta(E_3, H_3, U_3) \cdot\equiv\cdot (\exists s)[i^\frown r^\frown s \in \beta(E_2, H_2, U_2]$$

and therefore by Lemma 2,

(3) $$i^\frown r \in \beta(E_3, H_3, \tilde{U}_3) \equiv (\exists s)[i^\frown r^\frown s \in \beta(E_2, H_2, U_2)].$$

Because the complementation destroys the expanded form we are forced to use Lemma 3 again to obtain $E_4{}^\frown H_4{}^\frown U_4$ in expanded form and such that

(4) $$\beta(E_4, H_4, U_4) = \beta(E_3, H_3, \tilde{U}_3).$$

By repeatedly applying Lemma 6 we next construct $E_5{}^\frown H_5{}^\frown U_5$ such that for every special $i$, $i \in \beta(E_5, H_5, U_5) \cdot\equiv\cdot (\exists r) \cdot i^\frown r \in \beta(E_4, H_4, U_4)$ and therefore by Lemma 2,

(5) $$i \in \beta(E_5, H_5, \tilde{U}_5) \cdot\equiv\cdot \sim (\exists r)\, i^\frown r \in \beta(E_4, H_4, U_4).$$

From (1), ..., (5) it clearly follows that $i \in \beta(E_5, H_5, U_5) \cdot\equiv\cdot \mathfrak{C}^*(i)$. Because $\mathfrak{C}^*$ is equivalent to $\mathfrak{C}$ this shows that the behavior of $E_5{}^\frown H_5{}^\frown U_5$ is $\hat{i}\,\mathfrak{C}(i)$.

We next obtain a rather strong converse to Theorem 1.

Theorem 2 (Analysis). *To every $n/m$-automaton with special output $E^m{}^\frown H^m{}^\frown U$ one can construct a formula $\mathfrak{C}(i^n)$ of* W. 2. A. *such that $\hat{i}\,\mathfrak{C}(i) = \beta(E, H, U)$, and such that furthermore $\mathfrak{C}(i)$ is of the form $(\exists j^p) \cdot K[j(0)] \wedge (\forall t)\, B[i(t), j(t), j(t')]$.*

Proof. By definition 3 it is clear that for every special $i$,

$$i \in \beta(E, H, U) \cdot\equiv\cdot (\exists r)[r(0) \equiv E \wedge (\forall t)[r(t') \equiv H[r(t), i(t)]] \wedge (\exists x)(\forall t)_x^\infty U[r(t)]].$$

However, the range of $(\exists r)$ in this formula may not be restricted to special predicates. On the other hand, because $U$ is assumed to be special, the formula may be slightly modified so that the range of $(\exists r)$ can be restricted to special predicates. Namely, it is clear that for any special $i$,

(1) $$i \in \beta(E, H, U) \cdot\cdot\equiv\cdot\cdot (\exists r)(\exists x) \cdot r(0) \equiv E \wedge (\forall t)_x^0 [r(t') \equiv H(t)] \wedge (\forall t)_x^\infty [i(t) \equiv \mathbf{\Lambda} \wedge U[r(t)]].$$

It remains to change this definition of $\beta$ in W. 2. A. to one of the simple form required in Theorem 2. This is accomplished by using the following device for changing the individual quantification $(\exists x)$ to a quantification $(\exists j)$ over restricted predicates,

(2) $$i \in \beta(E, H, U) \cdot\cdot\equiv\cdot\cdot (\exists r)(\exists j) \cdot (\forall t)[j(t') \supset j(t)] \wedge r(0) \equiv E \wedge (\forall t)[j(t) \supset [r(t') \equiv H(t)]] \wedge (\forall t)[\tilde{j}(t) \supset [i(t) \equiv \mathbf{\Lambda} \wedge U[r(t)]]].$$

To see that (2) is correct in case $(\exists r)(\exists j)$ is interpreted as ranging over special predicates we observe that the right sides of (1) and (2) are equivalent because of an obvious one-to-one relationship between numbers $x$ and restricted predicates $j$ satisfying $(\forall t)[j(t') \supset j(t)]$.

Thus the right side of (2) is a formula $\mathfrak{C}(i)$ as required in Theorem 2.

## § 5. Definability in W. 2. A.

We will use the notation "$[0, x+1, \exists_0 i]$" to denote W. 2. A. This is intended to indicate that we are dealing with an interpreted system which besides first-order quantification over natural numbers contains $0$, the function $x+1$, and quantification over special predicates. Similar notations for other interpreted systems, all containing first-order quantification over natural numbers, are used below. "$\exists_0 \zeta$" indicates quantification over eventually constant monadic functions from natural numbers to natural numbers.

The following systems are known to be very strong in the sense that all recursively enumerable predicates are definable in each.

| | |
|---|---|
| $[0, =, x+1, \exists_0 \zeta]$ | Gödel [5], |
| $[0, =, x+y, x \cdot y]$ | Gödel [5], |
| $[0, x+1, 2x, \exists_0 i]$ | Robinson [14]. |

(In fact these systems are equivalent in the sense that the same predicates on natural numbers can be defined in each.) In contrast we will now show that $[0, x+1, \exists_0 i]$ is much weaker, in particular the only monadic predicates on natural numbers definable in W. 2. A. are those which are ultimately periodic.

Definition 5. A formula of W. 2. A. is said to be in *normal form* if it is of the following type, $\mathfrak{C}(\boldsymbol{i}^m, x_1, \ldots, x_p)$:

$$(\exists \boldsymbol{j}^m)\, [K[\boldsymbol{j}(0)] \wedge (\forall t)\, B[\boldsymbol{i}(t), \boldsymbol{j}(t), \boldsymbol{j}(t')] \wedge A_1[\boldsymbol{j}(x_1)] \wedge \cdots \wedge A_p[\boldsymbol{j}(x_p)].$$

Theorem 3. *For every formula* $\mathfrak{C}(\boldsymbol{i}^n, x_1, \ldots, x_p)$ *of* W. 2. A. *one can construct an equivalent formula* $\mathfrak{C}^*(\boldsymbol{i}^n, x_1, \ldots, x_p)$ *which is in normal form, i. e., every predicate on numbers and special predicates definable in* W. 2. A. *is definable by a formula in normal form.*

Proof. Suppose first that $\mathfrak{A}(\boldsymbol{i}^n)$ is a formula without free individual variables. Then by Theorem 1 one can construct an automaton $\boldsymbol{E}^\frown \boldsymbol{H}$ with special output $U$ such that $\beta(\boldsymbol{E}, \boldsymbol{H}, U) = \hat{\boldsymbol{i}}\, \mathfrak{A}(\boldsymbol{i})$. Next, by Theorem 2 one can construct a formula $\mathfrak{A}^*(\boldsymbol{i})$ in normal form such that $\hat{\boldsymbol{i}}\, \mathfrak{A}^*(\boldsymbol{i}) = \beta(\boldsymbol{E}, \boldsymbol{H}, U)$. It follows, $\hat{\boldsymbol{i}}\, \mathfrak{A}(\boldsymbol{i}) = \hat{\boldsymbol{i}}\, \mathfrak{A}^*(\boldsymbol{i})$. Thus,

(1) If $\mathfrak{A}(\boldsymbol{i})$ does not contain free individual variables one can construct a normal $\mathfrak{A}^*(\boldsymbol{i})$ equivalent to $\mathfrak{A}(\boldsymbol{i})$.

Let us next start with a formula $\mathfrak{C}$ which contains free individual variables, say for example $\mathfrak{C}(\boldsymbol{i}, x_1, x_2)$. Let $\mathfrak{A}$ be defined thus, $\mathfrak{A}(\boldsymbol{i}, s_1, s_2)$:

$$s_1(0) \wedge s_2(0) \wedge (\forall t)\, [s_1(t') \supset s_1(t)] \wedge (\forall t)\, [s_2(t') \supset s_2(t)] \wedge$$
$$\wedge\, (\forall t_1 t_2)\, [s_1(t_1)\, \bar{s}_1(t_1')\, s_2(t_2)\, \bar{s}_2(t_2') \supset \mathfrak{C}(\boldsymbol{i}, t_1, t_2)].$$

It is then easy to see that $\mathfrak{A}(\boldsymbol{i}, s_1, s_2) \supset \mathfrak{C}(\boldsymbol{i}, x_1, x_2)$, in case $x_1 + 1$, $x_2 + 1$ are respectively the length of the special predicates $s_1, s_2$. Restating this we obtain

$$\mathfrak{C}(\boldsymbol{i}, x_1, x_2) \cdot\!\equiv\!\cdot\, (\exists s_1 s_2)\, [\mathfrak{A}(\boldsymbol{i}, s_1, s_2) \wedge s_1(x_1)\, \bar{s}_1(x_1') \wedge s_2(x_2)\, \bar{s}_2(x_2')],$$

and therefore,

(2) $\mathfrak{C}(\boldsymbol{i}, x_1, x_2) \cdot\cdot\equiv\cdot\cdot (\exists s_1 s_2 r_1 r_2) \cdot \mathfrak{A}(\boldsymbol{i}, s_1, s_2) \wedge (\forall t)[r_1(t) \equiv s_1(t')] \wedge$

$\wedge (\forall t)[r_2(t) \equiv s_2(t')] \wedge s_1(x_1)\,\tilde{r}_1(x_1) \wedge s_2(x_2)\,\tilde{r}_2(x_2).$

Finally we use (1) to obtain $\mathfrak{A}^*(\boldsymbol{i}, s_1, s_2)$ in normal form and equivalent to $\mathfrak{A}$. Replacing in the right side of (2) $\mathfrak{A}$ by $\mathfrak{A}^*$ and performing some obvious shifts of quantifiers will then yield a $\mathfrak{C}^*$ equivalent to $\mathfrak{C}$ such that $\mathfrak{C}^*$ is in normal form.

Corollary 1. *There is a procedure for deciding whether or not a sentence of* W. 2. A. *is true.*[1])

Proof. Because of Theorem 3 it is sufficient to indicate a procedure which decides the truth of sentences $\mathfrak{C}$ of form $(\exists \boldsymbol{j}) \cdot K[\boldsymbol{j}(0)] \wedge (\forall t)\, B[\boldsymbol{j}(t), \boldsymbol{j}(t')]$. We may furthermore assume that $B[\boldsymbol{\Lambda}, \boldsymbol{\Lambda}]$, because otherwise the sentence $\mathfrak{C}$ is clearly false in W. 2. A. $\mathfrak{C}$ then is equivalent to the assertion,

(1) There is an $x$ and a sequence of states $\boldsymbol{Y}_0, \ldots, \boldsymbol{Y}_x$ such that $K[\boldsymbol{Y}_0], B[\boldsymbol{Y}_0, \boldsymbol{Y}_1], \ldots, B[\boldsymbol{Y}_{x-1}, \boldsymbol{Y}_x], \boldsymbol{Y}_x \equiv \boldsymbol{\Lambda}$.

Note that in case $x \geqq k =$ number of states of $\boldsymbol{j}$ there must occur a repetition in the sequence $\boldsymbol{Y}_0, \boldsymbol{Y}_1, \ldots, \boldsymbol{Y}_x$, say $\boldsymbol{Y}_y \equiv \boldsymbol{Y}_z$ for some $0 \leqq y < z \leqq x$. Then if $\boldsymbol{Y}_0, \ldots, \boldsymbol{Y}_y, \ldots, \boldsymbol{Y}_z, \ldots, \boldsymbol{Y}_x$ is a $B$-sequence, the shorter sequence $\boldsymbol{Y}_0, \ldots, \boldsymbol{Y}_y, \boldsymbol{Y}_{z+1}, \ldots, \boldsymbol{Y}_x$ is still a $B$-sequence. Consequently the assertion (1) is equivalent to,

(2) There is an $x \leqq k =$ number of states of $\boldsymbol{j}$, and there is a sequence of states $\boldsymbol{Y}_0, \ldots, \boldsymbol{Y}_x$ such that $K[\boldsymbol{Y}_0], B[\boldsymbol{Y}_0, \boldsymbol{Y}_1], \ldots, B[\boldsymbol{Y}_{x-1}, \boldsymbol{Y}_x], \boldsymbol{Y}_x \equiv \boldsymbol{\Lambda}$.

Because there are only a finite number of sequences $\boldsymbol{Y}_0, \ldots, \boldsymbol{Y}_x$, $x \leqq k$, it is clear that one can effectively check whether or not (2) holds. Because $\mathfrak{C}$ is equivalent to (1), and (1) is equivalent to (2), this establishes Corollary 1.

Corollary 2. *Every formula* $\mathfrak{C}(x)$ *of* W. 2. A. *defines an ultimately periodic set* $\hat{x}\,\mathfrak{C}(x)$ *of natural numbers, i. e., there are numbers $l$ (phase) and $p$ (period) such that* $(\forall t)[\mathfrak{C}(l + t + p) \equiv \mathfrak{C}(l + t)]$.

*Every ultimately periodic set of natural numbers is definable in* W. 2. A. *by a formula* $\mathfrak{C}(x)$.

Proof. By theorem 3 we may assume that $\mathfrak{C}(x)$ is of the form $\mathfrak{C}(x)$:

$$(\exists \boldsymbol{j}) \cdot K[\boldsymbol{j}(0)] \wedge (\forall t)\, B[\boldsymbol{j}(t), \boldsymbol{j}(t')] \wedge A[\boldsymbol{j}(x)].$$

Now, let $\boldsymbol{Y}_1, \ldots, \boldsymbol{Y}_a$ be those states $\boldsymbol{Y}$ of $\boldsymbol{j}$ for which $A[\boldsymbol{Y}]$, and for $\nu = 1, \ldots, a$ let, $\mathfrak{C}_\nu(x)$ stand for $(\exists \boldsymbol{j})[K(0) \wedge (\forall t)\, B(t) \wedge \boldsymbol{j}(x) \equiv \boldsymbol{Y}_\nu]$. Then clearly,

$$\mathfrak{C}(x) \equiv \mathfrak{C}_1(x) \vee \cdots \vee \mathfrak{C}_a(x),$$

$$\mathfrak{C}_\nu(x) \equiv (\exists \boldsymbol{j})[K(0) \wedge (\forall t)_0^x\, B(t) \wedge \boldsymbol{j}(x) \equiv \boldsymbol{Y}_\nu] \wedge (\exists \boldsymbol{j})[\boldsymbol{j}(0) \equiv \boldsymbol{Y}_\nu \wedge (\forall t)\, B(t)]$$

hold in W. 2. A. Therefore, if $\boldsymbol{Y}_1, \ldots, \boldsymbol{Y}_b$ $(b \leqq a)$ are those states among $\boldsymbol{Y}_1, \ldots, \boldsymbol{Y}_a$ for which $(\exists \boldsymbol{j})[\boldsymbol{j}(0) \equiv \boldsymbol{Y} \wedge (\forall t)\, B(t)]$, then

$$\mathfrak{C}(x) \equiv (\exists \boldsymbol{j}) \cdot K(0) \wedge (\forall t)_0^x\, B(t) \wedge [\boldsymbol{j}(x) \equiv \boldsymbol{Y}_1 \vee \cdots \vee \boldsymbol{j}(x) \equiv \boldsymbol{Y}_b]$$

holds in W. 2. A.

---

[1]) As R. L. Vaught remarks, this result can be obtained from a theorem of A. Ehrenfeucht; see Robinson [14].

Next let $k$ = number of states of $\boldsymbol{j}$, and let $\boldsymbol{Y}_1, \ldots, \boldsymbol{Y}_b, \ldots, \boldsymbol{Y}_k$ be the states of $\boldsymbol{j}$, and let $r_1, \ldots, r_k$ be defined by the recursion

$$(1) \quad \begin{cases} r_\nu(0) \equiv K[\boldsymbol{Y}_\nu], & \nu = 1, \ldots, k \\ r_\nu(t') \equiv r_1(t)\, B[\boldsymbol{Y}_1, \boldsymbol{Y}_\nu] \vee \cdots \vee r_k(t)\, B[\boldsymbol{Y}_k, \boldsymbol{Y}_\nu]. & \nu = 1, \ldots, k \end{cases}$$

One then easily shows that

$$(\exists \boldsymbol{j})\,[K(0) \wedge (\forall\, t)_0^x\, B(t) \wedge [\boldsymbol{j}(x) \equiv \boldsymbol{Y}_1 \vee \cdots \vee \boldsymbol{j}(x) \equiv \boldsymbol{Y}_b]]$$

holds if and only if $[r_1(x) \vee \cdots \vee r_b(x)]$. Consequently, $\hat{x}\mathfrak{C}(x) = \hat{x}[r_1(x) \vee \cdots \vee r_b(x)]$. Now $\boldsymbol{r}$ has $2^k$ states, therefore a repetition must occur in $\boldsymbol{r}(0), \ldots, \boldsymbol{r}(2^k)$, say $\boldsymbol{r}(l) \equiv \boldsymbol{r}(l+p)$ whereby $l + p \leqq 2^k$ and $0 < p$. By (1) it then follows that $\boldsymbol{r}(l+t) \equiv \boldsymbol{r}(l+p+t)$, for all $t$, so that $\hat{x}[r_1(x) \vee \cdots \vee r_b(x)] = x\,\mathfrak{C}(x)$ is ultimately periodic with phase $l$ and period $p$.

The second part of Corollary 2 is best shown by first obtaining definitions in W. 2. A. of the relations $x = y$, $x < y$, $x \equiv y \pmod{p}$, for fixed $p$.

Note also that the selection of $\boldsymbol{Y}_1, \ldots, \boldsymbol{Y}_b$ from $\boldsymbol{Y}_1, \ldots, \boldsymbol{Y}_a$ in the proof of Corollary 2 can be effectively made (by Corollary 1). As a result one can effectively find the phase and period of the set $x\,\mathfrak{C}(x)$.

By using similar methods to those employed in the proof of Corollary 2 one shows that every relation $R(x, y)$ definable in W. 2. A. must be of the form

$$R(x, y) \equiv \bigvee_{\nu=1,\ldots,a} [x \leqq y \wedge A_\nu(x) \wedge B_\nu(y-x)] \vee \bigvee_{\nu=1,\ldots,c} [y \leqq x \wedge C_\nu(y) \wedge D_\nu(x-y)]$$

whereby $A_\nu$, $B_\nu$, $C_\nu$, are ultimately periodic. In particular $y = f(x)$ is definable in W. 2. A. if and only if it is ultimately periodic, i. e., satisfies $f(l + x + p) = f(l + x) + q$, for some $l$, $p$, $q$ (compare this with ROBINSON's [14] result on $[0, x+1, 2x, \exists_0 i]$). However, the following result seems more informative.

Corollary 3. *If $R(x, y)$ well-orders a subset of natural numbers and is definable by a formula $\mathfrak{C}(x, y)$ of* W. 2. A., *then the type $\alpha$ of $R$ is less than $\omega^2$. Conversely, if $\alpha$ is an ordinal less than $\omega^2$ one can find a formula $\mathfrak{C}(x, y)$ of* W. 2. A. *such that $\hat{x}\hat{y}\mathfrak{C}(x, y)$ is an $\alpha$-well-ordering of all natural numbers.*

Proof. Suppose $R(x, y)$ is a well-ordering of natural numbers of type $\alpha$, and for $\eta \leqq \alpha$ let $A\,\eta$ be the initial segment relative to $R$ of type $\eta$. Then from a definition $\mathfrak{C}(x, y)$ of $R$ in W. 2. A. one can easily obtain definitions of the sets $A\,\eta$, for $\eta < \omega^2$; so that by Corollary 2 the sets $A\,\eta$, $\eta < \omega^2$ are ultimately periodic. Now it is easy to see that there is no strictly increasing $\omega^2$-sequence of ultimately periodic sets of natural numbers. Consequently if $R(x, y)$ is definable in W. 2. A. then its type $\alpha$ must be less than $\omega^2$. To indicate a proof of the second part of Corollary 3, let us consider the case $\alpha = \omega\, 3 + 2$. Because one can define $x \equiv y \pmod 3$ and $x < y$, and $x = y$ in W. 2. A., it is clear that one can also define a relation $R(x, y)$ which well-orders the natural numbers in sequence, 6, 9, 12, ..., 1, 4, 7, ..., 2, 5, 8, ..., 0, 3.

It seems clear that one could use Theorem 3 also to investigate the nature of sets $\hat{i}\,\mathfrak{C}(i)$ definable in W. 2. A. by formulas with free predicate variables. However, a rather concrete characterization of these sets of finite sets is given by Theorems 1 and 2, namely

Corollary 4. *The sets of n-tuples of finite sets definable in* W. 2. A. *are exactly the behaviors of special outputs of finite automata with n-ary input.*

We will terminate this section with the presentation of a first-order theory which is equivalent to W. 2. A. in a very strong sense. For this purpose note that

$$f(i) = \sum_{i(x)} (2^x)$$

defines a one-to-one mapping $f$ from all special predicates (finite sets) of natural numbers onto all natural numbers. $f(i) = y$ simply means that for $l$ = length of $i$, $i(0)\,i(1)\ldots i(l)$ is the binary expansion of $y$. This mapping $f$ induces a natural one-to-one correspondence between relations on special predicates and relations on natural numbers, let us say that $R(i_1, \ldots, i_n)$ and $S(x_1, \ldots, x_n)$ are *adjoined* to each other in case $R(i_1, \ldots, i_n) \equiv S(f(i_1), \ldots, f(i_n))$, for all special $i_1, \ldots, i_n$. Let furthermore "$Pw_2(x)$" stand for "$x$ is a power of 2".

Theorem 4. *The first-order theory* $[=, +, Pw_2]$ *is equivalent to the weak second-order theory* $[0, ', \exists_0 i]$ *in the sense that a relation* $S(x_1, \ldots, x_n)$ *on natural numbers is definable in* $[=, +, Pw_2]$ *if and only if its adjoined relation* $R(i_1, \ldots, i_n)$ *is definable in* $[0, ', \exists_0 i]$. *Moreover, from a definition* $\mathfrak{C}(x_1, \ldots, x_n)$ *of S one can effectively obtain a definition* $\mathfrak{D}(i_1, \ldots, i_n)$ *of R, and conversely.*

Proof. Let $S(i, j, s)$ be the adjoined to $x + y = z$, i. e., "$S(i, j, s)$" stands for "$f(i) + f(j) = f(s)$". Then by formalizing the procedure of adding natural numbers in binary expansion one obtains the following definition of $S$ in the weak second-order theory $[0, ', \exists_0 i]$,

$$S(i, j, s) : (\exists r) \cdot \tilde{r}(0) \wedge (\forall t)\,[r(t') \equiv [r(t)\,i(t) \vee r(t)\,j(t) \vee i(t)\,j(t)]]$$
$$\wedge\, (\forall t)\,[s(t) \equiv [r(t) \bigtriangledown i(t) \bigtriangledown j(t)]]$$

whereby "$X \bigtriangledown Y$" stands for "$X\tilde{Y} \vee Y\tilde{X}$". Furthermore, if $U(i)$ is the adjoined to $Pw_2(x)$; then $U(i) : (\exists x)\,(\forall t)\,[i(t) \equiv (t = x)]$ is a definition of $U$ in $[0, ', \exists_0 i]$. To prove Theorem 4 in one direction it therefore is sufficient to indicate a translation of formulas $\mathfrak{C}$ of $[=, +, Pw_2]$ into formulas $\mathfrak{C}^*$ of $[0, ', U, S, \exists_0 i]$ such that $\mathfrak{C}^*(i_1, \ldots, i_n)$ and $\mathfrak{C}(x_1, \ldots, x_n)$ define adjoined relations. Because every formula $\mathfrak{C}$ of $[=, +, Pw_2]$ is easily changed to an equivalent one in which no iteration of $+$ occurs, the following specifications yield the required translation.

1. $[x_\nu + x_\mu = x_\varrho]^*$ is $S(i_\nu, i_\mu, i_\varrho)$,
2. $[Pw(x_\nu)]^*$ is $U(i_\nu)$,
3. $[\mathfrak{C} \wedge \mathfrak{D}]^*$ is $\mathfrak{C}^* \wedge \mathfrak{D}^*$,
4. $[\sim\mathfrak{C}]^*$ is $\sim\mathfrak{C}^*$,
5. $[(\exists x_\nu)\,\mathfrak{C}]^*$ is $(\exists i_\nu)\,\mathfrak{C}^*$.

To prove Theorem 4 in the other direction we note that

$$E(x, y) : Pw_2(x) \wedge (\exists uv)[(y = u + x + v) \wedge (u < v) \wedge [v = 0 \vee 2x \leqq v]]$$

is a definition in $[=, +, Pw_2]$ of "$x$ is a power of 2, and $x$ occurs in the representation of $y$ as a sum of powers of 2". It therefore is sufficient to indicate a translation of formulas $\mathfrak{C}$ of $[0, ', \exists_0 i]$ into formulas $\mathfrak{C}^\circ$ of $[=, +, Pw_2, E]$ such that $\mathfrak{C}^\circ(x_1, \ldots, x_n)$ and $\mathfrak{C}(i_1, \ldots, i_n)$ define adjoined relations. If one notes that $i(x)$ means the same as $E(x, f(i))$, and that every formula $\mathfrak{C}(i_1, \ldots, i_n)$ can be changed to an equivalent one in which 0 does not occur and $'$ only occurs in parts of form $(\forall t)[j(t) \equiv i(t')]$, then it is clear that the following specifications yield such a translation. We single out the individual variable $t$ and divide the remaining ones in $x_1, x_2, \ldots; y_1, y_2, \ldots$. In $[0, ', \exists_0 i]$ we make use of $y_1, y_2, \ldots,$ and $t$ only (while it is intended that in $[=, +, Pw_2, E]$ the range of the $y$'s is restricted to $Pw_2$).

1. $[i_\nu(y_\mu)]^\circ$ is $E(y_\mu, x_\nu)$,
2. $[(\forall t)[i_\mu(t) \equiv i_\nu(t')]]^\circ$ is $[x_\nu = x_\mu + x_\mu \vee x_\nu = x_\mu + x_\mu + 1]$,
3. $[\mathfrak{C} \wedge \mathfrak{D}]^\circ$ is $\mathfrak{C}^\circ \wedge \mathfrak{D}^\circ$,
4. $[\sim\mathfrak{C}]^\circ$ is $\sim\mathfrak{C}^\circ$,
5. $[(\exists i_\nu)\mathfrak{C}]^\circ$ is $(\exists x_\nu)\mathfrak{C}^\circ$,
6. $[(\exists y_\nu)\mathfrak{C}]^\circ$ is $(\exists y_\nu)[Pw_2(y_\nu) \wedge \mathfrak{C}^\circ]$.

By Theorem 4, Corollary 1, Corollary 4 one clearly obtains,

Corollary 5. *The first-order theory $[=, +, Pw_2]$ is decidable. A relation $R(x_1, \ldots, x_n)$ is definable in this theory if and only if its adjoined $S(i_1, \ldots, i_n)$ is the behavior of a finite automaton with special output.*

One might attempt to prove Corollary 5 directly by extending PRESBURGER's [10] method for $[=, +]$. Corollary 1 would then follow by Theorem 4.

Let $Q_2(x)$ stand for "$x$ is a square". Then by PUTNAM [11], the first-order theory $[=, +, Q_2]$ is undecidable. It is interesting to compare this with Corollary 5, and one easily extends the results as follows,

Theorem 5.

*The first-order theory $[=, +, P]$ is*

a) *undecidable in case $P(x)$ is ultimately a hyper-arithmetic-progression, i. e., in case $P$ ultimately consists of the values of a polynomial of degree $\geqq 2$;*

b) *decidable in case $P(x)$ is ultimately a geometric progression.*

To better understand the jump in strength from $[=, +]$ to $[=, +, \cdot]$ it would be interesting to know whether there is a recursive predicate $P(x)$ such that $[=, +, P]$ is undecidable without admitting definitions for all recursive predicates.

## 6. Other concepts of finite automata

Clearly our concept of finite automata is very closely related to that of CHURCH [3]. What we have called the transition-recursion and output-recursion of an automaton (with output) are restricted recursions in his sense. While CHURCH allows instantaneous action (the input at time $t$ directly affects the transit and output at the same time $t$), it is just a matter of convenience that we have presented our theory in terms of automata with delayed action only; also our restriction to special outputs is inessential (see Section 7). The behavior-concept used implicitly by CHURCH is that of the operator $u = \psi(i)$ defined by the output-recursion. We hope to show elsewhere how his synthesis procedure (Case II) can be extended to condition-formulas containing predicate-quantifiers.

Recursions like our output-recursions were first used by BURKS and WRIGHT [2]. Their concept of well-formed logical net is equivalent to our finite automata, except that instantaneous action occurs in logical nets.

In the following discussion it is intended that a $n/m$-automaton (whose input has $n$ binary components) represents in coded form a $2^n/2^m$-automaton (whose input has one $2^n$-ary component. Let $\Sigma_2 = \{\wedge, \vee\}$, $\Sigma_n = \Sigma_2 \times \cdots \times \Sigma_2$ ($n$ factors). Under the conventional interpretation of propositional formulas $\boldsymbol{E}^m$ denotes an element $\sigma \in \Sigma_m$, while $\boldsymbol{H}^m[\boldsymbol{Y}^m, \boldsymbol{X}^n]$ may be interpreted to denote a function $\tau$: $\Sigma_m \times \Sigma_n \to \Sigma_m$. Thus, with an $n/m$-automaton $\boldsymbol{E}^m \frown \boldsymbol{H}^m[\boldsymbol{Y}^m, \boldsymbol{X}^n]$ is associated a system $\langle \Sigma_n, \Sigma_m, \sigma, \tau \rangle$ called its *transition system*. The elements of $\Sigma_n$ and $\Sigma_m$ are respectively called *input states* and *transit-states*, $\sigma$ is the *initial-state* and $\tau$ the *transit-function* of the automaton. An output $U[\boldsymbol{Y}^m]$ may be interpreted to denote a subset of transit-states. These remarks are intended to indicate how our notion of finite automata (with output) is related to those used by MOORE [9], MYHILL [8], RABIN and SCOTT [13], and others. While our notion is syntactic these authors take automata to be what we called transition systems (with output set). The latter definition is not correct if the mathematical concept "finite automaton" is to correspond to physical structures. Two physical systems (two syntactic systems) can be quite different structurally (syntactically) and still display the same transition-system. Also from a purely mathematical point of view it is not advisable to identify automata with their transition-structure, as it then becomes impossible to rigorously state existence of algorithms as, for example, in our Theorems 1 and 2. The point is that an algorithm is better thought of as applying to and yielding syntactic entities.

We will now indicate how our theory of behavior can be extended to automata with many inputs, each of which may have any number of binary components. For this purpose it is convenient to extend propositional calculus so as to contain $\omega$ lists, each list consisting of $\omega$ propositional variables. A notation like "$H[\boldsymbol{X}^n; \boldsymbol{Y}^k; \boldsymbol{Z}^h]$" will be used to stand for a formula of (extended) propositional calculus in which the variables $\boldsymbol{X}$, $\boldsymbol{Y}$, $\boldsymbol{Z}$ and only these may occur, and furthermore the variables $\boldsymbol{X}$ belong to a first list, which precedes a second list to which the variables $\boldsymbol{Y}$ belong, which precedes a third list to which the variables $\boldsymbol{Z}$ belong.

Definition 1'. A (*finite*) $\boldsymbol{n}_1; \ldots; \boldsymbol{n}_k/\boldsymbol{m}$*-automaton* with *inputs* $\boldsymbol{X}_1^{\boldsymbol{n}_1}, \ldots, \boldsymbol{X}_k^{\boldsymbol{n}_k}$ and transit $\boldsymbol{Y}^m$ is a $2m$-tuple $\boldsymbol{E}^m{}^\frown \boldsymbol{H}^m[\boldsymbol{Y}^m; \boldsymbol{X}_1^{\boldsymbol{n}_1}; \ldots; \boldsymbol{X}_k^{\boldsymbol{n}_k}]$ of propositional formulas (no variables occur in $\boldsymbol{E}^m$).

A (*binary*) *output* of such an automaton is a propositional formula $U[\boldsymbol{Y}^m]$, where $\boldsymbol{Y}^m$ is the transit of the automaton.

It is clear how one extends the notion of transition (output)-recursion, introduced in Section 3, to $\boldsymbol{n}_1; \ldots; \boldsymbol{n}_k/\boldsymbol{m}$-automata (with output). Also for the extension of the concept of special output we refer to Definition 2 in Section 3.

Definition 3'. The *behavior* $\beta(\boldsymbol{E}^m, \boldsymbol{H}^m, U)$ of a $\boldsymbol{n}_1; \ldots; \boldsymbol{n}_k/\boldsymbol{m}$-automaton with special output is the $k$-ary relation which holds for the special $\boldsymbol{n}_1$-predicate $\boldsymbol{i}_1, \ldots, \boldsymbol{n}_k$-predicate $\boldsymbol{i}_k$, just in case $(\exists x)(\forall t)_x^\infty U[\boldsymbol{r}(t)]$, whereby $\boldsymbol{r}^m = \zeta(\boldsymbol{E}, \boldsymbol{H}, \boldsymbol{i}_1; \ldots; \boldsymbol{i}_k)$ is obtained by the transition-recursion of $\boldsymbol{E}^\frown \boldsymbol{H}$.

To obtain an extension of the analysis and synthesis Theorems 1 and 2, we extend W. 2. A. to contain $\omega$ lists, each list containing $\omega$ predicate variables. A notation like "$\boldsymbol{C}(\boldsymbol{i}; \boldsymbol{j})$" will be used to denote a formula of (extended) W. 2. A. in which the variables $\boldsymbol{i}, \boldsymbol{j}$ and only these may have free occurrences, and such that the variables $\boldsymbol{i}$ all belong to the same list which precedes a list containing all the variables $\boldsymbol{j}$. A formula $\mathfrak{C}(\boldsymbol{i}_1^{\boldsymbol{n}_2}; \boldsymbol{i}_2^{\boldsymbol{n}_3}; \boldsymbol{i}_3^{\boldsymbol{n}_1})$ of W. 2. A. defines a ternary relation $\hat{\boldsymbol{i}}_1; \hat{\boldsymbol{i}}_2; \hat{\boldsymbol{i}}_3 \mathfrak{C}(\boldsymbol{i}_1; \boldsymbol{i}_2; \boldsymbol{i}_3)$ on special $\boldsymbol{n}_1$-predicates, $\boldsymbol{n}_2$-predicates, $\boldsymbol{n}_3$-predicates.

Theorem 1' (Synthesis). *To every formula* $\mathfrak{C}(\boldsymbol{i}_1^{\boldsymbol{n}_1}; \ldots; \boldsymbol{i}_k^{\boldsymbol{n}_k})$ *of (extended)* W. 2. A. *one can construct a* $\boldsymbol{n}_1; \ldots; \boldsymbol{n}_k/\boldsymbol{m}$*-automaton* $\boldsymbol{E}^m{}^\frown \boldsymbol{H}^m$ *with special output* $U$ *such that the behavior* $\beta(\boldsymbol{E}, \boldsymbol{H}, U)$ *is the k-ary relation* $\hat{\boldsymbol{i}}_1; \ldots; \hat{\boldsymbol{i}}_k \mathfrak{C}(\boldsymbol{i}_1; \ldots; \boldsymbol{i}_k)$.

Theorem 2' (Analysis). *To every* $\boldsymbol{n}_1; \ldots; \boldsymbol{n}_k/\boldsymbol{m}$*-automaton* $\boldsymbol{E}^m{}^\frown \boldsymbol{H}^m$ *with special output* $U$ *one can construct a formula* $\mathfrak{C}(\boldsymbol{i}_1^{\boldsymbol{n}_1}; \ldots; \boldsymbol{i}_k^{\boldsymbol{n}_k})$ *of* W. 2. A. *such that the k-ary relation* $\hat{\boldsymbol{i}}_1; \ldots; \boldsymbol{i}_k \mathfrak{C}(\boldsymbol{i}_1; \ldots; \hat{\boldsymbol{i}}_k)$ *is the behavior* $\beta(\boldsymbol{E}, \boldsymbol{H}, U)$.

To prove Theorem 1' one at first simply disregards the grouping of the $q = n_1 \cdot n_2 \cdots n_k$ variables in the formula $\mathfrak{C}(\boldsymbol{i}_1^{\boldsymbol{n}_1}; \ldots; \boldsymbol{i}_k^{\boldsymbol{n}_k})$ and constructs the $\boldsymbol{q}/\boldsymbol{m}$-automaton with special output according to Theorem 1. By properly grouping the $q$ components of the input of this automaton one obtains a $\boldsymbol{n}_1; \ldots; \boldsymbol{n}_k/\boldsymbol{m}$-automaton with special output as required in Theorem 1'. Similarly one proves Theorem 2'.

It thus appears that the theory of behavior for automata with many inputs does not present any additional difficulties, in case one works with special predicates rather than finite strings of input states. The reason for this is that a $k$-tuple $(\boldsymbol{i}_1^{\boldsymbol{n}_1}; \ldots; \boldsymbol{i}_k^{\boldsymbol{n}_k})$ of special $\boldsymbol{n}$-predicates may be reinterpreted as one special $\boldsymbol{n}_1 \cdot \boldsymbol{n}_2 \cdots \boldsymbol{n}_k$-predicate. In contrast a $k$-tuple $(\mathfrak{x}_1; \ldots; \mathfrak{x}_k)$ of finite strings of input states can not readily be reinterpreted as one finite string of states. This seems to be the reason that no theory of behavior of automata with many inputs occurs in the literature (see section 7).

Our concept of finite $\boldsymbol{n}_1; \ldots; \boldsymbol{n}_k/\boldsymbol{m}$-automaton still is somewhat specialized in that the components of each input and the transit are binary (i. e., can take values in $\Sigma_2 = \{\wedge, \vee\}$ only). Because we allow many components to occur in an input

6*

(transit) there is of course the possibility of indirectly dealing with automata whose sets of input states (transit states) have any finite cardinality, by way of binary coding. However, to obtain an entirely satisfactory theory of $\boldsymbol{n}_1; \ldots; \boldsymbol{n}_k/\boldsymbol{m}$-automata (for any numbers $n_1, \ldots, n_k, m$) one would have to replace $\Sigma_2$ by $\Sigma_2, \Sigma_3, \ldots$ ($\Sigma_k$ being a set of $k$ elements), and one would have to set up a calculus $PC_\omega$ to replace propositional-calculus ($PC_2$), such that $PC_\omega$ contains variables $X_k$ ranging over $\Sigma_k$ and provides names $H_k[X_{k_1}, \ldots, X_{k_a}]$ for all functions $\tau: \Sigma_{n_1} \times \cdots \times \Sigma_{k_a} \to \Sigma_{k_a}$. To generalize the results obtained in Section 4 one would extend W. 2. A. to contain variables $\boldsymbol{i}_k$ ranging over monadic functions from natural numbers to $\Sigma_k$ which ultimately take the fixed value $\Lambda_k \in \Sigma_k$. It seems clear that this program could be carried out, and moreover the form of the presentation in Section 4 would remain essentially unchanged. We did not present our results in this general version because no calculus $PC_\omega$ is available in the literature.

## 7. Finite strings versus special predicates; regular relations

According to our definition the behavior $\beta$ of an $\boldsymbol{n}/\boldsymbol{m}$-automaton with special output is a set of special $\boldsymbol{n}$-predicates (infinite strings of input-states which are ultimately $\boldsymbol{\Lambda}^n$). Simplifying KLEENE's [6] theory of regularity, MYHILL [8], RABIN and SCOTT [13], and COPI, ELGOT and WRIGHT [4] work instead with the set $\beta_{sg}$ consisting of all finite strings of input states which turn on the output. In this approach the restriction to special outputs is not needed, so that it may seem at first that our concept of behavior is too specialized. However we will see in this section that on the contrary our theory of behavior is more general in that it yields a rather natural extension of the concept of regular set of finite strings to regular relations on finite strings.

We begin by observing that there is a one-to-one mapping $f$ of all finite strings of elements of $\Sigma_n$ onto all special $\boldsymbol{n}$-predicates $\boldsymbol{i}^n$ which have the property $\boldsymbol{i}(l) \equiv \mathbf{V}^n$, if $l$ is the length of $\boldsymbol{i}$. Namely,

$$\boldsymbol{i} = f(\boldsymbol{Y}_0{}^\frown \cdots {}^\frown \boldsymbol{Y}_h) \cdot\equiv\cdot (\forall t)_{t<h} [\boldsymbol{i}(t) \equiv \boldsymbol{Y}_t] \wedge [\boldsymbol{i}(h') \equiv \mathbf{V}] \wedge (\forall t)_{t>h'} [\boldsymbol{i}(t) \equiv \boldsymbol{\Lambda}].$$

In particular the empty string corresponds to the predicate $\boldsymbol{i}(0) \equiv \mathbf{V}$, $\boldsymbol{i}(t') \equiv \boldsymbol{\Lambda}$. To simplify the presentation we will identify finite strings with the corresponding $\boldsymbol{n}$-predicate, i. e.,

(**) *A finite $\boldsymbol{n}$-string of length $l$* is a special $\boldsymbol{n}$-predicate $\boldsymbol{i}^n$ of length $l + 1$ which takes the value $\boldsymbol{i}(l') \equiv \mathbf{V}^n$.

The definition of finite string behavior which occurs in the literature can, in our notation, be formulated thus,

Definition 6. The sg-*behavior* of a $\boldsymbol{n}/\boldsymbol{m}$-automaton $\boldsymbol{E}^\frown \boldsymbol{H}$ with arbitrary output $U$ is the set $\beta_{sg}(\boldsymbol{I}, \boldsymbol{H}, U)$ consisting of all $\boldsymbol{n}$-strings $\boldsymbol{i}$ such that $U[\boldsymbol{r}(l+1)]$, if $l$ is the length of the string $\boldsymbol{i}$ and $\boldsymbol{r} = \zeta(\boldsymbol{I}, \boldsymbol{H}, \boldsymbol{i})$ is the $\boldsymbol{m}$-predicate determined by the transition-recursion of $\boldsymbol{E}^\frown \boldsymbol{H}$.

Theorem 6. *A set $\beta$ of $\boldsymbol{n}$-strings is the* sg-*behavior $\beta_{sg}(\boldsymbol{E}, \boldsymbol{H}, U)$ of some $\boldsymbol{n}/\boldsymbol{m}$-automaton with arbitrary output if and only if it is the behavior $\beta(\boldsymbol{I}, \boldsymbol{G}, W)$ of some $\boldsymbol{n}/\boldsymbol{q}$-automaton with special output.*

Proof. Suppose first that $\beta = \beta_{sg}(\boldsymbol{E}, \boldsymbol{H}, U)$, whereby $U$ is an arbitrary output of the $\boldsymbol{n}/\boldsymbol{m}$-automaton $\boldsymbol{E}^\frown\boldsymbol{H}$. Let the $\boldsymbol{n}/\boldsymbol{m}^\frown 2$-automaton $\boldsymbol{I}^\frown\boldsymbol{G}$ and its output $W$ be defined by,

$$I_\nu : E_\nu \quad G_\nu[\boldsymbol{Y}^m{}^\frown Z;\ \boldsymbol{X}^n] : H_\nu[\boldsymbol{Y}, \boldsymbol{X}), \qquad \nu = 1, \ldots, m$$

$$I_{m+1} : \boldsymbol{\wedge} \quad G_{m+1}[\boldsymbol{Y}^m{}^\frown Z;\ \boldsymbol{X}^n] : [\boldsymbol{X} \equiv \boldsymbol{\vee}]\, U[\boldsymbol{Y}] \vee [\boldsymbol{X} \equiv \boldsymbol{\wedge}]\, Z,$$

$$W[\boldsymbol{Y}^m{}^\frown Z] : Z.$$

Then clearly $G_{m+1}[\boldsymbol{Y}^\frown Z';\ \boldsymbol{\wedge}] \equiv Z$, and therefore $W[\boldsymbol{G}[\boldsymbol{Y}^\frown Z;\ \boldsymbol{\wedge}]] \equiv W[\boldsymbol{Y}^\frown Z]$, which by definition 2 means that $W$ is special output of $\boldsymbol{I}^\frown\boldsymbol{G}$. Furthermore note that the transition-recursions of the two automata are,

$$(1)\quad \begin{cases} \boldsymbol{r}(0) \equiv \boldsymbol{E}, \\ \boldsymbol{r}(t') \equiv \boldsymbol{H}[\boldsymbol{r}(t),\ \boldsymbol{i}(t)], \end{cases} \qquad (2)\quad \begin{cases} \boldsymbol{r}(0) \equiv \boldsymbol{E},\ s(0) \equiv \boldsymbol{\wedge}, \\ \boldsymbol{r}(t') \equiv \boldsymbol{H}[\boldsymbol{r}(t),\ \boldsymbol{i}(t)], \\ \boldsymbol{r}(t') \equiv [\boldsymbol{i}(t) \equiv \boldsymbol{\vee}]\, U[\boldsymbol{r}(t)] \vee [\boldsymbol{i}(t) \equiv \boldsymbol{\wedge}]\, s(t) \end{cases}$$

Now let $\boldsymbol{i}$ be any special predicate of length $l$. If $\boldsymbol{i} \in \beta(\boldsymbol{I}, \boldsymbol{G}, W)$ then by (*) of Section 3 and (2), $W[\boldsymbol{r}(l+1)^\frown s(l+1)]$, i. e., $s(l+1)$. Therefore by (2), $[\boldsymbol{i}(l) \equiv \boldsymbol{\vee}]\, U[\boldsymbol{r}(l)] \vee [\boldsymbol{i}(l) \equiv \boldsymbol{\wedge}]\, s(l)$. Because $l$ is the length of $\boldsymbol{i}$, $\boldsymbol{i}(l) \equiv \boldsymbol{\wedge}$ is not the case, so that $[\boldsymbol{i}(l) \equiv \boldsymbol{\vee}]\, U[\boldsymbol{r}(l)]$. Consequently by definition 6, $\boldsymbol{i} \in \beta_{sg}(\boldsymbol{E}, \boldsymbol{H}, U)$. These steps are reversible, so that also $\boldsymbol{i} \in \beta_{sg}(\boldsymbol{E}, \boldsymbol{H}, U)$ implies $\boldsymbol{i} \in \beta(\boldsymbol{I}, \boldsymbol{G}, W)$. Thus we have constructed an automaton $\boldsymbol{I}^\frown\boldsymbol{G}$ with special output $W$ such that also $\beta(\boldsymbol{I}, \boldsymbol{G}, W) = \beta$. This proves Theorem 6 in one direction.

Suppose next that $\beta = \beta(\boldsymbol{I}, \boldsymbol{H}, W)$, whereby $\boldsymbol{I}^\frown\boldsymbol{H}$ is an $\boldsymbol{n}/\boldsymbol{m}$-automaton with special output. Define $U[\boldsymbol{Y}^m] : W[\boldsymbol{H}[\boldsymbol{Y}^m, \boldsymbol{\vee}^n]]$. Noting that $\beta$ consists of finite $\boldsymbol{n}$-strings only, and using Definition 6 and (*) of Section 3, one easily shows that $\beta_{sg}(\boldsymbol{I}, \boldsymbol{H}, U) = \beta(\boldsymbol{I}, \boldsymbol{H}, W) = \beta$. This establishes Theorem 6 in the other direction.

Theorem 6 shows that the restriction to special outputs in our analysis and synthesis Theorems 1 and 2 (Section 4) is not a limitation in generality, but rather a natural feature which appears when one works with special predicates instead of finite strings. In fact it now is easy to obtain synthesis and analysis theorems in terms of sg-behavior and without restriction on the outputs. For this purpose we remark that the set of all finite $\boldsymbol{n}$-strings can be defined by a formula $Sg_n(\boldsymbol{i}^n)$ of W. 2. A.,

$$Sg_n^1(\boldsymbol{i}^n): \qquad (\exists j) \cdot j(0) \wedge (\forall t)\,[j(t') \supset j(t)]\,[j(t)\,\tilde{j}(t') \supset \boldsymbol{i}(t) \equiv \boldsymbol{\vee}]\,[\tilde{j}(t) \supset \boldsymbol{i}(t) \equiv \boldsymbol{\wedge}]$$

and we define a formula $\mathfrak{C}(\boldsymbol{i}^n)$ to be an sg-*formula* in case $(\forall \boldsymbol{i})\, \mathfrak{C}(\boldsymbol{i}) \supset Sg_n(\boldsymbol{i})$ holds in W. 2. A. (by Corollary 1 one can effectively decide whether a formula of W. 2. A. is a sg-formula).

Theorem 7.

Analysis. *To every $\boldsymbol{n}/\boldsymbol{m}$-automaton $\boldsymbol{E}^\frown\boldsymbol{H}$ with arbitrary output $U$ one can construct a* sg-*formula* $\mathfrak{C}(\boldsymbol{i}^n)$ *of* W. 2. A. *such that* $\mathfrak{C}(\boldsymbol{i})$ *defines the* sg-*behavior* $\beta_{sg}(\boldsymbol{E}^\frown\boldsymbol{H}, U)$.

Synthesis. *To every* sg-*formula* $\mathfrak{C}(\boldsymbol{i}^n)$ *of* W. 2. A. *one can construct a $\boldsymbol{n}/\boldsymbol{m}$-automaton $\boldsymbol{E}^\frown\boldsymbol{H}$ with output $U$ such that* $\beta_{sg}(\boldsymbol{E}, \boldsymbol{H}, U) = \hat{\boldsymbol{i}}\, \mathfrak{C}(\boldsymbol{i})$.

Proof. By Theorems 1, 2, and 6. Note that we actually established an effective version of Theorem 6.

In short form Theorem 7 states that the sg-behaviors of $n/m$-automata with arbitrary (binary) output are exactly the sets of finite $n$-strings definable (by sg-formulas) in W. 2. A. It has been shown that the sg-behaviors are the regular sets of finite strings (references at the beginning of this section). Thus,

Theorem 8. *For any set $\beta$ of finite $n$-strings the following are equivalent conditions,*

(1) *$\beta$ is the* sg-*behavior of an $n/m$-automaton with arbitrary output.*

(2) *$\beta$ is the behavior of an $n/m$-automaton with special output.*

(3) *$\beta$ is definable in* W. 2. A. *(by a* sg-*formula).*

(4) *$\beta$ is regular.*

In the literature sg-behavior has been studied only for automata with one input. Consequently the concept of regularity has been limited to sets. On the other hand we have seen in Section 6 that our theory of behavior can easily be extended to automata with many inputs. On the basis of Theorems 1′, 2′ (Section 6) and Theorem 8 the following definition suggests itself as a natural extension of the concept of regularity.

A formula $\mathfrak{C}(i_1^{n_1}; \ldots; i_k^{n_k})$ of (extended) W. 2. A. is called an sg-formula if $(\forall i_1 \cdots i_k) \cdot \mathfrak{C}(i_1; \ldots; i_k) \supset Sg_{n_1}(i_1) \cdots Sg_{n_k}(i_k)$ holds in W. 2. A.

Definition 7. A relation $R(i_1^{n_1}; \ldots; i_k^{n_k})$ on special $n_\nu$-predicates (finite $n_\nu$-strings) is *regular* if it is definable by a formula (sg-formula) $\mathfrak{C}(i_1; \ldots; i_k)$ of extended W. 2. A.

Theorems 1′ and 2′ can now be restated in the following short form,

Theorem 8′. *A relation $R(i_1^{n_1}; \ldots; i_k^{n_k})$ on special $n_\nu$-predicates (finite $n_\nu$-strings) is the behavior $\beta(E, H, U)$ of a $n_1; \ldots; n_k/m$-automaton with special output, if and only if $R$ is regular.*

In case one prefers to work with finite strings one could extend Definition 6 to a definition of sg-behavior of many-input-automata with arbitrary output, so as to obtain a corresponding extension of Theorem 6, and Theorem 7.

A binary relation on strings which naturally arises in connection with a $n/m$-automaton $E^m{}^\frown H^m$ is its *equi-response relation* $i^n \backsim j^n$ (mod $E^\frown H$). Two finite strings $i, j$ of input-states are in this relation if they produce the same transit-state, i. e., if $r(l') \equiv s(h')$ in case $r = \zeta(E, H, i)$, $s = \zeta(E, H, j)$, $l$ = length of string $i$, and $h$ = length of string $j$. It is easy to construct a $n; n/q$-automaton with special output whose behavior is the relation $i^n \backsim j^n$ (mod $E^\frown H$). Therefore, by Theorem 8′, the equi-response relation of any $n/m$-automaton is a regular relation on finite $n$-strings. In $k$-ary number theory (i. e., the theory of finite strings on the alphabet $1_1, \ldots, 1_k$) the equi-response relations module $k/q$-automata take the place of conventional congruence relations in ordinary (1-ary) number theory. Correspondingly the concept of regularity is the natural analogue in $k$-ary number theory to ultimate periodicity in ordinary number theory.

Another example of a regular relation on special predicates is the adjoined $S(i, j, s)$ under binary expansion (see Section 5) to the relation $x + y = z$. Moreover Theorem 4 may be restated as follows. A relation $R(x_1, \ldots, x_k)$ is definable in the first-order theory $[=, +, Pw_2]$ if and only if it is the adjoined under binary expansion to a regular relation $S(i_1, \ldots, i_k)$ on special predicates.

In case one prefers to deal with finite strings the following modified concept of binary expansion is appropriate. If $a_0, \ldots, a_h$ are either 1 or 2 then $a_0 a_1 \cdots a_h$ represents $a_0 2^0 + a_1 2^1 + \cdots + a_h 2^h$. In particular the empty string represents the number 0. It is easy to see that this yields a one-to-one mapping $x = g(\mathfrak{x})$ from all finite 2-strings to all natural numbers. Furthermore one can modify the proof of Theorem 4 to obtain; a relation $R(x_1, \ldots, x_n)$ is definable in $[=, +, Pw_2]$ if and only if its adjoined $S(\mathfrak{x}_1, \ldots, \mathfrak{x}_n)$ under modified binary expansion is a regular relation on finite 2-strings.

Consequently the following two conditions on a relation $R(x_1, \ldots, x_n)$ on natural numbers are equivalent.

(1) *The adjoined $S(i_1, \ldots, i_k)$ of $R(x_1, \ldots, x_k)$ under binary expansion is a regular relation on special predicates.*

(2) *The adjoined $S_{sg}(\mathfrak{x}_1, \ldots, \mathfrak{x}_k)$ of $R(x_1, \ldots, x_k)$ under modified binary expansion is a regular relation on finite 2-strings.*

Let us call such a relation $R$ a *2-regular relation on natural numbers* so that,

Theorem 9. *The relations definable in the first-order theory $[=, +, Pw_2]$ are exactly the 2-regular relations on natural numbers.*

For appropriate definition of $k$-regular relation (via adjoined under (modified) $k$-ary expansion) Theorem 9 holds if "2-regular" and "$Pw_2$" are replaced by "$k$-regular" and "$Pw_k$" ($Pw_k(x)$ means $x$ is a power of $k$). It thus appears that the first-order theory $[=, +, Pw_k]$ is suitable for a study of regularity on the alphabet $1_1, \ldots, 1_k$ consisting of $k$ letters. In this connection the following remarks are of interest.

1. Let $N_k$ denote the set of all finite strings on the alphabet $1_1, \ldots, 1_k$, and let the function $g_k : N_k \to N_1$ ($N_1$ = set of natural numbers) be defined by

$$g_k(0) = 0,$$

$$g_k(1_{\nu_0} 1_{\nu_1} \cdots 1_{\nu_h}) = \nu_0 k^0 + \nu_1 k^1 + \cdots + \nu_h k^h.$$

I.e., $g_k$ is that one-to-one enumeration of $N_k$ in which every string of length $h$ precedes every string of length $l > h$, while strings of equal length are enumerated in lexicographic order reading from right to left. The inverse $g_k^{-1}$ yields the modified $k$-ary expansion of natural numbers. Let $+^k$ denote the adjoined to $+$, i. e.,

$$\mathfrak{x} +^k \mathfrak{y} = g_k^{-1}(g_k(\mathfrak{x}) + g_k(\mathfrak{y})), \quad \text{for} \quad \mathfrak{x}, \mathfrak{y} \in N_k,$$

and let $Pw^k$ denote the adjoined to $Pw_k$, i. e.,

$$Pw^k(\mathfrak{x}) \equiv Pw_k(g_k(\mathfrak{x})), \quad \text{for} \quad \mathfrak{x} \in N_k.$$

We remark that it is easy to set up an algorithm for calculating $\mathfrak{x} +^k \mathfrak{y}$ (similar to the conventional algorithm for addition in ordinary expansion), and that $Pw^k(\mathfrak{x})$ simply means $\mathfrak{x}$ is $1_1$ or $\mathfrak{x}$ is $1_k$ followed by zero or more occurrences of $1_{k-1}$. Clearly $\langle N_1, +, Pw_k\rangle$ is isomorphic to $\langle N_k, +^k, Pw^k\rangle$, so that $[=, +, Pw_k]$ may be interpreted as first-order theory of either of these systems. Theorem 9 as well as the following remarks ought to be interpreted accordingly, and we will often speak of a relation when we actually mean its adjoined under $g$.

**2.** Let $x <_k y$ and $x >_k y$ denote respectively the initial segment relation and terminal segment relation on $N_k$ (or their adjoined). It is easy to construct a $\boldsymbol{k}$; $\boldsymbol{k/m}$-automaton with special output whose behavior is the relation $x <_k y$, so that by Theorem 9 (2 replaced by $k$), this relation is definable in $[=, +, Pw_k]$. Thus, by Theorem 4,

*The first-order theory of* $\langle N_k, <_k\rangle$ *is decidable.*

In contrast we will show elsewhere that the first-order theory of $\langle N_k, <_k, >_k\rangle$ is not decidable. Therefore, $x >_k y$ is not definable in $[=, +, Pw_k]$.

**3.** Let $x{}^\frown{}^k y$ denote the operation of concatenation on $N_k$ (or its adjoined on $N$). It is easy to see that

$$x{}^\frown{}^k y = x + k^{lg_k(x)} \cdot y$$

whereby $lg_k(x)$ is the length of the string $x$, i. e.,

$$lg_k(x) = a \cdot\equiv\cdot k^a \leqq x + 1 < k^{a+1}.$$

However, there is no definition of $x{}^\frown{}^k y$ in $[=, +, Pw_k]$, because this theory is decidable while it was shown by QUINE [12] that $[=, {}^\frown{}^k]$ is undecidable. It can be shown that $[=, {}^\frown{}^k]$ and $[=, +, \cdot]$ are equivalent in the sense that the primitives of one of these theories are definable in the other.

**4.** We have shown (see also RABIN and SCOTT [13]) that the relations $x \backsim y$ which are equi-response relations modulo automata with one $k$-ary input are exactly those right-congruence relations of the operation $x{}^\frown{}^k y$ which divide $N_k$ into finitely many classes. This emphasizes the analogy (remark following Theorem 8′) of these relations to ordinary congruence relations of $x + y = y{}^\frown{}^1 y$. We remark that while $x{}^\frown{}^k y$ is not definable in $[=, +, Pw_k]$, the equi-response relations are definable.

**5.** The study of congruences is fundamental in elementary number theory. It would be of great importance for a better understanding of regularity (behavior of finite automata) to develop analogously a theory of those (right, left) congruence relations of $x{}^\frown{}^k y$ whose partition is finite.

**6.** Every ultimately periodic set of natural numbers is definable in $[=, +]$. By Theorem 9 (extension to $k$) it follows,

*Every ultimately periodic* (i. e., 1-*regular*) *set of natural numbers is* $k$-*regular, for every* $k$.

**7.** Let $A_{n,k}(x)$ stand for $lg_k(x) \equiv 0 \pmod n$. It then is easy to see that $A_{n,k}(x)$ is $k$-regular, and therefore definable in $[=, +, Pw_k]$. Because

$$Pw_{k^n}(x) \cdot\equiv\cdot [Pw_k(x) \wedge A_{n,k}(x)] \vee [x = 1]$$

it follows that $Pw_{k^n}$ is definable in $[=, +, Pw_k]$. Conversely, for $n \geqq 2$, $Pw_k$ is definable in $[=, +, Pw_{k^n}]$, because

$$Pw_k(x) \cdot\equiv\cdot Pw_{k^n}(x) \vee Pw_{k^n}(k\,x) \vee \cdots \vee Pw_{k^n}(k^{n-1}x).$$

Thus, for any $k$ and $n \neq 0$ the theories $[=, +, Pw_k]$ and $[=, +, Pw_{k^n}]$ are equivalent with respect to definability. By theorem 9 (2 replaced by $k$) it follows,

*For any $k$ and $n \neq 0$, the $k^n$-regular relations on natural numbers are exactly the $k$-regular relations.*

8. Suppose that $n \geqq k$ and $Pw_n$ is $k$-regular. Let $A(x)$ stand for

$$(\exists t)\,[k^x \leqq n^t + 1 < k^{x+1}].$$

Then $A$ is the set of lengths of elements of $N_k$ which belong to the $k$-adjoined to $Pw_n$. Therefore, because $Pw_n$ is $k$-regular it follows that $A$ must be ultimately periodic. Let $h$ = phase, $p$ = period of $A$, and let $q$ be the number of elements of $A$ occurring in an interval $(x, x+p-1)$, where $x > h$. Take $a > h$ such that $A(a)$. Then also $A(a+p)$, $A(a+2p), \ldots, A(a+\nu p), \ldots$, i. e., there are $b, b_1, b_2, \ldots$ such that

$$(1) \qquad k^a \leqq n^b + 1 < k^{a+1}, \quad k^{a+\nu p} \leqq n^{b_\nu} + 1 < k^{a+\nu p+1}, \; \nu = 1, 2, 3, \ldots.$$

Note that there are exactly $q \leqq p$ members $a, a_2, \ldots, a_q$ of $A$ occurring in the interval $(a, a+p-1)$, and that, because $n \geqq k$, no $n^{t_1} \neq n^{t_2}$ can have the same length $x(n^{t_1}+1$ and $n^{t_2}+1$ cannot belong to the same interval $(k^x, k^{x+1}-1))$. It follows that $a, a_2, \ldots, a_q, a+p$ are in order the lengths of $n^b, n^{b+1}, \ldots, n^{b+q-1}$, $n^{b+q} = n^{b_1}$. Therefore $b_1 = b + q$, and similarly one shows $b_\nu = b + \nu q$, $\nu = 1$, $2, 3, \ldots$. I follows by (1) that

$$\frac{k^{a+\nu p}-1}{k^{a+1}} < n^{\nu q} < \frac{k^{a+\nu p+1}}{k^a - 1}. \qquad \nu = 1, 2, 3, \ldots$$

Therefore,

$$\sqrt[\nu]{k^{\nu p} - k^{-a}} \cdot \sqrt[\nu]{k^{-1}} < n^q < \sqrt[\nu]{\frac{k^{a+1}}{k^a - 1}}\, k^p. \qquad \nu = 1, 2, 3, \ldots$$

Letting $\nu$ approach infinity this clearly yields $n^q = k^p$.

Thus we have shown that for $n \leqq k$, if $Pw_n$ is $k$-regular then $n^q = k^p$ for some $p \neq 0$ and $q \neq 0$. This assertion extends to all $n \geqq 2$; because if $n \geqq 2$ there is an a such that $n^a \geqq k$, and by 7 if $Pw_n$ is $k$-regular then also $Pw_{n^a}$ is $k$-regular, so that there are $p \neq 0$ and $q \neq 0$ such that $n^{aq} = k^p$. Consequently, if one defines

(a) $x \approx y : (\exists\, uv)\,[u \neq 0 \wedge v \neq 0 \wedge x^u = y^v]$ one obtains by 7.

(b) If $n \geqq 2$ then $Pw_n$ is $k$-regular (definable in $[=, +, Pw_k]$) if and only if $n \approx k$.

(c) The $k$-regular relations are exactly the $n$-regular relations if and only if $n \approx k$.

(d) If $n \geqq 2$ and $Pw_n$ is $k$-regular (definable in $[=, +, Pw_k]$) then also $Pw_k$ is $n$-regular (definable in $[=, +, Pw_n]$), and $k$-regularity and $n$-regularity have the same meaning.

These facts can, of course, be stated in terms of sequences. For example, the one-to-one mapping $g_n^{-1} g_k$ of $N_k$ takes regular relations on $N_k$ onto regular relations on $N_n$ just in case $n \approx k$.

To better understand the equivalence relation $\approx$, let us say that a number $x$ is *simple* if $[x = 0 \vee x = 1 \vee \sim (\exists uv)\,(v \neq 1 \wedge x = u^v)]$, and let $spl(y)$ stand for that simple number $x$ of which $y$ is a power. Then by elementary number theoretic considerations (prime-decomposition) one shows,

(e) $x \approx u \cdot\!\equiv\!\cdot\, spl(x) = spl(y)$.

Furthermore, the simple numbers constitute a system of representatives for $\approx$, and the equivalence class of a simple number $x$ is $\{x^n \mid n \neq 0\}$.

Because of (b), (c), (d) it follows that it is sufficient to deal with $k$-regularity for simple $k$. The theory $[=, +, Pw_n]$ is equivalent to $[=, +, Pw_k]$, for $k = spl(n)$. (Note that a number $k$ is simple just in case g.c.d. $(a_1, \ldots, a_n) = 1$ if $k = p_1^{a_1} \cdots p_n^{a_n}$ is the prime-decomposition of $k$.)

By (b) it follows that $Pw_2$ is not 3-regular. This seems to mean that there is no finite automation with binary input and ternary output which translates binary expansions of numbers into the corresponding ternary expansions.

## 8. Remarks on feedback

The essentially new element added to pure switching circuits to obtain sequential circuits is feedback. The presence of feedback in our finite automata is reflected in the recursiveness of the definition of the input to transit operation $\boldsymbol{r} = \psi(\boldsymbol{i})$ of a finite automaton. In contrast, for a pure switching circuit this operation is introduced by explicit definition. The relationship of feedback to recursions was first clearly realized by BURKS and WRIGHT [2] and KLEENE [6], and is emphasized by CHURCH's contribution to automata theory [3].

A formalism, sufficiently rich to provide notations for all behaviors of finite automata, must contain an operator corresponding to feedback. In KLEENE's [6] formalism of regular expressions, this operator is $*$. COPI, ELGOT, and WRIGHT [4] and MYHILL [8], in their modified version of KLEENE's theory use a related monadic operator $*$. In W. 2. A. $= [0, ', \exists_0 i]$ it is *the finite-set-quantifier* $\exists_0 i$ *which corresponds to feedback*. This is best shown by indicating how the primitive operators $\cup$, $\cdot$, and $*$ on regular sets can be obtained in W. 2. A.

For this purpose to every $\mathrm{Sg}_n$-formula $\mathfrak{C}(\boldsymbol{i})$ of W. 2. A. we define the formulas $\mathfrak{C}_x^y(\boldsymbol{i})$ and $\mathfrak{C}_x^\omega(\boldsymbol{i})$ as follows. First obtain a normal form of $\mathfrak{C}(\boldsymbol{i})$, say

$\mathfrak{C}(\boldsymbol{i})$: $(\exists \boldsymbol{j}) \cdot K[\boldsymbol{j}(0)] \wedge (\forall t)\, B[\boldsymbol{i}(t), \boldsymbol{j}(t), \boldsymbol{j}(t')]$.

Then define

$\mathfrak{C}_x^y(\boldsymbol{i})$: $(\exists \boldsymbol{j}) \cdot K[\boldsymbol{j}(x)] \wedge (\forall t)_x^y\, B(t) \wedge B[\mathbf{V}, \boldsymbol{j}(y), \boldsymbol{j}(y')] \wedge (\forall t)_{y'}^\omega\, B[\mathbf{\Lambda}, \boldsymbol{j}(t), \boldsymbol{j}(t')]$,

$\mathfrak{C}_x^\omega(\boldsymbol{i})$: $(\exists \boldsymbol{j}) \cdot K[\boldsymbol{j}(x)] \wedge (\forall t)_x^\omega\, B[\boldsymbol{i}(t), \boldsymbol{j}(t), \boldsymbol{j}(t')]$.

Next for any $\mathrm{Sg}_n$-formulas $\mathfrak{C}(\boldsymbol{i})$, $\mathfrak{D}(\boldsymbol{i})$ of W. 2. A. we define $\mathfrak{C} \cup \mathfrak{D}$, $\mathfrak{C} \cdot \mathfrak{D}$, $\mathfrak{C}^*$ by

$[\mathfrak{C} \cup \mathfrak{D}](\boldsymbol{i})\colon\ \mathfrak{C}(\boldsymbol{i}) \vee \mathfrak{D}(\boldsymbol{i})$,

$[\mathfrak{C} \cdot \mathfrak{D}](\boldsymbol{i})\colon\ (\exists y) \cdot \mathfrak{C}_0^y(\boldsymbol{i}) \wedge \mathfrak{D}_y^\omega(\boldsymbol{i})$,

$[\mathfrak{C}^*](\boldsymbol{i})\colon\ (\exists s) \cdot (\forall x \acute{y})[s(x)s(y)(\forall t)_{x'}^{y}\, \tilde{s}(t) \supset \mathfrak{C}_x^y(\boldsymbol{i})] \wedge (\forall x)[\dot{s}(x)\,(\forall t)_{x'}^{\omega}\, \tilde{s}(t) \supset \mathfrak{C}_x^\omega(\boldsymbol{i})]$.

It is easy to see that these operators on $\mathrm{Sg}_n$-formulas of W. 2. A. define the operations $\cup$, $\cdot$, $*$ on sets of finite $\boldsymbol{n}$-strings. Note that $\cdot$ corresponds to individual quantification, while the feedback-operation $*$ is reflected as a finite-set quantification.

## 9. An unsolved problem

In essence our main result says that in $[0, ', \exists_0 i]$ one can define only very simple recursive operators $u = \psi(\boldsymbol{i})$, namely, input to output operators of finite automata. In contrast $[0, ', 2x, \exists_0 i]$ contains definitions for all recursive operators $u = \psi(\boldsymbol{i})$ (see ROBINSON [14]).

Problem. *Is the weak second-order theory* $[0, 2x+1, 2x+2, \exists_0 i]$ *decidable? Which are the recursive operators definable in this theory?*

These questions are of interest for automata theory, because via the enumeration of finite strings on two letters $1_1, 1_2$ (see § 7, 1). The theory $[0, 2x+1, 2x+2, \exists_0 i]$ is equal to the theory $[0, 1_1{}^\frown x, 1_2{}^\frown x, \exists_0 i]$ on the set $N_2$ of finite strings. Let $\exists_{00} i$ denote quantification over all finite subsets of $N_k$ which in addition are chains with respect to the terminal segment relation on $N_k$. Then by a straightforward extension of the methods used in this paper one obtains:

Theorem 10. *The "very-weak" second-order theory* $[0, 1_1{}^\frown x, \ldots, 1_k{}^\frown x, \exists_{00} i]$ *is decidable. All regular sets, and only regular sets, on the alphabet* $1_1, \ldots, 1_k$ *can be defined by formulas* $\mathfrak{C}(x)$ *of this theory.*

## Bibliography


[1] BEHMANN, HEINRICH, Beiträge zur Algebra der Logik, insbesondere zum Entscheidungsproblem. Math. Ann. **86**, 163—229 (1922).

[2] BURKS, A. W., and WRIGHT, J. B., Theory of Logical Nets. Proc. IRE, **41**, 1357—1365 (1953).

[3] CHURCH, ALONZO, Application of Recursive Arithmetic to the Problem of Circuit Synthesis, Proceedings of the Cornell Logic Conference, Cornell University, 1957. Also see „Application of Recursive Arithmetic in the Theory of Computing and Automata", Notes for a summer course, The University of Michigan, 1959.

[4] COPI, I. M., ELGOT, C. C., and WRIGHT, J. B., Realization of Events by Logical Nets. Journal Ass. Comp. Mach. **5**, pp. 181—196 (1958).

[5] GÖDEL, KURT, On undecidable propositions of formal mathematical systems. Notes by S. C. KLEENE and BARKLEY ROSSER on lectures at the Institute for Advanced Study, 1934. Mimeographed, Princeton, N. J., 30 pp.

[6] KLEENE, S. C., Representation of Events in Nerve Nets and Finite Automata. Automata Studies, Princeton University Press, 1956, 3—41.

[7] MEDVEDEV, I. T., On a Class of Events Representable in a Finite Automaton. MIT Lincoln Laboratory Group Report, 34—73, translated from the Russian by J. SCHORR-KON, June 30, 1958.

[8] MYHILL, JOHN, Finite Automata and Representation of Events. WADC Report TR 57—624, Fundamental Concepts in the Theory of Systems, October, 1957, 112—137.

[9] MOORE, E. F., Gedanken-Experiments on Sequential Machines. Automata Studies, Princeton, 1956, 129—153.

[10] PRESBURGER, M., Über die Vollständigkeit eines gewissen Systems der Arithmetik ganzer Zahlen, in welchem die Addition als einzige Operation hervortritt. Comptes-rendus du I Congres des Mathematiciens des Pays Slavs, Warsaw, 1930, 92—101, 395.

[11] PUTNAM, H., Decidability and Essential Undecidability. Journ. Symb. Log. **22**, 39—54. (1957)

[12] QUINE, W. V., Mathematical Logic. Harvard Univ. Press, Cambridge, 1947.

[13] RABIN, M., and SCOTT, D., Finite Automata and Their Decision Problems. IBM Journal, April, 1959, 114—125.

[14] ROBINSON, R. M., Restricted Set-Theoretical Definitions in Arithmetic. Proc. Am. Math. Soc. **9**, 238—242 (1958).

[15] SKOLEM, THORALF, Untersuchungen über die Axiome des Klassenkalküls und über Produktions- und Summationsprobleme, welche gewisse Klassen von Aussagen betreffen. Skrifter utgit av Videnskapsselskapet i Kristiania, I. Mat.-nat. kl. 1919, No. 3, 37 pp.

(Eingegangen am 14. September 1959)

SECTION

I

MATHEMATICAL LOGIC

# *Symposium on Decision Problems*

## ON A DECISION METHOD IN RESTRICTED SECOND ORDER ARITHMETIC

J. RICHARD BÜCHI

*University of Michigan, Ann Arbor, Michigan, U.S.A.*

Let SC be the interpreted formalism which makes use of individual variables t, x, y, z, . . . ranging over natural numbers, monadic predicate variables q( ), r( ), s( ), i( ), . . . ranging over arbitrary sets of natural numbers, the individual symbol 0 standing for zero, the function symbol ′ denoting the successor function, propositional connectives, and quantifiers for both types of variables. Thus SC is a fraction of the restricted second order theory of natural numbers, or of the first order theory of real numbers. In fact, if predicates on natural numbers are interpreted as binary expansions of real numbers, it is easy to see that SC is equivalent to the first order theory of [Re, +, Pw, Nn], whereby Re, Pw, Nn are, respectively, the sets of non-negative reals, integral powers of 2, and natural numbers.

The purpose of this paper is to obtain a rather complete understanding of definability in SC, and to outline an effective method for deciding truth

This work was done under a grant from the National Science Foundation to the Logic of Computers Group, and with additional assistance through contracts with the Office of Naval Research, Office of Ordnance Research, and the Army Signal Corps.

of sentences in SC. This answers a problem of A. Tarski's, which was discussed by R. M. Robinson [10].

A *congruence of finite rank* on words is a congruence with finite partition of concatenation; a *multi-periodic set* of words is a union of congruence classes of a congruence of finite rank. These concepts are intimately related to that of a finite automaton (Kleene [5], Myhill [6], Copi, Elgot, and Wright [3]), and turn out to be the key to an investigation of SC. Our results concerning SC may therefore be viewed as an application of the theory of finite automata to logic. In turn, SC arises quite naturally as a condition-language (Church [2]) on finite automata or sequential circuits, and "*sequential calculus*" is an appropriate name for SC. The significance of the decision method for SC is that it provides a method for deciding whether or not the input (**i**)-to-output (**u**) transformation of a proposed circuit $A(\mathbf{i}, \mathbf{r}, \mathbf{u})$ satisfies a condition $C(\mathbf{i}, \mathbf{u})$ stated in SC.

An important role in our theory of SC is played by Lemma 1, the *Sequential Lemma*. This is a combinatorial statement about $\omega$-sequences, which may well be of importance elsewhere. It turns out to be a simple consequence of Ramsey's Theorem A. The usefulness of the "Unendlichkeitslemma" of König (also known as the "fan-theorem" in its intuitionistic version) in related problems of automata theory was first observed by Jesse B. Wright. Because of its affinity to König's lemma the present application of Ramsey's theorem was suggested. The author wishes to thank Dr. Wright for his continued assistance in the work presented here.

## 1. Notations

**i** denotes an n-tuple of predicate variables. Expressions like $\mathrm{A}[\mathbf{i}(0)]$, $\mathrm{B}[\mathbf{i}(t), \mathbf{i}(t')]$ denote propositional formulas in the indicated constituents. $\Sigma_n$, $\Pi_n$, denote the classes of formulas of SC of the following type:

$$\begin{aligned}
\Sigma_1 &: (\exists\mathbf{r}) \cdot \mathrm{A}[\mathbf{r}(0)] \wedge (\forall t)\mathrm{B}[\mathbf{i}(t), \mathbf{r}(t), \mathbf{r}(t')] \wedge (\exists t)\, \mathrm{C}[\mathbf{r}(t)],\\
\Pi_1 &: (\forall\mathbf{r}) \cdot \mathrm{A}[\mathbf{r}(0)] \vee (\exists t)\, \mathrm{B}[\mathbf{i}(t), \mathbf{r}(t), \mathbf{r}(t')] \vee (\forall t)\, \mathrm{C}[\mathbf{r}(t)],\\
\Sigma_{n+1} &: (\exists\mathbf{r}) \cdot F(\mathbf{i}, \mathbf{r}), \text{ whereby } F \in \Pi_n,\\
\Pi_{n+1} &: (\forall\mathbf{r}) \cdot F(\mathbf{i}, \mathbf{r}), \text{ whereby } F \in \Sigma_n.
\end{aligned}$$

The quantifiers $(\exists t)_x^y A(t)$ for $(\exists t)\,[x \leqq t < y \wedge A(t)]$, $(\forall t)_x^y A(t)$ for $(\forall t)[x \leqq t < y \supset A(t)]$, $(\exists^\omega t)A(t)$ for $(\forall x)(\exists t)\,[x < t \wedge A(t)]$, $(\forall_\omega t)A(t)$ for $(\exists x)(\forall t)[x < t \supset A(t)]$, $(\exists j)_\omega A(j)$ for $(\exists j)\,[(\exists^\omega t)j(t) \wedge A(j)]$ can be defined in SC. The classes $\Sigma_1^\omega$ and $\Pi_1^\omega$ of formulas are defined as follows:

$$\begin{aligned}
\Sigma_1^\omega &: (\exists\mathbf{r}) \cdot \mathrm{A}[\mathbf{r}(0)] \wedge (\forall t)\, B[\mathbf{i}(t), \mathbf{r}(t), \mathbf{r}(t')] \wedge (\exists^\omega t)\, \mathrm{C}[\mathbf{r}(t)],\\
\Pi_1^\omega &: (\forall\mathbf{r}) \cdot \mathrm{A}[\mathbf{r}(0)] \vee (\exists t)\, \mathrm{B}[\mathbf{i}(t), \mathbf{r}(t), \mathbf{r}(t')] \vee (\forall_\omega t)\, \mathrm{C}[\mathbf{r}(t)].
\end{aligned}$$

Also the following classes of formulas will play an essential role:

$$\begin{aligned}
\Sigma^0 &: (\exists\mathbf{r}) \cdot \mathrm{A}[\mathbf{r}(x)] \wedge (\forall t)_x^y\, \mathrm{B}[\mathbf{i}(t), \mathbf{r}(t), \mathbf{r}(t')] \wedge \mathrm{C}[\mathbf{r}(y)],\\
\Pi^0 &: (\forall\mathbf{r}) \cdot \mathrm{A}[\mathbf{r}(x)] \vee (\exists t)_x^y\, \mathrm{B}[\mathbf{i}(t), \mathbf{r}(t), \mathbf{r}(t')] \vee \mathrm{C}[\mathbf{r}(y)].
\end{aligned}$$

These may be called *regular formulas*.

Let **i** be a k-tuple of predicates. The $2^k$ *states of* **i** are the k-tuples of truth-values. **i** may be viewed as an infinite sequence $\mathbf{i}(0)\mathbf{i}(1)\mathbf{i}(2)\ldots$ of states. The variables *u, v, w, ...* will be used for *words* (i.e., finite sequences) of states; *uv* denotes the result of juxtaposing the words *u* and *v*. A *congruence* is an equivalence relation $u \backsim v$ on words such that $u \backsim v$ implies $uw \backsim vw$ and $wu \backsim wv$; it is of *finite rank* n in case there are n equivalence classes. A set $\mathscr{S}$ of words is *multi-periodic* if $\mathscr{S} = \mathscr{E}_1 \cup \ldots \cup \mathscr{E}_m$, whereby $\mathscr{E}_1, \ldots \mathscr{E}_m$ are some of the congruence classes of a congruence of finite rank.

Note that the value of a regular formula $R(\mathbf{i}, x, y)$ depends only on the word $\mathbf{i}(x)\,\mathbf{i}(x+1)\ldots\mathbf{i}(y-1)$. If $\mathscr{R}$ is the set of all words $\mathbf{i}(0)\,\mathbf{i}(1)\ldots\mathbf{i}(h)$ such that $R(\mathbf{i}, 0, h+1)$, then the formula $R(\mathbf{i}, x, y)$ is said to *determine* the set $\mathscr{R}$ of words. The symbol "$\Sigma^0$" will be used also to denote the class of all sets $\mathscr{R}$ of words determined by formulas $R$ in $\Sigma^0$. Similarly, the symbol "$\Sigma_1^\omega$" is used also to denote the class of all sets $\hat{\mathbf{i}}F(\mathbf{i})$ defined by formulas $F(\mathbf{i})$ in $\Sigma_1^\omega$. Corresponding remarks hold for $\Pi^0$, $\Pi_1^\omega$, $\Sigma_1$, $\Pi_1$.

## 2. The Sequential Lemma

The working of the decision-method for SC is based on induction and a rather more sophisticated property of infinity, namely Theorem A of Ramsey [9]. Essential parts of this theorem can actually be formulated in SC, in the form of a surprising assertion about the division of infinite sequences into consecutive finite parts.

LEMMA 1. *Let* **i** *be any k-tuple of predicates, and let* $\mathscr{E}_0, \ldots, \mathscr{E}_n$ *be a partition of all words on states of* **i** *into finitely many classes. Then there exists a division* $\mathbf{i}(0)\,\mathbf{i}(1)\ldots\mathbf{i}(x_1-1),\ \mathbf{i}(x_1)\,\mathbf{i}(x_1+1)\ldots\mathbf{i}(x_2-1),\ \mathbf{i}(x_2)\,\mathbf{i}(x_2+1)\ldots\mathbf{i}(x_3-1),\ \ldots$ *of* **i** *such that all words* $\mathbf{i}(x_p)\,\mathbf{i}(x_p+1)\ldots\mathbf{i}(x_q-1)$ *belong to one and the same of the classes* $\mathscr{E}_0, \ldots, \mathscr{E}_n$.

PROOF. Assume $\mathbf{i}, \mathscr{E}_0, \ldots, \mathscr{E}_n$ are as supposed in Lemma 1. For $0 \leqq c \leqq n$ let $P_c$ consist of all $\{y_1, y_2\}$ such that $y_1 < y_2$ and $\mathbf{i}(y_1)\,\mathbf{i}(y_1+1)\ldots\mathbf{i}(y_2-1) \in \mathscr{E}_c$. Then $P_0, \ldots, P_n$ clearly is a partition of all 2-element sets of natural numbers. By Ramsey's Theorem A it follows that there is an infinite sequence $x_1 < x_2 < x_3 < \ldots$ and a $0 \leqq c \leqq n$ such that $\{x_p, x_q\} \in P_c$ for all $x_p < x_q$. By definition of $P_c$ this yields the conclusion of Lemma 1.

## 3. Finite Automata, Multi-periodic Sets, and $\Sigma^0$-formulas

The following methods and results are borrowed from the theory of finite automata, and play a very essential role in the study of SC. The reader is referred to Büchi [1], where some of the details are carried out in similar form, and where further references to the mathematical literature on finite automata are given. The basic result is

LEMMA 2. *The following are equivalent conditions on a set* $\mathscr{R}$ *of words:*

*(a)* $\mathscr{R}$ *is determined by a formula* $F(\mathbf{i}, x, y)$ *of* $\Sigma^0$.

*(b) There is a "finite automata recursion"* $\mathbf{r}(0) \equiv \mathbf{I}$, $\mathbf{r}(t') \equiv \mathbf{J}[\mathbf{i}(t), \mathbf{r}(t)]$, *and an "output"* $U[\mathbf{r}(t)]$ *such that a word* $\mathbf{i}(0)\,\mathbf{i}(1)\ldots\mathbf{i}(x-1)$ *belongs to* $\mathscr{R}$ *just in case the recursion yields an* $\mathbf{r}(x)$ *such that* $U[\mathbf{r}(x)]$ *holds.*

The implication a $\to$ b is shown in essence by Myhill's [6] "subset-construction"; nearly in the present form the details are in [1] Lemma 7. The implication b $\to$ a is trivial, $(\exists \mathbf{r}) \cdot \mathbf{r}(x) \equiv \mathbf{I} \wedge (\forall t)_x^y[\mathbf{r}(t') \equiv \mathbf{J}(t)] \wedge U[\mathbf{r}(y)]$ clearly determines $\mathscr{R}$.

The set $\mathscr{R}$ defined by (b) is sometimes called the *behavior* of $[\mathbf{I}, \mathbf{J}, U]$. In this terminology Lemma 2 says that $\Sigma^0$ is exactly the class of all behaviors of finite automata with outputs. It is easy to see that the class of behaviors is closed under disjunction and complementation. For example, if $\mathscr{R}$ is the behavior of $[\mathbf{I}, \mathbf{J}, U]$, then clearly $\tilde{\mathscr{R}}$ is the behavior of $[\mathbf{I}, \mathbf{J}, \tilde{U}]$. Therefore by Lemma 2,

LEMMA 3. *If the formulas* $R(\mathbf{i}, x, y)$, $S(\mathbf{i}, x, y)$ *determine* $\Sigma^0$*-sets of words, then so do the formulas* $R(\mathbf{i}, x, y) \wedge S(\mathbf{i}, x, y)$, $R(\mathbf{i}, x, y) \vee S(\mathbf{i}, x, y)$, *and* $\sim R(\mathbf{i}, x, y)$.

Suppose next that $R(\mathbf{i}, x, y)$ is the $\Sigma^0$-formula $(\exists \mathbf{r}) \cdot K(x) \wedge (\forall t)_x^y H(t) \wedge L(y)$. Then clearly $(\exists z)_x^y R(\mathbf{i}, x, y)$ is equivalent to $(\exists s\mathbf{r}) \cdot s(x) \wedge (\forall t)_x^y[(s(t') \supset s(t)) \wedge (s(t)\tilde{s}(t') \supset K(t)) \wedge (\tilde{s}(t) \supset H(t))] \wedge [\tilde{s}(y)L(y)]$, which is again in $\Sigma^0$. Therefore by Lemma 3,

LEMMA 4. *If the formula* $R(\mathbf{i}, x, y)$ *determines a* $\Sigma^0$*-set of words, then so do the formulas* $(\exists z)_x^y R(\mathbf{i}, x, y)$ *and* $(\forall z)_x^y R(\mathbf{i}, x, y)$.

Suppose again that $R(\mathbf{i}, x, y)$ is a $\Sigma^0$-formula. By Lemma 2 it follows that

$$(1) \qquad R(\mathbf{i}, 0, y) .\equiv. (\exists \mathbf{r}) \cdot \mathbf{r}(0) \equiv \mathbf{I} \wedge (\forall t)[r(t') \equiv \mathbf{J}(t)] \wedge U(y)$$

for properly chosen matrices $\mathbf{I}$, $\mathbf{J}[\mathbf{i}(t), \mathbf{r}(t)]$, and $U[\mathbf{r}(y)]$. It clearly follows that

$$(2) \qquad R(\mathbf{i}, 0, y) .\equiv. (\forall \mathbf{r}) \cdot [\mathbf{r}(0) \equiv \mathbf{I} \wedge (\forall t)[\mathbf{r}(t') \equiv \mathbf{J}(t)]] \supset U(y).$$

By (1) it follows that $(\exists y)\ R(\mathbf{i}, 0, y)$ is equivalent to a $\Sigma_1$-formula. By (2) it follows that $(\forall y)\ R(\mathbf{i}, 0, y)$ is equivalent to $(\forall \mathbf{r}).\ [\mathbf{r}(0) \equiv \mathbf{I} \wedge (\forall t)[r(t') \equiv \mathbf{J}(t)]] \supset (\forall t)\ U(t)$, and therefore to $(\exists \mathbf{r}).\ \mathbf{r}(0) \equiv \mathbf{I} \wedge (\forall t)[r(t') \equiv \mathbf{J}(t)]U(t)$ Thus,

LEMMA 5. *If* $R(\mathbf{i}, x, y)$ *determines a* $\Sigma^0$*-set of words, then* $(\exists t)\ R(\mathbf{i}, 0, t)$ *is equivalent to a* $\Sigma_1$*-formula, and* $(\forall t)\ R(\mathbf{i}, 0, t)$ *is equivalent to a* $\Sigma_1$*-formula of type* $(\exists \mathbf{r}).\ K(0) \wedge (\forall t)H(t)$.

As a consequence of Lemma 2, one thus obtains a rather clear picture of definability by $\Sigma^0$-formulas. However, a further characterization of behaviors is needed for the study of SC.

LEMMA 6. *A set* $\mathscr{R}$ *of words satisfies (b) of Lemma 2 (i.e., is the behavior of some finite automaton with output) if and only if it is multi-periodic.*

This fact has been observed by several authors; a proof can be found in Rabin and Scott [8]. By Lemma 2 it follows that $\Sigma^0$ consists exactly of the multi-periodic sets of words.

## 4. Definability by $\Sigma_1^\omega$-formulas

We will now show that also the class $\Sigma_1^\omega$, just like $\Sigma^0$, is closed under Boolean operations.

LEMMA 7. *If $F_1(\mathbf{i})$ and $F_2(\mathbf{i})$ are $\Sigma_1^\omega$-formulas, then also $F_1(\mathbf{i}) \vee F_2(\mathbf{i})$ is equivalent to a $\Sigma_1^\omega$-formula.*

PROOF. For $c = 1, 2$ let $F_c(\mathbf{i})$ be the formula

$$(\exists \mathbf{r}_c) \cdot K_c(0) \wedge (\forall t) H_c(t) \wedge (\exists^\omega t) L_c(t).$$

Then clearly the $\Sigma_1^\omega$-formula

$$(\exists s\, \mathbf{r}_1\mathbf{r}_2) \cdot [s(0)\, K_1(0) \vee \bar{s}(0)\, K_2(0)] \wedge (\forall t)[[s(t) \equiv s(t')] \wedge [s(t)H_1(t) \vee \bar{s}(t)H_2(t)]] \wedge (\exists^\omega t)[s(t)L_1(t) \vee \bar{s}(t)L_2(t)]$$

is equivalent to $F_1(\mathbf{i}) \vee F_2(\mathbf{i})$.

That $\Sigma_1^\omega$ also is closed under conjunction follows by

LEMMA 8. *A formula of form $(\exists \mathbf{r}) \cdot K(0) \wedge (\forall t) H(t) \wedge (\exists^\omega t) L_1(t) \wedge (\exists^\omega t) L_2(t)$ is equivalent to a $\Sigma_1^\omega$-formula.*

PROOF. If the predicate $s(t)$ is defined from $p_1(t)$ and $p_2(t)$ by the recursion $s(0) \equiv F$, $s(t') \equiv [\bar{s}(t)\, p_1(t) \vee s(t)\, \bar{p}_2(t)]$, then it is easy to see that $[(\exists^\omega t)\, p_1(t) \wedge (\exists^\omega t)\, p_2(t)] \equiv (\exists^\omega t)\, s(t)$. Using this device with $p_1$ and $p_2$ corresponding to $L_1$ and $L_2$, one obtains a $\Sigma_1^\omega$-formula as required in Lemma 8.

Using all previous lemmas, one can now establish the closure of $\Sigma_1^\omega$ under complementation.

LEMMA 9. *To every formula $A(\mathbf{i})$ in $\Sigma_1^\omega$ one can obtain a formula $B(\mathbf{i})$ in $\Sigma_1^\omega$ equivalent to $\sim A(\mathbf{i})$.*

PROOF. Suppose $A(\mathbf{i})$ is in $\Sigma_1^\omega$, say

$$(1) \qquad A(\mathbf{i}) : (\exists \mathbf{r}) \cdot K[\mathbf{r}(0)] \wedge (\forall t)\, H[\mathbf{i}(t), \mathbf{r}(t), \mathbf{r}(t')] \wedge (\exists^\omega t)\, L[\mathbf{r}(t)].$$

If V, W are states of $\mathbf{r}$ and if $x = X_0X_1 \ldots X_h$ is a word of states of $\mathbf{i}$, then define

$$[V,x,W]_1 : \bigvee_{U_1 \ldots U_h} \cdot H[X_0,V,U_1] \wedge H[X_1,U_1,U_2] \wedge H[X_2,U_2,U_3] \wedge \ldots \wedge H[X_h,U_h,W],$$

$$[V, x, W]_2 : \bigvee_{U_1 \ldots U_h} \cdot H[X_0,V,U] \wedge \ldots \wedge H[X_h,U_h,W] \wedge [L[U_1] \vee \ldots \vee L[U_h]].$$

(Read $[\ ]_1$ as "there is an H-transition from V by $x$ to W", and $[\ ]_2$ as "there is an H-transition through L from V by $x$ to W".) Next define the binary relation $\backsim$ on words of states of $\mathbf{i}$:

$$x \backsim y : \bigwedge_{VW} ([V,x,W]_1 \equiv [V,y,W]_1) \wedge \bigwedge_{VW} ([V,x,W]_2 \equiv [V,y,W]_2).$$

If m is the number of states of $\mathbf{r}$, then clearly $\backsim$ is the intersection of $m^2 + m^2$ dichotomies. Therefore, $\backsim$ is an equivalence relation of finite rank $a \leqq 2^{2m^2}$. Furthermore, using the definitions of $[\ ]_1$ and $[\ ]_2$, one obtains, $\backsim$ is a congruence relation on words. By Lemmas 2 and 6 it therefore follows that one can find formulas $E_1(\mathbf{i}, x, y), \ldots, E_a(\mathbf{i}, x, y)$ such that (2) $E_1, \ldots, E_a$ are $\Sigma^0$-formulas, and (3) $E_1, \ldots E_a$ determine the congruence classes of $\backsim$.

Next one applies Lemma 1 to the partition $E_1, \ldots, E_a$. It follows that for any $\mathbf{i}$

$$(4)\quad (\exists s)_\omega (\forall y)(\forall x)_0^y [s(x)s(y) \supset E_1(\mathbf{i}, x, y)] \vee \ldots \vee (\exists s)_\omega (\forall y)(\forall x)_0^y [s(x)s(y) \supset E_a(\mathbf{i}, x, y)].$$

If one defines for $1 \leqq c, d \leqq a$,

$$F_{c,d}(i) : (\exists s)_\omega \cdot (\exists x)[s(x) \wedge E_c(\mathbf{i}, 0, x)] \wedge (\forall y)(\forall x)_0^y [s(x)s(y) \supset E_d(\mathbf{i}, x, y)],$$

then clearly each disjunct of (4) is equivalent to a disjunction of $F_{c,d}$'s. Therefore,

$$(5)\qquad \bigvee_{1 \leqq c,d \leqq a} F_{c,d}(\mathbf{i})$$

holds for all $\mathbf{i}$.

Suppose now that $F_{c,d}(\mathbf{i}) \wedge F_{c,d}(\mathbf{j})$. Then, by definition of $F_{c,d}$ and by (3) there are $x_1 < x_2 < x_3 < \ldots$ and $y_1 < y_2 < y_3 \ldots$ such that

$$\mathbf{i}(0) \ldots \mathbf{i}(x_1 - 1) \backsim \mathbf{j}(0) \ldots \mathbf{j}(y_1 - 1),$$
$$\mathbf{i}(x_p) \ldots \mathbf{i}(x_{p+1} - 1) \backsim \mathbf{j}(y_p) \ldots \mathbf{j}(y_{p+1} - 1), \qquad p = 1, 2, 3 \ldots.$$

By definition of $\backsim$ and (1) it therefore follows that $A(\mathbf{i}) \equiv A(\mathbf{j})$. Thus if $F_{c,d}(\mathbf{i}) \wedge F_{c,d}(\mathbf{j})$, then $A(\mathbf{i}) \equiv A(\mathbf{j})$. Or restating this result,

$$(6)\qquad (\forall \mathbf{i})[F_{c,d}(\mathbf{i}) \supset A(\mathbf{i})] \vee (\forall \mathbf{i})[F_{c,d}(\mathbf{i}) \supset \sim A(\mathbf{i})], \qquad \text{for } 1 \leqq c,d \leqq a.$$

If now one defines the set $\Phi$ of pairs $(c, d)$ by

$$(7)\qquad \Phi(c,d) \equiv \sim (\exists \mathbf{j})[A(\mathbf{j}) \wedge F_{c,d}(\mathbf{j})], \qquad \text{for } 1 \leqq c,d \leqq a,$$

then it follows by (5) and (6) that,

$$(8)\qquad \sim A(\mathbf{i}) \equiv \bigvee_{\Phi(c,d)} F_{c,d}(\mathbf{i}).$$

By (2), definition of $F_{c,d}$, and Lemmas 3, 4, 5 it follows that $F_{c,d}$ is of form

$$F_{c,d}(\mathbf{i}) \equiv (\exists s\mathbf{pq}) \cdot I(0) \wedge (\forall t)\, J(t) \wedge (\exists t)\, M(t) \wedge (\exists^\omega t)\, s(t)$$

for some matrices $I[\mathbf{p}(0), \mathbf{q}(0)]$, $J[\mathbf{i}(t), s(t), \mathbf{q}(t), \mathbf{p}(t), \mathbf{q}(t'), \mathbf{p}(t')]$, $M[\mathbf{q}(t)]$. Note that $(\exists t)\, M(t) \wedge (\exists^\omega t)\, s(t)$ may be replaced by $(\exists^\omega t)[(\exists x)_0^t M(x) \wedge s(t)]$. Furthermore, $(\exists x)_0^t M(x)$ may be replaced by $r(t)$, if $r(0) \equiv F$, $r(t') \equiv [r(t) \vee M(t)]$ are conjoined to $I(0)$ and $J(t)$, respectively, and $(\exists r)$ is added to the prefix. Therefore, each $F_{c,d}(\mathbf{i})$ is equivalent to a $\Sigma_1^\omega$-formula. By (8) and Lemma 7 it follows that $\sim A(\mathbf{i})$ is equivalent to a $\Sigma_1^\omega$-formula, which concludes the proof of Lemma 9.

Note that by definition $F_{c,d} = E_c E_d E_d E_d \ldots$, and by (5) and (6) the set $A$ is the finite union of all $F_{c,d}$ such that $\sim \Phi(c, d)$. Furthermore, the sets of words $E_c$ are all multi-periodic. Thus our proof also yields

LEMMA 10. *Let $\mathscr{A}$ be a set of ω-sequences of states definable by a $\Sigma_1^\omega$-formula. Then $\mathscr{A} = \mathscr{S}_1 \cup \ldots \cup \mathscr{S}_m$, whereby each $\mathscr{S}_k$ is of form $\mathscr{C}\mathscr{D}\mathscr{D}\mathscr{D}\ldots$, for multi-periodic sets $\mathscr{C}$, $\mathscr{D}$ of words.*

This provides a rather clear understanding of $\Sigma_1^\omega$-definability, because multi-periodic sets of words have been investigated in automata theory.

## 5. Definability in SC

The following lemma may be proved by methods similar to those in [1] Lemma 1.

LEMMA 11. *To every formula $A(\mathbf{i})$ in* SC *one can obtain an equivalent formula $B(\mathbf{i})$ belonging to some $\Sigma_n$.*

Furthermore a $\Sigma_1$-formula can be transformed to an equivalent $\Sigma_1^\omega$-formula (see end of proof of Lemma 9). Repeated application of Lemma 9 now clearly yields an equivalent $\Sigma_1^\omega$-formula to every $\Sigma_n$-formula. Thus we obtain

THEOREM 1. *To every formula $A(\mathbf{i})$ of* SC *there is an equivalent formula $B(\mathbf{i})$ in $\Sigma_1^\omega$.*

We add the following remarks:

1. Because of Lemma 10 this theorem provides a clear understanding of which relations $A(i_1, \ldots, i_n)$ on predicates are definable in SC.
2. It is easy to see that $\Sigma_1^\omega$- and $\Sigma_2$-formulas define the same relations. Therefore, the hierarchy $\Sigma_n$, $\Pi_n$ collapses at $n = 2$. This result cannot be improved much; the set $\mathfrak{U}$ consisting of all infinite i's is definable by a $\Sigma_2$-(a $\Sigma_1^\omega$)formula, but not by a $\Pi_2$-formula.
3. Using Theorem 1, one easily shows that also formulas $A(\mathbf{i}, x_1, \ldots, x_m)$ of SC, containing free individual variables, have a normal form, namely,

$$(\exists \mathbf{r})\cdot K[\mathbf{r}(0)] \wedge (\forall t) H[\mathbf{i}(t), \mathbf{r}(t), \mathbf{r}(t')] \wedge (\exists^\omega t) L[\mathbf{r}(t)] \wedge U_1[\mathbf{r}(x_1)] \wedge \ldots \wedge U[\mathbf{r}(x_m)].$$

This yields rather complete information on definability in SC. For example,

4. A conjecture of Robinson [10]: A relation $\mathscr{R}(x_1, \ldots, x_m)$ on natural numbers is definable in SC if and only if it is definable in $SC_{fin}$, which is like SC except that the variables i, j, r, ... range over finite sets of natural numbers. This follows by remark 3 and methods similar to those in the proof of Lemma 12, Section 6. Similarly, one shows a relation $\mathscr{R}(i_1, \ldots, i_m)$ on finite sets of natural numbers is definable in SC if and only if it is definable in $SC_{fin}$. For a complete discussion of definability in $SC_{fin}$ see Büchi [1].
5. Theorem 1 holds in a stronger version: there is an algorithm which to any formula $A(\mathbf{i})$ in SC yields an equivalent formula $B(\mathbf{i})$ in $\Sigma_1^\omega$. See next section.

## 6. A Decision Method for SC

To obtain a method for deciding truth of sentences in SC we need a further lemma, whose proof again is typical for automata theory:

LEMMA 12. *There is an effective method for deciding truth of sentences A in $\Sigma_1^\omega$.*

PROOF. Let $C(\mathbf{r})$ be a formula of form $K[\mathbf{r}(0)] \wedge (\forall t)H[\mathbf{r}(t), \mathbf{r}(t')] \wedge (\exists^\omega t)\ L[\mathbf{r}(t)]$. Suppose $\mathbf{r}$ is a k-tuple of predicates such that $C(\mathbf{r})$ holds. Then there are $x_1 < x_2 < \ldots$ such that $L[\mathbf{r}(x_1)], L[\mathbf{r}(x_2)], \ldots$. Because $\mathbf{r}$ has but a finite number of states, there must be a repetition $\mathbf{r}(x_p) = \mathbf{r}(x_q)$ of some state U. Therefore, $(\exists\mathbf{r})\ C(\mathbf{r})$ implies the assertion

(1) There are words $x = X_0X_1 \ldots X_a$ and $y = Y_1Y_2 \ldots Y_b$ of states and a state U such that L[U], and $K[X_0] \wedge H[X_0, X_1] \wedge \ldots \wedge H[X_{a-1}, X_a] \wedge H[X_a, U]$, and $H[U, Y_1] \wedge H[Y_1, Y_2] \wedge \ldots \wedge H[Y_{b-1}, Y_b] \wedge H[Y_b, U]$.

Conversely (1) implies $(\exists\mathbf{r})\ C(\mathbf{r})$, because one has but to let $\mathbf{r} = xUyUyUy\ldots$. Thus, a method (I) which decides, for given propositional formulas K, H, L and given state U, whether or not (1) holds will also be a method for deciding truth of $\Sigma_1^\omega$-sentences $(\exists\mathbf{r})\ C(\mathbf{r})$. Clearly such a method (I) can be composed from a method (II) which, for given propositional formula H[X, Y] and given states V and W, decides whether or not

(2) There is a word $x = X_1X_2 \ldots X_a$ such that $H[V, X_1] \wedge H[X_1, X_2] \wedge \ldots \wedge H[X_{a-1}, X_a] \wedge H[X_a, W]$.

Let $n = 2^k$ be the number of states, and note that in a word $x = X_1X_2 \ldots X_a$ of length $a > n$ there must occur a repetition $X_p = X_q$, $p < q \leqq a$. Clearly if $x$ satisfies (2), then so does the shorter word $y = X_1X_2 \ldots X_pX_{q+1}X_{q+2} \ldots X_a$. Therefore, to establish whether or not (2) holds, it suffices to check among the finitely many words $x$ of length $\leqq n$. This remark clearly yields a method (II) for (2), whereby Lemma 12 is established.

Lemma 2 is proved in automata theory in a strong effective version. Also the proof of Lemma 6 actually yields the following result:

(a) Let $\backsim$ be a congruence of finite rank on words. Given a method for deciding $x \backsim y$ and a set of representatives $x_1, \ldots, x_a$ of the congruence classes of $\backsim$, one can construct a finite automaton $[\mathbf{I}, \mathbf{J}]$ and outputs $U_1, \ldots, U_a$ such that the congruence class of $x_c$ is equal to the behavior of $[\mathbf{I}, \mathbf{J}, U_c)$. Clearly also Lemmas 3, 4, 5, 7, 8, 11 hold effectively. This leaves only the following two critical steps in the proof of the crucial Lemma 9:

(b) The $\Sigma^0$-formulas $E_c(\mathbf{i}, x, y)$, $c = 1, \ldots, a$ can be effectively constructed from $A(\mathbf{i})$.

(c) The relation $\Phi(c, d)$ on the finite set $\{1, \ldots, a\}$ can be effectively constructed from $A(\mathbf{i})$ (so that the disjunction (8) can be effectively obtained).

To prove (b) note that given $A$ the definition of $\backsim$ in the proof of lemma 9 provides us with a method for deciding $x \backsim y$. Because we also have a bound $2^{2m^2}$ on the rank a of $\backsim$, it is possible to obtain a set of representatives $x_1, \ldots, x_a$ for the congruence classes. By (a) and Lemma 2 the assertion (b) follows.

To prove (c) we refer to the definition (7) of $\Phi$ in the proof of Lemma 9. By Lemma 8 one can actually construct a $\Sigma_1^\omega$-formula equivalent to $A(j) \wedge F_{c,d}(j)$. Lemma 12 therefore provides a method for deciding whether or not $\Phi(c, d)$ holds. This takes care of (c). Thus also Lemma 9 holds effectively.

It now follows that Theorem 1 holds effectively; in particular, to every sentence $A$ in SC one can construct an equivalent sentence in $\Sigma_1^\omega$. Applying Lemma 12 again, we have

THEOREM 2. *There is an effective method for deciding truth of sentences in* SC.

The strength of this result is best seen by noting some very special cases which occur in the literature and have been obtained by rather divergent methods:

1. The decidability of $\Sigma_2$-sentences of SC contains the result of Friedman [4], and implies the existence of various other algorithms of finite automata theory as programmed by Church [2]. It also implies some of the results of Wang [11].
2. In SC one can define $x = y$, $x < y$, $x \equiv y \pmod{k}$ (for $k = 1, 2, \ldots$). The decidability of SC therefore considerably improves a result of Putnam [7].
3. In SC one can define "i is finite". Theorem 2 therefore implies the decidability of $SC_{fin}$, which was also proved in Büchi [1], and according to Robinson [10] is due to A. Ehrenfeucht.
4. The decidability of the first order theory of [Nn, +, Pw] follows from Theorem 2 and improves the classical result of Presburger.
5. Theorem 2 is closely related to another classical result, namely, the decidability of the monadic predicate calculus of second order, proved first by Th. Skolem and later by H. Behmann. A modified form of Lemma 11 yields a rather simple solution to this problem.

## 7. Concluding Remarks: Unsolved Problems

A careful analysis of the decision method for SC would yield a complete axiom system for SC. The most interesting candidate for an axiom schema is that part of Lemma 1 which is used in the proof of Lemma 9, namely,

$$(\mathrm{Ax}) \qquad (\forall \mathbf{i})(\exists s)_\omega \cdot (\forall y)(\forall x)_0^y[s(x)s(y) \supset E(\mathbf{i}, x, y)] \vee (\forall y)(\forall x)_0^y[s(x)s(y) \supset \sim E(\mathbf{i}, x, y)]$$

for any formula $E(\mathbf{i}, x, y)$ in $\Sigma^0$.

Such an analysis also shows that the same method yields a decision about whether or not a sentence is true in $SC_{per}$, which is like SC except that the variables i, j, r, ... range over ultimately periodic sets of natural numbers. In particular it can be seen that (Ax) also holds in $SC_{per}$. However this is not shown by using Ramsey's theorem; rather one uses the fact that every element c of a finite semi-group has a power $c^n$ which is idempotent. These remarks outline a proof of

THEOREM 3. *A sentence A is true in* $SC_{per}$ *if and only if it is true in* SC.

Using predicates i as binary expansions of real numbers, one obtains as a corollary to Theorems 2 and 3

THEOREM 4. *The first order theories of* [Re, +, Pw, Nn] *and* [Ra, +, Pw, Nn] *are arithmetically equivalent, and decidable.*

Here Re, Ra, Nn, Pw stand for the sets of non-negative reals, rationals, integers, integral powers of 2, respectively.

It is interesting to note that SC becomes undecidable if the function 2x is added (Robinson [10]). Also in case monadic predicate quantification is replaced in SC by quantification over monadic functions, all recursive relations become definable (Gödel).

*Problem 1.* Let $SC^2$ be like SC, except that the functions 2x+1 and 2x+2 are taken as primitives in place of x+1. Is $SC^2$ decidable?

This is of some interest, because the functions 2x+1 and 2x+2 can be interpreted as the right-successor functions x1 and x2 on the set of all words on two generators 1 and 2.

*Problem 2.* Let SC($\alpha$) be like SC, except that the domain of individuals is the ordinal $\alpha$, and the well ordering on $\alpha$ is added as a primitive. Is SC($\omega^2$) decidable?

As outlined in the introduction, Theorem 2 may be interpreted as a method for deciding whether or not a given finite automaton satisfies a given condition in SC.

*Problem 3.* Is there a solvability algorithm for SC, i.e., is there a method which applies to any formula $C(\mathbf{i}, \mathbf{u})$ of SC and decides whether or not there is a finite automata recursion $A(\mathbf{i}, \mathbf{r}, \mathbf{u})$ which satisfies the condition $C$ (i.e., $A(\mathbf{i}, \mathbf{r}, \mathbf{u}) \supset C(\mathbf{i}, \mathbf{u})$)?

## REFERENCES


[1] BÜCHI, J. R. "Weak Second Order Arithmetic and Finite Automata", *Zeitschrift für Math. Log. und Grundl. der Math.*, 6 (1960), pp. 66–92.

[2] CHURCH, ALONZO. "Application of Recursive Arithmetic to the Problem of Circuit Synthesis", *Notes of the Summer Institute of Symbolic Logic*, Cornell, 1957, pp. 3–50, and "Application of Recursive Arithmetic in the Theory of Computing and Automata", Notes: *Advanced Theory of the Logical Design of Digital Computers*, U. of Michigan Summer Session, 1959.

[3] COPI, I. M., C. ELGOT, and J. B. WRIGHT. "Realization of Events by Logical Nets", *Journal of the Association for Computing Machinery*, Vol. 5, No. 2, April, 1958.

[4] FRIEDMAN, JOYCE. "Some Results in Church's Restricted Recursive Arithmetic", *Journal of Symbolic Logic*, 22, pp. 337–342 (1957).

[5] KLEENE, S. C. "Representation of Events in Nerve Nets and Finite Automata", *Automata Studies*, Princeton University Press, 1956, pp. 3–41.

[6] MYHILL, JOHN. "Finite Automata and Representation of Events", WADC Report TR 57–624, *Fundamental Concepts in the Theory of Systems*, October 1957, pp. 112–137.

[7] PUTNAM, HILLARY. "Decidability and Essential Undecidability", *Journal of Symbolic Logic*, 22 (1957), pp. 39–54.

[8] RABIN, M., and D. SCOTT. "Finite Automata and their Decision Problems", *IBM Journal*, April 1959, pp. 114–125.

[9] RAMSEY, F. P. "On a Problem of Formal Logic", *Proc. London Math. Soc.*, (2) 30 (1929), pp. 264–286.

[10] ROBINSON, R. M. "Restricted Set-theoretical Definitions in Arithmetic", *Proc. Am. Math. Soc.*, 9 (1958), pp. 238–242.

[11] WANG, HAO. "Circuit Synthesis by Solving Sequential Boolean Equations", *Zeitschrift für Math. Log. und Grundl. der Math.*, 5 (1959), pp. 291–322.

*Reprinted from*
PROCEEDINGS OF
THE 1964 INTERNATIONAL CONGRESS FOR LOGIC, METHODOLOGY AND PHILOSOPHY OF SCIENCE
Held in Jerusalem, August 26—September 2, 1964
*Published by* North-Holland Publishing Company, Amsterdam.

# TRANSFINITE AUTOMATA RECURSIONS AND WEAK SECOND ORDER THEORY OF ORDINALS*

J. RICHARD BÜCHI
*Purdue University, Lafayette, Indiana, U.S.A.*

**1. Introduction.** We identify the ordinal $\alpha$ with the set of all ordinals $x < \alpha$. The weak second order theory of $[\alpha, <]$ is the interpreted formalism WST $[\alpha, <]$ which makes use of: (a) the propositional connectives with usual interpretation; (b) a binary relation letter $<$ interpreted as ordering relation on $\alpha$; (c) individual variables $t, x, y, z, \cdots$, ranging over $\alpha$; (d) monadic predicate variables $i, j, s, r, \cdots$, ranging over finite subsets of $\alpha$; (e) quantifiers $\forall$, $\exists$ for both types of variables. The purpose of this paper is to provide, for any ordinal $\alpha$, a clear understanding of which relations $R$ on finite subsets of $\alpha$ can be defined by formulas $\Sigma(i_1, \cdots, i_n)$ of WST$[\alpha, <]$. In addition we obtain a decision method for truth of sentences $\Sigma$ in WST$[\alpha, <]$.

The binary expansion of natural numbers can be extended to ordinals. If $x < 2^\alpha$ the binary expansion $\phi x$ is a finite subset $i$ of $\alpha$, namely $\phi x = \{u_1, \cdots, u_n\}$ if $x = 2^{u_1} + \cdots + 2^{u_n}$, $u_n < \cdots < u_1 < \alpha$. $\phi$ is a one-to-one map of the ordinal $2^\alpha$ onto all finite subsets of $\alpha$, and it yields a simple translation of the first order theory FT$[2^\alpha, +, E]$ onto WST$[\alpha, <]$. Here $E$ stands for the binary relation "$x$ is a power of 2, and occurs in the representation of $y$ as decreasing sum of powers of 2," i.e., $(\exists u)[x = 2^u \wedge u \in \phi y]$. (Note that $2^\alpha$ is not necessarily closed under $+$; in this case $+$ is to be interpreted as a ternary relation.) This translation in fact is simply a change in the choice of primitive notions. Our main results can be restated: 1. For any $\alpha$ we obtain a clear understanding of which relations $R$ on $2^\alpha$ are definable by formulas $\Sigma(x_1, \cdots, x_n)$ in the first order theory of $[2^\alpha, +, E]$. 2. For any $\alpha$, there is a decision method for truth of first order sentences $[2^\alpha, +, E]$.

The key to the understanding of WST$[\alpha, <]$ is a simple sort of transfinite predicate recursions with finite predicates $i$ on $\alpha$ as inputs. In case $\alpha = \omega$ these are just the finite automata recursions, and the methods used in the present paper specialize to those used in [1]. We refer the reader

* This work was supported in part by grant PG-2754 from the National Science Foundation, and by a travel grant from Purdue University.

to this paper and its review [5], if he desires more explicit explanations of notations, or is interested in analyzing by-products of the main results.

Results on definability of individuals in $FT[\omega^\alpha, +]$ have been obtained earlier by A. Ehrenfeucht [6]. His methods are quite different from ours; we refer the reader to [3], for a lucid presentation of this work. In [3] and [4] it is stated that Ehrenfeucht also knew a decision method for $FT[\omega^\alpha, +]$. However, ours is the first published proof of this result.

Let $RST[\alpha, <]$, the restricted second order theory of $[\alpha, <]$, be like $WST[\alpha, <]$, except that predicate variables now range over all subsets of $\alpha$. That $RST[\omega, <]$ is decidable was shown in [2]. Whether this holds for $RST[\omega^2, <]$ is not known.*

**2. Transfinite automata recursions.** We show here how to extend the basic facts about the behavior of finite automata to a simple sort of transfinite recursion on monadic predicates, which we call "special recursions" for short. Lemma 4 says that the projection of the behavior of a special recursion is again the behavior of some special recursion. This is the crucial result of this paper, as it can be used to eliminate predicate quantifiers in formulas of $WST[\alpha, <]$.

We introduce the following abbreviations for limit ordinals of finite order $p$.

$$\begin{aligned}
&(\forall t)_x^y \Sigma &&: (\forall t)[x \leqq t < y \supset \Sigma] \\
&(\exists t)_x^y \Sigma &&: (\exists t)[x \leqq t < y \wedge \Sigma] \\
(1)\qquad &\mathrm{Lm}_0 x &&: T \\
&\mathrm{Lm}_{p+1} x &&: (\forall y)_0^x (\exists z)_{y+1}^x \mathrm{Lm}_p z \\
&\mathrm{lm}_p x &&: \mathrm{Lm}_p x \wedge \sim \mathrm{Lm}_{p+1} x
\end{aligned}$$

Thus $\mathrm{Lm}_0$ is the set of all ordinals. $\mathrm{Lm}_1$ consists of 0 and all limit ordinals. $\mathrm{Lm}_2$ consists of 0 and limits of limit ordinals, etc. While $\mathrm{lm}_p$ consists of all ordinals of form $r + \omega^p$. Let $x$ be a limit ordinal, $y < x$, and suppose that the predicate $j$ is constant between $y$ and $x$. Then $\lim_{t<x} jt$ stands for $jy$. I.e.

$$\begin{aligned}
&\lim_{t<x} jt \equiv T \quad \text{if } (\exists y)_0^x (\forall t)_y^x jt \\
(2)\qquad & \\
&\lim_{t<x} jt \equiv F \quad \text{if } (\exists y)_0^x (\forall t)_y^x \sim jt
\end{aligned}$$

Now consider the system of propositional formulas,

* Added in proof: $RST[\alpha, <]$ is decidable, for any countable $\alpha$.

$$
(3)\qquad
\begin{aligned}
r0 &\equiv E \\
r(x+1) &\equiv H[rx, ix] \\
rx &\equiv \lim_{t<x} U_1[rt], && \text{if } \mathrm{lm}_1 x \\
&\vdots \\
rx &\equiv \lim_{t<x} U_{p-1}[rt], && \text{if } \mathrm{lm}_{p-1} x \\
rx &\equiv \lim_{t<x} U_p[rt], && \text{if } \mathrm{Lm}_p x \text{ and } x \neq 0.
\end{aligned}
$$

We use here the vector notation explained in detail in [1]. Thus $\boldsymbol{i}$ stands for a vector $[i_1, \cdots, i_k]$ of predicate variables, $\boldsymbol{r}$ for a vector $[r_1, \cdots, r_n]$ of predicate variables, $\boldsymbol{E}$ for a vector $[E_1, \cdots, E_n]$ of truth values, $\boldsymbol{H}[Y, X]$ for a vector $[H_1, \cdots, H_n]$ of propositional expressions in $X_1, \cdots, X_k, Y_1, \cdots, Y_n$. Furthermore, each line of (3) abbreviates a conjunction of $n$ equivalences. $\boldsymbol{i}$ will be called the *input predicates* of (3), $k$-vectors $\boldsymbol{X} = [X_1, \cdots, X_k]$ of truth values are called the *input states* of (3). $\boldsymbol{r}$ is called the *transition predicates*, its states $\boldsymbol{Y} = [Y_1, \cdots, Y_n]$ are called *transit states*. In the sequel we will usually not explicitly mention the number of components of a vector $\boldsymbol{A}$ of propositional expressions; it can be recovered from the context. $\boldsymbol{F}$ always stands for a vector $[F, \cdots, F]$, all of whose components are the truth value $F$.

DEFINITION 1: *A special predicate recursion of order p*, or shortly a *p-recursion*, is a system $\mathfrak{R} = \boldsymbol{E}, \boldsymbol{H}[Y, X], \boldsymbol{U}_1[Y], \cdots, \boldsymbol{U}_p[Y]$ of propositional expressions satisfying

$$
(4)\qquad
\begin{aligned}
\boldsymbol{U}_a[\boldsymbol{U}_b[\boldsymbol{Y}]] &\equiv \boldsymbol{U}_a[\boldsymbol{Y}], \quad \text{for } 0 \leqq b < a \leqq p; \\
\boldsymbol{U}_p[\boldsymbol{U}_p[\boldsymbol{Y}]] &\equiv \boldsymbol{U}_p[\boldsymbol{Y}].
\end{aligned}
$$

Here $\boldsymbol{U}_0[\boldsymbol{Y}]$ stands for $\boldsymbol{H}[\boldsymbol{Y}, \boldsymbol{F}]$. An *output* to such a recursion is a propositional expression $O[\boldsymbol{Y}]$. A sequence consisting of a $p$-recursion and an output will also be called a *finite automaton of order p*, or a *p-automaton*.

Suppose now that $\boldsymbol{i}$ is a vector of specific predicates $i_1, \cdots, i_k$ on the ordinal $\alpha$. In some cases (3) will determine, in recursive fashion, a vector $\boldsymbol{r} = \zeta \boldsymbol{i}$ on $\alpha + 1$. Namely, just in case $\lim_{t<x} \boldsymbol{U}_a[\boldsymbol{r}t]$ becomes meaningful at each limit $x \leqq \alpha$, i.e., if $1 \leqq a < p$ and $\mathrm{lm}_a x$ or $a = p$ and $\mathrm{Lm}_a x$, there must exist $y < x$ such that $\boldsymbol{U}_a[\boldsymbol{r}t]$ remains constant for $y \leqq t < x$. By transfinite induction one easily shows that this really happens, in case $\boldsymbol{i}$ consists of finite predicates, and (4) holds. Just note that "$\boldsymbol{i}$ is finite" means $(\forall x)_{\mathrm{Lm}_1}(\exists y)_0^x(\forall t)_y^x\, \boldsymbol{i}t \equiv \boldsymbol{F}$. Thus we can state

LEMMA 1. *If* $\mathfrak{R} = \boldsymbol{E}, \boldsymbol{H}, \boldsymbol{U}_1, \cdots, \boldsymbol{U}_p$ *is a special predicate recursion of order p and* $\alpha$ *is any ordinal, then* (3) *uniquely determines a recursive operator* $\boldsymbol{r} = \zeta_{\mathfrak{R}} \boldsymbol{i}$, *defined for all vectors* $\boldsymbol{i}$ *of finite predicates on* $\alpha$, *and taking as values vectors of predicates on* $\alpha + 1$.

Note that the components of the values of $\zeta$ are not necessarily finite subsets of $\alpha + 1$. If $\boldsymbol{i}$ is finite and $\boldsymbol{r} = \zeta \boldsymbol{i}$ then $\boldsymbol{r}\alpha$ will be called the *terminal state* of $\mathfrak{R}$ under the *input signal* $\boldsymbol{i}$. If this terminal state belongs to the output $\boldsymbol{O}$, we will say that $\boldsymbol{i}$ belongs to the behavior of $\mathfrak{R}, \boldsymbol{O}$, i.e.

DEFINITION 2. Let $\mathfrak{A} = \boldsymbol{E}, \boldsymbol{H}, \boldsymbol{U}_1, \cdots, \boldsymbol{U}_p, \boldsymbol{O}$ be an automaton of order $p$. Let $\zeta = \zeta_{\mathfrak{A}}$ be the recursive operator determined by $\mathfrak{A}$. The *$\alpha$-behavior* of $\mathfrak{A}$ is the set $\mathrm{beh}_{\mathfrak{A}}^{\alpha}$ consisting of all vectors $\boldsymbol{i}$ of finite predicates on $\alpha$ such that $\boldsymbol{O}[\boldsymbol{r}\alpha]$ holds, if $\boldsymbol{r} = \zeta \boldsymbol{i}$.

The purpose of this section is to show how the well known facts about behavior of finite automata extend to $\alpha$-behaviors of $p$-automata. The following lemma is trivial.

LEMMA 2. *The complement (in the set of all finite subsets of* $\alpha$*) of an* $\alpha$*-behavior of a p-automaton is again the* $\alpha$*-behavior of some p-automaton. In fact* $\mathrm{beh}_{\mathfrak{R},\tilde{O}}^{\alpha} = \sim \mathrm{beh}_{\mathfrak{R},O}^{\alpha}$, *for any p-recursion* $\mathfrak{R}$.

Equally trivial is the closure under $\cup$ and $\cap$ of the class of $\alpha$-behaviors. Also the following lemma is quite obvious on intuitive grounds, even though its proof is somewhat cumbersome for typographic reasons.

DEFINITION 3. A $p$-recursion $\boldsymbol{E}, \boldsymbol{H}, \boldsymbol{U}_1, \cdots, \boldsymbol{U}_p$ is in *expanded form* if

(5)
$$\boldsymbol{H}[\boldsymbol{Y}, \boldsymbol{X}] \;.\equiv.\; \boldsymbol{A}_1[\boldsymbol{X}]Y_1 \vee \cdots \vee \boldsymbol{A}_n[\boldsymbol{X}]Y_n;$$
$$\boldsymbol{U}_a[\boldsymbol{Y}] \;.\equiv.\; \boldsymbol{B}_a Y_1 \vee \cdots \vee \boldsymbol{B}_a Y_n, \quad \text{for } a = 1, \cdots, p.$$

The output $\boldsymbol{O}[\boldsymbol{Y}]$ is in *expanded form* if

(5′)
$$O[\boldsymbol{Y}] \;.\equiv.\; C_1 Y_1 \vee \cdots \vee C_n Y_n.$$

LEMMA 3. *To every p-automaton* $\mathfrak{A} = \boldsymbol{E}, \boldsymbol{H}, \boldsymbol{U}_1, \cdots, \boldsymbol{U}_p, \boldsymbol{O}$ *one can construct a p-automaton* $\mathfrak{B} = \boldsymbol{G}, \boldsymbol{L}, \boldsymbol{W}_1, \cdots, \boldsymbol{W}_p, \boldsymbol{Q}$ *which is in expanded form and has the same* $\alpha$*-behavior as* $\mathfrak{A}$, *for any* $\alpha$.

PROOF. Given the automaton $\mathfrak{A}$ we define for $a = 1, \cdots, n$; $b = 1, \cdots, p$; $\mu, \nu_1, \cdots, \nu_n = T, F$:

(6)
$$A_{\mu,a,\nu_1,\ldots,\nu_n}[\boldsymbol{X}] \;.\equiv.\; H_a^{\mu}[\nu_1, \cdots, \nu_n, \boldsymbol{X}];$$
$$B_{b,\mu,a,\nu_1,\ldots,\nu_n} \;.\equiv.\; U_{b,a}^{\mu}[\nu_1, \cdots, \nu_n];$$
$$C_{\nu_1,\ldots,\nu_n} \;.\equiv.\; O[\nu_1, \cdots, \nu_n].$$

Here $Z^{v}$ stands for $Z$ or $\tilde{Z}$ according to whether $v = T$ or $v = F$. Now we define $\mathfrak{B} = G, L, W_1, \cdots, W_p, Q$ by

$$
(7)\quad
\begin{aligned}
G_{v_1,\ldots,v} &.\equiv.\ E_1^{v_1} \cdots E_n^{v_n};\\
L_{\sigma_1,\ldots,\sigma_n}[Z,X] &.\equiv.\ \bigvee_{v_1,\ldots,v_n} [\bigwedge_a A_{\sigma_a,a,v_1,\ldots,v_n}[X]] Z_{v_1,\ldots,v_n};\\
W_{b,\sigma_1,\ldots,\sigma_n}[Z] &.\equiv.\ \bigvee_{v_1,\ldots,v_n} [\bigwedge_a B_{b,\sigma_a,a,v_1,\ldots,v_n}] Z_{v_1,\ldots,v};\\
Q[Z] &.\equiv.\ \bigvee_{v_1,\ldots,v_n} C_{v_1,\ldots,v_n} Z_{v_1,\ldots,v_n}.
\end{aligned}
$$

Note that $\bigvee_{v_1,\ldots,v_n} D_{v_1,\ldots,v_n} Y_1^{v_1} \cdots Y_n^{v_n} .\equiv. D_{Y_1,\ldots,Y_n}$. Using this, it follows by (6) and (7) that

$$
(8)\quad
\begin{aligned}
L_{\sigma_1,\ldots,\sigma_n}[Z, X] &.\equiv.\ H_1^{\sigma_1}[Y, X] \cdots H_n^{\sigma_n}[Y, X];\\
W_{b,\sigma_1,\ldots,\sigma}[Z] &.\equiv.\ U_{b,1}^{\sigma_1}[Y] \cdots U_{b,n}^{\sigma_n}[Y], \text{ if } Z_{v_1,\ldots,v} \equiv Y_1^{v_1} \cdots Y_n^{v_n};\\
Q[Z] &.\equiv.\ O[Y].
\end{aligned}
$$

By assumption $E, H, U_1, \cdots, U_p$ is a special recursion. From Definition 3 and (8) one easily shows that,

$$
(9)\quad
\begin{aligned}
W_b[L[Z, F]] &= W_b[Z], \quad \text{if } 1 \leqq b \leqq p;\\
W_a[W_b[Z]] &= W_a[Z], \text{ if } 1 \leqq b < a \leqq p \text{ and } Z_{v_1,\ldots,v_n} \equiv Y_1^{v_1} \cdots Y_n^{v_n};\\
W_p[W_p[Z]] &= W_p[Z].
\end{aligned}
$$

For example, because $Z_{v_1,\ldots,v_n} \equiv Y_1^{v_1} \cdots Y_n^{v_n}$ it follows by (8) that also $L_{v_1\ldots,v_n}[Z, F] \equiv H_1^{v_1}[Y, F] \cdots H_n^{v_n}[Y, F]$. Therefore, $W_{b,\sigma_1,\ldots,\sigma_n}[L[Z, F]] \overset{(8)}{\equiv} U_1^{\sigma_1}[H[Y, F]] \cdots U_n^{\sigma_n}[H[Y, F]] \overset{Def.3}{\equiv} U_1^{\sigma_1}[Y] \cdots U_n^{\sigma_n}[Y] \overset{(8)}{\equiv} W_{b,\sigma_1,\ldots,\sigma_n}[Z]$, which establishes the first part of (9). The other parts are similar. Next we note that, by (7), $L, W_1, \cdots, W_p$ are expanded, and therefore $L[Z_1 \vee Z_2, X] \equiv L[Z_1, X] \vee L[Z_2, X]$ and $W_a[Z_1 \vee Z_2] \equiv W_a[Z_1] \vee W_a[Z_2]$, for $a = 1, \cdots, p$. Using these formulas and the fact that every $Z$ can be expressed in the form $Z = Z_1 \vee \cdots \vee Z_r$, whereby each of the $Z_1, \cdots, Z_r$ is of form $Y_1^{v_1} \cdots Y_n^{v_n}$, one sees that the assumption $Z_{v_1\ldots v_n} \equiv Y_1^{v_1} \cdots Y_n^{v}$ can be dropped in (9). Consequently $G, L, W_1, \cdots, W_p$ is a special recursion i.e. $\mathfrak{B}$ is a $p$-automaton. It is clearly in expanded form, see (7). Thus, it only remains to show that the behaviors $\text{beh}_{\mathfrak{A}}^{\alpha}$ and $\text{beh}_{\mathfrak{B}}^{\alpha}$ are equal, for any $\alpha$.

Let $r = \zeta_{\mathfrak{A}}(i)$ and let $s = \zeta_{\mathfrak{B}}(i)$ (see Lemma 1). Using the first two formulas of (8) one shows, by induction over $\alpha$, that $s_{\sigma_1\ldots\sigma} t \equiv (r_1 t)^{\sigma_1} \cdots (r_n t)^{\sigma_n}$, for all $t \leqq \alpha$. Therefore, by the third formula of (8), $Q[s\alpha] \equiv O[r\alpha]$. Thus by Definition 2, $\mathrm{beh}^{\alpha}_{\mathfrak{A}} = \mathrm{beh}^{\alpha}_{\mathfrak{B}}$. Q.e.d.

Based on Lemma 3 we now prove that the class of $\alpha$-behaviors is closed under finite-predicate quantification. This is the core of our argument; it will be used in Section 3 to eliminate predicate quantifiers in WST[$\alpha$, <]. We suggest that the reader carry out the proof of Lemma 4 in the special case $p = 1$. This will show clearly why the order $p$ of a recursion is stepped up by a quantification of part of the input predicates.

LEMMA 4 (*projection lemma*): *To every p-recursion* $\mathfrak{R} = \boldsymbol{E},\boldsymbol{H}[\boldsymbol{Y},\boldsymbol{X},\boldsymbol{Z}]$, $\boldsymbol{U}_1[\boldsymbol{Y}],\cdots,\boldsymbol{U}_p[\boldsymbol{Y}]$ *one can construct a* $(p+1)$*-recursion* $\mathfrak{S} = \boldsymbol{E},\boldsymbol{L}[\boldsymbol{Y},\boldsymbol{X}]$, $\boldsymbol{W}_1[\boldsymbol{Y}],\cdots,\boldsymbol{W}_{p+1}[\boldsymbol{Y}]$ *such that for any output* $O[\boldsymbol{Y}]$ *and any ordinal* $\alpha$, $(\exists \boldsymbol{j})\mathrm{beh}^{\alpha}_{\mathfrak{R},O}(\boldsymbol{i},\boldsymbol{j}) . \equiv . \mathrm{beh}^{\alpha}_{\mathfrak{S},O}(\boldsymbol{i})$.

PROOF. Given $\mathfrak{R}$ we let $\boldsymbol{U}_0[\boldsymbol{Y}]$ stand for $\boldsymbol{H}[\boldsymbol{Y},\boldsymbol{F},\boldsymbol{F}]$. By assumption on $\mathfrak{R}$ we have

$$(10) \qquad \boldsymbol{U}_p\boldsymbol{U}_p = \boldsymbol{U}_p, \quad \boldsymbol{U}_a\boldsymbol{U}_b = \boldsymbol{U}_a, \quad \text{for } 0 \leqq b < a \leqq p.$$

By Lemma 3 we may assume that $\mathfrak{R}$ is in expanded form, from which it follows that

$$(11) \qquad \begin{aligned} &\boldsymbol{Y} \supset \boldsymbol{Y}' . \supset . \ \boldsymbol{U}_a[\boldsymbol{Y} \supset \boldsymbol{U}_a[\boldsymbol{Y}'], \quad \text{for } 0 \leqq a \leqq p; \\ &\boldsymbol{Y} \supset \boldsymbol{Y}' . \supset . \ \boldsymbol{H}[\boldsymbol{Y},\boldsymbol{X},\boldsymbol{Z}] \supset \boldsymbol{H}[\boldsymbol{Y}',\boldsymbol{X},\boldsymbol{Z}]. \end{aligned}$$

Next we define $\boldsymbol{L}$ by

$$(12) \qquad \boldsymbol{L}[\boldsymbol{Y},\boldsymbol{X}] . \equiv . \bigvee_{\boldsymbol{Z}} \boldsymbol{H}[\boldsymbol{Y},\boldsymbol{X},\boldsymbol{Z}]$$

and we let $\boldsymbol{W}_0[\boldsymbol{Y}]$ stand for $\boldsymbol{L}[\boldsymbol{Y},\boldsymbol{F}]$. Then, clearly,

$$(\mathrm{u}_0) \qquad \boldsymbol{U}_0[\boldsymbol{Y}] \supset \boldsymbol{W}_0[\boldsymbol{Y}].$$

By (11) and $(\mathrm{u}_0)$ it follows that $\boldsymbol{U}_1\boldsymbol{U}_0[\boldsymbol{Y}] \supset \boldsymbol{U}_1\boldsymbol{W}_0[\boldsymbol{Y}]$. Therefore, by (10), $\boldsymbol{U}_1[\boldsymbol{Y}] \supset \boldsymbol{U}_1\boldsymbol{W}_0[\boldsymbol{Y}]$. Hence,

$$\boldsymbol{U}_1[\boldsymbol{Y}] \supset \boldsymbol{U}_1\boldsymbol{W}_0[\boldsymbol{Y}] \supset \boldsymbol{U}_1\boldsymbol{W}_0^2[\boldsymbol{Y}] \supset \cdots \supset \boldsymbol{U}_1\boldsymbol{W}_0^e[\boldsymbol{Y}] \supset \ldots..$$

Because there are but a finite number of propositional expressions in $\boldsymbol{Y}$, this sequence must break up at some place $e_0$ with $\boldsymbol{U}_1\boldsymbol{W}_0^{eo} = \boldsymbol{U}_1\boldsymbol{W}_0^{eo+1}$. Note that such an $e_0$ can be effectively found. We now define $\boldsymbol{W}_1$ by

$$(\mathrm{w}_1) \qquad \boldsymbol{W}_1[\boldsymbol{Y}] \equiv \boldsymbol{U}_1[\boldsymbol{W}_0^{eo}[\boldsymbol{Y}]]$$

and we note that

$$(\mathrm{u}_1) \qquad \boldsymbol{U}_1[\boldsymbol{Y}] \supset \boldsymbol{W}_1[\boldsymbol{Y}] \qquad\qquad (\mathrm{v}_1) \qquad \boldsymbol{W}_1\boldsymbol{W}_0 = \boldsymbol{W}_1$$

Iterating the argument which leads from ($u_0$) to the definition ($w_1$) and the formulas ($u_1, v_1$) we obtain numbers $e_0, \cdots, e_{p-1}$ and expressions $W_1, \cdots, W_p$ such that

(w) $$W_{a+1}[Y] \equiv U_{a+1}[W_a^{e_a}[Y]], \quad a = 0, \cdots, p-1;$$

(u) $$U_a[Y] \supset W_a[Y], \quad a = 0, \cdots, p;$$

(v) $$W_{a+1} W_a = W_{a+1}, \quad a = 0, \cdots, p-1.$$

From (w) and (v) one concludes

($\bar{\text{v}}$) $$W_a W_b = W_a, \quad 0 \leqq b < a \leqq p.$$

Using (10) and (w) one finds $W_p = U_p W_p$. This, together with (u), yields,

$$U_p[Y] \supset W_p[Y] \equiv U_p W_p[Y] \supset W_p^2[Y] \equiv U_p W_p^2[Y] \supset W_p^3[Y] \equiv \ldots..$$

It follows that one can find $e_p$ such that $W_p^{e_p} = W_p^{e_p+1} = U_p W_p^{e_p}$. We define $W_{p+1}$ by

(13) $$W_{p+1}[Y] \equiv U_p[W_p^{e_p}[Y]]$$

and we obtain

(14) $$U_p[Y] \supset W_{p+1}[Y], \; W_{p+1} W_p = W_{p+1}, \, W_{p+1} W_{p+1} = W_{p+1}.$$

By (14), ($\bar{\text{v}}$) it follows that $W_{p+1} W_b = W_{p+1}$, for $0 \leqq b \leqq p+1$. This, together with ($\bar{\text{v}}$), means that $\mathfrak{S} = E, L, W_1, \cdots, W_{p+1}$ is a special recursion. It remains to be shown that, for any expanded output $O[Y]$ and any $\alpha$,

(a) $$\mathrm{beh}_{\mathfrak{R},O}^{\alpha}(i,j) \supset \mathrm{beh}_{\mathfrak{S},O}^{\alpha}(i), \quad \text{for any finite } i,j;$$

(b) $$\mathrm{beh}_{\mathfrak{S},O}^{\alpha}(i) \supset (\exists j)\, \mathrm{beh}_{\mathfrak{R},O}^{\alpha}(i,j), \quad \text{for any finite } i.$$

PROOF OF (a). Let $i,j$ be any sequences of finite predicates on $\alpha$, let $r = \zeta_{\mathfrak{R}}(i,j)$ and $s = \zeta_{\mathfrak{S}}(i)$ (see Lemma 1). Then

(15)
$$\begin{aligned}
rO &\equiv E; \\
rx' &\equiv H[rx, ix, jx]; \\
rx &\equiv \lim_{t<x} U_1[rt], \text{ if } \mathrm{lm}_1 x; \\
&\vdots \\
rx &\equiv \lim_{t<x} U_{p-1}[rt], \text{ if } \mathrm{lm}_{p-1} x; \\
rx &\equiv \lim_{t<x} U_p[rt], \text{ if } \mathrm{Lm}_p x;
\end{aligned}$$

(16)
$$\begin{aligned}
sO &\equiv E; \\
sx' &\equiv L[sx, ix]; \\
sx &\equiv \lim_{t<x} W_1[st], \text{ if } \mathrm{lm}_1 x; \\
&\vdots \\
sx &\equiv \lim_{t<x} W_{p-1}[st], \text{ if } \mathrm{lm}_{p-1} x; \\
sx &\equiv \lim_{t<x} W_p[st], \text{ if } \mathrm{lm}_p x; \\
sx &\equiv \lim_{t<x} W_{p+1}[st], \text{ if } \mathrm{Lm}_{p+1} x.
\end{aligned}$$

By transfinite induction on $x \leqq \alpha$, and using (15), (16), (12), (u), (14), one shows $\boldsymbol{r}x \supset \boldsymbol{s}x$, for all $x \leqq \alpha$. In particular $\boldsymbol{r}\alpha \supset \boldsymbol{s}\alpha$. Because $O[\boldsymbol{Y}]$ is expanded it follows that $O[\boldsymbol{r}\alpha] \supset O[\boldsymbol{s}\alpha]$. Thus, if $(\boldsymbol{i},\boldsymbol{j})$ belongs to $\mathrm{beh}^{\alpha}_{\mathfrak{R},O}$, then $\boldsymbol{i}$ belongs to $\mathrm{beh}^{\alpha}_{\mathfrak{S},O}$, which establishes (a).

We use notations like $\boldsymbol{j}[y,z)$ to refer to predicates defined on the interval $[y,z) = \{t; y \leqq t < z\}$ of ordinals. $\boldsymbol{r}[y,z) = \zeta_C \boldsymbol{i}, \boldsymbol{j}[yz)$ means that $\boldsymbol{r}[y,z)$ is defined from $\boldsymbol{i}[y,z)$ and $\boldsymbol{j}[y,z)$ by the recursion (15), modified to start at $y$ with $\boldsymbol{r}y \equiv \boldsymbol{C}$. To prove (b) we need the following lemma.

(c) *Let $\boldsymbol{i}$ be a vector of finite predicates on $\alpha$, and let $\boldsymbol{s} = \zeta_{\mathfrak{S}}(\boldsymbol{i})$ be given by the recursion* (16). *If $z \neq 0$ and $s_h z$, then there exist $g, y < z, \boldsymbol{j}[y,z)$ such that $s_g y$, and if $C_g \equiv T$ and $\boldsymbol{r}[y,z) = \zeta_C \boldsymbol{i}, \boldsymbol{j}[y,z)$ then $r_h z$, and each component of $\boldsymbol{j}[y,z)$ is finite.*

The proof splits into 3 cases.

Case $z = t'$: By (16), $\boldsymbol{s}z \equiv \boldsymbol{L}[\boldsymbol{s}t, \boldsymbol{i}t]$, and because $s_h z$ it follows that $L_h[\boldsymbol{s}t, \boldsymbol{i}t]$. By (12) there is a $\boldsymbol{Z}$ such that $H_h[\boldsymbol{s}t, \boldsymbol{i}t, \boldsymbol{Z}]$, and because $H_h$ is expanded there must be a $g$ such that $s_g t$ and $H_g[\boldsymbol{C}, \boldsymbol{i}t, \boldsymbol{Z}]$, for arbitrary $\boldsymbol{C}$ with $C_g \equiv T$. Thus $g, y = t, \boldsymbol{j}[y,z) = \boldsymbol{Z}$ satisfy the requirements of (c).

Case $\mathrm{lm}_a z, 1 \leqq a \leqq p$: Since $\boldsymbol{i}$ consists of finite predicates there is an $x < z$ such that $(\forall t)^z_x(\boldsymbol{i}t \equiv \boldsymbol{F})$. We may also assume that $(\forall t)^z_x \sim \mathrm{Lm}_a t$. Then, by (16) and $(\bar{\mathrm{v}})$, $\boldsymbol{s}z \equiv \boldsymbol{W}_a[\boldsymbol{s}x]$, and by (w), $\boldsymbol{s}z \equiv \boldsymbol{U}_a \boldsymbol{W}^{e_{a-1}}_{a-1}[\boldsymbol{s}x]$. But, by (16) and $(\bar{\mathrm{v}})$, $\boldsymbol{W}^{e_{a-1}}_{a-1}[\boldsymbol{s}x] \equiv \boldsymbol{s}(x + \omega^{a-1} e_{a-1})$. Therefore, if $y = x + \omega^{a-1} e_{a-1}$, we have $\boldsymbol{s}z \equiv \boldsymbol{U}_a[\boldsymbol{s}y]$. Because $s_h z$ it follows that $U_{a,h}[\boldsymbol{s}y]$, and because $U_{a,h}$ is expanded there must be a $g$, such that $s_g y$ and $U_{a,g}[\boldsymbol{C}]$ in case $C_g \equiv T$. Now let $\boldsymbol{j}[y,z) = \boldsymbol{F}\boldsymbol{F}\cdots$. Then clearly $g, y, \boldsymbol{j}[y,z)$ satisfy the requirements of (c).

Case $\mathrm{Lm}_{p+1} z$: Since $\boldsymbol{i}$ consists of finite predicates there is an $x < z$ such that $(\forall t)^z_x(\boldsymbol{i}t \equiv \boldsymbol{F})$. Then, by (16) and $\boldsymbol{W}_{p+1}\boldsymbol{W}_a = \boldsymbol{W}_{p+1}$, $\boldsymbol{s}z \equiv \boldsymbol{W}_{p+1}[\boldsymbol{s}x]$, and by (13), $\boldsymbol{s}z \equiv \boldsymbol{U}_p \boldsymbol{W}^{e_p}_p[\boldsymbol{s}x]$. But, by (16) and $(\bar{\mathrm{v}})$, $\boldsymbol{W}^{e_p}_p[\boldsymbol{s}x] \equiv \boldsymbol{s}(x + \omega^p e_p)$. Therefore, if $y = x + \omega^p e_p$, we have $\boldsymbol{s}z \equiv \boldsymbol{U}_p[\boldsymbol{s}y]$. Because $s_h z$ it follows that $U_{p,h}[\boldsymbol{s}y]$, and because $U_{p,h}$ is expanded there must be a $g$, such that $s_g y$ and $U_{p,h}[\boldsymbol{C}]$ in case $C_g \equiv T$. Now let $\boldsymbol{j}[y,z) = \boldsymbol{F}\boldsymbol{F}\cdots$. Then clearly $g, y, \boldsymbol{j}[y,z)$ satisfy the requirements of (c).

This establishes (c) and there remains only to prove (b). Let $\boldsymbol{i}$ be any vector of finite predicates on $\alpha$. Let $\boldsymbol{s} = \zeta_{\mathfrak{S}}(\boldsymbol{i})$ be given by (16) and assume $\mathrm{beh}_{\mathfrak{S},O}(\boldsymbol{i})$. Then $O[\boldsymbol{s}\alpha]$, and because $O$ is expanded, there is an $h_0$ such that $s_{h_0}\alpha$ and

$$Y_{h_0} \supset O[\boldsymbol{Y}]. \tag{17}$$

By $s_{h_0}\alpha$ and (c) there exist $h_1, y_1, \boldsymbol{j}[y_1,\alpha)$ such that $y_1 < \alpha, s_{h_1} y_1$, and $r_{h_0}\alpha$ if $\boldsymbol{r}[y_1, \alpha) = \zeta_C \boldsymbol{i}, \boldsymbol{j}[y_1,\alpha)$ and $C_{h_1} \equiv T$. Because $s_{h_1} y_1$ we can use (c) again, and iterating this procedure we find the existence of $h_1, h_2, \cdots$; $\alpha = y_0 > y_1 > y_2 > \cdots$; $j[y_1, y_0), j[y_2, y_1), \cdots$, such that, for $v = 0, 1, 2 \cdots$,

(18) $s_{h_v} y_v$, every component of $j[y_{v+1}, y_v)$ is finite;

(19) if $r[y_{v+1}, y_v) = \zeta_C i, j[y_{v+1}, y_v)$ then $C_{h_v} \supset r_{h_v} y_v$.

Because $\alpha$ is an ordinal the sequence $y_0 > y_1 > y_2 > \cdots$ must end, say with $y_m = 0$. Now $j[y_m, y_{m-1}), \cdots, j[y_1, y_0)$ make up a $j$ defined on $[y_m, y_0) = [0, \alpha) = \alpha$, and this $j$ is finite, by (18). Let $r = \zeta_{\mathfrak{R}}(i, j)$ be defined by (15). Then we have $r0 \equiv s0 \equiv E$. Because $y_m = 0$, and (18), it follows that $r_{h_m} y_m$. But $r$ clearly satisfies $r[y_{v+1}, y_v) = \zeta_{r y_{v+1}} i, j[y_{v+1}, y_v)$, for $v + 1 = m, \cdots, 1$. Therefore, starting from $r_{h_m} y_m$ and using (19) we successively obtain $r_{h_{m-1}} y_{m-1}, \cdots, r_{h_0} y_0$. Put $y_0 = \alpha$, thus $r_{h_0}\alpha$, and by (17) it follows that $O[r\alpha]$, i.e., $\mathrm{beh}^{\alpha}_{\mathfrak{R},0}(i, j)$. Thus, for any finite $i$ in $\mathrm{beh}^{\alpha}_{\mathfrak{S},0}$ we have proved the existence of a finite $j$ such that $i, j$ belongs to $\mathrm{beh}^{\alpha}_{\mathfrak{R},0}$. This establishes (b). Q.e.d.

The crucial step in this proof is the proper construction of the $W$'s from the $U$'s; the step from ($\mathrm{u}_0$) to ($\mathrm{w}_1$), and from ($\bar{\mathrm{v}}$) to (13). A simple form of this idea was used in [1], and goes back to Church's synthesis algorithm for finite automata.

In analogy to $\alpha = \omega$, one might define a set $X$ of vectors $i$ of finite predicates on $\alpha$ to be an *$\alpha$-regular event of order $p$*, in case $X$ is the $\alpha$-behavior of a $p$-automaton. In this terminology we have established the closure of the class of $\alpha$-regular events under Boolean operations (not changing the order) and projection (stepping up the order by 1). We will now show that an $\alpha$-event $X$ is regular if and only if it is definable in WST$[\alpha, <]$.

**3. Definability in weak second order theory of** $[\alpha, <]$. Note that all the interpreted theories WST$[\alpha, <]$ have the same formulas; we will call them WS-formulas or just formulas. We will say that formulas $\Sigma$ and $\Gamma$ are *$\alpha$-equivalent* in case $\Sigma \equiv \Gamma$ is valid in WST$[\alpha, <]$, i.e., in case $\Sigma$ and $\Gamma$ define the same relation (or truth value) in $[\alpha, <]$. In case $\Sigma$ and $\Gamma$ are $\alpha$-equivalent, for all ordinals $\alpha$, we will also say that $\Sigma$ and $\Gamma$ are equivalent. Note that the bounded quantifier $(\forall x)_0^y \cdots$ may be introduced in WST as an abbreviation for $(\forall x)[x < y \supset \cdots]$. We introduce the following special sorts of formulas.

*Matrix.* A formula in which $\forall$, $\exists$, and $<$ do not occur. A notation like $M[ix_1, \cdots, ix_n]$ stands for a matrix in which only the indicated atomic formulas may occur.

*Kernel.* A formula of form $(\forall x_1)(\forall x_2)_0^{x_1} \cdots (\forall x_n)_0^{x_{n-1}} M[ix_1, \cdots, ix_n] \wedge (\exists x) Q_1[ix] \wedge \cdots \wedge (\exists x) Q_m[ix]$, where $M$ and the $Q$'s are matrices.

*Prenex formulas.* A formula $(\boldsymbol{j})\Gamma(\boldsymbol{i},\boldsymbol{j})$, where $\Gamma$ is a kernel and $(\boldsymbol{j})$ is a prefix consisting of predicate quantifiers only, and where the right-most quantifier is existential. Thus a prenex formula might look thus $(\forall j_1 j_2)(\exists j_3).(\forall x)(\forall y)_0^x M[ix, jx, iy, jy] \wedge (\exists x) j_3 x \wedge (\exists x) j_1 x$, where $M$ is a matrix. Note that in prenex formulas $<$ does not occur, except in the form $(\forall x)_0^y$. The *height* of prenex formulas is defined as follows. If $\Gamma$ is a kernel, then the height of $(\exists \boldsymbol{j})\Gamma$ is $1$. If $\Sigma$ is of odd (even) height $h$, then $(\forall \boldsymbol{j})\Sigma$ (respectively $(\exists \boldsymbol{j})\Sigma$) is of height $h+1$. Thus, odd (even) height means that the prefix of $\Sigma$ starts with $\exists$ (with $\forall$).

The predicate quantifiers $\overset{*}{\exists}$ and $\overset{*}{\forall}$ are defined as follows:

$$(\overset{*}{\exists} i)\Sigma\colon (\exists i)[(\exists x) ix \wedge \Sigma] \qquad (\overset{*}{\forall} i)\Sigma\colon (\forall i)\, [(\forall x)\widetilde{ix} \vee \Sigma]\,.$$

We make the following remarks.

Lemma 5. *Let $A$ be a quantifier free formula and let (prefix) be any prefix in which $\boldsymbol{x}$ does not occur. Then one can find a quantifier free formula $B$ such that $(\exists \boldsymbol{x})(\mathit{prefix})A. \equiv .(\overset{*}{\exists}\boldsymbol{i})(\forall \boldsymbol{x})(\mathit{prefix})B$ is a logical equivalence, if $\boldsymbol{i}$ does not occur in (prefix)$A$.*

Proof. Say, for example, $\boldsymbol{x} = x_1, x_2$. Then the following are all equivalent. $(\exists x_1 x_2)(\text{prefix})A$ and $(\overset{*}{\exists} i_1 i_2)(\forall x_1 x_2).[ix_1 \wedge ix_2] \supset (\text{prefix})A$ and $(\overset{*}{\exists} i_1 i_2)(\forall x_1 x_2)(\text{prefix})\cdot[ix_1 \wedge ix_2] \supset A$. Thus $B$ can be taken to be $[ix_1 \wedge ix_2)] \supset A$.

Lemma 6. *If $A(\boldsymbol{i}, x_1, \cdots, x_n)$ is a quantifier free formula then one can find a matrix $D[\boldsymbol{i}x_1, \cdots, \boldsymbol{i}x_n]$ such that $(\forall x_1 \cdots x_n)A$ is $\alpha$-equivalent to $(\forall x_1)(\forall x_2)_0^{x_1} \cdots (\forall x_n)_0^{x_{n-1}} D$, for every $\alpha$.*

Proof. Given the quantifier free formula $A(\boldsymbol{i}, x_1, \cdots, x_n)$. Let $B(\boldsymbol{i}, x_1, \cdots, x_n)$ be the conjunction of all formulas $A(\boldsymbol{i}, y_1, \cdots, y_n)$, where $(y_1, \cdots, y_n)$ is a combination (with or without repetitions) of the variables $(x_1, \cdots, x_n)$. Because $[\alpha, <]$ is a linear order it follows that $(\forall x_1 \cdots x_n)A$ is equivalent to $(\forall x_1 \cdots x_n)\cdot x_n < x_{n-1} < \cdots < x_1 \supset B$. Note that $B$ is a propositional expression in the atomic parts $\boldsymbol{i}x_1, \cdots, \boldsymbol{i}x_n$, and $x_\nu < x_\mu$ with $1 \leqq \nu, \mu \leqq n$. Let $D$ be obtained from $B$ by replacing the atomic parts $x_\nu < x_\mu$ by $T$ or $F$ according to whether $\mu < \nu$ or $\nu \leqq \mu$. Then clearly $(\forall x_1 \cdots x_n)\cdot x_n < \cdots < x_1 \supset B$ is equivalent to $(\forall x_1 \cdots x_n)\cdot x_n < \cdots < x_1 \supset D$. Thus we have found $D$ such that $(\forall x_1 \cdots x_n)A$ is equivalent to $(\forall x_1)(\forall x_2)_0^{x_1} \cdots (\forall x_n)_0^{x_{n-1}} D$. But $D$ is a propositional expression in the atomic parts $\boldsymbol{i}x_1, \cdots, \boldsymbol{i}x_n$ only. In other words $D$ is a matrix.

Lemma 7 (prenex form lemma): *To every WS-formula $\Sigma(\boldsymbol{i})$ one can construct a prenex formula $\Gamma(\boldsymbol{i})$ which is $\alpha$-equivalent to $\Sigma$, for every $\alpha$.*

PROOF. Starting from $\Sigma(\boldsymbol{i})$ we construct successively equivalent formulas $\Sigma_1(\boldsymbol{i}), \Sigma_2(\boldsymbol{i}), \Sigma_3(\boldsymbol{i}), \Sigma_4(\boldsymbol{i}), \Gamma(\boldsymbol{i})$ which are of the following forms

$\Sigma_1$: (ind-pred)$A$

$\Sigma_2$: (pred)*(ind)$B$

$\Sigma_3$: (pred)*$(\forall x_1 \cdots x_n)C$

$\Sigma_4$: (pred)$.(\forall x_1 \cdots x_n)D \wedge (\exists x)Q_1(x) \wedge \cdots \wedge (\exists x)Q_m(x)$

$\Gamma$: prenex form

Here $A, B, C, D$ are quantifier-free formulas, $Q_1, \cdots, Q_m$ are matrices, (ind) signifies a string of individual quantifiers, (pred) signifies a string of predicate quantifiers and (pred)* indicates that $\overset{*}{\forall}$ and $\overset{*}{\exists}$ may also occur in the string. We now indicate how $\Sigma_1, \cdots, \Gamma$ are obtained, each from its predecessor.

*Step* 1: In the usual manner move quantifiers, occurring in $\Sigma$, to the front, to obtain $\Sigma_1$.

*Step* 2: Use Lemma 5 and its dual to successively move the left-most individual quantifiers to the right, to eventually obtain $\Sigma_2$ starting from $\Sigma_1$. A single step in this procedure may look like this. (pred 1)*$(\exists \boldsymbol{x})(\forall \boldsymbol{s})$(pred-ind)$G$, using Lemma 5 pass to (pred 1)*$(\overset{*}{\exists}\boldsymbol{i})(\forall \boldsymbol{x})(\forall \boldsymbol{s})$(pred-ind)$H$, interchange $(\forall \boldsymbol{x})(\forall \boldsymbol{s})$, (pred 2)*$(\forall \boldsymbol{x})$(pred-ind)$H$.

*Step* 3: Use Lemma 5 and its dual to successively decrease the number of changes (from $\forall$ to $\exists$ and $\exists$ to $\forall$) in the part (ind) of $\Sigma_2$, to eventually arrive at $\Sigma_3$. A single step in this procedure may look like this (pred 1)*$(\exists \boldsymbol{x})(\forall \boldsymbol{y})$(ind)$G$, using Lemma 5 pass to (pred 2)*$(\forall \boldsymbol{x}\boldsymbol{y})$(ind)$H$.

*Step* 4: Let $(\overset{*}{\forall} j_1), \cdots, (\overset{*}{\forall} j_p)$ be all the quantifiers of form $(\overset{*}{\forall} j)$ occurring in $\Sigma_3$. Replace them by $(\forall j_1), \cdots, (\forall j_p)$ and replace $C$ by $[C \vee \widetilde{j_1 x_1} \vee \cdots \vee \widetilde{j_p x_1}]$ in $\Sigma_3$ to obtain an equivalent $\Sigma_3'$ of form (pred)*$(\forall x_1 \cdots x_n)C'$. Note that $\overset{*}{\forall}$ does not occur in $\Sigma_3'$. Let $(\overset{*}{\exists} s_1), \cdots, (\overset{*}{\exists} s_m)$ be all the quantifiers of form $(\overset{*}{\exists} s)$ occurring in $\Sigma_3'$. Replace them by $(\exists s_1), \cdots, (\exists s_m)$ and conjoin $(\exists x)s_1 x \wedge \cdots \wedge (\exists x)s_m x$ to $(\forall x_1 \cdots x_m)C'$ in $\Sigma_3'$. This yields the formula $\Sigma_4$ equivalent to $\Sigma_3$.

*Step* 5: Using Lemma 6 replace $(\forall x_1 \cdots x_n)D$ by $(\forall x_1)(\forall x_2)_0^{x_1} \cdots (\forall x_n)_0^{x_{n-1}} M$ in $\Sigma_4$. The resulting formula $\Gamma$ is prenex, because $M, Q_1, \cdots, Q_m$ are matrices, q.e.d.

Next consider the system of formulas

$$\text{(a)} \qquad \begin{aligned} rx &\equiv (\forall t)_0^x A[rt, st, it] \\ sx &\equiv (\exists t)_0^x \boldsymbol{B}[rt, st, it] \end{aligned}$$

It determines recursively an operator $\zeta$, defined for vectors $\boldsymbol{i}$ of predicates on ordinals and taking as values vectors $\boldsymbol{r}, \boldsymbol{s}$.

LEMMA 8. *From* $\boldsymbol{A}, \boldsymbol{B}$ *one can construct a special recursion* $\mathfrak{R}$ *of order* 1, *such that* $\zeta$ *given by* (a) *and* $\zeta_{\mathfrak{R}}$ *are identical for vectors* $\boldsymbol{i}$ *of finite predicates.*

PROOF. It is sufficient to deal with a recursion of form

$$\text{(b)} \qquad \boldsymbol{r}x \equiv (\exists t)_0^x \boldsymbol{A}[\boldsymbol{r}t, \boldsymbol{i}t],$$

because (a) can be reduced to this form by replacing in (a) $\boldsymbol{r}x$ by $\sim \boldsymbol{r}x$. Now (b) is equivalent to

$$\text{(c)} \qquad \begin{aligned} \boldsymbol{r}0 &\equiv \boldsymbol{F} \qquad \boldsymbol{r}(x+1) \,.\!\equiv\!.\, \boldsymbol{r}x \vee \boldsymbol{A}[\boldsymbol{r}x, \boldsymbol{i}x] \\ \boldsymbol{r}x &\equiv (\exists t)_0^x \boldsymbol{r}t, \quad \text{if } \mathrm{Lm}_1 x. \end{aligned}$$

Let $\boldsymbol{H}[\boldsymbol{Y}, \boldsymbol{X}]$ stand for $\boldsymbol{Y} \vee \boldsymbol{A}[\boldsymbol{Y}, \boldsymbol{X}]$, and let $\boldsymbol{U}_0[\boldsymbol{Y}]$ stand for $\boldsymbol{H}[\boldsymbol{Y}, \boldsymbol{F}]$. Then (c) can be restated in the following form,

$$\text{(c}'\text{)} \qquad \begin{aligned} \boldsymbol{r}0 &\equiv \boldsymbol{F} \qquad \boldsymbol{r}(x+1) \equiv \boldsymbol{H}[\boldsymbol{r}x, \boldsymbol{i}x] \\ \boldsymbol{r}x &\equiv \lim_{t<x} \boldsymbol{r}t, \quad \text{if } \mathrm{Lm}_1 x. \end{aligned}$$

This is not a special recursion. But note that $\boldsymbol{Y} \supset \boldsymbol{U}_0[\boldsymbol{Y}]$, and therefore $\boldsymbol{Y} \supset \boldsymbol{U}_0[\boldsymbol{Y}] \supset \boldsymbol{U}_0^2[\boldsymbol{Y}] \supset \cdots$. It follows that one can find $e$ such that $\boldsymbol{U}_0^e[\boldsymbol{Y}] \equiv \boldsymbol{U}_0^{e+1}[\boldsymbol{Y}]$. If we now define $\boldsymbol{U}_1$ by

$$\text{(d)} \qquad \boldsymbol{U}_1[\boldsymbol{Y}] \equiv \boldsymbol{U}_0^e[\boldsymbol{Y}]$$

we have

$$\text{(e)} \qquad \boldsymbol{U}_1\boldsymbol{U}_0 = \boldsymbol{U}_1, \quad \boldsymbol{U}_1^2 = \boldsymbol{U}_1, \quad \boldsymbol{Y} \supset \boldsymbol{U}_1[\boldsymbol{Y}].$$

Therefore $\mathfrak{R} = \boldsymbol{F}, \boldsymbol{H}, \boldsymbol{U}_1$ is a special recursion of order 1. Its operator $\zeta_{\mathfrak{R}}$ is given by

$$\text{(c}''\text{)} \qquad \begin{aligned} \boldsymbol{r}0 &\equiv \boldsymbol{F} \qquad \boldsymbol{r}(x+1) \equiv \boldsymbol{H}[\boldsymbol{r}x, \boldsymbol{i}x] \\ \boldsymbol{r}x &\equiv \lim_{t<x} \boldsymbol{U}_1[\boldsymbol{r}t] \qquad \text{if } \mathrm{Lm}_1 x. \end{aligned}$$

It remains to show that (c′) and (c″) are equivalent for all vectors $\boldsymbol{i}$ of finite predicates. This is done by transfinite induction and comes down to showing that $\lim_{t<x} \boldsymbol{r}t \equiv \lim_{t<x} \boldsymbol{U}_1[\boldsymbol{r}t]$, if $\boldsymbol{i}$ is finite, $\mathrm{Lm}_1 x$, and $\boldsymbol{r}[0, x)$ is given by (c″).

Suppose therefore that $\lim_{t<x} \boldsymbol{U}_1[\boldsymbol{r}t] = \boldsymbol{C}$. Then, because $\boldsymbol{i}$ is finite, there is a $y < x$ such that $\boldsymbol{i}t \equiv \boldsymbol{F}, \boldsymbol{U}_1[\boldsymbol{r}t] = \boldsymbol{C}$ for all $y \leqq t < x$. By (d) it follows $\boldsymbol{U}_0^e[\boldsymbol{r}t] = \boldsymbol{C}$. Because $\boldsymbol{i}t \equiv \boldsymbol{F}$ and (c″) we have $\boldsymbol{U}_0^e[\boldsymbol{r}t] \equiv \boldsymbol{r}(t+e)$, and therefore $\boldsymbol{r}(t+e) \equiv \boldsymbol{C}$, for all $y \leqq t < x$. Thus, $\lim_{t<x} \boldsymbol{r}t = \boldsymbol{C}$. Q.e.d.

LEMMA 9. *To every kernel* $\Gamma(\boldsymbol{i})$ *one can construct an automaton* $\mathfrak{A}$ *of order* 1 *such that for every* $\alpha$, *the behavior* $\mathrm{beh}_{\mathfrak{A}}^{\alpha}$ *is equal to the set* $\hat{\boldsymbol{i}}\,\Gamma(\boldsymbol{i})$ *defined by* $\Gamma$ *in* WST$[\alpha, <]$.

PROOF. For typographical reasons let us assume that the kernel $\Gamma(\boldsymbol{i})$ is of form $(\forall x_1)(\forall x_2)_0^{x_1}(\forall x_3)_0^{x_2}M[\boldsymbol{i}x_1, \boldsymbol{i}x_2, \boldsymbol{i}x_3] \wedge (\exists x)B[\boldsymbol{i}x]$, where $M$ and $B$ are matrices. (Longer kernels can be handled quite analogously.) Because $M$ contains only monadic atomic parts, its conjunctive normal form looks as follows:

$$M[\boldsymbol{i}x_1, \boldsymbol{i}x_2, \boldsymbol{i}x_3] .\equiv. A[\boldsymbol{i}x_1, \boldsymbol{i}x_2, \boldsymbol{i}x_3] \wedge \cdots \wedge A'[\boldsymbol{i}x_1, \boldsymbol{i}x_2, \boldsymbol{i}x_3]$$

where each $A$ is of the form $A_1[\boldsymbol{i}x_1] \vee A_2[\boldsymbol{i}x_2] \vee A_3[\boldsymbol{i}x_3]$. We now distribute $(\forall x_1)$ $(\forall x_2)_0^{x_1}(\forall x_3)_0^{x_2}$ in $\Gamma(\boldsymbol{i})$ over the conjuncts of $M$ to get

(1) $\Gamma(\boldsymbol{i}) .\equiv. (\forall x_1)(\forall x_2)_0^{x_1}(\forall x_3)_0^{x_2}[A_1[\boldsymbol{i}x_1] \vee A_2[\boldsymbol{i}x_2] \vee A_3[\boldsymbol{i}x_3]] \wedge \cdots \wedge (\exists x)B[\boldsymbol{i}x]$.

The $\cdots$ stands for zero or more conjuncts of the same form as the first. For typographical reasons, let us assume that no more occur; the general case can be handled analogously. Because the variables $x_1, x_2, x_3$ are isolated, we can move the universal quantifiers in (1) inside to get

(2) $\Gamma(\boldsymbol{i}) \equiv (\forall x_1)[A_1(x_1) \vee (\forall x_2)_0^{x_1}[A_2(x_2) \vee (\forall x_3)_0^{x_2} A_3(x_3)]] \wedge (\exists x)B(x)$.

Now consider the following system of formulas,

$$(3) \qquad \begin{aligned} r_3x &\equiv (\forall t)_0^x A_3[\boldsymbol{i}t] \\ r_2x &\equiv (\forall t)_0^x \,.\, A_2[\boldsymbol{i}t] \vee r_3t \\ r_1x &\equiv (\forall t)_0^x \,.\, A_1[\boldsymbol{i}t] \vee r_2t \\ r_0x &\equiv (\exists t)_0^x B[\boldsymbol{i}t]\,. \end{aligned}$$

It determines recursively an operator $\zeta$, defined for vectors $\boldsymbol{i}$ of finite predicates on $\alpha$ and taking as values vectors $\boldsymbol{r} = \zeta\boldsymbol{i}$ of predicates $r_0, r_1, r_2, r_3$ on $\alpha + 1$. Furthermore one easily verifies that the right side of (2) is equivalent to $(\forall \boldsymbol{r}) . \boldsymbol{r} = \zeta\boldsymbol{i} \supset [r_1\alpha \wedge r_0\alpha]$, for any particular $\alpha$. Thus, for any $\alpha$,

$$(4) \qquad \Gamma(\boldsymbol{i}) \equiv (\forall \boldsymbol{r}) . \boldsymbol{r} = \zeta\boldsymbol{i} \supset [r_1\alpha \wedge r_0\alpha].$$

But, by Lemma 8, the recursive operator $\zeta$ defined by (3) is also definable by a special recursion $\mathfrak{R}$ of order 1. Now, by Definition 2, the right side of (4) just says that $\boldsymbol{i}$ belongs to the behavior of $\mathfrak{R}, O$ where the output $O[\boldsymbol{r}\alpha]$ is given by $[r_1\alpha \wedge r_0\alpha]$. Thus, $\Gamma(\boldsymbol{i})$ defines $\mathrm{beh}_{\mathfrak{R},O}^{\alpha}$, for any $\alpha$. Furthermore we have actually given a procedure for constructing the 1-automaton $\mathfrak{R}, O$. Q.e.d.

The reader will find two ideas of Skolem's in the previous proofs. Namely "replacing bounded quantifiers by primitive recursions" in the proof of Lemma 8, and "isolation of individual quantifiers" in formulas containing monadic predicates only.

It is now easy to prove our first theorem.

THEOREM 1 (SYNTHESIS THEOREM): *To every* WS-*formula* $\Sigma(\boldsymbol{i})$ *one can construct a special automaton* $\mathfrak{A}$ *such that, for any ordinal* $\alpha$, *the set* $\hat{\imath}\Sigma(\boldsymbol{i})$ *defined in* WST$[\alpha, <]$ *by* $\Sigma$ *is just the* $\alpha$-*behavior of* $\mathfrak{A}$. *Moreover, if* $\Sigma$ *is a prenex formula of height* $h$ *then* $\mathfrak{A}$ *is of order* $(h+1)$.

PROOF. By Lemma 7 we may assume that $\Sigma(\boldsymbol{i})$ is a prenex formula, say for example of height $h = 3$. Then $\Sigma(\boldsymbol{i}) .\equiv. (\exists \boldsymbol{j}_3) \sim (\exists \boldsymbol{j}_2) \sim (\exists \boldsymbol{j}_1)\Gamma(\boldsymbol{i}, \boldsymbol{j}_1, \boldsymbol{j}_2, \boldsymbol{j}_3)$ where $\Gamma$ is a kernel. By Lemma 9 one can construct a 1-automaton $\mathfrak{A}_1$ whose $\alpha$-behavior satisfies

$$\mathrm{beh}^{\alpha}_{\mathfrak{A}}(\boldsymbol{i}, \boldsymbol{j}_1, \boldsymbol{j}_2, \boldsymbol{j}_3) \equiv \Gamma(\boldsymbol{i}, \boldsymbol{j}_1, \boldsymbol{j}_2, \boldsymbol{j}_3).$$

Using the Projection Lemma 4 three times, and Lemma 2 (twice) one obtains from $\mathfrak{A}_1$ an automaton $\mathfrak{A}$ of order $h + 1 = 4$, as required. Q.e.d.

The converse to Theorem 1 is rather trivial in comparison. None of the previous lemmas is needed for its proof.

THEOREM 2 (ANALYSIS THEOREM): *For every special automaton* $\mathfrak{A}$ *one can construct a formula* $\Sigma(\boldsymbol{i})$ *such that, for every ordinal* $\alpha$, $\Sigma$ *defines in* WST$[\alpha, <]$, *the* $\alpha$-*behavior of* $\mathfrak{A}$.

PROOF. Let $\mathfrak{A} = \boldsymbol{E}, \boldsymbol{H}, \boldsymbol{U}_1, \cdots, \boldsymbol{U}_p$ be the given $p$-automaton. By Definition 3, the $\alpha$-behavior of $\mathfrak{A}$ is given by

$$\text{(a)} \qquad \mathrm{beh}^{\alpha}(\boldsymbol{i}) .\equiv. (\exists \boldsymbol{r})[(\forall x)R(x, \boldsymbol{i}, \boldsymbol{r}) \wedge O[\boldsymbol{r}\alpha]].$$

Here, $R(x, \boldsymbol{i}, \boldsymbol{r})$ is the conjunction of the formulas occurring in (3) of Section 2. Thus the formula $\Gamma(\boldsymbol{i})$ on the right side of (a) defines $\mathrm{beh}^{\alpha}(\boldsymbol{i})$. Note however that $\Gamma(\boldsymbol{i})$ is not a WS-formula because (1) $R$ and $O[\boldsymbol{r}\alpha]$ are not WS-formulas (this could easily be corrected), and (2) the quantifiers $\exists \boldsymbol{r}$ in (a) must be interpreted to range also over non-finite predicates. To remedy this second point, we note that in calculating $\boldsymbol{r}\alpha$ from $\boldsymbol{i}$ (finite!) we actually make use of $\boldsymbol{r}x$ at finitely many places $x < \alpha$, only. These "significant places" make up a finite predicate $j$. For places $x$ outside of $j$ we then may assign the value $\boldsymbol{F}$ to $\boldsymbol{r}x$. With these hints in mind consider the conjunction $S(x, y, \boldsymbol{i}, \boldsymbol{j}, \boldsymbol{r})$ of the following formulas,

$ix \not\equiv F \supset j(x+1)$

$\mathrm{Lm}_a x \wedge j(x+\omega^a) . \supset . jx$, for $0 \leqq a \leqq (p-1)$

$\mathrm{Lm}_p y \wedge jy . \supset . (\exists x)_0^y [jx \wedge \mathrm{lm}_{p-1} x \wedge (\forall t)_{x+1}^y \tilde{j}t]$

$r0 \equiv E$

$jx . \supset . r(x+1) \equiv H[rx, ix]$

$\tilde{j}(x+\omega^{a-1}) \wedge \cdots \wedge \tilde{j}(x+1) \wedge jx . \subset . r(x+\omega^a) \equiv U_a[rx]$, for $1 \leqq a \leqq (p-1)$

$x < y \wedge \mathrm{Lm}_p y \wedge (\forall t)_{x+1}^y \tilde{j}t \wedge jx . \supset . ry \equiv U_p[rx]$.

Using $U_a U_b = U_a$, for $0 \leqq b \leqq a \leqq p$ and $U_p^2 = U_p$, it is easy to verify that for "significant places" $x \leqq \alpha$ (i.e. if $jx$) these formulas assign the same value to $rx$ as the formulas (3) of Section 2. Therefore we have

(b) $$\mathrm{beh}^\alpha(i) . \equiv . (\exists rj) . (\forall xy) S(x, y, i, j, r) \wedge j\alpha \wedge O[r\alpha].$$

Furthermore, in (b), the quantifiers $(\exists rj)$ may be interpreted to range over finite predicates on $\alpha + 1$. Next we note that in (b) one may equivalently replace $j\alpha \wedge O[r\alpha]$ by the disjunction $Q(i, j, r)$ of the following formulas

$$(\exists x) . (\forall t)[t \leqq x] \wedge jx \wedge O[H[rx, ix]]$$

$$(\forall x)(\exists t)_{x+1} \mathrm{Lm}_{a-1} t \wedge (\exists x) . (\forall t)_x \widetilde{\mathrm{Lm}}_a t \wedge jx \wedge \tilde{j}(x+1) \wedge \cdots \wedge \tilde{j}(x+\omega^a) \wedge O[U_a[rx]],\ 1 \leqq a \leqq p-1$$

$$(\forall x)(\exists t)_{x+1} \mathrm{Lm}_{p-1} t \wedge (\exists x) . jx \wedge (\forall t)_{x+1} \tilde{j}t \wedge O[U_p[rx]].$$

(Note that the $p+1$ disjuncts of $Q$ correspond to the cases $[\mathrm{lm}_0 \alpha \vee \alpha = 0]$, $\mathrm{lm}_1 \alpha, \cdots, \mathrm{lm}_{p-1} \alpha, \mathrm{Lm}_p \alpha$, describing various terminal characteristics of $\alpha$.) Finally we remark that $S$ and $Q$ can be expressed by WS-formulas, because for any $0 \leqq a \leqq p-1$ the expression $y = x + \omega^a$ is WS-definable by $x < y \wedge \mathrm{Lm}_a y \wedge (\forall t)[x < t < y \supset \sim \mathrm{Lm}_a t]$. Therefore the formula $(\exists rj) . (\forall xy) S \wedge Q$ defines $\mathrm{beh}^\alpha$ in $\mathrm{WST}[\alpha, <]$, for any $\alpha$. Q.e.d.

*Remark* 1. Theorems 1 and 2 provide a clear understanding of those relations $R(i_1, \cdots, i_k)$ between finite subsets of $\alpha$, which are definable in $\mathrm{WST}[\alpha, <]$. Namely, they are just the $\alpha$-behaviors of $p$-automata, and thus can be surveyed in a satisfactory manner. We will not engage here in a detailed study of $\alpha$-behaviors; the interested reader will find it easy to generalize from the discussion of $\omega$-behaviors given in [1]. Note, that for a $p$-recursion $\mathfrak{R}$, the equivalence relation $i \equiv j(\mathfrak{R})$ given by $\zeta_{\mathfrak{R}} i = \zeta_{\mathfrak{R}} j$ is of finite index and has the congruence property $i \equiv j(\mathfrak{R}) . \supset . i\hat{\ }s \equiv j\hat{\ }s(\mathfrak{R})$.

The behavior of an automaton $\mathfrak{R}, O$ is a (finite) union of congruence classes of $\equiv(\mathfrak{R})$.

*Remark* 2. An element $x$ of $\alpha$ can be represented by the finite predicate $it \equiv [x = t]$. Therefore, we could also survey the relations of type $R(i_1, \cdots, i_n, x_1, \cdots, x_m)$, definable in WST$[\alpha, <]$, and thus obtain a complete picture of definability in these theories. In the next section we will study, in some detail, the particular case $R(x)$.

*Remark* 3. The formula $\Sigma(\boldsymbol{i})$ constructed in the proof of Theorem 2 is of the form $(\exists \boldsymbol{j})(\forall x)\Gamma(\boldsymbol{i}, \boldsymbol{j}, x)$, where $\Gamma$ is a "generalized kernel" containing only propositional connectives, atomic parts $\boldsymbol{i}t$ and $\boldsymbol{j}t$, and bounded individual quantifiers. Note also the uniformity in $\alpha$ of both Theorem 1 and 2. Thus, to every WS-formula $\Sigma(\boldsymbol{i})$ we can construct a formula $\Sigma'(\boldsymbol{i})$ of form $(\exists \boldsymbol{j})(\forall x)$(generalized kernel), such that $\Sigma(\boldsymbol{i}) \equiv \Sigma'(\boldsymbol{i})$ holds in all WST$[\alpha, <]$. Furthermore, such a $\Sigma'(\boldsymbol{i})$ can also be constructed in the dual form $(\forall \boldsymbol{j})(\exists x)$(generalized kernel). Just note that the right side of (a), in the proof of Theorem 2, can be replaced by $(\forall \boldsymbol{r})[(\forall x)R(x, \boldsymbol{i}, \boldsymbol{r}) \supset O[\boldsymbol{r}\alpha]]$.

**4. Input free special recursions.** For any $p < \omega$ the ordinal $x$ can be represented in the form $x = y + \omega^{p-1}c_{p-1} + \cdots + \omega^0 c_0$, where $\mathrm{Lm}_p y$ (note that $y$ may be 0), and $0 \leqq c_0, \cdots, c_{p-1} < \omega$. We set $c = 0$ if $y = 0$, and $c = 1$ if $\omega^p \leqq y$, and call $(c, c_{p-1}, \cdots, c_0)$ the *p-terminal character* of $x$. Suppose now that $x = y + \omega^n c_n + \cdots + \omega^0 c_0$, where $\mathrm{Lm}_\omega y$, $c_n \neq 0$, $0 \leqq c_0, \cdots, c_n < \omega$. Set $c = 0$ if $y = 0$ and $c = 1$ if $y \neq 0$, and call $(c, c_n, \cdots, c_0)$ the *ω-terminal character* of $x$, $\omega / x = y$ the *ω-head* of $x$, and $x / \omega = \omega^n c_n + \cdots + \omega^0 c_0$ the *ω-tail* of $x$. Note that $n$ may be 0, namely in case $\mathrm{Lm}_\omega x$.

Suppose now that $\mathfrak{R} = \boldsymbol{E}, \boldsymbol{U}_0[\boldsymbol{Y}], \cdots, \boldsymbol{U}_p[\boldsymbol{Y}]$ is an input free $p$-recursion, i.e., $\boldsymbol{U}_a \boldsymbol{U}_b = \boldsymbol{U}_a$ for $0 \leqq b < a \leqq p$ and $\boldsymbol{U}_p^2 = \boldsymbol{U}_p$. Thus the predicate vector $\boldsymbol{r} = \zeta_{\mathfrak{R}}$ defined by the recursion $\mathfrak{R}$ is given by

(1) $\boldsymbol{r}x \equiv \boldsymbol{U}_0^{c_0} \cdots \boldsymbol{U}_{p-1}^{c_{p-1}} \boldsymbol{U}_p^{c}[\boldsymbol{E}]$ if $(c, c_{p-1}, \cdots, c_0)$ is the $p$-character of $x$.

Here $\boldsymbol{U}^0[\boldsymbol{Y}]$ stands for $\boldsymbol{Y}$, $\boldsymbol{U}^{n+1}[\boldsymbol{Y}]$ stands for $\boldsymbol{U}[\boldsymbol{U}^n[\boldsymbol{Y}]]$. It is now easy to prove the following theorem.

DEFINITION 4. $[\alpha, <]$ and $[\beta, <]$ are called WS-*equivalent* in case the same sentences $\Sigma$ are true in WST$[\alpha, <]$ and WST$[\beta, <]$.

THEOREM 3. *$[\alpha, <]$ and $[\beta, <]$ are* WS-*equivalent just in case their ω-terminal characters are equal. For any $\alpha$, the theory* WST$[\alpha, <]$ *is decidable. In fact there is a method M which, for a given sequence of* *numbers* $(c, c_n, \cdots, c_0)$ *and a given* WS-*sentence* $\Sigma$, *decides whether or not* $\Sigma$ *is true in* $[\alpha, <]$ *if* $\alpha$ *has* $\omega$-*terminal character* $(c, c_n, \cdots, c_0)$.

PROOF. Suppose first that $[\alpha, <]$ and $[\beta, <]$ are WS-equivalent and that $(c, c_n, \cdots, c_0)$ is the $\omega$-terminal character of $\alpha$. We show, by cases, that $(c, c_n, \cdots, c_0)$ must also be the $\omega$-terminal character of $\beta$.

*Case* 1. $c = 0$. Then $\alpha = \omega^n c_n + \cdots + \omega^0 c_0$. It is easy to write a WS-sentence $\Sigma$ which holds in $[\gamma, <]$ if and only if $\gamma = \omega^n c_n + \cdots + \omega^0 c_0$. Because $[\alpha, <]$ and $[\beta, <]$ are WS-equivalent, it follows that $\Sigma$ also holds in $[\beta, <]$ and therefore $\beta = \alpha$. Thus $(0, c_n, \cdots, c_0)$ is the $\omega$-terminal character of $\beta$.

*Case* 2. $c = 1$, $n = 0$. This means $\mathrm{Lm}_\omega \alpha$, i.e., the WS-sentences $(\forall x)(\exists y)_x \mathrm{Lm}_k y$, $k = 1, 2, 3, \cdots$ all hold in $[\alpha, <]$. Because $[\alpha, <]$ and $[\beta, <]$ are WS-equivalent, all these sentences also hold in $[\beta, <]$. Therefore (1) is also the $\omega$-terminal character of $\beta$.

*Case* 3. $c = 1$, $n \neq 0$. It is easy to write WS-sentences $\Sigma_k$, $k = 1, 2, 3, \cdots$ such that $\Sigma_k$ holds in $[\gamma, <]$ just in case $(\exists y)_1^\gamma [\mathrm{Lm}_k y \wedge \gamma = y + \omega^n c_n + \cdots + \omega^0 c_0]$. Because $(1, c_n, \cdots, c_0)$ is the $\omega$-terminal character of $\alpha$, all sentences $\Sigma_k$ hold in $[\alpha, <]$. Because $[\beta, <]$ is WS-equivalent to $[\alpha, <]$, these sentences also hold in $[\beta, <]$. It follows that $(1, c_n, \cdots, c_0)$ must also be the $\omega$-terminal character of $\beta$.

Suppose next that $\alpha$ and $\beta$ have the same $\omega$-terminal character. Note that $\alpha$ and $\beta$ then have the same $p$-terminal character for any $p$. Now, let $\Sigma$ be any WS-sentence. By Theorem 1 there is an input free $p$-recursion $\mathfrak{R}$ and an output $O$ such that, for $\boldsymbol{r} = \zeta_{\mathfrak{R}}$, $\Sigma$ holds in $[\gamma, <]$ if and only if $O[\boldsymbol{r}\gamma]$. But $\alpha$ and $\beta$ have the same $p$-terminal character. Therefore, by (1), $\boldsymbol{r}\alpha \equiv \boldsymbol{r}\beta$. Thus $O[\boldsymbol{r}\alpha] \equiv O[\boldsymbol{r}\beta]$, i.e., $\Sigma$ holds in $[\alpha, <]$ if and only if it holds in $[\beta, <]$. Thus $[\alpha, <]$ and $[\beta, <]$ are WS-equivalent.

This establishes the first part of Theorem 3. It remains to describe a decision method for WST$[\alpha, <]$ , if the $\omega$-terminal character of $\alpha$ is $(c, c_n, \cdots, c_0)$. Therefore let $\Sigma$ be any WS-sentence. Using Theorem 1 we construct an input free $p$-recursion $\mathfrak{R} = \boldsymbol{E}, \boldsymbol{U}_0[\boldsymbol{Y}], \cdots, \boldsymbol{U}_p[\boldsymbol{Y}]$ and an output $O[\boldsymbol{Y}]$ such that, for $\boldsymbol{r} = \zeta_{\mathfrak{R}}$, $\Sigma$ holds in $[\alpha, <]$ if and only if $O[\boldsymbol{r}\alpha]$. We may assume that $p \geqq n$, else modify $\mathfrak{R}$ letting $\boldsymbol{U}_{p+1} = \cdots = \boldsymbol{U}_n = \boldsymbol{U}_p$. Then the $p$-terminal character of $\alpha$ is $(c, 0, \cdots, 0, c_n, \cdots, c_0)$, and by (1), $\boldsymbol{r}\alpha \equiv \boldsymbol{U}_0^{c_0} \cdots \boldsymbol{U}_n^{c_n} \boldsymbol{U}_p^{c}[\boldsymbol{E}]$. Thus, $\Sigma$ holds in $[\alpha, <]$ if and only if $O\boldsymbol{U}_0^{c_0} \cdots \boldsymbol{U}_n^{c_n} \boldsymbol{U}_p^{c}[\boldsymbol{E}]$. Thus the truth of $\Sigma$ in WST $[\alpha, <]$ can be decided by evaluating the truth value of the propositional expression $O\boldsymbol{U}_0^{c_0} \cdots \boldsymbol{U}_n^{c_n} \boldsymbol{U}_p^{c}[\boldsymbol{E}]$. Q.e.d.

DEFINITION 5. Let $h$ be an isomorphism of $[\alpha, <]$ into $[\beta, <]$. $h$ is called a WS-*embedding* of $[\alpha, <]$ into $[\beta, <]$, if for any WS-formula $\Sigma(\boldsymbol{i})$, and for any vector $\boldsymbol{i}$ of finite subsets of $\alpha$, $\Sigma(\boldsymbol{i})$ holds in $[\alpha, <]$ if and only if $\Sigma(h\boldsymbol{i})$ holds in $[\beta, <]$. If there is a WS-embedding of $[\alpha, <]$ into $[\beta, <]$, we call $[\beta, <]$ a WS-*extension* of $[\alpha, <]$.

THEOREM 4. *For any $\alpha < \beta$, $[\beta, <]$ is a* WS-*extension of $[\alpha, <]$ if and only if $\alpha$ and $\beta$ have the same $\omega$-terminal character (i.e., if and only if $[\alpha, <]$ and $[\beta, <]$ are* WS-*equivalent). If this is the case, the* WS-*embedding $h$ of $[\alpha, <]$ into $[\beta, <]$ is given by*

$$(2) \qquad hx = \begin{cases} x, & \text{if } 0 \leqq x < (\omega/\alpha) \\ (\omega/\beta) + y, & \text{if } x = (\omega/\alpha) + y,\ y < (\alpha/\omega) \end{cases}$$

*In particular, if $\omega \leqq \alpha < \beta$ then $[\omega^\beta, <]$ is a* WS-*extension of $[\omega^\alpha, <]$, and the embedding is $hx = x$.*

PROOF. Suppose $[\beta, <]$ is a WS-extension of $[\alpha, <]$. Then $[\beta, <]$ and $[\alpha, <]$ are trivially WS-equivalent, and by Theorem 3, $\alpha$ and $\beta$ have the same $\omega$-terminal character.

Suppose next that $\alpha < \beta$ have the same $\omega$-terminal character $(c, c_n, \cdots, c_0)$, so that $\alpha/\omega = \beta/\omega = \omega^n c_n + \cdots + \omega^0 c_0$. Note that $c = 1$ (else $\alpha = \beta$), thus $0 < (\omega/\alpha) < (\omega/\beta)$ are in $\mathrm{Lm}_\omega$. We are to show that (2) defines a WS-embedding $h$ of $[\alpha, <]$ into $[\beta, <]$. Let therefore $\Sigma(\boldsymbol{i})$ be any WS-formula, and let $\boldsymbol{i}$ be any vector of finite subsets of $\alpha$. By Theorem 1 there is a $p$-recursion $\mathfrak{R} = \boldsymbol{E}, \boldsymbol{H}[\boldsymbol{Y}, \boldsymbol{X}], U_1[\boldsymbol{Y}], \cdots, U_p[\boldsymbol{Y}]$ and an output $O[\boldsymbol{Y}]$ such that for $\boldsymbol{r} = \zeta_{\mathfrak{R}}\boldsymbol{i}, \boldsymbol{s} = \zeta_{\mathfrak{R}}(h\boldsymbol{i})$,

$$(3) \qquad \begin{aligned} &\Sigma(\boldsymbol{i}) \text{ holds in } [\alpha, <] \text{ if and only if } O[\boldsymbol{r}\alpha] \\ &\Sigma(h\boldsymbol{i}) \text{ holds in } [\beta, <] \text{ if and only if } O[\boldsymbol{s}\beta]. \end{aligned}$$

Let $\boldsymbol{Y} \equiv \boldsymbol{r}(\omega/\alpha)$. Because $\mathrm{Lm}_p(\omega/\alpha)$ and $\omega/\alpha \neq 0$ it follows that $U_p[\boldsymbol{Y}] \equiv \boldsymbol{Y}$. By (2) we have $(\forall t)_0^{\omega/\alpha}[\boldsymbol{i}t \equiv (h\boldsymbol{i})t]$, therefore $\boldsymbol{r}(\omega/\alpha) \equiv \boldsymbol{s}(\omega/\alpha) \equiv \boldsymbol{Y}$. By (2) we have $(\forall t)_{\omega/\alpha}^{\omega/\beta}[(h\boldsymbol{i})t \equiv \boldsymbol{F}]$, therefore $\boldsymbol{s}(\omega/\beta) \equiv U_p[\boldsymbol{s}(\omega/\alpha)]$, thus $\boldsymbol{s}(\omega/\beta) \equiv \boldsymbol{Y}$. By (2) we have $(\forall t)_0^\gamma[\boldsymbol{i}(\omega/\alpha + t) \equiv (h\boldsymbol{i})(\omega/\beta + t)]$ if $\gamma = \alpha/\omega = \beta/\omega$; furthermore $\boldsymbol{r}(\omega/\alpha) \equiv \boldsymbol{s}(\omega/\beta) \equiv \boldsymbol{Y}$. Therefore $\boldsymbol{r}\alpha \equiv \boldsymbol{s}\beta$. By (3) it now follows that $\Sigma(\boldsymbol{i})$ holds in $[\alpha, <]$ if and only if $\Sigma(h\boldsymbol{i})$ holds in $[\beta, <]$. Q.e.d.

If $\alpha < \omega^\omega$, it is easy to see that every element of $\alpha$ is definable in $\mathrm{WST}[\alpha, <]$, and therefore, every finite predicate $\boldsymbol{i} \subseteq \alpha$ is definable in $\mathrm{WST}[\alpha, <]$. Also, if $\beta < \omega^\omega$, all finite predicates on $\omega^\omega + \beta$ are definable in $\mathrm{WST}[\omega^\omega + \beta, <]$. Consider now $[\alpha + \beta, <]$ where $\mathrm{Lm}_\omega\alpha, \alpha \neq 0, \beta < \omega^\omega$. By Theorem 4, this is a WS-extension of $[\omega^\omega + \beta, <]$. It follows that $x$ is definable in WST $[\alpha+\beta,<]$ if and only if either $x<\omega^\omega$ or $\alpha \leqq x < \alpha+\beta$. Therefore,

COROLLARY 1. *If* $\mathrm{Lm}_\omega \alpha, \alpha \neq 0, \beta < \omega^\omega$ *then the finite predicate i is definable in* WST$[\alpha+\beta,<]$ *if and only if* $i \subseteq \{x; x<\omega^\omega\} \cup \{x; \alpha \leqq x < \alpha+\beta\}$, *i.e., i does not enter into* $[\omega^\omega, \alpha]$.

To further test the strength of our definability criterion (Theorems 1 and 2) we discuss definability of subsets of $\alpha$ by WS-formulas $\Sigma(x)$. Consider the formula $\Sigma'(i):(\exists x).\Sigma(x) \wedge (\forall t)[it \equiv (t=x)]$. By Theorem 1 there is a $p$-recursion $\mathfrak{R} = E, H[Y,X], U_1[Y], \cdots, U_p[Y]$ and an output $O$ such that for any finite $i$ on $\alpha$, $\Sigma'(i)$ holds in $[\alpha,<]$ if and only if $O[r\alpha]$, if $r = \zeta_{\mathfrak{R}} i$. Let $U_0[Y]$ stand for $H[Y,F], V[Y]$ for $H[Y,T]$. By (1) and the relationship between $\Sigma(x)$ and $\Sigma'(i)$ it follows that, for any $x<\alpha$, $\Sigma(x)$ holds in $[\alpha,<]$ if and only if

$$OU_0^{a_0} \cdots U_{p-1}^{a_{p-1}} U_p^a V U_0^{b_0} \cdots U_{p-1}^{b_{p-1}} U_p^b [E] \tag{4}$$

where $(b, b_{p-1}, \cdots, b_0)$ is the $p$-terminal character of $x$ and $(a, a_{p-1}, \cdots, a_0)$ is the $p$-terminal character of $-(x+1)+\alpha$. Thus,

COROLLARY 2. *Let S be a subset of* $\alpha$ *which is definable in* WST$[\alpha,<]$. *There are propositional expressions* $E, U_0[Y], \cdots, U_p[Y], V[Y], O[Y]$, $U_c U_d = U_c$ *for* $c<d$, $U_d U_d = U_d$, *such that* $x \in S$ *if and only if* (4) *holds with* $(b, b_{p-1}, \cdots, b_0) =$ *the p-terminal character of x, and* $(a, a_{p-1}, \cdots, a_0)$ $=$ *the p-terminal character of* $-(x+1)+\alpha$.

Such a set $S$ might well be called *ultimately periodic of order p*. Let us for example take the case $\alpha = \omega^\omega$. Then any $x<\omega^\omega$ is of the form $\omega^n c_n + \cdots + \omega^0 c_0$, $c_n \neq 0$. The $p$-terminal character of $-(x+1)+\omega^\omega$ is (1). The $p$-terminal character of $x$ is $(0,0,\cdots,0,c_n,\cdots,c_0)$ if $n<p$, and $(1, c_{p-1}, \cdots, c_0)$ if $p \leqq n$. If we let $Q[Y]$ stand for $O[V[Y]]$ we therefore have that $x = \omega^n c_n + \cdots + \omega^0 c_0 (c_n \neq 0)$ belongs to $S$ just in case

$$\begin{aligned} &QU_0^{c_0} \cdots U_{n-1}^{c_{n-1}} U_n^{c_n}[E], && \text{if } n<p \\ &QU_0^{c_0} \cdots U_{p-1}^{c_{p-1}} U_p[E], && \text{if } p \leqq n. \end{aligned} \tag{4'}$$

It follows that, from $\omega^p$ on, $S$ repeats itself with period $\omega^p$; i.e., $S[\omega^p, \omega^p+\omega^p) = S[y, y+\omega^p)$ for all $0<y<\omega^\omega$, $\mathrm{lm}_p y$. Note furthermore that for each $0 \leqq d < p-1$ there are numbers $e, q$ such that $U_d^e = U_d^q U_d^e$. It follows that $S$ also becomes ultimately periodic in each segment $[y, y+\omega^{d+1}]$, $\mathrm{Lm}_d y$, for any $0 \leqq d < p-1$.

The subsets of $\alpha$, definable in WST$[\alpha,<]$, thus are closely related to certain right-congruences of finite index on $[\alpha,+]$. Similarly, the study of $\alpha$-behaviors of $p$-automata (i.e. sets $S$ of vectors of finite predicates on $\alpha$, definable in WST[$\alpha$, <]) is related to certain right-congruences of finite index on well-ordered sequences of letters from a finite alphabet. We do not want to elaborate here on this extension of automata theory to the transfinite.

**5. Elementary theory of addition of ordinals.** Every ordinal $x$ can be uniquely represented in the form $x = 2^{u_1} + \cdots + 2^{u_n}$, where $u_1 > \cdots > u_n$. The finite predicate $\phi x = \{u_n, \cdots, u_1\}$ is called the *binary expansion* of $x$. For any $\alpha$, $\phi$ is a one-to-one map of $2^\alpha$ onto the class $\mathrm{Fin}_\alpha$, consisting of all finite subsets of $\alpha$. In fact $\phi$ is an isomorphism between the systems

$$(1) \qquad \begin{matrix} 2^\alpha, & P, & E, & \prec \\ \mathrm{Fin}_\alpha, & \alpha, & A, & < \end{matrix}$$

where $P = \{2^x; x < \alpha\}$, $Exy$ stands for "$x$ is a power of 2 which occurs in the binary expansion of $y$" (i.e. $Exy . \equiv . (\exists u)[x = 2^u \wedge u \in \phi y]$), $\prec$ stands for the ordering relation on $P$, $Axi$ stands for $ix$, and $<$ is the ordering relation on $\alpha$. Thus

(2) The first order theory FT[$2^\alpha$, $P, E, \prec$] is identical with the weak second order theory WST[$\alpha$, <].

Let $i \oplus j = \phi(\phi^{-1} i + \phi^{-1} j)$, for any finite predicates $i,j$ on ordinals. We note next that the algorithm for addition of natural numbers in binary expansion can be extended to ordinals. Namely $i \oplus j = s$ holds if and only if there exist an ordinal $y$ and a (finite) predicate $r$ such that

$$(3) \qquad \begin{aligned} &y \text{ is the largest } \mathrm{Lm}_1 \text{ which is surpassed by members of } j. \\ &rx \equiv F, \text{ for all } \mathrm{Lm}_1 x \\ &r(x+1) . \equiv . [ix \wedge jx] \vee [ix \wedge rx] \vee [jx \wedge rx], \text{ for all } x \\ &sx . \equiv . ix \mathbin{\overline{\vee}} jx \mathbin{\overline{\vee}} rx, \quad \text{for all } x \geqq y \ (\overline{\vee} = \text{exclusive or}) \\ &sx \equiv jx, \text{ for all } x < y \end{aligned}$$

(to see this, note that $2^u + 2^v = 2^v$ if $u + \omega \leqq v$, $2^u + 2^v = 2^v + 2^u$ if $u - v$ is finite, $2^u + 2^u = 2^{u+1}$). It is easy to express (3) as a WS-formula $B(i,j,s,r,y)$. Thus, $i \oplus j = s$ is definable by a WS-formula $(\exists ry)B$. On the other hand $P$ is definable from $E$, and $\prec$ is definable from $+$ and $P$, by elementary formulas. Thus, the theory WST[$\alpha$, <] is but a reformulation of FT[$2^\alpha$, +, $E$] differing only in the choice of primitives. In particular, we have shown

LEMMA 10. a) *For every* WS-*formula* $\Sigma_<(i_1, \cdots, i_k)$ *one can effectively construct an elementary formula* $\Sigma'_{+,E}(x_1,\cdots,x_n)$ *such that, for any finite predicates* $i_1,\cdots,i_k$ *on* $\alpha$, $\Sigma(i_1,\cdots,i_k)$ *holds in* WST $[\alpha,<]$ *if and only if* $\Sigma'(\phi^{-1}i_1,\cdots,\phi^{-1}i_k)$ *holds in* $\mathrm{FT}[2^\alpha,+,E]$.

b) *For every elementary formula* $\Sigma_{+,E}(x_1,\cdots,x_k)$ *one can construct a* WS-*formula* $\Sigma^*_<(i_1,\cdots,i_k)$ *such that for any* $x_1,\cdots,x_n<2^\alpha$, $\Sigma(x_1,\cdots,x_k)$ *holds in* $\mathrm{FT}(2^\alpha,+,E]$ *if and only if* $\Sigma^*(\phi x_1,\cdots,\phi x_k)$ *holds in* $\mathrm{WST}[\alpha,<]$.

This lemma together with Theorems 1 and 2 yield complete information on those sets of ordinals which are elementarily definable in $[2^\alpha,+,E]$. In particular, we can restate Theorems 3, 4, and 5 in the following form.

THEOREM 3′. $[2^\alpha,+,E]$ *and* $[2^\beta,+,E]$ *are elementarily equivalent just in case* $\alpha$ *and* $\beta$ *have the same* $\omega$-*terminal character. For any* $\alpha$, *the elementary theory* $\mathrm{FT}[2^\alpha,+,E]$ *is decidable.*

THEOREM 4′. *For any* $\alpha<\beta$, $[2^\alpha,+,E]$ *is elementarily embeddable into* $[2^\beta,+,E]$ *if and only if* $\alpha$ *and* $\beta$ *have the same* $\omega$-*terminal character. If this is the case, the embedding h is given by*

$$h(2^{\omega/\alpha}x_1+x_2) = 2^{\omega/\beta}x_1+x_2$$

*for* $x_1<2^\gamma, x_2<2^{\omega/\alpha}$, $\gamma=\alpha/\omega=\beta/\omega$. *In particular, if* $\omega\leqq\alpha<\beta$ *then* $[\omega^{(\omega^\beta}, +,E]$ *is an elementary extension of* $[\omega^{(\omega^\alpha)},+,E]$; *the embedding is* $hx=x$.

In particular, if $Od$ is the class of all ordinals, $[Od,+,E]$ is an elementary extension of $[\omega^{(\omega^\omega)},+,E]$. As a corollary to this we get Ehrenfeucht's result [6]: $[Od,+]$ is an elementary extension of $[\omega^{(\omega^\omega)},+]$. (Note that $2^{(\omega^\omega)}=\omega^{(\omega^\omega)}$.)

COROLLARY 1′. *If* $\mathrm{Lm}_\omega\alpha$, $\alpha\neq 0$ $\beta<\omega^\omega$ *then x is elementarily definable in* $[2^{\alpha+\beta},+,E]$ *if and only if* $x=2^\alpha x_1+x_2$ *for* $x_1<2^\beta$ *and* $x_2<2^{(\omega^\omega)}$.

## REFERENCES


[1] J. R. BÜCHI, Weak second order arithmetic and finite automata. *Zeitschrift für Math. Log. und Grundl. der Math.* **6** (1960), pp. 66–92.

[2] ———, On a decision method in restricted second order arithmetic. *Logic, Method. and Phil. of Sc., Proc.* 1960 *Int. Congress*, Stanford Univ. Press, 1962.

[3] S. FEFERMAN, Some recent work of Ehrenfeucht and Fraïssé. Summer Institute for Symbolic Logic, Cornell Univ. 1957, Commun. Research Div., Institute for Defense Analysis, 1960, pp. 201–209.

[4] S. FEFERMAN and R. L. VAUGHT, The first order properties of products of algebraic systems. *Fund. Math.* **47** (1959), pp. 57–103.

[5] R. MCNAUGHTON, Review of [1, 2], *Journ. Symb. Logic* **28** (1963), pp. 100–102.

[6] A. EHRENFEUCHT, Application of games to some problems of mathematical logic. *Bull. de l'Acad. Pol. Sci.* **5** (1957), pp. 35–37.

# DECISION METHODS IN THE THEORY OF ORDINALS[1]

BY J. RICHARD BÜCHI

Communicated by D. Scott, May 21, 1965

For an ordinal $\alpha$, let RS($\alpha$), the restricted second order theory of $[\alpha, <]$, be the interpreted formalism containing the first order theory of $[\alpha, <]$ and quantification on monadic predicate variables, ranging over all subsets of $\alpha$. For a cardinal $\gamma$, RS($\alpha$, $\gamma$) is like RS($\alpha$), except that the predicate variables are now restricted to range over subsets of $\alpha$ of cardinality less than $\gamma$. $\omega = \omega_0$ and $\omega_1$ denote the first two infinite cardinals. In this note I will outline results concerning RS($\alpha$, $\omega_0$), which were obtained in the Spring of 1964 (detailed proofs will appear in [8]), and the corresponding stronger results about RS($\alpha$, $\omega_1$), which were obtained in the Fall of 1964.

The binary expansion of natural numbers can be extended to ordinals. If $x < 2^\alpha$, let $\phi x$ be the finite subset $\{u_1, \cdots, u_n\}$ of $\alpha$, given by $x = 2^{u_1} + \cdots + 2^{u_n}$, $u_n < \cdots < u_1$. $\phi$ is a one-to-one map of $2\alpha$ onto all finite subsets of $\alpha$. Let $Exy$ stand for $(\exists u)[x = 2^u \wedge u \in \phi y]$, and note that the algorithm $i + j = s$, for addition in binary notation can be expressed in RS($\alpha$, $\omega_0$). It now is easy to see that the first order theory FT$[2^\alpha, +, E]$ is equivalent to RS($\alpha$, $\omega_0$), in the strong sense that the two theories merely differ in the choice of primitive notions; the binary expansion $\phi$ yields the translation. Similarly, RS($\alpha$, $\gamma$) can be reinterpreted as a first order theory. We will state our results in one of the two forms, and leave it to the reader to translate.

THEOREM 1. *For any $\alpha$, there is a decision method for truth of sentences in $RS(\alpha, \omega_0)$. The same sentences are true in $RS(\alpha, \omega_0)$ and $RS(\beta, \omega_0)$, if and only if, $\alpha = \beta < \omega^\omega$ or else $\alpha, \beta \geqq \omega^\omega$ and have the same $\omega$-tail.*

If $\alpha = z + \omega^y + \omega^n c_n + \cdots + \omega^0 c_0$, $y \geqq \omega$, then $z + \omega^y$ is called the *$\omega$-head* of $\alpha$, and $\omega^n c_n + \cdots + \omega^0 c_0$ is called the $\omega$-tail of $\alpha$.

THEOREM 2. *For any ordinals $\beta > \alpha > \omega^\omega$, $[2^\beta, +, E]$ is an elementary extension of $[2^\alpha, +, E]$, if and only if, $\alpha$ and $\beta$ have the same $\omega$-tail. The elementary embedding is then given by $h(2^{\alpha_0}x + y) = 2^{\beta_0}x + y$, whereby $x < 2^\tau$, $y < 2^{\alpha_0}$, $\tau$ is the common $\omega$-tail of $\alpha$ and $\beta$, $\alpha_0$ and $\beta_0$ are respectively the $\omega$-heads of $\alpha$ and $\beta$.*

---

[1] This work was supported in part by grant GP-2754 from the National Science Foundation.

Let $\alpha=\alpha_0+\tau\geqq\omega^\omega$, where $\alpha_0$ is the $\omega$-head and $\tau$ is the $\omega$-tail of $\alpha$. From Theorem 2 one easily shows: the ordinals definable in FT$[2^\alpha, +, E]$ (in FT$[2^\alpha, +]$) are those of form $2^{\alpha_0}x+y$, whereby $x<2^\tau$ and $y<2^{(\omega^\omega)}$. Actually, Theorems 1 and 2 are but samples of corollaries to Theorem 3, which completely describes the relations on ordinals definable in FT$[2^\alpha, +, E]$.

The results on definability of individuals in FT$[\omega^\alpha, +]$ have been obtained earlier by A. Ehrenfeucht [6]. His methods are quite different; a lucid presentation of this work occurs in [3]. In [3] and [4] it is stated that Ehrenfeucht also knew a decision method for FT$[\omega^\alpha, +]$. However, it seems that nobody has checked out these ideas. The first published proof of the decidability of FT$[\omega, +, E]$, i.e., of RS$(\omega, \omega)$ occurs in [1], and a similar one in [7]. These are both based on my conjecture that RS$(\omega, \omega)$ is just strong enough to express the behavior of finite automata.

The key to the understanding of RS$(\alpha, \omega_0)$ is a natural extension of deterministic finite-state recursions to the transfinite. Let $I$ (input states) and $S$ (internal states) be finite sets. An automaton $\mathfrak{A}$ on $I$, $S$ consists of an element $A\in S$ (initial state) a map $H: S\times I\rightarrow S$, a map $U: 2^S\rightarrow S$, and a subset $0\subseteq S$ (the output). Let $\sup_{t<x}(rt)$ stand for the set of all values which the function $r$ takes on cofinal to $x$, i.e. $Y\in\sup_{t<x}(rt)\cdot\equiv\cdot(\forall z)_0^x(\exists t)_z^x[rt=Y]$. $[A, H, U]$ determines recursively an operator $s[o, \alpha]=\zeta i[o, \alpha)$ from $I^\alpha$ to $S^{\alpha+1}$, namely,

$$so = A,$$

$$s(x+1) = H[sx, ix],$$

$$sx = U\left[\sup_{t<x}(st)\right], \quad x \text{ a limit.}$$

An input sequence $i[o, \alpha)$ is said to be accepted by $\mathfrak{A}$, in case $s\alpha\in 0$. Extending the proofs given in [1], one now shows,

THEOREM 3. *Let $R(i_1, \cdots, i_n)$ be a relation on finite predicates on $\alpha$. $R$ is definable in $RS(\alpha, \omega_0)$ if and only if there is an automaton $\mathfrak{A}$ such that $R$ consists of those finite $(i_1, \cdots, i_n)$ on $\alpha$, for which the input signal $i[o, \alpha)$ is accepted by $\mathfrak{A}$.*

In fact there are effective methods, (1) for the construction of $\mathfrak{A}$ from a defining formula $\Sigma$ of $R$ (synthesis), and (2) for the construction of $\Sigma$ from $\mathfrak{A}$ (analysis). Theorems 1 and 2 now follow by investigating the behavior of input-free automata.

Let us now consider RS$(\alpha, \omega_1)$. The decidability of RS$(\omega_0, \omega_1)$, i.e., RS$(\omega_0)$ was proved in [2]. It is not difficult to extend the method used in [2], replacing ordinary automata recursions by transfinite automata. The result is,

THEOREM 1'. *For any countable ordinal* $\alpha$, $RS(\alpha)$ *is decidable. For* $\alpha<\beta<\omega_1$, $RS(\alpha)$ *and* $RS(\beta)$ *are equivalent if and only if either* $\alpha=\beta<\omega^\omega$ *or* $\alpha, \beta\geqq\omega^\omega$ *and have the same* $\omega$*-tail. Furthermore,* $RS(\alpha, \omega_1)$ *is decidable for any* $\alpha$.

As in [2] we actually obtain a complete survey over definability in $\mathrm{RS}(\alpha, \omega_1)$. In particular, the analog to Theorem 2 holds.

Define the $\alpha$-behavior of an automaton $\mathfrak{A}$ to be the set $\mathrm{Bh}(\mathfrak{A}, \alpha)$ consisting of all input-signals $i[,o\,\alpha)$ which are accepted by $\mathfrak{A}$. Thus, the $\omega$-behaviors are the ordinary regular sets of finite automata theory.

THEOREM 4. *To any automaton* $\mathfrak{A}$ *with input* $(i, j)$ *one can construct an automaton* $\mathfrak{C}$ *with input* $i$, *such that for any* $\alpha\leqq\omega_1$ *and any input-signal* $i$ *of length* $<\alpha$, $i\in Bh(\mathfrak{C}, \alpha)\cdot\equiv\cdot(\exists j)(i, j)\in Beh(\mathfrak{A}, \alpha)$.

For $\alpha=\omega$ this is the well-known projection-lemma for behaviors of finite automata. The case $\alpha=\omega+1$ constitutes a significant improvement of the crucial Lemma 9 of [2], and has recently been obtained by R. McNaughton. His construction is very ingenious, and his $\mathfrak{C}$'s are by far the most intricate finite automata this writer has seen in action. The extension to $\alpha\leqq\omega_1$ is an exercise in handling ordinals. Using this improved form of Lemma 9, the definability result of [2] extends as follows,

THEOREM 3'. *To every RS-formula* $\Sigma(i_1, \cdots, i_n)$ *one can construct an automaton* $\mathfrak{A}$, *and to every automaton* $\mathfrak{A}$ *with* $2^n$*-ary input* $(i_1, \cdots, i_n)$ *one can construct an RS-formula* $\Sigma(i_1, \cdots, i_n)$, *such that for any* $\alpha<\omega_1$ *the behavior* $Bh(\mathfrak{A}, \alpha)$ *is the relation defined by* $\Sigma$ *in* $RS(\alpha, \omega_1)$.

The following problem remains unsolved: Is $\mathrm{RS}(\omega_1)$ decidable?

## BIBLIOGRAPHY

1. J. R. Büchi, *Weak second order arithmetic and finite automata*, Z. Math. Logik Grundlagen Math. **6** (1960), 66–92.

2. ———, *On a decision method in restricted second order arithmetic*, Proc. Int. Cong. Logic, Method. and Philos. Sci., 1960, Stanford Univ. Press, Stanford, Calif., 1962.

3. S. Feferman, *Some recent work of Ehrenfeucht and Fraïssé*, Summer Institute for Symbolic Logic, Cornell Univ., 1957, Commun. Research Div., Institute for Defense Analyses, 1960, pp. 201–209.

4. S. Feferman and R. L. Vaught, *The first order properties of products of algebraic systems*, Fund. Math. **47** (1959), 57–103.

5. R. McNaughton, Reviews of *Weak second order arithmetic and finite automata*

and *On a decision method in restricted second order arithmetic* by J. R. Büchi, J. Symb. Logic **28** (1963), 100–102.

**6.** A. Ehrenfeucht, *Application of games to some problems of mathematical logic*, Bull. Acad. Polon. Sci. **5** (1957), 35–37.

**7.** C. C. Elgot, *Decision problems of finite automata design and related arithmetics*, Trans. Amer. Math. Soc. **98** (1961), 21–51.

**8.** J. R. Büchi, *Transfinite automata recursions and weak second order theory of ordinals*, Proc. Int. Cong. Logic, Method. and Philos. Sci., Jerusalem, 1964 (to appear).

OHIO STATE UNIVERSITY

Reprinted from the Bulletin of the American Mathematical Society, September, 1965, Vol. 71, No. 5, pp. 767–770.

THE JOURNAL OF SYMBOLIC LOGIC
Volume 34, Number 2, June 1969

# DEFINABILITY IN THE MONADIC SECOND-ORDER THEORY OF SUCCESSOR[1]

J. RICHARD BÜCHI and LAWRENCE H. LANDWEBER

**§1. Introduction.** Let $\mathscr{D} = \langle D, P_1, P_2, \cdots\rangle$ be a relational system whereby $D$ is a nonempty set and $P_i$ is an $m_i$-ary relation on $D$. With $\mathscr{D}$ we associate the (weak) monadic second-order theory $(W)MT[\mathscr{D}]$ consisting of the first-order predicate calculus with individual variables ranging over $D$; monadic predicate variables ranging over (finite) subsets of $D$; monadic predicate quantifiers; and constants corresponding to $P_1, P_2, \cdots$. We will often use $(W)MT[\mathscr{D}]$ ambiguously to mean also the set of true sentences of $(W)MT[\mathscr{D}]$.

In this note we study variants of the structure $\langle N, '\rangle$ where $N$ is the set of natural numbers and $'$ is the successor function on $N$. Our results are a consequence of McNaughton's [7] work on the $\omega$-behavior of finite automata and the decision procedure for $MT[N, ']$ given in [1]. The former is essential as we have been unable to obtain proofs which utilize only [1]'s characterization of $\omega$-behavior. In [2] we discuss related results.

§2 studies definability in $MT[N, ']$. For every formula $C(X)$ of $MT[N, ']$ where $X$ is a vector of unary predicate variables, the relation $C(X)$ is arithmetic and, in fact, is in the Boolean algebra over $\Pi_2$. In §3, we investigate the existence of decision procedures for $(W)MT[N, ', Q]$ where $Q$ is a subset of $N$. Such theories were previously studied by Elgot and Rabin [4]. For any recursive $Q$, the decision problem for $MT[N, ', Q]$ is in $\Sigma_3 \cap \Pi_3$. We also define a recursive $Q$ for which $(W)MT[N, ', Q]$ is undecidable. This provides a rather natural example of an undecidable theory which is still arithmetic.

**§2. Definability in $MT[N, ']$.** In this section we study definability in $MT[N, ']$ with respect to the arithmetic and classical Borel hierarchies. In particular we are interested in those relations definable by formulas $C(X)$, $X$ a vector of free monadic predicate variables, of $MT[N, ']$. The main result is that every such relation is in the Boolean algebra over $\Pi_2$ of the arithmetic hierarchy. In fact, Lemma 1 below also gives this result for a wider class of $C(X)$ than are definable in $MT[N, ']$. In the following $x, y, z, \cdots$ are individual variables ranging over $N$.

Let $\Pi_0$ be the class of recursive relations on $N^n \times P(N)^k$ where $P(N)$ is the power set of $N$. $\Pi_1$ ($\Pi_2$) is the class of relations presentable in the form $(\forall y)C(y, x_1, \cdots, x_n, X_1, \cdots, X_k)$ $((\exists z)(\forall y)C(z, y, x_1, \cdots, x_n, X_1, \cdots, X_k))$ where $C$ denotes a recursive relation. Relations in $\Pi_3, \Pi_4, \cdots$ are obtained by prefixing additional alternating quantifiers to relations in $\Pi_2$. The classes

Received October 10, 1967; revised July 22, 1968.

[1] This research was supported by the National Science Foundation (Contract 4730-50-395).

$\Pi_0, \Pi_1, \cdots$ comprise the *arithmetic hierarchy*. It is well known that $\Pi_{i+1} - \Pi_i \neq \varnothing$ for all $i$. Moreover, if $\Sigma_i$ is the class of relations whose complements are in $\Pi_i$, then for all $i$, $\Pi_i \subset \Pi_{i+1} \cap \Sigma_{i+1}$. We refer the reader to Kleene [6] and Rogers [9, Chapters 14–15] for a complete discussion of the properties of the arithmetic hierarchy.

A formula $C(x_1, \cdots, x_n, X_1, \cdots, X_k)$ of $MT[N, ']$ is in $\Pi_k(\Sigma_k)$ if the corresponding relation is in $\Pi_k(\Sigma_k)$. To simplify the notation we do not distinguish between formulas and the relations they define. $X$ is always used as an abbreviation for a vector of unary predicate variables. We implicitly use the obvious correspondence between $\omega$-sequences on $\{T, F\}^k$, $k$-tuples of unary predicates on $N$ and $k$-tuples of subsets of $N$. Let $I_n = \{T, F\}^n$. $I_n{}^*$ is the set of finite sequences on $I_n$. To simplify the notation we omit the subscript on $I_n$.

A recursive operator (RO) $Z = \mathscr{A}(X)$ is an operator mapping $\omega$-sequences over the finite set $I = \{T, F\}^n$ into $\omega$-sequences over a finite set $S$ which can be presented in the form

$$Zt = \Phi(\bar{X}\phi(t)) \tag{1}$$

whereby $\bar{X}t = X0 \cdots Xt$ and $\Phi$ and $\phi$ are recursive functions from $I^*$ into $S$ and from $N$ into $N$ respectively. Sup $Z$ is the set of members of $S$ appearing infinitely often in the $\omega$-sequence $Z = Z0, Z1, \cdots$.

LEMMA 1. *Let $Z = \mathscr{A}(X)$ be a RO and $U \subseteq 2^S$. Then the relation $F(X)$ given by*

$$(\exists Z)[Z = \mathscr{A}(X) \wedge \sup Z \in U] \tag{2}$$

*is in the Boolean algebra over $\Pi_2$ of the arithmetic hierarchy.*

PROOF. $F(X)$ can be written as

$$\bigvee_{B \in U} \centerdot (\exists x)(\forall y)[y \geq x \supset \Phi(\bar{X}\phi(y)) \in B] \wedge \bigwedge_{s \in B} (\forall x)(\exists y)[y \geq x \wedge \Phi(\bar{X}\phi(y)) = s].$$

The relations given by $[y \geq x \wedge \Phi(\bar{X}\phi(y)) = s]$ and $[y \geq x \supset \Phi(\bar{X}\phi(y)) \in B]$ are recursive because $\Phi$ and $\phi$ are recursive. Hence $F(X)$ is a Boolean combination of formulas of the form $(\forall y)(\exists x)M(X, x, y)$ where $M$ is recursive so $F(X)$ is in the Boolean algebra over $\Pi_2$. Q.E.D.

A finite automata operator (FAO) is a RO $Z = \mathscr{A}(X)$ which can be presented in the form

$$Z0 = c, \qquad Zt' = H[Xt, Zt] \tag{3}$$

whereby $H: I \times S \to S$ and $c \in S$. Let $C(X)$ be a formula of $MT[N, ']$. The main definability results of [1] and [7] (see [2] for more details) state that from $C$ we can effectively construct a presentation of a FAO $Z = \mathscr{E}(X)$ as in (3) (i.e., obtain $H$, $S$, and $c$) and a $U \subseteq 2^S$ such that

$$C(X) \centerdot\equiv\centerdot (\exists Z)[Z = \mathscr{E}(X) \wedge \sup Z \in U].$$

Hence by Lemma 1 we have

THEOREM 1. *Every relation between subsets of $N$ which is definable in $MT[N, ']$ is arithmetical, and in fact occurs in the Boolean algebra over $\Pi_2$. Furthermore, given a formula $C(X_1, \cdots, X_n)$ of $MT[N, ']$ one can construct an index of the relation $C$ in the Boolean algebra over $\Pi_2$.*

In contrast, all relations $R(y_1, \cdots, y_m, X_1, \cdots, X_n)$ appearing in the function-quantifier hierarchy over recursive relations are definable in $MT[N, ', 2x]$ (see [8]).

We can also consider $C(X)$ as defining a subset of the Cantor space of $\omega$-sequences over $I$, namely, the set of $\omega$-sequences over $I$ which satisfy $C$. Those sets that are both open and closed in the usual totally disconnected topology on this space are of the form $U_{w_1} \cup \cdots \cup U_{w_n}$ whereby $w_i \in I^*$ and $U_w = \{X \mid (\exists t)[\bar{X}t = w]\}$. A set is open if it is a denumerable union of sets which are both open and closed. $G_\delta(F_\sigma)$ is the class of sets which are denumerable intersections (unions) of open (closed) sets. $G_{\delta\sigma}, G_{\delta\sigma\delta}, \cdots$ and $F_{\sigma\delta}, F_{\sigma\delta\sigma}, \cdots$ sets are defined in the obvious manner. The *Borel hierarchy* is the increasing sequence of classes $G, G_\delta, G_{\delta\sigma}, \cdots$ (see [9, Chapter 15] for a comparison of the Borel and arithmetic hierarchies).

If $C$ is recursive, there is an effective procedure which decides whether $C(X)$ or $\sim C(X)$ is true after being given some finite portion $\bar{X}t = X0 \cdots Xt$ of $X$. Hence, if $X_0$ is such that $\bar{X}_0 t = \bar{X}t$, then $C(X) \equiv C(X_0)$. This implies that every recursive set of $X$'s is open and closed. But every $C(X)$ of $MT[N, ']$ is a Boolean combination of expressions of the form $(\forall x)(\exists y)M(x, y, X)$ where for fixed $x$ and $y$ $\hat{X}M(x, y, X)$ is open and closed (since $M$ is recursive). Thus by Theorem 1 we obtain

COROLLARY 1. *If $C(X)$ is a formula of $MT[N, ']$, then the relation $C(X)$ is in the Boolean algebra over $G_\delta$ of the Borel hierarchy.*

We conclude this section with an example of a $C(X)$ of $MT[N, ']$ which is neither a $G_\delta$ nor an $F_\sigma$ (and therefore neither a $\Sigma_2$ nor a $\Pi_2$). The following remark is observed in [3].

(1) A set $C(X)$ is a $G_\delta$, if and only if, there is a set $W$ of words over $I$ such that $C(X)$ holds if and only if $w < X$ for infinitely many $w \in W$.

Here $w < X$ ($w$ is initial segment of $X$) stands for $(\exists t)\bar{X}t = w$. Now define $C(X)$ by,

$$[X0 \wedge (\forall x)(\exists y)[x \leq y \wedge Xy]] \vee [\sim X0 \wedge (\exists x)(\forall y)[x \leq y \supset \sim Xy]]. \tag{2}$$

Suppose $C$ is a $G_\delta$. Then, by (1), there exists a $W \subseteq I^*$ such that

$$C(X) .\equiv. W \cap \{w \mid w < X\} \text{ is infinite.} \tag{3}$$

Define the sequence $w_0, w_1, w_2, \cdots$ by

$$\begin{aligned} w_0 &= \text{shortest } v,\ v \in W \wedge v \text{ of form } FF^k, \\ w_{n+1} &= \text{shortest } v,\ v \in W \wedge v \text{ of form } w_n TFF^k. \end{aligned} \tag{4}$$

By (2) $F^\omega$ belongs to $C$, therefore by (3) $w_0$ exists and $F \leq w_0$. Assume inductively that $w_n$ exists and $F \leq w_n$. Then by (2) $w_n TF^\omega$ belongs to $C$, therefore by (3) $w_{n+1}$ exists and $F \leq w_{n+1}$. Thus (4) really defines a sequence of words, and clearly $w_i \in W$, $F \leq w_0 < w_1 < w_2 \cdots$. Thus, by (3) and (2), the sequence $Y$ having all $w_i$'s as initial segments belong to $C$. But this is contradictory, as $Y$ starts with $F$ and has infinitely many $T$'s. Thus $C \notin G_\delta$, and similarly one shows $\sim C \notin G_\delta$. But $x \leq y$ is definable in $MT[N, ']$, and therefore $C$ is. Consequently, (2) provides an example of a set $C$, definable in $MT[N, ']$, but neither in $G_\delta$ nor $F_\sigma$.

**§3. Decision problems for extensions of** $MT[N, ']$. Elgot and Rabin [4] have studied the existence of decision procedures for extensions of $MT[N, ']$. In particular they have shown that $MT[N, ', Q]$ is decidable if $Q$ is either of $\{x^k \mid x \in N\}$, $\{k^x \mid x \in N\}$ or $\{x! \mid x \in N\}$ where $k$ is a fixed natural number. The results are obtained by reducing the decision problem for $MT[N, ', Q]$ to that for $MT[N, ']$ and then applying the procedure given in [1]. If $Q = \{(x, 2x) \mid x \in N\}$, then the corresponding weak monadic theory is undecidable [8].

Let $Q$ be a subset of $N$. If $WMT[N, ', Q]$ is undecidable, then so is $MT[N, ', Q]$. This follows from the definability of '$X$ is a finite set' in $MT[N, ']$, by the formula $(\exists x)(\forall t)[t \geq x \supset \sim Xt]$ where $t \geq x$ is an abbreviation of $(\forall Y) . Yt \wedge (\forall w)[Yw' \supset Yw] \supset Yx$.

If $Q$ is not recursive, then $WMT[N, ', Q]$ is undecidable (e.g., $0'^{\cdots'} \in Q$ can not be effectively decided). If $Q$ is recursive, the hierarchy result of §2 can be applied to give an upper bound to the complexity of decision problems for $MT[N, ', Q]$. $\psi(y, Z)$ is a universal predicate for $\Pi_2$ if for each $P(Z) \in \Pi_2$, there is an $e_p$ such that for all $Z$, $\psi(e_p, Z) \equiv P(Z)$.

THEOREM 2. *If $Q$ is recursive, then truth in $MT[N, ', Q]$ is in $\Sigma_3 \cap \Pi_3$.*

PROOF. Let $\Psi(e, Z)$ be a universal predicate for all predicates $P(Z)$ in $\Pi_2$, which is itself in $\Pi_2$ [6]. By Theorem 1, there is a recursive function $B$ which maps every formula $\Phi(Z)$ of $MT[N, ']$ into a Boolean expression $B_\Phi$, and a recursive function $f$ which maps every formula $\Phi(Z)$ of $MT[N, ']$ into a finite sequence $f_\Phi = \langle f_{\Phi,1}, \cdots, f_{\Phi,n}\rangle$ of numbers, such that for any $Z \subseteq N$,

(1) $\quad \Phi(Z)$ holds in $MT[N, '] .\equiv. B_\Phi[\Psi(f_{\Phi,1}, Z), \cdots, \Psi(f_{\Phi,n}, Z)]$.

Let $\chi(e)$ stand for $\Psi(e, Q)$, and note that because $\Psi \in \Pi_2$ and $Q$ is recursive it follows that $\chi \in \Pi_2$. Furthermore, (1) may be restated as,

(2) $\quad \Phi(Q)$ holds in $MT[N, ', Q] .\equiv. B_\Phi[\chi(f_{\Phi,1}), \cdots, \chi(f_{\Phi,n})]$.

Note that the functions $B, f$ are recursive, and all sentences of $MT[N, ', Q]$ are of form $\Phi(Q)$ where $\Phi(Z)$ is a formula of $MT[N, ']$. It follows that (2) provides for a recursive reduction of $\{\Sigma \mid \Sigma$ true in $MT[N, ', Q]\}$ to the set $\chi$ (i.e. a Turing machine can be built which, given a sentence $\Sigma$ of $MT[N, ', Q]$ and an oracle for membership in $\chi$, decides whether or not $\Sigma$ is true). Thus, truth in $MT[N, ', Q]$ is reducible to some $\chi \in \Pi_2$. It follows, by a well-known result of Post (see [9, p. 314]), that truth in $MT[N, ', Q]$ belongs to $\Sigma_3 \cap \Pi_3$. Q.E.D.

Theorem 2 shows that for no recursive $Q$ is it possible to prove $MT[N, ', Q]$ undecidable by the standard method of showing that all recursive relations are definable.

If $Q$ is the set of primes, then $(\forall x)(\exists y)[y > x \wedge Q(y) \wedge Q(y'')]$ states the twin prime problem in $MT[N, ', Q]$. Indeed, this sentence is in the first order theory of $\langle N, ', <, Q\rangle$. Hence, the problem as to whether $(W)MT[N, ',$ primes$]$ is decidable, would seem very difficult. Namely, a positive answer would settle the twin prime problem, while on the negative side, the standard methods of proving theories undecidable is not available.

THEOREM 3. *There is a recursive $Q$ such that $WMT[N, ', Q]$ is undecidable.*[2]

PROOF. Let $R$ be a recursively enumerable set of primes which is not recursive. Let $r_1, r_2, \cdots$ be a recursive enumeration of $R$ and let $Q_0 = \{r_i^2 p_i \mid i = 1, 2, \cdots\}$,

[2] Michael O. Rabin has obtained a similar result (personal correspondence).

whereby $p_i$ is the $i$th prime. $Q_0$ is obviously recursive. To prove that $WMT[N, ', Q_0]$ is undecidable it is sufficient to show that the first order theory ($FT$) of $\langle N, M_1, M_2, \cdots, Q_0\rangle$ is undecidable whereby $M_k$ stands for the set of multiples of $k$. Just note that each $M_k$ is definable in $WMT[N, ', Q_0]$ by the formula

$$M_k(w) : (\forall X) . Xw \wedge (\forall y)[X(y + k) \supset Xy] \supset X0.$$

From the definition of $R$ and $Q_0$ we obtain

$$(*) \qquad R(k) . \equiv . k \neq 1 \wedge (\exists y)[M_{k^2}(y) \wedge Q_0(y)].$$

Let $\Sigma_k$ be the sentence $k \neq 1 \wedge (\exists y)[M_{k^2}(y) \wedge Q_0(y)]$. By (*) $\Sigma_k$ is true in $FT[N, M_1, M_2, \cdots, Q_0]$ if and only if $k \in R$. But $R$ is not recursive so there is no effective procedure for deciding truth in $FT[N, M_1, M_2, \cdots, Q_0]$. Q.E.D.

PROBLEM 1. Is there an 'interesting' recursive $Q$ such that $(W)MT[N, ', Q]$ is undecidable? How about $Q$ = primes?

Although $WMT[N, ', Q_0]$ is undecidable, we have not classified its decision problem in the arithmetic hierarchy. This suggests

PROBLEM 2. Is there a recursive $Q$ such that the decision problem for $(W)MT[N, ', Q]$ is in $\Sigma_3 \cap \Pi_3$ but not in the Boolean algebra over $\Pi_2$?

Another interesting question is,

PROBLEM 3. Is there a recursive $Q$ such that $WMT[N, ', Q]$ is decidable but $MT[N, ', Q]$ is undecidable?

A negative answer to Problem 3 should imply the decidability of $MT[N, ']$ as a consequence of the decidability of $WMT[N, ']$ ($Q = \varnothing$). Hence, a negative answer might be quite difficult.

## BIBLIOGRAPHY

[1] J. R. BÜCHI, *On a decision procedure in restricted second order arithmetic*, ***Proceedings of the international congress on logic, methodology and the philosophy of science***, Stanford University Press, Stanford, California, 1962.

[2] J. R. BÜCHI and L. H. LANDWEBER, ***Solving sequential conditions by finite state operators***, Purdue Report CSD TR 14.

[3] M. DAVIS, *Infinitary games of perfect information*, ***Advances in game theory***, Princeton University Press, Princeton, New Jersey, 1964, pp. 85–101.

[4] C. C. ELGOT and M. O. RABIN, *Decidability and undecidability of extensions of second (first) order theories of (generalized) successor*, this JOURNAL, vol. 31 (1966), pp. 169–181.

[5] S. C. KLEENE, ***Introduction to metamathematics***, Van Nostrand, New York, Amsterdam and Noordhoff, Groningen, 1952.

[6] S. C. KLEENE, *Hierarchies of number theoretic predicates*, ***Bulletin of the American Mathematical Society***, vol. 61 (1955), pp. 193–213.

[7] R. MCNAUGHTON, *Testing and generating infinite sequences by a finite automaton*, ***Information and control***, vol. 9 (1966), pp. 521–530.

[8] R. M. ROBINSON, *Restricted set theoretical definitions in arithmetic*, ***Proceedings of the American Mathematical Society***, vol. 9 (1958), pp. 238–242.

[9] H. ROGERS, JR., ***Theory of recursive functions and effective computability***, McGraw-Hill, New York, 1967.

PURDUE UNIVERSITY
UNIVERSITY OF WISCONSIN

*Zeitschr. f. math. Logik und Grundlagen d. Math.*
*Bd. 29, S. 289–312 (1983)*

# THE COMPLETE EXTENSIONS OF THE MONADIC SECOND ORDER THEORY OF COUNTABLE ORDINALS[1])

by J. RICHARD BÜCHI in West Lafayette, Indiana (U.S.A.).
and DIRK SIEFKES in Berlin (West)[2])

## 0. Introduction

Let MT[co] be the Monadic second order Theory of Countable Ordinals. Thus MT[co] has individual variables and set variables, both quantifiable. The only primitive is the symbol $<$ for the order relation. Call this language $\mathscr{L}$. The true sentences of MT[co] are those of $\mathscr{L}$ which hold in all countable ordinals. Here "countable ordinal" refers to models of set theory, say ZERMELO-FRAENKEL, plus the axiom of choice. (For details compare the appendix, Section 7). From MT[co] we get ET[co], the Elementary Theory of Countable Ordinals, by cancelling the set variables. Let $\mathscr{L}_0$ be the language of ET[co]. Occasionally we will consider the Weak monadic second order Theory of Countable Ordinals, WT[co], which has the same language as MT[co], but the set variables are interpreted as ranging over finite sets only. Both ET[co] and WT[co] are the same as the corresponding theories of all ordinals (see proposition 1.5 below). Finally, let MT[$\alpha$] be the monadic second order theory of the ordinal $\alpha$.

The theories ET[co], WT[co], and MT[co] are shown to be decidable in the papers [8], [3], and [4], respectively. For the proof arbitrary sentences of the theory are reduced to certain basic types through quantifier elimination, in a completely different way though for the elementary and the monadic case. This method of proof allows to explore thoroughly the expressive power of the theory. Thus from the procedure of quantifier eliminiation the authors of [8] extract an axiom system for ET[co], and characterize the complete extensions of ET[co] by axiom systems and through the prime models of ET[co]. Similarly [5] contains axiom systems for MT[co] and for its standard extensions MT[$\alpha$], which are derived by formalizing the decision procedure of [4]. In this paper we present the general results characterizing all complete extensions

---

[1]) Editor's note. The present paper by J. R. BÜCHI and D. SIEFKES and the two following ones by J. R. BÜCHI and CH. ZAIONTZ were written in 1972–1974, containing material from 1971 and 1972. For reasons beyond the authors' control they were not published. Since there has been much recent progress in the area, and the problem of the decidability of the monadic theory of $\omega_2$ is (in a certain sense) settled by the results of Y. GUREVICH, M. MAGIDOR, and S. SHELAH [9], it seems appropriate to publish the three papers in their original form. The two papers on axiomatization might also stimulate work on the still unsolved problem of an axiom system for the monadic theory of two successors.

[2]) This paper was written while the second author was a guest at the „Forschungsinstitut für Mathematik der ETH Zürich". The results of the first three sections were obtained while the second author was visiting at Purdue University, and were presented there in a seminar. We are indebted to JOHN DONER for a copy of a report on unpublished work of MOSTOWSKI and TARSKI [14] which he prepared while working as a research assistant for ALFRED TARSKI in Berkeley. Upon receiving that copy we reformulated, partly corrected, and completed our results; so we present them here as an extension of the work of MOSTOWSKI and TARSKI on the elementary theory of all ordinals, although they were obtained independently. The content of [14] is contained in the later publication of [8].

of MT[co]. These results vastly generalize, partly in a surprising way, the results for the elementary case.

In the first two sections we collect the necessary syntactic and model theoretic notions, and state the results from [8], [3], and [4]. In Section 3 we characterize the complete extensions of MT[co] by axiom systems (Theorem 3.6). Any nonstandard extension of ET[co] splits into $2^{\aleph_0}$ many extensions of MT[co], depending on how the subsets of the domain are chosen. In Section 4 we introduce the principle of definable choice, and show that the same elements are definable in the monadic and in the elementary case (Lemma 4.3). We distinguish three types of extensions according to the order type of the definable elements. In Section 5 we show that only the extensions in $\mathfrak{Q}_0$ satisfy the principle of definable choice. In Section 6 we generalize the notion of ultimately periodic sets to the nonstandard ordinals of Section 4, and prove that these nonstandard ordinals together with the ultimately periodic subsets form the prime models of the extensions in $\mathfrak{Q}_0$ and $\mathfrak{Q}_1$, but no model at all for the extensions in $\mathfrak{Q}_2$. We note that these results can be obtained without using Section 5. In Section 7 (appendix) we discuss how these and other results depend on the underlying set theory.

## 1. Standard extensions

The (general) models of MT[co] are of the form

$$\mathfrak{D} = \langle D, \mathfrak{P}'D; \in, <\rangle$$

where $<$ is a linear ordering on $D$, $\mathfrak{P}'D$ is a subfield of the power set $\mathfrak{P}D$ of $D$, and $\in$ is the element-relation between $D$ and $\mathfrak{P}'D$. If $D = \alpha$ is well-ordered, and thus an ordinal, we write

$$[\alpha] =_{\mathrm{df}} \langle \alpha, \mathfrak{P}\alpha; \in, <\rangle.$$

We call these structures *standard models* of MT[co]. We write $\Phi[x/m, X/M] \succ\!\!- \mathfrak{T}$ if $m \in D$, $M \in \mathfrak{P}'D$ and the formula $\Phi(x, X)$ is satisfied by $\langle m, M\rangle$ in the structure $\mathfrak{T}$. For a set $\mathfrak{A}$ of sentences let cl($\mathfrak{A}$) denote the *logical closure* of $\mathfrak{A}$. This is the set of all sentences which are implied by $\mathfrak{A}$, that is, which hold in all models of $\mathfrak{A}$. We call two structures $\mathfrak{D}$ and $\mathfrak{D}'$ *equivalent* if they satisfy the same sentences in the appropriate language.

In [4] it is shown that MT[co] is decidable. [5] shows that the following set of sentences forms an axiom system for MT[co]:

(i) the domain is well-ordered,
(ii) all limit ordinals are $\omega$-cofinal,
(iii) the domain is $\omega$-cofinal,
(iv) every family of non-empty families has a choice family.

(For details see Section 2 of [5]; especially see there p. 153 for a precise formulation of (iv).) This axiom system is called $\mathfrak{C}_0$. *Axiom system* means that $\mathfrak{C}_0$ and MT[co] have the same general models. Or equivalently: exactly the true sentences of MT[co] are derivable from $\mathfrak{C}_0$, in a logic which contains, besides the axioms and rules of the two-sorted predicate calculus, the axioms of extensionality and comprehension.

A theory T is *complete* iff T is consistent and the negation of any sentence which is not true in T, is true in T. If $\mathfrak{C}$ is an axiom system for T, and T is complete, we say that $\mathfrak{C}$ is a *complete axiom system* for T. The complete extensions of MT[co] will be characterized in Section 3 by setting up sets of sentences which, together with $\mathfrak{C}_0$, yield complete axiom systems for the extensions in question.

Obviously, MT[$\alpha$] where $\alpha$ is a countable ordinal, is a complete extension of MT[co]. Theorem 4.9 of [4] shows that only countably many of them are different:

Proposition 1.1 ([4], Theorem 4.9, p. 93). *For any countable ordinals $\alpha$ and $\beta$, the structures $[\alpha]$ and $[\beta]$ are equivalent iff either*

(i) $\alpha = \beta < \omega^\omega$, *or* (ii) $\omega^\omega \leqq \alpha, \beta$ *and $\alpha$ and $\beta$ are congruent modulo $\omega^\omega$.*

We will call the theories MT[$\alpha$] where $\alpha < \omega^\omega \cdot 2$, the *standard extensions* of MT[co]. We have to take over from [4] some more detailed information. Recall that for any natural number $q$, any ordinal $\alpha$ has a unique representation

$$\alpha = \omega^q \cdot \nu + \omega^{q-1} \cdot n_{q-1} + \ldots + \omega^0 \cdot n_0,$$

where $\nu$ is an ordinal, and $n_{q-1}, \ldots, n_0$ are natural numbers. Set $(\nu) = 0$ if $\nu = 0$, and $(\nu) = 1$ otherwise. Call the $q$-tuple $\langle(\nu), n_{q-1}, \ldots, n_0\rangle$ the *$q$-character* of $\alpha$.

Proposition 1.2 ([4], Theorem 4.8, p. 89). *Let $\Sigma$ be any sentence in the language $\mathscr{L}$. There exists a natural number $q$, a finite set $K$, operators $F_0, \ldots, F_q$ on $K$, $a \in K$, $W \subseteqq K$ such that for any countable ordinal $\alpha$ of $q$-character $\langle n_q, \ldots, n_0\rangle$,*

$$\Sigma \succ [\alpha] \quad \textit{iff} \quad F_0^{n_0} \circ \ldots \circ F_q^{n_q}(a) \in W.$$

Corollary 1.3. *Let $\Sigma, q, K, F_0, \ldots, F_q$ be as in Proposition 1.2. For $i \leqq q$ let $\langle h_i, k_i\rangle$ be the first pair $h < k$ such that $F_i^h[c] = F_i^k[c]$ for all $c \in K$.*[1]) *Let $\alpha$ and $\beta$ have the $q$-character $\langle m_q, \ldots, m_0\rangle$ and $\langle n_q, \ldots, n_0\rangle$ respectively. For $i \leqq q$ let either $n_i = m_i$ or $h_i \leqq n_i, m_i$ and $n_i \equiv m_i$ $(k_i - h_i)$. Then $\Sigma \succ [\alpha]$ iff $\Sigma \succ [\beta]$.*

Proposition 1.1 follows from Corollary 1.3. We quote another, direct consequence of Proposition 1.2.

Proposition 1.4 ([4], p. 91/92). *Let $\Sigma$ be a sentence of $\mathscr{L}$ consistent with* MT[co]. *Then $\Sigma$ has a model $[\alpha]$ where $\alpha < \omega^\omega$.*

According to [8] and [2], the theories ET[co] and WT[co] reduce to the theories of all ordinals below $\omega^\omega$ in a similar way.

Proposition 1.5. a) ([8]) *Let $\Sigma$ be a sentence of $\mathscr{L}_0$ consistent with the elementary theory of all ordinals. Then $\Sigma$ has a model $\langle\alpha; <\rangle$ where $\alpha < \omega^\omega$.* b) ([2]) *Let $\Sigma$ be a sentence of $\mathscr{L}$ consistent with the weak monadic second order theory of all ordinals. Then $\Sigma$ has a model $\langle\alpha, \mathfrak{P}_{\text{fin}}\,\alpha; \in, <\rangle$ where $\alpha < \omega^\omega$.*[2])

Part b) follows from a modified Proposition 1.2; the proof of part a) uses quantifier elimination.

## 2. Non-standard extensions

We reproduce in short some notation of Section 2 of [5].

Let $\Sigma$ be a formula containing neither $X$ nor $x$. The *relativization of $\Sigma$ to the set $X$* is obtained by replacing all quantifiers of the form

$$\binom{\forall}{\exists} z \quad \text{by} \quad \left(\begin{smallmatrix}\forall\\ \exists\end{smallmatrix} z \in X\right), \quad \text{and} \quad \binom{\forall}{\exists} Z \quad \text{by} \quad \left(\begin{smallmatrix}\forall\\ \exists\end{smallmatrix} Z \subseteqq X\right).$$

The resulting formula is denoted by $\Sigma[X]$. The *relativization of $\Sigma$ to the upper bound $x$*

[1]) See [4], Lemma 4.10, p. 95. Remember that $K$ is finite.

[2]) Here $\mathfrak{P}_{\text{fin}}\,M$ is the set of all finite subsets of $M$.

19*

and *to the lower bound* $x$ are defined as

$$\Sigma[x] \equiv_{df} \Sigma[\{t; t < x\}], \quad \text{and} \quad \Sigma\{x\} \equiv_{df} \Sigma[\{t; x \leqq t\}]$$

respectively. Every ordinal is a *limit number of order* (at least) 0. A limit number of order $k+1$ is a limit of limit numbers of order $k$. The formal definitions are:

$$LM \equiv_{df} (\forall y)(\exists z)\, y < z,$$
$$LM_0 \equiv_{df} (\forall z)\, z = z,$$
$$Lm_k(x) \equiv_{df} LM_k[x],$$
$$LM_{k+1} \equiv_{df} LM[Lm_k],$$
$$lm_k \equiv_{df} LM_k \wedge \neg LM_{k+1}. \quad lm_k(x) \equiv_{df} lm_k[x].$$

An ordinal which satisfies $lm_k$ is called a *limit number of order exact* $k$. Note that $Lm_k(0)$ for every $k$; but $lm_k(0)$ only for $k = 0$.

Lemma 2.1. *For any* $k$, *the following is true in* MT[co]:

a) $LM_{k+1} \leftrightarrow (\forall y)(\exists z \in Lm_k)\, y < z$;

b) $LM_{k+1} \rightarrow LM_k$. $\quad Lm_{k+1}(x) \rightarrow Lm_k(x)$;

c) $\neg LM_{k+1} \rightarrow \bigvee_{i=0}^{k} lm_i$. $\quad \neg Lm_{k+1}(x) \rightarrow \bigvee_{i=0}^{k} lm_i(x)$.

A limit number of order $k$ is called *trailing* if all limit numbers of order $k+1$ are smaller: formally:

$$Lm_k(x) \wedge (\forall y \in Lm_{k+1})\, y < x.$$

Obviously, a trailing limit number of order $k$ is of order exact $k$.

Mostowski-Tarski [14] use the following abbreviations to describe the complete extensions of ET[co]. For $k, q \geqq 0$ let

$S_k \equiv_{df} (\exists x)\, lm_k(x)$ (there are limit numbers of order exact $k$);

$T_{k,q} \equiv_{df} (\exists x_1, \ldots, x_q \in Lm_k)\,[x_1 < \ldots < x_q \wedge (\forall y \in Lm_{k+1})\, y < x_1]$ (there are at least $q$ trailing limit numbers of order $k$).

Let $\alpha$ be an ordinal of $(k+1)$-character $\langle n, n_k, \ldots, n_0 \rangle$. Then

$S_k \succ\!\!- [\alpha]$ iff $\alpha > \omega^k$;

$T_{k,q} \succ\!\!- [\alpha]$ iff either $n_i = 0$ for all $i < k$ and $q < n_k$, or $n_i \neq 0$ for some $i < k$ and $q \leqq n_k$.[1])

The following lemma is easily proved:

Lemma 2.2 ([14]). *For any* $k, q \geqq 0$ *the following is true in* MT[co]:

a) $S_{k+1} \rightarrow S_k$;

b) $T_{k,q+1} \rightarrow S_k$;

c) $T_{k,q+1} \rightarrow T_{k,q}$;

d) $S_k \wedge \neg S_{k+1} \rightarrow T_{k,1}$.

The above notions are used in [5] to set up complete axiom systems $\mathfrak{A}_\alpha$ for MT[$\alpha$], for all countable ordinals $\alpha$. Mostowski-Tarski [14] show that even in the simple case ET[co], there are a lot more complete extensions than these standard ones. This is motivated by the following two examples.

---

[1]) Note that this corrects a mistaken statement on p. 151 of [5].

Let $T_0$ be the extension of ET[co] by the following sentences:

$$\neg LM, \neg S_i, T_{0,q} \quad \text{for } q = 0, 1, \ldots$$

These sentences say that: there is a last element, there are no limit numbers, there are infinitely many elements. $T_0$ is consistent, since each finite subset has a (finite) model. We will see in section 3 that $T_0$ is complete. Obviously, $T_0$ is different from all standard extensions of ET[co]. Let $\omega^*$ be the order type reverse to $\omega$. One shows by standard methods that $\omega + \omega^*$ is a model of $T_0$. We call a structure $\mathfrak{D}$ a *prime model of the theory* T if $\mathfrak{D}$ is a model of T, and $\mathfrak{D}$ is elementarily embeddable into every model of T which is equivalent to $\mathfrak{D}$. Then $\omega + \omega^*$ is the prime model of $T_0$. (Note that a complete theory has at most one prime model.)

Let $T_1$ be the extension of ET[co] by the sentences $T_{k,1} \wedge \neg T_{k,2}$ for $k = 0, 1, \ldots$ The models of $T_1$ have exactly one trailing limit number of order $k$, for each $k$. Therefore $T_1$ is not a standard extension of ET[co]. We will see later in this section that $T_1$ is complete. $T_1$ is consistent: its prime model is $\omega^\omega + \ldots + \omega^2 + \omega + 2$.
By combining the ideas of these two examples one gets all the complete extensions of ET[co].

In order to make our presentation independent from the limit type of the ordinals involved, we modify slightly the above notions of [14]. The remark in front of Lemma 2.2 suggests that actually we do not want to count the trailing limit numbers of order $k$, but the trailing intervals of order type $\omega^k$, which we will call *trailing limit types of order* (exact) $k$.

It is easy to define a trailing limit type of order $k$ as the smallest number which starts a trailing interval of order type $\omega^k$. But the following approach seems to be easier, and more convenient for Section 3. (For the first definition, define $T_{k,-1}$ arbitrarily, and compare Lemma 2.4.a) below.)

$$T^*_{k,q} \equiv_{df} T_{k,q} \vee [lm_k \wedge T_{k,q-1}]$$

(there are at least $q$ trailing limit types of order $k$);

$$t^*_{k,q} \equiv_{df} T^*_{k,q} \wedge \neg T^*_{k,q+1}$$

(there are exactly $q$ trailing limit types of order $k$).

Lemma 2.3. *Let $\alpha$ be of $(k+1)$-character $\langle n, n_k, \ldots, n_0\rangle$. Then*

$$T^*_{k,q} \succ [\alpha] \quad \text{iff} \quad n_k \geqq q.$$

Lemma 2.4. *The following are true in* MT[co]:

a) $lm_k \to T_{k,1}$;

b) $\neg T^*_{k,q+1} \wedge \neg LM_{k+1} \to \bigvee_{i=0}^{q} t^*_{k,i}$;

c) $LM_k \leftrightarrow \bigwedge_{i=0}^{k-1} t^*_{i,0}$;

d) $lm_k \leftrightarrow T^*_{k,1} \wedge LM_k$.

Define $\mathfrak{C}$ as the set of all functions $c\colon \omega + 1 \to \omega \cup \{\omega + \omega^*\}$ satisfying

(i) $c(\nu) \neq 0$ for at least one $\nu$,

(ii) $c(\omega) \in \{0, 1\}$,

(iii) $c(\omega) = 1$ if $c(i) \neq 0$ for infinitely many $i$.

For $c \in \mathfrak{Q}$ let $\mathfrak{B}_c$ be the following set of sentences:

$$t^*_{j,c(j)}, \quad \text{for each } j \text{ such that} \quad c(j) \in \omega:$$

$$T^*_{j,i}, \quad \text{for } i \in \omega, \quad \text{for each } j \text{ such that} \quad c(j) = \omega + \omega^*.$$

Now we define axiom systems $\mathfrak{A}_c$, $c \in \mathfrak{Q}$, as follows:

1. case. $c(\omega) = 0$, $c(k) \neq 0$, $c(i) = 0$ for $i > k$: $\mathfrak{A}_c =_{\mathrm{df}} \mathfrak{B}_c \cup \{\neg S_{k+1}\}$.

2. case. $c(\omega) = 1$: $\mathfrak{A}_c =_{\mathrm{df}} \mathfrak{B}_c \cup \{S_j;\ j \in \omega\}$.

The following proposition slightly generalizes the result of MOSTOWSKI-TARSKI, since in [14] only successor ordinals are taken into account.

Proposition 2.5 ([14]). ET[co] $\cup\ \mathfrak{A}_c$ *for* $c \in \mathfrak{Q}$ *yield exactly the complete extensions of* ET[co].

Actually the result of [14] is stronger, since there moreover a simple axiom system for ET[co] is furnished. The resulting axiomatic extensions are shown to be complete with the help of quantifier elimination.

Perhaps Proposition 2.5 is more easily understood in the following model theoretic version.

Proposition 2.6 ([14]). *For* $c \in \mathfrak{Q}$,

$$\alpha_c =_{\mathrm{df}} \sum_{i=\omega}^{0} \omega^i \cdot c(i)$$

*is the prime model of* ET[co] $\cup\ \mathfrak{A}_c$. *These are exactly the prime models of* ET[co].

The sum in the proposition is to be understood as a generalized Cantor representation. It is a possibly infinite polynomial in $\omega$; the powers and factors of $\omega$ are either $\omega$ and $\omega + \omega^*$ respectively, or else natural numbers. Thus in general $\alpha_c$ is not an ordinal.

For some $c \in \mathfrak{Q}$ Proposition 2.5 carries over to MT[co]. This will be proved below by the methods of [4]. For the remaining $c \in \mathfrak{Q}$, MT[co] $\cup\ \mathfrak{A}_c$ is not complete. The complete extensions in those cases will be exhibited in the next section.

To get the full axiomatic version of the result, we have to combine Proposition 1.4 with the completeness result of [5].

Lemma 2.7. *Let* $\mathfrak{C}_0$ *be the axiom system for* MT[co] *of* [5], *as described in Section* 1. *If the sentence* $\Sigma$ *is consistent with* $\mathfrak{C}_0$, *then there is an ordinal* $\alpha < \omega^\omega$ *such that* $\Sigma$ *holds in* $[\alpha]$

Theorem 2.8 *Let* $c \in \mathfrak{Q}$ *be such that* $c(i) \in \omega$ *for all* $i$. *Then* $\mathfrak{C}_0 \cup \mathfrak{A}_c$ *is a complete extension of* MT[co].

Proof. Let $c \in \mathfrak{Q}$ be such that

(1) $\quad c(i) \in \omega$ for all $i$.

$\mathfrak{C}_0 \cup \mathfrak{A}_c$ is consistent by a compactness argument (see e.g. Proposition 2.6 above). If there is $k \in \omega$ such that $c(i) = 0$ for all $k \leqq i < \omega$, then $\mathfrak{A}_c$ is identical with one of the axiom systems $\mathfrak{A}_\alpha$ of [5], p. 151. In fact, then case 1 and cases 2 and 3 of the definition given there, correspond respectively to the two cases above. $\mathfrak{C}_0 \cup \mathfrak{A}_\alpha$ is complete by Theorem 2.3 on p. 151 of [5]. Thus we may assume:

(2) $\quad c(i) \neq 0$ for infinitely many $i$.

Suppose that $\mathfrak{C}_0 \cup \mathfrak{A}_c$ is not complete. Then there is a sentence $\Sigma$ such that both $\Sigma$ and $\neg\Sigma$ are consistent with $\mathfrak{C}_0 \cup \mathfrak{A}_c$. Let $q = q(\Sigma)$ be the number associated with $\Sigma$ in Proposition 1.2. Define

$$\Psi \equiv_{\mathrm{df}} S_q \wedge \bigwedge_{j=0}^{q} t^*_{j,c(j)} .$$

$\Psi$ is well-defined by (1). By (2), $\mathfrak{A}_c$ implies $\Psi$. Therefore both $\Psi \wedge \Sigma$ and $\Psi \wedge \neg\Sigma$ are consistent with $\mathfrak{C}_0$. By Lemma 2.7 there are ordinals $\alpha, \beta < \omega^\omega$ such that

$$\Psi \wedge \Sigma \succ [\alpha], \qquad \Psi \wedge \neg\Sigma \succ [\beta].$$

By Lemma 2.3 and the remark in front of Lemma 2.2, $\alpha$ and $\beta$ have the same $q$-character $\langle 1, c(q-1), \ldots, c(0)\rangle$. By Corollary 1.3

$$\Sigma \succ [\alpha] \quad \text{iff} \quad \Sigma \succ [\beta],$$

contradiction. □

It should be noted that in [5] the completeness of $\mathfrak{C}_0 \cup \mathfrak{A}_\alpha$ is used to prove that $\mathfrak{C}_0$ is an axiom system for MT[co]. Therefore one cannot reprove the completeness of $\mathfrak{C}_0 \cup \mathfrak{A}_\alpha$ by a proof similar to the above.

## 3. The complete extensions of MT[co]

It is easy to see that MT[co] $\cup\, \mathfrak{A}_c$ is not complete if $c(i) = \omega + \omega^*$ for some $i$. Let e.g.

$$c(0) = \omega + \omega^*, \qquad c(\nu) = 0 \quad \text{for } \nu \neq 0 .$$

By Proposition 2.6, $\omega + \omega^*$ is the prime model of ET[co] $\cup\, \mathfrak{A}_c$. Obviously, $[\omega + \omega^*]$ is not a model of MT[co], since it violates the minimum principle. Can we extend $\omega + \omega^*$ to a general model of MT[co]? We know from SIEFKES [16], Theorem II.2.c.1, p. 115, that $\omega$ and its ultimately periodic subsets form the prime model of MT[$\omega$]. Here $M \subseteq \omega$ is *ultimately periodic* if there are numbers $q > 0$ and $h$ such that for all $n \geqq h$:

$$n \in M \quad \text{iff} \quad n + q \in M .$$

That is, $M$ is ultimately the union of some congruence classes modulo $q$. How do we extend the periodic sets to $\omega + \omega^*$? Let $\bar{0}$ be the last element of $\omega + \omega^*$. Is then, for example,

$$\bar{0} \equiv 0 \quad (2)$$

true or false? Since congruences are expressible in $\mathscr{L}$, these considerations suggest that sentences like the above are independent from MT[co] $\cup\, \mathfrak{A}_c$.

We will see that this is the case. We will show that the above $\mathfrak{C}_0 \cup \mathfrak{A}_c$ becomes complete, if we fix the sentence

$$\bar{0} \equiv d(p, k) \quad (p^k)$$

for each prime number $p$, and each $k$, by choosing an appropriate $d(p, k)$. Of course there are dependencies. E.g., $\bar{0} \equiv 0$ (2) implies $\bar{0} \equiv 0$ ($2^k$) for all $k > 0$.

We need some more abbreviations in the language $\mathscr{L}$. First we carry over the definition of congruence from the integers to the ordinals:

$y \equiv z \quad (n) \equiv_{\mathrm{df}} (\forall U)\,\{y \in U \wedge (\forall t)\,[t \in U \leftrightarrow t + n \in U] \rightarrow z \in U\}$

($y$ is *congruent* $z$ *modulo* $n$).

Note that for $n \neq 1$, $y \equiv z\ (n)$ implies that there is no limit number between $y$ and $z$. Therefore $\equiv (n)$ has the same properties as in the integers.

$LLm_k(x) \equiv_{\mathrm{df}} Lm_k(x) \wedge (\forall y \in Lm_k)\, y \leqq x$

($y$ is the *last limit number of order* (at least!) $k$).

Lemma 3.1. *The following are true in* MT[co]:

a) $\neg LM_{k+1} \leftrightarrow (\exists x)\, LLm_k(x)$; b) $LLm_k(x) \wedge LLm_k(y) \rightarrow x = y$.

Note that $LLm_k(x) \wedge LLm_j(x)$ does not imply $k = j$.

With the help of these abbreviations we define the sentences which will yield the wanted congruences. For a prime number $p$, and natural numbers $k, j, q$, put

$P_{k,p,j,q} \equiv_{\mathrm{df}} (\exists y \in LLm_{k+1})\,(\exists z \in LLm_k)\,(\{z \equiv y + q\ (p^j)\}\ [Lm_k])$

(the last and the $q$-th trailing limit number of order $k$ are congruent modulo $p^j$, relative to $Lm_k$).

Here $y + q$ is the $q$-th successor of $y$, a notion which for every $q$ is definable from $<$. Note, however, that in order to relativize a formule correctly, all abbreviations have to be replaced by their definitions. E.g., $(y + q)\,[Lm_k]$ is the $q$-th limit number of order $k$ following $y$. The reader is advised to write out the sentence defining $P_{k,p,j,q}$ in more detail. For the same reason as in Section 2, we modify the original definition. (Define $P_{k,p,j,-1}$ arbitrarily.)

$P^*_{k,p,j,q} \equiv_{\mathrm{df}} P_{k,p,j,q} \vee [lm_k \wedge P_{k,p,j,q-1}]$.

Lemma 3.2. *Let* $\alpha$ *be an ordinal of* $(k+1)$*-character* $\langle n, n_k, \ldots, n_0\rangle$. *Then*

$P^*_{k,p,j,q} \succ [\alpha]$ *iff* $\{\omega^k \cdot n_k \equiv \omega^k \cdot q\ (p^j)\}\ [Lm_k]$
*iff* $n_k \equiv q\ (p^j)$.

Lemma 3.3. *The following two sentences are true in* MT[co]:

a) $\neg LM_{k+1} \leftrightarrow \bigvee_{i=0}^{p^j - 1} P^*_{k,p,j,i}$;

b) $P^*_{k,p,j+1,q} \rightarrow P^*_{k,p,j,q}$.

c) $P^*_{k,p,j,q} \wedge P^*_{k,p,j,r}$ *is consistent with* MT[co] *iff* $q \equiv r\ (p^j)$.

Let $\boldsymbol{P}$ be the set of prime numbers. Let $\mathfrak{R}$ be the set of all functions $d: \boldsymbol{P} \times (\omega - \{0\}) \rightarrow \omega$ satisfying for all arguments $p$ and $j$

$d(p, j) < p^j, \qquad d(p, j+1) \equiv d(p, j) \quad (p^j)$.

Lemma 3.4. *Let* $d \in \mathfrak{R}$. *If* $n \equiv d(p, k) \quad (p^k)$, *then for all* $j \leqq k$

$n \equiv d(p, j) \quad (p^j)$.

Lemma 3.5 (Chinese remainder theorem). *Let* $m_1, \ldots, m_h$ *be relatively prime, let* $n_1, \ldots, n_h$ *be arbitrary. There is a number* $q$ *which can be choosen arbitrarily large, such that* $q \equiv n_i\ (m_i)$ *for* $i = 1, \ldots, h$.

Lemma 3.4 is easy. Lemma 3.5 is familiar; for a proof see e.g. DAVIS [7], Theorem 12 of the Appendix.

For any $d \in \mathfrak{R}$ and $k \in \omega$ define a set of sentences

$$\mathfrak{D}_{k,d} =_{\mathrm{df}} \{P_{k,p,j,d(p,j)};\ p \in \boldsymbol{P},\ j > 0\}.$$

Theorem 3.6. *Let $c \in \mathfrak{Q}$. For $k \in \omega$ such that $c(k) = \omega + \omega^*$ choose $d_k \in \mathfrak{R}$. Let $D$ be the set of these $d_k$. Then*

$$\mathfrak{A}_{c,D} =_{\mathrm{df}} \mathfrak{C}_0 \cup \mathfrak{A}_c \cup \bigcup_k \{\mathfrak{D}_{k,d_k}:\ d_k \in D\}$$

*is a complete extension of* MT[co]. *One gets all complete extensions of* MT[co] *in this way.*

Proof. Let $c \in \mathfrak{Q}$, $D \subseteqq \mathfrak{R}$, and $\mathfrak{A}_{c,D}$ be as in the theorem.

a) Consistency: Let $\mathfrak{F}$ be a finite, non-empty subset of $\mathfrak{A}_{c,D}$. If $c(\omega) = 0$ and $c(i) = 0$ for but finitely many $i$, let $k$ be such that $c(k) \neq 0$ and $c(i) = 0$ for $i > k$. Otherwise choose $k$ such that $c(k) \neq 0$ and, if $S_h$, or $T^*_{h,i}$ or $t^*_{h,i}$ for some $i$, or $P^*_{h,p,j,q}$ for some $p, j, q$ occur in $\mathfrak{F}$, then $h \leqq k$. Let $h \leqq k$ be given, let $c(h) = \omega + \omega^*$. For any prime number $p$ define

$$g(p, h) =_{\mathrm{df}} \max\{j;\ P^*_{h,p,j,d_h(p,j)} \in \mathfrak{F}\}.$$

(As usual, $\max \emptyset =_{\mathrm{df}} 0$.) With the help of the Chinese remainder theorem (Lemma 3.5 above) choose $b_h > \max\{i;\ T^*_{h,i} \in \mathfrak{F}\}$ such that

$$b_h \equiv d_h(p, g(p, h))\ (p^{g(p,h)}) \quad \text{for all } p \in \boldsymbol{P} \text{ such that } P^*_{h,p,j,d_h(p,j)} \in \mathfrak{F} \text{ for some } j.$$

By Lemma 3.4,

(1) $\qquad b_h \equiv d_h(p, j)\ (p^j) \quad$ for all $p, j$ such that $P^*_{h,p,j,d_h(p,j)} \in \mathfrak{F}$.

For $h \leqq k$ such that $c(h) \in \omega$ put $b_h =_{\mathrm{df}} c(h)$. Define $\alpha =_{\mathrm{df}} \omega^k \cdot b_k + \ldots + \omega^0 \cdot b_0$. We want to show that $[\alpha]$ is a model of $\mathfrak{F}$.

Let $h \leqq k$, $p, j$ be such that $P^*_{h,p,j,d_h(p,j)} \in \mathfrak{F}$. Then $c(h) = \omega + \omega^*$. Therefore by (1) and Lemma 3.2,

$$P^*_{h,p,j,d_h(p,j)} \succ\!\!- [\alpha].$$

(Note that in case $h = k$, $y = 0$ for the $y \in Lm_{h+1}$ which occurs in the definition of $P_{h,p,j,i}$.) Again let $h \leqq k$.

1. case. $c(h) \in \omega$: Then $b_h = c(h)$. Thus by Lemma 2.3 $t^*_{h,c(h)} \succ\!\!- [\alpha]$.

2. case. $c(h) = \omega + \omega^*$: Then there is no $t^*_{h,i}$ in $\mathfrak{F}$. By the definition of $b_h$ and Lemma 2.3,

$$T^*_{h,i} \succ\!\!- [\alpha] \quad \text{for all } i \text{ such that } T^*_{h,i} \in \mathfrak{F}.$$

Finally, $\neg S_{k+1} \succ\!\!- [\alpha]$. Since $b_k \neq 0$, $S_j \succ\!\!- [\alpha]$ if $S_j \in \mathfrak{F}$.

This shows that $[\alpha]$ satisfies all sentences in $\mathfrak{F}$. Thus $\mathfrak{F}$ is consistent. Therefore $\mathfrak{A}_{c,D}$ is consistent.

b) Completeness: Suppose that $\mathfrak{A}_{c,D}$ is not complete. Then there is a sentence $\Sigma$ such that both $\Sigma$ and $\neg\Sigma$ are consistent with $\mathfrak{A}_{c,D}$. Let $q$ and $h_j < k_j$ for $j = 0, \ldots, q$ be the numbers associated with $\Sigma$ by Corollary 1.3. For $j \leqq q$ write

$$k_j - h_j = \prod_{i=0}^{f(j)} p_i^{g(j,i)}$$

as a product of prime powers.

1. case. $c(\nu) \neq 0$ for some $\nu \geqq q$: Define the sentence $\Psi$ as the conjunction of the following sentences:

$$S_q;$$

$$t^*_{j,c(j)}, \quad \text{for each } j < q \text{ such that } c(j) \in \omega;$$

$$T^*_{j,h_j} \wedge \bigwedge_{i=0}^{f(j)} P^*_{j,p_i,g(j,i),d_j(p_i,g(j,i))}, \quad \text{for each } j < q \text{ such that } c(j) = \omega + \omega^*.$$

Then $\mathfrak{A}_{c,D}$ implies $\Psi$. Therefore both $\Psi \wedge \Sigma$ and $\Psi \wedge \neg\Sigma$ are consistent with $\mathfrak{C}_0$. By Lemma 2.7 there are ordinals $\alpha, \beta < \omega^\omega$ such that

$$\Psi \wedge \Sigma \succ [\alpha], \qquad \Psi \wedge \neg\Sigma \succ [\beta].$$

Let $\langle m_q, \ldots, m_0\rangle$ and $\langle n_q, \ldots, n_0\rangle$ be the $q$-character of $\alpha$ and $\beta$ respectively. Then $m_q = n_q = 1$. If $c(j) \in \omega$ for $j < q$, then $m_j = n_j = c(j)$ by lemma 2.3. Now let $c(j) = \omega + \omega^*$ for some $j < q$. Then by Lemma 2.3 $m_j, n_j \geqq h_j$. Further for all $i \leqq f(j)$,

$$m_j \equiv d_j(p_i, g(j, i)) \quad (p_i^{g(j,i)}), \qquad n_j \equiv d_j(p_i, g(j, i)) \quad (p_i^{g(j,i)}),$$

by Lemma 3.2. Therefore $m_j \equiv n_j$ $(k_j - h_j)$. Thus by Corollary 1.3,

$$\Sigma \succ [\alpha] \quad \text{iff} \quad \Sigma \succ [\beta],$$

contradiction.

2. case. $c(\nu) = 0$ for all $\nu \geqq q$: Then $\neg S_{k+1} \in \mathfrak{A}_{c,D}$ for some $k < q$. Replace $S_q$ by $\neg S_q$ in the above definition of $\Psi$. Then the proof of the 1. case works for this case; only now $m_q = n_q = 0$.

c) Let $\mathfrak{A}$ be any complete extension of MT[co]. We want to show that $\mathrm{cl}(\mathfrak{A}) = \mathrm{cl}(\mathfrak{A}_{c,D})$ for some $c$, $D$. Let

$$\mathfrak{M} = \langle M, \mathfrak{P}'M; \in, <\rangle$$

be a model of $\mathfrak{A}$. We are finished, if we exhibit $c$, $D$ such that $\mathfrak{M}$ is a model of $\mathfrak{A}_{c,D}$. Define $c: \omega + 1 \to \omega \cup \{\omega + \omega^*\}$ as follows. For $j \in \omega$,

$$c(j) =_{\mathrm{df}} \begin{cases} i, & \text{if } t^*_{j,i} \succ \mathfrak{M}; \\ \omega + \omega^*, & \text{if } T^*_{j,i} \succ \mathfrak{M} \quad \text{for all } i. \end{cases}$$

Put $c(\omega) = 1$ if $S_j \succ \mathfrak{M}$ for all $j$; $c(\omega) = 0$ otherwise Then $c \in \mathfrak{Q}$, and $\mathfrak{M}$ is a model of $\mathfrak{C}_0 \cup \mathfrak{A}_c$. For $k$ such that $c(k) = \omega + \omega^*$, define $d_k$: $\boldsymbol{P} \times (\omega - \{0\}) \to \omega$ by

$$d_k(p, j) = i \quad \text{iff} \quad P_{k,p,j,i} \succ \mathfrak{M} \text{ and } i < p^j.$$

By Lemma 2.4.c) and Lemma 3.3.a), $d_k$ is well-defined. By Lemma 3.3.b) and c)

$$d_k(p, j + 1) \equiv d_k(p, j) \quad (p^j).$$

Therefore $d_k \in \mathfrak{R}$. Let $D$ be the set of the $d_k$. Then $\mathfrak{M}$ is a model of $\mathfrak{A}_{c,D}$. □

Note that Theorem 2.8 and its proof, are a special case of the first half of Theorem 3.6 and parts a) and b) of its proof.

Corollary 3.7. *Let $c \in \mathfrak{Q}$ be such that $c(i) = \omega + \omega^*$ for some $i$. Then* MT[co] $\cup\, \mathfrak{A}_c$ *has $2^{\aleph_0}$ many complete extensions, whereas* ET[co] $\cup\, \mathfrak{A}_c$ *is complete.*

Corollary 3.8. MT[co] *has $2^{\aleph_0}$ many complete extensions which admit no standard model; it has $\aleph_0$ many standard extensions (i.e. extensions which admit standard models).*

## 4. Prime models

In analogy to Propositions 2.5 and 2.6 we will try now to indicate a model theoretic version of Theorem 3.6. Obviously, exactly the domains of the prime models of ET[co] (Proposition 2.6) yield the first domains of the prime models of MT[co]. (Recall that a monadic second order structure has two domains, which are the range of the individual variables and the set variables respectively. At places in this section we will regard such a structure as a first order one, suitable for a two-sorted first order language.)

Let T be a consistent theory formalized in higher order logic, with a choice operator (Hilbert $\varepsilon$-operator) for every level. By the completeness theorem for higher order logic the definable sets on every level form a (general) model of T. (See MONTAGUE-VAUGHT [12], Theorem 2.) If T is a monadic second order theory, it has to contain the following choice operators. (i) A choice operator $\varepsilon_U$ for the set variables. It picks an element from every non-empty set; thus it satisfies

$$\varepsilon_U \in U \leftrightarrow (\exists x)\, x \in U .$$

By the comprehension axiom this comes up to having a choice operator $\varepsilon_x \Phi(x)$ for every formula $\Phi$, satisfying

$$\Phi(\varepsilon_x \Phi(x)) \leftrightarrow (\exists x)\, \Phi(x) .$$

(ii) A choice operator $\varepsilon_X \Phi(X)$ for every formula $\Phi$. It picks a set from every definable set of sets.

Let T be one of the monadic second order theories of the preceding section (Theorem 3.6). By the axioms of well-order, the minimum operator $\mu_x \Phi(x)$ serves in case (i), picking the smallest element from every non-empty set. In case (ii) there is no such choice operator. Even in case (i), T does not contain the minimum operator. Rather $\mu_x \Phi(x)$ is defined by the formula $\Phi(x) \wedge (\forall y)\, [\Phi(y) \rightarrow x \leqq y]$.

Let us consider case (ii). We call the formula $\Delta$ *a choice formula for the formula $\Phi$ in the theory T with respect to the variables $Z = \langle Z_1, \ldots, Z_n \rangle$*, if the following are true in T.

(1) $(\forall Y)\, (\forall Z)\, [\Delta(Y) \wedge \Delta(Z) \rightarrow Y = Z]$;

(2) $(\forall Z)\, [\Delta(Z) \rightarrow \Phi(Z)]$;

(3) $(\exists Z)\, \Phi(Z) \rightarrow (\exists Z)\, \Delta(Z)$.

$\Phi$ may contain other free variables than $Z$. For each valuation of these other variables, $\Delta$ picks a unique valuation for $Z$ satisfying $\Phi$, if there is one. Occasionally we will write $\Delta(\Phi)\,(Z)$ for $\Delta$. We say that the theory T satisfies the *axiom of definable choice* if for every formula $\Phi$ in the language of T, and for any variables $Z$, there is a choice formula for $\Phi$ in T with respect to $Z$.

Let $\mathfrak{M} = \langle M, \mathfrak{P}'M; \ldots \rangle$ be any monadic second order structure. A subset $N \in \mathfrak{P}'M$ is called *definable in* $\mathfrak{M}$ if there is a formula $\Phi$ with the only free variable $X$, such that for all $L \in \mathfrak{P}'M$

$$\Phi[X/L] \succ\!\!- \mathfrak{M} \quad \text{iff} \quad L = N .$$

An element $n \in M$ is *definable in* $\mathfrak{M}$ if it satisfies an analoguous condition. It is easily seen that $N \in \mathfrak{P}'M$ is definable in $\mathfrak{M}$ iff there is a formula $\Phi$ with the only free variable $x$, such that

$$N = \{n \in M;\ \Phi[x/n] \succ \mathfrak{M}\}.$$

That is, exactly the elements of $N$ satisfy $\Phi$ in $\mathfrak{M}$. Let $\mathfrak{M}_{\mathrm{def}}$ and $\mathfrak{P}_{\mathrm{def}}\mathfrak{M}$ be the sets of elements and sets, respectively, which are definable in $\mathfrak{M}$. Put

$$\mathrm{DEF}(\mathfrak{M}) =_{\mathrm{df}} \langle \mathfrak{M}_{\mathrm{def}}, \mathfrak{P}_{\mathrm{def}}\mathfrak{M} : \in, < \rangle.$$

Remark 4.1. *If* $\mathfrak{M}$ *and* $\mathfrak{N}$ *are equivalent, then* DEF($\mathfrak{M}$) *and* DEF($\mathfrak{N}$) *are isomorphic.*

Proof. If $m \in \mathfrak{M}_{\mathrm{def}}$ is defined by the formula $\Phi$, then $\Phi$ defines $n \in \mathfrak{N}_{\mathrm{def}}$, since $\mathfrak{M}$ and $\mathfrak{N}$ are equivalent. Put $f(m) = n$. Analoguously define $f$ on $\mathfrak{P}_{\mathrm{def}}\mathfrak{M}$. Since $\mathfrak{M}$ and $\mathfrak{N}$ are equivalent, it is easily checked that $f$ is an isomorphism between DEF($\mathfrak{M}$) and DEF($\mathfrak{N}$). □

Let T be a complete theory. In view of Remark 4.1, we call an element or a set *definable in* T if it is definable in any model of T. We write DEF(T) for DEF($\mathfrak{M}$) where $\mathfrak{M}$ is any model of T.

Proposition 4.2. *If* T *is complete, and satisfies the axiom of definable choice, then* DEF(T) *is the prime model of* T.

Proof. Let T be a complete theory satisfying the axiom of definable choice, let $\mathfrak{M}$ be a model of T. Following the proof of Theorem 2 of MONTAGUE-VAUGHT [12], we will show that $\mathfrak{M}$ is an elementary extension of DEF($\mathfrak{M}$). Then it follows from Remark 4.1 that DEF($\mathfrak{M}$) is the prime model of T. Let $\Phi$ be a formula with the free variables $x_1, \ldots, x_n, X_0, \ldots, X_k$. Let $\beta_1, \ldots, \beta_n \in \mathfrak{M}_{\mathrm{def}}$, $M_1, \ldots, M_k \in \mathfrak{P}_{\mathrm{def}}\mathfrak{M}$. Let

(1) $\qquad (\exists X_0)\, \Phi[x_1/\beta_1, \ldots, x_n/\beta_n, \ldots, X_1/M_1, \ldots, X_k/M_k] \succ \mathfrak{M}.$

Let $\Phi_i$, $i = 1, \ldots, n$, and $\Psi_i$, $i = 1, \ldots, k$, be formulas which in $\mathfrak{M}$ define $\beta_1, \ldots, \beta_n$ and $M_1, \ldots, M_k$ respectively. Let $\Psi$ be the formula

$$(\exists x^n)\,(\exists X^k)\,[\Phi \wedge \bigwedge_{i=1}^{n} \Phi_i(x_i) \wedge \bigwedge_{i=1}^{k} \Psi_i(X_i)].$$

By (1)

(2) $\qquad (\exists X_0)\, \Psi(X_0) \succ \mathfrak{M}.$

Let $\Delta$ be a choice formula for $\Psi$ in T with respect to $X_0$. By (2), $\Delta$ defines a subset $M_0$ of $M$ in $\mathfrak{M}$. Therefore $M_0 \in \mathfrak{P}_{\mathrm{def}}\mathfrak{M}$, and

(3) $\qquad \Phi[x_1/\beta_1, \ldots, x_n/\beta_n, X_0/M_0, \ldots, X_k/M_k] \succ \mathfrak{M}.$

Similarly, if $\Phi$ contains the free variables $x_0, \ldots, x_n, X_1, \ldots, X_n$,

(1′) $\qquad (\exists x_0)\, \Phi[x_1/\beta_1, \ldots, x_n/\beta_n, \ldots, X_1/M_1, \ldots, X_k/M_k] \succ \mathfrak{M}$

implies

(3′) $\qquad \Phi[x_0/\beta_0, \ldots, x_n/\beta_n, X_1/M_1, \ldots, X_k/M_k] \succ \mathfrak{M}$

for some $\beta_0 \in \mathfrak{M}_{\mathrm{def}}$. (Here the minimum principle yields the choice formula.) Therefore by the TARSKI-VAUGHT criterion (see e.g. BELL-SLOMSON [1], Lemma 4.1.8), extended to many-sorted languages, DEF($\mathfrak{M}$) is an elementary substructure of $\mathfrak{M}$. □

It should be noted that the definable sets do not always serve as a prime model. For example, if T is the elementary theory of discretely ordered sets without first or last element, then T is complete and has the prime model $\langle \omega^* + \omega; < \rangle$. But DEF(T) is empty.

In applying the above remarks to our situation we have to distinguish three cases. Recall the definition of $\mathfrak{L}$ in Section 2. Define

$$\mathfrak{L}_0 =_{df} \{c \in \mathfrak{L}: c(\omega) = 0\};$$
$$\mathfrak{L}_1 =_{df} \{c \in \mathfrak{L}: c(i) \neq 0 \text{ for infinitely many } i\};$$
$$\mathfrak{L}_2 =_{df} \{c \in \mathfrak{L}: c(\omega) = 1, c(i) = 0 \text{ for all but finitely many } i\}.$$

Then $\mathfrak{L} = \mathfrak{L}_0 \cup \mathfrak{L}_1 \cup \mathfrak{L}_2$. The results of Sections 5 and 6 are the following. For $c \in \mathfrak{L}_0$, $\mathfrak{A}_{c,D}$ satisfies the axiom of definable choice, and thus DEF($\mathfrak{A}_{c,D}$) is the prime model of $\mathfrak{A}_{c,D}$. For $c \in \mathfrak{L}_2$, neither is true; thus very likely $\mathfrak{A}_{c,D}$ has no prime model. For $c \in \mathfrak{L}_1$, $\mathfrak{A}_{c,D}$ does not satisfy the axiom of choice, but has DEF($\mathfrak{A}_{c,D}$) as the prime model.

As in Proposition 2.6, for $c \in \mathfrak{L}$ we write

$$\alpha_c =_{df} \sum_{i=\omega}^{0} \omega^i \cdot c(i); \quad \alpha_c^\nu =_{df} \sum_{i=\omega}^{\nu} \omega^i \cdot c(i) \quad \text{for } \nu \leqq \omega + 1.$$

Note that for every $c$ there is $\mu \leqq \omega$ such that $\alpha_c^\mu \neq 0$ and $\alpha_c^\nu = 0$ for $\mu < \nu \leqq \omega + 1$. In analogy to the ordinals, we identify such a "generalized ordinal" $\alpha_c$ with the set of its "predecessors". A precise definition of this set is suggested by the definition of $\alpha_c$ and $\alpha_c^\nu$, and is left to the reader. For example,

$$\omega^k \cdot (\omega + \omega^*) = \omega^{k+1} \cup \{\omega^k \cdot (\omega + \omega^* - j) + \alpha: 0 < j < \omega, \alpha < \omega^k\}.$$

Note that $\omega^{k+1}$ is not an element of $\omega^k \cdot (\omega + \omega^*)$. Similarly, $\alpha_c^\omega \in \alpha_c$ iff $c \in \mathfrak{L}_0 \cup \mathfrak{L}_2$.

Motivated by Proposition 1.1, we will abuse the above notation as follows: in a model of $\mathfrak{A}_{c,D}$, $\alpha_c^\omega$ is the last limit number of arbitrary high order, and not the first one greater then 0. Thus in $[\omega^\omega \cdot 2]$, $\alpha_c^\omega = \omega^\omega$; whereas $\alpha_c^\omega = \omega^\omega \cdot 2$ in $[\omega^\omega \cdot 3]$. Consequently, $\alpha_c^j$ for $j \in \omega$, is the last limit number of order $j$ in $\alpha_c$ (if $\alpha_c^j \in \alpha_c$).

Since the elements of $\alpha_c$ are linearly ordered, we will use notations like

$$[\alpha, \beta) =_{df} \{\mu: \alpha \leqq \mu < \beta\}.$$

Lemma 4.3. *For $c \in \mathfrak{L}$ and any $D$, exactly the elements of $\alpha_c$ are definable in $\mathfrak{A}_{c,D}$.*

Proof. For any $\beta < \omega^\omega$ it is easy to make up a formula $\Phi_\beta(w)$ which defines $\beta$ in any model where it exists. (Hint: use the notations of Sections 2 and 3. See also [5], p. 151.) Now let $c \in \mathfrak{L}$ be given. $\alpha_c$ consists of elements of the form

$$\gamma = \alpha_c^{k+1} + \beta \quad \text{where} \quad \beta < \omega^{k+1}, \quad \text{and}$$
$$\gamma = \alpha_c^k - \omega^k \cdot j + \beta \quad \text{where} \quad \beta < \omega^k, \ 0 < j < \omega.$$

Then $\gamma$ is defined by the formula

$$(\exists v)\,[LLm_{k+1}(v) \wedge \Phi_\beta(w)\,\{v\}],$$

and

$$(\exists u, v)\,[LLm_k(u) \wedge Lm_k(v) \wedge (u = v + j)\,[Lm_k] \wedge \Phi_\beta(w)\,\{v\}],$$

respectively. (If $\alpha_c^k = \alpha_c$, replace $j$ by $j - 1$ in the second formula. For the notation $\Phi\{w\}$ see the beginning of Section 2.) For example, $\omega^\omega$ is the last limit number of order 2 in $[\omega^\omega + \omega + 1]$ and in any equivalent structure. This shows that all elements of $\alpha_c$ are definable in $\mathfrak{A}_{c,D}$.

Now let $\mathfrak{M}$ be any model of $\mathfrak{A}_{c,D}$ with domain $M$. By the first half of the proof, $\alpha_c \subseteqq M$. Let $\beta \in M - \alpha_c$. Assume $\beta$ is definable in $\mathfrak{M}$ by the formula $\Phi(w)$. Since $\Phi$ does not contain free variables besides $w$, there is an "input-free automaton"

(1) $\Gamma(Z) \equiv_{\mathrm{df}} Z\,0 = s_0 \wedge (\forall t)\, Zt' = F[Zt] \wedge (\forall x)\, Zx = \mathfrak{G}[\sup^x Z]$

such that

$$\Phi(w) \leftrightarrow (\exists Z)\,(\Gamma(Z) \wedge J[Zw]).$$

(See e.g. Remark 3 of [3], p. 7; Theorem II.2.a.1 of [16], p. 103; Theorems 4.8′ and 4.8″ of [4], p. 91/92. In general, for unexplained terminology and some of the details of this proof and similar proofs below, see Sections 2 and 4 of [4]; also Sections 3 and 5 of [5].) Let $N$ be the greatest interval of $M$ containing $\beta$ which does not intersect $\alpha_c$. That is

$$N =_{\mathrm{df}} \{\alpha \in M;\ (\alpha \leqq \beta \wedge [\alpha, \beta] \cap \alpha_c = \emptyset) \vee (\beta \leqq \alpha \wedge [\beta, \alpha] \cap \alpha_c = \emptyset)\}.$$

1. case. There is a maximal number $n$ such that there is $\gamma \in N$ satisfying

(2) $\beta \leqq \gamma \wedge \gamma \in lm_n$.

Choose $\gamma$ minimal satisfying (2). ($M$ need not be well-ordered; but the minimum principle holds for definable sets.) Thus $\gamma \in N$. Note that

(3) $y, z \in Lm_n \wedge y < z \wedge (\forall u)_{y'}^{z}\ u \notin Lm_n \rightarrow [y \in \alpha_c \leftrightarrow z \in \alpha_c]$

follows immediately from the definition of $\alpha_c$. Since $\gamma \notin \alpha_c$, $\gamma$ cannot be the last limit number of order $n$ in $M$. Therefore $\gamma$ has an immediate successor $\eta$ of order at least, and thus exact, $n$. By (3), $\eta \in N$. Since

$$lm_n(x) \rightarrow (\exists y < x)\,(Lm_n(y) \wedge (x = y')\ [Lm_n])$$

is true in MT[co], $\gamma$ has an immediate predecessor $\delta$ of order at least $n$. By (3), $\delta \in N$. If $\delta$ were of order $n+1$, then $\delta$ would have a successor of order $n+1$ in $N$, similar to above. Thus $\delta$ is of order exact $n$. This shows that there is a sequence $U$ of order type $\omega^* + \omega$ satisfying $U \subseteqq lm_n \cap N$, $\gamma \in U$. Since the run $Z$ of (1) has only finitely many states, there is a number $m$ such that $Zu = Zv$ for all $u, v \in U$, $(v = u + m)\ [U]$. (In fact, this would be true in all standard models of MT[co], and is thus true in $\mathfrak{M}$. Note that $m = k_n - h_n$ in the notation of Corollary 1.3.) This implies that $\{Zt;\ u \leqq t \leqq v\} = \{Zt;\ v \leqq t \leqq w\}$ for all $u, v, w \in U$, where

$$u < v < w \wedge (u \equiv v\ (m) \wedge v \equiv w\ (m))\ [U].$$

Since $\beta$ satisfies $\Phi$, $J[Z\beta]$ holds. Let $(\lambda = \delta + m)\ [U]$, $(\mu = \lambda + m)\ [U]$. Since $\beta \in (\delta, \lambda]$, there is $\nu \in (\lambda, \mu]$ such that $J[Z\nu]$ holds. Thus $\nu$ satisfies $\Phi$, contradiction.

2. case. Let $\gamma$ be minimal such that $\beta \leqq \gamma$, $\gamma \in N$, $\gamma \in Lm_n$ for all $n$. An analogous argument, involving the number $q$ of Corollary 1.3 besides $h_i$, $k_i$, leads to a contradiction as in the first case. □

The proof of Lemma 4.3 shows that the nonstandard models of $\mathfrak{A}_{c,D}$ arise from $\alpha_c$ by repeatedly inserting portions of the form $\omega^j \cdot (\omega^* + \omega)$, and of the form $\omega^\omega$ if $c(\omega) = 1$.

**5. The axiom of definable choice**

By Theorem 4.3 on p. 169 of [5], MT[$\alpha$] for $\alpha < \omega^\omega$ satisfies the axiom of definable choice. The problem stated there concerning MT[$\alpha$] for $\alpha \geqq \omega^\omega$, and MT[co], is solved in the negative as follows.

Theorem 5.1. MT[$\alpha$] *for* $\omega^\omega \leqq \alpha$, *and thus* MT[co], *do not satisfy the axiom of definable choice.*

Proof. Consider the formula

$$\Phi(U) \equiv_{\mathrm{df}} (\forall y)\,(\exists z \in U)\; y \leqq z \wedge \neg LM_2[U].$$

In a domain of limit type, $\Phi(U)$ says: $U$ is a cofinal $\omega$-sequence. Assume that $\Delta$ is a choice formula for $\Phi$ in MT[$\omega^\omega$] with respect to $U$. By Theorem 4.5 of [4] we may assume that $\Delta$ is in "deterministic automata normal form"

$$(\exists Z).\qquad Z\,0 = s_0 \wedge (\forall t)\; Zt' = H[Ut, Zt] \wedge (\forall x)\; Zx = \mathfrak{G}[\sup^x Z] \wedge \mathfrak{Y}[\sup Z].$$

Here

$$\sup Z =_{\mathrm{df}} \{s\colon (\forall y)\,(\exists t \geqq y)\; Zt = s\}, \qquad \sup^x Z =_{\mathrm{df}} \sup Z[x]$$

is the set of "states" of $Z$ which occur cofinal (in the domain, and below $x$, respectively). Since $\Delta$ is a choice formula for $\Phi$, there is a unique set $M$ satisfying $\Delta$ in $[\omega^\omega]$. By definition of $\Phi$,

(1) $\quad M$ is a cofinal $\omega$-sequence in $\omega^\omega$.

Now consider the recursion in $\omega^\omega$,

(2) $\quad Z\,0 = s_0 \wedge (\forall t)\; Zt' = H[Ut, Zt] \wedge (\forall x)\; Zx = \mathfrak{G}[\sup^x Z]$.

If in (2) we replace $Ut$ by $t \in M$, we get the "run" $Z_1$ of $\Delta$ over $M$ (in $\omega^\omega$). If in (2) we replace $Ut$ by $F$ (False), and moreover $s_0$ by any state $s$, we get the run $Z_s$ of $\Delta_s$ over the empty set. By Proposition 1.2 there is a number $q$ such that for all $s$ and all $\nu < \omega^\omega$, $\nu \neq 0$,

(3) $\quad Z_s\omega^q = Z_s\omega^q \cdot \nu$.

(See the proof of Theorem 4.8 of [4], especially Lemma 4.7 on p. 88 of [4].) By (1), there is $\alpha < \omega^\omega$ such that

(4) $\quad \sup^{\omega^\omega} Z_1 = \{Z_1 t;\, \alpha + \omega^q \leqq t < \omega^\omega\}$,

and

(5) $\quad M \cap [\alpha, \alpha + \omega^q) = \emptyset$.

By (1) and (4), there is $\beta > \alpha + \omega^q$, $\beta < \omega^\omega$ such that

(6) $\quad Z_1(\alpha + \omega^q) = Z_1\beta, \qquad M \cap [\alpha, \beta) \neq \emptyset$.

Let $c =_{\mathrm{df}} Z_1\alpha$, $d =_{\mathrm{df}} Z_1(\alpha + \omega^q)$. By (5), $Z_c\omega^q = d$. By (3) there is $\gamma < \omega^\omega$ such that $Z_c\gamma = d$ and

(7) $\quad \alpha + \gamma \geqq \beta$.

Define

$$N =_{\mathrm{df}} (M \cap [0, \alpha]) \cup \{\alpha + \gamma + t;\, \beta + t \in M,\, t < \omega^\omega\}.$$

By (7), $N \cap [\alpha, \beta) = \emptyset$. Thus by (6), $N \neq M$. Define $Z_2$ by

$$Z_2t = Z_1t \quad \text{for } t < \alpha; \quad Z_2(\alpha + t) = Z_ct \quad \text{for } t < \gamma:$$
$$Z_2(\alpha + \gamma + t) = Z_1(\beta + t) \quad \text{for } t < \omega^\omega.$$

Then $Z_2$ is the run of $\Delta$ over $N$. That is, (2) becomes true if we replace $Z$ and $Ut$ by $Z_1$ and $t \in N$ respectively. Since

$$\sup^{\omega^\omega} Z_2 = \sup^{\omega^\omega} Z_1.$$

$N$ satisfies $\Delta$ in $[\omega^\omega]$. This contradicts the fact that $\Delta$ is a choice formula. Therefore $\mathrm{MT}[\omega^\omega]$ does not satisfy the axiom of definable choice.

The same result holds for MT[co], which is a subtheory of $\mathrm{MT}[\omega^\omega]$.

Now let $\beta > \omega^\omega$. Assume that $\mathrm{MT}[\beta]$ satisfies the axiom of definable choice. Relativize the above formula $\Phi$ to $x$. $\Phi(U)\,[x]$ says that $U$ is an $\omega$-sequence cofinal in $x$. Let $\Delta(U, x)$ be a choice formula for $\Phi(U)\,[x]$. Since $(\forall x)\,(\exists U)\,\Phi(U)\,[x] \in \mathrm{MT}[\beta]$, the same holds for $\Delta$. Therefore there is a unique $\omega$-sequence $M$ such that

$$\Delta(U/M, x/\omega^\omega) \succ [\beta].$$

This yields the same contradiction as above. □

Corollary 5.2. *For $c \in \mathfrak{C}_1 \cup \mathfrak{C}_2$ and any $D$, the theory $\mathfrak{A}_{c,D}$ does not satisfy the axiom of definable choice.*

Proof. Any such theory has a model which contains $\omega^\omega$. (Choose $\alpha_c$ for $c \in \mathfrak{C}_2$, and $\omega^\omega + \alpha_c$ for $c \in \mathfrak{C}_1$.) Thus the above proof for $\mathrm{MT}[\beta]$, $\beta > \omega^\omega$, works here, too. (Note that $\omega^\omega$ is not definable in $\mathfrak{A}_{c,D}$ if $c \in \mathfrak{C}_1$.) □

We will show now that $\mathfrak{A}_{c,D}$ satisfies the axiom of definable choice if $c(\omega) = 0$. In order to extend the result of [5], we will present a proof for Theorem 4.3 of [5] which is omitted there.

Let $\Psi(X, Z)$ be a formula of the form

(1) $\quad E[Z\,0] \wedge (\forall t)\, H[Xt, Zt, Zt'] \wedge (\forall x)\, \mathfrak{K}[\sup^x Z, Zx] \wedge \mathfrak{Y}[\sup X, \sup Z]$.

$\Psi$ is the kernel of an "automata normal form" ([5], p. 160). That is, $X$ and $Z$ are strings of set variables: $E$, $H$, $\mathfrak{K}$, $\mathfrak{Y}$ are propositional formulas in at most the indicated components. Here and in similar context, $x$ ranges over limit numbers. If $X$ and $Z$ satisfying (1) exist, $Z$ is called a *run of $\Psi$ over $X$*. In $[\alpha]$ where $\alpha < \omega^\omega$, given $X$ we want to pick a unique run $Z$ over $X$ if it exists. For $\alpha = \omega$ this is done in [16], Section I.5.b: more explicitely in [5], Theorem 4.2, p. 168, and in [17], Section 4. The idea is: order the finite pieces of $Z$ lexicographically: define a set of "splicing points" by recursion: between each two neighbouring splicing points pick a smallest suitable finite piece of $Z$: get $Z$ by splicing these pieces together. This idea is extended to ordinals below $\omega^\omega$ as follows.

We identify an $n$-tuple $Z$ of subsets of a set $M$ with a function from $M$ into $\{0, 1\}^n$. The $n$-tuples of 0, 1 (or of $T$, $F$) are called the *states* of $Z$. For each $n$ let the states, and the sets of states, and the finite sequences of either (of fixed length), be ordered somehow. Let $Z\{u, v\} =_{\mathrm{df}} \{Zt\colon u \leqq t < v\}$, $Z[u, v] =_{\mathrm{df}} \langle Zt;\, u \leqq t \leqq v\rangle$ be the set and the sequence respectively, of states of $Z$ between $u$ and $v$. For $n$-tuples $Y$ and $Z$ define

$$Y =_u^v Z\,(k) \equiv_{\mathrm{df}} (\forall x \in Lm_k)_u^v\, [Yx = Zx \wedge Y\,\{x, x + \omega^k\} = Z\,\{x, x + \omega^k\}];$$
$$Y <_u^v Z\,(k) \equiv_{\mathrm{df}} (\exists x \in Lm_k)_u^v\, \{Y =_u^x Z\,(k) \wedge$$
$$\wedge\, [Yx < Zx \vee (Yx = Zx \wedge Y\{x, x + \omega^k\} < Z\,\{x, x + \omega^k\})]\};$$
$$Y \leqq_u^v Z\,(k) \equiv_{\mathrm{df}} Y =_u^v Z\,(k) \vee Y <_u^v Z\,(k).$$

In any ordinal let $u, v \in Lm_k$ such that

$$(\forall w \in Lm_{k+1})\,[w \leqq u \vee v < w].$$

Then for any $k$, $=^v_u(k)$ is an equivalence relation (on the sequences $Z[u, v]$) of finite index (that is, there are only finitely many equivalence classes). These equivalence classes are (well-)ordered by $<^v_u(k)$. Thus the above notions generalize Section I.5.b of [16]. In fact, for $k = 0$ they coincide with the notions introduced on p. 76 of [16].

For any $k$, the above notions are expressible in the language $\mathscr{L}$. We define for any formula $\Phi$

$$Min_k(\Phi)\,(Z; u, v) \equiv_{\mathrm{df}} \Phi(Z, u, v) \wedge (\forall Y)\,[\Phi(Y, u, v) \to Z \leqq^v_u Y\,(k)].$$

The following lemma is proved by relativizing to $Lm_k$ the proof of Theorem I.5.b.1 on p. 77 of [16].

Lemma 5.3. *For any number $q$, $Min_k(\Phi)\,(Z; 0, \omega^k \cdot q)$ is a choice formula for $\Phi$ in $[\omega^k \cdot q + 1]$ relative to $=^v_u(k)$. More generally, the following hold in* MT[ω]:

a) $$Min_k(\Phi)\,(Y; u, v) \wedge Min_k(\Phi)\,(Z; u, v) \to Y =^v_u Z\,(k);$$

b) $$Min_k(\Phi)\,(Z; u, v) \to \Phi(Z; u, v);$$

c) $$(\forall x \in Lm_{k+1})\,[x \leqq u \vee v < x] \to [(\exists Z)\,\Phi(Z; u, v) \to (\exists Z)\,Min_k(\Phi)\,(Z; u, v)].$$

Now recall the formula (1) above, below Corollary 5.2. For such a formula $\Psi$, for states $a$, $b$ of $Z$, and a set $C$ of states, we define

$$Run_{a,b,C}(Z, u, v) \equiv_{\mathrm{df}} Zu = a \wedge (\forall t)^v_u\, H[Xt, Zt, Zt'] \wedge$$
$$\wedge (\forall x)^{v'}_u\, \mathfrak{R}[sup^x Z, Zx] \wedge Zv = b \wedge Z\{u, v\} = C$$

($Z$ is a run of $\Psi$ over $X$ from state $a$ in $u$ to state $b$ in $v$ with set of states $C$).

Note that $Run_{a,b,C}$, as well as the notions defined below, contains the variables $X$, and depends on $\Psi$.

$$Sh\,Run^k_{a,b,C,d,D}(u, v) \equiv_{\mathrm{df}}$$
$$(\exists Z)\,[Run_{a,b,C}(u, u + \omega^{k+1}) \wedge \sup^{u+\omega^{k+1}} Z = D \wedge Z\{u, v\} = C \wedge Zv = d]$$
$$\wedge (\forall w \in Lm_k)^v_u \neg (\exists Z)\,[Run_{a,b,C}(u, u + \omega^{k+1}) \wedge \sup^{u+\omega^{k+1}} Z = D$$
$$\wedge Z\{u, w\} = C \wedge Zw = d]$$

(the shortest run of $\Psi$ over $X$ from $a$ in $u$ to $d$ in some $Lm_k$ with set of states $C$ which can be extended to a run to $b$ in $u + \omega^{k+1}$ with set of states $C$ and sup $D$, leads up to $v$).

Now we define by recursion the set of $Lm_k$-splicing points for $Run_{a,b,C}(y, y + \omega^{k+1})$.

$$Splice^k_{a,b,C,d,D}(U, y) \equiv_{\mathrm{df}} U \subseteq Lm_k \wedge (\forall v \in Lm_k)^{y+\omega^{k-1}}_y \{Uv \leftrightarrow Sh\,Run^k_{a,b,C,d,D}(y, v)$$
$$\vee (\exists u \in U)^v_y\,[(\forall t \in U)^v_y\, t \leqq u \wedge Sh\,Run_{d,b,D,d,D}(u, v)]\}.$$

To see what *Splice* does, note the following two remarks.

(2) $$Splice^k_{a,b,C,d,D}(U, y) \to (\forall u, v \in U)\,\{[(v = 0)\,[U] \to (\exists Z)\,Run_{a,d,C}(Z, y, v)]$$
$$\wedge [(v = u')\,[U] \to (\exists Z)\,Run_{d,d,D}(Z, u, v)]\};$$

(3) $$(\exists U)\,[Splice^k_{a,b,C,d,D}(U, y) \wedge (\forall u)^{y+\omega^{k+1}}_y (\exists v \in U)\, u \leqq v]$$
$$\leftrightarrow d \in D \wedge (\exists Z)\,[Run_{a,b,C}(Z, y, y + \omega^{k+1}) \wedge \sup^{y+\omega^{k+1}} Z = D].$$

We need a last auxiliary formula

$$\begin{aligned}&Min\ Set^k_{a,b,C,d,D}(y) \equiv_{df}\\ &\quad (\exists Z)\,[Run_{a,b,C}(Z, y, y+\omega^{k+1}) \wedge \sup^{y+\omega^{k+1}} Z = D]\\ &\quad \wedge \bigwedge_{D'<D} \neg(\exists Z)\,[Run_{a,b,C}(Z, y, y+\omega^{k+1}) \wedge \sup^{y+\omega^{k+1}} Z = D']\\ &\quad \wedge d \in D \wedge \bigvee_{d'<d} d' \notin D\end{aligned}$$

($D$ is a minimal set of states for which exists a run $Z$ from $a$ in $y$ to $b$ in $y + \omega^{k+1}$ with sup $D$, and $d$ is minimal in $D$).

Now we define choice formulas for $\omega^k \cdot q + 1$ by recursion.

$$\Delta_0(Run_{a,b,C})\,(Z;u,v) \equiv_{df} Min_0(Run_{a,b,C})\,(Z;u,v);$$

$$\begin{aligned}&\Delta^*_{k+1}(Run_{a,b,C})\,(Z;y) \equiv_{df} Run_{a,b,C}(Z, y, y+\omega^{k+1}) \wedge\\ &\quad \wedge \bigwedge_{d,D} (\forall U)\,(\forall u, v \in U)\,\{Min\ Set^k_{a,b,C,d,D}(y) \wedge Splice^k_{a,b,C,d,D}(U, y)\\ &\qquad \to [(v = 0)\,[U] \to \Delta_k(Run_{a,d,C})\,(Z; y, v)]\\ &\quad \wedge [(v = u')\,[U] \to \Delta_k(Run_{d,d,D})\,(Z; u, v)]\};\end{aligned}$$

$$\begin{aligned}&\Delta_{k+1}(Run_{a,b,C})\,(Z;u,v) \equiv_{df}\\ &\quad Min_{k+1}(Run_{a,b,C}(Z,u,v) \wedge (\forall y \in Lm_k)^v_u \bigvee_{c,e,B} \Delta^*_{k+1}(Run_{c,e,B})\,(Z;y))\,(Z;u,v)\end{aligned}$$

($Z$ is a minimal run of $\Psi$ over $X$ from $a$ in $y$ (in $u$ resp.) to $b$ in $y + \omega^{k+1}$ (in $v$ resp.) with set of states $C$).

Lemma 5.4. *$\Delta^*_k$ and $\Delta_k$ are choice formulas for $Run_{a,b,C}$ in $[\omega^k + 1]$ and in $[\omega^k \cdot q + 1]$ respectively. More generally, the following hold in* MT[co] (*where $k > 0$ in* a), c), e)):

a) $Lm_k(y) \wedge z = y + \omega^{k+1} \to [\Delta^*_k(Run_{a,b,C})\,(Y;y) \wedge \Delta^*_k(Run_{a,b,C})\,(Z;y) \to Y =^{z'}_y Z]$;

b) $Lm_k(u) \wedge Lm_k(v) \to [\Delta_k(Run_{a,b,C})\,(Y;u,v) \wedge \Delta_k(Run_{a,b,C})\,(Z;u,v) \to Y =^{v'}_u Z]$;

c) $\Delta^*_k(Run_{a,b,C})\,(Z;y) \to Run_{a,b,C}(Z, y, y+\omega^k)$;

d) $\Delta_k(Run_{a,b,C})\,(Z;u,v) \to Run_{a,b,C}(Z,u,v)$;

e) $(\exists Z)\,Run_{a,b,C}(Z,y,y+\omega^k) \to (\exists Z)\,\Delta^*_k(Run_{a,b,C})\,(Z;y)$;

f) $(\forall x \in Lm_{k+1})\,[x \leqq u \vee v < x]$
$\to [(\exists Z)\,Run_{a,b,C}(Z,u,v) \to (\exists Z)\,\Delta_k(Run_{a,b,C})\,(Z;u,v)]$.

Proof. c) and d) are trivial. Either of a) + b) and e) + f) is proved by simultaneous induction on $k$ in a straightforward way. These proofs are left to the reader. Hint: use Lemma 5.3, and (2) and (3) above. □

Theorem 5.5 ([5], Theorem 4.3, p. 169). *For $\alpha < \omega^\omega$,* MT[$\alpha$] *satisfies the axiom of definable choice.*

Proof. Let $\alpha < \omega^\omega$ be a successor ordinal. Then $\alpha$ can be written as

$$\alpha_c = \sum_{i=k}^{0} \omega^i \cdot c(i) \quad \text{where } c(0) > 0.$$

(See the definition preceding Lemma 4.3.) Define $\beta_0 =_{df} \alpha - 1$; $\beta_j =_{df} \alpha^j_c$ for $j = 1, \ldots, k+1$. Let $\Phi(X, U)$ be a formula containing no free variables besides the strings $X$ and $U$. By Theorem 4.5 of [4], p. 85, $\Phi$ is equivalent in MT[co] to a formula of the form

(4) $(\exists \tilde{Z}).\ E[\tilde{Z}0] \wedge (\forall t)\, H[Xt, Ut, \tilde{Z}t, \tilde{Z}t'] \wedge (\forall x)\, \mathfrak{K}[\sup^x \tilde{Z}, \tilde{Z}x] \wedge \mathfrak{Y}[\sup X, \sup U, \sup \tilde{Z}]$.

We put $Z = \langle U, \tilde{Z}\rangle$ and rewrite the kernel of (4) as

(5) $E[Z0] \wedge (\forall t)\, H[Xt, Zt, Zt'] \wedge (\forall x)\, \mathfrak{K}(\sup^x Z, Zx] \wedge (\forall w)\,[(\forall t)\, t \leqq w \to \mathfrak{Y}[Xw, Zw]]$.

By definition of sup, (4) is a formula $\Psi(X, Z)$ of type (1) above, below Corollary 5.2. We apply the above definitions to this $\Psi$, and write

$$Min\ Seq_{a_0, \ldots, a_{k+1}, C_0, \ldots, C_k} \equiv_{\mathrm{df}}$$

"$\langle a_0, \ldots, a_{k+1}, C_0, \ldots, C_k\rangle$ *is minimal such that:*

$$E[a_{k+1}] \wedge (\forall w)\,[(\forall t)\, t \leqq w \to \mathfrak{Y}[Xw, a_0]] \wedge \bigwedge_{j=0}^{k} (\exists Z)\, Run_{a_{j+1}, a_j, C_j}(Z, \beta_{j+1}, \beta_j)\text{''}.$$

Note that *Min Seq* is a formula of $\mathscr{L}$, which, as all the above notions, depends on $X$ (and $\Psi$). Define

$$\Delta_\alpha(\Psi)\,(Z) \equiv_{\mathrm{df}}$$

$$\Psi(Z) \wedge \bigvee_{\substack{a_0, \ldots, a_{k+1} \\ c_0, \ldots, c_k}} [Min\ Seq_{a_0, \ldots, a_{k+1}, C_0, \ldots, C_k} \to \bigwedge_{j=0}^{k} \Delta_j(Run_{a_{j+1}, a_j, C_j})\,(Z, \beta_{j+1}, \beta_j)].$$

It follows directly from Lemma 5.4 that $\Delta_\alpha(\Psi)$ is a choice formula for $\Psi$ in $\mathrm{MT}[\alpha]$ with respect to $Z$. Therefore $(\exists \tilde{Z})\, \Delta_\alpha(\Psi)\,(U, \tilde{Z})$ is a choice formula for $\Phi \equiv (\exists \tilde{Z})\, \Psi(X, U, \tilde{Z})$ in $\mathrm{MT}[\alpha]$ with respect to $U$.

If $\alpha$ is a limit ordinal, we have to define notions $Run_{a,b,C}(Z, u, \infty)$ etc., analoguous to the ones above, which take care of the last limit type in $\alpha$. This makes the choice formula $\Delta_\alpha(\Psi)$ slightly more complicated. Since no new idea is involved, we leave this case to the reader.

If $\Phi$ contains free individual variables, they can be eliminated with the help of set variables for singletons; for details see [17], Section 4. □

Corollary 5.6. *For $c \in \mathfrak{Q}_0$ and any $D$, the theory $\mathfrak{A}_{c,D}$ satisfies the axiom of definable choice.*

Proof. Let $c \in \mathfrak{Q}_0$ such that $c(0) > 0$. By definition of $\mathfrak{Q}$ there is a number $k$ such that $c(i) = 0$ for all $i > k$. Let $\beta_0, \ldots, \beta_k$ be as in the proof of Theorem 5.5. Now $c(i) = \omega + \omega^*$ is possible. Since $c(i) \in \omega$ was not used in the proof of Theorem 5.5, $\Delta_\alpha(\Psi)$ is a choice formula for $\Psi$ in $\mathfrak{A}_{c,D}$. More formally, in $\Delta_\alpha(\Psi)$ we replace $\beta_j$ by its explicit definition as the last limit number of order $j$.

$$\Delta^k(\Psi)\,(Z) \equiv_{\mathrm{df}} \Psi(Z) \wedge \bigwedge_{\substack{a_0, \ldots, a_{k+1} \\ C_0, \ldots, C_k}} (\forall x_0, \ldots, x_{k+1}) \bigwedge_{j=0}^{k+1} LLm_j(x_j)$$

$$\wedge\ Min\ Seq_{a_0, \ldots, a_{k+1}, C_0, \ldots, C_k} \to \bigwedge_{j=0}^{k} \Delta_j(Run_{a_{j+1}, a_j, C_j})\,(Z, x_{j+1}, x_j)].$$

Then

$$\Delta_\alpha(\Psi)\,(Z) \leftrightarrow \Delta^k(\Psi)\,(Z).$$

Thus $\Delta_\alpha(\Psi)$ depends only on $k$, not on $\alpha$. □

Corollary 5.7. *For $c \in \mathfrak{Q}_0$, $\mathrm{DEF}(\mathfrak{A}_{c,D})$ is the prime model of $\mathfrak{A}_{c,D}$.*

This follows directly from Proposition 4.2 by Theorem 3.6 and Corollary 5.6.

20*

## 6. Definable subsets

By Lemma 4.3, $\alpha_c$ is the first domain of DEF($\mathfrak{A}_{c,D}$), independent from $D$. The remark preceding Lemma 4.3 and the second half of the proof of Lemma 4.3, show that the definable subsets of $\alpha_c$ are those which occur as "output" of a deterministic finite automaton on $\alpha_c$. Such an automaton is nothing else than a recursion of the form (1) in the proof of Lemma 4.3. With the help of this fact we will now describe the second domain of DEF($\mathfrak{A}_{c,D}$), which will depend on $D$.

First recall the definition of $\mathfrak{R}$ in Section 3. Let $d \in \mathfrak{R}$. Define $d^*: \omega \to \omega$ as follows. Let $q \in \omega$ have the prime power representation

$$q = \prod_{i=0}^{n} p_i^{k_i}.$$

Put $d^*(q) = m$ iff $m < q$ and $m \equiv d(p_i, k_i)$ $(p_i^{k_i})$ for $i = 0, \ldots, n$. ($m$ exists by the Chinese remainder theorem, Lemma 3.5, and is uniquely determined.) $d^*$ extends $d$, since $d^*(p^k) = d(p, k)$. Let $\mathfrak{R}^* =_{\mathrm{df}} \{d^*: d \in \mathfrak{R}\}$.

Now let $\mathfrak{A}_{c,D}$ be any of theories of Theorem 3.6, let $\alpha_c$ be as defined in Section 4, let $q, h \in \omega$ where $q < 0$. The concept of an *ultimately D-periodic set* (sc. subset of $\alpha_c$) *of phase h and period q* is introduced as follows. (If $\alpha_c$ is an ordinal and thus $D = \emptyset$, we omit the prefix $D$. We write $d$-periodic instead of $\{d\}$-periodic. If phase and period are not specified, we omit them.)

The empty set and any 1-element set are ultimately periodic of any phase and period. Let $k \geqq 0$. For any $m > 1$, $N \subseteq \omega^k \cdot m$ is ultimately periodic if for all $j < m$

$$N \cap [\omega^k \cdot j, \omega^k \cdot (j + 1))$$

is ultimately periodic (regarded as a subset of $\omega^k$). $N \subseteq \omega^{k+1}$ is ultimately periodic of phase $h$ and period $q$ if

$$N \cap [0, \omega^k \cdot (h + q))$$

is ultimately periodic, and for all $m, n \geqq h$ such that $m \equiv n$ $(q)$,

$$N \cap [\omega^k \cdot m, \omega^k \cdot (m + 1)) = N \cap [\omega^k \cdot n, \omega^k \cdot (n + 1)).$$

$N \subseteq \omega^k \cdot (\omega + \omega^*)$ is ultimately $d$-periodic of phase $h$ and period $q$ if

$$N \cap [0, \omega^k \cdot (d^*(q) + q + h))$$

and

$$N \cap [\omega^k \cdot (\omega + \omega^* - h), \omega^k \cdot (\omega + \omega^*))$$

are ultimately periodic, and for all $m, n \geqq h$ such that $m \equiv -n$ $(q)$,

$$N \cap [\omega^k \cdot (d^*(q) + m), \omega^k \cdot (d^*(q) + m + 1))$$
$$= N \cap [\omega^k \cdot (\omega + \omega^* - (n + 1)), \omega^k \cdot (\omega + \omega^* - n)).$$

For $c \in \mathfrak{L}_0$, $N \subseteq \alpha_c$ is ultimately $D$-periodic if for all $j$, $N \cap [\alpha_c^{j+1}, \alpha_c^j)$ is ultimately periodic in case $c(j) \in \omega$, and is ultimately $d_j$-periodic otherwise. For $c \in \mathfrak{L}_1 \cup \mathfrak{L}_2$, $N \subseteq \alpha_c$ is ultimately $D$-periodic if in addition there is a number $k$ such that $N \cap [0, \omega^k \cdot 2)$ is ultimately periodic, and for all $\alpha \in Lm_k \cap \alpha_c$, $\alpha > 0$, $\alpha \neq \alpha_c^k$:

$$N \cap [\alpha, \alpha + \omega^k) = N \cap [\omega^k, \omega^k \cdot 2).$$

These definitions generalize the concept of an ultimately periodic subset of $\omega$, retaining its properties. For example, if $h$ is the greatest common divisor of $j$ and $l$, then $N \subseteq \omega^k \cdot (\omega + \omega^*)$ is ultimately $d$-periodic of period $h$ iff $N$ is ultimately $d$-periodic of period $j$ and of period $l$. Therefore if $N \subseteq \alpha_c$ is ultimately $D$-periodic, then $N \cap [\alpha_c^{j+1}, \alpha_c^j)$ is ultimately $d_j$-periodic of phase $h$ and period $q$ for a unique smallest $h$ and $q$. Similar for the intervals of $\alpha_c$ of order type $\omega^k$.

Let $\mathfrak{P}_{\mathrm{u.}\,D\text{-p.}}\alpha_c$ be the set of ultimately $D$-periodic subsets of $\alpha_c$.

Lemma 6.1. *Let $c \in \mathfrak{Q}$, let $\mathfrak{M} = \langle \alpha_c, \mathfrak{P}'\alpha_c; \in, < \rangle$ be a model of $\mathfrak{A}_{c,D}$. Then exactly the ultimately $D$-periodic subsets of $\alpha_c$ are definable in $\mathfrak{M}$. Thus*

$$\mathrm{DEF}(\mathfrak{A}_{c,D}) = \alpha_{c,D} =_{\mathrm{df}} \langle \alpha_c, \mathfrak{P}_{\mathrm{u.}\,D\text{-p.}}\alpha_c; \in, < \rangle.$$

The proof is similar to the proof of Lemma 4.3, and is left to the reader.

Corollary 6.2. *For $c \in \mathfrak{Q}_0$, $\alpha_{c,D}$ is the prime model of $\mathfrak{A}_{c,D}$.*

This follows immediately from Corollary 5.7 and Lemma 6.1.

Corollary 6.3. *For $c \in \mathfrak{Q}_2$, $\mathrm{DEF}(\mathfrak{A}_{c,D})$ is not a model of $\mathfrak{A}_{c,D}$.*

Proof. Let $c \in \mathfrak{Q}_2$ such that $\alpha_c \neq \omega^\omega$. Then $\alpha_c^\omega \in \alpha_c$, $\alpha_c^\omega > 0$, and $\alpha_c^\omega$ is a limit number of arbitrary high order. By Lemma 6.1 there is no definable $\omega$-sequence which is cofinal in $\alpha_c^\omega$. Therefore the definable sets form not a model for $\mathfrak{A}_{c,D}$. An analoguous argument works if $\alpha_c = \alpha_c^\omega = \omega^\omega$. □

Corollary 6.3 suggest that $\mathfrak{A}_{c,D}$, where $c \in \mathfrak{Q}_2$, has no prime model at all. Combined with Proposition 4.2, it yields another proof for Corollary 5.2.

Theorem 6.4. *For $c \in \mathfrak{Q}_1$, $\alpha_{c,D}$ is the prime model of $\mathfrak{A}_{c,D}$.*

Proof. As $\alpha_{c,D}$ is a general model, it satisfies the logical axioms. Also every set in $\alpha_{c,D}$ has a smallest element, thus $\alpha_{c,D}$ satisfies *WLO*. Obviously, $\alpha_{c,D}$ satisfies $\mathfrak{A}_{c,D} \wedge ACC$. Now let $\beta \in \alpha_c$ be any limit number. Since $\alpha_c^\omega \notin \alpha_c$ (see the remarks in Section 4, following the definition of $\alpha_c$), $\beta$ is of order exact $k$ for some $k \in \omega$. (I.e. either $\beta \in \omega^\omega$ or $\beta \in (\alpha_c - \omega^\omega)$.) Thus there is an $\omega$-sequence in $\mathfrak{P}_{\mathrm{u.}\,D\text{-p.}}\alpha_c$ which is cofinal in $\beta$. Therefore $\alpha_{c,D}$ is a model of $\mathfrak{A}_{c,D}$, and thus the prime model by Remark 4.1. □

Note that the same proof applies to $c \in \mathfrak{Q}_0$, thus avoiding the axiom of definable choice. We thus can get the results on prime models without using Section 5. The results on definable choice, however, are interesting on their own.

## 7. Appendix. Decidability of Monadic Second Order Theories of Ordinals and Underlying Set Theory

a) Decidability. Let $\mathscr{L}$ be the monadic second order language of order. For any model $\mathfrak{M}$ of some set theory T and any ordinal $\alpha$, let $\alpha^{\mathfrak{M}}$ be that ordinal in that model. Define

$$[\alpha]^{\mathfrak{M}} =_{\mathrm{df}} \langle \alpha^{\mathfrak{M}}, \mathfrak{P}^{\mathfrak{M}}(\alpha); \in_{\mathfrak{M}}, =_{\mathfrak{M}}, <_{\mathfrak{M}} \rangle$$

to be the standard structure of $\alpha$ for $\mathfrak{M}$. (Here $\mathfrak{P}^{\mathfrak{M}}(\alpha)$ is the power set of $\alpha$ in $\mathfrak{M}$.) Define

$$\mathrm{MT}[\alpha^{\mathfrak{M}}] =_{\mathrm{df}} \{\Psi \in \mathscr{L}; \Psi \succ\!- [\alpha]^{\mathfrak{M}}\};$$

$$\mathrm{MT}[\mathrm{co}^{\mathfrak{M}}] =_{\mathrm{df}} \bigcap \{\mathrm{MT}[\alpha^{\mathfrak{M}}]: [\alpha < \omega_1]^{\mathfrak{M}}\};$$

$$\mathrm{MT}[\alpha^T] =_{\mathrm{df}} \bigcap \{\mathrm{MT}[\alpha^{\mathfrak{M}}]: \mathfrak{M} \text{ model of T}\};$$

$$\mathrm{MT}[\mathrm{co}^T] =_{\mathrm{df}} \bigcap \{\mathrm{MT}[\mathrm{co}^{\mathfrak{M}}]: \mathfrak{M} \text{ model of T}\}.$$

In [4] it is shown that $MT[\alpha^{ZFC}]$ for $\alpha \leqq \omega_1$ and $MT[co^{ZFC}]$ are decidable. The proof shows (see [5]) that $MT[\alpha^{ZFC}]$ is complete; thus for any ZFC-models $\mathfrak{M}$ and $\mathfrak{M}'$ and any $\alpha \leqq \omega_1$

$$MT[\alpha^{\mathfrak{M}}] = MT[\alpha^{\mathfrak{M}'}] = MT[\alpha^{ZFC}];$$

therefore also $MT[co^{\mathfrak{M}}] = MT[co^{\mathfrak{M}'}] = MT[co^{ZFC}]$. LITMAN [11] uses BÜCHI's decision procedure to show that for $\alpha < \omega_1$

$$MT[\alpha^{ZF}] = MT[\alpha^{ZFC}],$$

thus also

$$MT[co^{ZF}] = MT[co^{ZFC}].$$

Therefore for $\alpha < \omega_1$, $MT[\alpha^{ZF}]$ is decidable and complete; also $MT[co^{ZF}]$ is decidable. Let $UD$ be the statement that there is a choice of fundamental sequences (uniform denumeration) on $\omega_1$. Let $\mathscr{I}_1$ be the filter generated by the closed cofinal subsets of $\omega_1$. Let $AT$ be the statement that the Boolean algebra $\mathfrak{P}\omega_1/\mathscr{I}_1$ is atomless. Let $ZFC^* =_{df} ZF + UD + AT$. It is known (cf. [5]) that $ZF \subset ZFC^* \subset ZFC$. LITMAN shows that

$$MT[\omega_1^{ZFC^*}] = MT[\omega_1^{ZFC}].$$

Therefore $MT[\omega_1^{ZFC^*}]$ is decidable and complete. Finally LITMAN shows that

$$ZF + UD \vdash \text{“}MT[\omega_1] \text{ is recursive”},$$

thus for any model $\mathfrak{M}$ of $ZF + UD$, $MT[\omega_1^{\mathfrak{M}}]$ is decidable. It remains open whether $MT[\omega_1^{ZF+UD}]$ is decidable and/or complete.

b) Standard Axioms. Let

$$\mathfrak{C}_0 =_{df} \{WLO, (\forall x)\, ACC[x], ACC\}, \qquad \mathfrak{C}_1 =_{df} \{WLO, (\forall x)\, ACC[x], \neg ACC\}.$$

Here $WLO$ and $ACC$ are $\mathscr{L}$-sentences stating that the domain is well-ordered and $\omega$-cofinal, resp. For $\alpha < \omega_1$ let $\mathfrak{A}_\alpha$ be the extension of $\mathfrak{C}_0$ describing the “tail” of $\alpha$. (See Section 2 of [5] for notation.) By [4], $\mathfrak{A}_\alpha$, $\mathfrak{C}_0$, and $\mathfrak{C}_1$ are (finite) ZFC-standard axiom systems for $MT[\alpha^{ZFC}]$, $MT[co^{ZFC}]$, and $MT[\omega_1^{ZFC}]$ resp. This means: in any ZFC-model $\mathfrak{M}$, $\mathfrak{C}_0$ and $MT[co^{ZFC}]$ have the same standard models, namely $[\alpha]^{\mathfrak{M}}$ for $\alpha < \omega_1$; also $[\alpha]^{\mathfrak{M}}$ and $[\omega_1]^{\mathfrak{M}}$ are the only standard models in $\mathfrak{M}$ of $\mathfrak{A}_\alpha$ and $\mathfrak{C}_1$ resp. Thus $\mathfrak{A}_\alpha$ and $\mathfrak{C}_1$ are ZFC-standard complete. In spite of the above result of LITMAN these results do not remain true if one replaces ZFC by ZF. Actually LITMAN shows that

$$\mathfrak{C}_1^* =_{df} \{WLO, (\forall x)\, (CD \wedge ACC)\, [x], \neg CD \vee \neg ACC\}$$

is ZF-standard categorical, and is thus a complete ZF-standard axiom system for $MT[\omega_1^{ZF}]$. (Here $CD$ is an $\mathscr{L}$-sentence such that, as LITMAN shows, for $\alpha \leqq \omega_1$, $CD \succ [\alpha]$ iff $UD[\alpha]$; see also Section 8 of [5].) Therefore $\mathfrak{C}_1$ is a $ZF + UD$-standard complete axiom system for $MT[\omega_1^{ZF+UD}]$. With $\mathfrak{A}_\alpha$ and $\mathfrak{C}_0$, however, the situation is different.

c) Axioms. Obtain $\bar{\mathfrak{A}}_\alpha$, $\alpha < \omega_1$, and $\bar{\mathfrak{C}}_0$ from $\mathfrak{A}_\alpha$ and $\mathfrak{C}_0$ resp. by adding $SPLICE$ ([5], p. 153). Let $\neg(\exists U)\, At(U)$ be the $\mathscr{L}$-sentence equivalent to $AT$. Let

$$\bar{\mathfrak{C}}_1 =_{df} \mathfrak{C}_1 + SPLICE + \neg(\exists U)\, At(U).$$

By [5], $\bar{\mathfrak{A}}_\alpha$ where $\alpha < \omega_1$, $\bar{\mathfrak{C}}_0$, and $\bar{\mathfrak{C}}_1$ are axiom systems for $MT[\alpha^{ZFC}]$, $MT[co^{ZFC}]$, and $MT[\omega_1^{ZFC}]$ resp. This means for example that in any ZF-model, $\bar{\mathfrak{C}}_0$ and $MT[co^{ZFC}]$ have the same general models. Equivalently, $MT[co^{ZFC}]$ is the set of sentences derivable from $\overline{\mathfrak{C}}_0$ in

$$\mathscr{P}_2 =_{\mathrm{df}} \text{two-sorted predicate logic} + \text{extensionality} + \text{comprehension}.$$

We write $\mathrm{cl}(\mathfrak{B})$ for the set of sentences derivable from $\mathfrak{B}$ in $\mathscr{P}_2$. It is further shown in [5] that *SPLICE* is independent from any of $\mathfrak{A}_\alpha$, $\omega^\omega \leqq \alpha < \omega_1$, $\overline{\mathfrak{C}}_0$, and $\overline{\mathfrak{C}}_1$. Therefore none of these axiom systems is complete. In fact there is a ZF-model $\mathfrak{M}$ such that for any $\alpha < \omega_1$

(1) $\quad [\omega_1 + \alpha]^{\mathfrak{M}} \models \mathfrak{A}_{\omega^\omega + \alpha} + \neg SPLICE.$

Here, by the result of LITMAN, for no ZF-model $\mathfrak{M}$ and no $\alpha < \omega_1$ we can have $\omega^\omega$ instead of $\omega_1$, since $SPLICE \in \mathrm{MT}[\mathrm{co}^{\mathrm{ZF}}]$. (1) proves that $\mathfrak{A}_\alpha$ where $\omega^\omega \leqq \alpha < \omega_1$, and $\mathfrak{C}_0$ are not even ZF-standard axiom systems for $\mathrm{MT}[\alpha^{\mathrm{ZF}}]$ and $\mathrm{MT}[\mathrm{co}^{\mathrm{ZF}}]$ resp.

Sections a) and c) together show that $\mathrm{MT}[\alpha]$ where $\alpha < \omega_1$, and $\mathrm{MT}[\mathrm{co}]$, and thus BÜCHI's decision procedures, are independent of the axiom of choice. The choice principle *SPLICE*, however, is needed for the axiomatization. This is not a contradiction: *SPLICE* is so weak that for countable ordinals it follows from ZF, namely from the well-orderedness of the domain and from the union axiom. LITMAN's proof shows that in $\overline{\mathfrak{A}}_\alpha$ and $\overline{\mathfrak{C}}_0$, *SPLICE* can be replaced by *CD*, if we consider ZF-standard consequence. Whether this is true for general consequences, is open. (See Problem 8.3 in [5].) It is also not known whether $\mathrm{cl}(\mathfrak{A}_\alpha)$ and $\mathrm{cl}(\mathfrak{C}_0)$ are decidable; similarly for the ZF-standard consequences of $\mathfrak{A}_\alpha$ and $\mathfrak{C}_0$. (See Problem 8.4 of [5].) For $\alpha < \omega^\omega$, by [5] *SPLICE* is derivable from $\mathfrak{A}_\alpha$. Therefore $\mathfrak{A}_\alpha$ is an axiom system for $\mathrm{MT}[\alpha^{\mathrm{ZFC}}]$ for $\alpha < \omega^\omega$. Now for $\alpha = \omega_1$, by [5] there is a ZF-model $\mathfrak{M}$ such that $\mathfrak{C}_1 + \neg SPLICE \succ [\omega_1]^{\mathfrak{M}}$. This implies that

$$\mathrm{MT}[\omega_1^{\mathrm{ZF}}] \subset \mathrm{MT}[\omega_1^{\mathrm{ZFC}}].$$

(See Section 8 and end of Section 1 of [5].) Therefore $\mathrm{MT}[\omega_1^{\mathrm{ZF}}]$ is not complete. It is open whether $\mathrm{MT}[\omega_1^{\mathrm{ZF}}]$ is decidable. Compare Problem 8.1 in [5]. See also Problem 2 in [11], quoted at the end of Section a) of this appendix.)

d) Definable Choice. Actually LITMAN proves that there is a choice function which chooses a run of an automaton on countable input if there is such a run. His proof does not translate into $\mathscr{L}$. In the present paper we show that $\mathrm{MT}[\alpha^T]$ for $\alpha \geqq \omega^\omega$ and any set theory T, does not satisfy the axiom of definable choice. This shows that LITMAN's choice function is not definable by a formula in $\mathscr{L}$. This explains better the independence of *SPLICE*. For $\alpha < \omega^\omega$, $\mathrm{MT}[\alpha^{\mathrm{ZF}}]$ satisfies the axiom of definable choice by Theorem 5.5 above. Both results extend to the appropriate non-standard extensions as stated in Corollaries 5.2 and 5.6 above.

e) The Situation at $\alpha < \omega_2$. In BÜCHI-ZAIONTZ [6] decision procedures for $\mathrm{MT}[\alpha^{\mathrm{ZFC}}]$ where $\omega_1 \leqq \alpha < \omega_2$, and for $\mathrm{MT}[<\omega_2^{\mathrm{ZFC}}] =_{\mathrm{df}} \bigcap\{\mathrm{MT}[\alpha^{\mathrm{ZFC}}];\ \alpha < \omega_2\}$, are given. It is shown that

$$\mathfrak{B}_2 =_{\mathrm{df}} \{WLO,\ Acc_{<2},\ (\forall t)\ Acc_{<2}[t]\}$$

is a standard axiom system for $\mathrm{MT}[<\omega_2^{\mathrm{ZFC}}]$. ZAIONTZ shows in [18] that

$$\overline{\mathfrak{B}}_2 =_{\mathrm{df}} \mathfrak{B}_2 + SPLICE + Atomless + (\forall t)\ Atomless\ [t]$$

is an axiom system for $\mathrm{MT}[<\omega_2^{\mathrm{ZFC}}]$. Corresponding results hold for $\mathrm{MT}[\alpha^{\mathrm{ZFC}}]$, $\alpha < \omega_2$; see [6] and [18] for details. Nothing is known if one drops *AC*.

f) The Case $\omega_2$. SHELAH [15] shows that in $\mathscr{L}$ one can express statements on $\omega_2$ which are independent of ZFC. Thus $\mathrm{MT}[\omega_2^{\mathrm{ZFC}}]$ is incomplete. Later, GUREVICH, MAGIDOR, and SHELAH show in [9] that $\mathrm{MT}[\omega_2^T]$ can be decidable, or undecidable of any given degree, depending on the chosen extension T of ZFC.

## Bibliography


[1] BELL, J. L., and A. B. SLOMSON, Models and Ultraproducts: An Introduction. North-Holland Publ. Comp., Amsterdam 1971.

[2] BÜCHI, J. R., On a decision method in restricted second order arithmetic. In: Proc. 1960 Int. Congr. for Logic, Method. and Philos. of Sci., Stanford Univ. Press, Stanford, Calif., 1962, pp. 1–11.

[3] BÜCHI, J. R., Transfinite automata recursions and weak second order theory of ordinals. In: Proc. 1964 Int. Congr. for Logic, Method. and Philos. of Sci., North-Holland Publ. Comp., Amsterdam 1965, pp. 3–23.

[4] BÜCHI, J. R., The monadic second order theory of $\omega_1$. In: Decidable Theories II: The monadic second order theory of all countable ordinals, Lecture Notes in Mathematics **328** (1973), Springer-Verlag, Berlin–Heidelberg–New York, pp. 1–127.

[5] BÜCHI, J. R., and D. SIEFKES, Axiomatization of the monadic second order theory of $\omega_1$. In: Decidable Theories II: The monadic second order theory of all countable ordinals, Lecture Notes in Mathematics **328** (1973), Springer-Verlag, Berlin–Heidelberg–New York, pp. 129–217.

[6] BÜCHI, J. R., and CH. ZAIONTZ, Deterministic automata and the monadic theory of ordinals $<\omega_2$. This Zeitschr. **29** (1983), 313–336.

[7] DAVIS, M., Computability and Unsolvability. McGraw-Hill, New York–Toronto–London 1958.

[8] DONER, J. E., MOSTOWSKI, A., and A. TARSKI, The elementary theory of well-ordering – a metamathematical study. In: Logic Colloquium 1977, North-Holland Publ. Comp., Amsterdam 1978, pp. 1–54.

[9] GUREVICH, Y., MAGIDOR, M., and S. SHELAH, The monadic theory of $\omega_2$. J. Symb. Logic **48** (1983), 387–398.

[10] HENKIN, L., Completeness in the theory of types. J. Symb. Logic **15** (1950), 81–91.

[11] LITMAN, A., On the monadic theory of $\omega_1$ without AC. Israel J. Math. **23** (1976), 251–266.

[12] MONTAGUE, R., and R. L. VAUGHT, A note on theories with selectors. Fund. Math. **47** (1959), 243–247.

[13] MOSTOWSKI, A., and A. TARSKI, Arithmetical classes and types of well-ordered systems. Bull. AMS **55** (1949), 65, abstract 55-1-78; Erratum, ibid., 1192.

[14] MOSTOWSKI, A., and A. TARSKI, The elementary theory of well-ordering. Unpublished.

[15] SHELAH, S., The monadic theory of order. Annals of Math. **102** (1975), 379–419.

[16] SIEFKES, D., Decidable Theories I: Büchi's monadic second order successor arithmetic. Lecture Notes in Mathematics **120** (1970), Springer-Verlag, Berlin–Heidelberg–New York.

[17] SIEFKES, D., The recursive sets in certain monadic second order fragments of arithmetic. Archiv Math. Logik und Grundl. **17** (1975), 71–80.

[18] ZAIONTZ, CH., Axiomatization of the monadic theory of ordinals $<\omega_2$. This Zeitschr. **29** (1983), 337–356.


(Eingegangen am 5. Oktober 1981)

J. Richard Büchi
Department Computer Sciences
Purdue University
West Lafayette, Indiana 47907 (U.S.A.)

Dirk Siefkes
Technische Universität Berlin
Fachbereich Informatik (20)
D-1000 Berlin 10
Franklinstr. 28/29

*Zeitschr. f. math. Logik und Grundlagen d. Math.*
*Bd. 29, S. 313–336 (1983)*

# DETERMINISTIC AUTOMATA AND THE MONADIC THEORY OF ORDINALS $<\omega_2$

by J. RICHARD BÜCHI in West Lafayette, Indiana (U.S.A.),
and CHARLES ZAIONTZ in Tampa, Florida (U.S.A.)[1])

## 1. Introduction

For any ordinal $\alpha$ let MT$[\alpha, <]$ be the monadic (second order) theory of $[\alpha, <]$ The language of MT$[\alpha, <]$ contains propositional connectives $\sim, \vee, \wedge, \supset, \equiv, T, F$; a binary predicate letter $<$ interpreted as the order relation on $\alpha$; individual variables $t, x, y, z, \ldots$, ranging over $\alpha$ and monadic predicate or set variables $X, Y, Z, \ldots$, ranging over subsets of $\alpha$ with quantification over both types of variables. MT$[\alpha, <]$ consists of all those sentences which are true in $[\alpha, <]$ with respect to the intended interpretation.

A method is given for deciding the truth of monadic sentences in $[\alpha, <]$ for any ordinal $\alpha < \omega_2$; also the monadic theory of all well-orders of cardinality $<\omega_2$ is decidable. These extend the results of [2, 3, 5] on the decidability of such theories for ordinals $\leqq \omega_1$ and are cited in [5]. A discussion of MT-definability in the $[\alpha, <]$, $\alpha < \omega_2$ is also included as well as a complete picture of MT-equivalence between these structures. In [11] axiom systems for the MT$[\alpha, <]$, $\alpha < \omega_2$, are presented.

Finite automata operating on transfinite strings are used and in fact the results presented may be viewed as contributions to automata theory. Using the transition system and techniques of [5], Section 6, and ideas of MCNAUGHTON [7], we produce a subset construction at $\omega_1$ (Section 3). However, it will be shown that for such transition systems there is no deterministic system of the same form which accepts the same input. To remedy the situation it is necessary to construct a deterministic system of equal behavior but with a more complicated terminal condition involving relativized filters of cofinal-closed sets. In Section 2 we exhibit the pertinent facts about such filters. Once the workings of the subset construction at $\omega_1$ are mastered the extension to all ordinals $<\omega_2$ is straightforward (Section 4). In order to prove a complementation lemma the relativized quantifiers must be removed. The result, however, is a non-deterministic system. As splicing and the fact that $\mathfrak{P}\alpha/\mathcal{J}_1^\alpha$ is atomless for any $\omega_1$-limit $\alpha$ is used, the axiom of choice is required.

Subsequent to our proof and using the ideas of [5], SHELAH [8] found a non-deterministic proof of the decidability of MT$[\alpha, <]$, $\alpha < \omega_2$. His proof avoids the complicated MCNAUGHTON construction, but since it is non-deterministic there is less information on definability.

---

[1]) This work has been supported by the National Science Foundation, grant number GJ-980.

## 2. The relativized filter of cofinal-closed sets

We begin by repeating some notation of [5]. $X, Y, Z, \ldots$ are used to denote $n$-tuples of set variables, $X = X_1 X_2, \ldots, X_n$. In this case $X$ is interpreted in $[\alpha, <]$ as an $\alpha$-sequence, $X: \alpha \to \{T, F\}^n$, where the elements of $\{T, F\}^n$ are referred to as the *states* of $X$. For $s \in \{T, F\}^n$, $Xt = s$ is an abbreviation for $\bigwedge_{1 \leqq i \leqq n} X_i t \equiv s_i$. Thus propositional formulas $E[Xt]$ range over sets of states. An extension of this idea is the use of expressions such as $Zt' = H[Xt, Zt]$ where $Z$ is an $n$-tuple and $X$ is an $m$-tuple to mean $\bigwedge_{1 \leqq i \leqq m} Z_i t' = H[Xt, Zt]$. Here $H[\cdot, \cdot]$ is actually an $n$-tuple of propositional formulas.

We also need the following relativized quantifiers:

$$(\exists t)^x \varphi: \ (\exists t).t < x \wedge \varphi, \qquad (\exists t)_x \varphi: \ (\exists t).x \leqq t \wedge \varphi,$$

$$(\exists t)^X \varphi: \ (\exists t).Xt \wedge \varphi, \qquad (\exists Y)^X \varphi: (\exists Y).Y \subseteq X \wedge \varphi,$$

$$(\exists X)^x \varphi: (\exists X).(\forall t)\ (Xt \supset t < x) \wedge \varphi.$$

Notations such as $(\exists t)_y^x \varphi$, $(\exists X)_x \varphi$ are defined similarly. Also there are various versions of relativized universal quantifiers, e.g. $(\forall t)^x \varphi$ abbreviates $(\forall t).t < x \supset \varphi$. Using these we may define the relativizations of formulas. These go by induction but basically the *relativization* $\varphi[X]$ *of* $\varphi$ *to* $X$ is obtained by relativizing all quantifiers to $X$. $\varphi[x]$ is obtained similarly—replace $(\exists t)$, $(\exists X)$ by $(\exists t)^x$ and $(\exists X)^x$ and similarly for $(\forall t)$.

Let $\mathscr{J}$ be any filter which is MT-definable in $[\alpha, <]$, i.e. $X \in \mathscr{J}$ is definable, and let $\mathscr{O}$ be the corresponding ideal, $\mathscr{O} = \{X; \tilde{X} \in \mathscr{J}\}$ ($\tilde{X}$ = complement of $X$). We can define the following quantifiers:

(1) $\quad (\forall_{\mathscr{J}} t)\ Xt: X \in \mathscr{J}, \qquad (\exists_{\mathscr{J}} t)\ Xt: X \notin \mathscr{O}.$

The fact that $\mathscr{J}$ is a filter is precisely what is needed to prove many of the properties of the usual quantifiers. Now let $Z$ be an $n$-tuple of set variables and define

(2) $\quad \sup Z = \{s \in \{T, F\}^n; (\exists_{\mathscr{J}} t).Zt = s\}.$

Here $(\exists_{\mathscr{J}} t).Zt = s$ means $\{t; Zt = s\} \notin \mathscr{O}$.

Lemma 2.1. ([5], Lemma 5.8). *Let* $\mathscr{J}$ *be a filter on* $\alpha$. *Conditions of the form* $\mathfrak{K}[\sup Z]$ *are exactly equivalent to propositional expressions in conditions of form* $(\exists_{\mathscr{J}} t)\ L[Zt]$ *(or of form* $(\forall_{\mathscr{J}} t)\ L[Zt]$*).*

The proof may be found in [5] and depends on the fact that $Z$ has but finitely many states. Note that if $\mathscr{J}$ is definable then so are conditions on $\sup Z$:

(3) $\quad \mathfrak{K}[\sup_{\mathscr{J}} Z]: \bigvee_{D \in \mathfrak{K}} . \bigwedge_{e \in D} (\exists_{\mathscr{J}} t)\ Zt = e \wedge \bigvee_{e \notin D} \sim(\exists_{\mathscr{J}} t)\ Zt = e.$

Of particular importance for our purposes is the *filter* $\mathscr{J}_0^\alpha$ *of terminal subsets* of a limit ordinal $\alpha$. $\mathscr{J}_0^\alpha$ is MT-definable by the formula:

(4) $\quad X \in \mathscr{J}_0: (\exists y)\ (\forall t).y < t \supset Xt.$

Note too that $X \notin \mathscr{O}_0^\alpha$ means that $X$ is cofinal $\alpha$ and is definable by $(\forall y)\ (\exists t).y < t \wedge Xt$. We replace the cumbersome notations $(\exists_{\mathscr{J}_0} t)$, etc. by $(\exists_0 t)$, $(\forall_0 t)$, $\sup_0 Z$. Also relativizations $(\exists_0 t)\ Xt[x]$, etc. are replaced by $(\exists_0^x t)\ Xt$, etc. As we only relativize these to limits we make the convention that henceforth $x$ ranges over limits, e.g. $(\forall x)\,\varphi$ means $(\forall x)^{Lm}\,\varphi$ where $Lm$ is the set of limits.

Let $\alpha$ be an $\omega_0$-inaccessible limit ordinal and let $X \subseteq \alpha$. The derivative $X'^{\alpha}$ of $X$ is the set of all limits of sequences in $X$ (not including $\alpha$). Then $\mathrm{cl}(X) = X \cup X'^{\alpha}$ is a closure operator, $X$ being *closed* if $X'^{\alpha} \subseteq X$. Let $\mathscr{J}_1^{\alpha}$ be the set of all those subsets of $\alpha$ which contain a cofinal-closed set. $\mathscr{J}_1^{\alpha}$ is then definable in $[\alpha, <]$ by

(5) $\quad X \in \mathscr{J}_1 : (\exists U)^X.(\forall x)\,[(\exists_0^x t)\; Ut \supset Ux] \wedge (\exists_0 t)\; Ut.$

Denote $\alpha$ is $\varkappa$-accessible by $\alpha \Leftarrow \varkappa$. We also assume the axiom of choice and so $\omega_1 = \varkappa_1$, the next inaccessible ordinal after $\omega_0$. The following are proved as in [5], Section 5.

Lemma 2.2 ([5], Lemma 5.1). *Suppose $\alpha$ is an $\omega_0$-inaccessible limit ordinal. If $X_i$, $i < \omega_0$, are cofinal-closed subsets of $\alpha$, then so is $\bigcap_{i<\omega_0} X_i$.*

Theorem 2.3 ([5], Theorem 5.2 (*AC*)). *For any $\omega_0$-inaccessible limit $\alpha$, $\mathscr{J}_1^{\alpha}$ is an $\omega_0$-filter on $\mathfrak{P}\alpha$ called the filter of cofinal-closed sets. Equivalently $\mathscr{O}_1^{\alpha} = \{\bar{X};\, X \in \mathscr{J}_1^{\alpha}\}$ is an $\omega_0$-ideal.*

Remark 2.4. *For any $\omega_0$-inaccessible limit $\alpha$: $X \notin \mathscr{O}_1^{\alpha} \wedge Y \in \mathscr{J}_1^{\alpha} \supset X \cap Y \notin \mathscr{O}_1^{\alpha}$.*

Remark 2.5 ([5], Remark 5.3). *If $\alpha$ is an $\omega_0$-inaccessible limit then*

$$X \text{ cofinal } \alpha \equiv X'^{\alpha} \text{ cofinal } \alpha \equiv X'^{\alpha} \text{ cofinal-closed } \alpha \equiv X'^{\alpha} \in \mathscr{J}_1^{\alpha}.$$

Remark 2.6. *If $\alpha \Leftarrow \omega_1$, there is an $\omega_1$-sequence $P$ of $\omega_0$-limits which is cofinal-closed $\alpha$.*

Proof. If $\alpha \Leftarrow \omega_1$, there is an $\omega_1$-sequence $P$ cofinal $\alpha$. Take $P = Q'^{\alpha}$ and the result follows by Remark 2.5. □

Remark 2.7 ([5], Remark 5.4). *Let $Z$ be a function from an $\omega_0$-inaccessible limit $\alpha$ to a finite set of states $K$. At most one state of $Z$ occurs cofinal-closed $\alpha$. Exactly one set $D$ of states occurs cofinal-closed as $\sup_0^x Z$, namely $D = \sup_0^{\alpha} Z$.*

In fact (using Remark 2.6) for $\alpha \Leftarrow \omega_1$ there is a cofinal-closed $Q$ such that $Qx \mathrel{.\supset.} x \Leftarrow \omega_0 \wedge \sup_0^x Z = \sup_0^{\alpha} Z$.

As before we use $(\exists_1 t)$, $(\forall_1 t)$, $\sup_1 Z$ for quantification with respect to $\mathscr{J}_1^{\alpha}$. This time though we will not only need relativizations to individuals, $(\exists_1^x t)$, etc. but also to sets.

Let $\alpha$ be an $\omega_0$-inaccessible limit and let $U$ be cofinal $\alpha$. We define $\mathscr{J}_1^{U,\alpha}$ to be the *filter of cofinal-closed sets relativized to* $U, \alpha$. That $\mathscr{J}_1^{U,\alpha}$ is a filter is a corollary to Theorem 2.3. As usual $\mathscr{O}_1^{U,\alpha}$ is the corresponding ideal and $(\forall_1^{U,x} t)$, etc. have their obvious meaning.

If $\{x_i;\, i < \lambda\}$ is a sequence in $U$, then the *$U$-limit* of the $x_i$'s is defined in the obvious way, $\lim^U_{i<\lambda} x_i = (\mu y).Uy \wedge y \geqq \lim_{i<\lambda} x_i$ (such a $y$ exists as $U$ is cofinal $\alpha$). If $V \subseteq U$, then the $U$-derivative of $V$ at $\alpha$, $V'^{U,\alpha}$, is the set of all $U$-limits of members of $V \cap \alpha$. Note that if $U$ is cofinal $\alpha$, then $\mathscr{J}_1^{U,\alpha}$ is the set of all those subsets of $U \cap \alpha$ which contain a set cofinal-closed in $U \cap \alpha$; i.e., cofinal $U$ and closed with respect to $U$-limits. We make the convention that if $U$ isn't cofinal $\alpha$, then we redefine $\mathscr{J}_1^{U,\alpha} = 0$·

Remark 2.8. *If $V \subseteq U$, $U$ cofinal $\alpha$, and $\alpha \not\Leftarrow \omega_0$, then*

(a) $V'^{\alpha} \cap U \subseteq V'^{U,\alpha}$, (b) $V'^{\alpha} \subseteq U \supset V'^{\alpha} = V'^{U,\alpha}$.

Lemma 2.9. *Let $\alpha$ be an $\omega_0$-inaccessible limit and suppose $X \subseteqq Y \in \mathscr{I}_1^\alpha$. Then*

$$X \in \mathscr{I}_1^{Y,\alpha} \equiv X \in \mathscr{I}_1^\alpha.$$

Proof. First suppose $X \in \mathscr{I}_1^{Y,\alpha}$. Thus, we have a $U \subseteqq X$ with $U$ cofinal-closed $Y \cap \alpha$. As $U$ is cofinal $Y$ and $Y$ is cofinal $\alpha$, it follows that $U$ is cofinal $\alpha$. Hence $U'^\alpha$ is cofinal closed $\alpha$, and since $Y \in \mathscr{I}_1^\alpha$ it follows that

(a) $\quad U'^\alpha \cap Y \in \mathscr{I}_1^\alpha$.

By Remark 2.8 we know that $U'^\alpha \cap Y \subseteqq U'^{Y,\alpha}$. As $U$ is closed in $Y \cap \alpha$ we get $U'^{Y,\alpha} \subseteqq U \subseteqq X$, and so

(b) $\quad U'^\alpha \cap Y \subseteqq X$.

By (a) and (b) we have $X \in \mathscr{I}_1^\alpha$. This argument shows that $X \in \mathscr{I}_1^{Y,\alpha} \supset X \in \mathscr{I}_1^\alpha$. In the next we complete the proof by showing $X \in \mathscr{I}_1^\alpha \supset X \in \mathscr{I}_1^{Y,\alpha}$. Suppose $X \in \mathscr{I}_1^\alpha$. Therefore, there is a $U \subseteqq X$ with $U$ cofinal-closed $\alpha$. As $U$ is cofinal $\alpha$ and $Y$ is cofinal $\alpha$ it follows that $U$ is cofinal $Y \cap \alpha$. Since $U$ is closed in $\alpha$ we know that (c) $U'^\alpha \subseteqq U$. As $U \subseteqq X \subseteqq Y$ we have $U'^\alpha \subseteqq Y$ and so by Remark 2.8, (d) $U'^\alpha = U'^{Y,\alpha}$. From (c) and (d) we get $U'^{Y,\alpha} \subseteqq U$, and consequently $U$ is cofinal-closed in $Y \cap \alpha$. Since $U \subseteqq X$, it follows that $X \in \mathscr{I}_1^{Y,\alpha}$. □

Corollary 2.10. *Let $\alpha$ be an $\omega_0$-inaccessible limit and suppose $Y \in \mathscr{I}_1^\alpha$. Then*

$$X \cap Y \notin \mathscr{O}_1^{Y,\alpha} \equiv X \notin \mathscr{O}_1^\alpha.$$

Corollary 2.11 *Let $\alpha$ be an $\omega_0$-inaccessible limit and suppose $Y$ is cofinal $\alpha$. For any $X$, define $V = \{v; (\exists t)_v^{X,Y} [v, t) \cap Y = 0\}$. Then*

$$\text{(a)}\quad X \in \mathscr{I}_1^{Y,\alpha} \equiv V \in \mathscr{I}_1^\alpha, \qquad \text{(b)}\quad X \notin \mathscr{O}_1^{Y,\alpha} \equiv V \notin \mathscr{O}_1^\alpha.$$

Proof. Let $X$, $Y$, $V$ be as above. For simplicity abbreviate $Y'^\alpha$ by $Y'$, and $Y'^{Y,\alpha}$ by $Y^+$. We define an isomorphism $f\colon [Y^+, <] \cong [Y', <]$ by $fy = (\mu x).x \leqq y \wedge x \in Y' \wedge \wedge [x, y) \cap Y = 0$. Now suppose $U \subseteqq Y^+$. Using the above isomorphism, $U \in \mathscr{I}_1^{Y^+} \equiv \equiv f[U] \in \mathscr{I}_1^{Y'}$. But as $Y' \in \mathscr{I}_1^\alpha$ by Lemma 2.9, $f[U] \in \mathscr{I}_1^{Y'} \equiv f[U] \in \mathscr{I}_1^\alpha$. Similarly, as $Y^+ \in \mathscr{I}_1^{Y,\alpha}$, by a relativized version of the same lemma we have $U \in \mathscr{I}_1^{Y^+} \equiv U \in \mathscr{I}_1^{Y,\alpha}$. Thus we have shown

(c) $\quad U \subseteqq Y^+ .\supset. U \in \mathscr{I}_1^{Y,\alpha} \equiv f[U] \in \mathscr{I}_1^\alpha$.

Using (c) we have the following sequence of equivalences

$$X \in Y_1^{Y,\alpha} \equiv X \cap Y^+ \in \mathscr{I}_1^{Y,\alpha} \equiv f[X \cap Y^+] = Y' \cap V \in \mathscr{I}_1^\alpha \equiv V \in \mathscr{I}_1^\alpha.$$

Thus (a) holds. The proof of (b) is almost identical and uses Corollary 2.10. □

The proof of the subset-construction for $\alpha < \omega_2$ will require the following extension of [5], Theorem 5.6.

Theorem 2.12 ($AC$). *Let $\alpha$ be an $\omega_1$-accessible limit. Every $U \notin \mathscr{O}_1^\alpha$ can be partitioned into $\omega_1$-many disjoint $U_i \notin \mathscr{O}_1^\alpha$. In particular $\mathfrak{P}\alpha/\mathscr{I}_1^\alpha$ is an atomless Boolean algebra.*

Proof. Using the Ulam argument [9], the proof of the theorem in the case $\alpha = \omega_1$ is given in [5] Theorem 5.6. Let $\alpha \Leftarrow \omega_1$ and $U \notin \mathscr{O}_1^\alpha$. By Remark 2.6 there is an $\omega_1$-sequence $P$ which is cofinal-closed $\alpha$. Then by Corollary 2.10, $U \cap P \notin \mathscr{O}_1^{P,\alpha}$. But $[P, <] \cong [\omega_1, <]$ and so by [5], Theorem 5.6, there are $\omega_1$-many disjoint $U_i$ in $U \cap P$ such that $U_i \notin \mathscr{O}_1^{P,\alpha}$. Since $U_i = U_i \cap P$, by Corollary 2.10 these $U_i \notin \mathscr{O}_1^\alpha$. □

### 3. The subset-construction at $\omega_1$

Throughout this section we will consider $\sup_1$-*transition systems* (at $\omega_1$) (also called *non-deterministic* $\sup_1$-*automata*), i.e. formulas of the form:

(1) $$\Gamma(X, Z)\colon E[Z0] \wedge (\forall t)\, H[Xt, Zt, Zt'] \wedge (\forall x)\, \mathfrak{K}[\sup_0^x Z, Zx] \wedge \mathfrak{L}[\sup_1 Z]$$

where $t'$ is the successor, $t+1$, of $t$ ($0, t'$ are clearly definable). $E, H, \mathfrak{K}, \mathfrak{L}$ are propositional expressions in the indicated parts. Intuitively, (1) states that the $\omega_1$-sequence $Z$ is a run of the non-deterministic finite automata $\Gamma = [E, H, \mathfrak{K}, \mathfrak{L}]$ with input $X$. If $X$ is an $m$-tuple and $Z$ an $n$-tuple then elements of $I = \{T, F\}^m$ are the *input states* and those of $K = \{T, F\}^n$ are the (*internal*) *states*. $E \subseteq K$ is the set of *initial states*, $H \subseteq I \times K \times K$ and $\mathfrak{K} \subseteq \mathfrak{P}K \times K$ are *transition relations*, and $\mathfrak{L} \subseteq \mathfrak{P}K$ is the *terminal condition*. If $E$ is a singleton and $H$ and $\mathfrak{K}$ are functions then $\Gamma$ is *deterministic*. $\Gamma$ *accepts* the input $X$ provided $(\exists Z)\, \Gamma(X, Z)$. The *behavior* of $\Gamma$ is the set of all those $\omega_1$-sequences accepted by $\Gamma$.

In [5], Section 6, a non-deterministic complementation lemma is given for $\sup_1$-transition systems, i.e., for every $\sup_1$-transition system $\Gamma$ a non-deterministic $\sup_1$-transition system $\Gamma'$ is constructed with complementary behavior. It is our purpose to present a subset-construction. Ideally we would like to show that for every $\sup_1$-transition system a deterministic $\sup_1$-transition system of equal behavior can be constructed. The following simple counterexample shows that this is impossible in a set theory in which $\mathscr{I}_1$ is not prime. In particular this is the case (Theorem 2.12) if $AC$ holds.

Remark 3.1. *$\mathscr{I}_1$ is definable by a $\sup_1$-transition system but not by a deterministic one.*

Proof. The first assertion is easy since

$$X \in \mathscr{I}_1 \;.\!\equiv\!.\; (\exists Z).(\forall t)\, [Zt \equiv Xt] \wedge (\forall_1 t)\, Zt.$$

Now suppose

$$X \in \mathscr{I}_1 \;.\!\equiv\!.\; (\exists Y).Y0 = a \wedge (\forall t)\, [Yt' = F[Xt, Yt]] \wedge (\forall x)\, [Yx = \mathfrak{G}[\sup_0^x Z]] \wedge \mathfrak{L}[\sup_1 Y].$$

For such a $Y$ let $D = \sup_0 Y$ and $d = \mathfrak{G}[D]$. By Remark 2.7, $\{x;\, \sup_0^x Y = D\} \in \mathscr{I}_1$ and so $\{x;\, Yx = d\} \in \mathscr{I}_1$. Consequently $\sup_1 Y = \{d\}$. From this we have, $X \notin \mathscr{I}_1 \;.\!\equiv\!.\; (\exists Y).\; Y0 \wedge (\forall t)\, [Yt' = F[Xt, Yt]] \wedge (\forall x)\, [Yx = \mathfrak{G}[\sup_0^x Z]] \wedge \mathfrak{L}'[\sup_0 Y]$ where $\mathfrak{L}'[\sup_0 Y]$ is $\sim \mathfrak{L}[\{\mathfrak{G}[\sup_0 Y]\}]$. Hence $\sim \mathscr{I}_1$ is definable by a $\sup_0$-transitionsystem. But by [5], Remark 5.5, this is impossible if we assume $\mathscr{I}_1$ is not prime. □

The subset construction which we present will use *relativized* $\sup_1$-*transition systems*. These are the same as ordinary $\sup_1$-transition systems, except that the terminal condition has form $\mathfrak{L}[\sup_1^{L_1} Z, \ldots, \sup_1^{L_k} Z]$ where $L_1[Zt], \ldots, L_k[Zt]$ are propositional formulas (we may suppose $k = 2^n$ and the $L_i$'s are essentially all such propositional formulas) and $\sup_1^L Z =_{\mathrm{def}} \sup_1^U Z$ where $U = \{t;\, L[Zt]\}$. The theorem which we prove is that for every $\sup_1$-transition system a deterministic relativized-$\sup_1$-transition can be constructed of equal behavior. Since for every relativized $\sup_1$-transition system a (non-deterministic) $\sup_1$-transition system of equal behavior can be constructed (see Lemma 4.3) we get another proof of the complementation lemma at $\omega_1$. As the techniques used easily extend to all $\omega_1$-accessible ordinals $<\omega_2$, a complementation lemma for all $\alpha < \omega_2$ is available.

For the remainder of this section we consider a fixed $\sup_1$-transition system $\Gamma = [E, H, \mathfrak{K}, \mathfrak{L}]$. A *C-exact* $X[u, v)$*-run of* $\Gamma$ *from c to d* is a sequence $Z[u, v]$ such that

$$Zu = c \wedge (\forall t)_u^v H[Xt, Zt, Zt'] \wedge (\forall x)_u^{v'} \mathfrak{K}[\sup_0^x Z, Zx] \wedge Z_u^v = C \wedge Zv = d$$

where $Z_u^v =_{\text{def}} \{Zt;\, u \leqq t < v\}$. For a given input $X$ and $u \leqq v$, define

(2) $\langle c, C, d\rangle \in S_u v \;.\equiv.\; (\exists Z).Z[u, v]$ is a $C$-exact $X$-run from $c$ to $d$.

Note that $S_u v$ depends upon the input $X$. If $Z$ is an $n$-tuple and so the set of states of $Z$ (and therefore of $\Gamma$) is $K = \{T, F\}^n$, then the states of $S_u$ are taken from the set $\mathfrak{P}(K \times \mathfrak{P}K \times K)$ and so $S_u$ is finite-state, and in fact has $g = 2^{r2^{2r}}$ $(r = 2^n)$. Often $\langle c, C, d\rangle \in S_u v$ is replaced by the more compact notation $d \in S_u^{c,C} v$, i.e. the projection onto the first two coordinates. Then $S_u^{c,C} v$ may be thought of as the set of all states $d$ which can be reached from $c$ by a $C$-exact run. The following merging relations are defined as in [5]:

$$u \approx v\ (t)\colon u, v \leqq t \wedge S_u t = S_v t \qquad (\textit{u and v merge by time t}),$$

(3) $$u \approx v\ (-t)\colon (\exists y)^t\, u \approx v\ (y) \qquad (\textit{u and v merge before t}),$$

$$u \sim v\ (t)\colon u \approx v\ (t) \wedge u \not\approx v\ (-t) \quad (\textit{u and v merge at t}).$$

Remark 3.2 ([5], Remark 3.2). *For any* $X$, $\approx(t)$ *and* $\approx(-t)$ *are equivalence relations of finite index* $\leqq g$. *Also* $u \approx v\ (t) \wedge t \leqq z \;.\supset.\; u \approx v\ (z)$ *and* $t \approx v\ (t) \supset v = t$.

Lemma 3.3. *For any input* $X$, $(\exists Z)\, \Gamma(X, Z) \equiv \Omega$, *where*

(4) $$\Omega\colon \bigvee_W . \bigwedge_{w \in W} (\exists_1 y)\, y \approx w(-\omega_1) \wedge (\forall_1 y) \bigvee_{w \in W} y \approx w(-\omega_1) \wedge (\exists_1 y) B_1[S_0 y, \langle S_w y; w \in W\rangle]$$

(5) $$B_1[s_0, T]\colon \bigvee_{c,C,D_0,D_1,Q} . E[c] \wedge \mathfrak{L}[D_1] \wedge \bigwedge_{e \in D_1} \mathfrak{K}[D_0, e] \wedge \bigvee_{e \in D_1} e \in s_0^{c,C}$$
$$\wedge\ Q \textit{ is a subdirect product of } D_1 \times T \wedge \bigwedge_{\langle e,s\rangle \in Q} D_1 \subseteqq s^{e,D_0}.$$

Here $W$ is a finite subset of $\omega_1$ and so $\bigvee_W$ abbreviates $(\exists w_1) \ldots (\exists w_n)$ where $W = \{w_1, \ldots, w_n\}$. $T$ is a finite sequence of states $s$ of the type $S_u$. Recall that $Q$ is a subdirect product of $D_1 \times T$ means that $Q \subseteqq D_1 \times T \wedge \bigwedge_{e \in D_1} \bigvee_{s \in T} \langle e, s\rangle \in Q \wedge \bigwedge_{s \in T} \bigvee_{e \in D_1} \langle e, s\rangle \in Q$. The lemma is an easy consequence of [5], Lemmas 6.1 and 6.3, and thus $AC$ is used (for splicing and so that $\omega_1 = \aleph_1$ and $\mathfrak{P}\omega_1/\mathscr{I}_1$ is atomless). A general version of the lemma is given in Section 4.

Now the subset construction for ordinals $<\omega_1$ ([5], Lemma 4.4) provides a finite-state recursion

(6) $$Y0 = e, \qquad Yt' = F[Xt, Yt], \qquad Yx = \mathfrak{G}[\sup_0^x Z],$$

such that $S_0$ is defined from some of the components of $Y$ (remember that $Y$ is a tuple of set variables). Thus the above recursion along with $\Omega$ yields a subset construction at $\omega_1$. All that we need do is put $\Omega$ in the form of a relativized $\sup_1$-transition system.

Because of the way the McNaughton construction works, the $(\exists_1 y)$-quantifier in (4) is undesirable. In the analogous situation in [5] the equivalence $(\exists_0 y)\, y \approx w\,(-\omega_0) \equiv (\exists_0 t)\,(\exists y)\, y \sim w\ (t)$ was used. But it is easy to see that $(\exists_1 y)\, y \approx w\,(-\omega_1) \not\equiv (\exists_1 t)\,(\exists y)\, y \sim w\ (t)$. It is here that the relativized filter of cofinal-closed sets enters the picture.

Let $V$ be a (finite) sequence of representatives of all the equivalence classes of $\approx (-\omega)$ in their natural order. It will often be convenient to purposely confuse $V$ (or any vector) with its set of components. Clearly,

$$(7) \qquad y < \omega_1 \supset \bigvee_{v \in V} y \approx v\,(-\omega_1),$$

$$v, w \in V \wedge v \neq w \,.\!\supset\!.\, v \not\approx w\,(-\omega_1).$$

For any such sequence $V$ we may define the sequences $y_i$, $t_i$ by induction on $i < \omega_1$:

$$(8) \qquad y_0 = (\mu y) \,.\, \bigwedge_{v \in V} v < y, \qquad t_i = (\mu t) \,.\, \bigvee_{v \in V} y_i \sim v\,(t) \quad \text{(exists by (7))},$$

$$y_{i+1} = t_i', \qquad y_\lambda = \lim_{i<\lambda} t_i, \qquad y_\lambda = \lim_{i<\lambda} y_i;$$

$$(9) \qquad Y_V = \{y_i;\, i < \omega_1\}, \qquad T_V = \{t_i;\, i < \omega_1\}.$$

The following is just what is needed to get rid of the $(\exists_1 y)$ quantifier from formula (4).

$$(10) \qquad (\exists_1 y)\, y \approx w\,(-\omega_1) \,.\!\equiv\!.\, (\exists_1^{T_V} t)\,(\exists y)\,[y \sim w\,(t) \wedge y \in T_V'].$$

Proof. We drop the subscript $V$. As $Y$ is cofinal-closed, by Corollary 2.10,

$$(\text{a}) \qquad \{y;\, y \approx w\,(-\omega_1)\} \notin \mathcal{O}_1 \equiv \{y_i;\, y_i \approx w\,(-\omega_1)\} \notin \mathcal{O}_1^Y.$$

Using (7) and (8) it is easy to see that

$$(\text{b}) \qquad y_1 \approx w\,(-\omega_1) \equiv y_i \sim w\,(t_i).$$

But $y_i \mapsto t_i$ defines an isomorphism $[Y, <] \cong [T, <]$. Thus, with (a), (b) we have

$$\{y;\, y \approx w\,(-\omega_1)\} \notin \mathcal{O}_1 \equiv \{t_i;\, y_i \sim w\,(t_i)\} \notin \mathcal{O}_1^T$$

and so

$$(\text{c}) \qquad \{y;\, y \approx w\,(-\omega_1)\} \notin \mathcal{O}_1 \equiv \{t_\lambda;\, y_\lambda \sim w\,(t_\lambda)\} \notin \mathcal{O}_1^T.$$

Assume $y \sim w\,(t_i)$ and $y \in T'$. Since $y \in T'$ there is a $\lambda$ such that $y = \lim_{i<\lambda} t_i$, and so $y = y_\lambda$. Using (8) there is a $v \in V$ with $y \sim v\,(t_\lambda)$. By the definition of $\sim$ we then have $w = v$ and $t_i = t_\lambda$. Thus $i = \lambda$ and so $y = y_i = y_\lambda$, $y_\lambda \sim w\,(t_\lambda)$. This shows that $\{t_i;\, (\exists y).y \sim w\,(t_i) \wedge y \in T'\} \subseteq \{t_\lambda;\, y_\lambda \sim w\,(t_\lambda)\}$. As the other inclusion is trivial we have by (c) that

$$\{y;\, y \approx w\,(-\omega_1)\} \notin \mathcal{O}_1 \equiv \{t;\, t \in T \wedge (\exists y).y \sim w\,(t) \wedge y \in T'\} \notin \mathcal{O}_1^T$$

which is the desired result. □

Using (10) it is not hard to put (4) into the more desirable form:

$$(11) \qquad \Omega \,.\!\equiv\!.\, \bigvee_{W \subsetneq V} \,.\, \bigwedge_{w \in W} (\exists_1^{T_V} t)\,(\exists y)\,[y \sim w\,(t) \wedge T_V' y]$$
$$\wedge\, (\forall_1^{T_V} t) \bigvee_{w \in W} (\exists y)\,[y \sim w\,(t) \wedge T_V' y] \wedge (\exists_1 t)\, B_1[S_0 t, \langle S_w t;\, w \in W\rangle].$$

In order for this new form of $\Omega$ to be useful it will be necessary to find recursions which define $T_V$, $Y_V$ and which are easily simulated by ones which are finite state. As a first step we remark that the $Y_V$, $T_V$ defined in (9) satisfy the following recursion:

$$Y_V \bar{v}', \quad \text{where } \bar{v} \text{ is the last member of } V,$$

$$(12) \qquad T_V t \,.\!\equiv\!.\, \bigvee_{v \in V} (\exists y)\,[Y_V y \wedge y \sim v\,(t)],$$

$$Y_V t' \,.\!\equiv\!.\, T_V t, \qquad Y_V x \,.\!\equiv\!.\, (\exists_0^x t)\, T_V t.$$

Note that in (11) there are references to both $t$ and $y$. We now introduce the piggyback device of McNaughton which enables the information at $y$ to be carried up to $t$. As in [5], we define,

| | | |
|---|---|---|
| | $r_y t$: $(\mu v)\, v \approx y\,(-t);\ y < t$ | *(the representation of $y$ at $t$),* |
| | $Ut$: The sequence in natural order of all $r_y t$, $y < t$ | *(the vector of indices active at time $t$),* |
| (13) | $u \leftarrow v\ (t)$: $v$ in $Ut$, and $u$ the first in $Ut$ such that $u \approx v\ (t)$ | *($u$ replaces $v$ at time $t$),* |
| | $v \in \bar{B}(t, u_0, \ldots, u_n)$: $(\exists y)^t . v = r_y t \wedge B(y, u_0, \ldots, u_n)$, $B$ any formula | *(representative at $t$ of past hit of $B$).* |

Lemma 3.4 (Piggyback lemma). *Suppose $w \leftarrow w\ (t)$. Then*

$$(\exists y)\,[w < y \wedge y \sim w\ (t) \wedge B(y)]\ .\equiv.\ (\exists v)\,[v \mp w \leftarrow v\ (t) \wedge v \in \bar{B}(t)].$$

Proof. Suppose $w \leftarrow w\ (t)$, $w < y$, $y \sim w\ (t)$, and $B(y)$. Clearly by the last statement in Remark 3.2, $y < t$. Therefore $v = r_y t$ exists and so $v \approx y\,(-t)$. Thus with $y \sim w\ (t)$ and $w \leftarrow w\ (t)$ it can be shown that $v \mp w \leftarrow v\ (t)$. As $y < t$ and $B(y)$, $v \in \bar{B}(t)$. This proves the $\supset$-case of the lemma. Conversely, suppose $v \mp w \leftarrow v\ (t)$, and $v \in \bar{B}(t)$. Thus $w < v$, and $w \sim v\ (t)$. Also there is a $y$ such that $y < t$, $v = r_y t$, and $B(y)$, and so $v \leqq y$ and $y \approx v\,(-t)$. Hence $w < y$ and $y \sim w\ (t)$, completing the proof of the $\subset$-case. □

Lemma 3.5. *The sequences $U$, $\bar{B}$ defined by* (13) *satisfy the following recursion:*

$$\begin{aligned}
&U0 = \textit{empty sequence}, \qquad \bar{B}(0) = 0,\\
&Ut' = \textit{remove from } Ut \textit{ all } v \nleftarrow v\ (t), \textit{ and add } t \textit{ at the end},\\
&Ux = \textit{all } v \textit{ such that } (\forall_0^x t)\ v \leftarrow v\ (t), \textit{ in natural order},\\
&v \in \bar{B}(t')\ .\equiv.\ (\exists p)\,[v \leftarrow p\ (t) \wedge p \in \bar{B}(t)] \vee [v = t \wedge B(t)],\\
&v \in \bar{B}(x)\ .\equiv.\ (\forall_0^x t)\ v \in \bar{B}(t).
\end{aligned}$$

The proof may be found in [5], Lemma 4.3. At first glance formula (11) is a bit nicer than the corresponding formula (12) of [5]. This is because the reference to the prepositional expression $B$ is isolated and therefore no piggybacking is needed. Unfortunately, piggybacking is needed to properly handle $V$, $Y_V$, $T_V$. Using (11), (12) and Lemma 3.4 we have:

Lemma 3.6. *For any input $X$ and any finite sequence $V$ in $\omega_1$, if*

$$\begin{aligned}
&Y_V \bar{v}', && \textit{where } \bar{v} \textit{ is the last member of } V,\\
&T_V t && .\equiv.\ \bigvee_{w \in V} (\exists v)\,[v \mp w \leftarrow v\ (t) \wedge P_V t],\\
&Y_V t' && .\equiv.\ T_V t, \qquad Y_V x\ .\equiv.\ (\exists_0^x t)\ T_V t,\\
&v \in P_V t\ .\equiv.\ (\exists y)^t\,[v = r_y t \wedge Y_V t], && \qquad v \in R_V t\ .\equiv.\ (\exists y)^t\,[v = r_y t \wedge T'_V t]
\end{aligned}$$

*where $P_V$, $R_V$ may be defined via a recursion as in Lemma 3.5, then*

$$\begin{aligned}
\Omega\ .\equiv.\ &\bigvee_{\substack{W, V\\ W \subsetneq V}} .\ \bigwedge_{v \in V} (\exists_0 t)\ v \leftarrow v\ (t) \wedge (\forall_0 t) \bigvee_{v \in V} v \leftarrow v\ (t)\\
&\wedge \bigwedge_{w \in W} (\exists_1^{T_V} t)\,(\exists v)\,[v \mp w \leftarrow v\ (t) \wedge v \in R_V t]\\
&\wedge (\forall_1^{T_V} t) \bigvee_{w \in W} (\exists v)\,[v \mp w \leftarrow v\ (t) \wedge v \in R_V t]\\
&\wedge (\exists_1 t)\ B_1[S_0 t, \langle S_w t;\ w \in W\rangle].
\end{aligned}$$

We now present a finite-state recursion for the $\omega_1$-sequences $S_i$, $Q_{i,k}$, $V$, $Y_J$, $P_J$, $R_J$ which simulate the sequences $S_u$, $Q_{u,w}$, $U$, $Y_V$, $P_V$, $R_V$. Here the $Q_{u,w}$ and also $B_0[s_0, s]$ which is used later are defined in [5], Section 4, and are used in piggybacking at $\omega_0$-limits.

$$S_it = \begin{cases} s_0, & \text{if } i \notin Vt \\ \mathfrak{S}_i(t), & \text{if } i \in Vt, \end{cases} \qquad Q_{i,k}t = \begin{cases} 0, & \text{if } i \notin Vt \vee k \notin Vt \\ \mathfrak{Q}_{i,k}(t), & \text{if } i, k \in Vt, \end{cases}$$

$$V0 = \text{empty sequence},$$

$$Vt' = \text{remove from } Vt \text{ all } i \nleftrightarrow i\ (t), \text{ add } \varrho(t) \text{ at end},$$

$$Vx = \text{all } i \text{ such that } (\forall_0^x t)\ i \leftarrow i\ (t),\ i \text{ before } j \text{ in case } (\forall_0^x t)\ [i \text{ before } j],$$

$$\tilde{Y}_J0, \qquad Y_Jt' .\equiv. T_Jt \vee \varrho(t) \in J, \qquad Y_Jx \equiv (\exists_0^x t)\ T_Jt,$$

$$P_J0 = 0, \qquad R_J0 = 0,$$

$$(14) \qquad P_Jt' = \begin{cases} 0, & \text{if } \varrho(t) \in J \\ \mathfrak{P}_J(t), & \text{if } \varrho(t) \notin J, \end{cases} \qquad R_Jt' = \begin{cases} 0 & \text{if } \varrho(t) \in J \\ \mathfrak{R}_J(t) & \text{if } \varrho(t) \notin J, \end{cases}$$

$$P_Jx = \{j;\ (\forall_0^x t)\ j \in P_Jt\}, \qquad R_Jx = \{j;\ (\forall_0^x t)\ j \in R_Jt\},$$

$$\bigvee_{\substack{J,K \\ K \subsetneqq J}} \cdot \bigwedge_{j \in J} (\exists_0 t)\ j \leftarrow j\ (t) \wedge (\forall_0 t) \bigvee_{j \in J} j \leftarrow j\ (t)$$

$$\wedge \bigwedge_{k \in K} (\exists_1^{T_J} t) \bigvee_h [h \neq k \leftarrow h\ (t) \wedge h \in R_Jt] \qquad \textit{terminal condition}$$

$$\wedge (\forall_1^{T_J} t) \bigvee_{k \in K} \bigvee_h [h \neq k \leftarrow h\ (t) \wedge h \in R_Jt] \wedge (\exists_1 t)\ B_1[S_0t, \langle S_kt;\ k \in K \rangle].$$

Here the indices $h, i, j, k$ range over $\{0, \ldots, g\}$ and $J$, $K$ range over finite sequences with distinct components in $\{0, \ldots, g\}$ ($g$ = number of states of the $S_u$'s), and the following abbreviations are used:

$$i \leftarrow j\ (t):\ j \text{ in } Vt,\ i \text{ first in } Vt \text{ such that } S_it = S_jt,$$

$$\varrho(t):\ \text{first } i \leqq g \text{ such that } i \notin Vt \text{ (used to simulate } t \text{ at time } t'),$$

$$s_0:\ \{\langle c, 0, c \rangle;\ c \text{ a state of } \Gamma\},$$

$$\mathfrak{S}_i(t'):\ \{\langle c, C, b \rangle;\ \bigvee_{a \in C} H[Xt, a, b] \wedge a \in S_i^{c,C}t \cup S_i^{c,C-\{a\}}t\},$$

$$\mathfrak{S}_i(x):\ \{\langle c, C, b \rangle;\ \bigvee_{k \in Vx} (\exists_0^x t) \bigvee_j [j \neq k \leftarrow j\ (t) \wedge j \in Q_{i,k}^{c,C,b}t]\},$$

$$\mathfrak{Q}_{i,k}(t'):\ \{\langle c, C, b, j \rangle;\ \bigvee_h [j \leftarrow h\ (t) \wedge h \in Q_{i,k}^{c,C,b}t] \vee [j = \varrho(t) \wedge b \in B_0^{c,C}[S_it, S_kt]]\},$$

$$\mathfrak{Q}_{i,k}(x):\ \{\langle c, C, b, j \rangle;\ (\forall_0^x t)\ j \in Q_{i,k}^{c,C,b}t\},$$

$$\mathfrak{P}_J(t):\ \{j;\ \bigvee_h [j \leftarrow h\ (t) \wedge h \in P_Jt] \vee [j = \varrho(t) \wedge Y_Jt]\},$$

$$\mathfrak{R}_J(t):\ \{j;\ \bigvee_h [j \leftarrow h\ (t) \wedge h \in R_Jt] \vee [j = \varrho(t) \wedge T'_Jt]\},$$

$$T_Jt:\ \bigvee_{j \in J} \bigvee_h .\ h \neq j \leftarrow h\ (t) \wedge h \in P_Jt.$$

Using Lemma 2.1 it is clear that (14) is a transition system of form,

$$(15) \qquad \Delta(X, Y):\ Y0 = a \wedge (\forall t)\ Yt' = F[Xt, Yt] \wedge (\forall x)\ Yx = \mathfrak{G}[\sup_0^x Z]$$
$$\wedge\ \mathfrak{L}[\sup_0 Y, \sup_1^{L_1} Y, \ldots, \sup_1^{L_n} Y].$$

where $Y$ is made up from the components $S_i$, $Q_{i,k}$, $V$, $Y_J$, $P_J$, $R_J$ and so has finitely many states. Using Remark 2.5 it is possible to express $\Delta$ as a deterministic relativized $\sup_1$-transition system $\Delta'$, i.e. $(\exists Y)\,\Delta(X, Y) \,.\!\equiv\!.\, (\exists U)\,\Delta'(X, U)$ (see Lemma 4.1).

Lemma 3.7 (Subset construction at $\omega_1$). *For every* $\sup_1$*-transition system* $\Gamma(X, Z)$, *a deterministic system* $\Delta(X, Y)$ *of form* (15) *can be constructed such that* $(\exists Z)\,\Gamma(X, Z) \equiv (\exists Y)\,\Delta(X, Y)$. *In fact* (14) *defines such a* $\Delta$, *i.e., for any input* $X$, $\Gamma$ *accepts* $X$ *just in case the recursion* $Y = A(X)$ *defined by* (14) *when applied to* $X$ *satisfies the terminal condition* $\mathfrak{L}$. *Using the remark after* (15) *there is even a deterministic relativized* $\sup_1$-*transition system* $\Delta'$ *such that* $(\exists Z)\,\Gamma(X, Z) \,.\!\equiv\!.\, (\exists U)\,\Delta'(X, U)$.

Proof. The reader is asked to make up the recursion $(\overline{14})$ obtained from (14) by making the following replacements: $h, i, j, k, J, K$, by $p, u, v, w, V, W$; $S, \mathfrak{S}, Q, \mathfrak{Q}, P, \mathfrak{P}, R, \mathfrak{R}, Y, T$ by $\bar{S}, \bar{\mathfrak{S}}, \bar{Q}, \bar{\mathfrak{Q}}, \bar{P}, \bar{\mathfrak{P}}, \bar{R}, \bar{\mathfrak{R}}, \bar{Y}, \bar{T}$; $V$ by $U$; $\leftarrow$ by $\Leftarrow$; $\varrho(t)$ by $t$. Note that $(\overline{14})$ contains the recursions of Lemmas 3.5 and 3.6 as well as those in $(\bar{8})$ of [5], Section 4. Also by Lemma 3.6, the terminal condition in $(\overline{14})$ is just $\Omega$. Thus recursion $(\overline{14})$ together with its terminal condition satisfies the conditions of the theorem but for the fact that $(\overline{14})$ isn't finite-state. Consequently the lemma is proved when we show how the finite state system (14) simulates the "quasi finite-state" system $(\overline{14})$. Toward this end we define the function $u_i t$:

(a) $\quad i = \varrho(t) \,.\!\supset\!.\, u_i t' = t, \qquad i \leftarrow i\,(t) \,.\!\supset\!.\, u_i t' = u_i t,$

$\quad (\forall_0^x t)\; i \leftarrow i\,(t) \,.\!\supset\!.\, u_i x = \text{that } u \text{ for which } (\forall_0^x t)\; u_i t = u.$

Exactly as in [5], Section 4, we may prove by induction on $t$:

(b$_1$) $\quad i \in Vt \,.\!\supset\!.\, u_i t$ defined and in $Ut$,

(b$_2$) $\quad u \in Ut \supset (\exists i)\,[i \in Vt \wedge u = u_i t]$,

(b$_3$) $\quad i$ before $j$ in $Vt \,.\!\supset\!.\, u_i t < u_j t$,

(b$_4$) $\quad i \in Vt \wedge u = u_i t \,.\!\supset\!.\, S_i t = \bar{S}_u t$,

(b$_5$) $\quad i, j \in Vt \,.\!\supset\!.\, i \leftarrow j\,(t) \equiv u_i t \Leftarrow u_j t\,(t)$,

(b$_6$) $\quad i, k \in Vt \wedge u = u_i t \wedge w = u_k t \,.\!\supset\!.\, j \in Q_{i,k}^{c,C,b} t \equiv u_j t \in \bar{Q}_{u,w}^{c,C,b} t$.

We next show how $Y, P, R$ simulate $\bar{Y}, \bar{P}, \bar{R}$.

Let $V = \{u;\, (\forall_0 t)\; u \Leftarrow u\,(t)\}$ and $J = \{i;\, (\forall_0 t)\; i \leftarrow i\,(t)\}$. Then

(c$_1$) $\quad (\forall_0 t).\, Y_J t \equiv \bar{Y}_V t$,

(c$_2$) $\quad (\forall_0 t).\, j \in P_J t \equiv u_j t \in \bar{P}_V t$,

(c$_3$) $\quad (\forall_0 t).\, j \in R_J t \equiv u_j t \in \bar{R}_V t$,

(c$_4$) $\quad (\forall_0 t).\, T_J t \equiv \bar{T}_V t$.

Let $\bar{v}$ be the last member of $V$. In order to prove (c) we first show the following:

(d$_1$) $\quad u = u_i y \wedge (\forall t)_y\; u \Leftarrow u\,(t) \,.\!\supset\!.\, (\forall t)_y\,[u = u_i t \wedge i \leftarrow i\,(t)]$,

(d$_2$) $\quad v \in V \supset \bigvee_{j \in J} (\forall t)_{\bar{v}'}\,[v = u_j t \wedge j \leftarrow j\,(t)]$,

(d$_3$) $\quad j \in J \supset \bigvee_{v \in V} (\forall t)_{\bar{v}'}\,[v = u_j t \wedge j \leftarrow j\,(t)]$,

($d_4$) $\bar{v} \in V \wedge \varrho(\bar{v}) \in J$,

($d_5$) $(\forall t)_{\bar{v}'} [t \notin V \wedge \varrho(t) \notin J]$.

($d_1$): The proof is by induction using (a) and ($b_5$).

($d_2$): Suppose $v \in V$. Then $(\forall t)_{\bar{v}'}\ v \Leftarrow v\ (t)$ and $v \in U\bar{v}'$. By ($b_2$) there is a $j \in V\bar{v}'$ such that $v = u_j\bar{v}'$, and so by ($d_1$), $j \in J \wedge (\forall t)_{\bar{v}'} [v = u_j t \wedge j \leftarrow j\ (t)]$.

($d_3$): Suppose $j \in J$. There is an $y > \bar{v}$ with $(\forall t)_y\ j \leftarrow j\ (t)$. Let $v = u_j y$. By (a), $(\forall t)_y\ v = u_j t$, and so by ($b_5$), $(\forall t)_y\ v \Leftarrow v\ (t)$. Thus $v \in V$ and therefore by ($d_2$) there is an $i \in J$ such that $(\forall t)_{\bar{v}'} [v = u_i t \wedge i \leftarrow i\ (t)]$. Therefore $u_j y = v = u_i y$ and so by ($b_3$), $i = j$. Hence $(\forall t)_{\bar{v}'} [v = u_j t \wedge j \leftarrow j\ (t)]$.

($d_4$): The first conjunct is trivial. Let $i = \varrho(\bar{v})$. By (a), $u_i\bar{v}' = \bar{v} \in V$. Thus $(\forall t)_y\ u_i\bar{v}' \Leftarrow u_i\bar{v}'\ (t)$. Using ($d_1$) we have $(\forall t)_y\ i \leftarrow i\ (t)$, and so $i \in J$.

($d_5$): The first conjunct is trivial. Let $\bar{v} < t$, $i = \varrho(t)$, and suppose $i \in J$. Then by ($d_3$), $i \leftarrow i\ (t)$ and so $i \in Vt$, contradicting $\varrho(t) \notin Vt$. Thus $i \notin J$.

We now show by induction on $t > \bar{v}$ that $Y_J t \equiv \bar{Y}_V t$, $j \in P_J t \equiv u_j t \in \bar{P}_V t$, $j \in R_J t \equiv u_j t \in \bar{R}_V t$, and $T_J t \equiv \bar{T}_V t$, thus proving (c).

($c_1$): By ($d_4$), (14), and ($\overline{14}$), we have $\bar{Y}_V\bar{v}'$ and $Y_J\bar{v}'$ and so $Y_J\bar{v}' \equiv \bar{Y}_V\bar{v}'$. We next give the inductive step from $t$ to $t'$. Let $\bar{v} < t$. By (14), ($\overline{14}$), ($d_5$), and the inductive assumption we have, $Y_J t' \equiv T_J t \equiv \bar{T}_V t \equiv \bar{Y}_V t'$. The inductive step at limits follows directly from (14), ($\overline{14}$), and the inductive assumption.

($c_2$): By ($d_4$), (14), ($\overline{14}$), it follows that $P_J\bar{v}' = 0 = \bar{P}_V\bar{v}'$. We next sketch the proof of the inductive step from $t$ to $t'$. Suppose $\bar{v} < t$, $j \in Vt'$, and $u = u_j t'$. By ($d_5$), $P_J t' = \mathfrak{P}_J(t)$ and $\bar{P}_V t' = \bar{\mathfrak{P}}_V(t)$, and so it remains to show $j \in \mathfrak{P}_J(t) \equiv u \in \bar{\mathfrak{P}}_V(t)$. By (14), (b), and the inductive assumption:

$$j \in \mathfrak{P}_J(t) \,.\!\equiv\!.\, \bigvee_h [u \Leftarrow u_h t\ (t) \wedge u_h t \in \bar{\mathfrak{P}}_V(t)] \vee [u = t \wedge \bar{Y}_V t].$$

From this equivalence we have by ($\overline{14}$) that $j \in \mathfrak{P}_J(t) \supset u \in \bar{\mathfrak{P}}_V(t)$. For the implication in the other direction note that from $u \Leftarrow p\ (t)$ we get by ($\overline{14}$) that $p \in Ut$ and so by ($b_2$) we know that $p = u_h t$ for some $h \in Vt$. The proof of the inductive step at limits is similar and will be omitted (see the proof of ($b_6$) in [5], Lemma 4.4).

As the proof of ($c_3$) is almost identical to that of ($b_8$) it too will be omitted.

($c_4$): By (14), (b), ($c_2$), and the inductive assumption:

$$T_J t \,.\!\equiv\!.\, \bigvee_{j \in J} \bigvee_h [u_h t + u_j t \Leftarrow u_h t\ (t) \wedge u_h t \in \bar{P}_V t].$$

Using ($d_2$), ($d_3$) it follows from the above equivalence that

$$T_J t \,.\!\equiv\!.\, \bigvee_{v \in V} \bigvee_h [u_h t + v \Leftarrow u_h t\ (t) \wedge u_h t \in \bar{P}_V t].$$

The rest of the proof follows as in ($c_2$).

The proof that the terminal conditions of (14) and ($\overline{14}$) are equivalent uses ($d_2$), ($d_3$), (14), ($\overline{14}$), ($c_3$) and ($c_4$). The proof is similar to that of ($c_4$) and will be omitted. □

21*

Lemma 3.8 (Complementation lemma at $\omega_1$). *For every* $\sup_1$- *or relativized* $\sup_1$-*transition system,* $\sup_1$- *and relativized* $\sup_1$-*transition systems of complementary behavior can be constructed.*

Proof. Let $\Gamma$ be a $\sup_1$-transition system. Using the subset construction Lemma 3.7, a deterministic relativized $\sup_1$-transition system $\Delta = [a, F, \mathfrak{G}, \mathfrak{L}']$ of equal behavior can be constructed. The relativized $\sup_1$-transition system $\Delta' = [a, F, \mathfrak{G}, \sim\mathfrak{L}']$ is then of complementary behavior. Using Remark 2.5 and Corollary 2.11 (see Lemma 4.3) for any relativized $\sup_1$-transition system a (non-deterministic) $\sup_1$-transition system of equal behavior can be constructed. In this way we obtain a $\sup_1$-transition system $\Gamma''$ with the same behavior as $\Delta'$, and thus of complementary behavior to that of $\Gamma$. If $\Gamma$ is a relativized $\sup_1$-transition system, first replace it by a $\sup_1$-transition system of equal behavior and proceed as above. □

Thus we have another proof of the complementation lemma at $\omega_1$. Although it is more difficult than the one found in [5], it can easily be generalized to handle all ordinals $<\omega_2$ (see Section 4). In [5], Lemma 1.5, it is shown that for every MT-formula $\Sigma(X)$ in the primitive $<$ one can construct a $\sup_0$-transition system $\Gamma$ such that $\Sigma(X)$ is MT-equivalent to a formula of form (predicate prefix $Y$) $(\exists Z)\,\Gamma(XY, Z)$. Using Remark 2.5 (see Corollary 4.2) we may suppose that $\Gamma$ is a $\sup_1$-transition system. Now using the complementation lemma we may find another $\sup_1$-transitive system $\Gamma'$ such that $\Sigma(X)$ is equivalent to (predicate prefix $Y'$) $(\exists Z')\,\Gamma'(XY', Z')$ where $Y'$ has less components than $Y$. Continuing in this way we obtain the following strong definability results.

Theorem 3.9 (MT-definability in $[\omega_1, <]$): *For every* MT-*formula* $\Sigma(X)$ *in the primitive* $<$ *one can construct a relativized* $\sup_1$-*automata* $A = [a, F, \mathfrak{G}, \mathfrak{L}]$ *such that for every input* $X$, $\Sigma(X)$ *holds in* $[\omega_1, <]$ *if and only if* $A$ *accepts* $X$.

Theorem 3.10 (Normal form of sentences): *For every* MT-*sentence* $\Sigma$ *an input-free relativized* $\sup_1$-*automata* $A$ *can be constructed such that* $\Sigma$ *holds in* $[\omega_1, <]$ *just in case (the unique run of)* $A$ *accepts* $X$.

## 4. The subset-construction extended up to $\omega_2$

Now we extend the subset-construction of the previous section up to $\omega_2$. We assume AC and so $\omega_1 = \aleph_1$ and $\mathfrak{P}\alpha/\mathcal{J}_1^\alpha$ is atomless for any $\alpha < \omega_2$, $\alpha \Leftarrow \omega_1$. Throughout this section individual variables range over $\omega_2$ and we consider $\sup_1$-transition systems of form

$$(1) \qquad \Gamma(X, Z)\colon E[Z0] \wedge (\forall t)\, H[Xt, Zt, Zt'] \wedge (\forall x)^{Acc}\, \mathfrak{R}_0[\sup_0^x Z, Zx] \wedge (\forall x)^{Acc_1}\, \mathfrak{R}_1[\sup_1^x Z, Zx] \wedge L[Z\alpha].$$

Here $Acc_0$ is the set of $\omega_0$-limits and $Acc_1$ is the set of $\omega_1$-limits. Both sets are easily seen to be definable (see [5]). We also use

$$\Gamma_u^v(X, Z)\colon (\forall t)_u^v\, H[Xt, Zt, Zt'] \wedge (\forall x)_u^{v', Acc_0}\, \mathfrak{R}_0[\sup_0^x Z, Zx] \wedge (\forall x)_u^{v', Acc_1}\, \mathfrak{R}_1[\sup_1^x Z, Zx].$$

$S_u v$, $u \approx v\,(-t)$, etc. are defined as in the last section. The $\omega_2$-behavior, $\mathrm{beh}_{\omega_2}\Gamma$, is the set of all input $X[0, \alpha)$ accepted by $\Gamma$, i.e. such that

$$(\exists Z).\, E[Z0] \wedge \Gamma_0^\alpha(X, Z) \wedge L[Z\alpha].$$

In [5], Lemma 1.5, it is shown that all MT-formulas can be put into the form (prefix $Y$) $(\exists Z)\,\Gamma(XY, Z)$ where $\Gamma$ is a $\sup_0$-transition system i.e., like (1) but with a single transition condition $\mathfrak{R}[\sup_0^x Z, Zx]$ at all limits $x$. The following lemma allows us to use $\sup_1$ rather than $\sup_0$ systems.

Lemma 4.1. *For every* $\sup_0$*-transition system* $\Gamma$ *one can effectively find a* $\sup_1$*-transition* $\Gamma'$ *which accepts the same* $\alpha$*-strings,* $\alpha < \omega_2$. *Furthermore if* $\Gamma$ *is deterministic then so is* $\Gamma'$.

Proof. Let $\Gamma = [E, H, \mathfrak{R}, L]$. By Lemma 2.1, there are $K_0, \ldots, K_r$ such that $\mathfrak{R}[\sup_0^x Z, Zx]$ may be restated as a propositional expression in $(\exists_0^x t)\, K_i[Zt]$, $i \leqq r$, and $Zx$. We show that

$$(\exists Z)\,\Gamma(X, Z) \equiv (\exists Y_0 \ldots Y_r Z).E[Z0] \wedge (\forall t)\, H[Xt, Zt, Zt'] \wedge (\forall x)^{Acc_0}\, \mathfrak{R}_0(x) \wedge (\forall x)^{Acc_0}\, \mathfrak{R}_1(x) \wedge L[Z\alpha]$$

where $\mathfrak{R}_0(x)$ is the conjunction of $\mathfrak{R}(x)$ and $Y_i x \equiv (\exists_0^x t)\, K_i[Zt]$, $i \leqq r$ and $\mathfrak{R}_1(x)$ is formed from $\mathfrak{R}(x)$ by replacing each occurence of $(\exists_0^x t)\, K_i[Zt]$ by $(\forall_1^x t)\, Y_i t$. Let $Z$ be an $X$-run of $\Gamma$, $U_i = \{t;\, K_i[Zt]\}$, and $Y_i = U_i'^{\omega_2}$. Then using Remark 2.5, $Y_0 \ldots Y_r Z$ is an $X$-run of $\Gamma' = [E, H, \mathfrak{R}_0, \mathfrak{R}_1, L]$. Conversely, suppose $Y_0 \ldots Y_r Z$ is an $X$-run of $\Gamma'$. Let $x \Leftarrow \omega_1$. By Remark 2.6 there is a $P \in \mathscr{J}_1^x$ all of whose members are $\omega_0$-limits. Thus if $U_i = \{t;\, K_i[Zt]\}$, $V_i = U_i'^{\omega_2}$ then $P \cap Y_i = P \cap V_i$ and so

$$Y_i \in \mathscr{J}_1^x \equiv P \cap Y_i \in \mathscr{J}_1^x \equiv P \cap V_i \in \mathscr{J}_1^x \equiv V_i \in \mathscr{J}_1^x.$$

Again using Remark 2.5 this implies $(\exists_0^x t)\, K_i[Zt] \equiv (\forall_1^x t)\, Y_i t$ and so $Z$ is an $X$-run of $\Gamma$. □

Corollary 4.2 (Prefix lemma). *For every* MT*-formula* $\Sigma(X)$ *in the primitive* $<$ *with only the free variables* $X = X_1 \ldots X_m$ *one can effectively find a formula which is equivalent to it in any structure* $[\alpha, <]$, $\alpha < \omega_2$, *and is of form (prefix* $Y$*)* $(\exists Z)\,\Gamma(XY, Z)$ *where* $\Gamma$ *is as in* (1).

Lemma 4.3. *For every relativized* $\sup_1$*-transition system* $\Gamma$ *one can effectively find a (non-deterministic)* $\sup_1$*-transition system* $\Gamma'$ *of equal behavior.*

Proof. Suppose $z \Leftarrow \omega_0$ and let

$$(*) \qquad (\forall t)\begin{bmatrix} Vt \wedge \sim Yt \,.\!\supset.\, Vt' \\ Vt \wedge \quad Yt \,.\!\supset.\, Ut \end{bmatrix} \wedge (\forall x).(\forall_0^x t)\; Vt \supset Vx.$$

Let $\bar{V} = \{v;\, (\exists t)_v^{x,y}\, [v, t) \cap Y = 0\}$. Then clearly $\bar{V}$ satisfies (*) and any $V$ satisfying (*) is a subset of $\bar{V}$. Then using Corollary 2.11 it is easy to see that

(a) $\quad (\forall_1^{Y,z} t)\; Ut \equiv (\exists V)\; [V \text{ satisfies } (*) \wedge (\exists_0^z t)\; Yt \wedge (\forall_1^z t)\; Vt]$,

(b) $\quad (\exists_1^{Y,z} t)\; Ut \equiv (\exists V)\; [V \text{ satisfies } (*) \wedge (\exists_0^z t)\; Yt \wedge (\exists_1^z t)\; Vt]$.

The lemma now follows from (a), (b) and Lemma 2.1. □

Our main task is to find a finite-state recursion which uniformly define the sequences $S_u$ for each $\omega_2$-sequence $X$. The recursion (8) of [5] Section 4 tells how to define $S_u t$ for $t = 0$, $t$ a successor, and $t \Leftarrow \omega_0$. We need to add to these recursions for $S_u t$ in case $t \Leftarrow \omega_1$. These will be modeled after the terminal condition of (14) of Section 3.

Lemma 4.4. *Let $\Gamma$ be a* $\sup_1$*-transition system with input $X$ and let $u < z \Leftarrow \omega_1$. There is a finite set $W$ such that $\bigwedge_{w \in W} (\exists_1^z y)\, y \approx w\,(-z)$, $(\forall_1^z y) \bigvee_{w \in W} y \approx w\,(-z)$, and $v, w \in W \wedge v \neq w \,.\supset.\, v \not\approx w\,(-z)$. In fact for any such $W$ there is an $\omega_1$-sequence $P$ of $\omega_0$-limits and sets $P_w$ such that $P = \bigcup_{w \in W} P_w$ is cofinal-closed in $z$, $w \in W \supset P_w \notin \mathcal{O}_1^z$, $v < y \wedge Py \wedge P_w v \,.\supset.\, v \approx w\,(y)$, and $v, w \in W \wedge v \neq w \,.\supset.\, P_v \cap P_w = 0$.*

Proof. As $\approx (-z)$ is an equivalence of finite index, some of the classes will occur non-null in the sense of $\mathcal{J}_1^z$, i.e. $\notin \mathcal{O}_1^z$. Let $W$ be a system of representatives of these classes. Clearly $W$ satisfies the first set of properties stated in the remark. Now for any such $W$ there is a set $P$ and for each $w \in W$ there are sets $P_w$ such that (a) $P_w \notin \mathcal{O}_1^z$, (b) $y \in P_w \supset y \approx w\,(-z)$, and (c) $P = \bigcup_{w \in W} P_w \in \mathcal{J}_1^z$. By (c) and Remark 2.6 we may assume (d) $P$ is a cofinal-closed $\omega_1$-sequence of $\omega_0$-limits. As $v, w \in W \wedge v \neq w \,.\supset.\, v \not\approx w\,(-z)$, by (b) we may further assume (e) $v, w \in W \wedge \wedge v \neq w \,.\supset.\, P_v \cap P_w = 0$. Now define $y_i, w_i, t_i$ by induction over $i < \omega_1$:

| | | |
|---|---|---|
| $y_0 = (\mu y) \cdot Py$ | (exists by (d)); | $Py_0$; |
| $w_i = (\mu w) \cdot w \in W \wedge P_w y_i$ | (exists as $Py_i$ and (c)); | $P_{w_i} y_i,\ w_i \in W$, |
| $t_i = (\mu t. y_i \approx w_i\,(t)$ | (exists as $y_i \approx w_i(-z)$); | by (b) $y_i \approx w_i\,(-z)$, $y_i \approx w_i\,(t_i)$; |
| $y_{i+1} = (\mu y). Py \wedge t_i < y$ | (exists by (d)); | $Py_{i+1}, t_i < y_{i+1}$ hence $y_i \approx w_i\,(y_{i+1})$; |
| $y_\lambda = \lim_{i<\lambda} y_i$ | (exists as $z \not\Leftarrow \omega_0$); | $Py_\lambda$ as $Py_i$ and (d). |

As $\{y_i : i < \omega_1\}$ is a cofinal closed subset of $P$ we may assume $P = \{y_i; i < \omega_1\}$. Suppose that $w \in W$, $v < y$, $P_w v$ and $Py$. So there are $i < j$ such that $v = y_i$ and $y = y_j$. As $y_i \approx w_i\,(y_{i+1})$ and $y_{i+1} \leqq y_j$, we have $v \approx w_i\,(y)$. Since $v = y_i \in P_{w_i}$ and $v \in P_w$, we have by (e) that $w_i = w$. Hence $v \approx w\,(y)$. □

Let $W$ be a finite set and let $s_0$, and for each $w \in W$, let $s_w$ be states of the $S_u$ type; we define the following set of states of $\Gamma$:

$$(2) \qquad b \in B_1^{c,C}[s_0, \langle s_w; w \in W\rangle] \,.\equiv.\, \bigvee_{D_0, D_1, Q} .D_0 \subseteq C \wedge \bigwedge_{e \in D_1} \mathfrak{R}_0[D_0, e] \wedge \mathfrak{R}_1[D_1, b]$$
$$\wedge\, D_1 \subseteq s_0^{c,C} \wedge Q \text{ is a subdirect product of } D_1 \times W \wedge \bigwedge_{\langle e, w\rangle \in Q} D_1 \subseteq s_w^{e, D_0}.$$

We now give the main lemma needed for the subset construction, the proof of which is modelled after [5], Lemma 6.3.

Lemma 4.5. *Let $\Gamma$ be a* $\sup_1$*-transition system with input $X$ and let $u < z \Leftarrow \omega_1$. If*

$$(3) \qquad (\forall w) \bigvee_s (\forall_1^z t)\, S_w t = s$$

*then*

$$(4) \qquad b \in S_u^{c,C_z} \,.\equiv.\, \bigvee_W .\, \bigwedge_{w \in W} (\exists_1^z y)\, y \approx w\,(-z) \wedge (\forall_1^z y) \bigvee_{w \in W} y \approx w\,(-z)$$
$$\wedge\, (\exists_1^z y)\, b \in B_1^{c,C}[S_u y, \langle S_w y; w \in W\rangle].$$

Proof. Suppose first that $b \in S_u^{c,C_z} z$. There is a run $Z[u, z]$ such that $Zu = c$, $\Gamma_u^z(X, Z)$, $Zz = b$, and $Z_u^z = C$. Let $D_0 = \sup_0^z Z$ and $D_1 = \sup_1^z Z$. Then clearly

$D_0 \subseteqq C$, and as $z \Leftarrow \omega_1$ and $Z$ is a run of $\Gamma$ ending in $b$, $\mathfrak{R}_1[D_1, b]$. So we have (a) $D_0 \subseteqq C$ and $\mathfrak{R}_1[D_1, b]$. By the definition of $D_1$, there are sets $P_e$ such that (b) $e \in D_1 \supset P_e \notin \mathcal{O}_1^z$, (c) $P = \bigcup_{e \in D_1} P_e \in \mathscr{J}_1^z$, and (d) $e \in D_1 \wedge P_e y \mathrel{.\supset.} Zy = e$. Using Remark 2.5 we may assume that $P$ is a cofinal-closed $\omega_1$-sequence of $\omega_0$-limits. Now we define $y_i$ by induction on $i < \omega_1$:

$$
\begin{aligned}
y_0 &= (\mu y).Py \wedge C \subseteqq Z_u^y \wedge Z_y^z \subseteqq D_0, && \text{(these exist as } P \text{ cofinal}\\
y_{i+1} &= (\mu y).Py \wedge y_i < y \wedge D_0 = Z_{y_i}^y, && \text{and } D_0 = \sup\nolimits_0^z Z)\\
y_\lambda &= \lim_{i<\lambda} y_i && \text{(exists as } z \not\Leftarrow \omega_0).
\end{aligned}
$$

As $P$ is closed, $\{y_\lambda; \lambda < \omega_1\}$ is a cofinal-closed subset of $P$, and so we may assume $P = \{y_\lambda; \lambda < \omega_1\}$ without losing any of the above properties of $P$; the new $P_e$'s being $\{y_\lambda; \lambda < \omega_1\} \cap P_e$. So we have the additional information (e) $v < y \wedge v, y \in P \mathrel{.\supset.} Z_v^y = D_0$, (f) $Py \supset Z_u^y = C$. Also $\supset Px$ $\sup_0^x Z = D_0$ and as $Z$ is a run of $\mathfrak{R}_0$ and $Px \supset x \Leftarrow \omega_0$, by (c), (d) we have (g) $e \in D_1 \supset \mathfrak{R}_0[D_0, e]$. By Lemma 4.4 there is a finite set $W$ and sets $\bar{P}$, $P_w$ such that (h) $w \in W \supset P_w \notin \mathcal{O}_1^z$, (i) $\bar{P} = \bigcup_{w \in W} P_w$ is cofinal-closed in $z$. (j) $v < y \wedge P_w v \wedge \bar{P} y \mathrel{.\supset.} v \approx w\,(y)$. By (3) we have an $s_0$, and for each $w \in W$ an $s_w$, such that $(\forall_1^z y)\, S_u y = s_0$ and $(\forall_1^z y)\, S_w = s_w$. As $W$ is finite, by (i) we may assume (k) $\bar{P} y \mathrel{.\supset.} S_u y = s_0 \wedge \bigwedge_{w \in W} S_w y = s_w$. Suppose $e \in D_1$. By (b), (i) we have $\bar{P} \cap P_e \notin \mathcal{O}_1^x$, and therefore there is a $y \in \bar{P} \cap P_e$. Then by (c), (d), (f) we know that $Z[u, y]$ is a $C$-exact run of $\Gamma$ from $c$ to $e$ and so with (k) we have $e \in S_u^{c,C} y = s_0^{c,C}$. This argument proves (l) $D_1 \subseteqq s_0^{c,C}$. Now define $\langle e, w\rangle \in Q \mathrel{.\equiv.} e \in D_1 \wedge w \in W \wedge P_e \cap P_w \neq 0$. Clearly $Q \subseteqq D_1 \times W$. Suppose $e \in D_1$. By (b), (i) we have a $w \in W$ such that $P_e \cap P_w \notin \mathcal{O}_1^z$. Hence $\langle e, w\rangle \in Q$. This argument shows that $Q$ projects onto $D_1$. Similarly, (c), (h) shows that $Q$ projects onto $W$. Thus, (m) $Q$ is a subdirect product of $D_1 \times W$. Assume $\langle e, w\rangle \in Q$ and $d \in D_1$. Then $e \in D_1$, $w \in W$, and there is a $v \in P_e \cap P_w$. By (b), (i) $\bar{P} \cap P_d \notin \mathcal{O}_1^z$. and hence there is a $y \in \bar{P} \cap P_d$ with $v < y$. Then by (c), (d), (e), $Z[v, y]$ is a $D_0$-exact run of $\Gamma$ from $e$ to $d$ and so with (j), (k) we have $d \in S_v^{e,D_0} y = S_w^{c,D_0} y = s_w^{e,D_0}$. This proves (n) $\langle e, w\rangle \in Q \supset D_1 \subseteqq s_w^{e,D_0}$. Hence by (a), (g), (l), (m), (n) it follows that $b \in B_1^{c,C}[s_0, \langle s_w; w \in W\rangle]$. Thus with (i), (k) we obtain (o) $(\exists_1^z y)\, b \in B_1^{c,C}[S_u y, \langle S_w y; w \in W\rangle]$. From $b \in S_u^{c,C} z$ we have now found $D_0, D_1$, $Q$, $W$ such that (h), (i), (o) and therefore the right side of (4) holds.

We now suppose the right side of (4) holds and show that $b \in S_u^{c,C} z$. First note that as $\mathcal{O}_1^z$ is an ideal, $(\exists_1^z t) \bigvee_i \varphi_i \equiv \bigvee_i (\exists_1^z t)\, \varphi_i$. Thus there is a finite $W$ and $\bar{P}, D_0, D_1, Q$ such that (p) $\bigwedge_{w \in W} (\exists_1^z y)\, y \approx w\,(-z) \wedge (\forall_1^z y) \bigvee_{w \in W} y \approx w\,(-z)$, (q) $\bar{P} \notin \mathcal{O}_1^z$, $D_0 \subseteqq C$, (r) $e \in D_1 \supset \mathfrak{R}_0[D_0, e]$, $\mathfrak{R}_1[D_1, b]$, (s) $y \in \bar{P} \supset D_1 \subseteqq S_u^{c,C} y$, $Q \subseteqq D_1 \times W$, $Q$ onto $D_1$, $Q$ onto $W$, and (t) $\langle e, w\rangle \in Q \wedge y \in \bar{P} \mathrel{.\supset.} D_1 \subseteqq S_w^{e,D_0} y$. Let $\overline{W} = \{w; w \in W \wedge (\forall v)^w. Wv \supset v \not\approx w\,(-z)\}$ and let $\bar{Q} = \{\langle e, w\rangle; w \in \overline{W} \wedge (\exists v).\langle e, v\rangle \in Q \wedge v \approx w\,(-z)\}$. Clearly $\overline{W}$ and $\bar{Q}$ satisfy all the above properties of $W$ and $Q$ respectively, and so we may assume (u) $v, w \in W \wedge v \neq w \mathrel{.\supset.} v \not\approx w\,(-z)$. From (p), (u) and the second part of Lemma 4.4 there is an $\omega_1$-sequence $P$ of $\omega_0$-limits and sets $P_w$ such that (v) $w \in W \supset P_w \notin \mathcal{O}_1^z$, (w) $P = \bigcup_{w \in W} P_w$ is cofinal closed in $z$, (x) $v < y \wedge Py \wedge P_w y \mathrel{.\supset.} v \approx w\,(y)$, and (y) $v, w \in W \wedge v \neq w \mathrel{.\supset.} P_v \cap P_w = 0$. By (q), (w) we may also assume (z) $\bar{P} \subseteq P$. Now we use the fact that $\mathfrak{P}z/\mathscr{I}_1^z$ has no atoms. Because of (v) we can split each $P_w$ into an arbitrary finite number of disjoint non-null parts. So for each $w \in W$ there are sets $P_{e,w}$ such that $P_w = \bigcup_{\langle e,w\rangle \in Q} P_{e,w}$ (a non-empty union since $Q$ projects onto $W$), (v') $\langle e, w\rangle \in Q \,.\supset.\, P_{e,w} \notin \mathcal{O}_0^z$, $\langle e, w\rangle \neq \langle d, w\rangle \in Q \,.\supset.\, P_{e,w} \cap P_{d,w} = 0$. By (w), (x), (y) we now have (y') $\langle e, w\rangle \neq \langle d, v\rangle \in Q \supset P_{e,w} \cap P_{d,v} = 0$, (w') $P = \bigcup_{\langle e,w\rangle \in Q} P_{e,w}$ is cofinal closed, and (x') $\langle e, w\rangle \in Q \wedge v < y \wedge P_{e,w}v \wedge Py \,.\supset.\, v \approx w(y)$.

By induction on $i < \omega_1$, we now define a subsequence of the $\omega_1$-sequence $P$ which is cofinal closed in $z$.

| | | |
|---|---|---|
| $y_0 = (\mu y)\, \bar{P}y$ | (exists by (q)); | $\bar{P}y_0$ (hence by (z), $Py_0$); |
| $y_{i+1} = (\mu y).\bar{P}y \wedge y > y_i$ | (exists by (q)); | $y_i < y_{i+1}$, $\bar{P}y_{i+1}$ (hence by (z), $Py_{i+1}$); |
| $y = \lim_{i<\lambda} y_i$ | (exists as $z \not\Leftarrow \omega_0$); | $Py_\lambda$ as $Py_i$ and (w'). |

As usual we may suppose $P = \{y_i;\, i < \omega_1\}$. By (w'), (y') each element of $P$ belongs to exactly one of the $P_{e,w}$. So for each $i < \omega_1$ there are $e_i, w_i$ such that (aa) $\langle e_i, w_i\rangle \in Q$, $P_{e_i,w_i}y_i$ and (bb) $\langle e, w\rangle \in Q \wedge P_{e,w}y_i \,.\supset.\, e = e_i$. As $Q \subseteq D_1 \times W$, by (aa) we also have (cc) $e_i \in D_1$. Since $y_0 \in \bar{P}$, by (x), (cc), $e_0 \in S_u^{c,C}y_0$ and so there is a $C$-exact run $Z[u, y_0]$ of $\Gamma$ from $c$ to $e_0$. Similarly for each $i < \omega_1$ as $y_{i+1} \in \bar{P}$, by (aa), (x') $e_{i+1} \in S_{w_i}^{e_i,D_0}y_{i+1}$ and so by (t), (cc), $e_{i+1} \in S_{y_i}^{e_i,D_0}y_{i+1}$. Thus there is a $D_0$-exact run $Z[y_i, y_{i+1}]$ from $e_i$ to $e_{i+1}$. Note that these runs have the same state where they overlap. As $D \subseteq C$ and the splicing points, $y_i$, form a $\omega_1$-sequence which is cofinal-closed in $z$ ($y_\lambda = \lim_{i<\lambda} y_i$) these partial runs may be spliced together to form a $C$-exact run $Z[u, z)$. Since $\sup_0^{y_\lambda} Z = D_0$ and $Zy_\lambda = e_\lambda$, by (cc), (r) it follows that $\mathfrak{R}_0[\sup_0^{y_\lambda} Z, Zy_\lambda]$. But $y_\lambda \Leftarrow \omega_0$ and so $Z$ satisfies the transition conditions at $y_\lambda$. As $Z$ trivially satisfies $\Gamma$ everywhere else we have that $Z[u, z)$ is an $X[u, z)$-run of $\Gamma$.

We now complete this run by defining $Zz = b$. It remains to show that $Z$ satisfies $\Gamma$ at $z$, i.e. $\mathfrak{R}_1[\sup_1^z Z, b]$. Since we already know that $\mathfrak{R}_1[D_1, b]$ it suffices to prove $\sup_1^z Z = D_1$. Suppose first that $e \in D_1$. Because $Q$ projects onto $D_1$, there is a $w \in W$ with $\langle e, w\rangle \in Q$. By (v'), $P_{e,w} \notin \mathcal{O}_1^z$ and so by (w') there is a $y_i \in P_{e,w}$. From (bb) it follows that $e = e_i$, and since $Zy_i = e_i$ we have $P_{e,w} \subseteq \{t;\, Zt = e\}$. Hence $\{t;\, Zt = e\} \notin \mathcal{O}_1^z$, i.e. $e \in \sup_1^z Z$. This shows that $D_1 \subseteq \sup_1^z Z$. Conversely, suppose $e \in \sup_1^z Z$. There is a $U \notin \mathcal{O}_1^z$ such that $Ut \supset Zt = e$. By (w'), $U \cap P \notin \mathcal{O}_1^z$ and so there is a $y_i \in U$. Therefore $Zy_i = e$. But we already know that $Zy_i = e_i$. Thus $e = e_i \in D$, by (cc). This proves that $\sup_1^z Z \subseteq D_1$. □

It turns out that (3) does hold for any $\omega_1$-accessible $z < \omega_2$. In [5] Lemma 6.1 (3) is proved in the case where $z = \omega_1$. The proof however required the subset-construction for countable ordinals. In a similar manner it is possible to prove (3) at a particular $\omega_1$-limit $z < \omega_2$ if we have a subset-construction for ordinals $<z$. Since we don't know of any direct proof it will be necessary to simultaneously prove (3) and the subset-construction for all ordinals $<\omega_2$ (see the proof of Lemma 4.7).

Exactly as in the previous section we can put (4) into a more desirable form as follows:

Lemma 4.6. *Let $z < \omega_2$, $z \Leftarrow \omega_1$, and $X[0, z)$ be any input such that* (3) *holds. Define $U$ as in Lemma* 3.5 *and for each finite sequence $V$ in $z$ define $Y_V$, $T_V$, $P_V$ and $R_V$ by recursion for $t < z$ as in Lemma* 3.6. *Then*

$$b \in S_u^{c,C} z \;.\equiv.\; \bigvee_{W \subsetneq Uz} \;.\; \bigwedge_{w \in W} (\exists_1^{T_{Uz},z} t)\,(\exists v)\,[v \neq w \leftarrow v\,(t) \wedge v \in R_{Uz} t]$$
$$\wedge\, (\forall_1^{T_{Uz},z} t) \bigvee_{w \in W} (\exists v)\,[v \neq w \leftarrow v\,(t) \wedge v \in R_{Uz} t]$$
$$\wedge\, (\exists_1^z t)\; b \in B_1^{c,C}[S_u t, \langle S_w t;\, w \in W\rangle].$$

Theorem 4.7 (Subset-construction). *For every* $\sup_1$*-transition system* $\Gamma = [E, H, \mathfrak{K}_0, \mathfrak{K}_1, L]$ *of form* (1), *a deterministic relativized* $\sup_1$*-transition recursion $A$ of form*

(5) $$Y0 = a \wedge (\forall t)\; Yt' = F[Xt, Yt] \wedge (\forall x)^{Acc_0} Yx = \mathfrak{G}_0[\sup_0^x Y]$$
$$\wedge\, (\forall x)^{Acc_1} Yx = \mathfrak{G}_1[\sup_1^{L_1,x}, \ldots, \sup_1^{L_n,x}]$$

*can be constructed with terminal condition $L'[\cdot]$ such that for any ordinal $\alpha < \omega_2$ and any input $X[0, \alpha)$, $\Gamma$ accepts $X$ if and only if $[A, L']$ accepts $X$, i.e.* $(\exists Z).E[Z0] \wedge \Gamma_0^\alpha(X, Z) \wedge L[Z\alpha]$ *holds just in case the recursion $A$ applied to $X$ yields a final state $Y\alpha$ with $L'[Y\alpha]$.*

Proof. Begin by constructing a recursion $Y = A(X)$ as in (14) of the previous section, dropping the terminal condition and adding

(6) $$\mathfrak{S}_i(x)\colon \{\langle c, C, b\rangle;\; \bigvee_{K \subsetneq Vx} \;.\; \bigwedge_{k \in K} (\exists_1^{T_{Vx},x} t) \bigvee_h [h \neq k \leftarrow h\,(t) \wedge h \in R_{Vx} t]$$
$$\wedge\, (\forall_1^{T_{Vx},x} t) \bigvee_{k \in K} \bigvee_h (h \neq k \leftarrow h\,(t) \wedge h \in R_{Vx} t]$$
$$\wedge\, (\exists_1^x t)\; b \in B_1^{c,C}[S_i t, \langle S_k t;\, k \in K\rangle]$$

for $x \Leftarrow \omega_1$ (the definition $\mathfrak{S}_i(x)$ for $x \Leftarrow \omega_0$ remains the same; $Vx$, $P_J x$, etc. are defined as in (14) for all limits $x$). $Y = A(X)$ is of form (5) except that the transition condition at $\omega_1$-limits is of form $Yx = \mathfrak{G}_1[\sup_0^x Y, \sup_1^{L_1,x}, \ldots, \sup_1^{L_n,x}]$. As in the proof of Lemma 4.1 we may omit the $\sup_0^x Y$. The theorem will be proved by showing that $Y = A(X)$ is the desired relativized $\sup_1$-recursion with terminal condition

$$L'[Y\alpha]\colon \bigvee_{c,C,b} \;.\; E[c] \wedge b \in L \cap S_0^{c,C}\alpha.$$

A recursion $\bar{Y} = \bar{A}(X)$ similar to $(\overline{14})$ of Section 3 is also constructed with changes that correspond to those made above. Now let $\alpha < \omega_2$, let $X[0, \alpha)$ be an arbitrary $\alpha$-sequence, and let $Y = A(X)$ and $\bar{Y} = \bar{A}(X)$. Note that $\bar{A}$ contains the recursions which occur in Lemmas 3.5, 4.6, and $(\bar{8})$ of [5], Section 4. Thus if $L[b]$

(*) $$z \leqq \alpha \wedge z \Leftarrow \omega_1 \;.\supset.\; (\forall w) \bigvee_s (\forall_1^z t)\; \bar{S}_w t = s$$

it follows that the $\bar{S}_0^{c,C}\alpha$ are just the sets defined in (2) of Section 3. In this case

$$b \in \bar{S}_0^{c,C}\alpha \equiv (\exists Z)\,[Z0 = c \wedge \Gamma_0^\alpha(X, Z) \wedge Z\alpha = b \wedge Z_0^\alpha = C]$$

and so the recursion $\bar{Y} = \bar{A}(X)$ together with the terminal condition $\bigvee_{c,c,b} . E[c] \wedge$ $\wedge\, b \in L \cap \bar{S}_0^{a,C}\alpha$ satisfies the conditions of the theorem but for the fact that $\bar{Y} = \bar{A}(X)$ isn't finite-state. Consequently the lemma will be proved if we can show (*) and $\bar{S}_0\alpha = S_0\alpha$. Toward this end we define the functions $u_i t$ as in (a) of the proof of

Lemma 3.7. We will also need (b) and versions of (c) and (d) at $\omega_1$-limits $z \leqq \alpha$, i.e. versions in which $J$ is replaced by $Vz$, $V$ by $Uz$, and $(\forall_0 t)$ by $(\forall_0^z t)$ for all $\omega_1$-limits $z \leqq \alpha$. Then the formulas of (b), (c), and (d) are proved simultaneously along with (*) by induction on $t \leqq \alpha$. The proofs are exactly as in Lemma 3.7 and will not be repeated here. As $S_0\alpha = \bar{S}_0\alpha$ will then follow from ($b_4$) all that remains is to give the inductive step for (*) at $z \leqq \alpha$, $z \Leftarrow \omega_1$. This is done in the next paragraph.

Let $Y$ be the unique run defined by the recursion $Y = A(X)$. Suppose $w < z$ and let $i$ be such that $w = u_i z$. Thus $S_i$ is one of the components of $Y$. By Remark 2.7 there is a set $Q \in \mathscr{J}_1^z$ and a set of states $D$ such that $Qx \,.\!\supset\!.\, x \Leftarrow \omega_0 \wedge \sup_0^x Y = D$. Letting $d = \mathfrak{G}_0[D]$ we have $(\forall_1^z t)\ Yt = d$. Thus if $s$ is the component of $d$ which corresponds to $S_i$ we have $(\forall_1^z t)\ S_i t = s$. But by induction hypothesis $\bar{S}_w t = S_i t$ for all $t.\ w \leqq t < z$. Hence $(\forall_1^z t)\ \bar{S}_0 t = s$. □

Lemma 4.8 (Complementation lemma for $\alpha < \omega_2$). *For every* $\sup_1$*-transition system of form* (1) *or relativized* $\sup_1$*-transition system* (of *the same form*), $\sup_1$- *and relativized* $\sup_1$*-transition systems of complementary behavior may be constructed.*

The proof uses Lemmas 4.3 and 4.7 and is similar to that of Lemma 3.8. Using Corollary 4.2 and Lemma 4.8 we also have as in Theorem 3.9

Theorem 4.9 (MT-definability for $\alpha < \omega_2$). *For every* MT-*formulas* $\Sigma(X)$ *in* $<$ *one can construct a relativized* $\sup_1$*-automata* $A = [a, F, \mathfrak{G}_0, \mathfrak{G}_1, M]$ *such that for every* $\alpha < \omega_2$ *and every input* $X[0, \alpha)$, $\Sigma(X)$ *holds in* $[\alpha, <]$ *if and only if* $A$ *accepts* $X[0, \alpha)$. *Also a* (*non-deterministic*) $\sup_1$*-transition* $\Gamma = [E, H, \mathfrak{R}_0, \mathfrak{R}_1, L]$ *can be constructed so that* $\Sigma(X)$ *holds just in case* $\Gamma$ *accepts* $X[0, \alpha)$, *i.e.* $\Sigma(X) \,.\!\equiv\!.\, (\exists Z),\ E[Z0] \wedge \Gamma_0^\alpha(X, Z) \wedge L[Z\alpha]$.

## 5. The decision method for $\alpha < \omega_2$

The decision procedure we now present is a non-deterministic version of that found in [5] Section 4. Define the *$\alpha$-spectrum*, $\operatorname{spec}_\alpha \Sigma$, of an MT-sentence $\Sigma$ to be the set of all those ordinals $\beta < \alpha$ such that $\Sigma$ holds in $[\beta, <]$. Since an input for an "input-free" transition system is an ordinal, the $\omega_2$-behavior of a transition system is just a set of ordinals. Thus by Theorem 4.9,

Remark 5.1. *For every* MT-*sentence* $\Sigma$, *an input-free* $\sup_1$*-transition system* $\Gamma$ *can be constructed such that* $\operatorname{spec}_{\omega_2} \Sigma = \operatorname{beh}_{\omega_2} \Gamma$.

In the sequel we consider a fixed input-free $\sup_1$-transition system $\Gamma = [H, \mathfrak{R}_0, \mathfrak{R}_1]$ (without initial or terminal conditions) where

(1) $\Gamma_u^v(z)\colon (\forall t)_u^v\, H[Zt, Zt'] \wedge (\forall x)_u^{v', Acc_0}\, \mathfrak{R}_0[\sup_0^x Z, Zx] \wedge (\forall x)_u^{v', Acc_1}\, \mathfrak{R}_1[\sup_1^x Z, Zx]$.

We also assume that $K$ is the set of states of $\Gamma$ (i.e. of $Z$) and $r = (K \times \mathfrak{P}K)$. For each $\alpha < \omega_2$ define the operator $G_\alpha\colon K \to \mathfrak{P}(\mathfrak{P}K \times K)$ by

(2) $\langle C, d\rangle \in G_\alpha[c] \,.\!\equiv\!.\, (\exists Z).Z[0, \alpha]$ is a $C$-exact run of $\Gamma$ from $c$ to $d$.

Let $G, G'\colon K \to \mathfrak{P}(\mathfrak{P}K \times K)$ and define their composition $G \circ G'\colon K \to \mathfrak{P}(\mathfrak{P}K \times K)$ by

(3) $\langle C, d\rangle \in G \circ G'[c] \,.\!\equiv\!.\, \bigvee_{D,E,e}.\ \langle D, e\rangle \in G'[c] \wedge \langle E, d\rangle \in G[e] \wedge C = D \cup E$.

Note then that

(4) $$G_{\alpha+\beta} = G_\beta \circ G_\alpha.$$

Using (3) the following is the natural definition of powers:

(5) $$G^0[c] = \{\langle 0, c\rangle\}, \qquad G^{h+1}[c] = G \circ G^h[c].$$

It is useful to note from (4) and (5) that for $\alpha < \omega_2$, $h < \omega$, $\langle C, d\rangle \in G^h_\alpha[c]$ just in case there is a $C$-exact $\alpha h$-run $Z$ of $\Gamma$ from $c$ to $d$. For any ordinals $\alpha < \beta$, there is a unique ordinal $\gamma$ such that $\beta = \alpha + \gamma$. We define $\beta - \alpha$ to be this $\gamma$.

Remark 5.2. *For all* $\alpha < \omega_2$ *and* $h < \omega$

(6) $$\langle C, d\rangle \in G^h_\alpha[c] \supset (\exists k)^{r+1}\langle C, d\rangle \in G^k_\alpha[c].$$

Proof. Let $h$ be the least number such that $\langle C, d\rangle \in G^h_\alpha[c]$, and suppose $h > r$. Thus, there is a $C$-exact run $Z[0, \alpha h]$ of $\Gamma$ from $c$ to $d$. For each $k \leqq h$ let $c_k = Z\alpha k$ and $C_k = Z_0^{\alpha k}$. As $h > r$, a repetition must occur among $\langle 0, c\rangle = \langle C_0, c_0\rangle$, $\langle C_1, c_1\rangle, \ldots, \langle C_h, c_h\rangle = \langle C, d\rangle$, say, $\langle C_m, c_m\rangle = \langle C_{m+n}, c_{m+n}\rangle$, $n > 0$. Then $Z[0, \alpha m)\, Z[\alpha(m+n), \alpha h]$ is a $C$-exact $\alpha(h-n)$-run of $\Gamma$ from $c$ to $d$. Thus $\langle C, d\rangle \in G^{h-n}_\alpha[c]$, which contradicts the assumption about $h$. This shows that $h \leqq r$. □

Lemma 5.3. *For each* $\alpha < \omega_2$

(7) $$\langle C, d\rangle \in G_{\alpha\omega}[c] \,.\!\equiv\!.\; \bigvee_{\substack{D,e\\ h,k \leqq r}} .\, D \subseteqq C \wedge \mathfrak{K}_0[D, d] \wedge \langle C, e\rangle \in G^h_\alpha[c] \wedge \langle D, e\rangle \in G^k_\alpha[e].$$

Proof. Suppose $\langle C, d\rangle \in G_{\alpha\omega}[c]$. There is a $C$-exact run $Z[0, \alpha\omega]$ from $c$ to $d$. Let $D = \sup_0^{\alpha\omega} Z$. Thus $D \subseteqq C$ and $\mathfrak{K}_0[D, d]$. As $\{\alpha n;\, n < \omega\}$ converges to $\alpha\omega$ there is a state $e$ and $p, m < \omega$, $0 < m$ such that $Z\alpha p = e = Z\alpha(p+m)$, $Z_0^{\alpha p} = C$, and $Z_{\alpha p}^{\alpha(p+m)} = D$. Then $\langle C, e\rangle \in G^p_\alpha[c]$ and $\langle D, e\rangle \in G^{p+m}_\alpha[e]$ and so by (6), there are $h, k \leqq r$ such that $\langle C, e\rangle \in G^h_\alpha[c]$ and $\langle D, e\rangle \in G^k_\alpha[e]$. This proves the right side of (7). Conversely, suppose $D \subseteqq C$, $\mathfrak{K}_0[D, d]$, $\langle C, e\rangle \in G^h_\alpha[c]$, and $\langle D, e\rangle \in G^k_\alpha[e]$. So there is a $C$-exact run $Z_1[0, \alpha h]$ from $c$ to $e$ and a $D$-exact run $Z_2[0, \alpha k]$ from $e$ to $e$. Thus $Z_1[0, \alpha h)\, Z_2[0, \alpha k)\, Z_2[0, \alpha k) \ldots d$ is a $C$-exact run of $\Gamma_0^{\alpha\omega}$ from $c$ to $d$ and so $\langle C, d\rangle \in G_{\alpha\omega}[c]$. □

Lemma 5.4. *Let* $\{\alpha_i;\, i < \omega\}$ *be a sequence which converges to* $\alpha$, $\alpha_0 = 0$, *and let* $G$ *be an operator such that* $i < j < \omega$ *implies* $G_{\alpha_j - \alpha_i} = G$. *Then*

(8) $$\langle C, d\rangle \in G_\alpha[c] \,.\!\equiv\!.\; \bigvee_{D,e} .\, D \subseteqq C \wedge \mathfrak{K}_0[D, d] \wedge \langle C, e\rangle \in G[c] \wedge \langle D, e\rangle \in G[e].$$

The proof of (8) is similar to that of (7) and will be omitted. We also need the following input-free version of Lemma 4.5.

Lemma 5.5. *Let* $\{\alpha_i;\, i < \omega_1\}$ *be a sequence cofinal-closed* $\alpha$, $\alpha_0 = 0$, *and let* $G$ *be an operator such that* $i < j < \omega_1$ *implies* $G_{\alpha_j - \alpha_i} = G$. *Then*

(9) $$\langle C, d\rangle \in G_\alpha[c] \,.\!\equiv\!.\; \bigvee_{D_0, D_1} .\, D_0 \subseteqq C \wedge \mathfrak{K}_1[D_1, d] \wedge \bigwedge_{e \in D_1} \mathfrak{K}_0[D_0, e] \wedge \bigwedge_{e \in D_1} \langle C, e\rangle \in G[c] \wedge \bigwedge_{b, e \in D_1} \langle D_0, e\rangle \in G[b].$$

For each $\mu < \omega_2$, $\nu < \omega_1$ define

(10) $$G_{\mu,\nu} = G_{\omega_1{}^\mu \omega_0{}^\nu}.$$

These $G_{\mu,\nu}$ will be taking the place of the simpler $F_\mu$ of [5]. Order $\omega_2 \times \omega_1$ lexigraphically by $<$, i.e. $\mu_0, \nu_0 < \mu_1, \nu_1 .\equiv. \mu_0 < \mu_1 \vee [\mu_0 = \mu_1 \wedge \nu_0 < \nu_1]$. Then the following absorption law is a consequence of (4):

$$(11) \qquad \mu_0, \nu_0 < \mu_1, \nu_1 .\supset. G_{\mu_1,\nu_1} \circ G_{\mu_0,\nu_0} = G_{\mu_1,\nu_1}.$$

We now show that for each $\mu < \omega_2$ the $G_{\mu,\nu}$ become equal from some $q < \omega$ on, thus extending [5], Lemma 4.7.

Lemma 5.6. *For each $\mu < \omega_2$ there is a finite $q_\mu$ such that $G_{\mu,q_\mu} = G_{\mu,q_\mu+1}$. Furthermore for such $q_\mu$*

$$(\forall \nu)_{q_\mu}^{\omega_1} (\forall h)_1^{\omega_0} G_{\mu,\nu}^h = G_{\mu,q_\mu}.$$

Proof. Fix $\mu < \omega_2$. Note that the $G_{\mu,\nu}$ are finite operators, i.e. $\langle G_{\mu,\nu}[c]; c \in K\rangle$ takes on only finitely many values. Thus there are finite $p, m, m > 0$, such that (a) $G_{\mu,p} = G_{\mu,p+m}$. If $m = 1$ we are done, so suppose (b) $m > 1$. For each $n < \omega$ let $\alpha_n = \omega_1^\mu \omega_0^p n$ and $\beta_n = \omega_1^\mu \omega_0^{p+1} n + \omega_1^\mu \omega_0^p$. Now by (a), (5), and (11), $G_{\mu,p}^2 = G_{\mu,p} \circ G_{\mu,p} = G_{\mu,p+m} \circ G_{\mu,p} = G_{\mu,p+m} = G_{\mu,p}$. Thus $G_{\mu,p}^2 = G_{\mu,p}$ and so by induction on $n$ using (5) we have (c) $0 < n < \omega \supset G_{\alpha_n} = G_{\mu,p}$. Also by (a), (b), (4) and (11) for $0 < n < \omega$, $G_{\beta_n} = G_{\mu,p} \circ G_{\mu,p+1}^n = G_{\mu,p+m} \circ G_{\mu,p+1}^n = G_{\mu,p+m} = G_{\mu,p}$. Thus, (d) $0 < n < \omega \supset G_{\beta_n} = G_{\mu,p}$. Now $\lim_n \alpha_n = \omega_1^\mu \omega_0^{p+1}$ and $\lim_n \beta_n = \omega_1^\mu \omega_2^{p+2}$ and thus by (c), (d) and (8), $G_{\mu,p+1} = G_{\mu,p+2}$, completing the proof of first part of the lemma. The second part follows from the first by induction on $\nu, h$ using (5) and (8). □

The above lemma shows that for any $\mu$ there is a $q_\mu$ such that $G_{\mu,q_\mu} = G_{\mu,q_\mu+1}$. As a matter of fact we have from the proof that $q_\mu \leqq 2^{m^2 \cdot 2^m}$ $(m = |K|)$. The next lemma shows that a similar fact is true of the $\mu$ part of the operator $G_{\mu,\nu}$.

Lemma 5.7. *There is a finite $p$ such that $G_{p+1,0} = G_{p,0}$. Furthermore for such $p$, we have $G_{p,1} = G_{p,2}$.*

Proof. $\langle G_{i,0}[c]; c \in K\rangle$ takes on finitely many values. Thus there are $p, m, m > 0$, such that (a) $G_{p,0} = G_{p+m,0}$. We may again suppose that (b) $m > 1$. For each $n < \omega$ let $\alpha_n = \omega_1^p n$ and $\beta_n = \omega_1^{p+1} n + \omega_1^p$. Using (a), (b), (4), (5), (11) we may show as in Lemma 5.6 that $0 < n < \omega$ implies $G_{\alpha_n} = G_{p,0} = G_{\beta_n}$. So by (8), as $\lim_n \alpha_n = \omega_1^p \omega_0$ and $\lim_n \beta_n = \omega_1^{p+1} \omega_0$ we have (c) $G_{p,1} = G_{p+1,1}$. By induction on $h, j$ using (5) and (7) we have $(\forall j)_1^\omega (\forall h)^\omega G_{p,j}^h = G_{p+1,j}^h$. Hence (d) $G_{p,q_p} = G_{p+1,q_{p+1}}$ where $q_p$ is as in Lemma 5.6. Let $q = \max\{q_p, q_{p+1}\}$. As $\{\omega_1^p \omega_0^\nu; q \leqq \nu < \omega_1\}$ is cofinal-closed $\omega_1^{p+1}$ and $\{\omega_1^{p+1} \omega_0^\nu; q \leqq \nu < \omega_1\}$ is cofinal-closed $\omega_1$, by (d), (9) and the second part of Lemma 5.6 it follows that $G_{p+1,0} = G_{p+2,0}$, proving the first part of the lemma. Suppose $G_{p+1,0} = G_{p,0}$. For each $n < \omega$ let $\alpha_n = \omega_1^p n$ and $\beta_n = \omega_1^p \omega n + \omega_1^p$. As usual we can show that $0 < n < \omega$ implies $G_{\alpha_n} = G_{p,0} = G_{\beta_n}$. But $\lim_n \alpha_n = \omega_1^p \omega_0$ and $\lim_n \beta_n = \omega_1^p \omega_0^2$ and so by (8), $G_{p,1} = G_{p,2}$. □

Lemma 5.8. *Let $G_{p,0} = G_{p+1,0}$ and for each $i < p$, $G_{i,q_i} = G_{i,q_i+1}$. Then for $1 \leqq h < \omega_0$, $\nu < \omega_1$, we have:*

$$i < p \wedge q_i \leqq \nu .\supset. G_{i,\nu}^h = G_{i,q_i},$$

$$(12) \qquad p \leqq \mu < \omega_2 .\supset. G_{\mu,\nu}^h = \begin{cases} G_{p,1} & \text{if } \mu \Leftarrow \omega_0 \vee \nu \neq 0 \\ G_{p,0} & \text{otherwise.} \end{cases}$$

Proof. The first implication is a restatement of the second part of Lemma 5.6. The second is proved by induction on $\mu$, $\nu$, $h$, using (5), (8), (9), and Lemma 5.7. □

For any $p$, every ordinal $\alpha$ has a unique representation $\alpha = \omega_1^p\mu + \omega_1^{p-1}\nu_{p-1} + \ldots + \omega_1^0\nu_0$ where each $\nu_i < \omega_1$; we call this the *p-expansion (base $\omega_1$) of* $\alpha$. For each $i < p$ and for any $q_i$, $\nu_i$ in turn has a unique representation $\omega_0^{q_i}\gamma_i + \omega_0^{q_i-1}m_{i,q_i-1} + \ldots + \omega_0^0 m_{i,0}$. This is the *$q_i$-expansion (base $\omega_0$) of $\nu_i$* that was discussed in [5] Section 4. The sequence $\langle(\gamma_i), m_{i,q_i-1}, \ldots, m_{i,0}\rangle$, where $(\gamma_i)$ is 0 if $\gamma_i = 0$ and 1 otherwise, is called the *$i, q_i$-character of $\alpha$*, denoted $\mathrm{char}_{i,q_i}\alpha$. Next define

$$\mathrm{head}_p\alpha = \begin{cases} \langle 0, 0\rangle, & \text{if } \mu = 0 \\ \langle 1, 0\rangle, & \text{if } \mu \Leftarrow \omega_0 \text{ (i.e. if } \omega_1^p\mu \Leftarrow \omega_0) \\ \langle 0, 1\rangle, & \text{if } \mu \text{ is a successor} \vee \mu \Leftarrow \omega_1 \text{ (i.e. if } \omega_1^p\mu \Leftarrow \omega_1). \end{cases}$$

We now define the *$p$; $q_{p-1}, \ldots, q_0$-character (base $\omega_1$) of $\alpha$* denoted $\mathrm{char}_{p;q_{p-1},\ldots,q_0}\alpha$ to be $\langle \mathrm{head}_p\alpha, \mathrm{char}_{p-1,q_{p-1}}\alpha, \mathrm{char}_{p-2,q_{p-2}}\alpha, \ldots, \mathrm{char}_{0,q_0}\alpha\rangle$. Note that associated with each $\alpha < \omega_2$ is a unique $p$; $q_{p-1}, \ldots, q_0$-character. We also have a unique representation $\alpha = \omega_1^\omega\mu + \omega_1^{n-1}\nu_{n-1} + \ldots + \omega_1^0\nu_0$, $n = 0 \vee \nu_{n-1} \neq 0$ and each $\nu_i < \omega_1$; we call this the *$\omega$-expansion (base $\omega_1$) of $\alpha$*. Each $\nu_i$ has in turn a unique representation $\omega_0^{n_i}\gamma_i + \omega_0^{n_i-1}m_{i,n_i-1} + \ldots + \omega_0^0 m_{i,0}$. $\langle(\gamma_i), m_{i,n_i-1}, \ldots, m_{i,0}\rangle$ is the *$i$, $\omega$-character of $\alpha$*, and is denoted $\mathrm{char}_{i,\omega}\alpha$. The *$\omega$-head of $\alpha$*, is defined as follows:

$$\mathrm{head}_\omega\alpha = \begin{cases} \langle 0, 0\rangle, & \text{if } \mu = 0 \\ \langle 1, 0\rangle, & \text{if } \mu \text{ successor} \vee \mu \Leftarrow \omega_0 \\ \langle 0, 1\rangle, & \text{if } \mu \Leftarrow \omega_1. \end{cases}$$

The *$\omega$-character (base $\omega_1$) of $\alpha$* denoted $\mathrm{char}_\omega\alpha$, is defined to be $\langle \mathrm{head}_\omega\alpha, \mathrm{char}_{n-1,\omega}\alpha, \ldots, \mathrm{char}_{0,\omega}\alpha\rangle$.

We now introduce some abbreviations. For each $\mu < \omega_2$ and $\nu < \omega_1$ the operator $F_{\mu,\nu}$: $\mathfrak{P}K \to \mathfrak{P}K$ is defined as follows:

$$F_{\mu,\nu}[D] = \{d;\ \bigvee_{c,C} .c \in D \wedge \langle C, d\rangle \in G_{\mu,\nu}[c]\}. \tag{13}$$

From (2) we see that $F_{\mu,\nu}[D]$ consists of all those states which are attainable via an $\omega_1^\mu\omega_0^\nu$-run of $\Gamma$ starting from some state in $D$. Let $\tau = \mathrm{char}_{p;q_{p-1},\ldots,p_0}\alpha$ and $\tau_i = \mathrm{char}_{i,q_i}\alpha$ be exactly as above and define (using ordinary composition):

$$\begin{aligned} F_{\tau_i}[D] &= F_{i,0}^{m_{i,0}}F_{i,1}^{m_{i,1}} \ldots F_{i,q_i-1}^{m_{i,q_i-1}}F_{i,q_i}^{(\gamma_i)}[D], \\ F_{\mathrm{head}_p\alpha}[D] &= F_{p,0}^h F_{p,1}^h[D], \quad \text{where } \mathrm{head}_p\alpha = \langle h, k\rangle, \\ F_\tau[D] &= F_{\tau_0}F_{\tau_1} \ldots F_{\tau_{p-1}}F_{\mathrm{head}_p\alpha}[D]. \end{aligned} \tag{14}$$

Theorem 5.9 (Normal form for sentences). *Let $\Sigma$ be an* MT-*sentence. There are finite numbers $p, q_p, \ldots, q_0$ ($q_p \leqq 1$), operators $\{G_{i,j};\ i \leqq p, j \leqq q_i\}$ on a finite set $K$, and $E$, $L \subseteq K$ such that for any $\alpha < \omega_2$, $\Sigma$ holds in $[\alpha, <]$ just in case $F_\tau[E] \cap L \neq 0$ where $\tau = \mathrm{char}_{p;q_{p-1},\ldots,q_0}\alpha$ and $F_\tau$ is defined by* (13) *and* (14).

Proof. Using Remark 5.1 we can construct an input-free $\sup_1$-transition system $\Gamma = [E, H, \mathfrak{R}_0, \mathfrak{R}_1, L]$ whose set of states is $K$ such that $\mathrm{beh}_{\omega_2}\Gamma = \mathrm{spec}_{\omega_2}\Sigma$. Let $G_{i,j}$ be defined from (10) and let $p, q_{p-1}, \ldots, q_0$ be defined as in Lemmas 5.6 and 5.9.

The proof of the theorem is straightforward and follows from Lemma 5.8 and (4), noting especially that $G_{p,1} \circ G_{p,0} = G_{p,1}$ by (11). □

Corollary 5.10. *For any ordinals* $\alpha, \beta < \omega_2$, $\mathrm{MT}[\alpha, <] = \mathrm{MT}[\beta, <]$ *just in case* $\mathrm{char}_\omega \alpha = \mathrm{char}_\omega \beta$.

Corollary 5.11. *For every* $\alpha < \omega_2$ *there is a* $\beta < \omega_1^{\omega+1} + \omega_1^\omega$ *such that* $\mathrm{MT}[\alpha, <] = \mathrm{MT}[\beta, <]$. *For every* $\alpha < \omega_1^{i+1}$ *there is a* $\beta < \omega_1^i \omega_0^{\omega_2}$ *such that* $\mathrm{MT}[\alpha, <] = \mathrm{MT}[\beta, <]$.

The system $A = [\mathfrak{P}K, E, \langle F_{i,j}; i \leqq p \wedge j \leqq q_i \rangle]$ is a finite transition algebra working on words over the input states $\{\langle i, j\rangle; i \leqq p \wedge j \leqq q_i\}$ in the usual sense of finite automata. We may also (see [5], Section 4) interpret it as a free algebra relative to the absorption laws:

$$(15) \qquad h, k < i, j \;.\supset.\; F_{i,j}F_{h,k} = F_{i,j}, \qquad F_{i,q_i}^2 = F_{i,q_i}, \qquad F_{p,0}^2 = F_{p,0}.$$

In this case an input is a $p; q_{p+1}, \ldots, q_0$-character. We call systems of the form $[A, L], p; q_{p-1}, \ldots, q_0$-*absorption automata* (with output $L$). Their behavior are sets of $p; q_{p-1}, \ldots, q_0$-characters, and Theorem 5.9 takes the form:

Theorem 5.12. *The* $\omega_2$-*spectrum of any sentence* $\Sigma$ *consists of all* $\alpha < \omega_2$ *whose* $p; q_{p-1}, \ldots, q_0$-*characters are accepted by an appropriately chosen* $p; q_{p-1}, \ldots, q_0$-*absorption automata.*

Let $\Gamma, p, q_p, q_{p-1}, \ldots, q_0$, and the operators $G_{i,j}$ be as in previous lemmas. For each $i < p$, $j < q_i$, clearly there are $h < h + k$ such that $G_{i,j}^h = G_{i,j}^{h+k}$ – call these $h_{i,j}, k_{i,j}$. Let $\tau_i = \langle q_i; h_{i,q_i-1}, \ldots, h_{i,0}; k_{i,q_i-1}, \ldots, k_{i,0}\rangle$ and $\tau = \langle p, q_p, \tau_{p-1}, \ldots, \tau_0\rangle$, $(q_p \leqq 1)$. Let $\alpha = \omega_1^p \mu + \sum_{i=p-1}^{0} [\omega_1^i \omega_0^{q_i} \nu_i + \sum_{j=q_i-1}^{0} \omega_1^i \omega_0^j m_{i,j}]$ We define the *reduction of* $\alpha$ *by* $\tau$, $\mathrm{red}_\tau \alpha$, to be $\omega_1^p \omega^{q_p} m + \omega_1^p n + \sum_{i=p-1}^{0} [\omega_1^i \omega_0^{q_i}(\nu_i) + \sum_{j=q_i-1}^{0} \omega_1^i \omega_0^j \bar{m}_{i,j}]$ where $\mathrm{head}_p \alpha = \langle m, n\rangle$ and for each $i < p$, $j < q_i$,

$$\bar{m}_{i,j} = \begin{cases} m_{i,j}, & \text{if } m_{i,j} < h_{i,j} \\ h_{i,j} + (\mu k)\, m_{i,j} - h_{i,j} \equiv k \bmod k_{i,j}, & \text{if } m_{i,j} \geqq h_{i,j}. \end{cases}$$

Clearly $\alpha \equiv_\tau \beta$ defined by $\mathrm{red}_\tau \alpha = \mathrm{red}_\tau \beta$ is an equivalence relation of finite index. If $\Sigma$ is an MT-sentence and $\Gamma$ is the input-free $\sup_1$-transition system constructed as in Remark 5 1 then we define $\mathrm{red}_\Sigma \alpha$ to be $\mathrm{red}_\tau \alpha$ and $\equiv_\Sigma$ to be $\equiv_\tau$ where $\tau$ is defined from $\Gamma$ as in the above The following Remark follows from Theorem 5.9.

Remark 5 13 *For any* MT-*sentence* $\Sigma$ *if* $\alpha \equiv_\Sigma \beta$ *then* $\Sigma \in \mathrm{MT}[\alpha, <] \equiv \Sigma \in \mathrm{MT}[\beta, <]$.

Clearly if $\Sigma$ has a model $[\alpha, <]$ where $\alpha < \omega_2$ then it has one where $\alpha < \omega_1^{p+1}$. Furthermore for each $i < p$ if $\Sigma$ has a model $[\alpha, <]$ where $\alpha < \omega_1^{i+1}$ then it has one where $\alpha < \omega_1^i \omega_0^{q_i+1}$. Consequently,

Theorem 5.14. *The* MT *of all well-orders of cardinality* $<\omega_2$ *is equal to the* MT *of all* $\alpha < \omega_1^\omega$. *The* MT *of all well-orders of type* $<\omega_1^{i+1}$ *is equal to the* MT *of all* $\alpha < \omega_1^i \omega_0^\omega$.

In [5] this observation was made in the case where $i = 0$. We define a subset $X$ of $\omega_2$ to be *ultimately periodic* (of type $p; q_p, q_{p-1}, \ldots, q_0$) if there is a $\tau$ as above such that $X$ is the union of some of the equivalence classes of $\equiv_\tau$. Then as in [5] we have

Theorem 5.15. *Let* $\alpha \leqq \omega_2$. *The following three classes are equal*: (a) *the class of definable subsets of* $\alpha$, (b) *the class of ultimately periodic subsets of* $\alpha$, (c) *the class of* $\alpha$-*spectra of sentences.*

Theorem 5.16 (Decision lemma) *There is a procedure which when applied to any* MT *sentence* $\Sigma$ *and any* $\omega$*-character (base* $\omega_1$*),* $\pi$*, decides whether or not* $\Sigma \in \mathrm{MT}[\alpha, <]$ *for any* $\alpha < \omega_2$ *such that* $\mathrm{char}_\omega \alpha = \pi$.

Proof. Let $\Sigma$ be a sentence. By Remark 5.1, one can construct an input free $\sup_1$-transition system $\Gamma = [E, H, \mathfrak{R}_0, \mathfrak{R}_1, L]$, with $K$ equals the set of states of $\Gamma$, $r = |\mathfrak{P}K \times K|$, such that $\mathrm{beh}_{\omega_2}\Gamma = \mathrm{spec}_{\omega_2}\Sigma$. Then by the results of this section we may define the operators $G_{i,j}$ for finite $i, j$ by the following recursion:

$$G_{0,0}[c] = \{\langle\{c\}, d\rangle; H[c, d]\},$$

$$(16) \qquad G_{i,j+1}[c] = \{\langle c, d\rangle; \bigvee_{\substack{D,e\\ h,k \leqq r}} . D \subseteqq C \wedge \mathfrak{R}_0[D, d] \wedge \langle C, e\rangle \in G^h_{i,j}[c] \wedge \langle D, e\rangle \in G^k_{i,j}[e]\}.$$

$$q_i = (\mu j). G_{i,j} = G_{i,j+1},$$

$$G_{i+1,0}[c] = \{\langle C, d\rangle; \bigvee_{D_0 D_1} . D_0 \subseteqq C \wedge \mathfrak{R}_1[D_0, e] \wedge \bigwedge_{e\in D_1} [\mathfrak{R}_0[D_0, e] \wedge \langle C, e\rangle \in G_{i,q_i}[c]] \wedge \bigwedge_{b,e\in D_1} \langle D_0, e\rangle \in G_{i,q_i}[b]\}.$$

Now use (16) to construct $G_{0,0}, G_{0,1}, \ldots$. This process is stopped when we reach a $q_0$ such that $G_{0,q_0} = G_{0,q_0+1}$ (which exists by Lemma 5.6). Next we construct $G_{1,0}, G_{1,1}, \ldots$, stopping when we reach a $q_1$ such that $G_{1,q_1} = G_{1,q_1+1}$. This above procedure is continued producing $G_{0,0}, G_{0,1}, \ldots, G_{1,0}, \ldots, G_{2,0}, \ldots$ until we reach a $p$ such that $G_{p,0} = G_{p+1,0}$ (which exists by Lemma 5.7). $E, K, L, G_{0,0}, \ldots, G_{p,0}, G_{p,1}, p, q_{p-1}, \ldots, q_0$ will then serve for Theorem 5.9. Now rewrite the $\omega$ character $\pi$ as a $p; q_{p-1}, \ldots, q_0$ character, $\tau$. From the $G_{i,j}$ above we can make the abbreviations of (13) and (14) and thus we can evaluate $F_\tau[E]$. By Theorem 5.9 $F_\tau[E] \cap K \neq 0$ just in case $\Sigma$ holds in $[\alpha, <]$, where $\alpha < \omega_2$ is such that $\mathrm{char}_\omega \alpha = \pi$. □

Corollary 5.17. *For any* $\alpha < \omega_2$, $\mathrm{MT}[\alpha, <]$ *is decidable, the decision algorithm depending only on the* $\omega$ *character (base* $\omega_1$*) of* $\alpha$. *Also* $\mathrm{MT}[\{[\alpha, <]; \alpha < \omega_2\}]$ *is decidable.*

Proof. The only thing remaining to be shown is a decision method for $\mathrm{T} = \mathrm{MT}[\{[\alpha, <]; \alpha < \omega_2\}]$. Let $\Sigma$ be an MT sentence. Let $Q = \{\mathrm{red}_\Sigma \alpha; \alpha < \omega_2\}$. Clearly $Q$ is a finite set and $\Sigma \in \mathrm{T} \equiv \bigwedge_{\alpha\in Q} \Sigma \in \mathrm{MT}[\alpha, <]$. Thus using the decision procedure of the previous lemma we can decide whether or not $\Sigma \in \mathrm{T}$. □

## Bibliography


[1] Büchi, J. R., Weak second order arithmetic and finite automata. This Zeitschr. **6** (1960), 66–92.

[2] Büchi, J. R., On a decision method in restricted second order arithmetik. In: Proc. 1960 Int. Congr. for Logic, Method. and Philos. of Sci., Stanford Univ. Press, Stanford, Calif. 1962, pp. 1–11.

[3] Büchi, J. R., Decision methods in the theory of ordinals. Bull. Amer. Math. Soc. **71** (1965), 767–770.

[4] Büchi, J. R., Transfinite automata recursions and weak second order theory of ordinals. In: Proc. 1964 Int. Congr. for Logic, Method. and Philos. of Sci., North-Holland Publishing Comp., Amsterdam 1965, pp. 3–23.

[5] Büchi, J. R., The monadic second-order theory of $\omega_1$. In: Decidable Theories II: The monadic second-order theory of all countable ordinals, Lecture Notes in Mathematics **328** (1973), Springer-Verlag, Berlin–Heidelberg–New York, pp. 1–127.

[6] BÜCHI, J. R., and D. SIEFKES, Axiomatization of the monadic second order theory of $\omega_1$. In: Decidable Theories II: The monadic second-order theory of all countable ordinals, Lecture Notes in Mathematics **328** (1973), Springer-Verlag, Berlin – Heidelberg – New York, pp. 129 – 217.
[7] MCNAUGHTON, R., Testing and generating infinite sequences by a finite automaton. Information and Control **9** (1966), 521 – 530.
[8] SHELAH, S., The monadic theory of order. Annals of Math. **102** (1975), 379 – 419.
[9] SIEFKES, D., Büchi's monadic second order successor arithmetic. Lecture Notes in Mathematics **120** (1970), Springer-Verlag, Berlin – Heidelberg – New York.
[10] ULAM, S., Zur Maßtheorie in der allgemeinen Mengenlehre. Fund. Math. **16** (1930), 140 – 150.
[11] ZAIONTZ, C., Axiomatization of the monadic theory of ordinals $<\omega_2$. This Zeitschr. **29** (1983), 337 – 356.

(Eingegangen am 5. Oktober 1981)

J. Richard Büchi
Department Computer Sciences
Purdue University
West Lafayette, Indiana 47907 (U.S.A.)

Charles Zaiontz
Department of Mathematics
University of South Florida
Tampa, Florida 33620 (U.S.A.)

# Section 6 Games and Determinacy

*With comments by Yuri Gurevich, University of Michigan, Ann Arbor*



*

# Games People Play

*Yuri Gurevich**

## Abstract

Büchi's approach to the determinacy of infinite games was original and not without a controversy. He resented determinacy proofs that did not live up to his standards of constructivity. We describe the infinite games in question, discuss the constructivity of determinacy proofs, and comment on Büchi's contributions to determinacy.

## 1. Games and Strategies

The infinite games of interest to us here were introduced by Gale and Stewart [8]. In this section we recall the definition of the games. To simplify the exposition, we restrict attention to a special case when the set of possible moves in any position is a fixed finite set $A$.

We view $A$ as an alphabet. As usual, $A^*$ is the set of strings over $A$. Borrowing the terminology of Muchnik [12] we call functions from the set $\omega$ of natural numbers to $A$ *superstrings*. Call a set $U$ of superstrings *open* if every superstring $X \in U$ has a prefix $x \in A^*$ such that the *cone*

$$[x] = \{Y: Y \text{ is a superstring with a prefix } x\}$$

is included into $U$; this gives the well-known Cantor topology on the set of superstrings.

Each set $W$ of superstrings gives rise to a *game* $G(W)$ between Player 0 and Player 1. In a course of the game, the two players build a superstring. Player 0 starts by choosing a letter $a_0$, then Player 1 chooses a letter $a_1$, then Player 0 chooses a letters $a_2$, and so on ad infinitum. If the resulting superstring (the *play*) $a_0 a_1 a_2 \ldots$ belongs to $W$ then Player 0 wins; otherwise Player 1 wins. The *winning set* $W_\varepsilon$ of a Player $\varepsilon$ is the collection of plays where he is victorious. If $\varepsilon = 0$ then $W_\varepsilon = W$, and if $\varepsilon = 1$ then $W_\varepsilon$ is the complement of $W$.

A game $G(W)$ is called *determined* if one of the players has a winning strategy. To formalize the notion of a strategy, let us notice that every string $x \in A^*$ can be viewed as a position in the game $G(W)$; if the length $|x|$ of $x$ is even then Player 0 makes a move in position $x$, and if $|x|$ is odd then Player 1

*Electrical Engineering and Computer Science Department, The University of Michigan, Ann Arbor, MI 48109-2122.
*Note*: The work was partially supported by NSF grant DCR 85-03275. Also, during the final stage of the work the author was with Stanford University and IBM Almaden Research Center (on a sabbatical leave from the University of Michigan).

does. For technical reasons, it is convenient to allow nondeterministic strategies. Formally, a *strategy* for a Player $\varepsilon$ is any function $F$ that assigns a nonempty subset $F(x)$ of $A$ (the set of recommended moves) to each position $x$ in the game where Player $\varepsilon$ makes a move. A strategy $F$ is *deterministic* if every $F(x)$ comprises a single letter (the recommended move). Consider the case when both players follow some strategies; if both strategies are deterministic then the play is uniquely determined, otherwise the number of plays consistent with the two strategies is in the range from one to the power of continuum. It should be clear what a *winning strategy* is; following a winning strategy, the player wins regardless of what the other player does. A winning strategy may be nondeterministic.

Each position $x$ determines a remainder of a game $G(W)$, called the *x-remainder* and denoted $G_x(W)$. The remainder is a game in its own right; if $|x|$ is even then Player 0 starts $G_x(W)$, otherwise Player 1 does. However, the initial position of $G_x(W)$ is $x$ rather than the empty string; the plays of $G_x(W)$ form the cone $[x]$. Player $\varepsilon$ wins a play $X$ of $G_x(W)$ if $X \in W_\varepsilon$. The definitions of strategies and winning strategies for Player $\varepsilon$ in the remainder game $G_x(W)$ should be clear.

## 2. Open Games

A game $G(W)$ is called *open*, $\mathscr{F}_\sigma$, *Borel*, etc. if $W$ is so (in the Cantor topology). If $U$ is an open set of superstrings, then a *support* for $U$ is any string set $S$ such that $U = \bigcup_{x \in S}[x]$. The collection of all strings $x$ with $[x] \subseteq U$ is the largest support for $U$.

**Theorem 1** (Gale and Stewart). *Every open or closed game $G(W)$ is determined.*

We give two proofs of the theorem. One is a proof by contradiction (essentially, the original proof); the other proof—also well known—is more constructive.

Proof 1. Suppose that a Player $\varepsilon$, whose winning set $W_\varepsilon$ is open, does not have a winning strategy (in the initial position). A winning strategy $F$ of his opponent Player $\delta$ is designed to ensure that Player $\varepsilon$ has no winning strategy in any position in the course of the game. If Player $\varepsilon$ does not have a winning strategy in $G_x(W)$ (and Player $\delta$ makes a move in $x$) then $F(x)$ comprises letters $a$ such that Player $\varepsilon$ has no winning strategy in $G_{xa}(W)$; otherwise we don't care what $F(x)$ is.

Any play $X$ consistent with $F$ belongs to $W_\delta$. For, if it belongs to $W_\varepsilon$ then it has a prefix $x$ such that $W_\varepsilon$ contains the whole cone $[x]$. But then Player $\varepsilon$ has a winning strategy in $G_x(W)$, which is impossible. □

To give a more constructive proof of Theorem 1, we define the *rank function* of a Player $\varepsilon$ corresponding to a given string set $S$. It is the unique partial function $\rho$ from $A^*$ to $\omega$ such that for every string $x$:

- $\rho(x) = 0$ if and only if $x$ has a prefix in $S$.
- $\rho(x) \leq n + 1$ if and only if either Player $\varepsilon$ makes a move in position $x$ and $\rho(xa) < n + 1$ for some letter $a$ or else his opponent makes a move in position $x$ and $\rho(xa) < n + 1$ for every letter $a$.

PROOF 2. Again, we suppose that $W_\varepsilon$ is open and $\delta = 1 - \varepsilon$. Let $S$ be any support for $W_\varepsilon$. Player $\varepsilon$ wins a play $X$ if and only if $X$ meets $S$, i.e., $X$ has a prefix that belongs to $S$. Let $\rho$ be the rank function of Player $\varepsilon$ corresponding to $S$.

If the rank of the position is defined then Player $\varepsilon$ has a winning strategy $F$ that can be called "decrease the rank": If $\rho(x)$ is defined and nonzero (and Player $\varepsilon$ makes a move in $x$) then $F(x)$ comprises all letters $a$ such that $\rho(xa) < \rho(x)$; otherwise we don't care what $F(x)$ is.

If the rank of the initial position is undefined then Player $\delta$ has a winning strategy $F'$ that can be called "keep off the ranked positions": If $\rho(x)$ is undefined (and Player $\delta$ makes a move in $x$) then $F'(x)$ comprises all letters $a$ such that $\rho(xa)$ is undefined; otherwise we don't care what $F'(x)$ is. □

The second proof actually presents winning strategies. (These winning strategies will be used in the next section.) If the rank function $\rho$ is computable then the second proof provides algorithms to decide who of the two players has a winning strategy and to execute the winning strategy.

## 3. $\mathscr{F}_\sigma$ Games

Recall that an $\mathscr{F}_\sigma$ set is the intersection of a countable collection of open sets, and a $\mathscr{G}_\delta$ set is the complement of an $\mathscr{F}_\sigma$ set.

**Theorem 2** (Wolfe). *Every $\mathscr{F}_\sigma$ or $\mathscr{G}_\delta$ game $G(W)$ is determined.*

Again we give two proofs. The first one is a proof by contradiction along the lines of the original proof. The second one, close to the proof of Wolfe's theorem in Moschovakis' book [11], is more constructive.

PROOF 1. Fix open sets $U_n$ such that the winning set of a Player $\varepsilon$ is the intersection of the sets $U_n$; without loss of generality, each $U_n$ includes $U_{n+1}$. Choose supports $S_n$ for $U_n$ in such a way that:

- Each $S_n$ is an antichain. In other words, if $x$ and $y$ belong to the same $S_n$ then the cones $[x]$ and $[y]$ are disjoint.
- Each $S_n$ is below $S_{n+1}$ (if $A^*$ is viewed as a tree that grows upward). More exactly, if $x \in S_n$, $y \in S_{n+1}$, and the cones $[x]$ and $[y]$ are not disjoint then $x$ is a proper prefix of $y$.

Let $T$ be the union of the antichains $S_n$. An arbitrary superstring $X$ belongs to $W_\varepsilon$ if and only if it meets $T$ infinitely many times, i.e., it has infinite many prefixes in $T$. For every string $x$, let $T_x$ be the set of nonempty strings $xy$ such that $xy \in T$. Let $D$ be the set of positions $x$ such that Player $\delta$, the opponent of Player $\varepsilon$, has a winning strategy in the $x$-remainder of the game.

For every position $x \notin D$, Player $\varepsilon$ has a strategy $F_x$ in the $x$-remainder $G_x(W)$ of the game $G(W)$ that allows him to reach a position in $T_x - D$. For, suppose that such $F_x$ does not exist. By Theorem 1, Player $\delta$ has a strategy to avoid $T_x - D$ in $G_x(W)$. Following such a strategy, Player $\delta$ either avoids $T_x$ and this way wins $G_x(W)$ or hits $D$ at some position $y$ where he can start playing a strategy winning $G_y(W)$. Thus, he has a winning strategy in $G_x(W)$, which is impossible.

We describe a winning strategy for Player $\varepsilon$ in any $G_{x_0}(W)$ such that $x_0 \notin D$: Use $F_{x_0}$ to reach a position $x_1$ in $T_{x_0} - D$, then use $F_{x_1}$ to reach a position $x_2$ in $T_{x_1} - D$, then use $F_{x_2}$ to reach a position $x_3$ in $T_{x_2} - D$, and so on. □

PROOF 2. Suppose again that $W_\delta$ is the intersection of open sets $U_n$ and $\delta = 1 - \varepsilon$. Let $S_n$ be the largest support for $U_n$. The goal of Player $\varepsilon$ is to hit every $S_n$. All rank functions in this proof are of Player $\varepsilon$.

If a position $x$ is unranked with respect to some $S_n$ then Player $\delta$ has an obvious winning strategy of keeping off $S_n$-ranked positions. Let $P_1$ be the collection of positions $x$ ranked with respect to all $S_n$. From a position in $P_1$, Player $\varepsilon$ can reach some position in any $S_n$, but that position may be unranked with respect to some other $S_m$. Let $P_2$ be the set of positions $x$ that are ranked with respect to all $P_1 \cap S_n$. From a position in $P_2$, Player $\varepsilon$ can hit any $P_1 \cap S_n$ from where he can hit any $S_m$; this is again insufficient for a winning strategy.

By induction on $k$, define $P_{k+1}$ to be the set of positions ranked with respect to all $P_k \cap S_n$. The index $k$ may be seen as a measure of "potential" to hit sets $S_n$. Starting in a position of potential $k$ and given any sequence $(n_1, \ldots, n_k)$ of natural numbers, Player $\varepsilon$ can hit $S_{n_1}$ in a position of potential $k-1$, from where he can hit $S_{n_2}$ in a position of potential $k-2$, and so on. A winning strategy seems to require an infinite potential.

Define $P_\omega = \bigcap_k R_k$. The potential of $\omega$ guarantees that Player $\varepsilon$ can hit any finite sequence of sets $S_n$, but it does not guarantee hitting an infinite sequence of sets $S_n$. This motivates the following definition by induction on a (countable) ordinal $\alpha$.

- $P_0 = A^*$.
- $P_{\alpha+1}$ is the set of positions ranked with respect to $P_\alpha \cap S_n$ for all $n$.
- If $\alpha$ is a limit ordinal then $P_\alpha = \bigcap_{\beta<\alpha} P_\beta$.
- $P$ is any $P_\alpha$ equal to $P_{\alpha+1}$.

It is easy to see that the sequence of $P_\alpha$ decreases; hence indeed some $P_\alpha = P_{\alpha+1}$. Notice that if $x \in P$ then it is ranked with respect to every $P \cap S_n$. This allows us to construct a strategy $F$ for Player $\varepsilon$ winning every $x$-remainder game such that $x \in P$. If $m = \min\{n : x \notin S_n\}$ (we don't care what $F(x)$ is in the case $x$ belongs to all $S_n$), $\rho$ is the rank function corresponding to $P \cap S_m$ and Player $\varepsilon$ makes a move in $x$ then

$$F(x) = \{a : \rho(xa) < \rho(x)\}.$$

Next we construct a strategy $F'$ for Player $\delta$ winning every $x$-remainder game such that $x \notin P$. We don't care what $F'(x)$ is when $x \in P$. If $x \notin P$, let $\alpha(x)$ be the least ordinal $\alpha$ such that $x \in P_\alpha - P_{\alpha+1}$ and $n(x)$ be the least number $n$ such that $x$ is not ranked with respect to $P_\alpha \cap S_n$. If Player $\delta$ makes a move in a position $x \notin P$, then

$$F'(x) = \{a : xa \text{ is not ranked with respect to } P_{\alpha(x)} \cap S_{n(x)}\}.$$

Say that a position $x \notin P$ is $\delta$-preferable to a position $y \notin P$ if either $\alpha(x) < \alpha(y)$ or else $\alpha(x) = \alpha(y)$ and $n(x) < n(y)$. Now suppose that $x \notin P$ and consider a play of $G_x(W)$ consistent with $F'$. Player $\delta$ steers clear of $P_{\alpha(x)} \cap S_{n(x)}$-ranked positions until a $\delta$-preferable position $y$ is reached. Then Player $\delta$ steers clear of $P_{\alpha(y)} \cap S_{n(y)}$-ranked positions until a $\delta$-preferable position $z$ is reached. And so on. From some moment on, Player $\delta$ sticks to steering clear of positions ranked with respect to a fixed $P_\alpha \cap S_n$. From that moment, all positions in the play belong to $P_\alpha$ and are not $P_\alpha \cap S_n$-ranked. This means that the play never hits $S_n$. □

## 4. Constructive Determinacy

In a pioneering paper [7], Büchi and Landweber proved a constructive determinacy theorem for games $G(W)$, where the winning set $W$ is a set of superstrings accepted by some finite automaton $A$. (The phenomenon of finite automata on superstrings is explained in McNaughton's article [10] in this volume.) Their proof yields an algorithm that, given an automaton $A$, decides which player has a winning strategy and constructs a finite automaton that executes a winning strategy. Later, Büchi tried to extend this result. In the case of more complicated winning sets, it was impossible to obtain strategies executable by finite automata, and Büchi sought strategies that can be played by finite automata, which have the game conditions given by an oracle. This seems to be the essence of Büchi's constructivism in the field of game determinacy.

By the way, the term "constructive" is used very informally here. Büchi was not an adherent of the school of constructive mathematics as far as I know. Problems of decidability of monadic second-order theories and some related issues led him to game determinacy. His intention was to give sharpened determinacy results. He was interested in winning strategies that can be defined or executed by certain restricted means; he was especially interested in winning strategies executable by finite automata. The constructivity we are talking about is more related to definability and complexity than to the philosophy of constructive mathematics. (Today, computational complexity folks improve Büchi's algorithms and complain that his papers are difficult to read.)

In [6], Büchi and Klein proved a constructive determinacy theorem for $\mathscr{F}_\sigma$ games. The paper was never published. "The referee said it was not worth the trouble," complained Büchi in [5]. It was an exciting time for the determinacy community [11]. People tried to extend determinacy to larger collections of sets and studied set-theoretic worlds where every Gale-Stewart game is determined. The complexity of winning strategies was very much in the center of attention. People have computed exactly the complexity (in the sense of descriptive set theory) of the simplest winning strategies for various classes of games, and constructive, substantially shorter proofs of the $\mathscr{F}_\sigma$ determinacy were well known to the community; one such proof is reproduced in the previous section. So, the referee's decision can be justified. Still, the special kind of constructivity of Büchi-Klein's proof is not found in other known proofs as far as I know.

An important application of determinacy is discussed in [4]. Let $\mathscr{B}$ be the collection of Boolean combinations of $\mathscr{F}_\sigma$ sets. Büchi stated in [4] (for the proof the reader is referred to [3]) his result on constructive $\mathscr{B}$ determinacy and then sketched how this result implies Rabin's complementation lemma, which is by far the hardest part in the famous proof by Rabin of the decidability of the monadic second-order theory of the standard binary tree. This application is further discussed in paper [5] whose declared goal is to prove a constructive $\mathscr{F}_{\sigma\delta} \cap \mathscr{G}_{\delta\sigma}$ determinacy. Unfortunately, the papers [4] and [5] are indeed very hard to understand. ($\mathscr{B}$ games are analyzed in [9]; simpler proofs of Rabin's complementation lemma can be found in [9] and [12].)

The writing of Büchi cannot be called boring. He held strong views and was not shy to express them. In particular, Büchi spoke sharply against "fancy model-theoretical proofs." In this connection, let me notice that a fancy and nonconstructive proof may be a shortcut to a constructive result. Examples of such phenomenon can be found in papers like [13] or [9] that seemed to irritate Büchi. It is most important however that—even though Büchi's papers are difficult to reconstruct or verify—their ideas are worth exploring.

*Acknowledgments.* The title of this article is stolen from [1]. I am thankful to Martín Abadi, Alekos Kechris, Saunders Mac Lane, Dirk Siefkes, Wolfgang Thomas, and Moshe Vardi for their reactions on a draft of this article.

## *References*

1. Berne, E. (1964). *Games people play; the psychology of human relationships*. Grove Press, New York.
2. Büchi, J.R. (1970). Algorithmisches Konstruieren von Automaten und die Herstellung von Gewinnstrategien nach Cantor-Bendixson. *Tagungsbericht Automatentheorie und Formale Sprachen, Math. Forschungsinst. Oberwolfach, 1969*. Mannheim, Germany, 385–398. Invited address.

3. Büchi, J.R. (1973). *The monadic second-order theory of* $\omega_1$. Lecture Notes in Mathematics, 328, Springer-Verlag, 1–127.
4. Büchi, J.R. (1977). *Using determinacy of games to eliminate quantifiers*. Lecture Notes in Computer Science, 56, Springer-Verlag, 367–378.
5. Büchi, J.R. (1983). State-strategies for games in $F_{\sigma\delta} \cap G_{\delta\sigma}$. *Journal of Symbolic Logic*, 48, 1171–1198.
6. Büchi, J.R. & Klein, S. (1972). On the presentation of winning strategies via the Cantor-Bendixson method. *Technical Report CSD TR-81*, Purdue University, 14 pp.
7. Büchi, J.R. & Landweber, L.H. (1969). Solving sequential conditions by finite state strategies. *Trans. AMS*, 138, 295–311.
8. Gale, D. & Stewart, F.M. (1953). Infinite games with perfect information. *Ann. of Math. Studies*, 28 (Contributions to the Theory of Games II). Princeton, N.J., 245–266.
9. Gurevich, Y. & Harrington, L. (1982). Trees, automata and games. *14th Symp. on Theory of Computations*, Association for Computing Machinery, 60–65.
10. McNaughton, R. (1989). Büchi's Sequential Calculus. This volume, 382–397.
11. Moschovakis, Y.N. (1980). *Descriptive set theory*. North-Holland, Amsterdam.
12. Muchnik, A.A. (1984). Games on infinite trees and automata with dead ends: A new proof of the decidability of monadic theory of two successor functions. *Semiotics and Information*, 24, 17–42 (in Russian).
13. Shelah, S. (1975). The monadic theory of order. *Annals of Math.*, 102, 379–419.
14. Wolfe, P. (1955). The strict determinateness of certain infinite games. *Pacific J. Math.*, 5, 841–847.

# SOLVING SEQUENTIAL CONDITIONS BY FINITE-STATE STRATEGIES([1])

BY

J. RICHARD BUCHI AND LAWRENCE H. LANDWEBER

Our main purpose is to present an algorithm which decides whether or not a condition $\mathfrak{C}(X, Y)$ stated in sequential calculus admits a finite automata solution, and produces one if it exists. This solves a problem stated in [4] and contains, as a very special case, the answer to Case 4 left open in [6]. In an equally appealing form the result can be restated in the terminology of [7], [10], [15]: Every $\omega$-game definable in sequential calculus is determined. Moreover the player who has a winning strategy, in fact, has a winning finite-state strategy, that is one which can effectively be played in a strong sense. The main proof, that of the central Theorem 1, will be presented at the end. We begin with a discussion of its consequences.

1. **Conditions on sequential operators.** Let $\mathfrak{C}(X, Y)$ be a condition (i.e., binary relation) on $\omega$-sequences $X = X0, X1, X2, \ldots$ and $Y = Y0, Y1, Y2, \ldots$ of members of the finite sets $I$ and $J$ respectively. Let $Y = \mathscr{A}(X)$ be an operator which maps $I$-sequences into $J$-sequences. We will say that the operator $\mathscr{A}$ *solves the condition* $\mathfrak{C}(X, Y)$ *for* $Y$ or that $\mathscr{A}$ is a *solution of* $\mathfrak{C}$ *for* $Y$, if $(\forall X)\mathfrak{C}(X, \mathscr{A}(X))$ or equivalently,

$$(1) \qquad (\forall XY) \cdot Y = \mathscr{A}(X) \supset \mathfrak{C}(X, Y).$$

If no further requirement is imposed on solutions, then the axiom of choice states: $(\forall X)(\exists Y)\mathfrak{C}(X, Y)$ is the solvability condition of $\mathfrak{C}$ for $Y$. The solvability question becomes more interesting if one requires the solution $\mathscr{A}$ to be continuous in the sense of the *natural Cantor topology* on the set of all $\omega$-sequences over the alphabets $I$ and $J$. Let $I^*$ denote the set of all finite sequences (words) over $I$. The members of $I^*$ form a tree if all words $wa$, $a \in I$ are taken as direct successors of $w \in I^*$. $\omega$-sequences over $I$ are represented by infinite paths through the tree. Let $U_w$ be the set of all those paths $X$ which contain $w$ (as an initial segment). The finite unions $U_{w_1} \cup \cdots \cup U_{w_n}$ are then the open-closed (clopen) sets of the totally disconnected space of all $I$-sequences. An operator $Y = \mathscr{A}(X)$ is *continuous* if it may be given in the form,

$$(2) \qquad Yt = \Phi(\bar{X}(\phi t))$$

whereby $\bar{X}z$ stands for the word $X0, \ldots, Xz$, $\phi$ is a map from $\omega$ into $\omega$ and $\Phi$ maps $I^*$ into $J$.

Received by the editors October 19, 1967 and, in revised form, April 15, 1968.

([1]) This research was sponsored by the National Science Foundation Grant No. GJ-120. The main result was announced in [13].

Among the continuous operators there are those for which the entries in the sequence $Y=\mathscr{A}(X)$ can in fact be computed, if sufficient information about the entries in $X$ is provided. The *recursive operators* (RO) are those presentable in form (2), whereby both $\phi$ and $\Phi$ are recursive.

A particularly simple class of recursive operators are the *finite automata operators* (FAO), that is those operators which may be presented in the form,

$$(3) \qquad Z0 = H[X0], \qquad Zt' = L[Xt', Zt], \qquad YT = W[Zt],$$

where $t'=t+1$. Here $Z$ varies over $\omega$-sequences from a finite set $K$. $H$, $L$ and $W$ are functions from $I$ into $K$, $I\times K$ into $K$, and $K$ into $J$. A system $\langle K, H, L, W\rangle$ is called a *finite automaton* with *input states* $I$, *output states* $J$, and (internal) *states* $K$. Finite automata were first studied by Kleene [12]. Also see [3], [5], and [16]. Besides being recursive, FAO's are deterministic in the sense that the state of $Y$ at time $t$ can be calculated without anticipating future states of the input $X$. More precisely, a continuous operator (2) is *deterministic* (DO) if $\phi t\leqq t$. I.e., if it can be given in the form,

$$(4) \qquad Yt = \Phi(\bar{X}t).$$

Thus we use the term deterministic in the sense familiar from physics.

Note that a DO is continuous but need not be recursive. A FAO is a recursive deterministic operator (RDO). Furthermore, one easily proves: The DO given by (4) is a FAO if and only if the right congruence $u\sim v$ on words, defined by $(\forall w)\Phi(uw)=\Phi(vw)$, has finite index. This explains in just which way a finite automaton is limited in its ability to memorize the input history $\bar{X}t$ at time $t$. To be a FAO is a very strong requirement on a RDO.

The operator (2) might be called *h-shift* in case $\phi t=0$ for $t<h$, $\phi t=t-h$ for $h\leqq t$. The deterministic operators now appear as 0-shift$\supseteq$1-shift$\supseteq$2-shift$\supseteq\cdots$. In particular (4) is a 0-shift operator and a 1-shift operator is one of form

$$(4') \qquad Yt = \Phi(\bar{X}(t-1))$$

whereby $\bar{X}(-1)$ stands for the empty word. Furthermore, the FAO defined by (3) is a 0-shift FAO, while a 1-shift FAO can always be presented in the form

$$(3') \qquad Z0 = c, \qquad Zt' = L[Xt, Zt], \qquad Yt = W[Zt],$$

whereby $c\in K$ is called the *initial state* of the 1-shift automaton $\langle K, c, L, W\rangle$.

DEFINITION 1. A condition $\mathfrak{C}$ is called *determined* if either there exists a 0-shift deterministic solution $Y=\mathscr{A}(X)$ of $\mathfrak{C}(X, Y)$ for $Y$ or else there exists a 1-shift deterministic solution $X=\mathscr{B}(Y)$ of $\sim\mathfrak{C}(X, Y)$ for $X$.

It is interesting to contemplate this notion in the context of the Cantor topology; say for example, if $\mathfrak{C}$ is a Borel set in the product of the two spaces. This is studied in a game theoretic context in [7], [10], and [15]. If $\mathfrak{C}$ is determined, it either contains the graph of a continuous function $Y = \mathscr{A}(X)$, or else $\sim\mathfrak{C}$ contains the graph of a continuous function $X=\mathscr{B}(Y)$.

LEMMA 1. *Let $\mathfrak{C}$ be an arbitrary condition. There cannot both exist a* 0-*shift deterministic solution* $Y=\mathscr{A}(X)$ *of* $\mathfrak{C}(X, Y)$ *for* $Y$ *and a* 1-*shift deterministic solution* $X=\mathscr{B}(Y)$ *of* $\sim\mathfrak{C}(X, Y)$ *for* $X$.

**Proof.** Suppose $Y=\mathscr{A}(X)$ is a 0-shift solution of $\mathfrak{C}$ for $Y$, given by $Yt=\Phi(\overline{X}t)$, and $X=\mathscr{B}(Y)$ is a 1-shift solution of $\sim\mathfrak{C}$ for $X$, given by $Xt=\Psi(\overline{Y}(t-1))$. The system of equations $Yt=\Phi(\overline{X}t)$, $Xt=\Psi(\overline{Y}(t-1))$ can be viewed as a simultaneous course-of-value induction, defining a pair $X_0$, $Y_0$ satisfying both equations, for all values of $t$. But then $Y_0=\mathscr{A}(X_0)$ and $X_0=\mathscr{B}(Y_0)$. Therefore, if $\mathscr{A}$ solves $\mathfrak{C}$ for $Y$ and $\mathscr{B}$ solves $\sim\mathfrak{C}$ for $X$, we have $\mathfrak{C}(X_0, Y_0)$ and $\sim\mathfrak{C}(X_0, Y_0)$, which is contradictory. Q.E.D.

2. **Finite-state conditions, the $\omega$-behavior of finite automata.** Let $Z=\mathscr{E}(X, Y)$ be a FAO from $\omega$-sequences on $I\times J$ into $\omega$-sequences on $S$, given by the recursions,

$$Z0 = s_0, \qquad Zt' = H[Xt, Yt, Zt]. \tag{5}$$

Here $s_0$ is a member of $S$ and $H$ maps $I\times J\times S$ into $S$. Furthermore, let $U$ be a class of subsets of $S$, called the *output condition*. Let $\sup Z$ denote the set of all states taken infinitely often by $Z$. I.e.,

$$s \in \sup Z \,.\!\equiv\!.\, (\forall x)(\exists t)[x \leqq t \wedge Zt = s]. \tag{6}$$

DEFINITION 2. The *$\omega$-behavior of* $\langle S, s_0, H, U\rangle$ (of the FAO $\mathscr{E}$ with output condition $U$) is the relation $\mathfrak{C}(X, Y)$ which holds for $X$ and $Y$ if $Z=\mathscr{E}(X, Y)$ satisfies $\sup Z \in U$. I.e.,

$$\mathfrak{C}(X, Y) \,.\!\equiv\!.\, (\exists Z)[Z0 = s_0 \wedge (\forall t)Zt' = H[Xt, Yt, Zt] \wedge \sup Z \in U]. \tag{7}$$

By a *finite-state condition* we mean one which is the $\omega$-behavior of some FAO with output condition. Our basic result may be stated thus:

THEOREM 1. *Every finite-state condition* $\mathfrak{C}(X, Y)$ *is determined. Moreover, either there is a* 0-*shift FAO which solves* $\mathfrak{C}$ *for* $Y$ *or else there is a* 1-*shift FAO which solves* $\sim\mathfrak{C}$ *for* $X$.

The proof is contained in §5. Actually we obtain there a constructive version of Theorem 1. In §3 we discuss a game theoretic form of this theorem which was conjectured by McNaughton. The purpose of §4 is to show that a surprisingly wide class of formulas $\mathfrak{C}$ in fact define finite-state conditions. We thereby extend the applicability of Theorem 1. An important step in this extension is provided by a recent result of McNaughton [14], which can truly be called the fundamental lemma of finite automata behavior. It can be stated as follows.

In place of the initial state $s_0$ we assume a set of initial states $K\subseteq S$. In place of the function $H$ from $I\times J\times S$ into $S$ we consider a relation $L$ on $I\times J\times S\times S$. An

$\omega$-sequence $Z$ on $S$ is called a *transition sequence* of the *transition system* (sometimes called nondeterministic finite automaton) $\langle S, K, L\rangle$ if

(5′) $$K[Z0], \quad L[Xt, Yt, Zt, Zt'], \quad \text{for all } t.$$

The notion of $\omega$-behavior naturally generalizes to transition systems. Namely,

DEFINITION 2′. The *$\omega$-behavior of the transition system $\langle S, K, L, U\rangle$ with output condition* is the relation $\mathfrak{C}(X, Y)$ which holds for $X$ and $Y$ if there is a transition sequence $Z$ such that $\sup Z \in U$. I.e.,

(7′) $$\mathfrak{C}(X, Y) .\equiv. (\exists Z)[K[Z0] \wedge (\forall t)L[Xt, Yt, Zt, Zt'] \wedge \sup Z \in U].$$

The fundamental lemma now states that $\omega$-behaviors of transition systems are still finite state conditions. More precisely,

FUNDAMENTAL LEMMA (MCNAUGHTON). *To every transition system with output condition $\mathbf{L}=\langle S, K, L, U\rangle$ on the input states $I\times J$ one can effectively construct a finite automaton with output condition $\boldsymbol{H}=\langle S', s_0, H, W\rangle$ on $I\times J$, such that $\mathbf{L}$ and $\boldsymbol{H}$ have the same $\omega$-behavior.*

Thus Theorem 1 remains true if $\mathfrak{C}(X, Y)$ is the $\omega$-behavior of a transition system. A further extension is discussed in §4.

3. **$\omega$-games and sequential conditions.** McNaughton has observed a close relationship between the notion of a deterministic solution of a condition $\mathfrak{C}(X, Y)$ and that of a winning strategy in purely combinatorial $\omega$-games studied in the literature [7], [10], and [15]. While this game terminology is not really needed for our purpose, it puts both the solvability problems of automata theory and game theory into a wider context, and adds appealing flavors to each. For example, the notion of determinateness (Definition 1) is very natural in terms of games, but did not arrive independently in automata theory. Indeed we could have avoided all reference to solutions of $\sim\mathfrak{C}(X, Y)$ for $X$, in a presentation of our solvability algorithm. But this would clearly be hiding important information.

A condition $\mathfrak{C}(X, Y)$ can be viewed as a game for two, player $I$ and player $J$. Intuitively, a play of the game $\mathfrak{C}(X, Y)$ goes as follows. At any time $t=0, 1, 2, \ldots$ player $I$ makes a move $Xt$ by selecting a member of $I$. Then player $J$ follows up with a move $Yt$ from $J$. The play $\langle X, Y\rangle$ is completed when all $\omega$ moves $X0$, $Y0$, $X1$, $Y1, \ldots$ have been made. Player $J$ wins if $\mathfrak{C}(X, Y)$, else player $I$ wins. It is intended that at time $t$, player $I$ has complete information about all previous moves $\bar{Y}(t-1)$ of his opponent, and player $J$ has complete information about all previous moves $\bar{X}t$ of his opponent. More rigorously this can be stated thus:

DEFINITION 3. *A strategy for player $I$ (for player $J$)*, in a game $\mathfrak{C}(X, Y)$, is a deterministic 1-shift operator $X=\mathscr{B}(Y)$ (deterministic 0-shift operator $Y=\mathscr{A}(X)$). If $\langle\mathscr{B}, \mathscr{A}\rangle$ is a pair of such strategies, then the *play* $\langle\mathscr{B}, \mathscr{A}\rangle$ produces the pair $\langle X_0, Y_0\rangle$ such that $\mathscr{A}(X_0)= Y_0$ and $\mathscr{B}(Y_0)=X_0$. The strategy $\mathscr{A}$ (of player $J$) *beats* the strategy $\mathscr{B}$ (of player $I$) in case $\mathfrak{C}(X_0, Y_0)$. Otherwise $\mathscr{B}$ beats $\mathscr{A}$. *A winning strategy* for either player is one which beats all strategies of the opponent.

That the play $\langle \mathscr{B}, \mathscr{A} \rangle$ exists has been pointed out in the proof of Lemma 1. We leave it to the reader to verify

LEMMA 2. *An operator* $Y = \mathscr{A}(X)$ $(X = \mathscr{B}(Y))$ *is a winning strategy for player J (player I) in the game* $\mathfrak{C}(X, Y)$ *if and only if it is a deterministic* 0-*shift* (1-*shift*) *solution of the condition* $\mathfrak{C}$ *for* $Y$ ($\sim\mathfrak{C}$ *for* $X$).

Thus Lemma 1 asserts the intuitively obvious fact that in no game $\mathfrak{C}$ can both players possess a winning strategy. Furthermore, the condition $\mathfrak{C}(X, Y)$ is determined (Definition 1) just in case the game $\mathfrak{C}$ is determined in the sense that one of the two players possesses a winning strategy.

If $\mathfrak{C}(X, Y)$ is called a *finite-state game* in case it is the $\omega$-behavior of a finite automata operator of form (5) with an output $U$, then Theorem 1 takes the following form.

THEOREM 1′. *Every finite-state game is determined. Moreover, the player who has a winning strategy in fact has one which can be executed by a finite automaton.*

We leave it to the reader to make up a particular finite-state game and to meditate about the sense in which such a game can actually be played. We also suggest that he review the results of §4 in game terminology.

We would like to emphasize here that the second stronger part of Theorem 1′ is critical for our solvability algorithm (§4). This second part is also a new kind of result in game theory. More generally, the following type of game problem is naturally suggested by automata theory. Given a class of games $G$: (1) can one effectively decide, for any $\mathfrak{C} \in G$, which player has a winning strategy? (2) Just how simple winning strategies do exist for games in $G$? For example, is there a recursive or even a finite automata winning strategy for $\mathfrak{C} \in G$? This general problem was considered in [17].

We suggest that the arithmetic hierarchy [11] provides more natural choices of $G$ (in connection with the above questions), than does the classical Borel hierarchy considered in the literature [7], [10] and [15]. To state a more concrete question we ask:

PROBLEM. For any $\forall_3$-game is there a winning strategy in the arithmetic hierarchy of operators? If yes, how high do they occur in the hierarchy?

Here $\forall_3$ stands for the class of all $\mathfrak{C}(X, Y)$ which are of the form

$$(\forall x)(\exists y)(\forall z)B(X, Y, x, y, z),$$

whereby $B$ is recursive. Note that $\forall_3$ is contained in $F_{\sigma\delta}$ of the Borel hierarchy over the product of the natural Cantor spaces of $I$-sequences and $J$-sequences (since $B(X, Y, x, y, z)$ is open and closed for fixed $x, y$ and $z$). Hence $\forall_3$ games are determined as a consequence of the following result of Davis [7].

(*) All $F_{\sigma\delta}$ games are determined.

It is easy to show, using the axiom of choice, that there is a $\mathfrak{C}(X, Y)$ which is not determined [10]. However, it is not known whether all $F_{\sigma\delta\sigma}$ or even all $\forall_4$ games are determined.

For comparison with our stronger proof of the full Theorem 1′, we end this digression into game theory with a proof, using (*), that all finite state games are determined. In fact Theorem 2 below is somewhat stronger.

Call $\mathfrak{C}(X, Y)$ a *continuous*-sup-*condition* (*recursive*-sup-*condition*) if it is of the form, $\sup Z \in U$ if $Z = \mathscr{E}(X, Y)$, whereby $\mathscr{E}$ is a continuous operator (recursive operator). I.e.,

$$\text{(8)} \qquad \mathfrak{C}(X, Y) .\equiv. (\exists Z)[(\forall t)Zt = \Phi(\bar{X}(\phi t), \bar{Y}(\phi t)) \wedge \sup Z \in U]$$

where $\Phi$ and $\phi$ are arbitrary (recursive) functions. Note that $\omega$-behaviors (i.e., finite-state games) are recursive-sup.

LEMMA 3. *Every continuous*-sup-*condition* (*recursive*-sup-*condition*) $\mathfrak{C}$ *is in the Boolean algebra over* $F_\sigma$ (*over* $\exists_2$).

**Proof.** Assume $\mathfrak{C}$ is given by (8), but drop the second argument $Y$ to avoid notational complexity. Using the definition of sup (6) it follows that

$$\mathfrak{C}(X) .\equiv. \bigvee_{B \in U} \Big[(\exists y)(\forall t)[y \leqq t \supset \Phi(\bar{X}(\phi t)) \in B] \wedge \bigwedge_{s \in B} (\forall y)(\exists t)[y \leqq t \wedge \Phi(\bar{X}(\phi t)) = s]\Big].$$

Note that $U$ and its members $B$ are finite sets, so that $\mathfrak{C}(X)$ is a Boolean combination of expressions of the form $(\exists y)(\forall t)M(X, y, t)$. The expressions $M(X, y, t)$, namely $[y > t \vee \Phi(\bar{X}(\phi t)) \in B]$ or $[y > t \vee \Phi(\bar{X}(\phi t)) \neq s]$ (for various values of $B$ and $s$), denote clopen sets for fixed $y$ and $t$ (recursive relations in case $\Phi$ and $\phi$ are recursive). This is true because $M(X, y, t)$ implies $M(X^*, y, t)$ whenever $\bar{X}^*(\phi(t)) = \bar{X}(\phi(t))$. Consequently each $(\exists y)(\forall t)M(X, y, t)$ denotes an $F_\sigma$ (an $\exists_2$) so that $\mathfrak{C}$ is a Boolean combination of $F_\sigma$'s (of $\exists_2$'s). Q.E.D.

THEOREM 2. *Every continuous*-sup-*game* (*recursive*-sup-*game*) $\mathfrak{C}(X, Y)$ *is determined.*

The proof is obvious from (*) and Lemma 3. We have not investigated whether Davis' proof of (*) can be analyzed to yield further information in case one assumes $\mathfrak{C}(X, Y)$ to be recursive-sup (or even an $\omega$-behavior). At any rate, if $\mathfrak{C}(X, Y)$ is an $\omega$-behavior our Theorem 1′ strengthens Theorem 2.

It seems unlikely that there is a presentation for recursive-sup-conditions which admits a method for deciding which of the players has a winning strategy. Note that our Theorem 6 states the existence for sequential conditions.

PROBLEM. Is it true that for every recursive-sup-game either of the players has a winning strategy which is arithmetical? If yes, how high does it occur?

**4. A solvability-synthesis algorithm for sequential calculus.** Our concern here is not so much to determine solutions for particular conditions. We rather ask for algorithms which for a class CL of conditions determine solvability questions with respect to a class OP of operators. Such algorithms are discussed in the literature [2], [5], [6], [8], [9], [18] and [19]. We will restate some known results and show what our basic Theorem 1 provides.

Let CL be an interpreted formalism (called the condition language) containing formulas $\mathfrak{C}(X, Y)$ denoting relations between $\omega$-sequences. Let OP be a class of operators. A *solvability algorithm for* CL with respect to OP is an effective procedure which applies to any $\mathfrak{C}(X, Y) \in$ CL and tells whether or not $\mathfrak{C}$ admits a solution $\mathscr{A} \in$ OP for $Y$. In case the members of OP are finitely presentable (as is a FAO by a finite automaton and a RDO by a Turing machine computing $\Phi$), one may ask for a *partial synthesis algorithm* which for any $\mathfrak{C}(X, Y) \in$ CL constructs a presentation of a solution $\mathscr{A} \in$ OP, if a solution exists, and a *solution algorithm* which, given a $\mathfrak{C}(X, Y) \in$ CL and a presentation of some $\mathscr{A} \in$ OP, decides whether or not $\mathscr{A}$ solves $\mathfrak{C}$ for $Y$.

In [4] *sequential calculus* (SC) is considered as a natural candidate for a condition language for FAO. SC is the monadic second-order theory of the successor function $'$ on natural numbers. That is SC is the interpreted formalism which includes the first order theory of $\langle \omega, 0, ' \rangle$ and quantification over monadic predicate variables ranging over sets of natural numbers. Note that a subset $X$ of $\omega$ (i.e., predicate on $\omega$) may also be interpreted as an $\omega$-sequence of members of $\{T, F\}$, and a finite sequence $X = \langle X_1, \ldots, X_k \rangle$ is an $\omega$-sequence of members of the set $\{T, F\}^k$. Thus, a formula $\mathfrak{C}(X, Y)$ of SC with free predicate variables $X = \langle X_1, \ldots, X_h \rangle$ and $Y = \langle Y_1, \ldots, Y_k \rangle$ denotes a condition on $\omega$-sequences over $I = \{T, F\}^h$ and $J = \{T, F\}^k$. Other finite $I$ and $J$ can be handled by coding their members as sequences of the truth-values $T$, $F$.

In [4] a method for deciding truth of sentences in SC was presented. Let $Y = \mathscr{A}(X)$ be a FAO given by (3). By appropriate coding of the automaton $\langle K, H, L, W \rangle$ one can construct a formula $\mathfrak{F}(X, Y) \in$ SC of form,

$$\begin{aligned}(\exists Z_1 \cdots Z_n) \,.\, Z_1 0 \equiv H_1(0) \wedge \cdots \wedge Z_n 0 \equiv H_n(0) \\ \wedge (\forall t)[Z_1 t' \equiv L_1(t) \wedge \cdots \wedge Z_n t' \equiv L_n(t')] \\ \wedge (\forall t)[Y_1 t \equiv W_1(t) \wedge \cdots \wedge Y_k t \equiv W_k(t)]\end{aligned}$$

(whereby the $H$, $L$, $W$'s are propositional formulas in the atomic parts $X_1 0, \ldots, X_h 0$, $Y_1 0, \ldots, Y_k 0$, $Z_1 0, \ldots, Z_n 0$, $X_1 t', \ldots, X_h t'$, $Y_1 t', \ldots, Y_k t'$, $Z_1 t, \ldots, Z_n t$), such that $\mathfrak{F}(X, Y)$ means $Y = \mathscr{A}(X)$. The assertion, $\mathscr{A}$ solves $\mathfrak{C}(X, Y) \in$ SC for $Y$, where $\mathscr{A}$ is a FAO can therefore be stated as a sentence of SC. Hence,

THEOREM 3. *There is a solution algorithm and a partial synthesis algorithm for SC with respect to FAO.*

A partial synthesis algorithm is available because all finite automata can be effectively enumerated, and one after the other checked as to whether it solves a proposed condition $\mathfrak{C}(X, Y)$ stated in SC. For a very small fragment of SC a solvability algorithm was found [5] and improved by Wright. This result can be extended to cover conditions of SC of the form

$$(\text{predicate prefix on } Z) \,.\, H[Z0] \wedge (\forall t)L[Xt,\ Yt, Zt, Zt']$$

(unpublished). It is easy to see [4] that addition of an existential conjunct $(\exists t)M[Zt]$ to such formulas yields all conditions $\mathfrak{C}(X, Y)$ expressible in SC. However, even for very special formulas including both kinds of individual quantifiers the problem of finding a solvability algorithm was left open in [6], and seemed rather hopeless at that time. It is only by using McNaughton's fundamental lemma and our Theorem 1 that we are now able to give a solvability algorithm for all of SC.

The main definability result of [4] can be restated thus: To every formula $\mathfrak{C}(X, Y)$ of SC one can construct a transition system $\langle S, K, L, U\rangle$ with output condition whose $\omega$-behavior is (the relation defined by) $\mathfrak{C}$. Thus by the fundamental lemma

THEOREM 4. *To every formula $\mathfrak{C}(X, Y)$ of SC one can effectively construct a finite automaton with output $\langle S, s_0, H, U\rangle$ whose $\omega$-behavior is (the relation defined by) $\mathfrak{C}$.*

Conversely, the $\omega$-behavior of every finite automaton can be defined in SC. In fact (7) with sup $Z$ replaced by its definition (6), up to coding, yields such a definition (as $x \leqq y$ is definable in SC by $(\forall Z)[Zy \wedge (\forall t)[Zt' \supset Zt] \supset Zx]$). Also note that the fundamental lemma yields another proof of the critical Lemma 9 of [4] (as explained in [14]), which does not make use of Ramsey's Theorem.

Because of Theorem 4 we can extend Theorem 1 to

THEOREM 5. *Every condition $\mathfrak{C}(X, Y)$ definable in SC is determined. In fact, either there is a* 0*-shift FAO which solves $\mathfrak{C}$ for $Y$ or else there is a* 1*-shift FAO which solves $\sim\mathfrak{C}$ for $X$.*

Because of Lemma 1 we have

COROLLARY. *If $\mathfrak{C}(X, Y)$ in SC has a deterministic solution for $Y$ then it has a FAO solution for $Y$.*

The corollary generalizes the statement: If $\mathfrak{C}(Y)$ in SC holds for some $Y$, then it holds for an ultimately periodic $Y$. Just note that a FAO solution of $\mathfrak{C}(Y)$ for $Y$ is an input free automaton.

THEOREM 6. *There is an algorithm which for any $\mathfrak{C}(X, Y)$ of SC, decides whether $\mathfrak{C}(X, Y)$ is deterministically solvable for $Y$, and either* (1) *produces a* 0*-shift FAO solution of $\mathfrak{C}(X, Y)$ for $Y$ if $\mathfrak{C}$ is deterministically solvable for $Y$ or* (2) *produces a* 1*-shift FAO solution of $\sim\mathfrak{C}(X, Y)$ for $X$ if $\mathfrak{C}$ is not deterministically solvable for $Y$.*

**Proof.** ALGORITHM 1. Systematically list all 0-shift FAO $Y=\mathscr{A}(X)$ and all 1-shift FAO $X=\mathscr{B}(Y)$. Check whether $\mathscr{A}$ solves $\mathfrak{C}$ for $Y$ or $\mathscr{B}$ solves $\sim\mathfrak{C}$ for $X$ using the algorithm of Theorem 3. By Theorem 5, eventually a solution of $\mathfrak{C}$ for $Y$ or a solution of $\sim\mathfrak{C}$ for $X$ will be found.

ALGORITHM 2. Use the algorithm of Theorem 4 to put $\mathfrak{C}(X, Y)$ in finite state form. Then use the method described in §5.

Note that there is a solvability algorithm for SC with respect to DO, which is also a solvability algorithm for SC with respect to FAO (Theorem 6). However, while there is a solution algorithm with respect to FAO (Theorem 3), there is no solution algorithm for SC with respect to RDO. For example, let $\mathfrak{C}(X, Y)$ be $(\exists y)(\forall z)[y<z \supset \sim Yz]$. Let $Y=\mathscr{A}_Q(X)$ be the RDO defined by

$$\begin{aligned} Yt &= T \quad \text{if } t \in Q, \\ &= F \quad \text{if } t \notin Q, \end{aligned}$$

whereby $Q$ is a recursive set. $\mathscr{A}_Q$ solves $\mathfrak{C}$ for $Y$ if and only if $Q$ is finite. Hence a solution algorithm for SC with respect to RDO would, given an index for $Q$, decide whether it is finite. It is well known that such a method does not exist.

We have not seriously investigated whcther the algorithms of Theorems 3 and 6 can be improved to a point of usefulness in the deisgn of sequential circuits. As they include conversion of propositional formulas into normal form, it seems that presently available computing equipment could not carry a significant part of our algorithms. Nevertheless, our solution automata of §5, like the construction of [14], provide examples of strictly finite devices which accomplish surprisingly intricate tasks.

The fundamental lemma can be extended to $\alpha$-behaviors, for any countable ordinal $\alpha$. This leads to a decision method for the monadic second-order theory of $\langle \alpha, < \rangle$ (see [1]). We hope to present elsewhere a corresponding extension of Theorem 6 from $\omega$ to any countable ordinal $\alpha$.

$\mathfrak{C}(X, Y)$ admits an $h$-shift solution for $Y$, if and only if,

$$\mathfrak{C}_h(X, Y) : (\exists Z) \,.\, \mathfrak{C}(Z, Y) \wedge (\forall t)Zt = X(t+h)$$

has a 0-shift solution for $Y$. Thus, for any fixed $h$, Theorem 6 yields a solvability algorithm for SC with respect to $h$-shift DO's and FAO's. Note that any $(h+1)$-shift recursion is also an $h$-shift recursion. This suggests

PROBLEM. Can one algorithmically determine whether or not for a condition $\mathfrak{C}(X, Y)$ stated in SC there exists an $h$ such that $\mathfrak{C}$ admits an $h$-shift, but no $(h+1)$-shift solution for $Y$?

5. **Solving finite state conditions.** We will present here our main proof, that of Theorem 1. Therefore, throughout this section $\mathfrak{C}(X, Y)$ will be the $\omega$-behavior, with respect to $U$, of the FAO given by

$$Z0 = s_0, \qquad Zt' = H[Xt, Yt, Zt]. \tag{9}$$

Let $I$, $J$, $S$ be the finite sets of states of $X$, $Y$, $Z$ respectively, so that $U$ is a class of subsets of $S$. We recall that $\mathfrak{C}(X, Y)$ stands for sup $Z \in U$, whereby $Z$ is given by (9).

Our proof is outlined as follows. In section (a) we will construct a subset $R_l[\ \ ]$ of the set of states $S$, (the significance of the brackets will be made clear below) such that if $s_0 \in R_l[\ \ ]$ then $\mathfrak{C}(X, Y)$ is solvable for $Y$ by a 0-shift FAO, and if $s_0 \notin R_l[\ \ ]$ then $\sim\mathfrak{C}(X, Y)$ is solvable for $X$ by a 1-shift FAO. Thus, $s_0 \in R_l[\ \ ]$ is the condition of solvability of $\mathfrak{C}(X, Y)$ for $Y$. The case $s_0 \in R_l[\ \ ]$ is treated in section (b), where we will present a 0-shift FAO $Y = \mathscr{A}(X)$ which solves $\mathfrak{C}$ for $Y$. The case $s_0 \notin R_l[\ \ ]$ is treated in section (c), where a 1-shift FAO $X = \mathscr{B}(Y)$ is presented which solves $\sim\mathfrak{C}$ for $X$.

(a) *Definition of* $R_l[\ \ ]$. For each $A \in U$ choose a cyclic permutation of its members. For simplicity of notation we denote the value of this permutation at $s \in A$ by $A(s)$. The crucial construction is that of the sets $R_k[A_1, s_1, \ldots, A_n, s_n]$, $P_k[A_1, s_1, \ldots, A_n, s_n]$, and $Q_k[A_1, s_1, \ldots, A_n, s_n]$, whereby $n \geqq 0$, $A_1 \supset \cdots \supset A_n$ range over strictly decreasing chains in $U$ and $s_1, \ldots, s_n$ range over members of $A_1, \ldots, A_n$, respectively. Note that $B \subset A$ will always mean "$B$ is properly contained in $A$." These sets are defined simultaneously by the following induction on $k = 0, 1, 2, \ldots$.

$$s \in R_0[A_1, s_1, \ldots, A_n, s_n] .\equiv. \text{false},$$

$$s \in P_k[A_1, s_1, \ldots, A_n, s_n] .\equiv. s \in R_k[\ \ ] \vee s \in A_1 \cap R_k[A_1, s_1] \vee \cdots \vee s \in A_n \cap R_k[A_1, s_1, \ldots, A_n, s_n],$$

$$(10) \quad s \in Q_k[A_1, s_1, \ldots, A_n, s_n] .\equiv. \bigvee_B . B \in U \wedge s \in B \subset A_n \wedge \bigwedge_{u \in B} u \in R_k[A_1, s_1, \ldots, A_n, s_n, B, B(u)],$$

$$s \in R_{k+1}[A_1, s_1, \ldots, A_n, s_n] .\equiv. \bigwedge_{x \in I} \bigvee_{y \in J} H[x, y, s] \in \{s_1, \ldots, s_n\} \cup P_k[A_1, s_1, \ldots, A_n, s_n] \cup Q_k[A_1, s_1, \ldots, A_n, s_n].$$

Note that $n$ is bounded by the length of maximal chains in $U$. If $n = 0$ we use the notations $R_k[\ \ ]$, $P_k[\ \ ]$, $Q_k[\ \ ]$.

*Caution*! In interpreting (10) for the case $n = 0$ the occurrence of $A_n$ (in the expression $s \in B \subset A_n$) is to be suppressed. A similar remark goes for all future occurrences of $A_0$. Also, the set $\{s_1, \ldots, s_n\}$ is the empty set if $n = 0$.

By induction on $k$, one easily shows that $R_k[v] \subseteq R_{k+1}[v]$, $P_k[v] \subseteq P_{k+1}[v]$, $Q_k[v] \subseteq Q_{k+1}[v]$, for all arguments $v = [A_1, s_1, \ldots, A_n, s_n]$. We include the induction step for the $R$ sets and leave it to the reader to complete the proof.

Assume $R_k[v] \subseteq R_{k+1}[v]$, $P_k[v] \subseteq P_{k+1}[v]$ and $Q_k[v] \subseteq Q_{k+1}[v]$ for all $v$. Let $v = [A_1, s_1, \ldots, A_n, s_n]$. Then

$$s \in R_{k+1}[v] .\equiv. \bigwedge_{x \in I} \bigvee_{y \in J} H[x, y, s] \in \{s_1, \ldots, s_n\} \cup P_k[v] \cup Q_k[v]$$

$$\text{(Ind. hyp.)} \quad .\supset. \bigwedge_{x \in I} \bigvee_{y \in J} H[x, y, s] \in \{s_1, \ldots, s_n\} \cup P_{k+1}[v] \cup Q_{k+1}[v]$$

$$.\supset. s \in R_{k+2}[v].$$

Because all $R_k[v]$, $P_k[v]$ and $Q_k[v]$ are subsets of the finite set $S$, and there are but a finite number of $v$'s, it follows that there is a number $k$ such that, for all $v$, $R_k[v]=R_{k+1}[v]$, $P_k[v]=P_{k+1}[v]$, and $Q_k[v]=Q_{k+1}[v]$. Accordingly we define $l$

(11) $l$ is the first number such that, $R_{l-1}[v]=R_l[v]$, $P_{l-1}[v]=P_l[v]$, and $Q_{l-1}[v]=Q_l[v]$, for all $v=[A_1, s_1, \ldots, A_n, s_n]$, $A_1 \supset \cdots \supset A_n$, $n \geqq 0$.

From (10) and (11) we obtain

$$s \notin R_l[A_1, s_1, \ldots, A_n, s_n] . \equiv . \bigvee_{x \in I} \bigwedge_{y \in J} H[x, y, s] \notin \{s_1, \ldots, s_n\} \cup P_l[A_1, s_1, \ldots, A_n, s_n] \cup Q_l[A_1, s_1, \ldots, A_n, s_n],$$

$$(12) \quad s \notin Q_l[A_1, s_1, \ldots, A_n, s_n] . \equiv . \bigwedge_B [B \in U \wedge s \in B \subset A_n] \supset \bigvee_{u \in B} u \notin R_l[A_1, s_1, \ldots, A_n, s_n, B, B(u)],$$

$$s \notin P_l[A_1, s_1, \ldots, A_n, s_n] . \equiv . s \notin R_l[\;\;] \wedge s \notin A_1 \cap R_1[A_1, s_1] \wedge \cdots \wedge s \notin A_n \cap R_l[A_1, s_1, \ldots, A_n, s_n].$$

(b) *The case* $s_0 \in R_l[\;\;]$. Choose a linear order of the members of $J$ and $U$. An expression $(\mu y)E(y)$ denotes the first member $y$ of $J$, in the chosen order, which satisfies $E(y)$, if it exists. We will now display a 0-shift FAO $Y=\mathscr{A}(X)$, and prove that it solves $\mathfrak{C}(X, Y)$ for $Y$.

In the sequel $X$, $Y$, $Z$, $k$ denote $\omega$-sequences over the sets $I$, $J$, $S$, $\{0, \ldots, l\}$, respectively. $V$ denotes $\omega$-sequences of elements of form $[A_1, s_1, h_1, \ldots, A_n, s_n, h_n]$, whereby $n \geqq 0$, $A_1 \supset \cdots \supset A_n$ is a chain of members of $U$, $s_1 \in A_1, \ldots, s_n \in A_n$ and $l > h_1 > \cdots > h_n$. Consider the following formulas (13), which define $Y$, $Z$, $V$ and $k$ recursively from $X$,

$$Z0 = s_0, \qquad V0 = [\;\;], \qquad k0 = l.$$

Assume now that $Vt=[A_1, s_1, h_1, \ldots, A_n, s_n, h_n]$. Then,

$$Yt = (\mu y) . H[Xt, y, Zt] \in \{s_1, \ldots, s_n\} \cup P_{kt-1}[A_1, s_1, \ldots, A_n, s_n] \cup Q_{kt-1}[A_1, s_1, \ldots, A_n, s_n],$$

$$Zt' = H[Xt, Yt, Zt].$$

($\alpha$) If $Zt' \in \{s_1, \ldots, s_n\}$, let $i$ be the first such that $Zt' = s_i$. Then,
$Vt' = [A_1, s_1, h_1, \ldots, A_i, A_i(s_i), h_i]$
$kt' = h_1$.

(13) ($\beta$) If $Zt' \in P_{kt-1}[A_1, s_1, \ldots, A_n, s_n]$ but not ($\alpha$), let $j$ be the first such that $Zt' \in A_j \cap R_{kt-1}[A_1, s_1, \ldots, A_j, s_j]$ ($Zt' \in R_{kt-1}[\;\;]$ if $j = 0$). Then,
$Vt' = [A_1, s_1, h_1, \ldots, A_j, s_j, h_j]$
$kt' = kt-1$.

($\gamma$) If $Zt' \in Q_{kt-1}[A_1, s_1, \ldots, A_n, s_n]$ but neither ($\alpha$) nor ($\beta$), let $B$ be the first in the chosen order of $U$ such that, $Zt' \in B \subset A_n$ and $\bigwedge_{u \in B} u \in R_{kt-1}[A_1, s_1, \ldots, A_n, s_n, B, B(u)]$. Then,
$Vt' = [A_1, s_1, h_1, \ldots, A_n, s_n, h_n, B, B(Zt'), kt-1]$
$kt' = kt-1$.

Note also the formulas (14).

$$\text{If } Vt=[A_1, s_1, h_1, \ldots, A_n, s_n, h_n], \text{ then}$$

$$Zt \in R_{kt}[A_1, s_1, \ldots, A_n, s_n], \quad Zt \in A_n \qquad (\text{if } n \neq 0),$$

$$(14) \qquad A_1 \supset \cdots \supset A_n, \quad s_i \in A_i \in U, \quad l > h_1 > \cdots > h_n \geqq kt > 0,$$

$$\bigwedge_{u \in A_i} u \in R_{h_i}[A_1, s_1, \ldots, A_i, A_i(u)].$$

Because we are dealing with the case $s_0 \in R_l[\ \ ]$, the values $Z0$, $V0$, $k0$ given by (13) clearly satisfy (14). Assume inductively that (14) holds for $t$, and $Xt$ is arbitrary. Using (10) it follows that there is a $y \in J$ such that

$$H[Xt, y, Zt] \in \{s_1, \ldots, s_n\} \cup P_{kt-1}[A_1, s_1, \ldots, A_n, s_n] \cup Q_{kt-1}[A_1, s_1, \ldots, A_n, s_n],$$

and therefore $Yt$ exists as described by (13), and so do $Zt'$, $Vt'$, $kt'$, in all cases ($\alpha$), ($\beta$), ($\gamma$). Furthermore, one easily checks that these values $Zt'$, $Vt'$, $kt'$ satisfy (14) with $t$ replaced by $t'$. Thus, the formulas (13) constitute a recursive definition of $Y$, $Z$, $V$, $k$ from $X$, and furthermore, (13) implies (14).

Let $Y=\mathscr{A}(X)$ be the operator from $I$-sequences to $J$-sequences, given by (13). Then $\mathscr{A}$ clearly is deterministic and recursive. Furthermore, because of (14), the auxiliaries $Z$, $V$, $k$ in the recursion (13) are finite valued. In fact it is easy to modify (13) so that it is of form (3). Therefore, $\mathscr{A}$ is a 0-shift FAO. It remains to show that $\mathscr{A}$ solves $\mathfrak{C}$ for $Y$, i.e., that (1) holds.

Note that a copy of (9) is built into the definition (13) of $\mathscr{A}$. As a consequence the assertion (1) is tantamount to the assertion: For any $X$, $Y$, $Z$, $V$, $k$, (13) implies $\sup Z \in U$. The remainder of section (b) constitutes a proof of this.

Assume that (13), and therefore (14), holds for $X$, $Y$, $Z$, $V$, $k$. From (13) one easily sees that $Vt_1=[\ \ ]$ and $t_1<t_2$ implies $kt_1>kt_2$. (Prove by induction on $t$ that if $V_t=[A_1, s_1, h_1, \ldots, A_n, s_n, h_n]$ for $t>t_1$, then (a) $kt<kt_1$ and (b) $h_i<kt_1$ for $i=1, \ldots, n$. This is obviously true for $t_1+1$. Assume it is true for $t$ and observe that ($\alpha$), ($\beta$) and ($\gamma$) of (13) preserve (a) and (b).) Therefore by (14), there can be but finitely many $t$ such that $Vt=[\ \ ]$. Accordingly there is a $t_1$ such that, $Vt\neq[\ \ ]$ for all $t\geqq t_1$, i.e., if $t\geqq t_1$ then $Vt$ is of form $[A_1, s_1, h_1, \ldots, A_n, s_n, h_n]$ with level $n\geqq 1$.

As the level $n$ of $Vt$ is bounded by the lengths of chains in $U$ (see (14)), some level $n\geqq 1$ must occur infinitely often. Let $m$ be the smallest of these. Then, $m\geqq 1$ and there is a $t_2$ such that for all $t\geqq t_2$ the level of $Vt$ is $\geqq m$. Thus we have,

(15) If $t\geqq t_2$ then $Vt=[A_1, s_1, k_1, \ldots, A_m, s_m, k_m, \ldots, A_n, s_n, k_n]$ whereby $n\geqq m$. Furthermore, $n=m$ occurs for infinitely many times $t$.

It follows from (15) that for $t'>t_2$ only the cases ($\alpha$) $i\geqq m$, ($\beta$) $j\geqq m$, and ($\gamma$) $n\geqq m$ of (13) can occur. Consequently, for $t\geqq t_2$ the entry $A_m$ (and all previous entries) in $Vt$ remains constant. By (14) it follows that $Zt \in A_m$ for $t\geqq t_2$, $A_m \in U$. Thus $\sup Z \subseteq A_m \in U$.

Suppose the case ($\alpha$) $i=m$ occurred for only finitely many $t$. Then there would be a $t_3 > t_2$ such that for $t \geqq t_3$ only the cases ($\alpha$) $i > m$, ($\beta$) $j \geqq m$, ($\gamma$) $n \geqq m$ could occur. Inspection of (13, 14) shows that then also the case ($\beta$) $j = m$ could occur only for finitely many $t \geqq t_3$ because each application of ($\beta$) $j = m$ followed by 0 or more steps ($\gamma$) $n \geqq m$, ($\alpha$) $i > m$, ($\beta$) $j > m$ followed by ($\beta$) $j = m$ lowers the value of $k$. (This is shown as follows: Assume ($\beta$) $j = m$ is used to obtain $Vt$ and $V(t+c)$, $c \geqq 1$ but no $V(t+\bar{c})$ for $\bar{c} < c$. Prove by induction on $\bar{c} \leqq c$ that if $V(t+\bar{c}) = [A_1, s_1, h_1, \ldots, A_m, s_m, h_m, A_{m+1}, \ldots, h_{m+n}]$, then (a) $h_{m+j} < kt$, $j = 1, \ldots, n$ and (b) $k(t+\bar{c}) < kt$.) Thus both cases ($\alpha$) $i = m$ and ($\beta$) $j = m$ would occur only finitely often. This contradicts the second part of (15). Therefore the case ($\alpha$) $i = m$ must occur infinitely often.

Let $t_3 < t_4 < t_5 < \cdots$ be the infinitely many consecutive places $t > t_2$ where ($\alpha$) $i = m$ is used. It clearly follows that $Zt_4 = A_m(Zt_3)$, $Zt_5 = A_m(Zt_4)$, $Zt_6 = A_m(Zt_5), \ldots$. Because $s \rightarrow A_m(s)$ was chosen to be a cyclic permutation of $A_m$, it follows that $Z$ will keep taking any value in $A_m$. Thus, $\sup Z \supseteq A_m$. Together with a former result, this yields $\sup Z = A_m \in U$. Q.E.D.

(c) *The case* $s_0 \notin R_l[\ ]$. Choose a linear order of $I$. The expression $(\mu x)E(x)$ denotes the first $x$ in $I$ such that $E(x)$, if it exists. We will now display a 1-shift FAO $X = \mathscr{B}(Y)$, and prove that it solves $\sim\mathfrak{C}(X, Y)$ for $X$.

In the sequel $X, Y, Z$ denote $\omega$-sequences over the sets $I, J, S$. $V$ denotes $\omega$-sequences over elements of form $[A_1, s_1, \ldots, A_n, s_n]$, whereby $A_1 \supset \cdots \supset A_n$ is a chain of members of $U$ and $s_1 \in A_1, \ldots, s_n \in A_n$. $W$ denotes $\omega$-sequences of chains of subsets of $S$. Consider the following formulas (16), which define $V, W, X$, and $Z$ recursively from $Y$.

$$Z0 = s_0, \qquad W0 = \{\{s_0\}\}, \qquad V0 = [\ ].$$

Assume that $Vt = [A_1, s_1, \ldots, A_n, s_n]$. Then,

$$Xt = (\mu x) \bigwedge_{y \in J} H[x, y, Zt] \notin \{s_1, \ldots, s_n\} U P_l[A_1, s_1, \ldots, A_n, s_n] \cup Q_l[A_1, s_1, \ldots, A_n, s_n].$$

$$Zt' = H[Xt, Yt, Zt].$$

$$Wt' = \{B \cup \{Zt'\};\ B \in Wt \vee B \text{ empty}\}.$$

(16) ($\alpha$) If $\bigvee_B .\ B \in U \cap Wt' \wedge Zt' \in B \wedge \bigvee_{0 \leqq i \leqq n} [[(0 < i < n \wedge A_i \supset B \supset A_{i+1})] \vee (i = n \wedge B \subset A_n) \vee (i = 0 \wedge B \supset A_1)] \wedge Zt' \notin R_l [A_1, s_1, \ldots, A_i, s_i, B, B(Zt')]$, let $B$ be the largest such. (Note that if $n = 0$, then the above reduces to $\bigvee_B B \in U \cap Wt' \wedge Zt' \in B \wedge Zt' \notin R_l[B, B(Zt')]$.)

$$Vt' = [A_1, s_1, \ldots, A_i, s_i, B, B(Zt')].$$

($\beta$) If not ($\alpha$), let $i$ be such that $Zt' \in A_i$, $Zt' \notin A_{i+1}$ ($A_{n+1}$ empty).

$$Vt' = [A_1, s_1, \ldots, A_i, s_i].$$

Note: $Vt' = [\ ]$, if $n = 0$.

Note also the formulas (17).

$$\text{If } Vt = [A_1, s_1, \ldots, A_n, s_n], \text{ then}$$

(17) $$\begin{aligned} &Zt \notin R_l[A_1, s_1, \ldots, A_n, s_n], \quad A_1 \supset \cdots \supset A_n \text{ all in } U \cap Wt, \\ &Zt \in A_n \text{ (if } n \neq 0), \quad s_1 \in A_1, \ldots, s_n \in A_n, \\ &Wt \text{ is a chain of subsets of } S. \end{aligned}$$

Because we are dealing with the case $s_0 \notin R_l[\ \ ]$, the values $Z0$, $P0$, $V0$ given by (16) satisfy (17) for $t=0$. Assume inductively that (17) holds for $t$ and $Yt$ is any member of $J$. By (12) it follows that $Xt$, $Zt'$, $Wt'$ as prescribed by (16) exists. Furthermore, $Wt'$ is still a chain of subsets of $S$, and $Zt' \notin P_l[A_1, s_1, \ldots, A_n, s_n]$. By (12) it therefore follows that, in cases ($\beta$), (17) holds for $t$ replaced by $t'$. The same can easily be checked in case $Vt'$ is calculated by ($\alpha$). The preceding argument shows that (16) constitutes a recursive definition of $Z$, $W$, $V$, $X$ from $Y$, and that (16) implies (17).

Let $X=\mathscr{B}(Y)$ be the operator from $I$-sequences to $J$-sequences, given by (16). Then $\mathscr{B}$ clearly is a 1-shift deterministic operator. Furthermore, the auxiliaries $Z$, $W$, $V$ take values in finite sets (see (17)). In fact, the recursion (16) is easily modified to the form (3'). Thus, $X=\mathscr{B}(Y)$ is a 1-shift FAO. To terminate the proof of Theorem 1, it remains to be shown that $\mathscr{B}$ solves $\sim\mathfrak{C}(X, Y)$ for $X$, i.e., that $X=\mathscr{B}(Y)$ and (9) imply $\sup Z \notin U$.

Note that the recursion (9) is built into the definition (16) of $X=\mathscr{B}(Y)$. As a consequence, "$\mathscr{B}$ solves $\sim\mathfrak{C}(X, Y)$ for $X$" is tantamount to the assertion: (16) implies $\sup Z \notin U$. The remainder of this section constitutes a proof of this, in the form: (16) and $\sup Z \in U$ yields a contradiction.

For the sequel assume that (16), and therefore (17), holds for $X$, $Y$, $Z$, $W$, $V$. Furthermore, assume $\sup Z = D \in U$. It follows that there is a $t_1$ such that

(18) $$\begin{aligned} &t \geqq t_1 \supset Zt \in D, \\ &u \in D \supset (\exists t)[t \geqq a \wedge Zt = u], \text{ for any time } a. \end{aligned}$$

From (16) one clearly sees that the chain $Wt$ consists of all sets $\{Z0, \ldots, Zt\}$, $\{Z1, \ldots, Zt\}, \ldots, \{Z(t-1), Zt\}, \{Zt\}$. It follows from (18), and $D \in U$ that there is a time $t_2 \geqq t_1$, such that

(19) $$t \geqq t_2 \supset D \in U \cap Wt.$$

Let $Vt=[A_1, s_1, \ldots, A_n, s_n]$. From (16), $Zt' \notin Q_l[A_1, s_1, \ldots, A_n, s_n]$. Because of (12) and $B \in Wt' \supset Zt' \in B$ this yields

$$[B \in U \cap Wt' \wedge (n = 0 \vee A_n \supset B)] \supset \bigvee_{u \in B} u \notin R_l[A_1, s_1, \ldots, A_n, s_n, B, B(u)].$$

This and (19) yield,

$$[t \geqq t_2 \wedge (n = 0 \vee A_n \supset D)] \supset \bigvee_{u \in D} u \notin R_l[A_1, s_1, \ldots, A_n, s_n, D, D(u)].$$

Because of (18) this yields,

(20) $$\begin{aligned} &[t \geqq t_2 \wedge Vt = [A_1, s_1, \ldots, A_n, s_n] \wedge (n = 0 \vee A_n \supset D)] \\ &\qquad \supset (\exists a)[a \geqq t \wedge Za' \notin R_l[A_1, s_1, \ldots, A_n, s_n, D, D(Za')]]. \end{aligned}$$

Define the quasi-order $\prec$ on chains $A_1 \supset \cdots \supset A_p$ ($p \geqq 0$) of members of $U$ by:

$$[B_1, \ldots, B_q] \prec [A_1, \ldots, A_p] . \equiv . \bigvee_{1 \leqq i \leqq q,p} \Big[ \bigwedge_{1 \leqq j \leqq i} A_j = B_j \wedge A_i \supset B_i \Big] \vee \Big[ \bigwedge_{1 \leqq j \leqq q} A_j = B_j \wedge p > q \Big].$$

By the *principal part* of the chain $A_1 \supset \cdots \supset A_n$ (of $[A_1, s_1, \ldots, A_n, s_n]$) we mean the chain $A_1 \supset \cdots \supset A_p$ (the sequence $[A_1, s_1, \ldots, A_p, s_p]$) whereby $p$ is the largest $i$ such that $A_i \supseteq D$, or $p=0$ if there is no such $i$. Note that $A_p \supseteq D \supset A_{p+1}$ if $1 \leqq p < n \neq 0$, $D \subseteq A_p$ if $p=n$, and $p=0$ if $n=0$.

Let $Vt=[A_1, s_1, \ldots, A_n, s_n]$, let $[A_1, s_1, \ldots, A_p, s_p]$ be the principal part of $Vt$. Inspection of (16) shows that the principal part of $Vt'$ will be equal or larger (in the sense of $\prec$), except if $(\beta)_{i<p}$ comes to use. If $t \geqq t_2$, so that by (18), $Zt' \in D \subseteq A_p$, $(\beta)_{i<p}$ cannot come to use. Therefore, for $t \geqq t_2$ the principal part of $Vt$ either stays equal or increases. Because $\prec$ is a quasi-order on a finite set, there must be a $t_3 \geqq t_2$ such that the principal part of $Vt$ remains constant from $t_3$ on. I.e., there are $m \geqq 0$, $\bar{A}_1, \bar{s}_1, \ldots, \bar{A}_m, \bar{s}_m$ such that $m=0$ or $\bar{A}_m \supseteq D$ and,

(21) if $t \geqq t_3$ then $Vt$ is of form $[\bar{A}_1, \bar{s}_1, \ldots, \bar{A}_m, \bar{s}_m, \ldots, A_n, s_n]$ whereby $n=m$ or $D \supset A_{m+1} \supset \cdots \supset A_n$.

Assume that, for all $t \geqq t_3$, $Vt$ were of form $[\bar{A}_1, \bar{s}_1, \ldots, \bar{A}_m, \bar{s}_m, A_{m+1}(t), s_{m+1}(t), \ldots]$. By (16), it follows that $A_{m+1}(t') \supseteq A_{m+1}(t)$, for all $t \geqq t_3$. Thus, there would have to be a $c \geqq t_3$ such that $A_{m+1}(t)$ remained constant, say $=A$, from $c$ on. By (21) $D \supset A$, so that by (18) there would exist a $d \geqq c$, $Zd' \notin A$. But $Vd= [\bar{A}_1, \bar{s}_1, \ldots, \bar{A}_m, \bar{s}_m, A, \ldots]$, so that the case $(\beta)_{i=m}$ of (16) would come to play. As a result, $Vd'=[\bar{A}_1, \bar{s}_1, \ldots, \bar{A}_m, \bar{s}_m]$. This is contradictory to the assumption, so that there must be a $t_4 \geqq t_3$ such that,

$$(22) \qquad Vt_4 = [\bar{A}_1, \bar{s}_1, \ldots, \bar{A}_m, \bar{s}_m].$$

Assume $\bar{A}_m \supset D$, or $m=0$. By (20) and (22) there would be an $a \geqq t_4$ such that $Za' \notin R_l[\bar{A}_1, \bar{s}_1, \ldots, \bar{A}_m, \bar{s}_m, D, D(Za')]$. By (21), $Va=[\bar{A}_1, s_1, \ldots, \bar{A}_m, \bar{s}_m]$, or $Va=[\bar{A}_1, s_1, \ldots, \bar{A}_m, \bar{s}_m, A, \ldots]$ and $D \supset A$. By (19), $D \in U \cap Wa'$. Thus, $D$ is a possible value for $B$ in $(\alpha)$ of (16). Therefore, $(\alpha)$ for some $B \supseteq D$ would come to use for calculating $Va'$. The result would be an entry $B \notin \{\bar{A}_1, \ldots, \bar{A}_m\}$, $B \supseteq D$ in $Va'$. This contradicts (21). Therefore,

$$(23) \qquad m \geqq 1 \wedge \bar{A}_m = D.$$

From (16) it clearly follows that $Zt' \neq \bar{s}_m$ if $Vt'=[\bar{A}_1, \bar{s}_1, \ldots, \bar{A}_m, \bar{s}_m, \ldots]$. Therefore, by (21) and (17), $Zt' \in \bar{A}_m$ and $Zt' \neq \bar{s}_m \in \bar{A}_m$ for any $t \geqq t_3$. It follows that $\bar{A}_m$ can not be $\sup Z$, i.e., $\bar{A}_m \neq D$. Together with (23) this yields the contradiction, ending the proof of Theorem 1.

The reader will easily find various modifications of our recursions (13) which also define DO's which solve $\mathfrak{C}(X, Y)$ for $Y$ in case $s_0 \in R_l[\ \ ]$. For example, such a recursion may keep $Zt' \in R_l[\ \ ]$ up to a fixed time $h$ and from there on act like

(13). More generally, the time $h$, at which the forcing of $\sup Z$ into $U$ is actually started, may be made to depend on the input $X$, and may be set and reset, deterministically depending on $X$. We have not investigated the following question.

PROBLEM. Modify the recursions (13) to a schema with parameters which, by proper additional specifications for the parameters, will yield any given deterministic operator which solves $\mathfrak{C}(X, Y)$ for $Y$. Do the same for (16) and solutions of $\mathfrak{C}(X, Y)$ for $X$.

To accomplish this it might be necessary to make (more basic) changes in the definition (10) of the sets $R_k[A_1, s_1, \ldots, A_n, s_n]$, which would make our proofs less intricate.

## BIBLIOGRAPHY


1. J. R. Buchi, *Decision methods in the theory of ordinals*, Bull. Amer. Math. Soc. **71** (1965), 767–770.

2. J. R. Buchi, C. E. Elgot and J. B. Wright, *The nonexistence of certain algorithms of finite automata theory*, Notices Amer. Math. Soc. **5** (1958), 98.

3. J. R. Buchi, *Weak second-order arithmetic and finite automata*, Z. Math. Logik Grundlagen Math. **6** (1960), 66–92.

4. ———, *On a decision method in restricted second order arithmetic*, Proc. Internat. Congr. Logic, Method. and Philos. Sci. 1960, Stanford Univ. Press, Stanford, Calif., 1962.

5. A. Church, *Application of recursive arithmetic to the problem of circuit synthesis*, Summaries of Talks Presented at the Summer Institute for Symbolic Logic, Cornell Univ. 1957, 2nd ed; Princeton, N. J., 1960, 3–50.

6. ———, *Logic arithmetic and automata*, Proc. Internat. Congr. Math. 1963, Almqvist and Wiksells, Uppsala, 1963.

7. M. Davis, "Infinite games of perfect information" in *Advances in game theory*, Princeton Univ. Press, Princeton, N. J., 1964, 85–101.

8. C. C. Elgot, *Decision problems of finite automata design and related arithmetics*, Trans. Amer. Math. Soc. **98** (1961), 21–51.

9. J. Friedman, *Some results in Church's restricted recursive arithmetic*, J. Symbolic Logic **22** (1957), 337–342.

10. D. Gale and F. M. Stewart, "Infinite games with perfect information" in *Contributions to the theory of games*, Vol. II, Princeton Univ. Press, Princeton, N. J., 1953, 245–266.

11. S. C. Kleene, *Arithmetical predicates and function quantifiers*, Trans. Amer. Math. Soc. **79** (1955), 312–340.

12. ———, "Representation of events in nerve nets and finite automata" in *Automata studies*, Princeton Univ. Press, Princeton, N. J., 1956, 3–41.

13. L. H. Landweber, *Finite state games—A solvability algorithm for restricted second-order arithmetic*, Notices Amer. Math. Soc. **14** (1967), 129–130.

14. R. McNaughton, *Testing and generating infinite sequences by a finite automaton*, Information and Control **9** (1966), 521–530.

15. J. Mycielski, S. Swierczkowski and A. Zieba, *On infinite positional games*, Bull. Acad. Polon. Sci. Cl. III **4** (1956), 485–488.

16. M. Rabin and D. Scott, *Finite automata and their decision problems*, IBM J. Res. Develop. **3** (1959), 114–125.

17. M. Rabin, "Effective computability of winning strategies" in *Contributions to the theory of games*, Vol. III, Princeton Univ. Press, Princeton, N. J., 1957, 147–157.

18. B. A. Trachtenbrot, *Synthesis of logic networks whose operators are described by means of single place predicate calculus*, Dokl. Akad. Nauk SSSR **118** (1958), 646–649.

19. ———, *Finite automata and the logic of single place predicates*, Dokl. Akad. Nauk SSSR **140** (1961), 320–323 = Soviet Math. Dokl. **2** (1961), 623–626.

20. J. R. Buchi and L. G. Landweber, *Definability in the monadic second-order theory of successor*, J. Symbolic Logic **34** (1969).

Purdue University,
Lafayette, Indiana
University of Wisconsin,
Madison, Wisconsin

Reprinted from Transactions of the American Mathematical Society, Vol. 138 (1969).

# ALGORITHMISCHES KONSTRUIEREN VON AUTOMATEN UND DIE HERSTELLUNG VON GEWINNSTRATEGIEN NACH CANTOR-BENDIXSON

von J. Richard Büchi in Lafayette, Indiana

Probleme der mechanischen Konstruktion von Schaltwerken mit Selbststeuerung wurden schon in den fünfziger Jahren von Church 5,6 und Buchi, Elgot, Wright 2 aufgeworfen. Diese Fragestellung hat sich, in verschiedener Hinsicht, als fruchtbar herausgestellt. Erstens liegen heute sehr starke Algorithmen für die Konstruktion von Automaten vor (Buchi 1, Buchi und Landweber 3). Zweitens wurden diese Methoden der Automatentheorie von Rabin 10 übernommen und haben zu Entscheidungsverfahren für erstaunlich reiche Sprachen geführt.

Wie R. McNaughton bemerkt hat ist das Lösen einer Konstruktionsbedingung durch einen endlichen Automaten identisch mit dem Aufstellen einer Gewinnstrategie in einem gewissen unendlichen Spiel. Dieses Problem wiederum ist verwandt mit der klassischen Frage nach der Existenz einer perfekten Teilmenge in einer vorgegebenen Menge und Cantors Ansätzen zur Lösung des Kontinuumproblems. Diese Zusammenhänge zwischen Automatentheorie, Spieltheorie und Cantorscher Mengenlehre sollen hier dargestellt werden.

1. Sequentialoperatoren: Die Figur zeigt eine Maschine, die eine Sequenz $X=X_0X_1X_2.....$ von Eingabedaten in eine Sequenz $Y=Y_0Y_1Y_2.....$ von Ausgabedaten verarbeitet. Die Eingabezustände werden einer endlichen Menge I entnommen, die Ausgabezustände gehören der

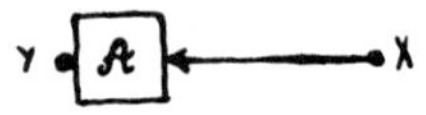


endlichen Menge J an. Der Operator $Y=\mathcal{A}(X)$ heisst: das Verhalten der Maschine und ist von der inneren Struktur der Maschine zu unterscheiden.

Es genügt den Fall I=J=2 zu betrachten, wobei 2 die Menge $\{F,T\}$ der Wahrheitswerte F=0=false und T=1=true ist. Die Menge $\omega=\{0,1,2,.....\}$ ist die Menge der Zeitpunkte (diskrete Zeit, digitale Maschine). Die Eingabe- und Ausgabesequenzen gehören dann der Menge 2 an, und können auch einfach als Teilmengen von $\omega$ aufgefasst werden. 2* ist die Menge aller Wörter $x, y, z, .....$ über dem Alphabet 2, und enthält auch das leere Wort e. Wenn $X\in 2^{\omega}$, $t\in\omega$ dann soll $\bar{X}_t$ das Wort $X_0...X_{t-1}$ bedeuten. Die nebenstehende Figur zeigt wie man die Wörter aus 2* als Eckpunkte des, und die Sequenzen aus $2^{\omega}$ als Wege durch den zweiverzweigten Baum auffassen kann.

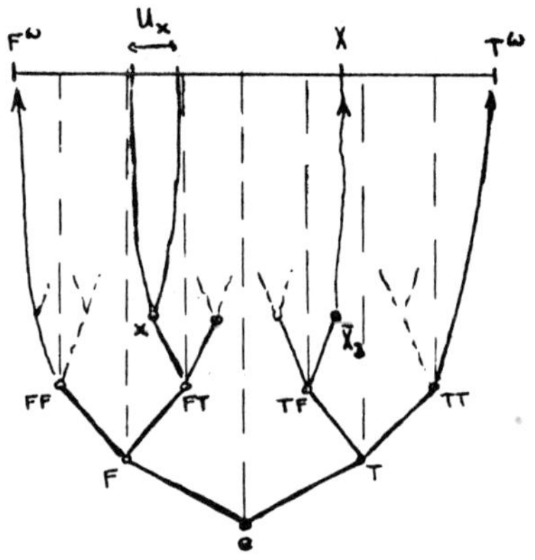

Es werde "$x$ ist Anfangsstück von Y" durch "$x \prec Y$" abgekürzt. Sodann sei $U_x = \{Y; x \prec Y\}$. Die Vereinigungen von Mengen $U_x$ sind dann genau die offenen Mengen des Cantorraumes über $2^\omega$, die endlichen Vereinigungen von Mengen $U_x$ sind genau die offen und abgeschlossenen Mengen dieser Topologie. Wird der Baum $2^*$ orientiert ($x$F=linker Nachfolger von $x$, $x$T=rechter Nachfolger von $x$) so wie in der Figur angedeutet, dann erhält man eine natürliche lineare Ordnung X<Y der Menge $2^\omega$. Die Mengen $U_x$ werden dann sowohl offene wie auch abgeschlossene Intervalle, und die schon erwähnte nulldimensionale Topologie über $2^\omega$ ist identisch mit der Ordnungstopologie. Aus dem Cantorraum $2^\omega$ geht übrigens die reelle Topologie und Ordnung durch schliessen der Sprünge $x$FTTT....., $x$TFFF..... hervor.

Es ist leicht zu zeigen, dass die <u>stetigen</u> <u>Operatoren</u> (SO) $Y = \mathcal{A}(X)$ genau diejenigen sind, die in der folgenden Form dargestellt werden können:

(1) $$Y_t = \Phi(X_0 \ldots X_{\varphi t}) \qquad \varphi : \omega \to \omega, \quad \Phi : 2^* \to 2$$

Ein <u>rekursiver</u> <u>Operator</u> (RO), also ein solcher, der als Verhalten einer Maschine realisierbar ist, muss von der Form (1), also stetig sein. Weiterhin müssen aber jetzt die Funktion $\varphi, \Phi$ rekursiv sein. Struktur der Maschine = paar von Algorithmen zur Berechnung von Funktionen $\varphi, \Phi$. Verhalten der Maschine = Operator $Y = \mathcal{A}(X)$ definiert durch (1). Struktur $< \varphi, \Phi <$ Verhalten.

Unter den SO kommen solche vor, die in der Form

(2) $$Y_t = \Phi(X_0 \ldots X_t) \qquad \Phi : 2^* \to 2$$

darstellbar sind. Dies sind die <u>deterministischen</u>

Operatoren (DO), genauer die 0-shift Operatoren. Die 1-shift Operatoren sind von der Form:

$$(3) \qquad Y_t = \Phi(X_0 \ldots X_{t-1}) \qquad\qquad \Phi : 2^* \to 2$$

Die rekursiv und deterministischen Operatoren (RDO) treten auf als Verhalten der real-time-computer. In diesen Maschinen erscheint die Ausgabe $Y_t$ zur Zeit t.

Eine spezielle Sorte von real-time-computer sind die endlichen Automaten. Die zugehörigen Operatoren heissen finite-state Operatoren (FSO). Sie sind darstellbar durch eine Rekursion der Form

$$(4) \qquad \begin{aligned} Z_0 &= S_0 \\ Z_{t'} &= F[Z_t, X_t] \\ Y_t &= G[Z_t, X_t] \end{aligned} \qquad\qquad \begin{aligned} &K \text{ endliche Menge (Zustände)} \\ &F : K \times 2 \to K \\ &G : K \times 2 \to 2 \end{aligned}$$

Dies ist ein 0-shift FSO. Beim 1-shift FSO ist die dritte Gleichung (4) von der Form $Y_t = G[Z_t]$. Das System $[K, S_0, F, G]$ ist der Automat; sein Verhalten ist der FSO $Y = \mathfrak{A}(X)$ definiert durch (4). Durch die Formel $x \backsim y \;.\!\equiv.\; (\forall u)\Phi(xu) = \Phi(yu)$ ist eine Rechtskongruenz über $2^*$ definiert. Der DO gegeben durch (2) ist ein FSO genau dann, wenn die Kongruenz $\backsim$ von endlichem Index ist.

2. Lösung von Konstruktionsbedingungen. Eine Forderung über das Verhalten einer Maschine ist eine Relation $\mathfrak{C}(X,Y)$. Entsprechend definieren wir "der Operator $\mathfrak{A}$ genügt der Bedingung $\mathfrak{C}$" oder "$\mathfrak{A}$ löst $\mathfrak{C}$ für Y" oder in Zeichen "$\mathfrak{A} \vdash \mathfrak{C}$" durch:

$$(5) \qquad \mathfrak{A} \vdash \mathfrak{C} \;.\!\equiv.\; (\forall XY)[Y = \mathfrak{A}(X) \supset \mathfrak{C}(X,Y)]$$

Die Borel-Hierarchie über dem Produktraum $2^{\omega}\times 2^{\omega}$ ergibt eine Klassifikation der Bedingungen. Zum Beispiel, sind

$$\mathfrak{C}(X,Y) \ .\equiv.\ (\forall t)M(\bar{X}_t,\bar{Y}_t)$$
$$\mathfrak{C}(X,Y) \ .\equiv.\ (\exists x)(\forall t)M_x(\bar{X}_t,\bar{Y}_t)$$
$$\mathfrak{C}(X,Y) \ .\equiv.\ (\forall y)(\exists x)(\forall t)M_{y,x}(\bar{X}_t,\bar{Y}_t)$$

respektive eine $\mathfrak{F}$-(abgeschlossene), eine $\mathfrak{F}_\sigma$-, und eine $\mathfrak{F}_{\sigma\delta}$-Bedingung.

Alle Borelschen Bedingungen sind Suslinsche, d.h. darstellbar in der Form:

$$\mathfrak{C}(X,Y) \ .\equiv.\ (\exists Z)\left[(\forall t)M_{\bar{Z}_t}(\bar{X}_t,\bar{Y}_t)\wedge(\exists^{\omega}t)Z_t\right.$$

"$(\exists^{\omega}t)A(t)$" ist Abkürzung für "$(\forall x)(\exists t)[x\leq t\wedge A(t)]$" und bedeutet: es gibt unendlich viele t, sodass A(t).

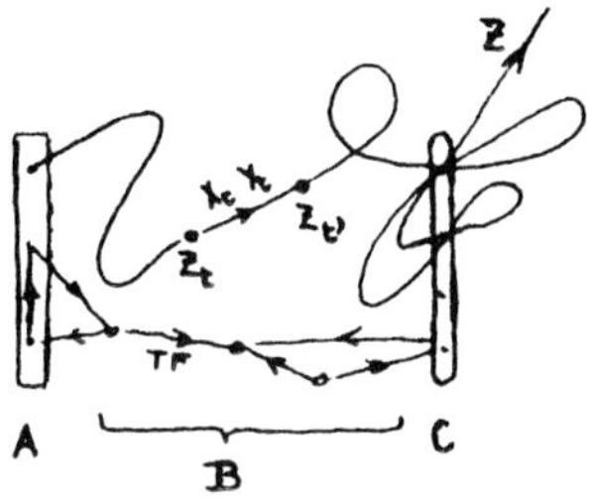


Es sei K eine endliche Menge, $A\subseteq K$, $B\subseteq K\times K\times 2\times 2$, $C\subseteq K$. Das System $[K,A,B,C]$ heisst ein <u>Transitionsystem</u> (nondeterministic finite automaton). Die Relation B kann als gerichteter Graph mit bewerteten Kanten veranschaulicht werden. Das Verhalten des Systems ist die Relation

(6) $\mathfrak{C}(X,Y) \ .\equiv.\ (\exists Z)[A[Z_0]\wedge(\forall t)B[Z_{t'},Z_t,Y_t]\wedge(\exists^{\omega}t)C[Z_t]]$

Also $\mathfrak{C}(X,Y)$ gilt, wenn es einen unendlichen Weg Z durch den Graph B gibt, der in A beginnt ($A[Z_0]$) unendlich oft nach C zurückkehrt ($(\exists^{\omega}t)C[Z_t]$) und, sodass auf den Kanten von Z die Werte $[X_0,Y_0]$, $[X_1,Y_1]$, $[X_2 Y_2]$, ..... erscheinen $((\forall t)B[Z_{t'},Z_t,X_t,Y_t])$. Eine Bedingung von der Form (6) heisse finite-state.

3. Konstruktionsalgorithmen. Es sei $\mathcal{S}$ eine Sprache mit Ausdrücken $\mathcal{C}(X,Y)$, die Relationen (Bedingungen) $\mathcal{C}(X,Y)$ bedeuten. Es sei $\mathcal{M}$ eine Klasse von Maschinen. Man kann dann nach der Existenz folgender Algorithmen fragen.

Lösungsalgorithmus: Für gegebene $\mathcal{A}\varepsilon\mathcal{M}$, $\mathcal{C}\varepsilon\mathcal{S}$ entscheide, ob $\mathcal{A}\vdash\mathcal{C}$.

Lösbarkeitsalgorithmus: Für gegebenes $\mathcal{C}\varepsilon\mathcal{S}$, entscheide, ob $(\exists\mathcal{A})[\mathcal{A}\varepsilon\mathcal{M}\wedge\mathcal{A}\vdash\mathcal{C}]$.

Synthesenalgorithmus: Für gegebenes $\mathcal{C}\varepsilon\mathcal{S}$, wenn $\mathcal{C}$ lösbar in $\mathcal{M}$, konstruiere ein $\mathcal{A}\varepsilon\mathcal{M}$ sodass $\mathcal{A}\vdash\mathcal{C}$.

Es soll im folgenden für $\mathcal{M}$ die Klasse der endlichen Automaten gewählt sein. Als besonders natürliche Bedingungssprache bietet sich dann der Sequentialkalkul (SC) an. Technisch ausgedrückt ist SC die monadische Theorie zweiter Ordnung der Struktur $[\omega,0,']$, enthält also die folgenden Ausdrucksmöglichkeiten:

| | |
|---|---|
| 0,' | Die Zahl 0 und die Nachfolgefunktion '. |
| t,x,y,..... | Variable über $\omega$ (Zeitpunkte) |
| X,Y,Z,..... | Variable über $2^\omega$ (Zustandssequenzen) |
| $\wedge,\vee,\sim$ | Logische Verknüpfungen |
| $\forall,\exists$ | Quantoren für beide Sorten von Variablen |

Beliebige endliche Mengen K können als Teilmengen von $2^n$ kodiert werden. Eine Zustandssequenz $Z\varepsilon K^\omega$ wird dadurch durch n zweiwertige Sequenzen $Z_1,\ldots,Z_n\varepsilon 2^\omega$ (die Komponenten von Z) repräsentiert. Daher gelten die folgenden Bemerkungen, die SC als geeignete Sprache zur Behandlung endlicher

Automaten ausweisen.

Bemerkung 1: Das Verhalten des endlichen Automaten (4) kann in SC durch eine Formel $\mathcal{C}(X,Y)$ ausgedrückt werden. Nämlich.

$$Y = \mathcal{A}(X) \;.\!\equiv.\; (\exists Z)[Z_0 = S_0 \wedge (\forall t) Z_{t'} = F[Z_t, X_t] \wedge (\forall t) Y_t = G[Z_t, X_t]]$$

In SC ist die Relation $x \leq y$ durch die Formel $(\forall Z)[(\exists t)[Z_{t'} \wedge \sim Z_t] \vee \sim Z_y \vee Z_x]$ definierbar. Daher besteht in SC auch die Möglichkeit $(\exists^{\omega} t)$ auszudrücken. Somit gilt,

Bemerkung 2: Jede finite-state Bedingung $\mathcal{C}(X,Y)$ kann in SC ausgedrückt werden. Siehe (6).

Schliesslich gilt wegen (5) und Bemerkung 1,

Bemerkung 3: Für gegebenen endlichen Automaten $\mathcal{A}$ und gegebenes $\mathcal{C}(X,Y) \in SC$ kann die Aussage " $\mathcal{A}$ löst $\mathcal{C}$ " ($\mathcal{A} \vdash \mathcal{C}$) in SC angeschrieben werden.

Das Hauptresultat aus Buchi 1 ist die folgende Umkehrung zur Bemerkung 2:

SATZ 1 (Normalform für SC): Jede Formel $\mathcal{C}(X,Y)$ aus SC definiert eine finite-state Bedingung; die Formel $\mathcal{C}(X,Y)$ kann mechanisch auf die Form (6) gebracht werden.

Es handelt sich also hier um ein Resultat über das Eliminieren von Quantoren. Der Satz gilt insbesondere, wenn in $\mathcal{C}$ die freien Variablen X,Y gar nicht auftreten. Daraus folgt leicht,

KOROLLAR 2: Die Sprache SC ist entscheidbar, d.h., es gibt einen Algorithmus, der über die Wahrheit von Aussagen in SC entscheidet.

Damit wurde ein Problem von Tarski mittels Methoden der Automatentheorie gelöst. Wegen Bemerkung 3 folgt nun leicht.

KORÖLLAR 3: Für endliche Automaten und relativ zur Bedingungssprache SC gibt es einen Lösbarkeits- und einen Synthesenalgorithmus.

Ein Synthesenalgorithmus besteht einfach im Durchprobieren aller Automaten $\mathcal{A}_0, \mathcal{A}_1, \mathcal{A}_2, \ldots$ . Jeder Automat wird mittels des Lösungsalgorithmus auf $\mathcal{A}_i \vdash \mathcal{C}$ getestet.

Im Gegensatz zur Bemerkung 3 kann die Aussage $(\exists \mathcal{A})[\mathcal{A}$ endlicher Automat $\wedge \mathcal{A} \vdash \mathcal{C}]$ nicht in SC angeschrieben werden. Daher führt das Entscheidungsverfahren für SC nicht direkt zum Lösbarkeitsalgorithmus für SC. Ein erster Schritt in dieser Richtung ist der folgende Satz von McNaughton 9.

SATZ 4: Jede finite-state Bedingung (6) kann effektiv auf die folgende Form gebracht werden:

$$(7) \qquad \mathcal{C}(X,Y) \ .\equiv.\ (\exists Z)[Z_0 = S_0 \wedge (\forall t) Z_{t'} = H[Z_t, X_t, Y_t] \wedge \sup Z \in W]$$

Hierbei ist W eine Menge von Teilmengen von K=Menge der Zustände von Z. supZ ist die Menge aller derjenigen Elemente von K, die von Z unendlich oft angenommen werden, d.h.,

$$(8) \qquad S \in \sup Z \ .\equiv.\ (\exists^{\omega} t) Z_t = S$$

Man beachte, dass in (6) $\mathcal{C}$ als Verhalten eines nicht determinierten Automaten [A,B,C] gegeben ist. Im Unterschied dazu wird $\mathcal{C}$ in (7) durch einen (determinierten) Automaten $[S_0, H, W]$ bestimmt. Allerdings tritt dabei in (7) eine etwas kompliziertere Ausgabebedingung, $\sup Z \in W$, auf. Die Stärke des Satzes 4 kann qualitativ auf folgende Weise eingesehen werden.

Der Automat $[S_0, H]$ aus (7) bestimme den Operator $Z = \mathfrak{F}(X, Y)$, und es sei $\mathfrak{D} = \{Z; \sup Z \in W\}$. Dann gilt

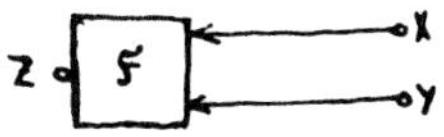


(7') $\mathfrak{C} = \mathfrak{F}^{-1}(\mathfrak{D})$ (8) $\mathfrak{D}(Z) \;.\!\equiv\!.\; \bigvee_{E \in W} [(\exists x)(\forall t)_x E[Z_t] \wedge \bigwedge_{S \in E} (\forall x)(\exists t)_x Z_t = S]$

Aus (8) sieht man, dass $\mathfrak{D}$ eine Boolesche Verbindung von $\mathfrak{F}_\sigma$-Mengen ist. Da $\mathfrak{F}$ stetig ist, gilt wegen (7'), auch $\mathfrak{C}$ ist eine Boolesche Verbindung von $\mathfrak{F}_\sigma$-Mengen. Wegen des Satzes 2 von McNaughton gilt also,

Bemerkung 4: Jede finite-state Bedingung $\mathfrak{C}(X, Y)$ ist Borelsch ja sogar eine Boolesche Verbindung von $\mathfrak{F}_\sigma$-Mengen, also sicher ein $\mathfrak{F}_{\sigma\delta}$ und auch ein $\mathfrak{G}_{\delta\sigma}$.

Diese Information ist nicht aus der Definition (6) der finite-state Bedingung ersichtlich. Aus (6) folgt zunächst nur die viel schwächere Aussage, dass $\mathfrak{C}$ Suslinsch ist.

Wegen Satz 1 und Bemerkung 4 folgt, dass sich in SC nur Boolesche Verbindungen von $\mathfrak{F}_\sigma$-Mengen definieren lassen (natürlich nicht alle diese, da ja SC nur abzählbar viele Ausdrücke enthält). Siehe Buchi und Landweber 4.

4. Gewinnstrategien als Lösungen von Bedingungen. Eine Bedingung $\mathfrak{C}(X, Y)$ lässt sich, auf folgende Weise, als

Spiel zwischen zwei Spielern I und J deuten. Spieler I beginnt zur Zeit 0. Zur Zeit t wählt der Spieler I ein Element $X_t$ aus $2=\{F,T\}$ , dann ist J am Zuge, er wählt ein $Y_t$ aus 2. Das Spiel geht weiter bis zum jüngsten Tage, d.h., bis alle Züge $X_0, Y_0, X_1, Y_1, X_2, Y_2, \ldots$ vorliegen. J gewinnt wenn $\mathfrak{C}(X,Y)$, I gewinnt $\sim\mathfrak{C}(X,Y)$.

Demnach ist eine Strategie für J eine Funktion $\Phi: 2^* \rightarrow 2$, sie bestimmt den Zug J zur Zeit t nach der Formel,

(9) $Y = \Phi(X_0 \ldots X_t)$ 0-shift DO, $Y = \mathfrak{A}(X)$

Eine Strategie für I dagegen ist von der Form,

(10) $X_t = \Psi(Y_0 \ldots Y_{t-1})$ 1-shift DO, $X = \mathfrak{B}(Y)$

Die Strategie (9) gewinnt das Spiel $\mathfrak{C}$ für J, genau dann, wenn der 0-shift Operator $\mathfrak{A}$ die Bedingung $\mathfrak{C}(X,Y)$ für Y löst. Die Strategie (10) gewinnt das Spiel $\mathfrak{C}$ für I, genau dann, wenn der 1-shift Operator $\mathfrak{B}$ die Bedingung $\sim\mathfrak{C}(X,Y)$ für X löst. Somit ist das Lösen von Bedingungen durch DO's identisch mit dem Auffinden von Gewinnstrategien.

Ein Spiel (eine Bedingung) $\mathfrak{C}(X,Y)$ heisst determiniert, wenn entweder der eine, oder dann der andere Spieler eine Gewinnstrategie besitzt. D.h., entweder soll $\mathfrak{C}$ für Y 0-shift-lösbar, oder dann $\sim\mathfrak{C}$ für X 1-shift-lösbar sein. Das folgende Hauptresultat über die Existenz eines Lösbarkeitsalgorithmus für SC wurde von McNaughton vermutet und von Büchi and Landweber 3 bewiesen:

SATZ 5: Jede finite-state Bedingung $\mathfrak{C}(X,Y)$ ist determiniert, und zwar besitzt derjenige Spieler der eine Gewinnstrategie hat, sogar eine gewinnende finite-state Strategie.

Im Beweis wird Satz 4 verwendet, der ja eine starke Vereinfachung in der Darstellung von finite-state Bedingungen, und damit ein besseres Verständnis von finite-state Spielen enthält. Danach besteht ein solches Spiel aus einem Automaten $Z=\mathfrak{f}(X,Y)$ mit zwei Eingabekanälen und einer Menge W von Teilmengen der Zustandsmenge K des Automaten. Zur Zeit t wählt I die Eingabe $X_t$, und dann J die Eingabe $Y_t$. Spieler J versucht dabei seine Züge $Y_t$ so zu bestimmen, dass die Zustände $Z_t$ schlussendlich zu einem $E\varepsilon W$ gehören, und jeden Wert $S\varepsilon E$ immer wieder durchlaufen. Spieler I versucht ihn daran zu hindern.

Wegen Satz 5 und Korollar 3 kann man nun für gegebene finite-state Bedingung $\mathfrak{C}(X,Y)$ auch wirklich entscheiden, welcher der Spieler die Gewinnstrategie besitzt (und übrigens ist ein solches Verfahren auch im Beweise von Satz 5 enthalten). Also, nach Satz 1,

KOROLLAR 6: Für endliche Automaten und relativ zur Bedingungssprache SC gibt es einen Lösbarkeitsalgorithmus.

5. Die Methode von Cantor und Bendixson. Cantor und Bendixson haben die von Cantor erfundene Theorie der Ordinalzahlen angewandt, um zu zeigen, dass für jede abgeschlossene Menge $\mathfrak{C}$ die folgende Alternative gilt:

(11) $\mathfrak{C}$ abzählbar v $\mathfrak{C}\ni$perfekte Menge

Da jede perfekte Menge die Mächtigkeit des Kontinuums besitzt, war damit die Kontinuumshypothese für abgeschlossene Menge bewiesen. Die Alternative (11) besagt gerade, dass das "lopsided game " determiniert ist (Davis 7). Die Methode von Cantor-Bendixson kann aber auch auf die gewöhnlichen Spiele angewandt werden. Der Beweis von

Satz 5 in Buchi und Landweber 3 ist ein Beispiel dafür. Es werde hier diese Methode, im Falle des abgeschlossenen Spieles $\mathfrak{C}(X,Y)$, vorgeführt.

(12) $$\mathfrak{C}(X,Y) \;.\!\equiv.\; (\forall t)M(\overline{X}_t,\overline{Y}_t)$$

Man definiere $M_i(\mathfrak{x},\mathfrak{y})$, für alle $i\leq\omega$, durch die Induktion:

(13) $$M_i(\mathfrak{x},\mathfrak{y}) \;.\!\equiv.\; M(\mathfrak{x},\mathfrak{y})\wedge \bigwedge_U \bigvee_V (\forall j)^i M_j(\mathfrak{x}U,\mathfrak{y}V)$$

wobei $(\forall j)^i-$ für $(\forall j)[j<i\wedge -]$ steht. Es ist nun leicht einzusehen, dass $M=M_0\supseteq M_1\supseteq M_2\supseteq\ldots\ldots\supseteq M_\omega$ , und daher $M_\omega(\mathfrak{x},\mathfrak{y}) \equiv (\forall i)^\omega M_i(\mathfrak{x},\mathfrak{y})$. Also,

(14) $$M_\omega(\mathfrak{x},\mathfrak{y}) \;.\!\supset.\; M(\mathfrak{x},\mathfrak{y})\wedge \bigwedge_U \bigvee_V M_\omega(\mathfrak{x}U,\mathfrak{y}V)$$

Fall $M_\omega(e,e)$: Nach (14) gibt es eine Funktion $\Phi(\mathfrak{x}U,\mathfrak{y})$, sodass $M_\omega(\mathfrak{x},\mathfrak{y})\supset M_\omega(\mathfrak{x}U,\mathfrak{y}\Phi(\mathfrak{x}U,\mathfrak{y}))$. Die Formel

(15) $$Y_t=\Phi(\overline{X}_t X_t,\overline{Y}_t)$$

bestimmt eine Strategie für Spieler J. Wenn J diese Strategie spielt, dann gilt induktiv, $M_\omega(\overline{X}_0,\overline{Y}_0), M_\omega(\overline{X}_1,\overline{Y}_1), M_\omega(\overline{X}_2,\overline{Y}_2),\ldots$ . Also nach (14) $(\forall t)M(\overline{X}_t,\overline{Y}_t)$, also $\mathfrak{C}(X,Y)$. Damit ist, im Falle $M_\omega(e,e)$ (15) eine Gewinnstrategie für J.

Fall $\sim M_\omega(e,e)$: Nach (13) gibt es eine Funktion $\Psi_i(\mathfrak{x},\mathfrak{y})$, sodass $\sim M_i(\mathfrak{x},\mathfrak{y})\wedge M(\mathfrak{x},\mathfrak{y}) .\!\supset.\; (\exists j)^i \sim M_j(\mathfrak{x}\Psi_i(\mathfrak{x},\mathfrak{y}),\mathfrak{y}V)$ . Die Formeln

(16) $$i_t = (\mu i)\sim M_i(\overline{X}_t,\overline{Y}_t) \qquad (\mu i)=\text{ erstes } i, \text{ sodass}$$
$$X_t = \Psi_{i_t}(\overline{X}_t,\overline{Y}_t)$$

bestimmen eine Strategie für Spieler I. Wenn I diese

Strategie spielt, dann gilt $[i_0 > \ldots > i_t] \vee M(\bar{X}_t, \bar{Y}_t)$. Da $i_0 > i_1 > i_2 \ldots\ldots$ nicht zutreffen kann folgt, $(\exists t) \sim M(\bar{X}_t, \bar{Y}_t)$, also $\sim \mathfrak{C}(X,Y)$. Damit ist, im Falle $\sim M_\omega(e,e)$, (16) eine Gewinnstrategie für I. Wir haben also das folgende Resultat:

SATZ 7: Jedes abgeschlossene Spiel (12) ist determiniert, und zwar hat J die Gewinnstrategie (15) wenn $M_\omega(e,e)$, und I die Gewinnstrategie (16) wenn $\sim M_\omega(e,e)$.

Dieser Satz ist nicht neu, und zwar hat Davis 10 bewiesen,

SATZ 8: Jedes $\mathfrak{F}_{\sigma\delta}$-Spiel ist determiniert.

Wegen Bemerkung 4 folgt daraus: Jedes finite-state Spiel ist determiniert. Dagegen liefert der Beweis von Davis nicht den stärkeren (und für den Lösbarkeitsalgorithmus, Korollar 6, essentiellen) Satz 5. Unser Beweis von Satz 7 deutet an, wie die Methode von Cantor-Bendixson, über die Determiniertheit hinaus, das effektive Ausschreiben von Gewinnstrategien erlaubt.

Für Suslinsche Mengen hat Hausdorf 8 gezeigt, dass die Alternative (11) immer noch gilt (lopsided Suslinsche Spiele sind determiniert). Sein Beweis ist eine Variante der Methode von Cantor-Bendixson. Dass auch gewöhnliche Suslinsche Spiele determiniert sind, wurde von D. A. Martin aus der Existenz von messbaren Kardinalzahlen $> \omega$ hergeleitet. Dieser Beweis ist, wie derjenige von Davis, nicht vom Cantor-Bendixsonschen Typus.

LITERATUR

1. J. R. Buchi, On a decision method in restricted second order arithmetic, Proc. Int. Congr. Logic, Method. and Philos. Sci. 1960, Stanford Univ. Press, Stanford, Cal. 1962

2. __________, Elgot and Wright, The nonexistence of certain algorithms of finite automata theory, Notices Am. Math. Soc. 5, 1958, 98.

3. __________ and Landweber, Solving sequential conditions by finite-state strategies. Trans. Am. Math. Soc. 138, 1969, 295-311.

4. _________________________, Definability in the monadic second order theory of successor, Journ. Symb. Logic 34, 1969, 166-170.

5. A. Church, Application of recursive arithmetic to the problem of circuit synthesis, Summaries of talks presented at the Summer Institute for Symbolic Logic, Cornell Univ. 1957; 2nd ed., Princeton N.J., 1960, 3-50.

6. ________, Logic arithmetic and automata, Proc. Int. Congr. Math. 1963, Almqvist and Wiksells, Uppsala 1963.

7. M. Davis, Infinite games of perfect information, Advances in game theory, Princeton Univ. Press, Princeton, N.J., 1964, 85-101.

8. F. Hausdorff, Grundzüge der Mengenlehre, New York, Chelsea Publ. Co. 1949.

9. R. McNaughton, Testing and generating infinite sequences by a finite automaton, Information and Control 9, 1966, 521-530.

10. M. Rabin, Decidability of second-order theories and automata on infinite trees, IBM Research RC 2012, 1968. Siehe auch Trans. Am. Math. Soc. 1969.

Reprinted from J. Dörr and G. Hotz (eds.), Tagungsbericht Automatentheorie und Formale Sprachen Math. Forschungsinst. Oberwolfach 1969. Mannheim: Bibliographisches Institut 1970, pp. 385–398.

**Algorithmisches Konstruieren von Automaten und die Herstellung von Gewinnstrategien nach Cantor-Bendixson**

von J. Richard Büchi in Lafayette, Indiana (U.S.A.)

Tagungsbericht Automatentheorie und Formale Sprachen
Math. Forschungsinst. Oberwolfach 1969, J. Dörr and G. Hotz (Eds.),
Mannheim: Bibliographisches Institut 1970, pp. 385-398

Translated by Sylvia Büchi, Saunders MacLane, Walter Schnyder, and Dirk Siefkes

# Algorithmic Construction of Automata and the Production of Winning Strategies According to Cantor-Bendixson.

by J. Richard Büchi in Lafayette, Indiana (U.S.A.)

Problems of the mechanical construction of electrical networks with self-control were already brought up by Church [5+6] and Büchi, Elgot, and Wright [2]. The questions proposed have proved fruitful in a variety of ways. First, there exist now very strong algorithms for the construction of automata (Büchi [1], Büchi-Landweber [3]). Secondly, these methods of automata theory were taken up by Rabin, and have led to decision methods for astonishingly rich languages. As R. MacNaughton has remarked, solving a construction condition through a finite automaton is identical with constructing a winning strategy in a particular infinite game. This problem is furthermore related to the classical question on the existence of a perfect subset in a given set, and thus to Cantor's approach to the solution of the continuum hypothesis. These relationships between automata theory, game theory, and Cantorian set theory will be set forth below.

## 1. Sequential operators

The figure shows a machine which transforms a sequence $X = X_0 X_1 X_2 \ldots$ of input data into a sequence $Y = Y_0 Y_1 Y_2 \ldots$ of output data.

$$Y \circ\!\boxed{A}\!\leftarrow\!\circ\, X$$

The input states will be taken from a finite set I. The output states belong to the finite set J. The operator $Y = \mathbf{A}(X)$ is the <u>behavior of the machine</u>, and is to be distinguished from the inner structure of the machine.

It suffices to consider the case I = J = 2, where 2 is the set {F, T} of the truth value F = 0 = false and T = 1 = true. The set $\omega = \{0, 1, 2, \ldots\}$ is the set of points of time (discrete time, digital machine). The input and output sequences belong then to the set $2^\omega$, and thus can be interpreted simply as subsets of $\omega$. $2^*$ is the set of all words x, y, z, ... over the alphabet 2, and contains also the empty word e. When $X \in 2^\omega$ and $t \in \omega$, then $\overline{X}_t$ means the word $X_0 \ldots X_{t-1}$. The adjoining figure shows how one can view the words of $2^*$ as vertices and the sequences $2^\omega$ as paths through a tree with two branches at each node.

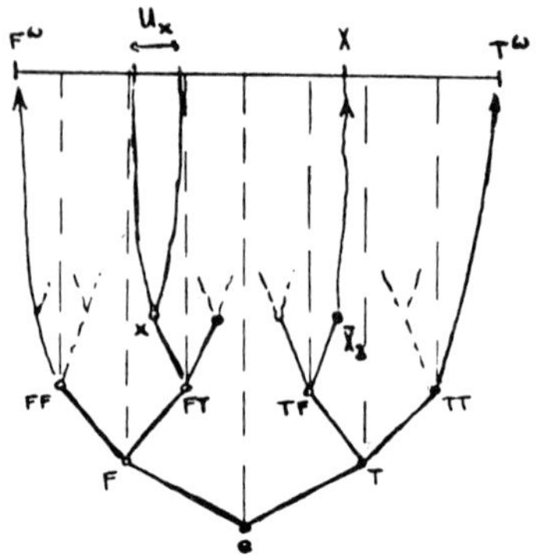


The phrase "x is an initial segment of $Y_o$" is abbreviated as $x \dashv Y$. Then set $U_x = \{Y; x \dashv Y\}$. The unions of sets $U_x$ are then exactly the open sets of the Cantor space over $2^\omega$, while the finite unions of sets $U_x$ are exactly the open and closed sets of this topology. If the tree $2^*$ is oriented (xF = left successor of x, xF = right successor of x) then, as shown in the figure, one obtains a natural linear order X < Y of the set $2^\omega$. The sets $U_x$ then become both open and closed intervals and the null dimensional topology over $2^\omega$ is identical with the order topology. By the way from the Cantor space $2^\omega$ one obtains the real topology and order by closing the gaps xFTTT..., xTFFF... .

It is easy to show that the <u>continuous operators</u> (CO) Y = **A**(X) are exactly those that can be represented in the following form:

(1) $\quad Y_t = \Phi(X_o \dots X_{\varphi t}), \quad \varphi : \omega \to \omega, \quad \Phi : 2^* \to 2$

A <u>recursive operator</u> (RO), that is, one which is realizable as the behavior of a machine, must be of the form (1), thus must be continuous; moreover, the functions $\varphi$, $\Phi$ must be recursive. Structure of the machine = pair of algorithms for computation of the functions $\varphi$, $\Phi$. Behavior of the machine = operator Y = **A**(X) defined by (1). Structure $\to$ $\langle\varphi, \Phi\rangle$ $\to$ behavior.

Among the CO there are those that can be represented in the form

(2) $\quad Y_t = \Phi(X_o \dots X_t), \quad \Phi : 2^* \to 2.$

These are the <u>deterministic operators</u> (DO), more exactly the <u>0-shift operators</u>. The <u>1-shift operators</u> are of the form

(3) $Y_t = \Phi(X_o \ldots X_{t-1}), \quad \Phi : 2^* \to 2.$

The recursive and deterministic operators (RDO) appear as the behavior of a real-time computer. In these machines the output $Y_t$ occurs at time t.

A special sort of real-time computers are the finite automata. The corresponding operators are called finite-state operators (FSO). They are represented by a recursion of the form

| | | |
|---|---|---|
| | $Z_o = S_o$ | K a finite set (of states) |
| (4) | $Z_{t'} = F[Z_t, X_t]$ | $F : K \times 2 \to K$ |
| | $Y_t = G[Z_t, X_t]$ | $G : K \times 2 \to 2$ |

This is a 0-shift FSO. For a 1-shift FSO the third equation of (4) is of the form $Y_t = G[Z_t]$. The system $<K, S_o, F, G>$ is the automaton; its behavior is the FSO $Y = \mathbf{A}(X)$ defined by (4). The formula $x \sim y. \equiv . (\forall u)\ \Phi(xu) = \Phi(yu)$ defines a right-congruence on 2*. The DO given by (2) is a FSO if and only if the congruence $\sim$ is of finite index.

## 2. **Solution of Construction Conditions**

A requirement on the behavior of a machine is a relation of the form C(X,Y). Correspondingly we define "the operator **A** satisfies the condition **C**" or "**A** solves $\Sigma$ for Y", or in symbols "$\mathbf{A} \vdash \Sigma$", by

(5) $\mathbf{A} \vdash \Sigma .\equiv. (\forall X\ Y)[Y = \mathbf{A}(X) \supset \mathbf{C}(X, Y)].$

The Borel-hierarchy on the product space $2^\omega \times 2^\omega$ gives a classification of the conditions. For example,

$$C(x, y) .\equiv. (\forall t)\ M(\overline{X}_t, \overline{Y}_t)$$
$$C(x, y) .\equiv. (\exists x)\ (\forall t)\ M_x(\overline{X}_t, \overline{Y}_t)$$
$$C(x, y) .\equiv. (\forall y)\ (\exists x)\ (\forall t)\ M_{y,x}(\overline{X}_t, \overline{Y}_t)$$

are respectively an **F**- (that is, closed), an $\mathbf{F}_\sigma$-, and an $\mathbf{F}_{\sigma\delta}$-condition.

All Borel conditions are Suslin, i.e., are representable in the form

$$C(X, Y) .\equiv. (\exists Z)\ [(\forall t)\ M_{\overline{Z}_t}(\overline{X}_t, \overline{Y}_t) \wedge (\exists^\omega t) Z_t].$$

Here "$(\exists^\omega t)\ A(t)$" is an abbreviation for "$(\forall x)\ (\exists t)\ [x \le t \wedge A(t)]$; it means there exist infinitely many t for which A(t) holds.

Let K be a finite set, let $A \subseteq K$, $B \subseteq K \times K \times 2 \times 2$, $C \subseteq K$. The system [K, A, B, C] is called a transition system (nondeterministic finite automaton). The relation B can be represented as a directed graph with labeled edges.

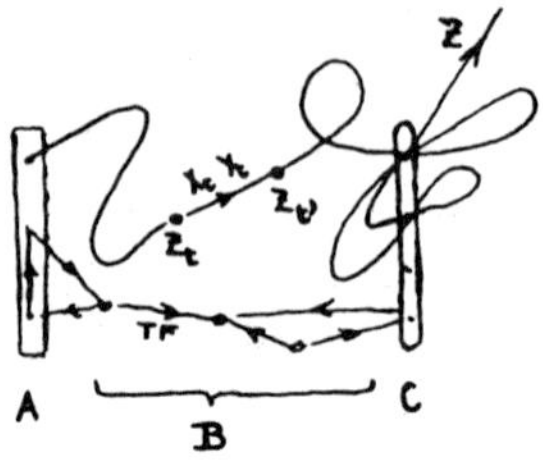


The behavior of the system is the relation

(6) $C(X, Y) .\equiv. (\exists Z) [A[Z_0] \wedge (\forall t) B[Z_{t'}, Z_t, X_t, Y_t] \wedge (\exists^{\omega} t) C[Z_t]]$

Thus **C**(X, Y) holds when there is an infinite path Z through the graph B that begins in A ($A[Z_0]$), infinitely often returns to C ($(\exists^{\omega} t)\ C[Z_t]$), and such that the labels $<X_0, Y_0>$, $<X_1, Y_1>$, $<X_2, Y_2>$, ... occur on the edges of B ($(\forall t)\ B[Z_{t'}, Z_t, X_t, Y_t]$). A condition of the form (6) is said to be finite-state.

## 3. Construction Algorithms

Let **S** be a language with expressions **C**(X, Y) that represent relations **C**(X, Y). Let **M** be a class of machines. One can ask then about the existence of the following algorithms.

Solution algorithm: For given **A** ε **M** and **C** ε **S** decide whether **A** ⊢ **C**.

Solvability algorithm: For given **C** ε **S** decide whether (∃**A**) [**A**ε**M** ∧ **A** ⊢ **C**].

Synthesis algorithm: For given **C** ε **S**, if **C** is solvable in **M**, construct an **A** ε **M** such that A ⊢ **C**.

In the sequel **M** will be the class of finite automata. The sequential calculus (SC) offers itself as an especially natural condition language. Expressed technically, SC is the monadic second order theory of the structure [ω, 0, '], and so contains the following types of expressions:

| | |
|---|---|
| 0, ' | the number 0 and the successor function ' |
| t, x, y, ... | variables over ω (time points) |

| | |
|---|---|
| X, Y, Z, ... | variables over $2^{\omega}$ (state sequences) |
| $\wedge$, $\vee$, $\sim$ | logical connectives |
| $\forall$, $\exists$ | quantifiers for both sorts of variables. |

Each finite set K can be encoded as a subset of $2^n$ for some n. A state sequence $Z \varepsilon K^{\omega}$ is thereby represented by n two-valued sequences $Z_1, \ldots, Z_n \varepsilon 2^{\omega}$ (the components of Z)[1]. Therefore the following remarks show that SC is a suitable language for the handling of automata.

**Remark 1:** The behavior of the finite automaton (4) can be expressed in SC through a formula **C**(X,Y); namely:

$$Y = \mathbf{A}(X) \;.\!\equiv\!.\; (\exists z)\, [Z_o = S_o \wedge (\forall t)\, Z_{t'} = F[Z_t, X_t] \wedge (\forall t)\, Y_t = G[Z_t, X_t]]$$

In SC the relation $x \leq y$ is definable by the formula $(\forall Z)\, [(\exists t)\, [Z_{t'} \wedge \sim Z_t] \vee \sim Z_y \vee Z_x]$. Thus in SC one can express $(\exists^{\omega} t)$. Therefore,

**Remark 2:** Every finite state condition **C**(X, Y) can be expressed in SC. See (6).

Finally by (5) and Remark 1,

**Remark 3:** For given finite automaton **A** and given **C**(X,Y) $\varepsilon$ SC, the formula "**A** solves **C**" ($\mathbf{A} \vdash \mathbf{C}$) can be expressed in SC.

The main result of Büchi [1] is the following converse of Remark 2.

**Theorem 1** (Normal form for SC): Every formula **C**(X, Y) of SC defines a finite-state condition; the formula **C**(X, Y) can be brought mechanically into the form (6).

This is thus a result on quantifier elimination. The theorem holds in particular when the free variables X, Y do not occur in **C** at all.

**Corollary 2**: The language SC is decidable, that is, there exists an algorithm that decides the truth of statements in SC.

In this way a problem of Tarski was solved by means of automata theory. From remark 3 one easily obtains:

---

[1]In SC-formulas the letters X, Y, Z are thus used for sequence variables, and for tuples of such variables as well. (The editors.)

**Corollary 3**: For finite automata and relative to the condition language SC there is a solution and synthesis algorithm.

A synthesis algorithm consists simply in testing through all automata $\mathbf{A}_0, \mathbf{A}_1, \mathbf{A}_2, \dots$ . Each automaton is tested for $\mathbf{A}_i \vdash \mathbf{C}$ by means of the solution algorithm.

In contrast to remark 3 the statement $(\exists \mathbf{A})[\mathbf{A}$ finite automaton $\wedge\ \mathbf{A} \vdash \mathbf{C}]$ cannot be written in SC. Therefore the decision method for SC does not lead directly to a solvability algorithm for SC. The final step in this direction is the following theorem of McNaughton [9].

**Theorem 4**: Every finite-state condition (6) can be brought effectively into the following form

(7) $\quad C(X, Y) \mathrel{.{\equiv}.} (\exists Z)\,[Z_0 = S_0 \wedge (\forall t)\, Z_{t'} \mathrel{.{=}.} H[Z_t, X_t, Y_t] \wedge \sup Z \in W]$

Here W is a set of subsets of K, the set of states of Z; supZ is the set of all those elements of K that are entered infinitely often by Z, i.e.

(8) $\quad S \in \sup Z \mathrel{.{\equiv}.} (\exists^{\omega} t)\, Z_t = S$

One should note that in (6) **C** is given as the behavior of a nondeterministic automaton [A, B, C]. In contrast, in (7) **C** is characterized by a (deterministic) automaton $[S_0, H, W]$. In (7), however, a somewhat more complicated output condition $\sup Z \in W$ occurs. The strength of theorem 4 can be seen qualitatively in the following way.

Let the automaton $[S_0, H]$ of (7) determine the operator $Z = \mathbf{F}(X, Y)$, and let $\mathbf{D} = \{Z; \sup Z \in W\}$. Then the following holds:

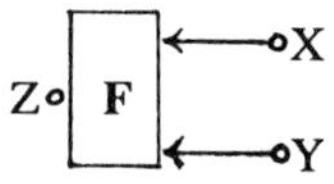


(7') $\quad C = \mathbf{F}^{-1}(\mathbf{D})$

(8') $\quad \mathbf{D}(Z) \mathrel{.{\equiv}.} \bigvee_{E \in W} [(\exists x)\,(\forall t)_x\, E[Z_t] \wedge \bigwedge_{S \in E} (\forall x)\,(\exists t)_x\, Z_t = S]$

From (8') one sees that **D** is a Boolean union of $\mathbf{F}_\sigma$-sets. By (7'), since **F** is continuous, also **C** is a Boolean union of $\mathbf{F}_\sigma$-sets. By theorem 4 of McNaughton one also has

**Remark 4**: Every finite-state condition **C**(X, Y) is Borel, in fact, even a Boolean union of $\mathbf{F}_\sigma$-sets, thus certainly an $\mathbf{F}_{\sigma\delta}$ and also a $\mathbf{G}_{\sigma\delta}$.

This information is not apparent in the definition (6) of the finite-state condition. From (6) at first only the much weaker statement follows that **C** is Suslin.

It follows from theorem 1 and remark 4 that in SC only Boolean unions of $\mathbf{F}_\sigma$-sets can be defined. (Naturally not all of these, as SC contains only countably many expressions.) See Büchi and Landweber [4].

## 4. Winning Strategies as Solutions of Conditions

A condition $\mathbf{C}(X,Y)$ can be viewed in the following way as a game between players I and J. Player I begins at time 0. At time t player I chooses an element $X_t$ from 2 = {F, T}, then player J has his turn, he chooses a $Y_t$ from 2. The game goes on till doomsday, that is until all moves $X_o, Y_o, X_1, Y_1, X_2, Y_2, \ldots$ are at hand. J wins if $\mathbf{C}(X, Y)$, I wins if $\sim \mathbf{C}(X, Y)$.

Accordingly a strategy for J is a function $\Phi : 2^* \to 2$. It determines the move J at time t by the formula

(9) $\quad Y_t = \Phi(X_o \ldots X_t)$ $\qquad$ 0-shift DO, $Y = \mathbf{A}(X)$

A strategy for I on the other hand is of the form,

(10) $\quad X_t = \Psi(Y_o \ldots Y_{t-1})$ $\qquad$ 1-shift DO, $X = \mathbf{B}(Y)$

The strategy (9) wins the game **C** for J if and only if the 0-shift operator **A** solves the condition $\mathbf{C}(X, Y)$ for Y. The strategy (10) wins the game **C** for I if and only if the 1-shift operator **B** solves the condition $\sim \mathbf{C}(X, Y)$ for X. Thus, solving the conditions by DO's is identical with obtaining winning strategies.

A game (a condition) $\mathbf{C}(X, Y)$ is called determined if either of the players has a winning strategy: That is, either **C** has to be 0-shift solvable for Y or else $\sim$ **C** has to be 1-shift solvable for X. The following main result on the existence of a solvability algorithm for SC was conjectured by McNaughton and proved by Büchi and Landweber [3].

**Theorem 5**: Every finite-state condition $\mathbf{C}(X,Y)$ is determined; and indeed the player who has a winning strategy even has a winning finite-state strategy.

In the proof theorem 4 is used, that indeed comprises a strong simplification in the representation of finite-state conditions, and thereby a better understanding of finite-state games. Accordingly such a game consists of an automaton $Z = \mathbf{F}(X,Y)$ with two input channels and a set W of subsets of the state-set K of the automaton. At time t player I chooses the input $X_t$, and thereafter J chooses the input $Y_t$. Player J tries to so choose his plays $Y_t$ that the states $Z_t$ finally belong to some $E \in W$, and every value $S \in E$ appears over and over again. Player I attempts to hinder him in this.

Due to theorem 5 and corollary 3 for a given finite-state condition C(X, Y) we can really decide which of the players possesses a winning strategy (besides such a method is also contained in the proof of theorem 5). Thus by theorem 1,

**Corollary 6**: There exists a solvability algorithm for finite automata and relative to the condition language SC.

## 5. The Cantor-Bendixson Method

Cantor and Bendixson used the theory of ordinal numbers discovered by Cantor in order to show that for every closed set **C**, the following alternative holds

(11) $\quad \mathbf{C}$ is countable $\vee$ $\mathbf{C} \supseteq$ a perfect set

As each perfect set has the cardinality of the continuum, this proves the continuum hypothesis for closed sets. The alternative (11) says precisely that the "lopsided game **C**" is determined (Davis [7]). The Cantor-Bendixson method can also be applied to ordinary games. The proof of theorem 5 in Büchi and Landweber [3] is an example of this. This method is presented below in the case of the closed game **C**(X, Y).

(12) $\quad \mathbf{C}(X, Y) \mathrel{.{\equiv}.} (\forall t)\, M(\overline{X}_t, \overline{Y}_t)$

One defines $M_i(x, y)$, for all $i \leq \omega$, by the induction:

(13) $$M_i(x, y) \mathrel{.{\equiv}.} M(x, y) \wedge \bigwedge_U \bigvee_V (\forall j)^i M_j(xU, yV)$$

here $(\forall j)^i$ - - stands for $(\forall j)\,[j < i \mathrel{.{\supset}.} \text{- -}]$. It is now easy to see that $M = M_0 \supseteq M_1 \supseteq M_2 \supseteq \ldots \supseteq M_\omega$, and therefore $M_\omega(x, y) \equiv (\forall i)^\omega M_i(x, y)$. Thus

(14) $$M_\omega(x, y) \mathrel{.{\supset}.} M(x, y) \wedge \bigwedge_U \bigvee_V M_\omega(xU, yV)$$

Case $M_\omega(e, e)$: According to (14) there is a function $\Phi(xU, y)$ so that

$$M_\omega(x, y) \supset M_\omega(xU, y\Phi(xU, y)).$$

The formula

(15) $\quad Y_t = \Phi(\overline{X}_t X_t, \overline{Y}_t)$

determines a strategy for player J. If J plays this strategy, then inductively $M_\omega(\overline{X}_o, \overline{Y}_o)$, $M_\omega(\overline{X}_1, \overline{Y}_1)$, $M_\omega(\overline{X}_2, \overline{Y}_2)$, ... holds. Thus, according to (14), $(\forall_t)\ M(\overline{X}_t, \overline{Y}_t)$, thus **C**(X, Y). Therefore, in case $M_\omega(e, e)$, (15) is a winning strategy for J.

**Case** $\sim M_\omega(e, e)$: According to (13) there is a function $\Psi_i(x, y)$ so that

$$\sim M_i(x, y) \wedge M(x, y) .\supset.(\exists j)^i \sim M_j(x\Psi_i(x, y), yV).$$

The formulas

(16) $i_t = (\mu i) \sim M_i(\overline{X}_t, \overline{Y}_t)$, $(\mu i)$ = the first i so that ...
$X_t = \Psi_{i_t}(\overline{X}_t, \overline{Y}_t)$

determine a strategy for player I. If I plays this strategy, then $[i_o > ... > i_t] \vee \sim M(\overline{X}_t, \overline{Y}_t)$. As $i_o > i_1 > i_2$ ... cannot happen, it follows that $(\exists t) \sim M(\overline{X}_t, \overline{Y}_t)$, thus $\sim$ **C**(X, Y). Thus in case $\sim M_\omega(e, e)$, (16) is a winning strategy for I. We thus have the following result:

**Theorem 7**: Every closed game (12) is determined. In fact, J has the winning strategy (15) if $M_\omega(e, e)$, and I has the winning strategy (16) if $\sim M_\omega(e, e)$.

This theorem is not new, and in fact Davis [7] has proved:

**Theorem 8**: Every $F_{\sigma\delta}$-game is determined.

By remark 4 follows: Every finite-state game is determined. However, the proof of Davis does not give the stronger (and for the solvability algorithm of corollary 6 essential) theorem 5. Our proof of theorem 7 indicates how the method of Cantor-Bendixson not only gives the determinacy, but also allows one to effectively write winning strategies.

Hausdorff [8] showed that for Suslin sets the alternative (11) still holds (lopsided Suslin games are determined). His proof is a variant of the Cantor-Bendixson method. That also ordinary Suslin games are determined was proved by D.A. Martin from the existence of measurable cardinal numbers. This proof, as that of Davis, is not of the Cantor-Bendixson type.

**Literature**

1. J.R. Büchi: *On a decision method in restricted second order arithmetic*, Proc. Int. Congr. Logic, Method. and Philos. Sci. 1960, Stanford Univ. Press, Stanford, Cal. 1962.

2. J.R. Büchi, Elgot and Wright: *The nonexistence of certain algorithms of finite automata theory*, Notices Am. Math. Soc. 5, 1958, p. 98.

3. J.R. Büchi and Landweber: *Solving sequential conditions by finite-state strategies.* Trans. Am. Math. Soc. 138, 1969, pp. 295-311

4. J.R. Büchi and Landweber: *Definability in the monadic second order theory of successor*, Journ. Symb. Logic 34, 1969, pp. 166-170

5. A. Church: *Application of recursive arithmetic to the problem of circuit synthesis*, Summaries of talks presented at the Summer Institute for Symbolic Logic, Cornell Univ. 1957; 2nd ed., Princeton N.J., 1960, pp. 3-50

6. J.R. Büchi: *Logic arithmetic and automata*, Proc. Int. Congr. Math. 1963, Almqvist and Wiksells, Uppsala 1963

7. M. Davis: *Infinite games of perfect information, Advances in game theory,* Princeton Univ. Press, Princeton, N.J., 1964, pp. 85-101

8. F. Hausdorff: *Grundzüge der Mengenlehre*, New York, Chelsea Publ. Co. 1949

9. R. McNaughton: *Testing and generating infinite sequences by a finite automaton,* Information and Control 9, 1966, pp. 521-530

10. M. Rabin: *Decidability of second-order theories and automata on infinite trees,* IBM Research RC 2012, 1968. See also Trans. Am. Math. Soc. 141, 1969, pp. 1-35

# On the Presentation of Winning Strategies via the Cantor-Bendixson Method

J. Richard Büchi and Stephen Klein

One of the first uses of Cantor's theory of ordinal numbers, if not the very reason Cantor invented ordinals, was in showing that for each closed set $\mathscr{C}$, the following alternative holds:

(1) $\mathscr{C}$ is countable $\vee$ $\mathscr{C} \supseteq$ perfect set.

Because every perfect set has the cardinality of the continuum, (1) implies the continuum hypothesis for closed sets. That (1) is true for closed $\mathscr{C}$ is a consequence of: There is an $\alpha$ less than $\omega_1$ such that the $\alpha$-derivative $\mathscr{C}^\alpha$ of $\mathscr{C}$ is either empty (in which case $\mathscr{C}$ is countable) or perfect.

A very nice way of presenting this idea is in the context of infinite games. A set $\mathscr{C} \subseteq 2^\omega = \{0,1\}^\omega$ defines the lopsided game $\mathscr{C}$ for two players. Player I first chooses a finite (possibly empty) sequence $x_0$ of 0's and 1's, and player II chooses $Y_0$ to be either 0 or 1. The players continue alternately, choosing $x_1$, $Y_1$, $x_2$ $Y_2$, etc., so that they form the infinite sequence $S = x_0Y_0x_1Y_1x_2Y_2\ldots$ . Player I wins if $S \in \mathscr{C}$, otherwise player II wins. A strategy for player I is a function $\Phi\colon \{0,1\}^* \to \{0,1\}^*$, where $A^*$ denotes the set of all finite sequences of elements of A. A strategy $\Phi$ for player I gives, for any sequence $Y \in \{0,1\}^\omega$ chosen by player II, the sequence $\underline{x}$, where $\underline{x}t = \Phi(\underline{x}_0Y_0\underline{x}_1Y_1\ldots\underline{x}(t-1)Y(t-1))$. $\Phi$ is a winning strategy if, for every $Y \in \{0,1\}^\omega$, $\Phi$ gives an

$\underline{x} \in (\{0,1\}^*)^{\omega}$ such that $\underline{x}_0 Y_0 \underline{x}_1 Y_1 \ldots \in \mathscr{L}$. By the following theorem, (1) says precisely, "The lopsided game $\mathscr{L}$ is determined".

<u>Theorem</u> (Davis [4]): Player II has a winning strategy in the lopsided game $\mathscr{L}$ iff $\mathscr{L}$ is countable. Player I has a winning strategy iff $\mathscr{L}$ contains a perfect set.

We will show that the Cantor-Bendixson method very naturally extends from F (the closed sets)(see [1]) to $F_\sigma$. This new proof of determinateness of $F_\sigma$ games has the advantage over other proofs [7] of displaying, in each case, the winning strategy in the sense of presenting a defining formula. This should be compared with the finite-state case, where the method was reinvented by Church and Wright (see [3]), and where the winning strategy is concretely displayed in the form of a finite-state recursion. (See also [2]).

<u>Notations</u>: Let A be any well-ordered set. If X is any function from the set $\omega$ of natural numbers into A, then $\overline{X}t$ is the finite sequence $X_0 X_1 X_2 \ldots X(t-1)$. The set of all finite sequences of elements of A is denoted by $A^*$. We use the notation $(\exists\eta)^{\xi}\Phi$ for $(\exists\eta)[\eta < \xi \wedge \Phi]$, $(\exists\eta)_{\xi}\Phi$ for $(\exists\eta)[\eta \geq \xi \wedge \Phi]$, $(\forall\eta)^{\xi}\Phi$ for $(\forall\eta)[\eta < \xi \supset \Phi]$, and $(\forall\eta)_{\xi}\Phi$ for $(\forall\eta)[\eta \geq \xi \supset \Phi]$. We write $\mathscr{L}(X,Y)$ in place of $\langle X,Y\rangle \in \mathscr{L}$,

and we call $\mathscr{C}$ an $F_\sigma$ condition if $\{\langle X,Y\rangle; \mathscr{C}(X,Y)\}$ is an $F_\sigma$ set. For $x$ and $x'$ members of $A^*$, $x \leq x'$ means $x$ is an initial segment of $x$. If $x$ ranges over a well-ordered set, $(\mu x)\Phi$ stands for "the least $x$ such that $\phi$" if $(\exists x)\Phi$, otherwise, $(\mu x)\Phi$ stands for the first $x$ in the implied range of $x$. The least ordinal with a cardinality greater than $\alpha$ is denoted by $\alpha^+$. The empty sequence is denoted by $e$.

Let $A$ and $B$ be well-ordered sets and let $\mathscr{C}(X,Y)$ be a condition on $A^\omega \times B^\omega$. The infinite game defined by $\mathscr{C}$ is the game in which players I and II alternately choose elements from $A$ and $B$, respectively, thus choosing sequences $X \in A^\omega$ and $Y \in B^\omega$. Player I wins if $\mathscr{C}(X,Y)$, otherwise player II wins.

Let $\mathscr{C}(X,Y)$ be given by

$$(2) \qquad \mathscr{C}(X,Y) \equiv (\exists u)(\forall t)M_u[\bar{X}t,\bar{Y}t],$$

where for each $u \in \omega$, $M_u$ is a condition on $A^* \times B^*$ and

$$(3) \qquad x' \leq x \wedge y' \leq y \supset. M_u[x,y] \supset M_u[x',y']$$

$$(4) \qquad v < u \supset. M_v[x,y] \supset M_u[x,y].$$

Every condition which can be so expressed is an $F_\sigma$ condition and, conversely, every $F_\sigma$ condition can be so expressed.

Let $\gamma = |\omega \times A^* \times B^*|^+$. For each $\xi \leq \gamma^+$ and for each $\mu \leq \gamma$ we define $M_\mu^\xi$ by:

$$(5) \quad M_\mu^\xi(u,x,y) \equiv (\exists\eta)^\xi\, M_\gamma^\eta(u+1,x,y) \vee [M_u[x,y] \wedge (\exists a)^A(\forall b)^B(\forall\nu)^\mu\, M_\nu^\xi(u,xa,yb)].$$

<u>Theorem</u>: There exists a $\zeta < \gamma^+$ such that $M_\mu^\zeta(u,x,y) \equiv M_\mu^{\zeta+1}(u,x,y)$.

a.) If $M_\gamma^\zeta(0,e,e)$, then player I has a winning strategy,

and

b.) If $\neg M_\gamma^\zeta(0,e,e)$, then player II has a winning strategy.

Thus, the game defined by $\mathscr{L}$ is determined.

<u>Proof</u>: We first note that

$$(6) \quad \sigma < \xi \supset. M_\mu^\sigma(u,x,y) \supset M_\mu^\xi(u,x,y).$$

This may be proved by induction on $\mu$ simultaneously for all u, x, y. By (6), the sets $\{\langle\mu,u,x,y\rangle;\, M_\mu^\xi(u,x,y)\}$ form a non-decreasing sequence, and each $\{\langle\mu,u,x,y\rangle;\, M_\mu^\xi(u,x,y)\}$ is contained in $\gamma \times \omega \times A^* \times B^*$, which has cardinality $\gamma$. Therefore, there exists a $\xi < \gamma^+$ such that $(\forall\mu uxy)M_\mu^\xi(u,x,y) \equiv M_\mu^{\xi+1}(u,x,y)$. Let $\zeta$ be the least such $\xi$. (Note that $(\forall\xi)_\zeta M_\mu^\zeta(u,x,y) \equiv M_\mu^\xi(u,x,y)$, but we will not need this fact.)

<u>Case a.)</u> $M_\gamma^\zeta(0,e,e)$. As an immediate consequence of (5), we have

$$(7) \quad \rho < \mu \supset. M_\mu^\xi(u,x,y) \supset M_\rho^\xi(u,x,y).$$

By (7), for each $\xi$, the sets $\{\langle u,x,y\rangle;\ M^{\xi}_{\mu}(u,x,y)\}$ form a non-increasing sequence, and $\{\langle u,x,y\rangle;\ M^{\xi}_{0}(u,x,y)\} \subseteq \omega \times A^* \times B^*$ has cardinality less than $\gamma$, so there is a $\mu < \gamma$ such that $M^{\xi}_{\mu}(u,x,y) \supset M^{\xi}_{\mu+1}(u,x,y)$. From this it follows by induction that $M^{\xi}_{\mu}(u,x,y) \supset (\forall\nu)_{\mu} M^{\xi}_{\nu}(u,x,y)$, so, in particular, $M^{\xi}_{\mu}(u,x,y) \supset M^{\xi}_{\gamma}(u,x,y)$. Thus, we have

$$(\forall\xi)(\exists\mu)^{\gamma}(\forall uxy)\cdot M^{\xi}_{\mu}(u,x,y) \supset M^{\xi}_{\gamma}(u,x,y),$$

and this implies

(8) $(\forall\xi)(\forall uxy)\cdot(\forall\mu)^{\gamma} M^{\xi}_{\mu}(u,x,y) \supset M^{\xi}_{\gamma}(u,x,y).$

From (8) and (5) we derive

(9) $M^{\xi}_{\gamma}(u,x,y) \supset (\exists\eta)^{\xi} M^{\eta}_{\gamma}(u+1,x,y) \vee [M_u[x,y] \wedge (\exists a)^{A}(\forall b)^{B} M^{\xi}_{\gamma}(u,xa,yb)].$

We now present a strategy for player I. Let $Y \in B^{\omega}$ be a play of player II. We will define sequences H, U and X by recursion such that for all t, $M^{Ht}_{\gamma}(Ut,\bar{X}t,\bar{Y}t)$. Player I first uses the disjunct $(\exists\eta)^{\xi} M^{\eta}_{\gamma}(u+1,x,y)$ repeatedly until he finds an $\eta < Ht$ and a $v > Ut$ such that $M^{\eta}_{\gamma}(v,\bar{X}t,\bar{Y}t)$ and $\neg(\exists\rho)^{\eta} M^{\rho}_{\gamma}(v+1,\bar{X}t,\bar{Y}t)$. Then let $Ht' = \eta$ and $Ut' = v$. In case $\neg(\exists\eta)^{Ht} M^{\eta}_{\gamma}(Ut+1,\bar{X}t,\bar{Y}t)$, then take $Ht' = Ht$, $Ut' = Ut$.

So we have $M_\gamma^{Ht'}(Ut',\bar{X}t,\bar{Y}t)$ and $\neg(\exists\eta)^{Ht'}M_\gamma^\eta(Ut'+1,\bar{X}t,\bar{Y}t)$. By (9) it follows that $(\exists a)^A(\forall b)^B M_\gamma^{Ht'}(Ut',(\bar{X}t)a,(\bar{Y}t)b)$. For Xt player I chooses an $a \in A$ (e.g., the smallest) such that $(\forall b)^B M_\gamma^{Ht'}(Ut',(\bar{X}t')a,(\bar{Y}t)b)$. We define H, U, X by the following recursions:

$$H0 = \zeta \qquad\qquad U0 = 0$$

$$\begin{aligned} Ht' &= (\mu\eta)(\exists v)_{Ut}[M_\gamma^\eta(v,\bar{X}t,\bar{Y}t) \wedge \neg(\exists\rho)^\eta M_\gamma^\rho(v+1,\bar{X}t,\bar{Y}t)] \\ Ut' &= (\mu v)[v \geq Ut \wedge M_\gamma^{Ht'}(v,\bar{X}t,\bar{Y}t) \wedge \neg(\exists\rho)^{Ht'}M_\gamma^\rho(v+1,\bar{X}t,\bar{Y}t)] \\ Xt &= (\mu a)[(\forall b)^B M_\gamma^{Ht'}(Ut',(\bar{X}t)a,(\bar{Y}t)b]. \end{aligned} \tag{10}$$

The recursions (10) clearly define a strategy for player I; we have yet to prove that it is a winning strategy. A simple application to (9) of the fact that every non-increasing sequence of ordinals is ultimately constant yields

$$M_\gamma^\xi(u,x,y) \supset (\exists\eta)^{\xi+1}(\exists v)_u[M_\gamma^\eta(v,x,y) \wedge \neg(\exists\rho)^\eta M_\gamma^\rho(v+1,x,y)]. \tag{11}$$

We now prove, for all t

$$M_\gamma^{Ht}(Ut,\bar{X}t,\bar{Y}t)$$

$$Ht \geq Ht'$$

$$Ht = Ht' \supset Ut = Ut' .$$

By the assumption $M_\gamma^\zeta(0,e,e)$ and (10), we have

$M_\gamma^{H0}(U0,\bar{X}0,\bar{Y}0)$.

Assume $M_\gamma^{Ht}(Ut,\bar{X}t,\bar{Y}t)$. By (11) we get

$$(\exists\eta)^{Ht+1}(\exists v)_{Ut}[M_\gamma^\eta(v,\bar{X}t,\bar{Y}t) \wedge \neg(\exists\rho)^\eta M_\gamma^\rho(v+1,\bar{X}t,\bar{Y}t)].$$

Therefore,

$$Ht \geq Ht' \wedge Ut' \geq Ut \wedge M_\gamma^{Ht'}(Ut',\bar{X}t,\bar{Y}t) \wedge \neg(\exists\rho)^{Ht'}M_\gamma^\rho(Ut'+1,\bar{X}t,\bar{Y}t).$$

From (9) it follows that

$$(\exists a)^A(\forall b)^B M_\gamma^{Ht'}(Ut',(\bar{X}t)a,(\bar{Y}t)b),$$

so by (10),

$$(\forall b)^B M_\gamma^{Ht'}(Ut',(\bar{X}t)Xt,(\bar{Y}t)b)$$

and in particular, $M_\gamma^{Ht'}(Ut',\bar{X}t',\bar{Y}t')$. It follows by induction that for all $t$,

$$Ht \geq Ht' \wedge M_\gamma^{Ht}(Ut,\bar{X}t,\bar{Y}t).$$

If $(\exists\eta)^\xi M_\gamma^\eta(u+1,x,y)$ and $(\exists\eta)^{\xi+1}(\exists v)_u[M_\gamma^\eta(v,x,y) \wedge \neg(\exists\rho)^\eta M_\gamma^\rho(v+1,x,y)]$, then $(\exists\eta)^{\xi}(\exists v)_u[M^{\eta}_{\gamma}(v,x,y) \wedge \neg(\exists\rho)^{\eta}M^{\rho}_{\gamma}(v+1,x,y)]$. From this it follows that if $(\exists\eta)^{Ht}M^{\eta}_{\gamma}(Ut+1,\overline{X}t,\overline{Y}t)$, then $Ht' < Ht$. Now, suppose $Ht = Ht'$. Then $M^{Ht'}_{\gamma}(Ut,\overline{X}t,\overline{Y}t)$ and $\neg(\exists\rho)^{Ht}M^{\rho}_{\gamma}(Ut+1,\overline{X}t,\overline{Y}t)$ so $Ut' \leq Ut$. Since we also have, by (10), $Ut' \geq Ut$, we get $Ut' = Ut$. Thus

(12) $Ht = Ht' \supset Ut = Ut'$.

As $H$ is a non-increasing sequence of ordinals, it must be ultimately constant. Let $\eta$ be the ultimate value of $H$ and let $t_o$ be such that $Ht_o = \eta$. By (12), $u = Ut_o$ is the ultimate value of $U$, i.e., $(\forall t)_{t_o} Ut = Ut_o = u$. So for all $t \geq t_o$ we have, by (10)

$$M^{\eta}_{\gamma}(u,Xt,Yt) \wedge \neg(\exists\rho)^{\eta}M^{\rho}_{\gamma}(u+1,\overline{X}t,\overline{Y}t)$$

and therefore, by (9), $M_u[\overline{X}t,\overline{Y}t]$. Finally, this yields, by (3), $(\forall t)M_u[\overline{X}t,\overline{Y}t]$, and thus $\mathscr{L}(X,Y)$, so player I wins. This proves that the strategy given by (10) is a winning strategy for player I.

Case b.) $\neg M^{\zeta}_{\gamma}(0,e,e)$. From $M^{\zeta}_{\mu}(u,x,y) \equiv M^{\zeta+1}_{\mu}(u,x,y)$ and (5) we get

(13) $\neg M_{\mu}^{\zeta}(u,x,y) \supset \neg M_{\gamma}^{\zeta}(u+1,x,y) \wedge [M_u[x,y] \supset (\forall a)^A (\exists b)^B (\exists \nu)^{\mu}\ \neg M_{\nu}^{\zeta}(u,xa,yb)]$.

Formula (13) indicates why $\neg M_{\gamma}^{\zeta}(0,e,e)$ assures that player II has a winning strategy, and it shows quite clearly what that strategy should be. Namely, player II chooses sequences N and U and X so that the following hold:

1. $(\forall t)\ \neg M_{Nt}^{\zeta}(Ut,\bar{X}t,\bar{Y}t)$
2. As long as $M_{Ut}[\bar{X}t,\bar{Y}t]$ continues to hold, U remains constant and N decreases.
3. When $\neg M_{Ut}[\bar{X}t,\bar{Y}t]$ holds, Ut' is chosen larger than Ut, and (4) assures that $\neg M_v[\bar{X}t,\bar{Y}t]$ for $Ut \le v < Ut'$.

We now present a strategy for player II via recursive definitions of sequences N, U and Y.

$$N0 = \gamma \qquad U0 = 0 \qquad Ut' = (\mu u)M_u[\bar{X}t,\bar{Y}t]$$

$$(14) \qquad Yt = \begin{cases} (\mu b)(\exists \nu)^{Nt}\ \neg M_{\nu}^{\zeta}(Ut',\bar{X}t',(\bar{Y}t)b) & \text{if } M_{Ut}[\bar{X}t,\bar{Y}t] \\ (\mu b)(\exists \nu)^{\gamma}\ \neg M_{\nu}^{\zeta}(Ut',\bar{X}t',(\bar{Y}t)b) & \text{if } \neg M_{Ut}[\bar{X}t,\bar{Y}t] \end{cases}$$

$$Nt' = (\mu\nu)\ \neg M_{\nu}^{\zeta}(Ut',\bar{X}t',\bar{Y}t').$$

Suppose that for some t, $(\forall u)\ \neg M_u[\bar{X}t,\bar{Y}t]$. This implies $(\forall u)(\exists t)\ \neg M_u[\bar{X}t,\bar{Y}t]$, so $\neg\mathscr{C}(X,Y)$ and player II wins.

For the remainder of the proof we will assume $(\forall t)(\exists u)M_u[\bar{X}t,\bar{Y}t]$, i.e., $(\forall t)M_{Ut'}[\bar{X}t,\bar{Y}t]$. We will prove by induction on $t$ that

$$(15) \qquad (\forall t).\ \neg M^{\zeta}_{Nt}(Ut,\bar{X}t,\bar{Y}t) \wedge \neg(\exists u)^{Ut}M_u[\bar{X}t,\bar{Y}t].$$

By the assumption $\neg M^{\zeta}_{\gamma}(0,e,e)$ and (14), we have

$$\neg M^{\zeta}_{N0}(U0,\bar{X}0,\bar{Y}0).$$

Assume $\neg M^{\zeta}_{Nt}(Ut,\bar{X}t,\bar{Y}t) \wedge \neg(\exists u)^{Ut}M_u[\bar{X}t,\bar{Y}t]$. We continue by cases, depending on whether $M_{Ut}[\bar{X}t,\bar{Y}t]$ holds or not.

Case 1, $M_{Ut}[\bar{X}t,\bar{Y}t]$. $M_{Ut}[\bar{X}t,\bar{Y}t]$ and $\neg(\exists u)^{Ut}M_u[\bar{X}t,\bar{Y}t]$ implies, by (14), $Ut = Ut'$. $\neg(\exists u)^{Ut'}M_u[\bar{X}t,\bar{Y}t]$ implies, by (3),

$$\neg(\exists u)^{Ut'}M_u[\bar{X}t',\bar{Y}t'].$$

Since $Ut' = Ut$, the induction hypothesis gives $\neg M^{\zeta}_{Nt}(Ut',\bar{X}t,\bar{Y}t)$, which, with (13), gives

$$(\forall a)^{A}(\exists b)^{B}(\exists v)^{Nt}\ \neg M^{\zeta}_{v}(Ut',(\bar{X}t)a,(\bar{Y}t)b).$$

Thus, by (14)

$$\neg M^{\zeta}_{Nt'}(Ut',\bar{X}t',\bar{Y}t')$$

and

$$Nt' < Nt.$$

Case 2, $\neg M_{Ut}[\bar{X}t,\bar{Y}t]$. By (14), $\neg(\exists u)^{Ut'} M_u[\bar{X}t,\bar{Y}t]$, and so, by (3),

$$\neg(\exists u)^{Ut'} M_u[\bar{X}t',\bar{Y}t'].$$

By (13), $\neg M^{\zeta}_{Nt}(Ut,\bar{X}t,\bar{Y}t)$ implies $(\forall v)_{Ut+1}\ \neg M^{\zeta}_{\gamma}(v,\bar{X}t,\bar{Y}t)$. Since $Ut' > Ut$, we get

$$\neg M^{\zeta}_{\gamma}(Ut',\bar{X}t,\bar{Y}t),$$

which, with $M_{Ut'}[\bar{X}t,\bar{Y}t]$, implies, by (13),

$$(\forall a)^{A}(\exists b)^{B}(\exists \nu)^{\gamma}\ \neg M^{\zeta}_{\nu}(Ut',(\bar{X}t)a,(\bar{Y}t)b).$$

Thus, $(\exists \nu)^{\gamma}\ \neg M^{\zeta}_{\nu}(Ut',\bar{X}t',\bar{Y}t')$, and therefore

$$\neg M^{\zeta}_{Nt'}(Ut',\bar{X}t',\bar{Y}t').$$

This completes the proof by induction of (15).

Note that the above proof also shows

$$(\forall t).[Nt' < Nt \wedge Ut' = Ut] \vee Ut' > Ut.$$

Thus, as long as U is constant, N must decrease, and therefore U can be constant only on a finite interval. Since U is also non-decreasing, we have

$$(\forall u)(\exists t)Ut > u.$$

Let $u$ be arbitrary. Pick $t_o$ so that $Ut_o > u$. Pick $t_1 \geq t_o$ such that $Ut_1 = Ut_o \wedge \neg M_{Ut_1}[\bar{X}t_1, \bar{Y}t_1]$. By (4), $\neg M_{Ut_1}[Xt_1, Yt_1]$ implies $\neg M_u[\bar{X}t_1, \bar{Y}t_1]$. Thus

$$(\forall u)(\exists t) \neg M_u(\bar{X}t, \bar{Y}t],$$

i.e., $\neg \mathscr{C}(X,Y)$ and player II wins.

Thus, the strategy given by (14) is a winning strategy for player II.

This completes the proof of the theorem.

The above proof does not depend on the axiom of choice. And it is general in that A and B may be any well-ordered sets. We cite some particular cases.

First, we may have $A = B = \{0,1\}$. In this case $A^\omega \times B^\omega$ (or $(A \times B)^\omega$) is the Cantor space. We may take $\gamma$ to be $\omega_o$ instead of $\omega_1$ and use an inaccessibility argument instead of a cardinality argument to prove (8). Then $\zeta$ will be less than $\omega_1$. More generally, we may take A and B to be any finite sets.

Second, we may take $A = B = \omega$, as does Martin [6]. Here, $\gamma = \omega_1$ and $\zeta$ is less than $\omega_2$.

Third, we may take $B = \{0,1\}$, $A = \{0,1\}^*$ and restrict our attention to those $\mathscr{C}(X,Y)$ which satisfy

$$X_oY_oX_1Y_1 \ldots = V_oW_oV_1W_1 \ldots \quad \mathscr{C}(X,Y) \supset \mathscr{C}(V,W)$$

and thus cover the lopsided games mentioned in the introduction. Thus, we have a new proof that every $F_\sigma$ $\mathscr{C}$ satisfies (1).

At this time we do not know how to extend our method to $F_{\sigma\delta}$ games. We feel that such an extension would yield important new insight into the matter of determinateness, and might eventually lead to a better result on Souslin games than Martin's result, which depends on the existence of a measurable cardinal. In the lopsided case, Hausdorff has proved that (1) holds for Souslin $\mathscr{C}$.

References

1. J. R. Büchi, Algorithmisches Konstruieren von Automaten und die Herstellung von Gewinnstrategien nach Cantor-Bendixson, Automateneheorie und formale Sprachen, Mannheim, 1970, 385-398.

2. ____________ and L. H. Landweber, Solving sequential conditions by finite-state strategies, Trans. Amer. Math. Soc. 138(1969), 295-311.

3. A. Church, Application of recursive arithmetic to the problem of circuit synthesis, Summaries of talks presented at the Summer Institute for Symbolic Logic, Cornell University, 1957, 2nd edition, 1960, 3-50.

4. Morton Davis, Infinite games of perfect information, Advances in game theory, Ann. of Math. Study No. 52 1964, 85-101.

5. F. Hausdorff, Grundzüge der Mengenlehre, Leipzig, 1914.

6. D. A. Martin, Measurable cardinals and analytic games, Fund. Math. 66 (1970), 287-291.

7. P. Wolfe, The strict determinateness of certain infinite games, Pacific J. of Math. 5 (1955), 891-897.

June, 1972

Report Purdue University CSD TR-81

This work was supported by Grant No. GJ-980 from the N.S.F.

# Using Determinancy of Games to Eliminate Quantifiers.

J. Richard Buchi, Purdue University

A. Combinatorial lemmas reduce quantifiers: To have recognized the fundamental role that quantifiers play, is one of Frege's contributions to mathematics. Elimination of quantifiers, however, was not invented by logicians. In fact, it is easily the most important thing that happens in any mathematical proof. Investigation would probably reveal a direct relation between the usefulness of a theorem, and its ability to simplify quantifications. (The same goes for notions, e.g., continuous everywhere versus uniformly continuous). In particular, (what some call) infinity lemmas, or (what others call) combinatorial lemmas, turn out to be simple instructions for replacing bad combinations $(\forall\exists)$ by more manageable ones $(\exists\forall)$. Here is a list of examples:

(1) Axiom of choice:

$$(\forall x)(\exists y)Rxy \;.\equiv.\; (\exists f)(\forall x)Rxfx$$

(2) Infinity lemma (compactness):

$$(\forall x)(\exists \bar{Z}x)(\forall t)^{x}M(\bar{X}t,\bar{Z}t) \;.\equiv.\; (\exists Z)(\forall t)M(\bar{X}t,\bar{Z}t)$$

(3) Ramsey's lemma:

$$(\forall Z)^{inf}(\exists y)(\exists x)^{y}[Zx\wedge Zy\wedge \tilde{R}xy] \;.\supset.\; (\exists Z)^{inf}(\forall y)(\forall x)^{y}[Zx\wedge Zy\supset Rxy]$$

Why, in the course of a proof, is the right side $\exists\forall$ more desirable? Having arrived at $(\exists x)(\forall y)Sxy$, I will simply say "let $b$ be one of these $x$, and so $(\forall y)Sby$". Such an "existentiation" permanently eliminates a quantifier. In contrast, having arrived at $(\forall x)(\exists y)Sxy$, I might say "let $b$ be any one of these $x$, and so $(\exists y)Sby$". This trick of "the generic element" eliminates the quantifier $(\forall x)$, but only temporarily. At any rate, in a later state I will have to recall that $b$ was generic.

In the course of this lecture we will add to the list some combinatorial lemmas from automata theory. They all come with nice proofs, which are just as non-trivial as you may desire, and they all say very real things about rationals and their relation to reals. Here we add some remarks about the history of infinity lemmas.

The compactness lemma generalizes to Tychonoff's theorem, and in this form becomes another version of AC. In 1949 R. Rado proved a useful combinatorial lemma (and uses it where a straight application of AC would do). Actually, his lemma is just a way of stating Tychonoff's. In 1964 A. Robinson announced a lemma (which he uses in model theory). This turned out to be a special case of Rado's, and actually just a way of stating the infinity lemma. Ramsey also used his lemma where nothing that refined is needed. (So did I in 1960). Moral: If you discover a combinatorial lemma, be careful, it may be old hat, or you might not actually need it.

The axiom of choice does, of course, not actually deserve to be called a combinatorial lemma. For one thing, a c.l. comes with a nice proof. Better think of AC as a schematic form, which indicates what is wanted, and which is to be proved for special R's. The infinity lemma is an example of this, and so are other combinatorial lemmas.

B. Systematic elimination of quantifiers: The idea of methodically eliminating quantifiers is due to Löwenheim 1915. He thus obtained the now famous countable models for elementary axiomsystems, and showed that the monadic theory MT[S] of any set S is decidable (called Behmann's theorem). These were contributions to the axiomatic method, and the foundations of mathematics, of a new caliber. So, nobody seems to have paid much attention, except for Skolem. He extended Löwenheim's theorem, about the countable model, to infinite axiomsystems. He then found that Zermelo's "Aussonderungs Axiom", and the induction axiom could be more rigorously formalized, as axiom schemata. And now the method of quantifier elimination gave these exiting results: 1. There are (if any) countable models for set-theory. 2. There are non-standard models of the axioms of number theory. Skolem also showed how to use the infinity lemma, in place of the more controversial AC. In this form quantifier elimination was used by Gödel, to show that the basic notion of axiomatics "$\Sigma$ is logical consequence of the axioms A" is semi-recursive.

Skolem also worked on Löwenheim's decision method, and showed how to eliminate quantifiers from the theory of $\langle N, + \rangle$ (Presburger's decision method). Both these theories are included in MT[$\omega$,o,'] (the monadic theory of one successor), and the early ideas can be recognized in the proofs we will discuss. Tarski's decision method for the ordered field of real numbers was the first to cover sentences which could (and I hear, did) occur as genuine mathematical problems. Of this same caliber is Rabin's method for the MT of two successors. We shall outline how, in this case, determinacy of games can be used, to systematically eliminate quantifiers.

Some have dreamed about implementing such decision methods on the beautiful modern machines. Some feel that the complexity boys have spoiled these dreams. But then, if one was to heed present complexity theory, how would he dare implement propositional calculus, and how else was he going to use the machine, if he was to believe it can't handle truth tables.

Others feel that the resulting decision method is just a nice way to summarize the work done. The crux of the matter are the ideas which go into the single step of quantifier elimination. I will take these ideas, and what they tell me about the nature of the theory under investigation; you can have the decision method. Apropo: such ideas do sometimes occur in papers which talk about model completeness. So you look up what this is all about; it's nice. But next time they will tell you about model companions. I am sure these are nice things too; but now you will have to work still more to find the place where the quantifier is being eliminated.

C. The two-splitting tree: Let $2 = \{0,1\} = \{F,T\}$ and let $\omega$ be the set of natural numbers. We use variables like $t,x,y$ to range over $\omega$, and $X,Y,Z$ to range over $2^\omega$. So $X$ may be viewed as a subset of $\omega$, or alternatively, as an $\omega$-sequence of members of 2. $\bar{X}t = Xo\ldots X(t-1)$ stands for the course of values of $X$, up to the place $t$. Thus $\bar{X}t$ is a member of $2^*$, the set of all words over the symbols F,T. Variables like $x$, $y$, $z$ are used to range over $2^*$.

The e items provide a very natural and efficient set of notations, for discussing the oriented two splitting tree $N_2$. Namely, $N_2$ has the vertices $2^*$, the empty word e is its root, and $xF$, $xT$ are the (left,right) successors of the vertex $x$.

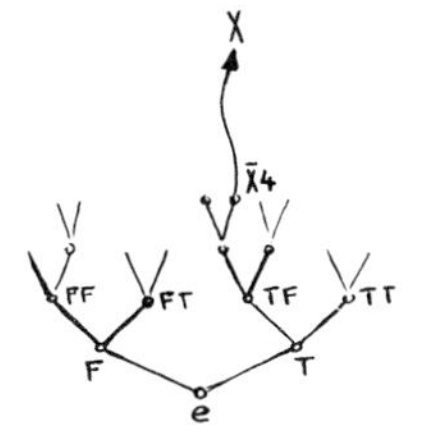


$\omega$-level $= 2^\omega$

4-level $= 2^4$

0-level $= 2^0$

The elements of $\omega$ mark the levels of the tree $N_2$, and the members of $2^\omega$ are just the infinite paths through $N_2$. Finally, $\bar{X}t$ is the vertex on the path $X$, which occurs at level $t$. Note that $N_2 = \langle 2^*,e,xF,xT\rangle \cong \langle \omega,0,2x+1,2x+2\rangle$ is the totally free algebra with one generator and two unary functions. It should be compared to the one-successor algebra $N_1 = \langle \omega,0,x+1\rangle$. In addition we use notations like $M,\Phi,\Psi$ for sets of vertices, i.e., functions of $2^*$ to 2. More generally $M$(or $Z$) may denote a function from $2^*$(from $\omega$) into any finite set $n$. Furthermore, $Z$ may be an m-tuple $Z_1,\ldots,Z_m$, which we identify with the sequence of tuples $Z_1t,\ldots,Z_mt$. The function $M:2^* \to n$ puts a marker $Mx$ onto every vertex $x$ of $N_2$. So $M$ may be visualized as a marked tree.

The brackets [,] are used so that you can spot propositional calculus from one mile off. So, D[-] denotes a truth-function, or an n-vector of truth-functions, or function from a finite set to a finite set. Occasionally we will violate this convention. Namely, when we are dealing with $\omega^*$ and $\omega^\omega$, and we are doing the analog to a thing which, in the case of $2^*$ and $2^\omega$, requires truth functions.

$(\exists t)^x$ and $(\forall t)^x$ are bounded quantifiers, made famous by Skolem. $(\exists t)_x$, $(\forall t)_x$ stand for $(\exists t)[x\le t\ \ldots]$ and its dual. From these one defines $(\exists^\omega t):(\forall x)(\exists t)_x$ and $(\forall^\omega t):(\exists x)(\forall t)_x$. They mean "for infinitely many t" and "for all but finitely many t". For $Z:\omega \to n$ we let $\sup Z = \sup_t(Zt) = \{a:(\exists^\omega t)Zt=a\}$. So $\sup Z$ is the set of values which $Z$ takes infinitely often. This provides for efficient and intuitively manageable handling of Boolean combinations of expressions, $(\exists^\omega t)L(t)$. Namely: Let $Z:\omega \to n$. Let $L(Z)$ be any Boolean combination of parts of form $(\exists^\omega t)C[Zt]$ and $(\forall^\omega t)C[Zt]$. Then $L(Z)$ is equivalent to $\sup Z \in \mathcal{U}$, whereby $\mathcal{U}$ is a set of subsets of $n$. Conversely, for any such $\mathcal{U}$, $\sup Z \in \mathcal{U}$ means $\bigvee_{D\in\mathcal{U}}[(\forall^\omega t)D[Zt] \wedge \bigwedge_{a\in D}(\exists^\omega t)Zt=a]$.

You might feel that all this makes a rather involved set of notations. I think that it is precisely fitted to talk about the very interesting structure $N_2$. This structure occurs in automata theory, and of course it also displays the basic (topological and algebraic) relationship between rational and real numbers.

| | | | | |
|---|---|---|---|---|
| t,x | nat. numbers | level of tree | pos. of approx. | time instance |
| $x,y$ | rationals | vertices | approximations | input signals |
| X,Y | sets of n.n. | paths | reals | infinite signals |
| M,Φ | sets of rationals | marked trees | $F,G,F_\sigma$ | strategies |
| $\bar{X}t$ | vertex at level t | on path X | t-th approximation of the real X | |

Actually $2^\omega$ more naturally carries the topology of the compact Cantor-set. The real topology is obtained by considering only the infinite subsets of $\omega$ (i.e., the infinitely right-turning path through the tree $N_2$). These are naturally identified with the members of $\omega^\omega$, for which we use notations like s,f. For $\omega^*$ we use the same notations as for 2*; it will be clear that in the expression $M\bar{s}t$ the M means a function from $\omega^\omega$ into 2 (or any finite set).

D. Descriptive set-theory is a beautiful collection of combinatorial lemmas: Cantor invented his set theory for the purpose of giving rigorous definitions to the basic concepts of analysis. For example, he defined reals to be Cauchy sequences of rationals. In particular he developed the fundamentals of topology, such as the theory of closed and open sets, and the set of limitpoints of a set. These investigations led to descriptive set-theory, a branch of analysis which flourished in the early 1900s, and has come back to life. Sets (and functions) of reals are here classified with regard to the form of defining expressions. The iterated occurrence of unions and intersections are recognized as the main ingredients in these expressions, and these are but another way of representing existential and universal quantifiers. A nice way of acquiring the ideas developed in the field, is to look at Hausdorff's book, and to do things in the special cases $2^\omega$ and $\omega^\omega$. Here are the very simplest Borel sets.

| | |
|---|---|
| $C(X) \equiv (\forall t)M\bar{X}t$ | closed, or $F$-set |
| $C(X) \equiv (\exists t)M\bar{X}t$ | open, or $G$-set |
| $C(X) \equiv (\forall^\omega t)M\bar{X}t$ | $F_\sigma$-set |
| $C(X) \equiv (\exists^\omega t)M\bar{X}t$ | $G_\delta$-set |
| $C(X) \equiv \sup_t M\bar{X}t \in U$ | $Bl(F_\sigma) = Bl(G_\delta)$ = Boolean over $F_\sigma$ |

You may prefer the notations $X \in C$ and $x \in M$; so do I at times. The class $B$ of all Borel-sets is obtained from $F \cup G$, by closing up under the operation $C(X) \equiv (\exists t)C_t(X)$ and its dual. This process of "closing up" can be analyzed into steps, using Cantor's countable ordinals. Souslin sets are the simplest which can be obtained by using quantification over reals:

$$C(X) \equiv \begin{array}{ll} (\exists Z)(\exists^{\omega} t)\, M(\bar{X}t,\bar{Z}t) & \text{equivalent forms for} \\ (\exists Z)\sup M(\bar{X}t,\bar{Z}t) \in U & \text{Souslin-, or } \mathcal{S}\text{-sets} \\ (\exists f)(\forall t)\, M(\bar{X}t,\bar{f}t) & \end{array}$$

Note that in the first line we can not put $(\forall t)$ in place of $(\exists^{\omega} t)$. Just look at our formulation of the infinity lemma! It says "what may look like a genuine Souslin set (right side of infinity lemma), actually is only a closed set (left side)". Consider this puzzle; the Souslin-like form is proof-technically more desirable than the closed form.

Souslin's Criterion: If both $C$ and $\tilde{C}$ are $\mathcal{S}$, then $C$ is $\mathcal{B}$.

This is one of the quantifier lemmas, that make up descriptive set-theory. (Souslin's proof you will find in Hausdorff, p. 191). Others take the form of determinacy results for games. To discuss these we need operators.

E. Presentation of deterministic operators: Let $f:\omega \to \omega$ and let $\Psi: I^* \to J^*$, where I and J are finite sets. These items can be used to define an operator $Y = \Psi fX$ from $2^I$ to $2^J$. Namely, $Yt = \Psi\bar{X}ft$. You may verify that this is a continuous map, and that every continuous map from $2^I$ to $2^J$ is of this form. We will be interested in the cases $ft = t$, and $ft = t'$:

| | |
|---|---|
| $Yt = \Psi\bar{X}t$ | the 1-delay operator $Y = (\Psi X)$ |
| $Yt = \Psi\bar{X}t'$ | the 0-delay operator $Y = [\Psi X]$ |

These are properly called deterministic operators, as the value $Yt$ depends only on values $Xv$ for times $v < t$ (1-delay) or $v \le t$ (0-delay). These same operators can be presented in a more general manner:

| | det. operator $X \to Y$, presented by |
|---|---|
| $so = c,\ st' = F[st,Xt],\ Yt = D[st]$ | an infinite-state induction |
| $Zo = c,\ Zt' = F[Zt,Xt],\ Yt = D[Zt]$ | a finite-state recursion |

As the type indicates, in the second line, $Z$ takes values in a finite set $K$, $F$ takes $K \times I$ to $K$ (the transition operator), $c$ is in $K$ (the initial state), and $D$ takes $K$ to $J$ (the output). In the first line $s$ takes values in $K = \omega$. Clearly $Yt$ depends on $Xo \ldots X(t-1)$ only. So these are 1-delay operators. The presentation of 0-delay operators are obtained by replacing the output by one of form $Yt = D[st,Xt]$. State inductions may be used to present Borel and Souslin sets in a more general form. For example:

$\mathcal{C}(X) \equiv (\exists s)[so=c \wedge (\forall t)st'=F[st,Xt] \wedge \sup_t D[st] \in \mathcal{U}]$

a $\mathcal{B}l(F_\sigma)$-set, presented by state-induction.

At first sight this may seem to be a Souslin set. However, the real quantification $(\exists s)$ is only apparent. Namely $s = gX$ is uniquely determined by the state-induction, and so $\mathcal{C}(X)$ also has the form $(\forall s)[s=gX \supset \sup \in \mathcal{U}]$. From this one sees (Souslin Criterium) that $\mathcal{C}$ is Borel. Actually $\mathcal{C}$ is a $\mathcal{B}l(F_\sigma)$. Namely, $\mathcal{C}=g^{-1}(\mathcal{D})$ whereby $\mathcal{D}(s) \equiv \sup_t D[st] \in \mathcal{U}$. So $\mathcal{D}$ is a $\mathcal{B}l(F_\sigma)$, and as the state-induction operator $g$ is continuous, $\mathcal{C}$ is also $\mathcal{B}l(F_\sigma)$. Here it is essential that D takes its values in a finite set. Otherwise the sup-condition is not a Boolean expression (in fact it would contain real-number quantification).

F. Determinacy is a schema for quantifier elimination: Let I and J be finite sets, and let $C(X,Y)$ be a condition for $X:\omega \to I$, $Y:\omega \to J$.

(4) C is determinate: $\sim(\exists\Psi)(\forall X)\mathcal{C}(X,(\Psi X)) \equiv (\exists\Phi)(\forall Y) \sim\mathcal{C}([\Phi Y],Y)$

$\Psi$ wins for J $\qquad$ $\Phi$ wins for I

Consider a game between players I and J. At any time t, first J makes a move $Yt \in J$, and then I makes a move $Xt \in I$. The winner is picked on doomsday, when all the moves have been made. Namely, J wins if $C(X,Y)$, and I wins if $\sim C(X,Y)$. Both players have complete information on previous moves of the opponent. I.e., a strategy for J is a function $Yt = \Psi\bar{X}t$, which tells J what his next move $Yt$ should be, given the previous moves $\bar{X}t$ of I. Thus, a strategy for J is a 1-delay operator $Y = (\Psi X)$. As I moves after J, his strategies are 0-delay operators $X = [\Phi Y]$. A winning strategy for J is one which, no matter how I picks his moves X, produces moves $Y = (\Psi X)$ for J, such that J wins, i.e., such that $C(X,Y)$. In turn, $\Phi$ wins for I if $C(X,Y)$, for every Y and $X = [\Phi Y]$. That C is determinate now takes this nice meaning: Either player J, or then player I has a winning strategy.

It has been suggested that determinacy should be an axiom, replacing AC. But then, setting up an axiom can be a very destructive affair. In the present case it would eliminate several interesting proofs of descriptive set theory. They use AC, but to me they still show clearly that determinacy is not evident. Furthermore, there is a proof of determinacy which does not use AC, and is very much non-trivial (see Buchi and Landweber 1969).

A lopsided game is one in which player J may make "long moves", picking a word $Yt \in J^*$ in place of just an element of J. The original Cantor-Bendixson proof, as it stands, shows that lopsided closed games are determinate. Countable ordinals are used here, and I think Cantor invented them for the purpose. W. H. Young extended the result to $G_\delta$. Hausdorff (see p. 179) gives a proof which is easily fixed up to show: every $S$ is determinate in the lopsided sense. Another way of stating "$C$ is lopsided determinate" is "$C$ is countable or contains a perfect set" (if the moves of both players are in 2). So these results say that closed, $G_\delta$, and even Souslin sets satsify the continuums hypothesis.

That $S$ games are determinate was shown by Martin 1970, using large cardinals. He now can handle $B$'s without the super cardinals. Hausdorff's proof and the modern results do not use the Cantor-Bendixson-type argument. Therefore (so it seems to me) they do not provide any presentation of the winner's strategy. But this is what is needed, if determinacy is to be used for systematic quantifier elimination.

G. Regularity: Let M be a subset of $2^*$, or more generally a map from $2^*$ into a finite set n. Think of M as being a marked tree, and consider the marked subtree $M_x$, which stands over the root $x$. Define,

$x \simeq y(M) \equiv (\forall u)M(xu) = M(yu)$       The congruence induced by M

This is actually a congruence relation on the free algebra $N_2$, and it simply means $M_x = M_y$. The index, rkM, of this congruence is called the rank of M. It tells how many of the marked trees $M_x$ are different from each other. The simplest possibility is that rkM is finite; we then say that M is regular.

Regular M's are exactly those which may be presented by a finite-state recursion. That is, in the form $M\bar{X}t \equiv (\exists Z)[Zo = c \wedge (\forall x)^t Zx' = F[Zx,Xx] \wedge D[Zt]]$. It now is clear what a regular closed (Borel, Souslin) set is. Here are examples:

| | |
|---|---|
| $C(X) \equiv (\exists Z)[Zo=c, Zt'=F[Zt,Xt], \sup Z \in U]$ | $Bl(F_\sigma)^{reg}$ given by f.s.r. |
| $C(X) \equiv (\exists Z)(\exists^\omega t)M(\bar{X}t,\bar{Z}t)$ | $S^{reg}$, if M is regular |

In these days of super-set-theory one finds those who sneer at sets of rationals, particularly at simple sets of rationals. Others like simple concepts, especially if they are finiteness notions, and very non-trivial things can be said about them. The fact is that regularity theory has produced a series of combinatorial lemmas, with the kind of proof which can be seen in any company. These ideas must be useful in other places.

H. Monadic second order theories: The formulas of these contain two sorts of variables, individual variables which are intended to range over a basic domain D, and set-variables which range over some or all subsets of D. The monadic theory, MT[S], of a structure S consists of all monadic sentences which are true in S, whereby the range of set-variables is taken to be all subsets of S. If S is just a set, MT[S] is decidable (Löwenheim). We will discuss $MT_1 = MT[\omega,o,']$ = the monadic theory of one successor, and $MT_2 = MT[2^*,e,xF,xT]$ = the monadic theory of two successors. The range of the variables, in both cases is of cardinality $2^\omega$. We must (by Skolem-Löwenheim) have equivalent countable structures. In fact these are obtained by restricting the set-variables to range over regular sets only. (Note that $\omega = 1^*$, and so its regular subsets are just the ultimately periodic sets). In other words, $MT_i = MT_i^{reg}$. Of course this is not obvious, but requires elimination of quantifiers, like every other proof of elementary equivalence of two structures.

$MT_2$ contains a significant fragment of the topology of reals. Namely: 1. In $MT_2$ one can define $Pth(\Phi) \equiv \Phi$ is a path through $2^*$. So the variables $X, Y, \ldots$ can be introduced in $MT_2$, by restricting its set-variables $\Phi, \Psi, \ldots$ to Pth. 2. We can use the set-variable $\Phi$ to simulate a variable over $F$, as $(\forall t)\Phi\overline{X}t$ is easily expressed as a formula $C(\Phi, X)$ of $MT_2$. The same goes for $F_\sigma$, and similarly one simulates a variable over larger and larger parts of $\mathcal{B}l(F_\sigma)$. Thus, $MT_2$ contains a good part of the elementary theory $ET$ of the structure $\langle \mathcal{B}l(F_\sigma), \subseteq, F_\sigma, F, \mathcal{B}_0 \rangle$, where $\mathcal{B}_0$ is the set of all clopen sets.

Problem: Does $MT_2$ contain all of ET? Are the two theories equivalent? If not, is ET decidable?

I. The basic facts about regularity are quantifier-eliminators: We are now ready to continue the list of infinity lemmas, started in section A.

(5) Büchi 1960: For every regular $M$ one can construct a regular $\overline{M}$ such that

$$\sim(\exists Z)(\exists^\omega t)M(\overline{X}t, \overline{Z}t) .\equiv. (\exists V)(\exists^\omega t)\overline{M}(\overline{X}t, \overline{V}t).$$

More carefully stated, the construction does of course apply to a given finite-state recursion for $M$, and yields one for $\overline{M}$. Lemmas of this type, where the new form is precisely the dual to the starting form, are called complementation lemmas. Such complementation lemmas are particularly suited for systematic elimination of quantifiers! One needs but iterate complementations to reduce a long prefix $\exists Z_1 \forall Z_2 \ldots$ $\ldots \forall Z_{n-1} \exists Z_n$ to the final $\exists Z$. Our lemma thus yields a decision method for $MT_1$. The same complementation also holds in $MT_1^{reg}$, and so $MT_1 = MT_1^{reg}$. Just as interesting is this interpretation: The lemma says that every regular Souslin set $C$ has a complement which is still regular Souslin. So, by Souslin's criterion: Every regular Souslin set is Borel; $\mathcal{S}^{reg} \subseteq \mathcal{B}$. A much stronger result is this:

(6) McNaughton 1966: For every regular $M$ one can construct a regular $\overline{M}$ and a $\mathcal{U}$ such that

$$(\exists Z)(\exists^\omega t)M(\overline{X}t, \overline{Z}t) .\equiv. \sup_t \overline{M}(\overline{X}t) \in \mathcal{U}.$$

In short, every regular Souslin set is actually in the Boolean algebra over $F_\sigma$; $\mathcal{S}^{reg} \subseteq \mathcal{B}l(F_\sigma)^{reg}$. McNaughton's is a quite ingenius proof. It shows how a bounded memory (finite automaton) can be made to keep track of essential information, accruing in an ever growing past. You will find a presentation of the matter in Büchi 1973, which also contains a version of the complementation lemma which works at $\omega_1$. We will use McNaughton's lemma in this form.

(6') $(\forall Y)(\exists Z)[Zo=c, Zt'=F[Zt,Xt,Yt], \sup Z \in \mathcal{U}\ .\equiv.$
$(\exists S)[So=d, St'=L[St,Xt], \sup S \in \mathcal{V}]$

What is meant here is that from $c,F,\mathcal{U}$ one can construct $a,G,\mathcal{V}$ such that the equivalence holds. McNaughton also conjectured this result:

(7) Landweber 1969: Regular $\mathcal{B}l(F_\sigma)$ games are determinate. One can tell who has the win, and the winner has a winning strategy which is finite-state (i.e., can be presented by a finite state recursion).

The classical Cantor-Bendixson process will give this result for $F^{reg}$ in place of $\mathcal{B}l(F_\sigma)^{reg}$. In fact, the process will terminate before $\omega$ (instead of before $\omega_1$). Landweber rediscovered a rather sophisticated version of this process. In the next section we will show how his result, and McNaughton's remove quantifiers in $MT_2{}^{reg}$.

(8) Rabin 1969: $\sim(\exists Z)\cdot Ze=c, A[Zx,ZxF,ZxT,\mathfrak{X}x], (\forall X)\sup(ZX) \in \mathcal{U}\ .\equiv.$
$(\exists H)\cdot He=d, B[Hx,HxF,HxT,\mathfrak{X}x], (\forall X)\sup(HX) \in \mathcal{V}$

This is the complementation lemma which eliminates quantifiers in $MT_2$. The same formula also holds in $MT_2^{reg}$, and so $MT_2 = MT_2^{reg}$. Rabin's proof also contains a Cantor-Bendixson process (now extending up to $\omega_1$), and it contains ideas from McNaughton (without actually using his lemma). We will show how a stronger version of Landweber yields Rabin.

K. Strong determinacy of Boolean $F_\sigma$-games: An important determinacy proof was Morton Davis (1964) for $F_{\sigma\delta}$. This implies the non-constructive part of Landweber's result, but does not seem to tell anything about the form of the winner's strategy. I think that our 1969 proof should be a first step in a very interesting and fruitful investigation of the forms of strategies. Here is the second step:

(9) Büchi 1973: $\mathcal{B}l(F_\sigma)$ games are determinate. The winner has a strategy which, up to finite-state, is no more complex as a given presentation of the game.

I have tried to fix up our 1969 proof; can you do it? A reasonably short proof only shows the second part of (9) and leaves the existence part to Davis. A much longer proof does both. Here is a precise statement of the result:

Given a presentation $so=c, st'=F[st,Xt,Yt], \sup_t D[st] \in \mathcal{U}$ of a game (D takes values in a finite set!) One can make up a finite state recursion $Vo=v_0, Vt'=G[Vt,D[st]]$ (in fact Vt is the order-vector of the values D[st]), such that the winner has a winning strategy presented by the joint inductions for s and V, and an output $Yt=B[st,Vt]$ (if J has the win) or $Xt=A[st,Vt,Yt]$ (if I has the win).

In the remaining sections we will show how this, and McNaughton's lemma yield Rabin's complementation lemma, and hence the decidability of $MT_2$. Actually this requires but a special case of (9). Can you tell me what more the full result will do? Here is what is needed for $MT_2$:

(9') Given the game $Zo=c_0, Zt'=F[Zt,Xt,Yt,\chi\overline{X}t], \sup Z \in \mathcal{U}$, whereby I = 2. J has no winning strategy, if and only if, I has a winning strategy presented by the recursion for Z, the ordervector recursion $Vo=v_0$, $Vt'=G[Vt,Zt]$ and an output $Xt=\Phi_{Zt,Vt,Yt}\overline{X}t$.

Note that I=2, J any finite set. As Z,V,Y all take values in finite sets, the strategy for I, in essence, consists of finitely many components $\Phi_{c,v,b} \subseteq 2^*$. To recognize this as a special case of (9), let $st=\langle Zt,\overline{X}t\rangle$, $D[st]=Zt$, and note that the game in (9'), is presented by a state-induction for s. In the sequel we will call these games <u>the special</u> $\mathcal{Bl}(F_\sigma)$ <u>games</u>.

<u>L. The monadic theory $MT_2$ is just right to handle special Boolean $F_\sigma$ games:</u>
Every sentence of $MT_2$, if put into prenex form and squeezed just a little more, will take the form $(\exists\chi_1)(\forall\chi_2)\ldots(\forall\chi_n)$ [J has a winning strategy in $G$]. Here $G$ is a special $\mathcal{Bl}(F_\sigma)$-game into which the $\chi_i$ enter. This will be easily established by one who knows monadic theories. Here is how the existential formula $C(\chi)$ in Rabin's lemma is transformed.

$(\exists Z)\cdot Ze=c, A[Zx,ZxF,ZxT,\chi x], (\forall X)\sup(ZX) \in \mathcal{U}$

let $\Psi x= ZxF,ZxT$

$(\exists\Psi)(\exists Z)\cdot Ze=c, ZxF=\Psi_1 x, ZxT=\Psi_2 x, A[Zx,\Psi x,\chi x], (\forall X)\sup(ZX) \in \mathcal{U}$

let $F[Zt,F,\Psi Xt]=\Psi_1\overline{X}t$, $F[Zt,T,\Psi\overline{X}t]=\Psi_2\overline{X}t$

$(\exists\Psi)(\forall X)(\exists Z)\cdot Zo=c, Zt'=F[Zt,Xt,\Psi Xt], A[Zt,\Psi\overline{X}t,\chi\overline{X}t], \sup Z \in \mathcal{U}$

let $Ut\equiv(\forall x)^t A[Zt,\Psi\overline{X}t,\chi\overline{X}t]$

$(\exists\Psi)(\forall X)(\exists ZU)\cdot Zo=c, Zt'=F[Zt,Xt,\Psi\overline{X}t], \sup Z \in \mathcal{U}$

$Uo\equiv T, Ut'\equiv Ut \wedge A[Zt,\Psi\overline{X}t,X\overline{X}t], \sup U = \{T\}$

The last formula $C'(X)$ is the statement "J has winning strategy in the special game....".

M. <u>The complementation lemma for $MT_2$ proved by determinacy:</u> As seen in section L, Rabin's (8) comes to put the negation of "J has win in the special game *G*" back into the same form. Here it goes, using the strong determinacy of $Bl(F_\sigma)$-games and McNaughton's lemma:

$\sim(\exists\Psi)(\forall X)(\exists Z)\cdot Zo=c_0, Zt'=F[Zt,Xt,\Psi\bar{X}t,\chi\bar{X}t],\ \sup Z \in U$

use (9'), $\Phi$ stands for all $\Phi_{c,v,b}$, c,v,b, the states of Z,V,Y

$(\exists\Phi)(\forall XY)(\exists Z)\cdot Zo=c_0, Zt'=F[Zt,Xt,Yt,\chi\bar{X}t]$
$Vo=v_0, Vt'=G[Vt,Zt]\quad ,\ \forall t[X_t=\Phi_{Zt,Vt,Yt}\bar{X}t] \supset \sup Z \notin U$

let $Wt\equiv(\forall x)^t Xx=\Phi_{Zt,Vt,Yt}\bar{X}t$

$(\exists\Phi)(\forall XY)(\exists ZW)\cdot Zo=c_0, Zt'=F[Zt,Xt,Yt,\chi\bar{X}t]$
$Vo=v_0, Vt'=G[Vt,Zt] \qquad (\exists^\omega t)Wt \vee \sup Z \in U$
$Wo\equiv T, Wt\equiv \bigwedge_{c,v,b}[Zt=c\wedge Vt=v\wedge yt=b \supset X_t=\Phi_{c,v,b}\bar{X}t]$

abbreviated, using new Z for Z,W

$(\exists\Phi)(\forall X)(\forall Y)(\exists Z)\cdot Zo=c, Zt'=H[Zt,Xt,Yt,\Phi\bar{X}t,\chi\bar{X}t], \sup Z \in V$

use (6'), to eliminate Y

$(\exists\Phi)(\forall X)(\exists S).\ So=d,\ St'=L[St,Xt,\Phi\bar{X}t,\chi\bar{X}t],\ \sup S \in W$

As the last formula is again in the form "J has win in the special game...", this proves our complementation lemma for $MT_2$. Precisely the same transformations will do, if we work in $MT_2^{reg}$ (only now (7) will be used to show correctness of the step where (9') was used). Hence we get $MT_2=MT_2^{reg}$. Having stepwise reduced the quantifiers of a sentence *C* it eventually takes the form $(\exists\Phi)[\Phi$ wins for J in *G*], where *G* is finite state. So now the truth of *C* can be decided using (7).

Bibliography:

R. Rado, Axiomatic treatment of rank in infinite sets, Canadian J. of Math. 1 (1949), 337-343.

F.P. Ramsey, On a problem of formal logic, Proc. London Math. Soc. 30 (1929), 264-286.

J.R. Büchi, The monadic second order theory of $\omega_1$, Lecture Notes in Math. 328, Springer (1973), 1-127.

J.R. Büchi, On a decision method in restricted second order arithmetic, Proc. 1960 Int. Cong. for Logic, Stanford Univ. Press (1962), 1-11.

L. Löwenheim, Über Möglichkeiten im Relativekalkul, Math. Ann. 76 (1915), 447-470.

F. Hausdorff, Mengenlehre, 3. Auflage, Dover N.Y. 1944.

R. McNaughton, Testing and generating infinite sequences by finite automata, Inform. and Control 9. (1966), 521-530.

Büchi and Landweber, Solving sequential conditions by finite state operators, Trans. Am. Math. Soc. 138 (1969) 295-311.

D.A. Martin, Measurable cardinals and analytic games, Fu. Math 66 (1970), 287-291.

M.O. Rabin, Decidability of second-order theories and automata on infinite trees, Trans. Am. Math. Soc. 141 (1969) 1-35.

M. Davis, Infinite games of perfect information, Advances in game theory, Ann. of Math. Study 52 (1964), 85-101.

---

This work was supported in part by a grant from the United States National Science Foundation.

Reprinted from Fundamentals of Computation Theory, Lecture Notes in Computer Science 56, Springer-Verlag, 1977, pp. 367–378.

THE JOURNAL OF SYMBOLIC LOGIC
Volume 48, Number 4, Dec. 1983

# STATE-STRATEGIES FOR GAMES IN $F_{\sigma\delta} \cap G_{\delta\sigma}$

J. RICHARD BÜCHI

The Cantor-Bendixson theorem says: For every closed set $\mathscr{C}$, $|\mathscr{C}| \leq \omega_0 \vee \mathscr{C} \supseteq$ perfect (and therefore $\mathscr{C}$ is countable or has the power of the continuum). W. H. Young proved the alternative for $G_\delta$-sets, and Hausdorff extended it to Souslin-sets. Cantor used his ordinals to prove the theorem (I like to think he invented them for this purpose). In Hausdorff's Mengenlehre ordinals do not officially enter in either of the three proofs. You should try to put them back. It is very important never to hide away $\omega_1$ when it actually is there, and it is there when the continuum problem is the subject.

In Davis [6] you find that Cantor's alternative is equivalent to another alternative: ($I$ wins $\mathscr{C}$) $\vee$ ($J$ wins $\mathscr{C}$). Here $\mathscr{C}$ means a lopsided game in which one of the players makes long moves on the game tree. So, the classical theorems can be restated as determinacy results for lopsided games, and now I make this observation: The original CB-proof "*actually presents*" a winning strategy. The HD-proofs do no such thing; all you know at the end is existence of a winning strategy (and Davis' remark is needed to gain this knowledge).

You understand now why I like CB-proofs, particularly for determinacy of games. They do exist also for games in which both players make short moves. Many years ago I tried to publish such a proof for determinacy of $F_\sigma$-games. The referee said it was not worth the trouble. I say you should remake this proof (it is a special case of the one presented in this paper), because it is the ideal of a determinacy proof.

The master-example of a non-CB-proof is Davis' [6] determinacy of $G_{\delta\sigma}$-games. Many have used his ideas; Martin [11] extends the method to work for all Borel-games.

The master-example of a CB-proof is given in Büchi-Landweber [5]. Here the "actually presents" hold in a very strong sense: A recursive method is given, which applies to the game automaton of any finite state game $\mathscr{C}$. It tells which of the players has the win, and it constructs a finite automaton which executes a winning strategy. These games are in $Bl(F_\sigma)$ (the Boolean algebra over $F_\sigma$). In Büchi [4] you find a discussion of this matter, and the statement of this theorem (Corollary 2, in the present paper): Let $\mathscr{C}$ be any game in $Bl(F_\sigma)$, and let $\mathscr{A}$ be a presentation of $\mathscr{C}$ by state-recursion. Then the winner has a winning strategy presented by $\mathscr{A}$ and the order-vector recursion for $\mathscr{A}$. So now you have an example of

Received July 2, 1981; revised June 8, 1982.


0022-4812/83/4804-0025/$03.80

what the "actually presents" may mean in a more sophisticated situation. My proof of this theorem on $Bl(F_\sigma)$ is over 10 years old and was communicated to many, privately and publicly, and on both sides of the Atlantic. In this paper you find an extension of the result to $F_{\sigma\delta} \cap G_{\delta\sigma}$. You see, there is no need for redoing old theorems. Better try to extend the result to Borel games.

A very interesting CB-proof occurs in Rabin [12]. It proves his complementation lemma for $MT_2$, my monadic second-order theory of two free successor-functions. The complementation-method was used in Büchi [2], to show decidability of $MT_1$. The automata used there very naturally suggested $MT_2$ and tree automata. I soon understood that $MT_2$ is not a plaything, it contains a significant fragment of analysis. In fact, $MT_2$ is basic analysis (rational order, real order, approximation). Among a host of possibilities Rabin found the correct version of the complementation lemma; the intuition was supplied by the tree automata. Given the statement of this lemma, and given McNaughton's handling of sup-conditions by order vectors, and given time, everybody can prove Rabin's theorem. He can also hide the CB-argument (à la Hausdorff), and he can hide the automata by using model-theory.

In Büchi [4] you find a short proof of Rabin's complementation lemma; it is reproduced at the end of this paper. Given my notions "special game" and "order-vector strategy", you can do it yourself: (1) Take the existential formula $F(\chi)$ which Rabin tells us to complement. (2) A bit of juggling puts $F(\chi)$ into the form $(\exists\Phi)[\Phi$ a winning strategy for $I$ in the special game $\mathscr{C}_\chi]$. (3) By determinacy $\sim F(\chi)$ takes the form $(\exists\Psi)[\Psi$ a winning strategy for $J$ in the game $\mathscr{C}_\chi]$. (4) Use the fact that the winner has a winning order-vector strategy. This almost puts $\sim F(\chi)$ back into $MT_2$. (5) Look for one application of McNaughton's lemma. It will put $\sim F(\chi)$ into the form $(\exists\Psi)[\Psi$ a winning strategy for $I'$ in the special game $\mathscr{C}'_\chi]$. So now $\sim F(\chi)$ is precisely in the same form as $F(\chi)$. This complementation lemma will transform any sentence $F$ in $MT_2$ into form $(\exists\Phi)[\Phi$ a strategy for $I$, in the special game $\mathscr{C}]$. But special games into which no $\chi$ enters are just finite-state games. So now you can use Büchi-Landweber [5] to decide the truth of $F$.

Automata naturally suggest $MT_2^{\mathrm{reg}}$, where the set-variables $\chi$ are interpreted to range over regular subsets of the two-successor tree $J^*$. Precisely the same algorithm decides truth in both $MT_2$ and $MT_2^{\mathrm{reg}}$ (and the regular-set model is the prime model of $MT_2 = MT_2^{\mathrm{reg}}$). In the proof of correctness for $MT_2^{\mathrm{reg}}$, Rabin's $BS$-argument goes over $\omega_0$ (in place of $\omega_1$, in the case of $MT$). Correspondingly, in step (4) of my algorithm, the verification for $MT_2^{\mathrm{reg}}$ is in [5]. So Büchi-Landweber [5] and McNaughton [10] give decidability of $MT_2^{\mathrm{reg}}$. This I knew soon after Rabin told us he used CB, and so $MT_2$ just had to be a game-language. It remained to verify step (4) in the case of $MT_2$. The special games $\mathscr{C}_\chi$ are obvious at this place. To understand how "ordervector-strategy" generalizes from finite-state to special games and $Bl(F_\sigma)$-games was much harder. In Büchi [4] you find this definition, and the statement of the theorem needed at (4), Corollary 2 and Lemma 5 in this paper. Finding the proof is much easier (it ought to be possible to generalize the proof in [5]). The one I give here happens to work for $F_{\sigma\delta} \cap G_{\delta\sigma}$.

I will not call this a new proof of decidability for $MT_2$. The old proof is quite good enough for me: statement of the complementation lemma; prove it by using McNaughton and a CB-argument over $\omega_1$. I will say how fortunate it was that Rabin did not hide CB, so now there is a first application of determinacy, and the very interesting concept of a state-strategy. These naturally suggest plenty of new problems. For example:

*Problem* 1. There must be a CB-proof of determinacy for Borel games. Find a kernel presentation for Borel-sets which admits a local universality criterion (such as my Lemma 2). The rest will come naturally.

A nice way of understanding what McNaughton's lemma does is this. The general Souslin-set admits this presentation $\mathscr{C}(X) \equiv (\exists Z)(\exists^\omega t)M[\bar{X}t, \bar{Z}t]$, $Z$ ranging over $2^\omega$. If $M$ is regular (i.e. is presented by a finite-state recursion), then $\mathscr{C}$ is in $Bl(F_\sigma)^{\text{reg}}$ (has a $Bl(F_\sigma)$ presentation with regular kernel). This gives evidence to *the thesis*: Every finitely presented set $\mathscr{C}$ is in $Bl(F_\sigma)^{\text{reg}}$.

*Problem* 2. Can you give more evidence for this thesis? Better yet, can you find an honestly finite presentation of a set $\mathscr{C}$ not in $Bl(F_\sigma)^{\text{reg}}$?

The work to be presented here started from the observation that $MT_2$ consists of sentences with a quantifier-prefix on variables $\chi_i$, followed by $[(\chi_1, \ldots, \chi_n)$ is a winning strategy for $I$, in the game $\mathscr{C}]$. Here $\mathscr{C}$ is finite-state.

You see now what "honestly finite" means in Problem 2: You are to find other games for which $[\cdots]$ can be spelled out in an extension $L$ of $MT_2$. You should then have an extension of $MT_2$ for which it is reasonable to ask whether it is decidable. (This would be what I would call a problem, as opposed to a frivolous question.)

More obvious extensions of $MT_2$ are obtained by adding special sets (of individuals or of sets) to the primitives. This has been done for $MT_1$, and the game-approach should make it feasible for $MT_2$. Try Theorem 4 on this:

*Problem* 3. Given $P: J^* \to E$ with finite $E$. Let GL($P$) be the extension of the game-language in §15, obtained by adding the primitive $P$. For which $P$ is GL($P$) decidable?

§1. **Notations.** Throughout this paper $I$ and $J$ are finite sets. The letter $a$ will usually mean a member of $I$, and $b$ will usually mean a member of $J$. $J^*$ denotes the set of all finite sequences over $J$, including the empty sequence. We will use variables like $y$, $v$ to range over $J^*$, and variables like $x$, $u$ will range over $I^*$. On $J^*$ we have the relation $y_1 \trianglelefteq y_2$, $y_1$ is initial segment of $y_2$, the right-successor operations $yb$, and concatenation $yv$. Think of $J^*$ as a tree with root $e$, and every vertex $y$ splitting into $J$ successors $yb$.

Variables like $Y$, $V$ range over $J^\omega$, while $X$, $U$ range over $I^\omega$. Think of $Y$ as an infinite sequence of members $Yt$ of $J$, or as an infinite path through the tree $J^*$. Think of $J^\omega$ as an extension of the tree $J^*$ to a level $\omega$. Read $y \triangleleft Y$ as the finite sequence $y$ is an initial segment of the infinite sequence $Y$, or as, $y$ approximates $Y$. $\bar{Y}t$ stands for the sequence $Y_0 \cdots Y_{t-1}$. Note $\bar{Y}0 = e$ and $\bar{Y}t' = Y_0 \cdots Y_t$. Think of $\bar{Y}t$ as the $t$th approximation of $Y$.

We use these quantifiers over $\omega$: $(\exists t)^z$ for "there is a $t$ less than $z$", $(\exists t)_z$ for "there is a $t$ larger or equal $z$", and $(\exists^\omega t)$ for $(\forall z)(\exists t)_z$. Each of these has a dual. Read $(\exists^\omega t)$ as "there are infinitely many $t$" and read $(\forall^\omega t)$ as "eventually all $t$." Closely related to these quantifiers is the notation $\sup Z = \sup_t Zt = \{c;\ (\exists^\omega t)\ Zt = c\}$.

§2. **State recursions.** Let $\mathscr{A} = \langle C, c_0, F\rangle$ consist of a finite or countable set $C$ (*the states*), a member $c_0$ of $C$ (*the initial state*), and a function $F$ from $C \times J$ to $C$ (*the transition operator*). The state recursion $\mathscr{A}$ is this:

$$\begin{aligned} S0 &= c_0 \\ St' &= F[St,\ Yt] \end{aligned} \qquad \text{for } Y \in J^\omega$$

$$\begin{aligned} Se &= c_0 \\ S(yb) &= F[Sy,\ b] \end{aligned} \qquad \text{for } y \in J^*,\ b \in J$$

The first version of the recursion defines a map $S = (\mathscr{A}Y)$ from $J^\omega$ to $C^\omega$. If $S = (\mathscr{A}Y)$ we have $St = S[\bar{Y}t]$, whereby $S[\bar{Y}t]$ is obtained by the second version. So this version also defines $(\mathscr{A}Y)$. Sometimes it will be convenient to use the one mode of defining $(\mathscr{A}Y)$, at other places the second mode is useful. If $C$ is finite, $\mathscr{A}$ is called a finite automaton. If $C$ is infinite and $F$ is not computable, "recursion" is to be taken in a wide sense. Note that $(\mathscr{A}Y)$ maps $J^\omega$ continuously into $C^\omega$, and so $\mathscr{A}^{-1}$ takes Borel-sets to Borel-sets.

Let $M$ be a subset of $J^*$, or more generally, a map $M\colon J^* \to D$. We will say that *$M$ is presented by the state recursion $\mathscr{A}$, with output $G$* if $My = GSy$, for any $y$ in $J^*$. So $G$ takes $C$ to $D$, and we have $M[\bar{Y}t] = G[St]$ for $S = (\mathscr{A}Y)$. For any $M$ there is the trivial presentation $\langle J^*, e, yb, M\rangle$. There are $M$'s for which one can do no better than this. The *regular* $M$'s are those which admit a finite presentation. An important conception is this:

$$y_1 \sim y_2(M) \cdot\!\equiv\!\cdot (\forall v) M[y_1 v] = M[y_2 v] \quad \textit{the right congruence induced by } M$$

Put $C = J^*/\!\sim$, $c_0 = \tilde{e}$, $F[\tilde{y}, b] = (yb)^\sim$, and $G\tilde{y} = My$. It is easy to show that this gives a presentation $M_{\min}$ for $M$, and that $M_{\min}$ is a homomorphic image of any other presentation $\mathscr{A}$, $G$ of $M$. As a consequence, $M$ is regular if and only if $\sim(M)$ has finite index.

§3. **Presentation of $\mathscr{C}$ by a kernel $M$; and by state recursion.** Over $J^\omega$ we have a Cantor topology. Its basic open sets are $\{Y;\ y \lhd Y\}$ for $y \in J^*$. Hence a set $\mathscr{C}$ is open (is a $G$) just in case it admits a presentation $\mathscr{C}(Y) \equiv (\exists t) M[\bar{Y}t]$. Accordingly the closed sets ($F$) are those of form $\mathscr{C}(Y) \equiv (\forall t) M[\bar{Y}t]$. The set $M \subseteq J^*$ is the kernel. An $F_\sigma$ set $\mathscr{C}(Y)$ may be presented in the form $(\exists i)(\forall t) M_i[\bar{Y}t]$, and here the kernel is the system of sets $M_i$, $i \in \omega$. Show that any $F_\sigma$ may also be presented in the form $\mathscr{C}(Y) \equiv (\forall^\omega t) M[\bar{Y}t]$ (it is an instructive exercise, see Davis [6] if you need help), and now the kernel is just a set $M$. For $\mathscr{C}$ in $Bl(F_\sigma)$ (the Boolean algebra over $F_\sigma$) there is this elegant presentation $\mathscr{C}(Y) \cdot\!\equiv\!\cdot \sup_t M[\bar{Y}t] \in \mathscr{U}$. Here the kernel $M$ maps $J^* \to C$, and $\mathscr{U}$ is a class of subsets of the finite set $C$ (find more about this in [3] and [4]). All these are examples of the following general setting:

$$\mathscr{C}(Y) \cdot\!\equiv\!\cdot (Qt) M[\bar{Y}t], \quad \text{kernel } M\colon J^* \to D \quad (\textit{positional presentation})$$

$D$ is a finite or countable set, and $(Qt)$ is an operator which, to any $Z$ in $D^\omega$, assigns a truth-value $(Qt)Zt$. Suppose that $\mathscr{A}$, $G$ is a presentation of the kernel $M$ by state recursion. So the set $\mathscr{C}$ can now be defined this way.

$$\mathscr{C}(Y) \cdot\!\equiv\!\cdot (Qt)G[St] \quad \text{for } S = (\mathscr{A}Y)$$

We call this a *presentation of* $\mathscr{C}$ *by state recursion.*

**§4. Positional strategy versus state-strategy.** A game between players $I$ and $J$ is a set $\mathscr{C}(X, Y)$ in $(I \times J)^\omega$. The *rules of the game* are these: At each time $t$, first player $I$ picks a move $Xt$ in $I$, and then player $J$ moves $Yt$ in $J$. The result, at $\omega$, is a match $(X, Y)$. $I$ wins this match if $\mathscr{C}(X, Y)$, $J$ wins if $\widetilde{\mathscr{C}}(X, Y)$. Each player's moves are known to the opponent; oracles telling future moves are forbidden.

From the rules one sees this: A *strategy for* $I$ is a map $\Phi$: $J^* \to I$. It tells $I$ to move $Xt = \Phi[\bar{Y}t]$ at any time $t$. A strategy for $J$ is a map $\Psi$: $I^* \to J$. It tells $J$'s moves $Yt = \Psi[\bar{X}t']$. Define $X = (\Phi Y)$ by $Xt = \Phi[\bar{Y}t]$, and define $Y = [\Psi X]$ by $Yt = \Psi[Xt']$. Hence, if $J$ plays $Y$ and $I$ uses the strategy $\Phi$, then $((\Phi Y), Y)$ is the resulting match. If $I$ plays $X$ and $J$ uses $\Psi$ the match is $(X, [\Psi X])$. Hence we define "*winning strategy in the game* $\mathscr{C}$" this way:

$$\Phi \text{ wins for } I \cdot\!\equiv\!\cdot (\forall Y)\mathscr{C}((\Phi Y), Y) \cdot\!\equiv\!\cdot (\forall XY)[X = (\Phi Y) \supset \mathscr{C}(X, Y)]$$

$$\Psi \text{ wins for } J \cdot\!\equiv\!\cdot (\forall X)\widetilde{\mathscr{C}}(X, [\Psi X]) \cdot\!\equiv\!\cdot (\forall XY)[Y = [\Psi X] \supset \mathscr{C}(X, Y)]$$

Note that these strategies ignore the player's own previous moves. More generally a strategy $\Phi$ for $I$ may look this way: $Xt = \Phi[\bar{X}t, \bar{Y}t]$. This $\Phi$ uses the full present *position* $(\bar{X}t, \bar{Y}t)$ *of the game.* Make up a strategy $Xt = \Phi'[\bar{Y}t]$ which simulates $\Phi$, in the sense that, for any play $Y$ of $J$, $\Phi$ and $\Phi'$ produce the same play $X$ of $I$. This seems to mean that $I$ may just as well forget his own previous moves, as long as he keeps track of the previous moves of the opponent.

We remark that strategies and the kernels used to present sets $\mathscr{C}$ are the same sort of objects. Hence a strategy $\Phi$ for $I$ may be presented by a state-recursion $\mathscr{A}$ over $J$, and an output $G$: $C \to I$. So we have $\Phi y = GSy$, with $S = (\mathscr{A}Y)$. Similarly a strategy $\Psi$ for $J$ may be presented as $\Psi[\bar{X}t] = H[St, Xt]$, where $H$: $C \times I \to J$ is an output on a state-recursion $\mathscr{B}$ over $I$. We are now ready to introduce the interesting notion of *state-strategy for a game* $\mathscr{C}$ *presented by state-recursion.* Find the history of its birth described in the introduction.

DEFINITION 1. Let $(Qt)$ be an operator (such as $(\forall^\omega t)$). Assume the game $\mathscr{C}(X, Y) = (Qt)M(St)$ presented by an output $M$: $C \to D$ on this state recursion $\mathscr{A}$:

$$\begin{aligned} S0 &= c_0 \\ St' &= F[St, Xt, Yt] \end{aligned} \quad \text{or} \quad \begin{aligned} S(e, e) &= c_0 \\ S(xa, yb) &= F[S(x, y), a, b] \end{aligned}$$

A state-strategy $\Phi$ for $I$ ($\Psi$ for $J$) is a strategy which may be presented by adding to the game-recursion $\mathscr{A}$ an output $Xt = G[St]$ (an output $Yt = G[St, Xt]$). Thus, in the case of $I$, a state-strategy $Xt = \Phi(\bar{Y}t)$ can be presented by this modified state recursion $\mathscr{A}$:

$$\begin{aligned} S0 &= c_0 \\ St' &= F[St, G[St], Yt] \\ Xt &= G[St] \end{aligned} \quad \text{or} \quad \begin{aligned} Se &= c_0 \\ S(yb) &= F[Sy, G[Sy], b] \\ \Phi y &= G[Sy] \end{aligned}$$

**§5. A normal form for members $\mathscr{C}$ of $E_{\sigma\delta} \cap G_{\delta\sigma}$.** We have already mentioned that $F_\sigma$ sets can be put into the form $(\exists^\omega t)A[\bar{Y}t]$. Therefore, any $\mathscr{C}$ in $F_{\sigma\delta} \cap G_{\delta\sigma}$ is of this form:

(a) $$\mathscr{C}(Y) \cdot\equiv\cdot (\exists i)(\exists^\omega t)A_i[\bar{Y}t] \qquad \tilde{\mathscr{C}}(Y) \cdot\equiv\cdot (\exists i)(\exists^\omega t)B_i[\bar{Y}t]$$

Here and at other places $i$ ranges over $\omega$. $A_i$ and $B_i$ are subsets of $J^*$. We may assume that $A_i \cap B_i = 0$; otherwise we replace $A_i$ by $A_i \cap \tilde{B}_i$ without destroying (a). Let $y_0, y_1, y_2, \ldots$ be an enumeration of those $y$ in $J^*$ which belong to none of the $A_i, B_i$. Define $A'_{2i} = A_i$, $B'_{2i} = B_i$, $A'_{2i+1} = \{y_i\}$, $B'_{2i+1} = 0$. It is easy to see that (a) still holds for these new sets $A'_i, B'_i$. Furthermore, every $y$ belongs to some $A'_i \cup B'_i$. So, without loss of generality, we may assume,

(b) $$A_i \cap B_i = 0 \qquad \bigcup_{i\in\omega}(A_i \cup B_i) = J^*$$

Next we define these items:

(c) $$hty = (\mu i)[A_i y \vee B_i y] \qquad My = A_{hty}y \qquad fY = (\mu i)(\exists^\omega t)[A_i[\bar{Y}t] \vee B_i[\bar{Y}t]]$$

Using the second part of (b) we see that $hty$ exists for every $y$ in $J^*$. Using the first part of (b) one shows:

(*) $$hty = i \supset [My \equiv A_i y \wedge \tilde{M}y \equiv B_i y]$$

Suppose now that $Y$ is any member of $J^\omega$. From (a) we see that there is an $i$ such that $(\exists^\omega t)A_i[\bar{Y}t] \vee (\exists^\omega t)B_i[\bar{Y}t]$. Hence $fY$ exists for every $Y$. Read $hty = i$ as "*y hits at i*".

LEMMA 1. *Let $\mathscr{C}$ be any $F_{\sigma\delta} \cap G_{\delta\sigma}$ over $J^\omega$. There is a kernel, consisting of $ht$: $J^* \to \omega$ and $M \subseteq J^*$, so that $\mathscr{C}$ is presented in either of these forms:*

(d) $$fY = i \cdot\supset\cdot \begin{cases} \mathscr{C}(Y) \equiv (\exists^\omega t)[ht[\bar{Y}t] = i \wedge M[\bar{Y}t]] \\ \tilde{\mathscr{C}}(Y) \equiv (\exists^\omega t)[ht[\bar{Y}t] = i \wedge \tilde{M}[\bar{Y}t]] \end{cases}$$
$$fY = i \cdot\supset\cdot \begin{cases} \mathscr{C}(Y) \equiv (\forall^\omega t)[ht[\bar{Y}t] = i \supset M[\bar{Y}t]] \\ \tilde{\mathscr{C}}(Y) \equiv (\forall^\omega t)[ht[\bar{Y}t] = i \supset \tilde{M}[\bar{Y}t]] \end{cases}$$

*Here $fY = (\mu i)(\exists^\omega t)ht\,\bar{Y}t = i$, the least infinite-hit position on $Y$. It exists for every $Y$.*

PROOF. We start from the presentation (a) of $\mathscr{C}$ and assume it is rectified so as to satisfy (b). Now (c) defines the new kernel $ht$, $M$ and the function $f$. For a given $Y$ abbreviate $ht[\bar{Y}t]$, $M[\bar{Y}t]$, $A_i[\bar{Y}t]$, $B_i[\bar{Y}t]$ as $h_t$, $Mt$, $A_i t$, $B_i t$.

Suppose $fY = i$. By definition (c) of $f$ we have $(\exists^\omega t)[A_i t \vee B_i t]$, and $\bigwedge_{j<i}(\forall^\omega t)[\tilde{A}_j t \wedge \tilde{B}_j t]$. In the second formula, the quantifier may be moved to the front. Using the definition (c) of $ht$, we see $(\exists^\omega t)h_t = i$ and $(\forall^\omega t)h_t \nless i$. This argument shows that $fY$ is the least infinite-hit position as claimed in the lemma.

Suppose $fY = i$. So we have $(\exists^\omega t)h_t = i$, and therefore by (c), $(\exists^\omega t)[[h_t = i \wedge A_i t] \vee [h_t = i \wedge B_i t]]$. As the quantifier distributes, this gives the alternative $(\exists^\omega t)[h_t = i \wedge A_i t] \vee (\exists^\omega t)[h_t = i \wedge B_i t]$. By (a) this implies the exclusive alternative $\mathscr{C}(Y) \vee \tilde{\mathscr{C}}(Y)$. Therefore, $\mathscr{C}(Y) \equiv (\exists^\omega t)[h_t = i \wedge A_i t]$ and $\tilde{\mathscr{C}}(Y) \equiv (\exists^\omega t)[h_t = i \wedge B_i t]$. Because of remark (*) following (c), we can replace $A_i$ by $M$, and whereby $S = (\mathscr{A}Y)$. So now the $\mathscr{C}$ in Lemma 2′ is in the hit-form $(\overline{ht}, \bar{M})$ as required in Lemma 2.

REMARK. In Lemma 2′, replace $\xi: J^* \to \Omega$ by $\xi: C \to \Omega$, and replace the condition on $\xi$ by $htc' = j \wedge Mc' \supset \xi c \geq_{j-1} \xi c'$ and $htc' = j \wedge \tilde{M}c \supset \xi c >_j \xi c'$ whereby $c'$ stands for $F[c, b]$. The result is still true. But now one has to modify the proof of Lemma 2, operating with state-versions of the items $\succ$, $H$, $H_i$. For example $c_1 \succ c_2$ is defined as $(\exists y)\,[c_1 \lhd c_2(y) \wedge \tilde{M}c_2 \wedge (\forall c)_{c_1 \lhd c \unlhd c_2(y)}\, htc_1 \leq htc]$. Here $c_1 \lhd c_2(y)$ means "$y$ takes $\mathscr{A}$ from the state $c_1$ to $c_2$," and $c_1 \lhd c \unlhd c_2(y)$ means that $y$ takes $\mathscr{A}$ from $c_1$ through $c$ to $c_2$. We do not use this remark.

Lemma 2 just about suffices to prove our Theorem 1. It does not enter into the second proof of the stronger Theorem 2. In this time of hyper set-theory I have met people who do not seem to understand a basic fact on ordinals and the notion of foundation. So I repeat: It is always possible to eliminate ordinals from a proof. Just find the founded relation of which they are the ranks. The next step (backwards) is to simply not use the word "founded". An infinite regress is such a commonplace in mathematics; the reader will probably not suspect that you have used one. The raison d'être for ordinals is that they measure one regress against another. In the case of our lemma the measure is $< \omega_1$, and in case $\mathscr{C}$ is presented by a finite-state recursion the measure becomes $< \omega_0$. What the measure of a proof tells me is to compare this proof with others of the same measure.

EXAMPLE. The CB-proof of Büchi-Landweber [5] has measure $\omega_0$. The CB-proof of Rabin [12] has measure $\omega_1$ (he did not hide it away by not using ordinals), and in the final step (deciding existential sentences) the measure becomes $\omega_0$. This indicates just where BL might fit into Rabin's proof, and that a stronger version of BL might do the full Rabin theorem.

Lemma 2 shows clearly what I mean by *local* universality criterion. Find one for Borel sets, and you will probably have the answer to Problem 1.

**§8. Ranking of winning strategies in a $F_{\sigma\delta} \cap G_{\delta\sigma}$ game.** Let $\Phi: J^* \to I$ be a strategy for $I$ in the game $\mathscr{C}(X, Y)$. That $\Phi$ wins for $I$ means $(\forall Y)\mathscr{C}((\Phi Y), Y)$. Hence we can use our criterion for universality as a criterion for winning strategies. You should first spell this out in detail, as it works for positional strategies. Here I will do it for the more general case of state-strategies.

Let $\mathscr{C}(X, Y)$ be an $F_{\sigma\delta} \cap G_{\delta\sigma}$ presented by state-recursion and in normal form. So we have given a state-recursion $\mathscr{A}$ consisting of the set of states $C$, an initial state $c_0$, and a transition operator $F: C \times I \times J \to C$. On $C$ we have given a hit-function $ht: C \to \omega$ and the set $M \subseteq C$. The state-recursion $\mathscr{A}$ defines the map $S = \mathscr{A}(X, Y)$ in either of these two modes:

$$(1) \qquad \begin{aligned} S_0 &= c_0 & \qquad S(e, e) &= C_0 \\ St' &= F[St, Xt, Yt] & \qquad S(xa, yb) &= F[S(x, y), a, b] \end{aligned}$$

The relation between the two modes is this: $St = S(\bar{X}t, \bar{Y}t)$ for any $X \in I^\omega$, $Y \in J^\omega$, $t \in \omega$. The hit-function $ht$ is known to be such that there is an infinite-hit position $i$ along any $S = \mathscr{A}(X, Y)$. Hence we have the smallest infinite-hit position $f(X, Y)$. It satisfies

$$\text{(2)} \qquad f(X, Y) = fS = i \mathrel{\cdot{\equiv}\cdot} (\exists^\omega t)ht[St] = i \wedge (\forall^\omega t)ht[St] \geq i$$

where $S$ is given by (1)

Finally, $\mathscr{C}$ is given by

$$\text{(3)} \quad fS = i \mathrel{\cdot{\supset}\cdot} \begin{cases} \mathscr{C}(X, Y) = (\exists^\omega t)[ht[St] = i \wedge M[St]] \\ \tilde{\mathscr{C}}(X, Y) = (\exists^\omega t)[ht[St] = i \wedge \tilde{M}[St]] \end{cases} \quad \text{where } S \text{ is given by (1)}$$

Let $\Phi$ be any strategy for $I$. Define $\mathscr{C}^\Phi(Y) \equiv ((\Phi Y), Y)$, and define the state-recursion $\mathscr{A}^\Phi$ by

$$\text{(4)} \quad \begin{aligned} S0 &= c_0 \\ St' &= F[St, \Phi[\bar{Y}t], Yt] \end{aligned} \quad \text{or} \quad \begin{aligned} Se &= c_0 \\ S(yb) &= F[Sy, \Phi y, b] \end{aligned}$$

More explicitly $\mathscr{A}^\Phi = \langle C \times J^*, (c_0, e), F^\Phi \rangle$ whereby $F^\Phi$ has the two components $F_1^\Phi[c, y, b] = F[c, \Phi\, y, b]$ and $F_2^\Phi\, [c, y, b] = yb$. From (3) we now see this:

$$fS = i \supset \begin{cases} \mathscr{C}^\Phi(Y) \equiv (\exists^\omega t)[ht[St] = i \wedge M[St]] \\ \tilde{\mathscr{C}}^\Phi(Y) \equiv (\exists^\omega t)[ht[St] = i \wedge \tilde{M}[St]] \end{cases} \text{whereby } S = \mathscr{A}^\Phi Y \text{ is given by (4)}$$

This shows that $\mathscr{C}^\Phi$ is in $F_{\sigma\delta} \cap G_{\delta\sigma}$ and is presented by the state-recursion $\mathscr{A}^\Phi$ in hit-form. From the definition of $\mathscr{C}^\Phi$ we see that $(\forall Y)\mathscr{C}^\Phi(Y)$ means $\Phi$ wins for $I$. Hence, the universality criterion, Lemma 2′, proves this:

LEMMA 3. *Let $\mathscr{C}$ be presented by* (3) *and the state-recursion* (1). *Let $\Phi$ be any strategy for I. Then $\Phi$ wins the game $\mathscr{C}$ for I if and only if there is a map $\xi: J^* \to \Omega$ which satisfies these conditions*:

$$\begin{aligned} ht[Syb] = j \wedge M[Syb] &\mathrel{\cdot{\supset}\cdot} \xi(y) \geq_{j-1} \xi(yb) \\ ht[Syb] = j \wedge \tilde{M}[Syb] &\mathrel{\cdot{\supset}\cdot} \xi(y) >_j \xi(yb) \end{aligned}$$

*whereby*

$$Se = c_0 \qquad S(yb) = F[Sy, \Phi y, b]$$

DEFINITION. The rank of a winning strategy for $I$ is the smallest starting value $\xi e$ for any $\xi: J^* \to \Omega$ which satisfies the conditions in Lemma 3.

You may want to spell out the dual version of this lemma. It is a criterion for $\Psi$ to be a winning strategy for $J$. So now $Y = [\Psi X]$ is the play for $J$ ordered by $\Psi$. Accordingly (4) contains the entry $\Psi[\bar{X}t']$ (or $\Psi xa$) in place of $\Phi[\bar{Y}t]$ (or $\Phi y$), and of course $Y$ is to be replaced by $X$.

**§9. The existence of winning state-strategies in $F_{\sigma\delta} \cap G_{\delta\sigma}$ games.** Let $\mathscr{C}$ be the game presented by (1) and (3) in §8 and assume that every state in $C$ is accessible from $c_0$ by $F$ (elimination of inaccessible states will not change $\mathscr{C}$). Suppose now that the recursion is started at $c$ in place of $c_0$. Using the accessibility of $c$ from $c_0$ shows that the resulting $S$ still hits some $i$ infinitely often and that (3) will consistently define a game $\mathscr{C}_c$. Note that $\mathscr{C} = \mathscr{C}_{c_0}$ and that the game $\mathscr{C}_c$ is presented by the same formulas (1), (3) with the only exception being that $c$ replaces $c_0$ in (1). We now define the *resolvent* $R_\xi$ for every $\xi \in \Omega$:

$$(5) \qquad c \in R_\xi \cdot\equiv\cdot (\exists\Phi)[\Phi \text{ wins } \mathscr{C}_c \text{ for } I \wedge \mathrm{rk}_c\Phi \le \xi]$$

Here $\mathrm{rk}_c\,\Phi$ is the rank of $\Phi$ in the game $\mathscr{C}_c$. From the definition we see $R_\eta \subseteq R_\xi$ for $\eta < \xi$. As $R_\xi \subseteq C$ is countable, and $\Omega$ is an $\omega_1$-order, we must have an $\alpha$ in $\Omega$ so that $R_\xi = R_\alpha$ for $\alpha \le \xi$. Note that $I$ has a win in the game $\mathscr{C}_c$ if and only if $c \in R_\alpha$.

Suppose now that $c \in R_\xi$. So there is a winning strategy $\Phi$ for $I$ in the game $\mathscr{C}_c$ with $\mathrm{rk}_c\Phi \le \xi$. By Lemma 3 we therefore have a map $\zeta\colon J^* \to \Omega$ such that $\zeta c \le \xi$ and

$$(\mathrm{i}) \quad \begin{array}{ll} ht[Syb] = j \wedge M[Syb] & \cdot\supset\cdot\ \zeta y \ge_{j-1} \zeta(yb) \\ ht[Syb] = j \wedge \tilde{M}[Syb] & \cdot\supset\cdot\ \zeta y >_j \zeta(yb) \end{array} \quad \text{whereby} \quad \begin{cases} Se = c \\ Syb = F[Sy, \Phi y, b] \end{cases}$$

Let $a_0 = \Phi e$ be the first move ordered by $\Phi$, and let $b_0$ be any move for $J$, and let $c' = F[c, a_0, b_0]$. In (i) put $y = e$, $b = b_0$ to get

$$(\mathrm{j}) \quad htc' = j \wedge Mc' \cdot\supset\cdot\ \xi \ge \zeta c \ge_j \zeta c' \qquad htc' = j \wedge \tilde{M}c' \cdot\supset\cdot\ \xi \ge \zeta c >_j \zeta c'$$

Let $S'v = Sb_0v$, $\zeta'v = \zeta b_0 v$, $\Phi'v = \Phi(b_0v)$. Use (i) for $y = b_0v$ to get

$$(\mathrm{k}) \quad \begin{array}{l} ht[S'vb] = j \wedge M[S'vb] \supset \zeta'v \ge_j \zeta'(vb) \\ ht[S'vb] = j \wedge \tilde{M}[S'vb] > \zeta'v \supset_j \zeta'(vb) \\ \text{whereby } S'e = c',\ S'vb = F[S'v, \Phi'v, b] \end{array}$$

From (k) and Lemma 3, we see that $I$ has a winning strategy $\Phi'$ in the game $\mathscr{C}_{c'}$, and that $\mathrm{rk}_{c'}\Phi' \le \zeta c'$. So by (5), $c' \in R_{\zeta c'}$. Starting from $c \in R_\xi$ we now show: there is an $a_0$ such that for all $b_0$ and $c' = F[c, a_0, b_0]$, we have (*j*) and $c' \in R_{\zeta c'}$. Spelled out, with $\eta = \zeta c'$, this gives:

$$(6) \quad c \in R_\xi \cdot\supset\cdot \bigvee_a \bigwedge_b \begin{cases} htc' = j \wedge Mc' \supset (\exists\eta)^{\le_{j-1}\xi}\ c' \in R_\eta \\ htc' = j \wedge \tilde{M}c' \supset (\exists\eta)^{<_j\xi}\ c' \in R_\eta \end{cases} \quad \text{where } c' = F[c, a, b]$$

This is a typical Cantor-Bendixson statement: If $c \in R_\xi$ we only know existence of a winning strategy $\Phi$ for $I$ in $\mathscr{C}_c$, while (6) clearly tells how $I$ has to move from $c$ if he desires to win. Here it is spelled out formally:

THEOREM 1. *Let $\mathscr{C}(X, Y)$ be the $F_{\sigma\delta} \cap G_{\delta\sigma}$ game presented by* (1) *and* (3) *of* §8. *Assume that I has a winning strategy in $\mathscr{C}$, i.e., $c_0 \in R_\alpha$. Then I has a winning state-strategy $\Phi$ in $\mathscr{C}$. Let $A[c, \xi]$ be a choice for the a in* (6), *and put $Gc = A[c, (\mu\xi)c \in R_\xi]$. Then the recursion consisting of* (1) *and the formula $Xt = G[St]$ presents a winning state-strategy for I.*

PROOF. By assumption $c_0 \in R_\alpha$. Put $\xi c = (\mu\xi)c \in R_\xi$. So $c \in R_{\xi c}$ if $c$ is in some $R_\xi$, and $\xi c = 0$ otherwise. Let $A$, $G$ be as stated in the theorem. Hence we have, by (6),

$$(6') \quad c \in R_{\xi c} \wedge a = Gc \wedge b \in J \cdot\supset\cdot \begin{cases} htc' = j \wedge Mc' \supset c' \in R_{\xi c'} \wedge \xi c \ge_{j-1} \xi c' \\ htc' = j \wedge \tilde{M}c' \supset c' \in R_{\xi c'} \wedge \xi c >_j \xi c' \end{cases}$$

Let $\Phi$ be the strategy presented by the state-recursion for $\mathscr{C}$ and the output $G$. That is,

$$(\mathrm{h}) \qquad Se = c_0, \quad Syb = F[Sy, G[Sy], b] \qquad \Phi y = G[Sy]$$

Suppose now that $S$ is given by this state-recursion (h). By assumption we have $Se \in R_\alpha$, and therefore $Se \in R_{\xi(Se)}$. Assume inductively that $Sy \in R_{\xi(Sy)}$, and let $b$ be any member of $J$. Note that $c' = Syb = F[Sy, QSy, b]$. So by (6′) we see $Syb \in R_{\xi(Syb)}$. Hence, by induction over the tree $J^*$, we have proved $Sy \in R_{\xi(Sy)}$ for all $y$ (the strategy $\Phi$ keeps the play in winning states for $I$). But now (6′) yields this further information:

$$ht[Syb] = j \wedge M[Syb] \supset \xi(Syb) \geq_{j-1} \xi(Sy)$$

and

$$ht[Syb] = j \wedge \tilde{M}[Syb] \supset \xi(Syb] >_j \xi(Sy)$$

By Lemma 3 it follows that $\Phi$ wins $\mathscr{C}$ for $I$. Q.E.D.

It is clear how the story goes for player $J$. The resolvent for him is this:

$$(\bar{5}) \qquad c \notin R^\rho \cdot\!\equiv\!\cdot (\exists\Psi)[\Psi \text{ wins } \mathscr{C}_c \text{ for } J \wedge \overline{\mathrm{rk}}_c\Psi \leq \rho]$$

To understand the meaning of $\overline{\mathrm{rk}}$ one has to make up the dual version of Lemma 3. It yields this dual to (6):

$$(\bar{6}) \qquad c \notin R^\rho \cdot\!\equiv\!\cdot \bigwedge_a \bigvee_b \begin{cases} htc' = j \wedge Mc' \cdot\!\supset\!\cdot (\exists\sigma)^{<_j\rho} c' \notin R^\sigma \\ htc' = j \wedge \tilde{M}c' \cdot\!\supset\!\cdot (\exists\sigma)^{\leq_{j-1}\rho} c' \notin R^\rho \end{cases}$$

Note that the countable sets $R^\rho$ decrease as $\rho$ gets larger. So there is a $\beta \in \Omega$, $R^\beta = R^{\beta+\rho}$. That $J$ has a win in $\mathscr{C}$ means $c_0 \notin R^\beta$. If so, $(\bar{6})$ tells a winning state-strategy for $J$. You will now spell out the dual to Theorem 1. Using Davis [6]; $G_{\delta\sigma}$ are determinate, we obtain

THEOREM 2. *All games $\mathscr{C}(X, Y)$ in $F_{\sigma\delta} \cap G_{\delta\sigma}$ are determined. Furthermore, if the game is presented by a state-recursion $\mathscr{A}$, and in the hit-form, the winner has a winning strategy given by an output on $\mathscr{A}$.*

Note that what Davis tells us is $R_\alpha = R^\beta$. In the next section we will show how this may be proved in the Cantor-Bendixson style. That is, we will present a second proof of Theorem 2, which only uses the trivial part of Lemma 3 and does not use Davis' theorem.

The proof of Theorem 1 is very natural: The localized universality criterion (Lemma 2) gives a criterion for "winning strategy" and puts a rank on this strategy. This yields the crucial formula (6), which tells what a winning strategy has to do. If you show me the right universality criterion for Borel sets, I am sure I can do Theorem 1 for Borel-games. To also do Theorem 2 for Borel-games (without using Martin), one would probably have to make some fine adjustments in your universality criterion. (In my Lemma 2, I first had a $\leq_j$ in place of $\leq_{j-1}$. This gives the same change in the important formula (6). This weaker (6) still gives Theorem 1, but it suggested the wrong thing when it came to proving Theorem 2; the relation (7) was not founded in the original version.)

**§10. Determinacy proved by the Cantor-Bendixson method.** Let $\mathscr{C}$ be the game presented by (1), (3) in §8. We assume that all states in $C$ are accessible. We define the relation $\succ_i$ by

(7) $c, \rho, \xi \succ_i c', \sigma, \eta \cdot\equiv\cdot$

$$\bigvee_{a,b} c' = F[c, a, b] \wedge htc' \geq i \wedge \begin{cases} htc' = i + j \wedge Mc' \supset \sigma <_{i+j} \rho \wedge \eta \leq_{i+j-1} \xi \\ htc' = i + j \wedge \tilde{M}c' \supset \eta <_{i+j} \xi \wedge \sigma \leq_{i+j-1} \rho \end{cases}$$

Suppose there is an infinite regress in $\succ_i$. Then we should have $X$, $Y$, $S$, $\xi$, $\eta$ so that $S0 = c$, $St' = F[St, Xt, Yt]$, $ht(St) \geq i$,

$$htSt' = i + j \wedge M(St') \supset \rho t' <_{i+j} \rho t \wedge \xi t' \leq_{i+j-1} \xi t$$

and

$$ht(St') = i + j \wedge \tilde{M}(St') \supset \xi t' <_{i+j} \xi t \wedge \rho t' \leq_{i+j-1} \rho t$$

Let $i + k = fS$ be the least infinite-hit position on $S$. Suppose $\mathscr{C}_c(X, Y)$, that is, for $t \geq t_0$ we have $ht(St') \geq i + k$, $ht(St') = i + k \supset M(St')$, and $ht(St') = i + k$ happens infinitely often. So, for $t \geq t_0$, if $h(St') > i + k$ we have $\xi t' \leq_{i+k} \xi t$, and if $ht(St') = i + k$ we have $\xi t' <_{i+k} \xi t$. This contradicts the fact that $<_{i+k}$ is founded. But $\tilde{\mathscr{C}}_c(X, Y)$ is equally impossible, as now we would have a regress in the $\rho t$. This argument shows that $\succ_i$ is founded. We now use this fact to define, by induction over $\succ_i$,

$c \in R[U_0 W_0 \cdots U_{i-1} W_{i-1}]^{\rho}_{\xi}$

$$(8) \qquad \cdot\equiv\cdot \bigvee_a \bigwedge_b \begin{cases} htc' = j < i \wedge Mc' \supset c' \in U_j \\ htc' = j < i \wedge \tilde{M}c' \supset c' \in W_j \\ htc' = i + j \wedge Mc' \\ \qquad \supset (\forall\sigma)^{<_{i+j}\rho}(\exists\eta)^{\leq_{i+j-1}\xi} c' \in R[U_0 W_0 \cdots U_{i-1} W_{i-1}]^{\sigma}_{\eta} \\ htc' = i + j \wedge \tilde{M}c' \\ \qquad \supset (\exists\eta)^{<_{i+j}\xi}(\forall\sigma)^{\leq_{i+j-1}\rho} c' \in R[U_0 W_0 \cdots U_{i-1} W_{i-1}]^{\sigma}_{\eta} \end{cases}$$

Here and in the sequel $c'$ stands for $F[c, a, b]$. At this point $U_j$, $W_j$ are arbitrary subsets of $C$. Consider for a moment the case $i = 0$: (8) defines the sets $R^{\rho}_{\xi}$ into which no parameters $U$, $V$ enter. On the right side of (8) two of the clauses disappear. Compare this (8) with the crucial last formulas (6), $(\bar{6})$ which prove Theorem 1 and its dual. You will see that (8) is a self-dual amalgamation of (6) and $(\bar{6})$. In fact, it will be quite easy to show that the sets $R_{\xi} = \bigcap_{\rho} R^{\rho}_{\xi}$ satisfy (6), and the dual sets $R^{\rho} = \bigcap_{\xi} R^{\rho}_{\xi}$ satisfy $(\bar{6})$. Hence, these newly defined resolvents will yield winning state-strategies. Among the $R_{\xi}$ there is a largest, $R_{\alpha}$, and among the $R^{\rho}$ there is a smallest, $R^{\beta}$. If $c_0 \in R_{\alpha}$ we know that (6) yields a win for $I$, and if $c_0 \notin R^{\beta}$ we know how $(\bar{6})$ yields a win for $J$. To prove determinacy we therefore only have to show $R^{\beta} \subseteq R_{\alpha}$. (The not used inclusion $R_{\alpha} \subseteq R^{\beta}$ happens to be quite obvious from the definitions. It also can be seen this way: $c \in R_{\alpha}$ and $c \notin R_{\beta}$ give a win for $I$ and a win for $J$ in the game $\mathscr{C}_c$, which is impossible.)

We will first prove $R^{\beta} \subseteq R_{\alpha}$. This is where I need the fancy $R$'s with the parameters $U$, $W$. Using duality we can omit half of the proofs. That (8) actually is self-dual may be seen this way. At the right side of (8) the variable $j$ is used only to avoid clumsy notations. What we have there are four exclusive and complementary cases. Hence, negating both sides of (8) gives a formula for $\tilde{R}^{\rho}_{\xi}$ which has, on its right, the same four cases, each with the dualized requirement. We begin with this definition:

$$(9)\quad R[U_0W_0 \cdots U_{i-1}W_{i-1}]_\xi = \bigcap_\rho R[U_0V_0 \cdots U_{i-1}V_{i-1}]^\rho_\xi$$
$$R[U_0W_0 \cdots U_{i-1}W_{i-1}]^\rho = \bigcup_\xi R[U_0W_0 \cdots U_{i-1}W_{i-1}]^\rho_\xi$$

From (8) we see that $R^\rho_\xi$ decreases with $\rho$ and increases with $\xi$. This carries over to $R_\xi$ and $R^\rho$. Thus,

$$(10)\quad R[U_1W_1 \cdots U_{i-1}W_{i-1}]_\xi \subseteq R[U_1W_1 \cdots U_{i-1}W_{i-1}]_{\xi+\zeta}$$
$$R[U_1W_1 \cdots U_{i-1}W_{i-1}]^\rho \supseteq R[U_1W_1 \cdots U_{i-1}W_{i-1}]^{\rho+\sigma}$$

$$(11)\quad U \subseteq U' \wedge W \subseteq W' \mathbin{\cdot\!\supset\!\cdot} R[U_0W_0 \cdots U_{i-1}W_{i-1}\, UW]^\rho_\xi \subseteq R[U_0W_0 \cdots U_{i-1}W_{i-1}\, U'W']^\rho_\xi$$

PROOF. Omit the arguments $U_j$, $W_j$. We have to show $c \in R[UW]^\rho_\xi$ implies $c \in R[U'W']^\rho_\xi$. Use (8) to spell out what this means, and you will see that it suffices to show this:

a. $htc' = i \wedge Mc' \wedge c' \in U \mathbin{\cdot\!\supset\!\cdot} c' \in U'$

$htc' = i \wedge Mc' \wedge c' \in W \mathbin{\cdot\!\supset\!\cdot} c' \in W'$

b. $htc' = i + 1 + j \wedge Mc' \wedge \sigma <_{i+1+j} \rho \wedge \eta \leq_{i+j} \xi \wedge c' \in R[UW]^\sigma_\eta \mathbin{\cdot\!\supset\!\cdot} c' \in R[U'W']^\sigma_\eta$

$htc' = i + 1 + j \wedge \tilde{M}c' \wedge \eta <_{i+1+j} \xi \wedge \sigma \leq_{i+j} \rho \wedge c' \in R[UW]^\sigma_\eta \mathbin{\cdot\!\supset\!\cdot} c' \in R[U'W']^\sigma_\eta$

Part a is trivial by the assumption in (11). From either of the assumptions in b we see that $c, \rho, \xi \succ_{i+1} c', \sigma, \eta$. As this relation is founded we may assume inductively that (11) holds for $\sigma, \eta$. This gives b. Q.E.D.

From here on $\bar{\rho}$ stands for $\rho_1\rho_2\rho_3 \cdots$, and $\varphi\xi$ is taken from §6.

$$(12)\quad R[U_0W_0 \cdots U_{i-1}W_{i-1}]^\rho_\xi \subseteq R[U_0W_0 \ldots U_{i-1}\, W_{i-1}\, UW]^{\bar{\rho}}_{\varphi\xi} \quad \text{whereby}$$
$$U = \bigcap_{\sigma_0<\rho_0} R[U_0W_0 \cdots U_{i-1}W_{i-1}]^\sigma \qquad W = \bigcup_{\eta_0<\xi_0} R[U_0W_0 \cdots U_{i-1}W_{i-1}]_\eta$$

PROOF. The proof goes by induction relative to the founded relation (7). Suppress the arguments $U_0W_0 \cdots U_{i-1}W_{i-1}$. To be shown is $c \in R^\rho_\xi$ implies $c \in R[UW]^{\bar{\rho}}_{\varphi\xi}$. Use (8) to spell out what this means, and you will see that it suffices to prove these items:

a. $htc' = i \wedge Mc' \wedge (\forall\sigma)^{<_0\rho}(\exists\eta)^{\leq_{-1}\xi} c' \in R^\sigma_\eta \mathbin{\cdot\!\supset\!\cdot} c' \in U$

$htc' = i \wedge \tilde{M}c' \wedge (\exists\eta)^{<_0\xi}(\forall\sigma)^{\leq_{-1}\rho} c' \in R^\sigma_\eta \mathbin{\cdot\!\supset\!\cdot} c' \in W$

b. $htc' = i + j + 1 \wedge Mc' \wedge (\forall\sigma)^{<_{j+1}\rho}(\exists\eta)^{\leq_j\xi} c' \in R^\sigma_\eta \mathbin{\cdot\!\supset\!\cdot} (\forall\tau)^{<_j\bar{\rho}}(\exists\zeta)^{\leq_{j-1}\varphi\xi} c' \in R[UW]^\tau_\zeta$

$htc' = i + j + 1 \wedge \tilde{M}c' \wedge (\exists\eta)^{<_{j+1}\xi}(\forall\sigma)^{\leq_j\rho} c' \in R^\sigma_\eta \mathbin{\cdot\!\supset\!\cdot} (\exists\zeta)^{<_j\varphi\xi} (\forall\tau)^{\leq_{j-1}\bar{\rho}} c' \in R[UW]^\tau_\zeta$

Note that $\leq_{-1}$ is the universal relation. Hence the assumption in the first part of

$\underset{\sim}{a}$ is $(\forall\sigma)^{\sigma_0 < \rho_0}(\exists\eta)c' \in R^\sigma_\eta$. By (9) this becomes $(\forall\sigma)^{\sigma_0 < \rho_0}c' \in R^\sigma$ and so $c' \in U$. The second claim in $\underset{\sim}{a}$ is proved similarly.

To prove the first claim $\underset{\sim}{b}$, assume $\tau <_j \bar{\rho}$. Then clearly $\rho_0\tau <_{j+1} \rho_0\bar{\rho} = \rho$. Letting $\sigma = \rho_0\tau$, in the assumption of $\underset{\sim}{b}$, we obtain an $\eta \leq_j \xi$ with $c' \in R^{\rho_0\tau}_\eta$. By (7) we see that $c, \rho, \xi \succ_i c', \rho_0\tau, \eta$. Therefore, by inductive assumption, $R^{\rho_0\tau}_\eta \subseteq R[UW']^\tau_{\varphi\eta}$, whereby $W' = \bigcup_{\zeta_0 < \eta_0} R_\zeta$. As $\eta \leq_j \xi$ we have $\eta_0 \leq \xi_0$, and therefore $W' \subseteq W$. Hence by (11), $R[UW']^\tau_{\varphi\eta} \subseteq R[UW]^\tau_{\varphi\eta}$. Using the two inclusions we now have $c' \in R[UW]^\tau_{\varphi\eta}$. Using (g) in §6, we have $\varphi\eta \leq_{j-1} \varphi\xi$. Hence for every $\tau <_j \bar{\rho}$ we have found a $\zeta = \varphi\eta$, as required on the right side of the first claim $\underset{\sim}{b}$. The second claim in $\underset{\sim}{b}$ is proved similarly. Q.E.D.

$$R[U_0W_0 \cdots U_{i-1}W_{i-1}]^\rho \subseteq R[U_0W_0 \cdots U_{i-1}W_{i-1}UW]^{\bar{\rho}} \quad \text{whereby}$$
$$(13) \qquad U = \bigcap_{\sigma_0 < \rho_0} R[U_0W_0 \cdots U_{i-1}W_{i-1}]^\sigma \qquad W = \bigcup_\xi R[U_0W_0 \cdots U_{i-1}W_{i-1}]_\xi$$

PROOF. Omit the arguments $U_j$, $W_j$. Let $W' = \bigcup_{\eta_0 < \xi_0} R_\eta$. Note that $W' \subseteq W$, and therefore by (11), $R[UW]^{\bar{\rho}}_{\varphi\xi} \subseteq R[UW]^{\bar{\rho}}_{\varphi\xi}$. From (12) we see $R^\rho_\xi \subseteq R[UW']^{\bar{\rho}}_{\varphi\xi}$. Using the two inclusions we have $R^\rho_\xi \subseteq R[UW]^{\bar{\rho}}_{\varphi\xi}$. Take the union over all $\xi$, note that $\varphi$ is onto, and use definition (9) to get $R^\rho \subseteq R[UW]^{\bar{\rho}}$. Q.E.D.

$$U \subseteq R[U_0W_0 \cdots U_{i-1}W_{i-1}UW]^\tau \quad \text{whereby}$$
$$(14) \qquad U = \bigcap_\rho R[U_0W_0 \cdots U_{i-1}W_{i-1}]^\rho \qquad W = \bigcup_\xi R[U_0W_0 \cdots U_{i-1}W_{i-1}]_\xi$$

PROOF. Omit the arguments $U_j$, $W_j$. The set $C$ of all states is countable, and therefore $R^\rho$ is. But $\Omega$ has type $\omega_1$. From (10) we therefore see that $R^\rho = R^\beta$ from some $\beta$ on. Hence $U = R^\beta$. Let $\delta = \beta_0'\tau$. Clearly $\beta_0 < \delta_0$, and so $U = \bigcap_{\sigma_0 < \delta_0} R^\sigma$. From (13) we now see $R^\rho \subseteq R[UW]^{\bar{\rho}}$. Use this for $\rho = \delta$, $\bar{\rho} = \tau$ to obtain $R^\delta \subseteq R[UW]^\tau$. Here we can put $U$ for $R^\delta$ because $\beta < \delta$ and therefore $U = U^\beta = U^\delta$. Q.E.D.

Note that the dual to (14) says $W \supseteq R[U_0W_0 \cdots U_{i-1}W_{i-1}\,UW]_\zeta$ with precisely the same $U$ and $W$. Up to this point the $U_j$, $W_j$ enter as parameters. We now define these sets by the following induction over $i$.

$$(15) \qquad U_i = \bigcap_\rho R[U_0W_0 \cdots U_{i-1}W_{i-1}]^\rho \qquad W_i = \bigcup_\xi R[U_0W_0 \cdots U_{i-1}W_{i-1}]_\xi$$

We need these properties of the $U$'s and $W$'s:

$$c \in U_i \supset \bigvee_a \bigwedge_b \begin{array}{l} htc' < i \wedge Mc' \supset c' \in U_{htc'} \\ htc' < i \wedge \tilde{M}c' \supset c' \in W_{htc'} \end{array}$$
$$(16) \qquad c \notin W_i \supset \bigwedge_a \bigvee_b \begin{array}{l} htc' < i \wedge Mc' \supset c \notin U_{htc'} \\ htc' < i \wedge \tilde{M}c' \supset c' \notin W_{htc'} \end{array}$$
$$U_0 \subseteq U_1 \subseteq U_2 \subseteq \cdots \qquad \tilde{W}_0 \subseteq \tilde{W}_1 \subseteq \tilde{W}_2 \subseteq \cdots$$

PROOF. From (14) and (15) we have $U_i \subseteq R[U_0W_0 \cdots U_iW_i]^\tau$ and so $U_i \subseteq U_{i+1}$. Using the dual to (14) and (15) we get $W_i \supseteq W_{i+1}$. It remains to prove the two dual statements in the first line of (16). We do the second. So assume $c \notin W_i$. By (15) this gives $c \notin R[U_0W_0 \cdots U_{i-1}W_{i-1}]_0$. So by (9), there is a $\rho$ such that

$c \notin R[U_0 W_0 \cdots U_{i-1} W_{i-1}]^{\rho}_0$. Using (8), this will imply the required claim (read the second paragraph following (8)). Q.E.D.

(17) $$c \in U_0 \;\cdot\!\supset\!\cdot\; \bigvee_a \bigwedge_b [(Mc' \supset c' \in U_{htc'}) \wedge (\tilde{M}c' \supset c' \in W_{htc'})] \;\cdot\!\supset\!\cdot\; c \in W_0$$

PROOF. Abbreviate the formula $[htc' < i \wedge Mc' \supset c' \in U_{htc'}] \wedge [htc' < i \wedge \tilde{M}c' \supset c' \in W_{htc'}]$ as $L(i, a, b)$, and abbreviate $[Mc' \supset c' \in U_{htc'}] \wedge [\tilde{M}c' \supset c' \in W_{htc'}]$ as $L(a, b)$. Note that $(\exists^\omega i)L(i, a, b)$ implies $L(a, b)$. Assume $c \in U_0$. From (16) we see first $(\forall i)c \in U_i$ and then $(\forall i) \bigvee_a \bigwedge_b L(i, a, b)$. As $a$ ranges over the finite set $I$, this gives $\bigvee_a (\exists^\omega i) \bigwedge_b L(i, a, b)$. Here we can move $(\exists^\omega i)$ to the right of $\bigwedge_b$. Using the note, this yields $\bigvee_a \bigwedge_b L(a, b)$. So now we have proved the first implication (17). The second is just the dual. (If you do not quite trust this duality, prove the second claim by starting with the assumption $c \notin W_0$. Use the second half of (16) to obtain a statement $(\forall i) \bigwedge_a \bigvee_b L'(i, a, b)$. Here the $\bigwedge_a$ can be moved to the front, and $(\forall i)\bigvee_b$ can be replaced by $\bigvee_b (\exists^\omega i)$ because $J$ is finite, and $(\exists^\omega i)L'(i, a, b)$ implies $\sim L(a, b)$. So now you have $\bigwedge_a \bigvee_b \sim L(a, b)$, which is wanted.) Q.E.D.

What we need from (17) is the inclusion $U_0 \subseteq W_0$. It will give determinacy! From here on we use only the simple $R$ without parameters. In the proof of (14) we used a countability argument to see that $R^\rho$ takes on a minimal value $R^\beta$. Dually, (10) shows that $R_\xi$ takes on a maximal value $R_\alpha$. Using definition (15) of $U_0$ and $W_0$, this gives:

(18) $$\alpha \leq \xi \supset R_\xi = W_0 \qquad \beta \leq \rho \supset R^\rho = U_0$$

We now use the same countability argument to show this:

(19) $$(\exists \delta)(\forall \xi) R^\eta_\xi = R_\xi \qquad (\exists \gamma)(\forall \rho) R^\rho_\gamma = R^\rho$$

PROOF. From (8) we see that $R^\rho_\xi$ decreases with $\rho$. As the $R$'s are countable, and as $\Omega$ is of type $\omega_1$, this gives a $\tau = \tau(\xi)$ such that $R^\rho_\xi \supseteq R^\tau_\xi$ for all $\rho$. Hence, by (9), $R_\xi = R^\rho_\xi$ for all $\rho \geq \tau(\xi)$. There are but countably many $\xi \leq \alpha$. Hence there is a $\delta$ in $\Omega$, such that $\beta \leq \delta$, and $\xi \leq \alpha \supset \tau(\xi) \leq \delta$. Therefore,

(a) $$\xi \leq \alpha \supset R_\xi = R^\delta_\xi$$

Suppose $\alpha \leq \xi$. Using (18) this gives $R_\xi = W_0$, and by (17), $U_0 \subseteq R_\xi$. By (18) and $\beta \leq \delta$ we therefore have $R^\delta \subseteq R_\xi$. From (9), $R^\delta \supseteq R^\delta_\xi \supseteq R_\xi$. So now $R^\delta = R^\delta_\xi = R_\xi$. Hence we have shown $\alpha \leq \xi \supset R_\xi = R^\delta_\xi$. Together with (a) this gives the first claim (19). The second is dual. Q.E.D.

THEOREM 3. *Let $\mathscr{C}$ be the game presented by* (1) *and* (3). *Let $R^\rho_\xi$ be the sets of states given by the $\omega_1$-recursion* (8), $i = 0$. *Define $R_\xi$, $R^\rho$ by* (9), *and pick the first $\alpha$, $\beta$ for which* (18) *holds. Then $R^\beta = R_\alpha$. Furthermore, $c_0 \in R_\alpha$ gives a winning state-strategy for I, and $c_0 \notin R^\beta$ gives a winning state-strategy for J. Hence, the game $\mathscr{C}$ is determinate, and membership of $c_0$ in $R_\alpha$ tells who has the win.*

PROOF. From (17) and (18) we see $R^\beta \subseteq R_\alpha$. The other inclusion is obvious, by (9). To prove the rest of the theorem we show that $R_\xi$ satisfies (6), and therefore the dual $\tilde{R}^\rho$ satisfies $(\bar{6})$. From these new versions of (6) and $(\bar{6})$ one obtains the claimed strategies, as in the proof of Theorem 1. We only show (6), as ($\bar{6}$) is the dual.

The assumption of (6) is $c \in R_\xi$. By (19) we have a $\delta$ with $(\forall\eta)R_\eta^\delta = R_\eta$. So now $c \in R_\xi^\delta$. Let $\delta^+ = \delta_0'000 \cdots$. Then $\delta < \delta^+$, and therefore $c \in R_\xi^{\delta^+}$. Look up in (8) what this means, and look up the conclusion in (6). You will see that we only need to prove this:

$$\underset{\sim}{a}: \quad htc' = j \wedge Mc' \wedge (\forall\sigma)^{<_j\delta^+}(\exists\eta)^{\leq_{j-1}\xi} c' \in R_\eta^\sigma \supset (\exists\eta)^{\leq_{j-1}\xi} c' \in R_\eta$$

$$\underset{\sim}{b}: \quad htc' = j \wedge \tilde{M}c' \wedge (\exists\eta)^{<_j\xi}(\forall\sigma)^{\leq_{j-1}\delta^+} c' \in R_\eta^\sigma \supset (\exists\eta)^{<_j\xi} c' \in R\eta$$

Clearly $\delta <_j \delta^+$. So the assumption in $\underset{\sim}{a}$ gives an $\eta \leq_{j-1} \xi$ with $c' \in R_\eta^\delta$. As $R_\eta^\delta = R_\eta$ this gives the conclusion in $\underset{\sim}{a}$. The assumption in $\underset{\sim}{b}$ gives an $\eta <_j \xi$ such that $(\forall\sigma)^{\leq_{j-1}\delta^+} c' \in R_\eta^\sigma$. As $\delta \leq_{j-1} \delta^+$ we therefore have $c' \in R_\eta^\delta = R_\eta$, which is the conclusion in $\underset{\sim}{b}$. Q.E.D.

This CB-proof does more than show determinacy, just in the existential sense. First, it yields a presentation of a winning strategy. Second, it yields a "method for deciding" which player has the win. Namely, the deciding set $R_\alpha$ of states is given by the $\omega_1$-recursion $(8)_{i=0}$. I think this is a point which deserves more thought: Is there a nice class of state-recursions $\mathscr{A}$ with ouput $ht$, $M$, for which $\alpha$, $\beta$ become recursive ordinals ($R_\alpha$ becomes computable, $\Omega$ can be replaced by the first nonrecursive ordinal)? We will now tell what happens if the set $C$ of states is finite.

**§11. Finite-state games.** These are the games $\mathscr{C}$ presented by (3) and a finite-state recursion (1). As $C$ is finite, the hit-function $ht: C \to \omega$ is bounded. So we may assume $ht: C \to r$, with a finite $r = \{0, \ldots, r - 1\}$ in place of $\omega$, and we may assume $M \subseteq r$. Hence, finite-state games are as concretely presentable, as you may expect for an infinite game.

The formula (3) gives $\mathscr{C}(X, Y)$ in the form $\bigvee_{i<r}[(\forall^\omega t)h(t) \geq i \wedge (\exists^\omega t)[h(t) = i \wedge M(t)]]$. Therefore finite-state games are in $Bl(F_\sigma)$, the Boolean algebra over $F_\sigma$. In [4] you find the more general remark: Every Souslin-set $\mathscr{C}(X)$, presented in kernel-form $(\exists Z)(\exists^\omega t)M[\bar{X}t, \bar{Z}t]$ and $M$ given by a finite-state recursion, is in $Bl(F_\sigma)$. This is not obvious, and is in fact McNaughton's lemma. I would like to much better understand this matter; look at my Problem 2, if you need a rest from doing super-set theory.

COROLLARY 1. *Let $\mathscr{C}$ be a game which admits a hit-presentation* (3) *over a finite-state recursion* (1). *Then $\mathscr{C}$ is determinate, and the winner has a winning strategy which can be executed by a finite-state recursion. Furthermore, given the finite automaton* (1) *and its output* ($ht$, $M$), *one can compute the finite set $R_\alpha = R^\beta$. Hence, one can effectively decide whether I wins ($c_0 \in R_\alpha$), or J wins ($c_0 \notin R_\alpha$). One can also construct the output, on the finite-state recursion, which gives a winning strategy.*

PROOF. Let $\Omega_0$ be like $\Omega$, except that only finite ordinals are permitted. In the proof of Theorem 3, replace $\Omega$ by $\Omega_0$. The sets $R_\xi^\rho$ in (8) are in the set of states $C$, which is now assumed finite. Hence, the cardinality arguments we used to show that the various sequences (such as $R_0 \subseteq \cdots \subseteq R_\xi \subseteq \cdots$) end with a constant value (such as $R_\alpha$), now work for $\Omega_0$. So we now find $R_\alpha = R^\beta$ for finite $\alpha$, $\beta$.

As (8) now is a recursive definition, in the strict sense, the proof actually gives instructions for computing $\alpha$, $\beta$, and all the finitely many $R_\xi$, $\xi \leq \alpha$, $R^\rho$, $\rho \leq \beta$. By checking $c_0 \in R_\alpha$ we can tell who has the win. As shown in the proof, the $R_\xi$ satisfy (6). As all $R_\xi$, $\xi \leq \alpha$, are computed, we can actually make the picks $A[c, \xi]$, occurring in the proof of Theorem 1, and this computes the output $G$ which gives the winning state-strategy for $I$ in case $c_0 \in R_\alpha$. If $c_0 \notin R_\alpha$ we can do the same for $J$. Q.E.D.

Corollary 1 is a very concrete result on quite concrete objects. Its proof is not simple, the one of Theorem 1 and just determinacy will not do. In [5] you find another proof which is a streamlined version of Landweber's original proof. The result was conjectured by McNaughton, who found the finite-state games.

**§12. Order-vector strategies for games in $Bl(F_\sigma)$.** Let $r$ stand for the finite set $\{0, \ldots, r-1\}$. An *order-vector* over $r$ is a member of $r!$, or a permutation $v = d_0 \cdots d_{r-1}$ of all members of $r$. As initial order-vector $v_0$ we select $0 \cdots r-1$. The hit $ht(d, v)$ is the place $i$, where $d$ occurs in $v$, and $A(d, v)$ is the set $\{d_i, \ldots, d_{r-1}\}$ which consists of those entries in $v$ at places to the right of the hit $i$. *The new order-vector* $G[d, v]$ is obtained by moving $d$, from its position in $v$, all the way to the right of $v$. For $Z\colon \omega \to r$ this gives *the order-vector recursion* $V0 = v_0$, $Vt' = G[Zt, Vt]$. Note that it is finite-state.

For given $Z$ in $r^\omega$ compute $V$ by the order-vector recursion, and let $At = A(Zt, Vt)$, $ht = ht(Zt, Vt)$. As $r$ is finite there obviously exists the least infinite-hit position $fZ = i$. Consider the first time $t_0$, from which on all hits $ht \geq i$. Clearly $At$ will take the constant value $U$ from $t_0$ on. Therefore $Zt$ is in $U$ from $t_0$ on. Hence $\sup Z \subseteq U$. As the position $i$ is hit infinitely often, every $d \in M$ will keep recurring at position $i$ and will eventually be hit at $i$. Hence $U \subseteq \sup Z$. Thus, $fZ = i$ implies that the statements $\sup Z \in \mathscr{U}$, $(\exists^\omega t)[ht = i \wedge At \in \mathscr{U}]$ are equivalent. Defining $M(c, v)$ by $A(c, v) \in \mathscr{U}$ we therefore have this:

$$fZ = i \cdot\supset\cdot \begin{cases} \sup Z \in \mathscr{U} = (\exists^\omega t)ht[Zt, Vt] = i \wedge M[Zt, Vt] \\ \sup Z \notin \mathscr{U} = (\exists^\omega t)ht[Zt, Vt] = i \wedge \bar{M}[Zt, Vt] \end{cases}$$

Suppose now that $\mathscr{A} = \langle C, c_0, F\rangle$ is a state-recursion over $I \times J$, $\mathscr{U}$ is a subset of $2^r$, and $L$ is a map from $C$ to $r$. Let $\mathscr{C}(X, Y)$ be the game presented by $\mathscr{A}$ and $\mathscr{C}(X, Y) \equiv \sup_t L[St] \in \mathscr{U}$. From $\mathscr{A}$ we obtain *the order-vector recursion* $\bar{\mathscr{A}}$ *over* $\mathscr{A}$ by adding to $\mathscr{A}$ the formulas $Zt = L[St]$, $V0 = v_0$, $Vt' = G[Zt, Vt]$. So now we have $\mathscr{C}(X, Y) = \sup Z \in \mathscr{U}$, whereby $S$, $Z$, $V$ are given by the order-vector recursion $\bar{\mathscr{A}}$. Therefore we have this hit-presentation for the game $\mathscr{C}$:

$$f(X, Y) = fZ = i \cdot\supset\cdot \begin{cases} \mathscr{C}(X, Y) \equiv (\exists^\omega t)ht[Zt, Vt] = i \wedge M[Zt, Vt] \\ \bar{\mathscr{C}}(X, Y) \equiv (\exists^\omega t)ht[Zt, Vt] = i \wedge \bar{M}[Zt, Vt] \end{cases}$$

$$\text{whereby } (S, Z, V) = \bar{\mathscr{A}}(X, Y)$$

We call this *the order-vector presentation for the sup-game* $\mathscr{C}$, originally presented by a state-recursion and a sup-condition $L$, $\mathscr{U}$ on the states. The order-vector presentation for $\mathscr{C}$ also is in our hit-form. So we have this corollary to Theorem 3:

COROLLARY 2. *Given a state-recursion $\mathscr{A}$ over $I \times J$, a subset $\mathscr{U}$ of $2^r$, and an $L: C \to r$, let $\mathscr{C}(X, Y)$ be the $Bl(F_\sigma)$-game $\mathscr{C}(X, Y) = \sup_t L[St] \in \mathscr{U}$ with $S = \mathscr{A}(X, Y)$. The game $\mathscr{C}$ is determinate and the winner has a winning order-vector strategy. That is, a strategy given by the order-vector recursion $\bar{\mathscr{A}}$ and an output $K$ on $\bar{\mathscr{A}}$ ($K: C \times r \times r! \to I$ in case the win is for $I$, $K: C \times r \times r! \times I \to J$ in case the win is for $J$).*

Note that the order-vector recursion keeps track of some of the information contained in the course of values $\bar{S}t$ of all earlier states. Therefore, in an extended sense, order-vector strategies still are state-strategies. Also in our more restricted sense, Definition 1, order-vector strategies are state-strategies, namely with respect to the recursion $\bar{\mathscr{A}}$ (with the extended set of states $C \times r \times r!$).

Note that the condition $\sup Z \in \mathscr{U}$ can be spelled out as $\bigvee_{U \in \mathscr{U}} [(\forall^\omega t) U[Zt] \wedge \bigwedge_{d \in \mathscr{U}} (\exists^\omega t) Zt = d]$. Therefore, a sup-condition is in $Bl(F_\sigma)$. Of course it is essential here that $Z$ takes its values in a finite set $r$. Conversely, every $Bl(F_\sigma)$ can be put into sup-form (by adding some new states), see [3]. So, Corollary 1 covers all games in $Bl(F_\sigma)$. Also note that every $\mathscr{C}$ given in the hit-form (Lemma 1) is in $Bl(F_\sigma)$ if the function $ht$ is bounded.

The order-vector recursion puts a finite factor on the original number of states. In particular, if $\mathscr{A}$ is finite-state, then $\bar{\mathscr{A}}$ is also. That additional states are essential is shown by this example of a solitary ($J = 1$) finite-state game: $I = 2$, $C = 3$, $F[c, a]$ given by $00 \to 1$, $01 \to 2$, $10 \to 1$, $11 \to 0$, $20 \to 2$, $21 \to 0$. $\mathscr{C}(X) \equiv \sup S = C$. It is easy to find a win $\Phi$ which looks at the two last states only ($\Phi(c1) = \Phi(c2) = \Phi(10) = 1$, $\Phi(20) = \Phi(00) = 0$). There is no win $\Phi$ which uses just the last state. Hence, we have an example of this situation:

REMARK. A game $\mathscr{C}$ may be presented in a form $(Q_1 t) M_1[\bar{X}t, \bar{Y}t]$, whereby the states of the kernel $M_1$ do not carry the information needed in a winning strategy. The same $\mathscr{C}$ may have another presentation $(Q_2 t) M_2[\bar{X}t, \bar{Y}t]$ so that the states of $M_2$ do carry that information.

Mind this remark if you work on Problem 1. There are many nice ways of presenting Borel-sets; the answer to the problem may depend on which presentation you select.

## §13. McNaughton's lemma.

The order-vector-hit method occurs in [5] and in [12]. It is due to McNaughton [10] and was used by him to prove the projection-lemma for finite-state recursions. I will use it in this form:

LEMMA 4. *Given a finite-state recursion $\mathscr{A} = \langle D, d_0, F \rangle$ with $F: D \times I \times K \to D$ and a $\mathscr{U} \subseteq 2^D$, one can construct another finite-state recursion $\mathscr{A}' = \langle D', d_0', F' \rangle$ with $F': D' \times K \to D'$ and a $\mathscr{U}' \subseteq 2^{D'}$ such that, for every $U$ in $K^\omega$, these two formulas are equivalent:*

$$(\forall X)(\forall Z)[Z0 = d_0 \wedge (\forall t) Zt' = F[Zt, Xt, Ut] \supset \sup Z \in \mathscr{U}]$$
$$(\exists Z')[Z'0 = d_0 \wedge (\forall t) Z't' = F'[Z't, Ut] \wedge \sup Z' \in \mathscr{U}']$$

The first formula defines a set $\mathscr{C}'(U) = (\forall X)\mathscr{C}(X, U)$, $\mathscr{C}(X, U)$ presented by $\sup Z \in \mathscr{U}$ and the finite-state recursion $Z = \mathscr{A}(X, U)$. The second formula presents the same $\mathscr{C}'$ in the finite-state-sup form. Hence, the lemma says the quantifier

$(\forall X)$ preserves $Bl(F_\sigma)^{\text{reg}}$. Replacing $\mathscr{C}$ and $\mathscr{C}'$ by their complement, you see that $(\exists X)$ also preserves $Bl(F_\sigma)^{\text{reg}}$, and now you have a projection-lemma. Another nice way of stating the projection-lemma is this: Given a Souslin-set $\mathscr{C}'(U) \equiv (\exists X)(\exists^\omega t)M[\bar{X}t, \bar{Y}t]$, if the kernel $M$ is regular, then $\mathscr{C}'$ is actually in $Bl(F_\sigma)^{\text{reg}}$.

There are those who want to see an error in McNaughton [10]. What I found there is the very interesting piggy-back method. It will quite nicely prove the projection lemma. So an error it was not.

*Problem.* If you look up my version of McNaughton's proof (see [3]) you will find the notation $S_u$, $u \in \omega$. The piggy-back method simulates the action of all these $S_u$ by the action of finitely many. I ask you to compare these $S_u$ with the games $\mathscr{C}_c$ in the proof of Theorem 1. I suspect that McNaughton's lemma is a result about solitary games ($J = 1$) which extends to 2-player games. The equation $S_u = \mathscr{C}_c$ might be the clue.

**§14. Special games and the complementation lemma.** A somewhat strange sort of game is this:

(1) $\quad \mathscr{C}_\chi(X, Y) \equiv \sup Z \in \mathscr{U}$ whereby $Z0 = d_0$, $Zt' = F[Zt, \chi[\bar{Y}t], Xt, Yt]$

Here $\chi$ is a parameter of type $J^* \to E$, $E$ a finite set. The states of $Z$ come from a finite set $D$ (say $r$), $F$ takes $D \times E \times I \times J$ to $D$, and $\mathscr{U} \subseteq 2^D$. The game $\mathscr{C}_\chi$ clearly belongs to $Bl(F_\sigma)$; we call it a *special game*.

LEMMA 5. *Let $\mathscr{C}_\chi$ be the special game presented by* (1). *For any $\chi$: $J^* \to E$ these two statements are complementary*:

(a) $\quad (\exists \Phi)(\forall Y)(\exists Z)$

$$\cdot\ Z0 = d_0 \wedge (\forall t)Zt' = F[Zt, \chi[\bar{Y}t], \Phi[\bar{Y}t], Yt] \wedge \sup Z \in \mathscr{U}$$

(b) $\quad (\exists \Psi)(\forall YXZV)$

$$\cdot \begin{cases} Z0 = d_0 \\ V0 = v_0 \end{cases} \wedge (\forall t) \begin{cases} Zt' = F[Zt, \chi[\bar{Y}t], Xt, Yt] \\ Vt' = G[Zt, Vt] \\ Yt = \Psi_{Zt,Vt,Xt}[\bar{Y}t] \end{cases} \supset \sup Z \notin \mathscr{U}$$

*Here $v_0$, $G$ is the order-vector recursion on $D$, and $\Psi$ consists of the components $\Psi_{d,v,a} \subseteq J^*$, $d \in D$ and $v \in D!$ and $a \in I$.*

PROOF. We note that (1) is a presentation of a sup-game by state-recursion. Namely, $C = D \times J^*$ is the set of states, $(d_0, e)$ is the initial state, and the transition operator $\bar{F}$ has these components: $\bar{F}_1[d, y, a, b] = F[d, \chi y, a, b]$ and $\bar{F}_2[d, y, a, b] = yb$. Therefore we can use Corollary 2. It says that these two statements are complementary: (A) $(\exists \Phi)[\Phi$ wins $\mathscr{C}_\chi$ for $I]$, and (B) $(\exists \Psi)[\Psi$ order-vector-strategy $\wedge\ \Psi$ wins $\mathscr{C}_\chi$ for $J]$. Clearly (a) just spells out (A). To prove the lemma we now show that (B) can be spelled out as (b).

"$\Psi$ wins $\mathscr{C}_\chi$ for $J$" may be spelled out as $(\forall XY)(\forall Z)$[if $I$ plays $X$, and $J$ plays $Y = [\Psi X]$, and $Z$ is given by (1), then $\sup Z \notin \mathscr{U}$]. Use now the additional informations in (B), $\Psi$ is order-vector strategy. That is, add $(\forall V)$ to $(\forall Z)$, add to the ifs "and $V$ is given by the order-vector recursion", and replace "$J$ plays $Y = [\Psi X]$"

by "$J$ picks his moves $Yt = \Psi_{Zt,Vt,Xt}\ \bar{Y}t$". Now you have (b) as just a formulation of (B). Note that $Yt = \Psi_{Zt,Vt,Xt}\ \bar{Y}t$ does depend on the state $St = (Zt,\ \bar{Y}t)$ of the game $\mathscr{C}_\chi$, the order-vector $Vt$, and the last move $Xt$ of $I$. This is just as $Yt$ is supposed to depend in an order-vector strategy for $J$. Q.E.D.

In the copycat version of my proof it will seem as if the basic ideas, such as state-strategy and special games, and the formulation of Lemma 5, came like manna from heaven. The fact is that it all started with the desire of putting the complement of the first statement (a) back into a similar existential form (b). Just determinacy does it in a first approximation. But now the $\Psi$ ranges over subsets of $I^*$, instead of $J^*$, as does the original $\Phi$. Suppose now that by brute force and good luck you have found the second formula (b). There now are many $\Psi$'s, but they do range over subsets of $J^*$, and they are finite in number. Obviously the new $\Psi$ means a special kind of strategy for $J$, the original $\Phi$ was an arbitrary strategy for $I$. So you are stuck with a conjecture (Lemma 5) stating a lopsided sort of determinacy for the strange games (1). To take the strangeness away, the general notion of order-vector strategy for the general sup-games had to be invented. Now the conjecture (namely Theorem 2) takes a much nicer (self-dual) form. Finding a proof (the one in §9) was easy (I still think the one in [5] could be made to work). The desire for Lemma 5 will be explained now.

LEMMA 6. *Given the finite object* $\mathscr{A} = \langle D, d_0, F, \mathscr{U}\rangle$, *which defines the special games* $\mathscr{C}_\chi(X, Y)$ *as in* (1), *one can construct another such object* $\mathscr{A}' = \langle D', d_0', F', \mathscr{U}'\rangle$, *which defines a special game* $\mathscr{C}'_\chi(X', Y)$ *between a new player* $I'$ *and the old player* $J$, *such that for all* $\chi$:

$$\sim(\exists\Phi)[\Phi \text{ wins } \mathscr{C}_\chi \text{ for } I] \equiv (\exists\Phi')[\Phi' \text{ wins } \mathscr{C}'_\chi \text{ for } I']$$

PROOF. Start with the expression $\sim(\exists\Phi)[\Phi$ wins $\mathscr{C}_\chi$ for $I]$. It is the negation of (a) in Lemma 5 and therefore is equivalent to (b). Therefore, our complementation lemma is proved, if we can put (b) into the new form $(\exists\Phi')[\Phi'$ wins $\mathscr{C}'_\chi$ for $I']$.

The formula (b) is $(\exists\Psi)(\forall Y)A$, whereby $A$ stands for this formula:

$$(\forall X)(\forall ZV)\cdot\begin{cases} Z0 = d_0 \\ V0 = v_0\end{cases} \wedge\ (\forall t)\begin{cases} Zt' = F[Zt,\ \chi[\bar{X}t],\ Xt,\ Yt] \\ Vt' = G[Zt,\ Vt] \\ Yt = \Psi_{Zt,Vt,Xt}[\bar{Y}t]\end{cases} \supset \sup Z \notin \mathscr{U}$$

Set $Wt \equiv (\forall u)^t YU = \Psi_{ZU,VU,XU}[\bar{Y}U]$. As Skolem said, the bounded quantifier $(\forall u)^t$ may be replaced by a recursion. This will put (A) into this form:

$$(\forall X)(\forall ZVW)\cdot\begin{cases} Z0 = d_0 \\ V0 = v_0 \\ W0 \equiv T\end{cases}$$

$$\wedge\ (\forall t)\begin{cases} Zt' = F[Zt,\ \chi[\bar{X}t],\ Xt,\ Yt] \\ Vt' = G[Zt,\ Vt] \\ Wt' \equiv Wt \wedge Yt = \Psi_{Zt,Vt,Xt}\,[\bar{Y}t]\end{cases} \supset [(\forall^\omega t)Wt \supset \sup Z \notin \mathscr{U}]$$

Note that $Yt = \Psi_{Zt,Vt,Xt}[\bar{Y}t]$ can be replaced by $\bigvee_{d,v,a}[Zt = d \wedge Vt = v \wedge Xt = a \wedge Yt = \Psi_{d,v,a}[\bar{Y}t]]$ and note that $(\forall^{\omega}t)Wt \supset \sup Z \notin \mathcal{U}$ can be replaced by an expression $\sup(Z, V, W) \in \mathcal{V}$. Putting $Z^*$ for $(Z, V, W)$, (A) now takes this form:

$$(\forall X)(\forall Z^*)\cdot Z^*0 = d_0^* \wedge (\forall t)Z^*t' = F^*[Z^*t, Xt, \chi[\bar{Y}t], \Psi[\bar{Y}t], Yt] \supset \sup Z^* \in \mathcal{V}$$

Here $Z^*$ takes its values in the finite set $D \times D! \times [F, T]$. The initial state is $d_0^* = (d_0, v_0T)$. Minding the note you will spell out the three components of the transition operator $F^*$. We now use Lemma 4, with $Ut = (\chi[\bar{Y}t], \Psi[\bar{Y}t], Yt)$ to give this new form to (A):

$$(\exists Z') \cdot Z'0 = d_0' \wedge (\forall t)Z't' = F'[Z't, \chi[\bar{Y}t], \Psi[\bar{Y}t], Yt] \wedge \sup Z' \in \mathcal{U}'$$

In (b) replace (A) by this new form of (A). So now (b) takes the new form

$$(\exists \Psi)(\forall Y)(\exists Z') \cdot Z'0 = d_0' \wedge (\forall t)Z't' = F'(t) \wedge \sup Z' \in \mathcal{U}'$$

Recall that $\Psi$ has the components $\Psi_{d,v,a}$, $d \in D$ and $v \in D!$ and $a \in I$. Let $I' = D \times D! \times I$, and let $\mathcal{C}_\chi'$ be the special game $\mathcal{C}_\chi'(X', Y) \equiv \sup Z' \in \mathcal{U}$ with $Z'0 = d_0'$, $Z't' = F'[Z't, \chi[\bar{Y}t], X't, Yt]$. Then the new form of (b) says $(\exists\Psi)[\Psi$ wins $\mathcal{C}'_\chi$ for $I']$.

Q.E.D.

§15. **Decidability of the game-language** GL. Let $J$ be a fixed finite set. Our language GL contains quantifiers for variables such as $\chi$, $\Phi$. Attached to each variable $\chi$ is a finite set $E$, which indicates the range $\chi: J^* \to E$ of $\chi$. The *basic formulas* of GL are these: $[(\chi_1, \ldots, \chi_n)$ wins $\mathcal{C}$ for $E_1 \times \cdots \times E_n]$. Here $\chi_i: J^* \to E_i$ and $\mathcal{C}$ is a finite-state game $\mathcal{C}(\chi_1, \ldots, \chi_n, Y)$ between the two players $(E_1 \times \cdots \times E_n)$ and $J$. (More carefully stated, $\mathcal{C}$ is a finite object $\langle D, c_0, F, \mathcal{U}\rangle$ which presents a finite-state game between $E_1 \times \cdots \times E_n$ and $J$. So now the basic formulas are proper linguistic entities!) From the basic formulas all other formulas of GL are obtained by prefixing quantifiers $(\exists\chi)$ and $(\forall\chi)$. We will often use $\chi$ for a sequence $(\chi_1, \ldots, \chi_n)$ of variables, and then the values of $\chi$ range over $E_1 \times \cdots \times E_n$, with those of $\chi_i$ ranging over $E_i$. Consider now these formulas:

$$\mathcal{C}(U, X, Y) \equiv \sup Z \in \mathcal{U} \quad \text{whereby} \quad Z0 = d_0,\ Zt' = F[Zt, Ut, Xt, Yt]$$
$$\mathcal{C}_\chi^+(X, Y) \equiv \sup Z \in \mathcal{U} \quad \text{whereby} \quad Z0 = d_0,\ Zt' = F[Zt, \chi[\bar{Y}t], Xt, Yt]$$

The first formula defines a finite-state game $\mathcal{C}$ between players $(E \times I)$ and $J$. So $Z$, $U$, $X$, $Y$ take values in the finite sets $D$, $E$, $I$, $J$, respectively. The second formula defines a family of special games $\mathcal{C}_\chi^+$ between the players $I$ and $J$. We say that $\mathcal{C}$ and $\mathcal{C}^+$ are *associated*. It is easy to verify this:

Remark 1. Let $\mathcal{C}(U, X, Y)$ and $\mathcal{C}_\chi^+(X, Y)$ be an associated pair of games. Then $(\chi, \Phi)$ wins the game $\mathcal{C}$ for $E \times I$ if and only if $\Phi$ wins the game $\mathcal{C}_\chi$ for $I$.

Using this remark, Lemma 6 may be restated as:

Lemma 6′ (Complementation). *Given the finite object $\mathcal{A} = \langle D, d_0, F, \mathcal{U}\rangle$ which defines the finite-state game $\mathcal{C}(U, X, Y)$ between $(E \times I)$ and $J$. One can construct $\mathcal{A}' = \langle D', d_0', F', \mathcal{U}'\rangle$, which defines a finite-state game $\mathcal{C}'(U, X, Y)$ between $(E \times I')$ and $J$ such that for all $\chi: J^* \to E$,*

$$\sim(\exists\Phi)[(\chi, \Phi) \text{ wins } \mathscr{C} \text{ for } E \times I] \cdot\!\equiv\!\cdot (\exists\Phi')[(\chi, \Phi') \text{ wins } \mathscr{C}' \text{ for } E \times I']$$

It is now easy to prove

THEOREM 4. (a) *Every formula* $\Delta(\chi)$ *of* GL($\chi$: $J^* \to E$) *is equivalent to a formula* $(\exists\Phi)[(\chi, \Phi)$ *wins* $\mathscr{C}$ *for* $E \times I]$. *From* $\Delta(\chi)$ *one can construct the (presentation of) the finite-state game* $\mathscr{C}$. (b) *There is a method which decides truth of sentences* $\Delta$ *in* GL.

PROOF. By adding void quantifiers we may put $\Delta(\chi)$ into the form

$$(\forall\chi_1)(\exists\Phi_1)\cdots(\forall\chi_{n+1})(\exists\Phi_{n+1})[(\chi, \chi_1, \Phi_1, \ldots, \chi_{n+1}, \Phi_{n+1}, ) \text{ wins } \mathscr{C} \text{ for } E \times E_1 \times I_1 \times \cdots \times E_n \times I_n]$$

In case $n = 0$ the conversion to existential form goes this way: Start from $\Delta(\chi) \equiv (\forall\chi_1)(\exists\Phi_1)[(\chi, \chi_1, \Phi_1)$ wins $\mathscr{C}$ for $E \times E_1 \times I_1]$. Replacing $(\forall\chi_1)$ by $\sim(\exists\chi_1)\sim$ and using Lemma 6′ yields $\Delta(\chi) \equiv \sim(\exists\chi_1)(\exists\Phi_1')[(\chi, \chi_1, \Phi_1')$ wins $\mathscr{C}'$ for $E \times E_1 \times I_1']$. Now Lemma 6′ can be used to replace $\sim(\exists\chi_1\Phi_1')$. The result is $\Delta(\chi) \equiv (\exists\Phi)[(\chi, \Phi)$ wins $\mathscr{C}''$ for $E \times I_1'']$. This shows (a) in case $n = 0$.

In case $n > 0$, we may assume inductively, that $(\forall\chi_2)(\exists\Phi_2)\cdots(\exists\chi_{n+1})(\exists\Phi_{n+1})$ $[\ldots]$ has already been put into form $(\exists\Phi)[\chi, \chi_1, \Phi_1, \Phi)$ wins $\mathscr{C}'$ for $E \times E_1 \times I_1 \times I]$. So now

$$\Delta(\chi) \equiv (\forall\chi_1)(\exists\Phi_1\Phi)[(\chi, \chi_1, \Phi_1, \Phi) \text{ wins } \mathscr{C}' \text{ for } E \times E_1 \times I_1 \times I]$$

This is a case $n = 0$; so two more steps will put $\Delta(\chi)$ into the desired form. Hence we have proved part (a) of the theorem. Part (b) goes this way:

Applied to a sentence $\Delta$, the algorithm (a) will give $\Delta \equiv (\exists\Phi)[\Phi$ wins $\mathscr{C}$ for $I]$. As $\mathscr{C}$ is a finite-state game we can now use Corollary 1 to decide the truth of $\Delta$. Q.E.D.

Our game language GL may be modified to $\text{GL}^{\text{reg}}$, where the strategy variables $\chi$ are interpreted to range over regular $\chi: J^* \to E$. Lemma 6′ remains true in $\text{GL}^{\text{reg}}$, in fact Corollary 1 and Lemma 4 will now suffice for the proof. Hence, Theorem 4 remains true for $\text{GL}^{\text{reg}}$, and precisely the same algorithms work for GL and $\text{GL}^{\text{reg}}$. Therefore we see that $\text{GL} = \text{GL}^{\text{reg}}$.

Suppose $(\exists\chi)\Delta(\chi)$. Put $\Delta(\chi)$ into existential form $(\exists\Phi)[(\chi, \Phi)$ wins $\mathscr{C}$ for $E \times I]$. So there is a winning strategy $(\chi, \Phi)$ for $E \times I$ in $\mathscr{C}$. By Corollary 1 this strategy may be assumed to be presented by the order-vector recursion for $\mathscr{C}$. This recursion is finite-state, and so $(\chi, \Phi)$ is regular. Hence $\chi$ is regular and $\Delta(\chi)$. This argument proves

$$(\exists\chi)\Delta(\chi) \supset (\exists\chi)^{\text{reg}}\Delta(\chi) \quad \text{for any } \Delta \text{ in GL}$$

DEFINITION. $\boldsymbol{P}$: $J^* \to I$ is *definable in* GL if it is the unique solution of a formula $\Delta(\chi)$ in GL.

The remark just stated implies: Every $\boldsymbol{P}$ definable in GL is regular. This obviously holds the other way around. Therefore, the regular model of $\text{GL} = \text{GL}^{\text{reg}}$ is the prime model of the theory.

A further application of the definability theorem is this. Given a $\boldsymbol{P}$: $J^* \to E$, let GL $[\boldsymbol{P}]$ be the extension of GL obtained by adding $\boldsymbol{P}$ as a new primitive. By Theorem 4(a), every sentence $\Delta(\boldsymbol{P})$ of GL $[\boldsymbol{P}]$ can be put into the form $(\exists\Phi)$ $[(\boldsymbol{P}, \Phi)$

wins $\mathscr{C}$ for $E \times I$]. Hence, the truth of $\Delta(\boldsymbol{P})$ can be decided by telling who has the win in the special game $\mathscr{C}_{\boldsymbol{P}}$. This puts the decision problem for GL[$\boldsymbol{P}$] into this very clear form: Given the special game $\mathscr{C}_{\boldsymbol{P}}(X, Y)$, to decide who has the win. Using Theorem 3, this reads: Given the special game $\mathscr{C}_{\boldsymbol{P}}$, to decide whether or not $(d_0, v_0, e) \in R^{\alpha}$. We will now show that GL is the monadic second-order theory of the successor-functions on $J^*$.

**§16. $MT_2$ is the game-language** GL. Let $MT_J$ stand for the monadic second-order theory of the structure consisting of $J^*$, the empty word $e$, and the successor operations $yb$, $b \in J$. We will deal with the case $J = 2 = \{F, T\}$; nothing more happens for larger $J$. $MT_2$ contains the primitive notations $e$, $yF$, $yT$, and $\chi y$ (the membership relation $y \in \chi$, for $y \in 2^*$ and $\chi \subseteq 2^*$). It also contains propositional connectives and quantifiers for both types of variables. Here are some of the concepts definable in $MT_2$:

Equality $y = v$ by the Leibnitz formula $(\forall\chi)[\chi y \supset \chi v]$; equality $\chi = \Phi$ by extensionality; $\chi$ is $\uparrow$ hereditary by $(\forall y)[\chi y \supset \chi yF \wedge \chi yT]$; $\chi$ is hereditary by $(\forall y)[\chi yF \wedge \chi yT \supset \chi y]$; $\chi$ is a path through $2^*$ by $\chi e \wedge (\forall y)\ [\chi y \equiv \chi yF \nabla \chi yT]$; the initial segment relation $y \trianglelefteq v$ by $(\forall\chi)[\chi \downarrow \text{hereditary} \wedge \chi v \supset \chi y]$; the rational order $y \prec v$ by $yT \trianglelefteq v \vee (\exists u)[uF \trianglelefteq y \wedge uT \trianglelefteq v] \vee vF \trianglelefteq y$. A path $\chi$ can be used to simulate a member $Y$ of $2^{\omega}$, so in $MT_2$ one can express $y \triangleleft Y$ which means "the rational $y$ approximates the real $Y$", and one can quantify over reals.

You see now that $MT_2$ is not just a plaything. It contains a singificant fragment of basic analysis. Furthermore it does this in a very natural way; just think of the $\chi$'s as fans and spreads. These very same $\chi$'s have served us well, as strategies for games, and as kernels for Borel sets. We will now show that $MT_2$ may also be interpreted as the game-language GL over finite-state games.

Individual variables can be made to play a very restricted role in any $MT$. We will show how a sentence $\Delta$ of $MT_2$ is put into normal form. The first step is this:

$$\text{(predicate prefix)} \cdot (\forall y_1 \cdots y_n)B \wedge (\exists y)C_1 \wedge \cdots \wedge (\exists y)C_m$$

Here $B$ is a propositional formula with atomic parts $\chi y_i$, $\chi y_i F$, $\chi y_i T$ and each $C_i$ is a $\chi y$, with $\chi$ occurring in the prefix. You can find this proved in [3]. Put $B$ into conjunctive form $\bigwedge_i B_i$, and note that each $B_i$ is a disjunction of form $B^1(y_1) \vee \cdots \vee B^n(y_n)$. Each quantifier $(\forall y_i)$ can therefore be pushed inside (as Skolem said), and $(\forall y_i)B(y_i)$ can be replaced by $(\forall y)B(y)$. We now have this prenex form:

$$\text{(predicate prefix)} \cdot \bigwedge_i [(\forall y)B_i^1(y) \vee \cdots \vee (\forall y)B_i^n(y) \wedge \bigwedge_j (\exists y)C_j(y)]$$

Next we use this trick: $(\forall y)A(y) \vee (\forall y)B(y)$ is equivalent to $(\exists\chi\Phi)[[\chi e \vee \Phi e] \wedge (\forall y)[\chi y \supset \chi yF \wedge \chi yT \wedge A(y)] \wedge (\forall y)[\Phi y \supset \Phi yF \wedge \Phi yT \wedge B(y)]]$. The result is this form of $\Delta$:

$$\text{(predicate prefix)} \cdot A(e) \wedge \bigwedge_i (\forall y)B_i(y) \wedge \bigwedge_j (\exists y)C_j(y)$$

Let $\chi$ stand for the sequence consisting of all variables in the prefix. Similarly, $Zy$ may stand for a sequence $Z_1 y, \ldots, Z_n y$. Expressions like $Zy = \Phi y$, $Ze = d$

will stand for conjunctions $\bigwedge_i Z_i y = \Phi_i y$, $\bigwedge_i Z_i e = d_i$; $d$ here is a sequence of truth-values. We will now show that each of the conjuncts in the last form of our sentence $\Delta$ can be replaced by an expression

$$(\exists Z)\cdot Ze = d \wedge (\forall y)B[Zy, \chi y, ZyF, ZyT] \wedge (\forall Y)\sup_t Z[\bar{Y}t] \in \mathcal{U}$$

As a conjunction of such expressions can be put into the same form, we will then have this form for $\Delta$:

$$(*) \qquad (\text{prefix } \chi_1 \cdots \chi_n)(\exists Z)\cdot Ze = d_0 \wedge (\forall y)B[Zy, \chi y, ZyF, ZyT] \wedge (\forall Y)\sup_t Z[\bar{Y}t] \in \mathcal{U}$$

*Case* $A(e)$. This is of form $A[\chi e]$, which can be replaced by

$$(\exists Z)\cdot Ze \equiv T \wedge (\forall y)[Zy \supset A[\chi y]]$$

*Case* $(\forall y)B(y)$. Here $B$ is of form $B[\chi y, \chi yF, \chi yT]$. We add the conjunct $(\forall y)$ $\chi yT = \Phi y$, and replace $\chi yT$ in $B$ by $\Phi y$, and add $(\exists \Phi)$ to the prefix. So now we have a new $\chi$ (namely $\chi$, $\Phi$) in the prefix, and the new $B_i$ will either have the form $B[\chi y, \chi yF]$ or $B[\chi y, \chi yT]$. You will see that, for example, $(\forall y)B[\chi y, \chi yF]$ may be replaced by $(\exists Z_1 Z_2)\cdot Z_1 e \equiv F \wedge Z_2 e = d \wedge (\forall y)[Z_1 yF \wedge \tilde{Z}_1 yT \wedge Z_2 yF = \chi y \wedge [Z_1 y \supset B[Z_2 y, \chi y]]]$. Here $d$ is a sequence of $F$'s, and the new formula is as required.

We now know that every sentence $\Delta$ of $MT_2$ can be put into the form $(*)$. We have used tricks which Skolem invented for $MT$ of a set, and we earlier used his ideas on bounded quantifiers. That the notation sup ought to enter was clear when Rabin started to work on $MT_2$. Just how it should enter was not at all obvious; just ask Rabin. Given $(*)$, and the complementation method, and the fact that it will work, almost anybody can find a decision method for $MT_2$. This is how I do it:

In $(*)$ put $(\exists \Phi_1 \Phi_2)$ in front of $(\exists Z)$, replace $ZyF$ by $\Phi_1 y$, replace $ZyT$ by $\Phi_2 y$, and conjoin $(\forall y)[ZyF \equiv \Phi_1 y \wedge ZyT \equiv \Phi_2 y]$. Now we have a recursion for $Z$, namely $Ze = d_0$, $Zyb = F[Zy, \Phi_1 y, \Phi_2 y, b]$ where $F[d, a_1, a_2, F]$ stands for $a_1$, and $F[d, a_1, a_2, T]$ stands for $a_2$. In the other mode, this recursion is $Z0 = d_0$, $Zt' = F[Zt, \Phi_1 \bar{Y}t, \Phi_2 \bar{Y}t, Yt]$. You will see now that $(*)$ can be put into this form:

$$(\text{prefix } \chi)(\exists \Phi_1 \Phi_2)(\forall Y)(\exists Z)\cdot Z0 = d_0 \wedge (\forall t)Zt' = F[Zt, \Phi_1 \bar{Y}t, \Phi_2 \bar{Y}t, Yt] \wedge (\forall t)B[Zt, \chi \bar{Y}t, \Phi_1 \bar{Y}t, \Phi_2 \bar{Y}t] \wedge \sup Z \in \mathcal{U}$$

Add the components $\Phi_1$, $\Phi_2$ to the $\chi$, and now abbreviate this expression for $\Delta$ as:

$$(\text{prefix } \chi)(\forall Y)(\exists Z)\cdot Z0 = d_0 \wedge (\forall t)Zt' = F[Zt, \chi \bar{Y}t, Yt] \wedge (\forall t)B[Zt, \chi \bar{Y}t] \wedge \sup Z \in \mathcal{U}$$

The part $(\forall t)B(t)$ may be replaced by $(\forall^\omega t)Wt$, if we add $(\exists W)$ to $(\exists Z)$, and add the recursion $W0 \equiv T$, $Wt' \equiv [Wt \wedge B(t)]$. The conjunction $(\forall^\omega t)Wt \wedge \sup Z \in \mathcal{U}$ can be replaced by an expression $\sup(W, Z) \in \mathcal{V}$. Put $Z'$ for $(Z, W)$ to arrive at this final form for $\Delta$:

$$(\text{prefix } \chi)(\forall Y)(\exists Z')\cdot Z'0 = d_0' \wedge (\forall t)Zt' = F'[Z't, \chi \bar{Y}t, Yt] \wedge \sup Z' \in \mathcal{V}$$

Let $I = 2^n$, where $n$ is the number of components in the final $\chi$, and following the prefix you will find the assertion [$\chi$ wins $\mathscr{C}$ for $I$], whereby $\mathscr{C}$ is the finite-state game $d_0$, $F'$, $\mathscr{V}$. Hence we now know this:

LEMMA 7. *Every sentence $\Delta$ in $MT_2$ is equivalent to a sentence $\Delta'$ of the game language* GL. *In fact, one can construct $\Delta$ from $\Delta'$. Furthermore, the same statement holds for formulas $\Delta(\chi)$ in $MT_2$.*

Using this lemma we now can restate Theorem 4. In particular we have proved Rabin's theorem: $MT_2$ is decidable.

**§17. The history of monadic theories.** It starts in 1915 with Löwenheim [9] proving decidability of $MT$, the monadic theory over a set. He understood that an $MT$ sentence which has any model must have a finite model. I like to think that this made him think of the famous countable models for elementary sentences; they occur in the same paper. Today, originality of this magnitude, and by an obscure author, would probably be rejected by the referee. Luckily there was none, and luckily there was another original thinker to pay attention. Skolem showed an easy way to the countable models (replacing quantifiers by function letters), and he simplified the quantifier elimination for $MT$, the Klassenkalkul of [15].

In $MT$ replace set quantifiers by quantifiers over monadic functions to get $ST$. Gödel's incompleteness theory (1932) begins with this remark: Dedekind's explicit definitions for functions given by recursion are available in $ST_1 = ST[\omega, 0, ']$. In particular, one can define a pairing function. It can be used to simulate quantification over $n$-ary functions, and so $ST_1$ is a full second order theory. Tarski asked the very natural question: what happens in $MT_1 = MT[\omega, 0, ']$?

In 1958 Robinson [13] showed that $MT_1$ becomes undecidable if $x + x$ is adjoined. In Feferman and Vaught [8] you find a general product of structures which can be used to pep up $MT_1$ to richer but recursively equivalent theories. It is also mentioned that Ehrenfeucht knew that $WMT_1$ (set variables range over finite sets) was decidable. In 1960 the problem for $MT_1$ remained unsolved.

Note that at that time one knew fancy model-theoretic decidability proofs. I call these *outside proofs*, as opposed to old fashioned quantifier elimination, where you look at the *inside* of formulas. Working from inside you will more quickly understand the combinatorial lemma CL which is at the heart of your problem. The decision method DM is but a pepped up version of CL. The same CL also occurs in the outside proof, but now it will look as if the fancy talk about model companions or games was the heart of the matter. If it is a nice CL it will be used to do more important things.

EXAMPLES. Ramsey found his lemma by working on decidable prefix types. McNaughton's lemma is more useful than the decidability of $MT_1$. I hope that state-strategies will be useful not just to decide $MT_2$.

In the inside of $MT_1$ sentences I found formulas $\Delta(X)$: $(\exists Z)[A[Z0] \wedge (\forall t)B[Xt, Zt, Zt'] \wedge (\exists^{\omega} t)C[Zt]]$. To understand what this says, think of $Z$ as an $\omega$-sequence of states from a finite set. $A$ becomes an initial condition, $B$ a transition relation, and $C$ a terminal condition. Hence, the basic combinatorial lemma needed was about finite automata working on infinite inputs $X$. In the case of $WMT_1$ the input $X$ is finite, and the needed lemma was already proved by Myhill: The projection onto $J^*$ of a regular set in $(I \times J)^*$ is regular. I talked to Elgot about this, and by 1960 we separately worked out the decidability for $WMT_1$, see [1] and [7]. The obvious thing to try now was to prove the analogous projection lemma for infinite input $X$: In the formula $\Delta(X)$ one may assume $Z$ given by a finite-state recursion. It was obvious then, just as it is now, that the terminal condition for the deterministic $Z$ had to be self-dual, i.e., in the Boolean algebra over $(\exists^{\omega}t)$. But I could not prove projection, and hence I found the complementation method. In the 1962 paper [2] I used Ramsey's lemma to prove complementation for expressions $\Delta(X)$, and this readily gives decidability of $MT_1$.

As the historians have it all mixed up, I repeat: 1. Whether you like it or not, the automata are there, in the inside of $MT_1$ formulas. 2. The deterministic method (projection) was obvious, and complementation had to be invented. 3. Terminal conditions in $Bl(F_\sigma)$ were there first; it so happened that $G_\delta$ would do for complementation. 4. I did not use automata theory, rather the complementation lemma and decidability of $MT_1$ are contributions to automata theory. 5. The inside look very naturally suggested $MT_2$ and the feasibility of a complementation lemma for tree automata.

$MT_1$ is a natural language for stating conditions $\mathscr{C}(X, Y)$, to be solved by a finite-state recursion $Y = \mathscr{A}X$. In fact, "$\mathscr{A}$ solves $\mathscr{C}$" is a sentence in $MT_1$, and this is how I became interested in Tarski's problem. McNaughton [10] eventually proved the projection lemma in 1966. He also invented finite-state games, noting that "$\mathscr{A}$ solves $\mathscr{C}$" means the same thing as "$\mathscr{A}$ is a winning strategy of $J$ in the game $\mathscr{C}$". In his doctoral thesis, Purdue 1967, Landweber solved the problem about finite-state games. My version of his proof is in [5]. Rabin's [12] complementation lemma for $MT_2$ came in 1968. You now know how these matters are logically related, they were closely related in time and space.

In 1962 Shoenfield had a proof of the complementation lemma for $MT_1$ using only the infinity lemma. This is when I realized that the additive case was only a weak part of Ramsey's lemma (Shelah says I had a hint from Rabin, while actually the learning process was quite reversed, at that time). In McNaughton's proof I soon found my additive coloring $AR$ and, of course, the much more intricate piggy-back method $PB$. You find all this analyzed in Büchi [3]. The matter became essential at $\omega_1$, where $AR$ still works and Ramsey does not. The inside approach quite naturally suggests that cofinal closed sets (the filter $J(\omega_1)$) play an important role in $MT[\omega_1, <]$. I used $AR$, $PB$, and "$P(\omega_1)/J(\omega_1)$ has no atom" to show decidability at $\omega_1$. Shelah [14] copied my $AR$ (as Theorem 1.1), and my use of cofinal closed (as Conclusion 1.2), and now he can splice theories $MT[x, y)$ $x < y < \omega_1$, $x$ and $y$ from a cofinal closed set, where I splice runs $Z[x, y)$; he works on the $\omega_0$-level of Fraïsse tree while I work on the 0-level. He says he does not use automata; I say he is joking, the automata are right there in the combinatorial lemma. He does show that $PB$ (McNaughton's lemma) is not needed for $\omega_1$.

Of course, Shelah also tells what super sets will do at $\omega_2$. Recently Gurevich, Magidor and Shelah continued this work, showing that $MT[\omega_2, <]$ is almost completely arbitrary, if the universe is allowed to have large cardinals. They say this is what made me stop at $\omega_2$: but the fact is I stopped because sometimes one needs a rest to think about other things. Obviously we badly need a new Löwenheim with new thoughts about countable models and how to stop forcing. The complementation problem at $\omega_2$ is precisely right to stimulate such thoughts, only you have to forget about super sets and model theory. Just naively go about trying to understand the matter from inside. This method worked at $\omega_0$ in 1960. It just may work again.

Shelah [14] found the analogs to "Theorem 1.1 and Conclusion 1.2", which are needed to handle the much more important $MT$ of dense order. The usual axioms for the continuum are naturally stated in $MT[\lambda, <]$. More recently Gurevich continued this work on the reals. I would like to know how $PB$ enters in the case of dense order. It seems to be essential for $MT_2$ which contains a fragment of $MT[\lambda, <]$.

## REFERENCES

[1] J.R. Büchi, *Weak second order arithmetic and finite automata*, ***Zeitschrift fur Mathematische Logik und Grundlagen der Mathematik***, vol. 6 (1960), pp. 66–92.

[2] ———, *On a decision method in restricted second order arithmetic*, ***Proceedings of 1960 International Congress for Logic, Methodology and the Philosophy of Science***, Stanford University Press, Stanford, Calif., 1962, pp. 1–11.

[3] ———, *The monadic second order theory of* $\omega_1$, ***Lecture Notes in Mathematics***, vol. 328, Springer-Verlag, Berlin and New York, 1973, pp. 1–127.

[4] ———, *Using determinancy of games to eliminate quantifiers*, ***Fundamentals of Computation Theory, Lecture Notes in Mathematics***, vol. 56, Springer-Verlag, Berlin and New York, 1977, pp. 367–378.

[5] J.R. Büchi and L.H. Landweber, *Solving sequential conditions by finite state operators*, ***Transactions of the American Mathematical Society***, vol. 138 (1969), pp. 295–311.

[6] M. Davis, *Infinite games of perfect information*, ***Advances in Game Theory, Annals of Mathematical Studies***, no. 52 (1964), pp. 85–101.

[7] C.C.Elgot, *Decision problems of finite automata design and related arithmetics*, ***Transactions of the American Mathematical Society***, vol. 98 (1961), pp. 21–51.

[8] S. Feferman and R.L. Vaught, *The first order properties of products of algebraic systems*, ***Fundamenta Mathematicae***, vol. 47 (1959), pp. 57–103.

[9] L. Löwenheim, *Über Möglichkeiten im Relativkalkul*, ***Mathematische Annalen***, Band 76 (1915), pp. 447–470.

[10] R.McNaughton, *Testing and generating infinite sequences by finite automata*, ***Information and Control***, vol. 9 (1966), pp. 521–530.

[11] D.A. Martin, *Borel determinacy*, ***Annals of Mathematics*** (2), vol. 102 (1975), pp. 363–371.

[12] M.O. Rabin, *Decidability of second-order theories and automata on infinite trees*, ***Transactions of the American Mathematical Society***, vol. 141 (1969), pp. 1–35.

[13] R.M. Robinson, *Restricted set theoretical definitions in arithmetic*, ***Proceedings of the American Mathematical Society***, vol. 9 (1958), pp. 238–242.

[14] S. Shelah, *The monadic theory of order*, ***Annals of Mathematics*** (2), vol. 102 (1975), pp. 379–419.

[15] Th. Skolem, *Untersuchungen über die Axiome des Klassenkalkuls und über Produktations- und Summationsprobleme welche gewisse Klassen von Aussagen betreffen*, ***Skrifter utgit av Videnskapsselskapet i Kristiana. I: Matematisk-naturvidenskalbelig Klasse*** 1919, no. 3.

PURDUE UNIVERSITY
WEST LAFAYETTE, INDIANA 47907

# Section 7 Computability

*With comments by Martin Davis, The Courant Institute*



*

# Büchi's Work on Computability


*Martin Davis**


[9] Büchi J.R., (1962). Turing-Machines and the Entscheidungsproblem. *Mathematische Annalen*, 148, 201–213.

In 1928, David Hilbert and Wilhelm Ackermann proposed as the fundamental problem of mathematical logic what has become known as *the Entscheidungs-*


* The Courant Institute, New York, N.Y.

*problem*: find an algorithm for deciding of a given formula of the predicate calculus (otherwise known as the first-order functional calculus, or simply as first-order logic) whether or not it is universally valid (i.e., true, no matter what domain the variables range over and no matter how the predicate symbols are interpreted over that domain). In fact, work on the Entscheidungsproblem has usually focused on the equivalent dual problem of determining whether a given formula is satisfiable in some domain. Kurt Gödel, in his dissertation written in 1929, proved a conjecture also posed by Hilbert in 1928, showing that a formula was universally valid if and only if it could be deduced by means of the ordinary axioms and rules of inference for the predicate calculus, i.e., that the rules of elementary logic are complete. Thus a positive solution to the Entscheidungsproblem would have provided an algorithm for deciding whether any given mathematical proposition could be deduced from a given set of axioms using the ordinary rules of logic. In retrospect, we can say that such a result, which would have essentially reduced all of mathematics to an algorithm, was highly unlikely. This unlikelihood was underlined by Gödel in his famous 1931 paper on arithmetic undecidability; he showed that a very large class of propositions of elementary number theory (including in particular Goldbach's conjecture and Fermat's last theorem) were each equivalent to the universal validity of some particular formula.

In the mid-1930s work by Church, Kleene, Post, Rosser, and Turing had provided a precise mathematical notion that corresponded to the intuitive notion of algorithmic computability, and had exhibited specific mathematical problems that were unsolvable with respect to this notion. Proofs of the unsolvability of the Entscheidungsproblem quickly followed. Surprisingly, no one remarked on the easy path from Gödel's result mentioned earlier to the unsolvability result. Church proceeded in terms of a finite set of recursion equations as axioms. It then followed from known undecidability results that there was no algorithm for testing whether a given equation was a consequence of these axioms by the rules of logic. But, as Church remarked, this argument leads to the unsolvability of Hilbert's Entscheidungsproblem only via Gödel's completeness theorem. In fact Church's constructivist scruples were such that he expressed doubt that his result could be regarded as definitively settling the question.

Turing, working in England and entirely ignorant of the work going on in the United States, also found a proof of the unsolvability of the Entscheidungsproblem. In fact Turing developed his now famous concept of the "Turing machine," an extremely simple combinatorial device that could perform any conceivable computation, precisely in order to have at hand a tool for dealing with the Entscheidungsproblem. Turing's proof then amounted to associating with any given Turing machine a formula of the predicate calculus which is satisfiable if and only if that Turing machine will ever produce some specified symbol. Since Turing had shown that there was no algorithmic way to determine this in general, the result followed.

By means of some simple transformations, any formula of the predicate calculus is seen to be equivalent to a so-called *prenex* formula, i.e., one in which a string of quantifiers precedes a quantifer-free formula. Before 1936, research had focused on *prefix classes* of prenex formulas, defined as having particular quantificational prefixes. Decision algorithms were found for certain prefix classes called *decidable*; for other prefix classes, called *reduction classes*, algorithms were found that "reduce" the satisfiabilty of an arbitrary formula to that of a formula that belongs to the class. The goal (which of course we now know was unattainable) was to find a prefix class that was both decidable and a reduction class. An analysis of Turing's proof yields the result that the class of formulas with prefix $\exists\forall\exists\forall\forall\forall$ is a reduction class. But this was a very weak result; better results were already known in 1936.

Büchi's contribution was to reopen the issue of modeling the behavior of Turing machines by formulas of logic. He saw that a great deal of the complexity of description could be peeled away by relying on the Skolem-Herbrand analysis of satisfiability in terms of "Skolem functions." The best result that Büchi was able to obtain by direct use of his methods was that the class of formulas of the form $\exists \wedge \forall\exists\forall$ is a reduction class. At the time he was doing this work, the critical open question was the status of the prenex class $\forall\exists\forall$, and Büchi evidently believed that his methods had finally made this problem approachable. Indeed he was correct! In September 1961, Büchi discussed the matter with Hao Wang and laid out the technical obstacles to extending his methods to deal with the $\forall\exists\forall$ case. In the March 1962 issue of the *Proceedings of the National Academy of Sciences* an article appeared by Kahr, Moore, and Wang in which they showed that $\forall\exists\forall$ is indeed a reduction class. This very ingenious paper is solidly based on Büchi's new techniques, and the authors emphasize their indebtedness: "Our proof... depends heavily on [the] remark of Büchi's.... Büchi's application ... is a new step that leads to a better understanding of the undecidability phenomenon in the predicate calculus. In this sense, Büchi may be said to have invented a new approach to the Entscheidungsproblem." And they conclude: "Many of the methods and techniques used in this paper are derived from developments made by Richard Büchi in connection with his work on closely related problems.... We are heavily indebted to Büchi for an extended discussion...." Settling the $\forall\exists\forall$ case was the final link that made possible a complete classification of the decision problem for classes of prenex formulas defined by their quantificational prefix. Although Büchi was not part of the final thrust that led to this key result, he provided the basic point of view that made success possible.

It should finally be noted that Büchi's methods enabled him to improve a result of Trachtenbrot showing that even for prenex formulas belonging to the class $\exists \wedge \forall\exists\forall$ the set of formulas that are unsatisfiable and those satisfiable on a finite domain are recursively inseparable.

## *References*

1. Church, A. (1936). "A note on the Entscheidungsproblem," *Journal of Symbolic Logic*, 1, 40–41. Correction: pp. 101–102. Reprinted in [2].
2. Davis, M. (Ed.) (1965). *The undecidable*. Raven Press, New York.

3. Gödel, K. (1931). Über formal unentscheidbare Sätze der Principia Mathematica und verwandter Systeme I. *Monatshefte für Mathematik und Physik*, 38, 173–198. Reprinted with an English translation by Jean van Heijenoort in Solomon Feferman et al. (Ed.) (1986). *Kurt Gödel, Collected Works, I: Publications 1929–1936*. Oxford University Press.
4. Hilbert, D., & Ackermann, W. (1928). *Grundzüge der Theoretischen Logik*. Julius Springer, Berlin.
5. Kahr, A.S., Moore, E.F., & Wang, H. (1962). Entscheidungsproblem reduced to the $\forall\exists\forall$ case. *Proceedings of the National Academy of Sciences*, 48, 365–377.
6. Suranyi, J. (1959). *Reduktionstheorie des Entscheidungsproblems im Prädikatenkalkül der ersten Stufe*. Budapest Ungar. Akad. d. Wiss.
7. Turing, A. (1936–37). On computable numbers with an application to the Entscheidungsproblem. *Proceedings of the London Mathematical Society (2)*, 42, 230–267. Correction: vol. 43 (1937), 544–546. Reprinted in [2].

*

[32] Büchi J.R. & Senger. S. (1986–87). Coding in the existential theory of concatenation. *Archiv für Mathematische Logik und Grundlagenforschung*, 26, 101–106.
[33] Büchi, J.R. & Senger, S. (1988). Definability in the Existential Theory of Concatenation and Undecidable Extensions of This Theory. *Zeitschrift für Mathematische Logik und Grundlagen der Mathematik*, 34, 337–342.

These papers are derived from the doctoral dissertation of the second author written under Büchi's direction. Büchi's untimely death occurred while the papers were still in preparation.

The first proofs of the undecidability of first-order arithmetic proceeded by showing how to express "syntax" arithmetically. This amounted to interpreting the theory of string concatenation in arithmetic, for example, by Gödel numbering. In the case of both theories, there was a great deal of interest in the special case of the existential theory of single equations, that is, in classical terms, the problem of solvability of equations. In the case of arithmetic, it is a question of polynomial Diophantine equations and the problem is Hilbert's tenth problem, which has turned out to be undecidable; (see the reference below). In the case of concatenation, the problem was studied by Löb and Markov among others. In fact, Markov had observed (before Hilbert's tenth problem was settled) that the undecidability of Hilbert's tenth problem would follow from that of the problem of solvability of equations in the concatenation operation. However, Makanin was able to show, by a rather difficult argument, in effect providing a superexponential bound on the size of "least" solutions, that this latter problem is decidable.

It was long noted that (as in the case of Diophantine equations) the logical conjunction of a pair of equations in concatenation is equivalent to a single equation of this type. In the first of these papers, it is shown rather simply that the same is true for the disjunction of a pair of equations as well as for the negation of an equation. It follows that Makanin's algorithm can be extended to provide a decision algorithm for the class of purely existential prenex sentences of the theory of concatenation. These results were also found by Taimanov and Hmelevskii in the paper cited by the authors. The present simple proof was apparently found by Büchi many years ago but never published. The remainder of this first paper is concerned with finding codes for an $n$ letter alphabet as strings on $\{1,2\}$ such that replacing the letters by their codes in an equation permits control of the extraneous solutions thus introduced. It is pointed out that to completely eliminate these extraneous solutions for a code

$$C = \{u_1, u_2, \ldots, u_n\},$$

it would suffice to provide an existential definition of the set $C^*$. But the question of whether this is possible is left open.

In the second paper, a careful analysis of Makanin's proof is given including an explicit expression for the superexponential bound. Using this analysis, it is shown that various sets of strings are not existentially definable; that is, they are not definable by any purely existential prenex formula in the first-order theory of concatenation. In particular, it is shown that neither the context-free language

$$\{1^n 2^n \mid n = 0, 1, 2, \ldots\},$$

nor the relation of two strings having the same length is existentially definable.

The paper concludes with some undecidability results obtained from the unsolvability of Hilbert's tenth problem. In particular, an operation $Sp$ is defined on $\{1, 2\}^*$, where $Sp(u)$ is the string obtained by permuting the symbols in $u$ so that all 1s precede all 2s. It is shown that there is no algorithm for determining the solvability of equations in concatenation and $Sp$ over $\{1, 2\}^*$. Finally, considering the familiar correspondence between the natural numbers and the strings belonging to $\{1, 2\}^*$ determined by the lexicographic ordering of the latter, the authors are led to consider the variant of elementary arithmetic in which multiplication is replaced by the operation:

$$P(x, y) = x2^y.$$

It is shown that the operation of squaring is existentially definable in terms of $+$ and $P$. A familiar easy trick (the $n = 2$ case of the binomial theorem) then shows the same to be true of multiplication. This same problem was very recently also considered, quite independently, by Richard Statman who provided a similar, but somewhat simpler proof (private communication).

## *Reference*

1. Davis, M., Matijasevic, Y., & Robinson, J. (1976). Hilbert's tenth problem: Diophantine equations: Positive aspects of a negative solution. *Proceedings of Symposia in Pure Mathematics*, 28, 323–378.

*

[30] (with Bernd Mahr and Dirk Siefkes) Manual on REC—A language for use and cost analysis of recursion over arbitrary data structures. *Bericht Nr. 84-06 Technische Universität Berlin*, Fb. Informatik (1984), 79 pp.
[31] (with Bernd Mahr and Dirk Siefkes) Recursive definition and complexity of functions over arbitrary data structures. In G. Wechsung (Ed.), *Proc. 2 Frege Conference 1984*, Akademie-Verlag, Berlin (1984), 303–308.

The second item is an extended abstract in which a novel formalism for computable functions over arbitrary data structures is described, which may be thought of as a LISP-like extension of the familiar Herbrand-Gödel-Kleene equation calculus. The first item contains all of the details. The formalism developed out of courses taught by Büchi's two co-authors. The formalism in general form is nondeterministic, but provision is made for various more restricted "evaluation strategies." These may be deterministic and yet complete. The familiar "call-by-value" and "call-by-name" strategies are supported. A cost function is defined and asymptotic relations are found between the costs of computation in these terms (with a deterministic strategy) and that for a deterministic one tape Turing machine. (In Theorem 2 of the second item, the word "prize" should be "price.")

BÜCHI, R.
Math. Annalen 148, 201—213 (1962)

# Turing-Machines and the Entscheidungsproblem*

By

J. RICHARD BÜCHI in Ann Arbor and Mainz

Let $Q$ be the set of all sentences of elementary quantification theory (without equality). In its semantic version Hilbert's Entscheidungsproblem for a class $X \subseteq Q$ of sentences is,

$[X]$: To find a method which for every $S \in X$ yields a decision as to whether or not $S$ is satisfiable.

CHURCH [3] showed that $[Q]$ is recursively unsolvable. Shortly thereafter TURING [6] obtained this result more directly by reducing to $[Q]$ an unsolvable problem on Turing-machines. This reduction however is rather involved, and requires much detailed attention of the kind which does not add to one's overall understanding of the situation. We will show in this paper that the connection between Turing-machines and quantification theory is really a rather simple one. The key to it is lemma 3, which is closely related to the Skolem-Gödel-Herbrand work on quantification theory. As a result we obtain the first really elegant proof of unsolvability of $[Q]$. It can be outlined thus:

**Lemma 1:** *The set* Hlt, *consisting of all Turing-machines which eventually halt, if started on the empty tape, is not recursive.*

**Lemma 2:** *To any Turing-machine $M$ one can construct a matrix $\underline{M}(x,u,y)$, with individual variables $x$, $u$, $y$, monadic predicate letters, and 3 binary predicate letters, such that $M \notin$ Hlt if and only if $Zo \wedge (\forall xy)\ \underline{M}(x, x', y)$ is satisfiable in the natural number system $\langle N, o, '\rangle$*[1].

**Lemma 3:** *For any matrices $\underline{Z}(x)$ and $\underline{M}(x, u, y)$, the sentence $(\exists x)\ \underline{Z}(x) \wedge \wedge (\forall x)(\exists u)(\forall y)\ \underline{M}(x, u, y)$ is satisfiable if and only if $\underline{Z}(o) \wedge (\forall xy)\ \underline{M}(x,x',y)$ is satisfiable in $\langle N, o, '\rangle$.*

By lemmas 2 and 3, to any Turing-machine $M$ one can construct a sentence $S$ of form $\exists \wedge \forall\exists\forall$, such that $M \notin$ Hlt if and only if $S$ is satisfiable. Therefore, by lemma 1,

**Theorem 1:** *The problem $[\exists \wedge \forall\exists\forall]$ is not recursively solvable, even if restricted to sentences in which, besides monadic letters, only three binary predicate letters occur.*

* The results of this paper were announced in the Notices, Am. Math. Soc. 8, 354 (1961), and appeared as University of Michigan Technical Report, December, 1961. The work was supported by grants from the National Science Foundation and the U. S. Army Signal Corps, and by a contract with the Office of Naval Research.

[1]) The expression "satisfiable in $\langle N, o, '\rangle$" is not to be confused with "satisfiable in $N$". Here as in other places we prefer a somewhat abbreviated terminology, so as not to drown the main ideas in a formalistic flood. With a bit of good will the reader will find it possible to supply the details.

To stress the simplicity of this argument we wish to claim that the following hints suffice to prove the three lemmas: Lemma 1 is well known and immediately follows from the basic unsolvability result on Turing-machines (machine halting on the tape carrying its description). Proving lemma 2 is just an exercise on Turing-machines. One first describes the operation of a machine $M$ on the empty tape in the form of a matrix $\underline{C}(Q, S, K, o, ', x, y)$, whereby $Qx$ = state at time $x$, $Kyx$ = position $y$ is scanned at time $x$, $Syx$ = tape-symbol at position $y$ and time $x$. It is important to note that the functions $Q$ and $S$ are finite-valued, and therefore can be considered to be vectors of monadic and binary predicates on natural numbers. $M \in$ Hlt now means that no $Q$, $K$ and $S$ exist such that for all places $y$ and all times $x$, $\underline{C}(Q, S, K, o, ', x, y)$. It remains to invent some tricks to put the matrix $\underline{C}(o, ', x, y)$ into the form $Z(o) \wedge \underline{M}(x, x', y)$, by adding auxiliary predicates. Lemma 3 is trivial in the "if-direction". In the other direction one might use the axiom of choice to introduce a monadic Skolem-function $fx$ for $(\exists u)$ and a constant $c$ for $(\exists x)$. It remains to note that $\underline{Z}(c) \wedge (\forall xy)\ \underline{M}(x, fx, y)$ is satisfiable if and only if it is satisfiable by predicates in $N$ and $c = o$, $fx = x'$.

This proof has the further advantage of directly yielding unsolvability for the very simple type $\exists \wedge \forall\exists\forall$. The unsolvability of $[\exists \wedge \forall\exists \wedge \forall\forall\forall]$ follows, by replacing the part $\forall\exists\forall$ by its Skolem-form $\forall\exists \wedge \forall\forall\forall$. (Compare these results with Bernays' (1958) analysis of Turing's proof.) More important is, that our method of using lemma 3, for the first time provides real hope of settling the decision problem for the only remaining prefix-types $\forall\exists\forall$ and $\forall\exists \wedge \forall\forall\forall$. Furthermore, with slight modifications we obtain an improved version of TRACHTÉNBROT's (1950) result about satisfiability in finite domains:

**Theorem 2:** *There is no recursive set which separates the not-satisfiable sentences from those satisfiable in a finite domain; even if only sentences of form* $\exists \wedge \forall\exists\forall$ *are considered.*

**Corollary:** *The set of* $\exists \wedge \forall\exists\forall$ *sentences, which are finitely satisfiable, is not recursive.*

Clearly theorem 2 follows from the following stronger versions of lemmas 1, 2, and 3:

**Lemma I:** *Let* Cyl *be the set of all Turing-machines which eventually cycle, if started on the empty tape. The sets* Hlt *and* Cyl *are not recursively separable.*

**Lemma II:** *The construction* $M \to \underline{M}$ *of lemma 2 has the further property:* $M \in$ Cyl *if and only if* $Zo \wedge (\forall xy)\ \underline{M}(x, x', y)$ *is satisfiable in* $\langle N, o, '\rangle$ *by periodic predicates.*

**Lemma III:** *Add to lemma 3: The sentence* $(\exists x)\ \underline{Z}(x) \wedge (\forall x)\ (\exists u)\ (\forall y)\ \underline{M}(x, u, y)$ *is satisfiable in a finite domain, if and only if,* $\underline{Z}(o) \wedge (\forall xy)\ \underline{M}(x, x', y)$ *is satisfiable in* $\langle N, o, '\rangle$ *by periodic predicates.*

We will now indicate the proofs of the lemmas, and add some additional discussion at the end of the paper.

**Proof of lemmas 2 and II:** A (Turing-) *machine* we define to be a system $M = \langle \underline{D}, A, \underline{L}, \underline{R}, \underline{P}, \underline{Q}, \underline{S}\rangle$ consisting of a finite set $\underline{D}$ of elements called *states;*

an $A \in \underline{D}$ called the *initial state;* three binary predicates $\underline{L}[X, Y]$, $\underline{R}[Y, X]$, $\underline{P}[X, Y]$, called *commands* of *left-move, right-move,* and *print;* a binary function $\underline{Q}[X, Y]$ with values in $\underline{D}$, called the *new-state-function;* and a function $\underline{S}[X, Y]$ with values in $\{T, F\}$, called the *print-function.* All these predicates and functions have arguments $X \in \underline{D}$ and $Y \in \{T, F\}$, furthermore $\underline{L}$, $\underline{R}$, $\underline{P}$ are to be exclusive and complementary. The *tape-symbols* are $T$ and $F$, a *tape* is a one-way infinite sequence of tape-symbols, i.e. a predicate $Ix$ on $N$. The tape $Ix \equiv F$ is called the *empty tape.*

The *operation of a machine* $M$, set to work on a tape $I$, is as follows. $M$ is started in its initial state $A$, scanning the zero-position of the tape $I$. If at any time $x$ it is in state $X$ and scans position $y$ of the tape, which now carries the symbol $Y$, then,

if $\underline{L}[X, Y]$ it moves to scan position $y - 1$

if $\underline{R}[X, Y]$ it moves to scan position $y + 1$

if $\underline{P}[X, Y]$ it prints $\underline{S}[X, Y]$ in place of $Y$ at position $y$.

In all cases it goes into the new state $\underline{Q}[X, Y]$. Note that if $y = o$ and the command is $\underline{L}[X, Y]$, then $M$ will next scan position $-1$, i.e., it *runs off the tape.* In this case we say that the machine *halts* at time $x + 1$, and that the tape $I$ is *accepted* by $M$; in symbols $\mathrm{Hl}(M, I)$. The set Hlt consists of all machines $M$ which eventually halt if put to work on the empty tape.

Now let $M = \langle \underline{D}, A, \underline{L}, \underline{R}, \underline{P}, \underline{Q}, \underline{S} \rangle$ be any machine. From it we construct the formula $\underline{C}(Q, S, K, L, R, P, x, y)$ as conjunction of the following parts,

$$\text{(i)} \qquad \begin{array}{ll}
Qo = A \wedge \sim Syo & Koo \wedge \sim Ky'o \\
Kyx \supset \cdot\, Qx' = \underline{Q}[Qx, Syx] & Lx \supset \cdot\, Kyx' \equiv Ky'x \\
Kyx \wedge \underline{L}[Qx, Syx] \cdot \supset \cdot\, Lx \wedge \sim Px \wedge \sim Rx & Px \supset \cdot\, Kyx' \equiv Kyx \\
Kyx \wedge \underline{P}[Qx, Syx] \cdot \supset \cdot\, Px \wedge \sim Lx \wedge \sim Rx & Rx \supset \cdot\, Ky'x' \equiv Kyx \\
Kyx \wedge \underline{R}[Qx, Syx] \cdot \supset \cdot\, Rx \wedge \sim Lx \wedge \sim Px & Rx \supset \cdot \sim Kox' \\
\sim Kyx \vee Lx \vee Rx \quad \cdot \supset \cdot\, Syx' \equiv Syx & \sim [Kox \wedge Lx] \\
\qquad Kyx \wedge Px \quad \cdot \supset \cdot\, Syx' \equiv \underline{S}[Qx, Syx] &
\end{array}$$

and we claim that[1]),

$$\text{(1)} \qquad M \notin \mathrm{Hlt} \cdot \equiv \cdot (\forall xy)\underline{C} \quad \text{is satisfiable in } \langle N, o, ' \rangle.$$

Suppose first that $M \notin \mathrm{Hlt}$, i.e., $M$ put to work on the empty tape does never halt. Then clearly the functions and predicates

$$\text{(j)} \qquad \begin{array}{rl}
Qx = & \text{state of } M \text{ at time } x \\
Syx \equiv & \text{tape-symbol at time } x \text{ and position } y \text{ is } T \\
Kyx \equiv & \text{at time } x \text{ position } y \text{ is scanned by } M \\
Lx \equiv & \text{at time } x \text{ the command is left-move} \\
Rx \equiv & \text{at time } x \text{ the command is right-move} \\
Px \equiv & \text{at time } x \text{ the command is print}
\end{array}$$

are defined for all $x$ and $y$. Furthermore, because $M$ is started on the empty tape and never runs off the tape, these $Q$, $S$, $K$, $L$, $R$, $P$ will satisfy $Qo = A$, $Koo \wedge \sim Ky'o$, $\sim Syo$ and $\sim [Kox \wedge Lx]$, for any $x$ and $y$. Referring to our description of the operation of a machine, one easily verifies that also the remaining formulas of (i) are satisfied for all $x$ and $y$. In other words, $(\forall xy)\underline{C}$ becomes true in $\langle N, o, ' \rangle$, if its variables are given the values (j). Thus, (1) is established in the left-right direction.

Suppose next that $(\forall xy)\underline{C}$ is satisfiable in $\langle N, o, ' \rangle$. Then there are a function $Q : N \rightarrow \underline{D}$ and predicates $S$, $K$, $L$, $R$, $P$, such that for all $x$ and $y$ in $N$ the formulas (i) hold. Now suppose $M$ is put to work on the empty tape, and refer back to our description of the operation of a machine. By induction on $x$ and the use of (i) one then shows that $Q$, $S$, $K$, $L$, $R$, $P$ must satisfy (j), for all $x$ and $y$[2]). It follows that at no time $x$, the machine $M$ halts, if started on the empty tape, i.e., $M \notin$ Hlt. This completes the proof of (1).

Note that the states of a machine $M = \langle \underline{D}, A, \underline{L}, \underline{R}, \underline{Q}, \underline{S} \rangle$ may be coded as vectors of truth-values, so that $A$ stands for a vector $(A_1, \ldots, A_n)$ of truth-values, and $\underline{Q}$ for a vector $(\underline{Q}_1, \ldots, \underline{Q}_n)$ of formulas of propositional calculus ($n$ depending on the number of states of $M$). Furthermore, the expressions $Qo = A$ and $Qx' = \underline{Q}[Qx, Syx]$ stand for conjunctions of terms of the form $Q_i o \equiv A_i$ and $Q_i x' \equiv \underline{Q}_i[Q_1 x, \ldots, Q_n x, Syx]$. $\underline{C}$ therefore is a matrix of quantification-theory. With the construction $M \rightarrow \underline{C}$, and the established (1), we thus are pretty close to a proof of lemma 2. All that remains to be done is to note that the following modifications of $\underline{C}$ do not affect the validity of (1).

1. Introduce an additional binary predicate-letter $H$. On the right side of $\underline{C}$ replace $Ky'o$ by $Hyo$, $Ky'x$ by $Hyx$, $Ky'x'$ by $Hyx'$, and conjoin $Hxy \equiv Kx'y$. The resulting matrix is of form $\underline{C}^*(o, x, x', y)$.

2. Introduce an additional monadic predicate-letter $Z$. To $\underline{C}^*$ conjoin $Zo \wedge \sim Zx'$ and replace $[Qo=A] \wedge \sim Syo$ by $Zx \supset \cdot [Qx=A] \wedge \sim Syx$, $Koo \wedge \wedge \sim Hyo$ by $Zx \supset \cdot Kxx \wedge \sim Hyx$, $\sim Kox'$ by $Zy \supset \sim Kyx'$, and $\sim [Kox \wedge Lx]$ by $Zy \supset \sim [Kyx \wedge Lx]$. The resulting formula is of form $Zo \wedge \underline{M}(x, x', y)$. $M \rightarrow \underline{M}$ is the construction required in lemma 2. That the same construction also satisfies lemma II, will now be shown.

The machine $M$, if started on the tape $I$, *goes into a p-cycle* at time $l$, if at the later time $(l + p)$ it is faced with an identical situation, i.e., at times $l$ and $(l + p)$ $M$ is in the same state and scans identical tapes in the same position. We will say that $M$ (eventually) *p-cycles on I*, in symbols $\mathrm{Cy}_p(M, I)$, if at some time it goes into a $p$-cycle. (It may be shown that for some $p$, $\mathrm{Cy}_p(M, I)$ holds if and only if $M$ never runs off the tape and scans only in a bounded part of the tape.) The set Cyl is defined to consist of all those machines $M$ which eventually cycle, if put to work on the empty tape.

A predicate $Ux$ on $N$ is called *periodic* with *phase* $l$ and *period* $p > o$, if $U(x + p) \equiv Ux$ for all $x \geqq l$. A relation $Vxy$ on $N$ is called *periodic* with *phase* $l$ and period $p > o$, if $V(x + p)y \equiv Vxy$ for all $x \geqq l$ and all $y$, and

[2]) The details are best left to the reader. It should be noted that, by our definition of machines, $\underline{L}$, $\underline{R}$, $\underline{P}$ are mutually exclusive.

$Vx(y+p) \equiv Vxy$ for all $x$ and all $y \geqq l$. (Note that the entire line $Vxl$ has to repeat at $Vx(l+p)$, and not just from $x = l$ on!)

Suppose now that $M \in \mathrm{Cyl}$, that is, $M$ goes into a $p$-cycle at time $l$, if started on the empty tape. Then clearly the functions $Q$, $L$, $R$, $P$ are periodic and $Syx$ and $Kyx$ are periodic in the time-argument $x$, all of them with phase $l$ and period $p$. Furthermore, if $d$ is the maximum of all positions scanned by $M$ before time $(l+p)$, then at no time $t$ a position $y > d$ will be scanned. It follows that $Kyx \equiv F$, for all $x$ and all $y > d$, and because $Syo \equiv F$ and $M$ never scans beyond $d$, also $Syx \equiv F$ for all $x$ and all $y > d$. Thus, also $K$ and $S$ are periodic (with phase $\geqq l, d$ and period $p$). In other words, the implication from left to right of (2) is valid.

(2) $M \in \mathrm{Cyl} \cdot\equiv\cdot (\forall yx)\underline{C}$ is satisfiable by periodic predicates in $\langle N, o, {}' \rangle$.

In the other direction (2) is trivially valid because the only solution of $(\forall yx)\underline{C}$ are the predicates $Q$, $S$, $K$, $L$, $R$, $P$ describing the operation of $M$; and to say that $Q$, $S$, $K$, are periodic (in the time argument) just means that $M$ eventually cycles. Because the modifications 1. and 2. of $\underline{C}$ to $C^*$ to $Zo \wedge \underline{M}$ clearly do not affect the validity of (2), this ends the proof of lemma II.

**Proof of lemma I**: We omit giving a direct proof of lemma 1, as it is well known from the literature. Of the proof of lemma I we present an intuitive sketch:

The *block* of length $l$ is the tape given by $Iy \equiv (y < l)$. It is clear that one can effectively set up a *coding function* $\mathrm{cd}(M)$ which maps one-to-one all machines onto all blocks. A set $X$ of blocks is called *recursive* if there are machines $M_1$ and $M_2$ such that for all blocks $I$,

$$\text{(a)} \qquad \begin{aligned} I \in X &\cdot\equiv\cdot \mathrm{Hl}\,(M_1, I) \\ I \notin X &\cdot\equiv\cdot \mathrm{Hl}\,(M_2, I)\,. \end{aligned}$$

A set $Y$ of machines is called *recursive* if the set $\mathrm{cd}(Y)$ of blocks is recursive. As to the equivalence of this to other definitions of "recursive sets" we remark:

1. It is well known that the restriction to two tape-symbols (one of them the blank) and one-way infinite tapes is not a serious one, 2. Machines which print and move at the same time can, by adding new states, easily be modified so as to either only print or only move in each atomic act, 3. We might have added another command predicate $H[X, Y]$, to obtain halt-situations in addition to "running off the tape". But these can be eliminated by adding an additional state $B$, such that $H[X, Y]$ implies that the next state is $B$, and $B$ requires $M$ to stay in $B$ and move left.

Note that for a machine $M$ to 1-cycle simply means that it keeps scanning at the same position. Thus, one can find a predicate $\underline{C}_1[X, Y]$ such that $M$ at time $t$ goes into a 1-cycle just in case $\underline{C}_1[X, Y]$ holds for the state $X$ and scanned symbol $Y$. Let now $M'$ be obtained by adding a new state $B$ with $\underline{Q}[B, Y] = B$, and conjoining $\underline{C}_1$ to the right-move condition $\underline{R}$. Then clearly

$\mathrm{Hl}(M, I) \equiv \mathrm{Hl}(M', I)$, but $M'$ never 1-cycles. Similarly one can modify a machine $M$ to $M'$ such that $\mathrm{Hl}(M, I) \equiv \mathrm{Cy}_1(M', I)$ and $M'$ does not halt on any block $I$. Thus in the definition of recursive sets of blocks one may replace (a) by

(b)
$$I \in X \cdot\equiv\cdot \mathrm{Cy}_1(M_1, I) \text{ and } M_1 \text{ does not halt on blocks}$$
$$I \notin X \cdot\equiv\cdot \mathrm{Hl}(M_2, I) \text{ and } M_2 \text{ never 1-cycles}.$$

It now is possible to combine $M_1$ and $M_2$ into one machine $M$ which 1-cycles on $I \in X$ and halts on $I \notin X$. Thus, to every recursive set $X$ of blocks there is a machine $M_0$ such that for all blocks $I$,

(c)
$$I \in X \cdot\equiv\cdot \mathrm{Cy}_1(M_0, I)$$
$$I \notin X \cdot\equiv\cdot \mathrm{Hl}(M_0, I).$$

By the usual diagonal-argument we now can prove lemma I:

Suppose that $Y$ is a recursive set of machines and separates the sets of machines for which $\mathrm{Hl}(M, \mathrm{cd}\,M)$ respectively $\mathrm{Cy}_1(M, \mathrm{cd}M)$, i.e.,

$$\mathrm{Hl}(M, \mathrm{cd}\,M) \supset M \in Y$$
$$\mathrm{Cy}_1(M, \mathrm{cd}M) \supset M \notin Y.$$

By (c) there is a machine $M_0$, such that

$$M \in Y \supset \mathrm{Cy}_1(M_0, \mathrm{cd}M)$$
$$M \notin Y \supset \mathrm{Hl}(M_0, \mathrm{cd}M).$$

Now $M_0 \in Y$ implies $\mathrm{Cy}_1(M_0, \mathrm{cd}M_0)$ implies $M_0 \notin Y$, and $M_0 \notin Y$ implies $\mathrm{Hl}(M_0, \mathrm{cd}M)$ implies $M_0 \in Y$. This is contradictory, and therefore,

(d) The sets $A = \{M; \mathrm{Hl}(M, \mathrm{cd}\,M)\}$ and $B = \{M; \mathrm{Cy}_1(M, \mathrm{cd}M)\}$ are not separable by a recursive set.

Because there is a recursive mapping $f$ from machines to machines such that

$$\mathrm{Hl}(M, \mathrm{cd}\,M) \supset fM \in \mathrm{Hlt}$$
$$\mathrm{Cy}_1(M, \mathrm{cd}\,M) \supset fM \in \mathrm{Cyl}_1$$

it follows from (d) that also Hlt and $\mathrm{Cyl}_1$ are not separable. Finally, because $\mathrm{Cyl}_1 \subsetneq \mathrm{Cyl}$, we conclude that Hlt and Cyl are inseparable.

**Proof of lemma 3 and III:** We will assume that the matrices $\underline{Z}$ and $\underline{M}$ contain only one predicate-letter $R$, which is binary. The general case does not present any new problems. Let $\Sigma(R)$ stand for the sentence $(\exists x)\, \underline{Z}(x) \wedge \wedge (\forall x)\, (\exists u)\, (\forall y)\, \underline{M}(x, u, y)$, and $\Sigma^*(a, f, R)$ for its Skolem-transform $\underline{Z}(a) \wedge \wedge (\forall xy)\, \underline{M}(x, fx, y)$.

Now suppose that $\Sigma(R)$ is satisfiable, i.e., has a model $\langle D_1, R_1\rangle$. By the axiom of choice it follows that there is an $a_1 \in D_1$ and a function $f_1\colon D_1 \to D_1$, such that $\underline{D}_1 = \langle D_1, a_1, f_1, R_1\rangle$ is a model of $\Sigma^*(a, f, R)$. Let $D_2$ be the smallest subset of $D_1$ which contains $a_1$ and is closed under $f_1$, let $a_2 = a_1$, $f_2 =$ restriction of $f_1$ to $D_2$, $R_2 =$ restriction of $R_1$ to $D_2$. Because $\Sigma^*$ is a universal sentence it follows that $\underline{D}_2 = \langle D_2, a_2, f_2, R_2\rangle$ still is a model of $\Sigma^*$. Next we note that

$\langle N, o, '\rangle$ is the free algebra with one generator and one monadic function. Because $\langle D_2, a_2, f_2\rangle$ is generated by $a_2$ and $f_2$, it follows that there is a homomorphism $h$ from $\langle N, o, '\rangle$ onto $\langle D_2, a_2, f_2\rangle$. If we now define $R_3xy \equiv R_2(hx)(hy)$, then it is clear that $\underline{D}_2$ is strong homomorphic image of $\langle N, o, ', R_3\rangle$. Again because $\Sigma^*$ is universal it therefore follows that $\langle N, o, ', R_3\rangle$ is still a model of $\Sigma^*(a, f, R)$[3]). Thus,

(1) $\Sigma$ is satisfiable $\cdot\supset\cdot$ $\Sigma^*$ is satisfiable in $\langle N, o, '\rangle$.

Suppose now further, that the model $\langle D_1, R_1\rangle$ of $\Sigma$ is finite. Then clearly the algebra $\langle D_2, a_2, f_2\rangle$ is finite. It follows that the congruence relation $hx = hy$ on $\langle N, o, '\rangle$ is of finite index, and therefore must be of form

$$hx = hy \cdot\equiv\cdot x = y \vee [x \geqq l \wedge y \geqq l \wedge x \equiv y \,(\mathrm{mod}\, p)]\,,$$

for some $l$ and $p > o$. It follows that the relation $R_3xy$ is periodic with phase $l$ and period $p$. Thus we have shown,

(2) $\Sigma$ finitely satisfiable $\cdot\supset\cdot$ $\Sigma^*$ periodically satisfiable in $\langle N, o, '\rangle$.

The converse to (1) is trivial, so that lemma 3 is established. To establish lemma III it remains only to prove the converse to (2). This goes as follows:

Suppose $\underline{D} = \langle N, o, ', R_1\rangle$ is a model of $\Sigma^*(a, f, R)$, whereby $R_1$ is periodic, say of phase $l$ and period $p > o$. The relation

$$x \sim y \cdot\equiv\cdot x = y \vee [x \geqq l \wedge y \geqq l \wedge x \equiv y \,(\mathrm{mod}\, p)]$$

is clearly a congruence relation of $\langle N, o, '\rangle$, and because $R_1$ has phase $l$ and period $p$, it is also a congruence relation of $R_1$. Consequently one can form the factor $\underline{D}/\sim$ of the relational system $\underline{D}$. Because $\Sigma^*$ is universal and $\underline{D}/\sim$ is homomorphic image of $\underline{D}$, it follows that $\underline{D}/\sim$ is still a model of $\Sigma^*$. Furthermore, $\underline{D}/\sim$ is finite, because $\sim$ is of finite index. But from any model of $\Sigma^*(a, f, R)$ one obtains a model of $\Sigma(R)$, if one just omits the interpretations of $a$ and $f$. Thus $\Sigma$ has a finite model.

This concludes the proof of the lemmas. We add some further discussion of the results.

**General form of lemma III:** Without any essential change in the presented proof, one can establish the result for general sentences of $Q$. In place of $\langle N, o, '\rangle$ appear the *totally free* algebras $\underline{F}_n^{m_1, \ldots, m_k} = \langle N, o_1, \ldots, o_n, f_1, \ldots, f_k\rangle$ with $n$ generators and $k$ operations, $f_i$ having $m_i$ arguments. A *periodic relation on* $\underline{F}$ is one which admits a congruence of $\underline{F}$ of finite index. A *Skolem-transform* $\Sigma^*(o_1, \ldots, o_n, f_1, \ldots, f_k, R)$ of an arbitrary sentence $\Sigma(R)$ in $Q$ is obtained by first writing $\Sigma$ as a conjunction of prenex sentences, and next replacing existential quantifiers by individual-letters and function letters, in the well

[3]) A strong homomorphism $h$ of $\underline{D} = \langle D, f, R\rangle$ onto $\underline{D}^* = \langle D^*, f^*, R^*\rangle$ is characterized by $h(fxy) = f^*(hx)(hy)$ and $Rxy \equiv R^*(hx)(hy)$. There seems to be a widespread prejudice that $h^{-1}$ does not preserve the validity of universal sentences $S$. Of course, this is justified in case $S$ contains the equality-sign, and one demands that it must be interpreted as equality. In the other case, a bit of reflection will show the prejudice to be faulty.

known manner (suggested by the axiom of choice). The general form of lemma III now is,

**Lemma III:** *Let* $\Sigma(R_1, \ldots, R_s)$ *be any sentence of* $Q$, *let* $\Sigma^*(o_1, \ldots, o_n, f_1, \ldots, f_k, R_1, \ldots, R_s)$ *be a Skolem-transform of* $\Sigma$.

(a) $\Sigma$ *is satisfiable, if and only if,* $\Sigma^*$ *is satisfiable in the totally free algebra* $\underline{F}_n^{m_1, \ldots, m_k}$.

(b) $\Sigma$ *is satisfiable in a finite domain, if and only if,* $\Sigma^*$ *is satisfiable in* $\underline{F}_n^{m_1, \ldots, m_k}$ *by periodic relations.*

The proof we gave (using the axiom of choice) simply carries one step farther Skolem's first proof of Löwenheim's theorem. A more elementary proof of part (a) actually is contained in Skolem's second proof. It may be outlined thus.

The free algebra $\underline{F}$ can be built up by levels: $L_0 = \{o_1, \ldots, o_n\}$, $L_{k+1}$ is obtained by adding to $L_k$ the elements $fx \ldots y$ whereby $x, \ldots, y \in L_k$ and $f$ is one of $f_1, \ldots, f_m$. Let $\Sigma$ be a sentence. Its Skolem-transform $\Sigma^*$ is a universal sentence, say $(\forall x \ldots y)\, A(x, \ldots, y)$. For any $k$ we define $\Sigma_k$ to be the conjunction of all $A(u, \ldots, v)$ whereby $u, \ldots, v$ range over $L_k$. By a quite elementary argument one shows,

(c) If $\Sigma$ is satisfiable, then for every $k$, $\Sigma_k$ is satisfiable in $L_k$.

Furthermore, by König's infinity lemma,

(d) If for every $k$, $\Sigma_k$ is satisfiable in $L_k$, then $\Sigma^*$ is satisfiable in $\underline{F}$.

Because the "if-part" is trivial, this yields another proof of (a). It makes use of the infinity lemma, while the first proof uses the axiom of choice! We have not analyzed whether (b) also can be obtained in this second way.

**Syntactic version of the Entscheidungsproblem:** If in the statement of problem $[X]$ one replaces "satisfiable" by "formally consistent" one obtains the syntactic version $[X]_0$. By GÖDEL's completeness theorem it follows that $[X]$ and $[X]_0$ are equivalent, so that also $[\exists \wedge \forall\exists\forall]_0$ is not recursively solvable. However, one can prove this more directly by using HERBRAND's theorem. It can be stated thus,

(c′) $\Sigma$ is formally consistent, if and only if, for any $k$, $\Sigma_k$ is satisfiable in $L_k$.

Now (c′) and the infinity lemma (d) yield,

(a′) $\Sigma$ is formally consistent, if and only if, $\Sigma^*$ is satisfiable in $\underline{F}$.

From lemmas 1, 2, and (a′) the unsolvability of $[\exists \wedge \forall\exists\forall]_0$ follows.

**Reduction:** To the one who does not accept CHURCH's thesis, theorem 1 is of less interest. But our method also yields that $\exists \wedge \forall\exists\forall$ is a reduction-type, i.e., the problem $[Q]$ is effectively reducible (in fact $1-1$-reducible) to the problem $[\exists \wedge \forall\exists\forall]$. This can be seen by using a theorem of MYHILL's, because our proof clearly shows that $[\exists \wedge \forall\exists\forall]$ is of unsolvability degree 1. More directly one can obtain a reduction from $[Q]$ to $[\exists \wedge \forall\exists\forall]$ as follows.

To the sentence $\Sigma$ in $Q$ construct a Turing-machine $M$ which, if started on the empty tape, begins by checking $\Sigma_0$ for satisfiability in $L_0$. $M$ halts if it finds $\Sigma_k$ not to be satisfiable in $L_k$, and it proceeds to $\Sigma_{k+1}$ in case it has found a model of $\Sigma_k$. Thus by (c),

$$\Sigma \text{ satisfiable} \cdot \equiv \cdot M \notin \text{Hlt} .$$

The construction of lemma 2 now yields a matrix $\underline{M}(x, u, y)$, which by lemma 2 and 3 is such that,

$$M \notin \text{Hlt} \cdot \equiv \cdot \Delta \text{ satisfiable} ,$$

whereby $\Delta$ is the sentence $(\exists x)Zx \wedge (\forall x)(\exists u)(\forall y)\underline{M}$. Thus, the effective construction $\Sigma \rightarrow \Delta$, reduces $[Q]$ to $[\exists \wedge \forall \exists \forall]$.

**Prefix of length four:** The unsolvability of $[\forall \exists \forall \forall]$ does not follow from theorem 1, but it can be proved by the same method. (However, to obtain the necessary modified version of lemma 2, the author had to make use of ternary predicate letters.) We note that all prenex-types with prefix of length 4 are now settled; all except $\forall \exists \forall \forall$ and those falling under $\exists \wedge \forall \exists \forall$ and Suranyi's (1959) $\forall \forall \exists \wedge \forall \forall \forall$, have a solvable decision problem. There remains the question whether $[\exists \wedge \forall \exists \forall]$ is unsolvable if one admits, besides monadic letters, only two (only one) binary predicate letters.

**The prefix $\forall \exists \forall$:** The really important outstanding question is to prove $[\forall \exists \forall]$ unsolvable. For the first time there now is hope of obtaining this result. All that is missing is the following stronger form of lemma 2,

*Problem:* To any Turing-machine $M$ to construct a matrix $\underline{M}(x, u, y)$, with individual variables $x, u, y$, monadic predicate letters, and binary predicate letters, such that $M \notin$ Hlt if and only if $(\forall xy)\ \underline{M}(x, x', y)$ is satisfiable in the natural number system $\langle N, ' \rangle$.

In our proof of lemma 2 we fell short of obtaining this stronger result, because, in describing the action of $M$ on the empty tape we used special constraints on two axes, namely the tape-axis and the time-axis. It is important to realize that also in conditions of form $(\forall xy)\ \underline{M}(x, x', y)$ one still has use of one axis, namely one can formulate special restraints on the diagonal! In September 61 the author explained this situation to Hao Wang. He now claims, in collaboration with A. S. Kahr and E. F. Moore, to have found a construction $M \rightarrow \underline{M}$ as required in the above problem. Thus, even $[\forall \exists \forall]$ seems to be unsolvable, and the unsolvability of $[\forall \exists \wedge \forall \forall \forall]$ follows by passing to Skolem-form. These results settle (up to detailed questions, like number of binary predicate-letters) the Entscheidungsproblem for all generalized prenex types (conjunctions of prenex sentences). It is easy to see that for every generalized prenex type $X$ either one of the following four alternatives holds:

$A_1$: Every conjunct of $X$ is of form $\exists^n \forall^m$

$A_2$: Every conjunct of $X$ is of form $\exists^n \forall^2 \exists^m$

$B_1$: There is a conjunct of form $\ldots \forall \ldots \exists \ldots \forall \ldots$ in $X$.

$B_2$: There are conjuncts $\ldots \forall \ldots \exists \ldots$ and $\ldots \forall \ldots \forall \ldots \forall \ldots$ in $X$.

Thus, there are just two cases,

A: $[X]$ trivially reduces to $[\exists^n \forall^m]$ or $[\exists^n \forall^2 \exists^m]$

B: Either $[\forall \exists \forall]$ or $[\forall \exists \wedge \forall \forall \forall]$ trivially reduces to $[X]$.

It is well known (see ACKERMANN [1]) that $[\exists^n \forall^m]$ and $[\exists^n \forall^2 \exists^m]$ are solvable. Thus, in case A, $[X]$ is solvable, while in case B, depending on the result of KAHR, MOORE, and WANG, $[X]$ is unsolvable.

Let us say that a set $X$ of sentences has property $\Phi$, if every sentence of $X$ which is satisfiable also is finitely satisfiable. In other words $\sim\Phi X$ means that $X$ contains an "infinity-axiom". It is well known (see ACKERMANN [1]) that $\exists^n \forall^2 \exists^m$ and $\exists^n \forall^m$ both have property $\Phi$, while $\forall \exists \forall$ and $\forall \exists \wedge \forall \forall \forall$ do not. The trivial reductions mentioned preserve property $\Phi$, so that the cases A and B also divide those generalized prenex types $X$ having property $\Phi$ from those which do not.

**Matrices of special form:** We will show now that if in theorem 1 one drops the remark concerning the number of binary predicate-letters, one can in turn add very strong restrictions on the form of the matrices in the $\exists \wedge \forall \exists \forall$-sentences. We note that monadic letters can be eliminated (replace $Sv$ by $Svv$) without changing satisfiability of a sentence. In the following discussion $R, S, \ldots$ will stand for vectors of binary predicate-letters. By lemmas 1 and 2 the following is an undecidable problem,

(D) For any matrices $\underline{Z}[Roo]$ and $\underline{M}[Rxx, Rxx', Rxy; Rx'x, Rx'x', Rx'y; Ryx, Ryx', Ryy]$ to decide whether $\underline{Z}(o) \wedge (\forall xy)\, \underline{M}(x, y)$ is satisfiable in $\langle N, o, {}' \rangle$.

The following modifications of $\underline{M}$ do clearly not affect satisfiability of $\underline{Z}(o) \wedge (\forall xy)\, \underline{M}(x, y)$:

1. To $\underline{M}$ conjoin $Sxy \equiv Ryx$ and make the proper substitutions in $\underline{M}$ to obtain a new matrix of form $\underline{A}[Rxx, Rx'x, Rx'x', Rxy, Rx'y, Ryy] \wedge \wedge \underline{B}[Rxy, Ryx]$.

2. To $\underline{A} \wedge \underline{B}$ conjoin $Sxy \equiv Rx'y$ and make the proper substitutions in $\underline{A}$ to obtain a new matrix of form $\underline{A}[Rxx, Rxx', Rxy, Ryy] \wedge \underline{B}[Rxy, Ryx] \wedge \wedge \underline{C}[Rxy, Rx'y]$.

3. To $\underline{A} \wedge \underline{B} \wedge \underline{C}$ conjoin $[Rxy \equiv Syx] \wedge [Sx'y \equiv Rxy]$ and make the proper substitutions in $A$ to obtain the new matrix of form $\underline{A}[Rxx, Rxy, Ryy] \wedge \wedge \underline{B}[Rxy, Ryx] \wedge \underline{C}[Rxy, Rx'y]$.

4. To $\underline{A} \wedge \underline{B} \wedge \underline{C}$ conjoin $[Sxx \equiv Rxx] \wedge [Sx'y \equiv Sxy]$ and in $A$ substitute $Syx$ for $Rxx$, and $Sxy$ for $Ryy$ to obtain a new matrix of form $\underline{W}[Rxx] \wedge \wedge \underline{B}[Rxy, Ryx] \wedge \underline{C}[Rxy, Rx'y]$.

It therefore follows that the following is still an undecidable problem,

(D′) For any matrices $\underline{Z}[Roo]$, $\underline{W}[Rxx]$, $\underline{B}[Rxy, Ryx]$ and $\underline{C}[Rxy, Rx'y]$ to decide whether $\underline{Z}(o) \wedge (\forall x)\ \underline{W}(x) \wedge (\forall xy)\ \underline{B}(x, y) \wedge (\forall xy)\ \underline{C}(x, x', y)$ is satisfiable in $\langle N, o, ' \rangle$.

By lemma 3 one now obtains,

**Theorem 1′:** *There is no recursive method for deciding satisfiability of sentences of form* $(\exists v)\ \underline{Z}[Rvv] \wedge (\forall x)\ \underline{W}[Rxx] \wedge (\forall xy)\ \underline{B}[Rxy, Ryx] \wedge (\forall x)(\exists u)(\forall y)\ \underline{C}[Rxy, Ruy]$, *whereby* $R$ *is a vector of binary predicate letters.*

Depending on the mentioned result of Kahr, Moore, and Wang the initial condition $\underline{Z}[Roo]$ may be dropped without making (D) decidable. Correspondingly theorem 1′ remains true if "$(\exists v)\ \underline{Z}[Rvv]$" is dropped. The question, whether in addition the axial restraint $\underline{W}[Rxx]$ can be avoided, remains unanswered.

**Domino problems:** The reduction (D) to (D′) discussed in the previous section can be carried one step further by the following observation. Suppose that the predicates $Rxy$ on $N$ satisfy

(1) $$(\forall xy)\ \underline{D}[Rxy, Ryx, Rx'y].$$

Then clearly the predicates $Pxy \equiv Rxy$ and $Qxy \equiv Ryx$ satisfy

(2) $$\begin{aligned}(\forall x)\ [Qxx \equiv Pxx] \wedge (\forall xy)_{x \geqq y}\ \underline{D}[Pxy, Qxy, Px'y] \wedge \\ \wedge (\forall xy)_{x \geqq y}\ \underline{D}[Qxy, Pxy, Qxy']\end{aligned}$$

Conversely, if $Q$ and $P$ satisfy (2) then the predicates $R$ defined by $Rxy \equiv Pxy$, if $x \geqq y$ and $Rxy \equiv Qyx$, if $x \leqq y$ satisfy (1). Thus, (1) is satisfiable if and only if (2) is satisfiable. Note furthermore that the following formula uniquely defines the predicates $x \geqq y$ and $x > y$:

(3) $$x \geqq x \cdot\wedge\cdot \sim [x > x] \cdot\wedge\cdot [x' > y] \equiv [x \geqq y] \cdot\wedge\cdot [x > y] \supset [x \geqq y].$$

Thus, in case $\underline{D}[Rxy, Ryx, Rx'y]$ is of form $B[Rxy, Ryx] \wedge \underline{C}[Rxy, Rx'y]$ one obtains an (as to satisfiability) equivalent formula of form $\underline{W}[Rxx] \wedge \underline{U}[Rxy, Rx'y] \wedge \underline{V}[Rxy, Rxy']$, by conjoining (3) to (2). Consequently the problem (D′) reduces to the following,

(D″) For any matrices $\underline{Z}[Roo]$, $\underline{W}[Rxx]$, $\underline{U}[Rxy, Rx'y]$, $\underline{V}[Rxy, Rxy']$ to decide whether $\underline{Z}(o) \wedge (\forall x)\ \underline{W}(x) \wedge (\forall xy) \cdot \underline{U}(x, x', y) \wedge \underline{V}(x, y, y')$ is satisfiable in $\langle N, o, ' \rangle$.

It therefore follows that also this problem is unsolvable, and by lemma 3,

**Theorem 1″:** *There is no recursive method for deciding satisfiability of sentences of form* $(\exists v)\ \underline{Z}[Rvv] \wedge (\forall x)\ \underline{W}[Rxx] \wedge (\forall x)(\exists u)(\forall y) \cdot \underline{U}[Ruy, Ruy] \wedge \underline{V}[Ryx, Ryu]$.

Depending on the result of Kahr, Moore and Wang the conjunct $(\exists v)\underline{Z}$ can be dropped. However, the question whether in this theorem both restraints $(\exists v)\ \underline{Z}(v)$ and $(\forall x)\ \underline{W}(x)$ may be dropped, is a challenging unsolved problem. It can be stated thus,

*Problem* 1: Is there an effective method which applies to any $\langle \underline{S}, \underline{U}, \underline{V} \rangle$, $\underline{U}$ and $\underline{V}$ binary relations on the finite set $\underline{S}$, and decides whether or not there is a valuation $R: N \times N \rightarrow \underline{S}$ which satisfies the condition $(\forall xy) \cdot \underline{U}[Rxy, Rx'y] \wedge \wedge \underline{V}[Rxy, Rxy']$.

In a slightly different form this was first stated by WANG (1961), and called the domino problem. $\underline{U}$ and $\underline{V}$ may be interpreted as sets of bars of length one, whose ends are marked with colors from a finite set $\underline{S}$. The problem then takes the following rather appealing form: To decide whether the lattice $N \times N$ can be filled with bars from $\underline{U}$ along the $x$-direction and bars from $\underline{V}$ along the $y$-direction, such that the ends of all bars meeting at any lattice point carry the same color.

The domino problem 1 is distinguished from other decision-problems by the complete lack of "initial restraints". This seems to make it very hard to reduce to it any one of the standard unsolvable problems, which all contain initial conditions of one kind or another (empty or finite initial tapes, initial states, axioms = initial theorems). In contrast, such a reduction was possible in the case (D''), which is the domino problem 1 with initial restraints $\underline{Z}[Roo]$ and $W[Rxx]$ added. In fact, the claim of KAHR, MOORE and WANG is that the domino problem becomes unsolvable even in case only the axial-restraint $\underline{W}[Rxx]$ is added[4]).

Related to the domino problem is the following, also unanswered, question:

*Problem* 2: Is there a finite set $\underline{S}$ and binary relations $\underline{U}$ and $\underline{V}$ on $\underline{S}$ such that $(\forall xy)$. $\underline{U}[Rxy, Rx'y] \wedge \underline{V}[Rxy, Rxy']$ has a solution $R$, but none which is periodic ?

By lemma III this is simply the question whether or not there still is an infinity-axiom of form $(\forall x)(\exists u)(\forall y)$. $\underline{U}[Rxy, Ruy] \wedge \underline{V}[Ryx, Ryu]$, i.e., whether the set $\forall\exists\forall_0$, consisting of all these sentences, has property $\Phi$. As noted by WANG (1961), a negative answer to problem 2 would mean solvability of the domino problem 1. (This corresponds to the well known fact that $\Phi X$ implies solvability of $[X]$.) However, we rather expect problem 1 to be unsolvable (possibly not of maximal degree 1, which would explain the mentioned difficulties in setting up reductions of standard unsolvable problems).

**An unsolved problem on Turing-machines:** We will now present a very natural halting problem on Turing-machines. It came up in connection with $[\forall\exists\forall]$, but seems to be of interest in its own right.

($T_2$) To find an effective method, which for every Turing-machine $M$ decides whether or not, for all tapes $I$ (finite and infinite) and all states $B$, $M$ will eventually halt if started in state $B$ on the tape $I$.

This problem also displays the feature of lack of initial restraints. STANLEY TENNENBAUM has shown to the author that ($T_2$) becomes unsolvable if either

[4]) While the present paper was in print the author was informed that the mentioned results of KARR, MOORE and WANG are to be published in the February issue of the Proc. Nat. Ac. of Sc. U.S.A. 1962.

one of the following initial restraints is added: 1. Distinguished initial state A, 2. Initially the tape is empty.

## Bibliography

[1] ACKERMANN, W.: Solvable cases of the decision problem. North-Holland, Amsterdam 1954.

[2] BERNAYS, P.: Remarques sur les problème de la décision en logique élémentaire. Edition de centre nat. de la rech. scient. **13**, 39—44 (1958), Paris.

[3] CHURCH, A.: A note on the Entscheidungsproblem. J. Symbolic Logic **1**, 40—41, 101—102 (1936).

[4] SURANYI, J.: Reduktionstheorie des Entscheidungsproblems. Budapest 1959.

[5] TRACHTÉNBROT, B. A.: Névozmožnost' algorifma dlá problemy razrésimosti na konéčnyh klassah. Doklady Akadémii Nauk SSSR, n. s., **70**, 569—572 (1950).

[6] TURING, A. M.: On computable numbers, with an application to the Entscheidungsproblem. Proc. London Math. Soc. ser. 2, **42**, 230—265 (1936—7); **43**, 544—546 (1937).

[7] WANG, H.: Proving theorems by pattern recognition II. Bell System Technical J. **40**, 1—41 (1961).

*(Received November 30, 1961)*

# Recursive Definition and Complexity of Functions Over Arbitrary Data Structures

J. Richard Büchi [1)]

Bernd Mahr [2)]

Dirk Siefkes [2) 3)]

October 1983

## Abstract

We present a language REC in which we define the computable functions over arbitrary data structures through recursive definitions in familiar mathematical notation. We describe computations as derivations in production systems, independent of the representation of data. We measure the cost of computations through book-keeping of weighted evaluation steps. By employing familiar evaluation strategies we model different types of complexity.

---

1) Mathematics and Computer Sciences Department, Purdue University, USA

2) FB Informatik, Technische Universität Berlin

3) Supported by travel grants of the Deutsche Forschungsgemeinschaft and the Stiftung Volkswagenwerk in visiting Purdue University

# 1. Yet Another Algorithmic Language?

We present an algorithmic language REC over arbitrary data structures, and within it we model different types of cost of computations. REC-programs are recursive definitions of functions on a data structure, employing the familiar mathematical notation; its computations are sequences or trees of data operations, defined through evaluation rules, not depending on data representations; its cost definition yields machine cost or arithmetic cost as one wishes, through summing up appropriate step costs. In this section we describe the features of REC in a non-technical way, and explain why we present it here. In the rest of the paper we list the main definitions and basic results; more details can be found in the paper Büchi-Mahr-Siefkes /1983/. For earlier work on REC see section 1.4; further results will be published separately.

1.1 The language REC grew out of a course on the **Theory of Algorithms for Computer Science and Mathematics students**, taught by the last two authors. The students have some background in higher programming languages and software engineering; they value machine languages only as basic implements for working directly with the computer; and they are used from mathematics to figure with paper and pencil. Thus they have encountered in some way or other the basic problems in designing or analyzing an algorithm: to evaluate the problem and one's intention; to get straight the language and the data structure; to show that the algorithm works correctly; to estimate its cost; and finally again to evaluate the work and its consequences. We want them to see these problems purely, not coded into the confines of a particular programming language, or data structure, or data representation, or machine model; and still drawing directly on their experience mentioned above.

For these reasons we were not satisfied with the classical formalisms used in computability and complexity theory. Machines, as Turing or random access machines - or languages derived thereof - are particularly inappropriate, since machine computations get their meaning from machine steps and not from human actions; Turing machines moreover require coding of data into strings. (In spite of this argument we teach Turing machines in the course, too, for the sake of contrast and comparison.) Similarly the

$\lambda$-calculus treats computing as the manipulating of symbols, the reading of data for example is fixed into the language definition. One can formulate algorithms only through heavy doding, which hinders in practical analysis. By its control operator the language of $\mu$ -recursive functions is bound to the natural numbers; again other data have to be coded. The same holds for the LOOP languages in Brainerd-Landweber /1974/ and similar languages. The language used in Kfoury-Moll-Arbib /1982/ is condensed expressedly out of the programming language PASCAL, and thus carries its ballast. The languages of recursive definitions with a fixed point semantics (Scott-Strachey, e.g. /1971/, Cadiou-Manna /1972/, Manna-Ness-Vuillemin /1972/; see the books of Bird /1976/ and Manna /1974/) are syntactically similar to REC, but their concept of computation (if any) is not of practical use, and in particular does not allow to consider costs; their extensive mathematical formalisms aim mainly at constructing a single universe for everything involved in computing.

Therefore we developed the formalism REC as an aid in understanding - not hindered by unnecessary coding or ballast - what computing is. Working with REC should thus influence the students' practical work in computer sciences, and provide a basis for further studying computability and complexity theory as well. Although much of the detail will be known to the experts in the field in some form or other, we feel that the use of REC and the intentions behind it are worth further spreading.

1.2 Historically REC roots in the primitive recursive definitions of functions on the natural numbers by Dedekind,Peano, and Skolem. Prompted by the more general recursions of Ackermann and Péter, and modifying an idea of Herbrand, Gödel /1934/ proposed a general form of recursive definitions over the natural numbers: functions are defined by systems of equations of the form $h(f_1(x_1),\dots,f_n(x_n))=t$ where h is a function variable, the $f_i$ are names of "given" functions, the $x_i$ are strings of number variables, and t is a term composed of function variables; terms are evaluated using the equations and substitution; and a total function is "recursive" if it is defined uniquely within this formalism by a system of equations. Gödel's formalism extends the primitive recursion, and according to Church's thesis defines just the total computable functions; however, the "program property" (whether a set of equations defines a functions) is undecidable. Kleene /1936, 1943, 1952/ adapted Gödel's definition to partial functions, and made the program property decidable by force: the result of the shortest computation is taken as the value of the function; thus every set of equations uniquely defines a partial function.

We feel it is more natural to change slightly the syntax of Gödel's formalism. In the primitive recursion

$$(1) \qquad \begin{aligned} &h(x,0):=p(x)\\ &h(x,y+1):=q(x,y)\end{aligned}$$

one needs two equations to define h(x,y) for the cases y=0 and $y \neq 0$; similarly with three equations for the Ackermann function. If we write (1) in the form

$$(2) \quad h(x,y):=\begin{cases} p(x); & \text{if } y=0\\ q(x,y-1); & \text{if } y\neq 0\end{cases}$$

we are down to a single definition. Further we see that the predecessor function and the zero-predicate are hidden in the notation of (1). Therefore we change Gödel's definition as follows: a dyadic partial function h(x,y) is recursive if its values can be computed from an equation

$$(3) \quad h(x,y):=\begin{cases} p_1; & \text{if } x=0,y=0\\ p_2(x); & \text{if } x\neq 0,y=0\\ p_3(y); & \text{if } x=0,y\neq 0\\ p_4(x,y); & \text{if } x\neq 0,y\neq 0\end{cases}$$

where the $p_i$ are terms built from the indicated variables, zero, successor, predecessor, and names of already defined recursive functions; similarly for other arities of h. This definition again produces exactly the partial recursive functions on the natural numbers. It is moreover easily generalized to other data structures if data are represented by (normal forms of) terms. The syntax of this formalism, however, though decidable, is rather complicated, and depends on the data structure and even on the data representation. We therefore change it again by adding truth values and using definitions of the form

$$(4) \quad h(x):=\begin{cases} p_T(x); & \text{if } q(x)\\ p_F(x); & \text{if } \quad q(x)\end{cases}$$

where q is a predicate, i.e. a function into truth values, and x is a string of data variables. This is the formalism REC described below and defined precisely in section 2. If we prefer the linear notation now in vogue in computer sciences, we may employ the test operator if-then-else in building terms, and simply write definitions of the form

(5) h(x):=t(x).

We start with this formalism, RECT, in section 2.2, since it is easier to describe, although clumsier to apply, than REC, which we present in 2.3. In one form or the other the formalism RECT is investigated several times (see the literature quoted above and below), although not for cost analysis, and in general not for arbitrary data structures. The version most similar to RECT is in the book of Bird /1976/.

1.3 We discuss now the characteristics of RECT and REC in more detail. The reader interested in the exact formulations might read section 2 in parallel.

(1) We use arbitrary many-sorted partial algebras as data structures. In their signatures we collect the symbols used for sort and function names in both, data structures and programs. For function symbols we allow geometrical configurations consisting of an operation symbol and indicating the I/O-type. Thus terms are two-dimensional, too, and we can stick to the familiar mathematical notation; we need not decide between prefix and infix and postfix notation.

(2) We write programs as recursive definitions of new functions by terms built from the signature, function variables and if-then-else, one definition per function. Thus in RECT we can easily design and analyze arbitrary algorithms, many in their natural form. In this way we provide a general concept of recursive functions over arbitrary data structures; it yields the same functions as the known definitions on the natural numbers (Gödel-Kleene-Church-Post), on words (Turing), or on other data structures (through coding, see e.g. Bergstra-Broy-Tucker-Wirsing /1981/. RECT can be viewed as a restriction of the technique of algebraic specification where one allows arbitrary equations to describe the properties of functions; see e.g. Goguen-Thatcher-Wagner-Wright /1977/. RECT avoids the problems of correctness and completeness encountered there; compare the investigations in

Padawitz /1983/. Kapur-Mandayam /1979/ look at the expressive power of data abstraction restricted in this way.

(3) We separate programs and data. Therefore through a program we enrich the data structure by new functions, but - contrasting to the situation in algebraic specification - we cannot alter it otherwise. Also programs cannot modify themselves, as do programs in machine languages or e.g. in ALGOL 68. For processing programs through programs we have to use an appropriate data structure, e.g. words; we did this for example with the universal RECT-program used in section 4 of Mahr-Siefkes /1981/. If wanted, also in RECT we can represent data through terms and work on normal forms. This is done by Loeckx /1981/ in his formalism for supporting automated correctness proofs, which but for the fixed point semantics is similar to RECT otherwise. Klaeren /1980/ extends primitive recursion to term structures in this way.

(4) We describe computations as stepwise evaluations of programs on data. We specify the steps through production systems in the sense of Post /1921, 1936, 1943/: within a term we can either evaluate an operation of the data structure (basic step), or eliminate an if-then-else, or replace a function variable with terms as arguments by its definition (control steps). Thus in a computation we perform a sequence of data operations, intermixed with and directed by control steps; this is what we do when computing by hand or in a higher programming language, or when working on a functional design. Each step is meaningful in itself, as opposed to the symbol manipulations on the machine level; this concept supports correctness considerations better than automatization of design and/or proof. The evaluation system is nondeterministic; still computations are much more restricted than in algebraic specification where equations may have any form and may be applied in either direction (cf. e.g. O'Donnell /1977/, Hoffmann-O'Donnell /1979/).

(5) We may evaluate independently occurring subterms simulteneously. This provides for nondeterminism and parallel computations as much as possible without complications. It facilitates the use of RECT and simplifies the proof of Church-Rosser properties which guarantee that programs do define functions; cf. Raoult-Vuillemin /1980/.

(6) We change the formalism into REC by eliminating if-then-else from the program terms and using definition by cases instead; otherwise REC keeps the characteristics of RECT discussed in (1)-(5). Since now we evaluate test conditions separately, computations become trees instead of sequences. This further facilitates the use of REC, but complicates the semantics definition.

(7) We reduce (but do not eliminate) the nondeterminism of the evaluation system by using strategies, which tell what terms to evaluate next. All familiar strategies like outermost, call-by-name, call-by-value, innermost allow to compute all recursive functions over any data structure.

(8) We define the cost of a computation as the sum of the costs of its steps. The cost of different evaluations of the same term might differ arbitrarily; within the above strategies, however, all evaluations of the same term cost the same. We can thus define the cost of computing a function under a strategy. For functions not actually depending on some argument call-by-value may be arbitrarily more expensive than call-by-name; in general it is the other way around. For example multiplying natural numbers is quadratic with call-by-value, but may be exponential with call-by-name. By choosing appropriate step costs we can represent in REC any of the familiar types of costs, like arithmetic cost, machine (or unit) cost, call cost (cf. e.g. Vuillemin /1974/) or logarithmic cost; under certain assumptions the first three costs are of the same order of magnitude.

1.4 Earlier versions of REC can be found in the lecture notes Fleischmann-Mahr-Siefkes /1977 ff./ and in the papers Siefkes /1979, 1980/; see the latter for details on the development. In Kirchner-Röhrich-Siefkes /1979/ REC is enlarged to synchronize parallel computations; in Mahr-Siefkes /1981/ we relate the uniform REC-cost to the non-uniform term cost. Further work on REC includes results on recursive and recursively enumerable data structures and on the relation to partial specification; investigation of space cost is under way.

One can view the permission of parallel steps and the change from RECT to REC, as steps towards implementation. They facilitate calculations by hand, and the resulting REC-derivation trees describe fairly closely the familiar evaluation of recursive definitions in the machine using pointers, or more technically, stacks. A further step towards practical use would be a mode for storing results and fetching them for later use. Backus /1978/ does this step in extending his system FP, which is similar to REC (see section 6.10 of Siefkes /1980/ for a comparison), into the system FFP by adding data and control structures which allow to store and fetch coded data. In REC we could do a similar step by allowing to use (the result of) a side derivation again later in a derivation. (This is general enough, since all REC-terms are function terms.) Derivations would then be "collapsed trees" as treated e.g. in Ehrig-Rosen /1980/. An implementation following this line, however, would hardly be feasible. It seems more appropriate to extend REC by assignment statements so that one can store data and terms in variables. A formalism which enhances the flexibility of REC in this way and keeps many of its features, is under investigation.

# 2. Definitions and Results

## 2.1 DATA STRUCTURES AND TERMS

Definition: A data structure A consists of

(1) sets $A_s$, $s \in S$, where S is a given set of sorts;

(2) partial functions $f: A_v \rightarrow A_s$ where $s \in S$, v is a word $s_1 \ldots s_n \in S^*$, and $A_v := A_{s1} \times \ldots \times A_{sn}$.

Definition: A signature $\Sigma$ is a set of function symbols. A function symbol is a geometrical configuration W, e.g. a word over some alphabet, which we can decompose uniquely into an operational symbol f and sort symbols $s_1, \ldots, s_n, s$ where $n \geq 0$. We call the words $s_1 \ldots s_n s$ and $s_1 \ldots s_n$ and the symbol s the type, the I-type and the sort of W resp. (Note that we can read the order $s_1, \ldots, s_n, s$ uniquely from W.) We then write W as $f: s_1 .. s_n \rightarrow s$, or as Vs (or $V \rightarrow s$) if we want to indicate just the sort. If the I-type is empty, i.e. n=0, then W is an individual constant. —

We use the notation $f: v \rightarrow s$ only as a convenient form for the general discussion. Actually our definition allows infix, prefix, and postfix notation as well as the two-dimensional patterns used in mathematics. Examples are $(\mathbb{R} + \mathbb{R}) \rightarrow \mathbb{R}$ and $\lfloor \mathbb{R} + \mathbb{R} \rfloor \rightarrow \mathbb{N}$ for addition and truncated addition on the reals which use the operation symbols $(+)$ and $\lfloor + \rfloor$ and are of type $\mathbb{R}\mathbb{R}\mathbb{R}$ and $\mathbb{R}\mathbb{R}\mathbb{N}$ resp., $\sum(\mathbb{R}_1, \ldots \mathbb{R}_{\mathbb{N}}) \rightarrow \mathbb{R}$ or $\sum_{i=1}^{\mathbb{N}} \mathbb{R}_i \rightarrow \mathbb{R}$ for iterated addition of type $\mathbb{N}\mathbb{R}^*\mathbb{R}$, and

$$\mathbb{N} \begin{pmatrix} A & \ldots & A \\ A & \ldots & A \end{pmatrix}^{\mathbb{N}} \longrightarrow \mathbb{M} \quad \text{of type } \mathbb{N}\mathbb{N}A^*\mathbb{M}$$

for a matrix over A. If the type is clear from the context we omit it, thus identifying the function symbol with the operation symbol.

We assume that data are represented by expression of any kind - not necessarily as strings or terms - in such a way that we can use the representations as symbols in expressions. This allows the following:

Notation: If W is a function symbol of the form Vs of I-type $s_1 \ldots s_n$ and operation symbol f, and if $d_1, \ldots, d_n$ are any expressions (like data representations, or terms, see below) of sort $s_1, \ldots, s_n$, then we write

$V[d_1 \ldots d_n]$ and $V[d_1 \ldots d_n]s$ for the result of replacing $s_i$ by $d_i$ for $i=1,\ldots,n$ in V and W resp. (This is not substitution, since the $s_i$ need not be different! ) Instead of $V[d_1 \ldots d_n]$ we also write $fd_1 \ldots d_n$.

Sometimes we use the same signature $\Sigma$ for different data structures; examples are arrays over any set, or matrices over any ring. We call such data structures similar, and the class of all data structures with signature $\Sigma$ the similarity type of $\Sigma$. In this spirit one considers data structures as interpretations of signatures. One then calls A a data structure of signature $\Sigma$, or a $\Sigma$-structure, if it contains domains $A_s$ (or rather $s_A$) exactly for each sort s of $\Sigma$ and functions $f_A : A_v \longrightarrow A_s$ exactly for each function symbol $f:v \longrightarrow s$ in $\Sigma$. Thus $s_A$ and $f_A$ serve as names for the sorts and functions in A. Again we will use this notation for the general discussion, but in actual examples we will write fd=b in A instead of $f_A(d)=b$, or just fd=b since normally A will be clear from the context; for example "$\lfloor 1.3 + 2.8 \rfloor = 4$ (in $\mathbb{R}$)".

EXAMPLE: For a data structure A with addition and multiplication the data structure $\mathbb{M}_A$ of matrices over A consists of A and $\mathbb{N}$ plus the sort $\mathbb{M}$ (matrix) and the operations

$_{\mathbb{N}}(0)^{\mathbb{N}} \longrightarrow \mathbb{M}$ ($_m(0)^n$ is the "initial" or "zero" matrix of m rows and n columns)

$\downarrow \mathbb{M} \longrightarrow \mathbb{N}$, $\overrightarrow{\mathbb{M}} \longrightarrow \mathbb{N}$ (row and column dimension),

$\mathbb{M}_{\mathbb{N},\mathbb{N}} \longrightarrow A$ ($X_{i,j}$ is the entry of X at location (i,j)),

$(\mathbb{M} \rangle_{\overline{\mathbb{N},\mathbb{N}}}\ A) \longrightarrow \mathbb{M}$ ($X\rangle_{\overline{i,j}}\, a$ is the matrix X with $X_{i,j}$ changed into a).

Definition: $\Sigma$-terms are built from the operation symbols in $\Sigma$ in the usual way. By our choice of function symbols each term t decomposes uniquely into $ft_1 \ldots t_n$ where $f:s_1 \ldots s_n \longrightarrow s$ is in $\Sigma$ and the $t_i$ are terms of sort $s_i$; s is the sort of t. If t and q are terms and q occurs somewhere in t, then q is a subterm of t. Subterms p and q of t (possibly identical) occur independently in t if they occur at non-overlapping locations.

Definition: Let A be a data structure of signature $\Sigma$.

A data term (over A) is an expression which results when in a $\Sigma$-term we replace some ($\geq 0$ many) subterms by data from A of the same sort. A basic term is a data term fd where $f: v \longrightarrow s$ is in $\Sigma$ and $d \in A_v$. By $\underline{T_A}$ we denote the set of A-data $\Sigma$-terms.

## 2.2 THE LANGUAGE RECT

As data variables we use function symbols of the form xs of I-type $\lambda$ and sort $s \in S$; they are distinct from the individual constants of $\Sigma$. If $x_i$ are data variables of sort $s_i$ for $i=1,\dots,n$ we call $x := x_1 \dots x_n$ data variables of type $v := s_1 \dots s_n$. For the following we fix a set $\underline{\Phi}$ of data variables, and write $\underline{\Phi_s}$ for the data variables of sort s, $\underline{\Phi_v}$ for $\Phi_{s1} \times \dots \times \Phi_{sn}$ if $v = s_1 \dots s_n$.

Definition: Terms and data terms with variables $\Phi$ are terms and data terms, respectively, over the signature $\Sigma \cup \Phi$. For a (data) term t with variables let $\underline{\Phi_t}$ be the set of variables occuring in t. Call t variable-free if $\Phi_t$ is empty.

Definition: Let $x := x_1 \dots x_n$ and $p := p_1 \dots p_n$ be data variables and terms resp., of the same type; let t be a data term with variables. By substituting p for x in t, i.e. by replacing $x_i$ by $p_i$ for $i=1,\dots,n$ throughout t, we get the data term

$$\underline{t\begin{bmatrix} x_1 \dots x_n \\ \\ p_1 \dots p_n \end{bmatrix}}, \text{ or } \underline{t\begin{bmatrix} x \\ \\ p \end{bmatrix}}.$$

As names for functions which we want to define, we use sets $\underline{\Pi}$ of function symbols, called function variables, with I-types in $S^*$ and sorts in S. Finally we use the set $\underline{\Lambda}$ of test symbols

if $\mathbb{B}$ then s else $s \longrightarrow s$ for $s \in S$ where if c then d else e :=

$$:= \begin{cases} d; & c=T \\ e; & c=F \end{cases} \text{ in A}$$

The sets $\Sigma, \Phi, \Pi, \Lambda$ are disjoint from each other and normally from A. All notions to follow depend on some fixed A (or $\Sigma$).

Definition:

1. Program terms and program data terms are (sequences of) terms, and data terms resp., over the signature $\Sigma \cup \Phi \cup \Pi \cup \Lambda$; we often call them just terms (of type $v \in S^*$). Thus a term is either a variable, or a datum, or an operation term ft where $f \in \Sigma$, or a function term ht where $h \in \Pi$, or a test term if r then p else q where we call r,p,q the test condition, the T-term, and the F-term resp.

2. A function definition (for h) is of the form

hx:=t

where $h:v \longrightarrow s$ is a function variable in $\Pi$, $x=x_1 \ldots x_n \in \Phi_v$ are variables, and t is a program term of sort s with variables $\Phi_t \subseteq \{x_1,\ldots,x_n\}$.

3. If $\Pi$ is finite, then a program (over A, for $\Pi$) is a set P of function definitions, for each $h \in \Pi$ exactly one. We write P-term (or just term) and $T_P$ short for variable-free(!) program data terms and the corresponding set or $\Sigma \cup \Pi \cup \Lambda$-structure, assuming that we can recover A and the signature from P.

Definition:

1. The program evaluation system (E) consists of the following rules:

(E1) $fd \longrightarrow b$ for each basic term fd and $b \in A$ s.t. fd=b in A,
(evaluate operations on data)

(E2) if c then $p_T$ else $p_F \longrightarrow p_c$ for all P-terms $p_T$, $p_F$ and truth values c,
(branch on test)

(E3) $hp \rightarrow t\begin{bmatrix}x\\p\end{bmatrix}$ for each hx:=t in P and all P-terms p.
(call function definition)

2. In a simple step $p \xrightarrow[r]{} q$ in (E) we apply the rule $r \longrightarrow t$ to replace some occurrence of r in the data term p by t, getting the data term q; r is the handle of the simple step. In an (n-fold) step $p \xrightarrow[v]{} q$ in (E) with handle $v=r_1 \ldots r_n$ we apply to p simultaneously n simple steps (the components of $p \xrightarrow[v]{} q$) with the pairwise independently occurring handles $r_1,\ldots,r_n$, getting q. If the handle is of no concern, we simply write $p \longrightarrow q$.

3. Let p and q be data terms. A derivation of q from p in (E) is a finite sequence D of steps in (E) leading from p to q; we write $p \xrightarrow[D]{} q$ (in(E)). The number of steps in D is its length. q is derivable from p in (E) if there is a derivation of q from p in (E); we write $p \vdash q$ (in (E)).

4. If a derivation D of q from p cannot be continued at q since no rule applies to it, we call q *terminal*. Thus q is terminal iff it contains no function variables, and either it contains no operation symbols, and thus is a datum, or for all basic terms fd in q, $f_A(d)$ is undefined. In the second case we call D *terminating* (*at* q), in the first case we call D *successful* (*with* q).

*Fact:* Any two terminating derivations from the same P-term yield the same result. Furthermore if either is successful so is the other, and both yield the same datum.

*Definition:* Let P be a program.

1. For a data structure A the *program evaluation function*

   $VAL_P : T_P \longrightarrow A$

is the partial function on P-terms defined by

   $VAL_P(t) = b$ iff $t \vdash b$ in (E).

2. For each function variable $h: v \longrightarrow s$ in $\Pi$ the program P *defines* the function

   $h_P : A_v \longrightarrow A_s$ where $h_P(d) := VAL_P(hd)$.

3. Let *RECT* (*REC with test*) be the formalism of programs and their evaluation. We call the functions on A defined by programs *RECT-computable*.

*Example:* The following program over the data structure $\mathbb{M}_A$ of 1.5.4 defines the multiplication of two matrices with elements from A, again using the school method:

$$X \bullet Y := \underline{if}\ \vec{X} \neq \downarrow Y\ \underline{then}\ \text{error}\ \underline{else}\ {}_{\downarrow X}\boxed{X \bullet Y}^{\vec{Y}}$$

(matrix multiplication $\mathbb{M} \bullet \mathbb{M} \longrightarrow \mathbb{M}$)

$$_{x}\boxed{X \bullet Y}^{z} := \underline{if}\ z=0\ \underline{then}\ \underline{if}\ x=1\ \underline{then}\ {}_{\downarrow X}(0)^{\vec{Y}}\ \underline{else}\ {}_{x^-}\boxed{X \bullet Y}^{\vec{Y}}$$

$$\underline{else}\ {}_{x}\boxed{X \bullet Y}^{z^-} \succ_{x,z} \sum_{i=1}^{\vec{X}} X_{x,j} \bullet Y_{j,z}$$

(partial matrix multiplication ${}_{\mathbb{N}}\boxed{\mathbb{M}\cdot\mathbb{M}}^{\mathbb{N}} \longrightarrow \mathbb{M}$ where for an $l\times m$-matrix A, $m\times n$-matrix B, and $1\leq p\leq l$, $0\leq q\leq n$

${}_{p}\boxed{A\cdot B}^{q}$ is the $l\times n$-matrix 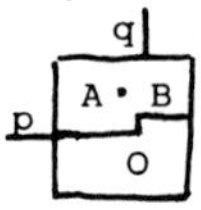 , which means that for

$1\leq i\leq l$, $1\leq k\leq n$

$$({}_{p}\boxed{A\cdot B}^{q})_{i,k} := \begin{cases} \sum_{j=1}^{m} A_{i,j}\cdot B_{j,k}; & \text{if } i<p \text{ or } (i=p \text{ and } k\leq q) \\ 0 & ;\ \text{otherwise} \end{cases} \quad )$$

$$\sum_{j=1}^{y} X_{x,j}\cdot Y_{j,z} := \underline{\text{if}}\ y=0\ \underline{\text{then}}\ 0\ \underline{\text{else}}\Big(\sum_{j=1}^{y^-} X_{x,j}\cdot Y_{j,z}\Big)+(X_{x,y}\cdot Y_{y,z})$$

(multiplication of the x-th row of X by the z-th column of Y, up the y-th entry

$$\sum_{j=1}^{\mathbb{N}} \mathbb{M}_{\mathbb{N},j}\cdot\mathbb{M}_{j,\mathbb{N}} \longrightarrow A)$$

## 2.3 THE LANGUAGE REC

In the following definitions we will define the formalism REC, using the same notations as for RECT in 2.2; only when we compare the two formalisms we will add the qualifiers "in REC" or "in RECT" resp. Throughout the chapter we fix a data structure A of signature $\Sigma$, and sets $\Phi$ and $\Pi$ of data variables and function variables resp., all disjoint; as in 1.2 we suppress the dependence on A and the symbols chosen.

Definition:

1. Program terms and program data terms are (finite sequences of) terms and data terms resp., over the signature $\Sigma\cup\Phi\cup\Pi$; we often call them just terms (of type $v\in S^*$). Thus a term is either a variable, or a datum, or an operation term, or a function term, or a finite sequence of those.

2. A function definition (for h) is of the form

$hx:=t$ (explicit definition), or

$$hx:=\begin{cases} t_T; & \text{if } q \\ t_F; & \text{if } \neg q \end{cases} \qquad \text{(definition by cases)}$$

where $h:v \rightarrow s$ is a function variable in $\Pi$, $x=x_1 \ldots x_n$ are variables in $\Phi_v$, and $t, t_T, t_F$ are program terms of sorts s with variables contained in $\{x_1, \ldots, x_n\}$. The terms t, or $t_T$, $t_F$, are the defining (T-,F-) term, and q is the test condition for h.

3. If $\Pi$ is finite, then a program (over A, for $\Pi$) is a set P of function definitions, exactly one for each $h \in \Pi$. We write P-term (or just term) and $T_P$ short for variable-free(!) program (data) terms and the corresponding set or $\Sigma \cup \Pi$-structure, assuming that we can recover A and the signature from P.

Definition:

1. We simultaneously define rules, steps, and derivations in the program evaluation system (H):

a) For each P-term p the empty derivation of p from p, $p \overset{\sigma}{\underset{p}{\triangle}} p$, is the tree with no edges and with the single node p as starting node and as root.

b) The rules of (H) are:

(H1) $fd \rightarrow b$ for each basic term fd and $b \in A$ s.t. fd=b in A (evaluate operations on data)

(H2) $hp \rightarrow t\begin{bmatrix} x \\ p \end{bmatrix}$ for each explicit definition hx:=t in P, and function term hp (call an explicit function definition)

(H3) $hp, C \rightarrow t_c\begin{bmatrix} x \\ p \end{bmatrix}$ for each definition by cases

$$hx:=\begin{cases} t_T; & \text{if } q \\ t_F; & \text{if } \neg q \end{cases}$$

in P, function term hp, truth value c, and derivation $q\begin{bmatrix} x \\ p \end{bmatrix} \underset{C}{\triangle} c$. (call a function definition by cases)

c) The notions of simple step and (n-fold) step ($p \underset{U}{\longrightarrow} q$) with premise(s) U, and of its components, are defined as in 2.2 for (E). We say "premises" instead of "handle", since - through the use of rule (H3) - U may contain names of derivations besides the replaced subterms. The subterms $p_1,\dots,p_n$ of p in the premises U of an n-fold step $p \underset{U}{\longrightarrow} q$ constitute the handle of the step. The different(!) derivations $C_1,\dots,C_m$, $0 \leq m \leq n$, which are named in U, are the side derivations of the step; for m=0 the step is straight, otherwise branching.

d) If $r \underset{B}{\triangleright} p$ is a derivation of p from r and $p \underset{U}{\longrightarrow} q$ is an n-fold step with side derivations $C_1,\dots,C_m$, then the tree

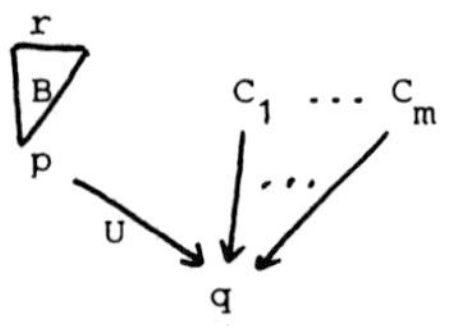


with starting node r and root q is a derivation D of q from r, denoted as $p \underset{D}{\triangleright} q$.

2. The number of steps (not edges) in a derivation $p \underset{B}{\triangleright} q$ is its size; the branch from p to q is its main derivation.

3. q is derivable from p, $p \triangleright q$, if there is a derivation of q from p.

4. The notions of terminal term, and of terminating and of successful derivation are defined as in 2.2. $D_P$ is the set of all derivations of P.

Proposition:

If $r \underset{B}{\triangleright} p$ and $r \underset{C}{\triangleright} q$ are two terminating derivations, then p=q. Especially any two successful derivations from the same P-term yield the same datum.

Definition:

1. The program evaluation function

$VAL_P : T_P \longrightarrow A$

is defined by

$VAL_P(t)=b$ iff $t \triangleright b$ in (H).

2. For any $h:v \longrightarrow s$ the program P defines the function

$h_P : A_v \longrightarrow A_s$ where $h_P(d) := VAL_P(hd)$.

3. Let REC be the above formalism of programs and their evaluation. Call the functions defined by programs REC-computable.

Theorem: REC- and RECT-programs can be effectively transformed into each other so that in both formalisms the same functions are computable.

## 2.4 STRATEGIES

Definition: An (evaluation) strategy is a rule R which for any P-term specifies which steps (if there are any) one is allowed to perform next. A strategy is deterministic if it specifies always at most one step. An R-derivation is a derivation following the strategy R.

Examples: The following are familiar strategies:

1) Call by value (V): Evaluate some operation; if none is possible, call an innermost function variable, i.e. call some hw where w contains no function variable.

2) Call by name (N): Evaluate some operation; if none is possible, call an outermost function variable, i.e. call some hw which is not proper subterm of a function term.

3) Innermost (I): Evaluate some innermost function, be it function variable or operation, i.e. some gw where no subterm of w (including w) can be evaluated.

4) Outermost (O): Evaluate some outermost function, be it function variable or operation, i.e. some gw where no term containing it properly can be evaluated.

5) Parallel call by value, call by name, innermost, outermost (PV,PN,PI,PO): The above strategies with as many steps as possible done in parallel, except that in PV and PN evaluation of operations and call of functions stay separated.

Note: The parallel strategies are deterministic. The other strategies allow only simple steps, on independently occurring subterms.

Fact: The call-by-name and the outermost strategies might lead to termination where the call-by-value and the innermost strategies do not. Nevertheless by all eight strategies the same functions are computable.

## 2.5 COST OF REC-COMPUTATIONS

Definition: The step costs for REC-programs are functions of the following type:

$C_f : A_v \rightarrow \mathbb{N}$ for each $f : v \rightarrow s$ in $\Sigma$ (cost of evaluating f)

$C_h : T_{P,v} \rightarrow \mathbb{N}$ for each $h : v \rightarrow s$ in $\Pi$ where $hx := t$ in P (cost of explicit call)

$C_h^d : T_{P,v} \rightarrow \mathbb{N}$ for each $h : v \rightarrow s$ in $\Pi$ where $hx := \begin{cases} t_T ; & \text{if } q \\ t_F ; & \text{if } \neg q \end{cases}$ in P (cost of call by cases)

We require that f and $C_f$ have the same domain (subset of $A_v$), whereas for $h : v \rightarrow s$ in $\Pi$, $C_h$ is defined on all of $T_{P,v}$, independent of the domain of $h_A$.

Examples:

1. Arithmetic, or algebraic cost: $C_f = 1, C_h = 0$ for $f \in \Sigma, h \in \Pi$.
2. Unit cost: $C_g = 1$ for $g \in \Sigma \cup \Pi$.
3. Call cost: $C_f = 0, C_h = 1$ for $f \in \Sigma, h \in \Pi$.
4. Logarithmic cost: $C_g(w) = \text{size}(w)$ for $g \in \Sigma \cup \Pi$ where "size" for example is the number of symbols in w.

Definition: The cost of derivations (for B) is the function

$COST := COST_P : D_P \rightarrow \mathbb{N}$

defined recursively through the following clauses:

a) The empty derivations are free: $COST(O_p) = 0$.

b) The cost of applying a rule is:

(1) $COST(fd \rightarrow b) := C_f(d)$ for each basic step;

(2) $COST(hp \rightarrow t\begin{bmatrix} x \\ p \end{bmatrix}) := C_h(p)$ for each explicit function call;

(3) $COST(hp, C \rightarrow t_c\begin{bmatrix} x \\ p \end{bmatrix}) := C_h^c(p)$ for each function call by cases (independent of C!)

c) The cost of the k-fold step $p \xrightarrow[U]{} q$ with components $U_i \rightarrow V_i$ for $i = 1, \ldots, k$ is:

$$COST(p \xrightarrow[U]{} q) := \sum_{i=1}^{k} COST(U_i \rightarrow V_i).$$

d) If D is the derivation

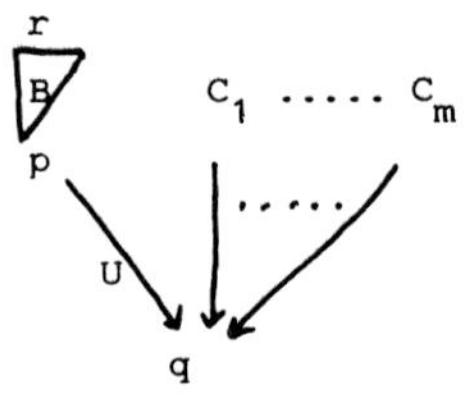

then $COST(D):=COST(B)+COST(p \xrightarrow[U]{} q)+\sum_{i=1}^{m} COST(C_i)$.

<u>Definition:</u> A strategy R is <u>stable</u> (sc. with respect to cost) if all terminating R-derivations starting from the same term have the same cost.

<u>Proposition:</u> The eight strategies in 2.4 are stable.

<u>Definition:</u> Let R be a stable strategy. Then the <u>cost of evaluating terms</u> (<u>through</u> P) <u>under</u> R is the function

$$COST_P^R : T_P \longrightarrow \mathbb{N}$$

defined as

$$COST_P^R(t)=COST_P(D) \qquad \text{for some terminating R-derivation } t \vdash_D d.$$

For a function variable $h\in\Pi$ the <u>cost of computing</u> h (<u>through</u> P) <u>under</u> R is the function

$$COST_h^R : A_V \longrightarrow \mathbb{N}$$

defined by

$$COST_h^R(d) \begin{cases} :=COST_P^R(hd); & \text{if } h \text{ is defined on } d \text{ in } A \\ \text{undefined}; & \text{otherwise} \end{cases}$$

<u>Example:</u> The arithmetic cost of multiplying an $l\otimes m$- and an $m\otimes n$-matrix using the algorithm of 2.2 is

$$l\cdot n\cdot(4m+5)+4\cdot l+9=O(l\cdot m\cdot n)$$

under the call-by-value or the innermost strategy.

<u>Proposition:</u> Call-by-value cost and innermost cost are the same; call-by-name cost might be higher than, equal to, or lower than outermost cost. Parallel cost is less or equal to, and might be less than, the corresponding simple cost. Call-by-value cost can be arbitrarily higher than call-by-name cost. Call-by-name cost can be exponentially higher than call-by-value cost.

<u>Definition:</u> Let $size:A^* \longrightarrow \mathbb{N}^k$ be a size function for the program P. Then the <u>size cost of</u> $h\in\Pi$ (<u>through</u> P) <u>under</u> R is for $n\in\mathbb{N}^k$

$$scost_h^R(n):=\begin{cases} \max\{cost_h^R(b);\ b\in A_V,\ size(b)=n\}; & \text{if this set is finite and nonempty} \\ \text{undefined}; & \text{otherwise} \end{cases}$$

<u>Proposition:</u>

1. Let $scost^c$ and $scost^u$ be the size cost of some function variable under some stable strategy using call cost and unit cost resp. Then $scost^u \in \Theta(scost^c)$, in fact $scost^c \leq scost^u \leq k \cdot scost^c$.

2. Let $scost^c$ and $scost^a$ be the size cost of some function variable under call-by-value or call-by-name using call cost and arithmetic cost, resp. Then

$$scost^a \in \Theta(scost^c), \text{ in fact } scost^c \leq scost^a \leq m \cdot scost^c.$$

# 3. References

1. Backus, J., 1978: Can Programming be Liberated from the von-Neumann Style? A Functional Style and Its Algebra of Programs. Comm. ACM vol.21, pp. 613-641

2. J.A. Bergstra, M. Broy, J.V. Tucker, M. Wirsing, 1981: On the power of algebraic specification. Proceedings 10th MFCS, Springer-Verlag, pp. 193-204

3. Bird, R., 1976: Programs and machines. Wiley

4. Brainerd, W., Landweber, L., 1974: Theory of Computation. Wiley

5. J. R. Büchi, B. Mahr, D. Siefkes, 1983: Manual on REC - a language for use and cost analysis of recursion over arbitrary data structures. Technical Report, FB Informatik, Techn. Universität Berlin

6. J. M. Cadiou, Z. Manna, 1972: Recursive definitions of partial functions and their computations. Proc. of an ACM conference on proving assertions about programs, SIGPLAN Notices 7.1 and SIGACT News 14, pp. 58-65

7. H. Ehrig, B.K. Rosen, 1980: Parallelism and Concurrency in Graph Manipulations. Theoret. Comp. Sci. 11, pp. 247-275

8. Fleischmann, K., Mahr, B., Siefkes, D., 1977: Algorithmentheorie. Lect. Notes FB. Informatik, Techn. Univ. Berlin, 1977 ff.

9. Fleischmann, K., Müller, F., Siefkes, D., 1975: Theorie der Berechenbarkeit. Lect. Notes FB Informatik, Techn. Universität Berlin, 1975/76 and 76/77

10. Gödel, K., 1934: On Undecidable Propositions of Formal Mathematical Systems. Lect. Notes Institute Advanced Study, Princeton, 28 pp.

11. Goguen, J.A., Thatcher, J.W., Wagner, E.G., Wright, J.B., 1977: Initial Algebra Semantics and Continuous Algebras. Journal ACM vol. 24, pp. 68-95.

12. Hoffmann, C.M., O'Donnell, M.J., 1979: Programming with Equations. Techn. Report Comp. Sci. Dept., Purdue University, 36 pp.

13. D. Kapur, S. Mandayam, 1979: Expressions of the Operation Set of a Data Abstraction. Computation Structures Group Memo 179-1, M.I.T., Laboratory for Computer Science

14. A.J. Kfoury, R.N. Moll, M.A. Arbib, 1982: A Programming Approach to Computability Theory. Springer-Verlag

**REFERENCES (cont'd)**


15. Kirchner, W., Röhrich, T., Siefkes, D., 1979: Synchronizing Computations by Conditions. Extended Abstract, FB Informatik, Techn. Univ. Berlin, 9 pp.

16. H. A. Klaeren, 1980: A simple class of algorithmic specifications for abstract software modules. Proc. 9th Symp. on Math. Found. of Comp. Sci., Lect. Notes in CS 88, Springer-Verlag, pp. 362-374

17. Kleene, S.C., 1936: General Recursive Functions of Natural Numbers. Math. Ann. vol. 112, pp. 727-742

17.a) --, 1943: Recursive predicates and quantifiers. Transactions AMS vol. 53, pp. 41-73

17.b) --, 1952: Introductions to Metamathematics. Wolters-Noordhoff and North-Holland

18. J. Loeckx, 1981: Implementations of Abstract Data Types and their Verification. Proc. GI-11. Jahrestagung München, pp.96-108

19. B. Mahr, D. Siefkes, 1981: Relating uniform and nonuniform models of computations. Proc. GI-11. Jahrestagung München, pp. 41-48

20. Z. Manna, 1974: Mathematical Theory of Computation. McGraw Hill

21. Z. Manna, S. Ness, J. Vuillemin, 1972: Inductive methods for proving properties of programs. Proc. of an ACM conference on proving assertions about programs, SIGPLAN Notices 7.1 and SIGACT News 14, pp. 27-50

22. M. J. O'Donnell, 1977: Computing in Systems Described by Equations. Lect. Notes in Comp. Sci. 58, Springer-Verlag

23. P. Padawitz, 1983: Correctness, Completeness and Consistency of Equational Data Type Specifications. Bericht-Nr. 83-15, FB Informatik, Techn. Univ. Berlin

24. Post, E.L., 1921: Introduction to a general theory of elementary propositions. Am. Journ. Math. vol. 43, pp. 163-185

24.a) --, 1936: Finite combinatory processes - formulation I. Journ. Symb. Logic vol. 1, pp. 103-105

24.b) --, 1943: Formal reductions of the general combinatorial decision problem. Am. Journ. Math. vol. 65, pp. 197-215

25. J.-C. Raoult, J. Vuillemin, 1980: Operational and Semantic Equivalence Between Recursive Programs. Journ. ACM 27,pp.772-796

**REFERENCES (cont'd)**

26. Scott, D., Strachey, C., 1971: Toward a mathematical semantics for computer languages. PRG-6, Oxford University

27. Siefkes, D., 1979: Recursive Equations as a Programming System. In: "Begriffsschrift" Jenaer Frege-Konferenz, Universität Jena, pp. 433-448

27.a) --, 1980: Programming with Recursion. Techn. Report CSD-TR 331, Purdue University, 72 pp.

28. J. Vuillemin, 1974: Correct and Optimal Implementations of Recursion in a Simple Programming Language. Journ. of Comp. System Sci., vol. 9, pp. 332-354

Reprinted from Wechsung (Ed.), Proceedings of the Second Frege Conference, Akademie Verlag Berlin, 1984, pp. 303-308.

Arch. math. Logik **26** (1986/7), 101–106

# CODING IN THE EXISTENTIAL THEORY OF CONCATENATION*, [1]

J. Richard Büchi [2] and Steven Senger

We will show (Theorem I) that every existentially definable relation over concatenation is diophantine. That is, it can be defined by an existential formula whose matrix is a single equation between terms in concatenation. We show that a) disjunction of diophantine relations is diophantine, and b) that the relation "not equals" is diophantine. These results were discovered by the first author about 20 years ago but were never published. We publish them now because they are interesting in view of Makanin's positive result on the decidability of solvability of equations in concatenation [3]. A diophantine definition for the relation "not equals" has recently appeared in Taimanov and Hmelevskii [5]. Since [5] is in Russian, and our definition is very simple we include it here. For the question of decidability a) is insignificant, but b) does extend Makanin's work and shows that solvability of systems of equations and inequations is decidable. For questions of definability both are significant, and will be used in a subsequent paper to present the main results of [4] on definability and decidability of extensions of the theory of concatenation.

In the second half of the paper we define a strong type of prefix code which we call a Thue [3] code. We use these codes to prove (Theorem II) that given an equation over an arbitrary alphabet one can construct, using a Thue code, an equation over the two letter alphabet such that their solutions are closely related via the coding. The constructed equation may however have extraneous solutions. To eliminate these solutions it would be sufficient to existentially define membership in the set of words over the code alphabet. To prove that this is not possible or to give a definition should lead to a better understanding of Makanin's proof.

---

* Eingegangen am 2. 12. 1985.

[1] The results in this paper are contained in the Ph. D. thesis [4] of the second author, written under the direction of the first author.

[2] Professor Büchi died suddenly in April 1984 while this paper was still in preparation. He will be greatly missed.

[3] So named because they are closely related to the work of Thue on infinite number sequences in [6].

**Section I.**

Let $I$ denote a finite set called the alphabet. As letters of $I$ we use consecutive numerals. $I$ is usually the two letter alphabet $\{1, 2\}$. $I^*$ is all words over $I$. Among these words is the empty word $e$. $I^+$ is $I^*$ without $e$. The binary operation of concatenation on words is denoted by $\hat{}$. We often abbreviate $x\hat{}y$ by $xy$. The structure $\langle I^*, e, 1, 2, \hat{}\rangle$ is a free monoid on two elements. Let $\Xi$ be a set of variables, then $(I \cup \Xi)^*$ is the set of terms for this system. We denote terms by $\alpha$ or $\beta$. We define the coefficient set of a term $\alpha$ by writing $\alpha = a_0\xi_0 a_1 \ldots a_{n-1}\xi_{n-1}a_n$, where $a_i \in I^*$ and $\xi_i \in \Xi$ and defining $\mathrm{cof}(\alpha) = \{a_i \neq e | 0 \leqq i \leqq n\}$. $\Sigma$ will denote an equation between terms. The existential theory of concatenation, that is, the set of true existential statements involving the conjunction, disjunction, and negation of equations in concatenation is denoted by $\exists\langle I^*, e, 1, 2, \hat{}\rangle$.
A relation $R(\vec{x})$ is diophantine if there is an equation $\Sigma(\vec{x}, \vec{y})$ such that $R(\vec{x}) \equiv (\exists\vec{y})\,[\Sigma(\vec{x}, \vec{y})]$. Several authors have shown that the conjunction of diophantine relations is diophantine. The pairing function $g(x, y) = x1yx2y$ and the definition $x = y \wedge u = v \equiv g(x, u) = g(y, v)$ suffices to show this. Since $g$ introduces no new variables, given equations $\Sigma_1(\vec{x})$ and $\Sigma_2(\vec{x})$ we can construct a new equation $\Sigma_3(\vec{x})$ such that $\Sigma_1(\vec{x}) \wedge \Sigma_2(\vec{x}) \equiv \Sigma_3(\vec{x})$. This fact was known to Makanin [2] and implies that his algorithm for deciding solvability of equations in concatenation, will in fact decide solvability for a conjunction and disjunction of equations.
Further it can be shown that the full existential theory of concatenation is decidable by showing that the relation "not equal" may be replaced by an existential statement involving only conjunction and disjunction of equations. We require the following relations: $x \leqq y \equiv (\exists z)\,[x\hat{}z = y]$ (read "$x$ is a prefix of $y$"), $x \neq e \equiv 1 \leqq x \vee 2 \leqq x$ (read "$x$ is not empty"), and $x < y \equiv x1 \leqq y \vee x2 \leqq y$ (read "$x$ is a strict prefix of $y$"). These last two are of course dependent upon the alphabet. We define a rational order on $I^*$ in terms of $\leqq$ by

1) $x \prec y \equiv x2 \leqq y \vee y1 \leqq x \vee (\exists u)\,[u1 \leqq x \wedge u2 \leqq y]$.

The relation "not equal" is defined by

2) $$x \neq y \equiv x \prec y \vee y \prec x$$
$$\equiv x < y \vee y < x \vee (\exists u)\,[(u1 \leqq x \wedge u2 \leqq y) \vee (u2 \leqq x \wedge u1 \leqq y)].$$

Decidability of the full existential theory of concatenation now follows. Inequations and conjunctions of equations may be reduced to a disjunction of equations and each disjunct may be decided separately. We can show the relation "not equal" to be diophantine by proving that the disjunction of diophantine relations is diophantine. First,

3) $$x \leqq y \vee u \leqq v \equiv (\exists p, p', p'', q, q', q'')\,[x = pp' \wedge u = qq' \wedge y = pp'' \wedge v = qq'' \wedge (p' = e \vee q' = e)].$$

The last conjunct is diophantine as is shown by

4) $x=e \vee y=e \equiv xy=yx \wedge (\exists u, u', v, v')[uu'=1 \wedge vv'=2 \wedge xuyu'=yu'xu \wedge xvyv'=yv'xv]$.

Finally disjunction is defined by

5) $x=y \vee u=v \equiv (x \leqq y \vee u \leqq v) \wedge (x \leqq y \vee v \leqq u) \wedge (y \leqq x \vee u \leqq v) \wedge (y \leqq x \vee v \leqq u)$.

Consequently, given equations $\Sigma_1(\vec{x})$, $\Sigma_2(\vec{x})$ we can use 3) through 5) to construct an equation $\Sigma_3(\vec{x}, \vec{y})$ such that $\Sigma_1(\vec{x}) \vee \Sigma_2(\vec{x}) \equiv (\exists \vec{y})[\Sigma_3(\vec{x}, \vec{y})]$. This definition introduces a considerable number of new variables $\vec{y}$. An alternate definition which, using some of the strong properties of the equations $xy=yx$ and $xy=yz$, introduces only three new variables, is given in [4]. The relation "not equal" is diophantine since given an equation $\Sigma_1(\vec{x})$, using 1) through 5) we can construct an equation $\Sigma_2(\vec{x}, \vec{y})$ such that $\sim \Sigma_1(\vec{x}) \equiv (\exists \vec{y})[\Sigma_2(\vec{x}, \vec{y})]$.

**Theorem I.** *Relations formed from equations and inequations in concatenation using conjunction, and disjunction are diophantine. Consequently every existential formula in concatenation is diophantine.*

We can also give an existential definition of a universally quantified equation [4] which shows that equations with prefix $\exists\forall$ are also decidable.

**Section II.**

We introduce the partial order of proper occurence defined by $x \propto y \equiv (\exists uv) \cdot [uxv=y \wedge (u \neq e \vee v \neq e)]$ and $x \underline{\propto} y \equiv (x \propto y \vee x=y)$. The words $u$, $v$ are called flankers of, and the pair of words $\langle u, v \rangle$ is called the position of, $x$ in $y$. Let $C \subseteq I^+$. Define the set $C^{\uparrow\propto} = \{y | (\exists c)^C [c \underline{\propto} y]\}$. That is, $C^{\uparrow\propto}$ is the set of all words above some element of $C$ in the partial order $\propto$. Let $x \in I^*$. A form of $x$ relative to $C$ is

6) $x=u_0 v_0 \dots v_{n-1} u_n$ where $u_i \in I^* - C^{\uparrow\propto}, v_i \in C$.

The form is determined by the tuples $sp(x)=\langle u_0 \dots u_n \rangle$ called the spacers of $x$, and $cn(x)=\langle v_0 \dots v_{n-1} \rangle$ the content of $x$. Given a function which determines a form relative to $C$ for each word $x$, there is a corresponding function, called the deletion function, defined by $x^d = v_0 \dots v_{n-1}$.
We require a set $C$ and a deletion function $d$ satisfying $x^d y^d = (xy)^d$ for all $x, y$ in some diophantine subset of $I^*$ containing $C^*$.
If $C$ satisfies $(\forall x, y)^C [x \not< y]$, (i.e. is disordered in $<$) then if $\langle u, v \rangle$, $\langle u', v' \rangle$ are positions of distinct occurences of members of $C$ in a word $x$, then $u \neq u'$. In this situation (using the notation of 6)) we are able to define the left form of $x$ relative to $C$, by the condition: if $\langle u, v \rangle$ is a position of $c \in C$ in $x$ then for some $i$, $u_0 v_0 \dots u_i \leqq u < u_0 v_0 \dots u_i v_i$. In words, each $v_i$ is the first occurence of an element of $C$ since $v_{i-1}$. The left deletion function relative to $C$ is the deletion function associated with the left form relative to $C$.

Define $x \mathrel{\bar{\frown}} y \equiv (\exists u, v, w)[x = uv \wedge y = vw \wedge v \neq e \wedge (u \neq e \vee w \neq e)]$ (read "$x$ overlaps $y$"). If a set $C$ is disordered in either $\propto$ or $\mathrel{\bar{\frown}}$ then it is disordered in $<$, and the left deletion function exists. We call a set $C$ disordered in both $\propto$ and $\mathrel{\bar{\frown}}$ a Thue code. While $<$ and $\propto$ are irreflexive, $x \mathrel{\bar{\frown}} x$ may be true if $x$ overlaps itself in some non-trivial way. Consequently for $C$ to be disordered in $\mathrel{\bar{\frown}}$ no member of $C$ may overlap itself except identically.

**Lemma I.** *Let $C \subseteqq I^+$ be a Thue code, and let $d$ be the left deletion function for $C$, then $x \in I^*$, $y \in C\hat{}I^* \cup \{e\}$ imply $x^d y^d = (xy)^d$.*

*Proof.* The case $y = e$ is trivial so suppose $y \in C\hat{}I^*$. Let $x = u_0 v_0 \ldots v_{n-1} u_n$, $y = q_0 p_0 \ldots p_{m-1} q_m$, and $xy = s_0 t_0 \ldots t_{k-1} s_k$, be the left forms of $x$, $y$, and $xy$. Since $y \in C\hat{}I^*$, $q_0$ is empty. Hence $x^d = v_0 \ldots v_{n-1}$, $y^d = p_0 \ldots p_{m-1}$, and $(xy)^d = t_0 \ldots t_{k-1}$. Suppose $x^d y^d \neq (xy)^d$. Let $t_i$ contain the first place where $(xy)^d$ disagrees with $x^d y^d$ (i.e. $t_i \neq v_i$ if $i \leqq n-1$ or $t_i \neq q_{i-n-1}$ otherwise). If $t_i$ occurs completely in $x$ i.e. $s_0 t_0 \ldots t_i \leqq x$ then the definition of the left form is contradicted as either $t_i$ or $v_i$ must occur first. If $t_i$ occurs completely in $y$, i.e. $x \leqq s_0 t_0 \ldots s_i$ then similarly the definition of the left form is contradicted. Therefore $t_i$ must overlap the boundary between $x$ and $y$, that is $s_0 t_0 \ldots s_i < x < s_0 t_0 \ldots s_i t_i$. Clearly $t_i \mathrel{\bar{\frown}} y$ therefore either $t_i \mathrel{\bar{\frown}} p_0$ or $p_0 \propto t_i$. As $t_i$ and $p_0$ are elements of $C$, either possibility contradicts the assumptions of the lemma. Therefore $x^d y^d = (xy)^d$.

One can verify that $\{221^i 21 | 1 \leqq i \leqq n\}$ is an $n$-element Thue code.

**Lemma II.** *Given an equation $\Sigma(\vec{x})$ one can construct a system of equations $\Sigma'(\vec{x}, \vec{y}, z)$, $\Sigma'$ consisting of a conjunction of equations of the form $uv = w$, where $v$, $w$ are variables and $u$ is either a variable or $u \in cof(\Sigma)$, such that $\Sigma(\vec{x})$ is equivalent to $(\exists \vec{y}, z)[\Sigma'(\vec{x}, \vec{y}, z)]$. Further the variables $\vec{y}$, $z$ either have an original variable $\vec{x}$, or an element of $cof(\Sigma)$ as a prefix.*

*Proof.* Let $\Sigma(\vec{x})$ be $\alpha(\vec{x}) = \beta(\vec{x})$, $cof(\Sigma) = \{c_1, \ldots, c_n\}$. Let $\alpha$ be written as $\alpha = u_0 \ldots u_m$ where either $u_i \in cof(\Sigma)$ or $u_i$ is a variable and if $u_i \in cof(\Sigma)$ then $u_{i+1}$ must be a variable. Introduce the variables $z, y_0, \ldots, y_{m-2}$ and form the equations $u_{m-1} u_m = y_{m-2}$, $u_{m-2} y_{m-2} = y_{m-3}, \ldots, u_1 y_1 = y_0$, and $u_0 y_0 = z$. By construction $\alpha = z$ and each $y_i$ is a tail segment of $z$. Similarly decompose the term $\beta$ reusing the variable $z$. $\Sigma'$ is the conjunction of these equations. It is clear from the form of these equations that a solution $\vec{x}$ to $\Sigma$ uniquely determines values for $\vec{y}$, and $z$, and thus $\vec{x}, \vec{y}, z$ solves $\Sigma'$. Conversely if $\vec{x}, \vec{y}, z$ solves $\Sigma'$ then $\vec{x}$ solves $\Sigma$. The last claim of the lemma is also clear from the form of the equations.

For convenience we now rename all variables, introducing $n$ new variables to be identified with the coefficients and present $\Sigma'$ in the form

$$\Sigma'': \begin{array}{ll} c_i = y_i & 1 \leqq i \leqq n \\ y_i y_j = y_k & ijRk \end{array},$$

where the relation $R$ is determined by the renaming.

**Lemma III.** *Let C be a Thue code, Σ an equation, and* $cof(\Sigma) \subseteqq C^+$. *Then* $\Sigma(\vec{x})$ *and* $\vec{x} \in C\hat{}I^* \cup \{e\}$ *implies* $\Sigma(\vec{x}^d)$.

*Proof.* From $\Sigma(\vec{x})$ derive the system $\Sigma''(\vec{y})$ as above. Since $\vec{x}$ is a solution to $\Sigma$ it determines a solution $\vec{y}$ to $\Sigma''$. Since $\vec{x} \in C\hat{}I^* \cup \{e\}$, $\vec{y} \in C\hat{}I^* \cup \{e\}$. Therefore $\vec{y}^d$ is a solution to $\Sigma''$, since by lemma I, $y_i^d y_j^d = (y_i y_j)^d = y_k^d$ for $ijRk$ and $y_i^d = c_i^d = c_i$ for $1 \leqq i \leqq n$. From $\vec{y}^d$ we can recover a solution to $\Sigma$ by separating the original variables from the introduced variables. This solution is $\vec{x}^d$.

We conclude with the following theorem.

**Theorem II.** *Let Σ be an equation in the n letter alphabet J. Let* $C \subseteqq I^+$ *be an n element Thue code, d the left deletion function relative to C, and f an isomorphism between J and C. Then an equation Σ′ (with additional variables) over I can be constructed such that:*

(1) $\Sigma(\vec{x}) \Rightarrow (\exists \vec{y})[\Sigma'(\vec{x}^f, \vec{y})]$
(2) $\Sigma'(\vec{x}, \vec{y}) \Rightarrow \Sigma(\vec{x}^{df^{-1}})$.

*Proof.* Let $\Sigma^f$ be $\Sigma$ with all letters of $J$ replaced by the corresponding members of $C$. If $\vec{x}$ is a solution to $\Sigma$ then $\vec{x}^f$ is a solution to $\Sigma^f$. Conversely, since $C$ is a code, if $\vec{x} \in C^*$ is a solution to $\Sigma^f$, then $\vec{x}^{f^{-1}}$ is a solution to $\Sigma$.

By theorem I we can find an equation $\Sigma'(\vec{x}, \vec{y})$ such that $(\exists \vec{y})[\Sigma'(\vec{x}, \vec{y})]$ is equivalent to

$$\Sigma^f \wedge \bigwedge_i (c_0 \leqq x_i \vee \ldots \vee c_n \leqq x_i \vee x_i = e),$$

the new variables $\vec{y}$ being introduced in the definitions of $\leqq$ and disjunction. If $\vec{x}$ solves $\Sigma$ then $\vec{x}^f$ solves $\Sigma^f$ and since $\vec{x}^f \in C^*$ there exist values for $\vec{y}$ satisfying the conjoined conditions. Thus $(\vec{x}^f, \vec{y})$ solves $\Sigma'$. This proves (1). For (2), if $(\vec{x}, \vec{y})$ solves $\Sigma'$ then $\vec{x}$ solves $\Sigma^f$ and $\vec{x} \in C\hat{}I^* \cup \{e\}$, by the conjoined conditions. Therefore by Lemma III, $\vec{x}^d$ solves $\Sigma^f$. Since $\vec{x}^d \in C^*$, $\vec{x}^{df^{-1}}$ solves $\Sigma$.
There may be extraneous solutions to $\Sigma'$ in the sense that $\vec{x} \neq \vec{z}$ may both solve $\Sigma'$, yet $\vec{x}^d = \vec{z}^d$. Clearly if one could existentially define membership in the submonoid $C^*$, one could adjoin to $\Sigma^f$ this definition instead of the prefix requirements we have used and thus obtain an exact correspondence between solutions. That is, we would have

$$\Sigma(\vec{x}) \Rightarrow (\exists \vec{y})[\Sigma'(\vec{x}^f, \vec{y})]$$
$$\Sigma'(\vec{x}, \vec{y}) \Rightarrow \Sigma(\vec{x}^{f^{-1}}).$$

Conversely if we had this correspondence then for $\Sigma(x): x = x$, $\Sigma'$ would define membership in the submonoid.

REFERENCES

[1] Hmelevskii, Ju.I.: Equations in Free Semigroups. Trudy Mat. Inst. Steklov. Vol. 107 (1971)
[2] Makanin, G.S.: The Problem of Solvability of Equations in a Free Semigroup. Math. USSR Sbornik Vol. 32 Number 2 (1977) pg. 129–198
[3] Makanin, G.S.: The Problem of Solvability of Equations in a Free Semigroup. Soviet Math. Dokl. Vol. 18 Number 2 (1977) pg. 330–334
[4] Senger, S.: The Existential Theory of Concatenation. Ph. D. Dissertation, Purdue University (1982)
[5] Taimanov, A.D., Hmelevskii, Ju.I.: Decidability of the Universal Theory of a Free Semigroup. (Russian) Sibirsk. Mat. Z. 21 (1980), no. 1, pg. 228–230
[6] Thue, A.: Über unendliche Zeichenreihen. Videnskabs-Selskabets Skrifter. Math.-Naturw. Klasse. No. 6 (1906)

J. Richard Büchi†
Purdue University
W. Lafayette, Indiana 47906
USA

Steven Senger
Univ. of Wisconsin LaCrosse
LaCrosse, Wisconsin 54601
USA

*Zeitschr. f. math. Logik und Grundlagen d. Math.*
*Bd. 34, S. 337–342 (1988)*

# DEFINABILITY IN THE EXISTENTIAL THEORY OF CONCATENATION AND UNDECIDABLE EXTENSIONS OF THIS THEORY

by J. RICHARD BÜCHI† and STEVEN SENGER in LaCrosse, Wisconsin (U.S.A.)[1])

We study extensions of the decidable existential theory of concatenation over words. We show that two equal-length predicates are equivalent and are not existentially definable by equations in concatenation. By adding a further length predicate we get an undecidable existential theory. Whether the existential theory of concatenation and equal length is decidable remains unknown.

In [4] MAKANIN gave a decision procedure for the solvability of equations in the theory of concatenation. From this decision procedure we obtain several definability results. Specifically MAKANIN uses a refinement of a lemma originally due to BULITKO [2], which gives a bound on a measure of periodicity of a minimal solution to an equation. We will analyze this lemma and find that it gives information about the form of the solutions and we will prove that there is no equation in concatenation defining the predicate "equal length" and consequently (since the conjunction, disjunction and negation of equations can be replaced by a single equation) there is no existential definition of "equal length". We further show that if stronger length predicates are added, the existential theory becomes undecidable.

Let $I$ stand for a finite alphabet containing at least the two letters 1 and 2. Let $e$ stand for the empty word. Lower case roman letters will stand for words over the alphabet. Terms in concatenation using the variables $\vec{x}$ will be denoted by $\alpha(\vec{x})$ and $\beta(\vec{x})$. $\Sigma$ will denote an equation between terms. The *prefix, suffix* and *occurrence* relations are defined by

$$x \leqq y\colon (\exists z)\,[x^\frown z = y], \quad y \geqq x\colon (\exists z)\,[y = z^\frown x] \text{ and } x \sqsubseteq y\colon (\exists u, v)\,[u^\frown x^\frown v = y].$$

The strict prefix and suffix relations are denoted by $x < y$ and $y > x$. If $x \sqsubseteq y$ then the pair of words $\langle u, v\rangle$ such that $u^\frown x^\frown v = y$ are called the *flankers* of the occurrence. Note that the prefix (suffix) notation is not symmetric; $x \leqq y$ means that $x$ is a prefix of $y$, while $y \geqq x$ means that $x$ is a suffix of $y$. A word $a$, which is not a proper power of some word, is called an *atom*. Atomic words have the following properties: If $a$ is an atom, then $a \sqsubseteq a^n$ with flankers $\langle u, v\rangle$ implies that $u, v \in a^*$: If $au = ua$, then $u \in a^*$. MAKANIN calls an occurrence of $a$ in $x$ with flankers $ua$ and $av$ a *stable* occurrence of $a$ in $x$.

We say that a set $S$ is *existentially definable from concatenation* if there is an equation $\Sigma$ such that $x \in S \equiv (\exists \vec{y})\,[\Sigma(x, \vec{y})]$. Since the conjunction, disjunction and negation of equations can be replaced by a single equation [7] this definition in fact allows definition by an arbitrary existential formula. We use the notation $S \mathsf{E}^\frown$ to indicate that $S$ is existentially definable from concatenation.

[1]) The results in this paper are contained in the Ph. d. thesis [6] of the second author written under the direction of the first author. Professor BÜCHI died in April 1984.

For any atom $a$ and word $w$ we define a *presentation* of $w$ relative to $a$ using a sequence of nonempty words $Sp_a(w) = \langle u_0, \ldots, u_{n+1}\rangle$ called the *spacers* of $w$ relative to $a$, and a sequence of positive integers $Pw_a(w) = \langle p_0, \ldots, p_n\rangle$ called the *stable powers* of $w$ relative to $a$, satisfying $w = u_0 a^{p_0} u_1 a^{p_1} \ldots u_n a^{p_n} u_{n+1}$, where

1) $a < u_i$ for $1 \leqq i \leqq n$, $u_i > a$ for $1 \leqq i \leqq n$, $a \leqq u_{n+1}$, $u_0 \geqq a$,

2) $\sim(a^2 \leqq u_i)$ for $1 \leqq i \leqq n+1$, $\sim(u_i \geqq a^2)$ for $0 \leqq i \leqq n$,

3) $\sim(a^3 \sqsubseteq u_i)$ for $0 \leqq i \leqq n+1$.

The degenerate case is when there are no stable occurrences of $a$ in $w$, in which case the stable powers are empty and the spacers consist only of the word $w$ which does not contain an occurrence of $a^3$. The maximum $Pw_a(w)$ is called the *exponent of periodicity* of the word $w$ relative to the atom $a$, denoted by $Exp_a(w)$.

Lemma 1. *The presentation of a word $w$ relative to an atom $a$ is unique.*

Proof. Suppose that $\langle u_0, \ldots, u_{n+1}\rangle$, $\langle p_0, \ldots, p_n\rangle$ and $\langle v_0, \ldots, v_{m+1}\rangle$, $\langle q_0, \ldots, q_m\rangle$ both determine nondegenerate presentations of the word $w$. That is

$$w = u_0 a^{p_0} u_1 a^{p_1} \ldots u_n a^{p_n} u_{n+1} = v_0 a^{q_0} v_1 a^{q_1} \ldots v_m a^{q_m} v_{m+1}.$$

Let $k$ be the first subscript where the two presentations disagree. That is $u_i = v_i$ and $p_i = q_i$ for $i < k$. We must have that $k < n+1$ and $k < m+1$. If $u_k = v_k$ and $p_k \neq q_k$, then either $u_{k+1}$ or $v_{k+1}$ has $a^2$ as a prefix in contradiction of condition 2. Therefore $u_k \neq v_k$. Consequently, we may suppose that $u_k \leqq v_k$. Then since $p_k, q_k \geqq 1$ and $a \leqq u_{k+1}, v_{k+1}$ we have $u_k a^2 \leqq v_k a^2$. By condition 1, $u_k = ua$, $v_k = va$ for some words $u$ and $v$. By condition 3, $va < ua^3$ so $v < ua^2$. Let $w \neq e$ be such that $vw = ua^2$, then $ua^3 = vwa \leqq va^3$. This means that $w = a$ or $w = a^2$. If $w = a$, then $v = ua$ so $v_k = ua^2$ which contradicts condition 2. Therefore $w = a^2$ which means that $u_k = v_k$. □

The lemma used in MAKANIN's decision procedure makes a connection between a solution to an equation in concatenation and a constructed system of diophantine equations. We state this lemma as follows.

Lemma 2 (MAKANIN [4]). *Given an equation $\Sigma$: $\alpha(\vec{x}) = \beta(\vec{x})$, $l$ symbols in length, an atom $a$, and for each variable $x_i$ in $\Sigma$ spacers $\vec{b}_i$ satisfying the required properties with respect to the given atom $a$, we can construct,* i) *a system of linear equations $\Gamma$ with variables ranging over the positive integers satisfying $p \leqq l^2$, $n \leqq 2l^2$, $m \leqq 3l$ (where $p$, $n$, $m$ are respectively, the number of equations in the system, the number of variables in the system, and the maximum absolute value of the coefficients), and* ii) *a list $E$: $c_i = d_i$ of equalities between words over the alphabet, such that the following hold:*

A) *If $\vec{x}$ is a solution to $\Sigma$ with $Sp_a(x_i) = \vec{b}_i$ for each component of the solution, then $Pw_a(x_i)$ for every component form a solution to $\Gamma$ and the list $E$ of word equalities are true.*

B) *If the equalities $E$ are true and $\vec{z}$ is a solution to $\Gamma$, then defining $\vec{x}$ to be the words such that $Sp_a(x_i) = \vec{b}_i$ and $Pw_a(\vec{x}) = \vec{z}$ yields a solution to $\Sigma$.*

Proof. We present an outline of the proof. Introducing the variables $\vec{z}$ (ranging over the positive integers) and using the spacers $\langle b_0, \ldots, b_{k+1}\rangle$ we write each variable $x$ of $\Sigma$ as

(i) $\quad x = b_0 a^{z_0} b_1 a^{z_1} \ldots b_k a^{z_k} b_{k+1}.$

That is each variable $x$ is now determined by the parameters $\vec{z}$. We take the parameterized words of (i) and substitute them into the equation $\Sigma$. We thus obtain $\alpha$ and $\beta$ as words parameterized by the variables $\vec{z}$. A stable occurrence of the atom $a$ in $\alpha$ and $\beta$ must be a) an original occurrence $a^z$, or b) an occurrence contained in the constants of $\Sigma$, or c) a new occurrence formed by the concatenation of variables or constants in $\Sigma$. If $c$ and $d$ have no stable occurrences of $a$, then $c^\frown d$ has at most three stable occurrences of $a$. Since there are only $l$ symbols in $\Sigma$ there can be at most $3l$ occurrences of variety c). Consequently we can uniquely present $\alpha$ and $\beta$ relative to $a$ with $Pw_a(\alpha)$ and $Pw_a(\beta)$ consisting of the variables $\vec{z}$ and constants arising from occurrences of types b) and c). Let $Pw_a(\alpha) = \vec{L}$, $Sp_a(\alpha) = \vec{c}$ and $Pw_a(\beta) = \vec{M}$, $Sp_a(\beta) = \vec{d}$. Both of $\vec{L}$ and $\vec{M}$ are linear terms consisting of the variables $\vec{z}$ and constants. The equalities $c_i = d_i$ form the list $E$. The system $\Gamma$ consists of the equations $L_i = M_i$, where we eliminate all equations of the form $z_i = z_j$, or the pair of equations $L = z_i$ and $L = z_j$, by identifying variables. If a term contains more than one variable, then it must come from a point of concatenation in the equation $\Sigma$. Consequently there are no more than $l$ terms which contain more than one variable $z$. Further there are no more than $l$ terms which consist only of a constant. Consequently the number of equations is bounded by $l^2$, the number of variables is bounded by $2l^2$ and the maximum absolute value of the coefficients is bounded by $3l$.

A) If $\vec{x}$ is a solution to $\Sigma$ such that $Sp_a(x_i) = \vec{b}_i$, then, since the presentation of a word relative the atom $a$ is unique, $Sp_a(\alpha(\vec{x})) = Sp_a(\beta(\vec{x}))$, which means the list $E$ of equalities $c_i = d_i$ are true. Also $Pw_a(\alpha(\vec{x})) = Pw_a(\beta(\vec{x}))$. Since the values of the variables in $L_i$ and $M_i$ are given by $Pw_a(\vec{x})$ we have $L_i = M_i$. Therefore $Pw_a(\vec{x})$ provide a solution to $\Gamma$.

B) If $\vec{z}$ is a solution to $\Gamma$ and the equalities of the list $E$ are true, then the words defined by equation (i) form a solution to $\Sigma$. □

The *index* of a system of equations $\Gamma$ of the above form, $I(\Gamma)$, is $q + (p - 1)\,n$, where $q$ is the number of non-zero coefficients in the first row and $p$ and $n$ are as before. We order vectors of positive integers by defining $\vec{x} < \vec{y}$ to mean that $x_i \leqq y_i$ for all $i$, and $x_i < y_i$ for some $i$. MAKANIN uses the following lemma to provide a bound on the size of the minimal solution to a solvable system $\Gamma$ where $i$ is the index and $m$ is as before. The bounding function is $B(i, m) = m(2m)^{(2^i-2)}$.

Lemma 3 (MAKANIN [4]). *If a system $\Gamma$: $A_{p,n}z_n = B_p$ has a solution $\vec{z}$ such that* $\max(\vec{z}) > B(I(\Gamma), m)$, *then there is a solution $\vec{z}' < \vec{z}$ of $\Gamma$.*

Lemmas 2 and 3 provide us with the following bound on the exponent of periodicity of a solution to an equation in terms of the length of the equation. The bounding function is

$$B(l) = 6l^{2^{2l^4}}.$$

We collect together several consequences of this bound in the following lemma.

Lemma 4. *If $\vec{x}$ is a solution to $\Sigma$, and $a$ is an atom such that $Exp_a(\vec{x}) > B(l)$, where $l$ is the length in symbols of $\Sigma$, then we have:*

A) *There is a solution $\vec{y}$ to $\Sigma$ such that $Sp_a(\vec{y}) = Sp_a(\vec{x})$ and $Pw_a(\vec{y}) < Pw_a(\vec{x})$.*

B) *There is a solution $\vec{y}$ to $\Sigma$ such that $Exp_a(\vec{y}) \leqq B(l)$ and $Sp_a(\vec{y}) = Sp_a(\vec{x})$.*

C) *$\Sigma$ has an infinite number of solutions with the same spacers as $\vec{x}$.*

22*

Proof. If $\vec{x}$ is a solution to $\Sigma$ such that $Exp_a(\vec{x}) > B(l)$, then $Pw_a(\vec{x})$ are a solution to the corresponding diophantine system $\Gamma$ which exceeds the bound $B(I(\Gamma), m)$. By lemma 3 we can find a smaller solution to the system $\Gamma$ which by lemma 2 yields a smaller solution $\vec{y}$ to $\Sigma$. Furthermore this solution uses the same spacers as $\vec{x}$. By repeating this process we can eventually obtain a solution $\vec{z}$ such that $Exp_a(\vec{z}) \leqq B(l)$. Finally, as the system $\Gamma$ has a strictly ordered pair of solutions, it has infinitely many. □

We now give three corollaries on definability. The first is a direct consequence of part B).

Corollary 1. *No infinite set $S$, where $(\forall w \in S)\,[Exp_a(w) > B(l)]$, is definable by an equation in concatenation of length less than $l$.*

The second is a consequence of part C).

Corollary 2. *If $S$ is a set such that for some atom $a$ there is a $w \in S$ such that $Exp_a(w) < B(l)$ and there are only finitely many other elements of $S$ with the same spacers relative to $a$ as $w$, then $S$ is not existentially definable in concatenation.*

As a consequence of this we have:

Corollary 3. *The set $\{a^n b^n \mid n = 0, 1, \ldots\}$ is not existentially definable in concatenation, where $a$ and $b$ are atoms such that $\sim(a \leqq b)$ and $\sim(b \leqq a)$. In particular $\{1^n 2^n \mid n = 0, 1, \ldots\}$ is not existentially definable.*

Proof. Suppose $S$ were definable by an equation $\Sigma$. Let $l$ be the length of $\Sigma$, and let $i > B(l)$. We then have that $a^i b^i$ is a solution to $\Sigma$, such that $Exp_a(a^i b^i) > B(l)$. The spacers of $a^i b^i$ relative to the atom $a$ are $\langle a, ab^i \rangle$, and this is the only word in $S$ with these spacers. This set is therefore undefinable by the previous corollary. □

The condition on the atoms $a$ and $b$ can be relaxed in several ways.

Define $Elg(x, y)$ to mean that the words $x$ and $y$ have the same length. Define $Lg(x)$ to be that element of $1^*$ having the same length as $x$, or equivalently, $x$ with all occurrences of 2 replaced by 1. These two notions of length are equivalent since

$$Elg(x, y)\colon Lg(x) = Lg(y) \quad \text{and} \quad Lg(x) = y\colon 1y = y1 \wedge Elg(x, y).$$

Theorem 1. *The predicate Elg is not definable in the existential theory of concatenation.*

Proof. From $Elg$ we can define the set $\{1^n 2^n\}$ by

$$w \in \{1^n 2^n\}\colon (\exists u, v)\,[1u = u1 \wedge 2v = v2 \wedge Elg(u, v) \wedge uv = w].$$

Therefore $Elg$ is not existentially definable in concatenation. □

We now consider several extensions to the theory of concatenation by introducing various versions of length. Define $Lg_1(x)$ to be $x$ with all occurrences of 2 removed. Define $Lg_2(x)$ to be $x$ with all occurrences of 1 removed. Define $Sp(x) = Lg_1(x)^\frown Lg_2(x)$. Clearly $Sp \mathrel{\mathsf{E}} Lg_1, Lg_2$. We also have $Lg_1 \mathrel{\mathsf{E}} Sp$ since

$$Lg_1(x) = y \equiv 1y = y1 \wedge (\exists z)\,[2z = z2 \wedge yz = Sp(x)],$$

and $Lg_1 \mathrel{\mathsf{E}} Lg, Lg_2$ since

$$Lg_1(x) = y \equiv 1y = y1 \wedge (\exists u, v)\,[Lg_2(x) = u \wedge Lg(u) = v \wedge Lg(x) = yv].$$

Similarly we define $Lg_2$ from $Sp$ or $Lg, Lg_1$. Finally $Lg \mathrel{\mathsf{E}} Lg_1, Lg_2$ since

$$Lg(x) = y \equiv (\exists u, v, w, z)\,[Lg_1(x) = u \wedge Lg_2(x) = v \wedge 12w = w12 \wedge Lg_2(w) = v \wedge Lg_1(w) = z \wedge y = uz].$$

Any pair of these functions or $Sp$ alone allows the matching of the number of occurrences of certain subsequences and this ability allows us to code arithmetic into an extension of the theory of concatenation.

Theorem 2. *In $\langle I^*, e, {}^\frown, Sp\rangle$ we can existentially define the operations of addition and multiplication on an existentially definable subset of $I^*$.*

Proof. As the natural numbers we take the set of words $N = \{e\} \cup \{1^*2\}$. These are essentially the tally sequences except that we use atoms. The length of a word is the number that it represents. Membership in this set is defined by

$$n \in N\colon n = e \vee (\exists x)\,[1x = x1 \wedge n = x2].$$

For elements $x$, $y$ and $z$ in this set we define addition and multiplication by

$$x + y = z\colon Lg(xy) = Lg(z) \vee (y = e \wedge x = z),$$

$$x \cdot y = z\colon \quad ((x \neq e) \wedge (\exists u)\,[xu = ux \wedge Lg(Lg_2(u)) = Lg(y) \wedge Lg(u) = Lg(z)]) \vee (x = e \wedge z = e).$$

The definition for addition is clear. For multiplication, if $x$ is not empty, the first condition means that $u$ is a repetition of $x$. The second condition means that $u$ consists of $Lg(y)$ copies of $x$. Finally, the last condition ensures that $z$ has the same length as $u$. □

Corollary 4. *Solvability of equations in the theory $\langle I^*, e, {}^\frown, Sp\rangle$ is undecidable.*

A direct proof of this undecidability, not resorting to arithmetic, would undoubtedly yield interesting information concerning the theory of words. In $\langle I^*, e, {}^\frown, Lg\rangle$ one can define addition and the predicate "divides". These give a decidable existential theory [1, 3] over the natural numbers, but the authors do not know whether the existential theory of concatenation and length over words is decidable.

The words over $\{1, 2\}$ can be viewed as binary representations of natural numbers if we assign to the word $w = w_l \ldots w_1 w_0$, where the $w_i$ are single letters, the number $w_l 2^l + \ldots + w_1 2 + w_0$. This close connection between the binary tree and arithmetic suggests that instead of considering extensions to the theory of concatenation we can consider fragments of arithmetic in which concatenation and length are definable. In particular we consider the structure $\langle \mathsf{N}, 0, 1, +, P\rangle$ where $P$ is a fragment of multiplication and exponentiation defined by: $P(x, y) = x2^y$. Using this product we can define:

$$m \in Pow_2 \equiv (\exists n)\,[m = P(1, n)], \quad \text{the set of powers of } 2,$$

$$Lg(m) = n \equiv P(1, n) \leqq m + 1 < P(1, n) + P(1, n), \quad \text{the length of } m.$$

In this context $Lg(m)$ is in fact the length of $m$ and not just the element of $1^*$ with the same length. We can define concatenation by $m^\frown n = P(m, Lg(n)) + n$.

We will need the following number theoretic result due to Julia Robinson.

Lemma 5. *If $m \leqq n$, $l \gg n$ and $l + m, l - m \mid l^2 - n$, then $m^2 = n$.* (It suffices to let $l \gg n$ mean that $l > 2n^2$.)

Lemma 6. *In* $\langle \mathsf{N}, 0, 1, +, P\rangle$ *we can existentially define the operations of addition and multiplication on* $\mathsf{N}$.

Proof. We define addition and multiplication on the positive powers of 2, which correspond to $\{1^n 2 \mid n \geqq 0\}$ (cf. proof of theorem 2). For $m$, a power of 2, addition is defined by

$$m \oplus n = 2^{Lg(m)+Lg(n)} = P(m, Lg(n)).$$

This corresponds to the definition of addition on the set $\{1^*2\}$. Similarly $m \cdot n$ should be $2^{Lg(m)Lg(n)}$. Since multiplication can be obtained from addition and squaring we need only define $2^{Lg(m)Lg(m)}$. If $Lg(m)$ is itself a power of 2, we can define

$$2^{Lg(m)Lg(m)} = 2^{P(Lg(m), Lg(Lg(m)))} = P(1, P(Lg(m), Lg(Lg(m)))).$$

Since concatenation and $Lg$ are definable, we can define the predicate "divides" on the powers of 2 by $m \mid n \equiv (\exists p)\, [m^\frown p = p^\frown m \wedge Lg(n) = Lg(p)]$. Lemma 5 now provides us with full squaring.

$$\begin{aligned} m^2 = n \equiv (\exists l, l', p, q, r)\, [m &\leqq n \wedge l' \in Pow_2 \wedge l = P(1, l') \wedge l \gg n \\ &\wedge p = l \oplus m \wedge q \oplus m = l \\ &\wedge r \oplus n = P(1, P(Lg(l), Lg(Lg(l)))) \wedge p \mid r \wedge q \mid r]. \end{aligned}$$

We can define $l \gg n$ by $l > P(2, P(n, n))$. This definition consists of the conditions of lemma 5 except that we require $l$ to be a power of a power of 2 in order to calculate $l^2$. Since $Lg(m)$ is the actual length of $m$, we can transfer these definitions from the powers of 2 to addition and multiplication on the natural numbers. □

Corollary 5. *The existential theory of* $\langle \mathsf{N}, 0, 1, +, P\rangle$ *is undecidable.*

## References

[1] Bel'tjukov, A. D., Decidability of the universal theory of the natural numbers with addition and divisibility. Zap. Nauchn. Sem. Leningrad. Odtel. Mat. Inst. Steklov (LOMI) **60** (1976), 15–28.

[2] Bulitko, V. K., Equations and inequalities in a free group and a free semigroup. Tul. Gos. Ped. Inst. Uchen. Zap. Mat. Kafedr Vyp. 2 Geometr. i Algebra (1970), 242–252.

[3] Lipshitz, L., The Diophantine problem for addition and divisibility. Trans. Amer. Math. Soc. **235** (1978), 271–283.

[4] Makanin, G. S., The problem of solvability of equations in a free semigroup. Math. USSR Sbornik **32** (1977), 129.

[5] Makanin, G. S., The problem of solvability of equations in a free semigroup. Soviet Math. Dokl. **18** (1977), 330.

[6] Senger, S., The Existential Theory of Concatenation. Ph. D. Dissertation, Purdue University, 1982.

[7] Senger, S., Coding in the existential theory of concatenation. Arch. Math. Logik Grundlag. **26** (1986/87), 101–106.

S. Senger
Departments of Computer Science
and of Mathematics
University of Wisconsin–LaCrosse
LaCrosse, Wisconsin 54601
U.S.A.

(Eingegangen am 9. Juni 1987)

# Section 8 Quadratic Forms, the Five Squares Problem, and Diophantine Equations

*With comments by Leonard Lipshitz, Purdue University*

No publications.

*

# Quadratic Forms, the Five Squares Problem, and Diophantine Equations

*Leonard Lipshitz**

Richard Büchi had an abiding interest in problems related to Hilbert's tenth problem, both before and after its solution. A variant that attracted his attention in the middle 1970s is the following:

1. Is there an algorithm for deciding the truth of formulas of the form

$$(*) \qquad \exists x_1 \ldots \exists x_{m \in \mathbb{N}} \left(\bigwedge_{i=1}^{N} \sum_{j=1}^{m} a_{ij} x_j^2 = b_i\right)$$

where the $a_{ij}$, $b_j \in \mathbb{Z}$?

A negative answer to 1 of course immediately implies the negative solution of Hilbert's tenth problem. The reverse implication seems far less clear. Problem 1 is equivalent to the following.

2. Is the positive existential theory of $\mathbb{N}$ in the language $\mathscr{L} = \{0, 1, +, <, Sq\}$ decidable, where $Sq(x) \leftrightarrow x$ is a square?

*Purdue University, West Lafayette, Indiana.
*Note*: I have drawn heavily from reference [7] below.

Every formula of the form (*) can easily be rewritten as a positive existential formula in $\mathscr{L}$. The other direction is also easy, using Lagrange's theorem to replace variables over $\mathbb{N}$ by sums of four squares of new variables.

The natural approach to a negative answer to 1 is to try to show how to associate with every polynomial $P(y_1, \ldots, y_k)$ over $\mathbb{Z}$ a positive existential formula $\varphi$ in $\mathscr{L}$ such that

$$\exists y_1, \ldots \exists y_{k \in \mathbb{N}}(P(y_1, \ldots, y_k) = 0) \leftrightarrow \varphi.$$

It is well known that every formula of the form $\exists y_1, \ldots \exists y_{k \in \mathbb{N}} P(y_1, \ldots, y_k) = 0$ is equivalent to a formula of the form $\exists z_1 \ldots \exists z_{l \in \mathbb{N}}(\bigwedge_{i=1}^{N} \varphi_i(z))$ where each $\varphi_i$ is of one the forms $z_i + z_j = z_k$, $z_i + j = z_k$ or $z_i^2 = z_j$. (Note that $xy = x \leftrightarrow (x + y)^2 = x^2 + y^2 + 2z$). Hence, to carry out this program it would be enough to show how to define the squaring function $y = x^2$ by a positive existential formula of $\mathscr{L}$. Since the difference between $(x + 1)^2$ and $x^2$ is $2x + 1$ we would be able to do this if we could define the set of successive squares by such a formula.

Notice that if $x_i = x_0 + i$ for $i = 1, \ldots, n - 1$, then the $n - 2$ second differences $x_{i+2}^2 - 2x_{i+1}^2 + x_i^2 = 2$ for $i = 0, \ldots, n - 3$. This led Büchi to ask the following.

*The n-Square Problem.* Is there an $n(\geq 5)$ such that whenever $x_0, \ldots, x_{n-1} \in \mathbb{N}$ satisfy $x_0^2 < x_1^2 < \cdots < x_{n-1}^2$ and $x_{i+2}^2 - 2x_{i+1}^2 + x_i^2 = 2$ for $i = 0, \ldots, n - 3$ then $x_i = x_0 + i$ for $i = 1, \ldots, n - 1$. (For $n = 4$ there are sequences of nonsuccessive squares with second differences 2; see later.)

**Theorem.** *A positive answer to the n-squares problem implies a negative answer to problems 1 and 2.*

Büchi became more interested in the $n$-square problem (especially for $n = 5$), than in problems 1 and 2. From time to time, he mentioned partial results to colleagues and students, but left nothing resembling a write-up of what he knew. I presented the problem at a Western Number Theory Conference (Asilomar, Dec. 1985), Problem 85:13. [1] and [2] contain results connected with his problem. [2] in particular considers a problem closely related to the five-squares problem. Duncan Buell [3] has partial results on the five-square problem, and a complete characterization of the sequences of four squares with second differences 2. Douglas Hensley has an unpublished manuscript with results related to the five-squares problem dating back to about 1978, when I mentioned the problem to him.

Büchi's interest in the five-squares problem soon expanded into an interest in the theory of (binary) quadratic forms. Beginning in the middle 1970s he reworked the classical theory of Gauss, Lagrange, and Legendre [4] in his own way.

A fundamental notion in the theory of binary quadratic forms is that of equivalence. The forms $ax^2 + bxy + cy^2$ and $AX^2 + BXY + CY^2$ are equivalent if one can be transformed into the other by a change of variables of the form $X = px + qy$, $Y = rx + sy$ for some integer matrix $\begin{pmatrix} p & q \\ r & s \end{pmatrix}$ with $\begin{vmatrix} p & q \\ r & s \end{vmatrix} = ps - qr = 1$. The group of these unimodular matrices plays a central role in the theory. Gauss uses the generators $L = \begin{pmatrix} 0 & 1 \\ -1 & 0 \end{pmatrix}$ and $\begin{pmatrix} 1 & 1 \\ 0 & 1 \end{pmatrix}$. Büchi preferred to work with the generators $1 = \begin{pmatrix} 1 & 0 \\ 1 & 1 \end{pmatrix}$ and $2 = \begin{pmatrix} 1 & 1 \\ 0 & 1 \end{pmatrix}$. Their inverses are $\dot{1} = \begin{pmatrix} 1 & 0 \\ -1 & 1 \end{pmatrix}$ and $\dot{2} = \begin{pmatrix} 1 & -1 \\ 0 & 1 \end{pmatrix}$. He let $\mathcal{U}$ be the Cayley graph of this group with respect to these generators. The vertices of the graph are the unimodular matrices and each matrix $A$ has the four neighbors $A1$, $A2$, $A\dot{1}$, and $A\dot{2}$ via directed edges labeled 1, 2, $\dot{1}$, and $\dot{2}$, respectively. The group is not free, as $L = 2\dot{1}2$ satisfies $L^4 = e = \begin{pmatrix} 1 & 0 \\ 0 & 1 \end{pmatrix}$. However the submonoid generated by 1 and 2 is free and is exactly the set of unimodular matrices with nonnegative entries. (The free monoid on two generators occurred repeatedly in Büchi's work.) This freeness is an advantage of Büchi's generators over the traditional ones.

A closely related structure is the structure that Büchi called NIC or NIC$(a, b, c)$ (for Nicomachus; see [6]). Starting with a triple $(a, b, c)$ of integers one generates the directed graph of triples, where each triple has four-neighbors via the operations $(a, b, c) \to (a + b + c, b + 2c, c)$; $(a, b, c) \to (a, b + 2a, a + b + c)$; $(a, b, c) \to (a, b - 2a, a - b + c)$ and $(a, b, c) \to (a - b + c, b - 2c, c)$. If we represent the form $ax^2 + bxy + cy^2$ by the triple $F = (a, b, c)$ or equivalently by the matrix $F = \begin{pmatrix} a & b/2 \\ b/2 & c \end{pmatrix}$ then the triples occurring in the graph NIC$(a, b, c)$ are exactly the forms equivalent to the form $(a, b, c)$. In fact NIC$(a, b, c)$ is homomorphic with $\mathcal{U}$ via the mapping $\mathcal{U} \ni A \to A^T F A$. (For example, $1^T(a, b, c)1 = (a + b + c, b + 2c, c)$). A third structure isomorphic to $\mathcal{U}$ and NIC is the graph of all conjugates $\dot{A}XA$ of a given matrix $X$ under the unimodular matrices $A$. A matrix $Z$ in this conjugacy structure has the four neighbors $\dot{1}Z1$, $\dot{2}Z2$, $1Z\dot{1}$, and $2Z\dot{2}$ via conjugation by 1, 2, $\dot{1}$, and $\dot{2}$, respectively. In fact if we map the form $F = (a, b, c)$ into the matrix $X = \begin{pmatrix} \frac{1}{2}(\imath - b) & -c \\ a & \frac{1}{2}(\imath + b) \end{pmatrix}$, where $\imath = 0, 1$ is the residue mod 4 of the discriminant $b^2 - 4ac$ of $F$, then the conjugacy graph is homomorphic to NIC.

Büchi worked out all the usual theorems, and his versions of the usual algorithms, on these structures. He felt strongly that his algorithms were easier and clearer than the classical ones. Among his papers were many, many pages of computations and notes, but no coherent exposition of his treatment. In the fall of 1979 he lectured on his treatment of the theory of binary quadratic forms. Dirk Siefkes and Steve Senger have notes from these lectures.

## *References*


1. Allison, D. (1986). On square values of quadratics. *Math. Proc. Camb. Phil. Soc.*, 99, 381–383.
2. Barbeau, E.J. (1985). Numbers differing from consecutive squares by squares. *Canadian Math Bull.*, 28, 337–342.
3. Buell, D.A. (1987). Integer squares with constant second difference. *Math. Comp.*, 49.
4. Gauss, C.F. (1801). Disquisitiones Arithmeticae. New Haven, Conn.: Yale University Press, 1960.
5. Hensley, D. Sequences of squares with second difference 2 and a problem of logic. Unpublished manuscript.
6. Nicomachus, Arithmetic, Life of Pythagoras, Harvard series.
7. Siefkes, D. (1989). The work of J. Richard Büchi. This volume, pp. 7–17.

# Section 9 Graphs and Matroids

*With comments by William E. Fenton, Bellarmine College and Saunders Mac Lane, The University of Chicago*



*

# Work with Büchi on Graphs and Matroids

*William E. Fenton**

During the late 1960s, Büchi was working on combining linear independence with the notion of orientation, to better understand convex sets. With Mei (dissertation, 1971), he developed what he at first called the paramatroid and later the convexity space; others have independently devised the same structure, now known as the oriented matroid. The dissertation of Fenton (1982) revised the axioms for convexity spaces and explored their implications. The paper [34] presents some of the results of that work, focusing on the convex sets containing at least half the elements of the space.

If graphs are to be models of oriented matroids, an orientation of the edges must be included. Büchi decided that a natural approach was to view each edge as a pair of opposing "half-edges"; each vertex, then, is a set of half-edges. Beginning with this viewpoint, the paper [35] uses Edmonds' [2] idea of rotation and Gustin's [3] idea of circulation to define (directed) circuits as the orbits of certain permutations of the half-edges. Whitney's [4] concepts of

* Bellarmine College, Louisville, Kentucky.

connectedness and nonseparability are rephrased in these terms. The principal result is this: for a graph $G$ with source and sink vertices $s$ and $t$ let $G^+$ be the graph formed from $G$ by adding a half-edge $e$ from $s$ to $t$ and its reverse $-e$ from $t$ to $s$. Then $G$ is series-parallel if and only if in $G^+$ the union of the circuits containing $e$ is disjoint from the union of the circuits containing $-e$ if and only if $G^+$ contains no subgraph homeomorphic from $K_4$. We discovered while writing that this result had already appeared in a paper by Duffin [1].

## *References*


1. Duffin, R. (1965). Topology of series-parallel networks. *J. Math. Anal. Appl.*, 10, 303–318.
2. Edmonds, J.R. (1960). A combinatorial representation for polyhedral surfaces. *A.M.S. Notices*, 7, 646.
3. Gustin, W. (1963). Orientable embeddings of Cayley graphs. *Bulletin A.M.S.*, 69, 272–275.
4. Whitney, H. (1932). Nonseparable and planar graphs. *Trans. A.M.S.*, 34, 339–362.


*

# Büchi's Work on Graphs and Matroids

*Saunders Mac Lane*

Since the comments by William Fenton cover the subject matter of these two papers by Büchi, I take this occasion to speak about the general thrust and scope of Richard's work. In the process of editing this volume, I have become more and more impressed with the depth and the scope of Büchi's activity. The depth of his insights has been set forth, for example, in the description of his work on automata, their background in ideas and problems of logic and their interest for computer science. There is also a considerable development of the range of his work, starting with the first penetrating papers on Boolean algebra. Some of this I knew in advance, but just now I was pleased to find still another of his effective interests: this time in some of the combinatorial problems that arise from geometry; explicitly, for graphs and matroids—subjects with which I had some early experience.

In these topics, much of the original impetus came from Hassler Whitney [5]. After Kuratowski had shown that a nonplanar graph must contain one of two special such (a five-point and a six-point graph), Whitney [5] went on to formulate a much more geometrical condition for planarity: A graph is planar if there is a suitably defined combinatorial dual. This led to my first paper [3] on the question, describing planarity in terms of a suitable basis of circuits—which, with one added circuit, contains each edge exactly twice, just as in the boundaries of regions in the plane.

Whitney also initiated the study of abstract linear independence in his paper on matroids—where he introduced that name. At the time, I added a small paper [2] exhibiting the connection of Hassler's notion with standard ideas of geometry. In my view at the time, the essential idea involved in independence was the Steinitz exchange condition—which I illustrated with a subsequent and more consequential paper [4] using this notion for a lattice that arises for fields of prime characteristic.

More recently, the study of planar graphs has expanded, with somewhat elaborate consideration of lists of non-toral graphs. Matroids have grown far beyond my early expectations, with several texts and a recent source-book, edited by J.P.S. Kung [1]. This is but one example of the way in which mathematical research has proliferated.

In these topics, as elsewhere, Richard Büchi took a vital part.

## *References*

1. Kung, J.P.S. (1986). *A source book in matroid theory.* Boston: Birkhauser Boston, 413 pp.
2. Mac Lane, S. (1936). Some interpretations of abstract linear dependence in terms of projective geometry. *Amer J. Math.*, 58, 236–240.
3. Mac Lane, S. (1937). A combinatorial condition for planar graphs. *Fundamenta Math.*, 28, 22–32.
4. Mac Lane, S. (1938). A lattice formulation for transcendence degrees and p-bases. *Duke Math. J.*, 4, 455–468.
5. Whitney, H. (1932). Non-separable and planar graphs. *Trans. Amer. Math. Soc.*, 34, 339–362.

Reprinted from JOURNAL OF COMBINATORIAL THEORY, Series B
All Rights Reserved by Academic Press, New York and London
Vol. 45, No. 3, December 1988
*Printed in Belgium*

# Large Convex Sets in Oriented Matroids

J. RICHARD BUCHI*

*Department of Mathematics, Purdue University, West Lafayette, Indiana 47907*

AND

WILLIAM E. FENTON

*Department of Mathematics and Computer Science, Bellarmine College, Louisville, Kentucky 40205*

*Communicated by the Managing Editors*

Received July 31, 1985

After a brief discussion of the axiom systems for oriented matroids we consider two basic types of convex sets, arising from the opposition, and establish some fundamental properties. These are used to investigate large convex sets, those containing at least one element from each opposite pair. A characterization is given for large convex sets, three types are examined, and the relationships between these three types are shown. These relationships are particularly strong in a dense space. 

The fundamental fact of linear algebra (geometry without order) is the Basis Theorem: Independent sets which span a space $U$ have the same cardinality, namely the rank of $U$. Steinitz [11] discovered that this is a consequence of an exchange property: If $x$ depends on $V$ and $y$ but not solely on $V$ then $y$ depends on $V$ and $x$. It is easily shown that algebraic closure over a field satisfies his exchange property, so Steinitz has proven the Basis Theorem for transcendentals. This is a nice application of the axiomatic method.

Whitney [12] continued the axiomatic study of the exchange property. He introduced the notion of circuit (minimal dependent set) and found an elegant axiomatization of linearity based on it. This is the matroid. Whitney also applied this notion to graphs, where the circuits are sets of edges forming a minimal closed path. In this situation the Basis Theorem becomes: All spanning trees of a graph have the same cardinality. This is a second success for the axiomatic method.

* Deceased.



0095-8956/88 $3.00

Whitney's axioms for linear dependence govern basic facts in geometry, algebra, and graph theory. However, linear dependence is only the skeleton of geometry, for it ignores the order of points on a line and the orientation of a plane. If these concepts are incorporated we have geometry with order, or convexity, a richer theory. This is the oriented matroid, which combines the dependence with an opposition $x^*$ of each element $x$. For example, the convex version of the exchange property is: If $x^*$ depends on $V$ and $y$ but not solely on $V$ then $y^*$ depends on $V$ and $x$. Further, a linear circuit $X$ is a convex circuit if $x^*$ rather than $x$ depends minimally on $X-x$.

Oriented matroids were developed independently in the 1960s and early 1970s by Novoa [10], Bland [1, 2], Las Vergnas [2, 8], Folkman and Lawrence [5], and by Buchi, whose axiom systems were first worked out in a doctoral dissertation by Mei [9] and were revised and expanded upon by Fenton [4]. The axioms presented here are a mixture of Buchi's and those of Folkman and Lawrence, slightly simplifying both systems. Infinite structures are allowed by a discreteness assumption.

We begin by displaying two equivalent sets of axioms, based on the notions of closure and circuit, and developing some basic properties. The connection to linearity is mentioned to provide the idea of rank and then the Carathéodory Theorem on the number of generators required for convex closure is proven. Section B is concerned with two types of convex set which arise naturally from the opposition operator and explores their relation to convexity, including a proof from the axioms of the Stone separation lemma. Section C focuses on the large convex sets, those containing at least one element from each opposite pair $\{x, x^*\}$. We examine three types of large convex sets, give a characterization of large convex sets, and show how these three types are related in a dense space.

## A. The Axioms and Some Basic Properties

The system of Folkman and Lawrence assumes a finite set. We remove this restriction by a discreteness assumption (Axiom 3).

Definition. An *oriented matroid* is a system $\langle E, *, cx\rangle$ where $E$ is a set with maps $*: E \to E$ and $cx: P(E) \to P(E)$ such that

(1) $x^{**} = x$,

(2) $cx$ is a closure operator,

(3) if $x \in cx(X)$ there exists a finite set $Y \subseteq X$ so that $x \in cx(Y)$,

(4) $cx(X^*) = cx(X)^*$,

(5) $x \in cx(X \cup x^*)$ implies $x \in cx(X)$,

(6) $y \in cx(X \cup x^*)$ and $y \notin cx(X)$ imply $x \in cx(X \cup y^* - x)$.

Note that $x \in cx(x^*)$ implies $x \in cx(\emptyset)$. We denote $cx(\emptyset)=0$; this set includes, but is not limited to, the elements $x = x^*$. The set 0 is often a nuisance, however, so we will work in the reduced space $E-0$ with $cx(X) \cap (E-0) = cx(X) - 0$ as the closure operator. This is not necessary for what follows but unclutters the definitions and proofs.

The following pair of results illustrates the "one-sided" aspect of an oriented matroid.

PROPOSITION 1. *If $y \in cx(U)$ then either $y \in cx(U-x)$ or $y \in cx(U-x^*)$.*

*Proof.* Assume $U$ contains both $x$ and $x^*$ (obvious otherwise). If $y \in U$ then $y \in cx(y)$ and the conclusion holds. If $y \notin cx(U-x^*)$ then by Axiom 6 $x \in cx(U \cup y^* - x)$ and Axiom 5 gives $x \in cx((U-x-x^*) \cup y^*)$. Let $V = U - x - x^*$. If $x \in cx(V)$ then $y \in cx(V \cup x^*) = cx(U-x)$. If $x \notin cx(V)$ then $y \in cx(V \cup x^* - y) = cx(V \cup x^*) \subseteq cx(U-x)$.

PROPOSITION 2. *If $y \in cx(U \cup x)$ and $y \in cx(U \cup x^*)$ then $y \in cx(U)$.*

*Proof.* Assume $y \notin cx(U)$. Then $x^* \in cx(U \cup y^* - x^*) \subseteq cx(U \cup y^*)$, which implies $y \in cx(U \cup y^*)$. By Axiom 5, $y \in cx(U)$.

One model for the oriented matroid is the vector space $V^n(F)$ over an ordered field $F$. There is a natural opposition: $x^* = -x$. The closure operator can be defined by

$$cx(X) = \left\{ \sum_1^m \alpha_i x_i \,\middle|\, \alpha_i \in F^+ \text{ and } x_i \in X \right\},$$

i.e., all positive linear combinations of finitely many vectors from $X$. The closed sets are the convex cones of the vector space (this motivated the notation $cx$). The set 0 consists of the zero vector, so $V^n(F)$ is not reduced.

Vector spaces provide a link between oriented matroids and convexity theory. Consider the oriented matroid of unit vectors from $V^{n+1}(F)$; this can be interpreted as the sphere $S_n$. The half-sphere of vectors with $x_1 < 0$ can be centrally projected onto the hyperplane $x_1 = -1$ (see Fig. 1). Convexity is preserved by this projection, so convex sets in the hyperplane, an

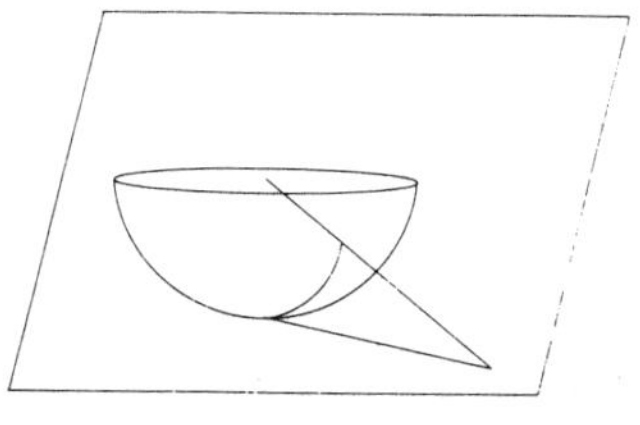

FIG. 1. The unit sphere $S_2$ projected to an affine space $A_2(F)$.

affine space, correspond to convex sets on the sphere. These are given by the $cx$ operator. Notice that the hyperplane is not an oriented matroid, for it contains no opposites.

Another formulation of oriented matroids can be given in terms of circuits. The closure system defined above is equivalent to this under the following translations.

(T1) $X$ is a circuit, $X \in \text{Cir}$: $X$ is a minimal nonempty finite set such that $X^* \subseteq cx(X)$,

(T2) $cx(U) = \{x \mid (\exists V \subseteq U)(V \cup x^*) \in \text{Cir}\}$.

THEOREM 3. *An oriented matroid can also be defined as a system* $\langle E, *, \text{Cir}\rangle$ *where* $E$ *is a set,* $*$ *is a map from* $E$ *to* $E$, *and* Cir *is a collection of subsets of* $E$ *satisfying*

(C1) $x^{**} = x$,

(C2) $X \in \text{Cir}$ *implies* $X$ *is nonempty and finite,*

(C3) $X, Y \in \text{Cir}$ *and* $X \subseteq Y$ *imply* $X = Y$,

(C4) $X \in \text{Cir}$ *implies* $X^* \in \text{Cir}$,

(C5) *If* $X, Y \in \text{Cir}$ *with* $a \in X \cap Y$ *and* $b \in X - Y$, *there is a* $Z \in \text{Cir}$ *such that* $b \in Z \subseteq (X - a) \cup (Y - a)^*$.

Notice that a set of the form $\{x, x^*\}$ satisfies these axioms, except perhaps C3; if $\{x\}$ is a circuit, $\{x, x^*\}$ cannot be a circuit. Circuits of the form $\{x, x^*\}$ will be designated *improper* and the other circuits, with $X \cap X^*$ empty, will be *proper*. This slightly relaxes the axioms of Folkman and Lawrence but permits cleaner translations. Furthermore, the vector set $\{(0, 1), (0, -1)\}$ seems as legitimate as $\{(0, 2), (0, -1)\}$ for a circuit. Note that all circuit axioms hold for the set of proper circuits.

*Proof of the Theorem.* It is straightforward to show that the circuit axioms imply the closure axioms, after noting that $X \subseteq cx(X)$. This holds because for every $x$, either $\{x\}$ or $\{x, x^*\}$ is a circuit. The converse is also straightforward, except for the proof of C5, which comes from two lemmas important in themselves.

LEMMA. *If* $X$ *is a circuit, either* $X \cap X^*$ *is empty or* $X = \{x, x^*\}$, *i.e., every circuit is either proper or improper.*

LEMMA. *Suppose a set* $X$ *with an element* $a \in X$ *is minimal for* $a \in X - X^*$ *and* $a^* \in cx(X)$. *Then* $X$ *is a circuit.*

The first lemma follows from the minimality of circuits. The second is more complicated; the proof given in [5] holds after one notices that $\langle (X \cup X^*), *, cx_X\rangle$ with $cx_X(U) = cx(U) \cap X$, is a finite oriented matroid.

To finish the proof that the closure axioms and T1 imply C5, let $X$, $Y$ be circuits with $a \in X \cap Y$ and $b \in X - Y$. The cases where one or both circuits are improper are easily dealt with; assume $X$ and $Y$ are proper. Then $a^* \in cx(Y - a)$, so $b^* \in cx((X - b - a) \cup (Y - a)^*)$. Pick a minimal finite $V \subseteq (X - b - a) \cup (Y - a)^*$ with $b^* \in cx(V)$. Then $(V \cup b)$ is the desired circuit.

The notion of circuit comes from graph theory and graphs provide additional models for oriented matroids. To have an opposition each edge on vertices $v_1$ and $v_2$ must be viewed as a pair of directed edges $(v_1, v_2)$ and $(v_2, v_1)$. A circuit $X$ is a minimal nonempty set of directed edges for which every vertex incident on $X$ is incident on $X^*$. The circuit properties of Theorem 3 are easily verified, except for C5; see Mei [9] or Fenton [4] for full details.

There is a nice "one-sidedness" result for circuits also.

PROPOSITION 4. *If $X$ and $Y$ are circuits, $X$ is proper, and $X \subseteq Y \cup Y^*$ then either $X = Y$ or $X = Y^*$. (This useful result appears in [5].)*

*Proof.* The circuit $Y$ is also proper. Assume the conclusion is false and pick an $X$ so that $|X \cap Y^*|$ is as small as possible. Since $X \nsubseteq Y$ and $X \nsubseteq Y^*$ there is an $a \in X \cap Y^*$ and $b \in X - Y^*$ and by C5 there is a circuit $Z$ such that $b \in Z \subseteq (X - a) \cup (Y - a^*)$. Hence $Z \cap Y^* \subseteq (X - a) \cap Y^*$ and $|Z \cap Y^*| < |X \cap Y^*|$, a contradiction.

In [12] Whitney used circuits as a primitive notion for matroids. In addition he gave an axiom system based on rank, the cardinality of a maximal independent subset. This is not a viable approach for convex dependence; in $R^2$ the sets $\{(1, 0), (0, 1), (-1, -1)\}$ and $\{(1, 0), (0, 1), (-1, 0), (0, -1)\}$ are both maximal convexly independent sets. Rank is available, however, from the underlying matroid $\langle E, cl \rangle$ where $cl(X) = cx(X \cup X^*)$. The Fundamental Theorem of linear algebra holds in $\langle E, cl \rangle$ and thus the rank of a set is well defined as the cardinality of a maximal linearly independent subset. Rank provides a bound on the discreteness of $cx$, as was first shown by Carathéodory [3].

THEOREM 5 (Carathéodory). *If $x \in cx(X)$ there is a $Y \subseteq X$ such that $x \in cx(Y)$ and $|Y| \leqslant rk(X)$.*

*Proof.* If $x \in X$ let $Y = \{x\}$. If $x \notin X$ pick a minimal finite $Y \subseteq X$ so that $x \in cx(Y)$. The set $(Y \cup x^*)$ is then a circuit. It only remains to show that $Y$ is linearly independent.

Assume not, so there is a $y \in Y$ for which $y \in cl(Y - y) = cx((Y - y) \cup (Y - y)^*)$. There is a minimal finite $V \subseteq (Y - y) \cup (Y - y)^*$ such that $y \in cx(V)$. By Proposition 1, $V \cap V^*$ is empty so $(V \cup y^*)$ is a circuit and

$(V \cup y^*) \subseteq (Y \cup x^*) \cup (Y \cup x^*)^*$. By Proposition 4, either $V \cup y^* = Y \cup x^*$ or $V \cup y^* = Y^* \cup x$. But neither $x$ nor $x^*$ is in $V$, a contradiction. Therefore $|Y| \leqslant rk(X)$.

This proof shows that the cardinality of a circuit is one greater than the rank of the subspace it generates, just as in the linear case.

## B. Flats, Sharps, and Hemispaces

We are interested in the sets $U$ for which $U = cx(U)$, to be called the *convex* sets of the oriented matroid. The opposition * suggests two important types of convex sets, the self-opposite and the opposite-disjoint.

Definition. $U$ is a *flat* if $U = cx(U)$ and $U = U^*$ (equivalently, if $U = cx(U^*)$). $U$ is a *sharp* if $U = cx(U)$ and $U \cap U^*$ is empty.

The flats are the subspaces, for they are closed under both operators. The sharps correspond to convex sets of an affine space, as in Fig. 1.

*Some notation.* $X^\flat = X \cap X^*$ and $X^\# = X - X^*$. These operators can be applied to any set, not just the convex ones. There are some easy corollaries.

$(X^*)^\flat = (X^\flat)^* = X^\flat$,

$(X^*)^\# = (X^\#)^*$,

$X \subseteq Y$ implies $X^\flat \subseteq Y^\flat$,

if $U$ is convex then $U^\flat$ is a flat and $U^\#$ is a sharp.

Notice that $X \subseteq Y$ does not imply $X^\# \subseteq Y^\#$, even for convex sets. For example, $U^\# \subseteq cx(U \cup U^*)^\# = \varnothing$ but $(U^\#)^\# = U^\#$ and $cx(U \cup U^*) = \varnothing$.

Lemma 6. *If $U$ is convex and $X \subseteq U$ then $cx(X) \cap U^\flat = cx(X \cap U^\flat)$. In particular, if $U = cx(X)$ then $U^\flat$ is generated by the elements of $X$ actually in $U$.*

*Proof.* Because $X \cap U^\flat \subseteq cx(X)$ we have $cx(X \cap U^\flat) \subseteq cx(X) \cap U^\flat$. Let $x \in cx(X) \cap U^\flat$ and pick a minimal finite $V \subseteq X$ for which $x \in cx(V)$. Then $(V \cup x^*)$ is a circuit, so $(V \cup x^*)^* \subseteq cx(V \cup x^*)$, implying $V \subseteq U^\flat$ and $x \in cx(X \cap U^\flat)$.

A flat $F$ can be enlarged by picking an $x$ not in $F$ and forming $cx(F \cup x \cup x^*)$, a linear operation. A sharp $S$ can be enlarged by picking an $x$ not in $S$ and forming $cx(S \cup x^*)$. Lemma 6 shows that this is a sharp. In a finite oriented matroid this allows the construction of sharps containing exactly one element from each opposite pair, which will be maximal among the sharps. Maximal sharps occur in every oriented matroid but in general a stronger argument is needed.

THEOREM 7. *If $S$ is a sharp and $x \notin S$ there exists a maximal sharp containing $S$ but not $x$.*

*Proof* (Using the Axiom of Choice). The set $cx(S \cup x^*)$ is a sharp, perhaps equal to $S$, so there are sharps containing $S$ but not $x$. Let $S_1 \subseteq S_2 \subseteq \cdots$ be a chain of such sharps and denote $T = \bigcup S_i$. If $y \in cx(T)$ there is a finite $V \subseteq T$ so that $y \in cx(V)$. Then for some $j$, $V \subseteq S_j$ so $y \in cx(S_j) \subseteq T$. Hence $T$ is convex.

If $z, z^* \in T$ there are $m, n$ such that $z \in S_m$ and $z^* \in S_n$. Let $k$ be the larger of $m, n$; then $z, z^* \in S_k$ (a contradiction). Thus $T$ is a sharp and $x \notin T$, so Zorn's Lemma applies.

The sharps containing exactly one element from each opposite pair $\{x, x^*\}$ will be called hemispaces; more formally, $H$ is a *hemispace* if $H = cx(H)$ and $H = E - H^*$. Hemispaces are clearly maximal sharps and the converse is also true.

PROPOSITION 8. *$H$ is a hemispace iff $H$ is a maximal sharp.*

*Proof.* Hemispaces are maximal among the sharps. Let $H$ be a maximal sharp, hence convex. If $x^* \notin H$ then $cx(H \cup x)$ is a sharp, implying $x \in H$. Also, if $x \in H$ then $x^* \notin H$. Therefore $H = E - H^*$.

With Theorem 7 this establishes the existence of hemispaces. Notice that the opposite $H^*$ is also a hemispace. The hemispaces are the smallest convex sets $U$ for which $U \cup U^* = E$ and are the only sharps for which this holds.

(Hammer [6] has studied semispaces, which are the projections of hemispaces onto a hyperplane, as in Fig. 1. His basic properties are easily verified when rephrased in the terminology presented here.)

A familiar topic in convexity theory is separation properties and a famous one is the Stone Separation Lemma (see Holmes [7]) which states that for disjoint convex sets $A$, $B$ in a linear (affine) space there are complementary convex sets $C$, $D$ such that $A \subseteq C$ and $B \subseteq D$. Recall that an affine space is "one side" of an oriented matroid, so its convex sets are sharps. Oriented matroids allow a nice generalization of Stone's result under an additional assumption.

DEFINITION. An oriented matroid $E$ of rank at least 3 is *full* if whenever $(U \cup V)$ is a circuit with $U$, $V$ disjoint and nonempty then $cx(U) \cap cx(V^*)$

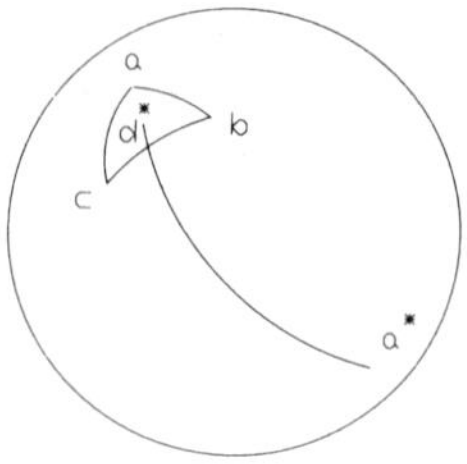


FIG. 2. Fullness on $S_2$.

is nonempty. When $\text{rk}(E) < 3$, $E$ is full if $E$ is a subspace (flat) of a full space of higher rank.

Fullness is the Pasch Axiom from geometry generalized to spaces of any rank. Figure 2 illustrates this on the sphere $S_2$, a space of rank 3. For the circuit $\{a, b, c, d\}$ we have $d^* \in cx(a, b, c)$ and $cx(b, c) \cap cx(a, d)$ is nonempty.

THEOREM 9. *Let $E$ be a full oriented matroid of finite rank. If $A$, $B$ are disjoint sharps there exists a hemispace $H$ such that $A \subseteq H$ and $B \subseteq H^*$.*

*Proof.* It suffices to show that $cx(A \cup B^*)$ is a sharp, for any maximal sharp containing $cx(A \cup B^*)$ contains $A$ and is disjoint from $B$.

*Case* 1. $\text{rk}(E) = 1$. For any $x \in A$, $A = cx(x)$ and $B = cx(x^*)$. Hence $cx(A \cup B^*) = cx(x) = A$.

*Case* 2. $\text{rk}(E) = 2$. Denote $(cx(A \cup B^*))^\flat$ by $F$. Let $x \in F$. If $x \in A$ then $x^* \notin A \cup B^*$ so there exists $a \in A$, $b \in B$ for which $x^* \in cx(a, b^*)$. (The Carathéodory Theorem comes into play here.) Then $b \in cx(a, x) \subseteq A$ (a contradiction). Hence $x \notin A$ and similarly $x \notin B$, so $(A \cup B^*) \cap F$ is empty. Then by Lemma 6, $F$ is empty.

*Case* 3. $\text{rk}(E) > 3$. Again denote $(cx(A \cup B^*))^\flat$ by $F$ and assume $F$ is nonempty. Pick $X \subseteq A \cup B^*$ so that $cx(X) = F$ (Lemma 6 ensures that $(A \cup B^*) \cap F$ is nonempty). Because $\text{rk}(E)$ is finite $X$ can be finite. Note that $X$ is disjoint from $X^*$. Pick a minimal $Z \subseteq X$ for which $cx(Z)$ is a flat; $Z$ will be a circuit. Since $Z \subseteq A \cup B^*$, $(Z - A)^* \subseteq B$. By the fullness of $E$ there exists an $x \in cx(Z \cap A) \cap cx((Z - A)^*) \subseteq A \cap B$, a contradiction.

The proofs for the rank 1 and 2 cases did not use fullness and thus hold in any oriented matroid. In higher ranks, however, fullness is necessary. Figure 3 shows a nonfull space with sharps $A$, $B$ for which $cx(A \cup B^*) = E$.

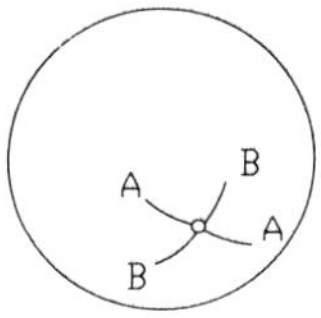


FIG. 3. The sphere $S_2$ minus an opposite pair.

## C. LARGE CONVEX SETS

A set $X$ in an oriented matroid is *large* if $X \cup X^* = E$, i.e., if $X$ contains at least one element from each opposite pair. For such an $X$, $E = X^{\#} \cup X^{\flat} \cup (X^*)^{\#}$ and these three sets are pairwise disjoint. Hemispaces are large by definition and by Proposition 8 are minimal among the large convex sets.

PROPOSITION 10. *Let $U$ be a large convex set.*

a. *$U$ contains a hemispace; furthermore, $U$ is the union of the hemispaces it contains. These are precisely the hemispaces which contain $U^{\#}$.*

b. *If $V$ is also a large convex set, $U \subseteq V$ iff $V^{\#} \subseteq U^{\#}$.*

This is straightforward to prove.

A particularly important type of large convex set is the *maximal convex set* (or simply *maximal*), a convex set $M$ such that if $M \subseteq U \subset E$ for a convex $U$ then $M = U$. The set of such maximals will be denoted $Mcx$. These are large because assuming $x, x^* \notin M$ would imply $x^* \in cx(M \cup x)$ and thus $x^* \in M$. We can use the maximals to describe any large convex set in the following manner.

*Notation.* $Mcx(F)$ will be the set of maximals in the subspace $F$.

THEOREM 11. *Suppose $E$ is an oriented matroid of finite rank $r$ with $U \subset E$. The set $U$ is a large convex set iff there exist $M_1 \supset M_2 \supset \cdots \supset M_n$ with $n \leqslant r$ such that*

$$U = M_1^{\#} \cup M_2^{\#} \cup \cdots \cup M_n^{\#} \cup M_n^{\flat}$$

*with $M_1 \in Mcx$ and $M_{i+1} \in Mcx(M_i^{\flat})$.*

(*If* $\text{rk}(E)$ *is infinite the sequence $M_\nu$ may need to be extended to a transfinite ordinal. At limit ordinals we would have $M_\lambda \in Mcx(\bigcup_{\nu<\lambda} M_\nu)$.*)

*Proof* ($\rightarrow$). The large convex set $U$ contains a hemispace $H$ and is contained in a maximal $M_1$. Hence $M_1^{\#} \subseteq H \subseteq U \subseteq M_1$. Note that $(H \cap F)$ will be a hemispace in a subspace $F$; in particular $(H \cap M_1^\flat)$ is a hemispace in the subspace $M_1^\flat$. If $M_1^\flat \not\subseteq U$ there is an $M_2 \in Mcx(M_1^\flat)$ containing the convex set $U \cap M_1^\flat$, so as before $M_2^\# \subseteq H \cap M_1^\flat \subseteq U \cap M_1^\flat \subseteq M_2$. Repeat this until some $M_n^\flat$ is contained in $U$; then $M_n = M_n^\# \cup M_n^\flat \subseteq U \cap M_{n-1}^\flat \subseteq M_n$. Since $\text{rk}(M_{i+1}) < \text{rk}(M_i)$, not only must this process stop at some $M_n^\flat$, but $n \leqslant r$.

For $(\leftarrow)$ we need a preliminary result.

LEMMA. *If $A$, $B$ are convex sets with $B \subseteq A^\flat$ then $(A^\# \cup B)$ is convex.*

*Proof of the Lemma.* Assume $z \in cx(A^\# \cup B) - (A^\# \cup B)$, so $z \in A^\flat$. Pick a minimal finite $X \subseteq A^\#$ so that $z \in cx(X \cup B)$. Since $z \notin B$, $X$ cannot be empty. Then for any $x \in X$ we get $x^* \in cx((X - x) \cup B \cup z^*) \subseteq A$, implying $x \in A^\# \cap A^\flat$, a contradiction.

*Proof of Theorem 11, continued.* By repeated application of the lemma, $U$ is convex. Since $E = M_1^\# \cup M_1^\flat \cup (M_1^*)^\#$, $M_1^\flat = M_2^* \cup M_2^\flat \cup (M_2^*)^\#$, etc., $U$ contains at least one element from each opposite pair.

Notice that the parts $M_i^\#$, $M_n$ are pairwise disjoint (up to 0; remember, we are working in a reduced space) and that $U^\# = M_1^\# \cup M_2^\# \cup \cdots \cup M_n^\#$ and $U^\flat = M_n^\flat$. The hemispaces have $n = r$ and $M_n^\flat = \varnothing$ in this decomposition. Notice also that, since $E - U = (M_1^*)^\# \cup \cdots \cup (M_n^*)^\#$, the large convex-closed sets are convex-open as well. There are, however, convex-clopen sets which are not large; for example a maximal $M$ gives the convex-clopen set $(M^*)^\# = E - M$.

An argument simular to that of Theorem 7 (with $AC$) shows the existence of convex sets which are maximal for not containing a given element $a$. These are the *relatively maximal convex sets* (or simply *relative maximals*), convex sets $R$ such that for some $a$ and any convex $U$, $R \subset U$ implies $a \in U - R$. These are large, for if $x$, $x^* \notin R$ then either $a \notin cx(R \cup x)$ or $a \notin cx(R \cup x^*)$, by Proposition 1.

Clearly a maximal $M$ is relatively maximal for any $a \in (M^*)^\#$, for the only larger convex set is $E$ itself. Hemispaces can be relatively maximal; when $H$ is written as in Theorem 11, the element $a$ would be in $(M_n^*)^\#$. In general, a relative maximal written as in Theorem 11 can use any element of $(M_n^*)^\#$ as the excluded element.

Some strong statements can be made about relative maximals if the space is assumed to be *dense*, meaning that whenever $a \notin cx(b, b^*)$ then $cx(a, b) \neq cx(a) \cup cx(b)$. This says that there exists an $x$ so that $\{a, b, x\}$ is a circuit. Dense spaces of rank $>1$ are of course infinite.

PROPOSITION 12. *If $R$ is relatively maximal in a dense oriented matroid there is a unique maximal containing $R$.*

*Proof.* Assume $M_1$, $M_2$ are maximals with $R \subseteq M_1 \cap M_2$. Let $x \in M_1 - M_2$ and $y \in M_2 - M_1$. Since $R$ contains a hemispace $H$ we have $M_1^\# \subseteq H \subseteq R \subseteq M_2$, implying $M_1 - M_2 \subseteq M_1^\flat$. Note that $x^* \in M_2$ and $y^* \in M_1$.

Because $E$ is dense there is a $z \in cx(x, y^*) - cx(x) - cx(y^*)$. Then $z \in cx(x, y^*) \subseteq M_1$ and $z^* \in cx(x^*, y) \subseteq M_2$. By the exchange property $y \in cx(x, z^*)$ and $x \in cx(y, z)$. Since $y \notin M_1$ we have $z^* \notin M_1$. However, because $x \notin M_2$ we have $z \notin M_2$. Hence $z \in M_1 - M_2 \subseteq M_1^\flat$, a contradiction.

PROPOSITION 13. *In a dense oriented matroid, $U$ is a large convex set iff $U$ is relatively maximal.*

LEMMA. *Let $E$ be dense. If $M \in Mcx$ and $x \notin M$ then $cx(M^\# \cup x) = E$.*

*Proof of the Lemma.* It suffices to show that $M^\flat \subseteq cx(M^\# \cup x)$, for then $E = cx(M \cup x) \subseteq cx(M^\# \cup x)$. Let $y \in M - M^\# = M^\flat$. $E$ is dense so there is a $z \in cx(x, y^*) - cx(x) - cx(y^*)$. Then $x^* \in cx(y^*, z^*)$ and $y \in cx(z^*, x)$. The former implies $x \in cx(y, z)$ so $z$ is not in $M$. But then $z^* \in M^\#$ so $y \in cx(M^\# \cup x)$.

*Proof of Proposition* 13. As noted earlier, relative maximals are large in any oriented matroid. For a large convex $U$ use Theorem 11 to write $U$ as $M_1^\# \cup \cdots \cup M_n^\# \cup M_n^\flat$. Pick any $a \in (M_n^*)^\#$ and consider $x \notin U$. For some $i$ we have $x \in (M_i^*)^\#$. This implies $a \in (M_n^*) \subseteq M_{n-1}^\flat \subseteq M_{n-2}^\flat \subseteq \cdots \subseteq M_{i-1}^\flat = cx(M_i^\# \cup x) \subseteq cx(U \cup x)$, so $U$ is relatively maximal for $a$ (the set $M_0^\flat$ is to be interpreted as $E$).

## D. AFTERWORD

The subject of oriented matroids has a strong flavor of geometry and topology due to the closure operator $cx$. Fullness and density, for example, are powerful assumptions on an oriented matroid. These notions require an infinite space, of course, so the results derived from them have no bearing on graphs and other finite structures. The flats are valuable as the subspaces in finite or infinite spaces but the value of sharps is unresolved. Affine convexity involves sharps but their usefullness in finite situations is unclear. Large convex sets occur in spaces of any size and their characterization in Theorem 11 is always valid.

The proofs from the axioms of the Carathéodory and Stone theorems and the multitude of intuitive notions suggest that much fruitful work remains to be done on oriented matroids.

## Acknowledgment

We thank Matt Kaufman, formerly of Purdue University, for his careful reading of and helpful comments on [4].

## References

1. R. G. Bland, A combinatorial abstraction of linear programming, *J. Combin. Theory Ser. B* **23** (1977), 33–57.
2. R. G. Bland and M. Las Vergnas, Orientability of matroids, *J. Combin. Theory Ser. B* **24** (1978), 94–123.
3. C. Carathéodory, Uber den Variabilitätsbereich der Koeffizienten von Potenzreihen, de gegebene Werte nicht annehmen, *Math. Ann.* **64** (1907), 95–115.
4. W. E. Fenton, "Axiomatic Convexity Theory," Ph.D. dissertation, Purdue University, 1982.
5. J. Folkman and J. Lawrence, Oriented matroids, *J. Combin. Theory Ser. B* **25** (1978), 199–236.
6. P. Hammer, Maximal convex sets, *Duke Math. J.* **22** (1955), 103–106.
7. R. B. Holmes, "Geometric Functional Analysis and Its Applications," Springer-Verlag, New York/Berlin, 1975.
8. M. Las Vergnas, Matroides orientables, *C. R. Acad. Sci. Paris Sér. I Math.* **20** (1975), 61–64.
9. P. Mei, "Axiomatic Theory of Linear and Convex Closure," Ph.D. dissertation, Purdue University, 1971.
10. L. G. Novoa, On *n*-ordered sets and order completeness, *Pacific J. Math.* **15**, No. 4 (1965), 1337–1345.
11. E. Steinitz, "Algebraische Theorie der Korper," de Gruyter, Berlin, 1930.
12. H. Whitney, On the abstract properties of linear dependence, *Amer. J. Math.* **57** (1935), 509–533.

Printed by Catherine Press, Ltd., Tempelhof 41, B-8000 Brugge, Belgium